Differentiation Rules

$f(t)$	$\dfrac{df}{dt} = f'(t)$
cu	$c\dot{u}$
$u + v$	$\dot{u} + \dot{v}$
uv	$u\dot{v} + v\dot{u}$
$\dfrac{u}{v}$	$\dfrac{v\dot{u} - u\dot{v}}{v^2}$
$u[v(t)]$	$\dfrac{du}{dv}[v(t)]\dot{v}(t)$

$$\frac{d}{dt}\,f[x(t), y(t)] = \frac{\partial f}{\partial x}\,\dot{x} + \frac{\partial f}{\partial y}\,\dot{y}.$$

Differentiation Formulas

$f(t)$	$\dfrac{df}{dt} = f'(t)$
c	0
t^p	pt^{p-1}
e^t	e^t
a^t	$(\ln a)a^t$
$\ln t$	$1/t$
$\sin t$	$\cos t$
$\cos t$	$-\sin t$
$\tan t$	$\sec^2 t$
$\cot t$	$-\csc^2 t$
$\sec t$	$\tan t \sec t$
$\csc t$	$-\cot t \csc t$
$\arcsin t$	$\dfrac{1}{\sqrt{1 - t^2}}$
$\arctan t$	$\dfrac{1}{1 + t^2}$
$\sinh t$	$\cosh t$
$\cosh t$	$\sinh t$
$\tanh t$	$\operatorname{sech}^2 t$

Polar Coordinates

$$\begin{cases} x = r \cos \theta \\ y = r \sin \theta \end{cases}$$

$$dx\, dy = r\, dr\, d\theta$$

Spherical Coordinates

$$\begin{cases} x = \rho \sin \phi \cos \theta \\ y = \rho \sin \phi \sin \theta \\ z = \rho \cos \phi \end{cases}$$

$$dx\, dy\, dz = \rho^2 \sin \phi\, d\rho\, d\phi\, d\theta$$

8.
$$\iint\limits_{\substack{x \geq 0,\, y \geq 0, \\ x+y \leq 1}} x^m y^n (1 - x - y)^p \, dx\, dy = \frac{m!\, n!\, p!}{(m + n + p + 2)!}$$

9.
$$\iiint\limits_{\substack{x \geq 0,\, y \geq 0,\, z \geq 0 \\ x+y+z \leq 1}} x^p y^q z^r (1 - x - y - z)^s \, dx\, dy\, dz$$
$$= \frac{p!\, q!\, r!\, s!}{(p + q + r + s + 3)!}$$

Integrals (constant of integration omitted)

1. $\displaystyle\int u\, dv = uv - \int v\, du$

2. $\displaystyle\int \frac{dx}{x} = \ln |x|$

3. $\displaystyle\int \frac{dx}{a^2 + x^2} = \frac{1}{a}\arctan\left(\frac{x}{a}\right)$

4. $\displaystyle\int \frac{dx}{a^2 - x^2} = \frac{1}{2a}\ln\left|\frac{x + a}{x - a}\right|$

5. $\displaystyle\int \frac{dx}{(a^2 + x^2)^n} = \frac{x}{(2n - 2)a^2(a^2 + x^2)^{n-1}}$
$$+ \frac{2n - 3}{(2n - 2)a^2}\int \frac{dx}{(a^2 + x^2)^{n-1}} \qquad (n > 1)$$

6. $\displaystyle\int \frac{dx}{\sqrt{a^2 - x^2}} = \arcsin\left(\frac{x}{a}\right)$

7. $\displaystyle\int \sqrt{a^2 - x^2}\, dx = \frac{x}{2}\sqrt{a^2 - x^2} + \frac{a^2}{2}\arcsin\left(\frac{x}{a}\right)$

(continued inside back cover)

Vectors

$$\mathbf{a} \cdot \mathbf{b} = a_1 b_1 + a_2 b_2 + a_3 b_3$$

$$\mathbf{a} \times \mathbf{b} = \begin{vmatrix} \mathbf{i} & \mathbf{j} & \mathbf{k} \\ a_1 & a_2 & a_3 \\ b_1 & b_2 & b_3 \end{vmatrix}$$

$$\nabla = \mathbf{i}\frac{\partial}{\partial x} + \mathbf{j}\frac{\partial}{\partial y} + \mathbf{k}\frac{\partial}{\partial z}$$

Calculus with Analytic Geometry

Calculus with Analytic Geometry

Harley Flanders
Justin J. Price

ACADEMIC PRESS
NEW YORK
SAN FRANCISCO
LONDON
A Subsidiary of Harcourt Brace Jovanovich, Publishers

ACADEMIC PRESS, INC.
111 FIFTH AVENUE, NEW YORK, NEW YORK 10003

UNITED KINGDOM EDITION PUBLISHED BY
ACADEMIC PRESS, INC. (LONDON) LTD.
24/28 OVAL ROAD, LONDON NW1

ISBN: 0-12-259672-2
Library of Congress Catalog Card Number: 77-90891

SET BY SANTYPE INTERNATIONAL LTD. IN
MONOPHOTO TIMES MATHEMATICS 569B,
UNIVERS BOLD 693, UNIVERS EXTRA BOLD 696,
AND MONOTYPE UNIVERS MEDIUM 689

PRINTED IN THE UNITED STATES OF AMERICA

Preface

This book is partly based on *Calculus* by H. Flanders, R. Korfhage, and J. J. Price, Academic Press, 1970, and on the spin-offs of that text: *A First Course in Calculus with Analytic Geometry*, Academic Press, 1973, and *A Second Course in Calculus*, Academic Press, 1974. However, it is essentially a new text rather than a second edition. We have rethought our basic approach to calculus in general and to many topics in particular; we have rewritten virtually everything taken from *Calculus*, and we have made many additions and subtractions of topics, examples, exercises, and figures.

Our basic objective is teaching the student how to set up and solve calculus problems—in short, how to apply calculus. Our initial approach to each topic is intuitive, numerical, and motivated by examples, with theory kept to a bare minimum. Later, after much experience in the use of the topic, we present an appropriate amount of theory.

We have included more than enough accurate definitions, theorems, and proofs, but they are definitely of secondary importance in this text. We believe that intuitive derivations students can remember are far more valuable than precise formal proofs they memorize for tests and promptly forget.

Organization Some basic pre-calculus algebra and analytic geometry is reviewed in Chapter 1, enough analytic geometry to hold us until Chapter 9, the big chapter on plane analytic geometry—which concludes with some applications of calculus. There is also a brief review of the trigonometric functions in Chapter 4.

Chapters 2—4 and 7 include differentiation, some usual and unusual applications, and the basic transcendental functions and their inverses. Chapters 5, 6, and 8 cover integration and applications. Chapters 10–12 form a unit on approximation, infinite series and integrals, and power series, winding up one-variable calculus. There is no separate chapter on differential equations as such, but there is a good deal of material in examples and exercises.

Chapers 13 and 14 contain solid analytic geometry, vectors, and curves. Chapters 15 and 16 contain the differential calculus of several variables, and the final Chapters 17 and 18, double and triple integrals.

We tried to think of each section as a teaching unit and to keep the time required per section constant. We did not entirely succeed, but still we hope this will have a positive effect on the teachability of the text. Similarly, we have tried to keep chapter lengths more or less equal, thinking of each as a unit for a test.

Applications Calculus was invented to solve real world problems and has proved indispensable in applications. We think the subject should be presented with this in mind, not as an abstract discipline. Therefore we have tried to include a variety of realistic and interesting applications.

Examples and Exercises The worked examples are the core of this text. We have tried to select appropriate ones that illustrate how each topic in calculus is used, and to grade their difficulties. There are about 480 formal examples, many with two or three parts, and about 170 informal examples.

Give or take a few, there are 5,010 exercises in the text, maybe 1,500 new, many unusual. About 35% are easy and routine drill, about 40% are middle level, and about 25% are hard. (Very hard exercises are *-ed.) There is a high correlation between topics in the examples and in the exercises, particularly in the easy and middle level exercises.

Each chapter ends with a set of 20–40 miscellaneous exercises. In the early chapters, they are mostly review exercises. Later when we have more material to work with, we use some of these exercises for interesting and off-beat material and applications.

As far as possible, we have tried to include interesting exercises, exercises that arouse curiosity and make the student want to know their answers. We have also tried for a reasonable balance between exercises that come out with clean solutions and those that do not.

Many exercises call for numerical answers. Today almost all calculus students have access to a scientific pocket calculator. This allows them to put their time into the set-up and solution of numerical exercises rather than into the drudgery of computations with tables. We recommend that those students without a calculator work just to slide rule accuracy, and we include absolutely minimal tables for this purpose.

In many examples and exercises, the correct answer is an arithmetic expression, but we usually include a numerical estimate to reinforce the reality of the answer. For instance, an exercise on vectors calls for the dihedral angle between faces of a regular tetrahedron. The answer is arc $\cos \frac{1}{3}$, but we add $\approx 70.529°$.

A short, but very practical integral table (plus some useful formulas) is printed inside the front and back covers. It includes some one- to three-dimensional definite integrals that are frequently needed in exercises.

Figures The drawings are an intimate part of this text and have been designed to convey a maximum of information. Virtually every figure from *Calculus* has been redrawn or replaced. A guiding principle in all figures is clarity and lack of clutter. Another is accuracy. For contrast and emphasis we have introduced a second color. Sometimes we use different scales on the axes to achieve a more reasonable graph. Altogether there are about 880 figures in the text and exercises and about 165 in the Answer Section.

Space Geometry Inability to visualize in space is a key difficulty for students of calculus. We believe that lack of adequate figures is a major cause of this difficulty. We have put as much work into the space figures for this book as into the accompanying text.

In our drawings, we have tried to choose projections that make the spatial aspects of the figures clear, usually not the standard oblique projections. To simplify our figures, we have transferred as much information as possible to legends. We have used a small computer to plot accurately the projections of curved figures. Finally, we have used a variety of techniques, for instance, varied line weights, color, degrees of shading, soft edges, overlap of black and color screens, and always breaking lines that pass behind others, all intended for greater clarity.

Choice of Topics   Because of the impact of computer science, numerical analysis, and new areas of application, there is considerable interest in having new topics and a shift in emphasis in the calculus course. We have chosen subject matter accordingly, and in most subjects we have included (possibly optional) material in the hope of satisfying the growing demand for new and diversified topics.

Although nothing is marked optional as such, no one will want to cover everything. Besides the common core of standard material, there is wide variety of additional topics for the instructor to choose from, according to the needs of his or her classes. An instructor may want to cover more theory or less theory, more or less applications (more biological, less physical, more social science, etc.), and similarly for numerical analysis-computing oriented material, routine exercises versus harder ones, and so on.

Notation   We have taken special pains to keep the notation simple, clean, and visually pleasing. For instance, we prefer c rather than x_0 for a special value of x because subscripts are always hard on students. We prefer $f[g(x)]$ rather than the more syntactically correct $f(g(x))$ because the double parentheses make one needless additional hurdle to a student's reading. There are many other instances of this sort.

At the same time, we do not deviate from the standard notation for calculus that students will meet in later courses. And in notes we alert students to alternative notations they may encounter in life.

Design   The design of this book, a natural evolution from several previous texts, places emphasis where it belongs, and we note the following features: (1) Each term that is defined is printed in **bold face**. That way, a student flipping pages to locate a half-forgotten definition finds the key words easily—it jumps out of the page. The usual italic type for definitions doesn't do this at all, and we reserve italics for *stress* in the text. (2) Main statements (theorems, rules, definitions, etc.) are boxed, usually with a bold face title in the box. (3) Formal examples begin and end with a solid square ■ in color. That way, a student is never left guessing where the solution ends. (4) Remarks, notes, warnings, etc., are set off from the text and used to give insights into what is really going on. (5) Vectors are printed in extra bold face sans serif type (**v**, **w**, etc.). This is by far the best method of emphasizing the distinction between scalar and vector quantities, and is attractive to read. (6) In each section, the examples are numbered simply Example 1, Example 2, etc., and the figures, Fig. 1, Fig. 2, etc. Very few formulas at all are numbered and then only for immediate reference. We feel that the elaborate decimal system numbering of theorems, formulas, etc., of research writing has no place in a lower division text. (7) We have simplified the punctuation by deleting the usual (unnecessary) periods in titles, captions, exercises' numbers, etc. (8) We have made a conscious effort to keep our English simple and to write in a plain, active style, without fuss. (9) Finally, we hope our detailed index will be useful.

History of Calculus   Our experience is that most students are uninterested and bored by material on the historical development of the calculus. They are generally motivated to learn the use of calculus and they resent as wasted time, such cultural material. Therefore, we have included extremely little history and few names and dates.

Acknowledgments   We and our publisher acknowledge with pleasure the assistance rendered by the following individuals and organizations. First, the Department of Mathematics of University College, London where the authors were guests for the academic year 1974–75. Next, our co-author Bob Korfhage of *Calculus*, who was overcommitted and could not join us on this project, but whose influence we feel keenly. We are very grateful to Charles C. Alexander, University of Mississippi, Arthur Copeland, Jr., University of New Hampshire, Charles Miller, Foothill Community College, and William Ritter, Rose-Hulman Institute of Technology who reviewed the manuscript in several drafts and contributed numerous valuable suggestions. Our technical artist, Rino Dussi, did a superb job of executing our sketches. And finally, our typesetter, Santype, Ltd., produced remarkably accurate work from our handwritten manuscript.

Contents

16. HIGHER PARTIALS AND APPLICATIONS

17. DOUBLE INTEGRALS

18. MULTIPLE INTEGRALS

NUMERICAL TABLES

ANSWERS TO ODD-NUMBERED EXERCISES

Functions and Graphs █

1. INTRODUCTION

Everyone is familiar with the use of graphs to summarize data (Fig. 1). The figure shows three typical graphs. There are many others; one sees graphs concerning length, time, speed, voltage, blood pressure, supply, demand, etc.

All graphs have an essential common feature; they illustrate visually the way one numerical quantity depends on (or varies with) another. In Fig. 1, (a) shows how the price of a certain manufactured item depends on the quantity available, (b) shows a "doomsday" estimate of world population versus time, and (c) shows how Fahrenheit readings depend on (are related to) centigrade readings.

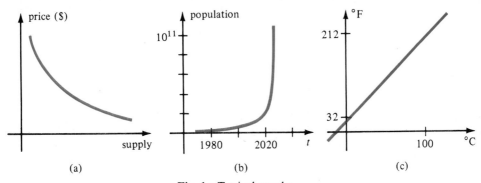

Fig. 1 Typical graphs

Graphs are pictures of **functions**. Roughly speaking, a function describes the dependence of one quantity on another or the way in which one quantity varies with another. We say, for instance, that price is a function of supply, or that population is a function of time, etc.

Functions lurk everywhere; they are the basic idea in almost every application of mathematics. Therefore, a great deal of study is devoted to their nature and properties. As Fig. 1 illustrates, a graph is an excellent tool in understanding the nature of a function. For it is a kind of "life history" of a function, to be seen at a glance.

The functions and graphs we shall deal with concern quantities measured in the **real number system.** This consists of the familiar numbers of our experience. Before starting our study of functions and graphs, we shall devote a brief section to the real number system itself.

2. REAL NUMBERS

The real numbers are the common numbers of everyday life. Everyone is familiar with their arithmetic. In more advanced courses they are defined and developed rigorously. However, that is a deep and lengthy project, not in the spirit of this book. We shall be content simply to quote their properties as we need them.

First of all, we list the basic rules of arithmetic that we use automatically when computing with real numbers.

Arithmetic of Real Numbers		
Associative laws	$a + (b + c) = (a + b) + c$	$a(bc) = (ab)c.$
Commutative laws	$a + b = b + a$	$ab = ba.$
Zero and unity laws	$a + 0 = a$	$a \cdot 1 = a.$
Distributive laws	$a(b + c) = ab + ac$	$(a + b)c = ac + bc.$
Inverse laws		

If a is any real number, then there is a unique real number $-a$ such that

$$a + (-a) = 0.$$

If a is any real number different from 0, then there is a unique real number a^{-1} such that

$$a \cdot a^{-1} = 1.$$

We write $a - b = a + (-b)$, and $a/b = ab^{-1}$ if $b \neq 0$.

Order Besides satisfying the rules of arithmetic, the real number system is endowed with an **order relation**; we can say that one number is greater or less than another. Recall the notation:

$a < b$ a is less than b,

$a \leq b$ a is less than or equal to b,

$a > b$ a is greater than b,

$a \geq b$ a is greater than or equal to b.

The three most basic properties of the order relation are these:

Properties of Order	
Reflexivity	$a \leq a.$
Anti-symmetry	If $a \leq b$ and $b \leq a$, then $a = b.$
Transitivity	If $a \leq b$ and $b \leq c$, then $a \leq c.$

Picturing the real number system as a number line (Fig. 1) helps us visualize the order relation. Thus $a < b$ means that a is to the left of b.

Fig. 1   The number line

The set of all numbers between two fixed numbers is called an **interval** on the number line. An interval may include one or both of its end points, or neither. For example, $-2 \le x \le 1$ describes the **closed interval** of all numbers between -2 and 1, *including* the end points, while $3 < x < 7$ describes the interval *strictly* between 3 and 7, that is, *excluding* the end points (Fig. 2).

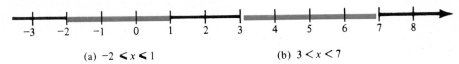

(a) $-2 \le x \le 1$ (b) $3 < x < 7$

Fig. 2   Examples of intervals

In working with the order relation, we use the following rules:

> ### Rules for Order
> (1)   If $a < b$ and if c is any real number, then $a \pm c < b \pm c$.
> (2)   If $a < b$ and $c > 0$, then $ac < bc$.
> (3)   If $a < b$ and $c < 0$, then $bc < ac$.
> (4)   If $0 < a < b$ or $a < b < 0$, then $1/b < 1/a$.
> (5)   The statement $a < b$ is equivalent to the statement $b - a > 0$.

These rules apply just as well with $\le$ in place of $<$, except that in (4) you must have

$$0 < a \le b \quad \text{or} \quad a \le b < 0$$

because division by 0 is excluded.

Examples   From $\sqrt{2} < \frac{3}{2}$ follow

(a) $1 + \sqrt{2} < \frac{5}{2}$ (b) $2\sqrt{2} < 3$ (c) $-2\sqrt{2} > -3$ (d) $1/\sqrt{2} > \frac{2}{3}$.

Several natural extensions of the rules for order are practical in computations.

> (6)   If $a < A$ and $b < B$, then $a + b < A + B$.
> (7)   If $0 < a < A$ and $0 < b < B$, then $ab < AB$.

Rules (6) and (7) apply also to three or more inequalities. For instance, if $a < A$, $b < B$, and $c < C$, then

$$a + b + c < A + B + C.$$

Examples   From $\sqrt{2} < \frac{3}{2}$ and $\sqrt{3} < \frac{7}{4}$ follow

(a) $\sqrt{2} + \sqrt{3} < \frac{13}{4}$ (b) $\sqrt{6} < \frac{21}{8}$.

■ **EXAMPLE 1**   Solve $\dfrac{1}{x + 1} > 2$.

Solution   The left-hand side must be positive, so we must have $x + 1 > 0$, that is, $x > -1$. Multiply both sides by the (positive) number $x + 1$; the result is $1 > 2(x + 1) = 2x + 2$. Add -2 to both sides: $-1 > 2x$. Multiply both sides by $\frac{1}{2}$ to obtain $x < -\frac{1}{2}$.

Answer $-1 < x < -\frac{1}{2}$ ∎

Absolute Values   The **absolute value** of a real number a, written $|a|$, is a measure of the size of a, regardless of its sign. We define

$$|a| = \begin{cases} a & \text{if} \quad a \geq 0 \\ -a & \text{if} \quad a \leq 0. \end{cases}$$

For example, $|-7| = 7$, $|-5.2| = 5.2$, $|3.6| = 3.6$, $|0| = 0$. Every real number except 0 has a positive absolute value. While it is false to say $-6 > 4$, it is correct to say $|-6| > |4|$.

Rules for Absolute Values

(1) $|a| = \pm a$, $|a| > 0$ if $a \neq 0$, $|0| = 0$.

(2) $|-a| = |a|$.

(3) $|ab| = |a| \cdot |b|$.

(4) $\left|\dfrac{a}{b}\right| = \dfrac{|a|}{|b|}$ $(b \neq 0)$.

(5) $|a + b| \leq |a| + |b|$ (triangle inequality).

Examples

Rule (3): $|(-6)(-5)| = |30| = 30 = |-6| \cdot |-5|$.

Rule (4): $|-3/4| = |3/4| = 3/4 = |-3|/|4|$.

Rule (5): $|3 + 4| = |7| = 7 \leq |3| + |4|$.

$|-3 + (-4)| = |-7| = 7 \leq |-3| + |-4|$.

$|5 - 2| = |3| = 3 < |5| + |-2|$.

Rules (3) and (5) extend to three or more numbers. For instance

$$|abc| = |a| \cdot |b| \cdot |c|, \qquad |a + b + c| \leq |a| + |b| + |c|.$$

Rule (5), the triangle inequality, is an equality if a and b have the same sign and is a strict inequality if they have opposite signs. The triangle inequality has two useful corollaries:

(6) $|a - b| \leq |a| + |b|$.

(7) $|a| - |b| \leq |a - b|$.

To obtain (6), replace b by $-b$ in (5). To obtain (7), apply (5) to the sum $(a - b) + b = a$:

$$|a| = |(a - b) + b| \leq |a - b| + |b|, \qquad \text{hence} \quad |a| - |b| \leq |a - b|.$$

Geometric Interpretation By its very definition, $|a|$ is the distance on the number line from a to 0. Thus the "size" of a is measured by its *distance* from 0, a non-negative quantity no matter whether a is right or left of 0.

Given distinct points a and b on the number line, the distance between them is $|a - b|$. See Fig. 3. Thus the distance between 5 and 9 is $|5 - 9| = |-4| = 4$. Since $|a - b| = |b - a|$, it doesn't matter which you subtract from which; the answer comes out the same.

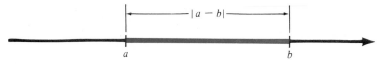

Fig. 3   Length of a segment: $L = |a - b| = |b - a|$

Using absolute values and inequalities, we can develop a nice shorthand to express geometric facts involving distances. For example, $|x - 4| < 1$ describes the set of points x whose distance from the point 4 is less than 1; in other words the interval $3 < x < 5$. By the same token, $|x - 4| \le 1$ describes the closed interval $3 \le x \le 5$.

> The inequality $|x - a| < r$ describes the interval (excluding end points)
>
> $$a - r < x < a + r.$$
>
> The inequality $|x - a| \le r$ describes the closed interval (including end points)
>
> $$a - r \le x \le a + r.$$

We can think of $|x - a| < r$ as representing the interval with center at a and "radius" equal to r. See Fig. 4.

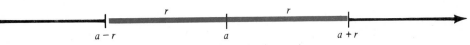

Fig. 4   The interval $|x - a| < r$

Of course $|x| < r$ represents the interval with center 0 and radius r, since $|x| = |x - 0|$. In calculus, the letter ε generally denotes a small positive number, so $|x - a| < \varepsilon$ describes a small interval centered at a. Here is a simple, yet important principle:

> If $|x| < \varepsilon$ for each $\varepsilon > 0$, then $x = 0$.

In other words, if $|x|$ is smaller than every positive number, then $x = 0$. For the only point contained in every interval centered at 0, no matter how small, is the point 0.

The Triangle Inequality This inequality, Rule (5), is more subtle than Rules (1)–(4), yet is extremely valuable in calculus. Here is a typical application. Suppose a is close to x and b is close to y. It seems that $a + b$ should be close to $x + y$. How do you prove that?

Well, *a close to x and b close to y* mean

$$|x - a| < \varepsilon_1 \quad \text{and} \quad |y - b| < \varepsilon_2,$$

where ε_1 and ε_2 are small positive numbers. To show that $|(x + y) - (a + b)|$ is small, use the triangle inequality:

$$|(x + y) - (a + b)| = |(x - a) + (y - b)| \leq |x - a| + |y - b| < \varepsilon_1 + \varepsilon_2.$$

Therefore $a + b$ approximates $x + y$ to within $\varepsilon_1 + \varepsilon_2$.

Example If $|x - 1| < 0.01$ and $|y - 5| < 0.02$, then $3x + 2y$ should be close to 13. By the triangle inequality,

$$|(3x + 2y) - 13| = |(3x - 3) + (2y - 10)| \leq |3x - 3| + |2y - 10|$$
$$= |3| \cdot |x - 1| + |2| \cdot |y - 5| < 3(0.01) + 2(0.02) = 0.07.$$

EXERCISES

Prove, using the rules of arithmetic and inequalities

1 if $ab \neq 0$, then $a \neq 0$ and $b \neq 0$ **2** if $ab = 0$, then $a = 0$ or $b = 0$

3 if $a \neq 0$ and $b \neq 0$, then $(ab)^{-1} = a^{-1}b^{-1}$

4 if $b \neq 0$ and $d \neq 0$, then $(a/b)(c/d) = (ac)/(bd)$

5 if $b \neq 0$, $c \neq 0$, and $d \neq 0$, then $(a/b)/(c/d) = (ad)/(bc)$

6 if $b \neq 0$ and $d \neq 0$, then $a/b + c/d = (ad + bc)/bd$

7 if $a \neq 0$ and $b \neq 0$, then $(a/b)^{-1} = b/a$

8 if $0 < a < b$, then $0 < a/b < 1$

9 if a, b, c, d are not all zero, then $a^2 + b^2 + c^2 + d^2 > 0$

10 $\dfrac{1}{a^2 + b^2 + c^2 + d^2} = \left(\dfrac{a}{a^2 + b^2 + c^2 + d^2}\right)^2 + \cdots + \left(\dfrac{d}{a^2 + b^2 + c^2 + d^2}\right)^2$

11 if $a \leq b$ and $-a \leq b$, then $|a| \leq b$

12 if a and b have the same sign, or if either is zero, then $|a + b| = |a| + |b|$

13 if a and b have opposite signs, then $|a + b| < |a| + |b|$

14 $||a| - |b|| \leq |a - b|$.

Express, using absolute values

15 x is either 2 or -2

16 x is farther from a than from b

17 x is at least as close to a as to b

18 x is either to the left of -3 or to the right of 3

19 x is between 16 and 18

20 x is within distance 2 of 7.

Express in terms of intervals, without using absolute values, all numbers x such that

21 $|3x| \leq 12$ **22** $|\frac{1}{2}x| \leq 5$ **23** $|x| \leq 0.01$

24 $0 < |x - 3| < 10^{-4}$ **25** $|x - 3| \leq 1$ **26** $|x + 4| \leq 2$

27 $|-5x| < 10$ **28** $0 < |x + 5| < 5 \times 10^{-3}$ **29** $|x - 4| < 5$

30 $|2x - 1| < 4$ **31** $|3x + 1| > 1$ **32** $|7 - x| < 6$

33 $|x^2 - 1| < 8$ **34** $|x| < |x + 5|$ **35** $0 < |x + 2| < 0.1$

36 $0 < |x - 9| < 2$ **37** $|x| + |x - 4| \leq 4$ **38** $|x| + |x - 2| < 3$.

Solve the inequalities

39 $6x - 1 < 3$ **40** $2x + 10 < 70$ **41** $4x - 5 > 8x + 1$

42 $3(x - 4) > \frac{1}{2}x - 6$ **43** $-4 < 2x + 6 < 16$ **44** $6 < \frac{1}{2}(x + 3) < 10$

45 $\dfrac{1}{x+5} > \dfrac{1}{8}$ **46** $\dfrac{1}{2} > \dfrac{4}{3-x}$ **47** $\dfrac{2x+1}{4x+1} < 1$

48 $\dfrac{1}{x+1} > \dfrac{2}{x}$ **49** $\dfrac{x}{x-3} < 0$ **50** $\dfrac{x}{8x-3} > 0.$

51 Find all points that are 3 times as far from 5 as from 1.

52 Find all points that are 10 times as far from 5 as from 1.

53 Explain how you can tell at a glance that there is no x for which both $|x-1| < 2$ and $|x-12| < 3$.

54 Explain why $|a| + |b| + |c| > 0$ is algebraic shorthand for "at least one of the numbers a, b, c is different from 0."

55 Suppose $|x-a| < 10^{-6}$. Show that $|7x - 7a| < 10^{-5}$.

56 Suppose $|x-7| < 10^{-6}$ and $|y-5| < 10^{-6}$. Show that $|(x+y) - 12| < 10^{-5}$. [*Hint* Use the triangle inequality.]

57 Suppose $|x-5| < \frac{1}{10}$ and $|y-7| < \frac{1}{10}$. Prove that $|xy - 35| < 1.3$. [*Hint* $xy - 35 = x(y-7) + 7(x-5)$.]

58 (cont.) Suppose $|x-5| < 10^{-6}$ and $|y-7| < 10^{-6}$. Prove that $|xy - 35| < 2 \times 10^{-5}$.

59 Suppose $|x-3| < 10^{-6}$. Prove that $|x^2 - 9| < 10^{-5}$. [*Hint* Factor $x^2 - 9$.]

60 (cont.) Prove $|x^3 - 27| < 5 \times 10^{-5}$.

3. COORDINATES

The real numbers provide labels for the points on a line. First, choose a point and mark it 0. Then choose a point to the right of 0 and mark it 1. In other words, choose a starting point, a unit length, and a positive direction (the direction from 0 toward 1). Then mark the points 2, 3, 4, $\cdots$ to the right and -1, -2, -3, $\cdots$ to the left. See Fig. 1. (It is perfectly possible to take the positive direction to the left; perhaps that is the convention on some planet in some galaxy.)

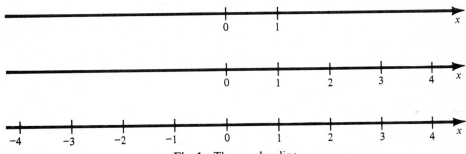

Fig. 1 The number line

Here we must make a fundamental assumption. We take it as an axiom that there is a perfect one-to-one correspondence between the points on the line and the system of real numbers. That means each point is assigned a unique real number label, and each real number labels exactly one point.

Because of this close association of the real number system and the set of points on a line, it is common to refer to a line as the real number system and to the real number system as a line. For instance, in a mathematical discussion, the real number 5.2 and the point labeled 5.2 might be considered the same. Although this is not correct logically, it almost never causes confusion; in fact it often sharpens our feeling for a problem.

Once the identification between real numbers and points on the line has been made, many arithmetic statements can be translated into geometric statements, and vice versa. Here are a few examples:

ARITHMETIC STATEMENT	GEOMETRIC STATEMENT
a is positive.	The point a lies to the right of the point 0.
$a > b$.	a lies to the right of b.
$a - b = c > 0$.	a lies c units to the right of b.
$a < b < c$.	b lies between a and c.
$\|3 - a\| < \frac{1}{2}$.	The point a is within $\frac{1}{2}$ unit of the point 3.
$\|a\| < \|b\|$.	The point a is closer to the origin than the point b is.

This close relationship between arithmetic and geometry is extremely important; often we can use arithmetical reasoning to solve geometrical problems or geometrical reasoning to solve arithmetical problems. Thus we may have two different ways of looking at a problem and, hence, increased chances of solving it.

If we denote a typical real number by x, we call the corresponding line the x-axis and draw Fig. 2.

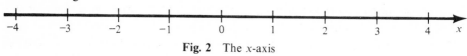

Fig. 2 The x-axis

When measuring time, we generally use t in place of x and call the corresponding line the t-axis. Usually 0 on the t-axis represents the time when an experiment begins; negative numbers represent past time, positive numbers future time.

Coordinates in the Plane When the points of a line are specified by real numbers, we say that the line is **coordinatized**: each point has a label or **coordinate**. It is possible also to label, or coordinatize, the points of a plane.

Draw two perpendicular lines in the plane. Mark their intersection **0** and coordinatize each line as shown in Fig. 3a. By convention, call one line horizontal and name it the x-axis; call the other line vertical and name it the y-axis.

Consider all lines parallel to the x-axis and all lines parallel to the y-axis (Fig. 3b). These two systems of parallel lines impose a rectangular grid on the whole plane. We use this grid to coordinatize the points of the plane.

Take any point P of the plane. Through P pass one vertical line and one horizontal line (Fig. 3c). They meet the axes in points x and y, respectively. Associate with P the ordered pair (x, y); it completely describes the location of P.

Conversely, take any ordered pair (x, y) of real numbers. The vertical line through x on the x-axis and the horizontal line through y on the y-axis meet in a point P whose coordinates are precisely (x, y). Thus there is a one-to-one correspondence,

$$P \longleftrightarrow (x, y),$$

between the set of points of the plane and the set of all ordered pairs of real numbers. The numbers x and y are the **coordinates** of P. The point $(0, 0)$ is called the **origin**.

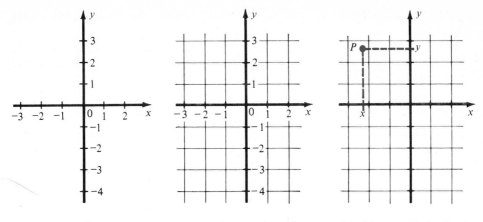

(a) Coordinate axes in the plane   (b) Rectangular grid   (c) P has coordinates (x, y).

Fig. 3

Remarks The coordinate system we have introduced is called a **rectangular** or **Cartesian** coordinate system.

Some writers refer to the horizontal coordinate of a point as its **abscissa** and the vertical coordinate as its **ordinate**.

Sometimes the pair (x, y) is called (ungrammatically) the coordinates of the corresponding point.

The coordinate axes divide the plane into four quadrants which are numbered as in Fig. 4.

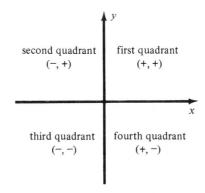

second quadrant
$(-, +)$

first quadrant
$(+, +)$

third quadrant
$(-, -)$

fourth quadrant
$(+, -)$

Fig. 4 The four quadrants

Sometimes the two coordinate axes are used to represent incompatible physical quantities. When this is the case, there is no reason whatsoever for choosing equal unit lengths on the two axes; on the contrary, it is usually best to take different unit lengths, or scales. For example, Fig. 5 shows the distance y in miles covered by a car in t seconds moving in city traffic.

If we are interested in the car's progress for about one minute, a reasonable choice of unit on the t-axis is 10 sec. Since we expect the car's speed to be at most 40 mph

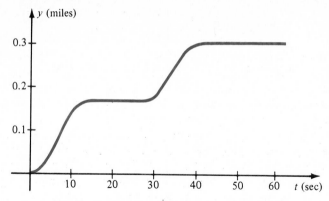

Fig. 5   Example of different scales on the axes

(about 0.1 mi per 10 sec) a reasonable choice for the unit on the y-axis is 0.1 mi. If we choose 1 sec and 1 mi for units, the graph will be silly and impractical. Try it!

If, however, we wish to plot the car's progress for 10 or 15 min, then 10 sec would probably be too small as a unit of time. A more practical choice might be 1 min as the time unit and 0.5 mi as the distance unit.

EXERCISES

Plot and label the points on one graph

1 $(-4, 1)$, $(3, 2)$, $(5, -3)$, $(1, 4)$
2 $(0, -2)$, $(3, 0)$, $(-2, 2)$, $(1, -3)$
3 $(0.2, -0.5)$, $(-0.3, 0)$, $(-1.0, -0.1)$
4 $(75, -10)$, $(-15, 60)$, $(95, 40)$.

Choose suitable scales on the axes and label the points

5 $(150, 0.3)$, $(50, 0.6)$ **6** $(-0.02, 5)$, $(0.03, 12)$
7 $(0.1, -0.003)$, $(-0.3, 0.007)$ **8** $(-0.02, 35)$, $(0.00, -60)$.

Indicate on a suitable diagram all points (x, y) in the plane for which

9 $x = -3$ **10** $y = 2$
11 x and y are positive   **12** either x or y (or both) is zero
13 $1 \leq x \leq 3$ **14** $-1 \leq y \leq 2$
15 $-2 \leq x \leq 2$ and $-2 \leq y \leq 2$ **16** $x > 2$ and $y < 3$
17 both x and y are integers   **18** $x^2 > 4$
19 $|x| \geq 1$ and $|y| \leq 2$ **20** $|x| \geq 2$ and $|y| \geq 2$
21 $xy > 0$ and $|x| \leq 3$ **22** $|x| + |y| > 0$.

Write the coordinates (x, y) of the

23 vertices of a square centered at $(0, 0)$, sides of length 2 and parallel to the axes
24 vertices of a square centered at $(1, 3)$, sides of length 2, at $45°$ angles with the axes
25 vertices of a 3-4-5 right triangle in the first quadrant, right angle at $(0, 0)$, hypotenuse of length 15
26 vertices of an equilateral triangle, sides of length 2, base on the x-axis, vertex on the positive y-axis.

4. FUNCTIONS AND GRAPHS

Let the symbol x represent a real number, taken from a certain set D of real numbers. Suppose there is a rule that associates with each such x a real number y. Then this rule is called a **function** whose **domain** is D.

For instance, suppose that to each real x is assigned a number y by the rule $y = x^2$. Then this assignment is a function whose domain is the set of all real numbers. As another example, take the assignment of $+\sqrt{x}$ to each real number x that has a square root. This assignment is a function whose domain is the set of non-negative numbers.

The set of all numbers y that a function assigns to the numbers x in its domain is called the **range** of the function. For example, the range of the function given by the rule $y = 2x$ is the set of all real numbers; the range of the function given by $y = x^2$ is the set of all *non-negative* real numbers. We sometimes say that a function **maps** or **carries** its domain onto its range.

Notation   The symbol used to denote a typical real number in the domain of a function is sometimes called the **independent variable**. The symbol used to denote the typical real number in the range is called the **dependent variable**.

Generally, but not always, variables are denoted by lowercase letters such as t, x, y, z. Functions are denoted by f, g, h and by capital letters.

If f denotes a function, x the independent variable, and y the dependent variable, then it is common practice to write $y = f(x)$, read " y equals f of x " or " y equals f at x." This means that the function f assigns to each x in its domain a number $f(x)$ which is abbreviated by y.

There are several common variations of this notation. For instance, if f is the function that assigns to each real number its square, then we write $f(x) = x^2$ or $y = x^2$.

Warning 1   It is logically incorrect to say "the function $f(x)$," or "the function x^2," or the function "$y = f(x)$." The symbols $f(x)$, y, x^2 represent numbers, the numbers assigned by the function f to the numbers x. A function is not a number, but an assignment of a number y or $f(x)$ to each number x in a certain domain. Nevertheless, these slight inaccuracies are so universal, we shall not try to avoid them.

Warning 2   A function is not a formula, and need not be specified by a formula. It is true that in practice most functions are indeed *computed* by formulas. For instance, f may assign to each real number x the real number y computed by formulas such as $y = x^2$, or $y = (\sqrt{x^2 + 1})/(1 + 7x^4)$, etc. Yet there are perfectly good functions not given by formulas. Here are a few examples:

(a) $f(x) =$ the largest integer (whole number) y for which $y \leq x$.

(b) $f(x) = \begin{cases} 1 & \text{if } x > 0 \\ 0 & \text{if } x = 0 \\ -1 & \text{if } x < 0. \end{cases}$

(c) $f(x) = 1$ if x is an integer, $f(x) = -1$ if x is not an integer.

(d) $f(x) =$ the number of letters in the English spelling of the rational number x in lowest terms. For example, $f(\frac{1}{2}) = 7$, $f(3) = 5$.

Keep in mind that $f(x)$ is the *number* assigned to x by the function f. If, for instance, $f(x) = x^2 + 3$, then $f(1) = 4$, $f(2) = 7$, $f(3) = 12$. By the same token $f(x + 1) = (x + 1)^2 + 3$, $f(x^2) = (x^2)^2 + 3 = x^4 + 3$, etc. For this particular function, you must boldly square and add 3 to whatever appears in the window, no matter what it is called:

$$f(x + y) = (x + y)^2 + 3, \qquad f\left(\frac{1}{x}\right) = \left(\frac{1}{x}\right)^2 + 3,$$

$$f[f(x)] = [f(x)]^2 + 3 = (x^2 + 3)^2 + 3 = x^4 + 6x^2 + 12.$$

Most functions arising in practice have simple domains. The most common domains are the whole line, an interval (segment) $a \le x \le b$, a "half-line" such as $x \ge 0$ or $x < 2$ or some simple combination of these. Examples:

FUNCTION	DOMAIN
$f(x) = 2x + 1$	all real x (the whole line)
$f(x) = \sqrt{x + 2}$	$x \ge -2$ (half line)
$f(x) = \sqrt{1 - x^2}$	$-1 \le x \le 1$ (interval)
$f(x) = \dfrac{1}{x}$	all x except $x = 0$ (union of two half-lines)

Construction of Functions   There are several standard methods for building new functions out of old ones. We shall list the most common of these constructions.

1. *Addition of functions.*   If f and g are functions of x defined on the same domain, then their **sum** $f + g$ is a function defined on the same domain by

$$[f + g](x) = f(x) + g(x).$$

For example, let $f(x) = 2x - 3$ and $g(x) = x^2 - x - 1$. Then

$$[f + g](x) = (2x - 3) + (x^2 - x - 1) = x^2 + x - 4.$$

2. *Multiplication of a function by a constant.*   If c is a constant and f is a function, the function cf is defined by

$$[cf](x) = cf(x).$$

For example, if $f(x) = x^2 - 2x - 1$, then

$$[-5f](x) = (-5)(x^2 - 2x - 1) = -5x^2 + 10x + 5.$$

3. *Multiplication of functions.*   If f and g are functions of x defined on the same domain, then their **product** fg is defined by

$$[fg](x) = f(x)g(x).$$

For example, if $f(x) = 2x - 1$ and $g(x) = 3x + 4$, then

$$[fg](x) = (2x - 1)(3x + 4) = 6x^2 + 5x - 4.$$

4. *Composition of functions.*   If g is a function whose range lies in the domain of a second function f, then the **composite** $f \circ g$ of f and g is defined by the formula

$$[f \circ g](x) = f[g(x)].$$

Think of substituting one function into the other, or replacing the variable of f by the function g. Here are some examples:

1. $f(y) = y^2 + 2y,\quad g(x) = -3x.$ Replace y by $g(x)$:

$$[f \circ g](x) = f[g(x)] = [g(x)]^2 + 2[g(x)] = (-3x)^2 + 2(-3x) = 9x^2 - 6x.$$

The domain of f is all real numbers; hence the range of g certainly lies in the domain of f.

2. $f(y) = 3y - 4,\quad g(x) = 2x^2 - x + 1.$

$$[f \circ g](x) = f[g(x)] = 3g(x) - 4 = 3(2x^2 - x + 1) - 4 = 6x^2 - 3x - 1.$$

Again the domain of f is all real numbers.

3. $f(y) = \sqrt{y - 1},\quad g(x) = -x^2.$

The domain of f is the set of real numbers y with $y \geq 1$. But $g(x) \leq 0$. Therefore the composition $f[g(x)]$ is not defined. Stated briefly, $\sqrt{-x^2 - 1}$ makes no sense.

If, however, $g(x) = 4x^2$, then $g(x) \geq 1$ provided $|x| \geq \frac{1}{2}$. Hence

$$[f \circ g](x) = \sqrt{4x^2 - 1}$$

is defined for $|x| \geq \frac{1}{2}$.

Graphs of Functions Given a function f, we construct its graph, a geometric picture of the function. For each number x in the domain of f, we find the associated number $y = f(x)$ in the range and plot the point (x, y). The locus (totality) of all such points is called the **graph** of $f(x)$. For example, if $f(x) = x$, then for each real number x, the associated number is $y = x$; we plot all points of the form (x, x). The locus (Fig. 1a) is obviously a straight line. If $f(x) = x + 1$, we plot all points of the form $(x, x + 1)$. The locus is again a straight line parallel to the graph of $y = x$ and one unit above it (Fig. 1b).

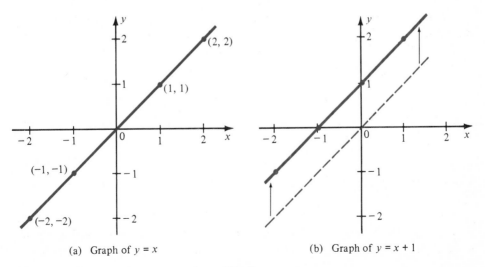

(a) Graph of $y = x$ (b) Graph of $y = x + 1$

Fig. 1

Terminology   Instead of referring to the "graph of $f(x) = x + 1$," we often say "the graph of $y = x + 1$," or when we are especially lazy, "the graph of $x + 1$."

■ **EXAMPLE 1**   Graph the function $f(x) = |x|$.

Solution   By definition, $|x| = x$ when $x \geq 0$, and $|x| = -x$ when $x \leq 0$. Consider the two cases separately. If $x \geq 0$, then $f(x) = x$, hence the graph is identical for $x \geq 0$ with the one shown in Fig. 1a. However, if $x \leq 0$, then $f(x) = -x$. Therefore this portion of the graph consists of all points $(x, -x)$, for instance, $(-1, 1)$, $(-2, 2)$, $(-3, 3)$, etc. Plot a few points; obviously the graph for $x \leq 0$ is a half line (ray) at angle $135°$ to the positive x-axis (Fig. 2).

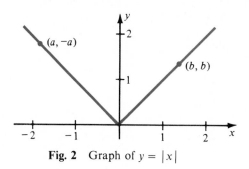

Fig. 2   Graph of $y = |x|$ ■

■ **EXAMPLE 2**   Graph $y = \dfrac{1}{x}$.

Solution   The domain of $f(x) = 1/x$ consists of all real numbers except $x = 0$, so the graph is not defined at $x = 0$. Plot a few points to get a general picture: $(1, 1)$, $(2, \frac{1}{2})$, $(3, \frac{1}{3})$, etc. As x increases, y decreases, so the graph approaches the x-axis from above (Fig. 3a). As x decreases from 1 toward 0, the curve rises steeply as is seen from plotting $(\frac{1}{2}, 2)$, $(\frac{1}{3}, 3)$, $(\frac{1}{4}, 4)$, etc. Hence the graph approaches the positive y-axis (Fig. 3b).

Finally, plot a few points for $x < 0$; for instance, $(-\frac{1}{2}, -2)$, $(-1, -1)$, $(-2, -\frac{1}{2})$, $(-3, -\frac{1}{3})$, etc. This part of the curve is obviously symmetric to the part for $x > 0$, but *below* the y-axis. Combine all this information to sketch the complete graph (Fig. 3c).

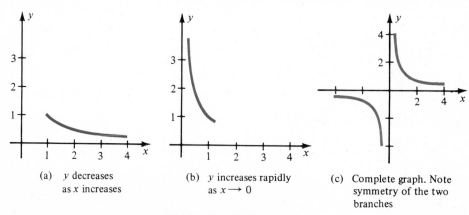

(a) y decreases
as x increases

(b) y increases rapidly
as $x \rightarrow 0$

(c)   Complete graph. Note
symmetry of the two
branches

Fig. 3   Graph of $y = 1/x$ ■

Graphical Definition of Functions   Not only does each function have a graph, but each graph defines a function. By a graph, we mean here, a collection of points (x, y) in the plane such that no two of the points have the same first coordinate (only one point can lie above a point on the x-axis). Such a graph automatically defines a function: to each x that occurs as a first coordinate of a point (x, y), it assigns the second coordinate y. In other words, $f(x)$ is the "height" of the graph above x.

Graphical definition of functions is standard procedure in science. For instance, a scientific instrument recording temperature or blood pressure on a graph is defining a function of time. There is hardly ever an explicit formula for such a function.

The graph of a function provides a picture of its domain and range. The vertical projection of the graph onto the x-axis is the domain; the horizontal projection onto the y-axis is the range (Fig. 4).

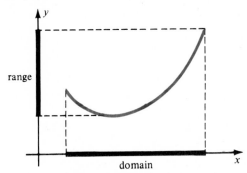

Fig. 4   Reading the domain and range of a function from its graph

EXERCISES

1 Let $f(x) = 2x + 5$. Compute
   (a) $f(0)$ (b) $f(2)$ (c) $f(\frac{1}{2})$ (d) $f(1/x)$ (e) $f(x - 3)$.

2 Let $f(x) = x^2 + x + 1$. Compute
   (a) $f(0)$ (b) $f(-x)$ (c) $f(x^2)$ (d) $f(\sqrt{x})$ (e) $f(x + h) - f(x)$.

Graph

3 $f(x) = x + 2$ **4** $f(x) = x - 1$ **5** $f(x) = -x$

6 $f(x) = -x + 1$ **7** $f(x) = -17$ **8** $f(x) = 0.03$

9 $f(x) = x + 0.01$ **10** $f(x) = -x - 2.5$ **11** $f(x) = |x|$

12 $f(x) = |x - 1|$ **13** $f(x) = \begin{cases} 0, & x \le 0 \\ 2x, & x > 0 \end{cases}$ **14** $f(x) = \begin{cases} x - 1, & x \le 3 \\ 2, & x > 3 \end{cases}$

15 $f(x) = \begin{cases} 1, & x > 0 \\ 0, & x = 0 \\ -1, & x < 0 \end{cases}$ **16** $f(x) = \begin{cases} 1 & \text{if } x \text{ is an integer} \\ -1 & \text{if } x \text{ is not an integer.} \end{cases}$

Find the domain and the range of $f(x)$

17 $f(x) = 3x - 2$ **18** $f(x) = -7x + 6$ **19** $f(x) = 4x - 5$

20 $f(x) = 7 - x$ **21** $f(x) = 1/(2x - 3)$ **22** $f(x) = x/(x + 2)$

23 $f(x) = x/(3x - 5)$ **24** $f(x) = 1/\sqrt{1 - x}$ **25** $f(x) = \sqrt{x - 6}$

26 $f(x) = \sqrt{5 - 2x}$ **27** $f(x) = \sqrt{4 - 9x^2}$ **28** $f(x) = \sqrt{15x^2 + 11}$

29 $f(x) = \sqrt{2x - 3}$ **30** $f(x) = \dfrac{1}{\sqrt{x + 4}}$ **31** $f(x) = \sqrt{\frac{1}{4} - x^2}$

32 $f(x) = \sqrt{x^2 - 1}$ **33** $f(x) = \sqrt{(x-1)(x-4)}$ **34** $f(x) = \sqrt{x^3 + 1}$.

Find $[f + g](x)$, and $[fg](x)$, where

35 $f(x) = 3x + 1$, $g(x) = -2$ **36** $f(x) = 2x - 1$, $g(x) = 2x + 3$
37 $f(x) = x^2$, $g(x) = -2x + 1$ **38** $f(x) = x^2 + 1$, $g(x) = -x^2 + x$.

39 Does it make sense to add the functions $y = \sqrt{1 - x}$ and $y = \sqrt{x - 2}$?
40 A function f is called **strictly increasing** if whenever $x_1 < x_2$, then $f(x_1) < f(x_2)$. Show that the sum of two strictly increasing functions is strictly increasing.

Find $f \circ g$ and $g \circ f$, where

41 $f(x) = 3x + 1$, $g(x) = x - 2$ **42** $f(x) = 2x - 1$, $g(x) = -x^2 + 3x$
43 $f(x) = 2x^2$, $g(x) = -x - 1$ **44** $f(x) = x + 1$, $g(x) = -x + 1$
45 $f(x) = 2x$, $g(x) = -2x$ **46** $f(x) = x + 3$, $g(x) = -x + 1$
47 $f(x) = x^2$, $g(x) = 3$ **48** $f(x) = \pi x^2$, $g(x) = 2x + 5$.
49 If $f(x) = x$ and $g(x)$ is any function, find $f \circ g$.
50 If $g(x) = x$ and $f(x)$ is any function, find $f \circ g$.
51 Let $f(x) = 1 - x$. Compute $[f \circ f](x)$.
52 Let $f(x) = 1/x$ for $x \neq 0$. Compute $[f \circ f](x)$.
53 Find an example of a function $f(x)$ such that $f(x^2) \neq [f(x)]^2$.
54 Find an example of a function $f(x)$ such that $f(1/x) \neq 1/f(x)$.
55 Does it make sense to form $f \circ g$ if $f(x) = \sqrt{2x - 5}$ and $g(x) = 1 - x^2$?
56 Prove that if $f(x) = 3x - 5$, then $f\left(\dfrac{x_0 + x_1}{2}\right) = \dfrac{f(x_0) + f(x_1)}{2}$.

57 (cont.) Is the same true for $f(x) = ax + b$?
58 (cont.) Is the same true for $f(x) = x^2$?

59 If $f(x) = 1/x$, show that $f\left(\dfrac{x_0 + x_1}{2}\right) = 2f(x_0 + x_1)$.
60 If $f(x) = 1/x^2$, show that $f(x_0 x_1) = f(x_0)f(x_1)$.

5. LINEAR FUNCTIONS

A function $f(x)$ is called **linear** if

$$f(x) = ax + b$$

for all real values of x, where a and b are constants. In the special case that $a = 0$, $f(x) = b$, a **constant** function. It assigns the constant value b to each x.

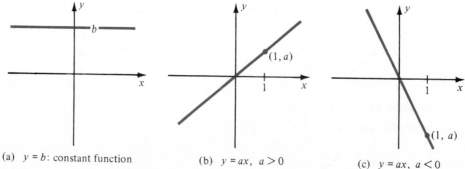

(a) $y = b$: constant function   (b) $y = ax$, $a > 0$ (c) $y = ax$, $a < 0$
Fig. 1 Graphs of certain linear functions.

The graph of the constant function $f(x) = b$ is obviously the horizontal line (Fig. 1a) at level $y = b$.

The graph of $f(x) = ax$ is the straight line through $(0, 0)$ and $(1, a)$. (This can be proved easily by means of similar triangles.) If $a > 0$, the line lies in the first and third quadrants (Fig. 1b); if $a < 0$, it lies in the second and fourth quadrants (Fig. 1c).

The graph of the general linear function $y = ax + b$ is just the graph of $y = ax$ moved up or down $|b|$ units, depending on whether $b > 0$ or $b < 0$. Figure 2 shows the four possibilities.

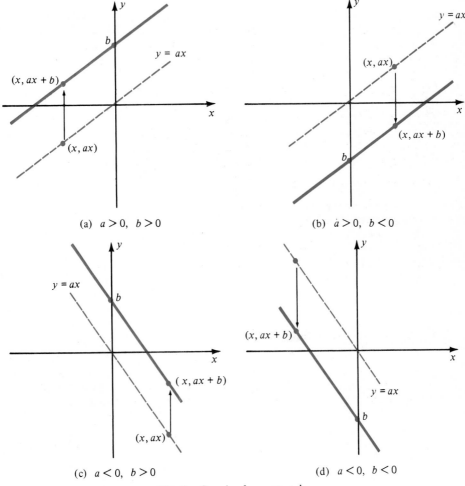

(a) $a > 0$, $b > 0$

(b) $a > 0$, $b < 0$

(c) $a < 0$, $b > 0$

(d) $a < 0$, $b < 0$

Fig. 2   Graph of $y = ax + b$

The graph of a linear function is a non-vertical straight line.

Conversely, each non-vertical straight line is the graph of a linear function.

Note that a vertical straight line is represented by an equation, $x = c$, but that it is not the graph of a function.

Slope   We define **slope**, a measure of the steepness of a non-vertical line. Choose two points on the line (x_0, y_0) and (x_1, y_1). As x advances from x_0 to x_1, the variable y changes from y_0 to y_1, so the change in y is $y_1 - y_0$. See Fig. 3. The slope is the ratio of the change in y to the change in x:

$$\text{Slope} = \frac{y_1 - y_0}{x_1 - x_0}.$$

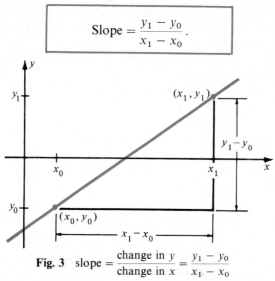

Fig. 3 $\text{slope} = \dfrac{\text{change in } y}{\text{change in } x} = \dfrac{y_1 - y_0}{x_1 - x_0}$

If the line rises as x increases, then both $x_1 - x_0$ and $y_1 - y_0$ have the same sign; hence the slope is positive. If the line falls as x increases, then $x_1 - x_0$ and $y_1 - y_0$ have opposite signs; hence the slope is negative.

The more steeply a line rises, the greater is the change in y compared to the change in x; hence the greater is the slope. For a line making a $45°$ angle with the positive x-axis, $y_1 - y_0 = x_1 - x_0$ so its slope is 1. For a horizontal line, $y_1 - y_0 = 0$ so its slope is 0.

Remark   The slope formula is valid whether (x_1, y_1) is to the right of (x_0, y_0) or to the left of (x_0, y_0). That is because

$$\frac{y_1 - y_0}{x_1 - x_0} = \frac{y_0 - y_1}{x_0 - x_1}.$$

To compute the slope of the line $y = ax + b$, use any two points on the line:

$$\frac{y_1 - y_0}{x_1 - x_0} = \frac{(ax_1 + b) - (ax_0 + b)}{x_1 - x_0} = \frac{a(x_1 - x_0)}{x_1 - x_0} = a.$$

Because the line $y = ax + b$ cuts the y-axis at $(0, b)$, the number b is called the y-**intercept** of the line.

The line $y = ax + b$ has slope a and y-intercept b.

Equations of Lines   Given that a line has slope a and passes through $(0, b)$, the line is the graph of $f(x) = ax + b$, that is, $y = ax + b$ is an equation for the line.

Suppose a line has slope a and passes through a point (x_0, y_0), not necessarily on the y-axis. How can we find its equation? Well, by the slope formula, a point $(x, y) \neq$

(x_0, y_0) lies on the line precisely when

$$\frac{y - y_0}{x - x_0} = a.$$

This is the desired equation. It is usually written in the form $y - y_0 = a(x - x_0)$; in this form we may legally substitute $x = x_0$ (without dividing by 0).

How about the case in which we are given two points on a line, (x_0, y_0) and (x_1, y_1), where $x_0 \neq x_1$. What is the equation of the line? We use the preceding result with a computed by

$$a = \frac{y_1 - y_0}{x_1 - x_0}.$$

Generally the equation is written in the form $\dfrac{y - y_0}{x - x_0} = \dfrac{y_1 - y_0}{x_1 - x_0}$.

A special case of this formula is sometimes handy. If the x-intercept is c and the y-intercept is d, that is, the line passes through $(c, 0)$ and $(0, d)$, then the equation becomes

$$\frac{x}{c} + \frac{y}{d} = 1.$$

We leave the derivation as an exercise.

Equations of lines

Slope-intercept form $y = ax + b$ *Point-slope form* $y - y_0 = a(x - x_0)$

Two-point form $\dfrac{y - y_0}{x - x_0} = \dfrac{y_1 - y_0}{x_1 - x_0}$ *Two-intercept form* $\dfrac{x}{c} + \dfrac{y}{d} = 1$

These formulas are used to obtain the equation of a line from data. Conversely, given an equation in one of these forms, it represents a line that can easily be identified. For example, $y - 3 = 2(x - 1)$ is the equation of the line through $(1, 3)$ with slope 2.

Bear in mind that an equation $cx + dy + e = 0$ can be put in the slope-intercept form if $d \neq 0$. For example, $3x - 2y + 8 = 0$ can be written as

$$y = \tfrac{3}{2}x + 4;$$

hence it represents a line of slope $\frac{3}{2}$ and y-intercept 4.

Angle between Lines   A non-vertical line makes an angle ϕ with the positive x-axis (Fig. 4a) called the **inclination angle** of the line. The slope m is the tangent of this angle: $m = \tan \phi$.

Two non-vertical lines with distinct inclination angles ϕ_1 and ϕ_2 intersect at angle $\theta = \phi_2 - \phi_1$. See Fig. 4b. By a formula from trigonometry,

$$\tan \theta = \tan(\phi_2 - \phi_1) = \frac{\tan \phi_2 - \tan \phi_1}{1 + \tan \phi_1 \tan \phi_2} = \frac{m_2 - m_1}{1 + m_1 m_2}.$$

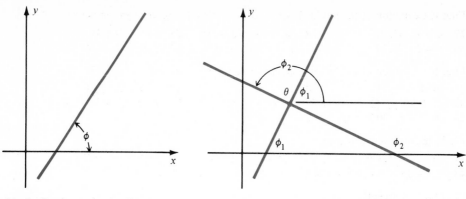

(a) Inclination angle: ϕ. Slope: $\tan \phi$ (b) Angle of intersection: $\theta = \phi_2 - \phi_1$

Fig. 4

In particular, the lines are perpendicular when $\theta = 90°$. This happens when $\tan \theta$ is undefined, that is, when the denominator $1 + m_1 m_2 = 0$.

The angle θ between two non-vertical lines of slopes m_1 and m_2 is given by

$$\tan \theta = \frac{m_2 - m_1}{1 + m_1 m_2}.$$

The lines are perpendicular when their slopes are negative reciprocals:

$$m_2 = -\frac{1}{m_1}.$$

Remark   Clearly $\tan \theta$ is only determined up to sign. That is because there is no way to decide between θ and its supplementary angle $\pi - \theta$.

EXERCISES

Graph

1 $y = 2x - 3, \quad 0 \le x \le 4$ **2** $y = 2x - 3, \quad -2 \le x \le 0$
3 $y = 2x + 9, \quad 1 \le x \le 2$ **4** $y = -3x + 1, \quad 0 \le x \le 1$
5 $y = -3x + 1, \quad -5 \le x \le 5$ **6** $y = -2x + 1, \quad -20 \le x \le -10$
7 $y = 3x + 40, \quad 25 \le x \le 50$ **8** $y = 9x - 50, \quad 100 \le x \le 200$
9 $y = 0.1x + 1.5, \quad 2 \le x \le 3$ **10** $y = -0.3x + 0.2, \quad -1 \le x \le 1.$

Graph; t in seconds, x in feet

11 $x = 0.2t - 1, \quad 0 \le t \le 5$ **12** $x = 25t + 15, \quad 50 \le t \le 100$
13 $x = 9t - 9, \quad 1 \le t \le 2$ **14** $x = -100t + 20, \quad -1 \le t \le 1$
15 $x = -t + 10, \quad 25 \le t \le 50$ **16** $x = 40t + 40, \quad 0 \le t \le 100.$

Find the slope of the line through the given points

17 $(0, 0), \quad (3, 4)$ **18** $(0, 0), \quad (2, 6)$ **19** $(-1, 2), \quad (1, 2)$
20 $(-1, 2), \quad (1, 0)$ **21** $(0, 1), \quad (1, 2)$ **22** $(0, -1), \quad (1, 2)$
23 $(-1, -1), \quad (1, 2)$ **24** $(-1, 2), \quad (2, -1)$ **25** $(-3, 1), \quad (-2, 2)$
26 $(-2, -2), \quad (3, -4).$

Find the equation of the line with given slope a and passing through the given point

27 $a = 1$, $(1, 2)$ **28** $a = -1$, $(2, -1)$ **29** $a = 0$, $(4, 3)$

30 $a = 2$, $(1, 3)$ **31** $a = \frac{1}{2}$, $(2, -2)$ **32** $a = \frac{2}{3}$, $(-1, 1)$.

Find the equation of the line through the two given points

33 $(0, 0)$, $(1, 2)$ **34** $(1, 0)$, $(3, 0)$ **35** $(-1, 0)$, $(2, 4)$

36 $(-1, -1)$, $(2, 6)$ **37** $(\frac{1}{2}, 1)$, $(\frac{3}{2}, 2)$ **38** $(-2, 0)$, $(-\frac{1}{2}, -1)$

39 $(0.1, 3.0)$, $(0.3, 2.0)$ **40** $(-2.01, 4.10)$, $(-2.00, 4.00)$.

Find the slope and y-intercept

41 $3x - y - 7 = 0$ **42** $x + 2y + 6 = 0$

43 $3(x - 2) + y + 5 = 2(x + 3)$ **44** $2(x + y + 1) = 3x - 5$.

Find both intercepts

45 $\dfrac{x}{2} + \dfrac{y}{3} = 1$ **46** $\dfrac{x}{a} + \dfrac{y}{b} = 1$

47 $2x + 3y = 1$ **48** $ax + by = 1$.

49 Derive the two-intercept form of a line.

50 Find the equation of the line through $(4, 5)$ that is perpendicular to $x - 3y - 4 = 0$.

Find the angle between the lines

51 $y = 2x + 1$, $y = -3x - 1$ **52** $y = \frac{2}{3}(x - \frac{1}{2})$, $y = -\frac{3}{2}x$

53 $y = \frac{1}{2}x - 2$, $2y + 4x = 1$ **54** $y = x\sqrt{3} - 1$, $y + x = 3$.

6. QUADRATIC FUNCTIONS

A function $f(x)$ is called **quadratic** if

$$f(x) = ax^2 + bx + c,$$

where a, b, and c are constants and $a \neq 0$. The domain of a quadratic function is all real x.

Let us graph the simplest quadratic, $y = x^2$. We consider first only $x \geq 0$. As x increases, y increases but very slowly at first. For example $(0.01)^2 = 0.0001$, and $(0.1)^2 = 0.01$. The curve passes through $(1, 1)$ and then begins to rise rapidly, passing through $(2, 4)$, $(3, 9)$, $(4, 16)$, etc. Plotting some of these points gives a rough idea of the graph for $x \geq 0$. See Fig. 1a. For $x < 0$, we note that $(-x)^2 = x^2$. Hence, for each point (x, y) on the curve, the point $(-x, y)$ is also on the curve. It follows that the graph is symmetric in the y-axis (Fig. 1b). The curve is called a **parabola**.

Next we graph $y = ax^2$, assuming first that $a > 0$. The graph of $y = ax^2$ can be obtained from the graph of $y = x^2$ in a simple way: Each point (x, y) on $y = x^2$ is changed to (x, ay), in other words, the graph $y = x^2$ is stretched (or shrunk) by the factor a in the y-direction only (Fig. 2).

If $a < 0$, then $-a > 0$, and the graph of $y = ax^2$ is obtained from the graph of $y = (-a)x^2$ by changing each y to $-y$, that is, by forming a mirror image in the x-axis (Fig. 3). Note that $(0, 0)$ is the lowest point on the graph of $y = ax^2$ if $a > 0$, and is the highest point on the graph if $a < 0$.

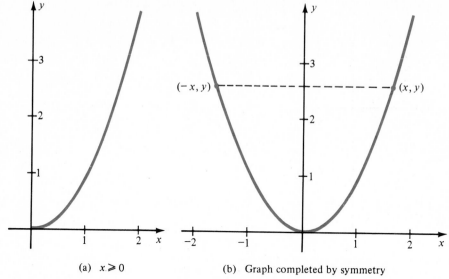

(a) $x \geqslant 0$ (b) Graph completed by symmetry

Fig. 1 Graph of $y = x^2$

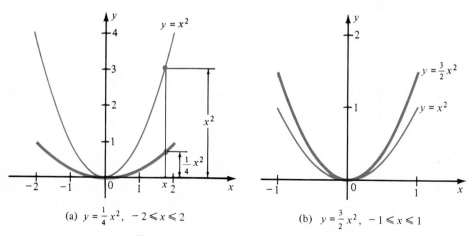

(a) $y = \frac{1}{4}x^2$, $-2 \leqslant x \leqslant 2$ (b) $y = \frac{3}{2}x^2$, $-1 \leqslant x \leqslant 1$

Fig. 2 Graphs of $y = ax^2$ for $a > 0$

The graph of $y = ax^2 + c$ is obtained by shifting the graph of $y = ax^2$ up or down by $|c|$ units (Fig. 4).

Next, we graph $y = a(x + h)^2$, assuming first that $h > 0$. For each point (x, y) on this curve, the point $(x + h, y)$ is on the curve $y = ax^2$. Thus, if we start with $y = a(x + h)^2$ and move each point h units to the right, we get the curve $y = x^2$. In other words, the curve $y = a(x + h)^2$ is the curve $y = x^2$ shifted h units to the left. If $k < 0$, similar reasoning shows that $y = a(x + k)^2$ is the curve $y = x^2$ shifted $|k|$ units to the right (Fig. 5).

Completing the Square To graph the most general quadratic function, we need an important technique called **completing the square**. (We can suppose $a \neq 0$,

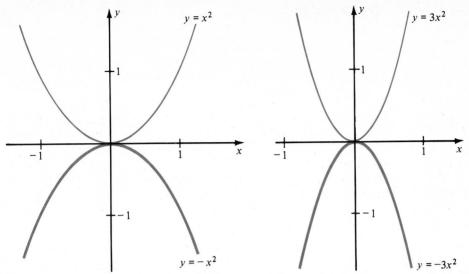

Fig. 3 Graphs of $y = ax^2$ for $a < 0$.

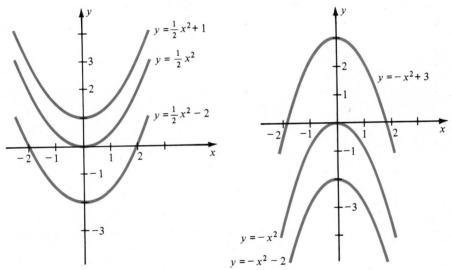

Fig. 4 Graphs of $y = ax^2 + c$.

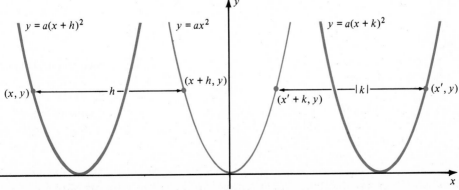

Fig. 5 Graph of $y = a(x + h)^2$ for $h > 0$ and graph of $y = a(x + k)^2$ for $k < 0$

otherwise the function is linear.) We write

$$y = ax^2 + bx + c = a\left(x^2 + \frac{b}{a}x + \frac{c}{a}\right)$$

and observe that the first two terms in the parentheses are part of a perfect square:

$$\left(x + \frac{b}{2a}\right)^2 = x^2 + \frac{b}{a}x + \frac{b^2}{4a^2}.$$

We "complete the square" by inserting $b^2/4a^2$ in the parentheses, and then compensate by subtracting the same quantity:

$$y = a\left(x^2 + \frac{b}{a}x + \frac{b^2}{4a^2} + \frac{c}{a} - \frac{b^2}{4a^2}\right)$$

$$= a\left(x^2 + \frac{b}{a}x + \frac{b^2}{4a^2}\right) + a\left(\frac{c}{a} - \frac{b^2}{4a^2}\right) = a\left(x + \frac{b}{2a}\right)^2 + \left(c - \frac{b^2}{4a}\right).$$

Hence for $a \neq 0$, the graph of $y = ax^2 + bx + c$ is the graph of

$$y = a\left(x + \frac{b}{2a}\right)^2 + \frac{4ac - b^2}{4a} = a(x + h)^2 + c',$$

where $h = b/2a$ and $c' = (4ac - b^2)/4a$. Therefore the graph of $y = ax^2 + bx + c$ is the graph of $y = ax^2$ shifted horizontally $|b/2a|$ units and vertically $|c'|$ units. The horizontal shift is left if $b/2a > 0$, right if $b/2a < 0$. The vertical shift is up if $c' > 0$, down if $c' < 0$. See Fig. 6.

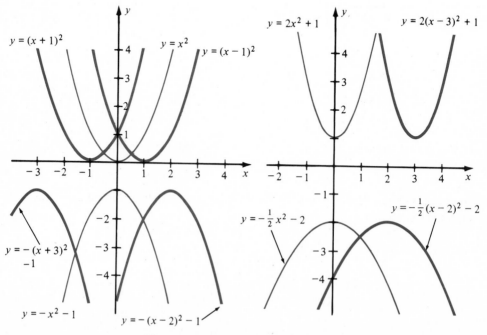

Fig. 6 Graphs of $y = a(x + h)^2 + c'$

Maxima and Minima   Completing the square enables us to write the quadratic $y = ax^2 + bx + c$ as

$$y = a(x + h)^2 + c'.$$

From this form, we can just read off some valuable information. We see that if $a > 0$, then $a(x + h)^2 \geq 0$ for all values of x. Hence the *smallest* value of y is c', and it occurs at $x = -h$. If $a < 0$, then $a(x + h)^2 \leq 0$ for all values of x. Hence the *largest* value of y is c'; again it occurs at $x = -h$. Thus the range of the function $f(x) = ax^2 + bx + c$ is $y \geq c'$ if $a > 0$, and $y \leq c'$ if $a < 0$.

■ **EXAMPLE 1** (a) Find the lowest point on the curve $y = x^2 - 6x$.
        (b) Find the highest point on the curve $y = -2x^2 - 4x + 1$.

Solution (a) Complete the square: $y = x^2 - 6x = x^2 - 6x + 9 - 9 = (x - 3)^2 - 9$.

The value of y is least at $x = 3$. The lowest point is $(3, -9)$. See Fig. 7a.

  (b) Complete the square: $y = -2x^2 - 4x + 1 = -2(x^2 + 2x) + 1$
              $= -2(x^2 + 2x + 1) + 2 + 1 = -2(x + 1)^2 + 3.$

The value of y is greatest at $x = -1$. The highest point is $(-1, 3)$. See Fig. 7b.

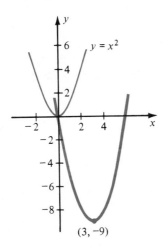

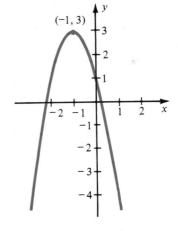

(a) Graph of $y = x^2 - 6x$ (b) Graph of $y = -2x^2 - 4x + 1$

Fig. 7   ■

■ **EXAMPLE 2** What is the largest possible area of a rectangular rug whose perimeter is 60 ft?

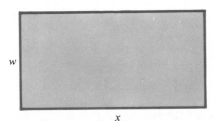

Fig. 8
perimeter $= 2x + 2w = 60$
$w = 30 - x$
area $= xw = x(30 - x)$

Solution   Let x be the length of the rug. By Fig. 8, its area is $A = x(30 - x)$. Since x is a length, $x > 0$. Since x is less than half the perimeter, $x < 30$. The problem, therefore, is to find the largest value of A for $0 < x < 30$. Complete the square:

$$A = 30x - x^2 = -(x^2 - 30x) = -(x^2 - 30x + 15^2) + 15^2 = -(x - 15)^2 + 225.$$

Obviously the largest value of A is 225, which occurs for $x = 15$, an acceptable value of x.

Answer   225 ft². ∎

In Chapters 2 and 3, we develop tools of calculus for finding maximum and minimum values of general functions. But for *quadratic* functions, no fancy methods are needed; completing the square does the trick.

EXERCISES

Graph

1 $y = 2x^2$	**2** $y = -2x^2$	**3** $y = -\frac{1}{2}x^2$
4 $y = \frac{1}{2}x^2$	**5** $y = x^2 + 3$	**6** $y = -x^2 - 3$
7 $y = 2x^2 - 1$	**8** $y = -2x^2 - 1$	**9** $y = -\frac{1}{4}x^2 + 2$
10 $y = -\frac{1}{4}x^2 - 2$.		

Graph on the indicated range (use different scales on the axes if necessary)

11 $y = 0.1x^2, \quad 0 \le x \le 100$ **12** $y = -x^2, \quad -0.1 \le x \le 0$.

Graph and locate the highest (or lowest) point

13 $y = x^2 - 4x + 1$	**14** $y = x^2 + 2x - 5$	**15** $y = x^2 + x + 1$
16 $y = x^2 - x + 1$	**17** $y = -x^2 - 2x$	**18** $y = -x^2 + 2x$
19 $y = -x^2 - 4x - 3$	**20** $y = -x^2 + 4x + 1$	**21** $y = 2x^2 - 6x + 1$
22 $y = 2x^2 + 4x$	**23** $y = 3x^2 + 12x - 8$	**24** $y = -3x^2 + 12x - 8$
25 $y = -2x^2 + 8x - 10$	**26** $y = -2x^2 + 12x$	**27** $y = -4x^2 + x$
28 $y = 2x^2 + 2x + 2$	**29** $y = 2x^2 - 3x$	**30** $y = x^2 - 6x + 2$
31 $y = x^2 + x - 4$	**32** $y = 3x^2 + 3x$	**33** $y = -x^2 + x - 2$
34 $y = -x^2 - 2x$	**35** $y = -2x^2 + x$	**36** $y = -2x^2 - 6x + 1$.

37   Show that the graph of $y = ax^2 + bx$ passes through the origin for all choices of a and b.

38   For what value of c does the lowest point of the graph of $y = x^2 + 6x + c$ fall on the x-axis?

39   Under what conditions is the lowest point of the graph of $y = x^2 + bx + c$ on the y-axis?

40   Show that the rectangle of given perimeter p and largest area is a square.

41   Show that for $0 \le x \le 1$, the product $x(1 - x)$ never exceeds $\frac{1}{4}$.

42   A farmer will make a rectangular pen with 100 ft of fencing, using part of a wall of his barn for one side of the pen. What is the largest area he can enclose?

43   A 4-ft line is drawn across a corner of a rectangular room, cutting off a triangular region. Show that its area cannot exceed 4 ft². [*Hint*   Use the Pythagorean theorem and work with A^2.]

44   A rectangular solid has a square base, and the sum of its 12 edges is 4 ft. Show that its total surface area (sum of the areas of its 6 faces) is largest if the solid is a cube.

7. MORE ON GRAPHING

This section starts with a kind of lazy man's guide to graphs, featuring techniques and shortcuts that can reduce the work in graphing. Some of the ideas popped up in Section 6 when we graphed quadratic functions. It is a good idea to spell them out.

Symmetry   Consider the graphs of $y = x^2$ and $y = x^3$ shown in Fig. 1. (The graph of $y = x^3$ will be discussed shortly.) These graphs possess certain symmetries. The one on the left is symmetric in the y-axis. The one on the right is symmetric in the origin, i.e., to each point of the graph corresponds an opposite point as seen through a peephole in the origin. In either case we need plot the curve only for $x \geq 0$; we obtain the rest by symmetry. Thus the work is "cut in half."

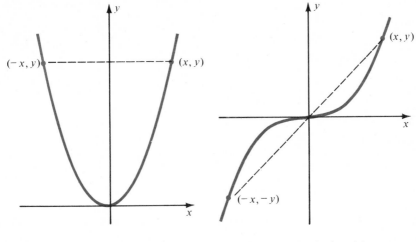

(a)   Symmetry in the y-axis   (b)   Symmetry in the origin

Fig. 1   Symmetry

When we plot $y = f(x)$, how can we recognize symmetry in advance? Look at Fig. 1a. The curve $y = f(x)$ is **symmetric in the y-axis** if for each x, the value of y at $-x$ is the same as at x; in mathematical notation, $f(-x) = f(x)$. If $f(x)$ satisfies this condition, it is called an even function.

Look at Fig. 1b. The curve $y = f(x)$ is **symmetric in the origin** if for each x, the value of y at $-x$ is the negative of the value at x, that is, $f(-x) = -f(x)$. If $f(x)$ satisfies this condition, it is called an odd function.

> An **even** function $f(x)$ is one for which $f(-x) = f(x)$.
>   The graph of an even function is symmetric in the y-axis.
> An **odd** function $f(x)$ is one for which $f(-x) = -f(x)$.
>   The graph of an odd function is symmetric in the origin.

Vertical and Horizontal Shifts   We know that if a positive constant c is added to or subtracted from $f(x)$, the graph of $y = f(x)$ is shifted up or down c units. Now let us consider horizontal shifts. How can we shift the graph of $y = f(x)$ three units to the right? More generally how can we find a function $g(x)$ for which the graph of $y = g(x)$ is precisely that of $y = f(x)$ shifted c units to the right?

Consider Fig. 2. For each point (x, y) on the curve $y = g(x)$, there corresponds a point $(x - c, y)$ on the curve $y = f(x)$. The values of y are the same. But on the first curve $y = g(x)$, on the second, $y = f(x - c)$. Conclusion: $g(x) = f(x - c)$. This makes sense. If x represents time, then the value of g "now" is the same as the value of f at c time units ago.

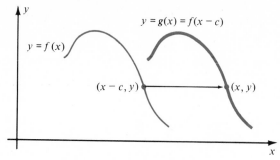

Fig. 2   Horizontal shift $(c > 0)$

The same reasoning shows that the graph of $y = f(x + c)$ is the graph of $y = f(x)$ shifted c units to the left.

Let $c > 0$. The graph of
$$\left.\begin{array}{l} y = f(x) + c \\ y = f(x) - c \\ y = f(x - c) \\ y = f(x + c) \end{array}\right\} \text{is the graph of } y = f(x) \text{ shifted } c \text{ units} \left\{\begin{array}{l} \text{up} \\ \text{down} \\ \text{right} \\ \text{left.} \end{array}\right.$$

Stretching and Reflecting   If $c > 0$, the graph of $y = cf(x)$ is obtained from that of $y = f(x)$ by stretching by a factor of c in the y-direction. Each point (x, y) is replaced by (x, cy). Note: "stretching" by a factor less than one is interpreted as shrinking (Fig. 3).

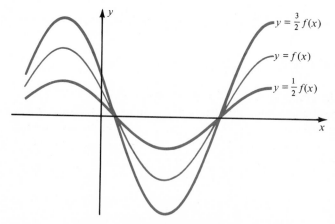

Fig. 3   Stretching in the y-direction

The graph of $y = -f(x)$ is obtained by reflecting the graph of $y = f(x)$ in the x-axis (turning it upside down). That is because each point (x, y) is replaced by the point $(x, -y)$. See Fig. 4.

Free Information   Very often you can get valuable information about a graph for free, just by looking at the equation involved. It is good practice not to start right

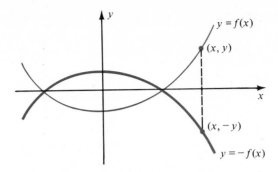

Fig. 4 Reflection in the x-axis

off plotting points, but to take a minute to think. Look for symmetry, shifts, and stretching and reflecting. In addition, there are some further points you can check.

Domain Is y defined for all real x or is there some restriction on x? For example, $y = \sqrt{1 - x^2}$ is defined only for $|x| \leq 1$ and $y = 1/(x - 1)(x - 4)$ is not defined for $x = 1$ or $x = 4$.

Range Is there some limitation on y? For example, if $y = 1/(1 + x^2)$, then by inspection, $0 < y \leq 1$. The graph does not extend above the level $y = 1$ or below the level $y = 0$.

Sign of y Can you tell where $y > 0$ or $y < 0$? For example $y = 1/x$ is positive for $x > 0$ and negative for $x < 0$. Also y is never 0.

Increasing or decreasing? For example, $1/x$ decreases as x increases through positive values.

You will not always be able to check all these points. At least see what you can find easily.

■ **EXAMPLE 1** Plot $y = \dfrac{1}{1 + x^4}$.

Solution The graph is defined for all x. Obviously $0 < y \leq 1$. In fact $y = 1$ only at $x = 0$, so the highest point on the curve is $(0, 1)$. As x increases through positive values, y decreases toward 0. The function is even, so the graph is symmetric in the y-axis. This free information alone is enough for a fairly good idea of the curve. Plotting a few points helps fix the shape (Fig. 5).

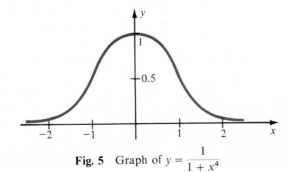

Fig. 5 Graph of $y = \dfrac{1}{1 + x^4}$ ■

Some Important Graphs Let us study the graphs of $y = x^n$, where n is any positive integer. We are already familiar with $y = x$ and $y = x^2$, so we start with $y = x^3$.

Since $f(x) = x^3$ is an odd function, its graph is symmetric in the origin. Therefore we concentrate on the right half of the graph, where $x \geq 0$. As x starts from 0 and increases, x^3 also starts from 0 and increases. Let us compute some values for $0 \leq x \leq 1$:

x	0.0	0.1	0.2	0.3	0.4	0.5	0.6	0.7	0.8	0.9	1.0
x^3	0.0	0.001	0.008	0.027	0.064	0.125	0.216	0.343	0.512	0.729	1.0

This table gives a pretty good idea of the graph from 0 to 1. See Fig. 6a. The curve is quite flat near $x = 0$, much flatter than the graph of $y = x^2$. (See Fig. 8 below.) Now consider some larger values of x:

x	0	1	2	3	4	5	6	7	8	9	10
x^3	0	1	8	27	64	125	216	343	512	729	1000

The graph rises very fast as x increases (Fig. 6b). We now have a good idea of the graph for $x \geq 0$.

By symmetry, we sketch the complete graph (Fig. 6c).

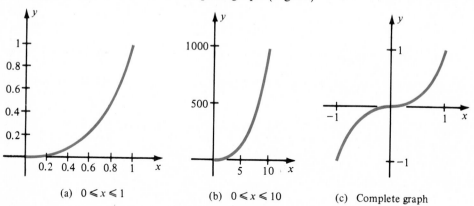

(a) $0 \leqslant x \leqslant 1$ (b) $0 \leqslant x \leqslant 10$ (c) Complete graph

Fig. 6   Graph of $y = x^3$

We obtain the graph of $y = x^4$ in a similar way. This time $f(x) = x^4$ is an even function, so its graph is symmetric in the y-axis (Fig. 7).

It is interesting to compare the graphs of x^2, x^3, and x^4 for small x and for large x. When x is small, x^2 is very small, x^3 is even smaller, and x^4 is even smaller yet (Fig. 8a). But when x is large, x^2 is very large, x^3 is even larger, and x^4 is even larger yet (Fig. 8b). The accompanying Tables 1 and 2 and graphs (Fig. 8) show this clearly.

The graphs of $y = x^5$, $y = x^7$, $y = x^9, \cdots, y = x^{2n+1}, \cdots$, where the exponent is odd, are all more or less like the graph of $y = x^3$. They are increasingly flat near $x = 0$ and grow increasingly rapidly for x large. They are all symmetric in the origin (Fig. 9). Similar remarks apply to the graphs of $y = x^4$, $y = x^6$, $y = x^8$, $y = x^{10}, \cdots$, $y = x^{2n}, \cdots$, where the exponent is even (Fig. 10).

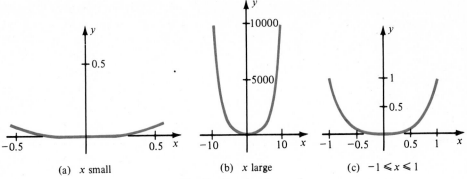

(a) *x* small   (b) *x* large   (c) $-1 \leqslant x \leqslant 1$

Fig. 7   Graph of $y = x^4$

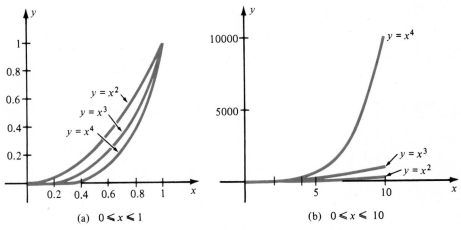

(a) $0 \leqslant x \leqslant 1$ (b) $0 \leqslant x \leqslant 10$

Fig. 8   Graphs of $y = x^2$, $y = x^3$, $y = x^4$

Table 1   Powers of small *x*

x	0.0	0.1	0.2	0.3	0.4	0.5	0.6	0.7	0.8	0.9	1.0
x^2	0.00	0.01	0.04	0.09	0.16	0.25	0.36	0.49	0.64	0.81	1.00
x^3	0.000	0.001	0.008	0.027	0.064	0.125	0.216	0.343	0.512	0.729	1.000
x^4	0.0000	0.0001	0.0016	0.0081	0.0256	0.0625	0.1296	0.2401	0.4096	0.6561	1.0000

Table 2   Powers of large *x*

x	0	1	2	3	4	5	6	7	8	9	10
x^2	0	1	4	9	16	25	36	49	64	81	100
x^3	0	1	8	27	64	125	216	343	512	729	1000
x^4	0	1	16	81	256	625	1296	2401	4096	6561	10000

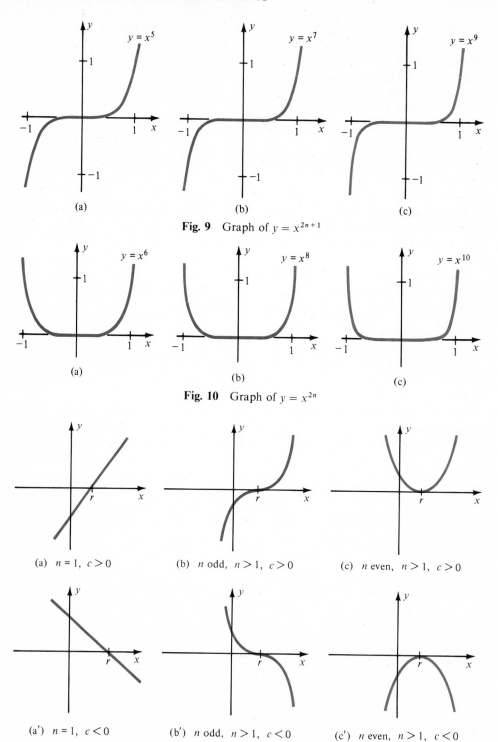

Fig. 9   Graph of $y = x^{2n+1}$

Fig. 10   Graph of $y = x^{2n}$

(a) $n = 1,\ c > 0$

(b) n odd, $n > 1,\ c > 0$

(c) n even, $n > 1,\ c > 0$

(a′) $n = 1,\ c < 0$

(b′) n odd, $n > 1,\ c < 0$

(c′) n even, $n > 1,\ c < 0$

Fig. 11   Graph of $y = c(x - r)^n$

It is important to remember that the graph of $y = x^n$ crosses the x-axis at $x = 0$ if n is odd, but does not cross if n is even. Algebraically this is obvious: if n is odd, x^n changes sign as x changes from negative to positive; if n is even, x^n is positive everywhere except at $x = 0$. Note that for $n = 1$, the graph of $y = x^n = x$ crosses the axis sharply, but for $n = 3, 5, 7, \cdots$ the graph of $y = x^n$ slithers across.

The graph of $y = c(x - r)^n$ is similar to that of $y = x^n$, except it touches the x-axis at r instead of 0, and is stretched or contracted by a factor $|c|$ in the y-direction (and reflected in the x-axis if $c < 0$). See Fig. 11.

Notation The graph of $y = x^n$ rises indefinitely (without limit) as x increases without limit. This behavior is commonly abbreviated by writing $x^n \longrightarrow \infty$ as $x \longrightarrow \infty$.

If n is even, the graph also rises indefinitely as x becomes more and more negative without limit. This is abbreviated by writing $x^n \longrightarrow \infty$ as $x \longrightarrow -\infty$. If n is odd, the graph falls indefinitely as x becomes more and more negative. We write $x^n \longrightarrow -\infty$ as $x \longrightarrow -\infty$.

We read " x approaches infinity" for $x \longrightarrow \infty$. If a is a real number, we shall also say " x approaches a " and write $x \longrightarrow a$ when x takes values nearer and nearer to a.

Graph of $1/x^n$ Finally, let us look at the graphs of $y = 1/x^n$. The function $1/x^n$ is even or odd, according as n is even or odd. Hence, in either case, it is enough to sketch the graph for $x > 0$.

All the curves pass through $(1, 1)$. Since $x^n \longrightarrow \infty$ as $x \longrightarrow \infty$, the reciprocal $1/x^n \longrightarrow 0$ as $x \longrightarrow \infty$. The larger n is, the faster $1/x^n \longrightarrow 0$. As $x \longrightarrow 0$ through positive values, $x^n \longrightarrow 0$, hence $1/x^n \longrightarrow \infty$. This gives the general shape of the graph for $x > 0$. We complete the graph using even or odd symmetry (Fig. 12). The graph of $y = 1/(x - r)^n$ is a shifted version of that of $y = 1/x^n$, centered about the vertical axis $x = r$. See Fig. 13.

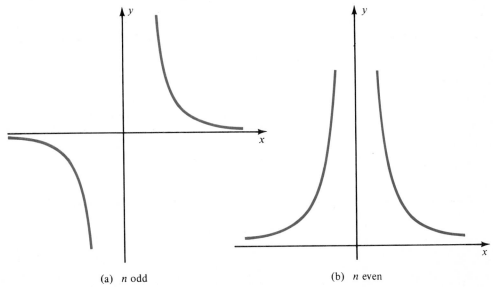

(a) n odd   (b) n even

Fig. 12   Graph of $y = \dfrac{1}{x^n}$

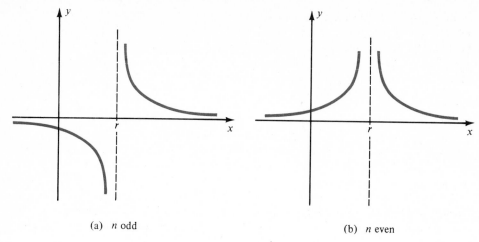

(a) n odd   (b) n even

Fig. 13   Graph of $y = \dfrac{1}{(x-r)^n}$

EXERCISES

1   Which of the functions are even?

$$x^2, \quad x^3, \quad x^4, \quad \frac{1}{x^2+1}, \quad \frac{1}{x^3+1}, \quad \frac{x}{x^2+1}, \quad \left(\frac{1}{x^3+x}\right)^5, \quad x^3+x^2+1.$$

2   (cont.) Which are odd?

3   Plot on the same graph

$$y = x, \quad y = 3x, \quad y = x - 1, \quad y = 3(x-1), \quad y = x+4, \quad y = -3(x+4).$$

4   Plot on the same graph

$$y = x^2, \quad y = 2x^2, \quad y = 2(x-1)^2, \quad y = 2(x+1)^2, \quad y = -2(x-3)^2.$$

5   Compute $\frac{1}{2}[f(x) + f(-x)]$ if $f(x)$ is $\quad x^3 + 1, \quad \dfrac{1}{x-3}, \quad \dfrac{x^2}{x+1}.$

Show in each case that the answer is an even function.

6   (cont.) Prove that for any function $f(x)$, the function $g(x) = \frac{1}{2}[f(x) + f(-x)]$ is even.

7   (cont.) Prove that for any function $f(x)$, the function $g(x) = \frac{1}{2}[f(x) - f(-x)]$ is odd.

8   (cont.) Prove that any function can be expressed as the sum of an odd function and an even function.

Graph

9 $y = \frac{1}{4}x^4$ **10** $y = -x^4$ **11** $y = \frac{1}{8}x^3$

12 $y = -x^3 + 1$ **13** $y = -x^6$ **14** $y = x^3 - 2$

15 $y = x^4 - 1$ **16** $y = x^5 + 1$ **17** $y = -x^5$

18 $y = \frac{1}{16}x^5$ **19** $y = (x+1)^3$ **20** $y = (x-1)^4$

21 $y = -(x+1)^3$ **22** $y = -(x-1)^4$ **23** $y = (x-1)^3 + 1$

24 $y = -(x+1)^4 - 1$ **25** $y = \frac{1}{3}(x+2)^4$ **26** $y = -\frac{1}{2}(x+2)^4 - 4$

27 $y = (x - \frac{1}{2})^5 - 1$ **28** $y = (x + \frac{1}{2})^6 - 2.$

Graph accurately

29 $y = (x-2)^3, \quad 1 \le x \le 3$ **30** $y = -(x+1)^3, \quad -2 \le x \le 0$

31 $y = -(x-1)^4, \quad 0 \le x \le 2$ **32** $y = \frac{1}{2}(x-2)^4, \quad 1 \le x \le 3.$

Graph roughly; plot for the given values of x

33 $y = x^3 - 4x$, $x = 0, \pm 1, \pm 2, \pm 3$
34 $y = x^3 + 2x^2 - x - 2$, $x = -3, -2, -1, 0, 1, 2$
35 $y = \frac{1}{3}x^3 - \frac{1}{2}x^2 - 6x + 1$, $x = -3, -2, -1, 0, 1, 2, 3, 4$
36 $y = x^3 - 3x + 3$, $x = -3, -2, -1, 0, 1, 2.$

Graph

37 $y = \dfrac{1}{x - 4}$ **38** $y = \dfrac{-1}{(x + 2)^2}$ **39** $y = \dfrac{1}{(x + 1)^2} - 2$ **40** $y = \dfrac{1}{4x - 3}.$

8. POLYNOMIALS AND RATIONAL FUNCTIONS

It is pretty hard to say much about the graph of a general polynomial

$$f(x) = a_n x^n + a_{n-1} x^{n-1} + \cdots + a_1 x + a_0, \qquad a_n \neq 0.$$

Given one, about the most we can do is plot a bunch of points and hope for the best. We shall show later (Chapter 10) that the graph of $f(x)$ changes direction at most $n - 1$ times, which is a little help. This means for instance that if Fig. 1 represents the graph of a polynomial, then its degree must be seven or more.

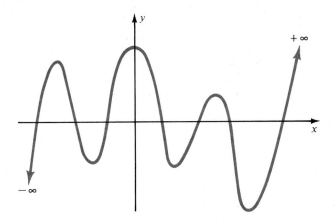

Fig. 1   The graph changes direction six times.

We can, however, say something about the behavior of the graph as $x \longrightarrow \infty$ or $x \longrightarrow -\infty$. We write

$$f(x) = a_n x^n + a_{n-1} x^{n-1} + \cdots + a_1 x + a_0$$

$$= a_n x^n \left(1 + \frac{a_{n-1}}{a_n x} + \cdots + \frac{a_1}{a_n x^{n-1}} + \frac{a_0}{a_n x^n} \right).$$

For $|x|$ very large, the quantity in parentheses is very close to 1, so $f(x)$ is about the size of $a_n x^n$. Hence for $|x|$ very large, the graph of $y = f(x)$ is like the graph of $y = a_n x^n$. As $x \longrightarrow \infty$ or $x \longrightarrow -\infty$, it either zooms up or down, depending on the sign of a_n and (for $x \longrightarrow -\infty$) whether n is even or odd.

Factored Polynomials   Polynomials of the form

$$f(x) = (x - r_1)(x - r_2) \cdots (x - r_n)$$

are particularly easy to graph. Each r_i is a **zero** of $f(x)$, that is, $f(r_i) = 0$. There are no other zeros of $f(x)$ because the product

$$(x - r_1)(x - r_2) \cdots (x - r_n)$$

can equal 0 only if one of the factors equals 0, that is, only if x is one of the numbers $r_1, r_2, \cdots, r_n$.

Let us study factored cubic polynomials ($n = 3$). There are three possible cases to consider.

Case 1 $f(x) = (x - r)^3$. We have already discussed this cubic. Its graph is a shifted version of the graph of $y = x^3$. See Fig. 11b, p. 32.

Case 2 $f(x) = (x - r)^2(x - s)$, where $r \neq s$. First of all, $f(x) = 0$ at $x = r$ and $x = s$. Now suppose $x \neq r$. Then $(x - r)^2 > 0$, so the sign of $f(x)$ is the same as the sign of $(x - s)$. We conclude that $f(x)$ changes sign as x passes through s. However the function does not change sign as x passes through r. We summarize the signs of $f(x)$ in Fig. 2a for the case $r < s$. Then we sketch the graph (Fig. 2b), using the additional information that $f(x) \longrightarrow \infty$ as $x \longrightarrow \infty$ and $f(x) \longrightarrow -\infty$ as $x \longrightarrow -\infty$. For $r > s$, the situation is similar; the graph is shown in Fig. 3.

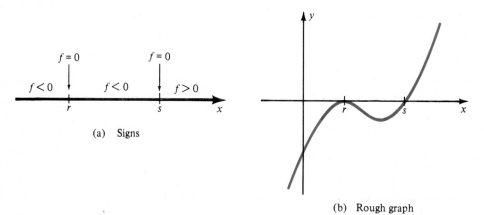

(a)   Signs

(b)   Rough graph

Fig. 2   Graph of $y = (x - r)^2(x - s)$, $r < s$

Case 3 $f(x) = (x - r)(x - s)(x - t)$ with $r < s < t$. As x passes through each zero, the sign of f changes. As before $f(x) \longrightarrow \infty$ as $x \longrightarrow \infty$ and $f(x) \longrightarrow -\infty$ as $x \longrightarrow -\infty$. The rough graph is now evident (Fig. 4).

We are ready to examine the general factored polynomial of the form

$$f(x) = (x - r_1) \cdots (x - r_n).$$

It is important how many times each factor $(x - r_i)$ occurs, so we write

$$f(x) = (x - r_1)^{m_1} \cdots (x - r_s)^{m_s}, \qquad r_1 < r_2 < \cdots < r_s,$$

to show clearly each zero r_i with its **multiplicity** m_i.

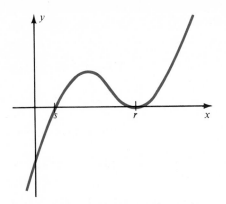

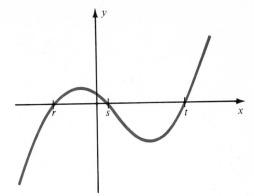

Fig. 3 Graph of $y = (x - r)^2(x - s)$, $r > s$

Fig. 4 Graph of $y = (x - r)(x - s)(x - t)$, $r < s < t$

Let us choose one of the zeros r of multiplicity m and study the graph of $y = f(x)$ near $x = r$. We write $f(x) = (x - r)^m h(x)$, lumping all the other factors together in $h(x)$. Now $h(r) \neq 0$ as can be seen from the factored form above, and near r the function $h(x)$ does not change sign. Thus, for x near r, the graph of $y = f(x)$ looks pretty much like the graph of $y = h(r)(x - r)^m$; it must be close to one of the six types shown in Fig. 11, p. 32. The graph crosses the x-axis if m is odd, but does not cross if m is even.

We shall show later (Chapter 10) by calculus that there is a single peak (or pit) between successive zeros of a completely factored polynomial.

■ **EXAMPLE 1** Sketch the graph of $f(x) = (x + 1)^3 x^2 (x - 2)(x - 3)^5$.

Solution Since the degree of $y = f(x)$ is 11, an odd number, $y \longrightarrow -\infty$ as $x \longrightarrow -\infty$ and $y \longrightarrow \infty$ as $x \longrightarrow \infty$. Taking the multiplicities into account, we find the various signs and sign changes (Fig. 5). Next we look closely at $f(x)$ near each zero (Fig. 6). Now we can sketch the rough shape of the graph (Fig. 7).

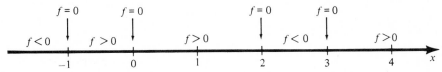

Fig. 5 Signs of $f(x) = (x + 1)^3 x^2 (x - 2)(x - 3)^5$

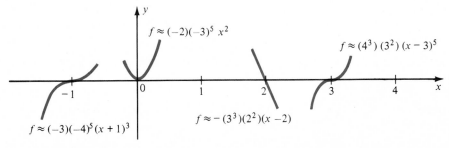

Fig. 6 Behavior of $f(x) = (x + 1)^3 x^2 (x - 2)(x - 3)^5$ near its zeros

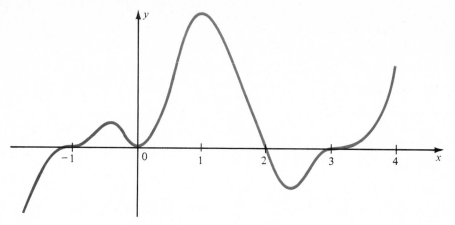

Fig. 7   Rough graph of $y = (x + 1)^3 x^2 (x - 2)(x - 3)^5$ ■

Rational Functions   A rational function is the quotient of two polynomials:

$$r(x) = \frac{p(x)}{q(x)}.$$

Let us agree that all rational functions will be expressed in lowest terms, that is, the numerator and denominator will have no common polynomial factors. For example, we shall write

$$r(x) = \frac{1}{x + 1}, \quad \text{not} \quad r(x) = \frac{x - 1}{x^2 - 1}.$$

Graphs of rational functions can be even more complicated than those of polynomials. Nevertheless many rational functions can be graphed without too much trouble. One important bit of information is that the behavior of $r(x)$ as $x \longrightarrow \pm\infty$ is predictable.

Suppose

$$r(x) = \frac{a_m x^m + a_{m-1} x^{m-1} + \cdots + a_1 x + a_0}{b_n x^n + b_{n-1} x^{n-1} + \cdots + b_1 x + b_0}, \quad a_m \neq 0, \quad b_n \neq 0.$$

Then as $|x| \longrightarrow \infty$,

$$|r(x)| \longrightarrow \infty \qquad \text{if} \quad m > n,$$
$$r(x) \longrightarrow 0 \qquad \text{if} \quad m < n,$$
$$r(x) \longrightarrow \frac{a_m}{b_n} \qquad \text{if} \quad m = n.$$

Thus, if the degree of the numerator exceeds the degree of the denominator (top-heavy case), then $|r(x)| \longrightarrow \infty$ as $|x| \longrightarrow \infty$. In the opposite (bottom-heavy) case, $r(x) \longrightarrow 0$ as $|x| \longrightarrow \infty$. If the degrees of the numerator and denominator are equal, then $r(x)$ tends to a finite non-zero number, the quotient of the leading coefficients.

To prove this assertion, we use the technique introduced at the beginning of this section. We write the numerator as $a_m x^m h(x)$ and the denominator as $b_n x^n k(x)$, where $h(x) \longrightarrow 1$ and $k(x) \longrightarrow 1$ as $|x| \longrightarrow \infty$. Then for $|x|$ large,

$$r(x) = \frac{a_m x^m h(x)}{b_n x^n k(x)} \approx \frac{a_m x^m}{b_n x^n} = \frac{a_m}{b_n} x^{m-n},$$

and the conclusion follows.

Notation   The symbol $\approx$ is shorthand for "approximately equals."

Examples   Assume $|x|$ is large.

1. $r(x) = \dfrac{x^5 + 3x}{x^2 + 12} \approx \dfrac{x^5}{x^2} = x^3,$

hence $r(x) \longrightarrow \infty$ as $x \longrightarrow \infty$, and $r(x) \longrightarrow -\infty$ as $x \longrightarrow -\infty$.

2. $r(x) = \dfrac{6x^2 + 7x - 3}{2x^3 + x^2 + 1} \approx \dfrac{6x^2}{2x^3} = \dfrac{3}{x},$

hence $r(x) \longrightarrow 0$ as $x \longrightarrow \pm\infty$.

3. $r(x) = \dfrac{2x^3 + 1}{5x^3 - 4x^2 - x - 7} \approx \dfrac{2x^3}{5x^3} = \dfrac{2}{5},$

hence $r(x) \longrightarrow \frac{2}{5}$ as $x \longrightarrow \pm\infty$.

■ **EXAMPLE 2**   Graph $y = \dfrac{x}{x^2 + 1}$.

Solution   The function is odd; we need only plot the graph for $x \geq 0$. Obviously $y = 0$ for $x = 0$ and $y > 0$ for $x > 0$. Furthermore, by the preceding criterion, $y \longrightarrow 0$ as $x \longrightarrow \infty$.

One other piece of information is helpful: near $x = 0$, the denominator $x^2 + 1 \approx 1$, hence $y \approx x$. This suggests that the graph passes through $(0, 0)$ at an angle of $45°$.

We sketch these clues in Fig. 8a. Apparently the graph starts upward from the origin at $45°$, but soon starts to decline, ultimately dying out toward 0. Plotting a few points confirms this (Fig. 8b).

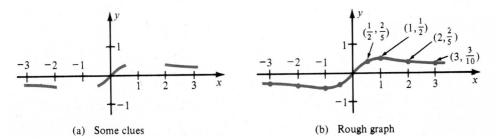

(a) Some clues   (b) Rough graph

Fig. 8   Graph of $y = \dfrac{x}{x^2 + 1}$

Factored Rational Functions   It is fairly easy to sketch the graph of a rational function that is completely factored into linear factors:

$$y = \frac{(x - r_1)^{m_1}(x - r_2)^{m_2} \cdots (x - r_h)^{m_h}}{(x - s_1)^{n_1}(x - s_2)^{n_2} \cdots (x - s_k)^{n_k}}.$$

We assume this expression is in lowest terms, hence none of the numbers r_i is the same as any of the numbers s_j.

Suppose r is one of the zeros of the numerator. Write

$$y = g(x)(x - r)^m,$$

where $g(x)$ is composed of all the other factors of the numerator and denominator lumped together. Note that $g(r) \neq 0$. If $g(r) = c$, then near $x = r$ the graph is like that of $y = c(x - r)^m$. Similarly, near a zero s of the denominator, the graph is like that of $y = d/(x - s)^n$.

We have further information too: we can find the behavior of y as $x \longrightarrow \pm\infty$, and we know that y changes sign at r_i or s_j if the corresponding exponent m_i or n_j is odd.

■ **EXAMPLE 3**   Graph $y = \dfrac{x - 1}{x^2}$.

Solution   First collect some free information. The graph is undefined at $x = 0$. Otherwise the sign of y is the same as that of $x - 1$:

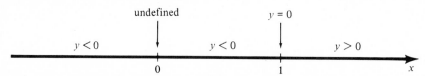

Further immediate information: the rational function is bottom-heavy, hence $y \longrightarrow 0$ as $x \longrightarrow \pm\infty$.

Now look at how the function behaves near the critical point $x = 0$. Write

$$y = \frac{g(x)}{x^2}, \quad \text{where} \quad g(x) = x - 1.$$

(a) Behavior near critical points   (b) Rough graph

Fig. 9   Graph of $y = \dfrac{x - 1}{x^2}$

Since $g(0) = -1$, $\qquad y \approx -\dfrac{1}{x^2}$, near $x = 0$.

Combine all the information (Fig. 9a); it suggests the desired graph (Fig. 9b). ∎

■ **EXAMPLE 4** Graph $y = \dfrac{x^2}{(x + 2)(x - 1)}$.

Solution The graph is undefined at $x = -2$ and $x = 1$. It meets the x-axis at $x = 0$ and nowhere else. The function is positive for large values of x and changes sign at $x = 1$ and $x = -2$, but not at $x = 0$:

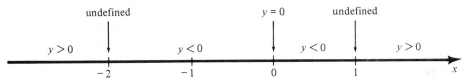

Next, look at the behavior of the function near the critical points $x = 0, -2$, and 1. Near $x = 0$, the curve is like

$$y = \frac{1}{(0 + 2)(0 - 1)} x^2 = -\frac{x^2}{2}.$$

Near, $x = -2$ the curve is like $\qquad y = \dfrac{(-2)^2}{(-2 - 1)}\dfrac{1}{x + 2} = -\dfrac{4}{3}\dfrac{1}{x + 2}.$

Near $x = 1$, it is like $\qquad y = \dfrac{(1)^2}{(1 + 2)}\dfrac{1}{x - 1} = \dfrac{1}{3}\dfrac{1}{x - 1}.$

Finally, since $\qquad y = \dfrac{x^2}{(x + 2)(x - 1)} = \dfrac{x^2}{x^2 + x - 2},$

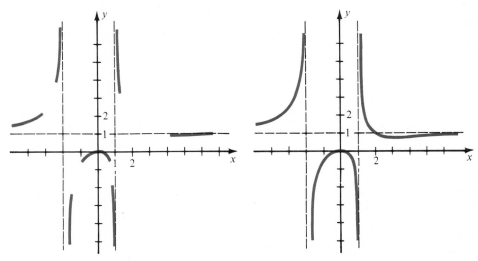

(a) Behavior near critical points (b) Rough graph

Fig. 10 Graph of $y = \dfrac{x^2}{(x + 2)(x - 1)}$

our criterion shows that $y \longrightarrow 1$ as $x \longrightarrow \pm\infty$. One thing is not quite obvious however: does $y \longrightarrow 1$ from above or from below? For large positive values of x, the denominator is greater than the numerator, so y is below 1. Hence $y \longrightarrow 1$ from below as $x \longrightarrow \infty$. For large negative values of x, it is just the other way around: $y \longrightarrow 1$ from above as $x \longrightarrow -\infty$.

Sketch all this information (Fig. 10a). A rough graph is suggested (Fig. 10b). ■

Two more rational functions are graphed in Fig. 11. A glance at Fig. 11a shows that the lines $x = 1$, $x = -1$, and $y = 1$ play a special role. These lines are called asymptotes of the graph. In general the line $x = a$ is called a **vertical asymptote** of the graph $y = f(x)$ if $|f(x)| \longrightarrow \infty$ as $x \longrightarrow a$. A non-vertical line L is called an **asymptote** of a graph if the vertical distance between the line and the graph approaches 0 as $x \longrightarrow \infty$ or $x \longrightarrow -\infty$ (or both), for example, the line $y = 1$ in Fig. 10b.

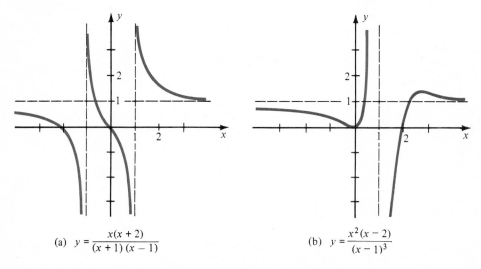

(a) $y = \dfrac{x(x + 2)}{(x + 1)(x - 1)}$

(b) $y = \dfrac{x^2(x - 2)}{(x - 1)^3}$

Fig. 11   Graphs of rational functions

EXERCISES

Draw a line graph of the signs and sign changes; then sketch the graph of $y = f(x)$

1 $f(x) = x(x - 1)(x - 2)$
2 $f(x) = (x + 2)x(x - 2)$
3 $f(x) = (x + 2)(x + 1)(x - 1)$
4 $f(x) = (x + 2)(x - 1)(x - 2)$
5 $f(x) = x^2(x - 1)$
6 $f(x) = (x - 1)(x - 2)^2$
7 $f(x) = -(x + 1)(x - 1)^2$
8 $f(x) = -x^2(x - 1)$
9 $f(x) = \frac{1}{6}(x - 1)(x - 2)(x - 3)(x - 4)$
10 $f(x) = \frac{1}{24}x(x - 2)(x - 3)(x - 4)$
11 $f(x) = x^2(x^2 - 1)$
12 $f(x) = \frac{1}{4}(x - 1)^2(x^2 - 4)$
13 $f(x) = -\frac{1}{4}(x + 1)^2(x^2 - 4)$
14 $f(x) = -\frac{1}{12}x^2(x - 2)(x - 3)$
15 $f(x) = x(x - 1)^3$
16 $f(x) = \frac{1}{8}(x^2 - 1)^2$.

Solve the inequality

17 $(x - 3)(x - 5)(x - 8) > 0$
18 $(x + 1)(x - 2)^2 < 0$
19 $x^4 - 5x^2 + 4 > 0$
20 $(x - 3)^2(x - 4)^5(x - 5)^6 < 0$.

Graph

21 $y = \dfrac{1}{x^2} - 1$ **22** $y = \dfrac{1}{x} + 3$ **23** $y = -\dfrac{1}{x} + 1$

24 $y = -\dfrac{1}{x^2} - 2$ **25** $y = \dfrac{x}{5x - 3}$ **26** $y = \dfrac{2}{1 + x^4}$

27 $y = \dfrac{-1}{4 + x^2}$ **28** $y = \dfrac{-x}{3x + 7}$ **29** $y = \dfrac{x^2}{x + 1}$

30 $y = \dfrac{x - 1}{x^2}$.

Describe the behavior of $r(x)$ as $x \longrightarrow \infty$

31 $r(x) = \dfrac{1}{x + 1} - \dfrac{1}{x - 1}$ **32** $r(x) = x^2 - \dfrac{1}{x^2}$.

Graph

33 $y = \dfrac{(x + 1)(x - 1)}{x^3}$ **34** $y = \dfrac{x^3}{(x + 1)(x - 1)}$ **35** $y = \dfrac{(x + 2)^2}{x^3}$

36 $y = \dfrac{x^3}{(x - 1)^2}$ **37** $y = \dfrac{-x^2}{(x + 1)^2}$ **38** $y = \dfrac{x^3}{(x + 1)^3}$

39 $y = \dfrac{x}{x + 3} + 2$ **40** $y = \dfrac{3x^2}{(x + 1)^2} - 2$ **41** $y = \dfrac{x(x - 1)}{(x + 1)(x - 2)}$

42 $y = \dfrac{(x + 2)(x - 3)}{(x + 1)(x - 2)}$ **43** $y = \dfrac{(x + 2)x(x - 2)}{(x + 1)(x - 1)}$ **44** $y = \dfrac{(x + 1)(x - 1)}{(x + 2)x(x - 2)}$

45 $y = \dfrac{x^4 + 2}{x(x - 1)(2x^2 + 5)}$ **46** $y = \dfrac{x^2 + 2}{(x - 1)(x + 1)(x^2 + 1)}$

47 $y = \dfrac{x^2 + 1}{x(x + 1)(x^2 + 4)}$ **48** $y = \dfrac{(x - 2)^3}{(x + 1)^2}$.

49 Suppose $r(x) = f(x)/g(x)$ is expressed in lowest terms and $\deg f(x) = 1 + \deg g(x)$. Why does the graph of $y = r(x)$ have an oblique asymptote?

50 Under what circumstances does the graph of a rational function have a horizontal asymptote? vertical asymptote?

9. DISTANCE FORMULA AND APPLICATIONS

Given two points (x_1, y_1) and (x_2, y_2) in the coordinate plane, what is the distance between them? We show several cases in Fig. 1. In each case we introduce the auxilliary point (x_2, y_1), forming a right triangle as shown. The legs have lengths $|x_2 - x_1|$ and $|y_2 - y_1|$, so by the Pythagorean theorem,

$$d^2 = |x_2 - x_1|^2 + |y_2 - y_1|^2 = (x_2 - x_1)^2 + (y_2 - y_1)^2.$$

Distance Formula The distance between (x_1, y_1) and (x_2, y_2) is
$$\sqrt{(x_2 - x_1)^2 + (y_2 - y_1)^2}.$$

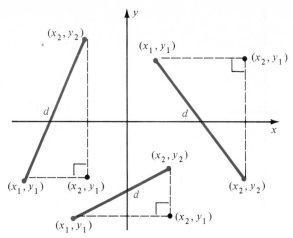

Fig. 1 Distance between points

If (x_1, y_1) and (x_2, y_2) lie on the same horizontal line or the same vertical line, the distance is $|x_1 - x_2|$ or $|y_1 - y_2|$; there is no need to introduce an auxiliary point. Nevertheless, the distance formula still yields the correct answer.

As applications of the distance formula, let us discuss a few **locus** problems, that is, problems of finding all points in the plane that satisfy certain geometric conditions.

Circles The locus of all points one unit from the origin is a circle of radius 1, called the **unit circle**. By the distance formula, a point (x, y) is on the circle if and only if $(x - 0)^2 + (y - 0)^2 = 1^2$,

$$x^2 + y^2 = 1.$$

This formula is the equation of the unit circle.

The unit circle consists of all points (x, y) in the plane that satisfy the condition

$$x^2 + y^2 = 1.$$

The equation is simply a restatement of the Pythagorean theorem for right triangles of hypotenuse 1. See Fig. 2a.

In general, the locus of all points at a fixed distance r from a point (a, b) is the circle with center (a, b) and radius r. See Fig. 2b. By the distance formula, the distance from (x, y) to (a, b) is r if and only if

$$(x - a)^2 + (y - b)^2 = r^2.$$

Equation of a Circle The circle with center (a, b) and radius r is the locus of all points satisfying the equation

$$(x - a)^2 + (y - b)^2 = r^2.$$

Parabolas A **parabola** is the locus of all points equidistant from a fixed line D and a fixed point P not on D. We call D the **directrix** and P the **focus** of the parabola. To

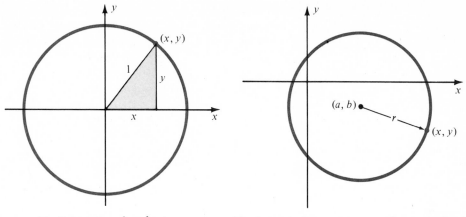

(a) Unit circle, $x^2 + y^2 = 1$ (b) Center (a, b), radius r: $(x - a)^2 + (y - b)^2 = r^2$

Fig. 2 Circles

find an equation for a parabola, set up the coordinate system (Fig. 3a) so that $P = (0, p)$ and D is the line $y = -p$. (This guarantees that $(0, 0)$ will be on the curve.)

By definition, a point (x, y) is on the parabola if and only if the distances d_1 and d_2 are equal, or equivalently, if and only if $d_1^2 = d_2^2$. See Fig. 3b. By the distance formula,

$$d_1^2 = (x - 0)^2 + (y - p)^2; \quad \text{also} \quad d_2^2 = (y + p)^2.$$

Hence $(x - 0)^2 + (y - p)^2 = (y + p)^2,$

$$x^2 + y^2 - 2py + p^2 = y^2 + 2py + p^2, \quad x^2 = 4py.$$

The steps can be read backward; therefore if $x^2 = 4py$, then (x, y) is a point of the parabola.

The equation of the parabola with focus $(0, p)$ and directrix $y = -p$ is

$$y = \frac{1}{4p} x^2.$$

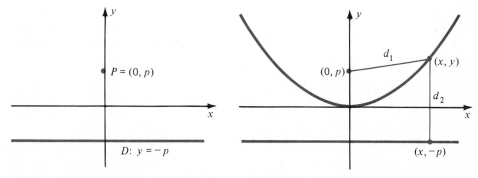

(a) Choose convenient axes   (b) The parabola is the locus of $d_1 = d_2$.

Fig. 3 The parabola.

With our choice of axes, the parabola is the graph of a quadratic function of the form $y = ax^2$. Conversely, given any quadratic function $y = ax^2$, its graph is the parabola with focus $(0, p)$ and directrix $y = -p$, where $a = 1/4p$. It follows that the graph of any quadratic function $y = ax^2 + bx + c$ is a parabola. For, by completing the square, we can write the quadratic in the form $y = a(x + h)^2 + c'$. Hence its graph is the parabola $y = ax^2$ shifted horizontally $|h|$ units and vertically $|c'|$ units, still a parabola.

Perpendicular Bisector of a Segment   The perpendicular bisector of a line segment $\overline{PQ}$ is the locus of all points equidistant from P and Q.

■ **EXAMPLE 1**   Find the perpendicular bisector of the segment $\overline{PQ}$, where $P = (1, 2)$ and $Q = (3, -5)$.

Solution   A point (x, y) is on the required bisector if and only if $d_1 = d_2$, where d_1 is the distance from (x, y) to $(1, 2)$ and d_2 is the distance from (x, y) to $(3, -5)$. It is easier to use the equivalent condition $d_1^2 = d_2^2$. By the distance formula, this condition is

$$(x - 1)^2 + (y - 2)^2 = (x - 3)^2 + (y + 5)^2,$$

$$x^2 - 2x + 1 + y^2 - 4y + 4 = x^2 - 6x + 9 + y^2 + 10y + 25,$$

$$-2x - 4y + 5 = -6x + 10y + 34, \qquad y = \tfrac{1}{14}(4x - 29).$$

Answer   The line $y = \tfrac{1}{14}(4x - 29)$.   ■

Midpoints   In Example 1, we found the perpendicular bisector of a line segment $\overline{PQ}$ by solving a locus problem. There is a different approach: If we knew the midpoint M of $\overline{PQ}$, we could write the equation of the line through M and perpendicular to $\overline{PQ}$. So, how do we find the midpoint of a line segment? The answer is easy:

Midpoint Formula   The midpoint of the line segment joining (x_0, y_0) and (x_1, y_1) is

$$(\tfrac{1}{2}(x_0 + x_1), \tfrac{1}{2}(y_0 + y_1)).$$

This formula seems reasonable: the coordinates of the midpoint are the averages of the coordinates. The proof requires a little geometry (Fig. 4).

If $(\bar{x}, \bar{y})$ is the midpoint, then $\bar{x}$ is the midpoint of the interval $x_0 \le x \le x_1$, and $\bar{y}$ is the midpoint of the interval $y_0 \le y \le y_1$. But for intervals on an axis, it is obvious that the midpoint is the average of the endpoints. Thus

$$\bar{x} = \tfrac{1}{2}(x_0 + x_1) \qquad \text{and} \qquad \bar{y} = \tfrac{1}{2}(y_0 + y_1).$$

■ **EXAMPLE 2**   Use the midpoint formula to find the perpendicular bisector of the segment joining $(1, 2)$ and $(3, -5)$.

Solution   The bisector is the line perpendicular to the segment and passing through its midpoint. The midpoint is

$$(\tfrac{1}{2}(1 + 3), \tfrac{1}{2}(2 - 5)) = (2, -\tfrac{3}{2}).$$

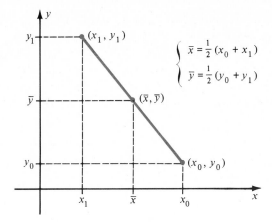

$$\begin{cases} \bar{x} = \frac{1}{2}(x_0 + x_1) \\ \bar{y} = \frac{1}{2}(y_0 + y_1) \end{cases}$$

Fig. 4   Midpoint of a segment.

The slope m of the bisector is the negative reciprocal of the slope of the segment:

$$m = -\frac{3-1}{-5-2} = \frac{2}{7}.$$

Therefore, by the point–slope form of a line, the equation of the bisector is

$$y - \left(-\tfrac{3}{2}\right) = \tfrac{2}{7}(x - 2), \qquad y = \tfrac{2}{7}x - \tfrac{4}{7} - \tfrac{3}{2} = \tfrac{1}{14}(4x - 29),$$

which agrees with the answer to Example 1.   ■

■ **EXAMPLE 3**   Show that any pair of points (x, y) and (y, x), with $x \neq y$, are mirror images of each other with respect to the line $y = x$.

Solution   "Mirror images" means that $y = x$ is the perpendicular bisector of the segment joining the two points (Fig. 5a). Check that it is. First of all the midpoint of (x, y) and (y, x) is

$$\left(\tfrac{1}{2}(x + y), \tfrac{1}{2}(y + x)\right) = (z, z) \qquad [z = \tfrac{1}{2}(x + y)],$$

which lies on the line $y = x$. Second the slope of $y = x$ is 1 and the slope of the segment is

$$\frac{x - y}{y - x} = -1,$$

the negative reciprocal. Hence $y = x$ is perpendicular to the segment and passes through its midpoint.   ■

The symmetry of the points (x, y) and (y, x) relative to the line $y = x$ is an important fact, which pops up often. For example, consider the graph of $y = 1/x$. For each point (x, y) on the graph, the reversed point (y, x) is also on the graph because x and y can be interchanged in the equation $y = 1/x$. Therefore the graph is symmetric with respect to the line $y = x$. See Fig. 5b.

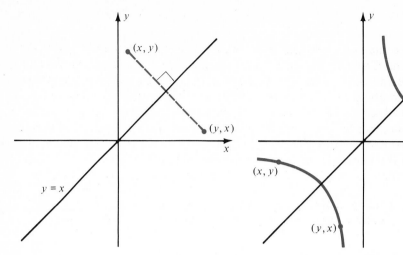

(a) (x, y) and (y, x) are mirror images in the line $y = x$.

(b) If x and y are interchangeable in the equation $y = f(x)$, then the graph is symmetric in the line $y = x$.

Fig. 5

EXERCISES

Write the equation of the circle

1 center $(1, 3)$, radius 6
2 center $(5, 12)$, radius 13
3 center $(-4, 3)$, radius 5
4 center $(-2, -1)$, radius 1
5 center $(1, 5)$, through $(0, 0)$
6 center $(3, 3)$, through $(-2, -4)$.
7 center $(-5, 2)$, tangent to y-axis
8 center $(1, 2)$, tangent to x-axis
9 diameter from $(0, 1)$ to $(3, 3)$
10 diameter from $(-2, -3)$ to $(4, 1)$.

Write the equation for the most general circle.

11 radius 3, tangent to x-axis
12 center in first quadrant, tangent to both axes
13 passing through $(0, 0)$
14 tangent to the line $y = 3$.

Do the circles intersect?

15 $x^2 + y^2 = 4$ and $(x - 3)^2 + (y + 2)^2 = 1$
16 $(x - 1)^2 + (y - 5)^2 = 9$ and $(x - 4)^2 + (y - 3)^2 = 4$.

17 Show that each point of the circle $(x - 4)^2 + y^2 = 4$ is twice as far from $(0, 0)$ as from $(3, 0)$.
18 Prove that the circle $(x - 1)^2 + (x + 2)^2 = 9$ lies inside the circle $x^2 + y^2 = 36$.

Plot

19 $x = y^2$
20 $x = y^2 - 2y + 7$.

21 Find the locus of the centers of all circles that are tangent to the x-axis and pass through the point $(0, 2)$.
22 Show that (x, y) and $(-y, -x)$ are symmetric in the line $x + y = 0$.
23 Show that $(5x, 5y)$ and $(3x + 4y, 4x - 3y)$ are symmetric in the line $2y = x$.
24 Find the mirror image of $(2, 0)$ in the line $y = 3x$.

10. TANGENTS

This section contains an introduction to one of the main problems of calculus, the tangent problem: Given a function $y = f(x)$ and a point (a, b) on its graph, find the equation of the line that is tangent to the graph at (a, b). In the following chapters we shall solve the problem completely by methods of calculus. In this section we shall solve it in a few special cases by methods of pure algebra alone.

Perhaps it appears a bit peculiar in a book that, after all, is called *Calculus* to solve calculus problems by a primitive, pre-calculus method. We offer two excuses for doing so. First, this material provides a motivation for differential calculus. Second, it contains the germ of an idea (dividing out zeros) that grows in importance later.

The Circle   Finding tangents to a circle is an easy problem in elementary geometry. The tangent is perpendicular to the radius, and that's all there is to it.

■ **EXAMPLE 1**   Find the tangent at (a, b) to the circle $x^2 + y^2 = r^2$.

Solution   We are assuming that (a, b) is a point of the circle, so $a^2 + b^2 = r^2$. The radius through (a, b) has slope b/a, so the tangent has slope $-a/b$, its negative reciprocal (Fig. 1). The tangent line has the equation (point–slope form)

$$y - b = \left(-\frac{a}{b}\right)(x - a).$$

Simplify:
$$a(x - a) + b(y - b) = 0, \qquad ax + by = r^2.$$

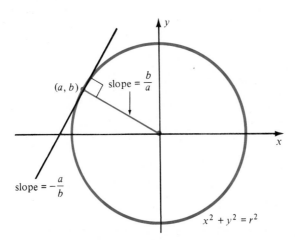

Fig. 1   Tangent to a circle   ■

Dividing out Zeros   Let us turn to the problem of finding the tangent at a given point to the graph of $y = f(x)$, where $f(x)$ is a polynomial. For example, we might consider finding the tangent at (a, a^2) to $y = x^2$.

To solve this and related problems, we recall an important tool from algebra, called the Factor Theorem.

Factor Theorem   Let $f(x)$ be a polynomial of degree n, and let $x = r$ be a zero of $f(x)$. Then

$$f(x) = (x - r)g(x),$$

where $g(x)$ is a polynomial of degree $n - 1$.

We can interpret the Factor Theorem as a statement about *dividing out zeros*: If $f(x)$ has degree n and $f(r) = 0$, then $f(x)$ is divisible by $(x - r)$; hence

$$g(x) = \frac{f(x)}{x - r}$$

is a polynomial of degree $n - 1$.

It may happen that the quotient $g(x)$ also has $x = r$ as a zero. In that case $x - r$ can be divided out of $g(x)$, so

$$\frac{g(x)}{x - r} = \frac{f(x)}{(x - r)^2}$$

is a polynomial. Then $x = r$ is called a **multiple zero** of $f(x)$.

The Method   Now for the tangent problem. We want the equation of the line tangent to the graph of the polynomial $y = f(x)$ at a given point (a, b). The most general (non-vertical) line through (a, b) is

$$y - b = m(x - a).$$

We must choose the slope m to make the line tangent.

In general the line intersects the graph not only at (a, b) but at other points $P_1, P_2, \cdots$. See Fig. 2a. If P_1 is very close to (a, b), then the line will be nearly tangent (Fig. 2b). Therefore we can think of the tangent as the line through (a, b) and P_1 when P_1 coincides with (a, b). In other words, (a, b) is a point of "multiple intersection."

Let us translate this approach into algebra. We shall show that a multiple intersection corresponds algebraically to a multiple zero.

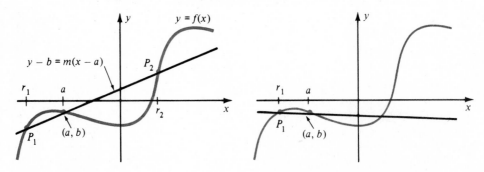

(a)   An arbitrary line through (a, b) meets   (b)   The closer P_1 is to (a, b), the nearer
       $y = f(x)$ in other points P_1, P_2, etc.   the line comes to being tangent.

Fig. 2   Method for finding tangents

We look again to Fig. 2a. The points of intersection have x-coordinates $a, r_1, r_2, \cdots$. To find these x-coordinates, we solve $y = f(x)$ and $y - b = m(x - a)$ simultaneously by eliminating y:

$$f(x) - b - m(x - a) = 0.$$

The polynomial

$$p(x) = f(x) - b - m(x - a)$$

has zeros $a, r_1, r_2, \cdots$, so by the Factor Theorem, $p(x)$ is divisible by $x - a, x - r_1,$ $x - r_2, \cdots$. To construct the tangent at (a, b), we choose m so that P_1 coincides with (a, b), or equivalently, so that $r_1 = a$. Therefore, the problem becomes purely algebraic: to choose m so that $p(x)$ has $x = a$ as a multiple zero; equivalently, so that $p(x)$ is divisible by $(x - a)^2$.

Now whatever m is, the polynomial $p(x)$ is divisible by $x - a$, so

$$\frac{p(x)}{x - a} = \frac{f(x) - b - m(x - a)}{x - a} = \frac{f(x) - b}{x - a} - m = g(x) - m$$

is a polynomial. We choose m so that *this polynomial also* is divisible by $x - a$, in other words, so that $x = a$ is a zero of $g(x) - m$:

$$g(a) - m = 0, \qquad m = g(a).$$

This proves that there is precisely one value of the slope m, namely, $m = g(a)$, for which the line through (a, b) has a multiple intersector with $y = f(x)$. The line with this slope is our solution to the tangent problem.

■ **EXAMPLE 2**   Find the tangent to $y = x^2$ at $x = a$.

Solution   The corresponding point on the graph is (a, a^2). The line with slope m through this point is

$$y - a^2 = m(x - a).$$

(a) The other intersection is
     $P = (m - a, (m - a)^2)$.

(b) It coincides with (a, a^2)
     when $m = 2a$.

Fig. 3   Tangent to $y = x^2$ at (a, a^2)

It meets $y = x^2$ where

$$x^2 - a^2 = m(x - a).$$

This quadratic equation has two solutions. One we know in advance is $x = a$; divide it out:

$$x + a = m.$$

The other is $x = m - a$. It also is equal to a if and only if $m = 2a$. So here is our desired slope. Consequently the tangent line (Fig. 3) is

$$y - a^2 = 2a(x - a), \qquad \text{that is,} \quad y = 2ax - a^2. \qquad \blacksquare$$

■ **EXAMPLE 3** Find the tangent to $y = x^3$ at $x = a$.

Solution The corresponding point on the graph is (a, a^3). The line with slope m through this point is

$$y - a^3 = m(x - a).$$

It meets $y = x^3$ where $x^3 - a^3 = m(x - a)$.

This cubic equation has at most three roots. One of them, $x = a$, is visible; divide it out:

$$\frac{x^3 - a^3}{x - a} = m, \qquad \text{that is,} \quad x^2 + ax + a^2 = m.$$

Now choose m so that $x = a$ is also a root of this last equation. The only possibility is

$$m = a^2 + a \cdot a + a^2 = 3a^2.$$

This is the required slope. Consequently the tangent line (Fig. 4) is

$$y - a^3 = 3a^2(x - a), \qquad \text{that is,} \quad y = 3a^2x - 2a^3. \qquad \blacksquare$$

Tangent to $y = 1/x$ With a little modification, our method can be applied to certain functions other than polynomials.

■ **EXAMPLE 4** Find the tangent to $y = 1/x$ at $x = a \neq 0$.

Solution The general line through $(a, 1/a)$ is

$$y - \frac{1}{a} = m(x - a).$$

It intersects $y = 1/x$ where

$$\frac{1}{x} - \frac{1}{a} = m(x - a), \qquad \text{that is,} \quad \frac{-(x - a)}{ax} = m(x - a).$$

Divide the root $x = a$ out of this equation; the result is

$$\frac{-1}{ax} = m.$$

This equation in turn has $x = a$ as a root if and only if $m = -1/a^2$. That is the slope of the tangent line (Fig. 5), so its equation is

$$y - \frac{1}{a} = -\frac{1}{a^2}(x - a), \qquad \text{that is,} \quad y = -\frac{1}{a^2}x + \frac{2}{a}. \qquad \blacksquare$$

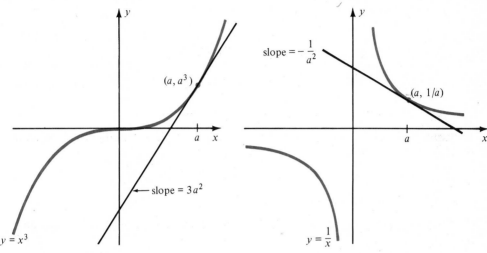

Fig. 4   Tangent to $y = x^3$ **Fig. 5**   Tangent to $y = 1/x$

We have shown that the slopes of the lines tangent to $y = x^2$, $y = x^3$, and $y = 1/x$ are $2a$, $3a^2$, and $-1/a^2$. These facts and others will be obtained in Chapter 2 using calculus. Still it is remarkable that these difficult problems can be solved by pure algebra.

EXERCISES

1 Find the locus of the centers of all circles that are tangent both to the x-axis and to the circle $x^2 + (y - 1)^2 = 1$.
2 Find the locus of the centers of all circles that are both tangent externally to the circle $x^2 + y^2 = 1$ and tangent to the line $y = -2$.
3 Show that 2 is a multiple zero of $f(x) = x^5 - 4x^4 + 3x^3 + 7x^2 - 16x + 12$.
4 Show that -1 is a multiple zero of $f(x) = 3x^4 + 7x^3 + 3x^2 - 3x - 2$.

Find the tangent to the curve

5 $y = 3x^2$ at $(-1, 3)$ 6 $y = x^2 - x$ at $(0, 0)$
7 $y = x^3 - x^2$ at $(1, 0)$ 8 $y = 2x^3 - 1$ at $(1, 1)$
9 $y = -3/x$ at $(1, -3)$ 10 $y = 1/x^2$ at $(a, 1/a^2)$.

11. MISCELLANEOUS EXERCISES

1 Given $f(x) = 3x + 1$:
   (a) Compute $f(0), f(-2), f[f(x)]$.   (b) Show that $f(a + b) = f(a) + f(b) - 1$.
2 Find a linear function whose graph passes through $(0, 3)$ and $(1, 5)$.
3 Graph and find the lowest point: $y = 2x^2 - 12x + 14$.

4 If $f(x) = \sqrt{x}$ and $g(x) = 3 - x$, compute   (a) $[f \circ g](x)$ (b) $[g \circ f](x)$.   In each case state the domain of the function.

5 Find a linear function $f(x) = ax + b$ whose graph passes through $(0, 6)$ and is parallel to the graph of $y = -x$.

6 For what numbers b is the value of $x^2 + bx + 1$ positive regardless of the choice of x?

7 Plot on the same set of coordinate axes: (a) $y = (x - 2)^2$ (b) $y = -3(x - 2)^2$.

8 Use the distance formula to find the perpendicular bisector of the segment from $(0, 0)$ to $(1, m)$. Hence give a proof of the formula relating the slopes of perpendicular lines.

9 Find the domain of $r(x) = \dfrac{1}{(x - 1)(x^2 + 1)} + \dfrac{2x - 3}{x^2 - 1}$.

10 Find a cubic polynomial whose graph crosses the x-axis at $x = -1$, is tangent to the x-axis at $x = 2$, and crosses the y-axis at $y = 8$.

11 Construct a rational function with vertical asymptotes $x = 0$ and $x = 4$, and horizontal asymptote $y = 3$.

12 What is the relation between the graph of $y = ax^2 + bx + c$ and that of $y = ax^2 - bx + c$?

Graph

13 $y = x^3 - 3x$

14 $y = \frac{1}{4}(x + 2)(x - 1)^3$

15 $y = x + \dfrac{1}{x}$

16 $y = \dfrac{x^3}{(x + 1)^2(x - 2)(x - 3)}$.

17 $y = x^3 - 2x^2$

18 $y = \dfrac{x^2}{(x - 3)^2}$

19 $y = \dfrac{x(x - 1)}{x^2 - 4}$

20 $y = \dfrac{(x - 2)^2}{x + 1}$

21 $y = \dfrac{x^2 + 4}{x(x^2 + 1)}$

22 $y = \frac{1}{4}x^2(x + 2)(x - 3)^2$

23 $y = \dfrac{x^2 + x}{x^2 - 9}$

24 $y = \dfrac{x^2}{(x - 2)(x^2 - 1)}$.

25 Find the locus of all points that are three times as far from $(0, 1)$ as from $(1, 0)$.

26* Show that $(0, 0)$, (a, b), $(-b, a)$,   and $(a - b, a + b)$ are the vertices of a square.

Derivatives 2

1. THE SLOPE PROBLEM

In the last section of Chapter 1 we constructed tangents to some simple curves, using special techniques that would be hard to apply to more complicated curves. In this chapter we shall solve the general problem of constructing tangents to graphs of functions $y = f(x)$. This problem is far more important than it appears to be at first glance; its solution leads to the *derivative*, one of the most applicable ideas in mathematics.

What shall we mean by *the tangent* to a curve $y = f(x)$ at a point P on the graph? We sidestep this question slightly and instead look for a reasonable definition of the *slope* of the graph at P. Then the tangent will be simply the line through P having that slope. What should the *slope* of $y = f(x)$ at P tell us? Well, it ought to convey somehow the idea of the "direction of the graph" at P.

Here is the method for solving this problem, for getting at the elusive slope. (This method is really the basic idea of differential calculus.) First we choose a point Q_h of the graph, very close to P. Then we compute the slope of the secant* through P and the nearby point Q_h. Generally, as Q_h moves closer and closer to P the slope of $\overline{PQ_h}$ will become closer and closer to some number, and this "limiting value" will be the slope of the graph at P. See Fig. 1.

This attack sounds find, but does it really work in practice? Let us try an example: $y = x^2$ at $P = (1, 1)$. See Fig. 2. A nearby point to $(1, 1)$ is $Q_h = (1 + h, (1 + h)^2)$, where h is a small number, positive or negative. By the slope formula,

$$\text{slope of } \overline{PQ_h} = \frac{(1 + h)^2 - 1}{(1 + h) - 1} = \frac{(1 + h)^2 - 1}{h}.$$

So far so good; now we must test some small values of h and see if a message comes through. For instance, if $h = 0.1$, then

$$\text{slope of } \overline{PQ_h} = \frac{(1.1)^2 - 1}{0.1} = \frac{1.21 - 1}{0.1} = \frac{0.21}{0.1} = 2.1.$$

---

* The word **secant** simply means a line through two or more distinct points of a curve.

55

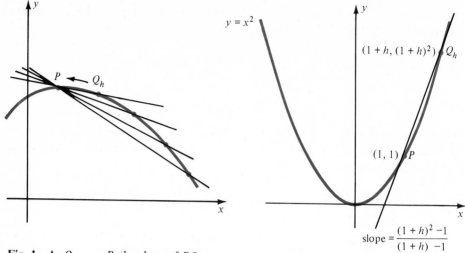

Fig. 1 As $Q_h \longrightarrow P$, the slope of $\overline{PQ_h}$ approaches the slope of the graph at P.

Fig. 2

Let us compute a few other values:

h	0.1	0.01	0.001	0.000001
slope of $\overline{PQ_h}$	2.1	2.01	2.001	2.000001

The message comes in loud and clear: As h gets smaller and smaller, so that Q_h moves closer and closer to P, the slope of the secant gets closer and closer to 2. Conclusion: the slope of $y = x^2$ at $(1, 1)$ is 2.

Now we shall test some other points on the same curve. The typical point P on the graph of $y = x^2$ can be written $P = (a, a^2)$. The typical nearby point Q_h is $Q_h = (a + h, (a + h)^2)$. Therefore

$$\text{slope of } \overline{PQ_h} = \frac{(a + h)^2 - a^2}{(a + h) - a} = \frac{(a + h)^2 - a^2}{h}.$$

Let us try $a = 2$, so $P = (2, 4)$, and various h:

h	0.1	-0.01	0.001	-0.000001
slope of $\overline{PQ_h}$	4.1	3.99	4.001	3.999999

Again the message is clear: the slope of $y = x^2$ at $(2, 4)$ is 4. Here are two further examples:

	$a = -7$			$a = 10$		
h	0.01	-0.001	0.000001	0.01	-0.001	-0.000001
slope of $\overline{PQ_h}$	-13.99	-14.001	-13.999999	20.01	19.999	19.999999

The message: the slope of $y = x^2$ at $(-7, 49)$ is -14; the slope at $(10, 100)$ is 20. This is our experimental evidence so far:

P	$(1, 1)$	$(2, 4)$	$(-7, 49)$	$(10, 100)$
slope	2	4	-14	20

For each point $P = (a, a^2)$ we have tried, we have found slope $2a$. Let us now prove that this is the correct answer in general. To do so, we simplify the formula for the slope of a secant:

$$\text{slope of } \overline{PQ_h} = \frac{(a + h)^2 - a^2}{h} = \frac{(a^2 + 2ah + h^2) - a^2}{h} = \frac{2ah + h^2}{h} = 2a + h.$$

Clearly, if h is very small, $2a + h$ is very close to $2a$, as close as we please if h is small enough. Conclusion: the slope of $y = x^2$ at (a, a^2) is $2a$. A glance at Fig. 3 shows we are in the right ball park: the predicted slope $2a$ is positive where $a > 0$, negative where $a < 0$, and zero for $a = 0$, which agrees with the figure. Furthermore, $2a$ increases as a increases, which agrees with graph's increasing steepness. Finally, the line through (a, a^2) with slope $2a$, hopefully the tangent line, is

$$y - a^2 = 2a(x - a), \qquad \text{that is,} \quad y = a^2 + 2a(x - a) = 2ax - a^2,$$

which agrees with the result found on p. 52 by an entirely different method.

General Curves   Let us return to the general graph $y = f(x)$. Its typical point is $P = (a, f(a))$, and the typical nearby point is $Q_h = (a + h, f(a + h))$. See Fig. 1 again.

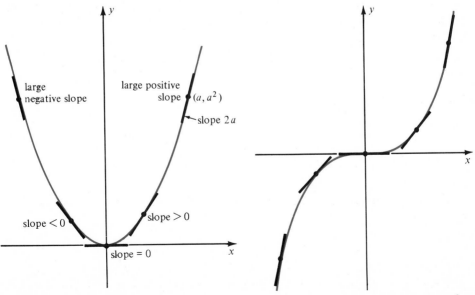

Fig. 3   Slopes at various points of $y = x^2$ **Fig. 4**   Slopes at various points of $y = x^3$

The slope of the secant through P and Q_h is

$$\text{slope of } \overline{PQ_h} = \frac{f(a+h) - f(a)}{(a+h) - a} = \frac{f(a+h) - f(a)}{h}.$$

This ratio $[f(a+h) - f(a)]/h$ is called the **difference quotient** of $f(x)$ at $x = a$. The figure suggests that the difference quotient is close to the slope of the graph at P provided Q_h is close to P, that is, provided h is small (but not 0). So we hope to predict the slope at P by studying the difference quotient for small values of h.

■ **EXAMPLE 1**   Compute the slope of $y = x^3$, first at $(-2, -8)$, then at the general point (a, a^3).

Solution   The difference quotient for $y = x^3$ at $x = -2$ is

$$\frac{f(-2+h) - f(-2)}{h} = \frac{(-2+h)^3 - (-2)^3}{h} = \frac{(-8 + 12h - 6h^2 + h^3) + 8}{h}$$

$$= 12 - 6h + h^2.$$

If h is small, then $-6h$ and h^2 are both small. Therefore, $12 - 6h + h^2$ is close to 12. The smaller h is, the closer $12 - 6h + h^2$ is to 12.

Answer   The slope of $y = x^3$ at $(-2, -8)$ is 12.

In general, the difference quotient at (a, a^3) is

$$\frac{(a+h)^3 - a^3}{h} = \frac{(a^3 + 3a^2h + 3ah^2 + h^3) - a^3}{h} = 3a^2 + 3ah + h^2.$$

When h is very small, both $3ah$ and h^2 are small. Thus as h approaches 0, both $3ah$ and h^2 die out, while $3a^2$ remains fixed.

Answer   The slope of $y = x^3$ at (a, a^3) is $3a^2$.   ■

The answer makes good sense geometrically. The slope $3a^2$ is always positive, except that it is zero at $a = 0$. As $|a|$ increases, $3a^2$ increases rapidly, so the curve becomes very steep (Fig. 4, previous p.).

EXERCISES

Notation   Given $y = f(x)$, a, and h, set

$P = (a, f(a))$ and $Q_h = (a + h, f(a + h))$.

Compute the slope of the secant through P and Q_h

$f(x) = 3x^2$ $a = -2$

1 $h = 0.1$ **2** $h = 0.01$ **3** $h = 10^{-6}$.
4 (cont.) Now compute the slope of $y = f(x)$ at $(-2, 12)$.

$f(x) = x^2 - x$ $a = 0$

5 $h = 0.1$ **6** $h = -0.03$ **7** $h = -10^{-4}$.
8 (cont.) Now compute the slope of $y = f(x)$ at $(0, 0)$.

$f(x) = 2x^2 + 3x$ $a = 2$

9 $h = -0.1$ **10** $h = 0.001$ **11** $h = -10^{-5}$.

12 (cont.) Now compute the slope of $y = f(x)$ at $(2, 14)$.

$f(x) = x^3 + x^2 \qquad a = 1$

13 $h = 0.1$ **14** $h = -0.002$ **15** $h = -10^{-10}$.
16 (cont.) Now compute the slope of $y = f(x)$ at $(1, 2)$.

$f(x) = x^3 - x + 1 \qquad a = 4$

17 $h = 0.01$ **18** $h = 0.001$ **19** $h = 0.0001$.
20 (cont.) Now compute the slope of $y = f(x)$ at $(4, 61)$.
21 Compute the slope of $y = 3x^2$ at $(a, 3a^2)$.
22 Compute the slope of $y = x^2 - x$ at $(a, a^2 - a)$.
23 Compute the slope of $y = 2x^2 + 3x$ at $(a, 2a^2 + 3a)$.
24 Compute the slope of $y = x^3 + x^2$ at $(a, a^3 + a^2)$.

2. LIMITS

In the last section we computed the slope of $y = x^2$ at (a, a^2) by examining the difference quotient for small values of h. It equals $2a + h$, which is as close as we like to $2a$ when h is small enough. Notation for this assertion is

$$\lim_{h \to 0}(2a + h) = 2a,$$

read, "The limit of $2a + h$ as h approaches 0 (tends to 0) equals (is) $2a$."

Similarly, when we computed the slope of $y = x^3$ at $(-2, -8)$, the crucial fact was that the difference quotient $12 - 6h + h^2$ is as close as we like to 12, provided h is close enough to 0. That is,

$$\lim_{h \to 0}(12 - 6h + h^2) = 12.$$

In general, suppose $F(h)$ is a function defined for all small values of h, both positive and negative, but not necessarily for $h = 0$ itself. Precisely, assume the domain of $F(h)$ includes all h satisfying $0 < |h| < A$, where A is a fixed positive number. Now suppose there is a number L such that the values of $F(h)$ are as close as we like to L, provided h is close enough to 0. Then we write

$$\lim_{h \to 0} F(h) = L.$$

An alternative and useful notation is

$$F(h) \longrightarrow L \qquad \text{as} \quad h \longrightarrow 0.$$

This is usually read, "$F(h)$ approaches L as h approaches 0."

Limits will be defined carefully in Section 10. For the present, let us try to get an intuitive feel for limits and learn how to work with them.

Examples   1. $F(h) = 6 - 7h, \quad \lim_{h \to 0} F(h) = 6.$

2. $F(h) = \dfrac{1}{4 + 3h}, \quad \lim_{h \to 0} F(h) = \dfrac{1}{4}.$ 3. $F(h) = (3 + h)(5 - h), \lim_{h \to 0} F(h) = 15.$

Warning It can perfectly well happen that $\lim_{h \to 0} F(h)$ does not exist. That is, there simply may be no number that $F(h)$ approaches as h approaches 0. For example, consider

$$F(h) = \begin{cases} 1 + h & \text{if} \quad h > 0 \\ 0 & \text{if} \quad h < 0. \end{cases}$$

Then $F(h)$ is near 1 if h is small and positive, and $F(h)$ is near 0 (actually is 0) if h is small and negative. Clearly there is no number L such that $F(h)$ is near L for *all* values of h near 0, both positive and negative.

In the three examples above it is easy to compute the limit; just set $h = 0$ and the limit $F(0)$ pops out. But things are usually not so simple. For example, consider

$$\lim_{h \to 0} \frac{\sin h}{h} \qquad (h \text{ in radians}).$$

(We cannot just set $h = 0$ because the meaningless quotient 0/0 results.) Observe how different this is from something like

$$\lim_{h \to 0} \frac{2h - h^2}{h} = \lim_{h \to 0} (2 - h) = 2,$$

because there is no apparent way to cancel h from $\sin h$. Instead, let us test some small values of h. Since $(\sin h)/h$ is an even function, we can restrict attention to $h > 0$. Using tables or a calculator to 5-place accuracy, we find

h	0.2	0.1	0.05	0.01
$\sin h$	0.198669	0.099833	0.049979	0.010000
$(\sin h)/h$	0.99335	0.99833	0.99958	1.00000

To slightly higher accuracy, the last entry is 0.999983. The numerical evidence strongly suggests

$$\lim_{h \to 0} \frac{\sin h}{h} = 1.$$

We shall meet this limit again in Chapter 4. Here it illustrates an important principle: that $\lim_{h \to 0} F(h)$ can exist even when $F(h)$ is undefined at 0. What matters are the values of $F(h)$ where h is very near to 0, but not 0 itself.

Limit as $h \longrightarrow a$ The idea of $\lim F(h)$ as $h \longrightarrow 0$ is important, as we know from studying slopes. Also important is the idea of $\lim F(h)$ as $h \longrightarrow a$. Here a is a fixed number and $F(h)$ is defined for all values of h near a, but not necessarily at $h = a$. That is, $F(h)$ is defined at least for $0 < |h - a| < A$ for some $A > 0$. Then we write

$$\lim_{h \to a} F(h) = L \qquad \text{or} \qquad F(h) \longrightarrow L \qquad \text{as} \quad h \longrightarrow a$$

if $F(h)$ is as close to L as we please, provided h is sufficiently close to a. Another interpretation is that $F(h)$ moves closer and closer to L as h moves closer and closer to a.

Examples $F(h) = h + 3h^2,$ $\lim\limits_{h \to -4} F(h) = -4 + 3 \cdot 16 = 44.$

$$F(h) = \frac{h^2 - 4}{h - 2}, \quad \lim_{h \to 2} F(h) = \lim_{h \to 2} \frac{(h + 2)(h - 2)}{h - 2} = \lim_{h \to 2} (h + 2) = 4.$$

Rules for Limits   In calculus we have to compute many limits. Some are obvious, such as

$$\lim_{h \to a} c = c \qquad \lim_{h \to a} h = a \qquad \lim_{h \to 0} h^2 = 0 \qquad \lim_{h \to 0} h^3 = 0.$$

(In the first of these, $F(h) = c$ is a constant function.) Others, such as

$$\frac{\sin h}{h} \longrightarrow 1 \qquad \text{as} \quad h \longrightarrow 0,$$

are not obvious and require special methods of proof.

Once we have mastered a few basic limits, we can compute many others by a batch of rules that express new limits in terms of known ones.

Rules for Limits   Suppose $\lim F(h)$ and $\lim G(h)$ exist as $h \longrightarrow a$. Then

1. $\lim\limits_{h \to a} cF(h) = c \lim\limits_{h \to a} F(h)$ (c constant);

2. $\lim\limits_{h \to a}[F(h) \pm G(h)] = \lim\limits_{h \to a} F(h) \pm \lim\limits_{h \to a} G(h);$

3. $\lim\limits_{h \to a} F(h)G(h) = \left[\lim\limits_{h \to a} F(h)\right]\left[\lim\limits_{h \to a} G(h)\right];$

4. $\lim\limits_{h \to a} \dfrac{F(h)}{G(h)} = \dfrac{\lim\limits_{h \to a} F(h)}{\lim\limits_{h \to a} G(h)},$ provided $\lim_{h \to a} G(h) \neq 0.$

We postpone the somewhat technical proofs of these rules until Section 10. The rules themselves, however, are quite natural. Rule 1 means that if $F(h)$ is near L when h is near a, then $cF(h)$ is near cL when h is near a. The other rules say that if $F(h)$ is near L and $G(h)$ is near M, then $F(h) \pm G(h)$ is near $L \pm M$ and $F(h)G(h)$ is near LM, while $F(h)/G(h)$ is near L/M, provided $M \neq 0$.

■ **EXAMPLE 1**   Use the rules to find

(a) $\lim\limits_{h \to 0} \dfrac{(3 + h) \sin h}{h}$ (b) $\lim\limits_{h \to 0} \dfrac{4 - h^2}{2 + 5h}$

(c) $\lim\limits_{h \to -1} [(1 + h^2)(2 - h) + (3 - h^3)(1 - h)].$

Solution

(a) $\lim\limits_{h \to 0} \dfrac{(3 + h) \sin h}{h} = \lim\limits_{h \to 0}\left[(3 + h) \dfrac{\sin h}{h}\right]$

$\qquad\qquad = \lim\limits_{h \to 0}(3 + h) \lim\limits_{h \to 0} \dfrac{\sin h}{h} = \left(\lim\limits_{h \to 0} 3 + \lim\limits_{h \to 0} h\right) \cdot 1$

$\qquad\qquad = (3 + 0) \cdot 1 = 3.$

(b)   As $h \longrightarrow 0$, we have $h^2 = h \cdot h \longrightarrow 0 \cdot 0 = 0$,

$$4 - h^2 \longrightarrow 4 - 0 = 4, \qquad \text{and} \qquad 2 + 5h \longrightarrow 2 + 5 \cdot 0 = 2 \neq 0.$$

Hence by Rule 4, $\dfrac{4 - h^2}{2 + 5h} \longrightarrow \dfrac{4}{2} = 2$.

(c)   As $h \longrightarrow -1$,

$$h^2 = h \cdot h \longrightarrow (-1)(-1) = 1 \qquad \text{and} \qquad h^3 = h \cdot h^2 \longrightarrow (-1)(1) = -1.$$

Hence

$$(1 + h^2)(2 - h) \longrightarrow (1 + 1)[2 - (-1)] = 6,$$

$$(3 - h^3)(1 - h) \longrightarrow [3 - (-1)][1 - (-1)] = 8,$$

$$(1 + h^2)(2 - h) + (3 - h^3)(1 - h) \longrightarrow 6 + 8 = 14. \qquad \blacksquare$$

In practice, limits like these are found by inspection, skipping obvious steps. For instance,

$$\lim_{h \to 2} \frac{3 + h^2}{(1 + h)(2 + h)} = \frac{3 + 4}{3 \cdot 4} = \frac{7}{12},$$

$$\frac{3h}{\sin h} = \frac{3}{(\sin h)/h} \longrightarrow \frac{3}{1} = 3 \qquad \text{as} \quad h \longrightarrow 0.$$

Rule 2 holds for any number of summands, not just two, and Rule 3 holds for any number of factors. These statements can be proved easily by mathematical induction from Rules 2 and 3.

Example $\displaystyle\lim_{h \to a}(a^3 + a^2 h + ah^2 + h^3)$

$$= \lim a^3 + a^2 \lim h + a \lim h^2 + \lim h^3$$

$$= a^3 + a^2 \cdot a + a \cdot a^2 + a^3 = 4a^3.$$

Notation   There is really nothing special about the letter h for the variable in limits; the variable can be any suitable letter. Thus

$$\lim_{h \to a} F(h) = \lim_{x \to a} F(x) = \lim_{y \to a} F(y), \quad \text{etc.}$$

Sometimes the variable is shifted, usually to 0. For instance x is near a if and only if $x = a + h$, where h is near 0, so

$$\lim_{x \to a} F(x) = \lim_{h \to 0} F(a + h).$$

For example,

$$\lim_{x \to a} 3x^2 = \lim_{h \to 0} 3(a + h)^2 = 3 \lim_{h \to 0}(a^2 + 2ah + h^2) = 3(a^2 + 0 + 0) = 3a^2.$$

EXERCISES

Find the limit

1 $\displaystyle\lim_{h \to 0}(3 + h^2)$ **2** $\displaystyle\lim_{h \to 0}\frac{1}{1 + h}$ **3** $\displaystyle\lim_{h \to 0}(2 + h)^3$

$\frac{t^2+h+t^2-1}{t} \quad \frac{}{t(1+t)}$

$d+2t(2)^2+3(t)^2 2-8 \quad \frac{}{t}$

4 $\lim\limits_{h\to 0} \dfrac{1+h}{1-h}$ /

5 $\lim\limits_{h\to 0} \dfrac{(1+\frac{1}{2}h)^2-1}{h} = 1.$

6 $\lim\limits_{h\to 0} \dfrac{(2+h)^3-8}{h}$ $=12$

7 $\lim\limits_{h\to 0} \dfrac{(-1+h)^4-1}{h}$

8 $\lim\limits_{h\to 0} \dfrac{(1-2h)^3-1}{h}$ -6

9 $\lim\limits_{h\to 0} \dfrac{(1+2h)^2-1}{h}$ 4

10 $\lim\limits_{h\to 0} \dfrac{(2-h)^5-32}{h}$

11 $\lim\limits_{h\to 0} \dfrac{1}{(h-1)^3}$

12 $\lim\limits_{h\to 0} \dfrac{h}{(1+2h)^3-1}.$ $\frac{1}{6}$

13 Compute to five places $(1-\cos h)/h$ for $h=0.1,\ 0.05,\ 0.01,\ 0.001$. Conclusion?

14 Compute to five places $(1-\cos h)/h^2$ for $h=0.1,\ 0.05,\ 0.01,\ 0.001$. Conclusion?

Find the limit

15 $\lim\limits_{h\to 2}(h^2-1)$

16 $\lim\limits_{h\to 3}(h^3+3)$

17 $\lim\limits_{k\to 1} \dfrac{1-2k}{1+k}$ $\frac{1}{2}$

18 $\lim\limits_{y\to -1} \dfrac{1+2y}{1-y}$

19 $\lim\limits_{h\to -2} \dfrac{(1+h)(3+h)}{5+h}$

20 $\lim\limits_{h\to 1}(1+h+h^2+h^3+\cdots+h^{12})$

21 $\lim\limits_{h\to 1} \dfrac{h-1}{h^4-1}$ $\frac{1}{4}$

22 $\lim\limits_{h\to 2} \dfrac{h^5-32}{h-2}$

23 $\lim\limits_{x\to -1} \dfrac{x^3-1}{x^7-1}$

24 $\lim\limits_{x\to 1} \dfrac{x^6-1}{x^4-1}$

25 $\lim\limits_{h\to 0} \dfrac{h^2}{\sin h}$

26 $\lim\limits_{h\to 0}\left(\dfrac{\sin h}{3h}\right)^2.$

27 Find $\lim\limits_{x\to 0} \dfrac{\sqrt{2+x}-\sqrt{2}}{x}$.

[*Hint* "Rationalize" the numerator, i.e., multiply and divide by $\sqrt{2+x}+\sqrt{2}$.]

28 Find $\lim\limits_{x\to 0} \dfrac{(2+x)^{1/3}-(2)^{1/3}}{x}$.

[*Hint* Proceed as in Ex. 27, using the identity $y^3-z^3=(y-z)(y^2+yz+z^2)$.]

3. THE DERIVATIVE

Given a function $f(x)$ and a point a of its domain, the **derivative** of $f(x)$ at a is

$$f'(a) = \lim\limits_{h\to 0} \frac{f(a+h)-f(a)}{h},$$

provided the limit exists. The derivative $f'(a)$ is a number, the slope of the graph $y=f(x)$ at its point $(a, f(a))$. The derivative was described geometrically in Section 1 in terms of slope, but now is expressed mathematically as a limit.

Examples   1. $f(x)=x^2,\qquad a=3.$

$$f'(a) = \lim\limits_{h\to 0} \frac{(3+h)^2-9}{h} = \lim\limits_{h\to 0} \frac{6h+h^2}{h} = \lim\limits_{h\to 0}(6+h) = 6.$$

2. $f(x)=x^3.$

$$f'(a) = \lim\limits_{h\to 0} \frac{(a+h)^3-a^3}{h} = \lim\limits_{h\to 0} \frac{3a^2h+3ah^2+h^3}{h} = \lim\limits_{h\to 0}(3a^2+3ah+h^2) = 3a^2.$$

There is a useful alternative form of the limit defining the derivative. Replace $a + h$ by x, and note that $x \longrightarrow a$ as $h \longrightarrow 0$:

$$f'(a) = \lim_{x \to a} \frac{f(x) - f(a)}{x - a}.$$

Example $f(x) = x^2$.

$$f'(a) = \lim_{x \to a} \frac{x^2 - a^2}{x - a} = \lim_{x \to a} \frac{(x - a)(x + a)}{x - a} = \lim_{x \to a}(x + a) = a + a = 2a.$$

The derivative $f'(a)$ is a number defined at each point a of the domain of $f(x)$ for which a certain limit exists. Therefore $f'(a)$ can be considered as a value of a new *function*, defined on part of the domain of $f(x)$, maybe on the whole domain. This point of view suggests writing $f'(x)$ for the slope of the tangent at $(x, f(x))$. The function $f'(x)$ is defined by

$$f'(x) = \lim_{h \to 0} \frac{f(x + h) - f(x)}{h}.$$

With this notation, we can restate Examples 1 and 2 as follows:

1. $f(x) = x^2$, $f'(x) = 2x$.   2. $f(x) = x^3$, $f'(x) = 3x^2$.

In different notation: $(x^2)' = 2x$, $(x^3)' = 3x^2$.

Notation and Terminology   A function whose derivative exists is called a **differentiable** function. You **differentiate** a function when you take its derivative; the process is called **differentiation**.

Unfortunately, lots of different notations are commonly used for the derivative, and it is important to recognize them. If $y = f(x)$, then the derivative may be denoted in any of these ways:

$$f'(x), \qquad y'(x), \qquad y', \qquad \frac{dy}{dx}, \qquad \frac{df}{dx}, \qquad \frac{df(x)}{dx}.$$

The first one you read "f prime of x"; the fourth, "*DYDX*." Note that the third, fourth, and fifth do not show the point where the derivative is to be evaluated. To show the value of the derivative at $x = a$, write

$$y'(a), \qquad \text{or} \qquad \frac{dy}{dx}\bigg|_{x=a}, \qquad \text{or} \qquad \frac{df}{dx}\bigg|_{a}, \qquad \text{or} \qquad \frac{df}{dx}(a).$$

Sometimes it pays to think of the derivative as an operation. The operator

$$\frac{d}{dx}$$

is applied to the function $y = f(x)$ to produce its derivative dy/dx.

The variable may be denoted by a letter other than x; that does not change anything. For instance,

$$\frac{d}{du}(u^2) = 2u, \qquad \frac{d}{dy}(y^3) = 3y^2.$$

The letter t is usually used for time. There is a special, and commonly used notation for time derivatives, a dot instead of a prime:

$$\frac{d}{dt}[x(t)] = \dot{x}(t).$$

Thus if $x(t) = t^2$, then $\dot{x} = 2t$.

Some Basic Derivatives   Derivatives are so important that sooner or later we must learn how to differentiate every function in sight. Let us first attack some simple, but basic ones.

The simplest is a constant function $f(x) = c$. Clearly its graph is a horizontal line, with slope 0, so $f'(x) = 0$. More formally,

$$\frac{f(x+h) - f(x)}{h} = \frac{c - c}{h} = \frac{0}{h} = 0 \longrightarrow 0$$

as $h \longrightarrow 0$, so $f'(x) = 0$.

Now consider a linear function $f(x) = mx + b$. The graph of $y = mx + b$ is a straight line of slope m, so $f'(x) = m$. Formal verification:

$$\frac{f(x+h) - f(x)}{h} = \frac{[m(x+h) + b] - [mx + b]}{h} = \frac{mh}{h} = m \longrightarrow m$$

as $h \longrightarrow 0$, so $f'(x) = m$.

Next we differentiate $f(x) = x^n$, where n is any positive integer. There are two ways to do this one, with factoring or with a binomial expansion.

Method 1

$$f(x) - f(a) = x^n - a^n = (x - a)(x^{n-1} + ax^{n-2} + a^2x^{n-3} + \cdots + a^{n-2}x + a^{n-1}),$$

hence $\quad f'(a) = \lim_{x \to a} \dfrac{f(x) - f(a)}{x - a} = \lim_{x \to a}(x^{n-1} + ax^{n-2} + a^2x^{n-3} + \cdots + a^{n-1}).$

The limit of the sum equals the sum of the limits. Clearly as $x \longrightarrow a$,

$$\lim ax^{n-2} = a \lim x^{n-2} = a \cdot a^{n-2} = a^{n-1},$$

$$\lim a^2x^{n-3} = a^2 \lim x^{n-3} = a^2 \cdot a^{n-3} = a^{n-1}, \quad \text{etc.}$$

Thus each term has the same limit, a^{n-1}. Since there are n terms in the sum, $f'(a) = na^{n-1}$.

Method 2   By the binomial theorem

$$(x + h)^n = x^n + nx^{n-1}h + b_2 x^{n-2}h^2 + \cdots + b_n h^n,$$

where $b_2, \cdots, b_n$ are binomial coefficients, whose exact values aren't needed. Hence

$$f'(x) = \lim_{h \to 0} \frac{(x+h)^n - x^n}{h} = \lim_{h \to 0}(nx^{n-1} + b_2 x^{n-2}h + \cdots + b_n h^{n-1}).$$

The first summand is constant as far as h is concerned; each of the others contains a power of h, hence has limit 0. Therefore $f'(x) = nx^{n-1}$.

Note that the special cases $n = 2$ and $n = 3$ agree with previous computations: $(x^2)' = 2x$ and $(x^3)' = 3x^2$.

Now let us differentiate $f(x) = 1/x$, defined for $x \neq 0$. According to the recipe,

$$f'(a) = \lim_{x \to a} \frac{1/x - 1/a}{x - a} = \lim_{x \to a} \frac{(a - x)/ax}{x - a} = \lim_{x \to a} \frac{-1}{ax} = \frac{-1}{a \cdot a} = \frac{-1}{a^2}.$$

Therefore $f'(x) = -1/x^2$.

Basic Differentiation Formulas

$$\frac{d}{dx}(c) = 0 \qquad\qquad \frac{d}{dx}(mx + b) = m$$

$$\frac{d}{dx}(x^n) = nx^{n-1}, \qquad n \text{ a positive integer}$$

$$\frac{d}{dx}\left(\frac{1}{x}\right) = -\frac{1}{x^2}, \qquad x \neq 0.$$

Rate of Change   The derivative has an important interpretation besides slope of a curve. Suppose, for instance, that $f'(a) = 3$. This means that

$$\frac{f(a + h) - f(a)}{h} \longrightarrow 3 \qquad \text{as} \quad h \longrightarrow 0.$$

Now $f(a + h) - f(a)$ is the change in $y = f(x)$ as x changes by amount h from a to $a + h$. When h is very small, the difference quotient, the ratio of these changes, is very close to 3. Thus, a change in x causes a change that is about 3 times as much in $f(x)$. We say that the rate of change of y with respect to x is 3 at $x = a$. Geometrically, the curve $y = f(x)$ is rising 3 times as fast at $(a, f(a))$ as it is moving to the right (Fig. 1).

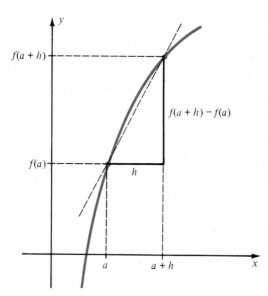

Fig. 1   The difference quotient approximates the rate of change of y with respect to x. Its limit, the derivative, *equals* the rate of change.

In general, $f'(a)$ is the **rate of change** of y with respect to x at $x = a$. For a linear function $f(x) = mx + b$, this interpretation of the derivative is natural. The graph is a straight line, and since $f'(a) = m$ at all points, the rate of change of y with respect to x is everywhere equal to the slope m. For other functions, the graph $y = f(x)$ is curved, and the rate of change of y with respect to x varies from point to point.

Take $y = x^2$ for instance. At $x = a$, a small change h in x produces a change in y that is about $2ah$. If x increases by $h = 0.01$ from 1.00 to 1.01, the change in y is

$$(1.01)^2 - 1.00 = 0.0201 \approx 2(0.01) = 2ah.$$

If x increases by $h = 0.01$ from 5.00 to 5.01, the change in y is

$$(5.01)^2 - (5.00)^2 = (5 + 0.01)^2 - 5^2 = 2(5)(0.01) + 0.0001 \approx 2(5)(0.01) = 2ah.$$

If x decreases from 3 to 2.99, so $h = -0.01$, the change in y is

$$(2.99)^2 - 3^2 = (3 - 0.01)^2 - 3^2 = 3^2 - 2(3)(0.01) + 0.0001 - 3^2$$
$$= -2(3)(0.01) + 0.0001 \approx -2(3)(0.01) = 2ah.$$

The symbol $\approx$ in these examples is read "approximately equal to."

EXERCISES

Find

1 $\dfrac{dy}{dx}$ for $y = x^2$

2 $\dfrac{dV}{dP}$ where $V = P^2$

3 $\dfrac{dy}{dx}$ for $y = -4(x - 5)$

4 $\dfrac{ds}{dt}$ where $s = -3(4 - 5t)$

5 $\dfrac{d}{dx}(x)$

6 $\dfrac{d}{dx}(x^2)$

7 $\dfrac{df}{dx}$ for $f(x) = 3x + 2$

8 $\dfrac{df}{dx}$ where $f(x) = 12x - 7$

9 $f'(x)$ where $f(x) = 8x$

10 $F'(z)$ for $F(z) = z^2$

11 $\dfrac{dx}{dy}$ for $x = y^3$

12 $\dfrac{dF}{dx}$ for $F(x) = x^3$

13 $\dfrac{d}{dx}(4^3)$

14 $\dfrac{d}{dt}(5)\Big|_{t=1}$

15 $\dfrac{dy}{dx}$ at $x = 0, 3, -3$, if $y = x^3$

16 $\dfrac{ds}{dt}$ for $t = 0, 1, 2, 3$ if $s = t^3$

17 $\dfrac{dG}{dz}\Big|_{6}$ for $G(z) = z^3$

18 $\dfrac{dy}{dx}\Big|_{-4}$, $y = x^3$

19 $V'(4)$, $V'(a)$, if $V(P) = P^3$

20 $s'(5)$, $s'(t_0)$, if $s(t) = t^3$.

Calculate

21 $f'(-6)$, $f'(12)$, $f'(1)$, where $f(x) = x^2$

22 $G'(-1)$, $G'(0)$, $G'(1)$, if $G(x) = x^2$

23 $f'(-1)$, $f'(1)$, $f'(a)$, $f'(-a)$, where $f(x) = 1/x$

24 $f'(-\tfrac{1}{2})$, $f'(\tfrac{1}{2})$, $f'(2)$, $f'(3)$, if $f(x) = 1/x$

25 $\dfrac{dy}{dx}\Big|_{x=a}$ and $\dfrac{dy}{dx}\Big|_{x=1/a}$, if $y = \dfrac{1}{x}$

26 $\dfrac{dV}{dP}\Big|_{P=1/4}$ and $\dfrac{dV}{dP}\Big|_{P=4}$, if $V = \dfrac{1}{P}$

27 $\dfrac{d}{ds}\left(\dfrac{1}{s}\right)\Big|_b$

28 $\dfrac{d}{dx}\left(\dfrac{1}{x}\right)\Big|_t$

29 $\dfrac{d}{dx}(x^2)$, at $x = 3, 7, 11$

30 $\dfrac{d}{dP}(P^2)$, at $P = -1, 1, 0$

31 $\dfrac{d}{dx}(13x + 5)$, at $x = 2$

32 $\dfrac{d}{dt}(-13t + v_0)$, for $t = 0$

33 $\dfrac{dy}{dx}\Big|_{x=3}$, $y = x^2$

34 $\dfrac{dR}{dI}\Big|_{I=-2}$, where $R = I^2$

35 $\dfrac{dv}{dt}\Big|_9$, $v = -32t + 200$

36 $\dfrac{ds}{dt}\Big|_{10}$, $s = t^2$.

37 Find all points on $y = x^2$ where the tangent has slope 6.   *(3,9)* ✗ *only.*   *slope of line*

38 Find a point on $y = x^2$ where the tangent is parallel to the line $x + 2y + 7 = 0$. ⟹ *x+2y+7=0* *= $-\frac{1}{2}$*

39 Find all points on the curve $y = x^3$ where the slope is 12. *$\frac{dy}{dx} = 2x$*

40 Find where the curve $y = 1/x$ has slope $-\frac{1}{2}$.   *$2x = -\frac{1}{2}$*   *$2x = -\frac{1}{2}$* ⟹ $\left(-\frac{1}{4}, \frac{1}{16}\right)$

41 Is the curve $y = 1/x$ ever horizontal?   *$x = -\frac{1}{4}$*

42 Do the curves $y = 1/x$ and $y = x^3$ ever have the same slope?

43 Is the graph of $y = x^3$ ever horizontal?

44 Show there do not exist points on the curve $y = x^3$ where the tangent is parallel to the line
   $x + y + 1 = 0$. ⟹ *slope = -1*.   *$\frac{dy}{dx} = 3x^2$ Since $3x^2 = -1$ $x^2 = -\frac{1}{3}$ impossible*

   for x^2 is not a negative no ∴ the pt is not find

45 Find which of the two curves $y = x^2$ and $y = x^3$ is steeper at $x = \frac{1}{2}$, at $x = 1$, at $x = 2$.

46 Find all positive values of x where $y = x^2$ is steeper than $y = x^3$.

47 Find all positive values of x where $y = x^3$ is steeper than $y = x^2$.

48 Give an interpretation to the identity

$$\frac{x^n - 1}{x - 1} = 1 + x + x^2 + \cdots + x^{n-1} \qquad \text{for } x = 1.$$

4. SUMS AND PRODUCTS

The definition of derivative in the last section is fine for theoretical purposes, but a nuisance to apply each time we differentiate a function. In practice, we use rules that express derivatives of new functions in terms of derivatives we already know. The two simplest rules are these:

Let $u(x)$ and $v(x)$ be differentiable functions and let c be a constant. Then $cu(x)$ and $u(x) \pm v(x)$ are differentiable and

Constant Factor Rule

$$\frac{d}{dx}[cu(x)] = c\frac{d}{dx}u(x), \qquad \text{that is,} \quad (cu)' = cu';$$

Sum Rule

$$\frac{d}{dx}[u(x) \pm v(x)] = \frac{d}{dx}u(x) \pm \frac{d}{dx}v(x), \qquad \text{that is,} \quad (u \pm v)' = u' \pm v'.$$

Examples $\dfrac{d}{dx}(3x) = 3\dfrac{d}{dx}(x) = 3 \cdot 1 = 3.$

$$\frac{d}{dx}\left(-\frac{2}{3}x^4\right) = -\frac{2}{3}\frac{d}{dx}(x^4) = -\frac{2}{3}(4x^3) = -\frac{8}{3}x^3.$$

$$\frac{d}{dx}(x^5 + x^3) = \frac{d}{dx}(x^5) + \frac{d}{dx}(x^3) = 5x^4 + 3x^2.$$

While these rules seem obvious, they really have to be proved. Differentiation is a new operation. Who says it must behave in the way we expect? For example,

$$[u(x) + v(x)]^2 \neq [u(x)]^2 + [v(x)]^2 \qquad \text{and} \qquad \log[u(x) + v(x)] \neq \log u(x) + \log v(x),$$

so why should

$$\frac{d}{dx}[u(x) + v(x)] = \frac{d}{dx}u(x) + \frac{d}{dx}v(x)\,?$$

Both rules hold because they reflect corresponding properties of limits. For the Constant Factor Rule, we use the definition of derivative and Rule 1 for limits (p. 61):

$$\frac{d}{dx}[cu(x)] = \lim_{h\to 0}\frac{cu(x+h) - cu(x)}{h} = \lim_{h\to 0}c\,\frac{u(x+h) - u(x)}{h}$$

$$= c\lim_{h\to 0}\frac{u(x+h) - u(x)}{h} = c\frac{d}{dx}u(x).$$

For the Sum Rule, we use Rule 2 for limits:

$$\frac{d}{dx}[u(x) \pm v(x)] = \lim_{h\to 0}\frac{[u(x+h) \pm v(x+h)] - [u(x) \pm v(x)]}{h}$$

$$= \lim_{h\to 0}\left[\frac{u(x+h) - u(x)}{h} \pm \frac{v(x+h) - v(x)}{h}\right]$$

$$= \lim_{h\to 0}\frac{u(x+h) - u(x)}{h} \pm \lim_{h\to 0}\frac{v(x+h) - v(x)}{h}$$

$$= \frac{d}{dx}u(x) \pm \frac{d}{dx}v(x).$$

By induction, the Sum Rule holds for any number of summands, not just two. For instance,

$$(u + v - w)' = u' + v' - w'.$$

■ **EXAMPLE 1** Differentiate   (a) $5x^3 - 8x^2 - 7$ (b) $\frac{1}{2}x^5 + \frac{3}{x}$.

Solution   (a)   We know the derivatives of x^3, x^2, and the constant 7. Apply the Sum Rule for three terms, then the Constant Factor Rule:

$$\frac{d}{dx}(5x^3 - 8x^2 - 7) = \frac{d}{dx}(5x^3) - \frac{d}{dx}(8x^2) - \frac{d}{dx}(7)$$

$$= 5\frac{d}{dx}(x^3) - 8\frac{d}{dx}(x^2) - 0 = 5 \cdot 3x^2 - 8 \cdot 2x = 15x^2 - 16x.$$

(b) Same procedure:

$$\frac{d}{dx}\left(\frac{1}{2}x^5 + \frac{3}{x}\right) = \frac{d}{dx}\left(\frac{1}{2}x^5\right) + \frac{d}{dx}\left(\frac{3}{x}\right)$$

$$= \frac{1}{2}\frac{d}{dx}(x^5) + 3\frac{d}{dx}\left(\frac{1}{x}\right) = \frac{1}{2}(5x^4) + 3\left(\frac{-1}{x^2}\right) = \frac{5}{2}x^4 - \frac{3}{x^2}. \qquad \blacksquare$$

Remark In practice, problems like these are usually done in one step, by inspection. It's just a matter of knowing the rules cold so you are confident of making no errors.

Products For the derivative of a product $u(x)v(x)$, a natural first guess is $u'(x)v'(x)$. This time intuition fails. The natural guess is just plain wrong. For instance, if $u(x) = x$ and $v(x) = x^2$, then $u'v' = 1 \cdot 2x = 2x$. But $uv = x^3$, so $(uv)' = 3x^2$, certainly not $2x$. Here is the correct formula.

Product Rule If $u(x)$ and $v(x)$ are differentiable functions, then $u(x)v(x)$ is differentiable and

$$\frac{d}{dx}[u(x)v(x)] = u(x)\frac{d}{dx}v(x) + v(x)\frac{d}{dx}u(x).$$

Briefly, $(uv)' = uv' + vu'.$

Example $$\frac{d}{dx}(x \cdot x^2) = x\frac{d}{dx}(x^2) + x^2\frac{d}{dx}(x)$$

$$= x \cdot 2x + x^2 \cdot 1 = 3x^2 = \frac{d}{dx}(x^3).$$

How does this unexpected formula arise? Here is an argument that makes it plausible. (We give a proof in Section 10.)

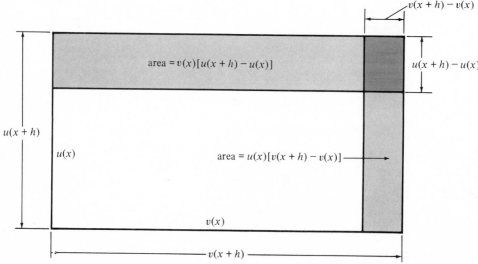

Fig. 1 Estimate of change in area

Assume $u(x)$ and $v(x)$ are positive, and interpret $u(x)v(x)$ as the area of a rectangle (Fig. 1). If x changes by a small amount to $x + h$, then $u(x)$ and $v(x)$ change to $u(x + h)$ and $v(x + h)$, and the area changes to $u(x + h)v(x + h)$. The change in area is the shaded region. If we ignore the tiny rectangle in the upper right corner (only a small part of the entire shaded region), then the change in area is

$$u(x + h)v(x + h) - u(x)v(x) \approx u(x)[v(x + h) - v(x)] + v(x)[u(x + h) - u(x)].$$

Divide by h and let $h \longrightarrow 0$. It seems probable that

$$\frac{d}{dx}[u(x)v(x)] = \lim_{h \to 0} \frac{u(x + h)v(x + h) - u(x)v(x)}{h}$$

$$= u(x) \lim_{h \to 0} \frac{v(x + h) - v(x)}{h} + v(x) \lim_{h \to 0} \frac{u(x + h) - u(x)}{h}$$

$$= u(x) \frac{d}{dx} v(x) + v(x) \frac{d}{dx} u(x).$$

■ **EXAMPLE 2** Differentiate

(a) $y = (3x - 2)(x^2 + 5x + 1)$ (b) $y = \dfrac{1}{x^2}$.

Solution (a) Apply the Product Rule with $u(x) = 3x - 2$ and $v(x) = x^2 + 5x + 1$:

$$\frac{dy}{dx} = (3x - 2)\frac{d}{dx}(x^2 + 5x + 1) + (x^2 + 5x + 1)\frac{d}{dx}(3x - 2)$$

$$= (3x - 2)(2x + 5) + (x^2 + 5x + 1)(3) = 9x^2 + 26x - 7.$$

Check Multiply out first, then differentiate:

$$y = 3x^3 + 13x^2 - 7x - 2, \qquad \frac{dy}{dx} = 9x^2 + 26x - 7.$$

(b) Write $y = \left(\dfrac{1}{x}\right)\left(\dfrac{1}{x}\right)$ and apply the Product Rule:

$$\frac{dy}{dx} = \frac{1}{x}\frac{d}{dx}\left(\frac{1}{x}\right) + \frac{1}{x}\frac{d}{dx}\left(\frac{1}{x}\right) = 2\left(\frac{1}{x}\right)\left(-\frac{1}{x^2}\right) = -\frac{2}{x^3}.$$ ∎

The last example suggests a formula for the derivative of the square of a function. By the Product Rule,

$$\frac{d}{dx}[u(x)]^2 = u(x)\frac{d}{dx}u(x) + u(x)\frac{d}{dx}u(x) = 2u(x)\frac{d}{dx}u(x).$$

Briefly,

$$\frac{d(u^2)}{dx} = 2u\frac{du}{dx}.$$

We can continue along this line with cubes and higher powers:

$$\frac{d(u^3)}{dx} = \frac{d}{dx}(u \cdot u^2) = u\frac{d(u^2)}{dx} + u^2\frac{du}{dx} = u\left(2u\frac{du}{dx}\right) + u^2\frac{du}{dx} = 3u^2\frac{du}{dx},$$

$$\frac{d(u^4)}{dx} = \frac{d}{dx}(u \cdot u^3) = u\frac{d(u^3)}{dx} + u^3\frac{du}{dx} = u\left(3u^2\frac{du}{dx}\right) + u^3\frac{du}{dx} = 4u^3\frac{du}{dx},$$

$$\frac{d(u^5)}{dx} = \frac{d}{dx}(u \cdot u^4) = u\frac{d(u^4)}{dx} + u^4\frac{du}{dx} = u\left(4u^3\frac{du}{dx}\right) + u^4\frac{du}{dx} = 5u^4\frac{du}{dx}.$$

Similarly, $\dfrac{d(u^6)}{dx} = 6u^5\dfrac{du}{dx}, \qquad \dfrac{d(u^7)}{dx} = 7u^6\dfrac{du}{dx}, \qquad$ etc.

If n is a positive integer, then $\dfrac{d(u^n)}{dx} = nu^{n-1}\dfrac{du}{dx}.$

Special case $u = x$.   Then the formula says that

$$\frac{d(x^n)}{dx} = nx^{n-1}\frac{dx}{dx} = nx^{n-1},$$

a familiar formula.

■ **EXAMPLE 3** Differentiate (a) $(x^2 + 1)^3$ (b) $(5x^3 - 17x + 12)^6$.

Solution   (a)   Take $u(x) = x^2 + 1$ and $n = 3$:

$$\frac{d(u^3)}{dx} = 3u^2\frac{du}{dx} = 3(x^2+1)^2\frac{d}{dx}(x^2+1) = 3(x^2+1)^2(2x) = 6x(x^2+1)^2.$$

Check   Expand $(x^2 + 1)^3$ by the binomial theorem and differentiate:

$$\frac{d}{dx}(x^2 + 1)^3 = \frac{d}{dx}(x^6 + 3x^4 + 3x^2 + 1) = 6x^5 + 12x^3 + 6x$$

$$= 6x(x^4 + 2x^2 + 1) = 6x(x^2 + 1)^2.$$

(b)   Take $u(x) = 5x^3 - 17x + 12$ and $n = 6$:

$$\frac{d(u^6)}{dx} = 6u^5\frac{du}{dx} = 6(5x^3 - 17x + 12)^5(15x^2 - 17). \qquad ■$$

Remark   Part (b) could be done by expanding u^6, then differentiating term-by-term. But the computation would be horrendous, which shows the great advantage of the formula.

EXERCISES

Find dy/dx for $y =$

1	$x^2 + 3x$	2	$5x^3 - x$	3	$-x^4 + 2x^2 - 1$
4	$3x^2 + 2x + 1$	5	$x + (1/x)$	6	$5x^5 - (7/x)$

7   $x^3 - 3x + 1$ at $x = -1$ 8 $x^2 + (2/x)$ at $x = 3$
9   $x^3 + x - (2/x)$ at $x = 2$ 10 $5x^4 - 4x^3 + x$ at $x = 1$
11   $(x + 1)(x^2 + 1)$ 12 $(x^2 - 1)(x^2 + 3)$
13   $(3x + 4)(x^2 - 2x - 3)$ 14 $(2x - 7)(x^3 + 1)$
15   $(x^5 - 2)(x^3 + x - 3)$ 16 $(x^4 - 2x)(2x^3 - 3x)$
17   $(x^2 - 1)^5$ 18 $(x^3 - 2)^4$ 19 $(x^2 - 2x + 1)^3$
20   $(\frac{1}{3}x - 1)^6$ 21 $(x^2 + 1)^3(x - 1)^2$ 22 $(x^2 + 1)^2(x - 1)^3$
23   $(x^3 + 6)/x$ 24 $(x + 2)^4/x$.   [*Hint*   Treat as a product.]

25   Find the derivative of $1/x^3$.   **26**   Find the derivative of $1/x^4$.

27   Find the derivative of $(mx + b)^n$.

28   Write $u^{mn} = (u^m)^n$ and differentiate: $mnu^{mn-1}\dfrac{du}{dx} = n(u^m)^{n-1}\dfrac{du^m}{dx}$.

Do the two sides agree?

29   Prove $(fgh)' = f'gh + fg'h + fgh'$.

30   Prove $(f_1 f_2 f_3 f_4)' = f'_1 f_2 f_3 f_4 + f_1 f'_2 f_3 f_4 + f_1 f_2 f'_3 f_4 + f_1 f_2 f_3 f'_4$.

Differentiate with respect to x

31 $(x + 1)(x + 2)(x + 3)$ **32** $(x + 1)(x^2 + 1)(x^3 + 1)$

33 $(2x + 1)(x - 3)(3x + 4)$ **34** $(3x^2 - 1)(x - 1)(2x + 3)$

35 $(2x + 1)^4(x - 3)(3x + 4)$ **36** $x(x + 1)^2(5x + 1)^3$

37 $(x^2 - 1)(x^2 - 2)(x^2 - 4)(x^2 - 8)$ **38** $(x - 1)(x + 1)(x^2 + 1)(x^4 + 1)$.

Find dy/dx at $x = 0$ for $y =$

39 $x^3(6x + 5)(2x - 7)$ **40** $x(x + 1)(x + 2)(x + 3)$.

5. QUOTIENTS AND SQUARE ROOTS

Next on the program is a rule for the derivative of a quotient. The natural guess

$$\frac{d}{dx}\left[\frac{u(x)}{v(x)}\right] = \frac{u'(x)}{v'(x)}$$

is again wrong. (Find some examples showing that this formula is false.) To derive the correct formula, suppose u/v has a derivative and apply the Product Rule to

$$u = v\,\frac{u}{v}: \qquad u' = \left(v\,\frac{u}{v}\right)' = v\left(\frac{u}{v}\right)' + \left(\frac{u}{v}\right)v'.$$

Now solve for $(u/v)'$:

$$v\left(\frac{u}{v}\right)' = u' - \left(\frac{u}{v}\right)v' = \frac{vu' - uv'}{v}, \qquad \left(\frac{u}{v}\right)' = \frac{vu' - uv'}{v^2}.$$

Quotient Rule   If $u(x)$ and $v(x)$ are differentiable and $v(x) \neq 0$, then $u(x)/v(x)$ is differentiable and

$$\frac{d}{dx}\left[\frac{u(x)}{v(x)}\right] = \frac{v(x)u'(x) - u(x)v'(x)}{[v(x)]^2}.$$

Briefly, $\dfrac{d}{dx}\left(\dfrac{u}{v}\right) = \dfrac{vu' - uv'}{v^2}.$

This is not just a restatement of the rule we derived above. There we *assumed* that u/v is differentiable, whereas the Quotient Rule says u/v *is* differentiable if u and v are (and $v \neq 0$). We postpone a proof of this technicality to Section 10.

■ **EXAMPLE 1**   Differentiate   (a) $\dfrac{x}{x^2 + 1}$ (b) $\dfrac{x^2 - 7}{x^3}$.

Solution (a) Take $u = x$ and $v = x^2 + 1$:

$$\frac{d}{dx}\left(\frac{x}{x^2 + 1}\right) = \frac{(x^2 + 1)(x)' - x(x^2 + 1)'}{(x^2 + 1)^2} = \frac{(x^2 + 1)(1) - x(2x)}{(x^2 + 1)^2} = \frac{1 - x^2}{(x^2 + 1)^2}.$$

(b) Take $u = x^2 - 7$ and $v = x^3$:

$$\frac{d}{dx}\left(\frac{x^2 - 7}{x^3}\right) = \frac{x^3(x^2 - 7)' - (x^2 - 7)(x^3)'}{x^6} = \frac{x^3(2x) - (x^2 - 7)(3x^2)}{x^6}$$

$$= \frac{-x^4 + 21x^2}{x^6} = \frac{-x^2 + 21}{x^4}. \qquad \blacksquare$$

Reciprocals   The special case $u = 1$ in the Quotient Rule yields a useful formula for the derivative of a reciprocal.

Reciprocal Rule   If $v(x) \neq 0$, then $\dfrac{d}{dx}\left(\dfrac{1}{v}\right) = -\dfrac{v'}{v^2}.$

This follows directly from the Quotient Rule, because if $u = 1$, then $u' = 0$. Therefore

$$\frac{d}{dx}\left(\frac{1}{v}\right) = \frac{v \cdot 0 - 1 \cdot v'}{v^2} = -\frac{v'}{v^2}.$$

Example $\dfrac{d}{dx}\left(\dfrac{1}{x^4 - 3x + 1}\right) = -\dfrac{(x^4 - 3x + 1)'}{(x^4 - 3x + 1)^2} = \dfrac{-4x^3 + 3}{(x^4 - 3x + 1)^2}.$

Derivatives of Powers   In the last section we derived the rule

$$\frac{d(u^n)}{dx} = nu^{n-1}\frac{du}{dx}$$

for positive integers n. The rule also holds for negative integers, provided $u(x) \neq 0$. For if $n < 0$, then $u^n = 1/u^{-n}$ and $-n > 0$. Hence by the Reciprocal Rule and the rule for positive powers,

$$\frac{d}{dx}(u^n) = \frac{d}{dx}\left(\frac{1}{u^{-n}}\right) = \frac{-1}{(u^{-n})^2}\frac{d}{dx}(u^{-n}) = (-u^{2n})\left(-nu^{-n-1}\frac{du}{dx}\right) = nu^{n-1}\frac{du}{dx}.$$

Power Rule   If $u(x)$ is differentiable and p is a positive or negative integer, then u^p is differentiable and

$$\frac{d(u^p)}{dx} = pu^{p-1}\frac{du}{dx}.$$

If $p < 0$, then $u(x) \neq 0$ is assumed.

$\blacksquare$ **EXAMPLE 2**   Differentiate $\dfrac{1}{(x^2 + 5x + 1)^3}.$

Solution   Take $u = x^2 + 5x + 1$ and $p = -3$:

$$\frac{d}{dx}(u^{-3}) = -3u^{-4}u' = -3(x^2 + 5x + 1)^{-4}(2x + 5) = \frac{-3(2x + 5)}{(x^2 + 5x + 1)^4}.$$

The result is valid for all x except for $x = \frac{1}{2}(-5 \pm \sqrt{21})$, the zeros of the quadratic $x^2 + 5x + 1$. Usually we shall not bother mentioning such obvious exceptions. ∎

The special case $u(x) = x$ of the Power Rule is important.

$$\frac{d}{dx}(x^p) = px^{p-1}, \qquad p \quad \text{any integer.}$$

Here the usual exception applies, $x \neq 0$ if $p < 0$. If $p = 0$, then $x^p = x^0 = 1$, so the formula only says $(1)' = 0$. The cases $p > 0$ and $p = -1$ agree with results we found in the last section.

Examples $\qquad \dfrac{d}{dx}\left(\dfrac{1}{x^6}\right) = -\dfrac{6}{x^7}, \qquad \dfrac{d}{dx}(x^{-5}) = -5x^{-6}.$

Square Roots   Suppose $u(x) > 0$. We want to derive a formula for the derivative of $v(x) = \sqrt{u(x)}$. If we assume that $v(x)$ is differentiable, then we can differentiate $u(x) = [v(x)]^2$ by the Power Rule and solve for v':

$$u = v^2, \qquad u' = 2vv', \qquad v' = \frac{u'}{2v} = \frac{u'}{2\sqrt{u}}.$$

Square Root Rule   If $u(x)$ is differentiable and $u(x) > 0$, then $\sqrt{u(x)}$ is differentiable and

$$\frac{d}{dx}\sqrt{u(x)} = \frac{u'(x)}{2\sqrt{u(x)}}.$$

In particular, $\qquad \dfrac{d}{dx}\sqrt{x} = \dfrac{1}{2\sqrt{x}} \qquad \text{for} \quad x > 0.$

In the derivation of the rule, we *assumed* $\sqrt{u}$ is differentiable. This will be proved in Section 10.

Remark   The Square Root Rule can be written

$$\frac{d}{dx}(u^{1/2}) = \frac{1}{2}u^{-1/2}u' \qquad (u > 0),$$

which is the Power Rule with $p = \frac{1}{2}$.

■ **EXAMPLE 3**   Differentiate

(a) $\sqrt{5x}$ (b) $\dfrac{\sqrt{x}}{6x + 1}$ (c) $(\sqrt{x^2 + 1} + 3)^4$.

Solution   (a)   Two possible solutions are

$$\frac{d}{dx}\sqrt{5x} = \frac{d}{dx}(\sqrt{5}\sqrt{x}) = \sqrt{5}\frac{d}{dx}\sqrt{x} = \frac{\sqrt{5}}{2\sqrt{x}}$$

and $\qquad \dfrac{d}{dx}\sqrt{5x} = \dfrac{(5x)'}{2\sqrt{5x}} = \dfrac{5}{2\sqrt{5x}} = \dfrac{\sqrt{5}}{2\sqrt{x}}.$

(b)   Use the Quotient and Square Root Rules:

$$\frac{d}{dx}\left(\frac{\sqrt{x}}{6x+1}\right) = \frac{(6x+1)\dfrac{d}{dx}\sqrt{x} - \sqrt{x}\dfrac{d}{dx}(6x+1)}{(6x+1)^2} = \frac{\dfrac{6x+1}{2\sqrt{x}} - 6\sqrt{x}}{(6x+1)^2}.$$

Multiply numerator and denominator by $2\sqrt{x}$:

$$\frac{d}{dx}\left(\frac{\sqrt{x}}{6x+1}\right) = \frac{6x+1-12x}{2\sqrt{x}\,(6x+1)^2} = \frac{-6x+1}{2\sqrt{x}\,(6x+1)^2}.$$

(c)   Use the Power and Square Root Rules:

$$\frac{d}{dx}(\sqrt{x^2+1}+3)^4 = 4(\sqrt{x^2+1}+3)^3 \frac{d}{dx}(\sqrt{x^2+1}+3)$$

$$= 4(\sqrt{x^2+1}+3)^3 \frac{(x^2+1)'}{2\sqrt{x^2+1}} = 4(\sqrt{x^2+1}+3)^3 \frac{2x}{2\sqrt{x^2+1}}$$

$$= \frac{4x(\sqrt{x^2+1}+3)^3}{\sqrt{x^2+1}}.\qquad \blacksquare$$

EXERCISES

Test the "formula" $(u/v)' = u'/v'$

1   $u = x$ $v = x$ 2 $u = x^2$ $v = x$

3   $u = x$ $v = x^2$ 4 $u = x^5$ $v = x^3$.

Differentiate with respect to x

5   $\dfrac{x}{x+1}$ 6 $\dfrac{x}{1-x}$ 7 $\dfrac{3x+1}{x+5}$

8   $\dfrac{x-2}{x+2}$ 9 $\dfrac{x^2+x+3}{x+4}$ 10 $\dfrac{x^2}{x^3+1}$

11   $\dfrac{1}{x^9}$ 12 $\dfrac{1}{(4x+3)^5}$ 13 $\dfrac{1}{(x^2+x)^4}$

14   $\dfrac{1}{(2x-3)^6}$ 15 $\left(\dfrac{x+1}{x+3}\right)^2$ 16 $\dfrac{x^2-1}{x^3-1}$

17   $\dfrac{1}{(x^2+1)^6}$ 18 $\dfrac{(x-1)^2}{(2x+3)^7}$ 19 $\dfrac{3x^2+1}{2x^3-1}$

20   $\dfrac{x}{(x^3+1)^2}$ 21 $\sqrt{x+3}$ 22 $\sqrt{x^2+x}$

23   $\sqrt{\dfrac{x}{x+1}}$ 24 $(1+2\sqrt{x})^3$ 25 $\dfrac{x}{\sqrt{1-x^2}}$

26   $\sqrt{\dfrac{x+a}{x-a}}$ 27 $x^2\sqrt{x^2-a^2}$ 28 $\sqrt{x+\sqrt{x}}$

29   $\dfrac{4-x}{\sqrt{8x-x^2}}$ 30 $\dfrac{1-\sqrt{3x}}{1+\sqrt{3x}}$ 31 $\left(\dfrac{x}{\sqrt{x}+1}\right)^3$

32   $\left(\dfrac{2}{x^2}+\dfrac{3}{x^5}\right)^{-2}$ 33 $\dfrac{1}{\sqrt{1+2x+3x^2}}.$

34   Show that the derivative of each rational function (quotient of polynomials) is a rational function.

35 Prove $\left(\dfrac{fg}{h}\right)' = \dfrac{(fgh)' - 2fgh'}{h^2}$.

36 Prove $\left(\dfrac{u}{v^n}\right)' = \dfrac{u'v - nuv'}{v^{n+1}}$.

6. THE CHAIN RULE

Composition of functions is an important way of combining functions to produce new ones. Recall that the composition of $y = f(u)$ and $u = g(x)$ is the function $y = f[g(x)]$. For instance, the composition of $y = u^5 - 3u^4$ and $u = 2x^2 + 1$ is

$$y = (2x^2 + 1)^5 - 3(2x^2 + 1)^4.$$

The Chain Rule allows us to differentiate composite functions.

Chain Rule If $y = f(u)$ and $u = g(x)$ are differentiable, then so is the composite function $y = f[g(x)]$ where defined, and

$$\frac{d}{dx}f[g(x)] = \left(\frac{df}{du}\bigg|_{u=g(x)}\right)\left(\frac{dg}{dx}\bigg|_{x}\right) = \frac{df}{du}[g(x)] \cdot \frac{dg}{dx}(x).$$

Briefly,

$$\frac{dy}{dx} = \frac{dy}{du} \cdot \frac{du}{dx}.$$

This rule will be proved in Section 11. In the meantime, it is easy to check that the rule is reasonable. Interpret dy/du as the rate of change of y with respect to u and du/dx as the rate of change of u with respect to x. According to the Chain Rule, the rate of change of y with respect to x is the product of these rates. For instance, if y changes five times as fast as u, and u in turn changes three times as fast as x, then y changes 15 times as fast as x.

Remark The brief version

$$\frac{dy}{dx} = \frac{dy}{du}\frac{du}{dx}$$

in the $DYDX$ notation contains a nice memory aid for the Chain Rule: "Cancel du." (However, do not think of du/dx as a fraction. So far the quantities du, dx, etc. have been given no meaning.)

■ **EXAMPLE 1** Differentiate

(a) $y = 5(x^2 + 1)^6 - 8(x^2 + 1)^2 + 9$ (b) $y = \sqrt{3x^4 + 2}$.

Solution (a) The function is the composite of $y = 5u^6 - 8u^2 + 9$ and $u = x^2 + 1$. By the Chain Rule,

$$\frac{dy}{dx} = \frac{dy}{du}\frac{du}{dx} = \frac{d}{du}(5u^6 - 8u^2 + 9)\frac{d}{dx}(x^2 + 1) = (30u^5 - 16u)(2x) = 4xu(15u^4 - 8).$$

Replace u by $x^2 + 1$. *Answer* $4x(x^2 + 1)[15(x^2 + 1)^4 - 8]$.

(b) Here $y(x)$ is the composite of $y = \sqrt{u}$ and $u = 3x^4 + 2$. By the Chain Rule,

$$\frac{dy}{dx} = \frac{dy}{du}\frac{du}{dx} = \frac{d}{du}(\sqrt{u})\frac{d}{dx}(3x^4 + 2) = \frac{1}{2\sqrt{u}}(12x^3) = \frac{6x^3}{\sqrt{u}} = \frac{6x^3}{\sqrt{3x^4 + 2}}. \quad ■$$

Special Cases   The Chain Rule implies other useful rules. For instance, the Power Rule for integers $p \neq 0$,

$$\frac{d}{dx}[u(x)]^p = p[u(x)]^{p-1}\frac{du}{dx},$$

follows from the Chain Rule and $du^p/du = pu^{p-1}$. The Square Root Rule

$$\frac{d}{dx}\sqrt{u(x)} = \frac{u'(x)}{2\sqrt{u(x)}}$$

follows from the Chain Rule and $d\sqrt{u}/du = 1/(2\sqrt{u})$.

Now we look at two further special cases.

Horizontal Shift Rule $\dfrac{d}{dx}f(x + c) = f'(x + c).$

Write $y = f(u)$ where $u = x + c$. By the Chain Rule,

$$\frac{dy}{dx} = \frac{d}{du}f(u)\frac{d}{dx}(x + c) = f'(u) \cdot 1 = f'(x + c).$$

For a geometric interpretation, recall that the graph of $y = f(x + c)$ is the graph of $y = f(x)$ shifted $|c|$ units; left if $c > 0$, right if $c < 0$. The Shift Rule says that the slope of $y = f(x + c)$ at $P = (x, f(x + c))$ equals the slope of $y = f(x)$ at $Q = (x + c, f(x + c))$. See Fig. 1.

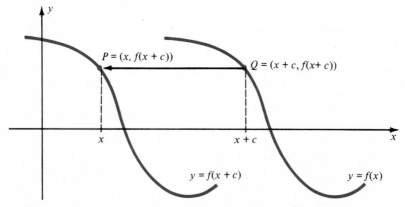

Fig. 1   Horizontal shift (with $c > 0$):   the slopes at P and Q are equal.

Change of Scale Rule $\dfrac{d}{dx}f(kx) = kf'(kx).$

Write $y = f(u)$ where $u = kx$. By the Chain Rule,

$$\frac{dy}{dx} = \frac{d}{du}f(u)\frac{d}{dx}(kx) = f'(u) \cdot k = kf'(kx).$$

Let us examine the graphs for the case $k = 2$. The curve $y = f(x)$ passes through all points $(x, f(x))$, hence through all points $(2x, f(2x))$. The curve $y = f(2x)$ passes through all points $(x, f(2x))$. Thus this second curve runs through the same values of y as the first, but twice as fast (Fig. 2). Therefore the slope of $y = f(2x)$ at $P = (x, f(2x))$ is twice the slope of $y = f(x)$ at $Q = (2x, f(2x))$. That is,

$$\frac{d}{dx} f(2x) = 2f'(2x).$$

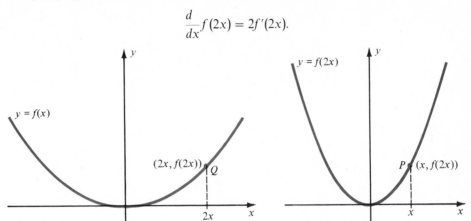

Fig. 2   Scale change: $k = 2$

Fractional Powers As an application of the Chain Rule, we shall find the derivative of the function $f(x) = x^r$, where r is rational, that is, a fraction. Here we shall assume that this function is defined for all $x > 0$ and is differentiable—these points will be discussed in Section 4, Chapter 7.

Write $r = p/q$, where p and q are integers. Then

$$f(x) = x^r = x^{p/q}, \qquad \text{so} \quad [f(x)]^q = x^p.$$

Differentiate, using the Chain Rule:

$$\frac{d}{dx}[f(x)]^q = \frac{d}{dx}(x^p), \qquad q[f(x)]^{q-1}f'(x) = px^{p-1}.$$

Solve for $f'(x)$ and simplify, using $[f(x)]^q = x^p$:

$$f'(x) = \frac{p}{q}\frac{x^{p-1}}{[f(x)]^{q-1}} = r\frac{x^p}{[f(x)]^q}\frac{f(x)}{x} = r\frac{x^p}{x^p}\frac{f(x)}{x} = r\frac{f(x)}{x} = r\frac{x^r}{x} = rx^{r-1}.$$

Power Rule   If r is rational, then $\dfrac{d}{dx}x^r = rx^{r-1}$ $(x > 0)$.

Notice the same pattern as found for integer powers:

$$\frac{d}{dx}(x^{\text{power}}) = (\text{power})x^{(\text{power})-1}.$$

■ **EXAMPLE 2**   Differentiate   (a) $y = \sqrt[3]{2x + 3}$ (b) $y = \dfrac{1}{\sqrt[4]{x^2 + 1}}$.

Solution (a)   Here $y = u^{1/3}$, where $u = 2x + 3$. By the Chain and Power Rules,

$$\frac{dy}{dx} = \frac{dy}{du}\frac{du}{dx} = \frac{1}{3}u^{-2/3} \cdot 2 = \frac{2}{3(2x + 3)^{2/3}}.$$

(b)   Here $y = u^{-1/4}$ and $u = x^2 + 1$, so

$$\frac{dy}{dx} = \frac{dy}{du}\frac{du}{dx} = -\frac{1}{4}u^{-5/4} \cdot 2x = \frac{-x}{2(x^2 + 1)^{5/4}}. \quad\blacksquare$$

Composition of Three or More Functions   In applications one frequently meets composite functions involving two or more successive compositions, such as

$$[(3x^2 + 1)^{1/3} + 1]^{1/4}.$$

Such functions are differentiated by applying the Chain Rule repeatedly. For instance if $y = y(u)$, $u = u(v)$, and $v = v(x)$, then

$$\frac{dy}{dx} = \frac{dy}{du}\frac{du}{dx}. \qquad \text{But} \qquad \frac{du}{dx} = \frac{du}{dv}\frac{dv}{dx}, \qquad \text{hence} \qquad \frac{dy}{dx} = \frac{dy}{du}\frac{du}{dv}\frac{dv}{dx}.$$

■ **EXAMPLE 3** Differentiate

(a) $y = [(3x^2 + 1)^{1/3} + 1]^{1/4}$ (b) $y = [(3x^2 + 1)^{1/3} + x]^{1/4}$.

Solution (a)   Write $y = u^{1/4}$, $u = v^{1/3} + 1$, and $v = 3x^2 + 1$. Then

$$\frac{dy}{dx} = \frac{dy}{du}\frac{du}{dv}\frac{dv}{dx} = \left(\frac{1}{4}u^{-3/4}\right)\left(\frac{1}{3}v^{-2/3}\right)(6x) = \frac{x}{2[(3x^2 + 1)^{1/3} + 1]^{3/4}(3x^2 + 1)^{2/3}}.$$

(b)   This is different because of the x term inside the square brackets. The method still works, but you must use the Sum Rule also. Write $y = u^{1/4}$, $u = v^{1/3} + x$, and $v = 3x^2 + 1$. Then

$$\frac{dy}{dx} = \frac{dy}{du}\frac{du}{dx} = \frac{1}{4}u^{-3/4}\frac{du}{dx},$$

$$\frac{du}{dx} = \frac{d}{dx}v^{1/3} + \frac{d}{dx}x = \frac{d}{dx}v^{1/3} + 1, \qquad \frac{d}{dx}v^{1/3} = \frac{d}{dv}v^{1/3}\frac{dv}{dx} = \frac{1}{3}v^{-2/3}(6x).$$

Now put the pieces together:

$$\frac{dy}{dx} = \frac{1}{4}u^{-3/4}\left[\frac{1}{3}v^{-2/3}(6x) + 1\right] = \frac{1}{4[(3x^2 + 1)^{1/3} + x]^{3/4}}\left[\frac{2x}{(3x^2 + 1)^{2/3}} + 1\right]. \quad\blacksquare$$

Remark   With some practice, you will soon do problems like these without using the intermediate variables u, v, etc., often by inspection. In numerical work, however, the extra variables can be useful because they split the calculations into several stages, for which you may have sub-routines. For instance, in Example 3(a), suppose you had to compute dy/dx for 25 values of x. Then it would make good sense to arrange the answer in the form

$$\frac{dy}{dx} = \frac{x}{2u^{3/4}v^{2/3}}, \qquad u = v^{1/3} + 1, \quad v = 3x^2 + 1,$$

and organize the computation accordingly.

A Friendly Tip Make a special effort to get the Chain Rule straight, because misuse of the Chain Rule is the cause of most mistakes in differentiation. Answers can end up rather long and messy, as Example 3 shows. Therefore keep cool, work systematically, and especially take care not to forget the innermost differentiation.

Examples

TYPICAL MISTAKE	CORRECT ANSWER
$\dfrac{d}{dx}(5x)^3 = 3(5x)^2$	$3(5x^2) \cdot 5$
$\dfrac{d}{dx}\left(\dfrac{1}{2x+7}\right) = \dfrac{-1}{(2x+7)^2}$	$\dfrac{-2}{(2x+7)^2}$
$\dfrac{d}{dx}\sqrt{3+4x^2} = \dfrac{1}{2\sqrt{3+4x^2}} \cdot 4$	$\dfrac{1}{2\sqrt{3+4x^2}} \cdot 8x$
$\dfrac{d}{dx}[1 + 5(1+4x)^2]^3$ $\qquad = 3[1 + 5(1+4x)^2]^2[10(1+4x)]$	$3[1 + 5(1+4x)^2]^2[10(1+4x)] \cdot 4$

EXERCISES

Differentiate with respect to x

1 $(x^3 + 1)^4 - 3(x^3 + 1)^2$

2 $(x^2 - x)^9 + 4(x^2 - x)^7$

3 $\left(x + \dfrac{1}{x}\right)^5 - \left(x + \dfrac{1}{x}\right)^3$

4 $5(2x - 1)^7 - 7(2x - 1)^5$

5 $-3\left(2x - \dfrac{1}{x}\right)^{10}$

6 $\dfrac{3x^6 + 1}{3x^6 - 1}$

7 $\dfrac{1}{\sqrt{2x + 1}}$

8 $[(3x^2 + 1)^2 + 1]^{1/5}$

9 $[(x^2 + 2x)^{1/3} - 1]^{1/2}$

10 $[(x^2 + 2x)^{1/3} - 3x]^{1/2}$

11 $\dfrac{\left(\dfrac{x+1}{x-1}\right)^{1/2} + 1}{\left(\dfrac{x+1}{x-1}\right)^{1/2} - 1}$

12 $(x^3 + 2x)^3 + 2(x^3 + 2x)$

13 $\dfrac{1}{2 + \sqrt[3]{x}}$

14 $\dfrac{1}{x\sqrt[3]{x^2 + a^2}} \cdot$

Find dy/dx for $y =$

15 $\dfrac{1}{1 - 2u}$ where $u = \dfrac{x - 1}{2x}$

16 $\dfrac{-u + 1}{u - 2}$ where $u = \dfrac{2x + 1}{x + 1}$

17 $u^2 + 1$ where $u = v^2 + 1$ and $v = x^2 + 1$

18 $u^2 + 1$ where $u = v^3 + 1$ and $v = x^4 + 1$

19 $\sqrt{u + 1}$ where $u = x^{2/3} + 1$

20 $u^{1/5}$ where $u = v^2 + 1$ and $v = 1/x$.

7. THE TANGENT LINE

In this section, we consider only differentiable functions. We sometimes refer to the graph of a differentiable function as a smooth curve.

Let $P = (a, f(a))$ be a point on the graph of $y = f(x)$. The **tangent** to the curve at P is defined as the line through P having slope $f'(a)$. By the point–slope form, the equation of this line is

$$y - f(a) = f'(a)(x - a), \qquad \text{that is,} \quad y = f(a) + f'(a)(x - a).$$

■ **EXAMPLE 1** Find the tangent to $y = x^2$ at $(-2, 4)$.

Solution The slope at $(-2, 4)$ is $\dfrac{dy}{dx} = 2x \Big|_{-2} = -4$.

The line through $(-2, 4)$ with slope -4 is

$$y - 4 = -4(x + 2), \qquad \text{that is,} \quad y = -4x - 4. \qquad \blacksquare$$

■ **EXAMPLE 2** Find the x-intercept of the tangent to $y = 1/x$ at $(3, \frac{1}{3})$.

Solution The equation of the tangent is

$$y = f(a) + f'(a)(x - a).$$

In this case, $f(x) = 1/x$ and $a = 3$, so

$$f(a) = \frac{1}{3}, \qquad f'(a) = \frac{-1}{x^2} \Big|_3 = -\frac{1}{9}.$$

Hence, the tangent is $y = \frac{1}{3} - \frac{1}{9}(x - 3) = -\frac{1}{9}x + \frac{2}{3}$.

To find the x-intercept, set $y = 0$ and solve for x:

$$-\tfrac{1}{9}x + \tfrac{2}{3} = 0, \qquad x = 6. \qquad \blacksquare$$

Approximating a Function Approximating complicated functions by simpler functions is one of the important ideas in calculus. Let us consider the simplest type of approximation, via linear functions.

Under a high-powered microscope, a smooth graph $y = f(x)$ appears nearly straight (Fig. 1). The tangent at $P = (a, f(a))$ is almost indistinguishable from the curve, at least very near P. See Fig. 2. Therefore, since the equation of the tangent is

$$y = f(a) + f'(a)(x - a),$$

we expect the linear function

$$y = t(x) = f(a) + f'(a)(x - a)$$

to be a good approximation to $f(x)$ provided x is near a.

To see how closely $t(x)$ approximates $f(x)$, we must measure the **error** in the approximation, that is, the difference between $f(x)$, the true value, and $t(x)$, the approximate value. So we set

$$e(x) = f(x) - t(x) = f(x) - f(a) - f'(a)(x - a).$$

In many cases, we can show by direct calculation that $|e(x)|$ is smaller than a

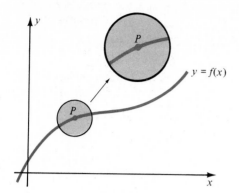

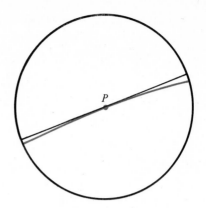

Fig. 1. Inspect the graph through a microscope.

Fig. 2. Higher magnification: the curve and its tangent are almost indistinguishable.

constant times $(x - a)^2$. That square is good news because $(x - a)^2$ is very small when x is near a, much smaller than $x - a$ itself.

Example $\qquad\qquad f(x) = x^2, \qquad a = -2.$
We found the tangent to $y = f(x)$ in Example 1:

$$y = t(x) = -4x - 4.$$

Hence $\qquad e(x) = f(x) - t(x) = x^2 - (-4x - 4) = (x + 2)^2.$

Accordingly, if x is within 0.1 of -2, that is, if $|x + 2| < 0.1$, then $e(x) \leq (0.1)^2 = 0.01$. If $|x + 2| < 0.01$, then $e(x) < 0.0001$, and in general, the error is the square of the distance from x to -2.

Example $\qquad\qquad f(x) = 1/x, \qquad a = 3.$
We found the tangent to $y = f(x)$ in Example 2:

$$y = t(x) = -\tfrac{1}{9}x + \tfrac{2}{3}.$$

The error is

$$e(x) = f(x) - t(x) = \frac{1}{x} - \left(-\frac{1}{9}x + \frac{2}{3} \right).$$

Let us tabulate to six places a few values of $e(x)$ for x near 3.

x	2.8	2.9	2.99	3.01	3.001
$1/x$	0.357143	0.344828	0.334448	0.332226	0.333222
$-\tfrac{1}{9}x + \tfrac{2}{3}$	0.355556	0.344444	0.334444	0.332222	0.333222
$e(x)$	0.001587	0.000384	0.000004	0.000004	0.000000

The approximation is extremely accurate. This is so because $e(x)$ is divisible by $(x - 3)^2$:

$$e(x) = \frac{1}{x} + \frac{1}{9}x - \frac{2}{3} = \frac{x^2 - 6x + 9}{9x} = \frac{(x - 3)^2}{9x}.$$

If $x \approx 3$, then $9x \approx 27$, so we can write

$$e(x) \approx \tfrac{1}{27}(x - 3)^2.$$

More precisely, if $x \geq 2.8$, then $9x > 25$, so

$$e(x) < \tfrac{1}{25}(x - 3)^2 \qquad \text{if} \quad x > 2.8.$$

Thus, for instance, if $|x - 3| \leq 0.2$, then

$$e(x) < \tfrac{1}{25}(0.2)^2 = \tfrac{1}{25}(0.04) = 0.0016 < 0.002.$$

Similarly, if $|x - 3| \leq 0.1$, then $e(x) \leq 0.00004$. If $|x - 3| \leq 0.01$, then $e(x) \leq 4 \times 10^{-6}$. Finally, if $|x - 3| \leq 0.001$, then $e(x) \leq 4 \times 10^{-8}$. These estimates compare very well to the exact values of $e(x)$ in the table above.

We shall return to this subject in Chapter 10.

EXERCISES

Find the equation of the tangent to the curve

1   $y = x^2$ through (2, 4)
2   $y = x^3$ through (2, 8)
3   $y = 1/x$ through $(-1, -1)$
4   $y = 1/x$ through $(2, \frac{1}{2})$
5   $y = x - \dfrac{1}{x}$ at $x = -1$
6   $y = \dfrac{x}{x^2 + 4}$ at $x = 0$
7   $y = \dfrac{1}{x^2}$ at $x = -3$
8   $y = (\frac{1}{2}x - 1)^5$ at $x = 4$.

Find the equation(s) of the tangent(s) to the curve

9   $y = x^2$ at all points where the slope is 10
10   $y = x^3$ at all points where the slope is 27
11   $y = x^2$ if the tangent crosses the y-axis at $y = -16$
12   $y = x^3$ if the tangent crosses the y-axis at $y = -128$.

Find the equation(s) of the tangent line(s) to the curve and determine where each crosses the coordinate axes

13   $y = x^2 + 3x - 1$ through (2, 9)
14   $y = x^3 - 8x^2$ through (10, 200)
15   $y = 1/x$ and the tangent has slope -81
16   $y = 1/x$ and the tangent has slope -6.

Find the area of the triangle bounded by the coordinate axes and the line tangent to $y = 1/x$

17   at $x = 2$
18   at $x = a$,   where $a \neq 0$.

19   Show that the tangents to the parabola $y = x^2$ at (3, 9) and $(-3, 9)$ cross on the y-axis. Where?

20   Exactly one of the lines $y = 3x + b$ is tangent to the parabola $y = x^2$. Which one?

Find the equation of the line tangent to the curve at the specified point; also find the error made in approximating the curve by its tangent

21   $y = 1 - x^2$ at $(0, 1)$ 22 $y = 2x^2 + 3$ at $(1, 5)$
23   $y = x^3$ at $(3, 27)$ 24 $y = x^2 + x + 1$ at $(-1, 1)$

25   $y = \dfrac{1}{3x + 4}$ at $(-1, 1)$ 26 $y = x^2 - x^3$ at $(1, 0)$.

27   Let $y = t(x)$ be the tangent to $y = \sqrt{x}$ at $(1, 1)$. Compute $t(x)$ and tabulate $\sqrt{x}$, $t(x)$, and $e(x) = \sqrt{x} - t(x)$ for $x = 0.9$, 0.98, 1.02, 1.1.
28   Let $y = t(x)$ be the tangent to $y = x^5 + 2x$ at $(0, 0)$. Compute $t(x)$ and tabulate $x^5 + 2x$, $t(x)$, and $e(x) = (x^5 + 2x) - t(x)$ for $x = -0.2$, -0.1, 0.1, 0.2.
29   Let $y = t(x)$ be the tangent to $y = x^3$ at $(2, 8)$.   Prove
$$|x^3 - t(x)| < 7|x - 2|^2 \qquad \text{for} \qquad |x - 2| < 1.$$
30   Let $y = t(x)$ be the tangent to $y = 1/x^2$ at $(1, 1)$.   Prove
$$\left| \frac{1}{x^2} - t(x) \right| < 16|x - 1|^2 \qquad \text{for} \qquad \frac{1}{2} < x < \frac{3}{2}.$$

8. ANTIDERIVATIVES

Basic Facts about Derivatives   Suppose $f(x)$ is a differentiable function whose derivative $f'(x)$ is zero at each point of an interval $a < x < b$. It seems almost obvious that $f(x)$ is a constant function because the tangent at each point of the curve $y = f(x)$ is horizontal. Surely then, the graph must be a horizontal straight line.

> If $f'(x) = 0$ for all x in the interval $a < x < b$, then $f(x) = c$, a constant function on the interval.

Here is another way of looking at it. If the rate of change of $f(x)$ is always 0, then $f(x)$ never changes, so $f(x)$ must be constant. (We shall give a rigorous proof on p. 150.)

Now suppose we are given two functions, $f(x)$ and $g(x)$, whose derivatives agree, $f'(x) = g'(x)$, at each point of an interval $a < x < b$. How are $f(x)$ and $g(x)$ related? Well, their difference $f(x) - g(x)$ has derivative 0:
$$[f(x) - g(x)]' = f'(x) - g'(x) = 0.$$
But a function with derivative 0 is constant, so $f(x) - g(x) = c$.

> If $f'(x) = g'(x)$ for all x in the interval $a < x < b$, then
> $$f(x) = g(x) + c.$$

The graphs of $f(x)$ and $g(x)$ are identical except for a vertical shift (Fig. 1). Conversely, the derivative cannot distinguish between vertically shifted curves, because if $f(x) = g(x) + c$, then $f'(x) = g'(x)$.

Example x^3, $x^3 + 1$, $x^3 - \frac{3}{2}$, $x^3 + 15$, etc. all have the same derivative, $3x^2$. Furthermore, any function with derivative $3x^2$ must be one of the functions $x^3 + c$.

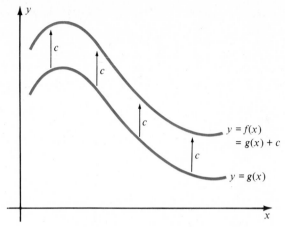

Fig. 1   Graphs of functions with equal derivatives

Antidifferentiation   Differentiation is the process of finding the derivative of a given function. **Antidifferentiation** is the reverse process: finding a function $f(x)$ having a given derivative $f'(x)$. There is not just one function whose derivative is $f'(x)$, but a family of functions which differ from each other by additive constants.

■ **EXAMPLE 1**   Find $f(x)$ for $x > 0$ if

(a) $f'(x) = x^2 + \dfrac{2}{x^2}$ (b) $f'(x) = 1 + \dfrac{3}{\sqrt{x}}$.

Solution   (a)   Determine an antiderivative for each term. From

$$\frac{d}{dx}x^3 = 3x^2 \quad \text{follows} \quad \frac{d}{dx}\left(\frac{1}{3}x^3\right) = x^2,$$

and from

$$\frac{d}{dx}\left(\frac{1}{x}\right) = -\frac{1}{x^2} \quad \text{follows} \quad \frac{d}{dx}\left(-\frac{2}{x}\right) = \frac{2}{x^2}.$$

Therefore $\frac{1}{3}x^3 - 2/x$ is a function whose derivative is $x^2 + 2/x^2$. Any other function with the same derivative must agree except for an additive constant.

Answer $f(x) = \frac{1}{3}x^3 - 2/x + c.$

(b)   Clearly $\dfrac{d}{dx}x = 1,$ and from

$$\frac{d}{dx}\sqrt{x} = \frac{1}{2\sqrt{x}} \quad \text{follows} \quad \frac{d}{dx}(6\sqrt{x}) = \frac{3}{\sqrt{x}}.$$

Therefore

$$\frac{d}{dx}(x + 6\sqrt{x}) = 1 + \frac{3}{\sqrt{x}}.$$

Answer $f(x) = x + 6\sqrt{x} + c.$ ■

Antidifferentiation produces a *family* of functions, of the form $f(x) + c$. The corresponding graphs $y = f(x) + c$ make up a family of parallel curves that differ from each other by vertical shifts. A particular choice of c singles out one curve of the family. Often c is chosen to make the curve pass through a given point.

■ **EXAMPLE 2** Graph the family of curves $y = f(x)$ where $f'(x) = x - 2$. Which of these curves passes through $(2, 1)$?

Solution $\frac{1}{2}x^2 - 2x$ is an antiderivative of $x - 2$. Therefore the family of curves consists of the parabolas $y = \frac{1}{2}x^2 - 2x + c$, where c is any constant (Fig. 2). To choose c so that $(2, 1)$ is a point of $y = \frac{1}{2}x^2 - 2x + c$, substitute $x = 2$, $y = 1$:

$$1 = \tfrac{1}{2}(2)^2 - 2 \cdot 2 + c = -2 + c, \qquad c = 3.$$

Answer $\quad y = \frac{1}{2}x^2 - 2x + 3.$

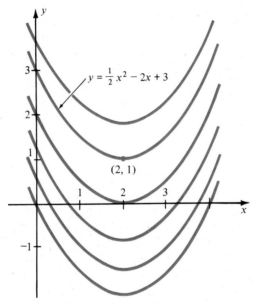

Fig. 2 The family $y = \frac{1}{2}x^2 - 2x + c$ of functions with $dy/dx = x - 2$ ■

Antidifferentiation Formulas The technique of antidifferentiation is a vital tool in integration; we shall study it in detail in Chapters 5 and 8. Here we shall note some general principles.

Each formula for the derivative of a function is also a formula for an antiderivative. For example, the differentiation formula

$$\frac{d}{dx}\left(\frac{1}{x}\right) = -\frac{1}{x^2}$$

becomes an antidifferentiation formula when read from *right to left*. It says that $1/x$ is an antiderivative of $-1/x^2$. Similarly, $d(\sqrt{x})/dx = 1/2\sqrt{x}$ says that $\sqrt{x}$ is an antiderivative of $1/2\sqrt{x}$.

We can adjust constant factors in antidifferentiation by means of the formula

$$\frac{d}{dx}[af(x)] = af'(x),$$

which says: If $f(x)$ is an antiderivative of $f'(x)$, then $af(x)$ is an antiderivative of

$af'(x)$. For example,

from $\dfrac{d}{dx}\left(\dfrac{1}{x}\right) = -\dfrac{1}{x^2}$ follows $\dfrac{5}{x^2} = \dfrac{d}{dx}\left(-\dfrac{5}{x}\right)$;

from $\dfrac{d}{dx}\sqrt{x} = \dfrac{1}{2\sqrt{x}}$ follows $\dfrac{1}{\sqrt{x}} = \dfrac{d}{dx}(2\sqrt{x})$.

Recall the formula

$$\frac{d}{dx}x^{r+1} = (r+1)x^r \qquad (r \text{ rational}).$$

If $r \neq -1$, we can divide by $r+1$ and obtain a basic antidifferentiation formula.

$$\boxed{\,x^r = \frac{d}{dx}\left(\frac{x^{r+1}}{r+1} + c\right), \qquad r \text{ rational}, \quad r \neq -1\,}$$

The formula is written so as to show all antiderivatives of x^r. It is valid on the domain of x^r except possibly at $x = 0$.

Remark   An antiderivative of $g(x)$ is often called a **primitive** of $g(x)$.

EXERCISES

Find all functions with derivative

1 $4x$	**2** -7	**3** $-x$
4 $5x + 6$	**5** $-2x + 1$	**6** $3(x+1)$
7 $x^2 - 2x$	**8** $3x^2 + 4$	**9** $x^3 + x - 1$
10 $8x^3 - 6x^2 + 1$	**11** $x + \dfrac{1}{x^2}$	**12** $x^2 - \dfrac{2}{x^3}$
13 $(x-2)(x-3)$	**14** $\left(x + \dfrac{1}{x}\right)^2$	**15** $\dfrac{1}{x^2} - \dfrac{3}{x^4}$
16 $(x-3)^5$	**17** $\dfrac{4}{(x-3)^3}$	**18** $(x+1)^3 - \dfrac{1}{2(x-1)^4}$
19 $(2x+1)^6$	**20** $(5x-3)^5$	**21** $(\tfrac{1}{2}x + \tfrac{1}{3})^7$
22 $\dfrac{1}{(2x-1)^3}$	**23** $x^2 + \dfrac{2}{\sqrt{x}}$	**24** $\dfrac{1}{\sqrt{x+5}}$
25 $\sqrt{x}$	**26** $\sqrt[3]{x}$	**27** $6x - \dfrac{1}{x^{3/4}}$
28 $\dfrac{x}{\sqrt{x^2+3}}$	**29** $3ax^2 + 2bx + c$	**30** $(ax+b)^n, \quad n \neq -1.$

Find the curve $y = f(x)$ such that

31 $f'(x) = x - 5$ and $f(1) = 2$

32 $f'(x) = -(x+3)$ and $f(0) = 4$

33 $\dfrac{dy}{dx} = 3x^2 - 4x + \dfrac{1}{x^2}$ and the point $(2, 9)$ is on the curve

34 $\dfrac{dy}{dx} = (3x-4)^2 + 5x$ and the point $(1, 2)$ is on the curve

[*Hint* $(3x-4)^2 = 9(x - \tfrac{4}{3})^2.$]

Find an antiderivative of

35 $f(x)f'(x)$

36 $\dfrac{f'(x)}{[f(x)]^2}$.

Suppose $F(x)$ is an antiderivative of $f(x)$. Find an antiderivative of

37 $[F(x)]^2 f(x)$

38 $f(x+3)$.

9. HIGHER DERIVATIVES

Second Derivatives Start with a differentiable function $f(x)$. Its derivative $f'(x)$ is itself a function. If $f'(x)$ in turn is differentiable, then

$$\frac{d}{dx}[f'(x)]$$

is called the **second derivative** of $y = f(x)$ and is written

$$\frac{d^2y}{dx^2} \quad \text{or} \quad f''(x) \quad \text{or} \quad y''.$$

A function that has a second derivative is called **twice differentiable**.

The symbol d^2y/dx^2 is usually read " D2YDX squared". This notation is natural in terms of operators. If we think of the second derivative operator as the compositve of the operator d/dx with itself, then

$$\frac{d^2}{dx^2} = \frac{d}{dx} \circ \frac{d}{dx} = \left(\frac{d}{dx}\right)^2.$$

Examples

$$f(x) = x^3 - 4x, \qquad f'(x) = 3x^2 - 4, \qquad f''(x) = 6x.$$

$$H(t) = 2(t-1)^4, \qquad \frac{dH}{dt} = 8(t-1)^3, \qquad \frac{d^2H}{dt^2} = 24(t-1)^2.$$

■ **EXAMPLE 1** Compute the second derivative of

(a) $y = \sqrt{x}$ \qquad (b) $y = (x^3 + 6)^5$.

Solution (a) $\dfrac{dy}{dx} = \dfrac{d}{dx}\sqrt{x} = \dfrac{1}{2\sqrt{x}} = \dfrac{1}{2}x^{-1/2}$.

$$\frac{d^2y}{dx^2} = \frac{d}{dx}\left(\frac{dy}{dx}\right) = \frac{d}{dx}\left(\frac{1}{2}x^{-1/2}\right) = \frac{1}{2}\frac{d}{dx}(x^{-1/2}) = -\frac{1}{4}x^{-3/2}.$$

(b) $\dfrac{dy}{dx} = \dfrac{d}{dx}(x^3 + 6)^5 = 5(x^3 + 6)^4(3x^2) = 15x^2(x^3 + 6)^4.$

By the Product and Chain Rules,

$$\frac{d^2y}{dx^2} = \frac{d}{dx}\left(\frac{dy}{dx}\right) = \frac{d}{dx}[15x^2(x^3 + 6)^4] = 15\left[x^2\frac{d}{dx}(x^3 + 6)^4 + (x^3 + 6)^4\frac{d}{dx}x^2\right]$$

$$= 15[x^2 \cdot 4(x^3 + 6)^3 \cdot 3x^2 + (x^3 + 6)^4 \cdot 2x]$$

$$= 30x(x^3 + 6)^3[6x^3 + (x^3 + 6)] = 30x(x^3 + 6)^3(7x^3 + 6). \qquad ■$$

■ **EXAMPLE 2** Find all functions $f(x)$ such that $f''(x) = 0$ for all x.

Solution $\dfrac{d}{dx}f'(x) = f''(x) = 0$, hence $f'(x)$ is a function with derivative zero.

Therefore $f'(x) = a$, a constant function. This means that $f(x)$ is an antiderivative of a. One antiderivative is ax, so the most general antiderivative is $ax + b$, where b is a constant.

$$\text{Answer}\quad f(x) = ax + b. \quad ■$$

The second derivative $f''(x)$ is the rate of change with respect to x of the first derivative $f'(x)$, hence $f''(x)$ measures how fast the slope of the curve $y = f(x)$ is changing. This important information will be applied in the next chapter.

Third and Higher Derivatives If the second derivative of $y = f(x)$ has a derivative, it is called the **third derivative** of $f(x)$ and is written

$$\frac{d^3y}{dx^3}\quad \text{or}\quad f'''(x)\quad \text{or}\quad y'''.$$

Examples
$$f(x) = x^{10},\quad f'(x) = 10x^9,\quad f''(x) = 90x^8,\quad f'''(x) = 720x^7.$$
$$y = x^4 - 5x^3 + 3x + 7,\quad y' = 4x^3 - 15x^2 + 3,$$
$$y'' = 12x^2 - 30x,\quad y''' = 24x - 30.$$

Fourth, fifth, and higher derivatives are defined similarly by repeated differentiation. The n-th derivative of $y = f(x)$ is written

$$\frac{d^ny}{dx^n}\quad \text{or}\quad f^{(n)}(x)\quad \text{or}\quad y^{(n)}.$$

Examples (continuing those above)
$$f^{(4)}(x) = 720(7x^6) = 5040x^6,\quad f^{(5)}(x) = 5040(6x^5) = 30{,}240x^5,$$
$$f^{(6)}(x) = 30{,}240(5x^4) = 151{,}200x^4,\quad \text{etc.}$$
$$y^{(4)} = 24,\quad y^{(5)} = 0,\quad y^{(6)} = y^{(7)} = \cdots = 0.$$

■ **EXAMPLE 3** Compute

(a) $y^{(4)}$ for $y = (\tfrac{1}{2}x - 3)^8$

(b) $y^{(5)}$ for $y = 19x^4 + 36x^3 + 499x^2 - 125x - 62.$

Solution (a) Differentiate carefully four times, using the Chain Rule. Don't forget the factor $\tfrac{1}{2}$ at each step:

$$y' = 8(\tfrac{1}{2}x - 3)^7(\tfrac{1}{2}),\quad y'' = 8 \cdot 7(\tfrac{1}{2}x - 3)^6(\tfrac{1}{2})^2,\quad y''' = 8 \cdot 7 \cdot 6(\tfrac{1}{2}x - 3)^5(\tfrac{1}{2})^3,$$
$$y^{(4)} = 8 \cdot 7 \cdot 6 \cdot 5(\tfrac{1}{2}x - 3)^4(\tfrac{1}{2})^4 = 105(\tfrac{1}{2}x - 3)^4.$$

(b) By inspection $y^{(5)} = 0$, because the derivative of a polynomial is a polynomial of one lower degree. Hence $y, y', y'', y''', y^{(4)}$ have degrees 4, 3, 2, 1, 0, respectively. Thus $y^{(4)}$ is a constant, so $y^{(5)} = 0.$ ■

■ **EXAMPLE 4**　Given　$y = 1/x$,　find　(a) $y^{(25)}$　(b) $y^{(n)}$.

Solution　(a)　Compute a few derivatives:

$$y' = -\frac{1}{x^2}, \qquad y'' = \frac{2}{x^3}, \qquad y''' = -\frac{2 \cdot 3}{x^4}, \qquad y^{(4)} = \frac{2 \cdot 3 \cdot 4}{x^5}.$$

A clear pattern emerges: The numerators are $1!$, $2!$, $3!$, $4!$, etc., and the denominators are x^2, x^3, x^4, x^5, etc. The signs alternate, minus for odd-order derivatives, plus for even. According to this pattern,

$$y^{(25)} = -\frac{25!}{x^{26}}.$$

(b)　By the same reasoning, $y^{(n)} = \pm \dfrac{\text{(factorial)}}{\text{(power of } x)}$.

The factorial must be $n!$ with the same n as in $y^{(n)}$. The exponent of x is one larger, that is, $n + 1$. For the sign, introduce the factor $(-1)^n$, an automatic sign changer. Its value is $+1$ for n even and -1 for n odd.

$$\textit{Answer} \quad y^{(n)} = (-1)^n \frac{n!}{x^{n+1}}. \quad ■$$

Remark　It is easy to make mistakes when finding general formulas as in Example 3(b). Be especially careful in case of an alternating sign; note that $(-1)^{n+1}$ is the sign changer that is $+1$ for n *odd* and -1 for n *even*. Check your answer for several low values of n and, if possible, prove it by induction.

■ **EXAMPLE 5**　Find the n-th derivative of　$y = \sqrt{x}$.

Solution　Write $y = x^{1/2}$ and compute a few derivatives:

$$y' = \tfrac{1}{2}x^{-1/2}, \qquad y'' = \left(\tfrac{1}{2}\right)\left(-\tfrac{1}{2}\right)x^{-3/2}, \qquad y''' = \left(\tfrac{1}{2}\right)\left(-\tfrac{1}{2}\right)\left(-\tfrac{3}{2}\right)x^{-5/2},$$

$$y^{(4)} = \left(\tfrac{1}{2}\right)\left(-\tfrac{1}{2}\right)\left(-\tfrac{3}{2}\right)\left(-\tfrac{5}{2}\right)x^{-7/2}.$$

The pattern is

$$y^{(n)} = \pm \frac{\text{(product of odd integers)}}{2^n} x^{\text{(exponent)}}.$$

Each differentiation lowers the exponent by one. Starting with exponent $\tfrac{1}{2}$, differentiation n times lowers the exponent to

$$\tfrac{1}{2} - n = \tfrac{1}{2}(1 - 2n) = -\tfrac{1}{2}(2n - 1).$$

The numerator is the product $1 \cdot 3 \cdot 5 \cdot 7 \cdots$, ending with the odd number just before $2n - 1$, that is, with $2n - 3$. Hence the numerator is $1 \cdot 3 \cdot 5 \cdots (2n - 3)$. Finally, the sign is $(-1)^{n+1}$, positive when $n = 1, 3, 5$, etc.

$$\textit{Answer} \quad y^{(n)} = (-1)^{n+1} \frac{1 \cdot 3 \cdot 5 \cdots (2n - 3)}{2^n} x^{-(2n-1)/2}.$$

Check it for $n = 2, 3,$ and 4.　　　　　　　　　　　■

Remark 1 Since there are tables for $n!$ but not for $1 \cdot 3 \cdot 5 \cdots (2n - 3)$, it is convenient to express the numerator in terms of factorials. We fill in the missing even factors and divide them out again:

$$1 \cdot 3 \cdot 5 \cdots (2n - 3) = \frac{1 \cdot 2 \cdot 3 \cdot 4 \cdots (2n - 2)}{2 \cdot 4 \cdot 6 \cdot 8 \cdots (2n - 2)}$$

$$= \frac{(2n - 2)!}{(2 \cdot 1)(2 \cdot 2)(2 \cdot 3) \cdots [2(n - 1)]} = \frac{(2n - 2)!}{2^{n-1}(n - 1)!}.$$

With this modification, the answer to Example 5 is

$$y^{(n)} = (-1)^{n+1} \frac{(2n - 2)!}{2^{2n-1}(n - 1)!} x^{-(2n-1)/2}.$$

This answer is correct for $n = 1$, but the previous answer is not. Why?

Remark 2 It may be difficult or impossible to find a formula for the n-th derivative of a given function. Just try computing four or five derivatives of $y = (x^2 - 2)/(x^3 + 1)$ for example and you will soon be convinced.

The Formula of Leibniz Suppose you want the second or third derivative of $y = x^3(2x - 1)^{10}$. You can expand $(2x - 1)^{10}$ by the binomial theorem, multiply by x^3, then differentiate. But that's a lot of work; it's much easier to treat y as the product of x^3 and $(2x - 1)^{10}$.

In general, if $y = uv$, then $y' = uv' + vu'$. Differentiate again, using the Product Rule on each term:

$$y'' = (uv')' + (vu')' = (uv'' + u'v') + (vu'' + u'v') = uv'' + 2u'v' + u''v.$$

For the third derivative differentiate again, using the Product Rule on each term. The result is

$$y''' = uv''' + 3u'v'' + 3u''v' + u'''v.$$

Similarly, the fourth derivative is

$$y^{(4)} = uv^{(4)} + 4u'v''' + 6u''v'' + 4u'''v' + u^{(4)}v.$$

Note the similarity of these formulas to the binomial expansions

$$(u + v)^2 = v^2 + 2uv + u^2, \qquad (u + v)^3 = v^3 + 3uv^2 + 3u^2v + u^3,$$

$$(u + v)^4 = v^4 + 4uv^3 + 6u^2v^2 + 4u^3v + u^4.$$

Leibniz Formula The n-th derivative of a product $y = uv$ is

$$y^{(n)} = uv^{(n)} + \binom{n}{1} u'v^{(n-1)} + \binom{n}{2} u''v^{(n-2)} + \cdots + \binom{n}{n-1} u^{(n-1)}v' + u^{(n)}v.$$

The coefficients are the binomial coefficients

$$\binom{n}{k} = \frac{n(n - 1)(n - 2) \cdots (n - k + 1)}{k!} = \frac{n!}{k!(n - k)!}.$$

■ **EXAMPLE 6** Compute

(a) y'', where $y = x^3(2x - 1)^{10}$ (b) $y^{(5)}$, where $y = x(x + 3)^8$.

Solution (a) Let $u = x^3$ and $v = (2x - 1)^{10}$. Then $y = uv$ and by the Leibniz Formula

$$y'' = uv'' + 2u'v' + u''v$$

$$= x^3[10 \cdot 9 \cdot 2^2(2x - 1)^8] + 2 \cdot 3x^2[10 \cdot 2(2x - 1)^9] + 6x(2x - 1)^{10}$$

$$= 6x(2x - 1)^8[60x^2 + 20x(2x - 1) + (2x - 1)^2]$$

$$= 6x(2x - 1)^8(104x^2 - 24x + 1).$$

(b) Let $u = x$ and $v = (x + 3)^8$. Use the Leibniz Formula for $y^{(5)}$ noting that $u'' = u''' = \cdots = 0$:

$$y^{(5)} = uv^{(5)} + 5u'v^{(4)} + 0 + 0 + 0 + 0$$

$$= x[8 \cdot 7 \cdot 6 \cdot 5 \cdot 4(x + 3)^3] + 5[8 \cdot 7 \cdot 6 \cdot 5(x + 3)^4] = 5040(x + 3)^3(3x + 5).$$

■

EXERCISES

Find d^2y/dx^2 for $y =$

1 $3x^2 - 2x + 1$	**2** $x^3 - 7x$	**3** $x^2(1 - x)$
4 $(x^2 - 1)(x^2 - 2)$	**5** $x^4(x + 1)^2$	**6** $x^9 - 8x^7$
7 $\dfrac{1}{1 + x^2}$	**8** $\dfrac{x}{x^2 + 4}$	**9** $\dfrac{x}{x - 2}$
10 $\dfrac{x - 1}{x + 1}$	**11** $\dfrac{x^2 + a}{x^2 - a}$	**12** $\dfrac{x^2}{x^3 + 2}$
13 $\dfrac{1}{\sqrt{1 - 3x}}$	**14** $\dfrac{1}{1 + \sqrt{x}}$	**15** $\sqrt{x^2 + 9}$
16 $\dfrac{x}{\sqrt{1 - x^2}}$	**17** $\sqrt[3]{x} - 5\sqrt[3]{1 + x}$	**18** $(x^{1/5} + 2)^6$.

Find $\left.\dfrac{d^2y}{dt^2}\right|_{t=a}$

19 $y = \frac{4}{3}\pi t^3$ $a = 2$

20 $y = \dfrac{1}{3t - 1}$ $a = 1$

21 $y = \frac{1}{2}t^{5/4}$ $a = 81$

22 $y = \dfrac{3}{t} - 8t^{10}$ $a = -1$

23 $y = t^2(t - 3)^2$ $a = 0$

24 $y = t^8(4t - 5)^2$ $a = 1$.

25 Find a formula for $\dfrac{d^n}{dx^n}(x^{1/3})$.

26 Find a formula for $\dfrac{d^n}{dx^n}\left(\dfrac{x + 1}{x - 1}\right)$.

27 For what values of x are the first and second derivatives of $x^3 - 2x^2 - x$ equal?

28 For what values of x is $x^3 + 6x + 1$ equal to its second derivative?

29 Find all functions $f(x)$ such that $f''(x) = 4x - 1$ for all x.

30 Find all functions satisfying $f'''(x) = 0$ for all x.

Use the formula of Leibniz to find

31 $\dfrac{d^2}{dx^2}[x^4(2x + 1)^6]$

32 $\dfrac{d^2}{dx^2}[\sqrt{x}\,(x - 10)^5]$

33　$\dfrac{d^5}{dx^5}[x(x-3)^8]$　　　　**34**　$\dfrac{d^3}{dx^3}[x^5(x+3)^6]$

35　$\dfrac{d^{10}}{dx^{10}}[(x+1)\sqrt{x}]$　　　　**36**　$\dfrac{d^2}{dx^2}[\sqrt{x-1}\,\sqrt[3]{x-2}\,]$

37　$\dfrac{d^5}{dx^5}[x^3(x-1)^{10}]$　at　$x=1$　　　　**38**　$\dfrac{d^5}{dx^5}[x^5\sqrt{x^3+16}\,]$　at　$x=0$.

Compute

39　$\dfrac{d}{dx}\left[x-\dfrac{f(x)}{f'(x)}\right]$　　　　**40**　$f''[f(x)]-\dfrac{d}{dx}\left(\dfrac{f'[f(x)]}{f'(x)}\right)$.

10. LIMITS AND CONTINUITY

The idea of a limit is basic in calculus. We gave a rough definition of limit in Section 2; now it is time to be precise.

We want to define

$$\lim_{x\to a} f(x) = L.$$

We want this batch of symbols to say that the values of $f(x)$ are as close to L as desired, *provided* x is close enough (but not equal) to a. By "as close to L as desired" we shall mean that for each positive number ε we have $|f(x)-L|<\varepsilon$, provided x is close enough to a. You should think of ε as *very* small: 10^{-10}, 10^{-100}, and even smaller. By "x is close to a" we shall mean that x lies in a small segment centered at a, but with a itself excluded. Such a segment (Fig. 1) is described by an inequality

$$0 < |x-a| < \delta, \qquad \text{where} \quad \delta > 0.$$

excluded

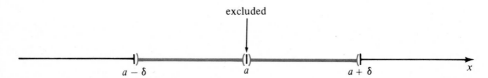

Fig. 1　The domain $0 < |x-a| < \delta$:
an interval $a - \delta < x < a + \delta$ with $x = a$ excluded

Let us summarize the discussion with a formal definition:

Definition of Limit　Suppose $f(x)$ is defined on an interval $x_0 \le x \le x_1$ and that a is a point of this interval. Then

$$\lim_{x\to a} f(x) = L$$

if for each $\varepsilon > 0$ there exists a number $\delta > 0$ such that

$$|f(x)-L|<\varepsilon \qquad \text{whenever} \quad 0 < |x-a| < \delta$$

and x is in the interval.

The expression $$\lim_{x \to a} f(x) = L$$

is read: "The limit as x approaches a of $f(x)$ is L." Often it is abbreviated

$$f(x) \longrightarrow L \quad \text{as} \quad x \longrightarrow a,$$

read: "$f(x)$ approaches L as x approaches a."

In actual practice, the definition works this way. It challenges you with an ε, and you must produce a suitable δ. You must be able to do so for *every* $\varepsilon > 0$, not just a particular ε.

Let us look at the definition of limit from a slightly different point of view. If $f(x) \longrightarrow L$ as $x \longrightarrow a$, then the difference between $f(x)$ and L becomes smaller and smaller as $x \longrightarrow a$. In other words, the difference approaches 0. Think of L as approximating $f(x)$. Then this difference is the error in the approximation; call it $E(x)$. Thus

$$f(x) = L + E(x), \quad \text{where} \quad E(x) = f(x) - L.$$

The assertion that $f(x) \longrightarrow L$ as $x \longrightarrow a$ is exactly the same as the assertion that $E(x) \longrightarrow 0$ as $x \longrightarrow a$, because $|f(x) - L| < \varepsilon$ is exactly the same as $|E(x)| < \varepsilon$.

Alternative Definition of Limit $f(x) \longrightarrow L$ as $x \longrightarrow a$ if and only if

$$f(x) = L + E(x),$$

where $E(x) \longrightarrow 0$ as $x \longrightarrow a$.

We shall soon prove some important basic properties of limits. In order to clarify the proofs, we first state a preliminary lemma about limit 0.

Lemma   Suppose $f(x) \longrightarrow 0$ and $g(x) \longrightarrow 0$ as $x \longrightarrow a$. Suppose also that c and $M \neq 0$ are constants. Then

(a) $\lim_{x \to a}[cf(x)] = 0$

(b) $\lim_{x \to a}[f(x) \pm g(x)] = 0$

(c) $\lim_{x \to a} f(x)g(x) = 0$

(d) $\lim_{x \to a} \dfrac{f(x)}{M + f(x)} = 0.$

Proof (a) This is the easiest to prove; we leave it as an exercise.

(b) Let $\varepsilon > 0$ be given. We must find $\delta > 0$ so that $|f(x) \pm g(x)| < \varepsilon$ whenever $0 < |x - a| < \delta$. By hypothesis, $\lim f(x) = 0$ and $\lim g(x) = 0$ as $x \longrightarrow a$. We apply the definition to each of these functions, but with the positive number $\frac{1}{2}\varepsilon$ rather than ε. (Note carefully this technique. We want to prove something about ε, but we apply something we already know to $\frac{1}{2}\varepsilon$ instead of ε. Watch to see why this pays off.) Since $f(x) \longrightarrow 0$, there is $\delta_1 > 0$ such that

$$|f(x)| < \tfrac{1}{2}\varepsilon \quad \text{whenever} \quad 0 < |x - a| < \delta_1 .$$

Since $g(x) \longrightarrow 0$, there is $\delta_2 > 0$ such that

$$|g(x)| < \tfrac{1}{2}\varepsilon \quad \text{whenever} \quad 0 < |x - a| < \delta_2 .$$

Now let δ be the smaller of δ_1 and δ_2, the minimum of δ_1 and δ_2. Then $\delta > 0$ and whenever $0 < |x - a| < \delta$, *both*

$$0 < |x - a| < \delta_1 \qquad \text{and} \qquad 0 < |x - a| < \delta_2.$$

Hence *both*

$$|f(x)| < \tfrac{1}{2}\varepsilon \qquad \text{and} \qquad |g(x)| < \tfrac{1}{2}\varepsilon,$$

so

$$|f(x) \pm g(x)| \le |f(x)| + |g(x)| < \tfrac{1}{2}\varepsilon + \tfrac{1}{2}\varepsilon = \varepsilon.$$

(We used the triangle inequality here.) To summarize, given $\varepsilon > 0$, we have found $\delta > 0$ such that

$$|f(x) \pm g(x)| < \varepsilon \qquad \text{whenever} \qquad |x - a| < \delta.$$

Therefore $f(x) \pm g(x) \longrightarrow 0$ as $x \longrightarrow a$, which completes the proof of (b). Please reread this proof carefully a couple of times until you are sure you really understand it. It is a model for other proofs, which will be presented with less detail.

(c)   Given $\varepsilon > 0$, choose $\delta_1 > 0$ and $\delta_2 > 0$ so that

$$|f(x)| < \varepsilon \qquad \text{whenever} \quad 0 < |x - a| < \delta_1,$$

$$|g(x)| < 1 \qquad \text{whenever} \quad 0 < |x - a| < \delta_2.$$

Let δ be the minimum of δ_1 and δ_2. Then $\delta > 0$ and

$$|f(x)| < \varepsilon \qquad \text{and} \qquad |g(x)| < 1 \qquad \text{whenever} \quad 0 < |x - a| < \delta.$$

Therefore $\quad |f(x)g(x)| = |f(x)| \, |g(x)| < \varepsilon \cdot 1 = \varepsilon \quad$ whenever $\quad 0 < |x - a| < \delta.$

(d)   The problem is to make the quotient

$$\frac{f(x)}{M + f(x)}$$

small. We can make the numerator small, but if the denominator becomes small at the same time, we may lose control of the quotient. This doesn't happen because $M \ne 0$. If $f(x)$ is very small, then $M + f(x)$ is very close to M, hence not so close to 0. For instance, if $|f(x)|$ is at most $\tfrac{1}{2}|M|$, then it stands to reason that $|M + f(x)|$ is at least $|M| - \tfrac{1}{2}|M| = \tfrac{1}{2}|M|$. Precisely, if $|f(x)| < \tfrac{1}{2}|M|$, then by the triangle inequality

$$|M| = |[M + f(x)] + [-f(x)]| \le |M + f(x)| + |f(x)| \le |M + f(x)| + \tfrac{1}{2}|M|.$$

Therefore $$|M + f(x)| \ge \tfrac{1}{2}|M|.$$

We are ready for (d). Let $\varepsilon > 0$ be given. Since $f(x) \longrightarrow 0$, there exists $\delta > 0$ such that

$$|f(x)| < \min\{\tfrac{1}{2}|M|, \tfrac{1}{2}|M|\varepsilon\}, \qquad \text{whenever} \quad 0 < |x - a| < \delta$$

($\min\{\cdots\}$ denotes the minimum of the numbers inside the braces). Since $|f(x)| < \tfrac{1}{2}|M|$, we have $|M + f(x)| > \tfrac{1}{2}|M|$, as just noted. Hence

$$\left| \frac{f(x)}{M + f(x)} \right| = |f(x)| \frac{1}{|M + f(x)|} \le \left(\tfrac{1}{2}|M|\varepsilon \right)\left(\frac{1}{\tfrac{1}{2}|M|} \right) = \varepsilon$$

whenever $0 < |x - a| < \delta$. This completes the proof.

Remark Statements (b) and (c) of the lemma extend easily to three or more terms. For instance, if $f(x) \longrightarrow 0$, $g(x) \longrightarrow 0$, and $h(x) \longrightarrow 0$, then $g(x) + h(x) \longrightarrow 0$, hence

$$f(x) + g(x) + h(x) = f(x) + [g(x) + h(x)] \longrightarrow 0, \quad \text{etc.}$$

The following theorem contains four basic properties of limits that are used time and time again.

Theorem 1 Basic Limit Rules

Assume $\quad \lim_{x \to a} f(x) = L \quad$ and $\quad \lim_{x \to a} g(x) = M$

and let c be a constant. Then

(1) $\quad \lim_{x \to a} cf(x) = cL$ 　　　　(2) $\quad \lim_{x \to a}[f(x) \pm g(x)] = L \pm M$

(3) $\quad \lim_{x \to a} f(x)g(x) = LM$ 　　(4) $\quad \lim_{x \to a} \dfrac{f(x)}{g(x)} = \dfrac{L}{M} \quad (M \neq 0)$.

Proof By hypothesis, $\quad f(x) = L + F(x) \quad$ and $\quad g(x) = M + G(x)$,

where $\quad F(x) \longrightarrow 0 \quad$ and $\quad G(x) \longrightarrow 0 \quad$ as $\quad x \longrightarrow a$.

(1) Clearly $cf(x) = cL + cF(x)$. By Lemma (a), we have $cF(x) \longrightarrow 0$ as $x \longrightarrow a$. By the alternative definition of limit, $cf(x) \longrightarrow cL$ as $x \longrightarrow a$.

(2) Clearly $f(x) \pm g(x) = (L \pm M) + [F(x) \pm G(x)]$. By Lemma (b), we have $F(x) \pm G(x) \longrightarrow 0$ as $x \longrightarrow a$. Hence $f(x) \pm g(x) \longrightarrow L \pm M$.

(3) We have

$$f(x)g(x) = [L + F(x)][M + G(x)] = LM + [LG(x) + MF(x) + F(x)G(x)],$$

so the error is a sum of three terms. By Lemma (b), the sum approaches 0 if each term does. Since L and M are constants,

$$LG(x) \longrightarrow 0 \quad \text{and} \quad MF(x) \longrightarrow 0 \quad \text{as} \quad x \longrightarrow a$$

by Lemma (a). Also $F(x)G(x) \longrightarrow 0$ by Lemma (c). Hence the error approaches 0; therefore $f(x)g(x) \longrightarrow LM$ as $x \longrightarrow a$.

(4) We split the proof into two parts. First we prove the special case $1/g(x) \longrightarrow 1/M$. The error is

$$\frac{1}{g(x)} - \frac{1}{M} = \frac{1}{M + G(x)} - \frac{1}{M} = \frac{-G(x)}{M[M + G(x)]} = \frac{-1}{M}\left(\frac{G(x)}{M + G(x)}\right),$$

so

$$\frac{1}{g(x)} = \frac{1}{M} + \frac{-1}{M}\left(\frac{G(x)}{M + G(x)}\right).$$

By Lemma (d), with an assist from (a), the error approaches 0. Hence $1/g(x) \longrightarrow 1/M$.

Now if $f(x) \longrightarrow L$ and $g(x) \longrightarrow M$, then $1/g(x) \longrightarrow 1/M$ and by (3),

$$\frac{f(x)}{g(x)} = f(x) \cdot \frac{1}{g(x)} \longrightarrow L \cdot \frac{1}{M} = \frac{L}{M}.$$

Remark   Parts (2) and (3) of Theorem 1 are easily extended to more than two terms.

One-Sided Limits   Sometimes we want a limit as $x \longrightarrow a$, where x is restricted to the left of a, or where x is restricted to the right of a. We abbreviate

$$x \longrightarrow a \qquad \text{and} \qquad x < a$$

by

$$x \longrightarrow a-.$$

Similarly, $x \longrightarrow a+$ means $x \longrightarrow a$ and $x > a$. For example

$$\lim_{x \to 1-} \sqrt{1 - x} = 0, \qquad \lim_{x \to 0+} \frac{|x|}{x} = 1, \qquad \lim_{x \to 0-} \frac{|x|}{x} = -1.$$

Continuous Functions   The concept of a continuous function is very important in calculus.

Definition of Continuous Function   Let $f(x)$ be defined on an interval $x_0 \le x \le x_1$ and let a be a point of this interval. Then $f(x)$ is **continuous** at a if

$$\lim_{x \to a} f(x) = f(a).$$

The function $f(x)$ is **continuous** if it is continuous at each point of its domain.

To say that $f(x)$ is continuous at a means that the value $f(a)$ is "predictable" from the values of $f(x)$ near $x = a$, namely, $\lim_{x \to a} f(x)$ must exist, *and* it must equal the "correct value" $f(a)$. Roughly speaking, a function is continuous if you can draw its graph without lifting your pencil. The word for continuous in some languages translates literally into "unbroken," a term that expresses the concept well.

Examples   (See Fig. 2.)

1. $f(x) = \dfrac{1}{x}$ is continuous for all $x \ne 0$.   It is undefined at $x = 0$.

2. $f(x) = \begin{cases} 0 & x < 0 \\ \frac{1}{2} & x = 0 \\ 1 & x > 0 \end{cases}$ is continuous for all $x \ne 0$.

The limit $\lim_{x \to 0} f(x)$ does not exist.

3. $f(x) = \begin{cases} 2x & x \ne 1 \\ -1 & x = 1 \end{cases}$ is continuous for all $x \ne 1$.

The limit $\lim_{x \to 1} f(x)$ exists and equals 2, not $f(1) = -1$. Knowing the values of $f(x)$ for x near 1, you still cannot predict $f(1)$.

4. $f(x) = |x|$ is continuous for all x.   It is continuous even at $x = 0$, despite the corner in its graph, because

$$\lim_{x \to 0} f(x) = \lim_{x \to 0} |x| = 0 = f(0).$$

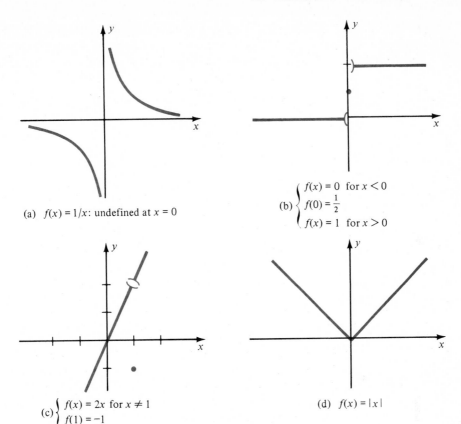

(a) $f(x) = 1/x$: undefined at $x = 0$

(b) $\begin{cases} f(x) = 0 & \text{for } x < 0 \\ f(0) = \frac{1}{2} \\ f(x) = 1 & \text{for } x > 0 \end{cases}$

(c) $\begin{cases} f(x) = 2x & \text{for } x \neq 1 \\ f(1) = -1 \end{cases}$

(d) $f(x) = |x|$

Fig. 2   Examples

We translate Theorem 1 into the language of continuity to obtain the following theorem.

Theorem 2   Let $f(x)$ and $g(x)$ be continuous on the same domain and let c be a constant. Then

(1) $cf(x)$ is continuous

(2) $f(x) \pm g(x)$ is continuous

(3) $f(x)g(x)$ is continuous

(4) $\dfrac{f(x)}{g(x)}$ is continuous wherever $g(x) \neq 0$.

It is pretty obvious that $f(x) = x$ is a continuous function. From this fact and Theorem 2, it follows easily that each polynomial is continuous and each rational function (quotient of polynomials) is continuous except where the denominator equals 0. We leave the details of the proof as exercises.

Theorem 3   Each polynomial

$$f(x) = a_n x^n + a_{n-1} x^{n-1} + \cdots + a_0$$

is continuous for all x. If $f(x)$ and $g(x)$ are polynomials, then the rational function $r(x) = f(x)/g(x)$ is continuous for all x such that $g(x) \neq 0$.

Let us add one more function to our repertory of continuous functions: the square root function. Recall that we needed its continuity for a detail in the proof of the Square Root Rule in Section 5.

Theorem 4 The function $f(x) = \sqrt{x}$ is continuous for all $x \geq 0$.

Proof We must prove for each $a \geq 0$ that

$$\lim_{x \to a} \sqrt{x} = \sqrt{a}.$$

First suppose $a > 0$. Then

$$\sqrt{x} - \sqrt{a} = (\sqrt{x} - \sqrt{a}) \frac{\sqrt{x} + \sqrt{a}}{\sqrt{x} + \sqrt{a}} = \frac{x - a}{\sqrt{x} + \sqrt{a}}.$$

Hence

$$\sqrt{x} = \sqrt{a} + E(x), \qquad E(x) = \frac{x - a}{\sqrt{x} + \sqrt{a}}.$$

Now $\sqrt{x} + \sqrt{a} \geq \sqrt{a}$ since $\sqrt{x} \geq 0$. Therefore

$$|E(x)| = \frac{|x - a|}{\sqrt{x} + \sqrt{a}} \leq \frac{1}{\sqrt{a}} |x - a|.$$

Clearly $|x - a|/\sqrt{a} \longrightarrow 0$ as $x \longrightarrow a$, and it follows that $E(x) \longrightarrow 0$ as $x \longrightarrow a$. Hence $\sqrt{x} \longrightarrow \sqrt{a}$ as $x \longrightarrow a$.

The case $a = 0$ is special because we are only allowed $x \geq 0$. Let $\varepsilon > 0$ be given. Set $\delta = \varepsilon^2 > 0$. If $0 < x < \delta$, then $0 < \sqrt{x} < \sqrt{\delta} = \varepsilon$. This proves

$$\lim_{x \to 0+} \sqrt{x} = 0 = \sqrt{0},$$

continuity at $x = 0$.

We finish this section by noting that the composite of continuous functions is continuous. This fact allows us to generate lots of new continuous functions out of old ones. For example, from the continuity of $\sqrt{x}$ follows the continuity of $\sqrt{u(x)}$ for any continuous function $u(x)$ taking non-negative values. Thus

$$\sqrt{x^2 + 1}, \qquad \sqrt{\frac{x}{x^4 + 6x^2 + 1}}$$

are continuous, the first for all x, the second for $x \geq 0$.

Theorem 5 If $g(x)$ is continuous at a and $f(x)$ is continuous at $g(a)$, then the composite function $f[g(x)]$ is continuous at a. Briefly, the composite of continuous functions is continuous.

The proof is very easy. By hypothesis, $g(x) \longrightarrow g(a)$ as $x \longrightarrow a$. Also $f[g(x)] \longrightarrow f[g(a)]$ as $g(x) \longrightarrow g(a)$. Therefore $f[g(x)] \longrightarrow f[g(a)]$ as $x \longrightarrow a$, which says that $f[g(x)]$ is continuous at a.

EXERCISES

Find the limit and give a reason for your conclusion

1 $\lim\limits_{x \to 2} \dfrac{1}{x}$

2 $\lim\limits_{x \to 11} \sqrt{2x + 3}$

3 $\lim\limits_{x \to -2} \dfrac{1}{x^2 + 1}$

4 $\lim\limits_{x \to 0} |x|$

5 $\lim\limits_{x \to 0} \dfrac{x^2}{x^2}$

6 $\lim\limits_{x \to 1} \dfrac{x^2 - 1}{x - 1}$

7 $\lim\limits_{x \to 0} \dfrac{x^2}{x}$

8 $\lim\limits_{x \to 1} (3x^5 - 7x^3 + 2x^2 + x - 5)$

9 $\lim\limits_{x \to 0} \sqrt{1 + x^3}$

10 $\lim\limits_{x \to 0} (\sqrt{x} - x)$

11 $\lim\limits_{h \to 0} \dfrac{3 + h + h^2}{5 - 4h + h^3}$

12 $\lim\limits_{h \to 0} \sqrt{1 + h^2}$.

13 Show that $\dfrac{1}{a + h} = \dfrac{1}{a} - \dfrac{h}{a^2} + \dfrac{h^2}{a^2(a + h)}$.

What information does this give about $\dfrac{d}{dx}\left(\dfrac{1}{x}\right)\Big|_{x=a}$?

14 Show that $\sqrt{a + h} = \sqrt{a} + \dfrac{h}{2\sqrt{a}} - \dfrac{h^2}{2\sqrt{a}(\sqrt{a + h} + \sqrt{a})^2}$.

What information does this give about $\dfrac{d}{dx}\left(\sqrt{x}\right)\Big|_{x=a}$?

15 Suppose $\lim_{x \to 4} f(x) = 10^{-6}$. Prove that there is a small segment centered at $x = 4$ on which $f(x) > 0$.

16 If $f(x) \geq 0$ and the limit exists, show that $\lim_{x \to a} f(x) \geq 0$.

17 Let $f(x) = 1$ for $x \geq 0$ and $f(x) = 0$ for $x < 0$. Show that $\lim_{x \to 0+} f(x)$ and $\lim_{x \to 0-} f(x)$ both exist, but $\lim_{x \to 0} f(x)$ does not exist.

18 Suppose both one-sided limits exist and $\lim_{x \to a-} f(x) = \lim_{x \to a+} f(x)$. Prove that $\lim_{x \to a} f(x)$ exists.

19 Prove (a) in the lemma, p. 95.

Let $f(x) \longrightarrow L$, $g(x) \longrightarrow M$, and $h(x) \longrightarrow N$ as $x \longrightarrow a$. Prove

20 $f(x)g(x)h(x) \longrightarrow LMN$

21 $f(x) + g(x) + h(x) \longrightarrow L + M + N$

22 $f(x)g(x)/h(x) \longrightarrow LM/N$ if $N \neq 0$.

23 Prove the first statement in Theorem 3, p. 99.

24 Prove the second statement in Theorem 3, p. 99.

Why is the function continuous for all x?

25 $x^3 - 4x + 6$

26 $x^2 + 3x$

27 $\dfrac{x}{x^2 + 3}$

28 $\dfrac{x^2 + 1}{x^4 + 1}$

29 $\sqrt{1 + x^2}$

30 $\sqrt{x^6 + 3x^2 + 7}$.

31 Prove that $\sqrt[3]{x}$ is continuous at $x = 0$.

32 Prove that $\sqrt[3]{x}$ is continuous at $x = a > 0$. [*Hint* $x - a = (\sqrt[3]{x} - \sqrt[3]{a}) \times (\sqrt[3]{x^2} + \sqrt[3]{a}\sqrt[3]{x} + \sqrt[3]{a^2})$.]

33 Suppose $f(x) \longrightarrow 0$ as $x \longrightarrow a$ and $|g(x)| \leq M$ for all $x \neq a$. Prove $f(x)g(x) \longrightarrow 0$ as $x \longrightarrow a$.

34 (cont.) Prove $x \sin(1/x) \longrightarrow 0$ as $x \longrightarrow 0$.

35 Suppose $f(x) \longrightarrow L$ as $x \longrightarrow a$. Prove $f(x)^3 \longrightarrow L^3$ as $x \longrightarrow a$.

36 Find $\delta > 0$ so $\left| \dfrac{h}{3 + h} \right| < 10^{-5}$ when $|h| < \delta$.

37 Find $\delta > 0$ so $\left| \dfrac{1 - h}{2 + h} - \dfrac{1}{2} \right| < 10^{-4}$ when $|h| < \delta$.

38 Where is $f(x) = |x^2 - 4|$ continuous?

39 Let A be the area of a triangle with sides a, b, c. Let $f(\varepsilon)$ be the area of a triangle with sides $a - \varepsilon, b - \varepsilon, c - \varepsilon$. Prove $\lim_{\varepsilon \to 0} f(\varepsilon) = A$. [*Hint* Heron's formula: $A^2 = s(s - a)(s - b) \times (s - c)$, where $2s = a + b + c$.]

40 Can the functions $f(x) = x/x$ and $g(x) = x/x^2$ be defined at $x = 0$ so as to be continuous there? If so, how?

11. DIFFERENTIABLE FUNCTIONS

In Section 3 we defined $f(x)$ to be **differentiable** at $x = a$ if the limit

$$\lim_{h \to 0} \frac{f(a + h) - f(a)}{h}$$

exists. If the limit does exist, it is called the **derivative** of $f(x)$ at $x = a$ and is denoted by $f'(a)$. Now according to the alternative definition of limit, p. 95, the number $f'(a)$ is the limit of the difference quotient if and only if

$$\frac{f(a + h) - f(a)}{h} = f'(a) + E(h), \qquad \text{where} \quad \lim_{h \to 0} E(h) = 0.$$

We multiply through by h and obtain the following useful alternative definition of derivative.

Alternative Definition of Derivative $f(x)$ is **differentiable** at $x = a$ if and only if there is a number $f'(a)$ and a function $E(h)$ such that

$$f(a + h) = f(a) + f'(a)h + E(h)h \qquad \text{and} \qquad \lim_{h \to 0} E(h) = 0.$$

The number $f'(a)$ is the **derivative** of $f(x)$ at $x = a$.

Example $f(x) = x^3$

$$f(a + h) = (a + h)^3 = a^3 + 3a^2h + 3ah^2 + h^3 = f(a) + (3a^2)h + hE(h),$$

where $$E(h) = 3ah + h^2 \longrightarrow 0 \qquad \text{as} \quad h \longrightarrow 0.$$

Hence $$f'(a) = 3a^2.$$

Differentiability implies continuity; we state this fact as a theorem.

Theorem   If $f(x)$ is differentiable at $x = a$, then $f(x)$ is continuous at $x = a$.

Proof By hypothesis,

$$f(a + h) = f(a) + f'(a)h + E(h)h, \qquad \text{where} \quad \lim_{h \to 0} E(h) = 0.$$

It follows that

$$\lim_{h \to 0} f(a + h) = \lim_{h \to 0} f(a) + \lim_{h \to 0}[f'(a)h] + \lim_{h \to 0}[E(h)h].$$

But

$$\lim_{h \to 0} f(a) = f(a), \qquad \lim_{h \to 0}[f'(a)h] = 0,$$

$$\lim_{h \to 0}[E(h)h] = \left(\lim_{h \to 0} E(h)\right)\left(\lim_{h \to 0} h\right) = 0.$$

Therefore

$$\lim_{h \to 0} f(a + h) = f(a), \qquad \text{that is,} \quad \lim_{x \to a} f(x) = f(a),$$

so $f(x)$ is continuous at $x = a$.

Warning   The converse of this theorem is false. A function may be continuous at a point, yet not have a derivative there. For example, $f(x) = |x|$ is continuous, but not differentiable at $x = 0$.

Roughly speaking, the graph of a continuous function is an unbroken curve. The graph of a differentiable function is even more, an unbroken curve rounded enough to have a tangent line at each point.

Proofs of Differentiation Rules   The proofs of the differentiation rules of Sections 4 and 5 were not complete. We are now in a position to fill the gaps.

The proof of the Sum Rule needed

$$\lim_{h \to 0}[f(h) + g(h)] = \lim_{h \to 0} f(h) + \lim_{h \to 0} g(h),$$

which is part of Theorem 1 of the last section. Similarly, the proof of the Constant Factor Rule needed $\lim cf(h) = c \lim f(h)$, another part of the same theorem.

What we lack are real proofs for the Product, Quotient, Square Root, and Chain Rules.

Product Rule   Assume $u(x)$ and $v(x)$ are differentiable at a. Write

$$u(a + h) = A + A'h + F(h)h, \qquad v(a + h) = B + B'h + G(h)h,$$

$$F(h) \longrightarrow 0 \qquad \text{and} \qquad G(h) \longrightarrow 0 \qquad \text{as} \quad h \longrightarrow 0,$$

where for brevity,

$$A = u(a), \qquad A' = u'(a), \qquad B = v(a), \qquad B' = v'(a).$$

Then

$$u(a + h)v(a + h) = [A + A'h + F(h)h][B + B'h + G(h)h]$$

$$= AB + (AB' + BA')h + E(h)h,$$

where

$$E(h) = AG(h) + BF(h) + h[A' + F(h)][B' + G(h)]$$

$$A \cdot 0 + B \cdot 0 + 0[A' + 0][B' + 0].$$

Clearly $E(h) \longrightarrow 0$ as $h \longrightarrow 0$, so $u(x)v(x)$ is differentiable at $x = a$, and its derivative is

$$AB' + BA' = u(a)v'(a) + v(a)u'(a).$$

Quotient Rule We use the same notation as above. We shall need $v(a + h) \longrightarrow B$ as $h \longrightarrow 0$, as given in the Theorem on p. 102, and the hypothesis $B \neq 0$. Let us compute the difference quotient:

$$\frac{1}{h}\left|\frac{u(a + h)}{v(a + h)} - \frac{A}{B}\right| = \frac{1}{h}\left[\frac{Bu(a + h) - Av(a + h)}{Bv(a + h)}\right]$$

$$= \frac{BA' - AB'}{Bv(a + h)} + \frac{BF(h) - AG(h)}{Bv(a + h)}$$

$$\downarrow \qquad\qquad \downarrow$$

$$\frac{BA' - AB'}{B^2} + \frac{B \cdot 0 - A \cdot 0}{B^2} = \frac{BA' - AB'}{B^2}.$$

Hence $u(x)/v(x)$ is differentiable at $x = a$, and the Quotient Rule follows.

Square Root Rule Some notation: $u(a) = A > 0$, and $u'(a) = A'$. We compute the difference quotient and use the continuity of $\sqrt{u(x)}$, proved in the last section:

$$\frac{\sqrt{u(a + h)} - \sqrt{A}}{h} = \frac{[\sqrt{u(a + h)} - \sqrt{A}][\sqrt{u(a + h)} + \sqrt{A}]}{h[\sqrt{u(a + h)} + \sqrt{A}]}$$

$$= \frac{u(a + h) - A}{h[\sqrt{u(a + h)} + \sqrt{A}]} = \frac{A' + F(h)}{\sqrt{u(a + h)} + \sqrt{A}} \longrightarrow \frac{A'}{2\sqrt{A}}$$

as $h \longrightarrow 0$. This proves $\sqrt{u(x)}$ is differentiable at $x = a$, and the Square Root Rule for the derivative.

Chain Rule We assume that $u = g(x)$ is differentiable at $x = a$ and $y = f(u)$ is differentiable at $u = g(a)$. We must prove that the composite function $y = f[g(x)]$ is differentiable at $x = a$, and that

$$\frac{dy}{dx}(a) = \left(\frac{df}{du}(b)\right)\left(\frac{dg}{dx}(a)\right), \qquad \text{where} \quad b = g(a).$$

Since $f(u)$ is differentiable at $u = b$, we have

$$f(b + k) = f(b) + B'k + E(k)k, \qquad \lim_{k \to 0} E(k) = 0,$$

where $B' = f'(b)$. We use this formula, with $k = g(a + h) - g(a) = g(a + h) - b$, in the difference quotient for $f[g(x)]$:

$$\frac{f[g(a + h)] - f[g(a)]}{h} = \frac{f(b + k) - f(b)}{h}$$

$$= \frac{B'k + E(k)k}{h} = [B' + E(k)]\left|\frac{g(a + h) - g(a)}{h}\right|$$

$$\downarrow \qquad \downarrow \qquad\qquad \downarrow$$

$$[B' + \quad 0\ ][\qquad g'(a) \qquad].$$

It follows that $f[g(x)]$ is differentiable at $x = a$ and that its derivative is $B'g'(a) = f'(b)g'(a)$. Note that $E(k) \longrightarrow 0$ as $h \longrightarrow 0$ because $k \longrightarrow 0$ as $h \longrightarrow 0$ (continuity of $g(x)$ at $x = a$).

EXERCISES

Find $E(h)$ satisfying $f(a + h) = f(a) + f'(a)h + E(h)h$

1 $f(x) = x^4$ $a = 1$

2 $f(x) = x^2 + x$ $a = -3$

3 $f(x) = 1/x$ $a = -2$

4 $f(x) = \dfrac{x^2}{x - 1}$ $a = 2.$ ▾

5 Let $f(x) = 1 + x$ for $x \geq 3$ and $f(x) = 7 - x$ for $x < 3$. Show that $f(x)$ is continuous at $x = 3$. Is it differentiable?

6 Let $f(x) = x^2$ for $x > 0$ and $f(x) = 0$ for $x \leq 0$. At what points is $f(x)$ differentiable?

7 Let $f(x) = x^2$ for $x \geq 0$ and $f(x) = -x^2$ for $x < 0$. At what points is $f(x)$ differentiable?

8 Let $f(x) = x^3$ for $x \geq 0$ and $f(x) = x^2$ for $x < 0$. At what points is $f(x)$ differentiable?

9 Suppose $f(x) = b + (x - a)g(x)$ and $g(x)$ is continuous at $x = a$. Show that $f(x)$ is differentiable at $x = a$ and find $f'(a)$.

10 Interpret the identity $\dfrac{x^n - 1}{x - 1} = x^{n-1} + x^{n-2} + \cdots + x + 1$

for $x = 1$ in terms of a derivative.

11 Let f be differentiable on (a, b) and suppose $a < c < b$. Define g by $g(c) = f'(c)$ and $g(x) = [f(x) - f(c)]/(x - c)$ if $x \neq c$. Prove g is continuous.

12 Suppose f is defined on (a, b) and $a < c < b$. Suppose f is differentiable at c. Prove that

$$\frac{f(c + h) - f(c - h)}{2h} \longrightarrow f'(c)$$

as $h \longrightarrow 0$. [*Hint*   Express $f(c + h)$ and $f(c - h)$, using the alternative definition of derivative, p. 102.]

12. MISCELLANEOUS EXERCISES

Find all values of x at which the curves have equal slope

1 $y = -\dfrac{27}{x}$, $y = x^3$

2 $y = 2x^3 - x^2$, $y = x^2 - 4$.

Differentiate

3 $(2 + \sqrt{3x})^5$

4 $\dfrac{1}{x\sqrt{9x^2 + 4}}$

5 $(x + 1)(2x + 1)^2(3x + 1)^3$

6 $\left[1 + \left(\dfrac{3x}{1 + x^2}\right)^{1/3}\right]^{1/2}.$

7 Show that $y = 1/x^2$ satisfies the relation $x^2 y'' + x y' - 4y = 0$.

8 Find $y(x)$ if $y'(x) = \dfrac{1}{\sqrt{5x + 4}}$ and $y(1) = 0$.

9 Show that the tangents to the curve $y = x^2$ at (a, a^2) and at $(a + 1, (a + 1)^2)$ intersect on the curve $y = x^2 - \frac{1}{4}$.

√10 Find where the tangent line to $y = 1/(2x - 5)$ at $x = 2$ crosses the line $x = -1$. $y \cdot \textcircled{3}$

Find

11 $\lim\limits_{x \to 0+} \dfrac{1 + \sqrt{x}}{3x^3 - 2}$

12 $\lim\limits_{x \to 2} \dfrac{x - 2}{x^3 - 8}$

13 $\lim\limits_{x \to 0} \dfrac{1 - \sqrt{1 + x}}{x}$

14 $\lim\limits_{x \to 1}(1 + \sqrt[3]{1 + 7x})(3 - x)$

15 $\dfrac{d^2}{dx^2}\left(\dfrac{3x+7}{x+2}\right)$ 　　　　　　　　　　　　 **16** $\dfrac{d^3}{dx^3}\left(\sqrt{3x-2}\right)\Big|_{x=1}$.

17 Where is $f(x) = |x+1| + |x+2| + |x+3|$ continuous? differentiable?

18 Let $f(x) = x^2$ for $x \le 1$ and $f(x) = 2x - 1$ for $x > 1$. Show that $f(x)$ is continuous and differentiable for all x.

19 Let $r(t)$ denote the larger of the two roots of the quadratic equation $x^2 + tx - 3 = 0$. Show that $r(t)$ is a continuous function of t for all t.

20 Compute $\dfrac{d^n}{dx^n}\left(\dfrac{x^2}{1-x}\right)$.

21 Let $h(x) = f[g(x)] - g[f(x)]$. Suppose $f(c) = g(c) = c$. Prove that $h'(c) = 0$.

22 Let $f(x) = x^n + \dbinom{n}{1}a_1 x^{n-1} + \dbinom{n}{2}a_2 x^{n-2} + \cdots + \dbinom{n}{n}a_n$, where $\dbinom{n}{j}$ are the binomial coefficients. Show that

$$\frac{1}{n}f'(x) = x^{n-1} + \binom{n-1}{1}a_1 x^{n-2} + \binom{n-1}{2}a_2 x^{n-3} + \cdots + \binom{n-1}{n-1}a_{n-1}.$$

Note the special case $[(x+a)^n]'/n = (x+a)^{n-1}$.

Applications of Differentiation **3**

1. CURVE SKETCHING

The derivative of a function is the slope of its graph at each point. Where the derivative is positive, the graph slopes upward. Where the derivative is negative, the graph slopes downward. Where the derivative is zero, the graph is horizontal (level). This information is of great help in sketching curves.

■ **EXAMPLE 1**   Sketch $y = x^3 - 3x + 1$.

Solution   The derivative is $y' = 3x^2 - 3 = 3(x^2 - 1)$. Clearly

$$y' > 0 \quad \text{if} \quad x^2 > 1, \qquad y' < 0 \quad \text{if} \quad x^2 < 1, \qquad y' = 0 \quad \text{if} \quad x^2 = 1.$$

The curve increases and decreases as indicated:

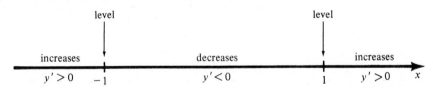

This information indicates a high point where $x = -1$ and a low point where $x = 1$. So compute the corresponding values of y, that is, $y(-1) = (-1)^3 - 3(-1) + 1 = 3$ and $y(1) = 1^3 - 3 + 1 = -1$. Then plot the high point $(-1, 3)$ and the low point $(1, -1)$. Also plot $(0, 1)$ because that point is obviously on the graph.

If $x \longrightarrow \infty$ or $x \longrightarrow -\infty$, then

$$y' = 3(x^2 - 1) \longrightarrow \infty,$$

so the graph is increasingly steep to the right and to the left. This is enough information for a reasonable sketch (Fig. 1).   ◼

■ **EXAMPLE 2**   Sketch $8y = 3 - x - x^3$.

Solution   Clearly

$$y' = -\tfrac{1}{8}(1 + 3x^2) < 0.$$

The derivative is always negative; the graph falls steadily from left to right. If $x \longrightarrow \infty$ or $x \longrightarrow -\infty$, then $y' \longrightarrow -\infty$, so the graph becomes steeper and

107

steeper. From the formula for y', the graph is least steep where $x = 0$, that is, at $(0, \frac{3}{8})$, and there the slope is $-\frac{1}{8}$. This is enough information for a reasonable sketch (Fig. 2). ∎

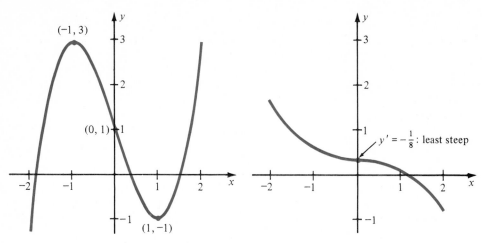

Fig. 1 $y = x^3 - 3x + 1$ **Fig. 2** $8y = 3 - x - x^3$

Convexity   The second derivative $f''(x)$ is the first derivative of $f'(x)$. Suppose $f''(x) > 0$. Then $f'(x)$ increases, hence the slope of the graph $y = f(x)$ increases. See Fig. 3 for the possibilities. The portion of the graph of $y = f(x)$ where $f''(x) > 0$ is called **convex from below**, or briefly, **convex**.

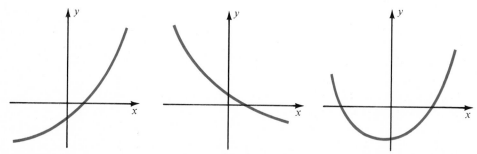

(a) Slope increases from small   (b) Slope increases from large   (c) Slope increases from negative
     positive to large positive.   negative to small negative.   to zero to positive.

Fig. 3   Convex graphs: $f''(x) > 0$

If $f''(x) < 0$, then $f'(x)$ decreases, and the graph $y = f(x)$ will have one of the shapes in Fig. 4. The portion of the graph of $y = f(x)$ where $f''(x) < 0$ is called **concave from below**, or briefly, **concave**.

A graph that is either convex or concave stays on the same side of its tangent at each point; the graph touches the tangent line, but does not cross it (Fig. 5).

A point at which the graph crosses its tangent is called an **inflection point**. Clearly $f''(x)$ must be zero at an inflection point. Furthermore, in the immediate neighborhood, $f''(x) > 0$ on one side of the inflection point and $f''(x) < 0$ on the other side. In

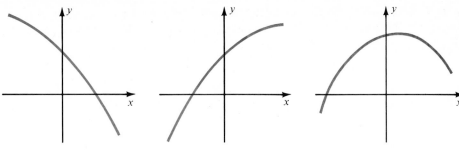

(a) Slope decreases from small negative to large negative.

(b) Slope decreases from large positive to small positive.

(c) Slope decreases from positive to zero to negative.

Fig. 4   Concave graphs: $f''(x) < 0$

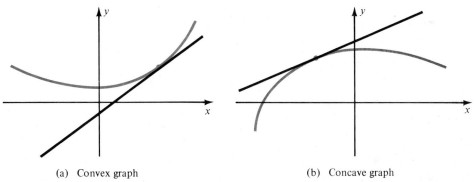

(a) Convex graph

(b) Concave graph

Fig. 5   Convex and concave graphs do not cross their tangents.

other words, the curve is convex on one side of the inflection point and concave on the other side (Fig. 6a). Note that $f''(x) = 0$ is not enough to guarantee an inflection point (Fig. 6b); $f''(x)$ must also change sign.

The graph of $y = f(x)$ has an inflection point at $(c, f(c))$ if and only if $f''(c) = 0$ and $f''(x)$ changes sign at $x = c$.

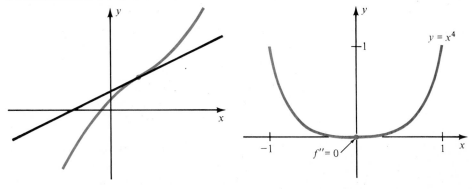

(a) Inflection point

(b) $f''(x) = 0$ does not guarantee an inflection point.

Fig. 6

Knowledge of convexity, concavity, and inflection points really helps in curve sketching. For instance, consider again the graph $y = x^3 - 3x + 1$ of Example 1. We used the derivative $y' = 3(x^2 - 1)$ to plot the curve. Now we can exploit the second derivative: $y'' = 6x$, so

$$y'' > 0 \quad \text{if} \quad x > 0, \qquad y'' < 0 \quad \text{if} \quad x < 0, \qquad y'' = 0 \quad \text{if} \quad x = 0.$$

The graph is convex for $x > 0$, concave for $x < 0$, and has an inflection at $x = 0$. This new information rules out a shape like that of Fig. 7a.

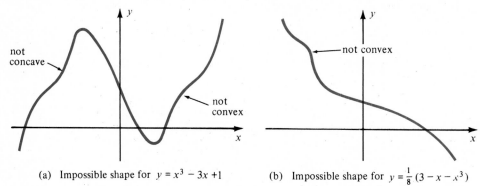

(a) Impossible shape for $y = x^3 - 3x + 1$ (b) Impossible shape for $y = \frac{1}{8}(3 - x - x^3)$

Fig. 7

Again, consider $y = \frac{1}{8}(3 - x - x^3)$ of Example 2. The second derivative is $y'' = -\frac{3}{4}x$, so the graph is convex for $x < 0$, concave for $x > 0$, and has an inflection point at $x = 0$. This rules out the shape of Fig. 7b.

■ **EXAMPLE 3** Sketch $y = (x - 2)^3 + 1$.

Solution   The first derivative is $y' = 3(x - 2)^2$, so $y' > 0$ if $x \neq 2$ and $y'(2) = 0$:

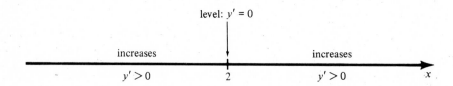

The second derivative is $y'' = 6(x - 2)$:

$x < 2$	$y'' < 0$	concave
$x = 2$	$y'' = 0$	inflection
$x > 2$	$y'' > 0$	convex

Thus $(2, 1)$ is an inflection point, and the tangent there is horizontal. This is enough information for a satisfactory sketch (Fig. 8).

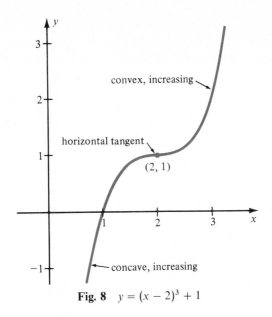

Fig. 8 $y = (x - 2)^3 + 1$ ■

Hints for Sketching Curves

(1)   Get as much "free" information as you can by inspection.

    (a)   If easily done, find where y is positive, negative, or zero.

    (b)   Look for symmetry. If $f(x)$ is an odd function, that is, if $f(-x) = -f(x)$, then the graph $y = f(x)$ is symmetric through the origin (Fig. 9a). If $f(x)$ is an even function, that is, if $f(-x) = f(x)$, then the graph $y = f(x)$ is symmetric about the y-axis (Fig. 9b).

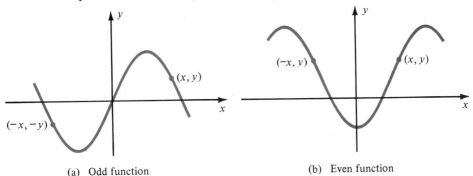

(a) Odd function   (b) Even function

Fig. 9

    (c)   Find the behavior of the curve as $x \longrightarrow \infty$ and as $x \longrightarrow -\infty$.

    (d)   Find values of x, if any, for which the curve is not defined.

       Examples $y = x + \dfrac{1}{x}$ not defined for $x = 0$,

                $y = \sqrt{1 - x}$ not defined for $x > 1$.

(2)   Take the derivative. Its sign will tell you where the curve is rising, falling, or

level. Plot all points where the tangent is horizontal, i.e., where $y' = 0$. If you can, find the behavior of y' as $x \longrightarrow \infty$ and as $x \longrightarrow -\infty$.

(3) Take the second derivative. Its sign will indicate convexity or concavity. Locate and plot all inflection points.

(4) For greater accuracy, plot a few points. Look for points that are easy to compute. Try $x = 0$, for example, or if not too hard, see where $y = 0$.

These hints are just suggestions; they are not sacred rules. Be flexible; there is no substitute for common sense.

■ **EXAMPLE 4** Sketch $y = x + \dfrac{1}{x}$.

Solution   There is some important free information: $y(-x) = -y(x)$, so the graph is symmetric through the origin. Once the curve is sketched for $x > 0$, it extends by symmetry to $x < 0$.

Further items of quick information: If $x > 0$, then $y > 0$. The curve is undefined at $x = 0$, and $y \longrightarrow \infty$ as $x \longrightarrow 0$. If x is large, $y = x + (1/x)$ is slightly larger than x. So as $x \longrightarrow \infty$, the graph is slightly above the line $y = x$.

Combining this information, you expect the graph to be something like Fig. 10a for $x > 0$. Check by inspecting the derivative $y' = 1 - (1/x^2)$:

$$y' < 0 \quad \text{if} \quad 0 < x < 1, \qquad y' > 0 \quad \text{if} \quad x > 1, \qquad y' = 0 \quad \text{if} \quad x = 1.$$

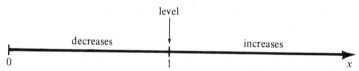

The curve indeed rises and falls as in Fig. 10a; also it has a lowest point at $(1, 2)$. Finally, extend the curve by symmetry (Fig. 10b) to $x < 0$.

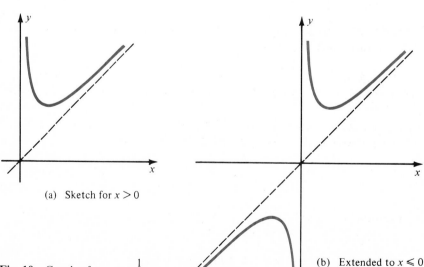

(a) Sketch for $x > 0$

(b) Extended to $x \leqslant 0$
by oddness

Fig. 10   Graph of $y = x + \dfrac{1}{x}$.

EXERCISES

Locate where $y = f(x)$ is convex, concave, and has inflection points

1 $f(x) = 4(x - 3)^2$ **2** $f(x) = -3(x + 7)^2$

3 $f(x) = (x - 2)^3 + 2x$ **4** $f(x) = x^3 + 3x^2 + 7x + 1$

5 $f(x) = -x^3 + 4x - 5$ **6** $f(x) = x^3 - 6x^2 + 3x + 1$

7 $f(x) = \dfrac{1}{\sqrt{x^2 + 5}}$ **8** $f(x) = x^4 - 6x^3 + 3x^2 + 2.$

Sketch the curve

9 $y = x^3 - x$ **10** $y = 3x^3 - 4x$ **11** $y = x^3 - 3x$

12 $y = 12x - x^3$ **13** $y = 3x^3 + x$ **14** $y = -12x^3 - 4x$

15 $y = 4x^3 - 2x^2$ **16** $y = 6x^2 - x^3$ **17** $y = -x^3 + 3x - 4$

18 $y = x^3 - 3x + 5$ **19** $y = 1 + \tfrac{1}{2}x + x^3$ **20** $y = \tfrac{1}{3}x - x^3$

21 $y = x^3 - x^2 - 8x + 4$ **22** $y = 7 - 3x^2 - 2x^3$ **23** $y = 2x + \dfrac{1}{x}$

24 $y = x - \dfrac{1}{x}$ **25** $y = \dfrac{x}{x + 1}$ **26** $y = \dfrac{x}{x - 1}$

27 $y = x + \dfrac{1}{x + 1}$ **28** $y = x + \dfrac{1}{x - 1}$ **29** $y = x + \dfrac{2}{x}$

30 $y = -x - \dfrac{16}{x}$ **31** $y = x^2(x - 3)$ **32** $y = x(x + 1)(x + 2)$

33 $y = x^2(x - 1)^2$ **34** $y = x^3(x + 2)$ **35** $y = \dfrac{1}{1 + x^2}$

36 $y = \dfrac{x}{1 + x^2}$ **37** $y = \dfrac{x - 1}{x^2}$ **38** $y = \dfrac{x}{x^2 - 1}$

39 $y = \sqrt{1 - x}$ **40** $y = \sqrt{x + 1} - \sqrt{x}$ **41** $y = x + \sqrt{x}$

42 $y = x - \sqrt{x}.$

2. RECTILINEAR MOTION

Rectilinear motion is motion along a straight line. From its study arise two important concepts: velocity and acceleration.

Velocity During the initial stage of flight, a rocket fired vertically reaches an elevation of $50t^2$ ft above the ground in t sec. How fast is the rocket rising 2 sec after it is fired?

This is a tricky question, because it is not really clear what is meant by velocity at an instant. Usually we compute **average velocity** by the formula

$$\text{average velocity} = \frac{\text{displacement}}{\text{time}}$$

applied to a certain time interval. To be more precise, for the time interval between t_1 and t_2,

$$\text{average velocity} = \frac{(\text{position at time } t_2) - (\text{position at time } t_1)}{t_2 - t_1}.$$

This formula does not apply to "instantaneous velocity." Nevertheless, we can compute the average velocity between $t = 2$ and, say, $t = 2.01$, or $t = 2.001$, or more generally, $t = 2 + h$.

The average velocity between $t = 2$ and $t = 2 + h$ is

$$\frac{50(2 + h)^2 - 50 \cdot 2^2}{(2 + h) - 2} = 50 \frac{(2 + h)^2 - 2^2}{h} = 50 \frac{4h + h^2}{h} = 50(4 + h).$$

The smaller h is, the closer this average velocity is to 200 ft/sec.

The average velocity between $t = t_0$ and $t = t_0 + h$ is

$$\frac{50(t_0 + h)^2 - 50t_0^2}{(t_0 + h) - t_0} = 50 \frac{(t_0 + h)^2 - t_0^2}{h} = 50 \frac{2t_0 h + h^2}{h} = 50(2t_0 + h).$$

The smaller h is, the stronger the message: At $t = t_0$ the instantaneous velocity is $100t_0$ ft/sec.

Here is an important observation. The preceding computation of velocity is exactly the same computation as that needed to find the slope of the curve $s = s(t) = 50t^2$. Thus the velocity of the rocket at time t_0 is numerically the same as the slope of $y = 50t^2$ at $t = t_0$.

This is no accident. The average velocity between $t = t_0$ and $t = t_0 + h$ is

$$\frac{\text{displacement}}{\text{time}} = \frac{s(t_0 + h) - s(t_0)}{h},$$

which is precisely the formula for the slope of the secant between two nearby points on $y = s(t)$. Thus the slope of the secant is like the "average speed" of $y = s(t)$ between t_0 and $t_0 + h$. As the interval gets smaller and smaller, the "average slope" approximates the "instantaneous slope" at t_0, that is, the derivative at t_0.

Definition   Let $s = s(t)$ be the position at time t of a particle moving on the s-axis. Its **velocity** is $v(t) = ds/dt$ and its **speed** is $|v| = |ds/dt|$.

Notice that ds/dt may be negative. This happens, for instance, in the case of a falling body whose position is measured above ground level. Then displacement in any time interval is negative, leading to a negative velocity. Speed ignores the sign of the derivative. It measures only the *rate* of motion, while velocity also takes into account the *direction*.

Notation   The use of a dot to indicate derivative with respect to *time* is common practice. Instead of ds/dt, you often see $\dot{s}(t)$ or just $\dot{s}$.

■ **EXAMPLE 1**   A ball is thrown straight up from a support 180 ft above the ground, which then moves aside. Assume the ball's height above ground after t sec to be $s = 180 + 64t - 16t^2$ ft. Compute (a) its velocity after 1 sec, (b) its maximum height, and (c) its velocity as it hits the ground.

Solution   The velocity of the ball after t sec is

$$v(t) = \dot{s}(t) = 64 - 32t = 32(2 - t).$$

(a)   After 1 sec, the velocity is $v(1) = 32(2 - 1) = 32$ ft/sec. (Since the velocity is positive, the ball is rising.)

(b)   We need the maximum value of $s(t)$. Examine $\dot{s}(t)$:

$$\dot{s}(t) > 0 \quad \text{when} \quad t < 2, \qquad \dot{s}(2) = 0, \qquad \dot{s}(t) < 0 \quad \text{when} \quad t > 2.$$

Therefore $s(t)$ increases for $t < 2$ and decreases for $t > 2$. It must take its maximum value when $t = 2$. The maximum is $s(2) = 244$ ft.

(c)   The ball hits the ground when $s(t) = 0$. Solve for t:

$$s(t) = 180 + 64t - 16t^2 = 0, \qquad 4t^2 - 16t - 45 = 0,$$

$$t = \frac{16 \pm \sqrt{16^2 + 4 \cdot 4 \cdot 45}}{8} = 2 \pm \frac{\sqrt{16 + 45}}{2}.$$

There is only one *positive* time t for which $s(t) = 0$; it is $t = 2 + \frac{1}{2}\sqrt{61}$. At this instant, the velocity is

$$v(2 + \tfrac{1}{2}\sqrt{61}) = 32[2 - (2 + \tfrac{1}{2}\sqrt{61})] = -16\sqrt{61} \approx -125.0 \text{ ft/sec.}$$

The velocity is negative because the ball is falling.   ∎

Acceleration   A falling weight moves faster and faster; its velocity increases; a car with brakes applied moves slower and slower; its velocity decreases. In many applications, it is important to know just how velocity is changing during motion.

Definition   If $v(t)$ is the velocity of a moving particle at time t, its **acceleration** is $a(t) = dv/dt = \dot{v}$.

Acceleration is the derivative of velocity. It measures the rate of change of velocity during motion. Positive acceleration indicates increasing velocity; negative acceleration, decreasing velocity; zero acceleration, constant velocity.

Remember that velocity itself is a derivative:

$$v(t) = \frac{ds}{dt} = \dot{s}(t),$$

where $s = s(t)$ is the position at time t. Therefore, acceleration is a second derivative, being the derivative of a derivative:

$$\text{acceleration} = \frac{d^2s}{dt^2} = \ddot{s}(t).$$

■ **EXAMPLE 2**   A ball is $s(t) = 180 + 64t - 16t^2$ ft above the ground at time t sec. Find its acceleration at time t.

Solution   Differentiate twice:

$$v(t) = \frac{ds}{dt} = 64 - 32t \text{ ft/sec}, \qquad a(t) = \frac{dv}{dt} = -32 \text{ ft/sec}^2.$$ ∎

Remark The *negative* acceleration means that the velocity is decreasing (from positive to negative to more negative).

■ **EXAMPLE 3** A bullet is shot straight up from the top of a hill s_0 meters high with an initial velocity of v_0 m/sec. Gravity causes a constant negative acceleration of $-g$ m/sec^2. After t sec, what are (a) its velocity, (b) its height above ground level, and (c) the distance it has traveled?

Solution (a) First find a formula for the velocity $v(t)$. Since acceleration is dv/dt, the data is

$$\frac{dv}{dt} = -g.$$

That means $v(t)$ is a function whose derivative is $-g$. In other words, $v(t)$ is an antiderivative of $-g$. One antiderivative is $-gt$, therefore all antiderivatives are of the form $-gt + c$, where c can be any constant. Hence,

$$v(t) = -gt + c.$$

To find the constant c that fits this problem, remember that the value of $v(t)$ is given for $t = 0$. Set $t = 0$:

$$v_0 = v(0) = -g \cdot 0 + c, \qquad c = v_0 .$$

Hence
$$v(t) = -gt + v_0$$

is the required formula.

  (b) Use the same sort of argument to find a formula for $s(t)$, the elevation at time t. Since

$$\frac{ds}{dt} = v(t) = -gt + v_0 ,$$

$s(t)$ is an antiderivative of $-gt + v_0$. Therefore

$$s(t) = -\tfrac{1}{2}gt^2 + v_0 t + k$$

for some appropriate constant k. To find the right value of k, remember that the value of $s(t)$ is given for $t = 0$. Set $t = 0$:

$$s_0 = s(0) = 0 + 0 + k, \qquad k = s_0 .$$

Hence
$$s(t) = -\tfrac{1}{2}gt^2 + v_0 t + s_0 \quad \text{meters.}$$

  (c) The bullet ascends until it reaches its maximum height above ground, when its velocity is zero, that is, when $t = v_0/g$. The maximum height is

$$s_{\max} = s\left(\frac{v_0}{g}\right) = \frac{1}{2}\frac{v_0^2}{g} + s_0 .$$

Then it descends until $t = 2v_0/g$, when it strikes the top of the hill ($s = s_0$). Let $D(t)$ be the distances traveled in t secs. There are two cases:

  (i) $0 \le t \le v_0/g$. Then $D(t) = s(t) - s_0 = -\tfrac{1}{2}gt^2 + v_0 t.$

(ii) $v_0/g \le t \le 2v_0/g$. Then (distance) = (up) + (down):

$$D(t) = [s_{max} - s_0] + [s_{max} - s(t)]$$

$$= \frac{1}{2}\frac{v_0^2}{g} + \left[\left(\frac{1}{2}\frac{v_0^2}{g} + s_0 \right) - \left(-\frac{1}{2}gt^2 + v_0 t + s_0 \right) \right]$$

$$= \frac{1}{2}gt^2 - v_0 t + \frac{v_0^2}{g}. \qquad\blacksquare$$

In Example 3 you solved a **differential equation**. That is an equation involving the derivatives of a function in which the function itself is the unknown. The data of Example 3 can be written:

$$\frac{d^2s}{dt^2} = -g, \qquad s(0) = s_0, \qquad \dot{s}(0) = v_0.$$

The first equation is the differential equation; the other equations are **initial conditions**. To find $s(t)$, antidifferentiate twice. First you get ds/dt, then $s(t)$ itself. Each antidifferentiation involves a constant to be determined. The two constants are obtained from the two initial conditions, $s(0) = s_0$ and $\dot{s}(0) = v_0$.

■ **EXAMPLE 4** An alpha particle enters a linear accelerator. It immediately undergoes a constant acceleration that changes its velocity from 1000 m/sec to 5000 m/sec in 10^{-3} sec. Compute its acceleration. How far does the particle move during this period of 10^{-3} sec?

Solution For convenience, assume the accelerator lies along the positive x-axis starting at the origin. Also assume the particle enters when $t = 0$, and t sec later reaches position $x(t)$. Then

$$\ddot{x}(t) = a, \qquad \dot{x}(0) = 1000, \qquad x(0) = 0,$$

where a is the unknown constant acceleration. This is the same problem as Example 3, with different numbers: a instead of $-g$, $v_0 = 1000$, and $x_0 = 0$.

By exactly the reasoning of Example 3,

$$v(t) = at + v_0 = at + 1000, \qquad x(t) = \tfrac{1}{2}at^2 + 1000t.$$

Use the first formula to find a. Since $v(10^{-3}) = 5000$,

$$5000 = 10^{-3}a + 1000, \qquad a = 4 \times 10^6 \quad \text{m/sec}^2.$$

From the second formula,

$$x(10^{-3}) = \tfrac{1}{2}(4)(10^6)(10^{-3})^2 + (1000)(10^{-3}) = 2 + 1 = 3 \quad \text{m.} \qquad\blacksquare$$

EXERCISES

1 A projectile shot straight up has height $s = -16t^2 + 980t$ ft after t sec. Compute its average velocity between $t = 2$ and $t = 3$, between $t = 2$ and $t = 2.1$. Compute its instantaneous velocity when $t = 2$.

2 During the initial stages of flight, a rocket reaches an elevation of $50t^2 + 500t$ ft in t sec. What is the average velocity between $t = 2$ and $t = 3$ sec? Find the instantaneous velocity when $t = 2$ and when $t = 3$. What is the average of the two instantaneous velocities?

3 A projectile launched from a plane has elevation $s = -5t^2 + 100t + 1500$ meters after

t sec. What is its maximum elevation? Find its vertical velocity after 15 sec, and upon striking the ground.

4 An object projected upward has height $s = -5t^2 + 30t$ m after t sec. Compute its velocity after 1.5 sec, its maximum height, and the speed with which it strikes the ground.

5 A ball is thrown straight up from the top of a 600-ft tower. After t sec, it is $s = -16t^2 + 24t + 600$ ft above ground. When does the ball begin to descend? What is its speed when 605 ft above ground, while going up, and while coming down?

6 A shell fired at angle 45° to the horizon with initial speed 300 m/sec has height $y = 150t\sqrt{2} - 5t^2$ m and horizontal distance $x = 150t\sqrt{2}$ m from its initial position t sec after firing. How far from its initial point does it strike the ground? What is its maximum elevation?

7 A body moves along a horizontal line according to the law $s = t^3 - 9t^2 + 24t$ ft. (a) When is s increasing and when decreasing? (b) When is the velocity increasing and when decreasing? (c) Find the total distance traveled between $t = 0$ and $t = 6$ sec.

8 Solve Ex. 7 if the law of motion is $s = t^3 - 3t^2 - 9t$ ft.

9 A ball is thrown straight up with an initial velocity of 48 ft/sec. Gravity causes a constant negative acceleration, -32 ft/sec². How high will the ball go if it is released from a height of 4 ft?

10 An object slides down a 200-ft inclined plane with acceleration 8 ft/sec². If the object starts from rest with zero velocity, when does it reach the end of the plane? How fast is it going?

11 Starting from rest, with what constant acceleration must a car proceed to go 75 ft in 5 sec?

12 The makers of a certain automobile advertise that it will accelerate from 0 to 100 mph in 1 min. If the acceleration is constant, how far will the car go in this time?

13 During the initial stages of flight after blast-off, a rocket shot straight up has acceleration $6t$ m/sec². The engine cuts out at $t = 10$ sec, after which only the gravitational acceleration, -10 m/sec², retards its motion. How high will the rocket go? How long does it take to reach its maximum height?

14 An airplane taking off from a landing field has a run of 1000 m. If it starts with speed 7 m/sec, moves with constant acceleration, and makes the run in 40 sec, with what speed does it take off?

15 A subway train starts from rest at a station and accelerates at the rate of 2 m/sec² for 10 sec. It then runs at constant speed for 60 sec, after which it decelerates at the rate of 3 m/sec² until it stops at the next station. Find the total distance it travels between the stations.

16 Gravitation on the moon is 0.165 times that on the earth. If a bullet shot straight up from the earth will rise 1 km, how far would it rise if shot on the moon?

Solve the differential equation

17 $dy/dx = -16x$, $y(0) = 12$

18 $dy/dt = 3t^2 + 4$, $y(1) = -3$

19 $d^2y/dt^2 = -32$, $y(1) = 48$, $\dot{y}(0) = 64$

20 $d^2y/dx^2 = 8$, $y(0) = 2$, $y'(0) = 1$

21 $d^2y/dt^2 = 2t - 1$, $y(0) = 5$, $\dot{y}(0) = 3$

22 $d^2y/dx^2 = 3 - 4x$, $y(0) = 2$, $y'(1) = 6$.

3. RELATED RATES

If two changing physical quantities are related, then their rates of change are also related. Quite a number of physical problems involve this idea.

■ **EXAMPLE 1** A large spherical rubber balloon is inflated by a pump that injects 10 ft³/sec of helium. At the instant when the balloon contains 972π ft³ of gas, how fast is its radius increasing?

Solution   Denote the radius and volume of the balloon at time t by $r(t)$ and $V(t)$. The derivative $dV/dt = 10$ ft³/sec is given. The derivative dr/dt is required at a specific time.

The formula for the volume V of a sphere of radius r is $V = \frac{4}{3}\pi r^3$. Hence

$$V(t) = \tfrac{4}{3}\pi[r(t)]^3.$$

To find a relation between dr/dt and dV/dt, differentiate by the Chain Rule:

$$\frac{dV}{dt} = \frac{dV}{dr}\frac{dr}{dt} = \frac{4}{3}\pi \cdot 3r^2 \cdot \frac{dr}{dt} = 4\pi r^2 \frac{dr}{dt}.$$

Solve for dr/dt:

$$\frac{dr}{dt} = \frac{1}{4\pi r^2}\frac{dV}{dt} = \frac{10}{4\pi r^2}.$$

This formula tells the rate of change of the radius at any instant, in terms of the radius. At the instant in question, the volume is 972π ft³, so the radius can be found:

$$\frac{4}{3}\pi r^3 = 972\pi, \qquad r^3 = \frac{3}{4\pi} \cdot 972\pi = 729, \qquad r = 9.$$

But when $r = 9$,

$$\frac{dr}{dt} = \frac{10}{4\pi \cdot 9^2} = \frac{10}{324\pi} \approx 9.82 \times 10^{-3} \quad \text{ft/sec.} \qquad \blacksquare$$

Example 1 is typical of related rate problems. You are given the time derivative of one physical quantity and asked for the time derivative of a related quantity at a certain instant. Usually you must find a relation between the two quantities, then differentiate it with respect to time to get a relation between their derivatives. Finally you substitute the data at the instant in question. (Finding this data may require some side computations.)

Recall the dot notation for derivatives with respect to time: $\dot{x} = dx/dt$, $\dot{y} = dy/dt$, etc. Since it is so common, we shall use this notation interchangeably with the notations dy/dx and y'.

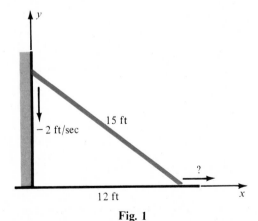

Fig. 1

■ **EXAMPLE 2**   A 15-ft ladder leans against a vertical wall. If the top slides downward at the rate of 2 ft/sec, find the speed of the lower end when it is 12 ft from the wall.

Solution Make a sketch placing axes as in Fig. 1. We are given $\dot{y} = -2$ and asked to find $\dot{x}$ at the instant when $x = 12$. There is an obvious relation between x and y:

$$x^2 + y^2 = 15^2.$$

We differentiate with respect to time:

$$2x\dot{x} + 2y\dot{y} = 0, \qquad \dot{x} = -y\dot{y}/x.$$

To find $\dot{x}$ at the instant in question, we need the values of y, $\dot{y}$, and x at that instant. We are given $\dot{y} = -2$ and $x = 12$. From the relation $x^2 + y^2 = 15^2$, we find $y = 9$. Therefore

$$\dot{x} = -\frac{9(-2)}{12} = \frac{3}{2} \quad \text{ft/sec.} \qquad \blacksquare$$

■ **EXAMPLE 3** If the volume of an expanding cube is increasing at the rate of 4 cm³/sec, how fast is its surface area increasing when the surface area is 24 cm² ?

Solution Let x be the edge of the cube. Then its area and volume are

$$A = 6x^2 \quad \text{and} \quad V = x^3.$$

Find a relation between A and V. Since $x = V^{1/3}$ and $A = 6x^2$,

$$A = 6V^{2/3}.$$

Differentiate carefully with respect to time:

$$\dot{A} = 6(\tfrac{2}{3})V^{-1/3}\dot{V} = 4V^{-1/3}\dot{V}.$$

Now find the value of V at the given instant, that is, when $A = 24$. From the relation $A = 6V^{2/3}$ follows

$$24 = 6V^{2/3}, \qquad V^{2/3} = 4, \qquad V = 8.$$

Use this value $V = 8$ and the given value $\dot{V} = 4$:

$$\dot{A} = 4(8)^{-1/3}(4) = 8 \quad \text{cm}^2/\text{sec.}$$

Alternative Solution Do not eliminate x. Differentiate the relations $A = 6x^2$ and $V = x^3$:

$$\dot{A} = 12x\dot{x}, \qquad \dot{V} = 3x^2\dot{x} = 4.$$

When $A = 24$ cm², $x = 2$ cm. From $3x^2\dot{x} = 4$ follows $\dot{x} = \tfrac{1}{3}$. Hence

$$\dot{A} = 12x\dot{x} = 12(2)(\tfrac{1}{3}) = 8 \quad \text{cm}^2/\text{sec.} \qquad \blacksquare$$

■ **EXAMPLE 4** A rectangular tank has a sliding panel S that divides it into two adjustable tanks of width 3 ft. See Fig. 2. Water is poured into the left compartment at the rate of 5 ft³/min. At the same time S is moved to the right at the rate of 3 ft/min. When the left compartment is 10 ft long it contains 70 ft³ of water. Is the water level rising or falling? How fast?

Solution Let x be the length of the left compartment; let y and V be the depth and

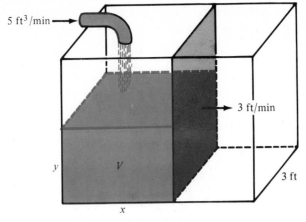

5 ft³/min →

3 ft/min →

3 ft

y

V

x

Fig. 2

the volume of the water in the left compartment. Then x, y, and V are all functions of time. Given:

$$\frac{dx}{dt} = 3 \quad \text{ft/min}, \qquad \frac{dV}{dt} = 5 \quad \text{ft}^3/\text{min}.$$

Compute: dy/dt when $x = 10$ and $V = 70$.

To do so, find a relation between x, y, and V. At any instant

$$V = 3xy.$$

Differentiate with respect to t:

$$\frac{dV}{dt} = 3x \frac{dy}{dt} + 3 \frac{dx}{dt} y.$$

Substitute the data at the instant in question:

$$5 = 3 \cdot 10 \cdot \frac{dy}{dt} + 3 \cdot 3 \cdot y, \qquad \frac{dy}{dt} = \frac{5 - 9y}{30}.$$

At the given instant, $V = 70$ and $x = 10$; hence $y = 7/3$. Therefore

$$\frac{dy}{dt} = \frac{5 - 9(7/3)}{30} = -\frac{16}{30}.$$

The water level is falling at the rate of 8/15 ft/min.

Alternative Solution Instead of differentiating the equation $V = 3xy$, solve for y first, then differentiate:

$$y = \frac{1}{3} \frac{V}{x}, \qquad \frac{dy}{dt} = \frac{1}{3} \frac{x \dfrac{dV}{dt} - V \dfrac{dx}{dt}}{x^2}.$$

At the given instant, $\dfrac{dy}{dt} = \dfrac{1}{3}\dfrac{10 \cdot 5 - 70 \cdot 3}{10^2} = -\dfrac{16}{30} = -\dfrac{8}{15}$ ft/min.   ∎

EXERCISES

Two functions $x = x(t)$ and $y = y(t)$ satisfy the given relation. Find $\dot{y}$ for the given data (and be sure to test the data for consistency).

1 $x^2 + 2 = y^3$ $x = 5,$ $y = 3,$ $\dot{x} = -1$

2 $y + \dfrac{1}{y} = x + 2$ $x = \frac{4}{3},$ $y = 3,$ $\dot{x} = 10$

3 $x = \dfrac{y + 1}{y - 1}$ $x = -1,$ $y = 0,$ $\dot{x} = 2$

4 $x^2 + xy + y^2 = 3$ $x = 1,$ $y = -2,$ $\dot{x} = -5$

5 $y + \sqrt{1 + y^2} = x$ $x = 3,$ $y = \frac{4}{3},$ $\dot{x} = 1$

6 $x^3(1 + y^2) = 2$ $x = 1,$ $y = 1,$ $\dot{x} = 2.$

7 A stone thrown into a pond produces a circular ripple which expands from the point of impact. When the radius is 8 ft, it is observed that the radius is increasing at the rate 1.5 ft/sec. How fast is the area increasing at that instant?

8 An inverted conical tank has height 4 m and radius 1 m at the top. When the depth is 2 m, oil flows in at the rate 2 m³/min. How fast is the level rising?

9 A 6-ft man walks away from a 15-ft lamp post. When he is 21 ft from the post, his walking rate is 5 ft/sec. How fast is his shadow lengthening at that instant?

10 Two cars leave an intersection P. After 60 sec, the one traveling north has speed 50 ft/sec and distance 2000 ft from P, and the one traveling west has speed 75 ft/sec and distance 2500 ft from P. At that instant, how fast are the cars separating from each other?

11 A point P moves along the curve $y = x^3 - 3x^2$. When P is at $(1, -2)$, its x-coordinate is increasing at rate 3. Find the rate of increase of the distance from the origin to P.

12 A point P moves along the curve $y = x^4 + x + 1$. When P is at $(1, 3)$, its y-coordinate increases at rate 1. Find the rate of increase of the distance from the origin to P.

13 Yarn of radius 2 mm is being wound on a ball at the rate of 60 cm/sec. Assume that the ball is a perfect sphere at each instant, and consists entirely of thread with no empty space. Find the rate of increase of the radius when the radius is 5 cm.

14 Two concentric circles are expanding. At a certain instant the outer radius is 10 ft and it is expanding at rate 2 ft/sec, while the inner radius is 3 ft and it is expanding at rate 5 ft/sec. Find the rate of change of the area between the circles.

15 Solve Ex. 14 for spheres and volume.

16 An elevated train on a track 30 ft above the ground crosses a (perpendicular) street at the rate of 50 ft/sec at the instant that an automobile, approaching at the rate of 30 ft/sec, is 40 ft up the street. Find how fast the train and the automobile are separating 2 sec later.

17 If a chord sweeps (without turning) across a circle of radius 10 ft at the rate of 6 ft/sec, how fast is the length of the chord decreasing when it is $\frac{3}{4}$ of the way across?

18 The power P in watts dissipated by an R-ohm resistor with V volts across it is $P = V^2/R$. At a certain instant V is 112 volts, R is 10,000 ohms, and V and R are changing at the rates of 3 volts/min and -200 ohms/min, respectively. Find the rate of change of P in watts/min.

19 The volume V and pressure P of a gas in a constant-temperature engine cylinder are related by $PV = k$, a constant. Express the time rate of change $\dot{P}$ in terms of P and $\dot{V}$.

20 Ship A sails due south toward a port P at 5 mph. Ship B sails due east away from P at 10 mph. At a given instant A is a miles from P and B is b miles from P. Show that the ships are getting closer together if $a > 2b$ and farther apart if $a < 2b$.

4. MAXIMA AND MINIMA

One of the most striking applications of calculus is in finding the largest or the smallest value of a function over a certain domain. In Chapter 1 we discussed this question for quadratic functions. To begin the general study of maximum and minimum values (briefly, maxima and minima, or extreme values), let us take a second look at quadratic functions.

Quadratic Functions   Our previous approach to extreme values of quadratic functions was based on completing the square:

$$f(x) = ax^2 + bx + c = a\left(x^2 + \frac{b}{a}x\right) + c = a\left(x + \frac{b}{2a}\right)^2 + \frac{4ac - b^2}{4a}.$$

The extreme value of $f(x)$ occurs at $x = -b/2a$. It is

$$f\left(\frac{-b}{2a}\right) = \frac{4ac - b^2}{4a},$$

a minimum if $a > 0$, a maximum if $a < 0$.

This formula is not easy to remember, so one usually rederives it for each particular example in hand. Another, more systematic, procedure is suggested by calculus. Wherever $f(x)$ takes its maximum or minimum value, the graph of $y = f(x)$ has a horizontal tangent (Fig. 1); hence $f'(x) = 0$. In other words, if we solve the equation $f'(x) = 0$, we find the x for which $f(x)$ is maximal or minimal.

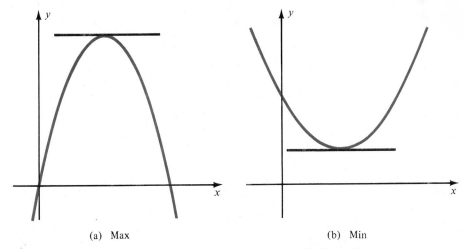

(a) Max   (b) Min

Fig. 1   At a max or a min, a quadratic has a horizontal tangent.

■ **EXAMPLE 1**   Find the maximum of $f(x) = 80x - 16x^2$.

Solution 1   (without calculus)

$$f(x) = -16(x^2 - 5x) = -16[(x - \tfrac{5}{2})^2 - \tfrac{25}{4}] = -16(x - \tfrac{5}{2})^2 + 100.$$

The maximum is 100, taken at $x = \tfrac{5}{2}$.

Solution 2 (with calculus) Graph $y = f(x) = 80x - 16x^2$ very roughly (Fig. 1a). At the maximum, the tangent is horizontal; hence $f'(x) = 0$. Since

$$f'(x) = 80 - 32x,$$

the condition is

$$80 - 32x = 0, \qquad x = \tfrac{5}{2}.$$

The maximum is

$$f(\tfrac{5}{2}) = 80(\tfrac{5}{2}) - 16(\tfrac{5}{2})^2 = 200 - 100 = 100. \qquad \blacksquare$$

General Functions Here is our basic problem: Find the maximal or minimal value of a differentiable function $f(x)$ in an interval $a \le x \le b$.
  In theory, it is easy to locate the maximum or the minimum:

The maximum of a differentiable function $f(x)$ in the interval $a \le x \le b$ occurs either at a value of x where $f'(x) = 0$, or at one of the end points, a or b. The same is true for the minimum of $f(x)$.

  This statement is easy to see graphically (Fig. 2). At points where $f'(x) > 0$, the graph is rising ($x = c_1$ for example). Neither the maximum nor the minimum can occur at such a point because the graph is higher to the right and lower to the left. At points where $f'(x) < 0$, the graph is falling ($x = c_2$ for example), and for a similar reason neither the maximum nor the minimum can occur there. Hence if $a < x_0 < b$ and $f(x_0)$ is the maximum or the minimum value of $f(x)$, then $f'(x_0) = 0$.

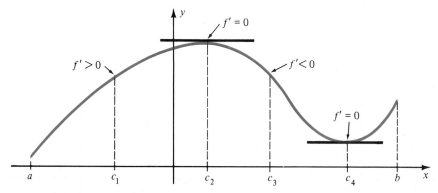

Fig. 2 Maxima and minima. Possible: a, c_2, c_4, b. Impossible: c_1, c_3.

  This argument does not apply at the end points a and b, however. (Why not?) The maximum or the minimum may occur at one of the end points without the derivative vanishing there (Fig. 3).
  This discussion suggests a procedure for solving the basic problem.

To find the maximum and minimum of $f(x)$ in the interval $a \le x \le b$, locate all points x where $f'(x) = 0$. Call these $x_1, x_2, \cdots, x_n$. The maximum is the largest of the numbers $f(a), f(x_1), f(x_2), \cdots, f(x_n), f(b)$. The minimum is the smallest of these.

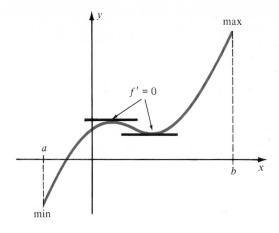

Fig. 3 The max of $f(x)$ occurs for $x = b$, *not* where $f'(x) = 0$; similarly, the min occurs for $x = a$.

■ **EXAMPLE 2** Find the max and min of

(a) $f(x) = \frac{1}{3}x^3 - 4x + 1$, $\qquad -3 \leq x \leq 3$,

(b) $f(x) = x^4 + 2x^3$, $\qquad -2 \leq x \leq 1$.

Solution (a) $f'(x) = x^2 - 4 = (x + 2)(x - 2)$, so $f'(x) = 0$ for $x = -2$ and $x = 2$. The corresponding values of $f(x)$ are

$$f(-2) = \frac{19}{3} \qquad \text{and} \qquad f(2) = -\frac{13}{3}.$$

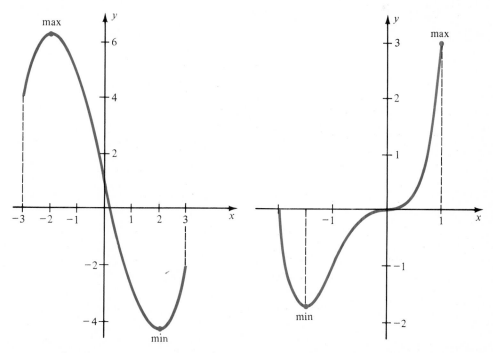

(a) $y = \frac{1}{3}x^3 - 4x + 1$, $-3 \leqslant x \leqslant 3$

(b) $y = x^4 + 2x^3$, $-2 \leqslant x \leqslant 1$

Fig. 4

These are candidates for the max and the min. The only other candidates are the values of $f(x)$ at the end points of the interval,

$$f(-3) = 4 \quad \text{and} \quad f(3) = -2.$$

The largest of these four numbers is $\frac{19}{3}$ and the smallest is $-\frac{13}{3}$, therefore

$$\text{max} = f(-2) = \tfrac{19}{3}, \quad \text{min} = f(2) = -\tfrac{13}{3}.$$

See Fig. 4a.

(b) $f'(x) = 4x^3 + 6x^2 = 2x^2(2x + 3)$, so $f'(x) = 0$ for $x = 0$ and for $x = -\frac{3}{2}$. The corresponding values of $f(x)$ are

$$f(0) = 0 \quad \text{and} \quad f(-\tfrac{3}{2}) = \tfrac{81}{16} - \tfrac{108}{16} = -\tfrac{27}{16}.$$

The only other candidates for max or min are the values of $f(x)$ at the end points,

$$f(-2) = 16 - 16 = 0 \quad \text{and} \quad f(1) = 3.$$

Therefore $\qquad\qquad \text{max} = f(1) = 3, \quad \text{min} = f(-\tfrac{3}{2}) = -\tfrac{27}{16}.$

See Fig. 4b.   ∎

Absence of End Points Some problems require the max or min of a function over an interval without end points, or with only one end point.

∎ **EXAMPLE 3** Find

(a) $\min(x^4 - 4x), \quad -\infty < x < \infty,$

(b) $\max\left(\dfrac{x}{2x^3 + 1}\right), \quad x \geq 0.$

Solution (a) Set $f(x) = x^4 - 4x$. Clearly

$$f(x) \longrightarrow \infty \quad \text{as} \quad x \longrightarrow \infty \quad \text{or} \quad x \longrightarrow -\infty,$$

so $f(x)$ *must* have a min someplace. At its min, $y = f(x)$ has a horizontal tangent, that is, $f'(x) = 0$. Consequently, we first solve the equation $f'(x) = 0$:

$$f'(x) = 4(x^3 - 1), \quad x^3 - 1 = 0, \quad x = 1.$$

There is only one value of x for which the tangent (Fig. 5a) is horizontal; it *must* give the minimum:

$$\text{min} = f(1) = 1 - 4 = -3.$$

(b) Set $f(x) = x/(2x^3 + 1)$. Obviously $f(x) > 0$ for $x > 0$, $f(0) = 0$, and

$$f(x) \longrightarrow 0+ \quad \text{as} \quad x \longrightarrow \infty.$$

The function increases starting from 0, but dies out to 0 as x increases indefinitely, so it must have a max someplace (Fig. 5b). Test for a horizontal tangent:

$$f'(x) = \frac{(2x^3 + 1) - 6x^3}{(2x^3 + 1)^2} = \frac{1 - 4x^3}{(2x^3 + 1)^2},$$

$$f'(x) = 0, \quad 1 - 4x^3 = 0, \quad x = \sqrt[3]{\tfrac{1}{4}} = \tfrac{1}{2}\sqrt[3]{2} \approx 0.629961.$$

There is only one horizontal tangent for $x > 0$; it must yield the max:

$$\max = f\left(\tfrac{1}{2}\sqrt[3]{2}\right) = \tfrac{1}{3}\sqrt[3]{2} \approx 0.419974.$$

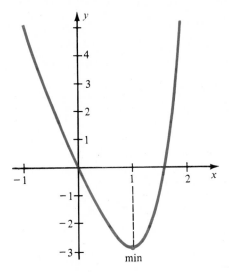

(a) $y = x^4 - 4x, \quad -\infty < x < \infty$

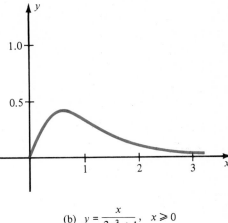

(b) $y = \dfrac{x}{2x^3 + 1}, \quad x \geqslant 0$

Fig. 5   ■

In general, a function $f(x)$ need not have a max or a min unless x is restricted to an interval with two end points. For example, take $f(x) = 1/(1 + x^2)$, where x is unrestricted. This function has maximum value 1, but no minimum value (it takes all values between 0 and 1, excluding zero). As another example, take $f(x) = 1/x$ for $x > 0$. This function has neither a maximum nor a minimum.

EXERCISES

Find the max and min

1 $x^2 - 4x + 6$

3 $-x^2 + 6x + 4$

5 $12x + \dfrac{1}{3x} \quad x > 0$

7 $2x^3 - 3x^2 - 12x + 1 \quad 0 \leq x \leq 3$

9 $\dfrac{1}{(x-1)(2-x)} \quad 1 < x < 2$

11 $\dfrac{x}{3} + \dfrac{1}{x} \quad 1 \leq x \leq 3$

13 $\dfrac{1}{x} - \dfrac{1}{x^2} - \dfrac{1}{x^3} \quad -3 \leq x \leq -1$

15 $4x^2 + \dfrac{1}{x} \quad \tfrac{1}{4} \leq x \leq 1$

17 $x^4 - 2x^2 + 1 \quad -1 \leq x \leq 2$

19 $-3x^4 - 16x^3 - 18x^2 - 12$
$\quad -3 \leq x \leq 0$

2 $2x^2 - 9x + 12$

4 $-3x^2 + 3x + 1$

6 $12x + \dfrac{1}{3x} \quad x < 0$

8 $2x^3 - 3x^2 - 12x + 1 \quad -2 \leq x \leq 2$

10 $\dfrac{x}{x^2 - x + 1}$

12 $\dfrac{1}{x} - \dfrac{1}{x^2} - \dfrac{1}{x^3} \quad 1 \leq x$

14 $x - x^4 \quad 0 \leq x \leq 1$

16 $\dfrac{x}{2 + x^3} \quad -1 \leq x$

18 $3x^4 - 16x^3 + 18x^2 + 12 \quad 0 \leq x \leq 4$

20 $x^3 + x + \dfrac{2}{x} \quad x > 0.$

21 $\dfrac{x}{x^2 - 4}$ $-2 < x < 2$ 22 $x^3 - 9x + 1.$

5. APPLICATIONS OF MAX AND MIN

■ **EXAMPLE 1** An open box is constructed by removing a small square from each corner of a tin sheet and then folding up the sides. If the sheet is L cm on each side, what is the largest possible volume of the box?

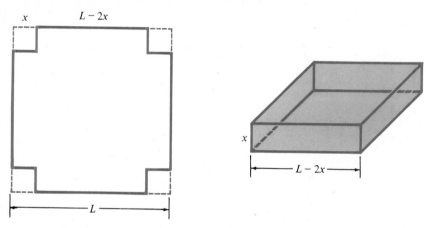

Fig. 1

Solution Let each cutout square have side x. See Fig. 1. Express the volume of the box as a function of x:

$$\text{Volume} = (\text{area of base}) \cdot (\text{height}).$$

The base is a square of side $L - 2x$, and the height is x. So the volume of the box is

$$V(x) = (L - 2x)^2 x = (L^2 - 4Lx + 4x^2)x = L^2 x - 4Lx^2 + 4x^3.$$

By the nature of the problem, x must be positive but less than $\frac{1}{2}L$, half the side of the sheet. The problem can now be stated in mathematical terms: Find the largest value of $V(x)$ in the domain $0 < x < \frac{1}{2}L$. Differentiate:

$$V'(x) = L^2 - 8Lx + 12x^2 = (L - 2x)(L - 6x).$$

The factor $L - 2x$ is positive because $x < \frac{1}{2}L$. Therefore the sign of $V'(x)$ is the same as the sign of $(L - 6x)$:

$$V'(x) > 0 \quad \text{for} \quad x < \tfrac{1}{6}L, \qquad V'(x) < 0 \quad \text{for} \quad x > \tfrac{1}{6}L,$$
$$V'(x) = 0 \quad \text{for} \quad x = \tfrac{1}{6}L.$$

This information clearly indicates a maximum for $x = \frac{1}{6}L$.

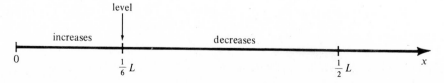

Therefore

$$V_{\max} = V(\tfrac{1}{6}L) = \left(L - \frac{2L}{6}\right)^2 \left(\frac{L}{6}\right) = \left(\frac{2L}{3}\right)^2 \left(\frac{L}{6}\right) = \frac{2L^3}{27} \quad \text{cm}^3.$$ ■

■ **EXAMPLE 2**   A length of wire 28 ft long is cut into two pieces. One piece is bent into a $3:4:5$ right triangle and the other piece is bent into a square. Show that the combined area is at least 18 ft^2.

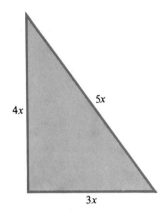

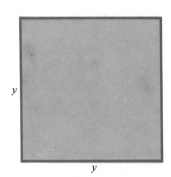

perimeter $= 12x$, area $= \frac{1}{2}(3x)(4x) = 6x^2$ perimeter $= 4y$, area $= y^2$

Fig. 2

Solution   This is just a disguised minimum problem: Find the minimum possible combined area and check that it is at least 18 ft^2.

To avoid fractions, name the sides of the right triangle $3x$, $4x$, and $5x$. Let y denote the side of the square (Fig. 2). The combined perimeter is

$$12x + 4y = 28, \quad \text{hence} \quad y = 7 - 3x.$$

The combined area is

$$A(x) = 6x^2 + y^2 = 6x^2 + (7 - 3x)^2 = 15x^2 - 42x + 49.$$

By the nature of the problem, x must be positive. But since the perimeter of the triangle is less than 28,

$$12x < 28, \quad \text{that is,} \quad x < \tfrac{28}{12} = \tfrac{7}{3}.$$

So the problem reduces to this: Find the least value of $A(x)$ in the domain $0 < x < \tfrac{7}{3}$. Take the derivative:

$$A'(x) = 30x - 42.$$

Hence $A'(x) < 0$ for $30x - 42 < 0$,   i.e., for $x < \tfrac{7}{5}$,

$\qquad\qquad A'(x) = 0$ for $x = \tfrac{7}{5}$,

$\qquad\qquad A'(x) > 0$ for $30x - 42 > 0$,   i.e., for $x > \tfrac{7}{5}$.

This information indicates a minimum for $x = \tfrac{7}{5}$.

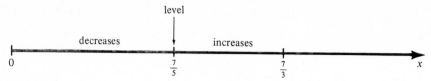

Therefore

$$A_{\min} = A(\tfrac{7}{5}) = 6(\tfrac{7}{5})^2 + (7 - \tfrac{21}{5})^2 = 6(\tfrac{49}{25}) + (\tfrac{14}{5})^2 = \tfrac{490}{25} = \tfrac{98}{5} = 19.6 \quad \text{ft}^2.$$

The combined area always exceeds 18 ft^2.　■

■ **EXAMPLE 3**   Ship A leaves a port at noon and sails due north at 10 mph. Ship B is 100 mi east of the port at noon and sailing due west at 6 mph. When will the ships be nearest each other?

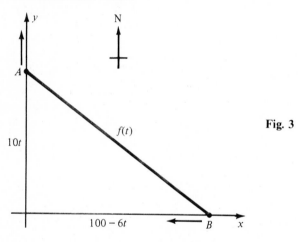

Fig. 3

Solution   Set up axes with the port at the origin and the y-axis pointing north. The relative position of the ships at t hr past noon is shown in Fig. 3. The distance between the ships is

$$f(t) = \sqrt{(100 - 6t)^2 + (10t)^2}.$$

The square root is annoying because when we differentiate, the resulting expression will be messy. A simple device for getting rid of the square root is squaring $f(t)$. Set

$$g(t) = [f(t)]^2 = (100 - 6t)^2 + (10t)^2.$$

The distance $f(t)$ is smallest precisely when its square $g(t)$ is smallest, and it is easier to minimize the quadratic function $g(t)$ than to minimize $f(t)$. We have

$$g(t) = (100 - 6t)^2 + (10t)^2 = 10{,}000 - 1200t + 36t^2 + 100t^2$$

$$= 10{,}000 - 1200t + 136t^2,$$

hence　　　　$$g'(t) = -1200 + 272t,$$

$$g'(t) = 0 \quad \text{for} \quad t = \tfrac{1200}{272} \approx 4.41 \approx 4 \text{ hr}, 25 \text{ min}.$$

Answer   The ships are closest at about $4:25$ PM.　■

■ **EXAMPLE 4** Compute the volume of the largest right circular cone inscribed in a sphere of radius R. (See Fig. 4a.)

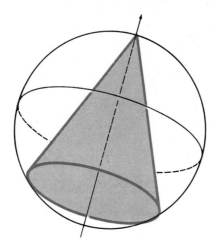

(a) Cone inscribed in sphere

(b) Cross-section

Fig. 4

Solution   The volume V of a cone is

$$V = \tfrac{1}{3}\pi r^2 h,$$

where r is the radius of its base and h is its height. You cannot maximize V directly, since V depends on *two* variables. In such cases, try to eliminate one of the variables, that is, express one variable in terms of the other. To do so, look for a relation between the variables. Often there is such a relation hidden in the conditions of the problem.

In this example, the cone is inscribed in a sphere, so there should be a relation between r, h, and R. Make a careful drawing of a cross-section (Fig. 4b). From the drawing,

$$r^2 + (h - R)^2 = R^2,$$
$$r^2 = R^2 - (h - R)^2 = 2Rh - h^2.$$

Substitute this expression and you eliminate the variable r:

$$V = \tfrac{1}{3}\pi r^2 h = \tfrac{1}{3}\pi(2Rh - h^2)h = \tfrac{1}{3}\pi(2Rh^2 - h^3).$$

By the physical nature of the problem, $0 < h < 2R$. Thus you must maximize

$$V(h) = \tfrac{1}{3}\pi(2Rh^2 - h^3)$$

in the interval $0 < h < 2R$.

There are no end points, hence the maximum occurs at a zero of the derivative:

$$\frac{dV}{dh} = \frac{\pi}{3}\left(4Rh - 3h^2\right) = \frac{\pi}{3}h(4R - 3h).$$

Therefore $\qquad \dfrac{dV}{dh} = 0 \qquad$ for $\quad h = 0 \quad$ or $\quad h = \dfrac{4}{3}R.$

But $h = 0$ is excluded; the maximum must occur at $4R/3$:

$$V_{\max} = V\!\left(\frac{4R}{3}\right) = \frac{\pi}{3}\left(\frac{4R}{3}\right)^{2}\!\left(\frac{2R}{3}\right) = \frac{32\pi R^{3}}{81}. \qquad\blacksquare$$

Remark The answer has the correct form; a volume should be a cubic expression. Since the sphere has volume $\frac{4}{3}\pi R^{3}$, it follows easily that the volume of the largest cone that can be inscribed in a sphere is $\frac{8}{27}$ the volume of the sphere.

■ **EXAMPLE 5** A 5-ft fence stands 4 ft from a high wall (Fig. 5). How long is the shortest ladder that can reach from the ground outside the fence to the wall?

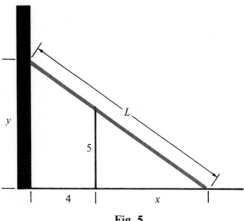

Fig. 5

Solution Take a moment to think. Note from Fig. 5 that if x is very small and positive, the ladder will be nearly vertical, certainly longer than is necessary. If x is large, the ladder will be nearly horizontal, again too long. The best choice of x seems to be somewhere around 5 or 6, surely between 2 and 10. In fact, as x increases starting near 0, it seems that L should decrease, reach a minimum, then increase thereafter.

To start the computation, note that

$$L^{2} = (x + 4)^{2} + y^{2}.$$

This is a function of two variables, but there is a relation between x and y: by similar triangles,

$$\frac{y}{x + 4} = \frac{5}{x}, \qquad \text{that is,} \quad y = \frac{5(x + 4)}{x}.$$

Hence, $\qquad L^{2} = (x + 4)^{2} + \dfrac{25(x + 4)^{2}}{x^{2}} = (x + 4)^{2}\!\left(1 + \dfrac{25}{x^{2}}\right).$

Rather than take the square root, minimize L^{2}. The range of x is all positive values; there are no end points in this problem.

Differentiate L^2:

$$\frac{d}{dx}(L^2) = 2(x+4)\left(1 + \frac{25}{x^2}\right) + (x+4)^2\left(\frac{-50}{x^3}\right) = 2(x+4)\left[1 + \frac{25}{x^2} - \frac{25(x+4)}{x^3}\right]$$

$$= 2(x+4)\left[\frac{x^3 + 25x - 25(x+4)}{x^3}\right] = \frac{2(x+4)(x^3 - 100)}{x^3}.$$

There is only one positive value of x for which the derivative is zero: $x = \sqrt[3]{100} \approx 4.64$. The derivative is negative for $x < \sqrt[3]{100}$, positive for $x > \sqrt[3]{100}$. Thus our physical intuition was correct: L^2 decreases, reaches a minimum near $x = 5$, then increases.

From the formula for L^2,

$$L = (x+4)\sqrt{1 + \frac{25}{x^2}} = \left(1 + \frac{4}{x}\right)\sqrt{x^2 + 25}.$$

Therefore $L_{\min} = L(\sqrt[3]{100}) = \left(1 + \frac{4}{\sqrt[3]{100}}\right)\sqrt{100^{2/3} + 25} \approx 12.7$ ft. ∎

∎ **EXAMPLE 6** The illumination of an object by a light source is directly proportional to the strength of the source and inversely proportional to the square of the distance between the source and the object. Two light sources, one five times as strong as the other, are 1 m apart (Fig. 6). At what point on the line between the sources should a screen be placed so that the illumination it receives is minimal?

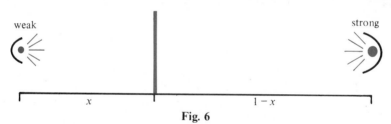

Fig. 6

Solution   Apparently the screen should be closer to the weaker source; $x < \frac{1}{2}$. Even though one source is five times as strong as the other, the screen cannot be too close to the weaker source because of the inverse square rule. A reasonable guess: x is around 0.3 or 0.4. The illumination from the weaker source is

$$I_1 = \frac{k}{x^2},$$

where the constant k depends on the units of measurement. The illumination from the stronger source is

$$I_2 = \frac{5k}{(1-x)^2}.$$

The problem is to minimize $I = \frac{k}{x^2} + \frac{5k}{(1-x)^2},$

for $0 < x < 1$. There are no end points in this problem since I is defined neither at $x = 0$ nor at $x = 1$. Differentiate:

$$\frac{dI}{dx} = -\frac{2k}{x^3} + \frac{10k}{(1-x)^3}.$$

This derivative is 0 for

$$\frac{2k}{x^3} = \frac{10k}{(1-x)^3}, \qquad \text{that is,} \quad 5x^3 = (1-x)^3.$$

Take cube roots: $(\sqrt[3]{5})x = 1 - x$, $\qquad x = \dfrac{1}{1 + \sqrt[3]{5}} \approx 0.369 \, \text{m}.$

Since $I \longrightarrow \infty$ as $x \longrightarrow 0+$ or $x \longrightarrow 1-$, this value of x must give the minimal I.

∎

■ **EXAMPLE 7** Light travels between two points along the path that requires the least time. In different substances (water, air, glass, etc.) light travels at different speeds. Assume the upper half of the x, y-plane is a substance in which the speed of light is v_1 and the lower half is another substance in which the speed of light is v_2. Describe the path of a light ray traveling between two points in opposite halves of the plane.

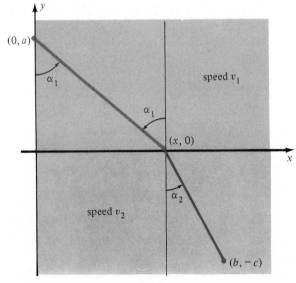

Fig. 7

Solution Let the two points be $(0, a)$ and $(b, -c)$ in Fig. 7. A ray will travel from $(0, a)$ along a straight line to some point $(x, 0)$ and then along another straight line to $(b, -c)$. A value x must be found so that the time of travel is a minimum. Obviously $0 \le x \le b$.

The time required for a ray to travel from $(0, a)$ to $(x, 0)$ is

$$t_1 = \frac{\text{distance}}{\text{speed}} = \frac{\sqrt{x^2 + a^2}}{v_1}.$$

The time required from $(x, 0)$ to $(b, -c)$ is

$$t_2 = \frac{\sqrt{(b - x)^2 + c^2}}{v_2}.$$

Hence you must minimize

$$t = \frac{\sqrt{x^2 + a^2}}{v_1} + \frac{\sqrt{(b - x)^2 + c^2}}{v_2}$$

in the interval $0 \leq x \leq b$. It is plausible physically that the minimum will not occur at either end point.

Compute dt/dx:

$$\frac{dt}{dx} = \frac{x}{v_1 \sqrt{x^2 + a^2}} - \frac{b - x}{v_2 \sqrt{(b - x)^2 + c^2}}.$$

This derivative looks complicated, but from Fig. 7,

$$\frac{x}{\sqrt{x^2 + a^2}} = \sin \alpha_1, \qquad \frac{b - x}{\sqrt{(b - x)^2 + c^2}} = \sin \alpha_2.$$

Hence the derivative has the simple form

$$\frac{dt}{dx} = \frac{\sin \alpha_1}{v_1} - \frac{\sin \alpha_2}{v_2}.$$

The derivative is zero if x is chosen to satisfy

$$\frac{\sin \alpha_1}{v_1} = \frac{\sin \alpha_2}{v_2}.$$

This equation is known as **Snell's Law of Refraction**. To see that it describes the path of least time, note that dt/dx is the difference of two terms. As x increases from 0 to b, the first term, $(\sin \alpha_1)/v_1$, increases steadily starting with 0. The second term, $(\sin \alpha_2)/v_2$, decreases steadily from some positive value to 0. Consequently, dt/dx starts negative at $x = 0$ and steadily increases to a positive value at $x = b$. Therefore, the minimum t occurs at the only x for which $dt/dx = 0$.

Answer   The path is the broken line for which $\dfrac{\sin \alpha_1}{v_1} = \dfrac{\sin \alpha_2}{v_2}.$ ∎

EXERCISES

1   A ball thrown straight up reaches a height of $3 + 40t - 16t^2$ ft in t sec. How high will it go?

2   Show that the rectangle of largest possible area, for a given perimeter, is a square.

3   Find the maximum slope of the curve $y = 6x^2 - x^3$.

4   The power output P of a battery is given by $P = EI - RI^2$, where E and R are constants and I is current. Find the current for which the power output is a maximum and find the maximum power.

5   A man with 300 m of fencing wishes to enclose a rectangular area and divide it into 5 pens with fences parallel to the short end of the rectangle. What dimensions of the enclosure make its area a maximum?

6   Find the dimensions of the rectangle of largest area that can be inscribed in an equilateral triangle of side s, if one side of the rectangle lies on the base of the triangle.

7   Find the dimensions of the rectangle of largest area that can be inscribed in a right triangle with legs of length a and b, if two sides of the rectangle lie along the legs of the triangle.

8   Find the maximal combined area in Example 2, p. 129.

9   Suppose in Example 3, p. 130, ship A is already 70 mi north of the port at noon, but otherwise the problem is the same. Now when are the ships closest?

10   An open rectangular box has volume 15 ft^3. The length of its base is 3 times its width. Materials for the sides and base cost 60¢ and 40¢ per ft^2, respectively. Find the dimensions of the cheapest such box.

11   What points on the curve $xy^2 = 1$ are nearest the origin?

12   Find the point on the graph of the equation $y = \sqrt{x}$ nearest to the point $(1, 0)$.

13   Two particles moving in the plane have coordinates $(2t, 8t^3 - 24t + 10)$ and $(2t + 1, 8t^3 + 6t + 1)$ at time t. How close do the particles come to each other?

14   Find the minimum vertical distance between the curves $y = 27x^3$ and $y = -1/x$ if $x \neq 0$.

15   As a man starts across a 200-ft bridge, a ship passes directly beneath the center of the bridge. If the ship is moving at the rate of 8 ft/sec and the man at the rate of 6 ft/sec, what is the shortest horizontal distance between them?

16   Suppose in Ex. 15, the bridge is 50 ft high. Find the shortest *distance* between the man and the ship.

17   Find the dimensions of the right circular cone having the greatest volume for a given slant height a.

18   A closed cylindrical can is to have volume 500 cm^3. For what dimensions will the total surface be a minimum?

A window of perimeter 16 ft has the form of a rectangle surmounted by a semicircle

19   For what radius of the semicircle is the window area greatest?

20   For what radius of the semicircle is the most light admitted, if the semicircle admits half as much light per unit area as the rectangle admits per unit area?

Suppose the cost of producing x units is $f(x)$ dollars, and the price of x units is $h(x)$ dollars. Calculate the maximum net revenue possible for a manufacturer if

21 $f(x) = 7.5x + 400$, $h(x) = 10x - 0.0005x^2$

22 $f(x) = 50x + 1200$, $h(x) = 65x - 0.001x^2$.

23   The cost per hour in dollars for fuel to operate a certain airliner is $0.012v^2$, where v is the speed in mph. If fixed charges amount to \$4000/hr, find the most economical speed for a 1500 mi trip.

24   During a typical 8-hr work day the quantity of gravel produced in a plant is $60t + 12t^2 - t^3$ tons, where t represents hours worked. When is the rate of production at a maximum?

25   A cylindrical tank (open top) is to hold V liters (1 liter = 1000 cm^3). How should it be made so as to use the least amount of material?

26   A page is to contain 27 in^2 of print. The margins at the top and bottom are 1.5 in, at the sides 1 in. Find the most economical dimensions of the page.

27   Find the largest possible area of an isosceles triangle whose equal legs have length L ft.

28   Find the dimensions of the rectangle of maximum area that can be inscribed in the region bounded by the parabola $y = -8x^2 + 16$ and the x-axis.

29   An athletic field of 500-m perimeter consists of a rectangle with a semicircle at each end. Find the dimensions of the field so that the area of the rectangular portion is the largest possible.

30   Two posts, 8 ft and 12 ft high, stand 15 ft apart. They are to be stayed by wires attached to a single stake at ground level, and running to the tops of the posts. Where should the stake be placed to use the least amount of wire?

31 A man in a rowboat 3 mi off a long straight shore wants to reach a point 5 mi up the shore. If he can row 2 mph and walk 4 mph, describe his fastest route.

32 Suppose in Ex. 31 the boat has a motor. How fast must the boat be able to go so that the fastest route is entirely by boat?

33 The strength of a beam of fixed length and rectangular cross-section is proportional to the width and to the square of the depth of the cross-section. Find the proportions of the beam of greatest strength that can be cut from a circular log.

34 The energy of a certain diatomic molecule is

$$U = \frac{a}{x^{12}} - \frac{b}{x^6},$$

where a and b are positive constants and x is the distance between the atoms. Find the **dissociation energy**, the maximum of $-U$.

35 The speed v of a surface wave depends on its wavelength λ according to

$$v = \sqrt{\frac{g}{2\pi}\lambda + \frac{2\pi\sigma}{\rho\lambda}},$$

where the constants are g, the force of gravity, σ, the surface tension of the fluid, and ρ, the density of the fluid. Find the minimum possible speed and the corresponding wavelength.

36 A one-port network (Fig. 8) at fixed frequency ω is to be terminated in a resistor x so the power

$$P = \frac{E^2 x}{(2\pi\omega L)^2 + R^2 + 2Rx + x^2}$$

dissipated by x (in heat) is maximal. Find x and $P_{\max}$.

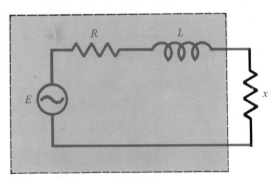

Fig. 8

37 Find the two positive numbers x and y for which $x + y = 1$, such that $x^3 y^4$ is maximum.

38 Given n numbers $a_1, a_2, \cdots, a_n$, show that

$$(x - a_1)^2 + (x - a_2)^2 + \cdots + (x - a_n)^2$$

is least when $x = \bar{a}$, the average of the numbers.

39 Find the maximum for $x \geq 0$ of $x/(1 + x)^2$.

40 Find the max and min of $(1 + x^2)/(1 + x^4)$.

The object of the next two examples is to prove an important inequality: if $a_1, a_2, \cdots, a_n$ are any positive numbers, then

$$\sqrt[n]{a_1 a_2 \cdots a_n} \leq \frac{a_1 + a_2 + \cdots + a_n}{n}.$$

In words, the **geometric mean** of a set of numbers does not exceed the **arithmetic mean** (average). We abbreviate the inequality by the notation

$$G_n \le A_n .$$

41* Show that the maximum value of the ratio

$$\frac{\sqrt[n+1]{a_1 a_2 \cdots a_n x}}{\dfrac{1}{n+1}(a_1 + a_2 + \cdots + a_n + x)}$$

occurs for $x = A_n$, and compute the maximum. Conclude that

$$\frac{G_{n+1}}{A_{n+1}} \le \left(\frac{G_n}{A_n}\right)^{n/(n+1)} .$$

42* By repeated application of Ex. 41, show that

$$\frac{G_n}{A_n} \le \left(\frac{G_1}{A_1}\right)^{1/n} ,$$

and therefore $G_n \le A_n$. Explain why $G_n = A_n$ if and only if $a_1 = a_2 = \cdots = a_n$.

6. SECOND DERIVATIVE TEST

Examine the function graphed in Fig. 1. At each of the points $x_1, x_2, \cdots, x_6$, its derivative equals 0. At the points x_1, x_3, x_6, the function is decreasing immediately to the left and increasing immediately to the right. The values $f(x_1)$, $f(x_3)$, $f(x_6)$ are accordingly called **local minima** of $f(x)$, also **relative minima**. Correspondingly, the values $f(x_2)$ and $f(x_5)$ are **local maxima**. The value $f(x_4)$ is different; it is neither a local max nor a local min, even though $f'(x_4) = 0$.

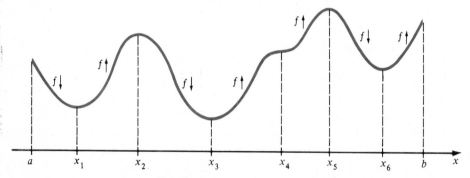

Fig. 1   Local min: x_1, x_3, x_6 neither: x_4
Local max: x_2, x_5

When we solve $f'(x) = 0$, we find all six of the x_i. Sometimes we can pick out the local maxs and mins by checking where $f' > 0$ and where $f' < 0$. But this may be tedious, so we seek another method.

Generally speaking, a curve $y = f(x)$ can have four shapes at a point where $f'(x) = 0$. Suppose $f'(c) = 0$, but otherwise $f'(x) \ne 0$ on an interval centered at c. The four possibilities are shown in Fig. 2. They are (a) local maximum, (b) local minimum, (c) and (d) horizontal inflection point.

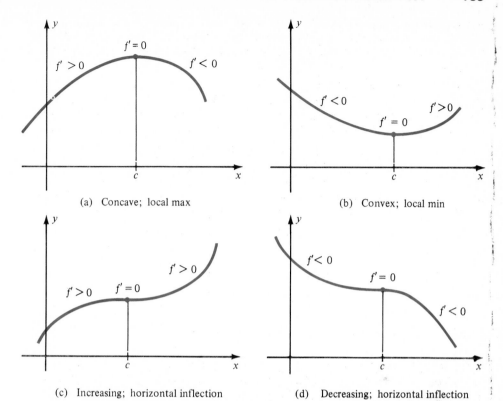

(a) Concave; local max

(b) Convex; local min

(c) Increasing; horizontal inflection

(d) Decreasing; horizontal inflection

Fig. 2   Possible shapes near a point where $f'(c) = 0$, provided $f'(x) \neq 0$ for $x \neq c$.

Let us summarize this discussion so far.

Let $f(x)$ be a differentiable function for $a \leq x \leq b$, and suppose $f'(c) = 0$, where $a < c < b$.

(1) If $f'(x)$ changes from positive to negative as x increases through c, then $f(c)$ is a local max.

(2) If $f'(x)$ changes from negative to positive as x increases through c, then $f(c)$ is a local min.

(3) If $f'(x)$ does not change sign as x passes through c, then $(c, f(c))$ is a horizontal inflection point of $y = f(x)$.

■ **EXAMPLE 1** Classify the points of $y = \frac{1}{5}x^5 - \frac{3}{4}x^4 + \frac{2}{3}x^3 - 1$ at which $f'(x) = 0$.

Solution   Examine the signs of $f'(x) = x^4 - 3x^3 + 2x^2 = x^2(x-1)(x-2)$:

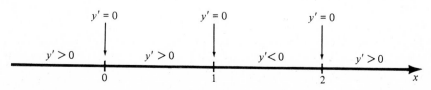

Accordingly:

$x = 0$	horizontal inflection
$x = 1$	local max
$x = 2$	local min

This information makes it quite easy to graph the function (Fig. 3).

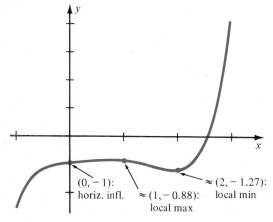

Fig. 3 $y = \frac{1}{5}x^5 - \frac{3}{4}x^4 + \frac{2}{3}x^3 - 1$

$(0, -1)$: horiz. infl. $\approx (1, -0.88)$: local max $\approx (2, -1.27)$: local min

Use of the Second Derivative   At a point $x = c$ where $f'(c) = 0$, the sign of the second derivative indicates whether the graph has a local min or local max. Suppose, for instance, that $f'(c) = 0$ and $f''(c) > 0$. Then $f'(x)$ is increasing near $x = c$; as x increases through c, then $f'(x)$ changes from negative to positive, so $f(c)$ is a local min (Fig. 4a). If $f'(c) = 0$ and $f''(c) < 0$, then $f'(x)$ is decreasing near $x = c$; hence $f'(x)$ changes from positive to negative, so $f(c)$ is a local max (Fig. 4b).

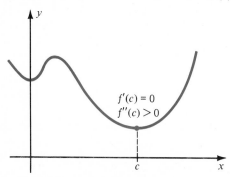

$f'(c) = 0$
$f''(c) > 0$

$f'(c) = 0$
$f''(c) < 0$

(a) Local min; $f'(x)$ increasing at $x = c$; graph convex near $x = c$

(b) Local max; $f'(x)$ decreasing at $x = c$; graph concave near $x = c$

Fig. 4

Geometrically speaking, the sign of $f''(c)$ indicates whether the graph is convex or concave. If $f''(c) > 0$, the curve is convex, hence $f(c)$ is a local min; if $f''(c) < 0$, the curve is concave, hence $f(c)$ is a local max.

Second Derivative Test Suppose that $f(x)$ is twice differentiable in an interval including $x = c$, and that $f'(c) = 0$.

(1) If $f''(c) > 0$, then $f(c)$ is a local min.

(2) If $f''(c) < 0$, then $f(c)$ is a local max.

Remark The case $f'(c) = 0$ *and* $f''(c) = 0$ is inconclusive. For example, the functions $f_4(x) = x^4$, $f_5(x) = x^5$, and $f_6(x) = -x^6$ all satisfy these conditions at $c = 0$. Yet $f_4(x)$ has a minimum at $x = 0$, and $f_6(x)$ has a maximum at $x = 0$, whereas $f_5(x)$ has neither (Fig. 5).

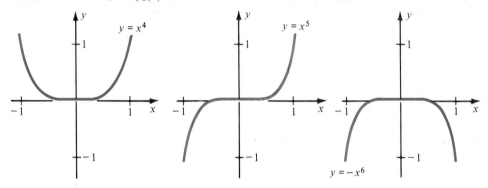

(a) Local min at $(0, 0)$ (b) Horizontal inflection at $(0, 0)$ (c) Local max at $(0, 0)$

Fig. 5 The second derivative test is inconclusive if $f''(c) = 0$.

■ **EXAMPLE 2** Find all local max and min of $f(x) = \dfrac{x}{1 + x^2}$.

Solution
$$f'(x) = \frac{(1 + x^2) - 2x \cdot x}{(1 + x^2)^2} = \frac{1 - x^2}{(1 + x^2)^2};$$

hence $f'(x) = 0$ only for $x = -1$ and $x = 1$. Next,

$$f''(x) = \frac{-2x(1 + x^2)^2 - 4x(1 + x^2)(1 - x^2)}{(1 + x^2)^4}$$

$$= \frac{-2x(1 + x^2) - 4x(1 - x^2)}{(1 + x^2)^3} = \frac{2x(x^2 - 3)}{(1 + x^2)^3}.$$

Test the two points where $f'(x) = 0$:

$x = -1$	$f''(-1) = \frac{1}{2} > 0$	$f(-1) = -\frac{1}{2}$, local min
$x = 1$	$f''(1) = -\frac{1}{2} < 0$	$f(1) = \frac{1}{2}$, local max

Continuation Let's graph $y = f(x)$. We have already located all local extrema. For further information we look at the sign of $f'(x)$:

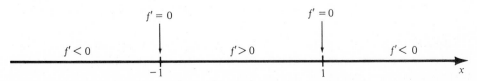

For information on convexity, concavity, and inflection points, we look at the sign of $f''(x)$:

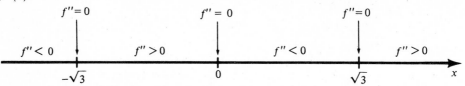

There are inflection points at $x = \pm\sqrt{3}$.

We note further that $f(x)$ is an odd function, that $f(x) \longrightarrow 0$ as $x \longrightarrow \infty$ or $x \longrightarrow -\infty$, and that

$$f(1) = \tfrac{1}{2}, \qquad f(\sqrt{3}) = \tfrac{1}{4}\sqrt{3}, \qquad f(0) = 0, \qquad f'(0) = 1.$$

We now have adequate information for a fairly accurate sketch (Fig. 6).

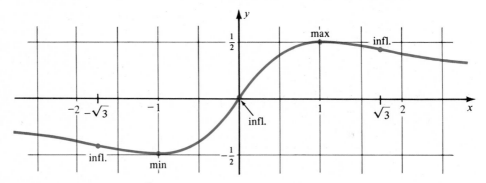

Fig. 6   Graph of $y = \dfrac{x}{1 + x^2}$

     Convex: $-\sqrt{3} < x < 0, \quad \sqrt{3} < x$ Concave: $-\sqrt{3} < x, \quad 0 < x < \sqrt{3}.$ ■

■ **EXAMPLE 3**   Find all local max and min of $f(x) = x^4(1 - x)^{10}$.

Solution

$$f'(x) = 4x^3(1 - x)^{10} - 10x^4(1 - x)^9$$
$$= 2x^3(1 - x)^9[2(1 - x) - 5x] = 2x^3(1 - x)^9(2 - 7x).$$

Therefore $f'(x) = 0$ for $x = 0$, $x = 1$, and $x = \tfrac{2}{7}$. Next,

$$f''(x) = 6x^2(1 - x)^9(2 - 7x) - 18x^3(1 - x)^8(2 - 7x) - 14x^3(1 - x)^9.$$

We want the signs of $f''(0), f''(1)$, and $f''(\tfrac{2}{7})$. Clearly, each term is zero for $x = 0$ and for $x = 1$, and the first two terms are zero for $x = \tfrac{2}{7}$. Hence,

$$f''(0) = 0, \qquad f''(1) = 0, \qquad f''(\tfrac{2}{7}) = -(14)(\tfrac{2}{7})^3(\tfrac{5}{7})^9 < 0.$$

Therefore $f(\tfrac{2}{7}) = (\tfrac{2}{7})^4(\tfrac{5}{7})^{10}$ is a local max, and the second derivative test is inconclusive for $x = 0$ and $x = 1$.

However, we see that $f(x) \geq 0$ for all x, while $f(0) = 0$ and $f(1) = 0$. Therefore $f(0) = 0$ and $f(1) = 0$ are local mins by inspection. A rough graph is useful here (Fig. 7).

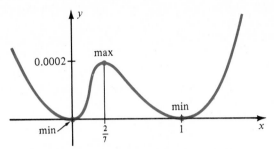

Fig. 7 Rough graph of $y = x^4(1 - x)^{10}$

Alternative Solution Rather than compute $f''(x)$, which is messy, examine the sign of

$$f'(x) = 2x^3(1 - x)^9(2 - 7x).$$

For very large positive values of x, this polynomial is obviously positive. Since it changes sign at $x = 1$, $\frac{2}{7}$, and 0, the whole picture is clear:

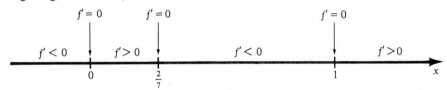

The behavior of $f'(x)$ indicates local mins at $x = 0$ and 1, a local max at $x = \frac{2}{7}$. ∎

Remarks on Finding Maxima and Minima In most maximum and minimum problems there are only one or two zeros of the derivative to consider, and possibly two end points. Often you can rule out the end points by physical considerations. Then you have to decide which zero of the derivative gives the maximum or the minimum. If it is easy to compute the second derivative, do so. If not, or if the second derivative is zero, try observing the sign of the derivative near the point in question. Better yet, graph the function if that is easy. Be flexible.

EXERCISES

Find all local max and min

1 $y = x^4 - 2x^2$

2 $y = x^5 - 20x - 3$

3 $y = 3x^4 - 4x^3 - 12x^2 + 2$

4 $y = x^3 + 4x^2 + 5x + 2$

5 $y = \dfrac{x}{\sqrt{x^4 + 16}}$

6 $y = \dfrac{\sqrt{x}}{5x + 4}$

7 $y = 2x - \dfrac{27}{x^2}$

8 $y = 2x + \dfrac{27}{x^2}$

9 $y = \dfrac{1}{x^2} - \dfrac{1}{x^3}$

10 $y = x^2 - \dfrac{1}{x}$

11 $y = \dfrac{x^2}{1 + x^4}$

12 $y = \dfrac{x^3}{1 + x^4}.$

Let $n \geq 2$. Find the maximum for $x \geq 0$

13 $y = \dfrac{x}{(x + 2)^n}.$

14 $y = \dfrac{x}{(1 + x^2)^n}.$

15* Assume $a > 0$ and $b > 0$. Find all local maxs and mins of $y = (a + x)\sqrt{b^2 + (a - x)^2}$. Consider various cases.

16* Show that the maximum of

$$y = \frac{(1 + x)^2}{1 + \sqrt{1 + x^4}}$$

is taken at $x = c$, where $c^3 - c - 2 = 0$. Conclude that $c \approx 1.5214$, and $y_{max} = (1 + c)^2/(2 + c) \approx 1.8054$.

17 Show that each inflection point of the graph $y = f(x)$ corresponds to a local max or min of $f'(x)$.

18 Given that $f''(c) = 0$ and $f'''(c) > 0$, draw a conclusion about the graph $y = f(x)$ near $x = c$.

19 Find $y(x)$ if $y(0) = y(1) = 0$ and $y''(x) = 0$ for all x.

20* Let $y = x^4 + a_1 x^3 + a_2 x^2 + a_3 x + a_4$. Assume $y'(c) = 0$ and $y''(c) < 0$. Prove that y has exactly two local mins. Give an explicit example of such a function.

7. ON PROBLEM SOLVING

A major part of your time in Calculus and other courses is devoted to solving problems. It is worth your while to develop sound techniques. Here are a few suggestions.

Think. Before plunging into a problem, take a moment to think. Read the problem again. Think about it. What are its essential features? Have you seen a problem like it before? What techniques are needed?

Try to make a rough estimate of the answer. It will help you understand the problem and will serve as a check against unreasonable answers. A car will *not* go 1000 mi in 3 hr; a weight dropped from 10,000 ft will *not* hit the earth at 5 mph; the volume of a tank is *not* −275 gal.

Examine the data. Be sure you understand what is given. Translate the data into mathematical language. Whenever possible, make a clear diagram and label it accurately. Place axes to simplify computations. If you get stuck, check that you are using *all* the data.

Avoid sloppiness.

(a) Avoid sloppiness in language. Mathematics is written in English sentences. A typical mathematical sentence is "$y = 4x + 1$." The equal sign is the verb in this sentence; it means "equals" or "is equal to." The equal sign is not to be used in place of "and," nor as a punctuation mark. *Quantities on opposite sides of an equal sign must be equal.*

Use short simple sentences. Avoid pronouns such as "it" and "which." Give names and use them. Otherwise you may write gibberish like the following:

"To find the minimum of it, differentiate it and set it equal to zero, then solve it which if you substitute it, it is the minimum."

Better: "To find the minimum of $f(x)$, set its derivative $f'(x)$ equal to zero. Let x_0 be the solution of the resulting equation. Then $f(x_0)$ is the minimum value of $f(x)$."

(b) Avoid sloppiness in computation. Do calculations in a sequence of neat, orderly steps. Include all steps except utterly trivial ones. This will help eliminate errors, or at least make errors easier to find. Check any numbers used; be sure that you have not dropped a minus sign or transposed digits.

(c)   Avoid sloppiness in units. If you start out measuring in feet, all lengths must be in feet, all areas in square feet, and all volumes in cubic feet. Do not mix feet and acres, seconds and years.

(d)   Avoid sloppiness in the answer. Be sure to answer the question that is asked. If the problem asks for the *maximum value* of $f(x)$, the answer is not the *point* where the maximum occurs. If the problem asks for a *formula*, the answer is not a *number*.

■ **EXAMPLE 1**   Find the minimum of $f(x) = x^2 - 2x + 1$.

Solution 1 $2x - 2, \quad x = 1, \quad 1^2 - 2 \cdot 1 + 1, \quad 0$

Unbearable. This is just a collection of marks on the paper. There is absolutely no indication of what these marks mean or of what they have to do with the problem. When you write, it is your responsibility to inform the reader of what you are doing. Assume he is intelligent, but not a mind reader.

Solution 2
$$\frac{df}{dx} = 2x - 2 = 0 = 2x = 2 = x = 1$$
$$= f(x) = 1^2 - 2 \cdot 1 + 1 = 0.$$

Poor. The equal sign is badly mauled. This solution contains such enlightening statements as "$0 = 2 = 1$," and it does not explain what the writer is doing.

Solution 3
$$\frac{df}{dx} = 2x - 2 = 0, \qquad 2x = 2, \qquad x = 1.$$

This is better than Solution 2, but contains two errors. Error 1: The first statement, "$df/dx = 2x - 2 = 0$," muddles two separate steps. First the derivative is computed, then the derivative is equated to zero. Error 2: The solution is incomplete because it does not give what the problem asks for, the minimum value of f. Instead, it gives the point x at which the minimum is assumed.

Solution 4   The derivative of f is
$$f' = 2x - 2.$$

At a minimum, $f' = 0$. Hence
$$2x - 2 = 0, \qquad x = 1.$$

The corresponding value of f is
$$f(1) = 1^2 - 2 \cdot 1 + 1 = 0.$$

If $x > 1$, then $f'(x) = 2(x - 1) > 0$, so f is increasing. If $x < 1$, then $f'(x) = 2(x - 1) < 0$, so f is decreasing. Hence f is minimal at $x = 1$, and the minimum value of f is 0.

This solution is absolutely correct, but long. For homework assignments the following may be satisfactory (check with your instructor):

Solution 5 $f'(x) = 2x - 2.$

At min, $f' = 0$, $2x - 2 = 0$, $x = 1$. For $x > 1$, $f'(x) = 2(x - 1) > 0, f\uparrow$; for $x < 1$, $f'(x) = 2(x - 1) < 0, f\downarrow$.

Hence $x = 1$ yields min, $f_{\min} = f(1) = 1^2 - 2 \cdot 1 + 1 = 0$.

The next solution was submitted by a student who took a moment to think.

Solution 6 $f(x) = x^2 - 2x + 1 = (x - 1)^2 \geq 0$.

But $f(1) = (1 - 1)^2 = 0$.

Hence the minimum value of $f(x)$ is 0.   ■

The Wyatt Earp Principle   That legendary gunfighter survived many a shoot-out in the old West. Yet Earp carried only one gun and used no fancy tricks. His secret? He took an extra split second *to aim*. While the bad guy blazed away wildly with two guns, Earp got his man on the first shot.

Try to face a calculus exam the way Wyatt Earp faced a gunfight. Instead of differentiating wildly with both hands, take a minute to think. You may find the problem is simpler than it looks at first. Certainly you will have a better chance of winning the showdown.

8. EXTREMA AND CONVEXITY

Our work on maxima and minima (extrema for short) has been based on the assumption that they exist. We are now going to state this assumption as a formal, precise theorem about continuous functions.

> **Existence of Extrema**   Let $f(x)$ be a continuous function with domain the closed interval $a \leq x \leq b$. Then there exist points c_0 and c_1 on this interval such that
>
> $$f(c_0) \leq f(x) \leq f(c_1)$$
>
> for all x on the interval.

In other words, the continuous function $f(x)$ *has* a minimum value $f(c_0)$ on the interval and *has* a maximum value $f(c_1)$. The minimum value may be taken at two or more points—the theorem says nothing about that. It does say that there is a minimum value, taken *at least once*, and likewise for a maximum value.

The existence of extrema is a deep property of continuous functions. Its proof is beyond the scope of this course, and we shall have to accept the theorem on faith.

An important corollary uses only part of this property, the part that says $f(x)$ stays between two fixed numbers.

> **Boundedness**   Let $f(x)$ be a continuous function with domain the closed interval $a \leq x \leq b$. Then $f(x)$ is **bounded**, that is, there exist constants A and B such that
>
> $$A \leq f(x) \leq B$$
>
> for all x on the interval. Alternatively, there exists a constant C such that
>
> $$|f(x)| \leq C$$
>
> for all x on the interval.

Proof By the existence of extrema, $f(c_0) \le f(x) \le f(c_1)$ for all x on the interval. Therefore the constants $A = f(c_0)$ and $B = f(c_1)$ do the trick.

Having A and B, set $C = \max\{|A|, |B|\}$. Then

$$f(x) \le B \le |B| \le C \quad \text{and} \quad -f(x) \le -A \le |A| \le C,$$

so $|f(x)| \le C$. Conversely, given C such that $|f(x)| \le C$, then $A = -C$ and $B = C$ satisfy $A \le f(x) \le B$.

Locally Increasing Functions The next result says that if $f'(c) > 0$, then $f(x)$ is larger than $f(c)$ slightly to the right of $x = c$ and smaller slightly to the left. A corresponding result holds for $f'(c) < 0$. Taken together, these results prove the first derivative test for extrema, as we shall see.

Theorem Let $f(x)$ be defined for $a < x < b$, and let $f(x)$ be differentiable at c, where $a < c < b$. If $f'(c) > 0$, then there exists $\delta > 0$ such that

$$\begin{cases} f(x) > f(c) & \text{for} \quad c < x < c + \delta \\ f(x) < f(c) & \text{for} \quad c - \delta < x < c. \end{cases}$$

Proof The proof is elementary, using only the definition of the derivative. Since

$$\lim_{x \to c} \frac{f(x) - f(c)}{x - c} = f'(c) > 0,$$

there exists $\delta > 0$ such that

$$\frac{f(x) - f(c)}{x - c} > 0 \quad \text{for} \quad 0 < |x - c| < \delta.$$

If $c < x < c + \delta$, then $x - c > 0$, so

$$f(x) - f(c) > 0, \quad \text{that is,} \quad f(x) > f(c).$$

If $c - \delta < x < c$, then $x - c < 0$, so

$$f(x) - f(c) < 0, \quad \text{that is,} \quad f(x) < f(c).$$

Remark Suppose $f'(c) < 0$. Then $-f'(c) > 0$, so the theorem applies to $-f(x)$. The conclusion is that $f(x) > f(c)$ for $c - \delta < x < c$ and $f(x) < f(c)$ for $c < x < c + \delta$.

The First Derivative Test We are now prepared to prove that the derivative must be zero at internal extrema.

First Derivative Test Let $f(x)$ be a differentiable function on the closed interval $a \le x \le b$. Suppose that $a < c < b$ and that $f(c)$ is either a local max or a local min of $f(x)$. Then $f'(c) = 0$.

Proof There are just three possibilities:

$$f'(c) > 0, \quad f'(c) < 0, \quad f'(c) = 0.$$

The last theorem rules out the first two. For if $f'(c) > 0$, then $f(x) > f(c)$ just to the right of $x = c$ and $f(x) < f(c)$ just to the left of $x = c$, so $f(c)$ is neither a local max nor min. Similarly, $f'(c) < 0$ is impossible. The only way out is $f'(c) = 0$.

Rolle's Theorem   We are preparing for a very fundamental theorem about differentiable functions called the Mean Value Theorem. It is precisely what is needed to prove that a function with everywhere zero derivative is constant, that a function with everywhere positive derivative is increasing, and many other results. Its proof depends on the following preliminary result.

> **Rolle's Theorem**   Let $f(x)$ be differentiable on the closed interval $a \leq x \leq b$ and suppose $f(a) = f(b)$. Then there exists a point c such that $a < c < b$ and $f'(c) = 0$.

Proof   Since $f(x)$ is differentiable, it certainly is continuous. Therefore $f(x)$ has a max and min:

$$f(c_0) \leq f(x) \leq f(c_1)$$

for all $a \leq x \leq b$, where $a \leq c_0 \leq b$ and $a \leq c_1 \leq b$. If $a < c_0 < b$, then $f'(c_0) = 0$ by the preceding theorem. If $a < c_1 < b$, then $f'(c_1) = 0$ for the same reason. If neither, then $c_0 = a$ or $c_0 = b$ *and* $c_1 = a$ or $c_1 = b$. By hypothesis, $f(a) = f(b)$. Hence

$$f(c_0) = f(a) = f(b) \qquad \text{and} \qquad f(c_1) = f(a) = f(b).$$

Therefore for *every* x on the interval

$$f(a) = f(c_0) \leq f(x) \leq f(c_1) = f(a),$$

so $f(x) = f(a)$. In other words, $f(x)$ is constant. This implies $f'(c) = 0$ for *every* $a < c < b$, more than we wanted! See Fig. 1 for the geometric meaning of Rolle's Theorem.

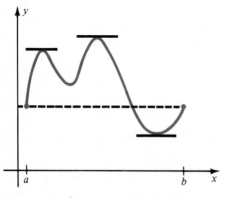

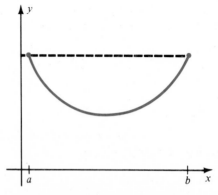

(a) $f'(c) = 0$ at each interior
local max or min

(b) Maybe there is no *interior* max
(min). Then there certainly is
an interior min (max)!

Fig. 1   Meaning of Rolle's Theorem

Mean Value Theorem The following result, known as the Mean Value Theorem or Law of the Mean, can be interpreted as an oblique form of Rolle's Theorem. Rolle's Theorem guarantees the existence of a horizontal tangent when $f(a) = f(b)$, that is, a tangent parallel to the (horizontal) chord joining $(a, f(a))$ and $(b, f(b))$. The Mean Value Theorem guarantees the existence of a tangent parallel to the chord joining $(a, f(a))$ and $(b, f(b))$ without the assumption $f(a) = f(b)$. See Fig. 2.

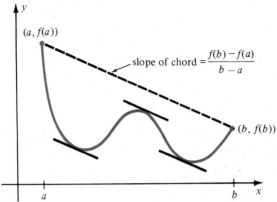

Fig. 2 Meaning of the Mean Value Theorem. there exist tangent(s) parallel to the chord.

Mean Value Theorem Let $f(x)$ be differentiable on the closed interval $a \leq x \leq b$. Then there exists a point c such that $a < c < b$ and

$$f'(c) = \frac{f(b) - f(a)}{b - a}.$$

The proof is not long or hard, but it involves an ingenious trick, subtracting a carefully chosen linear function from $f(x)$ and applying Rolle's Theorem to the difference.

Set $m = [f(b) - f(a)]/(b - a)$, the slope of the chord through $(a, f(a))$ and $(b, f(b))$. By the point–slope formula, the equation of the line containing this chord is

$$y = g(x) = f(a) + m(x - a).$$

Now set $h(x) = f(x) - g(x)$. Then $h(x)$ is a differentiable function for $a \leq x \leq b$ since it is the difference of two differentiable functions. What is more, $h(a) = 0$ and $h(b) = 0$, so $h(a) = h(b)$. Therefore $h(x)$ satisfies the hypotheses of Rolle's Theorem. Hence there exists a point c such that $a < c < b$ and $h'(c) = 0$. But

$$h'(x) = [f(x) - g(x)]' = f'(x) - g'(x) = f'(x) - m,$$

so $h'(c) = 0$ implies $f'(c) = m = [f(b) - f(a)]/(b - a)$. This completes the proof.

Remark Both Rolle's Theorem and the Mean Value Theorem are valid even if $f(x)$ is differentiable only for $a < x < b$. But in that case you must also *assume* that $f(x)$ is continuous at $x = a$ and at $x = b$. Then $f(x)$ is continuous on $a \leq x \leq b$ and the existence of extrema follows, etc. For instance, the Mean Value Theorem applies to $f(x) = \sqrt{x} + \sqrt{1 - x}$ on $0 \leq x \leq 1$. The function is differentiable on $0 < x < 1$ and continuous on $0 \leq x \leq 1$. It is *not* differentiable at $x = 0$ or at $x = 1$.

Constant Functions   Our first application of the MVT (Mean Value Theorem) will be to prove that a function with zero derivative is constant. This is unfinished business from p. 85.

> **Theorem**   Let $f(x)$ be a differentiable function on the closed interval $a \leq x \leq b$ and suppose $f'(x) = 0$ for all x on the interval. Then $f(x) = c$.

Proof   Suppose $a < x \leq b$. By the MVT, there exists x_1 such that $a < x_1 < x$ and

$$f(x) - f(a) = (x - a)f'(x_1).$$

But $f'(x_1) = 0$, hence $f(x) = f(a)$. Therefore $f(x) = f(a)$ for each x on the interval, so $f(x)$ is a constant function if you ever saw one.

Increasing Functions   Our next application of the MVT will be to prove that a function with everywhere positive derivative increases. Let us state this property precisely.

> **Theorem**   Let $f(x)$ be a differentiable function on the closed interval $a \leq x \leq b$.
>
> Suppose $f'(x) \geq 0$ for all x in the interval. Then $f(x)$ is an increasing function, that is,
>
> $$f(x_0) \leq f(x_1) \qquad \text{whenever} \quad a \leq x_0 < x_1 \leq b.$$
>
> Suppose $f'(x) > 0$ for all x inside the interval. Then $f(x)$ is a strictly increasing function, that is,
>
> $$f(x_0) < f(x_1) \qquad \text{whenever} \quad a \leq x_0 < x_1 \leq b.$$

Proof   Let $a \leq x_0 < x_1 \leq b$. Apply the MVT to the interval $x_0 \leq x \leq x_1$: There exists a point c such that $x_0 < c < x_1$ and

$$f'(c) = \frac{f(x_1) - f(x_0)}{x_1 - x_0}, \qquad \text{that is,} \quad f(x_1) - f(x_0) = f'(c)(x_1 - x_0).$$

But $x_1 - x_0 > 0$, so if $f'(c) \geq 0$, then $f(x_1) - f(x_0) \geq 0$, that is, $f(x_1) \geq f(x_0)$. And if $f'(c) > 0$, then $f(x_1) - f(x_0) > 0$, that is, $f(x_1) > f(x_0)$. This completes the proof.

Second Derivative Test

> **Theorem**   Let $f(x)$ be twice differentiable on $a \leq x \leq b$, and let $a < c < b$. If $f'(c) = 0$, and $f''(c) > 0$, then $f(c)$ is a local min of $f(x)$.

Proof   Since $f''(c) > 0$, it follows from the theorem on p. 147, applied to $f'(x)$, that there exists $\delta > 0$ such that

$$\begin{cases} f'(x) > f'(c) = 0 & \text{for} \qquad c < x < c + \delta \\ f'(x) < f'(c) = 0 & \text{for} \quad c - \delta < x < c. \end{cases}$$

Therefore $f(x)$ increases just to the right of $x = c$ and decreases just to the left:

$$\begin{vmatrix} f(x) > f(c) & \text{for} & c < x < c + \delta \\ f(x) > f(c) & \text{for} & c - \delta < x < c. \end{vmatrix}$$

It follows that $f(c)$ is a local min of $f(x)$.

Remark The corresponding statement for a local max follows easily.

Convex Functions   We begin with a precise definition of a convex function and a strictly convex function.

Definition   Let $f(x)$ be a twice differentiable function on an interval. Then $f(x)$ is **convex** if $f''(x) \geq 0$ and $f(x)$ is **strictly convex** if $f''(x) > 0$.

We shall state two basic properties of (strictly) convex functions. The graph of a convex function lies (1) above all of its tangent lines; (2) below all of its chords. These properties are what the word "convex" usually brings to mind, for instance if one thinks of a convex lens.

Tangent Theorem   Let $f(x)$ be a strictly convex function on the closed interval $a \leq x \leq b$. Let $a \leq c \leq b$. Then the graph of $y = f(x)$ for $x \neq c$ lies above its tangent line at $x = c$. See Fig. 3a.

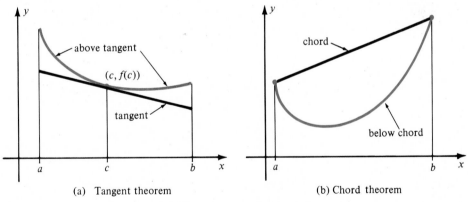

(a) Tangent theorem   (b) Chord theorem

Fig. 3   Basic properties of strictly convex functions $[f''(x) > 0]$

Chord Theorem   Let $f(x)$ be a strictly convex function on the closed interval $a \leq x \leq b$. Then the graph of $y = f(x)$ for $a < x < b$ lies below the chord joining $(a, f(a))$ to $(b, f(b))$. See Fig. 3b.

Since these results, although important, are off our main line of thought, we shall leave their proofs as exercises.

Remark 1   The Tangent and Chord Theorems have obvious modifications for convex functions rather than strictly convex functions.

Remark 2 They also have obvious counterparts for strictly concave ($f'' < 0$) and concave ($f'' \leq 0$) functions. The observation that $-f$ is (strictly) convex if and only if f is (strictly) concave makes everything routine.

EXERCISES

1. Use the first boxed theorem on p. 146 to prove that $f(x) = x^4 + 4x^3 - x - 1$ has a minimum for $x \geq 0$, and that this minimum is negative.

2. Prove that $y = x(1 - x)(x^{10} + x^8 + 1)$ has a positive maximum for $0 \leq x \leq 1$.

3. Give an example of a function that is bounded on a closed interval $a \leq x \leq b$, but does not have a maximum on this interval.

4. Suppose $f(x)$ is continuous and $f(x) > 0$ at each point of the closed interval $a \leq x \leq b$. Prove that $1/f(x)$ is bounded on this interval.

5. Show that the result of Ex. 4 is false on the domains $0 < x < 1$ and $-\infty < x < \infty$.

6. Is the function $f(x) = x^3/(1 + x^4)$ bounded for all x? If so, find a constant C such that $|f(x)| \leq C$ for all x.

7. Is $f(x) = (1/x) - (1/x^2)$ bounded for $x \geq 0$?

8. Is the function in Ex. 7 bounded for $x < -1$?

9. Suppose $f(0) = 0$ and $f'(0) = 1$. Prove that there exists $\delta > 0$ such that $f(x) < 2x$ for $0 < x < \delta$.

10. (cont.) Prove that $f(x) > 1.01x$ for $-\delta_1 < x < 0$, where δ_1 is some positive number.

11. Suppose $f'(x) = f(x)$ for all x and $f(0) = 1$. Prove that $f(x) > 1$ for $0 < x < \delta$ for some $\delta > 0$.

12. (cont.) Prove that $f(x)$ is an increasing function for $x \geq 0$. [*Hint* Differentiate $[f(x)]^2$. Conclude that $[f(x)]^2$ is increasing, so $f(x) \geq 1$, etc.]

13. Prove that the sum of two convex functions is convex.

14. If $f(x) \geq 0$ and $f(x)$ is convex, prove that $[f(x)]^2$ is convex.

15. (cont.) Is the converse true?

16. Let $f(x)$ and $g(x)$ be positive increasing convex functions. Prove that $f(x)g(x)$ is convex.

17. Let $f(x)$ and $g(x)$ be convex functions, $f(x)$ increasing. Prove that the composite function $f[g(x)]$ is convex.

18. Let $f(x)$ be convex. Prove the inequality

$$f(\tfrac{1}{2}x + \tfrac{1}{2}y) \leq \tfrac{1}{2}f(x) + \tfrac{1}{2}f(y).$$

[*Hint* Use the Chord Theorem.]

19. Let $f(x)$ be strictly convex and $f'(c) = 0$. Prove that $f(x) > f(c)$ for all $x \neq c$. [*Hint* Prove that $f' < 0$ for $x < c$ and $f' > 0$ for $x > c$, etc.]

20. (cont.) Now prove the Tangent theorem. [*Hint* Subtract a suitable linear function from $f(x)$.]

21. Suppose $f(x)$ is strictly convex and $f(a) = f(b) = 0$. Prove that $f(x) < 0$ for $a < x < b$. [*Hint* Locate max f and use the Second Derivative Test.]

22. (cont.) Now prove the Chord Theorem. [*Hint* Read the hint in Ex. 20.]

23. Let $f(x)$ be strictly convex, $x < y$, and $0 < t < 1$. Prove that

$$f[tx + (1 - t)y] < tf(x) + (1 - t)f(y).$$

24. (cont.) Let $x < z < y$. Prove that

$$(y - x)f(z) < (y - z)f(x) + (z - x)f(y).$$

25. Let $r > 1$ be rational. Apply Ex. 18 to prove that

$$(x + y)^r \leq 2^{r-1}(x^r + y^r)$$

for $x > 0$ and $y > 0$.

26. Prove that the reciprocal of the average of two positive numbers does not exceed the average of their reciprocals. [*Hint* Use Ex. 18.]

9. MISCELLANEOUS EXERCISES

1   Find a function $f(x)$ for which $f'(x) = x - 1/x^2$ and $f(1) = 3$.

2   A body moves along a horizontal line according to the law $s = 12t - t^3$, where $t \geq 0$. (a) When is s increasing and when decreasing? (b) What is the maximum velocity? (c) What is the acceleration at $t = 3$? The units are feet and seconds.

3   A train moving at 90 ft/sec slows up with a constant negative acceleration of 6 ft/sec^2. How long is it until the train stops? How far does it go?

4   If x is a number between 0 and 1, then $x > x^3$. Find the positive number that exceeds its cube by the greatest possible amount.

5   Find the largest value of $x^2 y$ if x and y are positive numbers whose sum is 15.

6   Graph $y = 2x^3 - 9x^2 + 12x - 5$. Plot all points where the tangent is flat, and plot all points of inflection.

7   A rectangular box with square bottom is to have volume 648 cubic inches. If the material on the top and bottom costs three times as much as that on the sides, find the most economical dimensions.

8   Compute the largest area of all rectangles inscribed in the ellipse $x^2/9 + y^2/4 = 1$ with sides parallel to the axes.

9   Compute the largest volume of all cones that can be generated by rotating a right triangle with hypotenuse c about one of its legs. [*Hint* $V = \frac{1}{3}\pi r^2 h$.]

10   A man M in a rowboat one km off shore wants to reach a point Q four km up the coast (Fig. 1). He rows to a point P, then walks the rest of the way. If he can row 4 km/hr and walk 6 km/hr, how should he choose P to minimize the total time required?

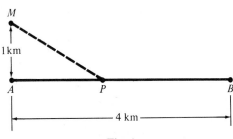

Fig. 1

11   Graph $y = \dfrac{2x + 1}{x^2 + 4}$.

12   Where is $y = (1 + x^3)^{-1}$ convex, concave, for $x > 0$?

13   A railroad will run a special train if at least 200 people subscribe. The fare will be $8 per person if 200 people go, but will decrease 1¢ for each additional person who goes. (For example, if 250 people go, the fare will be $7.50.) What number of passengers will bring the railroad maximum revenue?

14   Of all lines of negative slope through the point (a, b) in the first quadrant, find the one that cuts from the first quadrant a triangle of least area.

15   A wire 30 in. long is cut into two parts, one of which is bent into a circle, and the other into a square. How should the wire be cut so that the sum of the areas of the circle and the square is a minimum?

16   What is the maximum volume of the cylinder generated by rotating a rectangle of perimeter 48 cm about one of its sides?

17   Find the line tangent to the curve $y = 4 - x^2$ at a point of the first quadrant that cuts from the first quadrant a triangle of minimum area.

18   One corner of a long rectangular strip of width a is folded over and just reaches the opposite edge, as shown in Fig. 2. Find the largest possible area A.

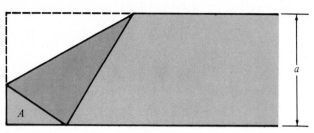

Fig. 2

19   A swampy region shares a long straight border with a region of farm land. A telephone cable is to be constructed connecting two locations, one in each region. Its cost is d_1 dollars per mile in the swampy region and d_2 dollars per mile in the farm land. What path should the cable take for its cost to be least?

20   The surface magnetic field strength B of a pulsar is related to its period P by $B^2 = kP\dot{P}$. Express $\dot{B}$ in terms of P and its time derivatives.

21*   Prove $\left(\dfrac{x^4 + 1}{2}\right)^{1/4} \geq \left(\dfrac{x^3 + 1}{2}\right)^{1/3}$ for all $x \geq 0$.

22*   An isosceles triangle is circumscribed about a circle of radius r. Find its altitude if its perimeter is to be as small as possible.

23   The drag on an airplane at (subsonic) speed v is

$$D = av^2 + \frac{b}{v^2}, \quad \text{where} \quad a > 0 \quad \text{and} \quad b > 0.$$

Find the speed at which D is least. (At this speed the range is greatest for a given fuel supply.)

24   (cont.) At a steady speed v, the thrust of the engine just balances the drag, and the power developed by the engine is

$$P = (\text{thrust})(\text{speed}) = Dv = av^3 + \frac{b}{v}.$$

Find the speed at which P is least. (Since P is proportional to the rate of consumption of fuel, the time airborne will be greatest at this speed.)

25   The distances x and y from a thin convex lens to an object on its axis and to its image, respectively, are related by

$$\frac{1}{x} + \frac{1}{y} = \frac{1}{f},$$

where f is the (constant) focal length. Minimize $x + y$.

26   The magnetic force at a point on the axis of a conducting loop at distance x from its center is $F = kx/(x^2 + a^2)^{5/2}$, where k and a are positive constants. Find the maximum of F and the value of x where F has an inflection.

Exponential and Trigonometric Functions   4

1. THE EXPONENTIAL FUNCTION

In the next four sections we study a function that is its own derivative. That such a function exists at all is not obvious. (We don't count the trivial function $y(x) = 0$.) To show that it does, we shall give a computational argument in this section and a geometric one in the next.

The desired function is going to be an exponential function like $y = 2^x$, $y = 3^x$, or more generally $y = a^x$, where $a > 1$. Let us first recall a few properties of the functions $y = a^x$, where $a > 1$. In Fig. 1 we show some examples. The function $y = a^x$ increases rapidly as x increases and decreases rapidly to 0 as x decreases. We can say

$$a^x \longrightarrow \infty \quad \text{as} \quad x \longrightarrow \infty;$$

$$a^x \longrightarrow 0 \quad \text{as} \quad x \longrightarrow -\infty.$$

In addition, $y = a^x$ satisfies the rules of exponents:

$$a^{u+v} = a^u a^v, \qquad a^0 = 1, \qquad a^{-x} = \frac{1}{a^x}.$$

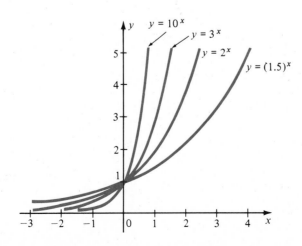

Fig. 1   Examples of $y = a^x$ for $a > 1$

155

We claim that one of the functions $y = a^x$ is its own derivative. To see why, differentiate $y = a^x$:

$$\frac{dy}{dx} = \lim_{h \to 0} \frac{a^{x+h} - a^x}{h} = \lim_{h \to 0} \frac{a^x(a^h - 1)}{h} = a^x \lim_{h \to 0} \frac{a^h - 1}{h}.$$

Suppose the limit on the right exists; call it k. Then

$$\frac{dy}{dx} = ka^x = ky, \qquad \text{where} \quad k = \lim_{h \to 0} \frac{a^h - 1}{h}.$$

The number k depends on a. If there is a choice of a for which k is 1, then $y = a^x$ will satisfy $dy/dx = y$, that is, $y = a^x$ will equal its own derivative.

Let us try to estimate k for a few values of a. (We used a pocket calculator in the computations that follow; tables or a slide rule would have served as well.)

$\underline{a = 2}$
$$k = \lim_{h \to 0} \frac{2^h - 1}{h}.$$

We tabulate $(2^h - 1)/h$ for several small h:

h	0.01	0.001	0.0001	-0.0001
$(2^h - 1)/h$	0.6956	0.6934	0.6932	0.6931

The numerical evidence suggests $k \approx 0.693$.

$\underline{a = 3}$
$$k = \lim_{h \to 0} \frac{3^h - 1}{h}.$$

h	0.01	0.001	0.0001	-0.0001
$(3^h - 1)/h$	1.1047	1.0992	1.0987	1.0986

Conclusion: $k \approx 1.099$.

Assuming always that the limit k exists, we have found

$$\frac{d}{dx} 2^x = k2^x, \quad k \approx 0.693; \qquad \frac{d}{dx} 3^x = k3^x, \quad k \approx 1.099.$$

This suggests that there is a number a in the interval $2 < a < 3$ for which $k = 1$. To get closer to a, we use linear interpolation (replacing 0.693 and 1.099 by 0.7 and 1.1 since the estimate is rough anyway). We find

$$a \approx 2 + \frac{1.0 - 0.7}{1.1 - 0.7} = 2.75,$$

so we suspect that $2.7 < a < 2.8$. It is; another calculation shows

$$\frac{d}{dx} (2.7)^x = k(2.7)^x, \quad k \approx 0.993; \qquad \frac{d}{dx} (2.8)^x = k(2.8)^x, \quad k \approx 1.030.$$

We narrow the gap further:

$$\frac{d}{dx}(2.71)^x = k(2.71)^x, \quad k \approx 0.9969; \qquad \frac{d}{dx}(2.72)^x = k(2.72)^x, \quad k \approx 1.0006.$$

Now interpolation suggests

$$a \approx 2.71 + \left(\frac{1.0000 - 0.9969}{1.0006 - 0.9969}\right)(0.01) \approx 2.7184.$$

(Actually 2.7183 is closer because of some round-off error.) The numerical evidence is pretty convincing that there is a number $a \approx 2.7183$ that does the trick. This number is always denoted by e:

The Exponential Function   There is a number $e \approx 2.7183$ such that

$$\frac{d}{dx}e^x = e^x.$$

Thus the function $y = e^x$ equals its own derivative (Fig. 2).

An alternative statement is that $y = e^x$ satisfies the equation

$$\frac{dy}{dx} = y.$$

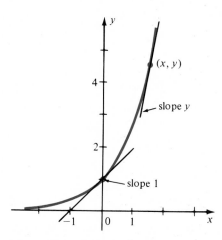

Fig. 2   Graph of $y = e^x$; the slope at (x, y) is y.

Every function $y = ce^x$ also satisfies the equation $dy/dx = y$ because

$$\frac{dy}{dx} = \frac{d}{dx}(ce^x) = c\frac{d}{dx}e^x = ce^x = y.$$

We assert that these are the *only* functions that equal their own derivatives. To prove this, let $y = f(x)$ be any function such that $dy/dx = y$. Then

$$\frac{d}{dx}\left[\frac{f(x)}{e^x}\right] = \frac{e^x f'(x) - f(x)(e^x)'}{(e^x)^2} = \frac{e^x f(x) - f(x)e^x}{(e^x)^2} = 0.$$

Therefore $f(x)/e^x = c$, a constant; that is, $f(x) = ce^x$.

> The solutions $y = f(x)$ of the equation
>
> $$\frac{dy}{dx} = y$$
>
> are precisely the functions $f(x) = ce^x$.

If $y = ce^x$, then $y(0) = ce^0 = c$, so the constant c is the value $y(0)$. Thus the value of y at $x = 0$ uniquely determines a solution of the equation $dy/dx = y$.

> There is a unique function $y = y(x)$ satisfying
>
> $$\frac{dy}{dx} = y \qquad \text{and} \qquad y(0) = c.$$
>
> It is $y = ce^x$.

Remark   The equation $dy/dx = y$ is an example of a **differential equation**. Other examples are

$$\frac{dy}{dx} = y^2, \qquad \frac{dy}{dx} = 2x - y, \qquad \frac{d^2y}{dx^2} - 3x\frac{dy}{dx} = xy^4 + 1.$$

Differentiation   We shall discuss another approach to the exponential function in Section 2 and another approach still in the last section of Chapter 7. Meanwhile let us take it for granted that this function exists. By the Chain Rule, if $u = u(x)$, then

$$\frac{d}{dx}e^u = e^u\frac{du}{dx}.$$

Examples $$\frac{d}{dx}e^{3x} = 3e^{3x}, \qquad \frac{d}{dx}e^{x^2} = 2xe^{x^2}, \qquad \frac{d}{dx}e^{-\sqrt{x}} = \frac{-e^{-\sqrt{x}}}{2\sqrt{x}}.$$

■ **EXAMPLE 1**   Find all values of k for which $y = e^{kx}$ satisfies the differential equation

$$y'' - 4y' + 3y = 0.$$

Solution $y' = ke^{kx}$ and $y'' = k^2e^{kx}$. Hence

$$y'' - 4y' + 3y = k^2e^{kx} - 4ke^{kx} + 3e^{kx} = (k^2 - 4k + 3)e^{kx}.$$

Consequently the differential equation is satisfied if and only if

$$(k^2 - 4k + 3)e^{kx} = 0.$$

But $e^{kx} \neq 0$, hence the condition is $k^2 - 4k + 3 = 0$. This quadratic has the solutions $k = 1$ and $k = 3$; they are the required values of k.   ■

EXERCISES

1   Use a calculator or tables to show

$$\frac{\log_{10} 2}{\log_{10} e} \approx 0.6931, \qquad \frac{\log_{10} 3}{\log_{10} e} \approx 1.0986, \qquad \frac{\log_{10} 10}{\log_{10} e} \approx 2.3026.$$

2   Compute $\dfrac{10^h - 1}{h}$ for $h = 10^{-3}, \pm 10^{-4}, \pm 10^{-5}$. Conclude that

$$\frac{d}{dx} 10^x = k \cdot 10^x, \qquad k \approx 2.3026.$$

3   On the basis of the numerical evidence in Exs. 1 and 2, guess a formula for k in terms of logs, where

$$\frac{d}{dx} a^x = k \cdot a^x.$$

4   (cont.) Test the formula numerically for $a = 5$.
5   Why is it reasonable to say $2^h \approx 1 + (0.6931)h$ when h is small?
6   (cont.) Estimate $\sqrt[16]{2}$. Check your estimate using tables or a calculator.
7   (cont.) Do the same for $1/\sqrt[10]{3}$.
8   Justify the formula $e^h \approx 1 + h$ for h small. Show that this approximation is consistent with $e^{0.01} e^{-0.01} = 1$.

Differentiate with respect to x

9 e^{3x^2} **10** $e^{1/x}$ **11** e^{x^4}
12 $e^x/(1 + e^{-x})$ **13** e^x/x **14** $x^2 e^{2x}$
15 $\dfrac{1}{1 + e^{-x}}$ **16** $\sqrt{x}\, e^{-\sqrt{x}}$ **17** $\left(\dfrac{x}{e^{2x} + 1}\right)^3$
18 $\dfrac{xe^x}{x^2 + 1}$ **19** $\dfrac{e^x + e^{-x}}{e^x - e^{-x}}$ **20** $e^{5x}(x^2 - 3x + 6).$

Find all k such that $y = e^{kx}$ satisfies

21 $y'' - 5y' + 4y = 0$ **22** $y'' - 4y' + 4y = 0$
23 $y'' - y' - 6y = 0$ **24** $y'' + 3y' + 2y = 0.$

25   Show that $y = e^x$ and $y = xe^x$ are solutions of $y'' - 2y' + y = 0$.
26   Given k, find A so $y = Ae^{kx}$ is a solution of $y'' - 4y' + 3y = e^{kx}$. For which values of k does this fail?

2. PROPERTIES OF EXPONENTIAL FUNCTIONS

Direction Fields   Let us examine the differential equation $y' = y$ geometrically. If a function $y(x)$ satisfies the equation, how must its graph behave? Well, the very meaning of $y' = y$ is that at each point (x, y) of its graph, the slope is y. For instance at any point where the graph touches the line $y = 1$, its slope is 1; at any point where the graph touches the line $y = 2$, its slope is 2. We are lead naturally to Fig. 1a. At each point of the line $y = 1$, we imagine a short segment of slope 1; at each point of $y = 2$, we imagine a short segment of slope 2, and in general, at each point of $y = k$, a short segment of slope k. In this way we assign a direction to each point of the plane. This configuration is called the **direction field** of the differential equation $y' = y$. A **solution** of the differential equation is a curve whose tangent at each point agrees with the direction field (Fig. 1b).

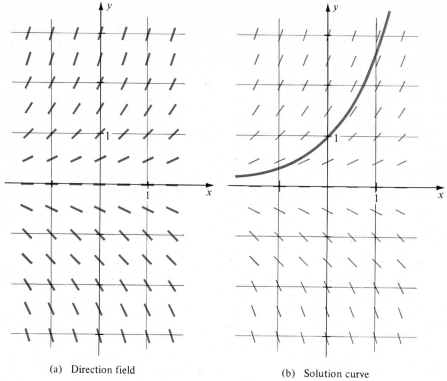

(a) Direction field   (b) Solution curve

Fig. 1   Geometric study of the differential equation $y' = y$

Our graph suggests certain properties of the solution curves:

(1) Each solution curve is the graph of a *function* $y = f(x)$, because it crosses every vertical line exactly once.

(2) Through each point of the plane passes a unique solution curve.

(3) Each solution curve is either completely above the x-axis, or completely below the x-axis, or is the x-axis itself.

Now we *define* a new function $y = E(x)$ to be the unique solution whose graph passes through $(0, 1)$. Thus $y = E(x)$ satisfies

$$\frac{dy}{dx} = y, \qquad y(0) = 1.$$

Each function $y = cE(x)$, where c is constant, also satisfies $y' = y$, so its graph is also a solution curve. Let us show that every solution curve is one of these.

Given any solution curve, it is the graph of some function $y = y(x)$. Set $c = y(0)$ and consider the function $z = z(x) = y(x) - cE(x)$. Then

$$z' = y' - cE' = y - cE = z$$

and

$$z(0) = y(0) - cE(0) = c - c \cdot 1 = 0.$$

By property (3), it follows that $z(x) = 0$; that is, $y(x) = cE(x)$, where $c = y(0)$.

> Each solution $y(x)$ of the differential equation $y' = y$ is given by
> $$y(x) = cE(x), \qquad \text{where} \quad c = y(0).$$

Remark   The notation $E(x)$ for e^x is makeshift for our purpose and is not usual. What is very common is the notation $e^x = \exp x$. This is particularly useful when the exponent is complicated, for example, $\exp x^3$ instead of e^{x^3}.

Laws of Exponents   In the last section we found that the exponential function $y = a^x$ satisfies a relation

$$\frac{dy}{dx} = ka^x, \qquad \text{where} \quad k = \left.\frac{dy}{dx}\right|_0 .$$

We defined e to be that number for which

$$\left.\frac{d}{dx} e^x\right|_0 = 1,$$

so that $de^x/dx = e^x$. Therefore this new function $E(x)$ that we have introduced from geometric considerations is nothing but the exponential function, that is,

$$E(x) = e^x.$$

Let us pretend, however, that we don't know the functions a^x at all, but we do know the function $E(x)$, defined by the two conditions

$$\frac{dE}{dx} = E, \qquad E(0) = 1.$$

We shall now *prove* that $E(x)$ satisfies the law of exponents

$$E(a + b) = E(a)E(b).$$

For consider $y(x) = E(a + x)$. By the Chain Rule,

$$\frac{dy}{dx} = \frac{d}{dx} E(a + x) = E(a + x) \cdot 1 = y.$$

Therefore y satisfies $y' = y$. Hence $y(x) = cE(x)$, where $c = y(0) = E(a + 0) = E(a)$. But $y = E(a + x)$, hence

$$E(a + x) = E(a)E(x).$$

Next, we shall prove that $E(x)$ behaves like powers of a fixed number. Set $e = E(1)$. Then

$$E(2) = E(1 + 1) = [E(1)]^2 = e^2,$$
$$E(3) = E(1 + 2) = E(1)E(2) = e \cdot e^2 = e^3.$$

Similarly, $E(n) = e^n$ for every positive integer n. Furthermore,

$$E(-n)E(n) = E(-n + n) = E(0) = 1,$$

hence

$$E(-n) = \frac{1}{E(n)} = \frac{1}{e^n} = e^{-n}.$$

Therefore $E(x) = e^x$ for all integers x (even 0).

Let us determine $E(x)$ for rational x. From

$$[E(\tfrac{1}{2})]^2 = E(\tfrac{1}{2})E(\tfrac{1}{2}) = E(\tfrac{1}{2} + \tfrac{1}{2}) = E(1) = e \qquad \text{follows} \quad E(\tfrac{1}{2}) = e^{1/2}.$$

Similarly, from

$$\left[E\left(\frac{1}{n}\right)\right]^n = E(1) = e \qquad \text{follows} \quad E\left(\frac{1}{n}\right) = e^{1/n}.$$

Now use m summands, each $1/n$:

$$E\left(\frac{m}{n}\right) = E\left(\frac{1}{n} + \cdots + \frac{1}{n}\right) = \left[E\left(\frac{1}{n}\right)\right]^m = (e^{1/n})^m = e^{m/n}.$$

Finally, $E(-m/n)E(m/n) = E(0) = 1$, so

$$E\left(-\frac{m}{n}\right) = \frac{1}{E(m/n)} = \frac{1}{e^{m/n}} = e^{-m/n}.$$

We have shown that $E(x) = e^x$ for each rational number x. The evidence is convincing: $E(x)$ really is an exponential function.

The Functions e^{kx} and Their Graphs   The functions $y = e^{kx}$ are called **exponential functions**. Let us start with $k > 0$. The graph of $y = e^{kx}$ is obtained from the graph of $y = e^x$ by a change of scale on the x-axis. For example $y = e^{2x}$ runs twice as fast over the same values as $y = e^x$. See Fig. 2a for several values of k.

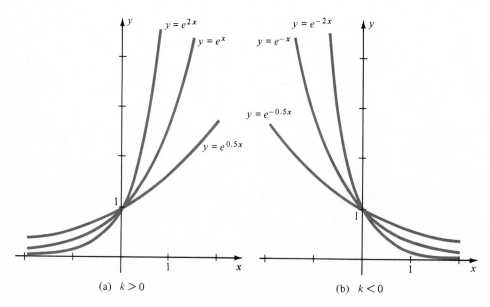

(a) $k > 0$ (b) $k < 0$

Fig. 2 Graphs of $y = e^{kx}$

The functions $y = e^{kx}$ share certain growth properties, suggested by Fig. 2.

Growth Properties   Let $k > 0$. Then

(1) $e^{kx} > 0$ for all x.

(2) e^{kx} is a strictly increasing function.

(3) $e^{kx} \longrightarrow \infty$ as $x \longrightarrow \infty$.

(4) $e^{kx} \longrightarrow 0$ as $x \longrightarrow -\infty$.

Property (1) holds because $e^u > 0$ for any u, in particular, for $u = kx$. Property (2) follows from $(e^{kx})' = ke^{kx} > 0$, which implies that $y = e^{kx}$ always has positive slope, hence is strictly increasing. Not only is $y' > 0$, but also

$$y'' = (ke^{kx})' = k^2 e^{kx} > 0;$$

hence $y = e^{kx}$ is strictly convex in addition to strictly increasing. Therefore (3) follows from the Tangent Theorem (p. 151). Finally, (4) follows from

$$e^{k(-x)} = e^{-kx} = \frac{1}{e^{kx}} \longrightarrow 0 \quad \text{as} \quad x \longrightarrow \infty.$$

Now we consider $y = e^{-kx}$, with $k > 0$. Since $e^{-kx} = e^{k(-x)}$, the graphs of $y = e^{-kx}$ and $y = e^{kx}$ are reflections of each other in the y-axis. Hence we obtain Fig. 2b from Fig. 2a by reflection. We also deduce the following properties of $y = e^{-kx}$ from the relation $e^{-kx} = 1/e^{kx}$:

Let $k > 0$. Then

(1) $e^{-kx} > 0$ for all x.

(2) e^{-kx} is a strictly decreasing function.

(3) $e^{-kx} \longrightarrow 0$ as $x \longrightarrow \infty$.

(4) $e^{-kx} \longrightarrow \infty$ as $x \longrightarrow -\infty$.

Differential Equation of e^{kx}   If $y = e^{kx}$, then $y' = ke^{kx} = ky$. Hence y is a solution of the differential equation $y' = ky$. So is each function $y = ce^{kx}$, where c is constant. Conversely, if y is any solution of $y' = ky$, then

$$(e^{-kx}y)' = (-ke^{-kx})y + e^{-kx}(ky) = 0;$$

hence $e^{-kx}y = c$, so $y = ce^{kx}$. Evidently $y(0) = c$.

Each solution $y(x)$ of the differential equation $\dfrac{dy}{dx} = ky$ is given by

$$y(x) = ce^{kx}, \qquad \text{where} \ \ c = y(0).$$

The Functions a^x   For each $a > 0$ there is an exponential function $y = a^x$. We choose k so that $a = e^k$, which we can do because e^x takes on all positive values. Then

$$a^x = (e^k)^x = e^{kx},$$

by the laws of exponents. Now we can differentiate:

$$\frac{d}{dx} a^x = \frac{d}{dx} e^{kx} = k e^{kx} = k a^x.$$

$$\boxed{\text{If } a > 0, \text{ then } \quad \frac{d}{dx} a^x = k a^x, \quad \text{where} \quad a = e^k.}$$

Example $\left. \dfrac{d}{dx} 2^x \right|_3 = k \cdot 2^x \Big|_3 = k \cdot 2^3 = 8k.$

To estimate k, we write $2 = e^k$ and solve for k by logarithms:

$$\log 2 = \log e^k = k \log e, \qquad k = \frac{\log 2}{\log e} \approx 0.69315.$$

Hence

$$\left. \frac{d}{dx} 2^x \right|_3 \approx 8(0.69315) = 5.5452.$$

EXERCISES

1   Plot the direction field of $dy/dx = \frac{1}{2}y$ for $-2 \le x \le 2$ and $-2 \le y \le 2$. Use it to sketch the graph of $y = e^{x/2}$.
2   Plot the direction field of $dy/dx = -y$ for $-1 \le x \le 1$ and $-1.5 \le y \le 2.5$. Use it to sketch the graph of $y = e^{-x}$.
3   Verify that $y = \frac{1}{2}(e^x + e^{-x})$ satisfies $y' + y = e^x$.
4   (cont.) Find the 78-th derivative of $y = \frac{1}{2}(e^x + e^{-x})$.

Sketch

5   $y = e^{x-1}$ 6 $y = e^{1-x}$
7   $y = \frac{1}{2}(e^x + e^{-x})$ 8 $y = \frac{1}{2}(e^x - e^{-x})$.

9   Show that $y = a e^{kx} + b e^{-kx}$ satisfies the differential equation $y'' = k^2 y$. Here a, b, and k are constants.
10 Show that $y = a e^{2x} + b e^{3x}$ satisfies the differential equation $y'' - 5y' + 6y = 0$.
11 Plot the direction field of $dy/dx = 2xy$ for $-1 \le x \le 1$ and $0 \le y \le 3$. Show that $y = cE(x^2)$ is a solution.
12 Plot the direction field of $dy/dx = x + y$ for $-1 \le x \le 1$ and $-2 \le y \le 2$. Show that $y = cE(x) - 1 - x$ is a solution.
13* Prove that $[E(x)]^y = E(xy)$ when y is an integer.
14* (cont.) Prove the same formula when y is rational.

Differentiate with respect to x

15 10^{3x} 16 10^{-4x} 17 5^{x-1}
18 5^{2x-1} 19 10^{4x-1} 20 10^{2-3x}
21 $(10^x + 10^{-x})^2$ 22 a^u where $u = b^x$.

Prove (using $a^x = e^{kx}$ where $a = e^k$)

23 $a^x b^x = (ab)^x$ **24** $(e^x)^y = e^{xy}$

25 $a^x a^y = a^{x+y}$ **26** $(a^x)^y = a^{xy}$.

Find an antiderivative

27 e^{2x} **28** $e^{-2x} + 1$ **29** 10^x

30 xe^{x^2} **31** xe^x **32** xe^{-x}.

Solve

33 $y' = e^x + e^{-x}$, $y(0) = 1$ **34** $y' = e^{-3x} + 1$, $y(1) = 0$.

3. APPROXIMATION AND GROWTH RATES

Approximate Value of *e* In our geometric approach to the exponential function, e is defined to be $E(1)$, where $y = E(x)$ is the unique function satisfying $y' = y$ and $y(0) = 1$. We can see in Fig. 1b, p. 160, that $2 < e < 3$. To estimate e more closely, we shall approximate the curve $y = E(x)$ by a polygonal line.

The slope of $y = E(x)$ at $(0, 1)$ is 1. Construct a line segment of slope 1, starting at $(0, 1)$. See Fig. 1a. This segment reaches $(\frac{1}{2}, \frac{3}{2})$. At level $\frac{3}{2}$, the slope of the direction field is $\frac{3}{2}$. Now change to a line segment of slope $\frac{3}{2}$. This segment reaches $(1, \frac{9}{4})$, so as a crude estimate, $e \approx \frac{9}{4} = 2.25$.

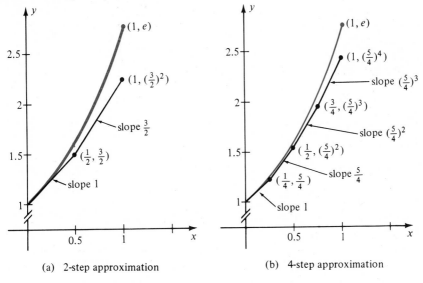

(a) 2-step approximation   (b) 4-step approximation

Fig. 1   Polygonal approximation to $y = e^x$

To refine the method, we approximate by a polygonal graph of n sides. At each corner, we correct the slope to conform with the direction field. For instance, let us take $n = 4$. Starting at $(0, 1)$ we construct the segment of slope 1 for $0 \leq x \leq \frac{1}{4}$. It terminates at $(\frac{1}{4}, \frac{5}{4})$, and from there we construct the segment of slope $\frac{5}{4}$ for $\frac{1}{4} \leq x \leq \frac{1}{2}$. It terminates at $(\frac{1}{2}, (\frac{5}{4})^2)$, because

$$\tfrac{5}{4} + \tfrac{1}{4} \cdot \tfrac{5}{4} = \tfrac{5}{4}(1 + \tfrac{1}{4}) = (\tfrac{5}{4})^2.$$

From there we construct the segment of slope $(\frac{5}{4})^2$ for $\frac{1}{2} \le x \le \frac{3}{4}$. It starts at height $(\frac{5}{4})^2$ and advances forward $\frac{1}{4}$ unit. Its slope is $(\frac{5}{4})^2$, so it reaches height

$$(\tfrac{5}{4})^2 + \tfrac{1}{4}(\tfrac{5}{4})^2 = (1 + \tfrac{1}{4})(\tfrac{5}{4})^2 = (\tfrac{5}{4})^3.$$

Hence the segment terminates at $(\frac{3}{4}, (\frac{5}{4})^3)$, and from there we construct the segment of slope $(\frac{5}{4})^3$ for $\frac{3}{4} \le x \le 1$. It terminates at $(1, (\frac{5}{4})^4)$. The result is the estimate

$$e \approx (\tfrac{5}{4})^4 \approx 2.44.$$

The y-coordinates of the vertices in Fig. 1b are

$$1, \quad \tfrac{5}{4}, \quad (\tfrac{5}{4})^2, \quad (\tfrac{5}{4})^3, \quad (\tfrac{5}{4})^4.$$

More generally, we construct the polygonal graph with vertices

$$(0, y_0) = (0, 1), \quad \left(\frac{1}{n}, y_1\right), \quad \left(\frac{2}{n}, y_2\right), \cdots, (1, y_n).$$

The segment from $(i/n, y_i)$ to $((i + 1)/n, y_{i+1})$ has slope y_i. By the slope formula

$$\frac{y_{i+1} - y_i}{\dfrac{i+1}{n} - \dfrac{i}{n}} = y_i, \qquad y_{i+1} - y_i = \frac{1}{n}\,y_i, \qquad y_{i+1} = \left(1 + \frac{1}{n}\right)y_i.$$

Therefore $\quad y_1 = \left(1 + \dfrac{1}{n}\right), \quad y_2 = \left(1 + \dfrac{1}{n}\right)y_1 = \left(1 + \dfrac{1}{n}\right)^2, \cdots, y_n = \left(1 + \dfrac{1}{n}\right)^n.$

But y_n is the height at $x = 1$ of the polygonal approximation to $y = E(x)$. Hence $y_n \approx e$. It seems believable that as n increases, the polygons approximate the curve more and more closely, so the approximation $y_n \approx e$ gets better and better. This argument is a rough justification for the following basic fact:

$$e = \lim_{n \to \infty} \left(1 + \frac{1}{n}\right)^n.$$

This is a fairly subtle limit. As $n \longrightarrow \infty$, the quantity $(1 + 1/n) \longrightarrow 1$. The large exponent n tries to make $(1 + 1/n)^n$ large, while $1 + 1/n$ itself pulls it down toward 1. The result is a compromise, neither 1 nor ∞. Here are some tabulated values:

n	10	10^2	10^3	10^4	10^5	10^6
$\left(1 + \dfrac{1}{n}\right)^n$	2.59374	2.70481	2.71692	2.71815	2.71827	2.71828

Thus the number e is the limit of the *sequence* $1 + 1$, $(1 + \frac{1}{2})^2$, $(1 + \frac{1}{3})^3, \cdots$. (The subject of limits of sequences is covered in Chapter 11.) To 15 places,

$$e \approx 2.71828 \; 18284 \; 59045.$$

The number e, like the number π, is a fundamental constant of nature, independent of units of measurement. It has been computed to great accuracy. Many remarkable properties of e have been discovered; for example, e is irrational, indeed, e is not a root of any equation $x^n + a_1 x^{n-1} + \cdots + a_n = 0$ with *rational* coefficients.

The polygon method used to approximate $e = E(1)$ works equally well for $e^x = E(x)$. We shall omit the computation and only state the result:

$$e^x = \lim_{x \to \infty} \left(1 + \frac{x}{n}\right)^n \qquad \text{for each} \quad x.$$

This formula has theoretical importance, but little practical value for computation where you want high accuracy for relatively small n. The preceding table shows that for $x = 1$, even $n = 10^5$ delivers only four-place accuracy. If $x > 1$, the situation is worse. Even if x is fairly small, the formula is impractical. For instance, take $x = 0.1$:

$$e^{0.1} \approx 1.10517, \qquad \left(1 + \frac{0.1}{10}\right)^{10} \approx 1.10462, \qquad \left(1 + \frac{0.1}{100}\right)^{100} \approx 1.10512.$$

Thus $n = 100$ gives four-place accuracy, but for hand estimates, 100 is a very large number.

Polynomial Approximation   We can approximate the curve $y = e^x$ by graphs of polynomials as well as by graphs of polygonal functions. Polynomials that do the trick are

$$p_1(x) = 1 + x, \quad p_2(x) = 1 + \frac{x}{1!} + \frac{x^2}{2!}, \cdots, p_n(x) = 1 + \frac{x}{1!} + \frac{x^2}{2!} + \cdots + \frac{x^n}{n!}.$$

To see why, note first that $p_n(0) = 1 = e^0$, and next that

$$\frac{d}{dx} p_n(x) = 1 + \frac{x}{1!} + \cdots + \frac{x^{n-1}}{(n-1)!} = p_n(x) - \frac{x^n}{n!},$$

because $(d/dx)(x^j/j!) = jx^{j-1}/j! = x^{j-1}/(j-1)!.$

Hence $p_n(x)$ is *almost* its own derivative, missing by the amount $x^n/n!$. This quantity is small if n is large relative to the size of x.

As an example take $n = 10$. If $|x| \le 1$, then

$$\left|\frac{x^{10}}{10!}\right| \le \frac{1}{10!} = \frac{1}{3628800} < 3 \times 10^{-7}.$$

Thus the two functions, e^x and $p_{10}(x)$ agree at $x = 0$, and furthermore, e^x equals its derivative, whereas $p_{10}(x)$ equals its derivative up to a small error if $|x| \le 1$. We conclude that $p_{10}(x)$ must be a very good approximation to e^x, at least for $|x| \le 1$.

In general, if x is not large, then $p_n(x)$ is a good approximation to e^x, even for modest values of n. Figure 2 illustrates graphically the approximation for $n = 2$, and the following data illustrate numerically:

x	0.1	0.5	1.0	2.0	−1.0	−2.0
$p_5(x)$	1.10517	1.64870	2.71667	7.26667	0.36667	0.06667
$p_{10}(x)$	1.10517	1.64872	2.71828	7.38899	0.36788	0.13538
e^x	1.10517	1.64872	2.71828	7.38906	0.36788	0.13534

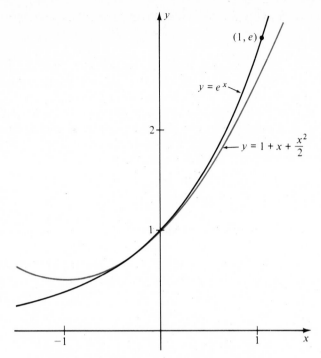

Fig. 2 Polynomial approximation to $y = e^x$

For $x = 1$, the numbers $p_n(1)$ give excellent approximations to e. We can write $e = \lim p_n(1)$, that is,

$$e = \lim_{n \to \infty}\left(1 + \frac{1}{1!} + \frac{1}{2!} + \cdots + \frac{1}{n!}\right).$$

Here is some numerical evidence, to be compared with $e \approx 2.71828\ 18285$:

n	5	6	8	10	15
$p_n(1)$	2.7167	2.71806	2.718279	2.71828 1801	2.71828 18285

With $n = 10$ we already get seven-place accuracy, something that requires $n \approx 2 \times 10^7$ when we use $e \approx (1 + 1/n)^n$.

Rate of Growth of e^x Certainly $e^x \longrightarrow \infty$ as $x \longrightarrow \infty$; but how rapidly? We tabulate some crude values:

x	1	2	3	4	5	6	7	8	9	10
e^x	2.7	7.4	20	55	150	400	1100	3000	8100	22000

x	20	30	40	50	100	1000
e^x	4.9×10^8	1.1×10^{13}	2.4×10^{17}	5.2×10^{21}	2.7×10^{43}	2.0×10^{434}

Evidently e^x increases very rapidly, much faster than x increases. It is a pretty safe bet that

$$\frac{e^x}{x} \longrightarrow \infty \qquad \text{as} \quad x \longrightarrow \infty.$$

In fact much more is true: e^x increases faster than x^2, than x^3, and even faster than any positive power of x. As an example, let us compare e^x with x^{10}, itself a rapidly increasing function:

x	10	20	30	40	50	100	1000
e^x/x^{10}	2.2×10^{-6}	4.7×10^{-5}	1.8×10^{-2}	22	5.3×10^4	2.7×10^{23}	2.0×10^{404}

After a slow start, e^x completely overwhelms x^{10}. The same holds for e^x versus any power x^n.

> **Rapid Growth of e^x**   For any positive n,
> $$\lim_{x \to \infty} \frac{e^x}{x^n} = \infty.$$

It suffices to prove this fact when n is an integer. For instance $x^{9.3} < x^{10}$ for $x > 1$, hence $e^x/x^{9.3} > e^x/x^{10}$. If we prove $e^x/x^{10} \longrightarrow \infty$, then certainly $e^x/x^{9.3} \longrightarrow \infty$ also.

Let $n \geq 0$ be an integer. We shall prove, by induction, that

$$e^x \geq p_n(x) = 1 + \frac{x}{1!} + \cdots + \frac{x^n}{n!} \qquad \text{for all} \quad x \geq 0.$$

The polynomials $p_n(x)$, which we have already met, have a useful property: the derivative of each one is the preceding one, that is,

$$\frac{d}{dx} p_n(x) = 1 + \frac{x}{1!} + \frac{x^2}{2!} + \cdots + \frac{x^{n-1}}{(n-1)!} = p_{n-1}(x).$$

We start the induction at $n = 0$:

$$e^x \geq 1 = p_0(x) \qquad \text{for all} \quad x \geq 0.$$

Now suppose we have proved

$$e^x \geq p_0(x), \quad e^x \geq p_1(x), \cdots, e^x \geq p_k(x)$$

for all $x \geq 0$. Set $y(x) = e^x - p_{k+1}(x)$. Then

$$y'(x) = e^x - p'_{k+1}(x) = e^x - p_k(x) \geq 0.$$

Therefore, $y(x)$ is an increasing function for $x \geq 0$, so

$$y(x) \geq y(0) = e^0 - p_{k+1}(0) = 1 - 1 = 0,$$

that is, $$e^x \geq p_{k+1}(x) \qquad \text{for all} \quad x \geq 0.$$

This completes the induction. We have shown that for each $n \geq 0$,

$$e^x \geq p_n(x) \qquad \text{for all} \quad x \geq 0.$$

Our statement about the growth rate of e^x follows easily. Given n, we have

$$e^x \geq p_{n+1}(x) = 1 + \frac{x}{1!} + \cdots + \frac{x^{n+1}}{(n+1)!} \geq \frac{x^{n+1}}{(n+1)!},$$

hence

$$\frac{e^x}{x^n} \geq \frac{x}{(n+1)!} \qquad \text{for all} \quad x > 0.$$

Clearly,

$$\frac{x}{(n+1)!} \longrightarrow \infty \qquad \text{as} \quad x \longrightarrow \infty,$$

since $1/(n+1)!$ is a constant. Therefore

$$\frac{e^x}{x^n} \longrightarrow \infty \qquad \text{as} \quad x \longrightarrow \infty,$$

because $e^x/x^n \geq x/(n+1)!$ and this latter quantity goes to infinity.

If $k > 0$, then e^{kx} has a similar rapid growth. For

$$\frac{e^{kx}}{x^n} = k^n \frac{e^{kx}}{(kx)^n} = k^n \frac{e^u}{u^n}, \qquad \text{where} \quad u = kx.$$

If $x \longrightarrow \infty$, then $u \longrightarrow \infty$; hence $e^u/u^n \longrightarrow \infty$ by the previous result. Therefore $e^{kx}/x^n \longrightarrow \infty$.

If $k > 0$ and n is positive, then $\displaystyle\lim_{x \to \infty} \frac{e^{kx}}{x^n} = \infty.$

For example, we are sure that $e^{x/100}$ will be much greater than x^{55}, provided x is large enough. In fact, $x^{55} > e^{x/100}$ for x up to about 60,500, but for larger values the exponential dominates, and as $x \longrightarrow \infty$ there is no contest.

Now consider e^{-kx} for $k > 0$. Since $e^{-kx} = 1/e^{kx}$, the last result implies

$$x^n e^{-kx} = \frac{x^n}{e^{kx}} \longrightarrow 0 \qquad \text{as} \quad x \longrightarrow \infty.$$

If $k > 0$, then for each $n \geq 0$, $\displaystyle\lim_{x \to \infty} x^n e^{-kx} = 0.$

EXERCISES

Find

1 $\displaystyle\lim_{x \to \infty} 2^x e^{-x}$ **2** $\displaystyle\lim_{x \to \infty} e^{\sqrt{x}}/x^3$ **3** $\displaystyle\lim_{n \to \infty} \left(1 - \frac{1}{n}\right)^n$ **4** $\displaystyle\lim_{n \to \infty} \left(1 + \frac{3}{n}\right)^n$

5 $\displaystyle\lim_{n \to \infty} \left(1 + \frac{1}{n}\right)^{5n}$ **6** $\displaystyle\lim_{n \to \infty} \left(1 - \frac{1}{n^2}\right)^n$ **7** $\displaystyle\lim_{n \to \infty} \frac{n^2}{2^n}$ **8** $\displaystyle\lim_{n \to \infty} \frac{(\sqrt{2})^n}{n^4}.$

9 Show that $\log_{10}(1 + 1/n) \approx 0.4343/n$ for n large.

10   Show that $p_2(x)p_2(y) = p_2(x + y) +$ (terms of degree at least three).

Show that y has a max for $x \geq 0$ and find it

11 $y = xe^{-x}$ **12** $y = x^2 e^{-x}$
13 $y = 3e^x - e^{2x}$ **14** $y = e^{-x} - e^x$.

15   Estimate e^x/x^x for $x = 10$, 100, 10^3, 10^6.
16   Prove $e^x/x^x \longrightarrow 0$ as $x \longrightarrow \infty$.

4. APPLICATIONS

   Exponential functions $y = ce^{kt}$ arise often in applications of mathematics because they are the solutions of the differential equation

$$\frac{dy}{dt} = ky.$$

For $k > 0$, this equation describes a quantity growing at a rate proportional to its own size; the larger y is, the faster it increases (the rich get richer). For $k < 0$, the equation describes a quantity decaying at a rate proportional to its own size; the smaller y is, the more slowly it decays.

Bacteria Growth   A colony of bacteria with unlimited food and no enemies grows at a rate proportional to its own size. We want a formula for $n(t)$, the number of bacteria in the colony at time t.

   To attack the problem we make an approximation. The function $n(t)$ is not continuous since it jumps by one each time a new bacterium is produced. However, since $n(t)$ is generally very large, and bacteria are produced at tiny time intervals, we simplify the problem by treating $n(t)$ as a continuous, even differentiable, function. In practice, this leads to satisfactory results.

   The growth law of $n(t)$ is

$$\frac{dn}{dt} = kn, \qquad k > 0.$$

If n_0 is the number of bacteria at time $t = 0$, then

$$n(t) = n_0 e^{kt}.$$

In a particular problem, k is found from additional data.

■ **EXAMPLE 1**   There are 10^5 bacteria in a culture at the start of an experiment and 10^6 after 5 hours. Find a formula for $n(t)$.

Solution   We are given $n_0 = 10^5$ and $n(5) = 10^6$. First we have

$$n(t) = n_0 e^{kt} = 10^5 e^{kt}.$$

Also, $$n(5) = 10^6 = 10^5 e^{5k}.$$

Therefore $e^{5k} = 10$. To find k, take logs:

$$5k \log e = \log 10 = 1, \qquad k = \frac{1}{5 \log e} \approx 0.46.$$

Hence $$n(t) \approx 10^5 e^{0.46t}, \qquad t \text{ in hours.}$$

There is an alternative form of the answer that doesn't require logs. Since $e^{5k} = 10$ we have

$$n(t) = 10^5 e^{kt} = 10^5 (e^{5k})^{t/5} = 10^5 \cdot 10^{t/5} = 10^{(25+t)/5}.$$ ■

Radioactive Decay   A radioactive element decays at a rate proportional to the amount present. Its **half-life** is the time in which a quantity decays to half of its original mass.

■ **EXAMPLE 2**   Carbon-14 has a half-life of 5568 years. Find its decay law.

Solution   Let $m(t)$ be the mass at time t, measured in years. Then

$$\frac{dm}{dt} = -\lambda m,$$

where the **decay constant** λ is positive. The solution is

$$m(t) = m_0 e^{-\lambda t},$$

where $m_0 = m(0)$, the initial mass. We are given

$$m(5568) = \tfrac{1}{2} m_0 = m_0 e^{-5568\lambda}.$$

Therefore $e^{-5568\lambda} = \tfrac{1}{2}$, that is, $e^{5568\lambda} = 2$. To find λ, take logs:

$$5568\lambda \log e = \log 2, \qquad \lambda = \frac{\log 2}{5568 \log e} \approx 1.245 \times 10^{-4}.$$

Hence $m(t) = m_0 e^{-\lambda t}, \qquad$ where $\quad \lambda \approx 1.245 \times 10^{-4}.$

Alternatively, $e^{5568\lambda} = 2$, so

$$m(t) = m_0 (e^{5568\lambda})^{-t/5568} = m_0 \, 2^{-t/5568}.$$ ■

Compound Interest

■ **EXAMPLE 3**   \$1000 is deposited in a savings account paying $5\tfrac{1}{2}\%$ annual interest, compounded daily. Estimate its value in 6 years.

Solution   There is an exact expression for the value. The daily interest rate is $i = 0.055/365$, and at each interest payment the current value is multiplied by $1 + i$. The number of payments in 6 years is $6 \times 365 = 2190$. Therefore the value after 6 years is

$$A = 1000(1 + i)^{2190} = 1000 \left(1 + \frac{0.055}{365} \right)^{2190}.$$

To the nearest cent, $A \approx \$1390.93$.

Two methods are available for estimating A. First, suppose interest is compounded continuously, not daily. Then the investment grows at a rate proportional to itself. If $A(t)$ is its value after t years, then

$$\frac{dA}{dt} = 0.055A;$$

hence $A(t) = 1000e^{0.055t}, \qquad A(6) = 1000e^{6(0.055)} \approx \$1390.97.$

The other approximation uses the formula

$$\lim_{n \to \infty} \left(1 + \frac{x}{n}\right)^n = e^x.$$

Choose $x = 0.055$ and $n = 365$: $\left(1 + \dfrac{0.055}{365}\right)^{365} \approx e^{0.055}.$

Therefore $1000\left(1 + \dfrac{0.055}{365}\right)^{6 \cdot 365} \approx 1000e^{6(0.055)} \approx \$1390.97.$ ∎

Approximate Value of $n!$ There is a useful estimate of $n!$ for large values of n. Its proof is too hard to present here; however, we do want to state the formula because of its importance and because of the way it involves the exponential function.

> **Stirling's Formula** $n! \approx \sqrt{2\pi n}\, n^n e^{-n}.$
>
> More precisely, $\lim_{n \to \infty} \left(\dfrac{n!}{\sqrt{2\pi n}\, n^n e^{-n}} \right) = 1.$

Here are some numerical examples, to 3 significant figures. We set

$$f(n) = \sqrt{2\pi n}\, n^n e^{-n}.$$

n	4	6	10	20	50
$n!$	24	720	3.63×10^6	2.43×10^{18}	3.04×10^{64}
$f(n)$	23.5	710	3.60×10^6	2.42×10^{18}	3.04×10^{64}

Do not assume that $n! - f(n) \longrightarrow 0$ as $n \longrightarrow \infty$. That is simply not true. For example,

$$50! - f(50) \approx 5 \times 10^{61}.$$

What *is* true is that the error $n! - f(n)$ is small compared to $n!$. For example,

$$\frac{50! - f(50)}{50!} \approx \frac{5 \times 10^{61}}{3 \times 10^{64}} \approx 0.0017.$$

Hence for $n = 50$, the formula is accurate to within $\frac{2}{10}$ of one percent.

Arms Race   Here is a simplified model of an arms race. Nation X has $x(t)$ weapons at time t and nation Y has $y(t)$ weapons. Each will increase its arsenal at a rate proportional to the arsenal of the other. Thus

$$\frac{dx}{dt} = ay \qquad \text{and} \qquad \frac{dy}{dt} = bx,$$

where a and b are positive constants.

We solve this system of differential equations by a little trick; we seek a constant c such that $z = x + cy$ is an exponential function, that is,

$$\frac{dz}{dt} = \lambda z$$

for a constant λ. Then $z = z_0 e^{\lambda t}$. Now

$$\frac{dz}{dt} = \frac{d}{dt}(x + cy) = \frac{dx}{dt} + c\frac{dy}{dt} = ay + bcx = bc\left(x + \frac{a}{bc}y\right).$$

Clearly we must choose c so that

$$\frac{a}{bc} = c, \qquad \text{that is,} \qquad c^2 = \frac{a}{b}, \qquad c = \pm\sqrt{\frac{a}{b}}.$$

Then $\lambda = bc = \pm\sqrt{ab}$.

To fix signs, let us set $\lambda = \sqrt{ab} > 0$ and $c = \sqrt{a/b} > 0$. Then the other possibility is $-\lambda$ and $-c$. We conclude that

$$\begin{cases} x + cy = (x_0 + cy_0)e^{\lambda t} \\ x - cy = (x_0 - cy_0)e^{-\lambda t}. \end{cases}$$

We solve for x and y by adding and subtracting these equations:

$$\begin{cases} x = x(t) = \dfrac{1}{2}(x_0 + cy_0)e^{\lambda t} + \dfrac{1}{2}(x_0 - cy_0)e^{-\lambda t} \\[2mm] y = y(t) = \dfrac{1}{2c}(x_0 + cy_0)e^{\lambda t} - \dfrac{1}{2c}(x_0 - cy_0)e^{-\lambda t}. \end{cases}$$

If either $x_0 > 0$ or $y_0 > 0$, then $x_0 + cy_0 > 0$, and we easily conclude that

$$x(t) \longrightarrow \infty, \qquad y(t) \longrightarrow \infty \qquad \text{as} \quad t \longrightarrow \infty.$$

An interpretation: war between X and Y is inevitable because at a certain time one nation can no longer bear the cost of the arms race.

EXERCISES

1   Thorium X has a half-life of 3.64 days. Find its decay law. How long will it take for $\frac{2}{3}$ of a quantity to disintegrate?

2   Two pounds of a certain radioactive substance loses $\frac{1}{5}$ of its original mass in 3 days. At what rate is the substance decaying after 4 days?

3   Money compounded continuously will double in a year at what annual rate of interest?

4   How long will it take a sum of money compounded continuously at 7.5% per annum to show a 50% return.

5   A colony of bacteria has a population of 3×10^6 initially, and 9×10^6 two hours later. What is the growth law? How long does it take the colony to double?

6   Assume that population grows at a rate proportional to the population itself. In 1950 the US population was 151 million, in 1970 it was 203 million. Make a prediction for the year 2050.

7   Under ideal conditions the rate of change of pressure above sea level is proportional to the pressure. If the barometer reads 30 in. at sea level, and 25 in. at 4000 ft, find the barometric pressure at 20,000 ft.

8 In a certain calculus course, it was found that the number of students dropping out each day was proportional to the number still enrolled. If 2000 started out and 10% dropped after 28 days, estimate the number left after 12 weeks.

9 A 5-lb sample of radioactive material contains 2 lb of radium F, which has a half-life of 138.3 days, and 3 lb of thorium X, which has a half-life of 3.64 days. When will the sample contain equal amounts of radium F and thorium X?

10 A salt in solution decomposes into other substances at a rate proportional to the amount still unchanged. If 10 kg of a salt reduces to 5 kg in $\frac{1}{2}$ hr, how much is left after 15 hr?

11 In the Arms Race example, find the ultimate weapons ratio, $\lim_{t \to \infty} x/y$.

12 A fixed region of area A is searched for oil. Let y denote the area of undiscovered oil, so $0 \leq y \leq A$. The probability of discovery upon drilling is y/A. Now let x denote the amount of exploratory drilling. Clearly y depends on x, and the larger x, the smaller y. The rate of discovery per unit of drilling is proportional to the probability of discovery;

hence $\quad \dfrac{d(A - y)}{dx} = k\,\dfrac{y}{A} \quad (k > 0)$. Express y in terms of x.

13 Suppose in a chemical reaction one molecule of A combines with one molecule of B to form a new substance X. The **law of mass action** says that the rate of increase of the amount $x(t)$ of X is proportional to the product $(a - x)(b - x)$, where a and b are the initial amounts of A and B. That is,

$$\frac{dx}{dt} = k(a - x)(b - x), \qquad k > 0.$$

Suppose $a \neq b$. Show that $\quad \dfrac{a(b - x)}{b(a - x)} = e^{k(b - a)t}$

provides a solution with $x(0) = 0$. [*Hint* Differentiate with respect to t by the Chain Rule.]

14 (cont.) Find $\lim x(t)$ as $t \longrightarrow \infty$, still assuming $a \neq b$.

15 Suppose a quantity of hot fluid is stirred so that at any time t its temperature $u(t)$ is uniform (the same) throughout the fluid. Let a denote the constant outside temperature. **Newton's law of cooling** says that the rate of decrease of u, due to heat loss at the surface, is proportional to $u - a$, that is,

$$\frac{du}{dt} = -k(u - a), \qquad k > 0.$$

Solve for u. Show that $u(t) \longrightarrow a$ as $t \longrightarrow \infty$. [*Hint* Set $v(t) = u(t) - a$.]

16 (cont.) Suppose $a = 0°C$, and the fluid cools from $100°C$ to $50°C$ in 5 minutes. How much longer will it take to cool to $5°C$?

17 The electric energy of a certain diatomic molecule is

$$U = k(e^{-2a(x - c)} - 2e^{-a(x - c)}),$$

where k, a, and c are positive constants and x is the distance between the atoms. Find the **dissociation energy**, the maximum of $-U$.

18 The rate of growth of the mass m of a falling raindrop is λm, where $\lambda > 0$. Show that $m = m_0\,e^{\lambda t}$.

19 (cont.) Newton's law of motion for the falling raindrop is

$$\frac{d}{dt}(mv) = mg,$$

where v is its downward speed and g is the gravitational constant. Express v in terms of t and find the **terminal velocity**, $\lim_{t \to \infty} v$. [*Hint* See the hint to Ex. 15.]

20 A rocket is traveling in a space region of negligible gravity. Let $m = m(t)$ denote its mass, $v = v(t)$ its speed, and u its constant speed of efflux (discharged hot gas). The rocket is propelled forward because hot gases are propelled backward (action equals reaction), and Newton's law of motion in this case says

$$\frac{d}{dt}(mv) = (v - u)\frac{dm}{dt}.$$

Show that

$$m\frac{dv}{dt} = -u\frac{dm}{dt}.$$

21 (cont.) Now use the Chain Rule to show that $\dfrac{dm}{dv} = -\dfrac{1}{u}m$. Deduce that $m = m_0\,e^{-v/u}$, assuming $v(0) = 0$ and $m(0) = m_0$.

22 (cont.) Let m_1 be the mass of the initial fuel supply and V the maximum speed. Show that

$$e^{-V/u} = \frac{m_0 - m_1}{m_0}.$$

23* (cont.) Suppose in addition that gravity is acting against the motion of the rocket. Then the equation of motion is

$$\frac{d}{dt}(mv) = (v - u)\frac{dm}{dt} - mg,$$

where g is the gravitational constant. Show that $m = m_0\,e^{-(v+gt)/u}$. [*Hint* Set $w = v + gt$ and find dm/dw.]

24* Suppose radioactive substance X decays into Y, which in turn decays into Z. Let $x(t)$ and $y(t)$ denote the amounts of X and Y at time t. Then

$$\frac{dx}{dt} = -\lambda x \qquad \text{and} \qquad \frac{dy}{dt} = \lambda x - \mu y,$$

where λ and μ are positive constants. Solve for x and y. Distinguish the cases $\lambda = \mu$ and $\lambda \neq \mu$. [*Hint* After you find $x(t)$, try $y(t) = e^{-\mu t}w(t)$ and solve for $w(t)$.]

25* Living wood absorbs atmospheric radiocarbon (^{14}C, half-life 5568 yr) and the rate of absorbtion exactly balances the rate of loss by radioactive decay. Therefore the amount of radiocarbon per gram of all living wood is a constant and has always been the same. When wood dies, it no longer absorbs fresh ^{14}C, and its ^{14}C content decays at the rate of 6.68 disintegrations per minute per gram (dpm). Wood excavated by archeologists in 1950 from a Babylonian city built in the time of King Hammurabi, the law-giver, measured 4.09 dpm. Show that Hammurabi lived about 1990 B.C.

26 (cont.) Wood from the Lascaux Caves in France found in 1950 measured 0.97 dpm. Estimate the age of the famous cave paintings as of 1950.

27 A bacteria colony grows in the presence of a toxin. Let $n = n(t)$ be the quantity of bacteria and $x = x(t)$ be that of the toxin. Then the toxin inhibits the growth of the colony by an amount proportional to the product of the amounts of toxin and bacteria. Hence

$$\frac{dn}{dt} = kn - anx, \qquad a > 0, \quad k > 0.$$

Suppose $x = ct$, where $c > 0$. Verify that $n(t) = n_0\exp(kt - act^2/2)$. Find $\lim_{t \to \infty} n(t)$.

28 Here is a model for the growth of a concentrated bacteria colony with limited space and food supply. The colony tends to squeeze itself out, so its concentration decreases at a rate proportional to itself. If nutrient is added, however, the concentration also increases at a rate proportional to the rate of nutrient addition. Let $x(t)$ be the concentration of cells, and assume that nutrient is added at a constant rate. Then

$$\frac{dx}{dt} = -ax + b, \qquad a > 0, \quad b > 0.$$

Solve for $x(t)$ and find the steady state concentration $x_\infty = \lim_{t\to\infty} x(t)$. [*Hint* Try $x = e^{-at}y$. Alternatively, see the hint to Ex. 19.]

29 Colonies of $p(t)$ parasites and $h(t)$ hosts live together. The hosts will tend to increase at a rate proportional to their number and tend to decrease at a rate proportional to the product ph. The parasites will tend to decrease at a rate proportional to their number and tend to increase at a rate proportional to the product ph. This leads to the **Lotka–Volterra system**

$$\frac{dh}{dt} = a_1 h - b_1 ph, \qquad \frac{dp}{dt} = -a_2 p + b_2 ph,$$

where the a's and b's are constants. It applies in a variety of predator–prey situations. Show that

$$h = \frac{a_2}{b_2}, \qquad p = \frac{a_1}{b_1}$$

is a solution (called the equilibrium solution).

30* (cont.) Prove for any solution that

$$(h^{a_2} e^{-b_2 h})(p^{a_1} e^{-b_1 p})$$

is a positive constant. Conclude that $h \longrightarrow \infty$ or $p \longrightarrow \infty$ as $t \longrightarrow \infty$ is impossible.

31 When a population is small, it grows at a rate proportional to itself. But as it gets larger, its members compete with each other for food and living space, and studies indicate that the population also tends to decline at a rate proportional to its square. This leads to a basic principle in ecology, the **Verhulst logistic equation** for population growth,

$$\frac{dp}{dt} = ap - bp^2,$$

where a and b are positive constants. Usually b is small relative to a. Show that

$$\frac{p}{a - bp} = ke^{at}, \qquad k = \frac{p_0}{a - bp_0},$$

and conclude that $\quad p = p(t) = \dfrac{ak}{bk + e^{-at}}.$

32 (cont.) The US population in 1790 was 3.93×10^6. Use $a = 3.054 \times 10^{-2}$ and $b = 1.189 \times 10^{-10}$. What does the Verhulst equation predict for 1850, 1900, and 1950? If the constants a and b are correct, what is the ultimate US population?

33 (cont.) For world population, some ecologists estimate $a = 0.0290$ and $b = 2.94 \times 10^{-12}$. The UN Statistical Office estimates mid-1970 world population as 3.63×10^9. Estimate world population for 2000, 2050, and 2100. What is the ultimate population?

34* "Double decay law" The problem is to solve the system

$$\frac{dx}{dt} = -ax, \qquad \frac{dy}{dt} = bxy, \qquad a > 0, \quad b > 0,$$

given x_0 and $y_\infty = \lim_{t\to\infty} y(t)$. [*Hint* You can find $x(t)$; to find $y(t)$, think of y as a function of x and find dy/dx by the Chain Rule.]

35* (cont.) Let $\rho = \rho(h)$ denote the density of the atmosphere at height h. Then the rate of decrease of ρ is proportional to ρ, that is, $d\rho/dh = -\rho/H$, where H is a positive constant. Let $R = R(h)$ be the intensity of solar radiation at height h. It is assumed that radiation is absorbed at a rate proportional to ρR, that is, $dR/dh = k\rho R$, where k is a positive constant. Express R in terms of h. Assume ρ_0 and R_∞ given. [*Hint* Translate the hint to Ex. 34 into this notation.]

36* (cont.) According to Chapman's theory of the formation of the ionosphere, ions are produced at a rate $J = J(h)$ proportional to the rate of absorption of radiation. Show that

$$J = bk\rho_0 R_\infty \exp[(-h/H) - kH\rho_0 e^{-h/H}].$$

Suppose J reaches its maximum at height h_1. Show that

$$e^{h_1/H} = Hk\rho_0 \quad\text{and}\quad J_{max} = bR_\infty/He.$$

5. TRIGONOMETRIC FUNCTIONS

In this section, we review the trigonometric functions, stressing those properties useful in calculus. We assume that the student is familiar with the elementary trigonometry of triangles.

Radian Measure   The unit of angle measurement in calculus is the radian. One **radian** is the angle which, placed at the center of a circle, subtends an arc whose length equals the radius (Fig. 1a). A central angle of θ radians subtends an arc of length θ times the radius (Fig. 1b).

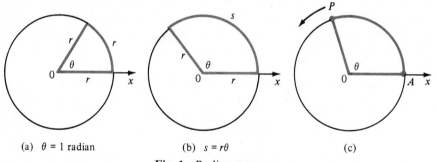

(a) $\theta = 1$ radian   (b) $s = r\theta$ (c)

Fig. 1   Radian measure

In a unit circle ($r = 1$), a central angle θ subtends an arc of length θ. A central angle of $360°$ subtends the entire circumference, whose length is 2π. Hence we have a conversion relation between radians and degrees: 2π radians $= 360°$.

$$\pi \text{ rad} = 180°, \quad 1 \text{ rad} = \frac{180}{\pi} \approx 57.2958°, \quad 1° = \frac{\pi}{180} \approx 0.0174533 \text{ rad}.$$

We shall follow the convention of omitting *rad* in radian measure. It is accepted procedure to write $\pi = 180°$ and to speak of the angle $\frac{1}{2}\pi$.

Look at Fig. 1c. If $\overline{OP}$ starts at position $\overline{OA}$ and swings counterclockwise, then θ starts at 0 and increases. After a complete revolution, $\theta = 2\pi$. After another revolution $\theta = 4\pi$, and so on. In general, starting from a given position, one counterclockwise revolution of $\overline{OP}$ increases θ by 2π. If $\overline{OP}$ starts at $\overline{OA}$ and swings clockwise, then θ is considered to be a negative angle. For example after a quarter revolution clockwise, $\theta = -\frac{1}{2}\pi$; after three full clockwise revolutions, $\theta = -3 \cdot 2\pi = -6\pi$.

According to this scheme, each real number θ determines a unique position of $\overline{OP}$ and a unique angle. However, each position of $\overline{OP}$ corresponds to infinitely many angles, differing from each other by integer multiples of 2π. For example, if $\overline{OP}$ points straight up, then the corresponding angles are $\frac{1}{2}\pi$, $\frac{1}{2}\pi \pm 2\pi$, $\frac{1}{2}\pi \pm 4\pi$, $\frac{1}{2}\pi \pm 6\pi$, etc.

Sine and Cosine   Each central angle θ determines a point (x, y) on the unit circle $x^2 + y^2 = 1$. See Fig. 2.

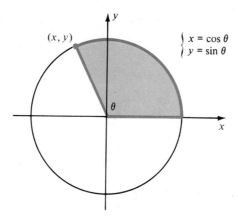

Fig. 2 θ determines (x, y)

Set $x = \cos \theta$ and $y = \sin \theta$. This defines two functions of θ for all real θ. Note that θ and $\theta + 2\pi$ determine the same point (x, y). Therefore,

$$\cos(\theta + 2\pi) = \cos \theta, \qquad \sin(\theta + 2\pi) = \sin \theta.$$

Thus the values of $\cos \theta$ and $\sin \theta$ repeat when θ increases by 2π. We say that these functions are **periodic** with **period** 2π.

Since $(\cos \theta, \sin \theta)$ is a point on the unit circle $x^2 + y^2 = 1$, the two functions satisfy the relation

$$\cos^2 \theta + \sin^2 \theta = 1.$$

We can see other basic properties of $\cos \theta$ and $\sin \theta$ from Fig. 3.
From Fig. 3a,

$$\cos(-\theta) = \cos \theta, \qquad \sin(-\theta) = -\sin \theta.$$

Thus $\cos \theta$ is an even function and $\sin \theta$ is an odd function. From Fig. 3b,

$$\cos(\theta + \pi) = -\cos \theta, \qquad \sin(\theta + \pi) = -\sin \theta.$$

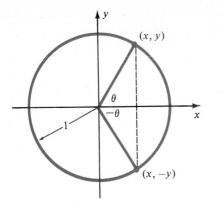

(a) Reflection in the x-axis:
  $(x, y) \longrightarrow (x, -y)$, $\theta \longrightarrow -\theta$

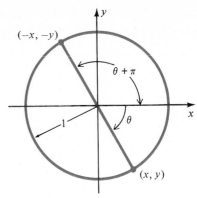

(b) Reflection in $(0, 0)$:
  $(x, y) \longrightarrow (-x, -y)$, $\theta \longrightarrow \theta + \pi$

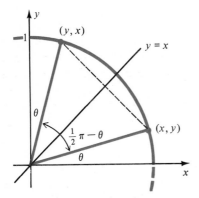

(c) Reflection in $y = x$:
  $(x, y) \longrightarrow (y, x)$, $\theta \longrightarrow \frac{1}{2}\pi - \theta$

Fig. 3 Properties of sine and cosine.

From Fig. 3c,

$$\boxed{\cos(\tfrac{1}{2}\pi - \theta) = \sin \theta, \qquad \sin(\tfrac{1}{2}\pi - \theta) = \cos \theta.}$$

Similar arguments yield the relations

$$\boxed{\begin{aligned} \cos(\pi - \theta) &= -\cos \theta, & \sin(\pi - \theta) &= \sin \theta, \\ \cos(\tfrac{1}{2}\pi + \theta) &= -\sin \theta, & \sin(\tfrac{1}{2}\pi + \theta) &= \cos \theta. \end{aligned}}$$

Identities   Among the most basic formulas in mathematics are the addition laws for sine and cosine.

$$\boxed{\begin{aligned} \sin(\alpha + \beta) &= \sin \alpha \cos \beta + \cos \alpha \sin \beta, \\ \cos(\alpha + \beta) &= \cos \alpha \cos \beta - \sin \alpha \sin \beta. \end{aligned}}$$

For proofs, see p. 353. We obtain alternative versions of these formulas by substituting $-\beta$ for β and using $\cos(-\beta) = \cos\beta$ and $\sin(-\beta) = -\sin\beta$:

$$\sin(\alpha - \beta) = \sin\alpha\cos\beta - \cos\alpha\sin\beta,$$
$$\cos(\alpha - \beta) = \cos\alpha\cos\beta + \sin\alpha\sin\beta.$$

By setting $\alpha = \beta = \theta$ in the addition laws, we obtain the double angle formulas for $\sin 2\theta$ and $\cos 2\theta$ in terms of $\sin\theta$ and $\cos\theta$:

$$\sin 2\theta = 2\sin\theta\cos\theta, \qquad \cos 2\theta = \cos^2\theta - \sin^2\theta.$$

The second formula has two alternative forms, both derived from $\cos^2\theta + \sin^2\theta = 1$:

$$\cos 2\theta = 1 - 2\sin^2\theta = 2\cos^2\theta - 1.$$

Graphs   Let us graph $y = \sin\theta$. Since $\sin\theta$ has period 2π, we need only plot $\sin\theta$ on the interval $-\pi \le \theta \le \pi$. We can then extend the graph indefinitely to the right and left, making it repeat every 2π. Actually, since $\sin\theta$ is an odd function, we need only plot $\sin\theta$ for $0 \le \theta \le \pi$; the part for $-\pi \le \theta \le 0$ is the reflection in the origin.

We can get a fairly accurate graph by plotting the points for $\theta = 0, 0.1\pi, 0.2\pi, \cdots, 1.0\pi$ with the aid of a sine table:

θ	0.0	0.1π	0.2π	0.3π	0.4π	0.5π	0.6π	0.7π	0.8π	0.9π	π
$\sin\theta$	0.00	0.31	0.59	0.81	0.95	1.00	0.95	0.81	0.59	0.31	0.00

We plot this data in Fig. 4a, extend it to $-\pi \le \theta \le 0$ in Fig. 4b, then obtain the complete graph in Fig. 5.

From the graph of $y = \sin\theta$, we obtain the graph of $y = \cos\theta$ free of charge. We use the relation $\cos\theta = \sin(\theta + \tfrac{1}{2}\pi)$, which shows that the graph of $y = \cos\theta$ is just the graph of $y = \sin\theta$ shifted $\tfrac{1}{2}\pi$ units to the left (Fig. 6).

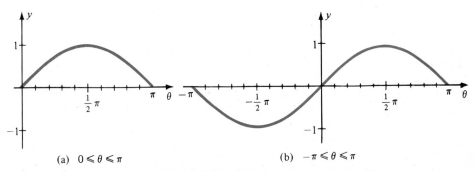

(a) $0 \le \theta \le \pi$ (b) $-\pi \le \theta \le \pi$

Fig. 4   Graph of $y = \sin\theta$

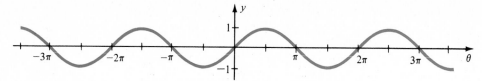

Fig. 5   Complete graph of $y = \sin \theta$

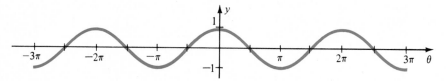

Fig. 6   Graph of $y = \cos \theta$

Multiple Angles   The graph of $y = \sin \theta$ makes one complete cycle on the interval $0 \leq \theta \leq 2\pi$. The graph of $y = \sin 2\theta$ makes one complete cycle on the interval $0 \leq \theta \leq \pi$ because 2θ runs from 0 to 2π as θ runs from 0 to π. See Fig. 7a. Therefore $\sin 2\theta$ oscillates twice as fast as $\sin \theta$. Similarly, $\sin \frac{1}{2}\theta$ oscillates half as fast as $\sin \theta$. See Fig. 7b.

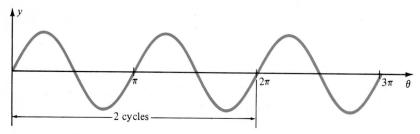

(a)   The graph of $y = \sin 2\theta$ oscillates twice as fast as the graph of $y = \sin \theta$.

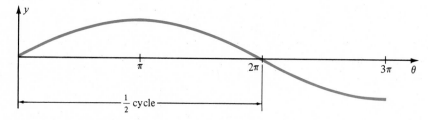

(b)   The graph of $y = \sin\frac{1}{2}\theta$ oscillates half as fast as the graph of $y = \sin \theta$

Fig. 7

In general, $\sin k\theta$ and $\cos k\theta$ make k cycles on $0 \leq \theta \leq 2\pi$, or equivalently, one full cycle on $0 \leq \theta \leq 2\pi/k$. It follows that $\sin k\theta$ and $\cos k\theta$ are periodic functions with period $2\pi/k$. To confirm this assertion, note that

$$\sin k(\theta + 2\pi/k) = \sin(k\theta + 2\pi) = \sin k\theta.$$

The same for $\cos k\theta$ can be confirmed similarly.

EXERCISES

Convert to radians (for example $30° = \frac{1}{6}\pi$)

1	$60°$	$150°$	$-240°$	$390°$	$-450°$	$900°$
2	$90°$	$-120°$	$210°$	$420°$	$-330°$	$480°$
3	$45°$	$-180°$	$270°$	$630°$	$-135°$	$495°$
4	$-90°$	$135°$	$-315°$	$405°$	$-270°$	$15°$.

Convert to degrees

5	$\frac{1}{2}\pi$	$-\frac{2}{3}\pi$	$\frac{5}{3}\pi$	3π	$-\frac{8}{3}\pi$	$\frac{50}{3}\pi$
6	$-\pi$	$-\frac{3}{2}\pi$	$\frac{7}{6}\pi$	5π	$-\frac{13}{6}\pi$	$\frac{17}{3}\pi$
7	$\frac{1}{4}\pi$	$\frac{2}{15}\pi$	$\frac{5}{12}\pi$	$\frac{1}{36}\pi$	$-\frac{11}{12}\pi$	$-\frac{1}{15}\pi$
8	$\frac{7}{2}\pi$	$-\frac{3}{4}\pi$	$\frac{7}{15}\pi$	$\frac{3}{8}\pi$	$\frac{1}{9}\pi$	$-\frac{1}{18}\pi$.

Find the point on the unit circle with central angle $\theta =$

9 $\frac{1}{4}\pi$ $-\frac{3}{4}\pi$ $\frac{11}{4}\pi$ **10** $\frac{3}{2}\pi$ $-\frac{1}{4}\pi$ $\frac{13}{4}\pi$

11 π $-\frac{2}{3}\pi$ $\frac{13}{6}\pi$ **12** $\frac{1}{6}\pi$ $\frac{5}{6}\pi$ $-\frac{2}{3}\pi$.

Find all θ in radians, $0 \le \theta < 2\pi$, such that

13 $\sin\theta = \frac{1}{2}\sqrt{2}$ **14** $\sin\theta = -\frac{1}{2}$ **15** $\cos\theta = -\frac{1}{2}\sqrt{3}$

16 $\cos\theta = \frac{1}{2}\sqrt{2}$ **17** $\sin(\theta + \pi) = 1$ **18** $\cos(\theta + \pi) = 1$

19 $\sin\theta = \cos\theta$ **20** $\sin(\theta - \pi) = -1$.

21 There are some simple rules of parity in the multiplication of integers:

$$\text{even} \cdot \text{even} = \text{even}, \quad \text{even} \cdot \text{odd} = \text{even}, \quad \text{odd} \cdot \text{odd} = \text{odd}.$$

Do the same rules hold in the multiplication of functions?

22 If $f(x)$ is an odd function, show that $f(0) = 0$.

Find the parity (odd or even?)

23 $\sin 2x$ **24** $\sin^4 x$ **25** $x \sin x$

26 $x \sin^3 x + 2\cos x$ **27** $\dfrac{\sin x}{x}$ **28** $e^x \cos x$.

Find the least period

29 $\sin 2\pi x$ **30** $\cos\frac{1}{3}x$ **31** $\sin x \sin 3x$

32 $\sin 3x + \cos 4x$ **33** $\cos^2 x$ **34** $\cos x \cos 2x$.

Show how the formulas (given on p. 180) are consequences of those in previous boxes

35 $\cos(\pi - \theta) = -\cos\theta, \quad \sin(\pi - \theta) = \sin\theta$

36 $\cos(\frac{1}{2}\pi + \theta) = -\sin\theta, \quad \sin(\frac{1}{2}\pi + \theta) = \cos\theta$.

What happens to the addition laws in the special cases

37 $\beta = \pm\pi$ **38** $\beta = \pm\frac{1}{2}\pi$?

Express in terms of $\sin\theta$ and $\cos\theta$

39 $\cos(\theta - \frac{1}{3}\pi)$ **40** $\sin(\theta + \frac{1}{6}\pi)$ **41** $\sin 3\theta$

42 $\cos 3\theta$ **43** $\sin 4\theta$ **44** $\cos 4\theta$.

45 Prove $\cos\alpha\cos\beta = \frac{1}{2}[\cos(\alpha + \beta) + \cos(\alpha - \beta)]$.

46 (cont.) Find similar formulas for $\sin\alpha\cos\beta$ and $\sin\alpha\sin\beta$.

47* Prove

$$\cos x \cos 2x \cos 4x \cos 8x = \frac{1}{8}(\cos x + \cos 3x + \cos 5x + \cdots + \cos 13x + \cos 15x).$$

48 Prove $\cos \frac{1}{2}\theta = \pm\sqrt{\frac{1}{2}(1 + \cos \theta)}$.   If $-2\pi \leq \theta \leq 2\pi$, when is the sign $+$ and when $-$?

49 Prove $\sin \frac{1}{2}\theta = \pm\sqrt{\frac{1}{2}(1 - \cos \theta)}$.   If $-2\pi \leq \theta \leq 2\pi$, when is the sign $+$ and when $-$?

50 (cont.) Compute $\sin 15°$ in two ways, using $15° = 45° - 30° = \frac{1}{2}(30°)$, to prove $\frac{1}{4}(\sqrt{6} - \sqrt{2}) = \frac{1}{2}\sqrt{2 - \sqrt{3}}$.

Graph

51 $y = \cos 3\theta$ **52** $y = \sin 4\theta$ **53** $y = 2 - \sin \theta$

54 $y = 1 + \cos \theta$ **55** $y = \cos 2\pi t$ **56** $x = \cos(\theta + \frac{1}{4}\pi)$.

6. ADDITIONAL TRIGONOMETRIC FUNCTIONS

In the last section we reviewed the properties of sine and cosine. Now let us recall the definitions of the other trig functions:

$$\tan x = \frac{\sin x}{\cos x}, \quad \cot x = \frac{\cos x}{\sin x}, \quad \sec x = \frac{1}{\cos x}, \quad \csc x = \frac{1}{\sin x}.$$

These four functions are defined in terms of $\sin x$ and $\cos x$. We can easily see some of their basic properties.

(1) **Domains.** Because of zeros in the denominators, $\tan x$ and $\sec x$ are not defined where $\cos x = 0$: at $\pm\frac{1}{2}\pi$, $\pm\frac{3}{2}\pi$, $\pm\frac{5}{2}\pi$, etc. Similarly, $\cot x$ and $\csc x$ are not defined where $\sin x = 0$: at 0, $\pm\pi$, $\pm 2\pi$, etc.

(2) **Periods.** The functions $\sin x$ and $\cos x$ are periodic with period 2π; the other four trigonometric functions inherit their periods. For example,

$$\sec(x + 2\pi) = \frac{1}{\cos(x + 2\pi)} = \frac{1}{\cos x} = \sec x,$$

and similarly, $\csc(x + 2\pi) = \csc x$. Even more can be said about $\tan x$ and $\cot x$. Recall that

$$\sin(x + \pi) = -\sin x, \qquad \cos(x + \pi) = -\cos x.$$

Therefore $\tan(x + \pi) = \dfrac{\sin(x + \pi)}{\cos(x + \pi)} = \dfrac{-\sin x}{-\cos x} = \tan x,$

and similarly for $\cot x$. Thus $\tan x$ and $\cot x$ have period π.

> The functions $\sec x$ and $\csc x$ have period 2π.
> The functions $\tan x$ and $\cot x$ have period π.

(3) **Parity.** Recall that $\sin(-x) = -\sin x$ and $\cos(-x) = \cos x$. Hence,

$$\tan(-x) = \frac{\sin(-x)}{\cos(-x)} = \frac{-\sin x}{\cos x} = -\tan x, \qquad \sec(-x) = \frac{1}{\cos(-x)} = \frac{1}{\cos x} = \sec x,$$

and similarly, $\cot(-x) = -\cot x, \qquad \csc(-x) = -\csc x.$

> The functions tan x, cot x, and csc x are odd functions.
> The function sec x is an even function

Graphs   Let us graph $y = \tan x$. Since tan x is an odd function and has period π, it suffices to sketch the part of the curve from 0 to $\frac{1}{2}\pi$. Then the complete graph can be obtained by symmetry and periodicity.

Since $\tan x = \sin x/\cos x$, we see that $\tan 0 = 0$. Furthermore, as x increases from 0 toward $\frac{1}{2}\pi$, the numerator increases from 0 toward 1, while the denominator decreases from 1 toward 0. Therefore tan x increases, and $\tan x \longrightarrow \infty$ as $x \longrightarrow \frac{1}{2}\pi$. See Fig. 1a.

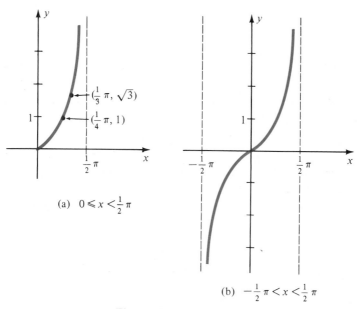

(a) $0 \leqslant x < \frac{1}{2}\pi$

(b) $-\frac{1}{2}\pi < x < \frac{1}{2}\pi$

Fig. 1   Graph of $y = \tan x$

By the oddness of tan x, we extend the graph (Fig. 1b) from 0 to $-\frac{1}{2}\pi$. Then we obtain the complete graph (Fig. 2) of $y = \tan x$ by extending the curve in Fig. 1b so as to have period π.

By similar reasoning, we obtain the graph (Fig. 3) of $y = \cot x$. Note that on each interval of length π, both tan x and cot x take every real value once.

Let us sketch $y = \csc x$. This function is odd and has period 2π, so by our usual argument, we need only plot the graph on the interval $0 < x < \pi$. The points $x = 0$ and $x = \pi$ are excluded because csc x is not defined there.

Assume $0 < x < \pi$. Then $0 < \sin x \leq 1$, hence $\csc x = 1/\sin x \geq 1$. Actually $\csc x = 1$ only at $x = \frac{1}{2}\pi$, where $\sin x = 1$, so the graph has a minimum point at $(\frac{1}{2}\pi, 1)$. If $x \longrightarrow 0+$ or $x \longrightarrow \frac{1}{2}\pi-$, then $1/\sin x \longrightarrow \infty$ since the denominator approaches 0. One other useful piece of information: the graph is symmetric about the line $x = \frac{1}{2}\pi$, a property it inherits from $y = \sin x$. (You should check that $\sin(\frac{1}{2}\pi - x) = \sin(\frac{1}{2}\pi + x)$.) We now have enough data for a reasonable sketch of

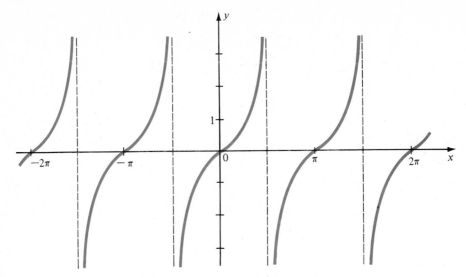

Fig. 2 Complete graph of $y = \tan x$

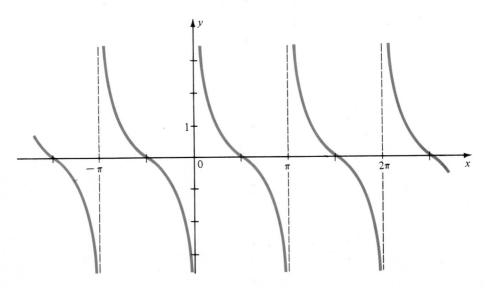

Fig. 3 Graph of $y = \cot x$

$y = \csc x$ on $0 < x < \pi$. We extend the curve to $-\pi < x < 0$ by oddness, then obtain the complete graph by periodicity (Fig. 4). There is another approach to the graph of $y = \csc x$: first sketch $y = \sin x$, then sketch its reciprocal.

We obtain the graph of $y = \sec x$ free of charge, just as we obtained the graph of $y = \cos x$ from that of $y = \sin x$. From $\cos x = \sin(x + \frac{1}{2}\pi)$ follows $\sec x = \csc(x + \frac{1}{2}\pi)$. Hence the graph of $y = \sec x$ is just the graph of $y = \csc x$ shifted $\frac{1}{2}\pi$ units to the left (Fig. 5).

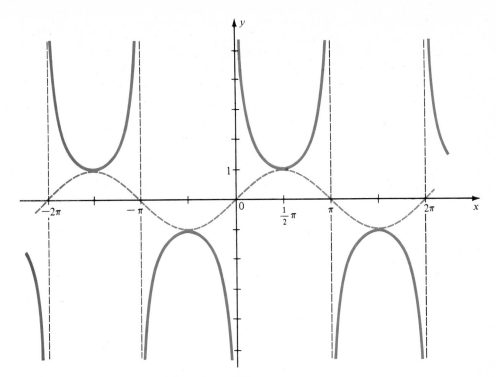

Fig. 4   Graph of $y = \csc x$

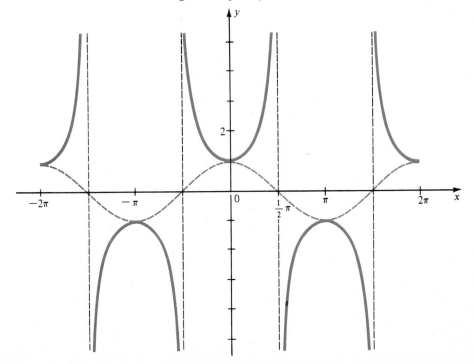

Fig. 5   Graph of $y = \sec x$

Identities   Start with the identity

$$\sin^2 x + \cos^2 x = 1,$$

and divide both sides first by $\cos^2 x$, then by $\sin^2 x$:

$$\tan^2 x + 1 = \sec^2 x, \qquad \cot^2 x + 1 = \csc^2 x.$$

These identities are helpful in expressing one trigonometric function in terms of another. For example,

$$\sin x = \pm\sqrt{1 - \cos^2 x}, \qquad \sec x = \pm\sqrt{\tan^2 x + 1},$$

$$\cot x = \pm\sqrt{\csc^2 x - 1} = \pm\frac{\sqrt{1 - \sin^2 x}}{\sin x}.$$

In each case, additional information is needed for the correct choice of sign.

■ **EXAMPLE 1**   Express $\sin x$ and $\cos x$ in terms of $\tan x$ for $0 < x < \frac{1}{2}\pi$.

Solution

$$\cos x = \frac{1}{\sec x} = \frac{1}{\pm\sqrt{\tan^2 x + 1}}, \qquad \sin x = \tan x \cdot \cos x = \frac{\tan x}{\pm\sqrt{\tan^2 x + 1}}.$$

Since $\sin x$, $\cos x$, and $\tan x$ are positive for $0 < x < \frac{1}{2}\pi$, choose the positive square root. ■

The addition laws for the sine and cosine yield an addition law for the tangent. Write

$$\tan(\alpha + \beta) = \frac{\sin(\alpha + \beta)}{\cos(\alpha + \beta)} = \frac{\sin \alpha \cos \beta + \cos \alpha \sin \beta}{\cos \alpha \cos \beta - \sin \alpha \sin \beta}.$$

Divide numerator and denominator by $\cos \alpha \cos \beta$:

$$\tan(\alpha + \beta) = \frac{\dfrac{\sin \alpha}{\cos \alpha} + \dfrac{\sin \beta}{\cos \beta}}{1 - \dfrac{\sin \alpha \sin \beta}{\cos \alpha \cos \beta}} = \frac{\tan \alpha + \tan \beta}{1 - \tan \alpha \tan \beta}.$$

$$\tan(\alpha + \beta) = \frac{\tan \alpha + \tan \beta}{1 - \tan \alpha \tan \beta}.$$

In particular, for $\alpha = \beta = \theta$, we have the double angle formula

$$\tan 2\theta = \frac{2 \tan \theta}{1 - \tan^2 \theta}.$$

EXERCISES

1   Express $\sin x$ in terms of $\cot x$.   **2**   Express $\sin x$ in terms of $\sec x$.

3 Express $\cot^2 x$ in terms of $\cos^2 x$.   **4** Express $\cot x$ in terms of $\sec x$.

Prove

5 $\cot(\alpha - \beta) = \dfrac{\cot \alpha \cot \beta + 1}{\cot \beta - \cot \alpha}$ **6** $\cot 2\theta = \dfrac{\cot^2 \theta - 1}{2 \cot \theta}$

7 $\tan \theta = \dfrac{\sin 2\theta}{1 + \cos 2\theta}$ **8** $\sin 2\theta = \dfrac{2 \tan \theta}{1 + \tan^2 \theta}.$

Find the least period

9 $\tan x - \cot x$ **10** $\sec \theta - \csc \theta.$

Graph

11 $y = \cot 2x$ **12** $y = 1 - \csc x$ **13** $y = \tan^2 x$
14 $y = \sec^2 x$ **15** $y = \tan(x - \frac{1}{4}\pi)$ **16** $y = -\sec x.$

7. DERIVATIVES

Let us start with the derivative of $\sin x$. Once we have that, the derivatives of the other trig functions follow easily. The derivative of $\sin x$ depends on two important limits:

$$\text{(a)} \quad \lim_{h \to 0} \frac{\sin h}{h} = 1 \qquad \text{(b)} \quad \lim_{h \to 0} \frac{\cos h - 1}{h} = 0.$$

These limits express simple geometric facts, as we can see from the equivalent statements:

$$\lim_{h \to 0} \frac{\sin(0 + h) - \sin 0}{h} = 1, \qquad \lim_{h \to 0} \frac{\cos(0 + h) - \cos 0}{h} = 0.$$

Thus (a) asserts that the curve $y = \sin x$ has slope 1 at $x = 0$, and (b) asserts that $y = \cos x$ has slope 0 at $x = 0$. The first statement is believable because on p. 60 we found strong numerical evidence that $\sin h/h \longrightarrow 1$ as $h \longrightarrow 0$. The second statement is nearly obvious geometrically because the curve $y = \cos x$ has a maximum, hence a horizontal tangent, at $x = 0$. Let us accept these two limits for the time being; they are proved at the end of this section.

It is not hard to *guess* the derivative of $y = \sin x$. Since the graph of $y = \sin x$ repeats after 2π, the slope of the graph also repeats after 2π. (See Fig. 5, p. 182.) Therefore the derivative of $\sin x$ is a periodic function with period 2π. According to limit (a), the slope of $y = \sin x$ is 1 at $x = 0$. Looking at the graph, we see that the slope decreases to 0 as x goes from 0 to $\frac{1}{2}\pi$, then becomes increasingly negative and, by symmetry, reaches -1 at $x = \pi$. Then the slope becomes less negative, returning to 0 at $\frac{3}{2}\pi$, and finally becomes positive, returning to the value 1 at 2π. But this is exactly what $\cos x$ does! So an educated guess is that the derivative of $\sin x$ is $\cos x$. A similar argument suggests that the derivative of $\cos x$ is $-\sin x$.

$$\frac{d}{dx} \sin x = \cos x, \qquad \frac{d}{dx} \cos x = -\sin x.$$

These formulas are correct. We prove the first one by writing the difference quotient for $\sin x$ and using the addition law:

$$\frac{\sin(x + h) - \sin x}{h} = \frac{(\sin x \cos h + \cos x \sin h) - \sin x}{h}$$

$$= \sin x \frac{\cos h - 1}{h} + \cos x \frac{\sin h}{h}.$$

This is where limits (a) and (b) come in. We let $h \longrightarrow 0$:

$$\frac{d}{dx} \sin x = \lim_{h \to 0} \frac{\sin(x + h) - \sin x}{h} = (\sin x)\left(\lim_{h \to 0} \frac{\cos h - 1}{h}\right) + (\cos x)\left(\lim_{h \to 0} \frac{\sin h}{h}\right)$$

$$= (\sin x) \cdot 0 + (\cos x) \cdot 1 = \cos x.$$

Thus $\sin x$ is differentiable and its derivative is $\cos x$.

We can differentiate $\cos x$ in a similar way, but it is quicker to use the identities

$$\cos x = \sin(x + \tfrac{1}{2}\pi), \qquad \sin x = -\cos(x + \tfrac{1}{2}\pi).$$

By the Chain Rule, $\quad \dfrac{d}{dx} \cos x = \dfrac{d}{dx} \sin\left(x + \dfrac{1}{2}\pi\right) = \cos\left(x + \dfrac{1}{2}\pi\right) \cdot 1 = -\sin x.$

Finally, if $u = u(x)$ is differentiable, then the Chain Rule implies the formulas

$$\boxed{\frac{d}{dx} \sin u = (\cos u)\frac{du}{dx}, \qquad \frac{d}{dx} \cos u = -(\sin u)\frac{du}{dx}.}$$

■ **EXAMPLE 1**   Differentiate

(a) $\sin \tfrac{1}{3}x$ (b) $\sqrt{8 - 7 \cos 2x}$ (c) $\cos^2 x$.

Solution   (a)   Let $u = \tfrac{1}{3}x$. By the Chain Rule,

$$\frac{d}{dx} \sin u = (\cos u)\frac{du}{dx} = \left(\cos \frac{1}{3}x\right) \cdot \frac{1}{3} = \frac{1}{3} \cos \frac{1}{3}x.$$

(b)   Let $u = 8 - 7 \cos 2x$. Then

$$\frac{d}{dx} \sqrt{u} = \frac{1}{2\sqrt{u}} \frac{du}{dx} = \frac{1}{2\sqrt{u}} \frac{d}{dx}(8 - 7 \cos 2x) = \frac{1}{2\sqrt{u}}\left[-7 \frac{d}{dx} \cos 2x\right]$$

$$= \frac{1}{2\sqrt{u}}[-7(-\sin 2x) \cdot 2] = \frac{7 \sin 2x}{\sqrt{8 - 7 \cos 2x}}.$$

Note that two applications of the Chain Rule are needed. Don't forget the second.

(c)   Let $u = \cos x$. Then

$$\frac{d}{dx} u^2 = 2u \frac{du}{dx} = (2 \cos x)(-\sin x) = -2 \cos x \sin x.$$

Alternative Solution   From the double angle formula, $\cos^2 x = \tfrac{1}{2}(\cos 2x + 1)$.

Therefore,

$$\frac{d}{dx}\cos^2 x = \frac{1}{2}\frac{d}{dx}(\cos 2x + 1) = \frac{1}{2}(-\sin 2x)(2) = -\sin 2x.$$

Note that the two answers are the same. Why?   ∎

■ **EXAMPLE 2** Show that $\dfrac{d}{dx}\left(\sin x - \dfrac{1}{3}\sin^3 x\right) = \cos^3 x.$

Solution $\dfrac{d}{dx}\left(\sin x - \dfrac{1}{3}\sin^3 x\right) = \dfrac{d}{dx}\sin x - \dfrac{1}{3}\dfrac{d}{dx}\sin^3 x$

$$= \cos x - \frac{1}{3}(3\sin^2 x)(\cos x) = (\cos x)(1 - \sin^2 x) = \cos^3 x. \quad ∎$$

■ **EXAMPLE 3** Sketch the curve $y = \cos 2x + 2\cos x.$

Solution   The curve is periodic with period 2π; it is enough to sketch it for $0 \le x \le 2\pi$, then extend the curve periodically. For a reasonable sketch without too much work, first find a few points on the curve that are easy to compute:

$$(0, 3), \quad (\tfrac{1}{2}\pi, -1), \quad (\pi, -1), \quad (\tfrac{3}{2}\pi, -1), \quad (2\pi, 3).$$

Next, locate the critical points (where the derivative vanishes). From $y = \cos 2x + 2\cos x,$

$$\frac{dy}{dx} = -2\sin 2x - 2\sin x = -2(2\sin x \cos x + \sin x) = -2\sin x \cdot (2\cos x + 1).$$

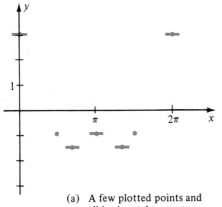

(a) A few plotted points and all horizontal tangents

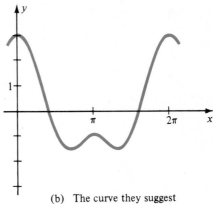

(b) The curve they suggest on one period

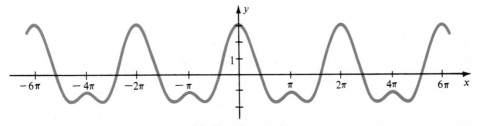

(c) Complete graph

Fig. 1   Graph of $y = \cos 2x + 2\cos x$

Therefore, $dy/dx = 0$ when either $\sin x = 0$ or $\cos x = -\frac{1}{2}$. The corresponding critical points are

$$(0, 3), \quad (\pi, -1), \quad (2\pi, 3) \quad \text{and} \quad (\tfrac{2}{3}\pi, -\tfrac{3}{2}), \quad (\tfrac{4}{3}\pi, -\tfrac{3}{2}).$$

Plot all of these points, showing the horizontal tangents at the critical points (Fig. 1a). Then fill in a smooth curve (Fig. 1b). Finally, extend the curve periodically (Fig. 1c). ∎

Other Derivatives   From the derivatives of $\sin x$ and $\cos x$, we easily find the derivatives of the other trig functions:

$$\frac{d}{dx} \tan x = \sec^2 x, \qquad \frac{d}{dx} \cot x = -\csc^2 x,$$

$$\frac{d}{dx} \sec x = \sec x \tan x, \qquad \frac{d}{dx} \csc x = -\csc x \cot x.$$

For example,

$$\frac{d}{dx} \tan x = \frac{d}{dx} \frac{\sin x}{\cos x} = \frac{\cos x \dfrac{d}{dx}(\sin x) - \sin x \dfrac{d}{dx}(\cos x)}{\cos^2 x}$$

$$= \frac{(\cos x)(\cos x) - (\sin x)(-\sin x)}{\cos^2 x} = \frac{\cos^2 x + \sin^2 x}{\cos^2 x} = \frac{1}{\cos^2 x} = \sec^2 x.$$

Similarly,

$$\frac{d}{dx} \sec x = \frac{d}{dx} \left(\frac{1}{\cos x} \right) = -\frac{\dfrac{d}{dx} \cos x}{\cos^2 x} = -\frac{-\sin x}{\cos^2 x} = \frac{1}{\cos x} \frac{\sin x}{\cos x} = \sec x \tan x.$$

■ **EXAMPLE 4** Differentiate (a) $\tan^4 3x$ (b) $e^{2x} \sec x$.

Solution (a) $\dfrac{d}{dx} (\tan 3x)^4 = 4(\tan 3x)^3 \dfrac{d}{dx} \tan 3x$

$$= 4(\tan^3 3x)(3 \sec^2 3x) = 12 \tan^3 3x \sec^2 3x.$$

(b) $\dfrac{d}{dx} (e^{2x} \sec x) = e^{2x} \dfrac{d}{dx} \sec x + \sec x \dfrac{d}{dx} e^{2x}$

$$= e^{2x} \sec x \tan x + (\sec x)(2e^{2x}) = e^{2x} \sec x \cdot (\tan x + 2). \quad ∎$$

■ **EXAMPLE 5** Show that $y = \sin kx$ and $y = \tan x$ satisfy, respectively, the differential equations

(a) $y'' + k^2 y = 0$ (b) $y' = 1 + y^2$.

Solution (a) $y' = k \cos kx$ and $y'' = k(-k \sin kx) = -k^2 \sin kx$.

Therefore $y'' + k^2 y = -k^2 \sin kx + k^2 \sin kx = 0.$

(b) $y' = \sec^2 x = 1 + \tan^2 x = 1 + y^2$.   ∎

Remark Our differentiation formulas for the trig functions require angles to be in radians. We can also differentiate these functions when expressed in degrees or in some other system, but the formulas do not turn out as simple. For example, suppose we want $d(\sin \theta)/d\theta$ when θ is in degrees. We convert to radians, θ deg $= (\pi/180)\theta$ rad, then differentiate using our standard formula:

$$\frac{d}{d\theta} \sin\left(\frac{\pi}{180}\theta\right) = \frac{\pi}{180} \cos\left(\frac{\pi}{180}\theta\right).$$

Therefore,

$$\frac{d}{d\theta} \sin \theta° = \frac{\pi}{180} \cos \theta°.$$

The awkward constant $\pi/180$ will appear in the derivatives of all trig functions if they are expressed in degrees.

Proofs of the Basic Limits Let us prove the limit statements (a) and (b) assumed at the beginning of this section. For (a) we need only consider $\sin h/h$ for $h > 0$ because $\sin(-h)/(-h) = \sin h/h$.

A central angle of h radians in a unit circle (Fig. 2a) subtends a chord of length l and an arc of length h. Clearly $\sin h < l < h$, hence

$$\frac{\sin h}{h} < 1.$$

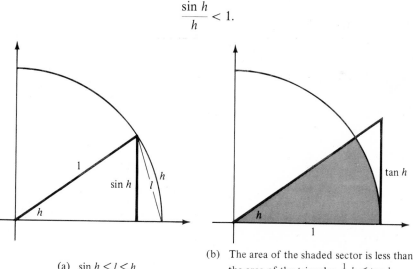

(a) $\sin h < l < h$

(b) The area of the shaded sector is less than the area of the triangle: $\frac{1}{2} h < \tan h$.

Fig. 2

The shaded sector in Fig. 2b has area $\frac{1}{2}h$ because

$$\frac{\text{shaded area}}{\pi} = \frac{\text{shaded area}}{\text{area of circle}} = \frac{h}{2\pi}.$$

The triangle has area $= \frac{1}{2}(\text{base})(\text{height}) = \frac{1}{2}(1)(\tan h) = \frac{1}{2} \tan h$. Since the area of the shaded sector is less than the area of the triangle,

$$\frac{h}{2} < \frac{1}{2} \tan h = \frac{1}{2} \frac{\sin h}{\cos h}, \qquad \text{which implies} \quad \cos h < \frac{\sin h}{h}.$$

Combining inequalities, we have

$$\cos h < \frac{\sin h}{h} < 1.$$

But $\cos h \longrightarrow 1$ as $h \longrightarrow 0$ and $\sin h/h$ is trapped between $\cos h$ and 1. Therefore

$$\lim_{h \to 0} \frac{\sin h}{h} = 1.$$

Limit (b) follows from (a). Write

$$\frac{\cos h - 1}{h} = \frac{\cos h - 1}{h}\frac{\cos h + 1}{\cos h + 1} = \frac{\cos^2 h - 1}{h(\cos h + 1)} = \frac{-\sin^2 h}{h(\cos h + 1)} = -\frac{\sin h}{h}\frac{\sin h}{\cos h + 1}.$$

Both factors on the right-hand side have limits as $h \longrightarrow 0$. Therefore

$$\lim_{h \to 0} \frac{\cos h - 1}{h} = -\left(\lim_{h \to 0} \frac{\sin h}{h}\right)\left(\lim_{h \to 0} \frac{\sin h}{\cos h + 1}\right) = -(1)(0) = 0.$$

EXERCISES

Differentiate with respect to x

1 $\cos x^2$ $-\sin x^2\,(2x)$

2 $e^{\sin x}$

3 $e^x \sin x - e^x \cos x$

4 $\cos^3 x$ $3\cos^2 x(-\sin x)$

5 $x \tan x$ $\tan x + x\sec^2 x$

6 $x^2 \sin(1/x)$ $(1+\tan x)\,6x\tan x - \sec x(\sec^2 x)$ $\frac{}{(1+\tan x)^2}$

7 $(\sin 2x)/x^2$

8 $\sin \sqrt{\tfrac{1}{3}x}$

9 $(\sec x)/(1 + \tan x)$

10 $\cot(\tan x)$

11 $\dfrac{\sin x + \cos x}{\sin x - \cos x}$

12 $\dfrac{\tan x}{\sqrt{x}}.$

13 Find the 82-nd derivative of $\cos x$.

14 Compute the derivative of $\cos^2 x - \sin^2 x$ in two different ways.

15 Verify the formulas given in the text for the derivatives of $\cot x$ and $\csc x$.

16 Find the 20-th derivative of $x \sin x$.

Prove

17 $\dfrac{d}{dx}(\sec^4 x - \tan^4 x) = 4 \sec^2 x \tan x$

18 $\dfrac{d}{dx}(x + \sec x - \tan x) = \dfrac{\sin x}{1 + \sin x}.$

Graph

19 $y = \sin 2x - \sin x$

20 $y = \sin x + \cos 2x$

21 $y = \sec x - \csc x$

22 $y = \tan x + \cot x$

23 $y = (\sin x)/x$

24 $y = e^{-x} \sin x$

25 $y = \sin(1/x)$

26 $y = x \sin(1/x).$

27 Find all points on $y = \sin x$ where the slope is $\frac{1}{2}$. $\frac{dy}{dx} = \cos x = \frac{1}{2}$

28 Find $k > 0$ so that the curves $y = \sin kx$ and $y = \cos kx$ intersect at right angles.

29 Show that $\sin x < x$ for all $x > 0$.

30 (cont.) Prove that $y = \cot x - (1/x)$ is a decreasing function for $0 < x < \pi$.

$x = 60° = \frac{\pi}{3}$ general form

$\frac{\pi}{3} + 2n\pi$ $(0, x)$

or $-\frac{\pi}{3} + 2n\pi$

8. APPLICATIONS

■ **EXAMPLE 1** A point P moves counterclockwise at constant speed 1 rpm around a circle of radius 50 ft. The tangent at P crosses the line OA at a point T. See Fig. 1. Compute the speed of T when $\theta = \frac{1}{4}\pi$.

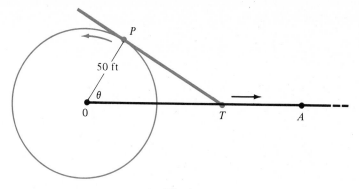

Fig. 1

Solution   We are given $d\theta/dt = 2\pi$ rad/min and are asked to find dx/dt, where $x = \overline{OT}$. The right triangle OPT suggests a relation between x and θ, namely, $x = 50 \sec \theta$. We differentiate with respect to t:

$$\frac{dx}{dt} = 50 \sec \theta \tan \theta \frac{d\theta}{dt}.$$

When $\theta = \frac{1}{4}\pi$, then $\sec \theta = \sqrt{2}$ and $\tan \theta = 1$. At that instant, $dx/dt = 50\sqrt{2} \cdot 2\pi$. *Answer*: $100\pi\sqrt{2} \approx 444.3$ ft/min.   ∎

■ **EXAMPLE 2**   A lighthouse stands 2 mi off a long straight shore, opposite a point P. Its light rotates counterclockwise at the constant rate of 1.5 rpm. How fast is the beam moving along the shore as it passes a point 3 mi to the right of P?

Solution   Set up axes with P at the origin and x-axis along the shore (Fig. 2a). The beam hits the shore at x. The rate of change of the angle θ is given: $d\theta/dt = 3\pi$ rad/min. The problem is to compute dx/dt at the instant when $x = 3$.

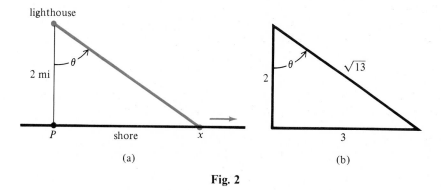

(a)   (b)

Fig. 2

The figure suggests a relation between x and θ:

$$\frac{x}{2} = \tan \theta, \qquad \text{that is,} \quad x = 2 \tan \theta.$$

Now differentiate with respect to t:

$$\frac{dx}{dt} = \frac{dx}{d\theta}\frac{d\theta}{dt} = (2\sec^2\theta)(3\pi) = 6\pi\sec^2\theta.$$

When $x = 3$, Fig. 2b shows that $\sec\theta = \frac{1}{2}\sqrt{13}$. Therefore at this instant,

$$\frac{dx}{dt} = 6\pi\left(\frac{1}{2}\sqrt{13}\right)^2 = \frac{39\pi}{2} \approx 61.26 \quad \text{mi/min.} \qquad \blacksquare$$

■ **EXAMPLE 3** A 5-ft fence stands 4 ft from a high wall (Fig. 3a). Find the angle of inclination of the shortest ladder that can lean against the wall from outside the fence.

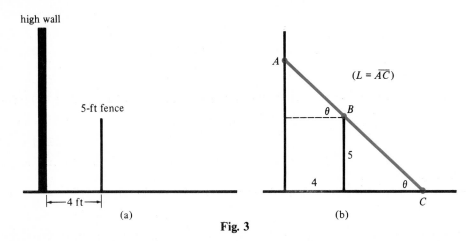

(a)   (b)

Fig. 3

Solution The shortest ladder must just touch the fence. If not, then a *shorter* ladder would reach from C to a point slightly below A on the wall (Fig. 3b). Let θ be the angle of inclination. Clearly, if θ is near 0 or near $\frac{1}{2}\pi$, the ladder will be very long.

Let L be the length of the ladder. Then

$$L = \overline{AB} + \overline{BC} = 4\sec\theta + 5\csc\theta.$$

The problem: find the angle θ in the range $0 < \theta < \frac{1}{2}\pi$ that minimizes $L = L(\theta)$. Now

$$\frac{dL}{d\theta} = 4\sec\theta\tan\theta - 5\csc\theta\cot\theta$$

$$= 4\frac{1}{\cos\theta}\frac{\sin\theta}{\cos\theta} - 5\frac{1}{\sin\theta}\frac{\cos\theta}{\sin\theta} = \frac{4\sin^3\theta - 5\cos^3\theta}{\sin^2\theta\cos^2\theta}$$

As θ increases, the numerator changes from negative to zero to positive. Therefore L is minimized for just one value of θ:

$$\frac{dL}{d\theta} = 0 \quad \text{if} \quad 4\sin^3\theta = 5\cos^3\theta, \quad \text{that is,} \quad \tan^3\theta = \frac{5}{4}.$$

Hence

$$\tan\theta = \sqrt[3]{\tfrac{5}{4}}, \quad \theta \approx 47.13°. \qquad \blacksquare$$

Note   Compare Example 5, p. 132, where we found $L_{min} \approx 12.7$ ft.

■ **EXAMPLE 4**   A man is in a rowboat 1 mi off shore. His home is 5 mi farther along the shore. If he can walk twice as fast as he can row, what is his quickest way home?

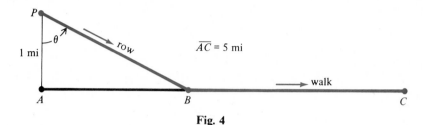

Fig. 4

Solution   Draw a diagram (Fig. 4). He should row to a point B, then walk to his home at C. Express everything in terms of the angle θ:

$$\overline{PB} = \sec\theta, \qquad \overline{BC} = 5 - \overline{AB} = 5 - \tan\theta.$$

Let v be his rowing speed and $2v$ his walking speed. The time required to reach home is

$$t = \frac{\overline{PB}}{v} + \frac{\overline{BC}}{2v} = \frac{\sec\theta}{v} + \frac{5 - \tan\theta}{2v}.$$

Since B must be between A and C, angle θ is at least 0 and at most the angle whose tangent is 5. Differentiate:

$$\frac{dt}{d\theta} = \frac{\sec\theta\tan\theta}{v} - \frac{\sec^2\theta}{2v} = \frac{\sin\theta}{v\cos^2\theta} - \frac{1}{2v\cos^2\theta} = \frac{2\sin\theta - 1}{2v\cos^2\theta}.$$

The derivative is 0 if $\sin\theta = \frac{1}{2}$, that is, $\theta = \frac{1}{6}\pi$. Its sign changes from minus to plus as θ increases through $\frac{1}{6}\pi$. Hence t has its minimum there. Notice that $\frac{1}{6}\pi$ falls within the permissible range of θ because $\tan\frac{1}{6}\pi < 5$.

Answer   Row to shore at an angle of $\frac{1}{6}\pi$, then walk the rest of the way.   ■

■ **EXAMPLE 5**   Describe the isosceles triangle of smallest area that circumscribes a circle of radius r.

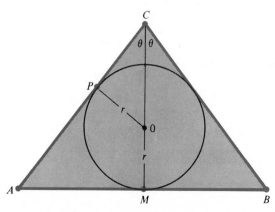

Fig. 5

Solution   Let 2θ be the apex angle, and express the area in terms of θ. See Fig. 5. Clearly

$$\overline{CM} = \overline{CO} + \overline{OM} = r \csc \theta + r.$$

From triangle ACM, observe that

$$\overline{AM} = \overline{CM} \tan \theta = (r \csc \theta + r) \tan \theta.$$

Therefore the area of the large triangle is

$$f(\theta) = \tfrac{1}{2}\overline{AB} \cdot \overline{CM} = \overline{AM} \cdot \overline{CM} = (r \csc \theta + r)^2 \tan \theta.$$

The problem is to minimize $f(\theta)$ for $0 < \theta < \tfrac{1}{2}\pi$. Clearly $f(\theta) > 0$ on this domain and $f(\theta) \longrightarrow +\infty$ if $\theta \longrightarrow 0+$ or $\theta \longrightarrow \tfrac{1}{2}\pi-$, so the minimum occurs where $f'(\theta) = 0$. Now

$$\frac{df}{d\theta} = r^2 \left[(\csc \theta + 1)^2 \frac{d}{d\theta}(\tan \theta) + \tan \theta \frac{d}{d\theta}(\csc \theta + 1)^2 \right]$$

$$= r^2[(\csc \theta + 1)^2 \sec^2 \theta + \tan \theta \cdot 2(\csc \theta + 1)(-\csc \theta \cot \theta)]$$

$$= r^2(\csc \theta + 1)[(\csc \theta + 1) \sec^2 \theta - 2 \csc \theta].$$

Since $\csc \theta + 1 \geq 2 > 0$ for $0 < \theta < \tfrac{1}{2}\pi$, the derivative $df/d\theta$ is 0 for

$$(\csc \theta + 1) \sec^2 \theta - 2 \csc \theta = 0.$$

To solve, multiply by $\sin \theta \cos^2 \theta$ and use $\cos^2 \theta = 1 - \sin^2 \theta$:

$$(1 + \sin \theta) - 2 \cos^2 \theta = 0, \qquad 2 \sin^2 \theta + \sin \theta - 1 = 0,$$

$$(2 \sin \theta - 1)(\sin \theta + 1) = 0, \qquad \sin \theta = \tfrac{1}{2} \quad \text{or} \quad -1.$$

Clearly $\sin \theta = -1$ is impossible, so $\sin \theta = \tfrac{1}{2}$, and $\theta = \tfrac{1}{6}\pi$. The triangle is equilateral,

$$\overline{CM} = r + r \csc \tfrac{1}{6}\pi = 3r,$$

and the side is

$$\overline{AB} = 2\overline{AM} = 2\overline{CM} \cdot \tan \tfrac{1}{6}\pi = 6r(\tfrac{1}{3}\sqrt{3}) = 2r\sqrt{3}. \qquad \blacksquare$$

Remark   Since $f(\theta) \longrightarrow +\infty$ at the end points and $f'(\theta) = 0$ for only *one* value of θ, that value must give us a minimum.

EXERCISES

1   Express the rate at which the chord x in Fig. 6 is lengthening in terms of the radius a, the central angle θ, and the angular speed $\omega = \theta^{\cdot}$. (Recall that $\theta^{\cdot}$ denotes $d\theta/dt$.)

2   Express the rate at which the segment x in Fig. 7 is lengthening in terms of a, b, θ, and $\omega = \dot{\theta}$.

3   A sector is cut from a circular piece of paper. The remaining paper is formed into a cone by joining together the edges of the sector without overlap. Find the sector angle that maximizes the volume of the cone.

4   An 8-ft ladder leans against the top of a 4-ft fence. Find the largest horizontal distance the ladder can reach beyond the fence.

5   A weight hangs at the end of an 8-m rope, rigged up by the pulley system shown in Fig. 8. The weight will seek equilibrium as far as possible below the level of the fixed pulleys (in order to minimize potential energy). How far is that? Ignore the small pulley diameters.

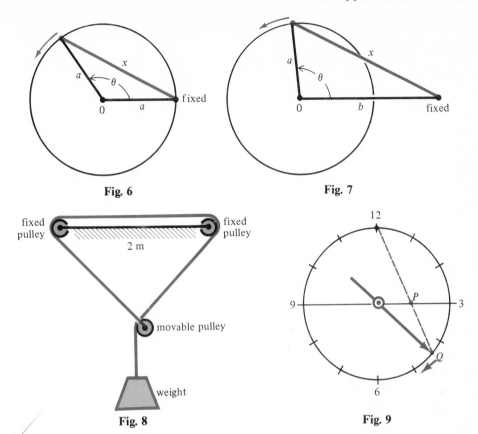

Fig. 6   Fig. 7

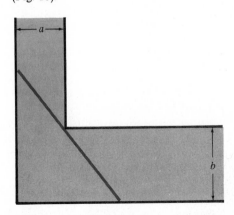

Fig. 8

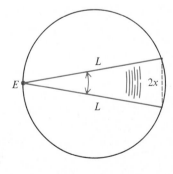

Fig. 9

6 A low-flying jet passes 450 ft directly over an observer on the ground. Shortly afterward its angle of elevation is 30° and decreasing at the rate of 20°/sec. Compute the plane's speed.

7 Suppose Fig. 9 represents an electric clock of radius 3 in. and that Q is the tip of its second hand. How fast is the point P moving at 20 sec past 3:47 p.m.?

8 Find the length of the longest pole that can be moved horizontally around the corner (Fig. 10).

Fig. 10   Fig. 11   Diameter unknown

9 (cont.) Suppose the pole can be tilted so its ends touch the floor and the ceiling, which is c ft from the floor. Now how long can the pole be?

10 In a tangent galvanometer (ammeter), current of I amps produces a deflection angle θ, where $I = k \tan \theta$. Compute the instrument's sensitivity,

$$S = \frac{I}{\theta} \bigg/ \frac{dI}{d\theta}.$$

Show that $S \approx 1$ for $\theta \approx 0$.

11 The hour hand of an electric clock has length a cm and the minute hand has length b cm. How fast are the tips of the hands separating at 3:00? How fast at 8:00?

12 To measure the diameter D of a circular hole (Fig. 11), a needle gauge, of length L, less than D, is rocked back and forth, with end E fixed. Let $2x$ denote the "rock". Express D in terms of x and show that the sensitivity of the gauge,

$$S = \frac{x}{D} \frac{dD}{dx} = \frac{x^2}{L^2 - x^2}.$$

(Therefore for x small, D is relatively insensitive to a relative change in x, so the gauge is highly accurate.)

13 If the target in Fig. 12 is 90 m from the range finder and running away at 11 m/sec, how fast is θ increasing in deg/sec?

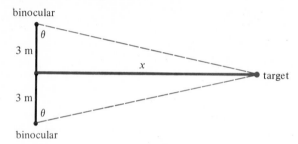

Fig. 12

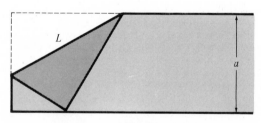

Fig. 13

14* One corner of a long rectangular strip of width a is folded over and just reaches the opposite edge (Fig. 13). Find the minimum length L of the crease.

15* (cont.) Find the minimum area of the heavily shaded triangle.

16* On p. 194, we proved the basic statement $(1 - \cos \theta)/\theta \longrightarrow 0$ as $\theta \longrightarrow 0$. Improve this proof to show that

$$\frac{1 - \cos \theta}{\theta^2} \longrightarrow \frac{1}{2}, \quad \text{hence} \quad \cos \theta \approx 1 - \frac{1}{2} \theta^2$$

for $\theta \approx 0$. Use this estimate to show that if a surveyor tilts his 12-ft vertical leveling rod as much as 2° out of plumb, then his error will be less than 0.01 ft.

9. MISCELLANEOUS EXERCISES

Differentiate with respect to x

1. $x^2 e^{-3/x}$
   2. $x^3 e^{-x}$
   3. x/e^x
   4. $e^{xu}, \quad u = e^x$.

5. Show that $e^x \geq 1 + x$ for all x.
   6. Find the maximum of $xe^{-\sqrt{x}}$.

Find

7. $\left. \dfrac{d}{dx} e^{-ex} \right|_{x=1/e}$
   8. $\displaystyle\lim_{t \to 0} \dfrac{e^{kt} - 1}{t}$
   9. $\displaystyle\lim_{x \to 0} \left(\dfrac{x}{e^x - 1} \right)^2$
   10. $\displaystyle\lim_{x \to 1} \dfrac{1 - x}{e^x - e}$.

11. Most pocket calculators handle numbers in the range $10^{-100} < x < 10^{100}$. How can you use a pocket calculator to estimate e^{1000} or $e^{1,000,000}$?

12. Show that $y = e^{-1/x}$ is strictly increasing for $x > 0$. Where is it convex and where concave?

13. Find the minimum of e^{-x/e^x} for $x \geq 0$.

14. Show that the derivative of $e^x p(x)$, where $p(x)$ is a polynomial, is a function of the same form.

15. Show, for $x \neq 0$, that

$$\frac{d^n}{dx^n} e^{-1/x} = \frac{P_n(x)e^{-1/x}}{x^{2n}},$$

where $P_n(x)$ is a polynomial of degree at most $n - 1$.

16*. Prove $\quad \dfrac{d^n}{dx^n}(x^{n-1}e^{1/x}) = (-1)^n \dfrac{e^{1/x}}{x^{n+1}}$.

17. In radiation theory, $f(x) = x^2 e^x/(e^x - 1)^2$ is known as an **Einstein function**. Show that $f(x) \longrightarrow 1$ as $x \longrightarrow 0+$, and $f(x) \longrightarrow 0+$ as $x \longrightarrow \infty$.

18*. (cont.) Show that $f(x)$ is strictly decreasing for $x > 0$.

19*. Define $\phi(x) = 0$ for $x \leq 0$ and $\phi(x) = e^{-1/x}$ for $x > 0$. Show that $\phi(x)$ has derivatives of all orders for every x. [*Hint* Use Ex. 15.]

20. (cont.) Graph $y = \psi(x) = e^2 \phi(1 + x)\phi(1 - x)$.

21. Suppose $a > 0$, $0 < c < b$, and $f(t) = t^a e^{-bt}$. Show that there is a constant $K > 0$ such that $f(t) \leq Ke^{-ct}$ for all $t \geq 0$.

22. Let $\alpha > 1$ and $y = \exp(-x^2)$ for $x > 0$. Find all x where the graph has an inflection point.

23. Assume that the population of a certain city grows at a rate proportional to the population itself. If the population was 100,000 in 1930 and 150,000 in 1970, predict what it will be in 1990.

24. The graph of $y = f(x)$ has the following property. The tangent line at each point (x, y) meets the x-axis at $(x - 1, 0)$. Find $f(x)$.

25. When the switch is closed in the circuit (Fig. 1) the current $I = I(t)$ satisfies the

conditions $\quad L\dfrac{dI}{dt} + RI = E, \quad I_0 = I(0) = 0. \quad$ Compute $\quad \dfrac{d}{dt}(e^{Rt/L}I)$.

Use the result to find a formula for I.

26. When the line from the boat (Fig. 2) is wrapped around the rough mooring post, friction causes the tension T in the rope, as a function of the length L of rope in contact with the post, to decrease at a rate proportional to itself. Precisely,

$$\frac{dT}{dL} = -\frac{\mu}{a} T, \quad \mu > 0.$$

Show that if n turns are taken, then a force of merely $T_0 e^{-2\pi n \mu}$ will hold the boat against the gale.

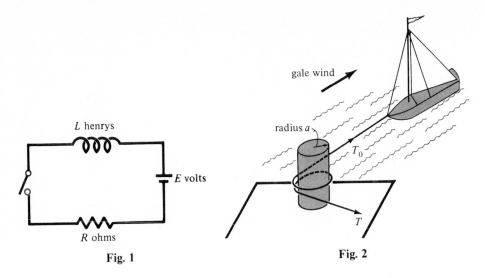

gale wind

L henrys

radius a

E volts

T_0

T

R ohms

Fig. 1   **Fig. 2**

Prove

27 $\cos 2x = \dfrac{1 - \tan^2 x}{1 + \tan^2 x}$ **28** $\cot x - \tan x = 2 \cot 2x$

29 $\sin \alpha - \sin \beta = 2 \sin \frac{1}{2}(\alpha - \beta) \cos \frac{1}{2}(\alpha + \beta)$

30 (cont.) $\cos \theta + \cos 3\theta + \cdots + \cos(2n - 1)\theta = \dfrac{\sin 2n\theta}{2 \sin \theta}$.

[*Hint* Find $2 \sin \theta \cos(2j - 1)\theta$ by Ex. 29.]

Find

31 $\displaystyle\lim_{x \to \infty} \dfrac{\cos x}{1 + 3x}$ **32** $\displaystyle\lim_{x \to \infty} \tan\left(\dfrac{\pi x^3}{2x^3 + 5}\right)$.

Find all local max and min

33 $y = \sin x \sin 2x$ **34** $y = e^x \sin x$.

35 A projectile is fired from the foot of a hill that rises at angle α with level ground. If the gun's elevation is θ, where $\alpha < \theta < \frac{1}{2}\pi$, and v_0 is the initial speed of the projectile, then the shell hits the hill at horizontal distance

$$x = \frac{v_0{}^2}{g} \left[\sin 2\theta - (\tan \alpha)(1 + \cos 2\theta)\right].$$

Find $x_{\max}$, and interpret geometrically the maximizing angle θ.

36 A line tangent to the unit circle at a point in the first quadrant meets the coordinate axes in points x and y. Minimize $x + y$.

37 A spring of length 1 ft is hung vertically. A weight attached to the free end stretches the spring 4 ft. If the weight is displaced 2 ft lower and released, then its distance (measured down from the ceiling) after t sec is $y = 5 + 2 \cos \omega t$, where $\omega^2 = g/4$ and $g = 32.2$ ft/sec^2. Describe the motion of the weight; give its velocity and acceleration.

38 A certain pendulum of length $2A$ swings out a circular arc when set into motion. If at time t the tip of the pendulum has horizontal position $x = A \sin 2\pi t$, describe this horizontal motion (shadow of the pendulum bob), giving velocity, acceleration, and position at critical values of t.

39 Suppose $x = x(t)$ satisfies the differential equation $\dot{x} = c \sin x$. If $y = 2x$, show that $\ddot{y} = c^2 \sin y$.

40 The power radiated in direction θ by an accelerated relativistic particle is

$$P = \frac{k \sin^2 \theta}{(1 - \beta \cos \theta)^5}, \qquad 0 < \beta < 1.$$

Find θ so P is maximal.

41 Find the maximum of $y = \frac{1}{2}x - \sin x$ for $0 \le x \le 4\pi$.

42 A balloon rises straight up from the ground at a constant rate of 5 ft/sec. At the instant it reaches an altitude of 100 ft, how fast is its angle of inclination changing as seen from the ground 100 ft from the point of release?

43 Prove $e^t(t - 1) + (1 - \frac{1}{2}t^2) > 0$ for all $t > 0$.

44 Prove $|\sin n\theta| \le |n \sin \theta|$.

Integration 5

1. THE AREA PROBLEM

The ancient Greek mathematician Archimedes used ingenious methods to compute the area bounded by a parabola and a chord (Fig. 1a). In this chapter we shall develop tools of Calculus that make the solution of this and similar problems routine. Although motivated originally by such area problems, these tools turn out to have a wide range of scientific and technical applications.

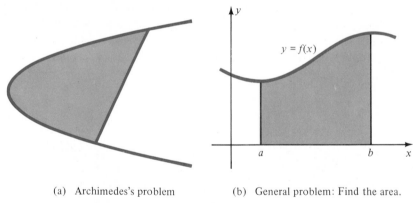

(a) Archimedes's problem   (b) General problem: Find the area.

Fig. 1

We start with a general problem: Find the area under the graph of a (non-negative) function $f(x)$ defined on an interval $a \le x \le b$. See Fig. 1b. Just as the slope problem led to the first major concept in calculus, the derivative, this area problem leads to the second major concept, the integral.

Part of the problem is the meaning of area. To be honest, we really know from elementary geometry only the areas of rectangles and figures derived from rectangles—triangles and other polygons. Therefore we need both a reasonable definition of area and, at the same time, a method for computing area.

Preliminaries   Most of the functions in this chapter will be continuous. Recall that $f(x)$ is a **continuous** function if

$$\lim_{x \to c} f(x) = f(c)$$

204

for each c in the domain of $f(x)$. Basic properties of continuous functions were given on pp. 98–100 and on pp. 146–147. In the discussion of maxima and minima, it was noted that a continuous function $f(x)$ on a closed interval $a \le x \le b$ is **bounded**, that is, there is a constant B such that $|f(x)| \le B$ for all x in the interval.

Intervals of the type $a \le x \le b$ occur frequently in this discussion, so we introduce a useful notation:

Closed Interval $[a, b]$ denotes the set of all x such that $a \le x \le b$.

When we write $[a, b]$, it is understood that $a < b$. The word "closed" means that the end points a and b are included, and rules out other kinds of intervals, like $a < x < b$ or $a \le x < b$, etc.

Analysis of the Problem   We return to Fig. 1b, assuming $f(x)$ is continuous on $[a, b]$. How shall we find the shaded area? Recall how we found the slope of a curve (and its definition). Assuming that the curve had a well-defined slope, we approximated what the slope ought to be (slope of secant), then found the limit of better and better approximations. This limit we *defined* to be the slope. We shall attack the area problem similarly.

Assume first that the shaded region in Fig. 1b really has a well-defined area. To approximate this area, slice the region into thin vertical strips (Fig. 2a) and approximate each strip by a rectangle (Fig. 2b). The sum of the areas of these rectangles ought to approximate the shaded area. The thinner the strips, the better the approximation should be.

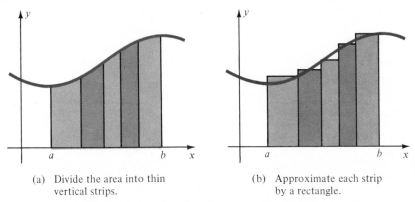

(a)   Divide the area into thin vertical strips.

(b)   Approximate each strip by a rectangle.

Fig. 2   Approximating the area under the curve

The base of a typical approximating rectangle is a subinterval $[c, d]$ of $[a, b]$; its height must be chosen between the highest and lowest values of $f(x)$ on $[c, d]$. See Fig. 3.

Now here is one of the most important facts in calculus: our approximations will converge to a limit. (We postpone the proof until Section 9.) This limit *defines* the area of the region under the graph of a continuous function. What we are saying is that there exists a number A, the limit value, with the following property: Every sum of areas of rectangles that can be produced by our approximation

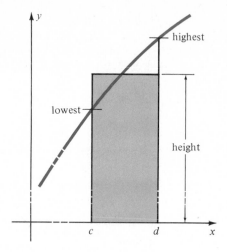

Fig. 3 The height of the approximating rectangle over $[c, d]$ is chosen between the highest and the lowest values of $f(x)$ on this subinterval.

process will be as close as we please to A, provided the widths of the approximating rectangle are *all* sufficiently small, that is, provided the thickest of these rectangles is sufficiently thin.

The limit value is usually written

$$\int_a^b f(x)\, dx$$

and read "the **integral** of f from a to b" or "the integral of $f(x)\, dx$ from a to b". The **integral sign** $\int$ is a distorted S, for sum, and it reminds us of the approximating sums of rectangles. The **limits of integration** a and b, **lower** and **upper**, respectively, remind us of the interval $[a, b]$. Of course $f(x)$ is the function whose area we want. Only dx appears unnatural; its usefulness will be explained later when we study change of variables in integrals. For the moment we shall think of dx as a reminder of the widths of the very thin approximating rectangles.

2. EXAMPLES OF INTEGRALS

We shall now compute several integrals. In each case we must find a computation process that leads to a numerical answer. This involves covering the region under $f(x)$ by thin strips, that is, decomposing $[a, b]$ into small subintervals, then choosing an appropriate height for each rectangle, then computing the area sum, and finally taking the limit. Some diabolically clever tricks, tricks that took centuries to discover, come into this work.

We shall need two formulas for sums. One is the formula

$$1 + 2 + \cdots + n = \tfrac{1}{2}n(n + 1)$$

for the sum of the first n positive integers. Recall its proof. For instance, if S_6

denotes $1 + 2 + \cdots + 6$, then

$$2S_6 = (1 + 2 + 3 + 4 + 5 + 6) + (6 + 5 + 4 + 3 + 2 + 1)$$
$$= (1 + 6) + (2 + 5) + (3 + 4) + (4 + 3) + (5 + 2) + (6 + 1)$$
$$= 7 + 7 + 7 + 7 + 7 + 7 = 6 \cdot 7.$$

Enough said! The other is the formula

$$1 + x + x^2 + \cdots + x^{n-1} = \frac{x^n - 1}{x - 1} \qquad (x \neq 1)$$

for the sum of a geometric progression. To prove it, multiply by $x - 1$ and watch the terms on the left-hand side telescope.

In summation notation, the two formulas are written

$$\sum_{j=1}^{n} j = \frac{1}{2}n(n+1), \qquad \sum_{j=0}^{n-1} x^j = \frac{x^n - 1}{x - 1} \qquad (x \neq 1).$$

■ **EXAMPLE 1** Find $\displaystyle\int_0^b cx\, dx.$

Solution The region under the graph of $y = cx$ is a triangle of base b and height cb. By elementary geometry the area is $\frac{1}{2}cb^2$. We shall show that the integration process leads to the same answer.

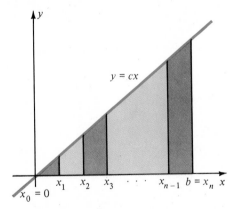 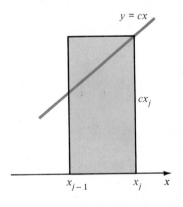

(a) Equal division into n strips of width b/n (b) Height of j-th rectangle is cx_j

Fig. 1 $\displaystyle\int_0^b cx\, dx$

Let us cover the region by n strips of *equal* width (Fig. 1a); later n will be taken larger and larger. The width of each strip is b/n. The points of subdivision of $[0, b]$ are

$$0 = x_0 < x_1 < x_2 < \cdots < x_{n-1} < x_n = b,$$

where $x_j = bj/n$. Now we look at the j-th rectangle. Its base is the subinterval

$[x_{j-1}, x_j]$, of width

$$x_j - x_{j-1} = \frac{bj}{n} - \frac{b(j-1)}{n} = \frac{b}{n},$$

as already noted. For its height, we choose the highest possible (Fig. 1b)

$$f(x_j) = cx_j = \frac{cbj}{n}.$$

Thus the area of the j-th rectangle is

$$\left(\frac{b}{n}\right)\left(\frac{cbj}{n}\right) = \frac{b^2c}{n^2}j.$$

We add the areas of these rectangles for $j = 1, 2, \cdots, n$:

$$\sum_{j=1}^{n} \frac{b^2c}{n^2}j = \frac{b^2c}{n^2}\sum_{j=1}^{n}j = \frac{b^2c}{n^2}\frac{n(n+1)}{2} = \frac{1}{2}b^2c\left(\frac{n+1}{n}\right).$$

The integral is the limit of this area sum as the widths approach 0, that is, as $1/n \longrightarrow 0$. Since

$$\frac{n+1}{n} = 1 + \frac{1}{n},$$

we can set $t = 1/n$ and write

$$\int_0^b cx \, dx = \lim_{t \to 0+}\left[\frac{1}{2}b^2c\frac{n+1}{n}\right] = \frac{1}{2}b^2c \lim_{t \to 0+}(1+t).$$

Clearly $1 + t \longrightarrow 1$ as $t \longrightarrow 0$, so our final result is

$$\int_0^b cx \, dx = \tfrac{1}{2}b^2c. \qquad \blacksquare$$

Remark   In working Example 1, we made two choices. We chose rectangles of equal widths, and for each rectangle, the greatest allowable height. We did so not because of any requirement, but to make the computation as easy as possible. The only requirement is that the widths all approach 0. Sometimes a clever choice of unequal widths can pay off, as the next example shows.

■ **EXAMPLE 2**   Find $\displaystyle\int_a^b x^3 \, dx$ for $0 < a < b$.

Solution   Divide the interval $[a, b]$ into n pieces by the points of a geometric progression:

This requires that $b = ar^n$, so the common ratio r of the progression must be $r = (b/a)^{1/n}$. Later we shall let $r \longrightarrow 1+$, or equivalently, $n \longrightarrow \infty$. Note that the *widest* strip (the one at the extreme right) has width

$$ar^n - ar^{n-1} = ar^n\left(1 - \frac{1}{r}\right) = b\left(1 - \frac{1}{r}\right),$$

which approaches 0 as $r \longrightarrow 1+$.

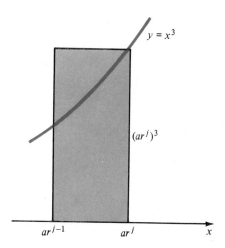

Fig. 2   Approximating rectangle for $\displaystyle\int_a^b x^3\,dx$

On the j-th strip construct a rectangle (Fig. 2). Its width is $ar^j - ar^{j-1}$; for its height choose $(ar^j)^3$. Thus the area of the j-th rectangle is

$$\left(ar^j - ar^{j-1}\right)\left(ar^j\right)^3 = a^4\left(r^{4j} - r^{4j-1}\right).$$

Hence the approximating sum is

$$A_n = \sum_{j=1}^{n} a^4\left(r^{4j} - r^{4j-1}\right) = a^4 \sum_{j=1}^{n}\left(r^{4j} - r^{4j-1}\right).$$

Now $r^{4j} - r^{4j-1} = r^4 r^{4(j-1)} - r^3 r^{4(j-1)} = \left(r^4 - r^3\right)r^{4(j-1)} = r^3(r-1)r^{4(j-1)},$

and $j - 1$ varies from 0 to $n - 1$ as j goes from 1 to n. Hence

$$A_n = a^4 r^3(r-1) \sum_{j=0}^{n-1} r^{4j} = a^4 r^3(r-1)\frac{r^{4n} - 1}{r^4 - 1}.$$

But $a^4\left(r^{4n} - 1\right) = (ar^n)^4 - a^4 = b^4 - a^4$,   so $A_n = (b^4 - a^4)r^3\dfrac{r-1}{r^4 - 1}.$

The integral is the limit of these approximating sums as n increases, that is, as $r \longrightarrow 1$. Hence

$$\int_a^b x^3\,dx = (b^4 - a^4)\lim_{r\to 1}\left(r^3\,\frac{r-1}{r^4 - 1}\right).$$

Clearly $\lim r^3 = 1$. The remaining limit can be evaluated as follows:

$$\lim_{r \to 1+} \left(\frac{r-1}{r^4-1}\right) = \frac{1}{\lim_{r \to 1+}\left(\frac{r^4-1}{r-1}\right)} = \frac{1}{\left.\frac{d}{dr}r^4\right|_{r=1}} = \frac{1}{4}.$$

Therefore

$$\int_a^b x^3 \, dx = \tfrac{1}{4}(b^4 - a^4).$$ ∎

Remark   What happens when $a = 0$? The method doesn't quite work, but from the answer, a reasonable (and correct) guess is

$$\int_0^b x^3 \, dx = \tfrac{1}{4}b^4.$$

We take this for granted now since another derivation is coming soon.

■ **EXAMPLE 3**   Find $\displaystyle\int_a^b e^x \, dx$.

Solution   Divide $[a, b]$ into n subintervals (Fig. 3) by equally spaced division points

$$x_j = a + \frac{b-a}{n}j, \qquad 0 \le j \le n.$$

Clearly $x_0 = a$, $x_n = b$, and $x_j - x_{j-1} = (b-a)/n$. Choose e^{x_j} for the height of the j-th rectangle. Because the typesetter of this page hates complicated exponents, write $e^x = \exp x$ for the exponential function. Then the height is

$$e^{x_j} = \exp x_j = \exp\left(a + \frac{b-a}{n}j\right) = (\exp a)\left[\exp\left(\frac{b-a}{n}j\right)\right].$$

Thus, the j-th rectangle has area

$$\frac{b-a}{n}e^a \exp\left(\frac{b-a}{n}j\right),$$

so the approximating sum is

$$A_n = \frac{b-a}{n}e^a \sum_{j=1}^{n} \exp\left(\frac{b-a}{n}j\right).$$

This sum is a disguised geometric progression. In fact, if $r = \exp[(b-a)/n]$, then

$$\sum_{j=1}^{n} \exp\left(\frac{b-a}{n}j\right) = \sum_{j=1}^{n} r^j = r\sum_{j=0}^{n-1} r^j = r\frac{r^n-1}{r-1},$$

so

$$A_n = \frac{b-a}{n} \cdot e^a r \frac{r^n - 1}{r-1}.$$

But

$$e^a(r^n - 1) = e^a(e^{b-a} - 1) = e^b - e^a,$$

so

$$A_n = (e^b - e^a)(b-a)\frac{r}{n(r-1)}.$$

Clearly $r = \exp[(b - a)/n] \longrightarrow 1$ as $1/n \longrightarrow 0$, that is, as $n \longrightarrow \infty$. Therefore,

$$\int_a^b e^x \, dx = (e^b - e^a)(b - a) \lim_{n \to \infty} \frac{1}{n\{\exp[(b - a)/n] - 1\}}.$$

The limit is easier than it looks. Set $t = 1/n$. Then

$$\lim_{n \to \infty} \frac{1}{n\{\exp[(b - a)/n] - 1\}} = \lim_{t \to 0+} \frac{t}{\exp[(b - a)t] - 1} = \left[\lim_{t \to 0+} \frac{e^{(b - a)t} - 1}{t} \right]^{-1}$$

$$= \left[\frac{d}{dt} e^{(b - a)t} \Big|_{t = 0} \right]^{-1} = (b - a)^{-1}.$$

Consequently, $\displaystyle\int_a^b e^x \, dx = (e^b - e^a)(b - a)(b - a)^{-1} = e^b - e^a,$

a surprisingly simple answer indeed. ∎

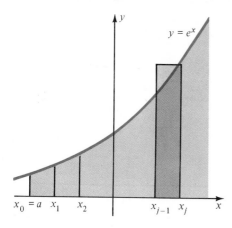

Fig. 3 $\displaystyle\int_a^b e^x \, dx$; equal subintervals

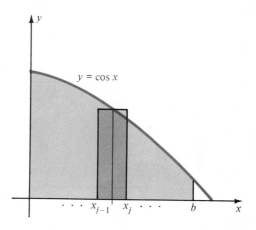

Fig. 4 $\displaystyle\int_0^b \cos x \, dx$;

equal subintervals, height at midpoint

For the next example we shall need a trigonometric identity:

$$\cos \theta + \cos 3\theta + \cdots + \cos(2n - 1)\theta = \frac{\sin 2n\theta}{2 \sin \theta}.$$

See Exercise 30, p. 202.

■ **EXAMPLE 4** Find $\displaystyle\int_0^b \cos x \, dx$ for $0 < b \leq \frac{1}{2}\pi$.

Solution Divide $[0, b]$ into n equal parts by the points

$$0 < \frac{b}{n} < \frac{2b}{n} < \cdots < \frac{(n - 1)b}{n} < \frac{nb}{n} = b.$$

On the j-th subinterval $[(j - 1)b/n, jb/n]$ construct a rectangle, computing its height at the *midpoint* of the interval (Fig. 4). The midpoint is $(j - \frac{1}{2})b/n = (2j - 1)b/2n$, the width is b/n, so the area of the j-th rectangle is

$$\frac{b}{n} \cos(2j - 1) \frac{b}{2n}.$$

Add these areas for $j = 1, 2, \cdots, n$ to obtain the area sum:

$$A_n = \frac{b}{n} \sum_{j=1}^{n} \cos(2j - 1) \frac{b}{2n} = \frac{b}{n} \frac{\sin 2n(b/2n)}{2 \sin(b/2n)} = \frac{b \sin b}{2n \sin(b/2n)}.$$

Therefore

$$\int_0^b \cos x \, dx = \lim_{n \to \infty} \frac{b \sin b}{2n \sin(b/2n)}.$$

Set $t = b/2n$ so $t \longrightarrow 0+$ as $n \longrightarrow \infty$. Then the fraction can be rewritten as

$$\frac{\sin b}{\left(\sin \dfrac{b}{2n}\right) / \left(\dfrac{b}{2n}\right)} = (\sin b) \left(\frac{\sin t}{t}\right)^{-1}.$$

But $\lim\limits_{t \to 0} \dfrac{\sin t}{t} = \dfrac{d}{dt} \sin t \Big|_{t=0} = \cos 0 = 1;$ hence $\displaystyle\int_0^b \cos x \, dx = \sin b.$ ∎

Discussion   Let us summarize the results of our four examples; they are all useful formulas:

$$\int_0^b cx \, dx = \tfrac{1}{2}b^2 c, \quad 0 < b, \qquad \int_a^b x^3 \, dx = \tfrac{1}{4}(b^4 - a^4), \quad 0 \le a < b,$$

$$\int_a^b e^x \, dx = e^b - e^a, \quad a < b, \qquad \int_0^b \cos x \, dx = \sin b, \quad 0 < b \le \tfrac{1}{2}\pi.$$

Their derivations involved some cunning tricks. We can hardly go through life expecting to find such tricks every time we have to integrate a function. Obviously we need a systematic method.

A curious thing occurred in Examples 2–4. The final step, leading from the area sum A_n to the integral, involved a certain limit. In each case, we could recognize the limit as a derivative! Apparently there is some subtle relation between integrals and derivatives, between the area problem and the slope problem.

EXERCISES

1   Find $\displaystyle\int_a^b x \, dx, \quad 0 < a < b.$ Use equal intervals and the left end points.

2   (cont.) Find the same integral, using equal intervals and the midpoints.

3   (cont.) Find the same integral, using division points in a geometric progression.

4   Find $\displaystyle\int_a^b x^2 \, dx, \quad 0 < a < b.$ [*Hint*   Geometric progression.]

5   (cont.) Find the same integral for $a < b < 0$.

6* Find $\displaystyle\int_a^b x^p\,dx$ for $p \neq -1,\quad 0 < a < b.$

7* Show that $\displaystyle\int_a^b \frac{1}{x}\,dx = \frac{d}{dt}\left(\frac{b}{a}\right)^t\bigg|_0$ $(0 < a < b).$

8 Find $\displaystyle\int_a^b e^{kx}\,dx$ $(a < b).$

9 Use your knowledge of the integral of cosine plus a geometric argument based on symmetry to find

$$\int_0^b \sin x\,dx, \qquad 0 < b \leq \pi.$$

10 Find $\displaystyle\int_a^b x^2\,dx$ for $0 < a < b,$ using *equal* divisions.

[*Hint* $1^2 + 2^2 + \cdots + n^2 = \frac{1}{6}n(n + 1)(2n + 1).$]

11 Find $\displaystyle\int_a^b x^3\,dx$ for $0 < a < b,$ using *equal* divisions.

[*Hint* $1^3 + 2^3 + \cdots + n^3 = \frac{1}{4}n^2(n + 1)^2.$]

12 Find $\displaystyle\int_1^2 5x^3\,dx.$

3. THE DEFINITE INTEGRAL AND
THE FUNDAMENTAL THEOREM

The process we discussed for computing the area under the graph of a positive continuous function applies to many questions, not just problems of area, and to a broad class of functions, not just positive continuous functions. Let us examine the process carefully.

We shall deal with a function $f(x)$ whose domain is a closed interval $[a, b]$. We assume that $f(x)$ is *bounded*, but assume nothing more, not even that $f(x)$ is continuous. Boundedness means, that $A \leq f(x) \leq B$ for some constants A and B and all x on the interval $[a, b]$.

Partition of the Interval   We shall be splitting the interval $[a, b]$ into subintervals. A **partition** Π of $[a, b]$ into n subintervals consists of an increasing sequence

$$a = x_0 < x_1 < x_2 < \cdots < x_{n-1} < x_n = b$$

of points of $[a, b]$, indexed as written. The **subintervals** of the partition are the closed intervals

$$[x_0, x_1],\quad [x_1, x_2], \cdots, [x_{n-2}, x_{n-1}],\quad [x_{n-1}, x_n].$$

Some partitions split $[a, b]$ into subintervals all of which are small; some do not. As a measure of the coarseness or fineness of a partition Π, we define its **mesh**:

$$\text{mesh}(\Pi) = \max\{x_1 - x_0, x_2 - x_1, \cdots, x_n - x_{n-1}\}.$$

A partition is fine if its mesh is small. For instance, $\text{mesh}(\Pi) = 0.001$ means that Π splits $[a, b]$ into subintervals, the largest of which has width 0.001.

Example   Let Π be the partition

$$0 < \tfrac{1}{2} < \tfrac{5}{6} < \tfrac{5}{4} < \tfrac{7}{4} < \tfrac{5}{2} < 3$$

of the interval $[0, 3]$. Then

$$\text{mesh}(\Pi) = \max\{\tfrac{1}{2} - 0, \tfrac{5}{6} - \tfrac{1}{2}, \tfrac{5}{4} - \tfrac{5}{6}, \tfrac{7}{4} - \tfrac{5}{4}, \tfrac{5}{2} - \tfrac{7}{4}, 3 - \tfrac{5}{2}\} = \max\{\tfrac{1}{2}, \tfrac{1}{3}, \tfrac{5}{12}, \tfrac{1}{2}, \tfrac{3}{4}, \tfrac{1}{2}\} = \tfrac{3}{4}.$$

Approximating Sums   Let us return to our bounded function $f(x)$ on $[a, b]$. Let Π be any partition of $[a, b]$. In each subinterval $[x_{j-1}, x_j]$ of Π, we choose a point $\bar{x}_j$. See Fig. 1. The choice of $\bar{x}_j$ is perfectly arbitrary. We use the symbol $\bar{x}$ to denote the whole choice of the points $\bar{x}_1, \bar{x}_2, \cdots, \bar{x}_n$.

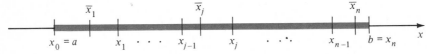

Fig. 1   Arbitrary choice of $\bar{x}_j$ in $[x_{j-1}, x_j]$

Now we evaluate $f(x)$ at $\bar{x}_j$ and multiply by $x_j - x_{j-1}$, the width of the j-th subinterval. (This imitates computing the area of the j-th rectangle.) We add these products for $j = 1, 2, \cdots, n$ and call the sum an **approximating sum** or a **Riemann sum** for f on $[a, b]$.

Approximating Sum

$$S(f, \Pi, \bar{x}) = f(\bar{x}_1)(x_1 - x_0) + f(\bar{x}_2)(x_2 - x_1) + \cdots + f(\bar{x}_n)(x_n - x_{n-1})$$

$$= \sum_{j=1}^{n} f(\bar{x}_j)(x_j - x_{j-1}).$$

The notation $S(f, \Pi, \bar{x})$ emphasizes that an approximating sum depends on (1) the function f, (2) the partition Π, and (3) the choice $\bar{x}$ of the points $\bar{x}_j$.

We are going to define the integral as a kind of limit of approximating sums. But first we note three properties of approximating sums because they imply corresponding properties of integrals.

Properties of Approximating Sums

   (1) If c is a constant, then $S(cf, \Pi, \bar{x}) = cS(f, \Pi, \bar{x})$.

   (2) $S(f + g, \Pi, \bar{x}) = S(f, \Pi, \bar{x}) + S(g, \Pi, \bar{x})$.

   (3) If $f(x) \geq 0$ on $[a, b]$,   then $S(f, \Pi, \bar{x}) \geq 0$.

The easy proofs are left as exercises.

The Integral   Suppose $f(x)$ is a function whose approximating sums $S(f, \Pi, \bar{x})$ approach a definite number L as the partitions get finer and finer, that is, as $\text{mesh}(\Pi)$ approaches 0. Such a function is called integrable, and the number L is called its definite integral (or just "integral") on $[a, b]$.

Definite Integral   A bounded function $f(x)$ on $[a, b]$ is called **integrable on** $[a, b]$ if there is a number L such that

$$S(f, \Pi, \bar{x}) \longrightarrow L \qquad \text{as} \quad \text{mesh}(\Pi) \longrightarrow 0.$$

Precisely, if $\varepsilon > 0$, there exists $\delta > 0$ such that

$$|S(f, \Pi, \bar{x}) - L| < \varepsilon \qquad \text{whenever} \quad \text{mesh}(\Pi) < \delta.$$

Then L is called the **integral** of $f(x)$ on $[a, b]$, and we write

$$L = \int_a^b f(x)\,dx = \lim_{\text{mesh}(\Pi) \to 0} S(f, \Pi, \bar{x}).$$

The definite integral is a complicated kind of limit, so complicated that it is not at all obvious whether a given function is integrable or not. About the only thing that is obvious is that constant functions are integrable:

$$\int_a^b c\,dx = c(b - a).$$

This is clear, because for each partition the approximating sum is the same:

$$S(c, \Pi, \bar{x}) = \sum_{j=1}^n c(x_j - x_{j-1}) = c \sum_{j=1}^n (x_j - x_{j-1}) = c(x_n - x_0) = c(b - a).$$

The sum telescopes.

Continuous Functions   The definition of integral is worthless unless the integral exists for a large class of useful functions. A basic theorem guarantees the existence of the integral for the important class of continuous functions. The proof will be discussed in Section 9.

Continuous Functions   If $f(x)$ is continuous on $[a, b]$, then $f(x)$ is integrable.

Thus we can be assured that the integrals

$$\int_1^5 (x^2 - 8x^3 + 2)\,dx, \quad \int_{3.1}^{7.5} \sqrt{x}\,dx, \quad \int_{-4}^1 \frac{x - 2}{x^2 + 1}\,dx, \quad \int_{-2}^0 e^{-x} \cos x\,dx$$

exist since each function being integrated is continuous on the interval of integration.

Terminology   The awkward phrase "function being integrated" is usually replaced by **integrand**.

Piecewise Continuous Functions   We hinted earlier that the integration process might apply to some discontinuous functions. It does, to bounded functions that are continuous *except at a finite number of points*. Such functions are called **piecewise continuous** (Fig. 2).

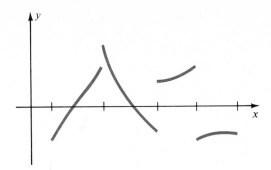

Fig. 2   Example of a piecewise continuous function

> **Theorem** If $f(x)$ is a bounded piecewise continuous function on $[a, b]$, then $f(x)$ is integrable.

Thus if

$$f(x) = \begin{cases} 1, & x < 0 \\ 4, & 0 \le x \le 3 \\ x^2, & x > 3, \end{cases}$$

then $f(x)$ is integrable on $[-10, 6]$.

Basic Properties In working with definite integrals, we frequently use the following properties.

> **Properties of Integrals**   (1) $\displaystyle\int_a^b cf(x)\,dx = c\int_a^b f(x)\,dx.$
>
> (2) $\displaystyle\int_a^b [f(x) + g(x)]\,dx = \int_a^b f(x)\,dx + \int_a^b g(x)\,dx.$
>
> (3)   If $f(x) \ge 0$ on $[a, b]$,   then $\displaystyle\int_a^b f(x)\,dx \ge 0.$
>
> (4)   If $a < c < b$,   then $\displaystyle\int_a^b f(x)\,dx = \int_a^c f(x)\,dx + \int_c^b f(x)\,dx.$

Properties (1), (2), and (3) follow easily from the corresponding properties of approximating sums (p. 214). We leave the proofs as exercises. A detailed proof of Property (4) is rather technical, and we shall omit it. The idea is to lump together a partition of $[a, c]$ with a partition of $[c, b]$, making a partition of $[a, b]$. Then corresponding approximating sums for

$$\int_a^c f(x)\,dx \qquad \text{and} \qquad \int_c^b f(x)\,dx$$

add up to an approximating sum for $\displaystyle\int_a^b f(x)\,dx.$

Geometrically, the result is reasonable for positive functions (Fig. 3).

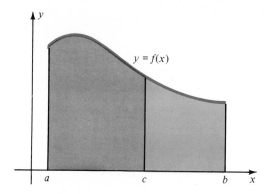

Fig. 3   The areas add up, so

$$\int_a^b f(x)\, dx = \int_a^c f(x)\, dx + \int_c^b f(x)\, dx.$$

EXAMPLE 1   Find $I = \displaystyle\int_0^{\pi/2} (3 \cos x - e^x + 8x^3)\, dx.$

Solution   By (1) and (2),

$$I = 3 \int_0^{\pi/2} \cos x\, dx - \int_0^{\pi/2} e^x\, dx + 8 \int_0^{\pi/2} x^3\, dx$$

$$= 3 \sin \tfrac{1}{2}\pi - (e^{\pi/2} - e^0) + 8(\tfrac{1}{4})(\tfrac{1}{2}\pi)^4 = 3 - e^{\pi/2} + 1 + \tfrac{1}{8}\pi^4 = 4 - e^{\pi/2} + \tfrac{1}{8}\pi^4. \quad \blacksquare$$

We have defined
$$\int_a^b f(x)\, dx$$

for $a < b$. Now we extend the definition to the cases $a = b$ and $a > b$. Property (4) in the box suggests how to proceed. Suppose (4) were true without any restrictions on a, b, c. For instance if $a = c < b$, then

$$\int_a^b f(x)\, dx = \int_a^a f(x)\, dx + \int_a^b f(x)\, dx,$$

so the first integral on the right must be 0. Now if $a = b < c$, then

$$0 = \int_a^a f(x)\, dx = \int_a^c f(x)\, dx + \int_c^a f(x)\, dx,$$

so the integrals on the right are negatives of each other.

Definitions $\displaystyle\int_a^a f(x)\, dx = 0, \qquad \int_a^b f(x)\, dx = -\int_b^a f(x)\, dx \quad \text{if } a > b.$

With these definitions, property (4) holds without any restriction on the order of a, b, and c:

Suppose $f(x)$ is integrable on a closed interval and a, b, c are any three points of this interval, not necessarily distinct. Then

$$\int_a^c f(x)\,dx = \int_a^b f(x)\,dx + \int_b^c f(x)\,dx.$$

We leave it for the reader to test the various possibilities.

The Fundamental Theorem of Calculus   So far in this section we have had a rather complicated definition of the integral and some of its formal properties. The definition is hardly an operational one; it does not give a practical method for integrating functions and getting answers. In the last section we needed a whole bag of tricks when working with approximating sums, so direct frontal attacks seem helplessly hard to use regularly. We did have a hint that integrals are somehow related to derivatives. Might that help? Where do we go from here?

If we don't see how to proceed, that's not surprising since it took mathematicians about 2000 years to find the right techniques. The ancient Greeks, particularly Archimedes, solved the area problem in a few special cases, always by ingenious geometric tricks. Nothing much happened then until after the Renaissance, when the types of tricks we used in Section 2 were discovered by Fermat and others.

The breakthrough came with the idea of changing the problem from a static one to a dynamic one. Instead of computing the area between two fixed lines, compute it between a fixed line and a second *moving* line. That is the key idea. At first, suppose $f(t)$ is continuous and $f(t) > 0$. Denote by $A(x)$ the area under the curve $y = f(t)$ between $t = a$ and $t = x$. See Fig. 4. Then

$$A(x) = \int_a^x f(t)\,dt.$$

Clearly $A(x)$ is a function of x. For instance, if $f(t) = t^3$ and $a = 0$, then

$$A(x) = \int_0^x t^3\,dt = \frac{x^4}{4}.$$

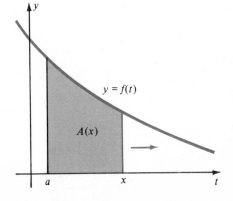

$y = f(t)$

$A(x)$

Fig. 4   Dynamics of area

Our aim is to find an explicit formula for $A(x)$. For this purpose, we study how $A(x)$ varies as x varies. Figure 4 shows that as x increases the area builds up, so $A(x)$ increases. But how fast? What change in $A(x)$ results from a small change h in x? Figure 5 provides a clue. When $f(x)$ is large, the corresponding change in $A(x)$ is large; when $f(x)$ is small, the change is small. Apparently the rate of change of $A(x)$ relative to x is very much like the value of $f(x)$. In the language of calculus, we suspect that the derivative dA/dx is proportional to $f(x)$. Let us test our hunch in the case of $f(t) = t^3$ and $a = 0$. Then

$$A(x) = \frac{x^4}{4}; \qquad \text{hence} \quad \frac{dA}{dx} = x^3.$$

Result: dA/dx is not only *proportional* to $f(x)$, but dA/dx is actually *equal* to $f(x)$.

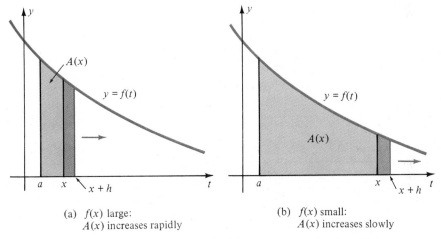

(a) $f(x)$ large:
      $A(x)$ increases rapidly

(b) $f(x)$ small:
      $A(x)$ increases slowly

Fig. 5   The rate of growth of $A(x)$ seems proportional to $f(x)$.

We have hit upon the crucial fact: that $dA/dx = f(x)$ in general. Let us sketch a proof. By definition,

$$\frac{dA}{dx} = \lim_{h \to 0} \frac{A(x + h) - A(x)}{h}.$$

Now $A(x + h)$ is the area under $y = f(t)$ between a and $x + h$, and $A(x)$ is the area between a and x. Hence $A(x + h) - A(x)$ is the area between x and $x + h$. (See the dark shaded region in Fig. 6a.) This area is approximately that of a rectangle of base h and height $f(x)$ as in Fig. 6b. Hence for small values of h,

$$A(x + h) - A(x) \approx hf(x), \qquad \text{that is,} \quad \frac{A(x + h) - A(x)}{h} \approx f(x).$$

Furthermore, Fig. 6 suggests that these approximations improve as $h \longrightarrow 0$. Our rough argument indicates that

$$\frac{dA}{dx} = \lim_{h \to 0} \frac{A(x + h) - A(x)}{h} = f(x).$$

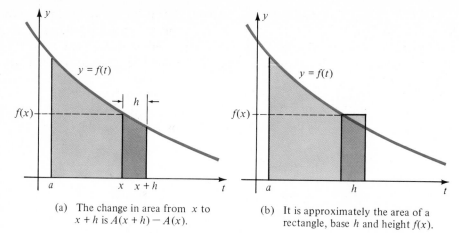

(a) The change in area from x to
     $x + h$ is $A(x + h) - A(x)$.

(b) It is approximately the area of a
     rectangle, base h and height $f(x)$.

Fig. 6

We shall show in Section 9 that this result holds for arbitrary continuous functions, not only positive ones. It is probably the most important theorem in all of calculus.

Fundamental Theorem of Calculus   Let $f(t)$ be continuous on the interval $[a, b]$. For each x in $[a, b]$, let

$$A(x) = \int_a^x f(t)\, dt.$$

Then $A(x)$ is differentiable and

$$\frac{dA}{dx} = \frac{d}{dx} \int_a^x f(t)\, dt = f(x).$$

Thus $A(x)$ is an antiderivative of $f(x)$.

We have almost achieved our goal of finding a formula for $A(x)$. We know that $A(x)$ is one of the antiderivatives of $f(x)$, but which one? Well, since they all differ by constants, it shouldn't be hard to sort out. Let $F(x)$ be *any* antiderivative of $f(x)$. Then $A(x) = F(x) + c$. To find c, note that the area is 0 where we start, that is, at $x = a$:

$$0 = A(a) = F(a) + c, \qquad c = -F(a).$$

Therefore $A(x) = F(x) - F(a)$, for *every* antiderivative $F(x)$ of $f(x)$.

In particular, $A(b) = F(b) - F(a)$. This gives us a supremely practical alternative version of the Fundamental Theorem:

Evaluation Rule   Let $f(x)$ be continuous on $[a, b]$ and let $F(x)$ be any antiderivative of $f(x)$. Then

$$\int_a^b f(x)\, dx = F(b) - F(a) = F(x)\Big|_a^b.$$

We have reached the end of a chain of thought that began by thinking of the definite integral as a function of its upper limit. The evaluation rule transforms a difficult problem, evaluating integrals, into a much easier one, finding antiderivatives. To compute $\int_a^b f(x)\, dx$, beg, borrow, or steal an antiderivative $F(x)$; then evaluate $F(b) - F(a)$.

For instance, in Section 2 we laboriously churned out the formulas

$$\int_a^b e^x\, dx = e^b - e^a \qquad \text{and} \qquad \int_0^b \cos x\, dx = \sin b.$$

Now they are obvious!—consequences of

$$e^x = \frac{d}{dx} e^x \qquad \text{and} \qquad \cos x = \frac{d}{dx} \sin x.$$

Because we know many derivatives, we know many antiderivatives, so many integrals are now within our grasp.

■ **EXAMPLE 2** Evaluate (a) $\displaystyle\int_1^2 x^5\, dx$ (b) $\displaystyle\int_3^5 \frac{dx}{x^2}$ (c) $\displaystyle\int_{\pi/4}^{\pi/3} \sec^2 x\, dx.$

Solution (a) $x^5 = \dfrac{d}{dx}\left(\dfrac{1}{6} x^6\right);$ hence $\displaystyle\int_1^2 x^5\, dx = \frac{1}{6} x^6 \Big|_1^2 = \frac{1}{6}(2^6 - 1^6) = \frac{63}{6} = \frac{21}{2}.$

(b) $\dfrac{1}{x^2} = \dfrac{d}{dx}\left(-\dfrac{1}{x}\right),$ hence $\displaystyle\int_3^5 \frac{dx}{x^2} = -\frac{1}{x} \Big|_3^5 = -\frac{1}{5} - \left(-\frac{1}{3}\right) = \frac{2}{15}.$

(c) $\sec^2 x = \dfrac{d}{dx} \tan x;$ hence

$$\int_{\pi/4}^{\pi/3} \sec^2 x\, dx = \tan x \Big|_{\pi/4}^{\pi/3} = \tan \frac{\pi}{3} - \tan \frac{\pi}{4} = \sqrt{3} - 1. \qquad ■$$

■ **EXAMPLE 3** Find $\displaystyle\int_{-2}^2 |x + 1|\, dx.$

Solution $|x + 1| = \begin{cases} -(x + 1) \\ x + 1 \end{cases} \quad \text{for} \quad \begin{matrix} x \le -1 \\ x \ge -1, \end{matrix}$

so we decompose the integral into two parts and deal with each separately. We know that

$$x + 1 = \frac{d}{dx} \frac{1}{2} (x + 1)^2,$$

so $\displaystyle\int_{-2}^2 |x + 1|\, dx$

$$= \int_{-2}^{-1} |x + 1|\, dx + \int_{-1}^2 |x + 1|\, dx = \int_{-2}^{-1} -(x + 1)\, dx + \int_{-1}^2 (x + 1)\, dx$$

$$= -\frac{1}{2}(x + 1)^2 \Big|_{-2}^{-1} + \frac{1}{2}(x + 1)^2 \Big|_{-1}^2 = -\frac{1}{2}(0 - 1) + \frac{1}{2}(9 - 0) = 5. \qquad ■$$

The next example shows another way to use the Fundamental Theorem.

■ **EXAMPLE 4**   Find $\dfrac{d}{dx} \displaystyle\int_0^{3x^2} te^{4t}\, dt$.

Solution   Set $F(u) = \displaystyle\int_0^u te^{4t}\, dt$.   We want $\dfrac{d}{dx} F(3x^2)$.

By the Chain Rule (with $u = 3x^2$) and the Fundamental Theorem,

$$\frac{d}{dx} F(u) = \frac{dF}{du} \cdot \frac{du}{dx} = \frac{d}{du}\left[\int_0^u te^{4t}\, dt\right]\left[\frac{d}{dx}(3x^2)\right]$$

$$= (ue^{4u})(6x) = (3x^2 e^{12x^2})(6x) = 18x^3 e^{12x^2}. \qquad ■$$

Remark   The word **primitive** is sometimes used as a synonym for antiderivative, especially in connection with integration.

Dummy Variable   The integral

$$\int_a^b f(x)\, dx$$

depends on three things: its lower limit a, its upper limit b, and its integrand f. We might write

$$\int_a^b f(x)\, dx = I(a, b; f)$$

to display this dependence. The integral does *not* depend on x; for that reason x is sometimes called a **dummy variable**. Thus

$$\int_a^b f(x)\, dx = \int_a^b f(y)\, dy = \int_a^b f(t)\, dt = \text{etc.}$$

EXERCISES

Evaluate

1   $\displaystyle\int_{-1}^{2} x\, dx$

2   $\displaystyle\int_{0}^{2} (1 - x)\, dx$

3   $\displaystyle\int_{0}^{\pi} \sin 2x\, dx$

4   $\displaystyle\int_{\pi/2}^{3\pi/2} \cos 3x\, dx$

5   $\displaystyle\int_{-2}^{-1} \frac{1}{x^2}\, dx$

6   $\displaystyle\int_{-4}^{-2} \left(\frac{1}{x^2} + x\right) dx$

7   $\displaystyle\int_{0}^{3\pi} (\cos x + \sin x)\, dx$

8   $\displaystyle\int_{0}^{\pi} (3\sin 3x + 2\cos 2x)\, dx$

9   $\displaystyle\int_{-2}^{2} (3 + 2x - x^2)\, dx$

10   $\displaystyle\int_{-1}^{2} (x - 1)(3x + 1)\, dx$

11   $\displaystyle\int_{1}^{2} \left(\frac{5}{x^2} + e^{1-x}\right) dx$

12   $\displaystyle\int_{0}^{1} (e^{3x} - x^2)\, dx$

13   $\displaystyle\int_{a}^{b} (b - x)(x - a)\, dx$

14   $\displaystyle\int_{-a}^{a} (x - a)^2\, dx$

15   $\displaystyle\int_{-1}^{-2} 5t^3\, dt$

16   $\displaystyle\int_{1}^{-1} -t^4\, dt$

17 $\displaystyle\int_0^1 (4u^3 - 3u^2)\, du$ **18** $\displaystyle\int_1^1 \sin(u^2)\, du$ **19** $\displaystyle\int_0^{\pi/4} \sec^2\theta\, d\theta$

20 $\displaystyle\int_{\pi/6}^{3\pi/4} \csc\theta \cot\theta\, d\theta.$

Find dy/dx

21 $y = \displaystyle\int_0^x \sec\theta\, d\theta$ **22** $y = \displaystyle\int_1^{x+1} \frac{1}{t}\, dt$

23 $y = \displaystyle\int_0^{\sin x} \sqrt{t}\, dt$ **24** $y = \displaystyle\int_{3x}^{10} u^4\, du.$

25 The point x moves to the right at the rate of 4 cm/sec. Let $A(x)$ be the area under the curve $y = \sin^2 u$ between $u = 0$ and $u = x$. Find dA/dt when $x = \frac{1}{3}\pi$. The unit of length on the u- and the y-axes is the centimeter.

26 Criticize: $\displaystyle\int_{-1}^{1} \frac{1}{x^2}\, dx = -\frac{1}{x}\Big|_{-1}^{1} = -1 - 1 = -2.$

4. APPLICATIONS

Area   As applications of the definite integral, let us compute the areas of some regions bounded by curves.

■ **EXAMPLE 1**   Find the area of the region bounded by the curve $y = 1 + x^3$, the two axes, and the line $x = 1$.

Solution   The region (Fig. 1) is simply the region under the graph of $y = 1 + x^3$ between $x = 0$ and $x = 1$. Its area is

$$\int_0^1 (1 + x^3)\, dx = \left(x + \frac{x^4}{4}\right)\Big|_0^1 = \left(1 + \frac{1}{4}\right) - 0 = \frac{5}{4}. \qquad ■$$

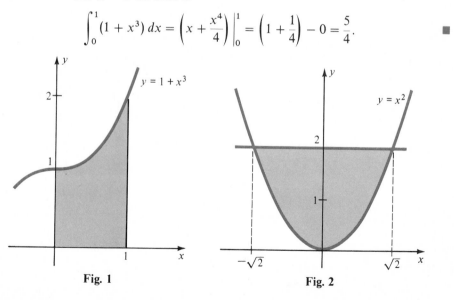

Fig. 1   Fig. 2

■ **EXAMPLE 2**   Find the area of the parabolic segment bounded by the curve $y = x^2$ and the line $y = 2$.

Solution   Sketch the region (Fig. 2). The line and the curve meet where $x^2 = y = 2$, that is, where $x = \pm\sqrt{2}$. To find the shaded area, compute the area *under* the curve and subtract it from the area of the rectangle:

$$\text{area} = (2\sqrt{2})(2) - \int_{-\sqrt{2}}^{\sqrt{2}} x^2\,dx = 4\sqrt{2} = \tfrac{1}{3}x^3\Big|_{-\sqrt{2}}^{\sqrt{2}}$$

$$= 4\sqrt{2} - \tfrac{1}{3}[(\sqrt{2})^3 - (-\sqrt{2})^3] = 4\sqrt{2} - \tfrac{4}{3}\sqrt{2} = \tfrac{8}{3}\sqrt{2}. \qquad \blacksquare$$

Algebraic Areas   When the integrand is positive, its definite integral is an area. How can we interpret an integral when the integrand is not always positive? First suppose $f(x) < 0$ on $[a, b]$. The integral of $f(x)$ is the limit of approximating sums,

$$\int_a^b f(x)\,dx = \lim S(f, \Pi, \bar{x}), \qquad \text{mesh}(\Pi)\longrightarrow 0,$$

where

$$S(f, \Pi, \bar{x}), = \sum_{j=1}^{n} f(\bar{x}_j)(x_j - x_{j-1}).$$

Each summand $f(\bar{x}_j)(x_j - x_{j-1})$ is the *negative* of the area of a rectangular strip (Fig. 3a) since $f(\bar{x}_j) < 0$. Therefore the limit is precisely the negative of the shaded area (Fig. 3b). In other words, the integral equals the negative of the shaded area:

$$\int_a^b f(x)\,dx = -A.$$

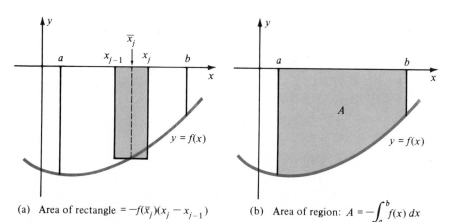

(a)   Area of rectangle $= -f(\bar{x}_j)(x_j - x_{j-1})$ (b)   Area of region: $A = -\int_a^b f(x)\,dx$

Fig. 3

In general, a continuous function on $[a, b]$ may change sign several times, so its graph and the x-axis bound several regions, some above the x-axis, some below. We claim that its integral equals the *algebraic sum* of these areas—each area above the x-axis is added, each area below is subtracted

For instance consider the function in Fig. 4a. We have

$$\int_a^b f(x)\,dx = \int_a^{c_1} f(x)\,dx + \int_{c_1}^{c_2} f(x)\,dx + \int_{c_2}^b f(x)\,dx.$$

The middle term equals the area A_2, but each of the two end terms equals the negative of an area, so

$$\int_a^b f(x)\, dx = -A_1 + A_2 - A_3.$$

So it goes in general. For example (Fig. 4b), we can see without computing that

$$\int_0^{2\pi} \sin x\, dx = 0,$$

since the areas above and below the x-axis cancel.

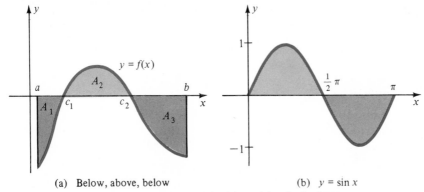

(a) Below, above, below   (b) $y = \sin x$

Fig. 4   Continuous functions taking both signs

■ **EXAMPLE 3**   One can of paint will cover 50 ft². How much paint is needed to cover the region bounded by $y = \sin x$ and the x-axis between $x = 0$ and $x = 2\pi$? Assume the unit on each axis is one yard.

Solution   If we treat the problem carelessly and integrate $\sin x$ between 0 and 2π, we reach a ridiculous conclusion: it takes no paint at all to cover the region! But remember, the integral

$$\int_a^b f(x)\, dx$$

gives the actual area under $y = f(x)$ only if $f(x) \geq 0$ between a and b.

Look at Fig. 4b. The two humps of the curve have equal area: compute the area of the hump between 0 and π, then double the result.

$$\int_0^{\pi} \sin x\, dx = (-\cos \pi) - (-\cos 0) = 1 - (-1) = 2.$$

Thus the area of one hump is 2 yd² $= 18$ ft²; the total area is 36 ft². This will require $\frac{36}{50} = 0.72$ cans of paint.   ■

Mean Value of a Function   Consider this problem: Find the mean value (average value) of a function $f(x)$ for $a \leq x \leq b$. Now the (arithmetic) mean, or average, of *numbers* $x_1, \cdots, x_n$ is defined by

$$\bar{x} = \frac{x_1 + \cdots + x_n}{n}.$$

This formula won't work here because there are infinitely many values of $f(x)$. Indeed, it is not at all clear what the "mean value" of a function is. So part of the problem is to define the problem, just as it was for slopes and areas.

One possible approach is this. Take the mean of a large sample of the values of $f(x)$:

$$M_n(f) = \frac{f(\bar{x}_1) + \cdots + f(\bar{x}_n)}{n} = \frac{1}{n} \sum_{j=1}^{n} f(\bar{x}_j),$$

and see what happens to $M_n(f)$ as n approaches ∞. For a fair sample, choose $\bar{x}_1, \ldots, \bar{x}_n$ well distributed throughout $[a, b]$, just as a political pollster samples opinions in all parts of the country.

This approach seems reasonable but hard to carry out. What saves the day is that $M_n(f)$ can be interpreted as an approximating sum of an integral. Partition $[a, b]$ into n *equal* subintervals

$$a_0 = x_0 < x_1 < \cdots < x_n = b, \quad \text{with} \quad x_j - x_{j-1} = \frac{b-a}{n} \quad (j = 1, 2, \cdots, n),$$

and let $\bar{x}_j$ be any point in the subinterval $[x_{j-1}, x_j]$. The sample $\bar{x}_1, \bar{x}_2, \ldots, \bar{x}_n$ is then fairly well distributed throughout $[a, b]$. The corresponding approximating sum is

$$S(f, \Pi, \bar{x}) = \sum_{j=1}^{n} f(\bar{x}_j)(x_j - x_{j-1}) = \frac{b-a}{n} \sum_{j=1}^{n} f(\bar{x}_j).$$

Except for the factor $b - a$, the right-hand side equals the mean $M_n(f)$ of the sample. Hence

$$M_n(f) = \frac{1}{b-a} S(f, \Pi, \bar{x}).$$

Now let $n \longrightarrow \infty$. The approximating sum approaches the integral; therefore

$$\lim_{n \to \infty} M_n(f) = \frac{1}{b-a} \int_a^b f(x)\,dx.$$

This reasoning motivates the following definition:

> **Mean Value** Let $f(x)$ be an integrable function on $[a, b]$. Its **mean value** is
>
> $$M(f) = \frac{1}{b-a} \int_a^b f(x)\,dx.$$

The definition has a simple geometric interpretation: The algebraic area under $f(x)$ between $x = a$ and $x = b$ equals the (algebraic) area of a rectangle with base $b - a$ and height $M(f)$. See Fig. 5.

■ **EXAMPLE 4** A 35-milligram (mg) sample of a certain radioactive substance decays in t days to $35e^{-t/10}$ mg. Compute the average mass $\bar{m}$ during the first three days.

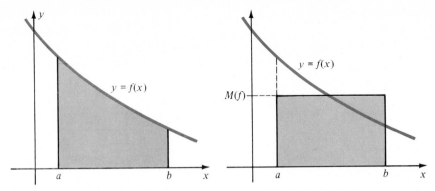

Fig. 5   Geometric interpretation of the mean $M(f)$: the shaded areas are equal.

Solution   The mean value of $f(t) = 35e^{-t/10}$ from $t = 0$ to $t = 3$ is

$$\bar{m} = \frac{1}{3 - 0}\int_0^3 f(t)\,dt = \frac{35}{3}\int_0^3 e^{-t/10}\,dt.$$

Now
$$e^{-t/10} = \frac{d}{dt}\left(-10e^{-t/10}\right),$$

so an antiderivative of $e^{-t/10}$ is $-10e^{-t/10}$. Therefore the average mass is

$$\bar{m} = \tfrac{35}{3}\left(-10e^{-t/10}\right)\Big|_0^3 = -\tfrac{350}{3}\left(e^{-t/10}\right)\Big|_0^3$$

$$= -\tfrac{350}{3}(e^{-0.3} - 1) \approx -\tfrac{350}{3}(0.7408 - 1) \approx 30.24 \quad \text{mg.} \qquad \blacksquare$$

Estimates of Sums   The Evaluation Rule makes the computation of integrals easier than the computation of many finite sums. In certain cases, we can obtain estimates of difficult sums by relating them to integrals. The next example illustrates the idea.

■ **EXAMPLE 5**   Find a quick estimate for

$$s_n = \sqrt{1} + \sqrt{2} + \cdots + \sqrt{100}.$$

Solution   There is a connection between s_n and the integral of $\sqrt{x}$ on $[0, 1]$. Partition the interval by n equally spaced points,

$$0 = x_0 < x_1 < x_2 < \cdots < x_n = 1,$$

where $x_j = j/n$. Choose $\bar{x}_j = x_j = j/n$. The approximating sum is

$$S(\sqrt{x}, \Pi, \bar{x}) = \sum_{j=1}^{n} f(\bar{x}_j)(x_j - x_{j-1}) = \sum_{j=1}^{n} f\left(\frac{j}{n}\right)\frac{1}{n}$$

$$= \frac{1}{n}\sum_{j=1}^{n}\sqrt{\frac{j}{n}} = \frac{1}{n^{3/2}}\sum_{j=1}^{n}\sqrt{j} = \frac{s_n}{n^{3/2}}.$$

Now let $n \longrightarrow \infty$:

$$\frac{S_n}{n^{3/2}} = S(\sqrt{x}, \Pi, \bar{x}) \longrightarrow \int_0^1 \sqrt{x}\, dx = \frac{2}{3} x^{3/2} \Big|_0^1 = \frac{2}{3}.$$

Hence for large values of n,

$$\frac{S_n}{n^{3/2}} \approx \frac{2}{3}, \qquad \text{that is,} \quad S_n \approx \frac{2}{3} n^{3/2}.$$

In particular, $\qquad S_{100} \approx \frac{2}{3}(100)^{3/2} = \frac{2}{3}(1000) \approx 667.$

(Actually, to six significant figures, $s_{100} \approx 671.463$.)   ∎

■ **EXAMPLE 6**   Use the technique of Example 5 to estimate $1^2 + 2^2 + \cdots + 300^2$.

Solution   Let $s_n = 1^2 + 2^2 + \cdots + n^2$. As in Example 5, partition $[0, 1]$, setting $\bar{x}_j = x_j = j/n$. The approximating sum to the integral of $f(x) = x^2$ is

$$S(x^2, \Pi, \bar{x}) = \sum_{j=1}^{n} (\bar{x}_j)^2 (x_j - x_{j-1}) = \sum_{j=1}^{n} \left(\frac{j}{n}\right)^2 \frac{1}{n} = \frac{1}{n^3} \sum_{j=1}^{n} j^2 = \frac{S_n}{n^3}.$$

Let $n \longrightarrow \infty$: $\qquad \dfrac{S_n}{n^3} = S(x^2, \Pi, \bar{x}) \longrightarrow \displaystyle\int_0^1 x^2\, dx = \frac{1}{3}.$

Hence for large values of n,

$$\frac{S_n}{n^3} \approx \frac{1}{3}, \qquad \text{that is,} \quad S_n \approx \frac{1}{3} n^3.$$

In particular, $s_{300} \approx \frac{1}{3}(300)^3 = 9 \times 10^6$. The precise value is 9,045,050. Our estimate is off by about $\frac{1}{2}$ of 1%.   ∎

Remark 1   An exact formula for s_n is known:

$$s_n = \frac{n(n+1)(2n+1)}{6} = \frac{2n^3 + 3n^2 + n}{6} = \frac{1}{3} n^3 + \frac{1}{2} n^2 + \frac{1}{6} n.$$

As n increases, the term $\frac{1}{3}n^3$ dominates the other two terms on the right. Thus $s_n \approx \frac{1}{3}n^3$ seems a reasonable estimate. But how can that be when the error, $\frac{1}{2}n^2 + \frac{1}{6}n$, increases as n does? The reason is simple: the *relative* error approaches 0 as $n \longrightarrow \infty$. For instance, when n is large enough, $\frac{1}{3}n^3$ will be more than 99.99% of the total $\frac{1}{3}n^3 + \frac{1}{2}n^2 + \frac{1}{6}n$.

Remark 2   Examples 5 and 6 illustrate **order of magnitude** estimates, which predict approximate values of large quantities. For example, $1^2 + 2^2 + \cdots + 300^2$ is certainly large, but about how large? Is it on the order of 10^4? 10^7? 10^{10}? 10^{20}?

EXERCISES

Compute the area under the graph $y = f(x)$ over the closed interval $[a, b]$

1	$y = (x - 1)^2$, $[1, 4]$	**2**	$y = 4 - x^2$, $[-2, 2]$
3	$y = 4(x - 3)^2$, $[0, 6]$	**4**	$y = 4x - x^2$, $[0, 4]$
5	$y = \cos x$, $[-\frac{1}{2}\pi, \frac{1}{2}\pi]$	**6**	$y = \sin 2x$, $[0, \frac{1}{2}\pi]$
7	$y = \sin x + 4 \cos x$, $[0, \frac{1}{2}\pi]$	**8**	$y = x + \sin x$, $[\pi, 2\pi]$
9	$y = (x^2 - 1)^2$, $[-1, 1]$	**10**	$y = e^{-2x}$, $[-1, 2]$.

Find the area of the region determined by

11   the y-axis, $y = x^3$,   and $y = 8$

12 $x \geq 0$, $y \geq 0$,   and $y = 1 - x^2$.

Find the *geometric* area of the region bounded by

13   the x-axis, $y = \cos x$, $x = 0$,   and $x = \pi$

14   the x-axis, $y = x - \dfrac{2}{x^2}$, $x = 1$,   and $x = 2$

15   the x-axis, $y = x^2 - 5x + 6$, $x = 0$,   and $x = 3$

16   the x-axis, $y = x^2 - 7x + 12$, $x = 1$,   and $x = 5$.

17   The graph $y = x^2$, $x \geq 0$,   may be looked at as the graph of $x = \sqrt{y}$, $y \geq 0$. Show that this implies geometrically the relation

$$\int_0^b x^2 \, dx + \int_0^{b^2} \sqrt{y} \, dy = b^3 \qquad (b > 0).$$

Now verify by computing the integrals.

18   (cont.) Find a similar formula for $y = x^3$.

19   Use your knowledge of the area of a circle to compute

$$\int_{-a}^a \sqrt{a^2 - x^2} \, dx.$$

20*   (cont.) Find the area enclosed by the ellipse

$$\frac{x^2}{a^2} + \frac{y^2}{b^2} = 1 \qquad (a > 0, \, b > 0).$$

Find the mean value $M(f)$ on $[a, b]$

21 $f(x) = x^5$, $[-2, 2]$ **22** $f(x) = \sin x$, $[-\frac{1}{3}\pi, \frac{1}{3}\pi]$

23 $f(x) = x^4$, $[0, 3]$ **24** $f(x) = 1/x^3$, $[1, 2]$

25 $f(x) = e^x$, $[-1, 1]$ **26** $f(x) = x^n$, $[0, b]$.

27   If x shares of a certain stock are sold, the market value in dollars per share is $V = 37 + [2.5 \times 10^6/(x + 500)^2]$. Find the average value per share on sales of 0 to 2000 shares.

28   Find the average area of circles with radius between 1 and 2 cm.

29   The rainfall per day in Erewhon, x days after the beginning of the year, is $R = (5.1 \times 10^{-5})(6511 + 366x - x^2)$ cm. Estimate the average daily rainfall for the first 100 days of the year.

30   A certain car, starting from rest, accelerates 11 ft/sec². Find its average speed during the first 10 sec.

31   Estimate $1^3 + 2^3 + 3^3 + \cdots + n^3$ for large n. For $n = 100$, what is the relative error, given the exact expression $\frac{1}{4}n^2(n + 1)^2$ for the sum?

32   Estimate $\sqrt[3]{1} + \sqrt[3]{2} + \sqrt[3]{3} + \cdots + \sqrt[3]{n}$ for large n.

5. APPROXIMATE INTEGRATION

In real-life situations, we must frequently approximate a definite integral rather than compute its exact value. This happens in two cases:

Case 1   The function to be integrated is given experimentally, say by a computer print-out. There is no formula for it.

Case 2   We have a formula for the function, but we do not know an antiderivative.

Let us consider these cases. In Case 1, we are given values $f(x_0), f(x_1), \ldots, f(x_n)$ at $n + 1$ points x_j. Usually the points are equally spaced on an interval:

$$x_0 = a, \quad x_1 = a + h, \quad x_2 = a + 2h, \cdots, x_n = a + nh = b.$$

Thus, the interval $[a, b]$ is divided into n equal subintervals, each of length

$$h = \frac{b - a}{n},$$

and we know the values of $f(x)$ only at the division points $x_0, \cdots, x_n$. We *assume* that there is a nice, smooth function $f(x)$, having domain $[a, b]$ and taking these values.

Example $[a, b] = [1, 3], \qquad n = 4, \quad h = 0.5,$

x_j	1.0	1.5	2.0	2.5	3.0
$f(x_j)$	1.00	0.67	0.50	0.40	0.33

Problem: From this data, find a reasonable estimate for $\int_1^3 f(x)\,dx$.

In Case 2, the problem is to estimate the integral of $f(x)$ without knowing an antiderivative of $f(x)$.

Examples $\int_1^3 \dfrac{dx}{x}, \qquad \int_0^{\sqrt{\pi}} \sin x^2 \, dx, \qquad \int_1^{10} \dfrac{e^x}{x} \, dx.$

Until now, we have not seen antiderivatives of the functions $1/x$, $\sin x^2$, e^x/x, so we cannot compute their integrals exactly. Still we should somehow be able to estimate them with reasonable accuracy.

We shall handle Case 2 by computing values of $f(x)$ at a number of equally spaced points and using these values to estimate the integral. Thus we shall treat both cases by the same method. It seems reasonable to expect better results in Case 2, however. Since we have a formula for $f(x)$, we can compute more and more data points for greater and greater accuracy. In addition, we can use further information about $f(x)$ to find limits on the possible error in estimating the integral.

Remark   It is standard to use the letter h for $(b - a)/n$. The symbol Δx is another standard notation. We prefer h here.

Rectangular and Trapezoidal Approximations   The simplest case of the problem is this: given $f(a)$ and $f(b)$—nothing else—estimate

$$\int_a^b f(x)\,dx.$$

Geometry (Fig. 1a) suggests two possible estimates by (algebraic) areas of rectangles:

$$\int_a^b f(x)\,dx \approx (b - a) f(a) \qquad \text{and} \qquad \int_a^b f(x)\,dx \approx (b - a)\, f(b).$$

Notice that each uses only part of the data. Which is the better estimate? That

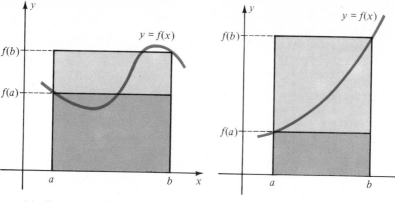

(a) Two rectangular approximations

(b) Over and under rect. approximations, curve increasing

Fig. 1 Approximations to $\displaystyle\int_a^b f(x)\,dx$

depends on $f(x)$. If $f(x)$ is an increasing (or decreasing) function, at least we can say the following:

If $f(x)$ increases on $[a, b]$, then $\displaystyle (b - a)f(a) \le \int_a^b f(x)\,dx \le (b - a)f(b).$

This assertion is obvious geometrically (Fig. 1b).

Of the two rectangular approximations, $(b - a)f(a)$ and $(b - a)f(b)$, one is an overestimate and the other is an underestimate, at least when $f(x)$ is increasing or decreasing on $[a, b]$. Common sense suggests *averaging* these estimates:

$$\int_a^b f(x)\,dx \approx (b - a)\,\frac{f(a) + f(b)}{2}.$$

Now we have used all the data—that is a good sign. Furthermore, the average $\frac{1}{2}(b - a)[f(a) + f(b)]$ of the two rectangular areas is itself an area, the (algebraic)

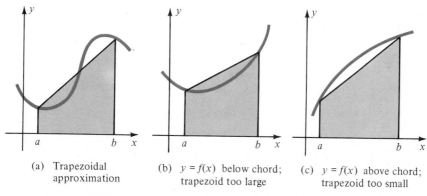

(a) Trapezoidal
      approximation

(b) $y = f(x)$ below chord;
      trapezoid too large

(c) $y = f(x)$ above chord;
      trapezoid too small

Fig. 2 Trapezoidal approximations to $\displaystyle\int_a^b f(x)\,dx$

area of a trapezoid (Fig. 2a). The figure suggests that the trapezoidal area may be a reasonable approximation to the area under the curve.

Can we draw any conclusion about the trapezoidal approximation, knowing that $f(x)$ increases? No. If $f(x)$ lies below the chord (Fig. 2b), the trapezoidal estimate is too large; if $f(x)$ lies above, it is too small (Fig. 2c). But if neither is the case (Fig. 2a), what happens is anybody's guess.

The Trapezoidal Rule   The trapezoidal estimate of an integral may be inaccurate if $[a, b]$ is a large interval (Fig. 3a). Hoping for a better estimate, we split $[a, b]$ into n equal parts. On each subinterval we estimate the integral of $f(x)$ by the area of a thin trapezoid, then add up these areas (Fig. 3b). The result is an approximation called the **Trapezoidal Rule**.

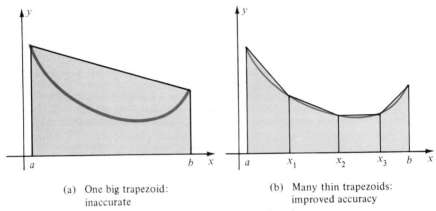

(a) One big trapezoid: (b) Many thin trapezoids:
     inaccurate   improved accuracy

Fig. 3   Trapezoidal approximation

To derive a workable formula, we write

$$a = x_0 < x_1 < \cdots < x_n = b, \qquad x_j - x_{j-1} = h = \frac{b-a}{n}.$$

We apply the trapezoidal estimate on each subinterval:

$$\int_a^b f(x)\,dx = \int_{x_0}^{x_1} + \int_{x_1}^{x_2} + \cdots + \int_{x_{n-1}}^{x_n} f(x)\,dx$$

$$\approx (x_1 - x_0)\frac{f(x_0) + f(x_1)}{2} + \cdots + (x_n - x_{n-1})\frac{f(x_{n-1}) + f(x_n)}{2}$$

$$= \frac{h}{2}[f(x_0) + f(x_1) + f(x_1) + f(x_2) + \cdots + f(x_{n-1}) + f(x_n)]$$

$$= \frac{h}{2}[f(x_0) + 2f(x_1) + 2f(x_2) + \cdots + 2f(x_{n-1}) + f(x_n)].$$

Trapezoidal Rule   Let $h = (b-a)/n$, $x_j = a + jh$,   and $f_j = f(x_j)$.

Then $\displaystyle\int_a^b f(x)\,dx \approx \frac{h}{2}[f_0 + 2f_1 + 2f_2 + \cdots + 2f_{n-1} + f_n]$.

■ **EXAMPLE 1** Estimate $\int_1^3 \frac{dx}{x}$ with $n = 2, 4, 10, 20, 100$.

Solution When $n = 2$, we have $x_0 = 1$, $x_1 = 2$, $x_2 = 3$, and $h = 1$, so

$$\int_1^3 \frac{dx}{x} \approx \frac{1}{2}[f_0 + 2f_1 + f_2] = \frac{1}{2}\left[\frac{1}{1} + \frac{2}{2} + \frac{1}{3}\right] = \frac{7}{6} \approx 1.17.$$

When $n = 4$, we have $x_0 = 1$, $x_1 = \frac{3}{2}$, $x_2 = 2$, $x_3 = \frac{5}{2}$, $x_4 = 3$, and $h = \frac{1}{2}$, so

$$\int_1^3 \frac{dx}{x} \approx \frac{1}{4}\left[1 + 2\left(\frac{2}{3} + \frac{1}{2} + \frac{2}{5}\right) + \frac{1}{3}\right] = \frac{67}{60} \approx 1.117.$$

For the larger values of n, we need some sort of computer facility. When $n = 10$, then $h = 0.2$ and

$$\int_1^3 \frac{dx}{x} \approx \frac{0.2}{2}\left[1 + 2\left(\frac{1}{1.2} + \frac{1}{1.4} + \cdots + \frac{1}{2.8}\right) + \frac{1}{3}\right] \approx 1.1016.$$

The corresponding results for $n = 20$ and $n = 100$ are

$$\int_1^3 \frac{dx}{x} \approx \frac{0.1}{2}\left[1 + 2\left(\frac{1}{1.1} + \frac{1}{1.2} + \cdots + \frac{1}{2.9}\right) + \frac{1}{3}\right] \approx 1.09935,$$

$$\int_1^3 \frac{dx}{x} \approx \frac{0.02}{2}\left[1 + 2\left(\frac{1}{1.02} + \cdots + \frac{1}{2.98}\right) + \frac{1}{3}\right] \approx 1.098642.$$

Summary:

n	2	4	10	20	100
approx.	1.17	1.117	1.1016	1.09935	1.098642

Because $f(x) = 1/x$ is convex on $[1, 3]$, each of the approximations is too large. Our computations yield approximations that decrease as n increases, a good sign. The actual value of the integral to nine places is

$$\int_1^3 \frac{dx}{x} \approx 1.098612289.$$ ■

Rectangular Rules We take the rectangle with base $x_j - x_{j-1}$ and height f_{j-1}, the value of $f(x)$ at the left end point of the interval $[x_{j-1}, x_j]$. This leads to a left rectangular rule

$$\int_a^b f(x)\, dx \approx h[f_0 + f_1 + \cdots + f_{n-1}].$$

See Fig. 4a. Similarly, we have a right rectangular rule (Fig. 4b)

$$\int_a^b f(x)\, dx \approx h[f_1 + f_2 + \cdots + f_n].$$

Since each of these ignores one datum, they are less desirable than the Trapezoidal Rule. It should be no surprise that averaging these two approximations leads again to the Trapezoidal Rule:

$$\int_a^b f(x)\,dx \approx \frac{h}{2}\left[(f_0 + f_1 + \cdots + f_{n-1}) + (f_1 + f_2 + \cdots + f_n)\right]$$

$$= \frac{h}{2}\left(f_0 + 2f_1 + 2f_2 + \cdots + 2f_{n-1} + f_n\right).$$

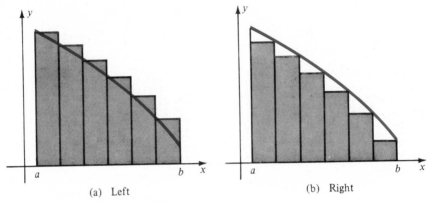

(a) Left   (b) Right

Fig. 4   Rectangular rules

Terminology   Formulas such as the Rectangular and Trapezoidal Rules for approximate integration are sometimes called **numerical quadrature** formulas. The word quadrature by itself is an older term for integration.

Error   If $f(x)$ is a linear function, then

$$\int_a^b f(x)\,dx = \frac{b-a}{2}\left[f(a) + f(b)\right],$$

as is easily checked. So the trapezoidal approximation is exact, which implies that the Trapezoidal Rule is exact for linear functions. Therefore the error in the Trapezoidal Rule for a general function $f(x)$ should depend on how far $f(x)$ deviates from being linear. Now the second derivative of a linear function is zero, so the second derivative of a general function measures its deviation from linearity. Therefore we expect an error estimate in the Trapezoidal Rule to involve the second derivative of $f(x)$. The following is such an estimate, but its proof is too hard to include. (However, see Ex. 18, p. 494.)

Error in the Trapezoidal Rule   Suppose $|f''(x)| \le M$ for $a \le x \le b$.

Then $$\int_a^b f(x)\,dx = \frac{h}{2}\left(f_0 + 2f_1 + \cdots + 2f_{n-1} + f_n\right) + \text{error},$$

where $$|\text{error}| \le \frac{Mnh^3}{12} = \frac{M(b-a)}{12}\,h^2.$$

■ **EXAMPLE 2**   Find an upper bound for the error in the Trapezoidal Rule with $n = 100$ for $\int_1^3 \dfrac{dx}{x}$.

Solution   Set $f(x) = 1/x$. Then $f''(x) = 2/x^3$, which has its largest value on $[1, 3]$ at $x = 1$. Hence take $M = f''(1) = 2$. Since $h = 0.02$, we have

$$|\text{error}| \le \frac{Mnh^3}{12} \le \frac{2(100)(0.02)^3}{12} = \frac{4}{3} \times 10^{-4} < 1.4 \times 10^{-4}.$$ ■

Remark   In Chapter 10 we shall discuss more accurate methods of approximate integration.

EXERCISES

Approximate by the Trapezoidal Rule with $n = 4$ (three significant figures), $n = 10$ (four significant figures), and if a programmable computer is available, $n = 50$ (six significant figures)

1   $\int_1^2 \dfrac{dx}{x}$ 2 $\int_0^3 \dfrac{dx}{x + 1}$ 3 $\int_1^3 \dfrac{x - 1}{x + 1}\, dx$

4   $\int_{-1}^1 \dfrac{dx}{1 + x^2}$ 5 $\int_1^2 \dfrac{x}{1 + x^2}\, dx$ 6 $\int_0^1 \sqrt{1 - x^2}\, dx$

7   $\int_0^4 \sqrt{1 + x^2}\, dx$ 8 $\int_0^2 xe^x\, dx$ 9 $\int_0^{10} xe^{-x}\, dx$

10   $\int_0^4 \exp(-x^2)\, dx$ 11 $\int_1^2 e^{-1/x}\, dx$ 12 $\int_0^{\pi} e^x \sin x\, dx$

13   $\int_1^2 \log_{10} x\, dx$ 14 $\int_1^2 x \log_{10} x\, dx$ 15 $\int_0^{\pi/4} \tan x\, dx$

16   $\int_0^{\pi} x \sin x\, dx$ 17 $\int_0^{\pi/2} x^2 \sin x\, dx$ 18 $\int_{-\pi/4}^{\pi/4} \sec x\, dx$

19   $\int_0^{\pi/2} \sin^2 x\, dx$ 20 $\int_0^{\pi/2} x \cos x\, dx$.

21   An automobile starting from rest accelerates for 15 sec. Velocity readings (in feet per second) taken at 1-sec intervals are: 0, 0.5, 2.1, 4.7, 8.3, 13.0, 18.7, 25.5, 33.3, 43.1, 53.0, 63.9, 75.9, 88.9, 102.9, 118.0. Estimate the distance the car traveled during this time.

22   Soundings in feet are taken across a 90-ft river at 5-ft intervals, resulting in the readings: 0, 1.0, 2.5, 5.0, 8.0, 10.5, 11.5, 12.0, 13.0, 12.5, 13.5, 16.0, 16.0, 14.0, 10.5, 9.0, 6.5, 4.0, 0. If the average current at this point in the river is 5 ft/sec, estimate, to the nearest million cubic feet, the daily flow of water.

The **midpoint approximation** is $\int_a^b f(x)\, dx \approx (b - a)f(\bar{x})$, $\bar{x} = \frac{1}{2}(a + b)$.

More generally the interval is split into n equal subintervals and this approximation is used in each. Compare the exact value, Trapezoidal Rule with $n = 2$, and midpoint approximation with $n = 2$

23   $\int_0^2 x^2\, dx$ 24 $\int_1^2 x^3\, dx$ 25 $\int_1^3 \dfrac{1}{x}\, dx$ 26 $\int_1^3 \dfrac{1}{x^2}\, dx$.

27   Show that the Trapezoidal Rule with $n = 100$ will yield at least four-place accuracy in approximating

$$\int_4^7 e^{-x} \sin 2x\, dx.$$

28 Show that the Trapezoidal Rule with $n = 100$ will approximate $\displaystyle\int_0^1 \frac{dx}{x^3 + 10}$ to within 10^{-6}.

6. INTEGRATION OF PRODUCTS

The formula

$$\int_a^b [f(x) + g(x)]\, dx = \int_a^b f(x)\, dx + \int_a^b g(x)\, dx$$

gives us the integral of a sum of functions, provided we know the integral of each summand. Can we also find the integral of a product,

$$\int_a^b f(x)g(x)\, dx,$$

knowing the integrals of $f(x)$ and $g(x)$? The answer in general is no. However, we can integrate many products of functions using two basic *differentiation* formulas, the Product Rule and the Chain Rule.

Integration by Parts Recall the product rule for differentiation:

$$\frac{d}{dx}[u(x)v(x)] = u(x)\frac{dv}{dx} + v(x)\frac{du}{dx}.$$

Turned around slightly, it reads

$$u(x)\frac{dv}{dx} = \frac{d}{dx}[u(x)v(x)] - v(x)\frac{du}{dx}.$$

Now suppose we want an antiderivative for uv'. If we *know* an antiderivative F for the product vu', then $uv - F$ is the desired antiderivative for uv'. Thus the problem of antidifferentiating (or integrating) the product uv' is transformed into the problem of antidifferentiating the product vu', which may be easier. Then again it may not be. These possibilities are illustrated in Example 1.

■ **EXAMPLE 1** Find an antiderivative of xe^x.

Solution 1 Write $xe^x = u(x)v'(x)$. There are two obvious choices: $u(x) = x$ and $v'(x) = e^x$, or the other way around. Try

$$u(x) = e^x, \qquad v'(x) = x.$$

For $v(x)$, take any antiderivative of x, say $v(x) = \frac{1}{2}x^2$. Since $u'(x) = e^x$, we have

$$uv' = (uv)' - vu', \qquad xe^x = \frac{d}{dx}\left(\frac{1}{2}x^2e^x\right) - \frac{1}{2}x^2e^x.$$

This "reduces" the problem of antidifferentiating xe^x to that of antidifferentiating x^2e^x, hardly a simplification!

Solution 2   This time write

$$xe^x = u(x)v'(x), \qquad u(x) = x, \qquad v'(x) = e^x.$$

Then $u' = 1$ and one choice of v is $v = e^x$. Now the formula $uv' = (uv)' - vu'$ reads

$$xe^x = \frac{d}{dx}(xe^x) - e^x,$$

which reduces antidifferentiation of xe^x to antidifferentiation of e^x. But we *know* an antiderivative of e^x, namely, e^x. Therefore

$$xe^x = \frac{d}{dx}(xe^x) - \frac{d}{dx}e^x = \frac{d}{dx}(xe^x - e^x).$$

We have integrated a new function!

$$\int_a^b xe^x \, dx = (xe^x - e^x)\Big|_a^b. \qquad\blacksquare$$

Because of the Evaluation Rule, this method of antidifferentiating uv' yields an important tool for computing integrals, called integration by parts.

Integration by Parts   If $u(x)$, $v(x)$, $u'(x)$, and $v'(x)$ are continuous on $[a, b]$, then

$$\int_a^b u(x)\frac{dv}{dx}\,dx = u(x)v(x)\Big|_a^b - \int_a^b v(x)\frac{du}{dx}\,dx.$$

Briefly,

$$\int_a^b uv'\,dx = uv\Big|_a^b - \int_a^b vu'\,dx.$$

■ **EXAMPLE 2**   Compute $\displaystyle\int_0^\pi x \sin x\, dx.$

Solution   After the last example, we won't fall into the trap of "reducing" the integrand to $x^2 \cos x$. Instead we write

$$x \sin x = uv', \qquad u = x, \qquad v' = \sin x.$$

Then $v = -\cos x$ is an antiderivative of $\sin x$ and $u' = 1$. Therefore

$$\int_0^\pi x \sin x\, dx = uv\Big|_0^\pi - \int_0^\pi vu'\, dx = -x \cos x\Big|_0^\pi - \int_0^\pi (-\cos x)\, dx$$

$$= \pi + \int_0^\pi \cos x\, dx = \pi - \sin x\Big|_0^\pi = \pi. \qquad\blacksquare$$

Successful application of integration by parts is tricky. It depends on "seeing" the integrand as a product uv', where you know both u and an antiderivative of vu'. We shall return to this topic in Chapter 8.

Change of Variable   We have exploited almost every differentiation formula we know to get integration formulas except one, the Chain Rule:

$$\frac{d}{dx} F[g(x)] = F'[g(x)]g'(x).$$

If we can "see" a product of two functions in the form $F'[g(x)]g'(x)$, then we immediately have an antiderivative, $F[g(x)]$. This is not as hard as it sounds; with a little practice we can get used to looking at the Chain Rule the other way around.

Examples

$$\sin^3 x \cos x = \frac{d}{dx}\left(\frac{1}{4}\sin^4 x\right), \qquad F(y) = y^4, \qquad g(x) = \sin x.$$

$$e^{x^2/2} \cdot x = \frac{d}{dx} e^{x^2/2}, \qquad\qquad F(y) = e^y, \qquad g(x) = \frac{1}{2} x^2.$$

$$\frac{3x^2}{2\sqrt{x^3 + 4}} = \frac{d}{dx}\sqrt{x^3 + 4}, \qquad F(y) = \sqrt{y}, \qquad g(x) = x^3 + 4.$$

By the Fundamental Theorem, the Chain Rule allows us to evaluate integrals of products of the form $F'[g(x)]g'(x)$:

$$\int_a^b F'[g(x)]g'(x)\,dx = F[g(x)]\,\Big|_a^b.$$

The right-hand side of this relation can be interpreted in another way, again by the Fundamental Theorem:

$$F[g(x)]\,\Big|_a^b = F[g(b)] - F[g(a)] = F(y)\,\Big|_{g(a)}^{g(b)} = \int_{g(a)}^{g(b)} F'(y)\,dy.$$

This procedure is called a change of variable, since the problem is stated in terms of a variable x, but the computation is done in terms of a new variable y.

> **Change of Variable Formula**   Assume $F(y)$ and $g(x)$ are differentiable functions with continuous derivatives, and that the composite function $F[g(x)]$ is defined for $a \le x \le b$. Then
>
> $$\int_a^b F'[g(x)]g'(x)\,dx = F[g(x)]\,\Big|_a^b = \int_{g(a)}^{g(b)} F'(y)\,dy.$$

This is also called the Substitution Rule because y is substituted for $g(x)$. We shall discuss the rule in more detail in Chapter 8, Sections 2 and 3.

■ **EXAMPLE 3**   Compute $\displaystyle\int_{\pi/4}^{\pi/2} \sin^3 x \cos x\,dx.$

Solution   As noted before,

$$\sin^3 x \cos x = F'[g(x)]g'(x), \qquad \text{where} \quad F(y) = \tfrac{1}{4}y^4 \quad \text{and} \quad g(x) = \sin x.$$

Therefore an antiderivative is $F[g(x)] = \frac{1}{4}\sin^4 x$. Hence

$$\int_{\pi/4}^{\pi/2} \sin^3 x \cos x \, dx = \frac{1}{4}\sin^4 x \Big|_{\pi/4}^{\pi/2} = \frac{1}{4}\left(1 - \frac{1}{4}\right) = \frac{3}{16}.$$

As a slightly different approach, substitute $y = \sin x$ into $F(y) = \frac{1}{4}y^4$. Then the formula in the box, with $g(x) = \sin x$, yields

$$\int_{\pi/4}^{\pi/2} \sin^3 x \cos x \, dx = \int_{g(\pi/4)}^{g(\pi/2)} F'(y) \, dy = \int_{\sqrt{2}/2}^{1} F'(y) \, dy$$

$$= F(y)\Big|_{\sqrt{2}/2}^{1} = \frac{1}{4}y^4 \Big|_{\sqrt{2}/2}^{1} = \frac{1}{4}\left(1 - \frac{1}{4}\right) = \frac{3}{16}. \qquad \blacksquare$$

In practice, the Change of Variable Formula is often applied in a slightly different form, obtained by substituting the letter f for the continuous function F':

$$\boxed{\int_a^b f[g(x)]g'(x) \, dx = \int_{g(a)}^{g(b)} f(y) \, dy.}$$

Translation   As an application, we prove the following rule for translating (shifting) the variable.

$$\boxed{\textbf{Translation Rule} \qquad \int_a^b f(c + x) \, dx = \int_{c+a}^{c+b} f(x) \, dx.}$$

Let $F(y)$ be any antiderivative of $f(y)$, and set $g(x) = c + x$. Then

$$F'(y) = f(y) \qquad \text{and} \qquad g'(x) = 1,$$

so

$$\int_a^b f(c + x) \, dx = \int_a^b F'[g(x)]g'(x) \, dx = \int_{g(a)}^{g(b)} F'(y) \, dy = \int_{c+a}^{c+b} f(y) \, dy = \int_{c+a}^{c+b} f(x) \, dx.$$

This translation formula has an obvious geometric interpretation (Fig. 1).

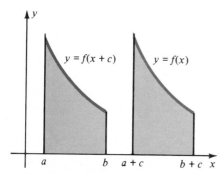

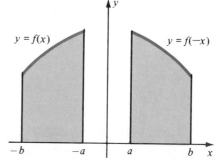

Fig. 1   Translation rule: the shaded areas are equal.

Fig. 2   Reflection rule: the shaded areas are equal.

Examples $$\int_{1}^{2}\sqrt{x+3}\ dx = \int_{3+1}^{3+2}\sqrt{x}\ dx = \int_{4}^{5}\sqrt{x}\ dx.$$

$$\int_{-\pi/2}^{0}\cos(x+\tfrac{1}{4}\pi)\ dx = \int_{\pi/4-\pi/2}^{\pi/4+0}\cos x\ dx = \int_{-\pi/4}^{\pi/4}\cos x\ dx.$$

Reflection   As a second application, we prove the following formula for reflection through the origin.

> **Reflection Rule** $\displaystyle \int_{a}^{b}f(-x)\ dx = \int_{-b}^{-a}f(x)\ dx.$

Again let $F(y)$ be any antiderivative of $f(y)$, but set $g(x) = -x$, so

$$F'(y) = f(y) \qquad \text{and} \qquad g'(x) = -1.$$

Then
$$\int_{a}^{b}f(-x)\ dx = -\int_{a}^{b}F'[g(x)]g'(x)\ dx = -\int_{g(a)}^{g(b)}F'(y)\ dy$$

$$= -\int_{-a}^{-b}f(y)\ dy = \int_{-b}^{-a}f(y)\ dy = \int_{-b}^{-a}f(x)\ dx.$$

This reflection formula, like the previous translation formula, also has a clear geometric interpretation (Fig. 2). We shall make good use of both formulas in the next section.

Examples $$\int_{3}^{4}\left(x^2 - x - \frac{1}{x}\right) dx = \int_{-4}^{-3}\left(x^2 + x + \frac{1}{x}\right) dx.$$

$$\int_{1}^{10}e^{-x}\ dx = \int_{-10}^{-1}e^{x}\ dx.$$

EXERCISES

Evaluate

1 $\displaystyle\int_{0}^{\pi}x\cos x\ dx$

2 $\displaystyle\int_{0}^{1}xe^{3x}\ dx$

3 $\displaystyle\int_{-1}^{1}xe^{-2x}\ dx$

4 $\displaystyle\int_{0}^{\pi/4}x\sin 2x\ dx.$

5 Suppose $F(x)$ is an antiderivative of $\tan x$. Find an antiderivative of $x\sec^2 x$.
6 Suppose $F(x)$ is an antiderivative of $1/x$. Find an antiderivative of $xF(x)$.

Evaluate

7 $\displaystyle\int_{0}^{1}2x(x^2+1)^9\ dx$

8 $\displaystyle\int_{0}^{1}\frac{4x^3+2x}{(x^4+x^2+1)^3}\ dx$

9 $\displaystyle\int_{1}^{2}\frac{9x^2}{(x^3+1)^4}\ dx$

10 $\displaystyle\int_{-1}^{0}\frac{x}{(x^2-4)^3}\ dx.$

Find an antiderivative

11 $x\exp(x^2)$

12 $e^x\sin(e^x)$

13 $(\sin x)\exp(\cos x)$

14 $x^2\sec^2(x^3)$

15 $(4x^3+1)\sqrt{x^4+x}$

16 $\dfrac{\exp\sqrt{x}}{\sqrt{x}}$

Prove by inspection

17 $\displaystyle\int_0^2\sqrt{2x+7}\,dx=\int_{-1}^1\sqrt{2x+5}\,dx$

18 $\displaystyle\int_0^3 e^{-x^2}\,dx=\int_{-3}^0 e^{-x^2}\,dx$

19 $\displaystyle\int_{-4}^{-1}\frac{dx}{e^x+1}=\int_1^4\frac{e^x}{e^x+1}\,dx$

20 $\displaystyle\int_0^2(x^2-1)^n\,dx=\int_{-1}^1 x^n(x+2)^n\,dx$

21 $\displaystyle\int_{2\pi}^{4\pi}e^x\sin x\,dx=e^{2\pi}\int_0^{2\pi}e^x\sin x\,dx$

22 $\displaystyle\int_{-2}^3(x^2+4x+5)^3\,dx=\int_0^5(x^2+1)^3\,dx.$

7. SYMMETRY

Sometimes a definite integral

$$\int_a^b f(x)\,dx$$

can be simplified because the integrand $f(x)$ has certain symmetry. In this section, we discuss some labor-saving devices for evaluating integrals with visible symmetry. We always assume, without further mention, that the integrals in question exist.

Even and Odd Functions

Even Functions   If $f(x)$ is an even function, that is, $f(-x)=f(x)$, then

$$\int_{-a}^a f(x)\,dx=2\int_0^a f(x)\,dx.$$

Since

$$\int_{-a}^a f(x)\,dx=\int_{-a}^0 f(x)\,dx+\int_0^a f(x)\,dx,$$

we must prove that

$$\int_{-a}^0 f(x)\,dx=\int_0^a f(x)\,dx.$$

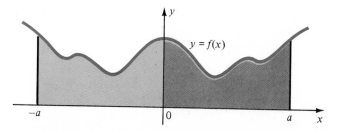

Fig. 1   Even function: left area = right area.

This is obvious from the graph (Fig. 1), and it follows easily by the reflection principle discussed in the preceding section:

$$\int_{-a}^{0} f(x)\, dx = \int_{0}^{a} f(-x)\, dx = \int_{0}^{a} f(x)\, dx.$$

■ **EXAMPLE 1**   Find $\displaystyle\int_{-\pi/2}^{\pi/2} \cos \tfrac{1}{3}x\, dx.$

Solution   The integrand is an even function:

$$\cos(-\tfrac{1}{3}x) = \cos(\tfrac{1}{3}x).$$

Therefore $\displaystyle\int_{-\pi/2}^{\pi/2} \cos \tfrac{1}{3}x\, dx = 2 \int_{0}^{\pi/2} \cos \tfrac{1}{3}x\, dx = 6 \sin \tfrac{1}{3}x \Big|_{0}^{\pi/2} = 6 \sin \tfrac{1}{6}\pi = 3.$ ■

Odd Functions   If $f(x)$ is an odd function, that is, $f(-x) = -f(x)$, then

$$\int_{-a}^{a} f(x)\, dx = 0.$$

This is obvious from a graph (Fig. 2), and it follows easily from the reflection principle:

$$\int_{-a}^{a} f(x)\, dx = \int_{-a}^{a} f(-x)\, dx = -\int_{-a}^{a} f(x)\, dx,$$

$$2\int_{-a}^{a} f(x)\, dx = 0, \qquad \int_{-a}^{a} f(x)\, dx = 0.$$

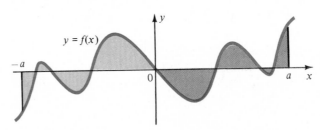

Fig. 2   Odd function: left area cancels right area.

Examples $\displaystyle\int_{-\pi/4}^{\pi/4} \sin^9 x\, dx = 0,$ $\displaystyle\int_{-6}^{6} \frac{x^5}{x^4 + 1}\, dx = 0.$

■ **EXAMPLE 2**   Show that

(a) $\displaystyle\int_{-3}^{3} (x \sin^2 x - 14x^3 + 1)\, dx = 6$ (b) $\displaystyle\int_{-8}^{10} \frac{x}{x^2 + 4}\, dx = \int_{8}^{10} \frac{x}{x^2 + 4}\, dx.$

Solution   (a)   Write the integral as

$$\int_{-3}^{3} (x \sin^2 x - 14x^3)\, dx + \int_{-3}^{3} dx.$$

The first integral is 0 because the integrand is odd; the second integral equals 6 by inspection.

(b)   The integrand is odd. Therefore

$$\int_{-8}^{10} \frac{x}{x^2 + 4}\, dx = \int_{-8}^{8} + \int_{8}^{10} = 0 + \int_{8}^{10} \frac{x}{x^2 + 4}\, dx. \qquad \blacksquare$$

Other Symmetry   The graph (Fig. 3a) of $y = (x - 3)^2 - 1$ is symmetric with respect to the line $x = 3$. The graph (Fig. 3b) of $y = \frac{1}{4}(x - 2)^3$ is symmetric with respect to the point $(2, 0)$. It is clear from the figures that

$$\int_{1}^{5} [(x - 3)^2 - 1]\, dx = 2 \int_{3}^{5} [(x - 3)^2 - 1]\, dx, \qquad \int_{0}^{4} \tfrac{1}{4}(x - 2)^3\, dx = 0.$$

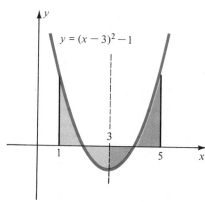

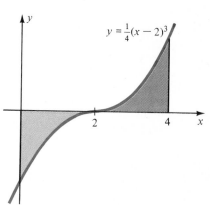

(a)   Symmetry with respect
to the line $x = 3$

(b)   Symmetry with respect
to the point $(2, 0)$

Fig. 3   Other symmetries

Symmetry with respect to the line $x = c$ means that the value of f at x units to the left of c equals the value of f at x units to the right of c; that is, $f(c - x) = f(c + x)$. Symmetry with respect to the point $(c, 0)$ means that the value of f at x units to the left of c equals the negative of the value of f at x units to the right of c, that is, $f(c - x) = -f(c + x)$.

Other Symmetries

(1) $y = f(x)$ is symmetric with respect to the line $x = c$ if $f(c - x) = f(c + x)$.
Then

$$\int_{c-a}^{c+a} f(x)\, dx = 2 \int_{c}^{c+a} f(x)\, dx.$$

(2) $y = f(x)$ is symmetric with respect to the point $(c, 0)$ if $f(c - x) = -f(c + x)$.
Then

$$\int_{c-a}^{c+a} f(x)\, dx = 0.$$

The integral formulas follow directly from the corresponding formulas for even and odd functions—via the translation formula discussed in the last section. For, set $g(x) = f(c + x)$. If $f(c - x) = f(c + x)$, then $g(-x) = g(x)$, so $g(x)$ is an even function. Then

$$\int_{c-a}^{c+a} f(x)\, dx = \int_{-a}^{a} f(c + x)\, dx = \int_{-a}^{a} g(x)\, dx$$

$$= 2 \int_{0}^{a} g(x)\, dx = 2 \int_{0}^{a} f(c + x)\, dx = 2 \int_{c}^{c+a} f(x)\, dx.$$

If $f(c - x) = -f(c + x)$, then $g(x) = f(c + x)$ is an odd function, and a similar argument applies.

■ **EXAMPLE 3**   Find $\displaystyle\int_{1}^{5} x(x - 1)(x - 2)(x - 3)(x - 4)(x - 5)(x - 6)\, dx.$

Solution $(3, 0)$ is a point of symmetry. To see this, set $f(x) = x(x - 1) \cdots (x - 6)$. Then

$$f(3 + x) = (3 + x)(2 + x)(1 + x)x(-1 + x)(-2 + x)(-3 + x)$$

is an odd function: $f(3 - x) = -f(3 + x)$. Therefore

$$\int_{1}^{5} f(x)\, dx = \int_{3-2}^{3+2} f(x)\, dx = 0. \qquad\qquad ■$$

Periodic Functions   A function $f(x)$ is **periodic** of **period** p if

$$f(x + p) = f(x).$$

The most familiar periodic functions are the trigonometric functions. We now note two basic properties of integrals of periodic functions:

Suppose $f(x)$ has period p. Then

(1) $\displaystyle\int_{a+p}^{b+p} f(x)\, dx = \int_{a}^{b} f(x)\, dx,$ (2) $\displaystyle\int_{a}^{a+p} f(x)\, dx = \int_{0}^{p} f(x)\, dx.$

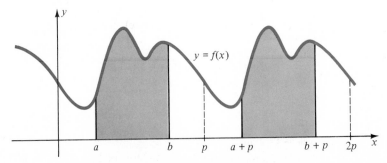

Fig. 4   Translation for a periodic function

The first relation is a special case of the translation formula. Its geometric content is clear from Fig. 4. The second relation is of a different nature; it says that all integrals of $f(x)$ over intervals of length p are equal. Its proof uses translation, but not directly (Fig. 5).

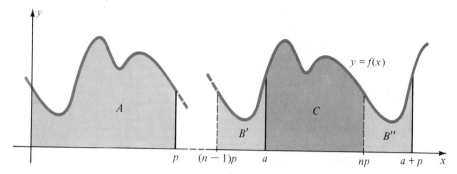

Fig. 5   area $B'' =$ area B'; hence $C + B'' = C + B' = B' + C = A$,

that is, $\displaystyle\int_a^{a+p} f(x)\,dx = \int_0^p f(x)\,dx$ for any a.

Choose n so that $(n-1)p \le a < np$. Then

$$\int_a^{a+p} f(x)\,dx = \int_a^{np} f(x)\,dx + \int_{np}^{a+p} f(x)\,dx.$$

By periodicity, $\displaystyle\int_{np}^{a+p} f(x)\,dx = \int_{(n-1)p}^{a} f(x)\,dx.$

Hence $\displaystyle\int_a^{a+p} f(x)\,dx = \int_a^{np} f(x)\,dx + \int_{(n-1)p}^{a} f(x)\,dx = \int_{(n-1)p}^{np} f(x)\,dx.$

Clearly $f(x) = f(x+p) = f(x+2p) = \cdots = f(x + (n-1)p)$, hence

$$\int_{(n-1)p}^{np} f(x)\,dx = \int_{(n-1)p-(n-1)p}^{np-(n-1)p} f(x)\,dx = \int_0^p f(x)\,dx.$$

This completes the proof of (2).

■ **EXAMPLE 4**   Let $f(y)$ be continuous on $[-1, 1]$. Prove

$$\int_0^{2\pi} f(\sin x)\,dx = \int_0^{2\pi} f(\cos x)\,dx.$$

Solution   The integrand has period $p = 2\pi$. Apply (2) with $a = \tfrac{1}{2}\pi$:

$$\int_0^{2\pi} f(\sin x)\,dx = \int_{\pi/2}^{2\pi+\pi/2} f(\sin x)\,dx.$$

By the translation formula,

$$\int_{\pi/2}^{2\pi+\pi/2} f(\sin x)\,dx = \int_0^{2\pi} f[\sin(x + \tfrac{1}{2}\pi)]\,dx.$$

But $\sin(x + \tfrac{1}{2}\pi) = \cos x$, so

$$\int_0^{2\pi} f(\sin x)\, dx = \int_0^{2\pi} f(\cos x)\, dx. \qquad \blacksquare$$

Example   Take $f(y) = \sqrt{1 + y^3}$. Then

$$\int_0^{2\pi} \sqrt{1 + \sin^3 x}\, dx = \int_0^{2\pi} \sqrt{1 + \cos^3 x}\, dx.$$

■ **EXAMPLE 5**   Use Example 4 to find $\displaystyle\int_0^{2\pi} \sin^2 9x\, dx$.

Solution   Set

$$A = \int_0^{2\pi} \cos^2 9x\, dx \qquad \text{and} \qquad B = \int_0^{2\pi} \sin^2 9x\, dx.$$

By Example 4, we have $A = B$, so $A + B = 2B$. But

$$A + B = \int_0^{2\pi} (\cos^2 9x + \sin^2 9x)\, dx = \int_0^{2\pi} dx = 2\pi.$$

Therefore, $A = B = \pi$. $\qquad \blacksquare$

EXERCISES

Determine all lines of symmetry $x = c$ and all points $(c, 0)$ of symmetry

1 $y = (\cos x) \exp(x^2)$ **2** $y = x^3 - 5x$

3 $y = (x^2 - 1)(x + 3)$ **4** $y = (x^2 + 2x)(x + 3)(x + 5)$

5 $y = \sin x + \cos x$ **6** $y = (x^2 + x + 1)^{-1}$

7 $y = [1 + (x + 1)^2]^{1/2}$ **8** $y = \exp(\sin x)$.

Reduce by symmetry (and periodicity) to an integral over a shorter interval, but do not evaluate

9 $\displaystyle\int_{-1}^{2} (x^3 - 5x)\, dx$ **10** $\displaystyle\int_{-4}^{4} \sin(x^2)\, dx$

11 $\displaystyle\int_{-2}^{3} \frac{2x - 1}{x^2 - x + 1}\, dx$ **12** $\displaystyle\int_{-4}^{0} \sqrt{x^2 + 4x + 6}\, dx$

13 $\displaystyle\int_0^{3\pi} (\sin x + \cos x)\, dx$ **14** $\displaystyle\int_0^{\pi} (\sin x + \cos x)\, dx$

15 $\displaystyle\int_0^{2\pi} \sin^4 x\, dx$ **16** $\displaystyle\int_0^{3\pi} \frac{\sin x}{2 + \cos x}\, dx$

17 $\displaystyle\int_{-100\pi}^{100\pi} \sin(\tfrac{1}{12}x - 4)\, dx$ **18** $\displaystyle\int_1^{4} \sin(\pi x + 3)\, dx$.

19 Find all functions $f(x)$ defined on $[-a, a]$ that are both even and odd.

20 Suppose $f(x)$ is defined for $-\infty < x < \infty$ and has both $x = 0$ and $x = c$ as axes of symmetry, where $c > 0$. Show that $f(x)$ is periodic.

A function $y = f(x)$, defined on the interval $[a - h, a + h]$, is **symmetric** in (a, b) if

$$\frac{f(a - x) + f(a + x)}{2} = b, \qquad 0 \le x \le h.$$

21   Show that $b = f(a)$ and find $\displaystyle\int_{a-h}^{a+h} f(x)\,dx$.

22* (cont.) Show that the Trapezoidal Rule on $[a - h,\ a + h]$, with any n, gives the exact integral.

8. INEQUALITIES AND ESTIMATES

Inequalities   The definite integral satisfies a number of inequalities that have useful applications.

Inequalities   Let $f(x)$ and $g(x)$ be integrable on $[a, b]$.

(1)   If $f(x) \geq 0$,   then $\displaystyle\int_a^b f(x)\,dx \geq 0$.

(2)   If $f(x) \leq g(x)$,   then $\displaystyle\int_a^b f(x)\,dx \leq \int_a^b g(x)\,dx$.

(3)   If $m \leq f(x) \leq M$,   then $m(b - a) \leq \displaystyle\int_a^b f(x)\,dx \leq M(b - a)$.

(4) $\left| \displaystyle\int_a^b f(x)\,dx \right| \leq \displaystyle\int_a^b |f(x)|\,dx$.

(5)   If $|f(x)| \leq M$,   then $\left| \displaystyle\int_a^b f(x)g(x)\,dx \right| \leq M \displaystyle\int_a^b |g(x)|\,dx$.

Inequality (1) was already given in Section 3. Inequality (2) follows easily, because if $f(x) \leq g(x)$, then $g(x) - f(x) \geq 0$ and (1) applies:

$$\int_a^b g(x)\,dx - \int_a^b f(x)\,dx = \int_a^b [g(x) - f(x)]\,dx \geq 0,$$

so $\int f \leq \int g$. Now (3) is a direct consequence of (2). For instance, from $m \leq f(x)$ follows

$$\int_a^b m\,dx \leq \int_a^b f(x)\,dx, \qquad m(b - a) \leq \int_a^b f(x)\,dx.$$

Inequality (4) also follows from (2) because $f(x) \leq |f(x)|$ and $-f(x) \leq |f(x)|$. If we take for granted the (non-obvious) fact that $|f(x)|$ is integrable, then

$$\int_a^b f(x)\,dx \leq \int_a^b |f(x)|\,dx \qquad \text{and} \qquad -\int_a^b f(x)\,dx \leq \int_a^b |f(x)|\,dx.$$

Therefore $\left| \displaystyle\int_a^b f(x)\,dx \right| = \max\!\left(\displaystyle\int_a^b f(x)\,dx,\ -\int_a^b f(x)\,dx \right) \leq \displaystyle\int_a^b |f(x)|\,dx$.

Inequality (5) follows easily:

$$\left| \int_a^b f(x)g(x)\,dx \right| \leq \int_a^b |f(x)g(x)|\,dx \leq \int_a^b M|g(x)|\,dx = M \int_a^b |g(x)|\,dx.$$

We are taking for granted the (again non-obvious) fact that the product $f(x)g(x)$ is integrable.

■ **EXAMPLE 1**   Prove

(a) $\displaystyle\int_0^{\pi/3} \sin^5 x \, dx \le \int_0^{\pi/3} \sin^4 x \, dx$

(b) $\displaystyle\sqrt{3}\int_1^4 x^2 \, dx \le \int_1^4 x^2\sqrt{3 + e^{-x}} \, dx \le 2\int_1^4 x^2 \, dx.$

Solution   (a)   If $0 \le x \le \frac{1}{3}\pi$, then

$$0 \le \sin x < 1, \qquad \text{hence} \quad \sin^5 x \le \sin^4 x.$$

Now integrate this inequality, that is, apply (2).

(b)   If $x \ge 0$, then $0 < e^{-x} \le 1$. Therefore

$$\sqrt{3 + 0} < \sqrt{3 + e^{-x}} \le \sqrt{3 + 1} = 2, \qquad \text{so} \quad x^2\sqrt{3} \le x^2\sqrt{3 + e^{-x}} \le 2x^2.$$

Now apply (2).   ■

■ **EXAMPLE 2**   Prove $\displaystyle 2\int_a^b f(x)g(x) \, dx \le \int_a^b f^2(x) \, dx + \int_a^b g^2(x) \, dx.$

Solution   From $[f(x) - g(x)]^2 \ge 0$ follows

$$f^2(x) + g^2(x) \ge 2f(x)g(x)$$

Now apply (2).   ■

Remark   Inequalities (1)–(5) can be sharpened somewhat because of the following fact, which we give without proof. If $f(x) \ge 0$ and continuous on $[a, b]$, and if $f(x) > 0$ at some point of the interval, then

$$\int_a^b f(x) \, dx > 0.$$

Thus for continuous functions, strict inequality implies strict inequality in (1). The same statement applies to (2), (3), and (5) as well. For instance, the result of Example 1(a) can be stated as

$$\int_0^{\pi/3} \sin^5 x \, dx < \int_0^{\pi/3} \sin^4 x \, dx.$$

The inequality is strict because $\sin^5 x < \sin^4 x$ for $0 < x \le \frac{1}{3}\pi$ and both functions are continuous.

Estimates   We have seen that accurate estimates of difficult integrals can be obtained by approximate integration. Sometimes, however, only rough estimates are needed. In such cases, quick estimates are usually possible via simple techniques based on inequalities.

■ **EXAMPLE 3**   Show that $\displaystyle\int_1^3 \frac{dx}{1 + x^4} < \frac{1}{3}.$

Solution $\dfrac{1}{1 + x^4} < \dfrac{1}{x^4}$ for $x > 0$, hence

$$\int_1^3 \frac{dx}{1 + x^4} < \int_1^3 \frac{dx}{x^4} = \left. \frac{-1}{3x^3} \right|_1^3 = \frac{1}{3} - \frac{1}{81} < \frac{1}{3}. \qquad \blacksquare$$

Remark   Example 3 illustrates an important technique: replacing the integrand by a slightly larger one that is easily integrated. Generally, this technique yields better results than does use of inequality (3). For instance, all we can deduce from (3) is

$$\int_1^3 \frac{dx}{1 + x^4} < 1.$$

(Here $a = 1$, $b = 3$, and $M = \frac{1}{2}$, the largest value of $1/(1 + x^4)$ in the interval $1 \le x \le 3$.) Nevertheless, (3) is effective when great accuracy is not needed.

■ **EXAMPLE 4**   Estimate $\displaystyle\int_6^8 \frac{dx}{x^3 + x + \sin 2x}$.

Solution   Find bounds for the integrand. Since $-1 \le \sin 2x \le 1$ and $x^3 + x$ is increasing (positive derivative),

$$221 = 6^3 + 6 - 1 < x^3 + x + \sin 2x < 8^3 + 8 + 1 = 521$$

in the range $6 \le x \le 8$. Take reciprocals (and reverse inequalities):

$$1.9 \times 10^{-3} < \frac{1}{521} < \frac{1}{x^3 + x + \sin 2x} < \frac{1}{221} < 4.6 \times 10^{-3} \qquad \text{for} \quad 6 \le x \le 8.$$

By inequality (3), $3.8 \times 10^{-3} < \displaystyle\int_6^8 \frac{dx}{x^3 + x + \sin 2x} < 9.2 \times 10^{-3}$. $\qquad \blacksquare$

■ **EXAMPLE 5**   Show that $\left| \displaystyle\int_2^5 \frac{\sin x}{(1 + x)^2} \, dx \right| \le \dfrac{1}{6}$.

Solution   Since $|\sin x| \le 1$, we use (5):

$$\left| \int_2^5 \frac{\sin x}{(1 + x)^2} \, dx \right| \le \int_2^5 \frac{dx}{|1 + x|^2} = \int_2^5 \frac{dx}{(1 + x)^2} = \left. \frac{-1}{1 + x} \right|_2^5 = \frac{1}{3} - \frac{1}{6} = \frac{1}{6}. \qquad \blacksquare$$

■ **EXAMPLE 6**   Show $\displaystyle\int_0^3 e^{-x^2} \, dx$ is a good approximation for $\displaystyle\int_0^{100} e^{-x^2} \, dx$. How good?

Solution   The error in this approximation is

$$\int_0^{100} e^{-x^2} \, dx - \int_0^3 e^{-x^2} \, dx = \int_3^{100} e^{-x^2} \, dx.$$

Since e^{-x^2} is extremely small even for moderate values of x, the integral on the right is small. Estimate the integral by (3). Because e^{-x^2} is a decreasing function, its largest value in the interval $3 \le x \le 100$ occurs at the left end point; this value is e^{-9}. By (3),

$$\int_3^{100} e^{-x^2} \, dx < (100 - 3)e^{-9} < (97)(1.3 \times 10^{-4}) < 1.3 \times 10^{-2}.$$

A better estimate can be obtained using (2). Note that $x^2 \geq 3x$ when $x \geq 3$. In
In this range therefore, $e^{-x^2} \leq e^{-3x}$ (which can be integrated). Hence

$$\int_3^{100} e^{-x^2} \, dx \leq \int_3^{100} e^{-3x} \, dx = \frac{e^{-3x}}{-3}\Big|_3^{100} = \frac{e^{-9}}{3} - \frac{e^{-300}}{3} < \frac{e^{-9}}{3} < \frac{1.3 \times 10^{-4}}{3}$$

$$< 5 \times 10^{-5}. \qquad \blacksquare$$

Remark   This example illustrates an important labor-saving device. If you intend to estimate

$$\int_0^{100} e^{-x^2} \, dx$$

by an approximate integration method, like the Trapezoidal Rule, you can save a great deal of
work by applying the method to

$$\int_0^3 e^{-x^2} \, dx.$$

Ignoring the rest of the integral introduces an error less than 5×10^{-5}. If that is not precise
enough, you might apply approximate integration to

$$\int_0^4 e^{-x^2} \, dx.$$

Then by the same argument as in the example, ignoring the rest of the integral introduces an
error less than $\frac{1}{4}e^{-16} < 3 \times 10^{-8}$.

EXERCISES

In the following exercises you are asked to find bounds for certain integrals. There are
generally several ways to estimate each integral. Try to obtain the bound given or to improve
on it. If you cannot, at least find *some* bound.

Show

1   $1 < \int_0^1 \exp(x^2) < e - 1$

2   $\int_0^1 \frac{dx}{4 + x} < \int_0^1 \frac{dx}{4 + x^3} < \frac{1}{4}$

3   $3 \int_3^5 \frac{dx}{x} < \int_3^5 \frac{\sqrt{3 + 2x}}{x} \, dx < 4 \int_3^5 \frac{dx}{x}$

4   $\frac{1}{15} < \int_1^2 \frac{dx}{x^3 + 3x + 1} < \frac{1}{5}$

5   $\int_0^{100} e^{-x} \sin^2 x \, dx < 1$

6   $\frac{3}{10} < \int_1^4 \frac{dx}{x^2 + x + 1} < \frac{3}{4}$

7   $6 < \int_5^{11} \frac{x}{\sqrt{5 + 4x}} \, dx < 10$

8   $2 < \int_0^4 \frac{dx}{1 + \sin^2 x} < 4$

9   $\frac{\pi}{2} < \int_0^{\pi/2} \frac{d\theta}{\sqrt{1 - k^2 \sin^2 \theta}} < \frac{\pi}{2\sqrt{1 - k^2}}$
$(0 < k < 1)$

10   $\int_0^{\pi/3} \sin 2\theta \cos \theta \, d\theta < \frac{3}{4}$

11   $\frac{15}{16} < \int_1^2 x^3 2^{-x} \, dx < \frac{15}{8}$

12   $\frac{9}{10}(1 - e^{-1}) < \int_1^{10} \frac{1 - e^{-x}}{x^2} \, dx < \frac{9}{10}$

13   $\int_0^1 \frac{\sin x + \cos x}{(1 + x)^2} \, dx < \frac{1}{2}\sqrt{2}$
[*Hint* $\sin x + \cos x = \sqrt{2} \cos(x - \frac{1}{4}\pi).$]

14* $\frac{2 - \sqrt{3}}{\sqrt{2}} < \int_0^{\pi/6} \sqrt{\sin x} \, dx < \frac{2}{3}\left(\frac{\pi}{6}\right)^{3/2}.$

15 Prove $\dfrac{7}{12} < \displaystyle\int_1^2 \dfrac{dx}{x} < \dfrac{5}{6}$ by splitting the interval $[1, 2]$ into two parts.

16 (cont.) Obtain bounds for $\displaystyle\int_1^2 \dfrac{dx}{1 + x^2}$ by a similar computation.

17 Prove $2 - \sqrt{2} < \displaystyle\int_1^2 \dfrac{dx}{x} < 2(\sqrt{2} - 1)$ by comparing the integrand to functions you

can integrate. [*Hint* $x^{1/2} < x < x^{3/2}$ for $x > 1$.]

18* (cont.) Prove the same inequality by splitting the interval into two cleverly chosen unequal parts.

19 Estimate how closely $\displaystyle\int_0^{10} e^{-x} \sin^2 x \, dx$ approximates $\displaystyle\int_0^{100} e^{-x} \sin^2 x \, dx$.

20 Find a as small as you can so that

$$\int_1^a \frac{1 + \sin x}{x^5} \, dx \qquad \text{approximates} \qquad \int_1^{200} \frac{1 + \sin x}{x^5} \, dx$$

within 5×10^{-5}.

21* Suppose $\displaystyle\int_a^b f(x)^2 \, dx = 0$. Prove that $\displaystyle\int_a^b f(x)g(x) \, dx = 0$ for each integrable $g(x)$.

[*Hint* Use Example 2; replace $g(x)$ by $\pm \varepsilon g(x)$.]

22* (cont.) For any integrable $f(x)$ and $g(x)$, prove

$$\left(\int_a^b f(x)g(x) \, dx \right)^2 \leq \left(\int_a^b f(x)^2 \, dx \right)\left(\int_a^b g(x)^2 \, dx \right).$$

[*Hint* Set $\int f^2 = A$, $\int fg = B$, $\int g^2 = C$ and consider $\int(\sqrt{C} f - \sqrt{A} g)^2$.]

23 (cont.) Prove $\displaystyle\sqrt{\int_a^b [f(x) + g(x)]^2 \, dx} \leq \sqrt{\int_a^b f(x)^2 \, dx} + \sqrt{\int_a^b g(x)^2 \, dx}$.

[*Hint* Write $(f + g)^2 = f(f + g) + g(f + g)$ and use Ex. 22.]

24* Prove $\displaystyle\int_a^b [f'(t)]^2 \, dt \geq \dfrac{[f(b) - f(a)]^2}{b - a}$. Assume $a < b$. [*Hint* Use Ex. 22.]

9. INSIGHTS INTO INTEGRATION

In this section we modify the approach to integration via approximating sums, and examine the reasons behind certain fundamental properties of the integral. To a large extent we shall start from scratch.

Step Functions   Our basic strategy will be to approximate integrable functions by functions of a very simple type, piecewise constant functions.

Definition   A **step function** on $[a, b]$ is a function $s(x)$ that is piecewise constant. That is, there is a partition Π :

$$a = x_0 < x_1 < \cdots < x_{n-1} < x_n = b$$

of $[a, b]$ into a finite number of subintervals such that $s(x)$ is constant on each *open* interval $x_{j-1} < x < x_j$ for $j = 1, \ldots, n$. The values $s(x_j)$ at the partition points can be anything.

See Fig. 1 for an illustration of a step function.

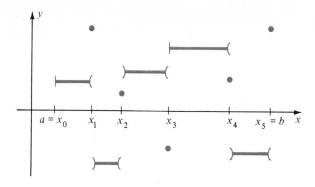

Fig. 1   Step function

Step functions can be combined in various ways to form new step functions.

If $s_1(x)$ and $s_2(x)$ are step functions on $[a, b]$ and k is a constant, then each of the following is also a step function:

$$s_1(x) + s_2(x), \qquad ks_1(x), \qquad |s_1(x)|, \qquad s_1(x)s_2(x).$$

The proof is routine. You lump together all of the partition points for $s_1(x)$ with all of those for $s_2(x)$ into one long string:

$$a = z_0 < z_1 < z_2 < \cdots < z_m = b.$$

Then $s_1(x)$ and $s_2(x)$ are both constant on $z_{j-1} < x < z_j$, etc.

We define the integral of a step function in a natural way, motivated by the area of rectangles.

Definition   Let $s(x)$ be a step function on $[a, b]$ with partition points

$$a = x_0 < x_1 < \cdots < x_m = b.$$

Suppose $s(x) = B_j$ for $x_{j-1} < x < x_j$. The **integral** of $s(x)$ on $[a, b]$ is

$$\int_a^b s(x)\, dx = \sum_{j=1}^n B_j(x_j - x_{j-1}).$$

Note that the values $s(x_j)$ at the partition points do not enter into the formula. Note also that if some additional partition points are inserted, then the integral does not change. If $s(x)$ is constant, then $s(x) = B$ on $a < x < b$ and

$$\int_a^b s(x)\, dx = B(b - a).$$

Here are the main properties of the integrals of step functions. The (easy) proofs are left as exercises.

Theorem   Let s, s_1, and s_2 be step functions on $[a, b]$. Then

(1) $\displaystyle\int_a^b (s_1 + s_2)(x)\, dx = \int_a^b s_1(x)\, dx + \int_a^b s_2(x)\, dx.$

(2) $\displaystyle\int_a^b ks(x)\, dx = k \int_a^b s(x)\, dx.$

(3) If $s(x) \geq 0$ for all $x \in [a, b]$,   then $\displaystyle\int_a^b s(x)\, dx \geq 0.$

(4) If $s_1(x) \leq s_2(x)$ for all $x \in [a, b]$,   then $\displaystyle\int_a^b s_1(x)\, dx \leq \int_a^b s_2(x)\, dx.$

(5) If $a < c < b$, then $\displaystyle\int_a^b s(x)\, dx = \int_a^c s(x)\, dx + \int_c^b s(x)\, dx.$

The Integral   We deal only with *bounded* functions $f(x)$ on $[a, b]$. If $|f(x)| \leq B$ for all $x \in [a, b]$, then there exist *some* step functions $s(x)$ and $S(x)$ such that

(*) $\qquad\qquad\qquad s(x) \leq f(x) \leq S(x) \qquad$ for all x.

For example $s(x) = -B$ and $S(x) = B$. We are interested in *all* step functions that satisfy (*). By studying the integrals of all such functions we hope to squeeze down on the "area" under the graph of f. See Fig. 2.

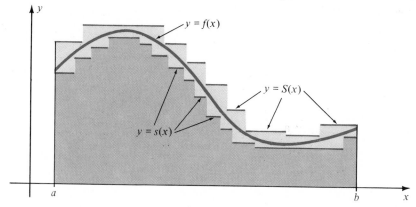

Fig. 2   Squeezing a function between two step functions: $s(x) \leq f(x) \leq S(x)$.

Definition   A bounded function $f(x)$ on $[a, b]$ is **integrable** if for each $\varepsilon > 0$, there exist step functions s and S such that

(1) $s(x) \leq f(x) \leq S(x)$ for all $x \in [a, b]$,

(2) $\displaystyle\int_a^b S(x)\, dx - \int_a^b s(x)\, dx < \varepsilon.$

Thus $f(x)$ is integrable provided its graph can be squeezed between the graphs of two step functions whose integrals are as close together as we please. If $f(x)$ is integrable on $[a, b]$, we can assign to $f(x)$ a number called its integral.

Theorem   Let $f(x)$ be integrable on $[a, b]$. Then there exists a *unique* real number I with the following property: If $s(x)$ and $S(x)$ are any step functions satisfying $s(x) \leq f(x) \leq S(x)$, then

$$\int_a^b s(x)\, dx \leq I \leq \int_a^b S(x)\, dx.$$

The number I is called the **integral** of $f(x)$ on $[a, b]$, and is written

$$I = \int_a^b f(x)\, dx.$$

The *existence* of at least one such number I follows from certain very basic properties of the real number system that are beyond the scope of this course. However, the *uniqueness* of I is not hard to prove from the integrability of $f(x)$. For suppose there were *two* numbers I_1 and I_2, both satisfying

$$\int_a^b s(x)\, dx \leq I \leq \int_a^b S(x)\, dx$$

for every pair of step functions such that $s(x) \leq f(x) \leq S(x)$. Since $f(x)$ is integrable, given any $\varepsilon > 0$, there are such step functions satisfying

$$\int_a^b S(x)\, dx - \int_a^b s(x)\, dx < \varepsilon.$$

This implies that I_1 and I_2 must be within ε of each other, that is, $|I_1 - I_2| < \varepsilon$. Since ε can be *any* positive number, it follows that $|I_1 - I_2| = 0$, $I_1 = I_2$.

Remark   The integral we have defined is called the **Riemann integral** and integrable functions are also called **R-integrable**. There are other integrals that apply to even more functions than does the R-integral. (Riemann is pronounced ree'-mahn.)

Continuous Functions   Our next goal is to justify the very important theorem, asserted in Section 3, that every continuous function on an interval $[a, b]$ is integrable. This fact depends on two properties of continuous functions that we shall have to accept since their proofs are too technical for this course. The first of these we have already met on p. 146:

Let $f(x)$ be continuous on a closed interval $[a, b]$. Then $f(x)$ is bounded and has both a maximum and a minimum on $[a, b]$. That is, there exist points z and w in $[a, b]$ such that

$$f(z) \leq f(x) \leq f(w) \qquad \text{for all} \quad x \quad \text{in} \quad [a, b].$$

The second property of continuous functions is called uniform continuity.

Definition   A function $f(x)$ with domain $[a, b]$ is **uniformly continuous** if for each $\varepsilon > 0$ there exists $\delta > 0$ such that

$$|f(z) - f(w)| < \varepsilon \qquad \text{whenever} \quad |z - w| < \delta.$$

This looks much like the definition of continuity, but there is a difference lurking in the word "uniformly". Remember that $f(x)$ is *continuous* on $[a, b]$ provided $f(x)$ is continuous at each point of $[a, b]$. Now $f(x)$ is continuous at c if for each $\varepsilon > 0$ there exists $\delta > 0$ such that $|f(x) - f(c)| < \varepsilon$ whenever $|x - c| < \delta$. Here δ *depends on* c. Change c and you may need a much smaller δ. Uniform continuity means that, given ε, there is one single δ that works at all points of the interval.

Now we can state the new basic property of continuous functions.

Theorem   Let $f(x)$ be continuous on a *closed* interval $[a, b]$. Then $f(x)$ is uniformly continuous on $[a, b]$.

The operating word in the theorem is "closed". Without it, the theorem is false. For example, on the open interval $0 < x < 1$, the function $f(x) = 1/x$ is continuous but not uniformly continuous. Given $\varepsilon > 0$, the closer x is to 0 the smaller δ must be. In fact $\delta \longrightarrow 0$ as $x \longrightarrow 0+$. Therefore, no single δ can work at all points of this interval.

We are nearly ready to prove that every continuous function is integrable. We shall use a corollary of the last theorem.

Corollary   Let $f(x)$ be continuous on $[a, b]$ and let $\varepsilon > 0$. Then there exist step functions $s(x)$ and $S(x)$ such that

$$s(x) \leq f(x) \leq S(x) \qquad \text{and} \qquad S(x) - s(x) < \varepsilon.$$

Proof   By the theorem, $f(x)$ is uniformly continuous on $[a, b]$. Hence there exists $\delta > 0$ such that

$$|f(z) - f(w)| < \varepsilon \qquad \text{whenever} \quad |z - w| < \delta.$$

Choose n so large that $(b - a)/n < \delta$. Then partition $[a, b]$ into n equal subintervals; each will have length less than δ:

$$\Pi: \quad a = x_0 < x_1 < x_2 < \cdots < x_n = b.$$

On the j-th (closed) subinterval $[x_{j-1}, x_j]$, the values of $f(x)$ vary very little. In fact, if z and w are any two points of this subinterval, then $|z - w| < \delta$, so

$$|f(z) - f(w)| < \varepsilon.$$

Define
$$M_j = \max\{f(x) \mid x_{j-1} \leq x \leq x_j\},$$
$$m_j = \min\{f(x) \mid x_{j-1} \leq x \leq x_j\}.$$

By the basic max and min property of continuous functions, $M_j = f(z_j)$ and

$m_j = f(w_j)$ for some points z_j and w_j of $[x_{j-1}, x_j]$. By uniform continuity,

$$M_j - m_j < \varepsilon \qquad (j = 1, 2, \cdots, n).$$

Now define step functions $S(x)$ and $s(x)$ by

$$S(x) = \begin{cases} M_j \\ f(x_j) \end{cases} \qquad \text{if} \qquad \begin{matrix} x_{j-1} < x < x_j \\ x = x_j, \end{matrix}$$

$$s(x) = \begin{cases} m_j \\ f(x_j) \end{cases} \qquad \text{if} \qquad \begin{matrix} x_{j-1} < x < x_j \\ x = x_j. \end{matrix}$$

Clearly, $\qquad s(x) \le f(x) \le S(x) \qquad \text{and} \qquad S(x) - s(x) < \varepsilon.$

This completes the proof of the corollary. The main theorem now follows easily.

Theorem   Each continuous function on $[a, b]$ is integrable.

Proof Let $f(x)$ be continuous on $[a, b]$. Given $\varepsilon > 0$, we must produce step functions $S(x) \ge f(x) \ge s(x)$ such that

$$\int_a^b S(x)\, dx - \int_a^b s(x)\, dx < \varepsilon.$$

Apply the corollary, but with ε replaced by $\varepsilon/(b - a)$: There exist step functions $S(x) \ge f(x) \ge s(x)$ such that $S(x) - s(x) < \varepsilon/(b - a)$. It follows that

$$\int_a^b S(x)\, dx - \int_a^b s(x)\, dx = \int_a^b [S(x) - s(x)]\, dx < \int_a^b \frac{\varepsilon}{b - a}\, dx = (b - a)\frac{\varepsilon}{b - a} = \varepsilon.$$

Therefore $f(x)$ is integrable.

Properties of Integrals The various formal properties of integrals hat we have studied, such as

$$\int_a^b [f(x) + g(x)]\, dx = \int_a^b f(x)\, dx + \int_a^b g(x)\, dx,$$

$$\int_a^b f(x)\, dx \ge 0 \qquad \text{if} \quad f(x) \ge 0, \qquad \int_a^b f(x)\, dx = \int_a^c f(x)\, dx + \int_c^b f(x)\, dx,$$

can be proved in detail via step functions and the definitions we have given. The proofs, however, are tedious and not too useful to us at this point.

The Fundamental Theorem of Calculus We now come to the most important thing in this section, a proof of the Fundamental Theorem, that important link between differential and integral calculus. Actually, there are two related theorems each of which is sometimes called the Fundamental Theorem. The first says that if $f(x)$ is continuous, then

$$F(x) = \int_a^x f(t)\, dt$$

is an antiderivative of $f(x)$ and

$$\int_a^b f(x)\,dx = F(b) - F(a).$$

The second says that if the derivative $F'(x)$ is integrable, then

$$\int_a^b F'(x)\,dt = F(b) - F(a).$$

We shall prove both versions.

Fundamental Theorem   Let $f(x)$ be continuous on $[a, b]$ and set

$$F(x) = \int_a^x f(t)\,dt.$$

Then $F(x)$ is differentiable on $[a, b]$ and

$$\frac{d}{dx} F(x) = f(x).$$

Proof   Given $\varepsilon > 0$, we must produce $\delta > 0$ such that

$$\left| \frac{F(x+h) - F(x)}{h} - f(x) \right| < \varepsilon \qquad \text{whenever} \quad |h| < \delta.$$

Since f is continuous, there exists $\delta > 0$ such that

$$|f(t) - f(x)| < \varepsilon \qquad \text{whenever} \quad |t - x| < \delta.$$

But

$$F(x+h) - F(x) = \left(\int_a^{x+h} - \int_a^x \right) f(t)\,dt = \int_x^{x+h} f(t)\,dt$$

and

$$f(x) = f(x)\left(\frac{1}{h} \int_x^{x+h} dt \right) = \frac{1}{h} \int_x^{x+h} f(x)\,dt,$$

so

$$\left| \frac{F(x+h) - F(x)}{h} - f(x) \right| = \left| \frac{1}{h} \int_x^{x+h} f(t)\,dt - \frac{1}{h} \int_x^{x+h} f(x)\,dt \right|$$

$$= \left| \frac{1}{h} \int_x^{x+h} [f(t) - f(x)]\,dt \right| \leq \left| \frac{1}{h} \int_x^{x+h} |f(t) - f(x)|\,dt \right| < \left| \frac{1}{h} \int_x^{x+h} \varepsilon\,dt \right| = \varepsilon,$$

provided $|h| < \delta$. (The extra bars in the next to last expression are to allow for h negative.) This completes the proof.

This theorem guarantees that every *continuous* function has an antiderivative. (But it is of little practical help for exhibiting an explicit antiderivative.)

■ **EXAMPLE 1**   (a)   Find an antiderivative of $f(x) = 1/(1 + x^2)$.

(b)   Use it to evaluate $\displaystyle\int_0^1 \frac{dx}{1 + x^2}.$

Solution   (a)   Since $f(x)$ is a continuous function, $F(x) = \displaystyle\int_0^x \frac{dt}{1 + t^2}$

is an antiderivative.

(b) $\displaystyle\int_0^1 \frac{dx}{1 + x^2} = F(x) \Big|_0^1 = \int_0^1 \frac{dt}{1 + t^2}.$ ∎

Remark   This example shows that merely knowing that an antiderivative exists and giving it a name are insufficient for evaluating an integral. (See Example 2, p. 334 for the integral in (b).)

Let us pass to the second version of the Fundamental Theorem. The proof will depend on the following lemma:

Lemma   Let $F(x)$ be continuous on $[a, b]$ and let $F'(x)$ exist for $a < x < b$. Suppose

$$n \le F'(x) \le M.$$

Then

$$m(b - a) \le F(b) - F(a) \le M(b - a).$$

Proof   Let us prove only the right-hand inequality. The other is proved similarly. We set

$$G(x) = M(x - a) - F(x).$$

Then $G(x)$ is continuous on $[a, b]$ and differentiable for $a < x < b$. What is more,

$$G'(x) = M - F'(x) \ge 0.$$

Therefore $G(x)$ is an increasing function (as proved on p. 150). Hence $G(b) \ge G(a)$, that is,

$$M(b - a) - F(b) \ge F(a), \qquad M(b - a) \ge F(b) - F(a).$$

Fundamental Theorem   Let $F(x)$ be differentiable on $[a, b]$, and assume that $F'(x)$ is integrable on $[a, b]$. Then

$$\int_a^b F'(x)\, dx = F(b) - F(a) = F(x) \Big|_a^b.$$

Proof   Let $s(x)$ and $S(x)$ be any pair of step functions on $[a, b]$ such that

$$s(x) \le F'(x) \le S(x).$$

Let

$$a = x_0 < x_1 < x_2 < \cdots < x_n = b$$

be a partition of $[a, b]$ such that both $S(x)$ and $s(x)$ are constant on each subinterval $x_{j-1} < x < x_j$. By the lemma,

$$\int_{x_{j-1}}^{x_j} s(x)\, dx \le F(x_j) - F(x_{j-1}) \le \int_{x_{j-1}}^{x_j} S(x)\, dx.$$

Sum these inequalities; the inside sum telescopes:

$$\int_a^b s(x)\,dx \le F(b) - F(a) \le \int_a^b S(x)\,dx.$$

But $F'(x)$ is integrable, so

$$I = \int_a^b F'(x)\,dx$$

is the *unique* real number such that $\int s \le I \le \int S$ for *all* step functions satisfying $s(x) \le F'(x) \le S(x)$. We have just seen that $F(b) - F(a)$ is such a number. Therefore

$$\int_a^b F'(x)\,dx = F(b) - F(a),$$

Remark   Note that $F'(x)$ is merely assumed *integrable*, not necessarily continuous, in this version of the Fundamental Theorem.

EXERCISES

Prove the following parts of the theorem on p. 253.

1   part 1   **2**   part 2   **3**   part 3
4   part 4   **5**   part 5.

6   Let $f(x) = x$ on $[0, 1]$. Find step functions $s(x) \le f(x) \le S(x)$ so that $\int_0^1 S - \int_0^1 s < 0.1$.

7   Let $f(x) = x^2$ on $[0, 1]$. Find step functions $s(x) \le f(x) \le S(x)$ so that $\int_0^1 S - \int_0^1 s < \frac{1}{2}$.

(Can you do it with two steps?)
8   Suppose $f(x)$ is continuous on $[a, b]$, $f(x) \ge 0$, and $\int_a^b f(x)\,dx = 0$.   Prove that $f(x) = 0$ on $[a, b]$.
9   Let $f(x)$ be a bounded function on $[a, b]$ that can be **uniformly approximated** by step functions. That is, given $\varepsilon > 0$, there exists a step function $s(x)$ such that $|f(x) - s(x)| < \varepsilon$ on $[a, b]$.   Prove that $f(x)$ is integrable.
10*   (cont.) Suppose $f(x)$ can be uniformly approximated on $[a, b]$ by *integrable* functions. Prove that $f(x)$ is integrable on $[a, b]$.
11   Prove that if $f(x)$ and $g(x)$ are integrable on $[a, b]$, then so is $f(x) + g(x)$, and its integral is the sum of the integrals of $f(x)$ and of $g(x)$.
12   Prove that if $f(x)$ is integrable and k is a constant, then $kf(x)$ is integrable and $\int_a^b kf = k \int_a^b f$.   (Be sure to take the sign of k into account.)
13   Let $f(x)$ be integrable on $[a, b]$ and $f(x) \ge 0$. Prove that $\int_a^b f \ge 0$.
14   Let $f(x)$ be any increasing function on $[a, b]$. Prove that $f(x)$ is integrable. [*Hint*   Partition the interval into n equal parts.]
15   Let $a < c < b$ and let $f(x)$ be a bounded function on $[a, b]$. Suppose f is integrable on $[a, c]$ and on $[c, b]$. Prove that f is integrable on $[a, b]$.
16   (cont.) Conversely, suppose f is integrable on $[a, b]$. Prove that f is integrable on $[a, c]$ and on $[c, b]$, and that

$$\int_a^b f(x)\,dx = \int_a^c f(x)\,dx + \int_c^b f(x)\,dx.$$

The next four exercises will prove that the product of integrable functions is integrable, a deeper property of the integral that is not at all obvious.

17 Let $s_1(x), \cdots, S_2(x)$ be step functions on $[a, b]$ such that

$$0 \le s_1(x) \le S_1(x) \le M, \qquad 0 \le s_2(x) \le S_2(x) \le M,$$

$$\int_a^b S_1 - \int_a^b s_1 < \varepsilon, \qquad \text{and} \qquad \int_a^b S_2 - \int_a^b s_2 < \varepsilon.$$

Prove
$$\int_a^b S_1 S_2 - \int_a^b s_1 s_2 < 2M\varepsilon.$$

[*Hint* $S_1 S_2 - s_1 s_2 = S_1(S_2 - s_2) + s_2(S_1 - s_1).$]

18 (cont.) Suppose $f(x)$ is integrable on $[a, b]$ and $0 \le f(x) \le M$. Given $\varepsilon > 0$, prove that there are step functions $s(x)$ and $S(x)$ such that

$$0 \le s(x) \le f(x) \le S(x) \le M \qquad \text{and} \qquad \int_a^b S - \int_a^b s < \varepsilon.$$

19* (cont.) Suppose $f(x)$ and $g(x)$ are integrable on $[a, b]$, $f(x) \ge 0$, and $g(x) \ge 0$. Prove that $f(x)g(x)$ is integrable.

20 (cont.) Suppose $f(x)$ and $g(x)$ are integrable on $[a, b]$. Prove that $f(x)g(x)$ is integrable. [*Hint* Add constants to $f(x)$ and to $g(x)$, and use Exs. 11 and 12.]

21* Suppose f is bounded on $[a, b]$. Suppose for each $\varepsilon > 0$, there exist *integrable* functions g and G such that (1) $g(x) \le f(x) \le G(x)$ for all x and (2) $\int_a^b G \, dx - \int_a^b g \, dx < \varepsilon$. Prove that f is integrable.

22* Let $f'(x)$ be integrable on $[a, b]$ and $|f'(x)| \le M$. Suppose also that $f(a) = f(b) = 0$. Prove

$$\left| \int_a^b f(x) \, dx \right| \le \tfrac{1}{4} M (b - a)^2.$$

23 Suppose $f(x)$ is twice differentiable and periodic with period p. Suppose $f''(x)$ is integrable. Prove

$$\int_a^{a+p} f''(x) \, dx = \int_a^{a+p} f'(x) \, dx = 0.$$

24* Suppose $f(x)$ is continuous on $0 \le x < \infty$ and

$$[f(x)]^2 = 2 \int_0^x f(t) \, dt \qquad \text{for all} \quad x \ge 0.$$

Find $f(x)$. [*Hint* Differentiate, then be careful.]

10. MISCELLANEOUS EXERCISES

Evaluate

1 $\displaystyle\int_{-2}^{2} (x^2 - 4)^2 \, dx$

2 $\displaystyle\int_0^{-a} 14(x - a)^6 \, dx$

3 $\displaystyle\int_1^2 \left(x - \frac{1}{x^2} \right) dx$

4 $\displaystyle\int_{5\pi/4}^{7\pi/4} \cos x \, dx$

5 $\displaystyle\int_{-2}^{2} \frac{x^3}{1 + x^8} \, dx$

6 $\displaystyle\int_0^{\pi/3} x \cos 3x \, dx$

7 $\displaystyle\left(\frac{d}{dx} \int_0^x \frac{dt}{\sqrt{t^4 + 3}} \right) \bigg|_{x=1}$

8 $\displaystyle\left(\frac{d}{dx} \int_0^{x^2} \frac{dt}{\sqrt{t^2 + 9}} \right) \bigg|_{x=2}$

9 $\displaystyle\frac{d}{dx} \int_x^1 \exp(-t^2) \, dt$

10 $\displaystyle\frac{d}{dx} \int_x^{x+1} t^3 e^{-t} \, dt.$

11 Find the area bounded by the x-axis, $x = -2$, $x = 0$, and the curve $y = x^3 + 4x + 5$.

12 Find the area bounded by $x = 1$, $x = 2$, $y = x^2$, and $y = x^4$.

Find the mean value of f

13 $f(x) = x \sin x$, $0 \le x \le \pi$ **14** $f(x) = e^{x/k}$, $-k \le x \le k$.

15 Let $c > 0$. Suppose $y = f(x)$ is symmetric in $(0, 0)$ and in $(c, 0)$. Prove $\displaystyle\int_{a-c}^{a+c} f(x)\, dx$

is independent of a.

16* Let $f(x)$ be strictly convex for $a \le x \le b$ and let $c = \frac{1}{2}(a + b)$. Prove

$$f\left(\frac{a+b}{2}\right) < M(f)\left[= \frac{1}{b-a}\int_a^b f(x)\, dx\right] < \frac{f(a) + f(b)}{2}.$$

17 Find $\displaystyle f(x) = \int_0^x |t|\, dt.$

18 Graph $\displaystyle y = \int_0^x f(t)\, dt$,   where $\begin{cases} f(t) = 1 & \text{for} \quad 4n - 1 \le t < 4n + 1 \\ f(t) = -1 & \text{for} \quad 4n + 1 \le t < 4n + 3. \end{cases}$

19 Estimate $\dfrac{1}{\sqrt{n+1}} + \dfrac{1}{\sqrt{n+2}} + \dfrac{1}{\sqrt{n+3}} + \cdots + \dfrac{1}{\sqrt{2n}}.$

20 Estimate $\dfrac{1}{\sqrt[3]{2n+1}} + \dfrac{1}{\sqrt[3]{2n+2}} + \cdots + \dfrac{1}{\sqrt[3]{3n}}.$

Applications of Integration **6**

1. INTRODUCTION

The definite integral

$$\int_a^b f(x)\, dx,$$

which was introduced in order to compute areas, turns out to be a powerful tool, not only in area problems, but in a surprisingly large number of other applications. The reason for its great versatility is basically this: the integral is a practical device for adding up lots of tiny quantities.

For example, take the problem of computing the area of a region under the curve $y = f(x)$. The region can be sliced vertically into a large number of thin pieces, each approximately a rectangle of area $f(\bar{x}_i)\, \Delta x$, where Δx is the width of the slice. The integral "adds up" these tiny areas. Even the notation is suggestive: $f(x)\, dx$ represents $f(\bar{x}_i)\, \Delta x$, and the symbol

$$\int_a^b f(x)\, dx$$

means "sum" the quantities $f(x)\, dx$ between $x = a$ and $x = b$. The integral sign $\int$ was originally an S for sum.

It happens in many applications other than area, that a complicated quantity can be divided into a large number of small parts, each given by an expression of the type $f(\bar{x}_i)\, \Delta x$. The integral "adds up" these parts just as it does for area. Here are three typical examples.

Volume of Revolution   A vase is being shaped on a potter's wheel (Fig. 1). For each x between a and b, its cross section is a circle of radius $f(x)$. What is the volume of the vase?

Slice the vase into thin slabs by cuts perpendicular to the x-axis. Each slab is nearly a cylindrical disk of volume

$$\text{(area of base)} \cdot \text{(thickness)} = [\pi f^2(\bar{x}_i)]\, \Delta x.$$

The integral

$$\int_a^b \pi f^2(x)\, dx$$

"adds up" these small volumes and gives the total volume of the vase.

262

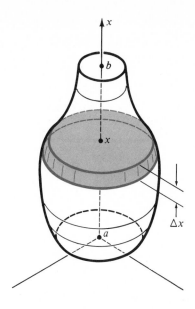

Fig. 1   Radius of slice $\approx f(x)$

Area of slice $\approx \pi f(x)^2$

Volume of slice $\approx \pi f(x)^2 \, \Delta x$

Volume of vase: $V = \int_a^b \pi f(x)^2 \, dx$

Work   Suppose at each point of the x-axis there is a force of magnitude $f(x)$ pulling a particle. How much work is done by the force in moving the particle from $x = a$ to $x = b$?

Slice the interval from a to b into a large number of small pieces of length Δx. In the i-th piece the force is nearly constant, so the work it does there is approximately

$$(\text{force}) \cdot (\text{distance}) = f(\bar{x}_i) \, \Delta x.$$

The integral $$\int_a^b f(x) \, dx$$

"adds up" these little bits of work and gives the total work done.

Distance   If a particle moves to the right along the x-axis with velocity $v(t)$ at time t, how far does it move between $t = a$ and $t = b$?

Divide the time interval into a large number of very short equal time intervals, each of duration Δt. In the i-th short time interval, the velocity is practically constant, so the distance traveled in this short period of time is approximately

$$(\text{velocity}) \cdot (\text{time}) = v(\bar{t}_i) \, \Delta t.$$

The integral $$\int_a^b v(t) \, dt$$

"adds up" all these little distances and gives the overall distance traveled.

Summary   The integral "adds up" many small quantities:

AREA = sum of thin rectangles of area $f(\bar{x}_i) \, \Delta x$;

VOLUME OF REVOLUTION = sum of thin cylindrical disks of volume $\pi f^2(\bar{x}_i) \, \Delta x$;

WORK = sum of small amounts of work, $f(\bar{x}_i)\,\Delta x$;

DISTANCE = sum of short distances, $v(\bar{t}_i)\,\Delta t$.

2. AREA

Suppose we want the area of the region (Fig. 1) bounded by two curves $y = f(x)$ and $y = g(x)$, where $g(x) \le f(x)$, and the lines $x = a$ and $x = b$. We think of the region as approximately a large number of thin rectangles. A typical one, shown in Fig. 1, has height $[f(\bar{x}_i) - g(\bar{x}_i)]$, width Δx, and area $[f(\bar{x}_i) - g(\bar{x}_i)]\,\Delta x$, which we shall abbreviate by $[f(x) - g(x)]\,dx$. The integral

$$A = \int_a^b [f(x) - g(x)]\,dx,$$

"adds up" these areas, so it is the required area.

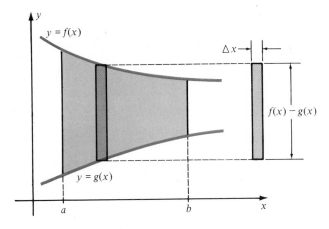

Fig. 1 Length of rectangle: $\approx f(x) - g(x)$

Area of rectangle: $\approx [f(x) - g(x)]\,\Delta x$

Area of region: $A = \int_a^b [f(x) - g(x)]\,dx$

It is important to keep straight which is the upper boundary and which is the lower boundary. If we get the upper and lower boundaries reversed, then we compute

$$\int_a^b [g(x) - f(x)]\,dx,$$

which is the negative of the area. If the curves cross, the upper and lower boundaries reverse (Fig. 2). Then we compute the shaded area by

$$\int_a^b [f(x) - g(x)]\,dx + \int_b^c [g(x) - f(x)]\,dx.$$

Under each integral sign, the upper curve comes first. If we compute just

$$\int_a^c [f(x) - g(x)]\, dx,$$

the two areas will be counted with opposite signs and the result will not be the geometric area.

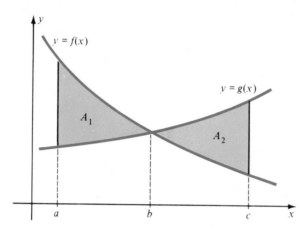

Fig. 2 The *geometric* area is

$$A = A_1 + A_2$$

$$= \int_a^b [f(x) - g(x)]\, dx + \int_b^c [g(x) - f(x)]\, dx.$$

In general, the absolute value $|f(x) - g(x)|$ always has the right sign for measuring geometric area, so we can state a principle:

The geometric area of the region bounded by the curves $y = f(x)$ and $y = g(x)$ and the lines $x = a$ and $x = b$ is

$$\int_a^b |f(x) - g(x)|\, dx.$$

In particular, if $g(x) \le f(x)$, then the geometric area is simply

$$\int_a^b [f(x) - g(x)]\, dx.$$

■ **EXAMPLE 1**   Compute the area of the region bounded by the curves $y = e^{x/2}$ and $y = 1/x^2$, and the lines $x = 2$ and $x = 3$.

Solution   Sketch the region (Fig. 3). Think of it as being composed of thin rectangular slabs. The area of a typical slab is

$$dA = \left(e^{x/2} - \frac{1}{x^2} \right) dx.$$

Hence
$$A = \int_{2}^{3} \left(e^{x/2} - \frac{1}{x^2} \right) dx = 2e^{x/2} + \frac{1}{x} \Big|_{2}^{3}$$
$$= (2e^{3/2} + \tfrac{1}{3}) - (2e + \tfrac{1}{2}) = 2(e^{3/2} - e) - \tfrac{1}{6}.$$ ■

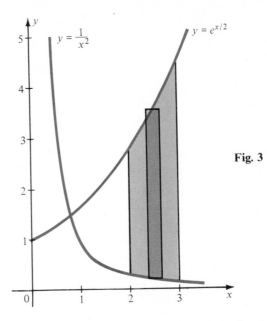

Fig. 3

Remark We sneaked a new notation into the solution: dA, the **element of area**. This is harmless, but it is a suggestive terminology for setting up applications of integration.

■ **EXAMPLE 2** Compute the area of the region bounded by $y = x^2 - 4$ and $y = \tfrac{1}{2}x + 1$.

Solution It is not clear from the statement of the problem what the region is; a graph is essential. A rough sketch (Fig. 4a) shows that region is a parabolic segment. But Fig. 4a is not accurate enough; the crucial points P and Q must be found. They can be computed by solving simultaneously
$$y = x^2 - 4 \quad \text{and} \quad y = \tfrac{1}{2}x + 1.$$

Eliminate y: $x^2 - 4 = \tfrac{1}{2}x + 1, \quad 2x^2 - x - 10 = 0, \quad (2x - 5)(x + 2) = 0.$

The solutions are $x = -2$ and $x = \tfrac{5}{2}$. The corresponding y values are 0 and $\tfrac{9}{4}$. Therefore, $P = (-2, 0)$ and $Q = (\tfrac{5}{2}, \tfrac{9}{4})$. Now Fig. 4a can be replaced by a more accurate sketch, Fig. 4b.

Slice the region between $x = -2$ and $x = \tfrac{5}{2}$ into thin vertical rectangles. The area of a typical rectangle, the element of area, is
$$dA = [(\tfrac{1}{2}x + 1) - (x^2 - 4)] \, dx = (5 + \tfrac{1}{2}x - x^2) \, dx.$$

Hence
$$A = \int_{-2}^{5/2} (5 + \tfrac{1}{2}x - x^2) \, dx = (5x + \tfrac{1}{4}x^2 - \tfrac{1}{3}x^3) \Big|_{-2}^{5/2}$$
$$= [5(\tfrac{5}{2}) + \tfrac{1}{4}(\tfrac{5}{2})^2 - \tfrac{1}{3}(\tfrac{5}{2})^3] - [5(-2) + \tfrac{1}{4}(-2)^2 - \tfrac{1}{3}(-2)^3] = \tfrac{729}{48} = \tfrac{243}{16}.$$

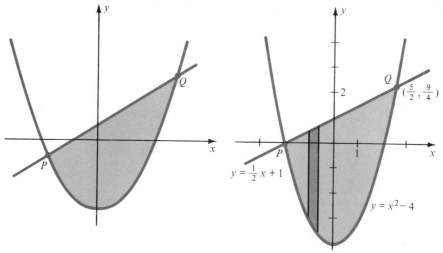

(a) Rough sketch

(b) More accurate sketch (showing positions of P and Q):

$$A = \int_{-2}^{5/2} [(\tfrac{1}{2}x + 1) - (x^2 - 4)] \, dx$$

Fig. 4

■ **EXAMPLE 3** Compute the (geometric) area of the region bounded by the graphs $y = f(x) = x^3 - 2x^2 - 5x + 6$ and $y = g(x) = -x^3 + 8x^2 - 9x - 10$.

Solution To sketch the region we need the intersections of the graphs. By trial and error we find $f(x) - g(x) = 0$ has the roots $x = -1, 2, 4$, so the intersections are $(-1, 8)$, $(2, -4)$, $(4, 18)$. This information plus our general knowledge of the shape of cubics is enough for an adaquate graph (Fig. 5). Clearly, $g(x) \le f(x)$ for $-1 \le x \le 2$ and $f(x) \le g(x)$ for $2 \le x \le 4$. Therefore

$$A = \int_{-1}^{4} |f(x) - g(x)| \, dx = \int_{-1}^{2} [f(x) - g(x)] \, dx + \int_{2}^{4} [g(x) - f(x)] \, dx.$$

Now $f(x) - g(x) = 2x^3 - 10x^2 + 4x + 16$, so

$$\int_{-1}^{2} [f(x) - g(x)] \, dx = \tfrac{1}{2}x^4 - \tfrac{10}{3}x^3 + 2x^2 + 16x \Big|_{-1}^{2} = \tfrac{63}{2}.$$

Similarly, $\displaystyle\int_{2}^{4} [g(x) - f(x)] \, dx = \tfrac{32}{3},$ so $A = \tfrac{63}{2} + \tfrac{32}{3} = \tfrac{253}{6}.$ ■

Remark   The *algebraic* area is

$$\int_{-1}^{4} [f(x) - g(x)] \, dx = \tfrac{63}{2} - \tfrac{32}{3} = \tfrac{125}{6}.$$

The ideas of this section apply word for word to regions bounded by curves $x = f(y)$ and $x = g(y)$ and lines $y = c$ and $y = d$. The area of such a region is given by the integral

$$\int_{c}^{d} |f(y) - g(y)| \, dy.$$

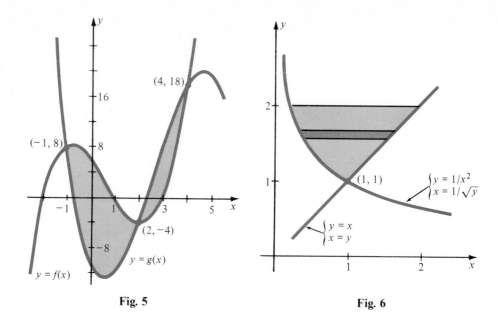

Fig. 5   Fig. 6

■ **EXAMPLE 4** Compute the area of the region bounded by the curves $y = x$, $y = 1/x^2$, and the line $y = 2$.

Solution   Sketch the region (Fig. 6). The curves obviously intersect at (1, 1). If you compute the area by vertical slicing, you will need two integrals because the lower boundary is made up of two different curves. So slice horizontally instead. Then the right-hand boundary is $x = y$ and the left-hand boundary is $y = 1/x^2$, that is, $x = 1/\sqrt{y}$. The typical horizontal slice has area

$$dA = \left(y - \frac{1}{\sqrt{y}} \right) dy.$$

Hence the area of the region is

$$\int_1^2 \left(y - \frac{1}{\sqrt{y}} \right) dy = \left(\frac{1}{2} y^2 - 2\sqrt{y} \right) \Big|_1^2 = \frac{7}{2} - 2\sqrt{2}.$$ ■

Practical hint   When you slice horizontally, you obtain $dA = $ (something) dy. That "something" must be expressed in terms of y before you can integrate. That's why in Example 4 we wrote the boundaries as $x = y$ and $x = 1/\sqrt{y}$.

EXERCISES

Compute the area of the region bounded by

1   $y = 8 - x^2$ and $y = -2x$
2   $y = x^2 + 5$ and $y = 6x$
3   $y = 3x^2$ and $y = -3x^2$,   and the lines $x = -1$ and $x = 1$
4   $y = 1 - x^2$ and $y = x^2 - 1$
5   $y = x^3 - 5x^2 + 6x$ and $y = x^3$
6   $y = x^2 - 8x$ and $y = x$

7 $y = \cos x$ and $y = \sin x$, between $x = \frac{1}{4}\pi$ and $x = \frac{5}{4}\pi$
8 $y = 2\sin(\frac{1}{4}\pi x)$ and $y = (x - 3)(x - 4)$, between $x = 2$ and $x = 4$
9 $y = e^x$, $y = e^{-x}$, and $y = e^2$
10 $y = \cos x - 1$ and $y = 1 - \cos 2x$, between $x = 0$ and $x = 2\pi$
11 $y = 1/x^2$, $y = x$, and $y = 8x$
12 $y = 1/x^2$, $y = 0$, $y = x^2$, and $x = 3$
13 $y^2 = 2x$ and $y = \frac{1}{2}x - 3$
14 $x = y^2$ and $x = 6 - y^4$.

15 Find a so that the area bounded by $y = x^2 - a^2$ and $y = a^2 - x^2$ is 9.
16 Find the fraction of the area of one hump of the curve $y = \sin x$ that lies above $y = \frac{1}{2}$.
17 Let $P = (a, ka^2)$ and $Q = (b, kb^2)$ be two points of the parabola $y = kx^2$. Find the area bounded by the parabola and the segment $\overline{PQ}$.
18 (cont.) Prove that this area is $\frac{4}{3}$ the area of the triangle PTQ, where T is the point of the parabola whose tangent is parallel to $\overline{PQ}$. (This is Archimedes' quadrature of the parabola.)

3. VOLUME

Before computing volumes, let us look at a simple area problem whose solution involves some useful ideas.

■ **EXAMPLE 1** Find the area of a circle of radius r. (Assume the formula $c = 2\pi r$ for circumference.)

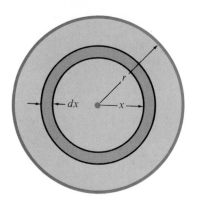

Fig. 1 Area of circle by concentric rings.

Solution Cut the circle into thin concentric rings (Fig. 1). Let x denote the radial distance of a ring from the center; $0 \leq x \leq r$. The typical ring has length $2\pi x$ and width dx, hence area $2\pi x\, dx$. Therefore, the area of the circle is

$$A = \int_0^r 2\pi x\, dx = \pi x^2 \Big|_0^r = \pi r^2.$$ ■

The strategy used in solving Example 1 is important and worth reviewing. First we sliced the circle into thin pieces, each having area $f(x) = 2\pi x\, dx$. Then we add up these small areas by integrating from 0 to r. Similar strategy applies to computing the volume of a solid. First we slice the solid into many thin pieces, each of which is approximately a familiar shape of known volume. The element of volume is

$f(x)\,dx$, the base area $f(x)$ of the typical slice multiplied by its thickness dx. Then we add these little volumes by integrating.

Thus the general plan of attack on a particular solid consists of four steps:

(1) choose a method of slicing the solid; (2) choose a variable x that locates the typical slice, and find the range $a \le x \le b$ that applies to the problem; (3) compute the volume $f(x)\,dx$ of the typical slice; (4) find an antiderivative of $f(x)$ and evaluate

$$V = \int_a^b f(x)\,dx.$$

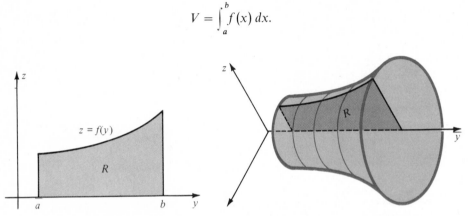

Fig. 2   The solid swept out when R revolves about the y-axis

Volume of Revolution   Let R be the region in the y, z-plane under the curve $z = f(y)$, where $a \le y \le b$. If R is revolved about the y-axis, a solid of revolution is swept out (Fig. 2). What is the volume of this solid?

We follow the strategy outlined above. First we slice R into thin rectangles (Fig. 3a). Each of these sweeps out a circular slab (Fig. 3b), so we have sliced the solid into slabs. The typical slab is identified by the variable y, where $a \le y \le b$. It has radius $f(y)$, base area $\pi f(y)^2$, thickness dy, hence volume $\pi f(y)^2\,dy$.

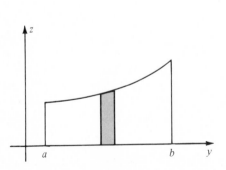

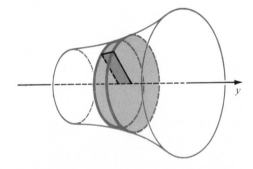

(b) Resulting slab has base area $\pi f(y)^2$ and thickness dy, hence volume $dV = \pi f(y)^2\,dy$.

(a) Thin "rectangle", dimensions $f(y) \times dy$

Fig. 3

Consequently, the volume of the solid of revolution is

$$V = \int_a^b \pi f(y)^2 \, dy.$$

■ **EXAMPLE 2** Find the volume of a sphere of radius a.

Solution The sphere is a solid of revolution obtained by rotating a semicircle about its diameter. Take the semicircle bounded by the y-axis, and $z = \sqrt{a^2 - y^2}$. See Fig. 4a.

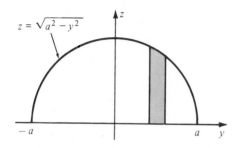

(a) Rotating the semi-circle about its diameter generates a sphere. The thin "rectangle" has dimensions $z \times dy$.

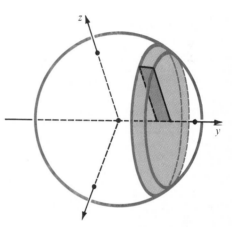

(b) Resulting slab has volume $dV = \pi z^2 \, dy = \pi(a^2 - y^2) \, dy$.

Fig. 4

Each slab in Fig. 4b has radius $z = \sqrt{a^2 - y^2}$ and thickness dy, so the element of volume is

$$dV = \pi z^2 \, dy = \pi(a^2 - y^2) \, dy.$$

Now integrate dV from $y = -a$ to $y = a$:

$$V = \int_{-a}^{a} \pi(a^2 - y^2) \, dy = 2\pi \int_{0}^{a} (a^2 - y^2) \, dy$$

$$= 2\pi\left(a^2 y - \tfrac{1}{3}y^3\right)\Big|_0^a = 2\pi(a^3 - \tfrac{1}{3}a^3) = \tfrac{4}{3}\pi a^3. \qquad ■$$

■ **EXAMPLE 3** A right circular cone of height h is constructed over a base of radius a. Compute its volume.

Solution The cone is a solid of revolution, obtained by revolving a right triangle about one leg. Choose the triangle indicated in Fig. 5a, and rotate it about the y-axis to generate the cone (Fig. 5b). Slice the cone into thin slabs, each of width dy. Note that the triangle is the region under the curve $z = (a/h)y$, where $0 \le y \le h$. Thus

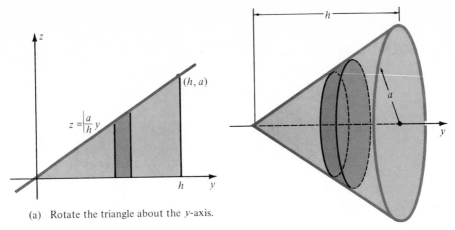

(a) Rotate the triangle about the y-axis.

(b) The result is a right circular cone.

Fig. 5   Volume of cone by disks

each slab has radius $(a/h)y$, hence volume

$$dV = \pi \left(\frac{a}{h} y\right)^2 dy.$$

Therefore the volume of the cone is

$$V = \int_0^h \pi \frac{a^2}{h^2} y^2 \, dy = \pi \frac{a^2}{h^2} \left(\frac{1}{3} y^3\right) \Big|_0^h = \pi \frac{a^2}{h^2} \left(\frac{1}{3} h^3\right) = \frac{1}{3} \pi a^2 h.$$

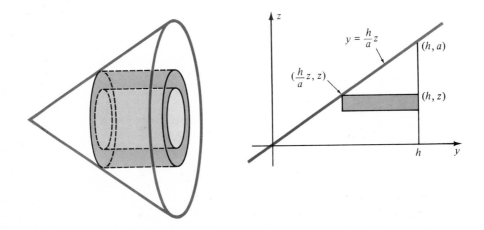

Fig. 6   Volume of cone by cylindrical shells

Alternative Solution   Slice the cone into cylindrical shells rather than slabs (Fig. 6a), and choose z as the variable. This corresponds to slicing the triangle into thin strips parallel to the y-axis (Fig. 6b). Each strip sweeps out a thin cylindrical shell with radius z, height $h - (h/a)z$, and thickness dz. The volume of this shell is the product of its three dimensions:

$$dV = (\text{circumference}) \cdot (\text{height}) \cdot (\text{thickness}) = (2\pi z)\left(h - \frac{h}{a}\right) dz.$$

Therefore $\quad V = \int_0^a (2\pi z)\left(h - \frac{h}{a}z\right) dz = 2\pi h \int_0^a \left(z - \frac{z^2}{a}\right) dz$

$$= 2\pi h\left(\frac{z^2}{2} - \frac{z^3}{3a}\right)\Big|_0^a = 2\pi h\left(\frac{a^2}{2} - \frac{a^2}{3}\right) = \frac{1}{3}\pi a^2 h. \qquad \blacksquare$$

■ **EXAMPLE 4** The region in the y, z-plane bounded by the parabola $z = y^2$, the line $y = a$, and the y-axis is revolved about the z-axis. (Assume $a > 0$.) What is the resulting volume?

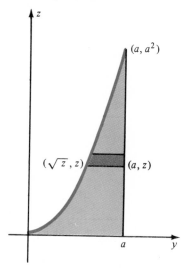
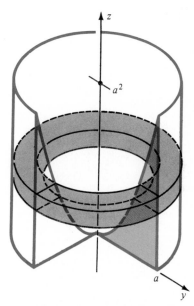

(a) Slice into strips parallel
to the y-axis.

(b) Cut-away view of the solid of
revolution, showing a typical "washer".

Fig. 7 Solution by washers

Solution Slice the plane region into thin rectangles parallel to the y-axis (Fig. 7a). When the typical resulting strip is revolved about the z-axis, it sweeps out a thin circular washer (Fig. 7b). The base of the washer is the region between two concentric circles of radii a and $\sqrt{z}$. Hence its area is

$$\pi a^2 - \pi(\sqrt{z})^2 = \pi(a^2 - z).$$

Since its thickness is dz, the element of volume is $dV = \pi(a^2 - z)\,dz$. Integrate these small volumes from the level $z = 0$ to $z = a^2$:

$$V = \int_0^{a^2} \pi(a^2 - z)\,dz = \pi(a^2 z - \tfrac{1}{2}z^2)\Big|_0^{a^2} = \pi(a^4 - \tfrac{1}{2}a^4) = \tfrac{1}{2}\pi a^4.$$

Alternative Solution   See Fig. 8. The region under the parabola is split into thin "rectangles" by parallels to the z-axis. Now y is the variable, and the solid of revolution is sliced into thin cylindrical shells. The typical shell has radius y, height y^2, and thickness dy, hence volume $dV = (2\pi y)y^2\, dy$. The volume of the solid is

$$V = \int_0^a (2\pi y)y^2\, dy = \int_0^a 2\pi y^3\, dy = 2\pi(\tfrac{1}{4}y^4)\,\Big|_0^a = \tfrac{1}{2}\pi a^4. \qquad \blacksquare$$

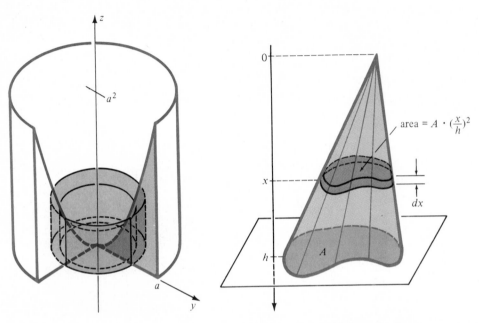

Fig. 8   Solution by cylindrical shells   **Fig. 9**   Cone with irregular base

   The volume of certain figures other than solids of revolution can also be found with the tools at our disposal.

■ **EXAMPLE 5**   A cone of height h has an irregular base of area A. Find the volume of the cone.

Solution   Let x denote distance measured from the apex towards the plane of the base (Fig. 9). The typical cross section of the cone by a plane parallel to the base, distance x from the apex, is a plane region similar to the base. This cross section has linear dimensions proportional to x, hence area proportional to x^2.

   Let $f(x)$ denote this area. Then $f(x) = cx^2$. To find the constant c, note that $f(h) = A$, hence $ch^2 = A$. Therefore $c = A/h^2$ and

$$f(x) = \frac{A}{h^2}\, x^2.$$

Slice the cone into slabs by planes parallel to the base. A typical slab has base area $f(x)$, thickness dx, and volume

$$dV = f(x)\, dx = \frac{A}{h^2}\, x^2\, dx.$$

Hence the volume of the cone is

$$V = \int_0^h \frac{A}{h^2} x^2 \, dx = \frac{A}{h^2} \left(\frac{1}{3} x^3 \right) \bigg|_0^h = \frac{1}{3} Ah.$$ ∎

Area of a Sphere   We know that the volume of a sphere of radius a is $\frac{4}{3}\pi a^3$. Let us use this information to find the surface area of the sphere. We don't yet know "officially" what surface area is. Nevertheless, let's assume that it exists alright and that, as with plane area, it has the property that if all linear dimensions are stretched by a factor k, then surface area is multiplied by the factor k^2.

■ **EXAMPLE 6**   Find the surface area A of a sphere of radius a.

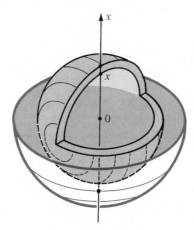

Fig. 10   Cutaway view of sphere and a concentric spherical shell

Solution   Find the volume of the sphere by slicing it into concentric spherical shells (Fig. 10). Let $f(x)$ by the surface area of the typical shell at distance x from the center $(0 \le x \le a)$. Then $f(x)$ is proportional x^2 and $f(a) = A$, hence $f(x) = Ax^2/a^2$. The volume of the shell is

$$dV = f(x)\, dx = \frac{A}{a^2} x^2 \, dx.$$

Hence

$$V = \int_0^a \frac{A}{a^2} x^2 \, dx = \frac{A}{a^2} \left(\frac{1}{3} x^3 \right) \bigg|_0^a = \frac{1}{3} Aa.$$

But

$$V = \tfrac{4}{3}\pi a^3,$$

therefore

$$\tfrac{4}{3}\pi a^3 = \tfrac{1}{3} Aa, \qquad \text{so} \quad A = 4\pi a^2.$$ ∎

EXERCISES

The region of the y, z-plane whose boundary curves are given is revolved about the y-axis. Find the volume of the resulting solid of revolution.

1   y-axis, $z = 2y + 3$, $y = 0$, $y = 3$
2   y-axis, $2z + y = 3$, $y = -1$, $y = 1$
3   y-axis, $(y + 1)z = 1$, $y = 0$, $y = 4$
4   y-axis, $(y + 4)z + 3 = 0$, $y = -1$, $y = 1$

5 y-axis, $z = 3\sqrt{y}$, $y = 0$, $y = 4$

6 $y = 16z^2$, $y = 2$, $y = 4$.

7 The region of the x, y-plane bounded by the x-axis, $y = x + 1$, $x = 1$, and $x = 4$ is revolved about the line $y = -3$. Find the volume of the resulting solid.

8 The region of the x, y-plane bounded by the lines $x = 1$, $y = x$, and $x = -2y + 6$ is revolved about the line $y = -2$. Find the volume of the resulting solid.

9 The region of the x, z-plane bounded by $z = -1$, $z = e^{2x}$, $x = 0$, and $x = 2$ is revolved about the line $z = -1$. Find the volume of the resulting solid.

10 The region of the y, z-plane bounded by the z-axis, $y^2 = \sin z$, $z = 0$, and $z = \pi$ is revolved about the z-axis. Find the volume of the resulting solid.

11 Find the volume of a frustum of a right circular cone with lower radius b, upper radius a, and height h.

12 Find the volume of a sphere of radius a with a hole of radius h drilled through its center.

13 A plane at distance h from the center of a sphere of radius a cuts off a spherical cap of height $a - h$. Find its volume.

14 Find the volume of the solid formed by revolving the triangle in the x, y-plane with vertices $(1, 1)$, $(0, 2)$, and $(2, 2)$ about the x-axis.

15 A circular hole is cut on center vertically through a sphere, leaving a ring of height h. Calculate the volume of the ring.

16 A circle of radius a is revolved about a line in the same plane at distance b from the center of the circle. Assume $b > a$. Show that the resulting torus has volume $2\pi^2 a^2 b$. [*Hint* Use washers; then identify the difficult integral as the area of a circle.]

17 The rectangle $-1 \le x \le 1$, $-2 \le y \le 2$, $z = 0$ moves upwards, its center always on the z-axis, and rotates counterclockwise at a uniform rate as it rises. Suppose it has turned $90°$ when it reaches $z = 1$. Find the volume swept out.

18 The region bounded by the x-axis, $y = f(x) = k/x$, $x = a$, and $x = b$ is revolved about the x-axis. Assume $k > 0$ and $0 < a < b$. Let $g(a)$ denote the limit as $b \longrightarrow \infty$ of the resulting volume. Find k so $g(x) = f(x)$ for all $x > 0$.

4. WORK

In elementary physics, we are taught that the work done by a constant force F in moving an object through distance D on a line is $W = FD$, that is, work equals force times distance.

When the force is variable, work is defined by an integral. The definition is motivated as follows. Suppose a continuous force $f(x)$ acts over an interval $[a, b]$ of the x-axis, and suppose an object is moved from a to b by the force. We divide $[a, b]$ into many small subintervals. On the typical one, of length Δx_j, the force is almost a constant $f(\bar{x}_j)$. Therefore, in any reasonable definition, the work over this interval must be approximately $\Delta W = f(\bar{x}_j) \Delta x_j$. Now we sum these approximations and take limits to *define* work. Thus the **element of work** is

$$dW = f(x)\, dx,$$

and the **work** done by the force in moving the object from $x = a$ to $x = b$ is

$$W = \int_a^b f(x)\, dx.$$

Units In the English system, work is measured in foot-pounds (ft-lb). In the CGS metric system, the unit of work is one **erg** = one dyne-centimeter; and in the MKS system, it is one **joule** (J) = one newton-meter. Here one joule = 10^7 ergs and one ft-lb $\approx$ 1.356 joules.*

■ **EXAMPLE 1** At each point of the x-axis (marked off in feet) there is a force of $5x^2 - x + 2$ pounds pulling an object. Compute the work done in moving it from $x = 1$ to $x = 4$.

Solution $W = \int_1^4 (5x^2 - x + 2)\, dx = \left(\tfrac{5}{3}x^3 - \tfrac{1}{2}x^2 + 2x\right)\Big|_1^4$

$$= \tfrac{5}{3}(64 - 1) - \tfrac{1}{2}(16 - 1) + 2(4 - 1) = 103.5 \text{ ft-lb.} \qquad ■$$

■ **EXAMPLE 2** According to Newton's Law of Gravitation, two bodies attract each other with a force F proportional to the product $m_1 m_2$ of their masses and inversely proportional to the square of the distance r between them:

$$F = G \frac{m_1 m_2}{r^2},$$

where G is the **gravitational constant**.† If one of the bodies is fixed at the origin, how much work is needed to move the other from $r = a$ to $r = b$, where a and b are positive?

Solution The element of work is $dW = F\, dr = G\dfrac{m_1 m_2}{r^2}\, dr,$

hence $W = \int_a^b G \dfrac{m_1 m_2}{r^2}\, dr = Gm_1 m_2 \left(-\dfrac{1}{r}\right)\Big|_a^b = Gm_1 m_2 \left(\dfrac{1}{a} - \dfrac{1}{b}\right).$ ■

Remark When $a < b$ the work is positive because you do work against the gravitational force. But when $a > b$ the free mass moves towards the fixed mass, *opposite* to your direction of pull. Hence you do negative work. Imagine moving the free mass from a to b and then back to a. The total work is zero Why?

■ **EXAMPLE 3** When a spring is stretched a small amount, there is a restoring force proportional to the distance of the moving end from its equilibrium point (Hooke's Law). Suppose 2 joules work are needed to stretch a certain spring 10 cm. How much work is needed to stretch it 25 cm?

Solution Let x denote the displacement of the free end from equilibrium (Fig. 1). The force needed at x to oppose the restoring force is kx, so the work in stretching the spring from $x = 0$ to $x = b$ is

$$W = \int_0^a kx\, dx = \tfrac{1}{2}kb^2.$$

When $b = 10$, then $W = 2$, hence $2 = \tfrac{1}{2}k(10)^2$, $k = \tfrac{1}{25}$. When $b = 25$,

$$W = \tfrac{1}{2}(\tfrac{1}{25})(25)^2 = 12.5 \text{ J.}$$

---

* The MKS force unit, the **newton** (N) is the force that applied to a one-kg mass will impart an acceleration of one m/sec². One kg of force = g newtons, where $g \approx 9.807$.

† In the MKS system, m_1 and m_2 are in kilograms, r in meters, F in newtons, W in joules, and

$$G \approx 6.670 \times 10^{-11} \text{ N-m}^2/\text{kg}^2.$$

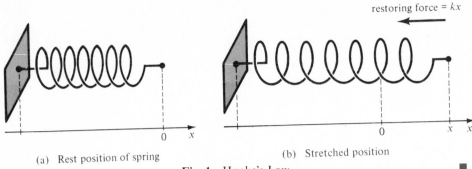

(a) Rest position of spring   (b) Stretched position

Fig. 1   Hooke's Law   ■

■ **EXAMPLE 4**   A sunken tank (Fig. 2a) has the shape of an inverted right circular cone. Compute the work done in pumping a tankful of water to ground level. (The density of water is 62.4 lb/ft^3.)

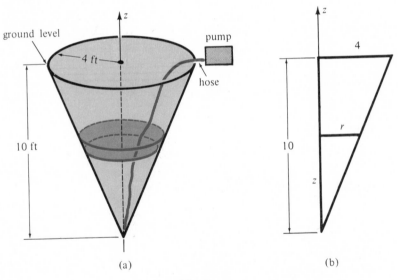

Fig. 2

Solution   Set up axes so the vertex of the cone is the origin and the axis of the cone is the z-axis. Imagine the tank sliced into thin slabs perpendicular to its axis. The idea is to compute the work done in raising each slab of water to the level $z = 10$, then to add up these elements of work by integrating.

Consider a typical slab at level z. It must be raised a distance $10 - z$. The upward force required equals its weight (to overcome the downward force of gravity). Since weight in pounds equals density δ times volume,

$$dW = (10 - z)\delta \, dV.$$

By similar triangles (Fig. 2b), the radius r of the slab satsifies

$$\frac{r}{z} = \frac{4}{10}, \quad \text{so} \quad r = \frac{2}{5} z.$$

Hence $\qquad dV = \pi(\frac{2}{5}z)^2\,dz, \qquad dW = \frac{4}{25}\pi\delta(10-z)z^2\,dz.$

Therefore the total work done is

$$W = \frac{4}{25}\pi\delta \int_0^{10} (10-z)z^2\,dz = \frac{4}{25}\pi\delta\left(\frac{10}{3}z^3 - \frac{1}{4}z^4\right)\Big|_0^{10}$$

$$= \frac{400}{3}\delta\pi = \frac{400}{3}(62.4)\pi \approx 26138 \text{ ft-lb.} \qquad\blacksquare$$

■ **EXAMPLE 5** A heavy buoy of weight w in the shape of a cone of revolution (Fig. 3a) floats in a lake with its lowest point at depth h. The buoy is raised by a winch until it just clears the water. How much work is done?

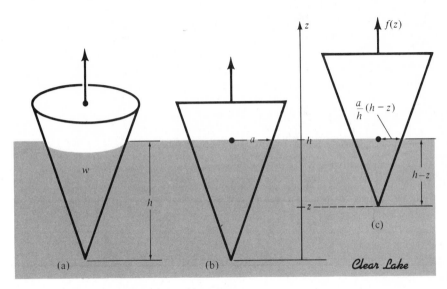

Fig. 3

Solution Fix the z-axis as in Fig. 3b, and denote by a the radius of the buoy *at the water level*. By Archimedes' principle for floating bodies, the buoy is acted on by an upward buoyant force of the water equal to the weight of the water displaced. When the buoy floats, this balances the downward force of gravity w. Thus,

$$w = \tfrac{1}{3}\pi a^2 h\delta g,$$

where δ is the density of the lake water and g is the gravitational force per unit mass.

Suppose the buoy is hoisted z units (Fig. 3c). The force $f(z)$ required to hold it in this position is the weight of the buoy minus the buoyant force of the water. But now the part submerged has radius $(a/h)(h-z)$ and height $h-z$. Hence

$$f(z) = w - \frac{1}{3}\pi\left[\frac{a}{h}(h-z)\right]^2 (h-z)\delta g.$$

Therefore the work done lifting the buoy out of the water is

$$W = \int_0^h f(z)\, dz = wh - \frac{\pi a^2 \delta g}{3h^2} \int_0^h (h - z)^3\, dz$$

$$= wh - \tfrac{1}{12}\pi a^2 \delta g h^2 = wh - \tfrac{1}{4}wh = \tfrac{3}{4}wh. \qquad \blacksquare$$

EXERCISES

✓**1** Find the work done by a force $f(x) = 3x + 2$ newtons in moving an object from $x = 1$ m to $x = 7$ m.

2 At each point of the x-axis (marked off in feet) there is a force of $x^2 - 5x + 6$ lb pushing to the right against an object. Compute the work done in moving it from $x = 1$ to $x = 5$.

✓**3** A 50-ft chain weighing 2 lb/ft is attached to a drum hung from the ceiling. The ceiling is high enough so that the free end of the chain does not touch the floor. How much work is required to wind the chain around the drum?

4 In the previous exercise, suppose that a 200-lb weight is attached to the free end. How much work is required to wind up the chain?

✓**5** The force in pounds required to stretch a certain spring x ft is $F = 8x$. How much work is required to stretch the spring 6 in., 1 ft, 2 ft?

6 A 3-lb force will stretch a spring 0.5 ft. How much work is required to stretch the spring 2 ft?

✓**7** A 100-lb bag of sand is hoisted 50 ft at a rate of 5 ft/sec. Because of a hole in the bag, 2 lb of sand is lost each sec. Compute the work done. *4600 lb-ft*

8 A 5-lb monkey is attached to the free end of a 20-ft hanging chain that weights 0.25 lb/ft. The monkey climbs the chain to the top. How much work does he do?

✓**9** How much work is required to lift a 500-kg payload from the surface of Earth to an orbit 500 km high? 1000 km high? You may assume $r_E \approx 6.37 \times 10^6$ m and $F = Gm_E m/x^2$, where $G \approx 6.67 \times 10^{-11}$ N-m^2/kg^2 and $m_E \approx 5.98 \times 10^{24}$ kg.

10 How much work is required to fill the tank in Fig. 4 with water pumped from the level of its base. The tank is in the shape of a paraboloid of revolution, obtained by revolving a parabola about its axis.

11 How much work is required to pump water from the bottom of the pipe and fill the spherical tank in Fig. 5?

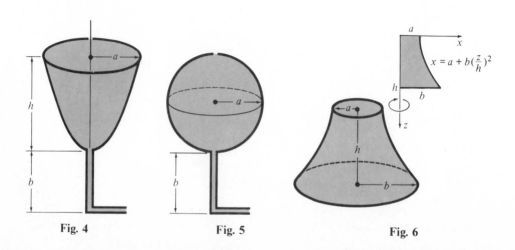

Fig. 4   Fig. 5   Fig. 6

12   A tank is obtained by revolving a parabolic segment as indicated in Fig. 6. How much work is required to pump a tankful of fluid of density δ to the level of its top?

13   Suppose in Example 5 that the top of the buoy is $\frac{1}{4}h$ from the water's surface. How much work is required to push the buoy down until its top is at water level?

14* A spherical mine has enough flotation so its average density equals 1030 kg/m³, that of the seawater it floats in—just touching the surface. Suppose its radius is 0.60 m. How much work is required to lift it so it just clears the water?

15   A particle of mass m is constrained to move on a vertical circle of radius a. (Think of a pendulum bob.) Suppose it is pushed from its downward rest position through a central angle ϕ. How much work is done? [*Hint*   Resolve the gravitational force into components parallel and perpendicular to the circle; only the former must be opposed.]

16* An open-top cylindrical tank of radius R and height H is filled with water. A cylindrical buoy of radius r floats on end in the tank, its base at depth h, where $r < R$ and $h < H$. How much work is required to raise the buoy until its base just clears the water. (Note that the water level goes down as the buoy comes up.)

5. FLUID PRESSURE

Fluid pressure is measured by the force it exerts on any piece of surface immersed in the fluid. At each point of the surface this force is exerted against the surface in the direction perpendicular to the surface.

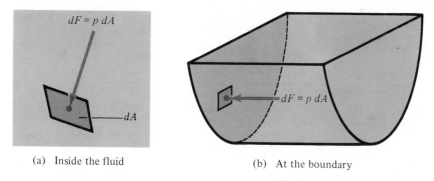

(a) Inside the fluid   (b) At the boundary

Fig. 1   Element of force due to pressure

Technically, **pressure** is force per unit area. In Fig. 1a the pressure p exerts a force dF against the element of area dA. Of course it also exerts an equal but oppositely directed force against the opposite surface (so nothing happens). However, if dD is part of the boundary of the fluid, i.e., part of the container wall, then the force is not opposed by fluid pressure on the other side (Fig. 1b). The wall's own strength must hold it up.

The pressure at depth y in a fluid is $p = \delta g y$, where δ is the density of the fluid and g the constant of gravity. (We often take $g = 1$ by measuring mass and force in the same units.) The **total pressure** against a plane surface submerged in the fluid is obtained by "summing" the elements of force $dF = p \, dA$ over the surface. Since p is constant at depth y, this can usually be done by an integration. The result is (the magnitude of) a force that is directed perpendicular to the plane surface. Total pressure is important in the design of tanks, dams, large buildings (wind pressure), and airfoils (lift). We are using the slightly inaccurate expression "total pressure" as an abbreviation for "magnitude of the total force due to pressure."

Units   The English system units are the $lb/in.^2$ and the lb/ft^2. The MKS unit is the N/m^2 (newton/meter squared), which sometimes is too small to be practical. The unit kg/cm^2 ($\approx 9.807 \times 10^4 \ N/m^2$) is common. Here kg means one kg of *force*, not mass. One kg is the force gravity at the surface of Earth exerts on one kg of mass. A tire guage in Europe might read $2.1 \ kg/cm^2$, equivalent to about $30 \ lb/in.^2$.*

Pressure Zero   We shall work at the surface of the Earth at "ordinary" pressures, so atmospheric pressure will be our reference point for zero. For instance $25 \ lb/in.^2$ pressure in a tire means $25 \ lb/in.^2$ *above* atmospheric pressure, which itself is about $14.7 \ lb/in.^2$ above vacuum, thus the tire pressure is about $39.7 \ lb/in.^2$ above vacuum.

For low pressure work, the reference point for zero is usually the true zero pressure of a vacuum. Thus $3 \ lb/in.^2$ pressure in a space ship en route to Mars means $3 \ lb/in.^2$ above the outside pressure.

■ **EXAMPLE 1**   The ends of the cylindrical tank of water† in Fig. 2a are parabolas with vertical axes. Find the total pressure the water exerts on each end of the tank.

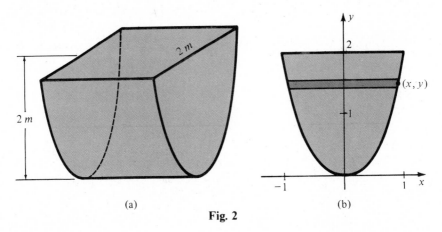

(a)   (b)

Fig. 2

Solution   Choose axes for one end as in Fig. 2b, the scale in meters. The strip shown, at height y from the bottom of the tank, has area $dA = 2x \ dy$ and lies at depth $2 - y$. The average pressure against the strip is $p = \delta(2 - y) = 1000(2 - y) \ kg/m^2$, so the element of force against the strip is

$$dF = 1000(2 - y)(2x \ dy) = 2000(2 - y)x \ dy.$$

The total pressure in kg against the end of the tank is

$$F = 2000 \int_0^2 (2 - y)x \ dy.$$

---

* Other common units: (1) the **atmosphere**, where $1 \ atm \approx 1.01325 \times 10^5 \ N/m^2 \approx 1.033 \ kg/cm^2 \approx$ $14.70 \ lb/in.^2$; (2) the **bar** $= 10^5 \ N/m^2$ (common in meteorology—just 1 atm rounded off); (3) the **torr** $\approx 1.316 \times 10^{-3} \ N/m^2$ (used in low pressure work).
† The density of water is $1000 \ kg/m^3 = 1 \ g/cm^3 \approx 62.4 \ lb/ft^3$.

To evaluate the integral, first we must express x in terms of y. The equation of the parabola has the form $y = kx^2$. Since $(1, 2)$ lies on it, $k = 2$ so

$$y = 2x^2, \qquad x = \tfrac{1}{2}\sqrt{2}\sqrt{y}.$$

Therefore

$$F = 1000\sqrt{2}\int_0^2 (2 - y)\sqrt{y}\, dy = 1000\sqrt{2}\int_0^2 (2y^{1/2} - y^{3/2})\, dy$$

$$= 1000\sqrt{2}\left(\tfrac{4}{3}y^{3/2} - \tfrac{2}{5}y^{5/2}\right)\Big|_0^2 = 1000\sqrt{2}\left(\tfrac{8}{3}\sqrt{2} - \tfrac{8}{5}\sqrt{2}\right)$$

$$= 16000(\tfrac{1}{3} - \tfrac{1}{5}) = \tfrac{1}{3}(6400) = 2133\tfrac{1}{3} \quad \text{kg.} \qquad \blacksquare$$

Vertical Slicing   Sometimes it is convenient to compute total pressure by slicing into vertical strips. To do so effectively we need a property of rectangles.

Suppose a vertical rectangle is submerged in a fluid (Fig. 3). Choose axes as indicated. Then the element of force against the horizontal strip at depth y is

$$dF = (\delta y)(a\, dy) = \delta a y\, dy \qquad (\delta = \text{density}).$$

Therefore $$F = \int_L^{L+b} \delta a y\, dy = \tfrac{1}{2}\delta a y^2 \Big|_L^{L+b} = \tfrac{1}{2}\delta a (2Lb + b^2) = (ab)[\delta(L + \tfrac{1}{2}b)].$$

But ab is the area of the rectangle and $L + \tfrac{1}{2}b$ is the depth of the midpoint of the rectangle.

> The total pressure against one side of a vertical rectangle submerged in a fluid is its area times the pressure at its midpoint.

Remark   The result is a special case of a much more general fact; the total pressure against one side of *any* submerged plane plate equals the area of the plate time the pressure at its center of gravity. This will become clear after our discussion of centers of gravity in a later chapter.

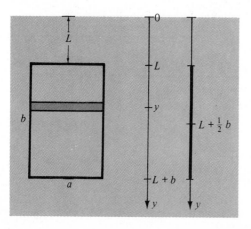

Fig. 3

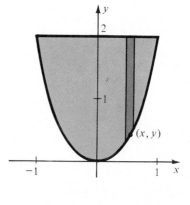

Fig. 4

EXAMPLE 2 Solve Example 1 by vertical slicing.

Solution The vertical strip in Fig. 4 has area $(2 - y)\,dx$, and its midpoint lies at depth $\frac{1}{2}(2 - y)$ so the pressure at the midpoint is $p = \frac{1}{2}(2 - y)\,\delta = 500(2 - y)$ kg. Therefore the element of force against this vertical strip is

$$dF = [500(2 - y)][(2 - y)\,dx] = 500(2 - y)^2\,dx.$$

But $y = 2x^2$ so $dF = 500(2 - 2x^2)^2\,dx = 2000(1 - x^2)^2\,dx$ and

$$F = \int_{-1}^{1} 2000(1 - x^2)^2\,dx = 4000 \int_{0}^{1} (1 - 2x^2 + x^4)\,dx$$

$$= 4000\left(x - \tfrac{2}{3}x^3 + \tfrac{1}{5}x^5\right)\bigg|_{0}^{1} = 4000\left(1 - \tfrac{2}{3} + \tfrac{1}{5}\right) = (4000)\left(\tfrac{8}{15}\right) = 2133\tfrac{1}{3} \text{ kg.} \qquad \blacksquare$$

EXERCISES

Each figure is the end of a tank filled with a fluid of density δ. Express the total pressure on the end of the tank in terms of the data.

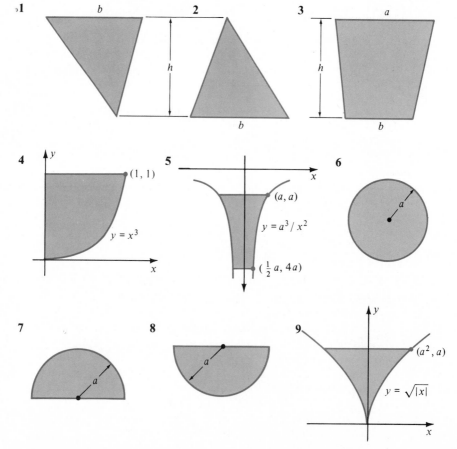

[*Hint for* **6–8** Don't overlook the area of a circle when you see it. Also differentiate $(a^2 - x^2)^{3/2}$ and see what you get.]

10   A tank of water has the shape of a high rectangular column with square base and height h meters and its volume is 1 m^3. Let F be the total pressure against one of the vertical sides and A be the area of the side. Find c so FA^c is independent of h.

11   Let F be the total pressure against one end of a tank. Suppose a second tank (filled with the same fluid) is exactly the same shape as the first tank except that each linear dimension is enlarged by the factor k. Find the total pressure against the corresponding end of the second tank.

12   One end of tank one has area A and height h, and the total pressure against it is F_1. Tank two is the same as tank one, except upside down, and the total pressure against the corresponding face is F_2. Find the relation between F_1, F_2, A, k, and the fluid density δ.

13   One end of a tank has area A and height h. Suppose the end is symmetric in the horizontal line at level $\frac{1}{2}h$. Express the total pressure against the end in terms of A, h, and the fluid density δ.

14   Air pressure at height h meters above sea level is $p = p_0 e^{-ah} \text{ kg/m}^2$, where $p_0 = 1.03 \times 10^4$ and $a = 1.25 \times 10^{-4}$. (See Exs. 16–17 below for a derivation.) A balloon made of 1 kg of reinforced plastic in the shape of a vertical cylinder of height 5 m and radius $\frac{3}{4}$ m is filled with helium (density 0.18 kg/m^3) and tethered near the ground; say its lower base is 50 m above sea level. What force is required to hold the balloon?

15   (cont.) Find the total pressure against a rectangular wall 40 m wide and 120 m high whose base is 200 m above sea level.

16*   In a gas, density is proportional to pressure. (This is a consequence of the gas law $pV = k$—at constant temperature.) Let $\delta = \delta(h)$ denote the density of the atmosphere at height h above ground. By computing the pressure in a long narrow cylinder of air, show that $d\delta/dh = -\delta/H$, where H is a constant. Conclude that $\delta = \delta_0 e^{-h/H}$. (See Ex. 35, p. 177.)

 The height of the cylinder may be limited to 100 km since $\delta(h)$ is known to be negligible above that height. This distance is small compared to the Earth's radius, so gravity may be assumed constant.

17   (cont.) At sea level (and 0°C) air pressure is $1.03 \times 10^4 \text{ kg/m}^2$ and air density is 1.29 kg/m^3. Use these data to derive the pressure formula in Ex. 14.

18   (cont.) At what height is the air pressure 10^{-2} kg/m^2? How many atm pressure on top of Mt. Everest (8848 m)?

6. MISCELLANEOUS APPLICATIONS

 The previous topics, area, volume, work, and pressure, should convince anyone of the enormous applicability of integration. As further evidence, we include some topics from a variety of directions: growth of money, suspension bridges, rotating fluids, shape of ice cubes, absorption of radiation, and escape velocities. A couple of additional applications are contained in the exercises.

Present Value of Future Income   We discussed compound interest briefly in Chapter 4, p. 172. Let us recall that if amount A is invested at annual interest rate r, compounded continuously, then its value after t years is Ae^{rt}.

 Now we consider compounding of funds invested not in one lump sum, but over a period of time. Suppose money is deposited into a certain account continuously at the rate of $f(t)$ dollars per year at time t. If the constant annual interest rate is r, what will be the value of the account after T years? Also, what is its fair present value, that is, the largest amount you would be willing to pay today for ownership of the account in T years?

To answer these questions, we divide the time interval $[0, T]$ into many small pieces of duration dt. In the typical short interval at time t, the "element of money" deposited is $f(t)\, dt$. This remains in the account for $T - t$ years, hence grows to $e^{r(T-t)}f(t)\, dt$. The total amount A of the account at time T is the total of all these small sums:

$$A = \int_0^T e^{r(T-t)}f(t)\, dt = e^{rT}\int_0^T e^{-rt}f(t)\, dt.$$

The present value of the account is the amount V which, if deposited today, will grow to amount A in T years. Therefore

$$Ve^{rT} = A = e^{rT}\int_0^T e^{-rt}f(t)\, dt, \qquad V = \int_0^T e^{-rt}f(t)\, dt.$$

Suppose that money is deposited continuously into an account at the rate of $f(t)$ dollars per year and that r is the constant annual interest rate. Then the **value** of the account after T years will be

$$A(T) = e^{rT}\int_0^T e^{-rt}f(t)\, dt.$$

Its **present value** is

$$V(T) = \int_0^T e^{-rt}f(t)\, dt.$$

For example, if funds are deposited at a continuous rate of \$10,000 per year, and the interest rate is 6%, then the value of the account after 10 years will be

$$A(10) = e^{10(0.06)}V(10) = e^{0.6}\int_0^{10} e^{-(0.06)t}\, 10{,}000 \; dt$$

$$= e^{0.6}\left(10{,}000\, \frac{e^{-(0.06)t}}{-0.06}\right)\Bigg|_0^{10} \approx (1.8221)(75{,}198) \approx \$137{,}020.$$

The present value is $V(10) \approx \$75198$. This is the lump sum which, deposited today at 6%, will grow to \$137,020 in 10 years.

Remark   In economics the function $V = V(T)$ is called the **capital value** of the **income stream** $f(t)$.

The Suspension Bridge   The problem is to find the shape of the cable that supports the roadway (Fig. 1a). Our model is based on the following simplifying assumptions:

(1)   The weight of the cable and of the suspension rods is negligible.
(2)   The weight of the roadway is uniform, δ per unit length.
(3)   The suspension rods are so close together that the horizontal loading of the cable is the uniform weight of the roadway: δ per unit length.

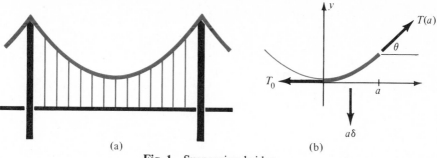

Fig. 1 Suspension bridge

We choose axes as in Fig. 1b and consider the portion of the cable for $0 \leq x \leq a$. Three forces act to hold it in equilibrium: the horizontal tension at the left end of magnitude T_0, the tension at the right end in the tangent direction with magnitude $T(a)$, and the downward weight $a\delta$.

We suppose the cable has the shape of a curve $y = f(x)$. If θ denotes the angle of the tangent at a, then clearly

$$\tan \theta = \frac{dy}{dx}\bigg|_a.$$

Since the three forces balance, their horizontal components balance, as do their vertical components. This gives us two relations:

$$T(a) \cos \theta = T_0, \qquad T(a) \sin \theta = a\delta.$$

Therefore

$$\tan \theta = \frac{T(a) \sin \theta}{T(a) \cos \theta} = \frac{\delta}{T_0} a.$$

Since this is true for any a, we deduce that

$$\frac{dy}{dx} = \frac{\delta}{T_0} x.$$

By integration,

$$y = \frac{\delta}{2T_0} x^2.$$

We conclude that the cable has the shape of a parabola. (Compare the hanging cable problem in Section 7 of the next chapter.)

Free Surface of a Rotating Fluid A partially filled bucket of water (Fig. 2a) is rotating at a steady speed. It was brought to this state gradually, so the water inside rotates with the bucket. The surface of the water is a surface of revolution, and our problem is to describe it.

Imagine that this surface is obtained by revolving $y = f(x)$ about the y-axis (Fig. 2b). Now consider a small particle of mass dm at the water's surface. As it rotates with the fluid, it keeps its relative position on the surface because of three forces: (1) a buoyant force (pressure) of the fluid acting perpendicular to the surface, (2) its weight acting downwards, (3) centrifugal force acting horizontally outwards. These forces are shown in Fig. 2c. The centrifugal force has magnitude $x\omega^2\, dm$, where ω is the angular speed in rad/sec.

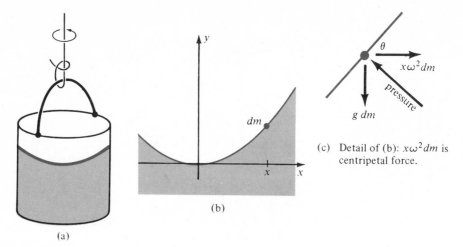

(c) Detail of (b): $x\omega^2 dm$ is centripetal force.

(b)

(a)

Fig. 2   Rotating fluid

These forces must balance; in particular their components in the direction tangent to the curve must balance, hence

$$(g\ dm)\cos(\tfrac{1}{2}\pi - \theta) = (x\omega^2\ dm)\cos\theta,$$

that is, $$g\sin\theta = x\omega^2\cos\theta, \qquad \tan\theta = \frac{\omega^2}{g}x.$$

(The pressure acts perpendicularly to the curve, hence makes no contribution to this tangential relation.) But $\tan\theta = dy/dx$, so we have obtained the relation

$$\frac{dy}{dx} = \frac{\omega^2}{g}x.$$

Here ω and g are constants. By integration,

$$y = \frac{\omega^2}{2g}x^2.$$

Therefore the rotating surface is a paraboloid of revolution.

Freezing of Liquids   The top of an ice cube is never flat; it is always curved a bit and often has a sharp peak. The following is a simplified model to explain this surface shape.

At the moment water freezes to ice, its volume expands by a factor $1 + \beta$, where $\beta > 0$. Actually $\beta = \frac{1}{8}$ for water, but we shall retain β since the model applies to other fluids as well.

Suppose that water is frozen in a cylindrical container having perfect radial symmetry, and that freezing takes place from the outside curved wall inwards. At an intermediate stage of the freezing, there is a flat cylinder of unfrozen water surrounded by a cylinder of ice (Fig. 3b). The water is higher than its initial level h because the expanding ice is squeezing it, forcing it up. Therefore, when freezing is complete, the top of the ice is peaked (Fig. 3a). The problem is to describe the

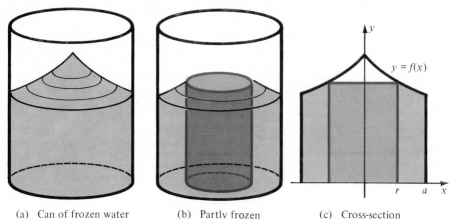

(a) Can of frozen water   (b) Partly frozen   (c) Cross-section
Fig. 3   Formation of cylindrical ice cube

shape of the frozen surface, that is, to identify the profile curve $y = f(x)$ in Fig. 3c.

At an intermediate stage, the central core of water is a cylinder of radius r, where $0 < r < a$, having volume $\pi r^2 f(r)$. Eventually this core will freeze into a solid of revolution whose upper surface is swept out by the curve $y = f(x)$. Its volume will be

$$\int_0^r 2\pi x f(x)\, dx.$$

But this volume of ice is $1 + \beta$ times the volume of the water it came from, hence

$$\int_0^r 2\pi x f(x)\, dx = (1 + \beta)\pi r^2 f(r), \qquad 0 < r < a.$$

The desired function $f(x)$ must satisfy this equation subject to the initial condition $f(a) = h$.

To find $f(x)$, we first differentiate with respect to r:

$$\frac{d}{dr}\int_0^r 2\pi x f(x)\, dx = (1 + \beta)\pi \frac{d}{dr}[r^2 f(r)],$$

$$2\pi r f(r) = (1 + \beta)\pi[2r f(r) + r^2 f'(r)].$$

We cancel πr and replace r by x:

$$2f(x) = (1 + \beta)[2f(x) + x f'(x)],$$

$$(1 + \beta)x f'(x) = -2\beta f(x),$$

$$x f'(x) = p f(x), \qquad \text{where} \quad p = \frac{-2\beta}{1 + \beta}.$$

We know a function that behaves this way under differentiation, $f(x) = x^p$, where p is rational.* Obviously, $f(x) = kx^p$ also satisfies the equation, and we can choose k

---

* We shall jump the gun a little and take it for granted that such functions exist also for any real p. The gap will be filled in the next chapter.

so that $f(a) = h$, that is, $ka^p = h$. Thus we arrive at the surface shape given by

$$y = h\left(\frac{x}{a}\right)^p, \qquad p = \frac{-2\beta}{1+\beta}.$$

For $0 < \beta < 1$, this curve is shown in Fig. 4a. Note that $p < 0$. If the liquid is a molten metal, then it contracts upon solidifying instead of expanding. Hence β is slightly less than 0. The same analysis applies, only this time $0 < p < 1$, yielding the shape shown in Fig. 4b.

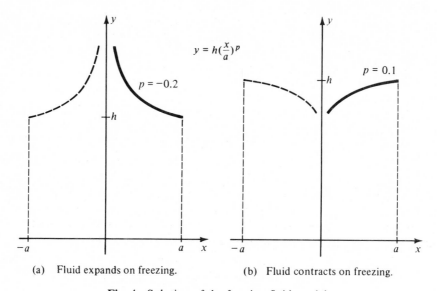

(a)   Fluid expands on freezing.   (b)   Fluid contracts on freezing.

Fig. 4   Solution of the freezing fluid model

Remark 1   The model fails when the cylinder is very thin because factors like surface tension, which we have ignored, become significant.

Remark 2   The mathematics in this example consisted of deriving a differential equation by differentiating an equation involving an integral with a variable (upper) limit. This is an important technique in physics, and the next topic provides another example.

Absorption of Radiation   The problem is to derive a differential equation for $R(h)$, the intensity of solar radiation in the atmosphere at height h above the Earth's surface. To get a feel for the meaning of "intensity of radiation" you might lie on the beach for an hour on a summer midday and measure your sunburn. More precisely, you hold a unit area of a perfect light absorber perpendicular to the Sun's rays for a unit time, then measure (by its temperature rise for instance) how much radiation it has absorbed. Thus "intensity of radiation" is measured by amount of (radiant) energy per unit area per unit time.

We shall make three simplifying physical assumptions:

(1)   We work in a "steady state," that is, everything is indpendent of time.
(2)   Energy from the Sun radiates directly downwards through the atmosphere.

(3)   The amount of radiant energy absorbed by the atmosphere per unit volume per unit time at height h is $c\delta(h)R(h)$, where $\delta(h)$ is the density of the atmosphere at height h and c is a positive constant.

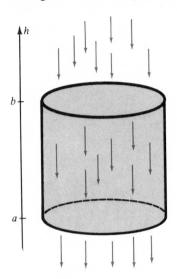

Fig. 5   Some of the radiation entering into the imaginary cylinder leaves; the rest is absorbed.

We imagine a vertical cylinder of air (Fig. 5) with base area A and extending between heights a and b. In a unit of time, the radiation passing into the cylinder of air through its upper base is $R(b)A$ and that passing out through the lower base is $R(a)A$. Hence, the net amount of radiant energy absorbed by the column of air per unit of time is

$$R(b)A - R(a)A = [R(b) - R(a)]A.$$

Let us compute this quantity by slicing. We slice the column of air into thin horizontal slabs. In the typical slab, lying at height h and with thickness dh, both $R(h)$ and $\delta(h)$ may be considered as constant. By assumption (3) the element of radiation absorbed by the slab of air per unit time is

$$c\delta(h)R(h)A \, dh.$$

Consequently, the total radiation absorbed by the cylinder of air per unit time is

$$\int_a^b c\delta(h)R(h)A \, dh = A \int_a^b c\delta(h)R(h) \, dh.$$

We equate the two expressions for this quantity, then cancel A:

$$R(b) - R(a) = \int_a^b c\delta(h)R(h) \, dh.$$

Finally, we let b vary and differentiate this equation with respect to b, then substitute h for b:

$$\frac{dR}{dh} = c\delta R.$$

This is the desired differential equation. For further details see Exs. 34–36, p. 177 and Ex. 16, p. 285.

Escape Velocity   A particle of mass m_0 is fixed at the origin on the x-axis. A second particle of mass m is initially ($t = 0$) at $x_0 > 0$ with initial velocity v_0. See Fig. 6. The only force F acting on m is the gravitational attraction of m_0; thus $F = -Gmm_0/x^2$ when m is at x. (See Example 2, p. 277.)

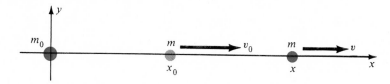

Fig. 6

By Newton's law of motion, the position $x = x(t)$ of m satisfies

$$m\frac{d^2x}{dt^2} = -G\frac{mm_0}{x^2}, \qquad x(0) = x_0, \quad \dot{x}(0) = v_0.$$

Our problem, only partially solved here, is to describe the motion.

It is convenient to set $c = \sqrt{2m_0 G}$ and $v = v(t) = \dot{x}$. Then the equation of motion and the initial conditions can be written as a system:

$$\begin{cases} \dfrac{dx}{dt} = v \\[2mm] \dfrac{dv}{dt} = \dfrac{-c^2}{2x^2} \end{cases} \qquad \begin{cases} x(0) = x_0 > 0 \\[2mm] v(0) = v_0. \end{cases}$$

Clearly $dv/dt < 0$, so $v(t)$ is a strictly decreasing function.

We derive a relation between v and x by multiplying the second differential equation by v:

$$v\frac{dv}{dt} = \frac{-c^2}{2x^2}v = \frac{-c^2}{2x^2}\frac{dx}{dt}, \qquad \text{that is,} \qquad \frac{d}{dt}\left(\frac{1}{2}v^2\right) = \frac{d}{dt}\left(\frac{c^2}{2x}\right).$$

Now we integrate:

$$\int_0^T \frac{d}{dt}\left(\frac{1}{2}v^2\right)dt = \int_0^T \frac{d}{dt}\left(\frac{c^2}{2x}\right)dt, \qquad \frac{1}{2}v(t)^2 - \frac{1}{2}v_0^2 = \frac{c^2}{2x(t)} - \frac{c^2}{2x_0}.$$

This relation can be written simply as

$$v^2 = \frac{c^2}{x} + b, \qquad \text{where} \quad b = v_0^2 - \frac{c^2}{x_0} = v_0^2 - \frac{2m_0 G}{x_0}.$$

We now examine cases according to the signs of v_0 and of b.

Case 1 $v_0 < 0$.   Since v is strictly decreasing, $v(t) < v_0$ for $t > 0$, that is, $dx/dt < v_0$ for $t > 0$. Hence, by integrating this inequality,

$$x(t) - x_0 = \int_0^t v\,dt < \int_0^t v_0\,dt = v_0 t,$$

so $x(t) < x_0 t$ for $t > 0$. The expression $x_0 + v_0 t$ is zero at time $t = -x_0/v_0 > 0$, so the particle collides with the fixed particle in at most this time.

Case 2 $v_0 = 0$.   Since v is strictly decreasing, $v(t_1) < 0$ for some $t_1 > 0$, and Case 1 applies.

Case 3 $v_0 > 0$, $b < 0$.   Then $\qquad \dfrac{c^2}{x} = v^2 - b \geq -b > 0$, $\qquad$ so $\qquad x \leq -\dfrac{c^2}{b}$.

(Note that $-c^2/b > 0$ since $b < 0$.) This implies an estimate for v. First,

$$x^2 \leq \frac{c^4}{b^2}, \qquad \frac{1}{x^2} \geq \frac{b^2}{c^4}, \qquad \frac{dv}{dt} = \frac{-c^2}{2x^2} \leq \frac{-b^2}{2c^2};$$

then by integration

$$v(t) - v_0 = \int_0^t \frac{dv}{dt}\, dt \leq \int_0^t \left(\frac{-b^2}{2c^2}\right) dt = \frac{-b^2 t}{2c^2}, \qquad v \leq v_0 - \frac{b^2}{2c^2}\, t.$$

Clearly, $v < 0$ for $t > 2c^2 v_0/b^2$, so we are back to Case 1 and a collision.

Case 4 $v_0 > 0$, $b > 0$.   Then

$$v^2 = \frac{c^2}{x^2} + b > b > 0, \qquad v > \sqrt{b}.$$

(The sign of $\sqrt{}$ is correct because $v_0 > 0$.)   By integrating, $x(t) - x_0 = \int_0^t v\, dt > t\sqrt{b}$ for $t > 0$, so

$$x(t) > x_0 + t\sqrt{b}.$$

Clearly, $x(t) \to \infty$ as $t \to \infty$ and

$$v^2 = \frac{c^2}{x} + b \to b, \qquad \text{so} \qquad v \to \sqrt{b}.$$

The particle moves off to ∞ (escapes) while its speed decreases towards its terminal speed $\sqrt{b}$.

Case 5 $v_0 > 0$, $b = 0$ (that is, $v_0^2 = 2m_0 G/x_0$).   This is the critical case because it separates the collision values of b from the escape values of b. What happens? Does the particle crash, escape, or something else?
    Since $b = 0$ and $v_0 > 0$, we have

$$\left(\frac{dx}{dt}\right)^2 = v^2 = \frac{c^2}{x}, \qquad \frac{dx}{dt} = \frac{c}{\sqrt{x}},$$

for as long as $dx/dt > 0$. We rewrite the last relation in the form

$$\sqrt{x}\,\frac{dx}{dt} = c, \qquad \text{that is,} \qquad \frac{d}{dt}\left(\frac{2}{3} x^{3/2}\right) = c.$$

By integrating, $\qquad\qquad \frac{2}{3} x^{3/2} - \frac{2}{3} x_0^{3/2} = ct,$

that is, $\qquad\qquad x = \left(\tfrac{3}{2} ct + x_0^{3/2}\right)^{2/3} = \left(\tfrac{3}{2} t \sqrt{2m_0 G} + x_0^{3/2}\right)^{2/3}.$

Clearly, $x \to \infty$ as $t \to \infty$, and $v = c/\sqrt{x} \to 0$. The particle decelerates towards its terminal speed 0, but goes off to ∞, thus escapes the gravitational attraction of m_0.

Remark   The value $v_0 = \sqrt{2m_0 G/x_0}$ is the **escape velocity** for a particle starting at x_0. Applied to the Earth it works as follows. It is known that the gravitational attraction of the Earth on an external particle is the same as the gravitational attraction of a particle at the center of the Earth with mass m_E, the Earth's mass, on the external particle. If we start at the Earth's surface, then $x_0 = r_E$, the radius of Earth, and the escape velocity is $v_0 = \sqrt{2m_E G/r_E}$.   Now $G \approx 6.67 \times 10^{-11}$ N-m^2/kg^2, $m_E \approx 5.98 \times 10^{24}$ kg,   and $r_E \approx 6.37 \times 10^6$ m, so $v_0 \approx 1.12 \times 10^4$ m/sec $\approx 25{,}100$ mile/hr.

EXERCISES

1   Find the capital value $V(T)$ of the constant income stream $f(t) = c$, interest rate r, time T years.

2   (cont.) A company deposits funds continuously at the constant rate of Y dollars per year in an account paying 8% interest, compounded continuously. Find Y so that the investment will be worth 10^7 dollars in 10 years.

3   Find the capital value $V(T)$ of the income stream $f(t) = bt$, interest rate r, T years. [*Hint*   Differentiate te^{-rt}.]

4   (cont.) A company deposits funds continuously in an 8% account, starting initially at the rate of $\$10^6$/yr and increasing the rate steadily to $\$2 \times 10^6$/yr at the end of 5 years. What is the value of the account at that time? [*Hint*   Differentiate re^{-rt}.]

In Exs. 5–10, the income stream $f(t)$ at annual interest rate r for T years has capital value $V = V(T)$.

5   Suppose money is steadily invested for one year, steadily withdrawn at the same rate the next year, then again steadily invested, etc. (cyclic behavior). Thus $f(t) = c$ for $2n \leq t < 2n + 1$ and $f(t) = -c$ for $2n + 1 \leq t < 2n + 2$. Find $V(T)$ for T an even integer, $T = 2N$.

6   (cont.) Find the ultimate capital value, $\lim V(2N)$, $N \to \infty$.

7   Suppose $f(t) = c \sin \pi t$, the simplest possible smooth periodic function in certain respects. Find $V(T)$. [*Hint* Differentiate $e^{-rt}(r \sin \pi t + \pi \cos \pi t)$.]

8   (cont.) Find $\lim_{N \to \infty} V(2N)$.

9   Find $f(T)$ if $V(T) = cT$.

10   Find $f(T)$ if $V(T) = cTe^{-kT}$.

11   Suppose that the interest rate $r = r(t)$ varies for $t \geq 0$. Show that an initial amount A_0 will grow in time t to

$$A(t) = A_0 e^{\phi(t)}, \qquad \text{where} \quad \phi(t) = \int_0^t r(u) \, du.$$

12*   (cont.) Find the capital value $V(T)$ of the income stream $f(t)$.

13   Express the tension $T(x)$ in the suspension bridge cable in terms of T_0, δ, and x.

14   Suppose the roadway of the suspension bridge is thicker at the ends than in the middle, say its linear density (weight per unit length) is $\delta(x) = A + Bx^2$. Find the shape of the cable.

15   Where does a teeter-totter (seesaw) balance? A thin, stiff, horizontal rod (Fig. 7) of varying linear density $\delta(x)$ is to be balanced on a fulcrum at $\bar{x}$. Each element of mass is acted on by gravity, causing a turning moment (torque) about $\bar{x}$ equal to the product of the mass by its distance from the fulcrum. Set up integrals for the moments on either side of the fulcrum. Equate these to find $\bar{x}$.

16   Solve the ice cube problem when the container is a thin rectangular slab (Fig. 8) and

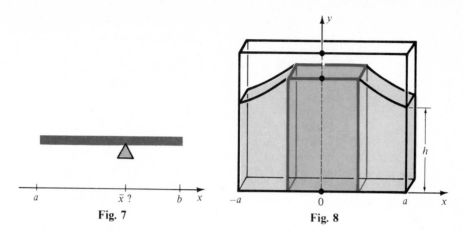

Fig. 7   Fig. 8

cooling is applied only at the ends (symmetrically). The flat faces and bottom are insulators.

17 Assume the mass of the Sun is $m_S \approx 1.99 \times 10^{30}$ kg and the (mean) radius of the Earth's orbit is $R_{SE} \approx 1.49 \times 10^{11}$ m. Find the escape velocity from a point on Earth's orbit to escape from the solar system.

The **kinetic energy** of a particle of mass m and speed v is $K = \frac{1}{2}mv^2$. For a finite system of moving particles, the **kinetic energy** of the system is the *sum* of their individual kinetic energies. For a moving rigid body (or fluid), $K = \frac{1}{2} \int v^2 \, dm$.

18 The mass m in Fig. 9 hangs at the end of a weightless spring with spring constant k. It is released from A. Write down the differential equation and initial conditions for its motion, and show that $x(t) = A \cos \omega t$, where $\omega^2 = k/m$, is the solution. (An example of **simple harmonic motion.**)

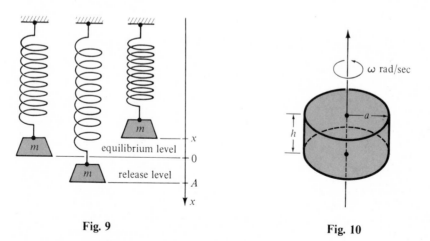

Fig. 9   Fig. 10

19 (cont.) Find the average kinetic energy over one period.
20 The cylindrical wheel (Fig. 10) rotates about its axis with steady angular speed ω rad/sec. It has uniform density (mass per unit volume) and total mass M. Find its kinetic energy. [*Hint* The cylindrical shell at distance x from the axis has speed $v = x\omega$ and thickness dx. Find its *element* of kinetic energy.]

21   (cont.) Solve the same problem for a right circular cone of base radius a and height h rotating about its axis.

22   Suppose a spherical planet of mass M and radius R has uniform density (mass per unit volume). A narrow tunnel is bored straight through the planet's center C from one side to the other. It is known that the gravitational attraction of the whole planet on a particle of mass m in the tunnel at distance x from C is the same as that due to a particle at C whose mass is that of the part of the planet within the sphere of radius x and center C. Express this force in terms of M, R, and x. Take the origin at C and the x-axis in the tunnel.

23   (cont.) At time 0 a particle of mass m is dropped into the tunnel from the planet's surface. Set up its differential equation of motion and initial conditions, and show that the solution is simple harmonic motion. (See Ex. 18.)

24   A mass m is towed, starting from rest, through a viscous fluid by a constant force F. Assume the opposing friction force of the fluid is kv, where $k > 0$. Set up the differential equation of motion and initial conditions, and show that

$$x = \frac{Fm}{k^2} (e^{-kt/m} - 1) + \frac{F}{k} t$$

is the solution. (This is a crude model of a tugboat towing a barge through an oil spill.)

25   (cont.) Find the terminal speed. Find the kinetic energy K at half terminal speed.

26* (cont.) Find the work W done by the towing force in bringing the mass to half terminal speed. Show that $W > K$ and explain the discrepancy. (This assumes you understand the conservation of energy principle.)

7. MISCELLANEOUS EXERCISES

1   Compute the area of the region bounded by the curves $y = x^3$ and $y = 2x^2$.

2   Compute the area of the region bounded by the curves $y = e^{2x}$, $y = -x^2$, $x = 0$, and $x = 3$.

3   Let R be the region under the curve $y = 1/x^2$ from $x = 1$ to $x = 10$. Find a vertical line that divides R into two regions of equal area.

4   In Fig. 1, we have $a < x < b$. Find the maximal area of the triangle.

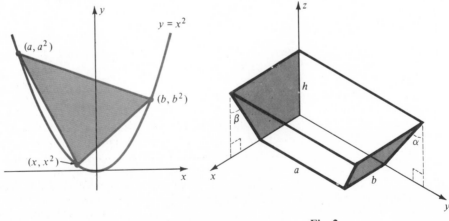

Fig. 1   Fig. 2

5 Find the area bounded by $y = x^2$, $y = 1/x^2$, and $y = 3$.

6 Find the area bounded by $y = x^2$, $y = 1/x^2$, and $x = 3$.

7 At time $t = 0$, a point is at the origin. It moves to the right along the x-axis, its speed at time t given by $1/\sqrt{3t + 1}$. Find its position at $t = 4$.

8 One particle is at $x = 0$ and another at $x = 5$. Each attracts a third particle p with a force $k/(\text{distance})^2$. Compute the work done moving p from $x = 10$ to $x = 15$.

9 Find the gravitational force of the uniform rod of linear density δ in Fig. 3 on the mass m_0.

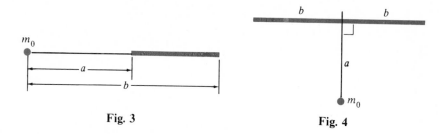

Fig. 3   **Fig. 4**

10 Do the same for Fig. 4. By symmetry, the horizontal component of the force is 0, so only the vertical component is required. [*Hint* The integration is easy if you use an angle as the variable.]

11 Compute the volume of a pinch-waist tank in the shape of the solid generated by rotating the region $-1 \le x \le \frac{1}{9}y^2$, $-3 \le y \le 3$ about the line $x = -1$. Take feet as the units.

12 (cont.) How much work is required to fill the tank with water from the level of its bottom? Assume water weighs 62.4 lb/ft^3.

The tank in Fig. 2 has its rectangular base on the horizontal x, y-plane. Its sides are trapezoids; the back sides are vertical, the two forward sides are oblique to the horizontal. The tank is filled with fluid of density δ.

13 Find the volume V of the tank. [*Hint* Slice by horizontal planes.]

14 How much work is required to fill the tank with fluid from the level of its base?

15 Compute the total pressure F_L on the (shaded) left end of the tank.

16 Find the total pressure F_R on the (shaded) right end of the tank. (Remember that pressure acts perpendicularly to any surface, even an oblique one.)

The tank in Fig. 5 consists of a standing right circular cylinder cut off by an oblique plane that cuts the horizontal x, y-plane at angle α. It is filled with fluid of density δ.

17 Use slabs perpendicular to the y-axis to set up the volume as an integral.

18 Evaluate this integral. Can you interpret the result geometrically?

19 Set up an integral for the work required to fill the tank with fluid from the level of its base.

20 The wedge in Fig. 6 has been cut from a right circular cylinder by a plane oblique to the horizontal base. Set up in two ways the volume as an integral and compute it.

21 In Ex. 17, p. 285 we show that the atmospheric pressure at height h is $p(h) = p_0 e^{-ah}$ By using a first order approximation for $p(h_1) - p(h_2)$, where, $h_1 < h_2$ and $h_2 - h_1$ is small, show that Archimedes' principle for the buoyancy of a fluid on a submerged object is a good approximation to the buoyant force on a balloon. Compare Ex. 14, p. 285.

22 A metal weight of density 5 gm/cm^3 has the shape of a truncated pyramid with square cross section. Its top measures 10×10 cm, its bottom 20×20 cm, and its height is 10 cm.

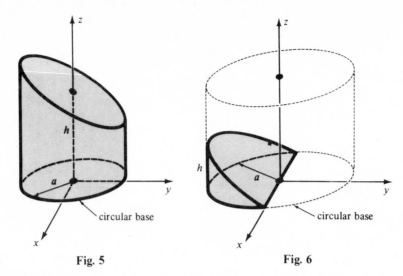

Fig. 5   **Fig. 6**

The weight is barely submerged in a lake of density 1 gm/cm^3. How much work (in gm-cm) is needed to lift the weight just clear of the water?

23 The piston of Fig. 7 has cross-sectional area A, and it is kept at constant temperature. When the piston is initially at x_0, the gas pressure inside the cylinder equals the outside atmospheric pressure p_0. Use the gas law (pressure)(volume) = const. to derive a formula for the pressure $p(x)$ when the piston is at x.

24 (cont.) Set up an integral for the work done by the external force F to move the piston from x_0 to x. (Don't forget that atmospheric pressure is helping you.)

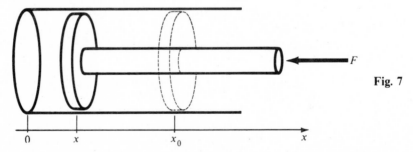

Fig. 7

25 The office space part of the Transamerica Building in San Francisco (Fig. 8) consists of a truncated pyramid. Its height is 621 ft, its base is a square of side 152 ft, and its top is a square of side 40 ft. Find its volume by integration, then use the Trapezoidal rule to estimate the total floor space of its 46 floors.

26 Find the area between the three semicircles in Fig. 9.

Fig. 8

Fig. 9

Inverse Functions 7

1. INVERSE FUNCTIONS AND THEIR DERIVATIVES

We shall create new functions by turning old functions inside out. Here is the idea. A function f assigns to each x in its domain a unique number y in its range (set of values). We ask whether each y in the range of f comes from a *unique* ancestor x. Given y, does the equation $y = f(x)$ determine a unique x? Equivalently, does f take on each of its values exactly once?

Suppose f has the property that each y in its range corresponds to exactly one x in its domain. Then y determines x, that is, x can be considered as a function of y. We think of solving the equation

$$f(x) = y$$

for x in terms of y.

For example, take $f(x) = 2x - 5$. Given any y, the equation $y = 2x - 5$ determines precisely one x. To find it, just solve the equation for x:

$$x = \tfrac{1}{2}(y + 5).$$

For another example, take $f(x) = x^3$. The equation $y = x^3$ determines a unique x, namely, $x = y^{1/3}$.

Not every function is so cooperative. The function $f(x) = x^2$ takes every positive value *twice*. Given $y > 0$, the equation $y = x^2$ determines *two* values of x, namely, $x = \sqrt{y}$ and $x = -\sqrt{y}$. Worse yet is $f(x) = \sin x$. This function takes every value in its range $-1 \le y \le 1$ *infinitely* many times. For instance, the equation $0 = \sin x$ has the solutions $x = 0, \ \pm \pi, \ \pm 2\pi, \ \pm 3\pi, \ \cdots$. For such functions we cannot expect $f(x) = y$ to determine x as a function of y unless we somehow restrict the set of values of x.

When can we be sure that x is determined as a function of y? There are two favorable situations:

(1) $f(x)$ strictly increasing, that is, $f(x_1) < f(x_2)$ whenever $x_1 < x_2$;
(2) $f(x)$ strictly decreasing, that is, $f(x_1) > f(x_2)$ whenever $x_1 < x_2$.

This is clear geometrically as illustrated in Fig. 1. In both examples, the graph intersects each horizontal line $y = c$ exactly once if c is in the range of the function.

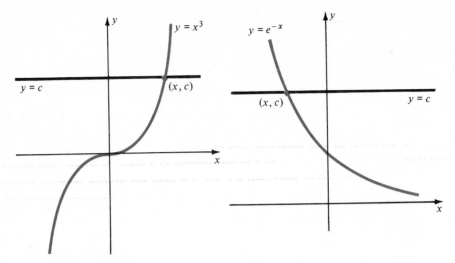

(a) Strictly increasing; range: all y (b) Strictly decreasing; range: $y > 0$

Fig. 1   The graph $y = f(x)$ crosses each horizontal line $y = c$ exactly once, where c is in the range of $f(x)$.

When a function $f(x)$ is not strictly increasing or strictly decreasing, the equation $y = f(x)$ may fail to have a *unique* solution x for some values of y in the range of $f(x)$. See Fig. 2 for examples.

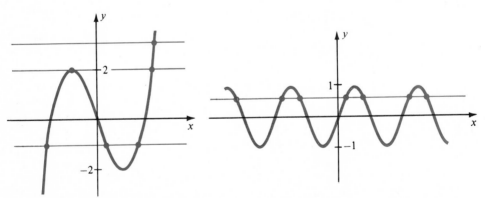

(a) $x^3 - 3x = y$ has a unique solution   (b) $\sin x = y$ has infinitely many solutions
     if $|y| > 2$, two solutions if $|y| = 2$,   if $|y| \leqslant 1$.
     and three solutions if $|y| < 2$.

Fig. 2   Graphs that cross some horizontal lines more than once

Definition of Inverse Functions   Suppose f is a function such that the equation $f(x) = y$ determines a unique x for each y in the range of f. Then we turn f inside out (invert f) and consider x as a function of y. We write $x = g(y)$ and call g the **inverse function** of f. By definition, the domain of g is the range of f.

Remark   Sometimes the inverse function is denoted f^{-1}. We shall avoid this notation because of possible confusion with the reciprocal $1/f$.

Examples (1) $y = f(x) = 2x - 5$, domain all x: $x = g(y) = \frac{1}{2}(x + 5)$.

(2) $y = f(x) = x^3$, domain all x: $x = g(y) = y^{1/3}$.

(3) $y = f(x) = x^2$, domain all x: no inverse function.

(4) $y = f(x) = \sin x$, domain all x: no inverse function.

It is pretty clear from Figs. 1 and 2 that a continuous function (defined on an interval) has an inverse function if and only if it is strictly increasing or strictly decreasing. (A proof of this assertion will be sketched in Section 9.) Hence if $f(x)$ is differentiable and $f'(x) > 0$ for all x in the domain of f, then f is strictly increasing so its inverse function exists. Likewise, if $f'(x) < 0$, then the inverse function exists. Let us summarize these facts, assuming $f(x)$ is defined on an interval, possibly an infinite interval.

Existence of Inverse Functions

(1) A continuous function $f(x)$ has an inverse function if and only if $f(x)$ is strictly increasing or strictly decreasing.

(2) A differentiable function $f(x)$ has an inverse function if either $f'(x) > 0$ for all x in the domain of f or $f'(x) < 0$ for all x in the domain of f.

Remark Inverting known functions sometimes produces brand new functions. Consider for example $f(x) = x^5 + x$ with domain all x. Then $f'(x) = 5x^4 + 1 \geq 1 > 0$, hence f has an inverse function g. But we cannot solve the quintic equation $x^5 + x = y$ explicitly for x, that is, we cannot express $g(y)$ by a formula involving familiar functions. So $x^5 + x = y$ defines a "new" function $x = g(y)$.

Properties of Inverse Functions Suppose f is a function having an inverse function g. By the very meaning of the inverse function,

$$y = f(x) \qquad \text{and} \qquad x = g(y)$$

are equivalent statements; they say the same thing. If f assigns the number y to x, then g assigns the number x to y. Thus the action of g undoes the action of f, and conversely f undoes the action of g:

$$g[f(x)] = g(y) = x, \qquad f[g(y)] = f(x) = y.$$

Examples (1) $f(x) = 2x - 5$, $g(y) = \frac{1}{2}(y + 5)$:

$$g[f(x)] = \tfrac{1}{2}[f(x) + 5] = \tfrac{1}{2}[(2x - 5) + 5] = x,$$

$$f[g(y)] = 2g(y) - 5 = 2[\tfrac{1}{2}(y + 5)] - 5 = y.$$

(2) $f(x) = x^3$, $g(y) = y^{1/3}$:

$$g[f(x)] = [f(x)]^{1/3} = [x^3]^{1/3} = x,$$

$$f[g(y)] = [g(y)]^3 = [y^{1/3}]^3 = y.$$

Obviously f and g play symmetric roles with respect to each other. Not only is g the inverse of f, but f is the inverse of g. Therefore we often speak of pairs of inverse functions.

Suppose $f(x)$ is a function having an inverse function $g(y)$. Then

(1) $y = f(x)$ and $x = g(y)$ are equivalent statements.

(2) $g[f(x)] = x$, $f[g(y)] = y$.

(3) f is the inverse of g.

Restricting the Domain   This is a technique for salvaging a kind of inverse for certain functions not having genuine inverses. The idea is simple: restrict the domain of the function to a smaller domain on which the graph is either increasing or decreasing.

For example, take $f(x) = x^2$ on the domain $-\infty < x < \infty$. This function has no inverse because its graph is neither increasing nor decreasing throughout the domain. But suppose you restrict the function to the smaller domain $x \geq 0$. There $y = x^2$ is strictly increasing (Fig. 3a), hence the restricted function $f(x) = x^2$ has an inverse.

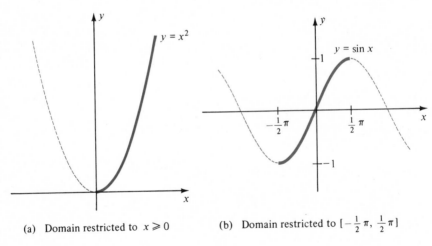

(a) Domain restricted to $x \geqslant 0$ (b) Domain restricted to $[-\frac{1}{2}\pi, \frac{1}{2}\pi]$

Fig. 3

All this makes good sense algebraically. The inverse of $y = x^2$ ought to be $x = \sqrt{y}$. But every positive number has *two* square roots, for instance $36 = (\pm 6)^2$. By insisting on $x \geq 0$, you accept only the unique positive square root of each positive number.

As another example, take $f(x) = \sin x$ on the domain $-\infty < x < \infty$. This function is far from having an inverse, as Fig. 2 shows. But on the subdomain $[-\frac{1}{2}\pi, \frac{1}{2}\pi]$, the sine function is strictly increasing (Fig. 3b). Therefore $f(x) = \sin x$, restricted to this domain, has an inverse function $g(y)$. Again this makes sense: if $-1 \leq y \leq 1$, then $g(y)$ is the unique angle x in the domain $[-\frac{1}{2}\pi, \frac{1}{2}\pi]$ whose sine is y.

Graphs   Suppose f has an inverse g. What is the relation between the graphs of these two functions? The points on the graph of $y = f(x)$ can be written in two ways, as

$$(x, f(x)) \quad \text{or as} \quad (g(y), y).$$

Hence the graph of $x = g(y)$ is the graph of $y = f(x)$.

That should be all there is to it. Yet the notation $x = g(y)$ seems awkward since we are accustomed to x as the independent variable and y as the dependent variable. So we reverse the roles of x and y, and study $y = g(x)$ instead. Then each point (x, y) on the graph of $y = g(x)$ corresponds to a point (y, x) on the graph of $x = g(y)$, that is, the graph of $y = f(x)$. See Fig. 4 (and compare Example 3, p. 47).

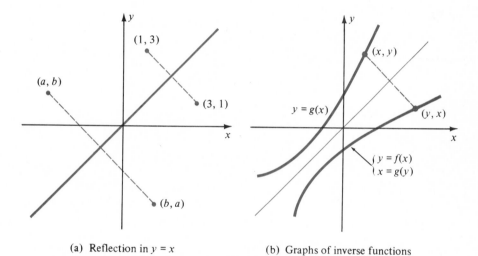

(a) Reflection in $y = x$ (b) Graphs of inverse functions

Fig. 4   Graphs of inverse functions are reflections of each other in $y = x$.

Graphs of Inverse Functions   Let f and g be a pair of inverse functions. Then the graphs of $y = f(x)$ and $y = g(x)$ are mirror images of each other in the line $y = x$.

Some explicit examples are shown in Fig. 5.

Remark   If two non-horizontal lines are reflections of each other in $y = x$, then their slopes are reciprocals. For example, in Fig. 5a the slopes are 2 and $\frac{1}{2}$. In general, suppose one of the lines passes through two distinct points, (a, b) and (c, d). By the slope formula, its slope m_1 is

$$m_1 = \frac{b - d}{a - c}.$$

The other line passes through the reflected points (b, a) and (d, c). Its slope m_2 is

$$m_2 = \frac{a - c}{b - d} = \frac{1}{m_1}.$$

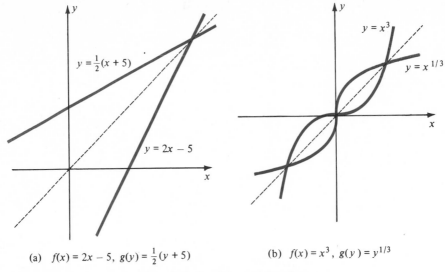

(a) $f(x) = 2x - 5$, $g(y) = \frac{1}{2}(y + 5)$ (b) $f(x) = x^3$, $g(y) = y^{1/3}$

Fig. 5   Pairs of inverse functions

Derivatives   Suppose $y = f(x)$ is differentiable and has inverse function $x = g(y)$. We suspect that $g(y)$ is also differentiable and that there is a rule for computing $g'(y)$ in terms of $f'(x)$. We can *guess* a rule, reasoning this way: if y is changing k times as fast as x, then x is changing $1/k$ times as fast as y. Therefore,

$$\frac{d}{dy}\, g(y) = 1 \left/ \frac{d}{dx}\, f(x) \right. ,$$

provided $f'(x) \neq 0$. We shall prove the following precise statement in Section 9.

Derivative of an Inverse Function   Let $y = f(x)$ be a differentiable function that has an inverse function $x = g(y)$. Then $g(y)$ is differentiable at each point $y = f(x)$ where $f'(x) \neq 0$, and

$$\left. \frac{d}{dy}\, g(y) \right|_{y = f(x)} = \frac{1}{f'(x)}.$$

Briefly, $$g'(y) = \frac{1}{f'(x)} \quad \text{or} \quad \frac{dx}{dy} = 1 \left/ \frac{dy}{dx} \right. .$$

Note that f' is evaluated at x, but g' is evaluated at $y = f(x)$.

The rule has a geometric interpretation (Fig. 6). The graphs of $y = f(x)$ and $y = g(x)$ are reflections of each other in the line $y = x$. At corresponding points (x, y) and (y, x), the tangents F and G are also reflections of each other, so their slopes are reciprocal:

$$(\text{slope of } G) = \frac{1}{(\text{slope of } F)}.$$

However, the slope of F is the derivative of f evaluated at x, and the slope of G

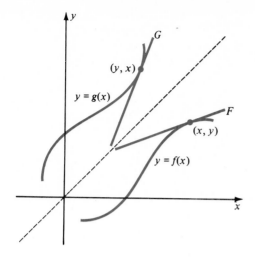

Fig. 6 The tangents at corresponding points have reciprocal slopes.

is the derivative of g evaluated at y. Therefore

$$g'(y) = \frac{1}{f'(x)}.$$

If F happens to be horizontal, then G is vertical. The slope of G is undefined, and $g(y)$ is not differentiable at the corresponding y.

■ **EXAMPLE 1** Differentiate $x^{1/3}$.

Solution The function $g(x) = x^{1/3}$ is the inverse function of $f(x) = x^3$. Therefore

$$g'(y)\Big|_{y=x^3} = \frac{1}{f'(x)} = \frac{1}{3x^2} \qquad (x \neq 0).$$

But $y = x^3$ is equivalent to $x = y^{1/3}$, so

$$g'(y) = \frac{1}{3x^2} = \frac{1}{3(y^{1/3})^2} = \frac{1}{3y^{2/3}} \qquad (y \neq 0).$$

Interchange x and y if you prefer the formula in terms of x:

$$\frac{d}{dx} x^{1/3} = \frac{1}{3x^{2/3}} \qquad (x \neq 0). \qquad\qquad ■$$

Remark At $x = 0$, the graph $y = x^3$ has a horizontal tangent. Therefore the reflected curve $y = x^{1/3}$ has a vertical tangent, hence no derivative (Fig. 5b). Nevertheless, it is reasonable to write $d(x^{1/3})/dx = \infty$ at $x = 0$, as justified by the limit calculation

$$\frac{g(x) - g(0)}{x - 0} = \frac{x^{1/3}}{x} = \frac{1}{x^{2/3}} \to \infty \qquad \text{as} \quad x \to 0.$$

■ **EXAMPLE 2** Show that $f(x) = x^5 + x$ has a differentiable inverse $g(y)$ and compute $g'(2)$.

Solution Since $f'(x) = 5x^4 + 1 > 0$ for all x, the graph of $y = x^5 + x$ is strictly

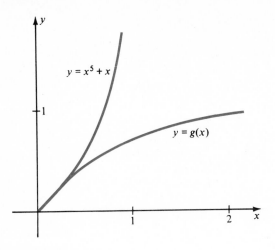

Fig. 7 Graphs of $y = x^5 + x$ and its inverse function $y = g(x)$

increasing (Fig. 7). Hence $f(x)$ has an inverse function $g(y)$. Since $f(x)$ is differentiable and $f'(x)$ is non-zero, $g(y)$ is also differentiable.

We cannot solve $y = x^5 + x$ for x, so we have no workable formula for $g(y)$. Still we can find $g'(y)$ by the rule

$$g'(y) = \frac{1}{f'(x)} = \frac{1}{5x^4 + 1},$$

where $y = x^5 + x$. Now $y = 2$ corresponds to $x = 1$ since $2 = 1^5 + 1$. Therefore,

$$g'(2) = \frac{1}{5(1)^4 + 1} = \frac{1}{6}.$$

EXERCISES

Give the inverse function in the form $x = g(y)$

1 $y = 3x - 7$

2 $y = -2x + 5$

3 $y = -\dfrac{1}{x}$ $(x \neq 0)$

4 $y = \dfrac{3}{10x - 7}$ $(x \neq \frac{7}{10})$

5 $y = \dfrac{2x - 7}{x + 4}$ $(x \neq -4)$

6 $y = \dfrac{x + 1}{x - 1}$ $(x \neq 1)$

7 $y = \dfrac{x + 2}{x + 3}$ $(x \neq -3)$

8 $y = \dfrac{3x + 4}{2x + 3}$ $(x \neq -\frac{3}{2})$

9 $y = \dfrac{x^3 + 2}{x^3 + 3}$ $(x^3 \neq -3)$

10 $y = \left(\dfrac{3x + 4}{2x + 3}\right)^3$ $(x \neq -\frac{3}{2})$

11 $y = \sqrt{2x - 8}$ $(x \geq 4)$

12 $y = x^5 + 1$

13 $y = x + \dfrac{1}{x}$ $(x \geq 1)$

14 $y = \dfrac{9}{x + 7} - 7$ $(x \neq -7)$.

Does $y = f(x)$ for $-\infty < x < \infty$ have an inverse function? (If so, you need not compute it.)

15 $y = e^{x^2}$

16 $y = \tan x$

17 $y = \frac{1}{5}x^5 + \frac{2}{3}x^3 + x + 12$ **18** $y = e^{-3x} + 1.$

19   Let $y = f(x)$ be an even function for $|x| < b$. Show that it has no inverse function.

20   Let $y = f(x)$ be an odd function for $|x| < b$ that has an inverse function $x = g(y)$. Show that the inverse function is odd.

21   Show that $y = x^3 + x$ has an inverse $x = g(y)$, and graph $y = g(x)$.

22   (cont.) Show that $g(x)^3 = -g(x) + x$.

Find $g'(c)$, where $x = g(y)$ is the inverse function of $y = f(x)$

23 $f(x) = x^3 + x, \quad c = 30$ **24** $f(x) = x^{11} + 2x^5 + x, \quad c = 4$

25 $f(x) = \dfrac{x^3}{1 + x^2}, \quad c = -\frac{8}{5}$ **26** $f(x) = -x^5 - 5x + 2, \quad c = 8$

27 $f(x) = \dfrac{1}{x^3} + \dfrac{1}{x^9}, \quad c = -2$ **28** $f(x) = 2x^3 - 9x^2 + 18x + 5, \quad c = 5.$

29   Find $\left. \dfrac{d}{dx} x^{4/3} \right|_0$ by use of $\left. \dfrac{d}{dx} f(x) \right|_0 = \lim\limits_{x \to 0} \dfrac{f(x) - f(0)}{x - 0}.$

30   (cont.) Generalize to $f(x) = x^{p/q}$ for $p/q > 1.$

31*   Suppose $y = f(x)$ has the inverse function $x = f^{-1}(y)$ and $z = g(y)$ has the inverse function $y = g^{-1}(z)$. Assuming the composites are defined, show that $(g \cdot f)^{-1} = f^{-1} \cdot g^{-1}.$

32   (cont.) Verify this result for $f(x) = 1/x$ and $g(y) = 1 - y.$

33*   Suppose $y = f(x)$ is differentiable at $x = a$ and $f(a) = b$. Suppose also $f(x)$ has the inverse function $x = g(y)$ and it is differentiable at $y = b$. Prove that $f'(a) \neq 0.$

34*   Suppose $f(x)$ is strictly increasing on $[0, 1]$ and equal to its own inverse. Prove $f(x) = x.$

2. THE LOGARITHM FUNCTION

The inverse of the strictly increasing function $y = e^x$ is denoted

$$y = \ln x$$

and is called the **natural logarithm** function. Figure 1 shows the graph of $y = \ln x$, the reflection in $y = x$ of the graph of $y = e^x$.

Certain basic properties are immediate from Fig. 1, next page.

> (1) $\ln x$ is defined only for $x > 0$;   (2) $\ln x$ is an increasing function;
> (3) $\ln x < 0$ for $0 < x < 1$;   (4) $\ln 1 = 0$;   (5) $\ln x > 0$ for $x > 1.$

Because $\ln x$ and e^x are inverse functions, two further statements are automatic:

> (6) $y = e^x$ is equivalent to $x = \ln y$;   (7) $\ln(e^x) = x$ and $e^{\ln y} = y.$

Examples $\ln e^4 = 4, \quad \ln e^{-2.3} = -2.3, \quad \ln e^\pi = \pi, \quad e^{\ln 7} = 7, \quad e^{\ln 10.6} = 10.6,$
$e^{\ln 1.2} = \frac{1}{2}.$

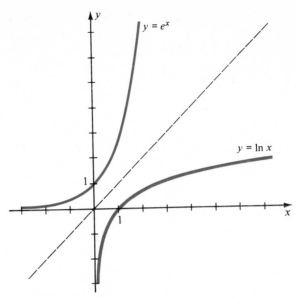

Fig. 1 ln x is the inverse function of e^x.

Be sure you understand these examples. They are easy, but unless you get the principle right, you are in for trouble. For instance, without any further information you should be able to differentiate $e^{\ln x}$. How?

■ **EXAMPLE 1** Given $2.5 < e < 3$, show that $2 < \ln 10 < 3$.

Solution Because e^x is strictly increasing, the desired inequality is equivalent to this one:

$$e^2 < e^{\ln 10} < e^3, \qquad \text{that is,} \qquad e^2 < 10 < e^3.$$

But $e < 3$, so $e^2 < 3^2 = 9 < 10$. Similarly, $e > 2.5$, so $e^3 > (2.5)^3 > 2 \times 2 \times 2.5 = 10$. Therefore, $e^2 < 10 < e^3$. ■

From Fig. 1 it is not clear whether the graph of $y = \ln x$ rises arbitrarily high or not. Actually, the graph reaches all levels because each $y > 0$ is the natural log of something, namely of e^y. For example,

$$10 = \ln e^{10}, \qquad 100 = \ln e^{100}, \qquad 1000 = \ln e^{1000}, \cdots.$$

Therefore the graph of $y = \ln x$ increases beyond all bound as x increases.
Similarly, each $y < 0$ is ln of something, $y = \ln e^y$. For example,

$$-10 \ln e^{-10}, \qquad -100 = \ln e^{-100}, \qquad -1000 = \ln e^{-1000}, \qquad \cdots$$

Therefore the graph of $y = \ln x$ decreases beyond all bound as $x \longrightarrow 0+$.

$$\boxed{\ln x \longrightarrow \infty \quad \text{as} \quad x \longrightarrow \infty; \qquad \ln x \longrightarrow -\infty \quad \text{as} \quad x \longrightarrow 0+.}$$

Algebraic Properties of ln x   Let us justify the name natural *logarithm* by showing that $\ln x$ behaves the way a logarithm should. Recall some algebraic properties of the exponential function e^x:

$$(1) \quad e^a e^b = e^{a+b} \qquad (2) \quad e^{-a} = \frac{1}{e^a} \qquad (3) \quad \frac{e^a}{e^b} = e^{a-b} \qquad (4) \quad (e^a)^b = e^{ab}.$$

Somehow, these properties ought to rub off onto the inverse function $\ln x$. They do: each of the four statements can be translated into a corresponding statement about natural logarithms. Set $e^a = x$ and $e^b = y$; that means $a = \ln x$ and $b = \ln y$. Then property (1) can be translated as follows:

$$e^a e^b = e^{a+b}, \qquad xy = e^{\ln x + \ln y},$$

hence

$$\ln(xy) = \ln x + \ln y.$$

Similarly, properties (2), (3), and (4) can be translated into properties of $\ln x$:

$$(1') \quad \ln(xy) = \ln x + \ln y \qquad (2') \quad \ln\left(\frac{1}{x}\right) = -\ln x$$

$$(3') \quad \ln\left(\frac{x}{y}\right) = \ln x - \ln y \qquad (4') \quad \ln x^b = b \ln x.$$

Warning   There are no nice formulas for $\ln(x + y)$ or $(\ln x)(\ln y)$.

■ **EXAMPLE 2**   Simplify
(a) $e^{(\ln 25)\,2}$ (b) $\ln 2 + \ln 4 + \ln 8 + \cdots + \ln 128$.

Solution   (a) $e^{(\ln 25)\,2} = e^{\ln\sqrt{25}} = e^{\ln 5} = 5$.

(b) $\ln 2 + \ln 4 + \cdots + \ln 128 = \ln 2 + \ln 2^2 + \cdots + \ln 2^7$

$$= \ln 2 + 2\ln 2 + \cdots + 7\ln 2 = (1 + 2 + 3 + \cdots + 7)\ln 2 = 28 \ln 2. \quad ■$$

Relation of Natural Logs to Common Logs   Natural logarithms are logarithms to the base e; common logarithms are logarithms to the base 10. There is a simple relation between the two logarithm systems. We express a positive number x in two ways,

$$x = e^{\ln x} = 10^{\log_{10} x},$$

then take logs to the base 10:

$$(\ln x)(\log_{10} e) = \log_{10} x.$$

Therefore

$$\ln x = \frac{1}{M}\log_{10} x, \qquad \text{where} \quad M = \log_{10} e \approx 0.43429.$$

Thus natural logs are proportional to common logs and can be computed from a table of common logs by this formula. Of course tables of natural logs are available. While natural logs are important for theoretical purposes, they are not nearly as well suited for practical computations as are common logs, which go so well with our decimal representation of numbers.

Derivative of ln x   The natural logarithm function inherits some useful algebraic properties from e^x. It also inherits a simple derivative:

$$\frac{d}{dx}(\ln x) = \frac{1}{x}, \qquad x > 0.$$

This follows easily from the rule for differentiating inverse functions. Let $f(x) = e^x$ and $g(x) = \ln x$, the inverse function. Then

$$g'(y) = \frac{1}{f'(x)}, \qquad \text{where} \quad y = f(x) = e^x.$$

Since $f'(x) = e^x$,

$$g'(y) = \frac{1}{e^x} = \frac{1}{y}, \qquad \text{that is,} \qquad \frac{d}{dy}(\ln y) = \frac{1}{y}.$$

Now replace y by x.

The formula agrees with the graph of $y = \ln x$ in Fig. 1. The slope $1/x$ is always positive, it becomes very large as $x \longrightarrow 0+$, and it dies out as $x \longrightarrow \infty$.

If $u(x)$ is differentiable and $u(x) > 0$, then the composite function $\ln u(x)$ is differentiable. By the Chain Rule,

$$\frac{d}{dx}\ln u(x) = \left(\frac{d}{du}\ln u\right)\left(\frac{du}{dx}\right) = \frac{1}{u(x)}\frac{du}{dx}.$$

$$\frac{d}{dx}\ln u = \frac{u'}{u} \qquad (u > 0).$$

■ **EXAMPLE 3**   Differentiate $y = \ln(x^2 + 1)$.

Solution $$\frac{dy}{dx} = \frac{u'}{u} = \frac{(x^2 + 1)'}{x^2 + 1} = \frac{2x}{x^2 + 1}. \qquad ■$$

■ **EXAMPLE 4**   Differentiate

$$y = \ln\left(\frac{1 + x}{1 - x}\right) \qquad \text{for} \quad -1 < x < 1.$$

Solution   Clearly, $1 + x > 0$ and $1 - x > 0$ for $-1 < x < 1$. By one of the rules for logs,

$$y = \ln(1 + x) - \ln(1 - x).$$

Hence $$\frac{dy}{dx} = \frac{(1 + x)'}{1 + x} - \frac{(1 - x)'}{1 - x} = \frac{1}{1 + x} - \frac{-1}{1 - x} = \frac{2}{1 - x^2}. \qquad ■$$

Note   An alternative, but more complicated, method is

$$\frac{dy}{dx} = \left(\frac{1 + x}{1 - x}\right)' \Big/ \frac{1 + x}{1 - x}, \quad \text{etc.}$$

Antiderivative of $1/x$ The formula

$$x^n = \frac{d}{dx}\left(\frac{x^{n+1}}{n + 1}\right) \quad (n \neq -1)$$

supplies an antiderivative for each integral power x^n except $x^{-1} = 1/x$. Now we can fill the gap at -1 because

$$\frac{1}{x} = \frac{d}{dx} \ln x \quad \text{for} \quad x > 0.$$

Thus $\ln x$ is an antiderivative of $1/x$, at least for $x > 0$. But what about x negative? The trouble is that $\ln x$ is not defined for $x < 0$; however $\ln(-x)$ is, because $-x > 0$. By the Chain Rule,

$$\frac{d}{dx} \ln(-x) = \frac{(-x)'}{-x} = \frac{-1}{-x} = \frac{1}{x} \quad \text{for} \quad x < 0.$$

An antiderivative of $\dfrac{1}{x}$ is $\begin{cases} \ln x \\ \ln(-x) \end{cases}$ for $\begin{matrix} x > 0 \\ x < 0. \end{matrix}$

In other words: An antiderivative of $\dfrac{1}{x}$ is $\ln |x|$ for $x \neq 0$.

■ **EXAMPLE 5**   Compute   (a) $\displaystyle\int_1^3 \frac{1}{x}\, dx$ (b) $\displaystyle\int_{-3}^{-1} \frac{1}{x}\, dx$.

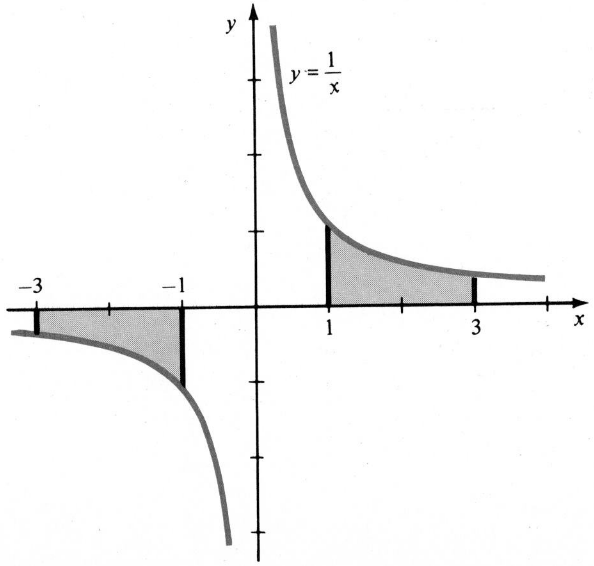

Fig. 2   The areas are equal; the integrals have opposite signs.

Solution Use the result above and the Fundamental Theorem:

(a) $\displaystyle\int_1^3 \frac{1}{x}\,dx = \ln x\,\Big|_1^3 = \ln 3 - \ln 1 = \ln 3.$

(b) $\displaystyle\int_{-3}^{-1} \frac{1}{x}\,dx = \ln(-x)\,\Big|_{-3}^{-1} = \ln 1 - \ln 3 = \ln \tfrac{1}{3}.$ ■

Remark The two answers are negatives of each other. A glance at Fig. 2 shows why; the integrals represent areas that are equal but of opposite sign.

Warning If $a < 0$ and $b > 0$, then

$$\int_a^b \frac{1}{x}\,dx$$

is undefined, because $1/x$ is undefined at $x = 0$.

ln x as an Area The function $\ln x$ can be expressed as an integral:

$$\ln x = \int_1^x \frac{1}{t}\,dt, \qquad x > 0.$$

Therefore $\ln x$ represents an area if $x > 1$, or the negative of an area if $0 < x < 1$. See Fig. 3.

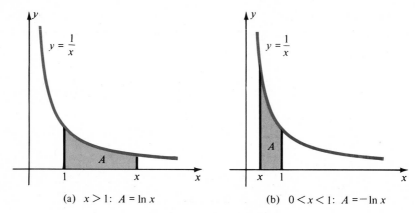

(a) $x > 1$: $A = \ln x$ 　　　　　 (b) $0 < x < 1$: $A = -\ln x$

Fig. 3 ln x as an area

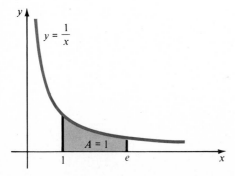

Fig. 4 e is defined by $\displaystyle\int_1^e \frac{dx}{x} = 1.$

From $\ln e = 1$, we obtain a geometric interpretation of the number e. It is the unique number greater than 1 such that the area under $y = 1/x$ from 1 to e equals one (Fig. 4).

From this viewpoint, we can give an easy geometric proof that $2 < e < 4$. See Fig. 5.

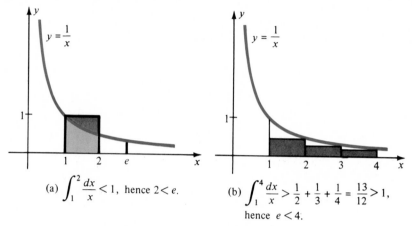

(a) $\displaystyle\int_1^2 \frac{dx}{x} < 1$, hence $2 < e$.

(b) $\displaystyle\int_1^4 \frac{dx}{x} > \frac{1}{2} + \frac{1}{3} + \frac{1}{4} = \frac{13}{12} > 1$, hence $e < 4$.

Fig. 5   Geometric proof that $2 < e < 4$

Antiderivative of ln x   We needed a new function, $\ln x$, to integrate $1/x$. Do we need another new function in order to integrate $\ln x$ in turn? We do not:

$$\frac{d}{dx}(x \ln x - x) = \ln x + \frac{x}{x} - 1 = \ln x.$$

How did we ever guess this? Well, that is a little secret until the next chapter.

EXERCISES

Find the inverse function

1 $y = e^{-x}$ **2** $y = e^{-x^3}$ **3** $y = 1/\ln x, \quad x > 1$

4 $y = 1/\ln x, \quad 0 < x < 1$ **5** $y = \ln(x + 5), \quad x > -5$ **6** $y = \ln(\ln x), \quad x > e$.

Simplify

7 $\ln e^{a+2}$ **8** $\exp(\ln x^2)$ **9** $e^{-\ln x}$

10 $e^{2 \ln x + 5 \ln y}$ **11** $\ln \sqrt{e}$ **12** $\ln(1/e^{3/2})$.

Differentiate with respect to x

13 $\ln 5x$ **14** $3 \ln 4x$ **15** $2 \ln x^2$ **16** $\ln x^4$

17 $\ln(1/x)$ **18** $\ln(x^2 + x)$ **19** $\ln(\sin x)$ **20** $\ln(\cos x)$

21 $\ln\left(\dfrac{x + 1}{x - 1}\right)$ **22** $\ln(\ln x)$ **23** $\sqrt{\ln x}$ **24** $(\ln x)^2$

25 $\dfrac{1}{\ln x}$ **26** $\ln\left(\dfrac{x + 2}{2x + 1}\right)$

27 $\ln(x + \sqrt{x^2 + 1})$ **28** $\ln(\sec x + \tan x)$ **29** $-\dfrac{1}{2}\ln\left(\dfrac{2 + \sqrt{x^2 + 4}}{x}\right)$

30 $\dfrac{1}{b}\ln(a + bx)$.   **31** $(\ln x)^n/n, \quad n \ne 0$ **32** $x^n(\ln x - 1/n)/n, \quad n \ne 0$.

Prove the boxed property on p. 309.

33 (2′)   **34** (3′)   **35** (4′)

36 Find the tangent line to $y = \ln x$ that passes through $(0, 1)$.

Find the integer n for which

37 $e^n < 1000 < e^{n+1}$ **38** $10^n < e^{100} < 10^{n+1}$.

39 Minimize $y = x \ln x$ for $x > 0$.
40 Show that $y = (\ln x)/x$ for $x > e$ has an inverse function $x = g(y)$, and compute $g'(3/e^3)$.
41 Show that $y = \ln x$ is strictly concave.   **42** Find $\lim_{r \to 1} [(r - 1)/\ln r]$.

43 Prove $\displaystyle\int_1^{100} \frac{dx}{x} = 2 \int_1^{10} \frac{dx}{x}$.   **44** Solve for x: $\displaystyle\int_{1/2}^x \frac{dt}{t} = 3 \int_1^x \frac{dt}{t}$.

45 Let $1 < x$. Prove $\ln x < x - 1$.   **46** Let $1 < x$. Prove $(x - 1)/x < \ln x$.
47 Use Exs. 45 and 46 to prove Napier's inequality: if $0 < x < y$, then

$$\frac{1}{y} < \frac{\ln y - \ln x}{y - x} < \frac{1}{x}.$$

48* Find $\displaystyle\lim_{n \to \infty} \left(\frac{1}{n + 1} + \frac{1}{n + 2} + \cdots + \frac{1}{2n}\right)$. [*Hint* Relate the sum to an integral.]

49 Use $1/t < 1/\sqrt{t}$ for $t > 1$ to prove $\ln x < 2(\sqrt{x} - 1)$ for $x > 1$.
50 (cont.) Similarly, prove $\ln x < n(x^{1/n} - 1)$ for $x > 1$ and $n \ge 1$.

51 (cont.) Prove $\left(1 + \dfrac{x}{n}\right)^n < e^x$ for $x > 0$ and $n \ge 1$.

52* Modify your solution to Ex. 50 to deduce that $n(1 - x^{-1/n}) < \ln x$ for $x > 1$ and $n \ge 1$.

53 (cont.) Prove $e^x < \left(1 - \dfrac{x}{n}\right)^{-n}$ for $x > 0$ and $n \ge 1$.

54 Use the results of Exs. 51 and 53 to prove $\left(1 + \dfrac{1}{n}\right)^n < e < \left(1 + \dfrac{1}{n}\right)^{n+1}$ for all $n \ge 1$.

55 Show that $1/x > \frac{1}{2} - \frac{1}{4}(x - 2)$ for $1 \le x < 2$ and $1/x > \frac{1}{3} - \frac{1}{9}(x - 3)$ for $2 \le x < 3$.
56 (cont.) Deduce that $\ln 3 > 1$, hence $e < 3$.

57 Show by algebra that $\dfrac{1}{x} < \dfrac{1}{2}\left(\dfrac{1}{x^{1/2}} + \dfrac{1}{x^{3/2}}\right)$ for $x > 1$.

58 (cont.) Show that $\ln x < \sqrt{x} - \dfrac{1}{\sqrt{x}}$ for $x > 1$. Deduce Kepler's inequality: if

$0 < x < y$, then $\dfrac{\ln y - \ln x}{y - x} < \dfrac{1}{\sqrt{xy}}$.

59 Compare the results of Exs. 45, 49, and 58 by completing the table to 5 places:

x	$\ln x$	$x - 1$	$2(\sqrt{x} - 1)$	$\sqrt{x} - 1/\sqrt{x}$
1.2				
1.5				
2.0				
e				
3.0				

60* Generalize Ex. 58: for $x > 1$ and $n \geq 2$, $\ln x < \dfrac{n}{2(n-1)}\left(x^{(n-1)/n} - \dfrac{1}{x^{(n-1)/n}}\right)$.

3. FURTHER PROPERTIES OF LOGARITHMS

Logarithmic Differentiation Differentiation of products, quotients, and powers can often be simplified by taking logs before differentiating. Be careful, however, to take logs of positive functions only.

■ **EXAMPLE 1** Find the derivative of $\quad y = \left(\dfrac{x^2}{x^4 + 1}\right)^{1/3} \quad$ for $\quad x \neq 0$.

Solution $\ln y = \frac{1}{3}[\ln x^2 - \ln(x^4 + 1)] = \frac{1}{3}[2 \ln x - \ln(x^4 + 1)]$. Therefore

$$\frac{y'}{y} = (\ln y)' = \frac{1}{3}\left(\frac{2}{x} - \frac{4x^3}{x^4 + 1}\right),$$

$$y' = \frac{y}{3}\left(\frac{2}{x} - \frac{4x^3}{x^4 + 1}\right) = \frac{1}{3}\left(\frac{x^2}{x^4 + 1}\right)^{1/3}\left(\frac{2}{x} - \frac{4x^3}{x^4 + 1}\right). \qquad ■$$

For the next example, we need a slight extension of the formula for the derivative of $\ln u(x)$. If $u(x)$ takes negative values, then $\ln |u(x)| = \ln[-u(x)]$ is defined. By the Chain Rule,

$$\frac{d}{dx} \ln |u| = \frac{d}{dx} \ln(-u) = \frac{(-u)'}{-u} = \frac{u'}{u}.$$

Again the derivative turns out to be u'/u. We combine the cases $u(x) > 0$ and $u(x) < 0$ into a single formula:

$$\boxed{\frac{d}{dx} \ln |u| = \frac{u'}{u}, \qquad u \neq 0.}$$

The next example concerns derivatives of factored polynomials and rational functions. If $f(x) = (x - a)(x - b)$, then

$$\frac{f'}{f} = \frac{(x - b) + (x - a)}{(x - a)(x - b)} = \frac{1}{x - a} + \frac{1}{x - b}.$$

If f has three or more factors, a similar formula holds, but it is clumsy to derive by direct differentiation. Logarithmic differentiation comes to the rescue.

■ **EXAMPLE 2**   Prove
  (1)   If $f(x) = (x - a_1) \cdots (x - a_p)$, then

$$\frac{f'(x)}{f(x)} = \sum_{i=1}^{p} \frac{1}{x - a_i} \qquad (x \neq a_i).$$

  (2)   If $f(x) = (x - a_1)^{m_1} \cdots (x - a_p)^{m_p}$, where the m_i are positive or negative integers, then

$$\frac{f'(x)}{f(x)} = \sum_{i=1}^{p} \frac{m_i}{x - a_i} \qquad (x \neq a_i).$$

Solution   It suffices to derive (2) since (1) is the special case $m_1 = m_2 = \cdots = m_p = 1$. The natural impulse is to take logs first, but to be safe take the log of $|f(x)|$:

$$|f(x)| = |x - a_1|^{m_1} |x - a_2|^{m_2} \cdots |x - a_p|^{m_p},$$

$$\ln |f(x)| = \sum_{i=1}^{p} m_i \ln |x - a_i|.$$

Now differentiate:

$$\frac{f'(x)}{f(x)} = \sum_{i=1}^{p} m_i \frac{d}{dx} \ln |x - a_i| = \sum_{i=1}^{p} \frac{m_i}{x - a_i}. \qquad ■$$

Rate of Growth of ln x   We know that $\ln x$ is strictly increasing and that $\ln x \longrightarrow \infty$ as $x \longrightarrow \infty$. Its graph (Fig. 1, p. 308) increases slowly, being the mirror image of the rapidly increasing graph of $y = e^x$.

  Actually, the rate of increase of $y = \ln x$ is agonizingly slow. The curve does not reach the level $y = 10$ until $x = e^{10} \approx 22,000$; it does not reach the level $y = 100$ until $x = e^{100} \approx 2.7 \times 10^{43}$. Obviously, the larger x is, the smaller $\ln x$ is by comparison. More precisely,

$$\boxed{\frac{\ln x}{x} \longrightarrow 0 \qquad \text{as} \quad x \longrightarrow \infty.}$$

  This assertion will be proved if we can show that for any positive integer n, no matter how large,

$$\frac{\ln x}{x} < \frac{1}{n},$$

for x sufficiently large. We use a fact proved on p. 169: given any positive integer n, then $e^x/x^n \longrightarrow \infty$ as $x \longrightarrow \infty$. Hence $x^n < e^x$ for x sufficiently large. Taking natural logs, we have

$$n \ln x < x, \qquad \frac{\ln x}{x} < \frac{1}{n},$$

which is what we wanted to prove.

Perhaps comparing $\ln x$ with x is unrealistic. Maybe we should compare $\ln x$ with a smaller function, for example with $\sqrt{x}$, which is much smaller than x. How does $\ln x$ compare with $\sqrt{x}$ or with $\sqrt[3]{x}$, which is smaller yet? The answer is that *any* positive power of x overwhelms $\ln x$.

$$\text{If } p > 0, \quad \text{then} \quad \frac{\ln x}{x^p} \longrightarrow 0 \quad \text{as} \quad x \longrightarrow \infty.$$

This follows from the behavior of $(\ln x)/x$. For if we set $x^p = y$, then $x = y^{1/p}$ and $y \longrightarrow \infty$ as $x \longrightarrow \infty$. Therefore

$$\frac{\ln x}{x^p} = \frac{\ln y^{1/p}}{y} = \frac{(1/p) \ln y}{y} = \frac{1}{p} \frac{\ln y}{y} \longrightarrow 0 \quad \text{as} \quad x \longrightarrow \infty.$$

As a check, let us tabulate $(\ln x)/x^{1/3}$ for some large values of x:

x	10^3	10^6	10^9	10^{30}	10^{300}
$\dfrac{\ln x}{x^{1/3}}$	0.691	0.138	0.0207	6.91×10^{-9}	6.91×10^{-98}

The function $\ln x$ increases so slowly that not only is $\ln x$ small compared to x but any positive power $(\ln x)^p$ is small compared to x:

$$\text{If } p > 0, \quad \text{then} \quad \frac{(\ln x)^p}{x} \longrightarrow 0 \quad \text{as} \quad x \longrightarrow \infty.$$

This follows from the last result: since $1/p > 0$,

$$\frac{\ln x}{x^{1/p}} \longrightarrow 0 \quad \text{as} \quad x \longrightarrow \infty.$$

Therefore

$$\frac{(\ln x)^p}{x} = \left[\frac{\ln x}{x^{1/p}} \right]^p \longrightarrow 0 \quad \text{as} \quad x \longrightarrow \infty.$$

As a check, let us tabulate $(\ln x)^5/x$ for some large values of x:

x	10	10^2	10^4	10^6	10^{20}	10^{100}
$\dfrac{(\ln x)^5}{x}$	6.47	20.7	6.63	0.503	2.07×10^{-12}	6.47×10^{-89}

Behavior as $x \longrightarrow 0+$ We have seen that $\ln x \longrightarrow -\infty$ as $x \longrightarrow 0+$. We ask how fast $\ln x$ approaches $-\infty$, for instance, fast enough that $x \ln x$ also approaches $-\infty$ as $x \longrightarrow 0+$, or less fast? Here is the precise answer:

$$\text{If } p > 0, \quad \text{then} \quad x^p \ln x \longrightarrow 0- \quad \text{as} \quad x \longrightarrow 0+.$$
$$\text{In particular,} \quad x \ln x \longrightarrow 0- \quad \text{as} \quad x \longrightarrow 0+.$$

To prove this assertion, replace x by $1/y$, so $y \longrightarrow \infty$ as $x \longrightarrow 0+$. Then

$$\lim_{x \to 0+} x^p \ln x = \lim_{y \to \infty} \frac{\ln(1/y)}{y^p} = \lim_{y \to \infty} \frac{-\ln y}{y^p} = 0.$$

Clearly $(-\ln y)/y^p < 0$ for large y, so the approach to 0 is from below, that is, through negative values.

Again, let us check this numerically, say for $x^{1/4} \ln x$. To simplify the tabulation, we shall change signs.

x	10^{-1}	10^{-2}	10^{-6}	10^{-20}	10^{-100}
$-x^{1/4} \ln x$	1.29	1.46	0.437	4.61×10^{-4}	2.30×10^{-23}

Limits   Logarithms are sometimes used to prove assertions about limits. The function $\ln x$ is continuous, being the inverse of the continuous function e^x. Therefore, if $a > 0$ and $b > 0$, and if $a \longrightarrow b$, then $\ln a \longrightarrow \ln b$. Conversely, by the continuity of e^x, if $\ln a \longrightarrow \ln b$, then $e^{\ln a} \longrightarrow e^{\ln b}$, that is, $a \longrightarrow b$. Combining these facts, we have the principle

$$a \longrightarrow b \quad \text{if and only if} \quad \ln a \longrightarrow \ln b, \qquad a > 0, \quad b > 0.$$

Here is an application of this principle.

$$\text{If } a \text{ is any positive number, then} \qquad \sqrt[n]{a} \longrightarrow 1 \qquad \text{as} \quad n \longrightarrow \infty.$$

It is enough to show that $\ln \sqrt[n]{a} \longrightarrow \ln 1$, that is,

$$\frac{1}{n} \ln a \longrightarrow 0 \qquad \text{as} \quad n \longrightarrow \infty.$$

But this is obvious since $\ln a$ is a fixed number and $1/n \longrightarrow 0$.

You can test this result on a pocket calculator with a square root key. Start with any number $a > 0$. Take repeated square roots, obtaining

$$\sqrt{a}, \quad \sqrt[4]{a}, \quad \sqrt[8]{a}, \quad \sqrt[16]{a}, \quad \cdots,$$

and watch the numbers approach 1. Starting with $a = 500$ for example, you obtain

22.36068	4.72871	2.17456	1.47464	1.21435	1.10197
1.04975	1.02457	1.01221	1.00609	1.00304	1.00152 etc.

An Inequality

■ **EXAMPLE 3**   (a)   Find the largest value of $y = (\ln x)/x$ for $x > 0$.
(b)   Show that $x^e \le e^x$ for all $x > 0$, with equality only for $x = e$.

Solution   (a) $\dfrac{dy}{dx} = \dfrac{x \cdot \dfrac{1}{x} - \ln x}{x^2} = \dfrac{1 - \ln x}{x^2}.$

It follows that $y' = 0$ if $x = e$, and

$$y' > 0 \quad \text{if} \quad 0 < x < e; \qquad y' < 0 \quad \text{if} \quad e < x.$$

Consequently, y increases steadily as x goes from 0 to e and decreases steadily as x continues beyond e. Therefore the max of y is $y(e) = (\ln e)/e = 1/e$.

(b) By (a) we have $\dfrac{\ln x}{x} \le \dfrac{1}{e}$

for all $x > 0$, with "$<$" unless $x = e$. Consequently, for $x > 0$ and $x \neq e$,

$$e \ln x < x, \qquad \ln x^e < x = \ln e^x.$$

It follows that $x^e < e^x$. ∎

Remark In particular, $\pi^e < e^\pi$.

EXERCISES

Differentiate where valid by logarithmic differentiation

1 $y = (x^3 + 2)^{1/2}$

2 $y = \left[\dfrac{x + 1}{(x + 2)^4} \right]^{1/3}$

3 $y = \left(\dfrac{x^2 + 4}{x + 7} \right)^6$

4 $y = \dfrac{(x + 1)(x + 2)}{(x + 4)(x + 5)}$

5 $y = (2x + 3)^{1/3} e^{-x^2}$

6 $y = \dfrac{x^2 e^{3x^3}}{(x + 3)^2}$

7 $y = \dfrac{e^x(x^3 - 1)}{\sqrt{2x + 1}}$

8 $y = \sqrt{\dfrac{x^2 - 1}{x^2 + 1}}.$

Compute y'/y

9 $y = x^{5/6}$

10 $y = (2x^4 + 1)^{-2/7}$

11 $y = (x - 3)(x - 5)$

12 $y = (x - 1)(x - 3)(x - 5)$

13 $y = (x + 2)(x + 7)^3$

14 $y = (x + 2)^2(x + 3)^3/(x + 4)^4.$

Find

15 $\lim\limits_{x \to \infty} \dfrac{x}{10^6 \ln x}$

16 $\lim\limits_{x \to \infty} \dfrac{(x + 1) \ln x}{x^2}$

17 $\lim\limits_{x \to \infty} \dfrac{(\ln x)^3}{x^{1/4}}$

18 $\lim\limits_{x \to \infty} \dfrac{\log_{10} x}{x}$

19 $\lim\limits_{x \to \infty} \dfrac{x^2 + 1}{x(\ln x)^2}$

20 $\lim\limits_{x \to 0+} \sqrt{x} \ln x$

21 $\lim\limits_{x \to 0+} x(\ln x)^{20}$

22 $\lim\limits_{x \to \infty} \dfrac{x \ln x}{x^p + 1} \quad (p > 1)$

23 $\lim\limits_{x \to \infty} \dfrac{3(\ln x)^2 + 1}{x}$

24 $\lim\limits_{x \to \infty} \dfrac{\ln(\ln x)}{\ln x}.$

25 Let $p > 0$ and $q > 0$. Prove $\lim\limits_{x \to \infty} \dfrac{(\ln x)^p}{x^q} = 0.$

26 Let $p > 0$ and $q > 0$. Maximize $y = \dfrac{(\ln x)^p}{x^q}$ for $x \ge 1$.

27 Graph $y = (\ln x)/x$ for $0 < x \le 6$.

28 Graph $y = x \ln x$ for $0 < x \le 1$.

4. APPLICATIONS OF LOGARITHMS

Exponential Functions With natural logarithms we shall define certain new functions involving exponents like $x^{\sqrt{2}}$ and x^x. The first question is precisely what we mean by a^b, where $a > 0$ and b are any real numbers, for example $(\sin 2)^{\sqrt{3}}$ or $\pi^{1/e}$. We *define* a^b by the formula

$$a^b = e^{b \ln a}.$$

This formula is really nothing new. On p. 164 we discussed a^x for any x and $a > 0$. Our definition of a^x was

$$a^x = e^{kx} \qquad \text{where} \quad a = e^k.$$

But $a = e^k$ means $k = \ln a$; therefore

$$a^x = e^{x \ln a},$$

which agrees with the given formula.

From the preceding expression for a^x, we can derive a neater formula for the derivative of a^x than we had before. By the Chain Rule,

$$\frac{d}{dx} a^x = \frac{d}{dx} e^{x \ln a} = e^{x \ln a} \frac{d}{dx} (x \ln a) = a^x (\ln a).$$

$$\frac{d}{dx} a^x = (\ln a)a^x.$$

■ **EXAMPLE 1** Find $\displaystyle\lim_{t \to 0} \frac{3^t - 1}{t}$.

Solution The limit is the derivative at 0 of $f(x) = 3^x$ because

$$f'(0) = \lim_{t \to 0} \frac{f(0 + t) - f(0)}{t} = \lim_{t \to 0} \frac{3^t - 1}{t}.$$

But $f'(x) = (\ln 3)3^x$, hence $f'(0) = \ln 3$.   *Answer* $\ln 3$.   ■

Remark By tables or a calculator, $\ln 3 \approx 1.09861$. Let us tabulate some values of the function in question for small t:

t	10^{-2}	10^{-3}	10^{-4}	10^{-6}
$\dfrac{3^t - 1}{t}$	1.10467	1.09922	1.09867	1.09861

Log to Base a We defined the function $\ln x$ as the inverse of the exponential function e^x. In exactly the same way, we now define the function $\log_a x$ for $a > 1$ as the inverse of the exponential function a^x. Since a^x is strictly increasing, the inverse exists; we call it the "log to the base a of x." (The common logarithm $\log x = \log_{10} x$ is a special case; the natural logarithm $\ln x = \ln_e x$ is another.)

By definition,

$$a^{\log_a x} = x \quad \text{and} \quad \log_a a^x = x.$$

Apply ln to the first of these relations:

$$(\log_a x)(\ln a) = \ln x,$$

$$\boxed{\log_a x = \frac{\ln x}{\ln a}.}$$

Thus $\log_a x$ is proportional to $\ln x$. To remember the constant of proportionality, we can reason as follows: $\log_a x = c \ln x$. Set $x = a$ and use $\log_a a = 1$. Result: $c = 1/\ln a$.

It follows that the derivative of $\log_a x$ is proportional to the derivative of $\ln x$:

$$\boxed{\frac{d}{dx} \log_a x = \frac{1}{x \ln a}.}$$

Power Functions   Let α be any real number, not necessarily rational. We define the power function x^α by

$$\boxed{x^\alpha = e^{\alpha \ln x} \qquad (x > 0).}$$

When $\alpha = n$, a positive integer, this agrees with our old power function

$$x^n = x \cdot x \cdots x$$

because

$$e^{n \ln x} = (e^{\ln x})^n = (x)^n.$$

It can also be shown, without much difficulty, that x^α agrees with the usual power function when α is a negative integer or a rational number.

The power functions x^α have three basic properties:

$$\boxed{\begin{array}{l} \text{If } \alpha \text{ and } \beta \text{ are any real numbers and } x > 0, \text{ then} \\[2mm] (1) \quad x^\alpha x^\beta = x^{\alpha + \beta} \qquad (2) \quad (x^\alpha)^\beta = x^{\alpha\beta} \qquad (3) \quad \dfrac{d}{dx} x^\alpha = \alpha x^{\alpha - 1}. \end{array}}$$

The first two are familiar laws of exponents. In the present context we derive them from corresponding properties of e^x. First (1):

$$x^\alpha x^\beta = e^{\alpha \ln x} e^{\beta \ln x} = e^{\alpha \ln x + \beta \ln x} = e^{(\alpha + \beta) \ln x} = x^{\alpha + \beta}.$$

Now set $y = x^\alpha$, so $\ln y = \alpha \ln x$. Then

$$(x^\alpha)^\beta = y^\beta = e^{\beta \ln y} = e^{\beta(\alpha \ln x)} = e^{(\alpha\beta) \ln x} = x^{\alpha\beta}.$$

We use the Chain Rule for Property (3):

$$\frac{d}{dx} x^\alpha = \frac{d}{dx} e^{\alpha \ln x} = (e^{\alpha \ln x}) \left[\frac{d}{dx} (\alpha \ln x) \right] = (x^\alpha)(\alpha x^{-1}) = \alpha(x^\alpha x^{-1}) = \alpha x^{\alpha - 1}.$$

Note how different this approach is from our earlier treatment of the derivative of x^n for $n = 2, 3, \cdots$.

Rational Powers The special case $\alpha = p/q$, a rational number, is worth a second look. Assume p/q is in lowest terms and $q > 0$.

First take $p = 1$, that is, $\alpha = 1/q$, where q is a positive integer. On p. 79, we found the derivative of $y = x^{1/q}$ without worrying about the meaning of this function. Now let us regard $y = x^{1/q}$ as the inverse function of $y = x^q$. If q is odd, x^q is strictly increasing for all x, so it has an inverse function. But if q is even, we consider only the restricted domain $x \geq 0$; there x^q is strictly increasing and has an inverse function (Fig. 1).

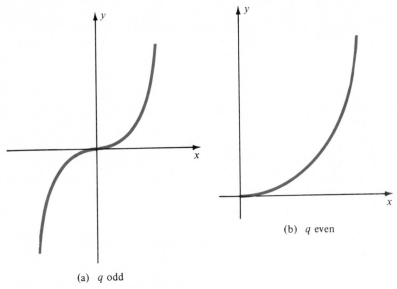

(a) q odd

(b) q even

Fig. 1   Graphs of $y = x^q$ for $q = 2, 3, 4, 5, \cdots$

Set $y = f(x) = x^q$. Then $f'(x) = qx^{q-1}$, so $f'(x) \neq 0$ if $x \neq 0$ (q odd) or $x > 0$ (q even). Therefore the inverse function $y^{1/q}$ is differentiable on its domain except where $x = 0$, that is, at $y = 0$. By the rule for differentiating inverse functions,

$$\frac{d}{dy} y^{1/q} = 1 \bigg/ \frac{d}{dx} x^q = \frac{1}{qx^{q-1}} = \frac{x}{qx^q} \qquad (x \neq 0).$$

But $x = y^{1/q}$ and $y = x^q$, so it follows that

$$\frac{d}{dy} y^{1/q} = \frac{y^{1/q}}{qy} = \frac{1}{q} y^{(1/q)-1}.$$

Therefore
$$\frac{d}{dx} x^\alpha = \alpha x^{\alpha-1} \qquad \text{if} \quad \alpha = \frac{1}{q}.$$

The formula for the derivative of x^α where $\alpha = p/q$ now follows from the Chain Rule applied to $x^{p/q} = (x^{1/q})^p$:

$$\frac{d}{dx}(x^{p/q}) = \frac{d}{dx}(x^{1/q})^p = p(x^{1/q})^{p-1}\left(\frac{1}{q}x^{(1/q)-1}\right) = \frac{p}{q}x^{(p/q)-1}.$$

Derivative of Rational Powers   Let $\alpha = p/q$ be a fraction in lowest terms with $q > 0$. Then

$$\frac{d}{dx}x^\alpha = \alpha x^{\alpha-1}.$$

If q is odd, this rule holds for all $x \neq 0$.   If q is even, it holds for $x > 0$.

Thus we have an independent verification of the rule for differentiating x^α when α is rational. Actually we have a little more: those cases in which $x < 0$ is allowed (odd denominator). They are not included in the pattern $x^\alpha = e^{\alpha \ln x}$ because $\ln x$ is undefined when $x < 0$. There is more than one way to skin a cat.

General Powers   Suppose f and g are functions such that $f(x) > 0$ for all x in their common domain. We define $f(x)^{g(x)}$ by the formula

$$f(x)^{g(x)} = e^{g(x) \ln f(x)}.$$

The compound function $f(x)^{g(x)}$ is a differentiable function if f and g are differentiable. While we could easily write down a formula for its derivative, the formula is not particularly memorable. Instead we shall concentrate on an example.

■ **EXAMPLE 2** Find (a) $\dfrac{d}{dx}x^x$ (b) $\lim\limits_{x \to 0+} x^x$.

Solution   (a)   Write $x^x = e^u$, where $u = x \ln x$. By the Chain Rule,

$$\frac{d}{dx}x^x = \frac{d}{dx}e^u = \frac{de^u}{du}\frac{du}{dx} = e^u\left(\frac{x}{x} + \ln x\right) = x^x(1 + \ln x).$$

Alternatively, set $v = x^x$ so $\ln v = x \ln x$. Differentiate:

$$\frac{v'}{v} = \frac{d}{dx}\ln v = \frac{d}{dx}(x \ln x) = 1 + \ln x, \qquad v' = (1 + \ln x)v = (1 + \ln x)x^x.$$

(b)   Write $x^x = e^u$, where $u = x \ln x$. Then

$$\lim_{x \to 0+} u = \lim_{x \to 0+}(x \ln x) = 0,$$

hence

$$\lim_{x \to 0+} x^x = \lim_{x \to 0+} e^u = e^{\lim u} = e^0 = 1.$$ ■

Remark   The following table illustrates the result of part (b):

x	0.1	0.01	0.001	0.0001	10^{-6}
x^x	0.79433	0.95499	0.99312	0.99908	0.99999

Applications to Economics We noted in Chapter 4 that the exponential function occurs in many diverse natural phenomena. It is not surprising that its inverse, the logarithm function, also occurs in nature as do the related power functions. For example, in the subject of psychophysics, Fechner's law says that a stimulus of magnitude x causes a sensation of magnitude $y = k \ln x$. In the study of winged animals, Greenewalt's equation for soaring birds is $W = 16.2446 \, H^{0.8787}$, where W is the wingspan and H the length of the humerus (bone from shoulder to elbow) in cm. Lawson's equation for the ancient flying reptile Pterodactylus is $W = 13.909 \, H^{1.0548}$.

Further examples: When a mole of an ideal gas undergoes a reversible isothermal expansion from volume V_0 to V_1, its entropy increase is $S_1 - S_0 = R \ln(V_1/V_0)$, where R is the universal gas constant. If $\pi(n)$ denotes the number of prime numbers from 2 to n, then $\pi(n) \approx n/\ln n$ for n large. (A prime number is a number like 19 or 67 that has no smaller factors. Thus $77 = 7 \times 11$ is not a prime.)

We now consider an important concept in economics. Suppose that x and y are positive variables and that y depends on x. The **elasticity** of y with respect to x is

$$\frac{Ey}{Ex} = \frac{d(\ln y)}{d(\ln x)} = \frac{x}{y}\frac{dy}{dx}.$$

Elasticity is the ratio of the percentage change in y to percentage change in x since these are dy/y and dx/x respectively.

■ **EXAMPLE 3** For what functions $y = y(x)$ is Ey/Ex constant?

Solution If $\dfrac{Ey}{Ex} = \dfrac{d(\ln y)}{d(\ln x)} = k,$

then $\ln y = k \ln x + \ln a = \ln ax^k,$

hence $y = ax^k.$ ■

In economics, a **good** is anything that is supplied and purchased, like bread, furnace repair, soybean futures, or life insurance. To each possible price p of the good corresponds a **demand** x, the amount of the good that would be purchased at price p. Thus we can write $x = x(p)$, or inversely, $p = p(x)$. We usually assume that x is a strictly decreasing differentiable function of p. The **revenue** is $R = xp$, the total money taken in by the sale of x units of the good at price p.

■ **EXAMPLE 4** Express the **marginal revenue** dR/dp in terms of the demand x and the elasticity η of demand with respect to price.

Solution First, $\eta = \dfrac{Ex}{Ep} = \dfrac{p}{x}\dfrac{dx}{dp}.$

Next, $\dfrac{dR}{dp} = \dfrac{d}{dp}(px) = x + p\dfrac{dx}{dp} = x + x\dfrac{p}{x}\dfrac{dx}{dp} = x + x\eta,$

hence $\dfrac{dR}{dp} = x(1 + \eta).$

The result is important in economics. ■

Poisson's Gas Equation Suppose a fixed quantity of a diatomic gas (such as O_2, N_2, or CO) is subject to a reversible adiabatic process.* Then the pressure p and the volume v of the gas satisfy

$$\frac{dp}{p} + \gamma \frac{dv}{v} = 0, \qquad \gamma = \frac{7}{5} = 1.40.$$

This relation says that dp/p, the percentage change in pressure, is negatively proportional to dv/v, the percentage change in volume.

■ **EXAMPLE 5** How are p and v related?

Solution The given differential relation can be written

$$d(\ln p) + \gamma \, d(\ln v) = 0,$$

that is,
$$d[\ln(pv^\gamma)] = 0.$$

Hence
$$pv^\gamma = c.$$

This result, known as Poisson's equation, is important in meteorology. ■

EXERCISES

Find

1   $\lim\limits_{x \to \infty} x^{1/x}$

2   $\lim\limits_{x \to 0+} x^{1/x}$

3   $\lim\limits_{x \to 0+} (x)^{x^x}$

4   $\lim\limits_{n \to \infty} n(a^{1/n} - 1) \quad (a > 0)$.

Differentiate with respect to x

5   $y = x^{x-1}$
6   $y = (x + 2)^{x+3}$
7   $y = x^{1/x}$
8   $y = 2^{2x}$

9   $y = 3^{\ln x}$
10   $y = \left(1 + \dfrac{1}{x}\right)^x$
11   $y = 10^{x^2}$
12   $y = x^{x^x}$

13   $y = (\ln x)^x$
14   $y = (2 + \sin x)^x$
15   $y = \log_x b$
16   $y = \log_x x$
17   $y = x^{\ln x}$
18   $y = x^{\sin x}$.

Graph

19   $y = x^{2/3}$
20   $y = x^{3/2}$
21   $y = x^{-2/3}$
22   $y = x^{-3/2}$
23   $y = x^x$
24   $y = x^{1/x}$.

Maximize for $x > 1$

25   $y = \dfrac{\ln x}{x^a}, \quad a > 0$
26   $y = \dfrac{x^{10}}{10^x}$
27   $y = x^{-1/\ln x}$.

28   Arrange these functions according to increasing size as $x \to \infty$:
$$y_1 = 2^x, \quad y_2 = x^{(\ln x)^3}, \quad y_3 = (\sqrt{x})^x, \quad y_4 = x^{\sqrt{x}}, \quad y_5 = e^{x^2}.$$

29   Let $f(x)$ and $g(x)$ be differentiable with $f(x) > 0$. Prove $f(x)^{g(x)}$ is differentiable.

---

* **Reversible** means that the process can go forward and then back to its original state and that no energy is dissipated. **Adiabatic** means that no heat is transferred to or from the gas. You can think of the process as taking place in a Thermos bottle.

30*   Let $a > 1$. Prove the inequality of James Bernoulli:

$$(1 + x)^a \geq 1 + ax \qquad \text{for} \quad x > -1$$

with equality only for $x = 0$.

31   Let $e < a < b$. Prove $a^b > b^a$.

32   Find all integers m and n such that $1 \leq m < n$ and $m^n = n^m$. [*Hint*   Refer to the graph of $y = (\ln x)/x$, Ex. 27 of Section 3.]

The functions in the next two exercises are called **Einstein functions** in radiation theory

33   Show that $y = \ln(1 - e^{-x})$ is strictly increasing.

34   Show that $y = \dfrac{x}{e^x - 1} - \ln(1 - e^{-x})$ is strictly decreasing for $x > 0$.

35   A projectile is shot straight up. If we assume air resistance is proportional to its velocity, then the equation of motion is

$$\frac{dv}{dt} = -kv - g, \qquad \text{where} \quad v = \frac{dx}{dt}$$

and k and g are positive constants. Show that

$$\frac{dx}{dv} = -\frac{1}{k} + \frac{g}{k^2}\left(\frac{1}{v + g/k}\right).$$

36   (cont.) Find the relation between x and v, taking x_0 and v_0 for their initial values.

37   A measure of the strength of a visible star S is its **flux** $\phi(S)$. This is the amount of radiant energy from the star falling on a unit area of Earth (perpendicular to the light from the star) per second. Another measure is the **magnitude** $m(S)$ of the star. This is defined so that if $\phi(S_2) = k\phi(S_1)$, then $m(S_1) - m(S_2)$ depends only on k. Guess this dependence, the relation between $m(S_1) - m(S_2)$ and $\phi(S_2)/\phi(S_1)$ if a flux ratio of 100 yields a magnitude difference of 5 (and $m(S)$ decreases as $\phi(S)$ increases).

38*   Find the most general continuously differentiable function $y = y(x)$ defined for all $x > 0$ and satisfying $y(uv) = y(u) + y(v)$.

39   Find the equation of the general straight line plotted on semi-log paper (Fig. 2).

40   Find the equation of the general straight line plotted on log–log paper (Fig. 3).

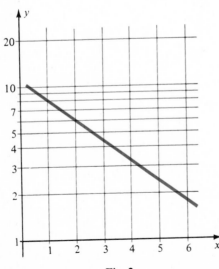

Fig. 2

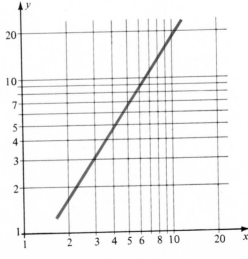

Fig. 3

Prove the following formulas for elasticity

41 $\dfrac{E(uv)}{Ex} = \dfrac{Eu}{Ex} + \dfrac{Ev}{Ex}$

42 $\dfrac{E}{Ex}\left(\dfrac{u}{v}\right) = \dfrac{Eu}{Ex} - \dfrac{Ev}{Ex}$

43 $\dfrac{Ey}{Ex} = \dfrac{Ey}{Eu}\dfrac{Eu}{Ex}$ (chain rule)

44 $\dfrac{E}{Ex}(y + c) = \dfrac{y}{y + c}\dfrac{Ey}{Ex}.$

Find

45 $\dfrac{E}{Ex}(cx^z)$

46 $\dfrac{E}{Ex}(ax + b)$

47 $\dfrac{E}{Ex}(ce^{kx})$

48 $\dfrac{E}{Ex}(xy).$

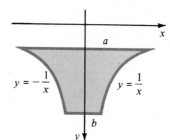

$y = -\dfrac{1}{x}$ a $y = \dfrac{1}{x}$ **Fig. 4** b

49 The region in Fig. 4 is one end of a tank that is filled with fluid of density δ. Find the total pressure on the end of the tank.

50 In a reactor, the uranium oxide fuel generates heat energy at the rate of $Q = 362$ watts/cm^3. This heat is extracted by a coolant passing over the surface of the fuel.
Suppose a long cylindrical fuel element has radius a cm. It is known that

$$Q = \frac{4}{a^2} \int_{T_1}^{T_0} \mu \, dT,$$

where T_0 is the temperature on the axis, T_1 is the surface temperature or temperature of the coolant, and μ is the thermal conductivity of uranium oxide. At high temperatures, $\mu = 31.7/T$ watts per cm per degree kelvin, where the temperature T is measured in °K. Given that $a = 0.610$ cm and $T_1 = 820$°K, compute T_0.

5. INVERSE TRIGONOMETRIC FUNCTIONS

Inverse of Sine   The sine of an angle does not completely determine the angle. For example, if $\sin x = \frac{1}{2}\sqrt{2}$, then x may be $\frac{1}{4}\pi$, or $\frac{3}{4}\pi$, or $\frac{9}{4}\pi$, or $\frac{11}{4}\pi$, etc. Figure 1 shows that the curve $y = \sin x$ intersects $y = c$ infinitely often if $-1 \le c \le 1$; hence there are infinitely many values of x for which $\sin x = c$.

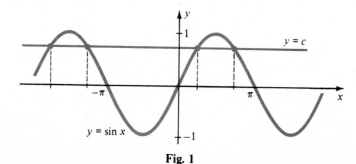

$y = c$

$y = \sin x$

Fig. 1

Certainly sin x, defined for all x, has no inverse function. To obtain an inverse, we restrict the domain to the interval $[-\frac{1}{2}\pi, \frac{1}{2}\pi]$. There the graph of $y = \sin x$ increases strictly and takes all values of y for $-1 \le y \le 1$. See Fig. 2a. Hence there exists an inverse function, called the **arc sine**. If $-1 \le y \le 1$, then arc sin y is the unique angle from $-\frac{1}{2}\pi$ to $\frac{1}{2}\pi$ whose sine is y.

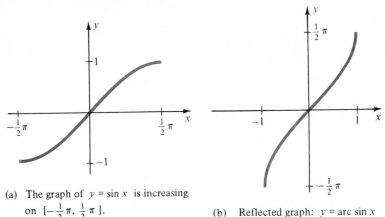

(a) The graph of $y = \sin x$ is increasing
on $[-\frac{1}{2}\pi, \frac{1}{2}\pi]$.

(b)   Reflected graph: $y = \text{arc sin } x$

Fig. 2

The graph of $y = \text{arc sin } x$ is the reflection of the curve in Fig. 2a across the line $y = x$. See Fig. 2b. The graph has vertical tangents at $(1, \frac{1}{2}\pi)$ and $(-1, -\frac{1}{2}\pi)$, reflecting the horizontal tangents of $y = \sin x$ at $(\frac{1}{2}\pi, 1)$ and $(-\frac{1}{2}\pi, -1)$.

The function sin x restricted to the domain $-\frac{1}{2}\pi \le x \le \frac{1}{2}\pi$ has an inverse function, called **arc sine**.

The function arc sin x is defined for $-1 \le x \le 1$, is strictly increasing, and its values range from $-\frac{1}{2}\pi$ to $\frac{1}{2}\pi$.

Examples

$$\text{arc sin } 1 = \tfrac{1}{2}\pi \qquad \text{arc sin}(-1) = -\tfrac{1}{2}\pi \qquad \text{arc sin } 0 = 0$$

$$\text{arc sin } \tfrac{1}{2} = \tfrac{1}{6}\pi \qquad \text{arc sin}(-\tfrac{1}{2}\sqrt{2}) = -\tfrac{1}{4}\pi \qquad \text{arc sin } \tfrac{1}{2}\sqrt{3} = \tfrac{1}{3}\pi.$$

Since the sine and arc sine are inverse functions, we have

$$\text{arc sin}(\sin x) = x, \quad -\tfrac{1}{2}\pi \le x \le \tfrac{1}{2}\pi, \quad \sin(\text{arc sin } x) = x, \quad -1 \le x \le 1.$$

Because sin x is an odd function, it follows that arc sin x is also odd:

$$\text{arc sin}(-x) = -\text{arc sin } x.$$

Inverse of Cosine   Just as for sin x, the inverse of cos x can be defined only if x is suitably restricted. A logical choice is to restrict x to the interval $0 \le x \le \pi$. See Fig. 3a. On this interval, the graph of $y = \cos x$ decreases strictly and takes

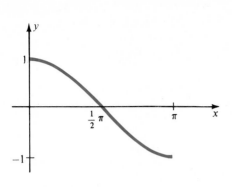
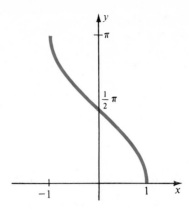

(a) The graph of $y = \cos x$ is decreasing on $[\,0, \pi\,]$.

(b) Reflected graph: $y = \text{arc } \cos x$

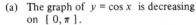
Fig. 3

all values of y for $-1 \le y \le 1$. Hence there exists an inverse function, called the **arc cosine**. If $0 \le y \le \pi$, then arc cos y is the unique angle from 0 to π whose cosine is y.

The graph of $y = \text{arc } \cos x$ is the reflection of the curve in Fig. 3a across the line $y = x$. See Fig. 3b.

The function $\cos x$, restricted to the domain $0 \le x \le \pi$, has an inverse function, called **arc cosine**.

The function arc cos x is defined for $-1 \le x \le 1$, is strictly decreasing, and its values range from 0 to π.

Examples

$$\text{arc } \cos 1 = 0 \qquad \text{arc } \cos(-1) = \pi \qquad \text{arc } \cos 0 = \tfrac{1}{2}\pi$$

$$\text{arc } \cos \tfrac{1}{2} = \tfrac{1}{3}\pi \qquad \text{arc } \cos \tfrac{1}{2}\sqrt{2} = \tfrac{1}{4}\pi \qquad \text{arc } \cos(-\tfrac{1}{2}\sqrt{2}) = \tfrac{3}{4}\pi.$$

Since the cosine and arc cosine are inverse functions, we have

---

$$\text{arc } \cos(\cos x) = x, \quad 0 \le x \le \pi, \qquad \cos(\text{arc } \cos x) = x, \quad -1 \le x \le 1.$$

We note two other useful relations:

$$\text{arc } \cos x + \text{arc } \cos(-x) = \pi, \qquad \text{arc } \sin x + \text{arc } \cos x = \tfrac{1}{2}\pi.$$

To prove the first, set $y = \pi - \text{arc } \cos(-x)$. Then

$$(1) \quad 0 \le y \le \pi \qquad \text{and} \qquad (2) \quad \cos y = x$$

since on the one hand, $0 \le \text{arc } \cos(-x) \le \pi$, and on the other hand,

$$\cos y = \cos[\pi - \text{arc } \cos(-x)] = -\cos[\text{arc } \cos(-x)] = -(-x) = x.$$

By (1) and (2), the number y is that unique number from 0 to π whose cosine equals x, that is, $y = \text{arc } \cos x$. The first relation follows.

For the second relation, set $y = \frac{1}{2}\pi - \text{arc cos } x$. Then by similar reasoning we have

(3) $-\frac{1}{2}\pi \leq y \leq \frac{1}{2}\pi$ and   (4) $\sin y = x$,

so $y = \text{arc sin } x$ follows.

Remark 1   It is very easy to make mistakes in arguments of this sort. You must always check that a number falls into the right range before concluding that it is an inverse anything of another number. For instance, if $0 \leq y \leq \pi$ and $\sin y = x$, we do *not* conclude that $y = \text{arc sin } x$.

Remark 2   The last two relations can be proved graphically by symmetries of the sine and cosine curves (Fig. 4).

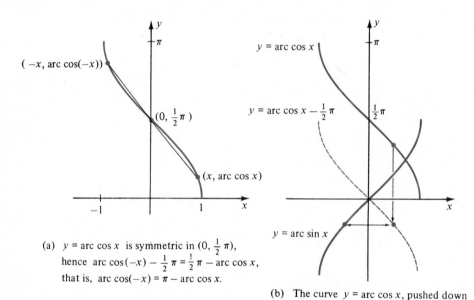

(a) $y = \text{arc cos } x$ is symmetric in $(0, \frac{1}{2}\pi)$,
hence $\text{arc cos}(-x) - \frac{1}{2}\pi = \frac{1}{2}\pi - \text{arc cos } x$,
that is, $\text{arc cos}(-x) = \pi - \text{arc cos } x$.

(b)   The curve $y = \text{arc cos } x$, pushed down $\frac{1}{2}\pi$ units is the curve $y = \text{arc sin}(-x)$, the reflection of $y = \text{arc sin } x$ in the x-axis. Therefore $\text{arc cos } x - \frac{1}{2}\pi = \text{arc sin}(-x) = -\text{arc sin } x$.

Fig. 4

Inverse of Tangent   Recall the graph of $\tan x$ (Fig. 5a). If we want an inverse, it is natural to consider just one branch of the graph. So we restrict the domain of $\tan x$ to $-\frac{1}{2}\pi < x < \frac{1}{2}\pi$. The graph increases strictly and takes all real values. Hence there exists an inverse function, called the **arc tangent**. If y is any real number, then arc tan y is the unique angle between $-\frac{1}{2}\pi$ and $\frac{1}{2}\pi$ whose tangent is y.

The graph of $y = \text{arc tan } x$ is the reflection of the heavy curve in Fig. 5a across the line $y = x$. See Fig. 5b. The graph approaches the height $\frac{1}{2}\pi$ as $x \longrightarrow \infty$ and the height $-\frac{1}{2}\pi$ as $x \longrightarrow -\infty$.

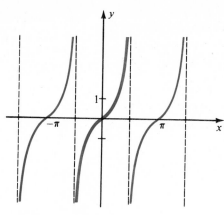

(a) The graph of $y = \tan x$ is increasing for $-\frac{1}{2}\pi < x < \frac{1}{2}\pi$.

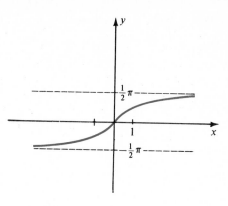

(b) Reflected graph: $y = \text{arc tan } x$

Fig. 5

The function $\tan x$ restricted to the domain $-\frac{1}{2}\pi < x < \frac{1}{2}\pi$ has an inverse function called the **arc tangent**.

The function $\text{arc tan } x$ is defined for all x, is strictly increasing, and its values range between $-\frac{1}{2}\pi$ and $\frac{1}{2}\pi$.

As $x \longrightarrow \infty$, $\text{arc tan } x \longrightarrow \frac{1}{2}\pi$;   as $x \longrightarrow -\infty$, $\text{arc tan } x \longrightarrow -\frac{1}{2}\pi$.

Examples

$$\text{arc tan } 0 = 0 \qquad \text{arc tan } 1 = \tfrac{1}{4}\pi \qquad \text{arc tan } \sqrt{3} = \tfrac{1}{3}\pi$$

$$\text{arc tan}(-\sqrt{3}) = -\tfrac{1}{3}\pi \qquad \text{arc tan } \tfrac{1}{3}\sqrt{3} = \tfrac{1}{6}\pi \qquad \text{arc tan } 100 \approx 1.5608 \approx 89.43°$$

$$\text{arc tan } 1000 \approx 1.5698 \approx 89.94° \qquad (\text{note: } \tfrac{1}{2}\pi \approx 1.5708).$$

Since the tangent and arc tangent are inverse functions, we have

$$\text{arc tan}(\tan x) = x, \quad -\tfrac{1}{2}\pi < x < \tfrac{1}{2}\pi, \qquad \tan(\text{arc tan } x) = x.$$

Since $\tan x$ is an odd function, it follows that $\text{arc tan } x$ is also odd:

$$\text{arc tan}(-x) = -\text{arc tan } x.$$

Notation   The following alternative notation is very common for inverse trigonometric functions:

$$\text{arc sin } x = \sin^{-1} x, \qquad \text{arc tan } x = \tan^{-1} x, \qquad \text{etc.}$$

Do not confuse

$$\sin^{-1} x \qquad \text{with} \qquad \frac{1}{\sin x}.$$

The notation $\sin^{-1} x$ is a bit awkward because we do write $\sin^n x$ for $(\sin x)^n$ when $n > 0$.

Inverses of the Other Trigonometric Functions   Inverses of the functions cot x, sec x, and csc x can be defined in a similar manner. Rather than discuss them in detail, we shall show their graphs (Figs 6–8) and list a few basic relations.

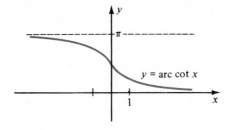

Fig. 6   Graph of $y = $ arc cot x; strictly decreasing for $-\infty < x < \infty$

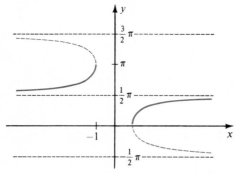

Fig. 7   Graph of $y = $ arc sec x; defined for $-\infty < x \le -1$ and for $1 \le x < \infty$. Strictly increasing.

Fig. 8   Graph of $y = $ arc sec x; defined for $-\infty < x \le -1$ and for $1 \le x < \infty$. Strictly decreasing.

The following relations hold among the various inverse functions:

For all x,   arc tan $x + $ arc cot $x = \frac{1}{2}\pi$.

For $|x| \ge 1$,   arc sec $x + $ arc csc $x = \frac{1}{2}\pi$,

$$\text{arc sec } x = \text{arc cos } \frac{1}{x}, \qquad \text{arc csc } x = \text{arc sin } \frac{1}{x}.$$

EXERCISES

Evaluate

1   arc sin$(\frac{1}{2}\sqrt{2})$

2   arc cos$(-\frac{1}{2}\sqrt{2})$

3   arc tan $\sqrt{3}$

4   arc sec 1

5   arc csc$(\frac{2}{3}\sqrt{3})$

6   arc cot(-1)

7   arc sin $\frac{1}{2} - $ arc sin$(-\frac{1}{2}\sqrt{3})$

8   arc cos $\frac{1}{2} - $ arc cos$(-\frac{1}{2})$

9   arc tan$(\tan \frac{4}{7}\pi)$

10   cos[arc cos(0.35)]

11   cos(arc sin $\frac{3}{5}$)

12   cot(arc tan 2)

13   arc sin $\frac{2}{9} + $ arc cos $\frac{2}{9}$

14   arc tan 6.2 $+ $ arc cot 6.2.

Find the inverse function $x = g(y)$ and its domain

15   $y = \ln \sin x$, $0 < x \le \frac{1}{2}\pi$

16   $y = ($arc tan $x)^3$.

17 Prove arc tan x + arc cot $x = \frac{1}{2}\pi$. [*Hint* See the corresponding proof for arc sine and arc cosine.]

18 Prove arc sec x = arc cos$(1/x)$ for $|x| \geq 1$. [*Hint* Split into two cases.]

19 Prove arc sec x + arc csc $x = \frac{1}{2}\pi$ for $|x| \geq 1$. [*Hint* See hint to Ex. 17.]

20 Prove arc sin x + arc cos $x = \frac{1}{2}\pi$ for $0 < x < 1$ by a right triangle argument.

21 Prove arc tan$(1/x)$ = arc cot x for $x > 0$.

22 (cont.) Find and prove the corresponding formula for $x < 0$.

23 Prove sin(arc cos x) $= \sqrt{1 - x^2}$ for $-1 \leq x \leq 1$.

24 For which x is y = arc tan(cot x) defined? For these x, express y in terms of x without using trigonometric functions.

25 Prove arc cos$(2x^2 - 1) = 2$ arc cos x for $0 \leq x \leq 1$. [*Hint* What trig identity does this suggest?]

26 (cont.) Find the corresponding formula for $-1 \leq x \leq 0$.

27 Prove arc tan x + arc tan y = arc tan $\dfrac{x + y}{1 - xy}$ provided $|$arc tan x + arc tan $y| < \frac{1}{2}\pi$.

28* (cont.) Find the corresponding formulas in the other cases.

Use Ex. 27 to prove

29 arc tan $\frac{1}{2}$ + arc tan $\frac{1}{3}$ = $\frac{1}{4}\pi$

30 2 arc tan $\frac{1}{3}$ + arc tan $\frac{1}{7}$ = $\frac{1}{4}\pi$

31 2 arc tan $\frac{1}{5}$ + arc tan $\frac{1}{7}$ + 2 arc tan $\frac{1}{8}$ = $\frac{1}{4}\pi$

32* 4 arc tan $\frac{1}{5}$ − arc tan $\frac{1}{239}$ = $\frac{1}{4}\pi$.

6. DERIVATIVES AND APPLICATIONS

The derivatives of the inverse trigonometric functions are:

$$\frac{d}{dx}(\text{arc sin } x) = \frac{1}{\sqrt{1 - x^2}} \qquad \frac{d}{dx}(\text{arc tan } x) = \frac{1}{1 + x^2}$$

$$\frac{d}{dx}(\text{arc cos } x) = \frac{-1}{\sqrt{1 - x^2}} \qquad \frac{d}{dx}(\text{arc cot } x) = \frac{-1}{1 + x^2}$$

$$\frac{d}{dx}(\text{arc sec } x) = -\frac{d}{dx}(\text{arc csc } x) = \begin{cases} \dfrac{1}{x\sqrt{x^2 - 1}}, & x > 1 \\[3mm] \dfrac{-1}{x\sqrt{x^2 - 1}}, & x < -1. \end{cases}$$

The top two formulas are by far the most important; let us justify them. First, if y = arc sin x, then $x = \sin y$. By the rule for differentiating inverse functions,

$$\frac{dy}{dx} = 1 \Big/ \frac{dx}{dy} = 1 \Big/ \frac{d}{dy} \sin y = \frac{1}{\cos y}.$$

Now $\cos y > 0$ because y is between $-\frac{1}{2}\pi$ and $\frac{1}{2}\pi$ by definition of the arc sine. Therefore,

$$\frac{d}{dx}(\text{arc sin } x) = \frac{1}{\cos y} = \frac{1}{+\sqrt{1 - \sin^2 y}} = \frac{1}{\sqrt{1 - x^2}}, \qquad -1 < x < 1.$$

Next if $y = \text{arc tan } x$, then $x = \tan y$; hence

$$\frac{dy}{dx} = 1 \bigg/ \frac{dx}{dy} = \frac{1}{\sec^2 y} = \frac{1}{1 + \tan^2 y} = \frac{1}{1 + x^2}.$$

The derivative of arc cos x follows from the derivative of arc sin x. Since arc cos $x = \frac{1}{2}\pi - \text{arc sin } x$, we have

$$\frac{d}{dx} (\text{arc cos } x) = -\frac{d}{dx} (\text{arc sin } x) = \frac{-1}{\sqrt{1 - x^2}}.$$

Derivations of the remaining formulas are left as exercises.

These new differentiation formulas may be used in conjunction with the Chain Rule. For instance, if $u = u(x)$, then

$$\boxed{\frac{d}{dx} \text{arc sin } u = \frac{1}{\sqrt{1 - u^2}} \frac{du}{dx}, \quad |u| < 1, \qquad \frac{d}{dx} \text{arc tan } u = \frac{1}{1 + u^2} \frac{du}{dx}.}$$

■ **EXAMPLE 1** Differentiate

(a) $\text{arc sin}(\sqrt{x}), \quad 0 < x < 1$ \qquad (b) $\text{arc tan}(5x^2 + 1)$.

Solution (a) Write $y = \text{arc sin } u$, where $u = \sqrt{x}$. Then

$$\frac{dy}{dx} = \frac{1}{\sqrt{1 - u^2}} \frac{du}{dx} = \frac{1}{\sqrt{1 - (\sqrt{x})^2}} \cdot \frac{1}{2\sqrt{x}} = \frac{1}{\sqrt{1 - x}} \cdot \frac{1}{2\sqrt{x}} = \frac{1}{2\sqrt{x - x^2}}.$$

(b) Write $y = \text{arc tan } u$, where $u = 5x^2 + 1$. Then

$$\frac{dy}{dx} = \frac{1}{1 + u^2} \frac{du}{dx} = \frac{1}{1 + u^2} \cdot 10x = \frac{10x}{1 + (5x^2 + 1)^2}.$$ ■

Applications An important use of inverse trigonometric functions is in evaluating integrals.

■ **EXAMPLE 2** Find the area under the curves

(a) $y = \dfrac{1}{\sqrt{1 - x^2}}$ from 0 to $\frac{1}{2}$, \qquad (b) $y = \dfrac{1}{1 + x^2}$ from 0 to 1.

Solution From the differentiation formulas for inverse trigonometric functions, we see that

$$\text{arc sin } x \qquad \text{is an antiderivative of} \qquad \frac{1}{\sqrt{1 - x^2}},$$

$$\text{arc tan } x \qquad \text{is an antiderivative of} \qquad \frac{1}{1 + x^2}.$$

Therefore

$$\int_0^{1/2} \frac{dx}{\sqrt{1 - x^2}} = \text{arc sin } x \bigg|_0^{1/2} = \frac{\pi}{6} - 0 = \frac{\pi}{6},$$

and
$$\int_0^1 \frac{dx}{1 + x^2} = \text{arc tan } x \Big|_0^1 = \frac{\pi}{4} - 0 = \frac{\pi}{4}.$$ ∎

■ **EXAMPLE 3** Show that $\int_0^t \frac{dx}{1 + x^2} < \frac{\pi}{2}$ no matter how large t is.

Solution $\int_0^t \frac{dx}{1 + x^2} = \text{arc tan } x \Big|_0^t = \text{arc tan } t.$

By definition, all values of arc tan t are between $\frac{1}{2}\pi$ and $-\frac{1}{2}\pi$ so the integral is less than $\frac{1}{2}\pi$. ∎

Remark   As $t \longrightarrow \infty$, the quantity arc tan $t \longrightarrow \frac{1}{2}\pi$. For this reason, we write

$$\int_0^\infty \frac{dx}{1 + x^2} = \frac{\pi}{2}.$$

In geometric terms, the area under the curve $y = 1/(1 + x^2)$ between 0 and t is close to $\frac{1}{2}\pi$ when t is large (Fig. 1). Furthermore, the larger t is, the closer the area is to $\frac{1}{2}\pi$. Integrals of this type will be studied in Chapter 11.

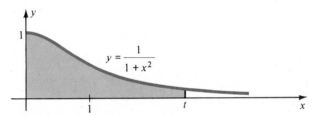

Fig. 1   The shaded area approaches $\frac{1}{2}\pi$ as $t \longrightarrow \infty$.

The next three examples illustrate further applications of inverse trig functions.

■ **EXAMPLE 4**   The circle shown in Fig. 2 has radius r meters. As the point P moves to the right at the rate of v m/sec, how fast is the length of the arc BQ increasing?

Solution   Express the arc length s in terms of the angle θ. From plane geometry, $s = 2r\theta$, where θ is measured in radians. From the triangle AOP,

$$\theta = \text{arc tan } \frac{x}{r}, \quad \text{hence} \quad s = 2r \text{ arc tan } \frac{x}{r}.$$

Differentiate with respect to time:

$$\dot{s} = 2r \frac{d}{dt}\left(\text{arc tan } \frac{x}{r}\right) = 2r \frac{1}{1 + (x/r)^2}\left(\frac{\dot{x}}{r}\right) = \frac{2r^2}{x^2 + r^2} \cdot v = \frac{2r^2 v}{x^2 + r^2} \quad \text{m/sec.} \quad ∎$$

Remark   As a rough check, we see from Fig. 2 that BQ should increase most rapidly when $x = 0$, then less rapidly as x increases. According to the answer, it does.

■ **EXAMPLE 5**   The Statue of Liberty is 150 ft tall and stands on a 150-ft pedestal. How far from the base should you stand so you can photograph the statue with largest possible angle? Assume camera level is 5 ft.

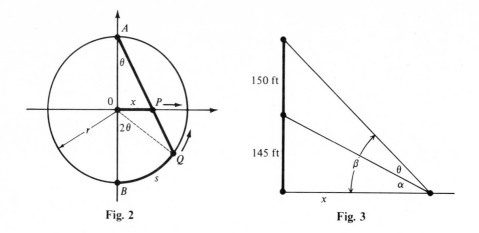

Fig. 2   Fig. 3

Solution Draw a diagram, labeling the various distances and angles as indicated (Fig. 3). The problem is to choose x in such a way that the angle θ is greatest. If x is very small or very large, θ will be small. Certainly the optimal value of x is between, say, 50 and 1000 ft.

Express θ as a function of x. From Fig. 3,

$$\theta = \beta - \alpha, \qquad \cot \alpha = \frac{x}{145}, \qquad \cot \beta = \frac{x}{295}.$$

Hence

$$\theta = \text{arc cot } \frac{x}{295} - \text{arc cot } \frac{x}{145}.$$

This is the function of x to be maximized. The domain of x is $x > 0$; there are no end points. Differentiate:

$$\frac{d\theta}{dx} = \frac{-\dfrac{1}{295}}{1 + \left(\dfrac{x}{295}\right)^2} + \frac{\dfrac{1}{145}}{1 + \left(\dfrac{x}{145}\right)^2} = \frac{-295}{(295)^2 + x^2} + \frac{145}{(145)^2 + x^2}.$$

Therefore

$$\frac{d\theta}{dx} = 0 \qquad \text{if} \qquad \frac{145}{(145)^2 + x^2} = \frac{295}{(295)^2 + x^2}.$$

Solve for x^2: $(145)(295)^2 + 145x^2 = (295)(145)^2 + 295x^2$,

$$x^2(295 - 145) = (145)(295)(295 - 145), \qquad x^2 = (145)(295).$$

The only positive root of this equation is

$$x = \sqrt{(145)(295)} = 5\sqrt{1711} \approx 5(41.36) = 206.8.$$

Answer Approximately 206.8 ft. ∎

■ **EXAMPLE 6** Maximize θ in Fig. 4a for $x > 0$.

Solution By Fig. 4b, we have $\theta = \text{arc cot } \frac{1}{2}(x - 2) - \text{arc cot}(x - 1).$

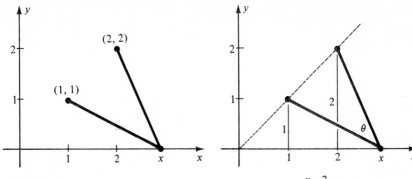

(a) Problem: maximize θ for $x > 0$

(b) $\theta = \text{arc cot}(\dfrac{x-2}{2}) - \text{arc cot}(x-1)$

Fig. 4

(You should check the validity of this formula for all positions of x.) Hence

$$\frac{d\theta}{dx} = \frac{-\frac{1}{2}}{1 + [\frac{1}{2}(x-2)]^2} - \frac{-1}{1 + (x-1)^2}.$$

Therefore $d\theta/dx = 0$ for $1 + \frac{1}{4}(x-2)^2 = \frac{1}{2} + \frac{1}{2}(x-1)^2$,

that is, for $4 + (x-2)^2 = 2 + 2(x-1)^2.$

This equation simplifies to $x^2 = 4$, hence $x = 2$ and

$$\theta = \text{arc cot } 0 - \text{arc cot } 1 = \tfrac{1}{2}\pi - \tfrac{1}{4}\pi = \tfrac{1}{4}\pi.$$

Since $\theta \longrightarrow 0$ if $x \longrightarrow 0+$ or $x \longrightarrow \infty$, this value of θ is the maximum.   ∎

  Query What does the solution $x = -2$ of $x^2 = 4$ mean?

EXERCISES

Differentiate

1   arc sin $\frac{1}{3}x$ **2**   arc cos $2x$ **3**   arc tan(x^2)

4 x arc sin$(2x+1)$ **5** $(\text{arc sin } 3x)^2$ **6**   arc tan $\sqrt{x}$

7   arc cot $\dfrac{1}{x}$ **8**   arc sin $\dfrac{x}{x+3}$ **9**   arc tan $\dfrac{x-1}{x+1}$

10   arc tan $\dfrac{1}{x}$ + arc cot x **11** $2x$ arc tan $2x - \ln \sqrt{1+4x^2}$

12 x arc cot $x + \ln \sqrt{1+x^2}$ **13** x arc sin $\frac{1}{4}x + \sqrt{16 - x^2}$

14 $\frac{1}{2}(x^2+1)$ arc tan $x - \frac{1}{2}x.$

Derive the formula on p. 333 for the derivative of

15   arc cot x **16**   arc sec x **17**   arc csc x.

18   Show that $\displaystyle\int_0^1 \frac{dx}{1+x^2} = 3 \int_1^{\sqrt{3}} \frac{dx}{1+x^2}.$

Compute

19 $\displaystyle\int_0^{\sqrt{3}/2} \frac{dx}{\sqrt{1-x^2}}$ **20** $\displaystyle\int_{2\sqrt{3}/3}^{\sqrt{2}} \frac{dx}{x\sqrt{x^2-1}}.$

21   A balloon is released from eye level and rises 10 ft/sec. According to an observer 100 ft from the point of release, how fast is the balloon's elevation angle increasing 4 sec later?

Express θ' in terms of x, $\dot{x}$, and the constant lengths a and b

22

23

24

25

26

27   (cont.) Show from Ex. 25 that θ increases as x increases.
28   Find the maximum of θ in Fig. 5. Here a and b are constants and $x \geq 0$.
29   (cont.) Find $\theta + 2\alpha$ when θ is maximal.

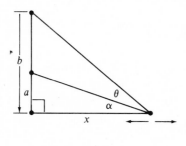

Fig. 5

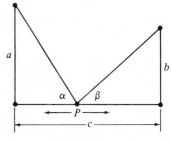

Fig. 6

30   In Fig. 6 the lengths a, b, and c are constant and P varies over the segment of length c. Prove that if $\alpha + \beta$ is minimal, then $b \sin^2 \alpha = a \sin^2 \beta$.
31   (cont.) Suppose $a = b$. Find $\min(\alpha + \beta)$.
32*   (cont.) Suppose $a > b$. Show that $\alpha + \beta$ achieves its min for P strictly inside the segment if and only if $c^2 > b(a - b)$.

7. HYPERBOLIC FUNCTIONS

The hyperbolic functions are certain combinations of exponential functions, with properties similar to those of the trigonometric functions. They are useful in solving differential equations and in evaluating integrals.

The three basic hyperbolic functions are the **hyperbolic sine**, the **hyperbolic cosine**, and the **hyperbolic tangent**:

$$\sinh x = \frac{e^x - e^{-x}}{2}, \qquad \cosh x = \frac{e^x + e^{-x}}{2}, \qquad \tanh x = \frac{\sinh x}{\cosh x} = \frac{e^x - e^{-x}}{e^x + e^{-x}}.$$

These functions are defined for all x. Tables of their values are available.
From the definitions,

$$\sinh(-x) = -\sinh x, \qquad \cosh(-x) = \cosh x, \qquad \tanh(-x) = -\tanh x.$$

Thus $\sinh x$ and $\tanh x$ are odd functions, while $\cosh x$ is an even function. Other immediate consequences of the definitions are that $\cosh x > 0$ for all x and that $|\tanh x| < 1$ for all x. (The numerator of $\tanh x$ is always a bit less in absolute value than the denominator.) Furthermore, since $e^{-x} \longrightarrow 0$ as $x \longrightarrow \infty$ and $e^x \longrightarrow 0$ as $x \longrightarrow -\infty$, we have

$$\sinh x \approx \tfrac{1}{2}e^x, \qquad \cosh x \approx \tfrac{1}{2}e^x, \qquad \tanh x \approx 1, \qquad \text{as} \quad x \longrightarrow \infty;$$
$$\sinh x \approx -\tfrac{1}{2}e^{-x}, \qquad \cosh x \approx \tfrac{1}{2}e^{-x}, \qquad \tanh x \approx -1, \qquad \text{as} \quad x \longrightarrow -\infty.$$

The less commonly used **hyperbolic cotangent**, **hyperbolic secant**, and **hyperbolic cosecant** are defined by

$$\coth x = \frac{1}{\tanh x}, \qquad \operatorname{sech} x = \frac{1}{\cosh x}, \qquad \operatorname{csch} x = \frac{1}{\sinh x}.$$

Derivatives Here are the derivatives of the hyperbolic functions.

$$\frac{d}{dx}\sinh x = \cosh x, \qquad \frac{d}{dx}\coth x = -\operatorname{csch}^2 x,$$

$$\frac{d}{dx}\cosh x = \sinh x, \qquad \frac{d}{dx}\operatorname{sech} x = -\operatorname{sech} x \tanh x,$$

$$\frac{d}{dx}\tanh x = \operatorname{sech}^2 x, \qquad \frac{d}{dx}\operatorname{csch} x = -\operatorname{csch} x \coth x.$$

These formulas are very easy to check. For example,

$$\frac{d}{dx}\cosh x = \frac{d}{dx}\frac{e^x + e^{-x}}{2} = \frac{e^x - e^{-x}}{2} = \sinh x.$$

■ **EXAMPLE 1** Show that $y = \sinh 3x$ satisfies the differential equation $y'' - 9y = 0$.

Solution Use the differentiation formulas for sinh x and cosh x, and the Chain Rule. If $y = \sinh 3x$, then

$$y' = 3 \cosh 3x, \qquad y'' = (3 \cosh 3x)' = 9 \sinh 3x.$$

Therefore $\qquad\qquad\qquad y'' - 9y = 9 \sinh 3x - 9 \sinh 3x = 0.$ ∎

Graphs Let us graph the three basic hyperbolic functions, using our knowledge of their derivatives. For sinh x and tanh x, the derivatives are positive; hence these functions are strictly increasing. For $y(x) = \cosh x$, we observe that

$$y'(0) = \sinh 0 = 0, \qquad y''(x) = \cosh x > 0.$$

Hence the graph of $y = \cosh x$ is convex upwards, with a minimum at $x = 0$. We now have plenty of information to sketch sinh x, cosh x, and tanh x. See Fig. 1.

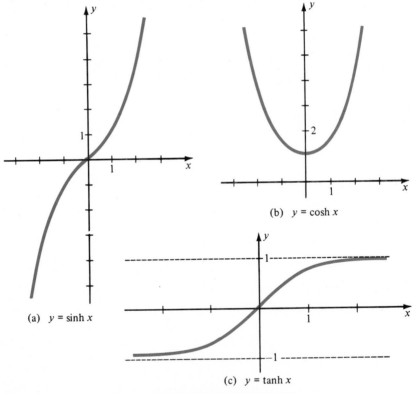

(a) $y = \sinh x$

(b) $y = \cosh x$

(c) $y = \tanh x$

Fig. 1   Graphs of hyperbolic functions

Identities The hyperbolic functions are related to each other by identities similar to trigonometric identities. For example, since

$$\cosh^2 x = \left(\frac{e^x + e^{-x}}{2}\right)^2 = \frac{1}{4}\left(e^{2x} + 2 + e^{-2x}\right),$$

and $\qquad\qquad \sinh^2 x = \left(\frac{e^x - e^{-x}}{2}\right)^2 = \frac{1}{4}\left(e^{2x} - 2 + e^{-2x}\right),$

it follows that

$$\cosh^2 x - \sinh^2 x = 1.$$

Easy consequences are the identities

$$\tanh^2 x + \operatorname{sech}^2 x = 1, \qquad \coth^2 x - \operatorname{csch}^2 x = 1.$$

Note the similarity to trig identities, except for signs. Virtually every trigonometric identity has a hyperbolic analogue. But you must be careful with signs.

■ **EXAMPLE 2** Prove the identity $\cosh(u + v) = \cosh u \cosh v + \sinh u \sinh v$.

Solution Express the right-hand side in terms of exponentials and simplify algebraically:

$$\cosh u \cosh v + \sinh u \sinh v$$

$$= \left(\frac{e^u + e^{-u}}{2}\right)\left(\frac{e^v + e^{-v}}{2}\right) + \left(\frac{e^u - e^{-u}}{2}\right)\left(\frac{e^v - e^{-v}}{2}\right)$$

$$= \frac{e^{u+v} + e^{u-v} + e^{-u+v} + e^{-u-v}}{4} + \frac{e^{u+v} - e^{u-v} - e^{-u+v} + e^{-u-v}}{4}$$

$$= \frac{2e^{u+v} + 2e^{-u-v}}{4} = \frac{e^{u+v} + e^{-(u+v)}}{2} = \cosh(u + v). \qquad ■$$

Inverse Hyperbolic Functions   The function $\sinh x$ increases strictly, taking each real value once. Hence $\sinh x$ has an inverse, written* $\sinh^{-1} x$ or arg sinh x. Thus the statements

$$y = \sinh^{-1} x \qquad \text{and} \qquad x = \sinh y$$

are equivalent. The graph of $y = \sinh^{-1} x$, shown in Fig. 2, is the reflection of $y = \sinh x$ in the line $y = x$.

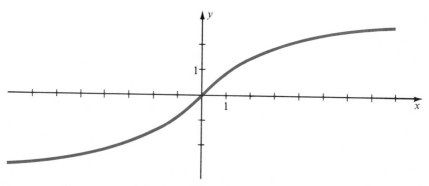

Fig. 2   Graph of $y = \sinh^{-1} x$

* When $y = \sinh x$, usually x is *not* an angle (arc), and x is called the **argument** of sinh x. Hence the name arg sinh.

Warning   Do not confuse $\sinh^{-1} x$ with $1/\sinh x$.

Since $\sinh x$ is defined in terms of exponentials, it seems reasonable that $\sinh^{-1} x$ should be expressible in terms of logarithms. Indeed, if $y = \sinh^{-1} x$, then

$$x = \sinh y = \frac{e^y - e^{-y}}{2}.$$

Hence $e^y - 2x - e^{-y} = 0,$ $e^{2y} - 2xe^y - 1 = 0.$

This is a quadratic equation for e^y. By the quadratic formula, $e^y = x \pm \sqrt{x^2 + 1}$. Since $e^y > 0$, the correct choice of sign is plus:

$$e^y = x + \sqrt{x^2 + 1}, \qquad \text{therefore} \qquad y = \ln(x + \sqrt{x^2 + 1}).$$

$$\boxed{\sinh^{-1} x = \ln(x + \sqrt{x^2 + 1}).}$$

To obtain an inverse for $\cosh x$, we restrict the domain to $x \geq 0$, where $\cosh x$ increases strictly, taking each value $y \geq 1$ once. Hence, there is an inverse function $\cosh^{-1} x$ defined for $x \geq 1$. It can be expressed in terms of logarithms by a derivation similar to the one given above for $\sinh^{-1}$:

$$\boxed{\cosh^{-1} x = \ln(x + \sqrt{x^2 - 1}), \qquad x \geq 1.}$$

The inverse hyperbolic tangent is defined for $-1 < x < 1$ and can be expressed by the formula

$$\boxed{\tanh^{-1} x = \frac{1}{2} \ln\left(\frac{1 + x}{1 - x}\right), \qquad -1 < x < 1.}$$

Graphs of $\cosh^{-1} x$ and $\tanh^{-1} x$ are shown in Fig. 3.

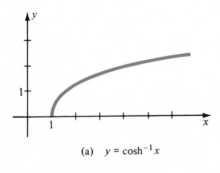

(a) $y = \cosh^{-1} x$

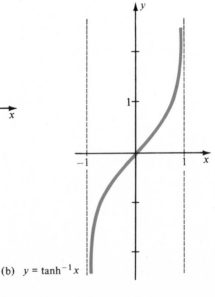

Fig. 3

(b) $y = \tanh^{-1} x$

Derivatives of Inverse Hyperbolic Functions If $y = \sinh^{-1} x$, then $x = \sinh y$. Therefore,

$$\frac{dy}{dx} = 1 \bigg/ \frac{dx}{dy} = \frac{1}{\cosh y} = \frac{1}{\sqrt{1 + \sinh^2 y}} = \frac{1}{\sqrt{1 + x^2}}.$$

In this way, we obtain the formulas

$$\frac{d}{dx} \sinh^{-1} x = \frac{1}{\sqrt{1 + x^2}}, \qquad \frac{d}{dx} \cosh^{-1} x = \frac{1}{\sqrt{x^2 - 1}}, \qquad \frac{d}{dx} \tanh^{-1} x = \frac{1}{1 - x^2}.$$

These formulas are useful for integration since they provide antiderivatives for the functions on the right-hand sides.

■ **EXAMPLE 3** Express $\displaystyle\int_0^5 \frac{dx}{\sqrt{1 + x^2}}$ in terms of natural logarithms.

Solution An antiderivative of $1/\sqrt{1 + x^2}$ is $\sinh^{-1} x = \ln(x + \sqrt{x^2 + 1})$. Therefore

$$\int_0^5 \frac{dx}{\sqrt{1 + x^2}} = \ln(x + \sqrt{x^2 + 1})\bigg|_0^5 = \ln(5 + \sqrt{26}) - \ln 1 = \ln(5 + \sqrt{26}). \quad ■$$

We shall mention an interesting consequence of the formula for the derivative of $\cosh^{-1}$. First we write the formula in Chain Rule form:

$$\frac{d}{dx} \cosh^{-1} u = \frac{1}{\sqrt{u^2 - 1}} \frac{du}{dx}.$$

The expression $\sqrt{u^2 - 1}$ suggests taking $u = \sec x$:

$$\frac{d}{dx} \cosh^{-1}(\sec x) = \frac{1}{\tan x} (\sec x \tan x).$$

Conclusion:

$$\frac{d}{dx} \cosh^{-1}(\sec x) = \sec x.$$

We have found an antiderivative for $\sec x$.

The Hanging Cable As an application, we shall find the slope of a uniform, heavy, flexible cable suspended between two points at the same level (Fig. 4a). Here "flexible" means that the only internal force in the cable is tension acting in the tangential direction. "Heavy" means that gravity must be taken into account. "Uniform" means that the density of the cable, in weight per unit length, is a constant, δ.

We choose axes as in Fig. 4b, so the lowest point of the cable is on the y-axis. The shape of the hanging cable is then some curve $y = f(x)$, to be found.

Let us look at the portion of the cable for $0 \leq x \leq a$. Three forces act on it: a horizontal tension of magnitude T_0 at the left end, a tangential tension of magnitude $T(a)$ at the right end, and a downward gravitational force $L\delta$, where L is the

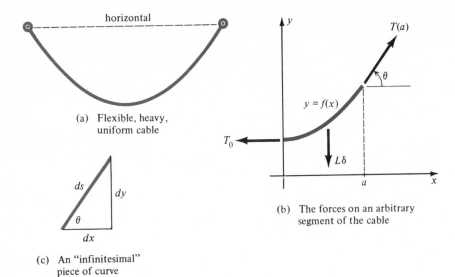

(a) Flexible, heavy, uniform cable

(b) The forces on an arbitrary segment of the cable

(c) An "infinitesimal" piece of curve

Fig. 4 Hanging cable

length. When the chain hangs in equilibrium, these forces balance; in particular the horizontal components balance, and similarly the vertical components.

Let $\theta = \theta(x)$ be the angle between the tangent and the positive x-axis. At the right end, the tension has components $T(a) \cos \theta$ and $T(a) \sin \theta$. Therefore,

$$T(a) \cos \theta = T_0, \qquad T(a) \sin \theta = L\delta.$$

Division yields

$$\tan \theta = \frac{\delta}{T_0} L.$$

It will help to express L in terms of θ. Figure 4c shows that the "element" of arc length* is $ds = \sec \theta \, dx$. Hence L is the sum of these elements:

$$L = \int_0^a \sec \theta \, dx.$$

Thus our equation for the balance of forces becomes

$$\tan \theta = \frac{\delta}{T_0} \int_0^a \sec \theta \, dx.$$

Now we consider a as a variable in this equation, differentiate both sides with respect to a, and apply the Fundamental Theorem of Calculus,

$$\frac{d}{da} \tan \theta = \frac{d}{da} \left[\frac{\delta}{T_0} \int_0^a \sec \theta \, dx \right] = \frac{\delta}{T_0} \sec \theta.$$

Hence

$$\sec^2 \theta \frac{d\theta}{da} = \frac{\delta}{T_0} \sec \theta, \qquad \frac{d\theta}{da} = \frac{\delta}{T_0} \frac{1}{\sec \theta}.$$

---

* This is all we need know about arc length here.

Next we replace a by x and take reciprocals:

$$\frac{dx}{d\theta} = \frac{T_0}{\delta} \sec \theta.$$

Here is where the antiderivative of $\sec \theta$ comes in; we can express x in terms of θ:

$$x = \frac{T_0}{\delta} \cosh^{-1}(\sec \theta) + C.$$

When $x = 0$, then $\theta = 0$ so $\cosh^{-1}(\sec \theta) = \cosh^{-1}(1) = 0$, hence $C = 0$ and

$$x = \frac{T_0}{\delta} \cosh^{-1}(\sec \theta), \qquad \sec \theta = \cosh\left(\frac{\delta}{T_0} x\right).$$

This is very good, but we still need to find $y = f(x)$. Now

$$\frac{dy}{dx} = \tan \theta = \sqrt{\sec^2 \theta - 1} = \sqrt{\cosh^2\left(\frac{\delta}{T_0} x\right) - 1} = \sinh\left(\frac{\delta}{T_0} x\right),$$

so we find y by antidifferentiating:

$$y = \frac{T_0}{\delta} \cosh\left(\frac{\delta}{T_0} x\right) + C.$$

We can make $C = 0$ by choosing $y(0) = T_0/\delta$. Then the equation of the hanging cable is simply

$$y = k \cosh\left(\frac{x}{k}\right), \qquad k = \frac{T_0}{\delta}.$$

The shape of a hanging cable is called a **catenary** after the Latin for chain. We now see that a catenary is simply a hyperbolic cosine curve.

EXERCISES

Prove

1 $\sinh(u + v) = \sinh u \cosh v + \cosh u \sinh v$

2 $\tanh(u + v) = \dfrac{\tanh u + \tanh v}{1 + \tanh u \tanh v}$

3 $\cosh 2x = \cosh^2 x + \sinh^2 x$

4 $\sinh(u + v) - \sinh(u - v) = 2 \cosh u \sinh v$

5 $\sinh 3x = 3 \sinh x + 4 \sinh^3 x$

6 $1 + 2 \displaystyle\sum_{k=1}^{n} \cosh kx = \dfrac{\sinh(n + \frac{1}{2})x}{\sinh \frac{1}{2}x}.$

Differentiate with respect to x

7 $\sinh 5x$

8 $\cosh \sqrt{x}$

9 $\tanh(x^2 + 1)^{1/2}$

10 $\tanh^3 x$

11 $\frac{1}{3}e^{2x}(2 \cosh x - \sinh x)$

12 $x \cosh x - \sinh x$

13 $\ln \cosh x$

14 $\sqrt{\cosh 4x}$

15 $x^2 \sinh x - 2x \cosh x + 2 \sinh x$

16 $\frac{1}{4}\sinh 2x - \frac{1}{2}x.$

Find $\lim_{x \to \infty}$ of

17 $e^{-x} \sinh x$ **18** $e^{-x} \cosh x$ **19** $e^x \operatorname{sech} x$

20 $e^{2x}(1 - \tanh x)$ **21** $\dfrac{\cosh x}{\cosh 2x}$ **22** $\dfrac{\ln \sinh x}{x}$

Prove

23 $\dfrac{d}{dx} \tanh x = \operatorname{sech}^2 x$ **24** $\dfrac{d}{dx} \operatorname{csch} x = -\operatorname{csch} x \coth x$

25 $\cosh^{-1} x = \ln(x + \sqrt{x^2 - 1})$ **26** $\tanh^{-1} x = \dfrac{1}{2} \ln\left(\dfrac{1+x}{1-x}\right)$

27 $\dfrac{d}{dx} \cosh^{-1} x = \dfrac{1}{\sqrt{x^2 - 1}}$ **28** $\dfrac{d}{dx} \tanh^{-1} x = \dfrac{1}{1 - x^2}$

29 $\sinh^{-1}(\tan \theta) = \cosh^{-1}(\sec \theta)$ $(-\tfrac{1}{2}\pi < \theta < \tfrac{1}{2}\pi)$

30 $\dfrac{d}{dx} \tanh^{-1}(\sin \theta) = \sec \theta.$

Express in terms of logarithms

31 $\displaystyle\int_0^1 \dfrac{dx}{\sqrt{1 + x^2}}$ **32** $\displaystyle\int_2^3 \dfrac{dx}{\sqrt{x^2 - 1}}$ **33** $\displaystyle\int_0^{1/4} \dfrac{dx}{1 - x^2}$ **34** $\displaystyle\int_0^{1/4} \dfrac{x^2}{1 - x^2}\, dx.$

35 Show that $y = a \sinh cx + b \cosh cx$ satisfies $y'' = c^2 y$.
36 (cont.) Find $y = y(x)$ such that $y'' = 4y$, $y(0) = 6$, and $y'(0) = -1$.
37 Prove $\sinh x > x$ for all $x > 0$.
38 Find a formula for the tension $T(x)$ in the hanging cable.
39 Let R be the region bounded by $y = \cosh x$, the y-axis, and $y = b$, where $b > 1$. Find the volume of the solid obtained by revolving R about the y-axis. [*Hint* $d/dx(x \sinh x - \cosh x) = ?$]
40 Let R be the region bounded by $y = \sinh x$, the x-axis, and $x = a$, where $a > 0$. Find the volume of the solid obtained by revolving R about the x-axis. [*Hint*   Use Ex. 16.]

8. BASIC PROPERTIES

The existence of inverse functions and many of their properties depend on certain basic facts about continuous functions. Like uniform continuity and the existence of maxima and minima, these facts are too hard to prove in a first calculus course. Nevertheless, they all follow from one basic principle, which is quite natural and easy to understand.

Intermediate Value Theorem   The idea is simple. To draw a continuous graph, you cannot lift your pencil from the paper. If the curve is below the x-axis at x_0 and above at x_1, then somewhere between it must cross the axis; it cannot jump over. Here is the general statement:

> **Intermediate Value Theorem**   Let $f(x)$ be a continuous function defined on an interval. Suppose that x_0 and x_1 are in this interval, and that y is a number between $f(x_0)$ and $f(x_1)$. Then $f(x) = y$ for some x between x_0 and x_1.

Rather than attempt the technical proof, let us accept this principle and give some examples of its use. One practical application is in finding roots of difficult equations such as $x^5 - x + 1 = 0$, or $\cos x = x$. A natural approach to root finding is a search technique based on a simple idea:

> Suppose that $f(x)$ is continuous on an interval $[a, b]$ and that $f(a)$ and $f(b)$ have opposite signs. Then $f(x) = 0$ for some x with $a < x < b$.

This is a direct corollary of the Intermediate value Theorem. The search technique squeezes down on a root of an equation $f(x) = 0$ by finding a sequence of smaller and smaller intervals, $[a_0, b_0]$, $[a_1, b_1]$, $[a_2, b_2]$, etc., each one inside the previous one, such that $f(a_i)$ and $f(b_i)$ have opposite sign.

■ **EXAMPLE 1** Estimate to six places a root of the equation

$$f(x) = x^5 - x + 1 = 0.$$

Solution By trial and error, $f(-2) = -29 < 0$ and $f(-1) = 1 > 0$, so there is a zero on the interval $[-2, -1]$ by the Intermediate Value Theorem. On this interval $f'(x) = 5x^4 - 1 > 0$, so $f(x)$ is strictly increasing; there is precisely one zero.

Now divide $[-2, -1]$ into 10 equal subintervals

$$[-2.0, -1.9], \quad [-1.9, -1.8], \cdots, [-1.1, -1.0],$$

and test the signs of $f(x)$ at the end points. We find $f(-1.2) = -0.29$ and $f(-1.1) = 0.49$, so there is a zero in $[-1.2, -1.1]$. Next, divide this interval into ten equal parts and repeat the process. Here is the table of results:

k	0	1	2	3	4	5	6
a_k	-2	-1.2	-1.17	-1.168	-1.1674	-1.16731	-1.167304
b_k	-1	-1.1	-1.16	-1.167	-1.1673	-1.16730	-1.167303
$f(a_k)$	-29	-0.29	-0.022	-5.8×10^{-3}	-8.0×10^{-4}	-5.0×10^{-5}	-1.8×10^{-7}
$f(b_k)$	1	0.49	0.060	2.5×10^{-3}	3.3×10^{-5}	3.3×10^{-5}	8.1×10^{-6}

Clearly, to six places, $x = -1.167304$ is a solution. ■

Of course, the calculations are laborious by hand, but are relatively easy on a scientific calculator.

There is another statement of the Intermediate Value Theorem that is useful in applications.

> If $f(x)$ is continuous on an interval and does not take the value 0, then either $f(x) > 0$ throughout the interval or $f(x) < 0$ throughout the interval.

For if $f(x_0)$ and $f(x_1)$ have opposite signs, then $f(x) = 0$ somewhere between x_0 and x_1.

■ **EXAMPLE 2** Suppose $f(x)$ is continuous on $[0, 1]$ and $f(0) = f(1)$. Show that the graph of $f(x)$ must have a horizontal chord of length $\frac{1}{3}$. See Fig. 1.

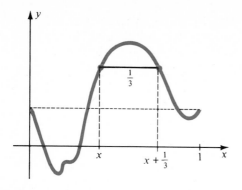

Fig. 1   Assumptions: $f(x)$ continuous and $f(0) = f(1)$. Conclusion: there is a horizontal chord of length $\frac{1}{3}$.

Solution   The problem is to prove that there is a point x with $0 \le x \le \frac{2}{3}$ and $f(x) = f(x + \frac{1}{3})$.
   To do so, consider

$$g(x) = f(x + \tfrac{1}{3}) - f(x).$$

It is a continuous function on $[0, \frac{2}{3}]$. We want to prove that $g(x) = 0$ for some x. If not, then either $g(x) > 0$ throughout the interval or $g(x) < 0$ throughout the interval. In either case, $g(0) + g(\frac{1}{3}) + g(\frac{2}{3}) \ne 0$. But

$$g(0) + g(\tfrac{1}{3}) + g(\tfrac{2}{3}) = [f(\tfrac{1}{3}) - f(0)] + [f(\tfrac{2}{3}) - f(\tfrac{1}{3})] + [f(1) - f(\tfrac{2}{3})] = f(1) - f(0) = 0,$$

a contradiction. Therefore $g(x) = 0$ for some x, that is, $f(x) = f(x + \frac{1}{3})$ for some x.

∎

Remark   The assumption $f(0) = f(1)$ is crucial. Without it, the graph could be increasing, for example, and not have any horizontal chords at all.

Inverse Functions

> Let $f(x)$ be a strictly increasing continuous function whose domain is a closed interval $[a, b]$. Then the range of $f(x)$ is the closed interval $[f(a), f(b)]$. For each y in $[f(a), f(b)]$ there is a unique x in $[a, b]$ for which $y = f(x)$.

   The proof consists of three parts. First, suppose $a \le x \le b$. Then $f(a) \le f(x) \le f(b)$ because $f(x)$ is strictly increasing. Therefore each point of the range of $f(x)$ lies in $[f(a), f(b)]$. Second, suppose $f(a) \le y \le f(b)$. By the Intermediate Value Theorem, there is an x in $[a, b]$ such that $f(x) = y$; hence y is in the range of $[a, b]$. These two conclusions show that the range of $f(x)$ is precisely the interval $[f(a), f(b)]$. Finally, suppose y is in the range of $f(x)$. Then $y = f(x)$ for exactly one x in $[a, b]$, because $f(x_1) \ne f(x_2)$ if $x_1 \ne x_2$ since $f(x)$ is strictly increasing. This completes the proof.

   A similar result holds for strictly decreasing continuous functions. For simplicity, we shall discuss only the increasing case.
   The last result says that f **maps** or **carries** the interval $[a, b]$ onto the interval $[f(a), f(b)]$ in a **one-to-one** manner. Each x in $[a, b]$ goes to one point y in $[f(a), f(b)]$, conversely, each y in $[f(a), f(b)]$ comes from exactly one x. Thus there is

a one-to-one correspondence between the two sets: each x corresponds to a unique y and each y corresponds to a unique x. When that happens, we can write $x = g(y)$. Then g is the inverse function of f. The domain of g is $[f(a), f(b)]$, the range of f. For example, $f(x) = x^3$ maps $[-1, 2]$ in a one-to-one manner onto $[-1, 8]$. The inverse function $g(y) = \sqrt[3]{y}$ maps $[-1, 8]$ onto $[-1, 2]$.

Theorem Let $f(x)$ be a strictly increasing continuous function with domain $[a, b]$. Then $f(x)$ has an inverse function $g(y)$, and $g(y)$ is a strictly increasing continuous function with domain $[f(a), f(b)]$.

The new part of this assertion is that $g(y)$ is continuous. To prove it, suppose $g(y_0) = x_0$. (We shall assume $a < x_0 < b$. The cases $x_0 = a$ and $x_0 = b$ require a slight modification of the argument that follows.) Take any $\varepsilon < 0$. We can assume

$$a \le x_0 - \varepsilon < x_0 < x_0 + \varepsilon \le b.$$

We must find a $\delta > 0$ so small that

$$|g(y) - x_0| < \varepsilon \qquad \text{whenever} \qquad |y - y_0| < \delta.$$

In geometric language, we must produce an interval of radius δ centered at y_0 that is carried by g to the inside of the interval of radius ε centered at x_0.

We concentrate on $f(x)$, restricted to the interval $[x_0 - \varepsilon, x_0 + \varepsilon]$. Since $f(x)$ is strictly increasing and continuous, it maps this interval onto the interval $[f(x_0 - \varepsilon), f(x_0 + \varepsilon)]$, containing y_0. See Fig. 2a. We simply choose $\delta > 0$ small enough that $[y_0 - \delta, y_0 + \delta]$ is contained in $[f(x_0 - \varepsilon), f(x_0 + \varepsilon)]$. See Fig. 2b. Then g maps $[y_0 - \delta, y_0 + \delta]$ into $[x_0 - \varepsilon, x_0 + \varepsilon]$. In other words, $|g(y) - x_0| < \varepsilon$ whenever $|y - y_0| < \delta$. This completes the proof.

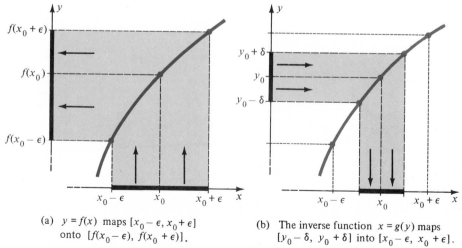

(a) $y = f(x)$ maps $[x_0 - \epsilon, x_0 + \epsilon]$ onto $[f(x_0 - \epsilon), f(x_0 + \epsilon)]$.

(b) The inverse function $x = g(y)$ maps $[y_0 - \delta, y_0 + \delta]$ into $[x_0 - \epsilon, x_0 + \epsilon]$.

Fig. 2 Proof of the continuity of the inverse function

Derivatives of Inverse Functions There remains one more basic fact whose proof we postponed earlier. Let us state the result in a precise form.

> **Theorem** Let $f(x)$ be a strictly increasing continuous function with domain $[a, b]$. Let $g(y)$ be the inverse function of $f(x)$. Suppose that $a < c < b$, that $f(x)$ is differentiable at $x = c$, and that $f'(c) > 0$. Then $g(y)$ is differentiable at $y = f(c)$ and
>
> $$g'[f(c)] = \frac{1}{f'(c)}.$$

Proof Consider the difference quotient of $g(y)$ at $f(c)$. Take $y \neq f(c)$, that is, $y = f(x)$, where $x = g(y) \neq c$. Then

$$\frac{g(y) - g[f(c)]}{y - f(c)} = \frac{x - c}{f(x) - f(c)} = \left[\frac{f(x) - f(c)}{x - c} \right]^{-1}$$

Let $y \longrightarrow f(c)$. Since $g(y)$ is continuous (previous theorem), we have $g(y) \longrightarrow g[f(c)]$, that is, $x \longrightarrow c$. Therefore

$$\frac{f(x) - f(c)}{x - c} \longrightarrow f'(c) > 0.$$

Recall that in general $\lim[1/h(t)] = 1/[\lim h(t)]$, provided $\lim h(t) \neq 0$. Applied here, this rule yields

$$\lim_{y \to f(c)} \frac{g(y) - g[f(c)]}{y - f(c)} = \frac{1}{f'(c)}.$$

Consequently $g(y)$ is indeed differentiable at $y = f(c)$, and its derivative has the specified value. For a geometric interpretation of this proof, see Fig. 3.

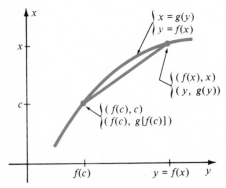

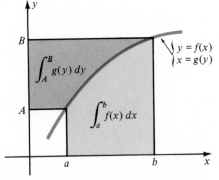

Fig. 3 Slope of the chord:

$$\frac{g(y) - g[f(c)]}{y - f(c)} = \frac{x - c}{f(x) - f(c)} = \left[\frac{f(x) - f(c)}{x - c} \right]^{-1}$$

Fig. 4 The sum of the two shaded areas plus the area Aa of the smaller rectangle equals the area Bb of the larger rectangle.

Integrals of Inverse Functions There is a useful relation between the integrals of a function and of its inverse function.

Let $f(x)$ be a strictly increasing continuous function with domain $[a, b]$. Let $g(y)$ be the inverse function of $f(x)$, so that the domain of $g(y)$ is $[A, B] = [f(a), f(b)]$. Then

$$\int_a^b f(x)\, dx + \int_A^B g(y)\, dy = Bb - Aa.$$

The proof is entirely geometrical (Fig. 4). We illustrate it in the simplest case, $0 < a < b$ and $0 < A < B$. Other cases can be reduced to this one by shifting.

■ **EXAMPLE 3** Show that $\displaystyle\int_1^b \ln x\, dx = b \ln b - b + 1$.

Solution Apply the formula with $f(x) = \ln x$, $g(y) = e^y$, $a = 1$, $A = 0$, $B = \ln b$:

$$\int_1^b \ln x\, dx + \int_0^{\ln b} e^y\, dy = b \ln b.$$

But

$$\int_0^{\ln b} e^y\, dy = e^y \Big|_0^{\ln b} = b - 1,$$

so the result follows. ■

Logarithm and Exponential When you get right down to it, our discussion of the exponential function in Chapter 4 was based on an assumption: that there exists a *differential* function $y = y(x)$ satisfying $y' = y$ and $y(0) = 1$. While we gave plausible reasons for the existence of such a function, we never really proved it. Undaunted, we went on to define the logarithm function as the inverse of the exponential function. Therefore we are not on solid ground with $\ln x$ either.

Clearly, we must start someplace. If we prove the existence of either the exponential or the logarithm function, we get the other one from the theory of inverse functions. There are several satisfactory approaches to the problem; here is the way one of these works. It is based on the theorem that every continuous function on a closed interval has an integral.

Definition $\quad \displaystyle \ln x = \int_1^x \frac{dt}{t}, \quad x > 0.$

This definition yields a function $y = \ln x$ having domain $x > 0$ and satisfying

$$\frac{dy}{dx} = \frac{1}{x}, \quad y(1) = 0.$$

Now the algebraic properties follow easily.

For instance, suppose $a > 0$ and $x > 0$. Then

$$\frac{d}{dx} \ln ax = a \cdot \frac{1}{ax} = \frac{1}{x} = \frac{d}{dx} \ln x.$$

Hence $\ln ax = \ln x + c$. To find the constant c, we set $x = 1$, obtaining $c = \ln a$. Therefore,

$$\ln ax = \ln x + \ln a.$$

Once the logarithm function is firmly established, we define e^x to be the inverse function and derive the usual properties; in particular,

$$e^x > 0 \quad \text{for all } x, \qquad \frac{d}{dx} e^x = e^x, \qquad e^0 = 1.$$

Let us now settle a point discussed also in Chapter 4. There we asserted that any differentiable function satisfying $y' = y$ and $y(a) = 0$ for some a, is identically zero. Suppose y is such a function. Then

$$\frac{d}{dx}\left(ye^{-x}\right) = y'e^{-x} - ye^{-x} = ye^{-x} - ye^{-x} = 0,$$

hence $\qquad\qquad\qquad ye^{-x} = c, \qquad \text{so} \qquad y = ce^x.$

But $\qquad\qquad 0 = y(a) = ce^a \qquad \text{and} \quad e^a \neq 0, \qquad \text{hence} \qquad c = 0.$

Therefore $\qquad\qquad\qquad y(x) = 0 \qquad \text{for all} \quad x.$

Trigonometric Functions   The trigonometric functions also can be put on a rigorous foundation along the same lines. One way is to start with arc tan x, defined by an integral.

Definition $\qquad \text{arc tan } x = \displaystyle\int_0^x \frac{dt}{1 + t^2}.$

This definition yields a function $y = $ arc tan x defined for all x and satisfying

$$\frac{dy}{dx} = \frac{1}{1 + x^2}, \qquad \text{arc tan}(-x) = -\text{arc tan } x.$$

Thus $dy/dx > 0$, so arc tan x is a strictly increasing function. Less obvious is that arc tan x is a bounded function. For, if $x > 1$, then

$$\text{arc tan } x = \int_0^x \frac{dt}{1 + t^2} = \int_0^1 \frac{dt}{1 + t^2} + \int_1^x \frac{dt}{1 + t^2}$$

$$< \int_0^1 \frac{dt}{1} + \int_1^x \frac{dt}{t^2} = 1 + \left(1 - \frac{1}{x}\right) = 2 - \frac{1}{x}.$$

Therefore arc tan $x < 2$ for all x. From this boundedness follows (for a theoretical reason we skip here) that arc tan x approaches a limit as $x \longrightarrow \infty$. We call this limit $\frac{1}{2}\pi$, that is, we define π by the relation

$$\frac{\pi}{2} = \lim_{x \to \infty} \int_0^x \frac{dt}{1 + t^2} = \int_0^\infty \frac{dt}{1 + t^2}.$$

Since arc tan $x < 2$, we have $\frac{1}{2}\pi \leq 2$, hence $\pi \leq 4$. Of course, better estimates of π

are available. Incidentally, we can also prove that

$$\frac{\pi}{4} = \text{arc tan } 1 = \int_0^1 \frac{dt}{1 + t^2},$$

but that requires a trick. See p. 558.

The function $y = \text{arc tan } x$ is defined for all x, is strictly increasing, and takes all values in the range $-\frac{1}{2}\pi < y < \frac{1}{2}\pi$. It has an inverse function $y = \tan x$ defined for $-\frac{1}{2}\pi < x < \frac{1}{2}\pi$, strictly increasing, taking all real values, and satisfying

$$\frac{d}{dx} \tan x = 1 + \tan^2 x.$$

We extend the domain of $\tan x$ by defining $\tan(x + \pi) = \tan x$. Finally, we define functions $\sin x$ and $\cos x$ motivated by the half-angle formulas:

$$\sin x = \frac{2t}{1 + t^2}, \qquad \cos x = \frac{1 - t^2}{1 + t^2},$$

where $t = \tan \frac{1}{2}x$. After several applications of the Chain Rule, we get the formulas

$$\frac{d}{dx} \sin x = \cos x, \qquad \frac{d}{dx} \cos x = -\sin x.$$

Previously (p. 190) we used the Addition Laws for the sine and cosine functions to derive the formulas for their derivatives. Also, we promised (p. 181) a proof of these Addition Laws. Obviously, up to now a proof based on the derivatives of sine and cosine would have been circular. Now, however, a valid proof is possible. Set

$$\begin{cases} S(x) = \sin(a + x) - \sin a \cos x - \cos a \sin x, \\ C(x) = \cos(a + x) - \cos a \cos x + \sin a \sin x. \end{cases}$$

Then $S(0) = C(0) = 0, \qquad S'(x) = C(x), \qquad$ and $\qquad C'(x) = -S(x).$

Set $$F(x) = S(x)^2 + C(x)^2.$$

Then $F(0) = 0$ and $F'(x) = 2SC - 2CS = 0$, so $F(x) = 0$ for all x. This implies $S(x) = 0$ and $C(x) = 0$ for all x, precisely the Addition Laws.

This has been only a sketch of a rigorous development of the trigonometric functions. Filling in the details is a big job.

EXERCISES

Show that $f(x)$ has exactly one zero

1 $f(x) = x^3 + px + q, \quad p > 0$ **2** $f(x) = x - a \sin x + b, \quad |a| < 1.$

Show that the equation has a solution.

3 $x^4 = 3 + e^{-x} \cos 7x$ **4** $\ln x = -x.$

Locate a solution within an interval of length 0.1.

5 $e^x = 2 \cos x$ **6** $e^x = -3x.$

7 Let $f(x)$ be continuous on $[a, b]$ and $f(a) = f(b)$. Let n be a positive integer. Prove that the graph $y = f(x)$ has a horizontal chord of length $(b - a)/n$. [*Hint* See Example 2.]

8*   (cont.) Assume that the temperature at the Earth's surface at any instant of time varies continuously. At a fixed instant of time, prove there exists a pair of antipodal points on the equator (opposite ends of a diameter) where the temperatures are the same.

9   Let $y = f(x)$ be twice continuously differentiable on $a < x < b$ and suppose $f'(x)$ is never 0. Let $x = g(y)$ be the inverse function. Express $g''(y)$ in terms of $f(x)$ and its derivatives.

10   (cont.) Suppose $y = f(x)$ is increasing and strictly convex. What can you say about $x = g(y)$?

11   Prove that $\ln x^p = p \ln x$ if $\ln x$ is defined by an integral as on p. 351.

12   Prove that $(\sin x)' = \cos x$ and $(\cos x)' = -\sin x$ if $\sin x$ and $\cos x$ are defined in terms of $\tan \frac{1}{2}x$ as on p. 353, line 10.

Define $f(x) = \displaystyle\int_0^x \frac{dt}{1 + t^4}$ for $x \geq 0$.

13   Show that $y = f(x)$ has an inverse function $x = g(y)$.

14   Prove $g'(y) = 1 + [g(y)]^4$.   **15**   Prove $f(x) < \frac{4}{3}$ for all $x \geq 0$.

16   Prove $f(x) = \arctan x$ has a solution $x > 0$.

9. MISCELLANEOUS EXERCISES

Differentiate with respect to x

1 $(x \ln x)^{4/3}$ **2** $x^{(x^2)}$ **3** $10^{\cdot x}$

4 $\arcsin\left(\dfrac{x + a}{x - a}\right)$ **5** $\sqrt{1 + x^2}\ \arctan x$ **6** $\sinh^3 2x$

7 $\dfrac{1}{6}\ln\left(\dfrac{1 + x + x^2}{(1 - x)^2}\right) + \dfrac{1}{\sqrt{3}}\arctan\left(\dfrac{2x + 1}{\sqrt{3}}\right)$ **8** $\dfrac{1}{4}\ln\left(\dfrac{1 + x}{1 - x}\right) + \dfrac{1}{2}\arctan x.$

By computing numerically, guess the value of

9 $20 \arctan \frac{1}{7} + 8 \arctan \frac{3}{79}$

10 $24 \arctan \frac{1}{18} + 16 \arctan \frac{1}{57} - 10 \arctan \frac{1}{239}.$

Find

11 $\displaystyle\lim_{x \to \infty} \frac{5x + 1}{2x + \ln x}$ **12** $\displaystyle\lim_{x \to \infty} \frac{\sinh x}{\sinh(x + 1)}.$

Solve for x to 3 significant figures

13 $\ln \ln x = 3$ **14** $\ln \ln x = 6.$

15   Find the point of intersection of the tangents to the graph $y = \ln x$ at $(e, 1)$ and $(1/e, -1)$.

16   Suppose $\theta = \arcsin x$ in *degrees*. Find $d\theta/dx$.

17   The "doomsday" equation of von Foerster *et al.*, predicts the population $N(t)$ in year t as

$$N(t) = \frac{1.79 \times 10^{11}}{(2026.87 - t)^{0.99}}   \quad (t < 2026.87).$$

(See Fig. 1b, p. 1.) Find the inverse function $t = t(N)$.

18   Prove that $e^{-x} = x^3$ has exactly one solution.

Prove

19 $\dfrac{99\pi}{400} < \displaystyle\int_{1}^{100} \dfrac{\arctan x}{x^2}\,dx < \dfrac{99\pi}{200}$ **20** $(1 - e^{-1})\ln 10 < \displaystyle\int_{1}^{10} \dfrac{1 - e^{-x}}{x}\,dx < \ln 10.$

21 Suppose $f(x) > 0$ and $y = f(x)$ is strictly concave for $a < x < b$. Show that $z = \ln f(x)$ is strictly concave.

22 Suppose $a > 0$. Find $\lim_{n \to \infty} n(\sqrt[n]{a} - 1)$.

23 Use the Trapezoidal Rule to prove

$$\tfrac{1}{2}[\ln 1 + 2\ln 2 + 2\ln 3 + \cdots + 2\ln(n - 1) + \ln n] < \int_{1}^{n} \ln x\,dx.$$

24 (cont.) Prove $n! < \sqrt{n}\,(n^n)e^{-n+1}$. [*Hint* $x\ln x - x$ is an antiderivative of $\ln x$.]

25 Prove $\tfrac{1}{2}\pi + \arctan x > 2\arctan 2x$.

26 Prove $\ln x > 2(x - 1)/(x + 1)$ for $x > 1$.

27 Some programming languages (Algol 60, Pascal, etc.) contain only sin, cos, and arc tan as standard trig functions, so it is necessary to express other trig functions in terms of these. Find a formula of the form $\arccos x = \arctan(?)$, and state where it is valid.

28 (cont.) Find a similar formula for arc sin x.

29 (cont.) Find a formula of the form $\arccos x = 2\arctan(?)$.

30* Express $y = \arctan\left(\dfrac{x - 1}{x + 1}\right) - \arctan x$ as simply as possible in terms of x.

Prove

31 $\cosh^4 x - \sinh^4 x = \cosh 2x$ **32** $2\cosh x \sinh x = \dfrac{2\tanh x}{1 - \tanh^2 x}$

33 $\tanh x = 1 - \dfrac{2}{\sinh 2x + \cosh 2x + 1}.$

34 Find a function of the form $y = a\sinh cx + b\sinh c(L - x)$ that solves the **boundary value** problem $y'' - k^2 y = 0$, $y(0) = y_0$, $y(L) = y_L$, where the constants satisfy $k > 0$ and $y_0 < y_L$.

35 A quantity of gas undergoes ionization at the constant rate ρ. Let $n = n(t)$ denote the number of positive ions at time t. At the same time, the ions tend to recombine with electrons at the rate αn^2, where α is the **coefficient of recombination**. Thus

$$\dfrac{dn}{dt} = \rho - \alpha n^2.$$

Solve for n in terms of t, given that initially $n(0) = 0$.

36 A model for the thermal breakdown of dialectrics leads to the initial value problem

$$\dfrac{d^2 u}{dx^2} + \beta e^u = 0, \qquad u(0) = u_0, \quad u'(0) = 0.$$

Find a solution of the form $u = a + b\ln\cosh cx$.

37* Graph $x^y = y^x$ and indicate by shading where $x^y > y^x$. [*Hint* Use the solution to Ex. 27 of Section 3.]

38* Let $y = f(x)$ be continuous and strictly increasing for $a \le x \le b$, where $0 \le a$. Let $A = f(a)$ and $B = f(b)$. Finally, let $x = g(y)$ be the inverse function. Prove geometrically that

$$2\int_{a}^{b} xf(x)\,dx + \int_{A}^{B} g(y)^2\,dy = Bb^2 - Aa^2.$$

39*   Find the most general linear substitution $\begin{cases} \bar{x} = a_{11}x + a_{12}y \\ \bar{y} = a_{21}x + a_{22}y \end{cases}$

such that $\bar{x}^2 - \bar{y}^2 = x^2 - y^2$ for all x and y. (If x is distance and $y = ct$, where t is time and c the speed of light in a vacuum, this is the **Lorentz transformation** of relativity.)

40*   Let $f(x) = x - \ln(1 + x)$. Show that $y = f(x)$ restricted to $-1 < x \leq 0$ has a strictly decreasing inverse function $x = g(y)$, and restricted to $0 \leq x < \infty$ has a strictly increasing inverse function $x = h(y)$. Both $g(y)$ and $h(y)$ have domain $0 \leq y < \infty$. Prove $g(y) + h(y) > 0$ for $y > 0$.

41   Let $0 < \alpha < 1$. Prove $(x + y)^\alpha < x^\alpha + y^\alpha$ for $0 < x, 0 < y$.

42   Let $0 < \alpha < 1$. Apply Ex. 23, p. 152, to $f(x) = -\ln x$ to prove Hölder's inequality

$$x^\alpha y^{1-\alpha} \leq \alpha x + (1 - \alpha)y \qquad \text{for} \quad x > 0, \quad y > 0.$$

Techniques of Integration 8

1. INDEFINITE INTEGRALS

We have developed rules for differentiation of most functions that arise in practice. The reverse process, antidifferentiation, is much harder. There is no systematic complete procedure for antidifferentiation, rather, a few important techniques and a large bag of miscellaneous tricks. What is worse, there are functions that are not derivatives of any common function, for example

$$e^{-x^2}, \quad \frac{\sin x}{x}, \quad \frac{1}{\sqrt{(1 - x^2)(1 - k^2 x^2)}}.$$

In this chapter, we discuss some basic techniques of antidifferentiation, a few of the more common tricks, and the use of integral tables.

First, some notation. The symbol

$$\int f(x)\,dx,$$

called the **indefinite integral** of $f(x)$, denotes the most general antiderivative of $f(x)$. For example,

$$\int x^2\,dx = \frac{x^3}{3} + C, \qquad \int \cos x\,dx = \sin x + C.$$

To each differentiation formula, there corresponds an indefinite integral formula. For instance, to

$$\frac{d}{dx}(\tan x) = \sec^2 x \qquad \text{corresponds} \qquad \int \sec^2 x\,dx = \tan x + C.$$

A Short Table of Integrals Let us list some of the indefinite integrals we know. All of these come from differentiation formulas derived so far.

$$\int x^n\,dx = \frac{x^{n+1}}{n+1} + C \quad (n \neq -1)$$

$$\int e^x\,dx = e^x + C$$

$$\int \frac{dx}{x} = \ln |x| + C$$

$$\int \sin x \, dx = -\cos x + C \qquad \int \sec x \tan x \, dx = \sec x + C$$

$$\int \cos x \, dx = \sin x + C \qquad \int \csc x \cot x \, dx = -\csc x + C$$

$$\int \sec^2 x \, dx = \tan x + C \qquad \int \csc^2 x \, dx = -\cot x + C$$

$$\int \frac{dx}{\sqrt{1 - x^2}} = \text{arc sin } x + C \quad (|x| < 1) \qquad \int \frac{dx}{1 + x^2} = \text{arc tan } x + C$$

$$\int \frac{dx}{x\sqrt{x^2 - 1}} = \text{arc sec } x + C \quad (x > 1)$$

$$\int \frac{dx}{\sqrt{x^2 + 1}} = \sinh^{-1} x = \ln(x + \sqrt{x^2 + 1}) + C$$

$$\int \frac{dx}{\sqrt{x^2 - 1}} = \cosh^{-1} x = \ln(x + \sqrt{x^2 - 1}) + C \quad (x > 1).$$

Our aim is to develop techniques for extending this table to cover a wide class of integrands.

Sums and Constant Factors   We can extend our formulas to sums and constant factors by two basic principles:

$$\int [f(x) + g(x)] \, dx = \int f(x) \, dx + \int g(x) \, dx, \qquad \int cf(x) \, dx = c \int f(x) \, dx.$$

The first formula splits an integration problem into two parts. The second allows constant factors to slide across the integral sign. Both are derived from simple properties of differentiation. If $F'(x) = f(x)$ and $G'(x) = g(x)$, then the two formulas are restatements of the differentiation formulas

$$[F(x) + G(x)]' = F'(x) + G'(x) = f(x) + g(x), \qquad [cF(x)]' = cF'(x) = cf(x).$$

Warning   Although constants can slide across the integral sign, variables cannot:

$$x \int x^2 \, dx \neq \int x^3 \, dx.$$

■ **EXAMPLE 1**   Find $\int (5e^x - 2x^3) \, dx$.

Solution $\int (5e^x - 2x^3) \, dx = 5 \int e^x \, dx - 2 \int x^3 \, dx.$

Both integrals on the right are known:

$$\int e^x \, dx = e^x + C, \qquad \int x^3 \, dx = \tfrac{1}{4}x^4 + C'.$$

Therefore
$$\int (5e^x - 2x^3) \, dx = 5 \int e^x \, dx - 2 \int x^3 \, dx = 5(e^x + C) - 2(\tfrac{1}{4}x^4 + C')$$

$$= 5e^x - \tfrac{1}{2}x^4 + (5C - 2C'). \qquad \blacksquare$$

Warning   Since both C and C' are arbitrary constants, so is $5C - 2C'$. Therefore the preceding answer can just as well be written $5e^x - \tfrac{1}{2}x^4 + C$. Henceforth, we shall lump together all arbitrary constants into one. So don't be surprised if $2C$, or $C + 1$, or $C_1 + 3C_2 + \ln 5$ end up as just plain C.

Reminder   In this subject you can always check your work. If you have a suspected indefinite integral of $f(x)$, take its derivative and see if that really equals $f(x)$.

2. SUBSTITUTIONS AND DIFFERENTIALS

Let us see how we might adapt known integration formulas to new situations. The first observation is that we must be careful. For example, the formula

$$\int x^3 \, dx = \tfrac{1}{4}x^4 + C$$

does not imply that

$$\int \sin^3 x \, dx = \tfrac{1}{4} \sin^4 x + C.$$

To check, we differentiate using the Chain Rule:

$$\frac{d}{dx} \, (\tfrac{1}{4} \sin^4 x) = \sin^3 x \cos x, \qquad \text{not} \quad \sin^3 x.$$

A *correct* formula is

$$\int \sin^3 x \cos x \, dx = \tfrac{1}{4} \sin^4 x + C.$$

Similarly, the formula

$$\int e^x \, dx = e^x + C \qquad \text{does not imply} \qquad \int e^{x^2} \, dx = e^{x^2} + C$$

because
$$\frac{d}{dx} e^{x^2} = e^{x^2} \cdot 2x, \qquad \text{not} \quad e^{x^2}.$$

A *correct* formula is

$$\int e^{x^2} \cdot 2x \, dx = e^{x^2} + C.$$

Obviously, this business involves careful use of the Chain Rule. Let us set things straight. (Actually we did so in Chapter 5, Section 6, but it is easy enough to do again from scratch.)

Suppose $F(x)$ is an antiderivative of $f(x)$ and $g(x)$ is a differentiable function. Then by the Chain Rule,

$$\frac{d}{dx} F[u(x)] = F'[u(x)]u'(x) = f[u(x)] \frac{du}{dx}.$$

We express this relation as an integration formula:

Suppose $F(x)$ is an antiderivative of $f(x)$ and $u(x)$ is differentiable. Then

$$\int f[u(x)] \frac{du}{dx} dx = F[u(x)] + C.$$

Use of this formula requires spotting an integrand in the form $f[u(x)]u'(x)$, a skill that comes with practice.

Examples

(1) $\displaystyle \int \sin^3 x \cos x \, dx = \int (\sin x)^3 \left(\frac{d}{dx} \sin x \right) dx = \frac{1}{4} \sin^4 x + C;$

$$f(x) = x^3, \qquad F(x) = \tfrac{1}{4}x^4, \qquad u(x) = \sin x.$$

(2) $\displaystyle \int e^{x^2} \cdot 2x \, dx = \int e^{x^2} \left(\frac{d}{dx} x^2 \right) dx = e^{x^2} + C;$

$$f(x) = e^x, \qquad F(x) = e^x, \qquad u(x) = x^2.$$

(3) $\displaystyle \int \cos(\sqrt{x}) \cdot \frac{1}{2\sqrt{x}} \, dx = \int \cos(\sqrt{x}) \left(\frac{d}{dx} \sqrt{x} \right) dx = \sin(\sqrt{x}) + C;$

$$f(x) = \cos x, \qquad F(x) = \sin x, \qquad u(x) = \sqrt{x}.$$

Differentials   The indefinite integral symbol $\int f(x) \, dx$ denotes the general antiderivative of $f(x)$. The "differential" dx is a formal notation that is not absolutely necessary, but helps simplify integration formulas.

For each function $u = u(x)$, we introduce the formal **differential**

$$du = u'(x) \, dx.$$

With this notation we replace

$$\int f[u(x)] \frac{du}{dx} dx \qquad \text{by} \qquad \int f(u) \, du.$$

Not only is the notation simpler, but also it reminds us of the important factor du/dx.

Properties of differential which will be used repeatedly:

$$d(u + v) = du + dv; \qquad d(cu) = c \, du, \qquad c \text{ constant}; \qquad df(u) = f'(u) \, du.$$

The last property follows from the Chain Rule:

$$df(u) = \frac{d}{dx}[f(u)]\, dx = f'(u)\frac{du}{dx}\, dx = f'(u)\, du.$$

Using the notation of differentials, we can abbreviate the basic integration formula of this section:

Suppose $F(x)$ is an antiderivative of $f(x)$ and $u(x)$ is differentiable. Then

$$\int f[u(x)]\frac{du}{dx}\, dx = \int f(u)\, du = F(u) + C.$$

■ **EXAMPLE 1**   Find $\int (e^x + x)^2(e^x + 1)\, dx.$

Solution   The integrand is of the form $u^2(du/dx)$, where $u = e^x + x$. So let $u = e^x + x$. Then

$$du = \frac{du}{dx}\, dx = (e^x + 1)\, dx.$$

Hence $\int (e^x + x)^2(e^x + 1)\, dx = \int u^2\, du = \tfrac{1}{3}u^3 + C = \tfrac{1}{3}(e^x + x)^3 + C.$ ■

Remark   The technique used in Example 1 is called the method of **substitution**, or **change of variable**. We simplified the integral by substituting $u = e^x + x$.

■ **EXAMPLE 2**   Find   (a) $\displaystyle\int \frac{\text{arc tan } x}{1 + x^2}\, dx.$ (b) $\displaystyle\int \frac{\ln x}{x}\, dx.$

Solution   (a)   Notice that $\dfrac{d}{dx}(\text{arc tan } x) = \dfrac{1}{1 + x^2}.$

Therefore set $u = \text{arc tan } x$. Then $du = \dfrac{du}{dx}\, dx = \dfrac{1}{1 + x^2}\, dx.$

Hence $\displaystyle\int \frac{\text{arc tan } x}{1 + x^2}\, dx = \int u\, du = \frac{1}{2}u^2 + C = \frac{1}{2}(\text{arc tan } x)^2 + C.$

(b)   Set $u = \ln x$. Then $du = dx/x$, so

$$\int \frac{\ln x}{x}\, dx = \int u\, du = \frac{1}{2}u^2 + C = \frac{1}{2}(\ln x)^2 + C.$$ ■

■ **EXAMPLE 3**   Find   (a) $\displaystyle\int \frac{4x - 5}{x^2 + 1}\, dx$ (b) $\displaystyle\int \frac{dx}{\sqrt{3x + 5}}.$

Solution   (a)   Split the integral into two:

$$\int \frac{4x - 5}{x^2 - 1}\, dx = 4\int \frac{x\, dx}{x^2 + 1} - 5\int \frac{dx}{x^2 + 1}.$$

In the first integral on the right, the numerator is nearly the differential of the denominator. Set $u = x^2 + 1$, so $du = 2x \, dx$, and throw in the needed factor 2:

$$4 \int \frac{x \, dx}{x^2 + 1} = 2 \int \frac{2x \, dx}{x^2 + 1} = 2 \int \frac{du}{u}$$

$$= 2 \ln|u| + C_1 = 2 \ln(x^2 + 1) + C_1 = \ln(x^2 + 1)^2 + C_1.$$

The second integral is arc tan $x + C_2$. Therefore

$$\int \frac{4x - 5}{x^2 + 1} \, dx = \ln(x^2 + 1)^2 - 5 \text{ arc tan } x + C.$$

(b)   Set $u = 3x + 5$. Then $du = 3 \, dx$ and

$$\int \frac{dx}{\sqrt{3x + 5}} = \int \frac{\frac{1}{3} du}{\sqrt{u}} = \frac{1}{3} \int \frac{du}{\sqrt{u}}$$

$$= \frac{1}{3} \int u^{-1/2} \, du = \frac{1}{3}(2u^{1/2}) + C = \frac{2}{3}\sqrt{3x + 5} + C.$$

Alternative solution   Set $u^2 = 3x + 5$. Then $2u \, du = 3 \, dx$ and

$$\int \frac{dx}{\sqrt{3x + 5}} = \frac{2}{3} \int \frac{u \, du}{u} = \frac{2}{3} u + C = \frac{2}{3}\sqrt{3x + 5} + C. \qquad \blacksquare$$

Definite Integrals   The method of substitution, or change of variables, applies to definite integrals as well as indefinite integrals. There is one crucial twist: the limits of integration must be suitably changed when a substitution is made.

Change of Variable Formula

$$\int_a^b f[u(x)] \frac{du}{dx} \, dx = \int_c^d f(u) \, du, \qquad \text{where} \quad c = u(a) \quad \text{and} \quad d = u(b).$$

As shown above, if $F(x)$ is an antiderivative of $f(x)$, then $F[u(x)]$ is an antiderivative of $f[u(x)]u'(x)$. Therefore, by two applications of the Evaluation Rule,

$$\int_a^b f[u(x)] \frac{du}{dx} \, dx = F[u(x)] \Big|_a^b = F[u(b)] - F[u(a)] = F(d) - F(c) = \int_c^d f(u) \, du.$$

Thus once the integral is changed into an integral in u, the computation can be done entirely in terms of u, *provided* the limits of the integral are changed correctly.

■ **EXAMPLE 4**   Compute $\displaystyle \int_0^4 \sqrt{x^2 + 9} \cdot 2x \, dx.$

Solution   (Old way)   First evaluate the indefinite integral

$$\int \sqrt{x^2 + 9} \cdot 2x \, dx.$$

Make the substitution $u = x^2 + 9$, $du = 2x\,dx$:

$$\int \sqrt{x^2 + 9} \cdot 2x\,dx = \int u^{1/2}\,du = \tfrac{2}{3}u^{3/2} + C.$$

Now change back to x:

$$\int \sqrt{x^2 + 9} \cdot 2x\,dx = \tfrac{2}{3}(x^2 + 9)^{3/2} + C.$$

Therefore $\displaystyle\int_0^4 \sqrt{x^2 + 9} \cdot 2x\,dx = \tfrac{2}{3}(x^2 + 9)^{3/2}\Big|_0^4 = \tfrac{2}{3}(5^3 - 3^3) = \tfrac{196}{3}.$

Solution (New way) Again substitute $u = x^2 + 9$. Note that

$$u = 25 \quad \text{for} \quad x = 4 \qquad \text{and} \qquad u = 9 \quad \text{for} \quad x = 0.$$

Therefore $\displaystyle\int_0^4 \sqrt{x^2 + 9} \cdot 2x\,dx = \int_9^{25} u^{1/2}\,du = \tfrac{2}{3}u^{3/2}\Big|_9^{25} = \tfrac{2}{3}(5^3 - 3^3) = \tfrac{196}{3}.$

Alternative solution (avoiding fractional exponents) Set $u^2 = x^2 + 9$. Then $2u\,du = 2x\,dx$. Now

$$u = 5 \quad \text{for} \quad x = 4 \qquad \text{and} \qquad u = 3 \quad \text{for} \quad x = 0.$$

Therefore $\displaystyle\int_0^4 \sqrt{x^2 + 9} \cdot 2x\,dx = \int_3^5 u \cdot 2u\,du = \tfrac{2}{3}u^3\Big|_3^5 = \tfrac{2}{3}(5^3 - 3^3) = \tfrac{196}{3}.$ ∎

■ **EXAMPLE 5** Evaluate $\displaystyle\int_{1/2}^{1/\sqrt{2}} \frac{\arcsin x}{\sqrt{1 - x^2}}\,dx.$

Solution Substitute $u = \arcsin x$, $du = \dfrac{dx}{\sqrt{1 - x^2}}$,

and note that $u(\tfrac{1}{2}) = \tfrac{1}{6}\pi$, $u(1/\sqrt{2}) = \tfrac{1}{4}\pi$.

Therefore $\displaystyle\int_{1/2}^{1/\sqrt{2}} \frac{\arcsin x}{\sqrt{1 - x^2}}\,dx = \int_{\pi/6}^{\pi/4} u\,du = \frac{u^2}{2}\Big|_{\pi/6}^{\pi/4} = \frac{1}{2}\left(\frac{\pi^2}{16} - \frac{\pi^2}{36}\right) = \frac{5\pi^2}{288}.$ ∎

EXERCISES

Find the indefinite integral. Make free use of the formulas on pp. 357–358.

1 $\displaystyle\int \sin x \cos x\,dx$

2 $\displaystyle\int \sin^4 x \cos x\,dx$

3 $\displaystyle\int 5e^{5x}\,dx$

4 $\displaystyle\int (e^x + 3x)(e^x + 3)\,dx$

5 $\displaystyle\int \frac{2x}{(1 + x^2)^3}\,dx$

6 $\displaystyle\int \frac{(\ln x)^2}{x}\,dx$

7 $\displaystyle\int \frac{-\sin x}{\cos^2 x}\,dx$

8 $\displaystyle\int (1 + \sin x)^3 \cos x\,dx$

9 $\displaystyle\int \frac{e^x + 2x}{e^x + x^2 + 1}\,dx$

10 $\displaystyle\int \frac{3x^2}{4 + x^3}\,dx$

11 $\displaystyle\int \tan^3 x \sec^2 x\,dx$

12 $\displaystyle\int \frac{-2x}{\sqrt{1 - x^2}}\,dx$

13 $\displaystyle\int \frac{e^{\sqrt{x}}}{2\sqrt{x}}\,dx$

14 $\displaystyle\int 8x(1 + 4x^2)^5\,dx$

15 $\displaystyle\int (3x + 1)^4\,dx$

16 $\int (x^2 + 1)^2 \, dx$ **17** $\int \cos 3x \, dx$ **18** $\int x^2 e^{-x^3} \, dx$

19 $\int \sec^4 x \tan x \, dx$ **20** $\int \cos^2 x \sin x \, dx$ **21** $\int \dfrac{dx}{\sqrt{1 + 5x}}$

22 $\int \dfrac{e^x \, dx}{1 + e^{2x}}$ **23** $\int \dfrac{\ln(2x + 7)}{2x + 7} \, dx$ **24** $\int \dfrac{e^x - e^{-x}}{e^x + e^{-x}} \, dx$

25 $\int \dfrac{dx}{(5 - 3x)^2}$ **26** $\int \dfrac{3x - 4}{1 + x^2} \, dx$ **27** $\int \left(x + \dfrac{1}{x} \right)^2 dx$

28 $\int \tan^2 x \, dx$ **29** $\int (ax + b)^n \, dx, \quad n \neq -1$ **30** $\int x\sqrt{cx^2 + d} \, dx$

31 $\int \dfrac{x^3}{1 + x^4} \, dx$ **32** $\int \dfrac{2x}{1 + x^4} \, dx$ **33** $\int \dfrac{dx}{x \ln x}$

34 $\int \tan^2 \frac{1}{2}x \sec^2 \frac{1}{2}x \, dx.$

Compute the definite integral by making an appropriate substitution and changing the limits of integration.

35 $\int_0^1 2x(x^2 - 2)^3 \, dx$ **36** $\int_0^1 \dfrac{e^x \, dx}{(e^x + 1)^2}$ **37** $\int_0^{\sqrt{\pi}} x \sin x^2 \, dx$

38 $\int_1^3 x e^{x^2} \, dx$ **39** $\int_0^4 \dfrac{x \, dx}{\sqrt{x^2 + 9}}$ **40** $\int_0^1 \dfrac{\arctan x}{1 + x^2} \, dx$

41 $\int_{\pi/6}^{\pi/4} \dfrac{\cos x}{\sin^2 x} \, dx$ **42** $\int_0^{\pi/4} (1 + \tan x)^3 \sec^2 x \, dx.$

3. OTHER SUBSTITUTIONS

Frequently an integral can be simplified by an appropriate substitution.

■ **EXAMPLE 1** Find $\int x\sqrt{x + 1} \, dx.$

Solution Set $u^2 = x + 1$. Then $\sqrt{x + 1} = u$ and $x = u^2 - 1$, so $dx = 2u \, du$. Therefore

$$\int x\sqrt{x + 1} \, dx = \int (u^2 - 1) \cdot u \cdot 2u \, du = 2 \int (u^4 - u^2) \, du$$

$$= 2 \left(\frac{u^5}{5} - \frac{u^3}{3} \right) + C = \frac{2u^3}{15} (3u^2 - 5) + C,$$

where $u = \sqrt{x + 1}$. Hence

$$\int x\sqrt{x + 1} \, dx = \frac{2(x + 1)^{3/2}}{15} [3(x + 1) - 5] + C = \frac{2}{15} (x + 1)^{3/2} (3x - 2) + C.$$

Alternative solution   Set $u = x + 1$. Then

$$\int x\sqrt{x+1}\, dx = \int (u-1)\sqrt{u}\, du = \int (u^{3/2} - u^{1/2})\, du$$

$$= 2\left(\frac{u^{5/2}}{5} - \frac{u^{3/2}}{3}\right) + C = \frac{2(x+1)^{3/2}}{15}(3x-2) + C$$

as before.

∎

■ **EXAMPLE 2**   Find $\displaystyle\int \frac{x\, dx}{(x-a)^3}.$

Solution   Set $u = x - a$. Then $x = u + a$ and $dx = du$:

$$\int \frac{x\, dx}{(x-a)^3} = \int \frac{(u+a)\, du}{u^3} = \int \left(\frac{1}{u^2} + \frac{a}{u^3}\right) du = -\frac{1}{u} - \frac{a}{2u^2} + C$$

$$= -\frac{1}{x-a} - \frac{a}{2(x-a)^2} + C = \frac{-2x+a}{2(x-a)^2} + C.$$

∎

■ **EXAMPLE 3**   Find $\displaystyle\int \frac{dx}{1 + \sqrt[3]{x}}.$

Solution   Substitute $x = u^3$ and $dx = 3u^2\, du$:

$$\int \frac{dx}{1 + \sqrt[3]{x}} = 3\int \frac{u^2\, du}{1 + u}.$$

By long division, $\displaystyle\frac{u^2}{1+u} = u - 1 + \frac{1}{1+u}.$

Hence

$$\int \frac{dx}{1 + \sqrt[3]{x}} = 3\int \left(u - 1 + \frac{1}{1+u}\right) du = 3\left(\frac{u^2}{2} - u + \ln|1+u|\right) + C$$

$$= 3\left(\frac{1}{2}x^{2/3} - x^{1/3} + \ln|1 + x^{1/3}|\right) + C = \frac{3}{2}x^{2/3} - 3x^{1/3} + \ln|1 + x^{1/3}|^3 + C.$$

∎

Often a suitable substitution can change an integral into one that is already known.

■ **EXAMPLE 4**   Find   (a) $\displaystyle\int \frac{dx}{a^2 + x^2}$ (b) $\displaystyle\int \frac{dx}{\sqrt{a^2 - x^2}}$ $a > 0.$

Solution   (a) We already have the formula $\displaystyle\int \frac{dx}{1 + x^2} = \text{arc tan } x + C.$

The given integral is so like this one, we try to change it into this form. We set $x = ay$; then $dx = a\, dy$:

$$\int \frac{dx}{a^2 + x^2} = \int \frac{a\, dy}{a^2 + (ay)^2} = \frac{1}{a}\int \frac{dy}{1 + y^2} = \frac{1}{a}\text{arc tan } y + C = \frac{1}{a}\text{arc tan }\frac{x}{a} + C.$$

(b)   This integral recalls the formula

$$\int \frac{dx}{\sqrt{1-x^2}} = \arcsin x + C.$$

Again we set $x = ay$: $\int \frac{dx}{\sqrt{a^2-x^2}} = \int \frac{a\,dy}{\sqrt{a^2-(ay)^2}} = \int \frac{dy}{\sqrt{1-y^2}}$

$$= \arcsin y + C = \arcsin \frac{x}{a} + C. \qquad \blacksquare$$

■ **EXAMPLE 5**   Find   (a) $\displaystyle\int \frac{dx}{\sqrt{x^2+a^2}}$ (b) $\displaystyle\int \frac{dx}{\sqrt{x^2-a^2}}$.

Solution   (a)   We already have the formula

$$\int \frac{dx}{\sqrt{x^2+1}} = \ln(x + \sqrt{x^2+1}) + C.$$

Set $x = ay$:

$$\int \frac{dx}{\sqrt{x^2+a^2}} = \int \frac{a\,dy}{\sqrt{(ay)^2+a^2}} = \int \frac{dy}{\sqrt{y^2+1}} = \ln(y + \sqrt{y^2+1}) + C$$

$$= \ln\left(\frac{x}{a} + \sqrt{\left(\frac{x}{a}\right)^2+1}\right) + C = \ln\left(\frac{1}{a}(x + \sqrt{x^2+a^2})\right) + C$$

$$= \ln(x + \sqrt{x^2+a^2}) - \ln a + C = \ln(x + \sqrt{x^2+a^2}) + C.$$

(b)   The solution is practically identical. We obtain

$$\int \frac{dx}{\sqrt{x^2-a^2}} = \ln(x + \sqrt{x^2-a^2}) + C, \qquad x > a > 0.$$

For $x < -a < 0$, a similar result holds, and generally

$$\int \frac{dx}{\sqrt{x^2-a^2}} = \ln|x + \sqrt{x^2-a^2}| + C, \qquad |x| > |a|.$$

This can be checked by differentiation.   ■

The integration formulas found in Examples 4 and 5 will be useful in the next section.

EXERCISES

Find the indefinite integral

1   $\displaystyle\int x\sqrt{x+3}\,dx$

2   $\displaystyle\int (\sin x)e^{\cos x}\,dx$

3   $\displaystyle\int \frac{x}{\sqrt{2x+5}}\,dx$

4   $\displaystyle\int \frac{dx}{1+b^2x^2}$

5   $\displaystyle\int \frac{x^2\,dx}{(x-1)^3}$

6   $\displaystyle\int \frac{dx}{\sqrt{1-4x^2}}$

7   $\displaystyle\int \frac{dx}{1+\sqrt{x}}$

8   $\displaystyle\int e^{2x}\sqrt{1+e^x}\,dx$

9   $\displaystyle\int \frac{dx}{x+\sqrt[4]{x}}$

10 $\int \dfrac{(x-1)\,dx}{\sqrt{1-x^2}}$ **11** $\int \dfrac{e^{2x}}{1+e^{4x}}\,dx$ **12** $\int x^3\sqrt{x^2+1}\,dx$

13 $\int \dfrac{dx}{1+(5x+2)^2}$ **14** $\int \dfrac{x+1}{a^2+b^2x^2}\,dx$ **15** $\int \dfrac{x^3\,dx}{\sqrt{1-x^2}}$

16 $\int \dfrac{x}{1+\sqrt{x}}\,dx$ **17** $\int \dfrac{\sin 2x\,dx}{3+\cos 2x}$ **18** $\int \dfrac{\sqrt{x+1}}{x+3}\,dx$

19 $\int x\sqrt[3]{2x+1}\,dx$ **20** $\int \dfrac{x^3-5}{(x+2)^2}\,dx$ **21** $\int \dfrac{x^3}{x^2+1}\,dx$

22 $\int x^2\sqrt{x+3}\,dx$ **23** $\int \dfrac{dx}{\sqrt{9x^2+1}}$ **24** $\int (x^2+x+1)\sqrt{x+1}\,dx.$

Compute the definite integral by making an appropriate substitution and changing the limits of integration

25 $\int_0^2 x^3\sqrt{x^4+9}\,dx$ **26** $\int_1^2 (x-1)^3(x+2)\,dx$ **27** $\int_4^5 \dfrac{x}{(x-2)^3}\,dx$

28 $\int_0^5 x\sqrt{x+4}\,dx$ **29** $\int_0^1 x(2x-1)^5\,dx$ **30** $\int_{-\ln 2}^{-(\ln 2)/2} \dfrac{e^x\,dx}{\sqrt{1-e^{2x}}}$

31 $\int_0^3 \dfrac{dx}{1+\sqrt{1+x}}$ **32** $\int_0^{12} \dfrac{x^2}{\sqrt{2x+1}}\,dx$ **33** $\int_0^1 \dfrac{x^3\,dx}{\sqrt{4-x^2}}$

34 $\int_0^2 \dfrac{dx}{(x+2)\sqrt{x+1}}$ **35** $\int_0^4 (x-1)(x-2)(x-3)\,dx$

36 $\int_0^1 [x^n-(1-x)^n]\,dx.$

4. USE OF IDENTITIES

When working integration problems, keep in mind the possibility of simplifying the integrand by algebraic manipulation. Such tactics as long division, factoring, combining fractions, and using trigonometric identities may convert a function into an equivalent form easier to integrate.

■ **EXAMPLE 1** Find $\int \dfrac{x^4}{1+x^2}\,dx.$

Solution By long division, $\dfrac{x^4}{1+x^2}=x^2-1+\dfrac{1}{1+x^2}.$

Hence, $\int \dfrac{x^4}{1+x^2}\,dx=\int\left(x^2-1+\dfrac{1}{1+x^2}\right)dx=\dfrac13 x^3-x+\arctan x+C.$ ■

■ **EXAMPLE 2** Find $\int \sqrt{\dfrac{1+x}{1-x}}\,dx.$

Solution   Simplify the integrand by writing

$$\sqrt{\frac{1+x}{1-x}} = \sqrt{\frac{1+x}{1-x}} \cdot \frac{\sqrt{1+x}}{\sqrt{1+x}} = \frac{1+x}{\sqrt{1-x^2}}$$

Then $\displaystyle\int \sqrt{\frac{1+x}{1-x}}\, dx = \int \frac{dx}{\sqrt{1-x^2}} + \int \frac{x\,dx}{\sqrt{1-x^2}} = \text{arc sin } x - \sqrt{1-x^2} + C.$ ∎

Integrals of sec x and csc x   The integral of sec x is done by a trick:

$$\sec x = \frac{\sec x(\sec x + \tan x)}{(\sec x + \tan x)} = \frac{\sec x \tan x + \sec^2 x}{\sec x + \tan x}$$

$$= \frac{1}{\sec x + \tan x} \frac{d}{dx}(\sec x + \tan x).$$

Hence $\displaystyle\int \sec x\, dx = \int \frac{d(\sec x + \tan x)}{\sec x + \tan x} = \ln|\sec x + \tan x| + C.$

In a similar manner, we may derive the formula

$$\int \csc x\, dx = -\ln|\csc x + \cot x| + C.$$

Trigonometric Identities

■ **EXAMPLE 3**   Find   (a) $\displaystyle\int \cos^3 x\, dx$ (b) $\displaystyle\int \cos^3 x \sin^2 x\, dx.$

Solution   (a)   Convert the integrand into powers of sin x, reserving a factor of cos x for the differential:

$$\cos^3 x = \cos^2 x \cos x = (1 - \sin^2 x)\cos x.$$

Hence $\displaystyle\int \cos^3 x\, dx = \int \cos x\, dx - \int \sin^2 x \cos x\, dx = \sin x - \frac{\sin^3 x}{3} + C.$

(b)   Same technique:

$$\cos^3 x \sin^2 x = \cos^2 x \sin^2 x \cos x = (1 - \sin^2 x)\sin^2 x \cos x,$$

$$\int \cos^3 x \sin^2 x\, dx = \int \sin^2 x \cos x\, dx - \int \sin^4 x \cos x\, dx = \frac{\sin^3 x}{3} - \frac{\sin^5 x}{5} + C.$$ ∎

■ **EXAMPLE 4**   Find   (a) $\displaystyle\int \sin^2 x\, dx$ (b) $\displaystyle\int \sin^4 x\, dx.$

Solution   (a)   Use the identity $\sin^2 x = \frac{1}{2}(1 - \cos 2x)$:

$$\int \sin^2 x\, dx = \frac{1}{2}\int dx - \frac{1}{2}\int \cos 2x\, dx = \frac{x}{2} - \frac{\sin 2x}{4} + C.$$

(b) $\displaystyle\int \sin^4 x \, dx = \int (\sin^2 x)^2 \, dx = \int \left[\frac{1}{2}(1 - \cos 2x)\right]^2 dx$

$\displaystyle = \frac{1}{4}\int dx - \frac{1}{2}\int \cos 2x \, dx + \frac{1}{4}\int \cos^2 2x \, dx = \frac{x}{4} - \frac{\sin 2x}{4} + \frac{1}{4}\int \cos^2 2x \, dx.$

Now use the identity $\cos^2 2x = \frac{1}{2}(1 + \cos 4x)$:

$$\int \cos^2 2x \, dx = \frac{1}{2}\int dx + \frac{1}{2}\int \cos 4x \, dx = \frac{x}{2} + \frac{\sin 4x}{8} + C.$$

Combine results:

$$\int \sin^4 x \, dx = \frac{x}{4} - \frac{\sin 2x}{4} + \frac{1}{4}\left(\frac{x}{2} + \frac{\sin 4x}{8} + C\right) = \frac{3x}{8} - \frac{\sin 2x}{4} + \frac{\sin 4x}{32} + C. \quad \blacksquare$$

■ **EXAMPLE 5** Find (a) $\displaystyle\int \tan x \, dx$ (b) $\displaystyle\int \tan^2 x \, dx$

(c) $\displaystyle\int \tan^3 x \, dx.$

Solution (a) By definition of $\tan x$,

$$\int \tan x \, dx = \int \frac{\sin x}{\cos x} \, dx = \int \frac{-du}{u},$$

where $u = \cos x$. Hence $\displaystyle\int \tan x \, dx = -\ln |\cos x| + C = \ln |\sec x| + C.$

(b) Use the identity $\tan^2 x = \sec^2 x - 1$:

$$\int \tan^2 x \, dx = \int (\sec^2 x - 1) \, dx = \int \sec^2 x \, dx - \int dx = \tan x - x + C.$$

(c) $\displaystyle\int \tan^3 x \, dx = \int \tan x(\sec^2 x - 1) \, dx = \int \tan x \sec^2 x \, dx - \int \tan x \, dx.$

The first integral is of the form $\int u \, du$, where $u = \tan x$; the second integral was done in (a). Hence

$$\int \tan^3 x \, dx = \frac{1}{2} \tan^2 x + \ln |\cos x| + C. \quad \blacksquare$$

Completing the Square In Examples 4 and 5 of the previous section, we derived the following formulas:

$$\int \frac{dx}{x^2 + a^2} = \frac{1}{a} \arctan \frac{x}{a} + C, \qquad a > 0,$$

$$\int \frac{dx}{\sqrt{a^2 - x^2}} = \arcsin \frac{x}{a} + C, \qquad |x| < a, \quad a > 0,$$

$$\int \frac{dx}{\sqrt{x^2 + a^2}} = \ln(x + \sqrt{x^2 + a^2}) + C, \qquad a > 0,$$

$$\int \frac{dx}{\sqrt{x^2 - a^2}} = \ln |x + \sqrt{x^2 - a^2}| + C, \qquad |x| > a > 0.$$

One other integral of this type will be obtained in the next section:

$$\int \frac{dx}{a^2 - x^2} = \frac{1}{2a} \ln \left| \frac{a + x}{a - x} \right| + C.$$

These formulas are useful when the integrand involves a quadratic polynomial or the square root of a quadratic polynomial. The basic trick is completing the square.

■ **EXAMPLE 6** Find $\displaystyle \int \frac{dx}{x^2 - 10x + 29}$.

Solution Complete the square:

$$x^2 - 10x + 29 = x^2 - 10x + 25 + 4 = (x - 5)^2 + 2^2 = u^2 + a^2,$$

where $u = x - 5$ and $a = 2$. Therefore

$$\int \frac{dx}{x^2 - 10x + 29} = \int \frac{dx}{2^2 + (x - 5)^2} = \int \frac{du}{a^2 + u^2}$$

$$= \frac{1}{a} \arctan \frac{u}{a} + C = \frac{1}{2} \arctan \frac{x - 5}{2} + C. \qquad ■$$

■ **EXAMPLE 7** Find $\displaystyle \int \frac{dx}{\sqrt{3 - x - x^2}}$.

Solution Complete the square:

$$3 - x - x^2 = 3 - (x^2 + x) = 3 - (x^2 + x + \tfrac{1}{4}) + \tfrac{1}{4} = \tfrac{13}{4} - (x + \tfrac{1}{2})^2 = a^2 - u^2,$$

where $u = x + \tfrac{1}{2}$ and $a = \tfrac{1}{2}\sqrt{13}$. Therefore

$$\int \frac{dx}{\sqrt{3 - x - x^2}} = \int \frac{dx}{\sqrt{\tfrac{13}{4} - (x + \tfrac{1}{2})^2}} = \int \frac{du}{\sqrt{a^2 - u^2}}$$

$$= \arcsin \frac{u}{a} + C = \arcsin \left(\frac{2x + 1}{\sqrt{13}} \right) + C. \qquad ■$$

■ **EXAMPLE 8** Find $\displaystyle \int \frac{dx}{\sqrt{5x^2 - 2x}}$.

Solution Complete the square:

$$5x^2 - 2x = 5(x^2 - \tfrac{2}{5}x + \tfrac{1}{25} - \tfrac{1}{25}) = 5[(x - \tfrac{1}{5})^2 - \tfrac{1}{25}] = 5(u^2 - a^2),$$

where $u = x - \tfrac{1}{5}$ and $a = \tfrac{1}{5}$. Therefore

$$\int \frac{dx}{\sqrt{5x^2 - 2x}} = \int \frac{dx}{\sqrt{5}\sqrt{(x - \tfrac{1}{5})^2 - \tfrac{1}{25}}}$$

$$= \frac{1}{\sqrt{5}} \int \frac{du}{\sqrt{u^2 - a^2}} = \frac{1}{\sqrt{5}} \ln|u + \sqrt{u^2 - a^2}| + C$$

$$= \frac{1}{\sqrt{5}} (\ln|5x - 1 + \sqrt{25x^2 - 10x}| - \ln 5) + C.$$

Hence $\displaystyle\int \frac{dx}{\sqrt{5x^2 - 2x}} = \frac{1}{\sqrt{5}} \ln|5x - 1 + \sqrt{25x^2 - 10x}| + C.$ ∎

EXERCISES

Compute

1 $\displaystyle\int \frac{1 + x^4}{9 + x^2} dx$ 2 $\displaystyle\int \frac{x^2 \, dx}{x^2 + 3}$ 3 $\displaystyle\int \frac{2x + 1}{x - 4} dx$ 4 $\displaystyle\int \frac{x^2}{x - 1} dx$

5 $\displaystyle\int \frac{(x + 1)^3}{x^2} dx$ 6 $\displaystyle\int \frac{x^8 + 1}{x^2 + 1} dx$ 7 $\displaystyle\int \sqrt{\frac{1 + ax}{1 - ax}} dx$ 8 $\displaystyle\int \frac{dx}{\sqrt{x + 5} - \sqrt{x}}$

9 $\displaystyle\int \sec x \csc x \, dx$ 10 $\displaystyle\int \cos x \csc x \, dx$ 11 $\displaystyle\int \cos^3 x \sin^4 x \, dx$

12 $\displaystyle\int (\cos x - \sin x)^2 \, dx$ 13 $\displaystyle\int \sin^3 x \cos^2 x \, dx$ 14 $\displaystyle\int \sin^3 ax \, dx$

15 $\displaystyle\int \cos^5 3x \, dx$ 16 $\displaystyle\int \tan^3 x \, dx$ 17 $\displaystyle\int \cos^4 x \, dx$

18 $\displaystyle\int \sin^2 \frac{x}{3} \cos^2 \frac{x}{3} dx$ 19 $\displaystyle\int \tan^4 x \, dx$ 20 $\displaystyle\int \sec^4 x \, dx$

21 $\displaystyle\int x \tan(x^2) \, dx$ 22 $\displaystyle\int \tan^2 x \sec^4 x \, dx$ 23 $\displaystyle\int \frac{dx}{1 - \sin x}$

24 $\displaystyle\int \sin x \sqrt{\frac{\sec x + 1}{\sec x - 1}} \, dx.$

Evaluate the definite integrals

25 $\displaystyle\int_{3\pi/4}^{\pi} \tan x \, dx$ 26 $\displaystyle\int_{-2}^{-1} \frac{dx}{\sqrt{4x^2 - 1}}$ 27 $\displaystyle\int_{0}^{\pi/2} \sin 2x \, dx$ 28 $\displaystyle\int_{0}^{1} \cos^2 \pi x \, dx$

29 $\displaystyle\int_{0}^{2\pi} \cos 3x \cos 4x \, dx$ $[(\cos A)(\cos B) = \tfrac{1}{2}(\cos ? + \cos ?).]$ 30 $\displaystyle\int_{0}^{2\pi} \sin x \cos 3x \, dx.$

Compute

31 $\displaystyle\int \frac{dx}{x^2 + 2x + 5}$ 32 $\displaystyle\int \frac{dx}{2x^2 + x + 6}$

33 $\displaystyle\int \frac{dx}{\sqrt{6x - x^2}}$ 34 $\displaystyle\int \frac{3x + 10}{\sqrt{x^2 + 2x + 5}} dx$

35 $\displaystyle\int \frac{x \, dx}{\sqrt{4x - x^2}}$ 36 $\displaystyle\int \frac{x^2 \, dx}{x^2 - 4x + 9}$ (long division)

37 $\displaystyle\int \frac{x \, dx}{\sqrt{3x^4 - 4x^2 + 1}}$ 38 $\displaystyle\int \frac{2x \, dx}{1 - x^2 - x^4}$

39 $\displaystyle\int \frac{dx}{bx - ax^2}$ $a > 0, \quad b > 0$ 40 $\displaystyle\int \frac{dx}{a^2 x^2 + x}.$

5. PARTIAL FRACTIONS

Any fraction of the form $\dfrac{cx + d}{(x - a)(x - b)}$

can be split into the sum of two simpler fractions: $\dfrac{A}{x - a} + \dfrac{B}{x - b}$.

This decomposition into **partial fractions** simplifies integration since each term is easy to integrate.

■ **EXAMPLE 1**   Decompose $\dfrac{2x + 1}{(x - 3)(x - 4)}$ into partial fractions.

Solution   Write $\dfrac{2x + 1}{(x - 3)(x - 4)} = \dfrac{A}{x - 3} + \dfrac{B}{x - 4}$,

where A and B are constants to be determined. Multiply through by $(x - 3)(x - 4)$:

$$2x + 1 = A(x - 4) + B(x - 3) = (A + B)x - (4A + 3B).$$

The coefficients of x on both sides of this identity must be equal, and so must the constant terms. Hence

$$A + B = 2, \qquad -4A - 3B = 1.$$

The unknowns A and B must satisfy these two equations simultaneously.

Solve: $A = -7$, $B = 9$. Therefore,

$$\frac{2x + 1}{(x - 3)(x - 4)} = \frac{-7}{x - 3} + \frac{9}{x - 4}.$$

Alternative solution   There is a different way to compute A and B. Return to the equation

$$2x + 1 = A(x - 4) + B(x - 3).$$

This must hold for every value of x, in particular for $x = 3$ and $x = 4$:

$$x = 3: \qquad 6 + 1 = A(3 - 4) + 0, \qquad A = -7;$$

$$x = 4: \qquad 8 + 1 = 0 + B(4 - 3), \qquad B = 9.$$

Therefore $\dfrac{2x + 1}{(x - 3)(x - 4)} = \dfrac{-7}{x - 3} + \dfrac{9}{x - 4}.$ ■

■ **EXAMPLE 2**   Find $\displaystyle\int \frac{dx}{a^2 - x^2}.$

Solution   Write $\dfrac{1}{a^2 - x^2} = \dfrac{1}{(a - x)(a + x)} = \dfrac{A}{a - x} + \dfrac{B}{a + x}.$

Multiply through by $(a - x)(a + x)$:

$$1 = A(a + x) + B(a - x).$$

Set $x = a$ to obtain $A = 1/2a$; then set $x = -a$ to obtain $B = 1/2a$. Hence

$$\frac{1}{a^2 - x^2} = \frac{1}{2a}\left(\frac{1}{a - x} + \frac{1}{a + x}\right).$$

Therefore

$$\int \frac{dx}{a^2 - x^2} = \frac{1}{2a}\left(\int \frac{dx}{a - x} + \int \frac{dx}{a + x}\right)$$

$$= \frac{1}{2a}\left(-\ln|a - x| + \ln|a + x|\right) + C = \frac{1}{2a}\ln\left|\frac{a + x}{a - x}\right| + C. \quad \blacksquare$$

Rational Functions A **rational function** is the quotient of two polynomials. To integrate a rational function $p(x)/q(x)$, use partial fractions. However, in case degree $[p(x)] \geq$ degree $[q(x)]$, first divide $p(x)$ by $q(x)$. This yields

$$\frac{p(x)}{q(x)} = r(x) + \frac{s(x)}{q(x)},$$

where $r(x)$ is a polynomial and $s(x)$ is a polynomial whose degree is less than that of $q(x)$.

■ **EXAMPLE 3** Find $\displaystyle\int \frac{x^3 + 4}{x^2 + x}\, dx.$

Solution Divide $x^3 + 4$ by $x^2 + x$: $\displaystyle\frac{x^3 + 4}{x^2 + x} = x - 1 + \frac{x + 4}{x^2 + x}.$

Hence $\displaystyle\int \frac{x^3 + 4}{x^2 + x}\, dx = \int (x - 1)\, dx + \int \frac{x + 4}{x^2 + x}\, dx = \frac{x^2}{2} - x + \int \frac{x + 4}{x^2 + x}\, dx.$

The problem is now reduced to evaluating the last integral. Write

$$\frac{x + 4}{x^2 + x} = \frac{x + 4}{x(x + 1)} = \frac{A}{x} + \frac{B}{x + 1}.$$

Multiply by $x(x + 1)$:

$$x + 4 = A(x + 1) + Bx.$$

Set $x = 0$ and $x = -1$ to obtain $A = 4$ and $B = -3$. Thus $\displaystyle\frac{x + 4}{x^2 + x} = \frac{4}{x} - \frac{3}{x + 1},$

$$\int \frac{x + 4}{x^2 + x}\, dx = 4\int \frac{dx}{x} - 3\int \frac{dx}{x + 1} = 4\ln|x| - 3\ln|x + 1| + C.$$

Therefore $\displaystyle\int \frac{x^3 + 4}{x^2 + x}\, dx = \frac{x^2}{2} - x + \ln\left|\frac{x^4}{(x + 1)^3}\right| + C. \quad \blacksquare$

Partial fractions are useful in the integration of rational functions $p(x)/q(x)$, provided the denominator can be completely factored into linear and quadratic factors. In practice, this is hard to do for polynomials of degree 3 or more, except in special cases.

Assume the degree of $q(x)$ exceeds that of $p(x)$, and assume that $q(x)$ is factored into linear and quadratic factors. Then for each factor $x - a$ there is a term

$$\frac{A}{x - a}.$$

If $(x - a)^2$ occurs, there are two terms: $\dfrac{A_1}{x - a} + \dfrac{A_2}{(x - a)^2}.$

If $(x - a)^3$ occurs, there are three terms: $\dfrac{A_1}{x - a} + \dfrac{A_2}{(x - a)^2} + \dfrac{A_3}{(x - a)^3}.$

For each quadratic factor $x^2 + ax + b$ there is a term $\dfrac{Ax + B}{x^2 + ax + b}.$

If $(x^2 + ax + b)^2$ occurs, there are two terms: $\dfrac{Ax + B}{x^2 + ax + b} + \dfrac{Cx + D}{(x^2 + ax + b)^2}$

and so on. For instance:

$$\frac{1}{(x - a)(x - b)(x - c)} = \frac{A}{x - a} + \frac{B}{x - b} + \frac{C}{x - c},$$

$$\frac{1}{(x - a)^2(x - b)} = \frac{A}{x - a} + \frac{B}{(x - a)^2} + \frac{C}{(x - b)},$$

$$\frac{1}{(x - a)(x^2 + bx + c)} = \frac{A}{x - a} + \frac{Bx + C}{x^2 + bx + c},$$

$$\frac{1}{(x - a)(x^2 + b^2)^2} = \frac{A}{x - a} + \frac{Bx + C}{x^2 + b^2} + \frac{Dx + E}{(x^2 + b^2)^2},$$

$$\frac{1}{x^4 - 1} = \frac{1}{(x - 1)(x + 1)(x^2 + 1)} = \frac{A}{x - 1} + \frac{B}{x + 1} + \frac{Cx + D}{x^2 + 1}.$$

■ **EXAMPLE 4** Find $\displaystyle \int \frac{dx}{x^4 - 1}.$

Solution Write $\dfrac{1}{x^4 - 1} = \dfrac{A}{x - 1} + \dfrac{B}{x + 1} + \dfrac{Cx + D}{x^2 + 1}.$

Multiply through by $(x - 1)(x + 1)(x^2 + 1)$:

$$1 = A(x + 1)(x^2 + 1) + B(x - 1)(x^2 + 1) + Cx(x - 1)(x + 1) + D(x - 1)(x + 1).$$

Set $x = 1$ and $x = -1$ to obtain $A = -B = \frac{1}{4}$. Set $x = 0$ to obtain $1 = A - B - D = \frac{1}{4} + \frac{1}{4} - D$. Hence $D = -\frac{1}{2}$. Choose any other value of x to find C. Try $x = 2$, for example:

$$1 = 15A + 5B + 6C + 3D = \tfrac{15}{4} - \tfrac{5}{4} + 6C - \tfrac{3}{2},$$

from which $C = 0$. Therefore

$$\frac{1}{x^4 - 1} = \frac{1}{4}\left(\frac{1}{x - 1}\right) - \frac{1}{4}\left(\frac{1}{x + 1}\right) - \frac{1}{2}\left(\frac{1}{x^2 + 1}\right),$$

$$\int \frac{dx}{x^4 - 1} = \frac{1}{4} \int \frac{dx}{x - 1} - \frac{1}{4} \int \frac{dx}{x + 1} - \frac{1}{2} \int \frac{dx}{x^2 + 1}$$

$$= \frac{1}{4} \ln |x - 1| - \frac{1}{4} \ln |x + 1| - \frac{1}{2} \arctan x + C$$

$$= \ln \left| \frac{x - 1}{x + 1} \right|^{1/4} - \frac{1}{2} \arctan x + C. \qquad \blacksquare$$

■ **EXAMPLE 5** Find $\displaystyle\int \frac{2x + 5}{(x - 1)(x + 3)^2} \, dx.$

Solution Write $\displaystyle\frac{2x + 5}{(x - 1)(x + 3)^2} = \frac{A}{x - 1} + \frac{B}{x + 3} + \frac{C}{(x + 3)^2}.$

Multiply through by $(x - 1)(x + 3)^2$:

$$2x + 5 = A(x + 3)^2 + B(x - 1)(x + 3) + C(x - 1).$$

Set $x = 1$ to obtain $A = \frac{7}{16}$; set $x = -3$ to obtain $C = \frac{1}{4}$. Choose any other value of x to find B, for example, $x = 0$:

$$5 = 9A - 3B - C = \frac{63}{16} - 3B - \frac{1}{4},$$

from which $B = -\frac{7}{16}$. Therefore,

$$\frac{2x + 5}{(x - 1)(x + 3)^2} = \frac{7}{16}\left(\frac{1}{x - 1}\right) - \frac{7}{16}\left(\frac{1}{x + 3}\right) + \frac{1}{4}\left(\frac{1}{(x + 3)^2}\right),$$

$$\int \frac{2x + 5}{(x - 1)(x + 3)^2} \, dx = \frac{7}{16} \int \frac{dx}{x - 1} - \frac{7}{16} \int \frac{dx}{x + 3} + \frac{1}{4} \int \frac{dx}{(x + 3)^2}$$

$$= \frac{7}{16} \ln \left| \frac{x - 1}{x + 3} \right| - \frac{1}{4(x + 3)} + C. \qquad \blacksquare$$

Rational Functions of Sine and Cosine In theory, partial fractions allow the integration of any rational function $r(t)$. An interesting substitution, based on the half-angle formulas of trigonometry, reduces any rational function of $\sin \theta$ and $\cos \theta$ to a rational function $r(t)$. Thus, at least in theory, integrals such as

$$\int \frac{d\theta}{3 + \cos \theta} \qquad \text{and} \qquad \int \frac{1 + \sin^3 \theta \cos \theta}{4 \cos^5 \theta - 3 \sin \theta} \, d\theta$$

can be computed explicitly. However, the computations may be quite formidable. When a *definite* integral is required, approximation methods like the Trapezoidal Rule and the rules to be discussed in Chapter 10 usually involve less work.

Now we state the half-angle formulas in the form needed for computing integrals.

Half-Angle Formulas Set $t = \tan \frac{1}{2}\theta$. Then

$$\sin \theta = \frac{2t}{1 + t^2}, \qquad \cos \theta = \frac{1 - t^2}{1 + t^2}, \qquad t = \frac{\sin \theta}{1 + \cos \theta}, \qquad d\theta = \frac{2\,dt}{1 + t^2}.$$

The first three formulas are usually given in trigonometry courses, and they follow easily from the standard double-angle formulas. We shall give a geometric derivation in Section 9 of Chapter 9.

The last formula, the relation between $d\theta$ and dt, is obtained by differentiating $t = \tan \frac{1}{2}\theta$:

$$dt = \tfrac{1}{2}(\sec^2 \tfrac{1}{2}\theta)\ d\theta = \tfrac{1}{2}(1 + \tan^2 \tfrac{1}{2}\theta)\ d\theta = \tfrac{1}{2}(1 + t^2)\ d\theta.$$

With this preparation we can state a rule for transforming integrals of functions of $\sin \theta$ and $\cos \theta$. Suppose $f(x, y)$ is a polynomial in two variables or a quotient of two polynomials. Then

$$\int f(\sin \theta, \cos \theta)\ d\theta = 2 \int f\left(\frac{2t}{1 + t^2}, \frac{1 - t^2}{1 + t^2}\right) \frac{dt}{1 + t^2}, \qquad \text{where} \quad t = \tan \tfrac{1}{2}\theta.$$

■ **EXAMPLE 6**   Compute $\displaystyle \int \frac{d\theta}{3 + \cos \theta}$.

Solution   Set $t = \tan \frac{1}{2}\theta$. Then

$$\int \frac{d\theta}{3 + \cos \theta} = 2 \int \frac{dt/(1 + t^2)}{3 + [(1 - t^2)/(1 + t^2)]} = 2 \int \frac{dt}{3(1 + t^2) + (1 - t^2)}$$

$$= 2 \int \frac{dt}{4 + 2t^2} = \int \frac{dt}{2 + t^2} = \frac{1}{\sqrt{2}} \arctan \frac{t}{\sqrt{2}} + C$$

$$= \frac{1}{\sqrt{2}} \arctan \left(\frac{1}{\sqrt{2}} \tan \frac{\theta}{2}\right) + C. \qquad\qquad ■$$

EXERCISES

Decompose into partial fractions

1   $\dfrac{1}{(x + 1)(x - 1)}$ 2 $\dfrac{x}{(x + 2)(x + 3)}$ 3 $\dfrac{x^2}{(x + 1)(x - 2)}$

4   $\dfrac{1}{(x + 1)(x + 2)(x + 3)}$ 5 $\dfrac{x}{(x + 1)(x + 2)(x + 3)}$ 6 $\dfrac{1}{(x + 1)(x^2 + 4)}$

7   $\dfrac{x^4}{(x^2 + 1)^2}$ 8 $\dfrac{x^3 - 1}{x(x^2 + 1)}$ 9 $\dfrac{x + 1}{(x - 1)(x^2 + 4)}$

10   $\dfrac{1}{x(x + 1)^2}$.

Compute

11   $\displaystyle \int \frac{dx}{x^2 - 3x + 2}$ 12 $\displaystyle \int \frac{dx}{(x - 2)(x + 4)}$ 13 $\displaystyle \int \frac{x + 3}{x^2 + x}\ dx$

14   $\displaystyle \int \frac{x^2 + 1}{x^2 - 5x + 6}\ dx$ 15 $\displaystyle \int \frac{2x + 3}{x^3 + x}\ dx$ 16 $\displaystyle \int \frac{x\ dx}{(x + 1)^2(x - 3)}$

17   $\displaystyle \int \frac{dx}{(x - 2)^2(x^2 + 9)}$ 18 $\displaystyle \int \frac{dx}{3x^2 - 13x + 4}$ 19 $\displaystyle \int \frac{x^4\ dx}{x^3 - 1}$

20 $\displaystyle\int \frac{x\,dx}{x^4-1}$ **21** $\displaystyle\int \frac{x^3\,dx}{x^2+3x+2}$ **22** $\displaystyle\int \frac{dx}{(x-1)(x-2)(x-3)}$

23 $\displaystyle\int \frac{x^2+x+1}{(x-3)(x^2+2x+2)}\,dx$ **24** $\displaystyle\int \frac{dx}{x(x-3)^2}.$

Evaluate the definite integral

25 $\displaystyle\int_0^1 \frac{dx}{x^2-5x+6}$ **26** $\displaystyle\int_1^2 \frac{dx}{x^2(x+1)}$ **27** $\displaystyle\int_1^4 \frac{dx}{x^3+8}$

28 $\displaystyle\int_0^3 \frac{x\,dx}{(x+1)(x^2+9)}$ **29** $\displaystyle\int_0^{\pi/2} \frac{\cos\theta\,d\theta}{\sin^2\theta+7\sin\theta+10}$

30 $\displaystyle\int_4^9 \frac{dx}{\sqrt{x}\,(1+\sqrt{x}\,)(2+\sqrt{x}\,)}.$

Compute

31 $\displaystyle\int \frac{\sin\theta\,d\theta}{2+\cos\theta}$ **32** $\displaystyle\int \frac{4-\sin\theta}{3-\cos\theta}\,d\theta.$

6. TRIGONOMETRIC SUBSTITUTIONS

Integrals involving a^2-x^2 or a^2+x^2 are often simplified by trigonometric substitutions. The substitution $x=a\sin\theta$ changes a^2-x^2 into $a^2\cos^2\theta$; the substitution $x=a\tan\theta$ changes a^2+x^2 into $a^2\sec^2\theta$.

■ **EXAMPLE 1**   Find $\displaystyle\int \frac{dx}{x^2\sqrt{4-x^2}}.$

Solution   Set $x=2\sin\theta$. Then

$$\int \frac{dx}{x^2\sqrt{4-x^2}} = \int \frac{2\cos\theta\,d\theta}{(2\sin\theta)^2\sqrt{4-4\sin^2\theta}} = \int \frac{2\cos\theta\,d\theta}{4\sin^2\theta\cdot 2\cos\theta} = \frac{1}{4}\int \frac{d\theta}{\sin^2\theta}.$$

Hence $$\int \frac{dx}{x^2\sqrt{4-x^2}} = \frac{1}{4}\int \csc^2\theta\,d\theta = -\frac{1}{4}\cot\theta + C.$$

As a final step, express $\cot\theta$ in terms of x. This can be done quickly by drawing a right triangle (Fig. 1) showing $x=2\sin\theta$. It follows that

$$\cot\theta = \frac{\sqrt{4-x^2}}{x},\qquad \int \frac{dx}{x^2\sqrt{4-x^2}} = -\frac{\sqrt{4-x^2}}{4x} + C.$$ ■

■ **EXAMPLE 2**   Find $\displaystyle\int \frac{dx}{\sqrt{a^2+x^2}}.$

Solution   Set $x=a\tan\theta$. Then

$$\int \frac{dx}{\sqrt{a^2+x^2}} = \int \frac{a\sec^2\theta\,d\theta}{\sqrt{a^2(1+\tan^2\theta)}} = \int \frac{a\sec^2\theta\,d\theta}{a\sec\theta}$$

$$= \int \sec\theta\,d\theta = \ln|\sec\theta+\tan\theta| + C.$$

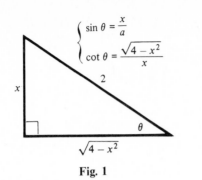

$$\begin{cases} \sin \theta = \dfrac{x}{a} \\[2mm] \cot \theta = \dfrac{\sqrt{4-x^2}}{x} \end{cases}$$

Fig. 1

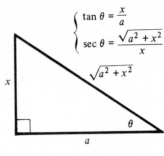

$$\begin{cases} \tan \theta = \dfrac{x}{a} \\[2mm] \sec \theta = \dfrac{\sqrt{a^2+x^2}}{x} \end{cases}$$

Fig. 2

Draw a right triangle (Fig. 2) showing $x = a \tan \theta$, and read off $\sec \theta$. Thus

$$\int \frac{dx}{\sqrt{a^2+x^2}} = \ln \left| \frac{\sqrt{a^2+x^2}}{a} + \frac{x}{a} \right| + C$$

$$= \ln \left| \sqrt{a^2+x^2} + x \right| - \ln |a| + C = \ln(x + \sqrt{a^2+x^2}) + C. \qquad \blacksquare$$

■ **EXAMPLE 3** Compute the definite integral $\displaystyle \int_0^1 \frac{dx}{(1+x^2)^2}.$

Solution Substitute $x = \tan \theta$. Then $\theta = 0$ for $x = 0$, and $\theta = \frac{1}{4}\pi$ for $x = 1$. Hence

$$\int_0^1 \frac{dx}{(1+x^2)^2} = \int_0^{\pi/4} \frac{\sec^2 \theta \, d\theta}{(1+\tan^2 \theta)^2} = \int_0^{\pi/4} \frac{d\theta}{\sec^2 \theta} = \int_0^{\pi/4} \cos^2 \theta \, d\theta.$$

Use the identity $\cos^2 \theta = \frac{1}{2}(1 + \cos 2\theta)$:

$$\int_0^1 \frac{dx}{(1+x^2)^2} = \int_0^{\pi/4} \left(\frac{1}{2} + \frac{1}{2} \cos 2\theta \right) d\theta = \left(\frac{1}{2}\theta + \frac{1}{4}\sin 2\theta \right) \Big|_0^{\pi/4}$$

$$= \left(\frac{1}{2} \cdot \frac{\pi}{4} + \frac{1}{4} \sin \frac{\pi}{2} \right) - \left(\frac{1}{2} \cdot 0 + \frac{1}{4} \sin 0 \right) = \frac{\pi}{8} + \frac{1}{4}. \qquad \blacksquare$$

■ **EXAMPLE 4** Find the indefinite integral $\displaystyle \int \frac{dx}{(1+x^2)^2}.$

Solution From the solution of the last example,

$$\int \frac{dx}{(1+x^2)^2} = \frac{\theta}{2} + \frac{\sin 2\theta}{4} + C, \qquad x = \tan \theta.$$

It remains to express this function of θ in terms of x. Obviously, $\theta = \arctan x$. To

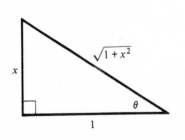

Fig. 3

$$\begin{cases} \tan \theta = \dfrac{x}{1} = x \\[2mm] \sin 2\theta = 2 \sin \theta \cos \theta \\[2mm] \qquad = 2\left(\dfrac{x}{\sqrt{1+x^2}} \right)\left(\dfrac{1}{\sqrt{1+x^2}} \right) = \dfrac{2x}{1+x^2} \end{cases}$$

express $\sin 2\theta$ in terms of x, draw a right triangle (Fig. 3) showing $x = \tan \theta$. It follows (via $\sin 2\theta = 2 \sin \theta \cos \theta$) that

$$\int \frac{dx}{(1 + x^2)^2} = \frac{1}{2} \arctan x + \frac{x}{2(1 + x^2)} + C. \qquad \blacksquare$$

Hyperbolic Substitutions In integrals involving $\sqrt{x^2 \pm a^2}$, a hyperbolic substitution is often useful. Because $\cosh^2 t - \sinh^2 t = 1$, the substitution $x = a \sinh t$ transforms $x^2 + a^2$ into $a^2 \cosh^2 t$; the substitution $x = a \cosh t$ transforms $x^2 - a^2$ into $a^2 \sinh^2 t$.

$\blacksquare$ **EXAMPLE 5** Find $\displaystyle\int \frac{dx}{\sqrt{x^2 - a^2}}$, $x > a > 0$.

Solution Let $x = a \cosh t$. Then $dx = a \sinh t \, dt$, and

$$\int \frac{dx}{\sqrt{x^2 - a^2}} = \int \frac{a \sinh t \, dt}{\sqrt{a^2 \cosh^2 t - a^2}} = \int \frac{\sinh t \, dt}{\sqrt{\sinh^2 t}} = \int dt = t + C = \cosh^{-1} \frac{x}{a} + C$$

$$= \ln\left(\frac{x}{a} + \sqrt{\left(\frac{x}{a}\right)^2 - 1}\right) + C = \ln(x + \sqrt{x^2 - a^2}) + C. \qquad \blacksquare$$

Some Tricks Integrals involving $\sqrt{x^2 \pm a^2}$ can also be handled by certain formal tricks. Make the substitution

$$y^2 = x^2 \pm a^2;$$

natural enough. Now fiddle with the differentials dx and dy:

$$y \, dy = x \, dx, \qquad \text{hence} \qquad \frac{dx}{y} = \frac{dy}{x}.$$

The first trick is to observe that if $\dfrac{a}{b} = \dfrac{c}{d}$, then $\dfrac{a}{b} = \dfrac{a + c}{b + d}$,

a fact from elementary ratios and proportions.

Hence $\dfrac{dx}{y} = \dfrac{dx + dy}{x + y} = \dfrac{d(x + y)}{x + y}$, $\displaystyle\int \frac{dx}{y} = \int \frac{d(x + y)}{x + y} = \ln|x + y| + C.$

Therefore $\displaystyle\int \frac{dx}{\sqrt{x^2 \pm a^2}} = \ln|x + \sqrt{x^2 \pm a^2}| + C.$

For the next trick write

$$y \, dx = \frac{y^2 \, dx}{y} = \frac{x^2 \pm a^2}{y} \, dx = \frac{x^2 \, dx}{y} \pm a^2 \frac{dx}{y} = x \, dy \pm a^2 \frac{dx}{y}.$$

But also $y \, dx = d(xy) - x \, dy$. Add: $2y \, dx = d(xy) \pm a^2 \dfrac{dx}{y}.$

Integrate, using the first integral: $\displaystyle\int y \, dx = \frac{1}{2} xy \pm \frac{a^2}{2} \ln|x + y| + C,$

$$\int \sqrt{x^2 \pm a^2}\, dx = \frac{1}{2} x\sqrt{x^2 \pm a^2} + \frac{a^2}{2} \ln |x + \sqrt{x^2 \pm a^2}| + C.$$

Similarly one can evaluate $\displaystyle \int \frac{\sqrt{x^2 \pm a^2}}{x}\, dx, \qquad \int \sqrt{a^2 - x^2}\, dx, \qquad$ etc.

EXERCISES

Use a trig substitution to find

1 $\displaystyle \int \frac{dx}{\sqrt{1 + x^2}}$ **2** $\displaystyle \int \frac{dx}{1 - x^2}$ **3** $\displaystyle \int \frac{x^3\, dx}{\sqrt{4 - x^2}}$

4 $\displaystyle \int \frac{\sqrt{9 - x^2}}{x^2}\, dx$ **5** $\displaystyle \int \frac{dx}{x^2\sqrt{16 - x^2}}$ **6** $\displaystyle \int \frac{x^2\, dx}{\sqrt{a^2 - x^2}}$

7 $\displaystyle \int \frac{\sqrt{x^2 + a^2}}{x^2}\, dx$ **8** $\displaystyle \int \frac{x^2\, dx}{(1 + x^2)^{3/2}}$ **9** $\displaystyle \int \frac{x^2\, dx}{(1 + x^2)^2}$

10 $\displaystyle \int \frac{dx}{x^2\sqrt{1 + x^2}}.$

Use a hyperbolic substitution to find

11 $\displaystyle \int \frac{dx}{\sqrt{x^2 + a^2}}$ **12** $\displaystyle \int \sqrt{x^2 + a^2}\, dx$ **13** $\displaystyle \int \frac{x^2\, dx}{\sqrt{x^2 + a^2}}$

14 $\displaystyle \int \frac{\sqrt{x^2 + a^2}}{x^2}\, dx$ **15** $\displaystyle \int \frac{dx}{a^2 - x^2}$ **16** $\displaystyle \int \frac{dx}{x\sqrt{a^2 - x^2}}.$

Use the "Some Tricks" method with $y^2 = x^2 \pm a^2$ so $y\, dy = x\, dx$ and (by the text)

$$(1) \quad \int \frac{dx}{y} = \int \frac{ay}{x} = \ln |x + y| \qquad (2) \quad \int y\, dx = \frac{1}{2} xy \pm \frac{1}{2} a^2 \ln |x + y|.$$

Reduce each integral to one that does not involve a radical, and evaluate it

17 $\displaystyle \int \frac{dx}{xy}$ **18** $\displaystyle \int \frac{y\, dx}{x}$ **19** $\displaystyle \int \frac{y\, dx}{x^2}$ **20*** $\displaystyle \int \frac{dx}{x^2 y}.$

Now let $y^2 = a^2 - x^2$ so $y\, dx = -x\, dx$. Hence $\displaystyle \int \frac{dx}{y} = -\int \frac{dy}{x} = \arcsin \frac{x}{a}$. Reduce the integral to one without a radical and evaluate it

21 $\displaystyle \int \frac{dx}{xy}$ **22** $\displaystyle \int \frac{y\, dx}{x^2}$ **23** $\displaystyle \int \frac{dx}{x^2 y}$ **24*** $\displaystyle \int y\, dx.$

7. INTEGRATION BY PARTS

This important technique, which we introduced in Chapter 5, p. 236, converts an integration problem into a different integration problem, hopefully an easier one. Let us recall how it goes.

Integration by parts comes from the Product Rule:

$$\frac{d}{dx}(uv) = u\frac{dv}{dx} + v\frac{du}{dx},$$

where u and v are functions of x.

This rule can be expressed in terms of differentials. Multiply through formally by the symbol dx:

$$\frac{d}{dx}(uv)\,dx = u\,\frac{dv}{dx}\,dx + v\,\frac{du}{dx}\,dx.$$

The term on the left is $d(uv)$. On the right we have

$$\frac{dv}{dx}\,dx = dv, \qquad \frac{du}{dx}\,dx = du.$$

Product Rule in Differential Form $d(uv) = u\,dv + v\,du.$

Rearranging the terms, we have

$$u\,dv = d(uv) - v\,du,$$

and consequently,

$$\int u\,dv = \int d(uv) - \int v\,du.$$

But

$$\int d(uv) = uv + C,$$

so we obtain the following formula:

Integration by Parts $\displaystyle\int u\,dv = uv - \int v\,du.$

The constant is absorbed into the second integral.

This formula converts the problem of integrating $u\,dv$ into that of integrating $v\,du$. With luck, the second integration is easier. There is no guarantee, however, that it need be.

■ **EXAMPLE 1**   Find $\displaystyle\int x\cos x\,dx.$

Solution   Interpret the integral as $\displaystyle\int x\,d(\sin x).$

Apply the formula for integration by parts with

$$u = x, \quad du = dx; \qquad dv = \cos x\,dx, \quad v = \sin x,$$

to obtain

$$\int x\cos x\,dx = uv - \int v\,du = x\sin x - \int \sin x\,dx.$$

This is a bit of luck; the integral on the right is easy. Conclusion:

$$\int x\cos x\,dx = x\sin x + \cos x + C. \qquad\blacksquare$$

Remark We could have chosen u and v differently:

$$\int x \cos x \, dx = \int u \, dv,$$

where $u = \cos x, \quad du = -\sin x \, dx; \quad dv = x \, dx, \quad v = \tfrac{1}{2}x^2.$

Then integration by parts yields

$$\int x \cos x \, dx = \cos x \cdot \frac{x^2}{2} + \int \frac{x^2}{2} \sin x \, dx,$$

which is true but does not help; the integral on the right is harder than the given one. Thus, it may be crucial how we factor the integrand into u and dv. It is also possible that no choice of u and dv helps.

■ **EXAMPLE 2** Find $\displaystyle\int \text{arc} \sin x \, dx.$

Solution Use integration by parts with

$$u = \text{arc} \sin x, \quad du = \frac{dx}{\sqrt{1 - x^2}}; \quad dv = dx, \quad v = x.$$

Then $\displaystyle\int \text{arc} \sin x \, dx = uv - \int v \, du = x \, \text{arc} \sin x - \int \frac{x \, dx}{\sqrt{1 - x^2}}.$

There are several ways to do the integral on the right. For example, substitute $y^2 = 1 - x^2$, $y \, dy = -x \, dx$:

$$-\int \frac{x \, dx}{\sqrt{1 - x^2}} = \int \frac{y \, dy}{y} = \int dy = y + C = \sqrt{1 - x^2} + C.$$

Therefore $\displaystyle\int \text{arc} \sin x \, dx = x \, \text{arc} \sin x + \sqrt{1 - x^2} + C.$ ■

Note "y" was used in the substitution step because "u" was used in the first step. Always take care not to confuse variables.

■ **EXAMPLE 3** Compute $\displaystyle\int_1^2 \ln x \, dx.$

Solution Use integration by parts with

$$u = \ln x, \quad du = \frac{dx}{x}; \quad dv = dx, \quad v = x.$$

Therefore $\displaystyle\int_1^2 \ln x \, dx = uv \Big|_1^2 - \int_1^2 v \, du = x \ln x \Big|_1^2 - \int_1^2 x \cdot \frac{dx}{x}$

$$= (2 \ln 2 - 1 \ln 1) - x \Big|_1^2 = 2 \ln 2 - 1.$$ ■

■ **EXAMPLE 4** Find $\displaystyle\int x^3 e^{x^2} \, dx.$

Solution   Substitute $y = x^2$, $dy = 2x \, dx$:

$$\int x^3 e^{x^2} \, dx = \int x^2 e^{x^2} \cdot x \, dx = \int y e^y \cdot \tfrac{1}{2} \, dy.$$

Now integrate by parts with $u = y$, $du = dy$; $dv = e^y \, dy$, $v = e^y$:

$$\int y e^y \, dy = y e^y - \int e^y \, dy = y e^y - e^y + C = (y - 1)e^y + C.$$

Therefore $\int x^3 e^{x^2} \, dx = \tfrac{1}{2}(x^2 - 1)e^{x^2} + C.$ ∎

Repeated Integration by Parts   Some problems require two or more integrations by parts.

■ **EXAMPLE 5**   Find $\int x(\ln x)^2 \, dx.$

Solution   Integrate by parts with

$$u = (\ln x)^2, \quad du = \frac{2 \ln x \, dx}{x}; \qquad dv = x \, dx, \quad v = \frac{x^2}{2}.$$

Therefore $\int x(\ln x)^2 \, dx = \dfrac{x^2(\ln x)^2}{2} - \displaystyle\int \frac{x^2}{2} \cdot \frac{2 \ln x}{x} \, dx = \dfrac{x^2(\ln x)^2}{2} - \int x \ln x \, dx.$

The problem now is to evaluate $\int x \ln x \, dx,$

which is similar to the original integral except that $\ln x$ appears only to the first power. Therefore another integration by parts should reduce the integral to

$$\int x \, dx.$$

Try it. Integrate by parts again with

$$u = \ln x, \quad du = \frac{dx}{x}; \qquad dv = x \, dx, \quad v = \frac{x^2}{2}.$$

Therefore

$$\int x \ln x \, dx = \frac{x^2 \ln x}{2} - \int \frac{x^2}{2} \cdot \frac{dx}{x} = \frac{x^2 \ln x}{2} - \frac{1}{2} \int x \, dx = \frac{x^2 \ln x}{2} - \frac{x^2}{4} + C.$$

Combine the results:

$$\int x(\ln x)^2 \, dx = \frac{x^2(\ln x)^2}{2} - \left(\frac{x^2 \ln x}{2} - \frac{x^2}{4} + C \right) = \frac{x^2}{4} [2(\ln x)^2 - 2 \ln x + 1] + C.$$

■

■ **EXAMPLE 6**   Find $\int x^3 e^x \, dx.$

Solution   Integrate by parts three times: $\displaystyle\int x^3 e^x\,dx = x^3 e^x - 3\int x^2 e^x\,dx,$

$$\int x^2 e^x\,dx = x^2 e^x - 2\int xe^x\,dx, \qquad \int xe^x\,dx = xe^x - \int e^x\,dx = xe^x - e^x + C.$$

Combine the results:

$$\int x^3 e^x\,dx = x^3 e^x - 3x^2 e^x + 6xe^x - 6e^x + C = e^x(x^3 - 3x^2 + 6x - 6) + C. \qquad \blacksquare$$

■ **EXAMPLE 7**   Find $\displaystyle\int e^x \cos x\,dx.$

Solution   Denote the integral by J. Integrate by parts with

$$u = e^x, \quad du = e^x\,dx; \qquad dv = \cos x\,dx, \quad v = \sin x.$$

Therefore $\displaystyle J = e^x \sin x - \int e^x \sin x\,dx.$

Integrate by parts again with

$$u = e^x, \quad du = e^x\,dx; \qquad dv = \sin x\,dx, \quad v = -\cos x.$$

Therefore $\displaystyle\int e^x \sin x\,dx = -e^x \cos x + \int e^x \cos x\,dx.$

The integral on the right is J again! Hence $\displaystyle\int e^x \sin x\,dx = -e^x \cos x + J.$

Have we gone in a circle? No, because substitution of this expression in the first equation for J yields

$$J = e^x \sin x + e^x \cos x - J.$$

The minus sign on the right saves us from disaster. Solve for J:

$$J = \tfrac{1}{2}(e^x \sin x + e^x \cos x).$$

Don't forget the constant of integration:

$$\int e^x \cos x\,dx = \tfrac{1}{2}e^x(\sin x + \cos x) + C. \qquad \blacksquare$$

EXERCISES

Find

1 $\displaystyle\int x \sin x\,dx$	**2** $\displaystyle\int x \cos 3x\,dx$	**3** $\displaystyle\int xe^{2x}\,dx$
4 $\displaystyle\int x \sec^2 x\,dx$	**5** $\displaystyle\int \sqrt{x}\,\ln x\,dx$	**6** $\displaystyle\int \text{arc} \cos x\,dx$
7 $\displaystyle\int \text{arc} \tan x\,dx$	**8** $\displaystyle\int \ln(x^2 + 1)\,dx$	**9** $\displaystyle\int x\,\text{arc} \tan x\,dx$
10 $\displaystyle\int x^5 e^{x^2}\,dx$	**11** $\displaystyle\int e^{2x} \sin 3x\,dx$	**12** $\displaystyle\int x^2 e^{-x}\,dx$

13 $\int x \cosh x \, dx$ **14** $\int x^2 \sinh 3x \, dx$ **15** $\int x^2 \cos ax \, dx$

16* $\int \sec^3 x \, dx.$

Compute

17 $\int_{\pi}^{2\pi} x \cos x \, dx$ **18** $\int_{1}^{e} (\ln x)^2 \, dx$ **19** $\int_{1}^{2} x^2 \ln x \, dx$

20 $\int_{0}^{1/2} \arcsin 2x \, dx$ **21** $\int_{0}^{1} e^x (x + 3)^2 \, dx$ **22** $\int_{0}^{\pi/2} e^{2x} \sin x \, dx$

23 $\int_{0}^{\pi/3} x \sin^2 x \cos x \, dx$ **24** $\int_{0}^{\pi/2} x \sin^2 x \, dx.$

25 Prove that $\int_{-1}^{1} x^2 e^{x^2} \, dx = e - \frac{1}{2} \int_{-1}^{1} e^{x^2} \, dx.$

26 (cont.) Find a relation between $\int_{-1}^{1} x^4 e^{x^2} \, dx$ and $\int_{-1}^{1} e^{x^2} \, dx.$

27 Show that $\int_{0}^{2\pi} f(x) \cos x \, dx = -\int_{0}^{2\pi} f'(x) \sin x \, dx.$

28 If $f(a) = f(b) = 0,$ show that $\int_{a}^{b} f(x) \, dx = -\frac{1}{2} \int_{a}^{b} (x - a)(b - x) f''(x) \, dx.$

29 Show that $\int_{0}^{2\pi} \cos x \cos 2x \, dx = 0$ by integrating twice by parts.

30 Let $P(x)$ be a polynomial. Show that $\int P(x) e^x \, dx = (P - P' + P'' - P''' + \cdots) e^x.$

31* Prove $\int_{0}^{x} \frac{\sin t}{t} \, dt > 0$ for all $x > 0.$

32* Suppose $f(x)$ has a continuous derivative on $[a, b].$ Prove $\lim_{x \to \infty} \int_{0}^{1} f(t) \sin(xt) \, dt = 0.$

8. REDUCTION FORMULAS

The integral

$$\int x^2 e^x \, dx$$

requires two integrations by parts. Each integration lowers the power of x by one until x disappears. In the same way

$$\int x^3 e^x \, dx \quad \text{and} \quad \int e^4 e^x \, dx$$

require three and four integrations by parts, respectively. It is convenient to have a **reduction formula**, a formula that reduces

$$\int x^n e^x \, dx \quad \text{to} \quad \int x^{n-1} e^x \, dx.$$

Repeated use of such a formula reduces

$$\int x^n e^x \, dx \quad \text{to} \quad \int e^x \, dx.$$

■ **EXAMPLE 1**   Derive a reduction formula for $\int x^n e^x \, dx$.

Solution   Integrate by parts with

$$u = x^n, \quad du = nx^{n-1} \, dx; \qquad dv = e^x \, dx, \quad v = e^x:$$

$$\int x^n e^x \, dx = x^n e^x - \int e^x \cdot nx^{n-1} \, dx.$$

Therefore $\int x^n e^x \, dx = x^n e^x - n \int x^{n-1} e^x \, dx.$ ■

Remark   For abbreviation, write $J_n = \int x^n e^x \, dx.$

Then the reduction formula is $J_n = x^n e^x - nJ_{n-1}.$

■ **EXAMPLE 2**   Find $\int x^5 e^x \, dx.$

Solution   Use the reduction formula just derived to find J_5. With $n = 5$, the reduction formula yields

$$J_5 = x^5 e^x - 5J_4.$$

With $n = 4$, it yields $J_4 = x^4 e^x - 4J_3.$

Hence $J_5 = x^5 e^x - 5(x^4 e^x - 4J_3) = x^5 e^x - 5x^4 e^x + 20J_3.$

By repeated use of the reduction formula,

$$J_3 = x^3 e^x - 3J_2 = x^3 e^x - 3(x^2 e^x - 2J_1)$$

$$= x^3 e^x - 3x^2 e^x + 6J_1 = x^3 e^x - 3x^2 e^x + 6(xe^x - J_0).$$

The integral J_0 is easy: $J_0 = \int x^0 e^x \, dx = e^x + C.$

Hence $J_3 = e^x(x^3 - 3x^2 + 6x - 6) + C,$

and consequently

$$J_5 = x^5 e^x - 5x^4 e^x + 20e^x(x^3 - 3x^2 + 6x - 6) + C.$$

$$= e^x(x^5 - 5x^4 + 20x^3 - 60x^2 + 120x - 120) + C.$$ ■

Question   Study the polynomial in the answer. How does each term follow from the preceding term? Can you write down the value of $\int x^6 e^x \, dx$ by inspection?

■ **EXAMPLE 3**   Derive a reduction formula for $\int \cos^n x \, dx.$

Solution   Write $J_n = \int \cos^n x \, dx = \int \cos^{n-1} x \cos x \, dx$

and integrate by parts with

$$u = \cos^{n-1} x, \quad du = -(n-1) \cos^{n-2} x \sin x \, dx; \qquad dv = \cos x \, dx, \quad v = \sin x:$$

$$\int \cos^n x \, dx = \cos^{n-1} x \sin x + \int (n-1) \cos^{n-2} x \sin^2 x \, dx$$

$$= \cos^{n-1} x \sin x + (n-1) \int \cos^{n-2} x (1 - \cos^2 x) \, dx.$$

Therefore, $\qquad J_n = \cos^{n-1} x \sin x + (n-1)J_{n-2} - (n-1)J_n .$

Combine the terms in J_n: $\qquad nJ_n = \cos^{n-1} x \sin x + (n-1)J_{n-2} .$

Now dividing by n gives the desired reduction formula:

$$\int \cos^n x \, dx = \frac{\cos^{n-1} x \sin x}{n} + \frac{n-1}{n} \int \cos^{n-2} x \, dx. \qquad \blacksquare$$

Remark This reduction formula lowers the power of $\cos x$ by two. Therefore, repeated application will ultimately reduce J_n to J_0 or J_1, according as n is even or odd. But both of these are easy:

$$J_0 = \int \cos^0 x \, dx = \int dx = x + C, \qquad J_1 = \int \cos x \, dx = \sin x + C.$$

■ **EXAMPLE 4** Compute $\displaystyle\int_0^{\pi/2} \cos^6 x \, dx.$

Solution Set $\qquad I_n = \displaystyle\int_0^{\pi/2} \cos^n x \, dx.$

Then by the reduction formula of the last example,

$$I_n = \frac{\cos^{n-1} x \sin x}{n} \Big|_0^{\pi/2} + \frac{n-1}{n} \int_0^{\pi/2} \cos^{n-2} x \, dx.$$

Hence, $\qquad I_n = 0 + \dfrac{n-1}{n} I_{n-2} .$

Apply this formula with $n = 6$, then repeat with $n = 4$ and $n = 2$:

$$I_6 = \frac{5}{6} I_4 = \frac{5}{6}\left(\frac{3}{4} I_2\right) = \frac{5}{6} \cdot \frac{3}{4}\left(\frac{1}{2} I_0\right).$$

Therefore, $\qquad \displaystyle\int_0^{\pi/2} \cos^6 x \, dx = \frac{5 \cdot 3 \cdot 1}{6 \cdot 4 \cdot 2} \int_0^{\pi/2} dx = \frac{5 \cdot 3 \cdot 1}{6 \cdot 4 \cdot 2} \cdot \frac{\pi}{2} = \frac{5}{32} \pi. \qquad \blacksquare$

EXERCISES

Find a formula reducing J_n to J_{n-1}

1 $J_n = \displaystyle\int (\ln x)^n \, dx$ $\qquad$ **2** $J_n = \displaystyle\int x^n e^{-2x} \, dx$ $\qquad$ **3** $J_n = \displaystyle\int x^2 (\ln x)^n \, dx$

4 $J_n = \displaystyle\int \frac{dx}{(a^2 + x^2)^n}$ $\qquad$ **5** $J_n = \displaystyle\int \frac{dx}{(x^2 - a^2)^n}$ $\qquad$ **6** $J_n = \displaystyle\int \frac{dx}{(a^3 \pm x^3)^n}.$

Find a formula reducing J_n to J_{n-2}

7 $J_n = \displaystyle\int \tan^n x \, dx$ $\qquad$ **8** $J_n = \displaystyle\int \frac{dx}{x^n \sqrt{x^2 \pm a^2}}$ $\qquad$ **9** $J_n = \displaystyle\int \sec^n x \, dx$

10* $J_n = \displaystyle\int e^{ax} \sin^n bx \, dx$ $\qquad$ **11** $J_n = \displaystyle\int \sin^n x \, dx$ $\qquad$ **12*** $J_n = \displaystyle\int (\arcsin x)^n \, dx.$

Compute by means of an appropriate reduction formula

13 $\displaystyle\int_0^{\pi/2} \sin^7 x \, dx$ $\qquad$ **14** $\displaystyle\int_0^{\pi/2} \sin^8 x \, dx$ $\qquad$ **15** $\displaystyle\int_0^{\pi/4} \tan^{10} x \, dx$

16 $\displaystyle\int_0^1 \frac{dx}{(1+x^2)^3}$ **17** $\displaystyle\int_1^2 (\ln x)^4 \, dx$ **18*** $\displaystyle\int_{\pi/2}^{\pi} x^4 \sin x \, dx.$

19   Set $J_n = \displaystyle\int e^{ax} \tan^n x \, dx.$ Express J_n in terms of J_{n-1} and J_{n-2}.

20* Prove $\displaystyle\int_a^b (x-a)^m (b-x)^n \, dx = \frac{m! \, n!}{(m+n+1)!} (b-a)^{m+n+1}.$

9. INTEGRAL TABLES

A busy scientist does not want to bother with various tricks each time he encounters an integral; he uses integral tables. Not only do they save time, but they help eliminate errors.

Inside the two covers of this book is a short table of indefinite integrals. Much longer tables are available, for example those in the C.R.C. Standard Mathematical Tables. We suggest that you get one of the more complete integral tables and spend some time browsing through it. Become familiar with the type of integral you can expect to find there. Not every integral is listed in a table, but many can be transformed into integrals that are listed.

■ **EXAMPLE 1**   Use integral tables to find $\displaystyle\int x^3 \sqrt{3 - 4x^2} \, dx.$

Solution   Most tables include a section on integrals involving $\sqrt{a^2 - x^2}$. A formula in the C.R.C. tables states that

$$\int x^3 \sqrt{a^2 - x^2} \, dx = -\left(\frac{1}{5} x^2 + \frac{2}{15} a^2 \right)(a^2 - x^2)^{3/2}.$$

This is very close to what is wanted, except that $\sqrt{3 - 4x^2}$ appears instead of $\sqrt{a^2 - x^2}$. There are two simple ways of modifying the integrand: write either

$$\sqrt{3 - 4x^2} = \sqrt{4\left(\frac{3}{4} - x^2\right)} = 2\sqrt{\frac{3}{4} - x^2},$$

or $\sqrt{3 - 4x^2} = \sqrt{3 - (2x)^2} = \sqrt{3 - u^2},$ where $u = 2x.$

By the first method with $a^2 = \frac{3}{4}$

$$\int x^3 \sqrt{3 - 4x^2} \, dx = 2\int x^3 \sqrt{\frac{3}{4} - x^2} \, dx = -2\left(\frac{1}{5} x^2 + \frac{2}{15} \cdot \frac{3}{4}\right)\left(\frac{3}{4} - x^2\right)^{3/2} + C.$$

By the second method with $u = 2x$ and $a^2 = 3$

$$\int x^3 \sqrt{3 - 4x^2} \, dx = \int \left(\frac{u}{2}\right)^3 \sqrt{3 - u^2} \cdot \frac{1}{2} \, du = \frac{1}{16} \int u^3 \sqrt{3 - u^2} \, du$$

$$= -\frac{1}{16}\left(\frac{1}{5} u^2 + \frac{2}{15} \cdot 3\right)(3 - u^2)^{3/2} + C = -\frac{1}{16}\left(\frac{1}{5} \cdot 4x^2 + \frac{2}{5}\right)(3 - 4x^2)^{3/2} + C.$$

A little algebra shows that both answers agree.   ■

■ **EXAMPLE 2**   Use integral tables to find $\displaystyle\int e^{2x}\sin^3 x\, dx.$

Solution   In the C.R.C. integral tables under Exponential Forms is the formula

$$\int e^{ax}\sin^n bx\, dx = \frac{1}{a^2 + n^2 b^2}\left[(a\sin bx - nb\cos bx)e^{ax}\sin^{n-1} bx\right.$$

$$\left. + n(n-1)b^2\int e^{ax}\sin^{n-2} bx\, dx\right].$$

This is a reduction formula which lowers the power of sin bx by two. Apply it with $a = 2,\ b = 1,\ n = 3$:

$$\int e^{2x}\sin^3 x\, dx = \frac{1}{4+9}\left[(2\sin x - 3\cos x)e^{2x}\sin^2 x + 6\int e^{2x}\sin x\, dx\right].$$

The integral on the right is also given in the C.R.C. tables. Its value is

$$\tfrac{1}{5}e^{2x}(2\sin x - \cos x).$$

It follows that

$$\int e^{2x}\sin^3 x\, dx = \frac{e^{2x}}{13}\left[(2\sin x - 3\cos x)\sin^2 x + \frac{6}{5}(2\sin x - \cos x)\right] + C. \quad ■$$

Integral tables use abbreviations for common expressions. For instance, one section of the C.R.C. tables contains formulas involving X and $\sqrt{X}$, where $X = a + bx + cx^2$.

■ **EXAMPLE 3**   Use integral tables to find $\displaystyle\int \frac{\sqrt{5x^2 + 2x + 3}}{x}\, dx.$

Solution   According to one of the formulas in the C.R.C. tables,

$$\int \frac{\sqrt{X}}{x}\, dx = \sqrt{X} + \frac{b}{2}\int \frac{dx}{\sqrt{X}} + a\int \frac{dx}{x\sqrt{X}}.$$

The integrals on the right are also given in the tables:

$$\int \frac{dx}{\sqrt{X}} = \frac{1}{\sqrt{c}}\ln\left(\sqrt{X} + x\sqrt{c} + \frac{b}{2\sqrt{c}}\right), \qquad c > 0,$$

$$\int \frac{dx}{x\sqrt{X}} = -\frac{1}{\sqrt{a}}\ln\left(\frac{\sqrt{X} + \sqrt{a}}{x} + \frac{b}{2\sqrt{a}}\right), \qquad a > 0.$$

Setting $X = 3 + 2x + 5x^2$, and $a = 3,\ b = 2,\ c = 5$, yields

$$\int \frac{\sqrt{5x^2 + 2x + 3}}{x}\, dx = \sqrt{X} + \frac{1}{\sqrt{5}}\ln\left|\sqrt{X} + x\sqrt{5} + \frac{1}{\sqrt{5}}\right|$$

$$-\sqrt{3}\ln\left|\frac{\sqrt{X} + \sqrt{3}}{x} + \frac{1}{\sqrt{3}}\right| + C,$$

where $X = 5x^2 + 2x + 3$. ■

EXERCISES

Find, using tables

1 $\displaystyle\int e^{-2x}\sin 5x\,dx$ **2** $\displaystyle\int \sqrt{4-x^2}\,dx$ **3** $\displaystyle\int x^2\sqrt{1-4x^2}\,dx$

4 $\displaystyle\int \frac{x^2\,dx}{\sqrt{4-3x^2}}$ **5** $\displaystyle\int \frac{x^2\,dx}{2+5x^2}$ **6** $\displaystyle\int (4-x^2)^{3/2}\,dx$

7 $\displaystyle\int x^3\sin\frac{x}{2}\,dx$ **8** $\displaystyle\int (\ln x)^4\,dx$ **9** $\displaystyle\int \frac{x^2-6x-2}{x\sqrt{10x^2+7}}\,dx$

10 $\displaystyle\int (x+3)^2\sqrt{x^2+2x+5}\,dx.$

Compute

11 $\displaystyle\int_0^1 \frac{x^4+2x^2-3}{x^4+2x^2+1}\,dx$ **12** $\displaystyle\int_0^1 \frac{x^2\,dx}{1+3x^2}$ **13** $\displaystyle\int_0^\pi e^{3x}\cos^6 x\,dx$

14 $\displaystyle\int_0^{2\pi}\cos^2 x\sin^8 x\,dx$ **15** $\displaystyle\int_0^1 \frac{dx}{(1+3x)^2(2+5x)}$ **16** $\displaystyle\int_1^2 (x\ln x)^3\,dx.$

10. MISCELLANEOUS EXERCISES

Compute

1 $\displaystyle\int \frac{dx}{(x-a)(x-b)}$ **2** $\displaystyle\int \frac{x^2\,dx}{\sqrt{x+1}}$ **3** $\displaystyle\int \frac{\sin 2x\,dx}{\sqrt{5+\cos 2x}}$

4 $\displaystyle\int \frac{dx}{(2-5x)^2}$ **5** $\displaystyle\int \sec^4 3x\tan^3 3x\,dx$ **6** $\displaystyle\int \tan^5 x\,dx$

7 $\displaystyle\int \sin^6 x\cos^3 x\,dx$ **8** $\displaystyle\int \sin^4 x\,dx$ **9** $\displaystyle\int x^5\sqrt{1+x^2}\,dx$

10 $\displaystyle\int \frac{x^3\,dx}{x^2+4x+13}$ **11** $\displaystyle\int \sin\sqrt{x}\,dx$ **12** $\displaystyle\int \frac{dx}{e^x+1}$

13 $\displaystyle\int \frac{x\,dx}{\sqrt{2x-x^2}}$ **14** $\displaystyle\int \frac{dx}{(1-x^2)^{3/2}}$ **15** $\displaystyle\int x^2\arctan x\,dx$

16 $\displaystyle\int \frac{(x+1)e^x}{xe^x+1}\,dx$ **17** $\displaystyle\int \frac{dx}{a^4x^2+b^2x^4}$ **18** $\displaystyle\int \frac{x^4+a^4}{x^4-a^4}\,dx$

19 $\displaystyle\int \frac{dx}{e^x+5+4e^{-x}}$ **20** $\displaystyle\int \frac{\ln(2+\sqrt{x})}{\sqrt{x}}\,dx$ **21** $\displaystyle\int \frac{dx}{x(\ln x)(\ln\ln x)}$

22 $\displaystyle\int \frac{x\,dx}{\sqrt{1-9x^4}}$ **23** $\displaystyle\int x^2\sin x\,dx$ **24** $\displaystyle\int \frac{e^x\,dx}{\sqrt{1+e^{2x}}}$

25 $\displaystyle\int \frac{dx}{x\sqrt{1+x^2}}$ **26** $\displaystyle\int \frac{\sqrt{x^2-a^2}}{x}\,dx.$

27   Evaluate $\displaystyle\int_{-a}^{a} \sqrt{a^2 - x^2}\ dx$ by inspection. [*Hint*   Interpret as an area.]

28 $\displaystyle\int_{-\alpha}^{\alpha} \sec x\ dx = \ln(\sec x + \tan x)\Big|_{-\alpha}^{\alpha} = \ln(\sec \alpha + \tan \alpha) - \ln(\sec \alpha - \tan \alpha)$

and by symmetry

$$\int_{-\alpha}^{\alpha} \sec x\ dx = 2\int_{0}^{\alpha} \sec x\ dx = 2\ln(\sec x + \tan x)\Big|_{0}^{\alpha} = 2\ln(\sec \alpha + \tan \alpha).$$

The answers appear different. Explain.

29* Suppose a quantity of hot fluid is stirred so that at any time t, its temperature $u(t)$ is uniform throughout the fluid. Suppose it loses heat to the outside only by *radiation*. Then it is known that

$$\frac{du}{dt} = -k(u^4 - a^4),$$

where $k > 0$ and a is the (constant) outside temperature. Express the *inverse* function $t = t(u)$ in terms of u. [*Hint*   See Example 4, p. 374.]

30* Find $\displaystyle\int \frac{x^2 + ax - 2a - 3}{(x - 1)^2}\ e^x\ dx.$

31   Find $\displaystyle\int_{0}^{b} x(b - x)^n\ dx.$ **32*** Find $\displaystyle\int \frac{dx}{\sqrt{e^a - e^x}}.$

Plane Analytic Geometry 9

1. TRANSLATION AND CIRCLES

René Descartes (1596–1650) slept till noon every day and invented analytic geometry. Analytic geometry is the study of geometry via algebra and coordinate systems. In honor of Descartes, rectangular coordinate systems are called *Cartesian*.

Translation of Axes A useful skill in working with coordinate systems is knowing where best to place the axes. While doing a problem in a given coordinate system, it may be convenient to introduce new coordinate axes parallel to the given ones. This operation is called **shifting** or **translating** the axes.

After a translation of axes, a point with coordinates (x, y) relative to the original axes acquires new coordinates $(\bar{x}, \bar{y})$ relative to the new axes. It is easy to express the relation between the old coordinates and the new ones. The translation sets up a new origin $\bar{\mathbf{0}}$ at a point (h, k) in the old coordinates (Fig. 1). From the figure, $x = \bar{x} + h$ and $y = \bar{y} + k$, no matter what quadrant $\bar{\mathbf{0}}$ lies in. Conversely, $\bar{x} = x - h$ and $\bar{y} = y - k$.

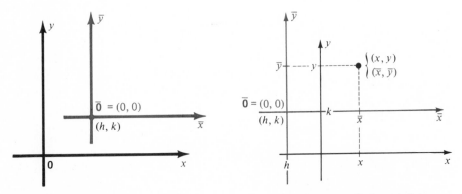

(a) Choice of new origin and axes

(b) By comparing horizontal distances (with due regard for sign): $x = \bar{x} + h$. Similarly, $y = \bar{y} + k$.

Fig. 1 Translation of axes

Translation of Axes   If the coordinate axes are translated so that the origin is moved to (h, k), then the new coordinates $(\bar{x}, \bar{y})$ and the old coordinates (x, y) of a point are related by the equations

$$\begin{cases} \bar{x} = x - h \\ \bar{y} = y - k \end{cases} \qquad \begin{cases} x = \bar{x} + h \\ y = \bar{y} + k. \end{cases}$$

To get the signs right, remember that $x = h$ and $y = k$ must imply $\bar{x} = 0$ and $\bar{y} = 0$, and vice versa.

Translation of axes is helpful in simplifying equations and computations. For example, take the equation $y + 3 = (x - 2)^2$. If we introduce a new coordinate system with origin at $(2, -3)$, then $x - 2 = \bar{x}$ and $y + 3 = \bar{y}$. Therefore the equation becomes $\bar{y} = \bar{x}^2$ in the new coordinates. Thus the graph of $y + 3 = (x - 2)^2$ is just the familiar quadratic curve $y = x^2$, but with its vertex shifted (Fig. 2).

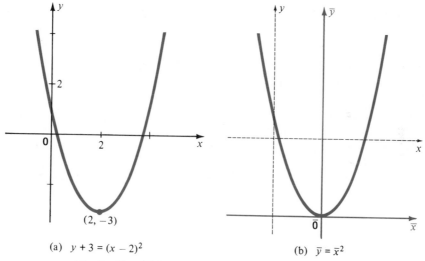

(a) $y + 3 = (x - 2)^2$ (b) $\bar{y} = \bar{x}^2$

Fig. 2 Simplifying an equation by translation of axes

In general, given an equation in x and y and new axes with origin at (h, k), it is easy to write the equation in $\bar{x}$ and $\bar{y}$ that describes the same locus. Simply substitute $x = \bar{x} + h$ and $y = \bar{y} + k$ for x and y in the given equation. For example, the equation of the line $ax + by = c$ is transformed into the equation $a(\bar{x} + h) + b(\bar{y} + k) = c$, that is

$$a\bar{x} + b\bar{y} = c - (ah + bk).$$

Note that the equation of the line is the same in either coordinate system except for the constant on the right side. Can you explain why geometrically?

Division of a Segment   Here is an example of a problem that is solved neatly using translation of axes. Given two points $\mathbf{p}_1 = (x_1, y_1)$ and $\mathbf{p}_2 = (x_2, y_2)$, find the point on the segment $\overline{\mathbf{p}_1\mathbf{p}_2}$ that is $\frac{2}{5}$ of the way from $\mathbf{p}_1$ to $\mathbf{p}_2$. More generally, if $0 \le r \le 1$, find the point r of the way from $\mathbf{p}_1$ to $\mathbf{p}_2$.

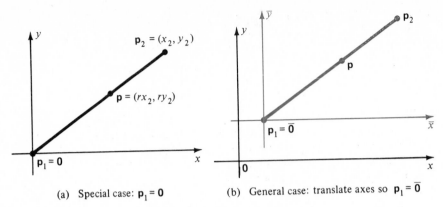

(a) Special case: $\mathbf{p}_1 = \mathbf{0}$ (b) General case: translate axes so $\mathbf{p}_1 = \overline{\mathbf{0}}$

Fig. 3   Division of a segment

In the special case that $\mathbf{p}_1 = \mathbf{0}$, the solution is obvious (Fig. 3a). The desired point is (rx_2, ry_2). In the general case (Fig. 3b), translate axes so the new origin is at $\mathbf{p}_1 = (x_1, y_1)$. Then the *new* coordinates of $\mathbf{p}_2$ are $(\bar{x}_2, \bar{y}_2) = (x_2 - x_1, y_2 - y_1)$. According to the special case, the *new* coordinates of the division point $\mathbf{p}$ are $(r\bar{x}_2, r\bar{y}_2) = (r(x_2 - x_1), r(y_2 - y_1))$. But the coordinate translation is $x = \bar{x} + x_1$ and $y = \bar{y} + y_1$, so the *old* coordinates of $\mathbf{p}$ are

$$(r(x_2 - x_1) + x_1, r(y_2 - y_1) + y_1) = ((1 - r)x_1 + rx_2, (1 - r)y_1 + ry_2).$$

Division of a Segment   If $0 \leq r \leq 1$, then the point that is r of the way from $\mathbf{p}_1 = (x_1, y_1)$ to $\mathbf{p}_2 = (x_2, y_2)$ is

$$\mathbf{p} = (x, y),$$

where $x = (1 - r)x_1 + rx_2$ and $y = (1 - r)y_1 + ry_2.$

Examples   The point that is $\frac{2}{5}$ of the way from $\mathbf{p}_1$ to $\mathbf{p}_2$ is

$$(\tfrac{3}{5}x_1 + \tfrac{2}{5}x_2, \tfrac{3}{5}y_1 + \tfrac{2}{5}y_2).$$

The midpoint of $\overline{\mathbf{p}_1\mathbf{p}_2}$ is $(\tfrac{1}{2}(x_1 + x_2), \tfrac{1}{2}(y_1 + y_2)).$

The Circle   Recall the equation of a circle, p. 44:

The equation of the circle with center (a, b) and radius r is

$$(x - a)^2 + (y - b)^2 = r^2.$$

This equation is the algebraic statement of the fact that the distance from each point (x, y) on the circle to the center equals the radius. If new axes are centered at (a, b), the equation becomes simply $\bar{x}^2 + \bar{y}^2 = r^2$.

Let us examine the general equation of a circle. In expanded form the equation is

$$(x^2 - 2ax + a^2) + (y^2 - 2by + b^2) = r^2,$$

$$x^2 + y^2 - 2ax - 2by + (a^2 + b^2 - r^2) = 0.$$

Now suppose we start with an equation of the form

$$x^2 + y^2 - 2ax - 2by + c = 0.$$

What is the locus of all points (x, y) that satisfy it? (Of course we suspect a circle.) We complete the squares for both the x and the y terms:

$$(x^2 - 2ax + a^2) + (y^2 - 2by + b^2) + c = a^2 + b^2,$$

$$(x - a)^2 + (y - b)^2 = a^2 + b^2 - c.$$

Now here is a hitch: the left side is a sum of squares, hence non-negative. But the right side is not guaranteed to be non-negative! We are forced to consider three cases:

(1) $a^2 + b^2 - c < 0$. Then no point (x, y) satisfies the equation; the locus is empty.

(2) $a^2 + b^2 - c = 0$. Then only $(x, y) = (a, b)$ satisfies the equation because the left side is positive for any other point; the locus is the single point (a, b).

(3) $a^2 + b^2 - c > 0$. Then we set $r = \sqrt{a^2 + b^2 - c}$, and the equation becomes

$$(x - a)^2 + (y - b)^2 = r^2.$$

The locus in this case is an honest circle.

■ **EXAMPLE 1** Describe the set of points (x, y) that satisfy

(a) $x^2 + y^2 - 2x + 4y - 4 = 0,$ (b) $x^2 + y^2 - 2x + 4y + 5 = 0,$
(c) $x^2 + y^2 - 2x + 4y + 6 = 0.$

Solution Complete the squares in x and y:

$$x^2 - 2x + y^2 + 4y = (x - 1)^2 + (y + 2)^2 - 5.$$

Hence the three cases become

(a) $(x - 1)^2 + (y + 2)^2 = 9 = 3^2$:   circle;   center $(1, -2)$,   radius 3.
(b) $(x - 1)^2 + (y + 2)^2 = 0$:   single point: $(1, -2)$.
(c) $(x - 1)^2 + (y + 2)^2 = -1$:   empty locus.   ▨

Intersection of Line and Circle Given a circle and a line, there are three possibilities: (1) they do not intersect, (2) they have exactly one common point— the line is **tangent** to the circle, (3) they intersect in two distinct points.

From the equations of the circle and line, we can find explicitly the intersections (if any). The algebra is simplest if the circle has its center at the origin. Otherwise, we make a preliminary translation to move the origin to the center.

Suppose the equations are

$$\begin{cases} x^2 + y^2 = r^2 \\ ax + by = c, \quad a^2 + b^2 > 0. \end{cases}$$

The points of intersection are the common solutions (x, y) of the two equations. Either $a \neq 0$ or $b \neq 0$; assume $a \neq 0$ and multiply the first equation by a^2:

$$a^2 x^2 + a^2 y^2 = a^2 r^2.$$

Substitute $c - by$ for ax, eliminating x:

$$(c - by)^2 + a^2y^2 = a^2r^2, \qquad (a^2 + b^2)y^2 - 2bcy + (c^2 - a^2r^2) = 0.$$

This is a quadratic equation for y; it has 0, 1, or 2 solutions, depending on the discriminant Δ.

If $\Delta < 0$, then there are no real solutions, hence no points of intersection. If $\Delta \geq 0$, solve for y by the quadratic formula, then find x from the relation $ax = c - by$. Note that if $\Delta = 0$, there is one point of intersection (tangency), and if $\Delta > 0$, there are two points of intersection. (In case $a = 0$, then $b \neq 0$ and a similar argument applies.)

■ **EXAMPLE 2**   Find the intersections of $x^2 + y^2 = 4$ and $x + 2y = 1$.

Solution   Replace x^2 by $(1 - 2y)^2$:

$$(1 - 2y)^2 + y^2 = 4, \qquad 1 - 4y + 5y^2 = 4, \qquad 5y^2 - 4y - 3 = 0.$$

The discriminant of the quadratic equation is

$$\Delta = (-4)^2 - 4(5)(-3) = 16 + 60 = 76 > 0.$$

The quadratic has two solutions:

$$y = \frac{4 \pm \sqrt{76}}{10} = \frac{4 \pm 2\sqrt{19}}{10} = \frac{2}{5} \pm \frac{1}{5}\sqrt{19}.$$

The corresponding values of x are

$$x = 1 - 2y = 1 - \tfrac{2}{5}(2 \pm \sqrt{19}) = \tfrac{1}{5} \mp \tfrac{2}{5}\sqrt{19}.$$

See Fig. 4. Hence the two intersection points are

$$(\tfrac{1}{5} - \tfrac{2}{5}\sqrt{19}, \tfrac{2}{5} + \tfrac{1}{5}\sqrt{19}) \quad \text{and} \quad (\tfrac{1}{5} + \tfrac{2}{5}\sqrt{19}, \tfrac{2}{5} - \tfrac{1}{5}\sqrt{19}). \qquad ■$$

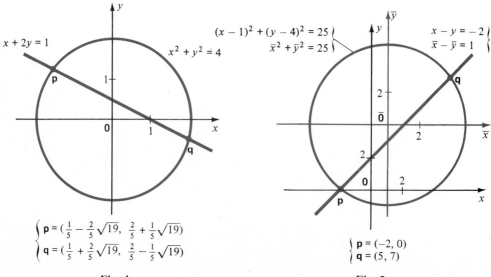

$$\begin{cases} \mathbf{p} = (\tfrac{1}{5} - \tfrac{2}{5}\sqrt{19}, \tfrac{2}{5} + \tfrac{1}{5}\sqrt{19}) \\ \mathbf{q} = (\tfrac{1}{5} + \tfrac{2}{5}\sqrt{19}, \tfrac{2}{5} - \tfrac{1}{5}\sqrt{19}) \end{cases}$$

Fig. 4

$$\begin{cases} \mathbf{p} = (-2, 0) \\ \mathbf{q} = (5, 7) \end{cases}$$

Fig. 5

■ **EXAMPLE 3**   Find the intersection of $(x - 1)^2 + (y - 4)^2 = 25$
and $x - y = -2$.

Solution   Translate the origin to $(1, 4)$. Then $\bar{x} = x - 1$, $\bar{y} = y - 4$. The new equation of the circle is $\bar{x}^2 + \bar{y}^2 = 25$, and the new equation of the line is

$$(\bar{x} + 1) - (\bar{y} + 4) = -2; \quad \text{that is,} \quad \bar{x} - \bar{y} = 1.$$

Now solve the system of equations

$$\begin{cases} \bar{x}^2 + \bar{y}^2 = 25 \\ \bar{x} - \bar{y} = 1. \end{cases}$$

Eliminate $\bar{x}$:

$$(\bar{y} + 1)^2 + \bar{y}^2 = 25, \quad 2\bar{y}^2 + 2\bar{y} - 24 = 0, \quad \bar{y}^2 + \bar{y} - 12 = 0,$$
$$(\bar{y} + 4)(\bar{y} - 3) = 0, \quad \bar{y} = -4, \quad \bar{y} = 3.$$

The corresponding values of $\bar{x} = \bar{y} + 1$ are $\bar{x} = -3$ and $\bar{x} = 4$, so the two points of intersection are $(\bar{x}, \bar{y}) = (-3, -4)$, $(4, 3)$. The (x, y) coordinates of these points are

$$(x, y) = (\bar{x} + 1, \bar{y} + 4) = (-2, 0), \quad (5, 7).$$

See Fig. 5.

     ■

Remark   Example 3 can be solved without shifting axes. For instance x can be eliminated from the original pair of equations and the resulting quadratic solved for y. The computations involved are a bit longer.

Intersection of Two Circles   Given two non-concentric circles, find their intersection. This problem can be reduced to the previous problem of a circle and a line. Let the equations of the circles be

(1) $$x^2 + y^2 - 2a_1 x - 2b_1 y = c_1,$$

(2) $$x^2 + y^2 - 2a_2 x - 2b_2 y = c_2.$$

The centers are (a_1, b_1) and (a_2, b_2) respectively, as can be seen by completing the squares. Subtract the two equations:

(3) $$2(a_1 - a_2)x + 2(b_1 - b_2)y = c_1 - c_2.$$

Because the circles are not concentric, $(a_1, b_1) \neq (a_2, b_2)$; hence $a_1 - a_2$ and $b_1 - b_2$ are not both 0, so (3) is the equation of a line L.
    This line L intersects either circle in precisely the same points where the circles intersect each other. Why? Because the intersections of the circles are the simultaneous solutions of the equations (1) and (2). The intersections of L and the first circle are the simultaneous solutions of equations (1) and (3). In either case the answers are the same. For if (x, y) satisfies both (1) and (2), then (x, y) also satisfies (3), the difference of (1) and (2). Likewise if (x, y) satisfies (1) and (3), then (x, y) also satisfies (2), the difference of (1) and (3). Similarly, the intersections of L and the second circle are the same points.
    Thus finding the intersection of two circles is equivalent to finding the intersection of a circle and a line. But we know how to do that!

■ **EXAMPLE 4** Find the intersections of the circles

$$x^2 + y^2 = 25, \qquad x^2 + y^2 + 2x + 2y = 31.$$

Solution Subtract the equations:

$$2x + 2y = 31 - 25 = 6, \qquad x + y = 3.$$

This is the equation of a line. To obtain its intersection with the first circle, solve the system

$$\begin{cases} x^2 + y^2 = 25 \\ \quad x + y = 3. \end{cases}$$

Eliminate y: $x^2 + (3 - x)^2 = 25,$ $2x^2 - 6x - 16 = 0,$

$$x^2 - 3x - 8 = 0, \qquad x = \tfrac{1}{2}(3 \pm \sqrt{41}\,).$$

The corresponding values of $y = 3 - x$ are $y = \tfrac{1}{2}(3 \mp \sqrt{41}\,)$. Therefore the two intersection points are

$$(\tfrac{3}{2} + \tfrac{1}{2}\sqrt{41}, \tfrac{3}{2} - \tfrac{1}{2}\sqrt{41}\,) \qquad \text{and} \qquad (\tfrac{3}{2} - \tfrac{1}{2}\sqrt{41}, \tfrac{3}{2} + \tfrac{1}{2}\sqrt{41}\,). \qquad ■$$

Tangents In Chapter 1, we found the tangent to a circle at one of its points (Example 1, p. 49.) The basic fact is that a line tangent to a circle is perpendicular to the radius at the point of contact. Thus, if (u, v) is a point on the circle $x^2 + y^2 = r^2$, the tangent at (u, v) is perpendicular to the radius [segment from $(0, 0)$ to (u, v)]. Since the radius has slope v/u, the tangent has slope $-u/v$. Therefore the equation of the tangent is

$$y - v = -\frac{u}{v}(x - u), \qquad ux + vy = u^2 + v^2 = r^2.$$

The problem of finding the two tangent lines to a circle from an external point is harder, but the basic idea is the same.

■ **EXAMPLE 5** Find the two tangents to the circle $x^2 + y^2 = 4$ from $(3, 0)$.

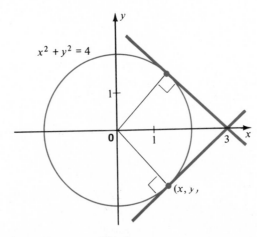

Fig. 6

Solution Suppose a tangent meets the circle at (x, y). Then the segments (Fig. 6)

$$\overline{(0, 0)(x, y)} \quad \text{and} \quad \overline{(3, 0)(x, y)}$$

are perpendicular. Therefore

$$\frac{y}{x} \cdot \frac{y}{x - 3} = -1, \quad \text{that is,} \quad x^2 + y^2 = 3x.$$

Therefore the points of contact satisfy two equations:

$$x^2 + y^2 = 4, \quad x^2 + y^2 = 3x.$$

Subtract: $3x = 4$, $x = \frac{4}{3}$. Hence

$$y^2 = 4 - \left(\frac{4}{3}\right)^2 = \frac{36}{9} - \frac{16}{9} = \frac{20}{9}, \quad y = \pm\frac{2}{3}\sqrt{5}.$$

Each tangent passes through $(3, 0)$ and one of the two points $(\frac{4}{3}, \pm\frac{2}{3}\sqrt{5})$. By the two-point form, the equations of the two tangents are

$$\frac{y - 0}{x - 3} = \frac{\pm\frac{2}{3}\sqrt{5}}{\frac{4}{3} - 3} = \frac{\pm\frac{2}{3}\sqrt{5}}{-\frac{5}{3}} = \mp\frac{2}{5}\sqrt{5}, \quad \text{that is,} \quad y = \pm\frac{2}{5}\sqrt{5}\,(x - 3).$$

Alternative solution The tangent lines are of the form $ux + vy = 4$, where (u, v) is the point of contact. Find the points (u, v) for which the tangent line passes through $(3, 0)$. The condition is simply $3u + 0 \cdot v = 4$, that is, $u = \frac{4}{3}$. But

$$u^2 + v^2 = 4, \quad v^2 = 4 - \left(\frac{4}{3}\right)^2 = \frac{20}{9}, \quad v = \pm\frac{2}{3}\sqrt{5}.$$

Therefore the tangent lines are

$$\frac{4}{3}x \pm \frac{2}{3}\sqrt{5}\,y = 4, \quad \text{that is,} \quad y = \pm\frac{2}{5}\sqrt{5}\,(x - 3). \quad \blacksquare$$

Remark The algebra in the solution of Example 5 turned out to be fairly simple because $(3, 0)$ is on the x-axis. For an external point off the axes, the algebra will be more complicated.

EXERCISES

An $\bar{x}, \bar{y}$-coordinate system is introduced with its origin at $(-7, 3)$. Find the $\bar{x}, \bar{y}$-coordinates of the point (x, y)

1 $(1, 6)$ **2** $(5, 4)$ **3** $(0, 0)$ **4** $(-8, 1)$ **5** $(-5, -6)$ **6** $(0, 2)$.

With the $\bar{x}, \bar{y}$-coordinate system as above, find the x, y-coordinates of the point:

7 $(\bar{x}, \bar{y}) = (1, 3)$ **8** $(\bar{x}, \bar{y}) = (4, -2)$ **9** $(\bar{x}, \bar{y}) = (-6, -7)$
10 $(\bar{x}, \bar{y}) = (0, 3)$ **11** $(\bar{x}, \bar{y}) = (-7, 3)$ **12** $(\bar{x}, \bar{y}) = (1, 6)$.

Describe a translation of axes that converts the first equation into the second

13 $3x - 2y = 1$, $3\bar{x} - 2\bar{y} = 0$ **14** $x + 2y = 5$, $\bar{x} + 2\bar{y} = 0$

15 $y = x^2 + 2x + 2$, $\bar{y} = \bar{x}^2 + 1$ **16** $y = 3 + \dfrac{x - 1}{1 + (x - 1)^2}$, $\bar{y} = \dfrac{\bar{x}}{1 + \bar{x}^2}$

17 $y = \sin(x - \frac{1}{6}\pi) - 1$, $\bar{y} = \sin \bar{x}$
18 $y = \cos x + \cos(x + \frac{1}{3}\pi) + \cos(x + \frac{2}{3}\pi)$, $\bar{y} = \cos(\bar{x} - \frac{1}{3}\pi) + \cos \bar{x} + \cos(\bar{x} + \frac{1}{3}\pi)$.

Find the point

19 $\frac{2}{3}$ of the way from $(1, 1)$ to $(4, -5)$.
20 $\frac{1}{5}$ of the way from $(4, 0)$ to $(0, 4)$.

21   Interpret $(x, y) = ((1 - r)x_1 + rx_2, (1 - r)y_1 + ry_2)$ when $r > 1$.
22   (cont.) Do the same when $r < 0$.

Describe the set of points that satisfy

23 $x^2 + y^2 - 4x - 4y = 0$

24 $x^2 + y^2 - 6x = 0$

25 $x^2 + y^2 + 2x + 6y = 26$

26 $x^2 + y^2 - x + y = \frac{33}{2}$

27 $x^2 + y^2 - x + 2y = 0$

28 $x^2 + y^2 + 6x - 8y = 25$

29 $2x^2 + 2y^2 - 3x - 5y + 1 = 0$

30 $3x^2 + 3y^2 - x - y = 0$.

Find the intersection of the circle and the line

31 $x^2 + y^2 = 9$, $y = x + 1$

32 $x^2 + y^2 = 10$, $x + y = 1$

33 $x^2 + y^2 = 5$, $x + 2y = 5$

34 $x^2 + y^2 = 6$, $y = 2x - 7$

35 $x^2 + y^2 - 4x - 2y + 4 = 0$, $2x - 5y = 6$ **36** $x^2 + y^2 + 4x + 5y = 0$, $x - 4y = 1$

37   circle with center $(1, 1)$ and radius 5, line through $(0, 2)$ and $(4, 0)$

38   circle with center at $(3, 4)$ and passing through $(0, 0)$, line through $(1, 1)$ parallel to $y = 3x$.

Find the point or points of intersection (if any) of the circles

39 $x^2 + y^2 = 9$, $x^2 + y^2 + 8x + 12 = 0$

40 $x^2 + y^2 - 2x - 2y = 0$, $(x - 2)^2 + (y - 3)^2 = 4$

41 $x^2 + y^2 - 10x + 6y + 33 = 0$, $x^2 + y^2 + 2x + 4y - 4 = 0$

42 $(x - 1)^2 + (y - 2)^2 = \frac{5}{4}$, $(x - 3)^2 + (y - 6)^2 = \frac{45}{4}$.

Find the tangent to $x^2 + y^2 = 1$ at

43 $(-1, 0)$

44 $(0, -1)$

45 $(\frac{1}{2}\sqrt{2}, -\frac{1}{2}\sqrt{2})$

46 $(-\frac{1}{2}\sqrt{2}, -\frac{1}{2}\sqrt{2})$

47 $(\frac{1}{2}\sqrt{3}, \frac{1}{2})$

48 $(-\frac{1}{2}, \frac{1}{2}\sqrt{3})$.

49   Find the tangents to $(x - 1)^2 + (y - 2)^2 = 1$ that pass through $(0, 0)$.

50   Find the tangents to $x^2 + y^2 = 1$ that pass through $(2, 2)$.

51   Find the tangents to $x^2 + y^2 = 13$ that pass through $(-5, 1)$.

52   Find the tangents to $(x - 1)^2 + y^2 = 1$ that pass through $(0, -2)$.

53   Show that the circles $x^2 + y^2 - 2x - 4y - 6\frac{1}{4} = 0$ and $x^2 + y^2 - 6x - 12y + 43\frac{3}{4} = 0$ are tangent. [*Hint*   Compute the distance between their centers.]

54*   Find all common tangents to the circles $x^2 - 2x + y^2 = 0$ and $x^2 + 4x + y^2 = 0$.

55   Show that the circles $x^2 + y^2 = 1$ and $25x^2 + 25y^2 - 8x - 6y = 15$ are tangent.

56   (cont.) Show that the tangent to $(x - a)^2 + (y - b)^2 = r^2$ at a point (u, v) on the circle is $(u - a)(x - a) + (v - b)(y - b) = r^2$.

57   Find the tangents to $x^2 + y^2 = 25$ with slope $\frac{3}{4}$.

58   Find the circle with center $(2, -1)$ tangent to $3x + y = 0$.

59   Find the circles through $(1, 8)$ tangent to both axes.

60   Find the circle tangent to the x-axis at $(a, 0)$ and passing through (b, c), $c \neq 0$.

61   Prove that the circles $x^2 + y^2 - 8x + 2y + 8 = 0$ and $x^2 + y^2 - 2x + 10y + 22 = 0$ are tangent and find their point of tangency.

62   Find the circle through $(2, 2)$, $(4, 1)$, and $(3, -1)$.

2. LOCUS

   A geometrical figure is called a **locus** (plural, **loci**) if it is described as the set of all points that satisfy a certain condition. For example, the locus of all points at a distance 3 from a fixed point **p** is a circle of radius 3 centered at **p**. The locus of all points equidistant from fixed points **p** and **q** is the perpendicular bisector of the segment $\overline{\mathbf{pq}}$. The locus of all points (x, y) for which $y = f(x)$ is the graph of the function $f(x)$.

■ **EXAMPLE 1**   Find the locus of all points that are twice as far from the point $(6, 0)$ as from the origin.

Solution   The condition is satisfied by a point (x, y) if and only if

$$\sqrt{(x - 6)^2 + y^2} = 2\sqrt{x^2 + y^2}.$$

Square both sides:

$$(x - 6)^2 + y^2 = 4x^2 + 4y^2, \qquad 3x^2 + 3y^2 + 12x = 36, \qquad x^2 + y^2 + 4x = 12.$$

This is the equation of a circle (Fig. 1). For more precise information, complete the square in x:

$$(x^2 + 4x + 4) + y^2 = 12 + 4, \qquad (x + 2)^2 + y^2 = 16.$$

The locus is the circle with center $(-2, 0)$, radius 4.

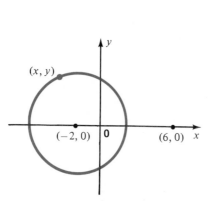

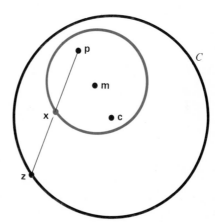

Fig. 1   Locus of all points twice as far from $(6, 0)$ as from **0**

Fig. 2   Locus of all midpoints **x** of segments $\overline{\mathbf{pz}}$, where **z** traces the circle C ■

■ **EXAMPLE 2**   Let **p** be a point and C a circle. Find the locus of the midpoint of the segment $\overline{\mathbf{pz}}$ as **z** traces C.

Solution   Choose coordinates so that C is $x^2 + y^2 = r^2$ and $\mathbf{p} = (a, 0)$. Let $\mathbf{z} = (u, v)$. Since **z** traces the circle, $u^2 + v^2 = r^2$. The midpoint of $\overline{\mathbf{pz}}$ is $(x, y) = (\frac{1}{2}(a + u), \frac{1}{2}v)$. Thus

$$x = \tfrac{1}{2}(a + u), \qquad y = \tfrac{1}{2}v, \qquad \text{where} \quad u^2 + v^2 = r^2.$$

The problem asks for the locus of (x, y), that is, a relation between x and y. Therefore, express u and v in terms of x and y and substitute into $u^2 + v^2 = r^2$:

$$u = 2x - a, \qquad v = 2y, \qquad (2x - a)^2 + (2y)^2 = r^2,$$

hence

$$(x - \tfrac{1}{2}a)^2 + y^2 = (\tfrac{1}{2}r)^2.$$

The locus is the circle with center $(\frac{1}{2}a, 0)$ and radius $\frac{1}{2}r$. In geometric language, it is the circle with center **m**, the midpoint of the segment from **p** to the center **c** of C, and radius half the radius of C. See Fig. 2.   ■

Remark   Example 2 illustrates a basic technique in locus problems. The point **x** that generates the locus is usually related to a point **z** that satisfies a known equation (locus). Express **z** in terms of **x**, then substitute into the equation for **z**. This yields an equation for **x**.

■ **EXAMPLE 3**   Let **z** trace the parabola $y = x^2$. Find the locus of the point **x** that is $\frac{1}{3}$ of the way from $(-6, 0)$ to **z**.

Solution   This is precisely the kind of situation discussed in the preceding remark. Here **z** traces the curve $y = x^2$, so we can write $\mathbf{z} = (z, w)$, where $w = z^2$. We write $\mathbf{x} = (x, y)$ and seek the relation between x and y. [Note carefully the change in notation. You cannot write *both* $\mathbf{x} = (x, y)$ *and* $\mathbf{z} = (x, y)$ because **x** and **z** are completely different points. So if you want the answer in x and y, you *must* give the coordinates of **z** some other names.] By the rule for division of a segment

$$x = \tfrac{2}{3}(-6) + \tfrac{1}{3}z = \tfrac{1}{3}z - 4 \qquad \text{and} \qquad y = \tfrac{2}{3} \cdot 0 + \tfrac{1}{3}w = \tfrac{1}{3}w,$$

hence $$z = 3x + 12 \qquad \text{and} \qquad w = 3y.$$

Substitute into $w = z^2$:

$$3y = (3x + 12)^2 = 9(x + 4)^2, \qquad \text{that is,} \qquad y = 3(x + 4)^2.$$

The result is another parabola (Fig. 3).

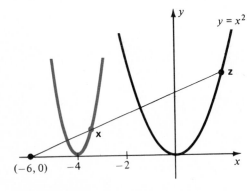

Fig. 3   Locus of all points **x** one-third of the way from $(-6, 0)$ to **z**

■

■ **EXAMPLE 4**   Let **p** lie inside a circle C. Find the locus of the midpoints of all chords through **p**.

Solution   Take axes so that **p** is at the origin and C is the circle $(x - c)^2 + y^2 = r^2$ with $0 < c < r$. The typical chord through **p** is $y = mx$. The line through the center $(c, 0)$ of C perpendicular to this chord is $y = (-1/m)(x - c)$. This line meets the chord precisely in the chord's midpoint **x**. Thus $\mathbf{x} = (x, y)$ is the solution of the system

$$y = mx, \qquad -my = x - c.$$

We easily find $$(x, y) = \left(\frac{c}{1 + m^2}, \frac{cm}{1 + m^2} \right).$$

To find the locus of (x, y) we eliminate the parameter m. The quickest way is to observe that

$$x^2 + y^2 = \frac{c^2 + (cm)^2}{(1 + m^2)^2} = \frac{c^2(1 + m^2)}{(1 + m^2)^2} = c\,\frac{c}{1 + m^2} = cx,$$

$$x^2 + y^2 - cx = 0.$$

Now we complete the square: $(x - \tfrac{1}{2}c)^2 + y^2 = (\tfrac{1}{2}c)^2.$

Therefore the locus is the circle with diameter $\overline{\mathbf{pc}}$, where $\mathbf{c}$ is the center of the given circle C. See Fig. 4.

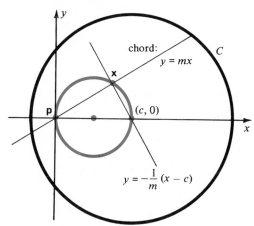

Fig. 4 Locus of midpoints $\mathbf{x}$ of all chords of C through $\mathbf{p}$ ($\mathbf{p}$ taken at $\mathbf{0}$)

■

EXERCISES

1   Let L be a line and $\mathbf{p}$ a point not on L. Find the locus of the midpoint of $\overline{\mathbf{pz}}$, where z varies over L.

2   A 10-ft ladder leans against a wall, and its foot slides along the floor as its top slides down the wall. Find the locus of its midpoint.

3   Find the locus of the midpoints of all chords of length 6 of a circle of radius 5.

4   At each point of a circle of radius 1 is drawn a segment of length 3, tangent at one end to the circle. Find the locus of the other ends of the segments.

5   Find the locus of all points whose distance from $(0, 0)$ is five times the distance from $(3, 4)$.

6   Do Example 2 by choosing the origin at $\mathbf{p}$.

7   Let $\mathbf{z}$ trace the parabola $y = x^2$ and let $\mathbf{a}$ be any point of the plane. Find the locus of the midpoint of $\overline{\mathbf{az}}$.

8   Find the locus in Example 4 if $\mathbf{p}$ lies on C, outside C.

9   Find the foot of the perpendicular dropped from the point (u, v) to the line $ax + by = c$.

10* (cont.) Given lines L_1, L_2, and L. For each $\mathbf{z}$ on L, let $\mathbf{p}_i$ be the foot of the perpendicular from $\mathbf{z}$ to L_i. Describe the locus of the midpoint of $\overline{\mathbf{p}_1\mathbf{p}_2}$.

11   Let $\mathbf{z} = (u, v)$ trace a circle C passing through $\mathbf{0}$. Find the locus of

$$\mathbf{x} = (x, y) = \left(\frac{u}{u^2 + v^2}, \frac{v}{u^2 + v^2}\right).$$

12   (cont.) Solve the same problem where $\mathbf{z}$ traces a line L *not* passing through $\mathbf{0}$.

13   Let C_1 and C_2 be two circles that are external to each other, and let $a > 0$. Describe the locus of all points $\mathbf{x}$ such that the length of the tangents from $\mathbf{x}$ to C_1 is a times the length of the tangents from $\mathbf{x}$ to C_2.

14*   Let C_1 and C_2 be circles through **0**. Each line L through **0** meets C_1 in a second point $\mathbf{x}_1$ and C_2 in another point $\mathbf{x}_2$. Find the locus of the midpoint of $\mathbf{x}_1$ and $\mathbf{x}_2$ as L varies.

15   Find the locus of the centers of all circles that are simultaneously tangent to the x-axis and to the circle $x^2 + y^2 = 2y$.

16   Let X and Y be fixed perpendicular lines and $\mathbf{a}$ a point on neither. A variable line L through $\mathbf{a}$ meets X in $\mathbf{u}$, and the line M through $\mathbf{a}$ perpendicular to L meets Y in $\mathbf{v}$. Find the locus of the midpoint of $\overline{\mathbf{uv}}$.

3. PARABOLA AND ELLIPSE

The ancient Greek geometers discovered that on cutting a right circular cone by various planes, they obtained three types of remarkable curves called **conic sections**, or **conics** (Fig. 1).

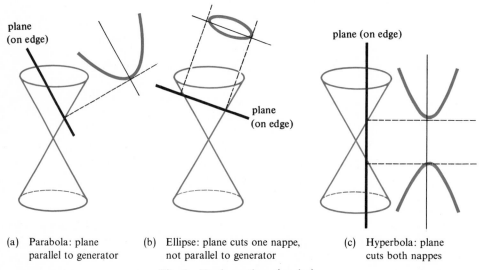

(a) Parabola: plane   (b) Ellipse: plane cuts one nappe,   (c) Hyperbola: plane
     parallel to generator   not parallel to generator   cuts both nappes

Fig. 1   Conic sections (conics)

A special case of the ellipse is a circle, obtained by cutting the cone by a plane parallel to its base.

A section of the cone by a plane through its apex is called a **degenerate conic** (Fig. 2).

It is possible to define the conic sections as certain geometric loci. We shall do so because this is less complicated than starting from plane sections of a cone. We shall discuss the parabola and the ellipse in this section and the hyperbola in Section 4.

The Parabola   A **parabola** is the locus of all points $\mathbf{x}$ equidistant from a fixed line D and a fixed point $\mathbf{p}$ not on D. See Fig. 3a. We call D the **directrix** and $\mathbf{p}$ the **focus** of the parabola. For convenience let us take $(0, p)$ with $p > 0$ as the focus and $y = -p$ as the directrix. Then the equation of the parabola follows easily, as was shown on p. 45, Chapter 1. The curve is shown in Fig. 3b.

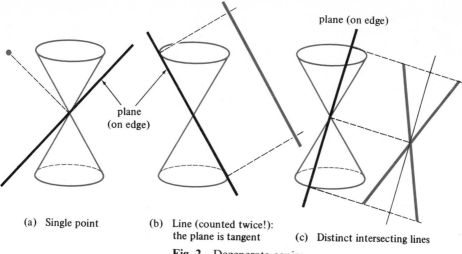

(a) Single point   (b) Line (counted twice!): the plane is tangent   (c) Distinct intersecting lines

Fig. 2 Degenerate conics

The equation of the parabola with focus $(0, p)$ and directrix $y = -p$ is

$$y = \frac{1}{4p} x^2.$$

For our choice of axes, the parabola is the graph of a quadratic polynomial. Conversely, the graph of any quadratic polynomial $y = ax^2 + bx + c$ (with $a \neq 0$) is a parabola. This is just a matter of completing the square:

$$y = a(x - h)^2 + k, \qquad y - k = a(x - h)^2,$$

where h and k are easily determined. Relative to new axes defined by $\bar{x} = x - h$, $\bar{y} = y - k$, the curve is $\bar{y} = a\bar{x}^2$, a parabola with $4p = 1/a$. If $a > 0$, the curve opens upward: if $a < 0$, it opens downward.

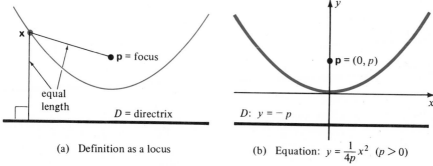

(a) Definition as a locus   (b) Equation: $y = \frac{1}{4p} x^2$ $(p > 0)$

Fig. 3 The parabola

The line through the focus perpendicular to the directrix is called the **axis** of the parabola. The point of intersection of the axis with the parabola is the **vertex** of the parabola (Fig. 4).

By interchanging the roles of x and y, we see that a parabola whose axis is parallel to the x-axis is the locus of an equation $x = ay^2 + by + c$. Some examples are shown in Fig. 5.

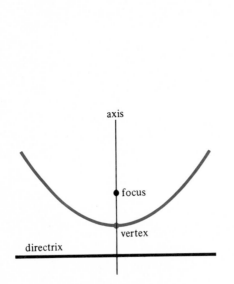

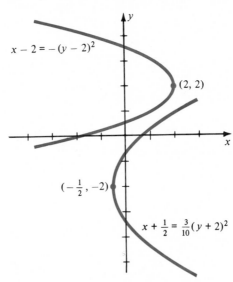

Fig. 4   Geometry of the parabola   **Fig. 5**   Parabolas with horizontal axes

---

The equation $$y - k = a(x - h)^2$$

represents a parabola with vertex at (h, k) and axis vertical. The parabola opens upward if $a > 0$, downward if $a < 0$.

The equation $$x - h = a(y - k)^2$$

represents a parabola with vertex at (h, k) and axis horizontal. The parabola opens to the right if $a > 0$, to the left if $a < 0$.

■ **EXAMPLE 1**   Sketch the curve $x = 2y^2 + 4y + 5$.   Find its vertex, axis, focus, and directrix.

Solution   Complete the square:

$$x = 2(y^2 + 2y + 1) - 2 + 5, \qquad x - 3 = 2(y + 1)^2.$$

The curve is a parabola, vertex $(3, -1)$, axis $y = -1$, horizontal opening to the right. To find the focus and the directrix, write the equation in the form $x - 3 = (1/4p)(y + 1)^2$, where $p = \frac{1}{8}$. The focus is $(3 + \frac{1}{8}, -1) = (\frac{25}{8}, -1)$; the directrix is $x = 3 - \frac{1}{8} = \frac{23}{8}$. See Fig. 6.   ■

■ **EXAMPLE 2**   Find the equation of a vertical parabola, vertex $(1, 2)$, passing through $(5, 0)$.

Solution   The equation is of the form

$$y - 2 = a(x - 1)^2.$$

To find a, set $x = 5$, $y = 0$:

$$-2 = a(5 - 1)^2, \qquad a = -\tfrac{1}{8}.$$

Hence the desired parabola is $y - 2 = -\tfrac{1}{8}(x - 1)^2$.   ■

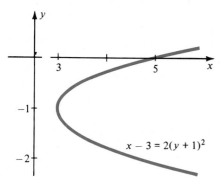

Fig. 6

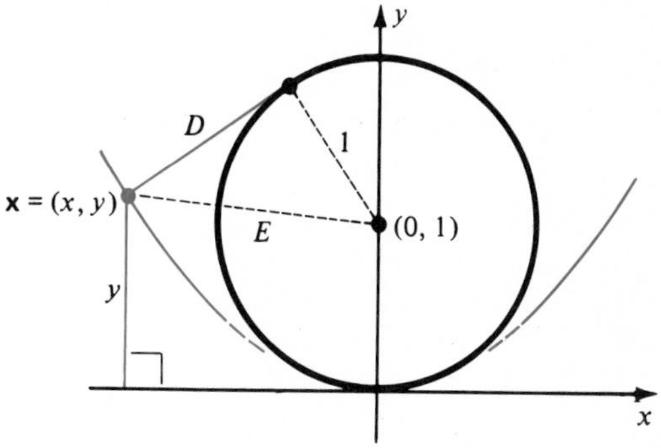

Fig. 7   From the right triangle,
$$D^2 = E^2 - 1 = [x^2 + (y - 1)^2] - 1$$

Parabolas arise in many locus problems. Here is an example.

■ **EXAMPLE 3**   Find the locus of all points **x** whose distance to the x-axis equals the length of the tangent(s) from **x** to the circle $x^2 + (y - 1)^2 = 1$.

Solution   Draw a careful figure (Fig. 7). The condition on the distance is

$$y^2 = D^2 = [x^2 + (y - 1)^2] - 1.$$

Simplify: $x^2 - 2y = 0, \qquad y = \tfrac{1}{2}x^2.$

The locus is a parabola.   ■

The Ellipse   An **ellipse** is the locus of all points **x** such that the sum of the distances of **x** from two fixed points **p** and **q** is a constant [greater than $d(\mathbf{p}, \mathbf{q})$, the distance from **p** to **q**]. The points **p** and **q** are the **foci** (plural of **focus**) of the ellipse.

We take the distance sum to be $2a$ and the distance between the foci to be $2c$. Thus $c < a$. See Fig. 8a. To find an equation for the ellipse, we choose coordinates so that the x-axis goes through the foci and the origin is their midpoint (Fig. 8b). A point (x, y) is on the ellipse if and only if its distance sum is $2a$, that is, if and only if

$$d_1 + d_2 = \sqrt{(x - c)^2 + y^2} + \sqrt{(x + c)^2 + y^2} = 2a.$$

This unpleasant expression is an equation for the ellipse. To derive an equivalent equation without radicals requires squaring the equation twice and doing some algebra. It also requires a tedious check that no undesired points creep in. [Squaring

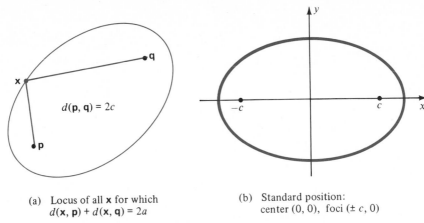

(a) Locus of all **x** for which
     $d(\mathbf{x}, \mathbf{p}) + d(\mathbf{x}, \mathbf{q}) = 2a$

(b) Standard position:
     center $(0, 0)$, foci $(\pm c, 0)$

Fig. 8   The ellipse

is tricky. For instance squaring $x = 2$ leads to $x^2 = 4$ which has a solution that is not a solution of $x = 2$.] We shall postpone the derivation until Section 6, where a much easier method saves a lot of blood, sweat, and tears. For the time being, we shall just state the result.

The equation of the ellipse with foci $(-c, 0)$ and $(c, 0)$ and length sum $2a$ is

$$\frac{x^2}{a^2} + \frac{y^2}{b^2} = 1,$$

where $b^2 = a^2 - c^2$.

Let us sketch this ellipse. Because both terms on the left-hand side of the equation are non-negative, it follows that

$$\frac{x^2}{a^2} \leq 1, \quad x^2 \leq a^2; \qquad \frac{y^2}{b^2} \leq 1, \quad y^2 \leq b^2.$$

Therefore the curve lies in the box $-a \leq x \leq a$, $-b \leq y \leq b$.

Next, we note symmetry: if (x, y) satisfies the equation then so do $(-x, y)$, $(x, -y)$ and $(-x, -y)$. Therefore the curve is symmetric in both axes and in the origin. We need plot it only in the first quadrant, then extend the curve to the other quadrants by symmetry.

We solve for y:

$$y = \frac{b}{a}\sqrt{a^2 - x^2}.$$

(The positive square root applies in the first quadrant.) If x starts at 0 and increases to a, then y starts at b and decreases to 0. The curve has a horizontal tangent at $(0, b)$, a vertical tangent at $(a, 0)$, and is strictly concave. We leave the proofs of these assertions as exercises. We now have enough information for a reasonable sketch (Fig. 9).

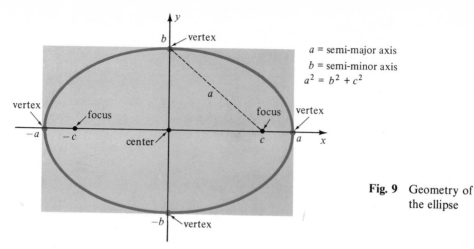

a = semi-major axis
b = semi-minor axis
$a^2 = b^2 + c^2$

Fig. 9 Geometry of the ellipse

The points $(\pm a, 0)$ and $(0, \pm b)$ are called the **vertices** of the ellipse. The numbers a and b are known (historically) by the names **semi-major axis** and **semi-minor axis**. The point halfway between the foci is called the **center** of the ellipse.

As Fig. 9 shows, the distance from a focus to one of the vertices $(0, \pm b)$ is a; that is because $a^2 = b^2 + c^2$.

If $a = b$, the equation of the ellipse becomes

$$\frac{x^2}{a^2} + \frac{y^2}{a^2} = 1, \qquad \text{that is,} \qquad x^2 + y^2 = a^2,$$

the equation of a circle of radius a. Thus a circle can be considered as a limiting (extreme) case of an ellipse, where $c = 0$. (The foci come together at one point, the center.)

If $a < b$, then

$$\frac{x^2}{a^2} + \frac{y^2}{b^2} = 1$$

is an ellipse with major axis along the y-axis instead of the x-axis and foci at $(0, \pm c)$. To see why, interchange the roles of x and y in the preceding discussion.

By a translation of axes, we get

$$\frac{(x - h)^2}{a^2} + \frac{(y - k)^2}{b^2} = 1,$$

the equation of an ellipse centered at (h, k).

Remark To construct an ellipse, tie a string of length $2a$ to two fixed pins $2c$ units apart $(a > c)$. Place your pencil against the string and move it so the string is taut. The locus generated is an ellipse, Why? If you move the pins closer and closer together $(c \longrightarrow 0)$, the ellipse becomes more and more like a circle.

■ **EXAMPLE 4** Show that

$$4x^2 + 9y^2 - 8x + 54y + 49 = 0$$

defines an ellipse. Locate its center, vertices, and foci.

Solution   Complete the squares in x and y:

$$4(x^2 - 2x + 1) + 9(y^2 + 6y + 9) + 49 - 4 - 81 = 0,$$

$$4(x - 1)^2 + 9(y + 3)^2 = 36, \qquad \frac{(x - 1)^2}{9} + \frac{(y + 3)^2}{4} = 1.$$

This is an ellipse with horizontal major axis, center at $(1, -3)$, semi-major axis $= a = 3$, semi-minor axis $= b = 2$. Hence its vertices are

$$(1 \pm a, -3) \qquad \text{and} \qquad (1, -3 \pm b),$$

that is, $(4, -3), (-2, -3)$ and $(1, -1), (1, -5)$.

The foci are $(1 \pm c, -3)$,   where $c^2 = a^2 - b^2 = 5, \quad c = \pm\sqrt{5}$.

Therefore the foci are the points $(1 \pm \sqrt{5}, -3)$.   ∎

■ **EXAMPLE 5**   Find the equation of the ellipse having foci at $(\pm 3, 1)$ and vertices at $(\pm 4, 1)$.

Solution   The center is $(0, 1)$, halfway between the foci. Therefore, the equation is of the form

$$\frac{(x - 0)^2}{a^2} + \frac{(y - 1)^2}{b^2} = 1.$$

Now $a = 4$, the distance from the center to either vertex on the major axis of the ellipse (through the foci). To find b, use $b^2 = a^2 - c^2$, where c is the distance from the center to either focus. Clearly, $c = 3$, so $b^2 = 4^2 - 3^2 = 7$. Hence the equation of the ellipse is

$$\frac{x^2}{16} + \frac{(y - 1)^2}{7} = 1.$$ ∎

Parameterization of the Ellipse   Consider again the ellipse

$$\frac{x^2}{a^2} + \frac{y^2}{b^2} = 1.$$

Since

$$\left(\frac{x}{a}\right)^2 + \left(\frac{y}{b}\right)^2 = 1,$$

there is an angle θ such that $x/a = \cos\theta$ and $y/b = \sin\theta$, so

$$x = a\cos\theta, \qquad y = b\sin\theta.$$

As θ runs from 0 to 2π, the point $(x, y) = (a\cos\theta, b\sin\theta)$ traverses the ellipse once in the counterclockwise sense, starting at $(a, 0)$. Note from Fig. 10 that the parameter θ is not the central angle of (x, y); rather θ is the central angle of two circles, one inscribed in the ellipse and the other circumscribed about the ellipse.

■ **EXAMPLE 6**   A 12-ft ladder leans against a wall. Its bottom is pulled along the floor away from the wall and its top slides down the wall. Find the locus of the point 4 ft from the top of the ladder.

Solution   Choose axes as indicated, and let θ be the angle the ladder makes with the floor (Fig. 11). Obviously, $x = 4\cos\theta$ and $y = 8\sin\theta$, that is, the point moves

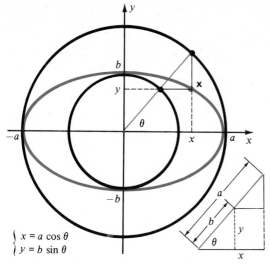

Fig. 10 Parameterization of the ellipse
$$\frac{x^2}{a^2} + \frac{y^2}{b^2} = 1$$

Fig. 11

$$\begin{cases} x = a\cos\theta \\ y = b\sin\theta \end{cases}$$

on an ellipse. As the ladder slides, θ runs from $\frac{1}{2}\pi$ to 0, so the locus is a quarter of the ellipse, traversed clockwise. The standard equation of this ellipse is

$$\frac{x^2}{4^2} + \frac{y^2}{8^2} = 1.$$

Note that its major axis is vertical. ∎

EXERCISES

Give the focus, directrix, axis, and vertex

1 $\quad x - 3 = 3y^2$	**2** $\quad x + 2 = -y^2$	**3** $\quad x = 2(y+1)^2$
4 $\quad 2y = -3(x-2)^2$	**5** $\quad x^2 + 4x - 6y = 0$	**6** $\quad 2y^2 - 4y + x + 2 = 0.$

Find the equation of the parabola

7   vertex $(0, 0)$,   through the point $(1, 3)$,   axis vertical
8   vertex $(0, 0)$,   through the point $(6, -1)$,   axis vertical
9   vertex $(1, 2)$,   through the point $(-3, 4)$,   axis horizontal
10   vertex $(-5, 0)$,   through the point $(0, 8)$,   axis horizontal
11   vertex $(2, -3)$ focus $(2, 1)$ **12**   vertex $(2, -3)$,   focus $(10, -3)$.

Find the locus of the centers of all circles that are simultaneously tangent to the y-axis and the circle

13 $\quad x^2 + y^2 = 2x$	**14** $\quad x^2 + y^2 = 1$
15 $\quad (x-1)^2 + y^2 = 4$	**16** $\quad (x-3)^2 + y^2 = 1.$

17   Let P be a parabola and L a line. Find the locus of the midpoints of all chords of P parallel to L.

18   Find the locus of all points $\mathbf{x}$ whose distance to the x-axis equals the length of the tangents from $\mathbf{x}$ to the circle $x^2 + (y - b)^2 = r^2$, where $b \geq 0, r > 0$.

19*   Let P be the parabola $x^2 = 4py$. Prove that all chords $\overline{\mathbf{x}_1\mathbf{x}_2}$ of P such that $\angle\mathbf{x}_1\mathbf{0}\mathbf{x}_2$ is a right angle pass through a common point.

20*   (cont.) Find the locus of the midpoints of all $\overline{\mathbf{x}_1\mathbf{x}_2}$.

Let P be the parabola $x^2 = 4py$, and let $\overline{\mathbf{x}_1\mathbf{x}_2}$ be an arbitrary chord of P through $\mathbf{p}$

21*   Find the locus of the midpoint of $\overline{\mathbf{x}_1\mathbf{x}_2}$.

22*   (cont.) Prove that the circle with diameter $\overline{\mathbf{x}_1\mathbf{x}_2}$ is tangent to the directrix of P.

23   Show that the circle with diameter $\overline{\mathbf{x}_1\mathbf{p}}$ is tangent to the x-axis.

24   Suppose a circle C meets $4py = x^2$ in four points (x_i, y_i), $i = 1, \cdots, 4$. Prove that $x_1 + x_2 + x_3 + x_4 = 0$.

Give the center, foci, major and minor semi-axes (a and b), and vertices

25 $x^2/25 + y^2/9 = 1$ **26** $x^2 + 4y^2 = 4$

27 $2(x + 1)^2 + (y - 2)^2 = 2$ **28** $4x^2 + y^2 - 2y = 0$

29 $2x^2 + y^2 - 12x - 4y = -21$ **30** $x^2 + 2y^2 + 8y = 0$.

Write the equation of the ellipse

31   center at $(1, 4)$,   vertices at $(10, 4)$ and $(1, 2)$

32   center at $(-2, -3)$,   vertices at $(7, -3)$ and $(-2, -7)$

33   foci at $(2, 0)$ and $(8, 0)$,   vertex at $(0, 0)$

34   foci at $(0, 3)$ and $(0, -3)$,   semi-major axis $= 10$

35   foci at $(-1, 0)$, $(3, 0)$, $c/a = \frac{1}{2}$

36   vertices at $(0, 2)$, $(0, 6)$, $c/a = \frac{3}{4}$, major axis vertical.

37   Show that $x^2/a^2 + y^2/b^2 = 1$ has horizontal tangents at $(0, \pm b)$ and vertical tangents at $(\pm a, 0)$.

38   Show that the upper half of the ellipse $x^2/a^2 + y^2/b^2 = 1$ is strictly concave.

39   Prove that the points on an ellipse farthest from its center are the two vertices on the major axis.

40   Prove that the points on an ellipse nearest to its center are the two vertices on the minor axis.

41   A rod moves with one end on the x-axis and the other on the y-axis. Describe the locus of any other point of the rod.

42*   A gadget (Fig. 12) consists of two disks glued together and a point P on their line of centers. The smaller disk can move only in the vertical tracks, and the larger disk can move only in the horizontal tracks. Describe the locus of P.

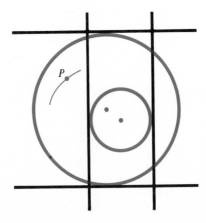

Fig. 12

43 Find the locus of the centers of all circles that are simultaneously tangent to the circles $x^2 + y^2 = 2ax$ and $x^2 + y^2 = 2bx$, where $0 < a < b$.

44 Find the locus of the midpoints of all chords of slope m of the ellipse $x^2/a^2 + y^2/b^2 = 1$.

45 Let D be a fixed diameter of a circle of radius r. At each point $\mathbf{p}$ of the circle drop a perpendicular to D. Find the locus of the midpoints of these perpendiculars.

46 (cont.) Extend each perpendicular beyond the circle so that its length increases by a factor $k > 1$. Find the locus of the end points of these segments.

4. HYPERBOLA

A **hyperbola** is the locus of all points $\mathbf{x}$ such that the difference of the distance of $\mathbf{x}$ from two fixed points $\mathbf{p}$ and $\mathbf{q}$ has constant absolute value, $2a$. The points $\mathbf{p}$ and $\mathbf{q}$ are the **foci** of the hyperbola. If $d(\mathbf{p}, \mathbf{q}) = 2c$, then $2a < 2c$ because $\mathbf{p}$, $\mathbf{q}$, and $\mathbf{x}$ form a triangle (Fig. 1a).

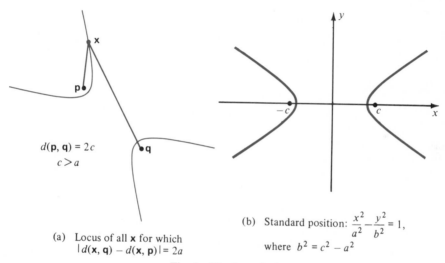

$d(\mathbf{p}, \mathbf{q}) = 2c$
$c > a$

(a) Locus of all $\mathbf{x}$ for which $|d(\mathbf{x}, \mathbf{q}) - d(\mathbf{x}, \mathbf{p})| = 2a$

(b) Standard position: $\dfrac{x^2}{a^2} - \dfrac{y^2}{b^2} = 1$,

where $b^2 = c^2 - a^2$

Fig. 1 The hyperbola

To obtain an equation for the hyperbola, we set up coordinate axes so the foci are $(\pm c, 0)$. See Fig. 1b. Then a point (x, y) is on the hyperbola if and only if

$$\left| \sqrt{(x - c)^2 + y^2} - \sqrt{(x + c)^2 + y^2} \right| = 2a.$$

As for the ellipse, we postpone the messy simplification of this equation until Section 6, and merely state the result:

The equation of the hyperbola with foci $(-c, 0)$ and $(c, 0)$ and absolute length difference $2a$ is

$$\frac{x^2}{a^2} - \frac{y^2}{b^2} = 1,$$

where $b^2 = c^2 - a^2$.

The size of b depends on the relative sizes of c and a; both $b \leq a$ and $b > a$ are possible for the hyperbola. For the ellipse with foci on the x-axis, however, only $b \leq a$ is possible since $b^2 = a^2 - c^2$.

Let us sketch the hyperbola

$$\frac{x^2}{a^2} - \frac{y^2}{b^2} = 1.$$

The curve is symmetric in both axes and in the origin; we need plot it only in the first quadrant, then extend the curve to the other quadrants by symmetry.

We solve for y:

$$y = \frac{b}{a}\sqrt{x^2 - a^2}.$$

(The positive square root applies for the first quadrant.) Since the quantity under the radical must be non-negative, the locus is defined only for $x \geq a$. Now when x starts at a and increases, y starts at 0 and increases. When x is very large, we suspect that y is slightly less than bx/a. To confirm this suspicion, we write

$$\frac{b}{a}x - y = \frac{b}{a}(x - \sqrt{x^2 - a^2}) = \frac{b}{a}(x - \sqrt{x^2 - a^2})\frac{x + \sqrt{x^2 - a^2}}{x + \sqrt{x^2 - a^2}}$$

$$= \left(\frac{b}{a}\right)\frac{x^2 - (x^2 - a^2)}{x + \sqrt{x^2 - a^2}} = \frac{ab}{x + \sqrt{x^2 - a^2}} < \frac{ab}{x}.$$

It follows that $(b/a)x - y$ is positive, but becomes smaller and smaller as x becomes larger and larger. This means the curve approaches the line $y = bx/a$ (from below) as x increases; the line is an asymptote of the hyperbola.

Further information: the hyperbola has a vertical tangent at $(a, 0)$ and is strictly concave in the first quadrant. (See Exs. 19–20.) We can now make a reasonable sketch of the curve (Fig. 2).

The lines $y = \pm bx/a$ are the **asymptotes** of the hyperbola. A neat way to remember this fact is to write

$$\frac{x^2}{a^2} - \frac{y^2}{b^2} = \left(\frac{x}{a} + \frac{y}{b}\right)\left(\frac{x}{a} - \frac{y}{b}\right).$$

The expression on the left is zero if and only if one of the factors on the right is zero, that is, if and only if $y = \pm bx/a$.

The asymptotes of the hyperbola $\dfrac{x^2}{a^2} - \dfrac{y^2}{b^2} = 1$ are the lines

$$y = \frac{b}{a}x \quad \text{and} \quad y = -\frac{b}{a}x,$$

or equivalently, the locus of the equation $\dfrac{x^2}{a^2} - \dfrac{y^2}{b^2} = 0.$

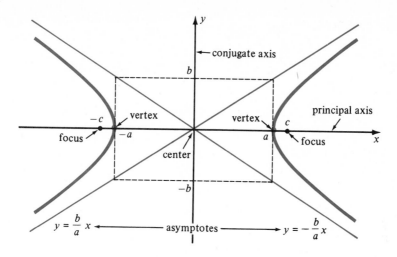

Fig. 2 Geometry of the hyperbola $\dfrac{x^2}{a^2} - \dfrac{y^2}{b^2} = 1$

Here is some of the official terminology for hyperbolas. A hyperbola consists of two **branches**: one where $d(\mathbf{x},\ \mathbf{p}) - d(\mathbf{x},\ \mathbf{q}) = 2a$, and the other one where $d(\mathbf{x},\ \mathbf{q}) - d(\mathbf{x},\ \mathbf{p}) = 2a$. The point halfway between the foci is the **center** of the hyperbola. The line through the foci is the **principal axis**, and the line through the center perpendicular to the principal axis is the **conjugate axis**. The points where the hyperbola meets its principal axes are its **vertices**.

A hyperbola is **rectangular** if its asymptotes are perpendicular. This happens when the slopes of the two asymptotes are negative reciprocals of each other:

$$\left(\frac{b}{a}\right)\left(-\frac{b}{a}\right) = -1, \qquad b^2 = a^2, \quad b = a.$$

Thus the locus of $\qquad\qquad\qquad x^2 - y^2 = a^2$

is a rectangular hyperbola.

By translation, $\qquad\qquad \dfrac{(x-h)^2}{a^2} - \dfrac{(y-k)^2}{b^2} = 1$

is the equation of a hyperbola with center at (h, k) and horizontal principal axis. Its asymptotes are the lines

$$y - k = \frac{b}{a}(x - h).$$

By interchanging the roles of x and y, we see that the equation

$$\frac{y^2}{a^2} - \frac{x^2}{b^2} = 1$$

defines a hyperbola with center at the origin, vertical principal axis, and foci at $(0, \pm c)$.

Remark The equations $\dfrac{x^2}{a^2} + \dfrac{y^2}{b^2} = 1$ and $\dfrac{x^2}{a^2} - \dfrac{y^2}{b^2} = 1$

differ by a little minus sign, but that makes all the difference in the world. The first equation, where the sign is plus, requires $x^2 \le a^2$ and $y^2 \le b^2$; the locus is confined. The second imposes no such restriction; x^2/a^2 and y^2/b^2 can both be enormous, yet differ by 1.

■ **EXAMPLE 1** Sketch the curve $3x^2 - 12x - 8y^2 = 12$. Locate its center, axes, foci, and asymptotes.

Solution Complete the square in x:

$$3(x - 2)^2 - 8y^2 = 12 + 12 = 24, \qquad \frac{(x - 2)^2}{8} - \frac{y^2}{3} = 1.$$

The curve is a hyperbola with center $(2, 0)$ and $a^2 = 8$, $b^2 = 3$. Its principal axis is the x-axis and its foci are

$$(2 \pm c, 0), \qquad \text{where} \quad c = \sqrt{a^2 + b^2} = \sqrt{11}.$$

Its asymptotes are

$$y = \pm \frac{b}{a}(x - 2) = \pm \sqrt{\frac{3}{8}}(x - 2).$$

See Fig. 3.

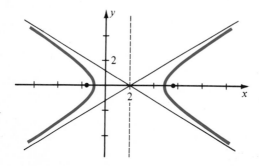

Fig. 3 $3x^2 - 12x - 8y^2 = 12$

■ **EXAMPLE 2** Find the equation of the hyperbola

  (a) vertices $(-1, 1)$, $(-1, 5)$, foci $(-1, 0)$, $(-1, 6)$;
  (b) foci $(0, 2)$, $(10, 2)$, asymptotes having slopes ± 3.

Solution (a) The principal axis is vertical and the center is $(-1, 3)$. Therefore the equation is of the form

$$\frac{(y - 3)^2}{a^2} - \frac{(x + 1)^2}{b^2} = 1.$$

Now $2a$ is the distance between the vertices, hence $2a = 4$, $a = 2$. Similarly, $2c$ is the distance between the foci, hence $2c = 6$, $c = 3$. Finally, $b^2 = c^2 - a^2$, so $b^2 = 9 - 4 = 5$. Hence the desired equation is

$$\frac{(y - 3)^2}{4} - \frac{(x + 1)^2}{5} = 1.$$

(b)   The principal axis is horizontal and the center is (5, 2). Therefore the equation is of the form

$$\frac{(x-5)^2}{a^2} - \frac{(y-2)^2}{b^2} = 1.$$

The distance between the foci is $10 = 2c$, hence $c = 5$. It follows that $a^2 + b^2 = c^2 = 25$. Furthermore, the slopes of the asymptotes are $\pm 3 = \pm b/a$, so $b^2 = 9a^2$. Therefore

$$25 = a^2 + b^2 = 10a^2, \qquad a^2 = \tfrac{5}{2}, \qquad b^2 = \tfrac{45}{2}.$$

Hence the equation of the hyperbola is

$$\frac{2(x-5)^2}{5} - \frac{2(y-2)^2}{45} = 1. \qquad \blacksquare$$

■ **EXAMPLE 3**   Find the locus of the centers of all circles that are tangent to the x-axis and cut off a segment of length $2k$ on the y-axis.

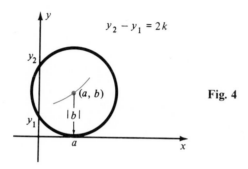

Fig. 4

Solution   Draw a picture (Fig. 4). If the center of a circle is (a, b) and the circle is tangent to the x-axis, then its radius is $|b|$ (absolute value in case $b < 0$). The equation of the circle is

$$(x-a)^2 + (y-b)^2 = b^2, \qquad \text{that is,} \qquad x^2 + y^2 - 2ax - 2by + a^2 = 0.$$

Suppose the circle intersects the y-axis in $(0, y_1)$ and $(0, y_2)$ with $y_1 < y_2$. Then the segment condition is $y_2 - y_1 = 2k$.
   To find y_1 and y_2, set $x = 0$ and solve for y:

$$y^2 - 2by + a^2 = 0, \qquad y = b \pm \sqrt{b^2 - a^2}.$$

Assume $b^2 > a^2$; otherwise the circle does not intersect the y-axis, or is tangent to it. Then

$$y_2 - y_1 = 2\sqrt{b^2 - a^2} = 2k, \qquad b^2 - a^2 = k^2.$$

Hence the centers (a, b) all lie on the rectangular hyperbola

$$y^2 - x^2 = k^2.$$

Conversely, each point on this hyperbola is a point of the locus, as is easily checked.

$$\blacksquare$$

■ **EXAMPLE 4**   Find all points that are 10 units from the origin and two units closer to $(3, 0)$ than to $(-3, 0)$.

Solution   The first condition means that the point (x, y) lies on the circle $x^2 + y^2 = 10^2$. The second condition means that it lies on one branch of the hyperbola with foci $(\pm 3, 0)$ and difference of the distances 2. This hyperbola is in standard position with principal axis horizontal, so its equation is of the form

$$\frac{x^2}{a^2} - \frac{y^2}{b^2} = 1.$$

From the data $2a = 2$, and the distance between the foci is $2c = 6$. Hence $a = 1$ and $c = 3$. Therefore $b^2 = c^2 - a^2 = 8$, so $1/b^2 = \frac{1}{8}$.

The desired points are the intersections of the circle and the right-hand branch of the hyperbola (where the points are closer to $(3, 0)$ than to $(-3, 0)$). Solve simultaneously:

$$x^2 + y^2 = 100, \qquad x^2 - \tfrac{1}{8}y^2 = 1.$$

Subtract: $\qquad\qquad \frac{9}{8}y^2 = 99, \qquad y^2 = 88, \quad y = \pm\sqrt{88},$

from which $\qquad x^2 = 100 - y^2 = 100 - 88 = 12, \qquad x = \pm\sqrt{12}.$

On the right-hand branch of the hyperbola, $x > 0$, so only the value $x = \sqrt{12}$ is acceptable. Therefore the desired points are $(\sqrt{12}, \pm\sqrt{88})$. ■

EXERCISES

Find the principal axis, center, foci, and asymptotes

1   $x^2/4 - y^2/9 = 1$ 2 $x^2/9 - y^2/4 = 1$
3   $-x^2/9 + y^2/4 = 1$ 4 $-x^2/4 + y^2/9 = 1$
5   $(x + 1)^2 - (y - 1)^2 = 1$ 6 $-(x - 2)^2 + 4(y + 1)^2 = 4$
7   $x^2 - 5y^2 + 4x - 20y = 0$ 8 $-x^2 + 2y^2 - 6x - 20y + 47 = 0$
9   $4x^2 - y^2 - 24x - 2y + 31 = 0$ 10 $3x^2 - 3y^2 - 3x - 2y = 31/12.$

Write the equation of the hyperbola

11   foci $(0, \pm 5)$, vertex $(0, -4)$ 12   vertices $(\pm 3, 0)$ focus $(-5, 0)$
13   asymptote $y = -2x$, vertices $(\pm 2, 0)$ 14   foci $(1, 7), (1, -3)$,   vertex $(1, 6)$
15   asymptotes $y = \pm(x - 1)$,   curve passes through $(3, 1)$
16   asymptotes $y = \pm 2x$,   curve passes through $(1, 1)$.

17   Show that $x^2 - y^2 + ax + by + c = 0$ represents a rectangular hyperbola.
18   Show that $3x^2 - y^2 + ax + by + c = 0$ represents a hyperbola whose asymptotes form a 60° angle.
19   Show that $x^2/a^2 - y^2/b^2 = 1$ has a vertical tangent at each vertex.
20   (cont.) Show that the right branch is convex with respect to the y-axis.
21   Let $0 < s < r$ and $r + s < 2a$. Describe the locus of the centers of all circles that are simultaneously tangent externally to the circles $(x + a)^2 + y^2 = r^2$ and $(x - a)^2 + y^2 = s^2$.
22   (cont.) Identify geometrically the other branch.
23* (cont.) Find another hyperbola lurking in this configuration.
24   A rifle at point **a** on level ground is shot at a target at point **b**. Find the locus of all observers who hear the shot and the impact of the shell simultaneously.

25 Three listening posts, A, B, and C record an explosion. Post A is 10 km west of post B, and post C is 8 km south of B. Posts B and C hear the explosion simultaneously; A hears it 6 sec later. Assuming the speed of sound in air is $\frac{1}{3}$ km/sec, locate the explosion.

26 A point moves in the plane starting at time $t = 0$. If its coordinates at time t are $x = a \cosh t$ and $y = b \sinh t$, where a, $b > 0$, describe its path.

5. POLAR COORDINATES

The usual rectangular coordinate system is fine in some situations but clumsy in others. Sometimes another coordinate system, called the **polar coordinate** system, fits much more naturally. The idea of polar coordinates is that you identify a point by telling how far it is from a given point **0** and in what direction. (This is the principle of the radar screen.)

In a rectangular coordinate system, two families of grid lines, $x = $ constant and $y = $ constant, fill the plane. Each point $\mathbf{x}$ is the intersection of two of these lines, $x = a$ and $y = b$, and receives the coordinates (a, b). See Fig. 1a.

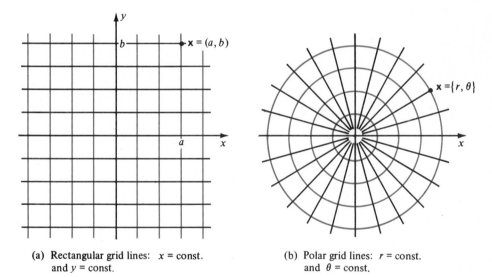

(a) Rectangular grid lines: $x = $ const. and $y = $ const.

(b) Polar grid lines: $r = $ const. and $\theta = $ const.

Fig. 1

Polar coordinates work on a similar principle. There are two families of grid lines: all circles centered at **0**, and all rays from **0**. See Fig. 1b. Each point $\mathbf{x}$ different from **0** is the intersection of one circle and one ray. The circle is identified by a positive number r, its radius, and the ray is identified by a real number θ, its angle in radians from the positive x-axis. Thus $\mathbf{x}$ is assigned the **polar coordinates** $\{r, \theta\}$. Since θ is determined only up to a multiple of 2π, we agree that

$$\{r, \theta + 2\pi n\} = \{r, \theta\} \qquad (n \text{ any integer}).$$

The point **0** does not determine an angle θ. Nonetheless, it is customary to say that *any* pair $\{0, \theta\}$ represents **0**.

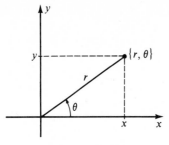
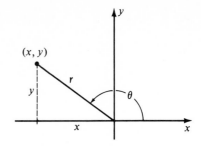

(a) Polar to rectangular:
$x = r \cos \theta, \; y = r \sin \theta$

(b) Rectangular to polar:
$r = \sqrt{x^2 + y^2}, \quad \cos \theta = \dfrac{x}{r}, \quad \sin \theta = \dfrac{y}{r}$

Fig. 2

Given the polar coordinates of a point, what are its rectangular coordinates? If its polar coordinates are $\{r, \theta\}$, then the point is r units from **0** in the direction θ. Hence $x = r \cos \theta$, $y = r \sin \theta$. See Fig. 2a.

Conversely, given the rectangular coordinates (x, y), what are the polar coordinates? Figure 2b shows that $r = \sqrt{x^2 + y^2}$, and that $\cos \theta = x/r$ and $\sin \theta = y/r$.

Polar to Rectangular	Rectangular to Polar
$\begin{cases} x = r \cos \theta. \\ y = r \sin \theta. \end{cases}$	$\begin{cases} r = \sqrt{x^2 + y^2} \\ \cos \theta = \dfrac{x}{\sqrt{x^2 + y^2}} = \dfrac{x}{r} \\ \sin \theta = \dfrac{y}{\sqrt{x^2 + y^2}} = \dfrac{y}{r}. \end{cases}$

■ **EXAMPLE 1**   (a)   Convert $(2, -2\sqrt{3})$ to polar coordinates.
(b)   Convert $\{3, \tfrac{1}{6}\pi\}$ to rectangular coordinates.

Solution   (a) $r^2 = 4 + 12 = 16, \quad r = 4$. Also $\cos \theta = \tfrac{2}{4} = \tfrac{1}{2}$ and $\sin \theta = \tfrac{1}{4}(-2\sqrt{3}) = -\tfrac{1}{2}\sqrt{3}$, so $\theta = \tfrac{5}{3}\pi$. *Answer* $\{4, \tfrac{5}{3}\pi\}$.

(b) $x = r \cos \theta = 3 \cos \tfrac{1}{6}\pi = \tfrac{3}{2}\sqrt{3}$, and $y = r \sin \theta = 3 \sin \tfrac{1}{6}\pi = \tfrac{3}{2}$.
Answer $(\tfrac{3}{2}\sqrt{3}, \tfrac{3}{2})$.   ■

Negative r   In applications it is convenient to allow points $\{r, \theta\}$ with $r < 0$. Consider a ray and a point $\{r, \theta\}$ on the ray (Fig. 3a). Suppose the point moves towards **0**, through **0**, and keeps on going! Then r decreases, becomes 0, but then what? So that θ won't jump abruptly to $\theta + \pi$, we agree that θ remains constant, but r becomes negative. This amounts to agreeing that

$$\{-r, \theta\} = \{r, \theta + \pi\}.$$

See Fig. 3b. For example, the point $(-1, -1)$ has polar coordinates $\{-\sqrt{2}, \tfrac{1}{4}\pi\}$ as well as $\{\sqrt{2}, \tfrac{5}{4}\pi\}$.

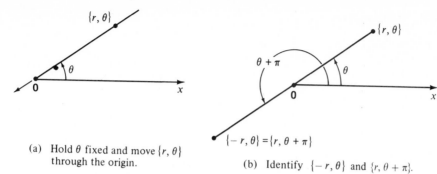

(a) Hold θ fixed and move $\{r, \theta\}$
      through the origin.

(b) Identify $\{-r, \theta\}$ and $\{r, \theta + \pi\}$.

Fig. 3

Polar Equation of a Line   In polar coordinates, the equation of a line through **0** is $\theta = \theta_0$, where θ_0 is the angle the line makes with the positive x-axis. What is the equation of a line L not through the origin? Drop a perpendicular to L from **0**. It has length $p > 0$ and polar angle α, See Fig. 4.
   The figure shows that for each point $\{r, \theta\}$ on the line, $r \cos(\theta - \alpha) = p$.

The polar equation of a line not passing through **0** is

$$r \cos(\theta - \alpha) = p, \qquad p > 0.$$

Here p is the distance from **0** to the line and the point $\{p, \alpha\}$ is the foot of the perpendicular from **0** to the line.

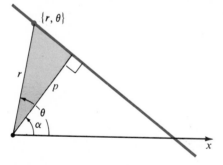

Fig. 4   From the right triangle,
$p = r \cos(\theta - \alpha)$.

The equation $r \cos(\theta - \alpha) = p$ has an analogue in rectangular coordinates. Use the trig identity for $\cos(\theta - \alpha)$, then replace $r \cos \theta$ by x and $r \sin \theta$ by y:

$$r \cos \theta \cos \alpha + r \sin \theta \sin \alpha = p, \qquad x \cos \alpha + y \sin \alpha = p.$$

The latter equation is called the **normal form** of the line. Note that even lines through the origin satisfy such an equation with $p = 0$.

Normal Form   Each line in the plane has an equation

$$x \cos \alpha + y \sin \alpha = p, \qquad p \geq 0.$$

If $p > 0$, then $(p \cos \alpha, p \sin \alpha)$ is the foot of the perpendicular from **0** to the line.

There is a trick for converting an equation $ax + by = c$ into normal form: just divide by $\pm\sqrt{a^2 + b^2}$, taking the sign the same as that of c:

$$\frac{a}{\pm\sqrt{a^2 + b^2}}\, x + \frac{b}{\pm\sqrt{a^2 + b^2}}\, y = \frac{c}{\pm\sqrt{a^2 + b^2}}.$$

Since

$$\left(\frac{a}{\pm\sqrt{a^2 + b^2}}\right)^2 + \left(\frac{b}{\pm\sqrt{a^2 + b^2}}\right)^2 = 1,$$

there is an angle α such that the equation can be written

$$(\cos \alpha)x + (\sin \alpha)y = p, \qquad p \geq 0.$$

Example   Given $x - 3y = 7$, divide by $+\sqrt{1^2 + 3^2} = \sqrt{10}$. The normal form is

$$\frac{x}{\sqrt{10}} - \frac{3y}{\sqrt{10}} = \frac{7}{\sqrt{10}}, \qquad \text{so} \qquad \cos\alpha = \frac{1}{\sqrt{10}}, \quad \sin\alpha = \frac{-3}{\sqrt{10}}, \quad p = \frac{7}{\sqrt{10}}.$$

The point on the line closest to **0** is $(p\cos\alpha,\, p\sin\alpha) = \left(\frac{7}{10},\, -\frac{21}{10}\right)$.

Polar Equation of a Circle   In polar coordinates, the equation of a circle of radius a, center **0**, is simply $r = a$.

Consider next the circle of radius a and center $(a, 0)$. Its Cartesian equation is

$$(x - a^2) + y^2 = a^2, \qquad \text{that is,} \qquad x^2 + y^2 = 2ax.$$

Substitute $x = r\cos\theta$ and $y = r\sin\theta$: $r^2 = 2ar\cos\theta$.

If $r \neq 0$, then $r = 2a\cos\theta$. But $r = 0$ represents only the point **0**, which is already on the locus $r = 2a\cos\theta$ for $\theta = \frac{1}{2}\pi$. Hence canceling r does not change the locus, so

$$r = 2a\cos\theta$$

is the polar equation of the given circle. The right triangle in Fig. 5 shows that the relation $r = 2a\cos\theta$ is satisfied by every point on the circle.

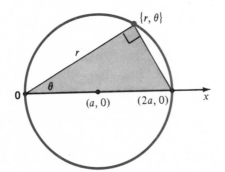

Fig. 5   Circle of radius a and center $(a, 0)$: by the right triangle, $r = 2a\cos\theta$.

Let us see how $\{r, \theta\}$ moves on the circle $r = 2a\cos\theta$ as θ makes a complete revolution. If θ starts at 0, then r starts at $2a$. If θ increases to $\frac{1}{2}\pi$, then r decreases to 0. (Think of an arm rotating counterclockwise and shrinking.) Hence $\{r, \theta\}$ traces the upper half of the circle (Fig. 6a).

If θ increases from $\frac{1}{2}\pi$ to π, then r decreases from 0 to $-2a$. Since r is negative, the point $\{r, \theta\}$ is measured "backward" and moves through the *fourth* quadrant, tracing the lower half of the circle (Fig. 6b).

Thus the full circle is described as θ runs from 0 to π. As θ runs from π to 2π, the same circle is traced again. For when θ is in the third quadrant, $r < 0$, so $\{r, \theta\}$ describes the semicircle in the first quadrant; when θ is in the fourth quadrant $r > 0$, so $\{r, \theta\}$ describes the semicircle in the fourth quadrant. Therefore, in one complete revolution of θ, the circle is traced twice.

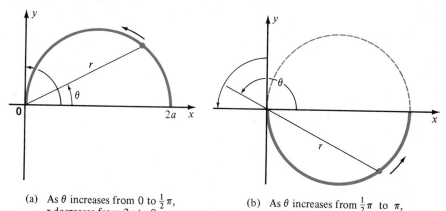

(a) As θ increases from 0 to $\frac{1}{2}\pi$,   (b) As θ increases from $\frac{1}{2}\pi$ to π,
      r decreases from $2a$ to 0. r decreases from 0 to $-2a$.

Fig. 6   The circle $r = 2a \cos \theta$ traced as $0 \le \theta \le \pi$

The graph of the equation

$$r = 2a \cos \theta$$

is a circle of radius a and center $\{a, 0\}$. The circle is traced twice as θ makes a complete revolution.

Distance Formula   In rectangular coordinates, the distance formula follows from the Pythagorean Theorem; in polar coordinates it follows from the Law of Cosines (which contains the Pythagorean Theorem as a special case.) See Fig. 7.

According to the Law of Cosines

$$d^2 = r_1{}^2 - 2r_1 r_2 \cos(\theta_2 - \theta_1) + r_2{}^2 .$$

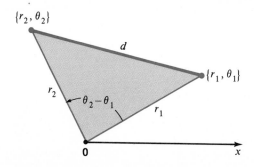

Fig. 7   Apply the Law of Cosines to express d in terms of the polar coordinates.

Distance Formula   If d is the distance between the points $\{r_1, \theta_1\}$ and $\{r_2, \theta_2\}$ then

$$d^2 = r_1{}^2 - 2r_1r_2 \cos(\theta_2 - \theta_1) + r_2{}^2.$$

An immediate corollary is the polar equation of the circle of radius a, center $\{p, \alpha\}$.

The graph of the equation

$$r^2 - 2rp \cos(\theta - \alpha) + p^2 = a^2$$

is a circle of radius a and center $\{p, \alpha\}$.

Note two special cases: If the center is $\mathbf{0}$, then $p = 0$, and the equation boils down to $r^2 = a^2$, that is, $r = a$, or $r = -a$. If the center is at the point $(a, 0)$, then $p = a$, $\alpha = 0$ and the equation becomes

$$r^2 - 2ra \cos \theta + a^2 = a^2, \qquad r - 2a \cos \theta = 0.$$

EXERCISES

Give the rectangular coordinates

1 $\{1, \frac{1}{2}\pi\}$ **2** $\{1, -\frac{1}{2}\pi\}$ **3** $\{1, -\frac{1}{6}\pi\}$ **4** $\{1, \frac{1}{3}\pi\}$ **5** $\{2, -\frac{3}{4}\pi\}$ **6** $\{2, \frac{5}{4}\pi\}$.

Give polar coordinates

7 $(1, 1)$ **8** $(0, -1)$ **9** $(-1, 1)$

10 $(-\frac{1}{2}, \frac{1}{2}\sqrt{3})$ **11** $(\sqrt{3}, -1)$ **12** $(\sqrt{2}, -\sqrt{2})$.

Find an equation in polar coordinates

13 line through $\mathbf{0}$ and $\{3, \frac{1}{4}\pi\}$ **14** circle, center $\mathbf{0}$, radius 5

15 line through $\{1, 0\}$ and $\{1, \frac{1}{2}\pi\}$ **16** line perpendicular to $\theta = \frac{3}{4}\pi$, tangent to circle $r = 1$

17 circle, center $\{a, \pi\}$, radius a **18** circle, center $\{a, \frac{1}{2}\pi\}$, radius a

19 circle through $\mathbf{0}$, center $\{1, -\frac{3}{4}\pi\}$ **20** circle through $\mathbf{0}$, center $\{2, \frac{1}{4}\pi\}$

21 line through $(1, 2)$, slope $-\frac{1}{2}$ **22** line through $(-2, 1)$, slope 3

23 circle, center $\{5, \frac{1}{6}\pi\}$, radius 4

24 circle, center $\{5, \frac{1}{6}\pi\}$, through $\{3, -\frac{1}{2}\pi\}$.

Give the line in normal form

25 $3x - 4y = 5$ **26** $-5x + 12y = -26$

27 $7x + 24y = -25$ **28** $5x - 3y = 2$

29 through $(1, 2)$ and $(2, 1)$ **30** through $(-1, -3)$, slope 2.

Suppose the normal form of a line is $x \cos \alpha + y \sin \alpha = p$.

31 If the origin is translated to (h, k), find the normal form in the new coordinates.

32 (cont.) Find all translations of coordinates that convert the equation to $\bar{x} \cos \alpha + \bar{y} \sin \alpha = 0$.

33 The circle with center $\{p, \alpha\}$ and radius a has polar equation $r^2 - 2pr \cos(\theta - \alpha) + p^2 = a^2$. Solve for r. Discuss the domain of θ and corresponding values of r for the case $a > p$.

34 (cont.) Now consider the cases $a = p$ and $a < p$.

6. POLAR GRAPHS

Graphing a function $r = f(\theta)$ in polar coordinates is tricky at first, because you must change your point of view. For $y = f(x)$, you think of x running along the horizontal axis, with the corresponding point (x, y) measured above or below. Basically your mental set is "left–right" and "up–down."

In polar coordinates, however, you must think of the angle θ swinging around (like a radar scope) and repeating after 2π. For each θ, you must measure forward from the origin a distance $f(\theta)$, or backward if $f(\theta) < 0$. Your mental set must be "round and round" and "in and out."

Because of the special nature of points $\{r, \theta\}$ where $r < 0$, pay close attention to the sign of $f(\theta)$ and be sure to plot points "backwards" if $f(\theta) < 0$.

Look for symmetries and periodicity. For example if $f(\theta + 2\pi) = f(\theta)$, the polar graph $r = f(\theta)$ will repeat after 2π. There are many symmetries possible; we mention only two, $f(\theta)$ even and $f(\theta)$ odd. If $f(\theta)$ is even, that is, $f(-\theta) = f(\theta)$, then the point $\{r, -\theta\}$ is on the graph whenever $\{r, \theta\}$ is; the curve is symmetric in the x-axis (Fig. 1a). If $f(\theta)$ is odd, $f(-\theta) = -f(\theta)$, then the point $\{-r, -\theta\}$ is on the graph whenever $\{r, \theta\}$ is; the curve is symmetric in the vertical axis (Fig. 1b).

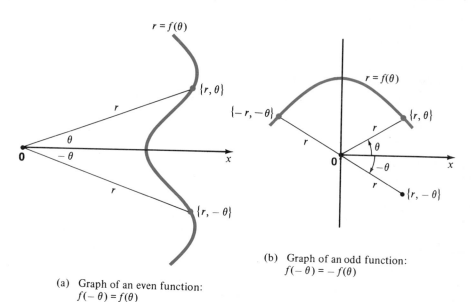

(a) Graph of an even function:
$f(-\theta) = f(\theta)$

(b) Graph of an odd function:
$f(-\theta) = -f(\theta)$

Fig. 1   Polar symmetry

■ **EXAMPLE 1**   Graph the "spiral of Archimedes" $r = \theta$.

Solution   If θ increases starting from 0, then r increases steadily, also starting from 0. Hence the locus goes round and round, its distance from **0** becoming greater and greater. The result is a spiral (Fig. 2a). Since $f(\theta) = \theta$ is an odd function, we obtain the locus for $\theta < 0$ by reflection in the vertical axis (Fig. 2b).

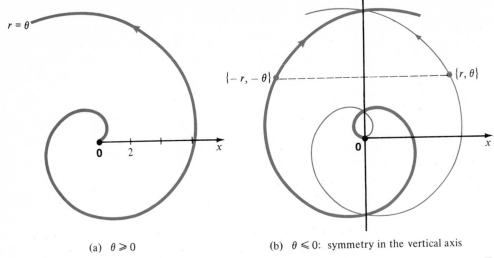

(a) $\theta \geqslant 0$ (b) $\theta \leqslant 0$: symmetry in the vertical axis

Fig. 2 Spiral of Archimedes: $r = \theta$ ■

■ **EXAMPLE 2** Graph the "rose" $r = a \cos 2\theta$, where $a > 0$.

Solution Since $\cos 2(\theta + 2\pi) = \cos 2\theta$, the curve repeats every 2π, so we need plot it only for $0 \leq \theta \leq 2\pi$.

The sign of $\cos 2\theta$ fluctuates, so we make a preliminary sketch showing the proper signs (Fig. 3a). If θ starts at 0 and increases to $\frac{1}{4}\pi$, then $\cos 2\theta$ starts at 1 and decreases to 0. Since $\cos 2\theta$ is an even function, this part of the graph is repeated below, forming a loop (Fig. 3b).

If θ increases from $\frac{1}{4}\pi$ to $\frac{1}{2}\pi$ to $\frac{3}{4}\pi$, then $\cos 2\theta$ is negative and goes from 0 to -1 and back to 0. Thus we get another loop, but between $\frac{5}{4}\pi$ and $\frac{7}{4}\pi$. See Fig. 4a. For θ going from $\frac{3}{4}\pi$ to $\frac{5}{4}\pi$, we get a third loop plotted forward, and from $\frac{5}{4}\pi$ to $\frac{7}{4}\pi$ a fourth loop plotted backwards. Figure 4b is the complete graph.

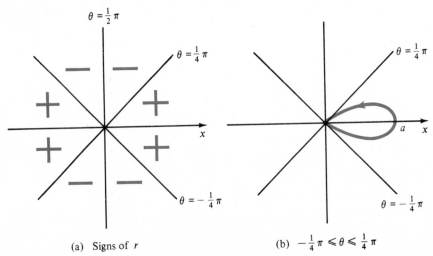

(a) Signs of r (b) $-\frac{1}{4}\pi \leqslant \theta \leqslant \frac{1}{4}\pi$

Fig. 3 Partial graph of $r = a \cos 2\theta$

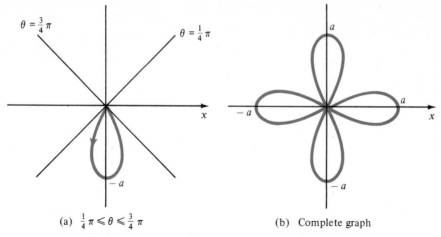

(a) $\frac{1}{4}\pi \leqslant \theta \leqslant \frac{3}{4}\pi$ (b) Complete graph

Fig. 4   Graph of $r = a \cos 2\theta$

Hindsight   It is necessary to plot only one of the petals in Fig. 4b. From $\cos(\theta + \pi) = -\cos\theta$ follows $\cos 2(\theta + \frac{1}{2}\pi) = -\cos 2\theta$. Hence $\cos 2\theta$ repeats itself negatively after $\frac{1}{2}\pi$. Therefore the first loop plotted (for $-\frac{1}{4}\pi \leq \theta \leq \frac{1}{4}\pi$) is repeated negatively as θ continues from $\frac{1}{4}\pi$ to $\frac{3}{4}\pi$. In other words, rotate the first loop *backward* by $\frac{1}{2}\pi$; the result is another loop of the curve. Rotate again, and once again, and you have generated the whole curve.

For an accurate picture of the petals, plot some points. One thing can be said without plotting; the petals are rounded at their ends, not pointed. That stems from a property of the cosine: for small angles, $\cos 2\theta$ is very close to 1. Hence for θ small (near the tip of the petal to the right) the curve $r = a \cos 2\theta$ looks like the circle $r = a$.   ■

Equation of the Ellipse   The orbit of a planet around a fixed star is an ellipse with the star at one focus. Because in astronomy one measures angles rather than distances, it is natural to study the polar equation of an ellipse with one focus at the origin.

Place the origin at one focus and take the polar axis through the other focus, $\{2c, 0\}$. See Fig. 5a. By definition of the ellipse, $r + d = 2a$, hence $d^2 = (2a - r)^2$. On the other hand, by the Law of Cosines,

$$d^2 = r^2 + (2c)^2 - 2r(2c)\cos\theta = r^2 + 4c^2 - 4rc\cos\theta.$$

Equate the two expressions for d^2:

$$(2a - r)^2 = r^2 + 4c^2 - 4rc\cos\theta, \qquad 4a^2 - 4ar = 4c^2 - 4rc\cos\theta.$$

Solve for r:

$$r(a - c\cos\theta) = a^2 - c^2 = b^2, \qquad \text{that is,} \qquad r\left(1 - \frac{c}{a}\cos\theta\right) = \frac{b^2}{a}.$$

Either form is the polar equation of the ellipse.

From the polar equation follows easily the rectangular equation of the ellipse,

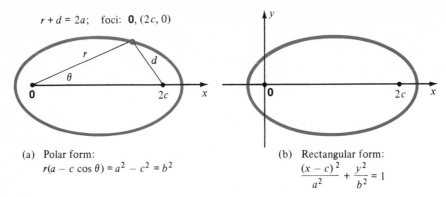

$r + d = 2a;$ foci: **0**, $(2c, 0)$

(a) Polar form:
$r(a - c \cos \theta) = a^2 - c^2 = b^2$

(b) Rectangular form:
$$\frac{(x - c)^2}{a^2} + \frac{y^2}{b^2} = 1$$

Fig. 5   Equation of the ellipse

as promised in Section 3. Substitute $r \cos \theta = x$ and $r^2 = x^2 + y^2$:

$$r(a - c \cos \theta) = b^2, \qquad ar - cx = b^2, \qquad a^2 r^2 = (cx + b^2)^2,$$

$$a^2(x^2 + y^2) = c^2 x^2 + 2cb^2 x + b^4, \qquad b^2 x^2 + a^2 y^2 = 2cb^2 x + b^4.$$

Now complete the square:

$$b^2(x^2 - 2cx) + a^2 y^2 = b^4, \qquad b^2(x - c)^2 + a^2 y^2 = b^4 + b^2 c^2 = a^2 b^2,$$

$$\frac{(x - c)^2}{a^2} + \frac{y^2}{b^2} = 1.$$

This is the rectangular form when the foci are $(0, 0)$ and $(2c, 0)$. See Fig. 5b. Translation of the foci to $(\pm c, 0)$ produces the standard form $x^2/a^2 + y^2/b^2 = 1$.

Eccentricity   Define the **eccentricity** of the ellipse to be the number $e = c/a$. Since $c < a$, it follows that $0 < e < 1$. Define also $p = b^2/a = b^2/ce$. In this notation, the polar equation of the ellipse becomes

> **Polar Equation for the Ellipse** $r(1 - e \cos \theta) = ep.$

The eccentricity determines the shape of the ellipse. If e is near zero, then c is small compared to a. That means the foci are close together relative to the semi-major axis, hence the ellipse is circlelike. If e is near 1, the foci are relatively far apart and the ellipse is cigar-shaped. See Fig. 6. Once e is given, the scale factor p determines the size of the ellipse (as the radius does for a circle).

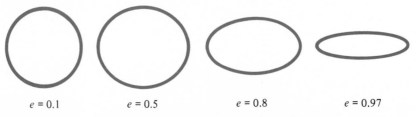

$e = 0.1$ $e = 0.5$ $e = 0.8$ $e = 0.97$

Fig. 6   Ellipses of various eccentricities

■ **EXAMPLE 3** Find the equation of the ellipse having eccentricity $\frac{3}{4}$ and foci at $(-5, -1)$ and $(1, -1)$.

Solution   The center is $(-2, -1)$ so the equation is of the form

$$\frac{(x + 2)^2}{a^2} + \frac{(y + 1)^2}{b^2} = 1,$$

where $a > b$ because the major axis is horizontal.

The distance between the foci is $6 = 2c$, hence $c = 3$. By the definition of eccentricity, $c = ae$, hence $a = c/e = 3/\frac{3}{4} = 4$. Finally, $b^2 = a^2 - c^2 = 16 - 9 = 7$. Therefore, the equation of the ellipse is

$$\frac{(x + 2)^2}{16} + \frac{(y + 1)^2}{7} = 1.$$ ■

■ **EXAMPLE 4**   An ellipse has semi-major axis a and eccentricity e. Find in terms of a and e the length of the chord through one of the foci perpendicular to the major axis.

Solution   Choose coordinates so that the ellipse is in standard position with foci at $(\pm c, 0)$. The length of the chord is $2y$, where $y > 0$ is evaluated at $x = c$. From

$$\frac{x^2}{a^2} + \frac{y^2}{b^2} = 1 \qquad \text{follows} \qquad y^2 = b^2\left(1 - \frac{x^2}{a^2}\right).$$

Set $x = c$: $y^2 = b^2\left(1 - \dfrac{c^2}{a^2}\right) = b^2(1 - e^2) = (a^2 - c^2)(1 - e^2)$

$$= (a^2 - a^2 e^2)(1 - e^2) = a^2(1 - e^2)^2.$$

Therefore the length of the chord is

$$2y = 2a(1 - e^2).$$ ■

Equation of the Hyperbola   Fix the polar coordinate system so the foci of a hyperbola are the origin and $\{2c, 0\}$, and let the absolute distance difference be $2a$, so $c > a$. See Fig. 7.

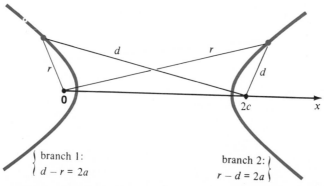

branch 1:
$d - r = 2a$

branch 2:
$r - d = 2a$

Fig. 7   Set-up for the hyperbola in polar coordinates

Branch 1 is defined by $d - r = 2a$ and branch 2 by $r - d = 2a$. Therefore both branches are defined by $d = r \pm 2a$, hence $d^2 = r^2 \pm 4ar + 4a^2$. However, by the law of cosines, $d^2 = r^2 + 4c^2 - 4cr \cos\theta$. Equate these expressions for d^2:

$$r^2 \pm 4ar + 4a^2 = r^2 + 4c^2 - 4cr \cos\theta, \qquad \pm 4ar + 4a^2 = 4c^2 - 4cr \cos\theta,$$

$$\boxed{r(c \cos\theta \pm a) = c^2 - a^2 = b^2.}$$

The "+" determines branch 1 in Fig. 7, the "−", branch 2.

The rectangular form is derived from this form exactly as for the ellipse:

$$cx \pm ar = b^2, \qquad \pm ar = b^2 - cx, \qquad a^2r^2 = (b^2 - cx)^2,$$

$$a^2(x^2 + y^2) = b^4 - 2b^2cx + c^2x^2, \qquad a^2y^2 = b^4 - 2b^2cx + (c^2 - a^2)x^2,$$

$$b^2x^2 - a^2y^2 - 2b^2cx = -b^4.$$

Complete the square:

$$b^2(x - c)^2 - a^2y^2 = b^2c^2 - b^4 = a^2b^2, \qquad \frac{(x - c)^2}{a^2} - \frac{y^2}{b^2} = 1.$$

The standard form, with the foci at $(\pm c, 0)$, follows by translation.

The **eccentricity** of the hyperbola is $e = c/a > 1$. If we set $p = b^2/a = b^2e/c$, the polar equation becomes

$$\boxed{\textbf{Polar Equation of the Hyperbola} \qquad r(e \cos\theta \pm 1) = ep.}$$

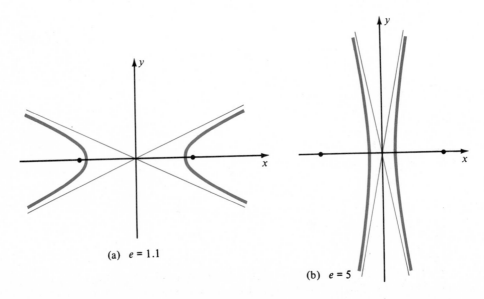

(a) $e = 1.1$

(b) $e = 5$

Fig. 8 Eccentricities near 1 and far from 1; note the relative positions of the foci.

The eccentricity determines the shape of the hyperbola. Suppose

$$\frac{x^2}{a^2} - \frac{y^2}{b^2} = 1$$

is a hyperbola in standard position with foci at $(\pm c, 0)$. Its asymptotes have slopes

$$\pm \frac{b}{a} = \pm \frac{\sqrt{c^2 - a^2}}{a} = \pm \sqrt{\frac{c^2}{a^2} - 1} = \pm \sqrt{e^2 - 1}.$$

Hence if e is near 1, the asymptotes have small slopes and the hyperbola is squeezed into a narrow angle. The larger the eccentricity, the broader the hyperbola (Fig. 8).

EXERCISES

Graph

1    $r = 2 \sin 2\theta$

2   the **rose** $r = \sin 5\theta$

3    $r = \cos 3\theta$

4    $r = -\cos 4\theta$

5    $r = \theta^2$

6   the **lemniscate** $r^2 = \cos 2\theta$

7   the **cissoid** $r = \sec \theta - \cos \theta$
         $= \sin \theta \tan \theta$

8   the **strophoid** $r = \cos 2\theta \sec \theta$

[*Hint for* 7 *and* 8   Use $x = r \cos \theta$ to find the vertical asymptote.]

9   the **cardioid** $r = 1 - \cos \theta$

10   the **limaçon** $r = 2 + \cos \theta$

11   the **limaçon** $r = 1 + 2 \cos \theta$

12   the **bifolium** $r = \sin \theta \cos^2 \theta$

13   the **conchoid** $r = \csc \theta - 2$.

14*   Graph $r = a + b \cos \theta$, $a > 0$, $b > 0$, in general. [*Hint*   Use Exs. 10 and 11.]

15   If an ellipse has eccentricity e and we write $e = \cos \alpha$ where $0 < \alpha < 90°$, then we call the ellipse an "α degree ellipse". Interpret α geometrically and express b/a in terms of α. (Templates for drawing ellipses go by degree.)

16   (cont.) Draw ellipses of $15°$, $30°$, $45°$, and $60°$.

17   The orbit of the Earth is approximately an ellipse with the Sun at one focus and semi-major and semi-minor axes 9.3×10^7 and 9.1×10^7 miles, respectively. Compute the eccentricity of the orbit.

18   (cont.) Find the distance from the Sun to the other focus of the ellipse.

19   Fix p and e with $0 < p$ and $0 < e < 1$. Find the locus of all points $\mathbf{x}$ whose distance from $\mathbf{0}$ is e times its distance from the line $x = -p$.

20   (cont.) Solve the problem for $e \geq 1$.

21   What is the eccentricity of a rectangular hyperbola?

22   Show that the distances of any point (x, y) of the ellipse $x^2/a^2 + y^2/b^2 = 1$ to its foci are $a \pm ex$.

7. ROTATION OF AXES

Suppose we start with a polar coordinate system and create a new system by rotating the polar axis forward through an angle α. See Fig. 1. A point with coordinates $\{r, \theta\}$ acquires new coordinates $\{\bar{r}, \bar{\theta}\}$. From the figure it is clear that $\bar{r} = r$ and $\bar{\theta} = \theta - \alpha$.

Rotation of Axes (Polar Coordinates)   If the polar axis is rotated by an angle α, a point $\{r, \theta\}$ acquires the new coordinates $\{\bar{r}, \bar{\theta}\}$, where

$$\bar{r} = r, \qquad \bar{\theta} = \theta - \alpha.$$

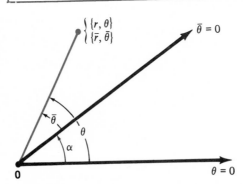

Fig. 1   Rotation of polar coordinates

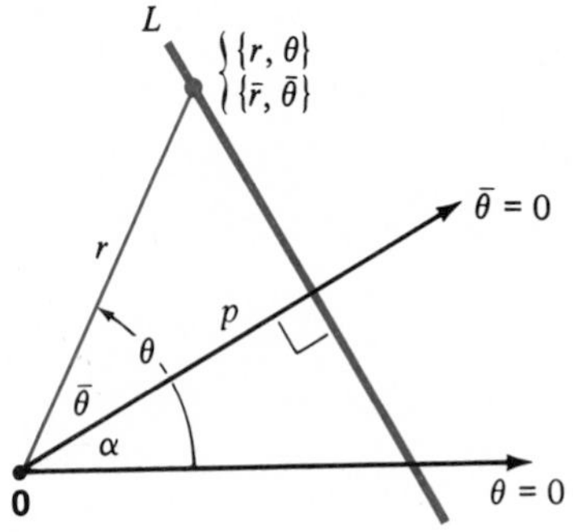

Fig. 2   Polar equation of line:
$$\bar{r} \cos \bar{\theta} = p, \qquad r \cos(\theta - \alpha) = p$$

As an application, let us find the polar equation of the line L that is p units from the origin, perpendicular to the ray $\theta = \alpha$. See Fig. 2. Relative to the tilted axis, the line has equation $\bar{r} \cos \bar{\theta} = p$. Its r, θ-equation therefore is $r \cos(\theta - \alpha) = p$. This is a quick derivation of the equation given on p. 421.

By the same reasoning $r = f(\theta - \alpha)$ represents the curve $r = f(\theta)$ rotated forward through angle α. For example, knowing that $r = 2a \cos \theta$ represents the circle of radius a and center $\{a, 0\}$, we can instantly write down the equation of the circle of radius a and center $\{a, \tfrac{1}{2}\pi\}$. The equation is

$$r = 2a \cos(\theta - \tfrac{1}{2}\pi) = 2a \cos(\tfrac{1}{2}\pi - \theta) = 2a \sin \theta.$$

Rotation of Rectangular Coordinate Systems   Now let us look at the effect on rectangular coordinates of a rotation of axes. Suppose we rotate the x- and y-axes through an angle α, obtaining new axes which we call the $\bar{x}$- and $\bar{y}$-axes (Fig. 3). They define a new rectangular coordinate system.

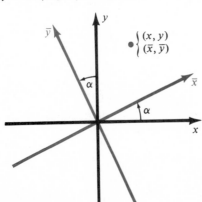

Fig. 3   Rotation of rectangular coordinates

A point **p** in the plane with coordinates (x, y) now acquires new coordinates $(\bar{x}, \bar{y})$. From Fig. 3 it is not obvious what relation exists between x, y and $\bar{x}, \bar{y}$. Still, knowing α, we should be able to express $\bar{x}, \bar{y}$ in terms of x, y and vice versa. The trick is passing through polar coordinates, for which the rotation rule is so easy: $\bar{r} = r$, $\bar{\theta} = \theta - \alpha$. Indeed,

$$\bar{x} = \bar{r}\cos\bar{\theta} = r\cos(\theta - \alpha) = r\cos\theta\cos\alpha + r\sin\theta\sin\alpha = x\cos\alpha + y\sin\alpha,$$

$$\bar{y} = \bar{r}\sin\bar{\theta} = r\sin(\theta - \alpha) = -r\cos\theta\sin\alpha + r\sin\theta\cos\alpha = -x\sin\alpha + y\cos\alpha.$$

Similarly,

$$x = r\cos\theta = \bar{r}\cos(\bar{\theta} + \alpha) = \bar{r}\cos\bar{\theta}\cos\alpha - \bar{r}\sin\bar{\theta}\sin\alpha = \bar{x}\cos\alpha - \bar{y}\sin\alpha,$$

$$y = r\sin\theta = \bar{r}\sin(\bar{\theta} + \alpha) = \bar{r}\cos\bar{\theta}\sin\alpha + \bar{r}\sin\bar{\theta}\cos\alpha = \bar{x}\sin\alpha + \bar{y}\cos\alpha.$$

Rotation of Axes (Rectangular Coordinates) Suppose the plane is rotated through angle α and the x- and y-axes, under this rotation, become the $\bar{x}$- and $\bar{y}$-axes. Then the x, y-coordinates and $\bar{x}, \bar{y}$-coordinates of any point are related by

$$\begin{cases} x = \bar{x}\cos\alpha - \bar{y}\sin\alpha \\ y = \bar{x}\sin\alpha + \bar{y}\cos\alpha \end{cases} \qquad \begin{cases} \bar{x} = x\cos\alpha + y\sin\alpha \\ \bar{y} = -x\sin\alpha + y\cos\alpha. \end{cases}$$

For example, if $\alpha = 45°$, then

$$x = \tfrac{1}{2}\sqrt{2}\,(\bar{x} - \bar{y}), \qquad y = \tfrac{1}{2}\sqrt{2}\,(\bar{x} + \bar{y}).$$

For another example, if $\alpha = -30°$, then $\cos\alpha = \tfrac{1}{2}$ and $\sin\alpha = -\tfrac{1}{2}\sqrt{3}$; hence

$$x = \tfrac{1}{2}(\bar{x} + \sqrt{3}\,\bar{y}), \qquad y = \tfrac{1}{2}(-\sqrt{3}\,\bar{x} + \bar{y}).$$

Conics We have learned how to graph quadratic equations of the form

$$ax^2 + cy^2 + dx + ey + f = 0 \qquad (a^2 + c^2 > 0).$$

By completing squares, we generally obtain one of the conic sections (sometimes a degenerate conic, or no locus at all). Now we tackle the most general quadratic equation

$$ax^2 + bxy + cy^2 + dx + ey + f = 0.$$

It is the term bxy that makes life difficult. Where does it come from and how can we get rid of it? We can learn a good deal from two experiments.

■ **EXAMPLE 1** Find the equation of the ellipse $\dfrac{x^2}{9} + \dfrac{y^2}{4} = 1$ in the $\bar{x}, \bar{y}$-coordinate system that results from a $\tfrac{1}{4}\pi$ rotation of the x, y-coordinate system.

Solution The rotation formulas are $x = \tfrac{1}{2}\sqrt{2}\,(\bar{x} - \bar{y})$, $\quad y = \tfrac{1}{2}\sqrt{2}\,(\bar{x} + \bar{y})$.

Substitute: $\dfrac{x^2}{9} + \dfrac{y^2}{4} = \dfrac{1}{2}\left[\dfrac{(\bar{x} - \bar{y})^2}{9} + \dfrac{(\bar{x} + \bar{y})^2}{4}\right] = \dfrac{1}{2}\left[\dfrac{13}{36}\bar{x}^2 + \dfrac{10}{36}\bar{x}\bar{y} + \dfrac{13}{36}\bar{y}^2\right].$

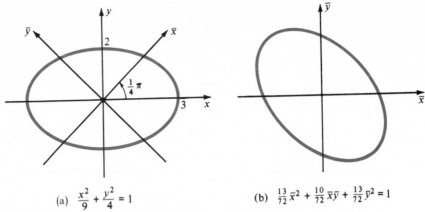

(a) $\dfrac{x^2}{9} + \dfrac{y^2}{4} = 1$ (b) $\dfrac{13}{72}\bar{x}^2 + \dfrac{10}{72}\bar{x}\bar{y} + \dfrac{13}{72}\bar{y}^2 = 1$

Fig. 4   Rotation of an ellipse

Therefore the $\bar{x}$, $\bar{y}$-equation of the ellipse (Fig. 4) is

$$\frac{13}{72}\bar{x}^2 + \frac{10}{72}\bar{x}\bar{y} + \frac{13}{72}\bar{y}^2 = 1.$$ ∎

The experiment suggests that the $\bar{x}\bar{y}$ term is due to the tilt of the coordinate axes relative to the axes of the ellipse. If that is so, then the same should be true for hyperbolas and parabolas. Let us try a parabola.

■ **EXAMPLE 2**   Apply the rotation of Example 1 to the parabola $y = x^2$.

Solution   Substitute $x = \frac{1}{2}\sqrt{2}\,(\bar{x} - \bar{y})$ and $y = \frac{1}{2}\sqrt{2}\,(\bar{x} + \bar{y})$:

$$\tfrac{1}{2}\sqrt{2}\,(\bar{x} + \bar{y}) = [\tfrac{1}{2}\sqrt{2}\,(\bar{x} - \bar{y})]^2 = \tfrac{1}{2}(\bar{x}^2 - 2\bar{x}\bar{y} + \bar{y}^2),$$
$$\sqrt{2}\,(\bar{x} + \bar{y}) = \bar{x}^2 - 2\bar{x}\bar{y} + \bar{y}^2.$$

Therefore the $\bar{x}$, $\bar{y}$-equation of the parabola (Fig. 5) is

$$\bar{x}^2 - 2\bar{x}\bar{y} + \bar{y}^2 - \sqrt{2}\,\bar{x} - \sqrt{2}\,\bar{y} = 0.$$

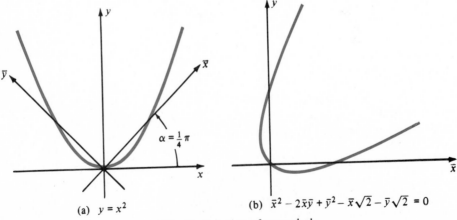

(a) $y = x^2$ (b) $\bar{x}^2 - 2\bar{x}\bar{y} + \bar{y}^2 - \bar{x}\sqrt{2} - \bar{y}\sqrt{2} = 0$

Fig. 5   Rotation of a parabola

Examples 1 and 2 suggest that an xy term occurs when the axes are "incorrectly" placed. Perhaps it can be eliminated by rotating the axes through a cleverly chosen angle. Let us look for a suitable angle. Now a rotation of coordinates,

$$\begin{cases} x = \bar{x} \cos \alpha - \bar{y} \sin \alpha \\ y = \bar{x} \sin \alpha + \bar{y} \cos \alpha, \end{cases}$$

changes a linear polynomial $dx + ey + f$ in x and y into a linear polynomial in $\bar{x}$ and $\bar{y}$. Of more interest to us is what it does to the quadratic $ax^2 + bxy + cy^2$. Substitute:

$$ax^2 + bxy + cy^2 = a(\bar{x} \cos \alpha - \bar{y} \sin \alpha)^2 + b(\bar{x} \cos \alpha - \bar{y} \sin \alpha)(\bar{x} \sin \alpha + \bar{y} \cos \alpha)$$
$$+ c(\bar{x} \sin \alpha + \bar{y} \cos \alpha)^2.$$

Multiply out and collect terms in $\bar{x}^2$, $\bar{x}\bar{y}$, and $\bar{y}^2$.

Under a rotation through an angle α, the quadratic polynomial

$$ax^2 + bxy + cy^2 + dx + ey + f$$

is changed to

$$\bar{a}\bar{x}^2 + \bar{b}\bar{x}\bar{y} + \bar{c}\bar{y}^2 + \bar{d}\bar{x} + \bar{e}\bar{y} + \bar{f},$$

where $\begin{cases} \bar{a} = a \cos^2 \alpha + b \cos \alpha \sin \alpha + c \sin^2 \alpha \\ \bar{b} = 2(c - a) \sin \alpha \cos \alpha + b(\cos^2 \alpha - \sin^2 \alpha) \\ \bar{c} = a \sin^2 \alpha - b \sin \alpha \cos \alpha + c \cos^2 \alpha. \end{cases}$

We are most concerned with the formula for $\bar{b}$, which we can write as

$$\bar{b} = (c - a) \sin 2\alpha + b \cos 2\alpha.$$

It is always possible to choose the rotation angle α so that $\bar{b} = 0$, that is, so that $(c - a) \sin 2\alpha + b \cos 2\alpha = 0$. For if $c = a$, we take $\alpha = \pm \frac{1}{4}\pi$; if $c \neq a$, we choose α so that

$$\tan 2\alpha = \frac{b}{a - c}.$$

A quadratic locus

$$ax^2 + bxy + cy^2 + dx + ey + f = 0$$

is changed into a quadratic locus

$$\bar{a}\bar{x}^2 + \bar{c}\bar{y}^2 + \bar{d}\bar{x} + \bar{e}\bar{y} + \bar{f} = 0$$

without an $\bar{x}\bar{y}$ term by rotating the axes through angle α, where

$$\tan 2\alpha = \frac{b}{a - c} \quad \text{if} \quad a \neq c; \qquad \alpha = \pm \frac{1}{4}\pi \quad \text{if} \quad a = c.$$

Because the tangent has period π, the angle 2α is determined up to a multiple of π, hence α is determined only up to a multiple of $\frac{1}{2}\pi$. Therefore we can always choose α in the first quadrant.

In numerical examples, we must compute $\bar{a}$ and $\bar{c}$, having a, b, c, and $\tan 2\alpha$. We write the formulas for $\bar{a}$ and $\bar{c}$ in the form

$$\begin{cases} \bar{a} = a \cos^2 \alpha + \frac{1}{2}b \sin 2\alpha + c \sin^2 \alpha \\ \bar{c} = a \sin^2 \alpha - \frac{1}{2}b \sin 2\alpha + c \cos^2 \alpha. \end{cases}$$

From $\tan 2\alpha$ we can find $\sin 2\alpha$ and $\cos 2\alpha$:

$$\sin 2\alpha = \frac{\pm \tan 2\alpha}{\sqrt{1 + \tan^2 2\alpha}}, \qquad \cos 2\alpha = \frac{\pm 1}{\sqrt{1 + \tan^2 2\alpha}}.$$

From $\cos 2\alpha$ we can find $\cos^2 \alpha$ and $\sin^2 \alpha$:

$$\cos^2 \alpha = \tfrac{1}{2}(1 + \cos 2\alpha), \qquad \sin^2 \alpha = \tfrac{1}{2}(1 - \cos 2\alpha).$$

Everything ties together neatly.

■ **EXAMPLE 3**   Describe the locus of $xy = 1$.

Solution   In this case $a = c = 0$, $b = 1$. Therefore we choose $\alpha = \frac{1}{4}\pi$ to make $\bar{b} = 0$. The rotation is

$$x = \tfrac{1}{2}\sqrt{2}\,(\bar{x} - \bar{y}), \qquad y = \tfrac{1}{2}\sqrt{2}\,(\bar{x} + \bar{y}),$$

so by direct computation,

$$xy = \tfrac{1}{2}(\bar{x} - \bar{y})(\bar{x} + \bar{y}) = \tfrac{1}{2}(\bar{x}^2 - \bar{y}^2).$$

The locus is $\tfrac{1}{2}(\bar{x}^2 - \bar{y}^2) = 1,$

a rectangular hyperbola (Fig. 6).

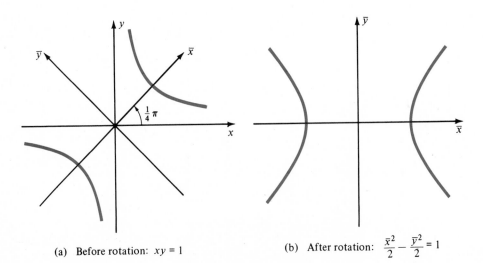

(a) Before rotation: $xy = 1$ (b) After rotation: $\dfrac{\bar{x}^2}{2} - \dfrac{\bar{y}^2}{2} = 1$

Fig. 6   Locus of $xy = 1$

■ **EXAMPLE 4**   Describe the locus of $x^2 - 2xy + 3y^2 = 1$.

Solution   Rotate the axes through angle α where

$$\tan 2\alpha = \frac{b}{a-c} = \frac{-2}{1-3} = 1, \qquad 2\alpha = \tfrac{1}{4}\pi, \quad \alpha = \tfrac{1}{8}\pi.$$

Hence $\sin 2\alpha = \cos 2\alpha = \tfrac{1}{2}\sqrt{2}$, and

$$\cos^2 \alpha = \tfrac{1}{2}(1 + \tfrac{1}{2}\sqrt{2}), \qquad \sin^2 \alpha = \tfrac{1}{2}(1 - \tfrac{1}{2}\sqrt{2}).$$

Substitute these values with $a = 1$, $b = -2$, $c = 3$ into the formulas for $\bar{a}$ and $\bar{c}$:

$$\begin{cases} \bar{a} = \tfrac{1}{2}(1 + \tfrac{1}{2}\sqrt{2}) - \tfrac{1}{2}\sqrt{2} + \tfrac{3}{2}(1 - \tfrac{1}{2}\sqrt{2}) = 2 - \sqrt{2}, \\ \bar{c} = \tfrac{1}{2}(1 - \tfrac{1}{2}\sqrt{2}) + \tfrac{1}{2}\sqrt{2} + \tfrac{3}{2}(1 + \tfrac{1}{2}\sqrt{2}) = 2 + \sqrt{2}. \end{cases}$$

Therefore, in the $\bar{x}$, $\bar{y}$-coordinate system, the locus is

$$(2 - \sqrt{2})\bar{x}^2 + (2 + \sqrt{2})\bar{y}^2 = 1.$$

Because $2 - \sqrt{2}$ and $2 + \sqrt{2}$ are both positive, this is an ellipse in standard form:

$$\frac{\bar{x}^2}{A^2} + \frac{\bar{y}^2}{B^2} = 1,$$

where $A^2 = 1/(2 - \sqrt{2})$ and $B^2 = 1/(2 + \sqrt{2})$. See Fig. 7.

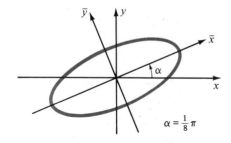

(a)   Before rotation:
$x^2 - 2xy + 3y^2 = 1$

$\alpha = \tfrac{1}{8}\pi$

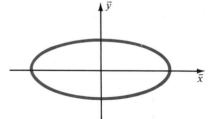

(b)   After rotation:

$$\frac{\bar{x}^2}{\tfrac{1}{2}(2 + \sqrt{2})} + \frac{\bar{y}^2}{\tfrac{1}{2}(2 - \sqrt{2})} = 1$$

Fig. 7   Locus of $x^2 - 2xy + 3y^2 = 1$ ■

EXERCISES

1   Solve the system of linear equations

$$x = \bar{x} \cos \alpha - \bar{y} \sin \alpha, \qquad y = \bar{x} \sin \alpha + \bar{y} \cos \alpha$$

for $\bar{x}$ and $\bar{y}$. Explain your answer.

2   Let $x = \bar{x} \cos \alpha - \bar{y} \sin \alpha$ and $y = \bar{x} \sin \alpha + \bar{y} \cos \alpha$. Compute $x^2 + y^2$. Explain your answer.

3   Let (x_1, y_1) and (x_2, y_2) be two points in the x, y-coordinate system. Let $(\bar{x}_1, \bar{y}_1)$ and $(\bar{x}_2, \bar{y}_2)$ be their coordinates in the $\bar{x}$, $\bar{y}$-coordinate system obtained by a rotation. Compute $x_1 x_2 + y_1 y_2$ in terms of $\bar{x}_1, \bar{x}_2, \bar{y}_1, \bar{y}_2$, and α, the angle of rotation. Explain your answer.

4   Follow a rotation through angle α by a rotation through angle β. The result is obviously a rotation through angle $\alpha + \beta$. Use this observation and the rotation formulas to verify the addition laws for sine and cosine.

5   Let $(x, y) = (\cos \theta, \sin \theta)$ be a point on the unit circle. Rotate the axes by an angle α and show geometrically that $(\bar{x}, \bar{y}) = (\cos(\theta - \alpha), \sin(\theta - \alpha))$.

6   (cont.) Combine this result with the rotation formulas to get a new verification of the addition laws for the sine and cosine.

Make a suitable rotation and write the $\bar{x}$, $\bar{y}$-equation (without an $\bar{x}\bar{y}$ term)

7 $x^2 - xy = 1$ **8** $xy - y^2 = 1$ **9** $xy + y^2 = 1$ **10** $2xy + y^2 = 1$.

Determine the type of the conic and the directions of its principal axes

11 $x^2 + xy + y^2 = 1$	**12** $x^2 - xy + y^2 = 1$	**13** $x^2 + xy - y^2 = 1$
14 $x^2 - xy - y^2 = 1$	**15** $x^2 + xy + 2y^2 = 1$	**16** $x^2 - xy + 2y^2 = 1$
17 $x^2 - 2xy + y^2 = 2y$	**18** $x^2 - 4xy + 4y^2 = x$	**19** $2x^2 - 6xy + y^2 = 1$
20 $x^2 + 3xy - y^2 = 1$.		

Suppose a rotation converts $ax^2 + bxy + cy^2$ into $\bar{a}\bar{x}^2 + \bar{b}\bar{x}\bar{y} + \bar{c}\bar{y}^2$. Prove

21 $a + c = \bar{a} + \bar{c}$ **22*** $4ac - b^2 = 4\bar{a}\bar{c} - \bar{b}^2$.

8. CALCULUS APPLIED TO CONICS

In this section we shall use calculus to learn more about conics.

Tangents to the Parabola   Let us find the tangents to the parabola $y = ax^2$, where $a > 0$, at one of its points (u, v). The slope of the tangent is

$$\left.\frac{dy}{dx}\right|_u = 2ax\Big|_u = 2au,$$

hence the equation of the tangent is

$$y - v = 2au(x - u), \qquad y - v + 2au^2 = 2aux.$$

But $au^2 = v$ since the point (u, v) is on the parabola, so the tangent can be written $y + v = 2aux$. See Fig. 1a.

> The tangent to the parabola $y = ax^2$ at (u, v) is $y + v = 2aux$.

The equation of the tangent has a remarkable symmetry property: the roles of (x, y) and (u, v) are interchangeable! Therefore $y + v = 2aux$ holds if a line through an outside point (x, y) is tangent to the parabola at (u, v) *and also* if a line through (u, v) is tangent at (x, y).

The equation of the tangent provides a tool for finding all tangents to a parabola from a given *exterior* point (u, v): the points of tangency (x, y) must satisfy $y + v = 2aux$. Hence they are the points of intersection (Fig. 1b) of the parabola $y = ax^2$ and the line $y + v = 2aux$, that is, solutions of the system

$$\begin{cases} y = ax^2 \\ y + v = 2aux. \end{cases}$$

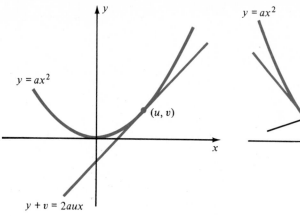

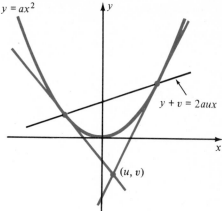

(a) (u, v) on the parabola: Then $y + v = 2aux$ is the equation of the tangent at (u, v).

(b) (u, v) outside the parabola: Then $y + v = 2aux$ is the line through the points of tangency of the tangent lines from (u, v).

Fig. 1 Tangents to the parabola

To solve the system, we eliminate y:

$$ax^2 + v = 2aux, \qquad ax^2 - 2aux + v = 0, \qquad x = \frac{au \pm \sqrt{a(au^2 - v)}}{a}.$$

The solution is valid if $au^2 - v \geq 0$. If equality, $au^2 = v$, then (u, v) is on the parabola and $x = u$; nothing new. If $au^2 - v > 0$, then (u, v) is outside of the parabola and there are two values of x, hence two tangents. In the remaining case, $au^2 - v < 0$, then (u, v) is inside of the parabola and there are no (real) solutions, hence no tangents.

Remark 1 The line $y + v = 2aux$ is called the **polar** of the point (u, v) with respect to the parabola $y = ax^2$.

Remark 2 This discussion can be worked out for any parabola, not just $y = ax^2$.

■ **EXAMPLE 1** Find the tangents to $y = x^2$ from $(-1, -3)$.

Solution Here $(u, v) = (-1, -3)$ and $a = 1$. The tangents through $(-1, -3)$ touch the parabola at (x, y), where

$$y = x^2 \qquad \text{and} \qquad y - 3 = -2x.$$

Eliminate y: $\qquad x^2 - 3 = -2x, \qquad x^2 + 2x - 3 = 0.$

There are two solutions, $x = 1, x = -3$. The corresponding values of y are $y = 1$ and $y = 9$, so the two points of tangency are $(1, 1)$ and $(-3, 9)$. Since the tangent to $y = ax^2$ at (u, v) is $y + v = 2aux$, the two tangent lines in this case are

$$y = 2x - 1 \qquad \text{and} \qquad y = -6x - 9. \qquad ■$$

Angle between Lines We shall shortly need a formula for the angle, measured counterclockwise, between two directed lines in terms of the slopes of the lines. With the notation chosen in Fig. 2 we have $\theta = \beta - \alpha$. By the addition formula for tangent,

$$\tan \theta = \tan(\beta - \alpha) = \frac{\tan \beta - \tan \alpha}{1 + \tan \alpha \tan \beta}.$$

Since $\tan \alpha$ and $\tan \beta$ are the respective slopes of the given lines, this is the required formula.

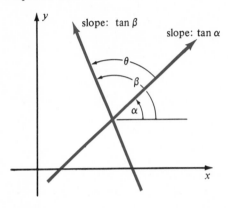

Fig. 2 $\tan \theta = \dfrac{\tan \beta - \tan \alpha}{1 + \tan \alpha \tan \beta}$

Reflection Property of the Parabola   The parabola has a remarkable geometric property with practical applications. Think of the inside of the parabola as a mirror.

If a point source of light is placed at the focus, the light rays striking the mirror at various places will all be reflected parallel to the axis of the parabola, forming a beam (Fig. 3a). This is the principle of the parabolic searchlight.

Conversely, light rays from infinity entering the parabola parallel to its axis will bounce off the mirror to the focus (Fig. 3b). Thus they are concentrated (focused) at this single point. This is the principle of the parabolic receiving antenna and the telescopic mirror.

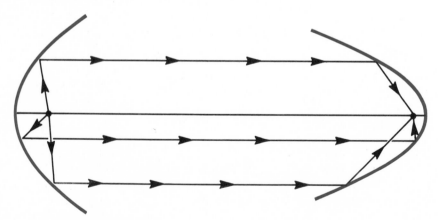

(a) Light from the focus is reflected in a parallel beam.

(b) Light from "infinity" is reflected to the focus.

Fig. 3   Reflection property of the parabola

Let us prove this property of the parabola. We take the standard parabola $4py = x^2$ with $p > 0$. A light ray from the focus $(0, p)$, striking the parabola at (u, v), is reflected so that "the angle of incidence equals the angle of reflection," in other words, so that $\alpha = \beta$ in Fig. 4.

We shall verify that β, the angle from the tangent to the *reflected* ray, equals the angle from the tangent to the *vertical* line through (u, v). This will show that the light ray is reflected vertically, parallel to the axis of the parabola.

First we compute $\tan \beta = \tan \alpha$. Now α is the angle between the ray, of slope $(v - p)/u$, and the tangent, of slope $dy/dx|_u = u/2p$. We use the formula we just learned (and the relation $4pv = u^2$):

$$\tan \alpha = \frac{\dfrac{u}{2p} - \dfrac{v - p}{u}}{1 + \left(\dfrac{v - p}{u}\right)\left(\dfrac{u}{2p}\right)} = \frac{u^2 - 2p(v - p)}{2up + u(v - p)} = \frac{2pv + 2p^2}{up + uv} = \frac{2p(v + p)}{u(v + p)} = \frac{2p}{u}.$$

Hence $\tan \beta = 2p/u$. (Note that $v + p \neq 0$ since $p > 0$ and $v \geq 0$.)

Now we find the angle β_1 from the tangent to the vertical line through (u, v). Clearly, $\beta_1 + \theta = \frac{1}{2}\pi$, where θ is the angle the tangent makes with the positive x-axis. Hence

$$\tan \beta_1 = \cot \theta = \frac{1}{\tan \theta} = \frac{1}{u/2p} = \frac{2p}{u} = \tan \beta.$$

Therefore $\beta_1 = \beta$, which completes the derivation.

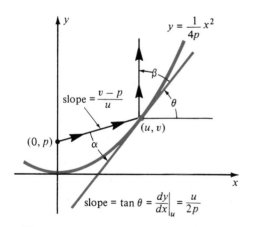

Fig. 4   Proof of the reflection property

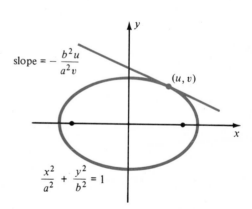

Fig. 5   Tangents to the ellipse

Tangents to the Ellipse   Let us find the tangent to the ellipse

$$\frac{x^2}{a^2} + \frac{y^2}{b^2} = 1 \qquad (a > 0, \quad b > 0)$$

at one of its points (u, v). See Fig. 5.

For the slope, we need dy/dx at $x = u$. The equation of the ellipse actually defines two functions of x, corresponding to $y > 0$ and $y < 0$:

$$y = b \sqrt{1 - \frac{x^2}{a^2}}, \qquad y = -b \sqrt{1 - \frac{x^2}{a^2}}.$$

However, both satisfy $\dfrac{x^2}{a^2} + \dfrac{y^2}{b^2} = 1$,

so we can simply differentiate this relation with respect to x and set $(x, y) = (u, v)$:

$$\frac{2x}{a^2} + \frac{2y}{b^2} \frac{dy}{dx} = 0, \qquad \frac{dy}{dx} \bigg|_{(u,v)} = -\frac{b^2 x}{a^2 y} \bigg|_{(u,v)} = -\frac{b^2 u}{a^2 v}.$$

(The sign takes care of itself.) Now we can write down the equation of the tangent:

$$y - v = -\frac{b^2 u}{a^2 v}(x - u), \qquad \frac{u}{a^2}(x - u) + \frac{v}{b^2}(y - v) = 0, \qquad \frac{ux}{a^2} + \frac{vy}{b^2} = \frac{u^2}{a^2} + \frac{v^2}{b^2} = 1.$$

> The tangent to the ellipse $\dfrac{x^2}{a^2} + \dfrac{y^2}{b^2} = 1$ at (u, v) is $\dfrac{ux}{a^2} + \dfrac{vy}{b^2} = 1.$

Remark   The reasoning fails if $v = 0$, but the result is still correct—just reverse the roles of x and y to prove it.

As in the case of the parabola, (u, v) and (x, y) are interchangeable in the tangent formula. It follows, as for the parabola, that if (u, v) lies outside of the ellipse and (x, y) is the point of contact with the ellipse of one of the two tangents through (u, v), then $ux/a^2 + vy/b^2 = 1$.

Remark   The line $ux/a^2 + vy/b^2 = 1$ is called the **polar** of (u, v) with respect to the ellipse $x^2/a^2 + y^2/b^2 = 1$.

■ **EXAMPLE 2**   Find the two tangents from $(6, 4)$ to the ellipse $\frac{1}{9}x^2 + \frac{1}{4}y^2 = 1$.

Solution   Here $(u, v) = (6, 4)$, $a^2 = 9$, and $b^2 = 4$. The tangents through $(6, 4)$ touch the ellipse at (x, y), where

$$\frac{x^2}{9} + \frac{y^2}{4} = 1 \quad \text{and} \quad \frac{2x}{3} + y = 1.$$

Eliminate y: $\dfrac{x^2}{9} + \dfrac{1}{4}\left(1 - \dfrac{2x}{3}\right)^2 = 1, \qquad 8x^2 - 12x - 27 = 0.$

Solve: $x = \dfrac{6 \pm \sqrt{6^2 + 8 \cdot 27}}{8} = \dfrac{3}{4} \pm \dfrac{3}{4}\sqrt{7},$

$$y = 1 - \tfrac{2}{3}x = 1 - \tfrac{2}{3}(\tfrac{3}{4} \pm \tfrac{3}{4}\sqrt{7}) = \tfrac{1}{2} \mp \tfrac{1}{2}\sqrt{7}.$$

Therefore the two points of tangency are $(\tfrac{3}{4}(1 \pm \sqrt{7}), \tfrac{1}{2}(1 \mp \sqrt{7}))$.

We know that the tangent to $x^2/a^2 + y^2/b^2 = 1$ at (u, v) is $ux/a^2 + vy/b^2 = 1$.

Hence the two tangent lines in this case are

$$\frac{\frac{3}{4}(1 \pm \sqrt{7})}{9}x + \frac{\frac{1}{2}(1 \mp \sqrt{7})}{4}y = 1.$$

Simplify: $2(1 \pm \sqrt{7})x + 3(1 \mp \sqrt{7})y = 24,$

$$y = \frac{-2(1 \pm \sqrt{7})x + 24}{3(1 \mp \sqrt{7})} = \frac{-2(1 \pm \sqrt{7})^2 x + 24(1 \pm \sqrt{7})}{3(1 - 7)}$$

$$= \frac{-2(8 \pm 2\sqrt{7})x + 24(1 \pm \sqrt{7})}{-18},$$

so the tangents are $y = \frac{1}{9}(8 \pm 2\sqrt{7})x - \frac{4}{3}(1 \pm \sqrt{7}).$ ∎

Remark 1   It is a bit simpler to start with one of the values of x, say $x = \frac{3}{4} + \frac{3}{4}\sqrt{7}$ and follow it through:

$$y = \frac{1}{2} - \frac{1}{2}\sqrt{7}, \qquad m = \frac{8}{9} + \frac{2}{9}\sqrt{7}, \qquad y = (\frac{8}{9} + \frac{2}{9}\sqrt{7})x + (-\frac{4}{3} - \frac{4}{3}\sqrt{7}).$$

Then replace $\sqrt{7}$ by $-\sqrt{7}$ for the second tangent line.

Remark 2   If (u, v) is inside the ellipse, that is, if $u^2/a^2 + v^2/b^2 < 1$, there are no tangents. When you eliminate y (or x), the resulting quadratic will have negative discriminant, hence no (real) roots.

Reflection Property of the Ellipse   Like the parabola, the ellipse has a remarkable reflection property. Think of an elliptical pool table. Then a ball cued from one focus will always pass through the other focus after one rebound off the side. This is the principle of whispering galleries. Sound waves emanating from one focus of an elliptical chamber will bounce off the walls and pass through the other focus. Hence a listener at one focus hears clearly a whisper from the other focus.

Let us prove this property of the ellipse. We take the standard ellipse $x^2/a^2 + y^2/b^2 = 1$ with foci at $(\pm c, 0)$. See Fig. 6.

We must verify that $\alpha = \beta$. For then, the ray from $(c, 0)$ will strike the ellipse and be reflected through the focus at $(-c, 0)$.

By the formula for the angle between two lines,

$$\tan \alpha = \frac{-\dfrac{b^2 u}{a^2 v} - \dfrac{v}{u - c}}{1 + \left(\dfrac{v}{u - c}\right)\left(-\dfrac{b^2 u}{a^2 v}\right)} = \frac{-b^2 u(u - c) - a^2 v^2}{a^2 v(u - c) - b^2 uv}.$$

To simplify this expression, we use the relations $c^2 = a^2 - b^2$ and $u^2/a^2 + v^2/b^2 = 1$ in the form $b^2 u^2 + a^2 v^2 = a^2 b^2$:

$$\tan \alpha = \frac{b^2 cu - (b^2 u^2 + a^2 v^2)}{(a^2 - b^2)uv - a^2 cv} = \frac{b^2 cu - a^2 b^2}{c^2 uv - a^2 cv} = \frac{b^2(cu - a^2)}{cv(cu - a^2)} = \frac{b^2}{cv}.$$

Note that $cu - a^2 \neq 0$ because $c < a$ and $|u| \leq a$, hence $|cu| < a^2$.

Now we look at the corresponding angle β with respect to the other focus $(-c, 0)$. We can use the preceding formula for $\tan \alpha$ with c replaced by $-c$. By Fig. 5, however, the angle *from* the ray *to* the tangent is $-\beta$, so the formula yields

$$\tan(-\beta) = \frac{b^2}{(-c)v}, \qquad \text{that is,} \qquad \tan \beta = \frac{b^2}{cv} = \tan \alpha.$$

Therefore $\beta = \alpha$.

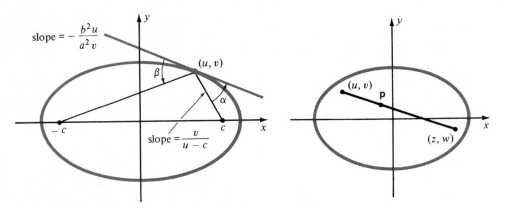

Fig. 6   Reflection property: $\alpha = \beta$ **Fig. 7**   Convexity property

Convexity of the Ellipse   The region inside the ellipse $x^2/a^2 + y^2/b^2 = 1$ is characterized by the inequality

$$\frac{x^2}{a^2} + \frac{y^2}{b^2} \le 1.$$

(This includes the boundary, the ellipse itself.) We propose to prove that this region is **convex** in the following sense: if (u, v) and (z, w) are any two points in the region, then the segment joining them is completely in the region.

The typical point **p** of this segment (Fig. 7) is t of the way from (u, v) to (z, w), where $0 \le t \le 1$. Hence (p. 394)

$$\mathbf{p} = ((1 - t)u + tz, \ (1 - t)v + tw).$$

What we must prove is that the assumptions

$$\frac{u^2}{a^2} + \frac{v^2}{b^2} \le 1, \qquad \frac{z^2}{a^2} + \frac{w^2}{b^2} \le 1, \qquad 0 \le t \le 1$$

imply

$$\frac{[(1 - t)u + tz]^2}{a^2} + \frac{[(1 - t)v + tw]^2}{b^2} \le 1.$$

Let us call the left-hand side $f(t)$. This function is defined for $0 \le t \le 1$, takes non-negative values, and satisfies

$$f(0) = \frac{u^2}{a^2} + \frac{v^2}{b^2} \le 1, \qquad f(1) = \frac{z^2}{a^2} + \frac{w^2}{b^2} \le 1.$$

We must prove that $f(t) < 1$ also for $0 < t < 1$.

We differentiate twice:

$$f'(t) = \frac{2}{a^2}[(1-t)u + tz](z-u) + \frac{2}{b^2}[(1-t)v + tw](w-v),$$

$$f''(t) = \frac{2}{a^2}(z-u)^2 + \frac{2}{b^2}(w-v)^2.$$

Hence $f''(t) > 0$, so the graph of $f(t)$ is strictly convex. By the Chord Theorem (p. 151) $f(t) < 1$ for $0 < t < 1$.

EXERCISES

Find the tangent to

1   $4y = x^2$ at $(-2, 1)$
3   $y = (x-1)^2$ at $(3, 4)$

2   $x = 2y^2$ at $(2, 1)$
4   $x = -3y^2$ at $(-3, -1)$.

Find the tangents to

5   $4y = x^2$ from $(-1, -1)$
7   $y = (x-1)^2$ from $(-1, 3)$

6   $x = 2y^2$ from $(0, -2)$
8   $x = -3y^2$ from $(3, 0)$.

9   Let (u, v) lie inside the parabola $y = ax^2$, $a > 0$, and let (x, y) lie on the polar of (u, v). Show that (x, y) lies outside the parabola.
10   Show that the region inside $y = ax^2$ is convex.

Find the tangent to

11   $x^2 + 2y^2 = 1$ at $(1, 0)$
13   $\frac{1}{15}x^2 + \frac{1}{40}y^2 = 1$ at $(3, 4)$

12   $\frac{1}{8}x^2 + \frac{1}{18}y^2 = 1$ at $(2, -3)$
14   $\frac{1}{6}x^2 + \frac{1}{3}y^2 = 1$ at $(-2, 1)$.

Find the tangents to

15   $x^2 + 2y^2 = 1$ from $(1, 3)$
17   $3x^2 + 2y^2 = 3$ from $(-1, 1)$

16   $x^2 + 2y^2 = 1$ from $(2, 0)$
18   $3x^2 + y^2 = 1$ from $(-2, -1)$.

19   Let (u, v) be a point of the hyperbola $x^2/a^2 - y^2/b^2 = 1$. Show that the tangent at (u, v) is $xu/a^2 - yv/b^2 = 1$.
20   (cont.) Find the tangents to $2x^2 - y^2 = 1$ from $(2, 3)$.
21   Find the (acute) angle of intersection of the two tangents to the parabola $y = x^2$ from the point $(2, 1)$.
22   Find all points on the coordinate axes from which the two tangents to the ellipse $x^2/a^2 + y^2/b^2 = 1$ are perpendicular.

9. MISCELLANEOUS EXERCISES

1   Let $(x - a_1)^2 + (y - b_1)^2 = r_1^2$ and $(x - a_2)^2 + (y - b_2)^2 = r_2^2$ be two non-concentric circles. The **radical axis** of the two circles is the locus of

$$(x - a_1)^2 + (y - b_1)^2 - (x - a_2)^2 - (y - b_2)^2 = r_1^2 - r_2^2.$$

Show that the radical axis is a line perpendicular to the line of centers.
2   (cont.) Suppose the circles intersect in two points. Prove that the radical axis is the line through these points.
3   (cont.) Suppose the circles are tangent. Prove that the radical axis is their common tangent line.

4   (cont.) Suppose the circles do not meet. Prove that the radical axis is the locus of all points **x** such that the tangents from **x** to the two circles all have the same length.

5   Show that $y = mx$ is tangent to the circle $x^2 + y^2 + 2ax + 2by + c = 0$ if and only if $(a + mb)^2 = c(1 + m^2)$.

6*   Show that if the circles $x^2 + y^2 - 2a_1x - 2b_1y + c_1 = 0$ and $x^2 + y^2 - 2a_2x - 2b_2y + c_2 = 0$ intersect at right angles, then $2(a_1a_2 + b_1b_2) = c_1 + c_2$, and conversely.

7   Let $0 < b < a$. Show algebraically that the ellipse $x^2/a^2 + y^2/b^2 = 1$ and the circle $x^2 + y^2 = r^2$ have four intersection points if $b < r < a$ and none if $0 < r < b$ or $r > a$.

8*   Find the locus of all points **x** such that the two tangents from **x** to the ellipse $x^2/a^2 + y^2/b^2 = 1$ are perpendicular.

9   Find all parabolas with focus **0** and axis the y-axis.

10   (cont.) Show that any two that open in opposite directions intersect at right angles.

11   Show for the ellipse in Fig. 1 that $uv/w^2 = $ const.

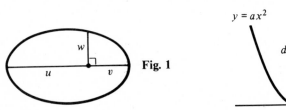

Fig. 1

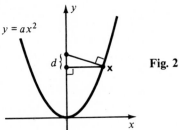

$y = ax^2$

Fig. 2

12   (cont.) Give a similar result for the hyperbola.

13   Find the equation of the rectangular hyperbola with foci (c, c) and $(-c, -c)$.

14   Show in Fig. 2 that d is independent of **x**.

15   For what values of m does $y = mx$ intersect the hyperbola $x^2/a^2 - y^2/b^2 = 1$?

16*   Prove that an ellipse and a hyperbola with the same foci always intersect at right angles.

17   Find the polar equation for the locus of all points, the product of whose distances from $(-a, 0)$ and $(a, 0)$ is a^2.

18   Show that the region $x \geq a$, $x^2/a^2 - y^2/b^2 \geq 1$ is convex.

19   Two firms are $2c$ km apart, and they both sell an item at the same price. However, the shipping cost per km (as the crow flies) for one firm is k times that for the other. Find the curve of equal cost.

20   (cont.) Suppose the shipping costs per km are equal, but one factory price is k times the other. Now solve the problem of equal cost.

Here are two drawing board and string constructions for curves. The string ends are attached at P and A (and the pencil point really touches the rule). Find the curve in

21

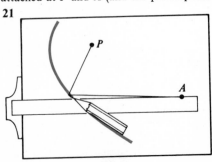

Fig. 3

22

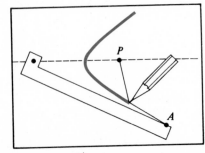

Fig. 4

Approximation **10**

1. INTRODUCTION

It is not always possible to solve a problem with an *exact* numerical answer, even though in theory a precise answer may exist. Examples: find *exactly* the solution of $\cos x = x$; find *exactly* the value of

$$\int_0^3 e^{-x^2}\, dx.$$

The next best thing is approximating the answer. In this chapter we shall learn some methods of approximating numbers and functions to as high an accuracy as desired.

From a practical point of view, approximate answers are just as good as exact answers because you cannot have greater accuracy than your data and your measuring tools. For instance, suppose a surveyor finds that the side of a square field measures 135.2 m. This means that the exact length is between 135.15 and 135.25 m. Hence it is useless to say that the diagonal is

$$135.2\sqrt{2} \approx 135.2 \times 1.414214 \approx 191.201674 \text{ m.}$$

Such accuracy is unjustified. All that can be said is that the true value is between

$$135.15\sqrt{2} \approx 191.13 \qquad \text{and} \qquad 135.25\sqrt{2} \approx 191.27.$$

This chapter is a brief introduction to a vast subject called Numerical Analysis. The main topic of the chapter is approximations of functions by polynomials, and their applications. Two kinds of polynomial approximations are discussed. The first uses data at a single point: values of the function and its successive derivatives. This approach leads to Taylor polynomials. The second uses values of the function at several points. This technique is called interpolation.

An approximation is useful only when we are able to tell how accurately it approximates, that is, to estimate its error. Estimation of error leads us to a second topic, a deeper study of Rolle's Theorem and the Mean Value Theorem than we made in Chapter 3.

Among the various applications of these theorems are the following three: (1) iteration methods for estimating zeros of functions, (2) Simpson's Rule for approximate integration, (3) Lhospital's Rule for limits of the form $\lim_{x \to a} f(x)/g(x)$, where $f(x) \longrightarrow 0$ and $g(x) \longrightarrow 0$ as $x \longrightarrow a$.

2. FIRST AND SECOND DEGREE APPROXIMATIONS

We begin with a basic application of differential calculus, approximating functions by linear functions. Graphically, this amounts to approximating a curve by its tangent.

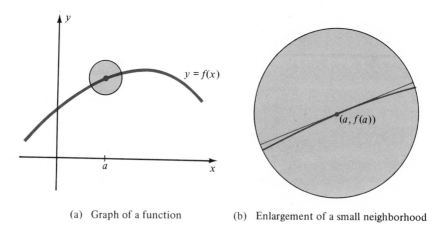

(a) Graph of a function   (b) Enlargement of a small neighborhood

Fig. 1

Under a microscope a smooth curve appears nearly straight. Its tangent at $(a, f(a))$ is quite close to the curve, at least in a small neighborhood of that point (Fig. 1). The slope of the tangent is $f'(a)$. Hence by the point–slope formula, the equation of the tangent line is

$$y - f(a) = f'(a)(x - a), \quad \text{that is,} \quad y = f(a) + f'(a)(x - a).$$

Since the tangent is close to the curve, it is reasonable to expect that the linear function $f(a) + f'(a)(x - a)$ is a good approximation of $f(x)$, at least for x near a. Just how good the approximation is depends on the error $e(x)$, where

$$e(x) = f(x) - [f(a) + f'(a)(x - a)].$$

■ **EXAMPLE 1**   How closely does the tangent to $y = \frac{1}{3}x^2$ at $(3, 3)$ approximate the function?

Solution   First check that $(3, 3)$ is on the curve: $\frac{1}{3}(3^2) = 3$. Next,

$$y' = \tfrac{2}{3}x, \quad y'(3) = 2,$$

so the tangent at $(3, 3)$ is

$$y = 3 + 2(x - 3) = 2x - 3.$$

We want to know how closely $2x - 3$ approximates $\frac{1}{3}x^2$ near $x = 3$. See Fig. 2. Let $e(x)$ denote the error in this approximation:

$$e(x) = \tfrac{1}{3}x^2 - (2x - 3) = \tfrac{1}{3}(x^2 - 6x + 9) = \tfrac{1}{3}(x - 3)^2.$$

This is good news: the error equals the *square* of the distance from x to 3 times a factor $\frac{1}{3}$. Therefore if $|x - 3|$ is small, $e(x)$ is very small. For instance, if $|x - 3| < 10^{-3}$, then $|e(x)| < \frac{1}{3}(10^{-3})^2 < 4 \times 10^{-7}$.   ■

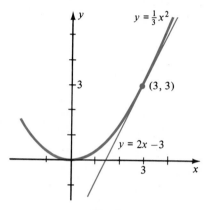

Fig. 2   Fig. 3

■ **EXAMPLE 2** How closely does the tangent to $y = (x + 1)(x - 2)^2$ at $(-1, 0)$ approximate the function?

Solution Clearly $y(-1) = 0$, so the given point lies on the graph (Fig. 3). Next,

$$y' = (x - 2)^2 + 2(x + 1)(x - 2), \qquad y'(-1) = 9.$$

The tangent at $(-1, 0)$ is $y = 9(x + 1)$.

Therefore, the error made in approximating the curve by its tangent is

$$e(x) = (x + 1)(x - 2)^2 - 9(x + 1) = (x + 1)[(x - 2)^2 - 9]$$
$$= (x + 1)(x^2 - 4x - 5) = (x + 1)^2(x - 5).$$

Suppose we limit attention to $|x + 1| < 1$, that is, $-2 < x < 0$. Then $|x - 5| < 7$, so

$$|e(x)| < 7|x + 1|^2.$$

For instance, if $|x + 1| < 10^{-4}$, then $|e(x)| < 7 \times 10^{-8} < 10^{-7}$.   ■

There is a lesson in these two examples. In each case the error

$$e(x) = f(x) - f(a) - f'(a)(x - a)$$

is more or less equal to a constant times $(x - a)^2$. In Example 1,

$$e(x) = \tfrac{1}{3}(x - 3)^2;$$

and in Example 2,

$$e(x) = (x - 5)(x + 1)^2 \approx -6(x + 1)^2$$

near $x = -1$. Later we shall prove that this is typical behavior provided $f(x)$ is smooth enough.

Second Degree Approximation The linear or **first degree approximation** to $f(x)$ at $x = a$ is

$$p_1(x) = f(a) + f'(a)(x - a).$$

Near $x = a$, the function $p_1(x)$ appears to be a good approximation to $f(x)$. This is not surprising since

$$p_1(a) = f(a), \qquad p_1'(a) = f'(a);$$

hence both $y = p_1(x)$ and $y = f(x)$ pass through the point $(a, f(a))$ with the same slope.

For greater accuracy, we shall approximate $f(x)$ by a polynomial whose graph passes through $(a, f(a))$ with the same slope as $y = f(x)$, and which also curves the same amount as $y = f(x)$. Now the curving of a graph is due to change in its slope. The slope is $f'(x)$, and the rate of change of the slope is the second derivative $f''(x)$. Therefore, we seek a polynomial $p_2(x)$ such that

$$p_2(a) = f(a), \qquad p_2'(a) = f'(a), \qquad \text{and} \qquad p_2''(a) = f''(a).$$

Let us try a quadratic

$$p_2(x) = A + B(x - a) + C(x - a)^2.$$

We note that $\qquad p_2(a) = A, \qquad p_2'(a) = B, \qquad p_2''(a) = 2C.$

(Verify these statements.) Since we want

$$p_2(a) = f(a), \qquad p_2'(a) = f'(a), \qquad p_2''(a) = f''(a),$$

there is no choice but

$$A = f(a), \qquad B = f'(a), \qquad C = \tfrac{1}{2}f''(a).$$

The **second degree approximation** to $f(x)$ at $x = a$ is

$$p_2(x) = f(a) + f'(a)(x - a) + \tfrac{1}{2}f''(a)(x - a)^2.$$

The polynomial $p_2(x)$ agrees with $f(x)$ at $x = a$, and its derivative and second derivative agree with those of $f(x)$ at $x = a$.

The first two terms of $p_2(x)$ are $f(a) + f'(a)(x - a)$, which is $p_1(x)$. Thus $p_2(x)$ consists of the linear approximation to $f(x)$ plus another term, $\tfrac{1}{2}f''(a)(x - a)^2$, which (we hope) corrects some of the error in linear approximation.

■ **EXAMPLE 3** Approximate e^x near $x = 0$ by the first and second degree polynomials $p_1(x)$ and $p_2(x)$. Test their accuracy.

Solution Use the formula

$$p_2(x) = f(a) + f'(a)(x - a) + \tfrac{1}{2}f''(a)(x - a)^2,$$

where $f(x) = e^x$ and $a = 0$. In this case $f(x) = f'(x) = f''(x) = e^x$, so $f(0) = f'(0) = f''(0) = 1$. Therefore

$$p_1(x) = 1 + x \qquad \text{and} \qquad p_2(x) = 1 + x + \tfrac{1}{2}x^2.$$

Table 1 compares e^x with $p_1(x)$ and $p_2(x)$ for various values of x.

Table 1

	Small values of x				Larger values of x		
x	e^x	$p_1(x)$	$p_2(x)$	x	e^x	$p_1(x)$	$p_2(x)$
-0.4	0.6703	0.6000	0.6800	-2.0	0.1353	-1.0000	1.0000
-0.3	0.7408	0.7000	0.7450	-1.5	0.2231	-0.5000	0.6250
-0.2	0.8187	0.8000	0.8200	-1.0	0.3679	0.0000	0.5000
-0.1	0.9048	0.9000	0.9050	-0.5	0.6065	0.5000	0.6250
0.0	1.0000	1.0000	1.0000	0.0	1.0000	1.0000	1.0000
0.1	1.1052	1.1000	1.1050	0.5	1.6487	1.5000	1.6250
0.2	1.2214	1.2000	1.2200	1.0	2.7183	2.0000	2.5000
0.3	1.3499	1.3000	1.3450	1.5	4.4817	2.5000	3.6250
0.4	1.4918	1.4000	1.4800	2.0	7.3891	3.0000	5.0000

From the table we see that $p_1(x)$ and $p_2(x)$ are good estimates of e^x for x near 0, but that $p_2(x)$ is considerably better than $p_1(x)$. Both estimates become poor as x moves away from 0, but $p_2(x)$ stays accurate in a wider range because its graph is curved like that of e^x near $x = 0$. See Fig. 4.

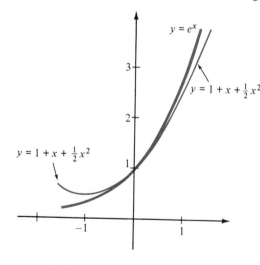

Fig. 4 Second degree approximation to $y = e^x$ at $x = 0$

■ **EXAMPLE 4** Estimate $y = 1/x$ near $x = 1$ by its first and second degree approximations. Test their accuracy.

Solution

$$y = \frac{1}{x}, \qquad y' = -\frac{1}{x^2}, \qquad y'' = \frac{2}{x^3},$$

hence $y(1) = 1$, $y'(1) = -1$, $y''(1) = 2$. Therefore

$$p_1(x) = 1 - (x - 1), \qquad p_2(x) = 1 - (x - 1) + (x - 1)^2.$$

Since we are dealing with numbers near 1, it is convenient to leave $p_1(x)$ and $p_2(x)$ in terms of $x - 1$.

Table 2 compares $1/x$ with $p_1(x)$ and $p_2(x)$.

Table 2

	x near 1				Other values of x		
x	$1/x$	$p_1(x)$	$p_2(x)$	x	$1/x$	$p_1(x)$	$p_2(x)$
0.85	1.1765	1.1500	1.1725	0.25	4.0000	1.7500	2.3125
0.90	1.1111	1.1000	1.1100	0.50	2.0000	1.5000	1.7500
0.95	1.0526	1.0500	1.0525	0.75	1.3333	1.2500	1.3125
1.00	1.0000	1.0000	1.0000	1.00	1.0000	1.0000	1.0000
1.05	0.9524	0.9500	0.9525	1.25	0.8000	0.7500	0.8125
1.10	0.9091	0.9000	0.9100	1.50	0.6667	0.5000	0.7500
1.15	0.8696	0.8500	0.8725	1.75	0.5714	0.2500	0.8125

From the table, we see that $p_1(x)$ is a good approximation to $1/x$ provided x is near 1, but that $p_2(x)$ is much better. Both estimates become poor as x moves away from 1, but $p_2(x)$ is accurate in a wider range.   ∎

Error Estimates   In our examples of linear approximations, the error had order of magnitude $|x - a|^2$. By analogy, we might expect the error in second degree approximations,

$$e(x) = f(x) - p_2(x),$$

to have order of magnitude $|x - a|^3$. By that we mean

$$e(x) \approx c(x - a)^3 \qquad \text{for} \quad x \text{ near } a,$$

where c is a suitable constant. We shall prove this conjecture in Section 4. Meanwhile, let us verify it in an example.

∎   **EXAMPLE 5** How closely does $p_2(x)$ approximate the function $y = 1/x$ near $x = 1$?

Solution   By Example 4,

$$e(x) = \frac{1}{x} - p_2(x) = \frac{1}{x} - 1 + (x - 1) - (x - 1)^2.$$

Hence $e(x) = \dfrac{1 - x}{x} + (x - 1) - (x - 1)^2 = (x - 1)\left[-\dfrac{1}{x} + 1 - (x - 1) \right]$

$$= (x - 1)\left[\frac{x - 1}{x} - (x - 1) \right] = (x - 1)^2\left(\frac{1}{x} - 1 \right) = -\frac{1}{x}(x - 1)^3.$$

For x near 1, $$e(x) \approx -(x-1)^3,$$

so the order of magnitude of the error is as expected.

For a more precise estimate, not involving the vague symbol $\approx$, let us restrict our attention to a neighborhood of $x = 1$, say $\frac{1}{2} < x < \frac{3}{2}$. On this interval $|1/x| < 2$, hence we get the specific estimate

$$|e(x)| = \left| -\frac{1}{x} \right| |x-1|^3 < 2|x-1|^3.$$

For instance, if $|x - 1| < 10^{-3}$, then $|e(x)| < 2 \times 10^{-9}$; the approximation is very close indeed.   ∎

EXERCISES

Given $f(x)$ and $x = a$; find $p_1(x)$ and the error in the form $e(x) = (x-a)^2 g(x)$

1	$1 - x^2$	$a = 0$	**2**	$2x^2 + 3$	1	**3**	x^3	-1
4	x^3	2	**5**	$1/x$	$\frac{1}{2}$	**6**	$x^2 - x^3$	1
7	x^4	1	**8**	x^5	-1	**9**	$1/x^2$	-1
10	$1/(x^3 + 1)$	1.						

Given $f(x)$ and $x = a$, find $p_1(x)$

11 $\cos x$ $a = 0$ **12** $\sin x$ $\frac{1}{4}\pi$
13 xe^x 0 **14** $1/(1 + e^x)$ 0.

Given $f(x)$ and $x = a$, find $p_2(x)$ and the error in the form $e(x) = (x - a)^3 g(x)$

15 $x + 1$ 0 **16** $(x-2)^3 + 3(x-2)^2 - 4(x-2) + 1$ 2
17 $1/x$ -2 **18** $1/(x^2 + 1)$ 0
19 $x^2/(x+1)$ 0 **20** $x^3/(x+1)$ 1.

Given $f(x)$ and $x = a$, find $p_2(x)$

21 e^{2x} 0 **22** $\cos x$ 0
23 $\tan x$ 0 **24** $\sinh x$ 0.

Complete the table (4 place accuracy) at $x = 0$

25

x	xe^x	$p_1(x)$	$p_2(x)$
0.1			
0.2			
0.3			
0.4			
0.5			

26

x	$\dfrac{1}{1-x^3}$	$p_1(x)$	$p_2(x)$
0.1			
0.2			
0.3			
0.4			
0.5			

27 Justify the approximation $\ln x \approx \dfrac{2(x-1)}{x+1}$ near $x = 1$. Test it numerically for $x = 0.5$, 0.8, 1.2, 1.5, and 2.0, and compare it to $p_2(x)$.

28 Find a, b, c so $e^{-x} \approx \dfrac{ax + b}{cx + 1}$ near $x = 0$ in the sense that both sides have the same $p_2(x)$. (This is called Padé approximation.)

3. TAYLOR APPROXIMATIONS

Let us extend the ideas of Section 2 to the approximation of functions by polynomials of arbitrary degree. We shall need an algebraic fact: every polynomial can be expressed not only in powers of x, but also in powers of $x - a$, where a is any number. This form of the polynomial is convenient for computations near $x = a$.

Examples (1) $p(x) = x^2 + x + 2$, $a = 1$.
Set $u = x - 1$ so $x = u + 1$:

$$p(x) = x^2 + x + 2 = (u + 1)^2 + (u + 1) + 2$$
$$= (u^2 + 2u + 1) + (u + 1) + 2 = u^2 + 3u + 4$$
$$= (x - 1)^2 + 3(x - 1) + 4.$$

(2) $p(x) = x^4$, $a = -1$.
$$p(x) = x^4 = [(x + 1) - 1]^4$$
$$= (x + 1)^4 - 4(x + 1)^3 + 6(x + 1)^2 - 4(x + 1) + 1.$$

In general, to express

$$p(x) = B_0 + B_1 x + B_2 x^2 + B_3 x^3 + \cdots + B_n x^n$$

in powers of $x - a$, we write $u = x - a$, then substitute $u + a$ for x:

$$p(x) = B_0 + B_1(u + a) + B_2(u + a)^2 + \cdots + B_n(u + a)^n.$$

Now we expand each power by the Binomial Formula and collect like powers of u. The result is a polynomial in $u = x - a$, as desired.

This method is laborious when the degree of $p(x)$ exceeds three or four. If

$$p(x) = A_0 + A_1(x - a) + A_2(x - a)^2 + \cdots + A_n(x - a)^n$$

is the desired expansion of $p(x)$ in terms of $x - a$, we would like a way of computing the coefficients A_0, A_1, etc. directly, without a lot of algebra.

Obviously $p(a) = A_0$, so finding A_0 is no problem at all. But how shall we get A_1? The trick is to differentiate $p(x)$, then set $x = a$:

$$p'(x) = A_1 + 2A_2(x - a) + \cdots + nA_n(x - a)^{n-1}, \qquad p'(a) = A_1.$$

This trick can be repeated:

$$p''(x) = 2A_2 + 3 \cdot 2A_3(x - a) + \cdots + n(n - 1)A_n(x - a)^{n-2}, \qquad p''(a) = 2A_2,$$
$$p'''(x) = 3 \cdot 2A_3 + \cdots + n(n - 1)(n - 2)A_n(x - a)^{n-3}, \qquad p'''(a) = 3 \cdot 2A_3 = 3! A_3$$

Continuing in this way, we find

$$p^{(4)}(a) = 4! A_4, \quad p^{(5)}(a) = 5! A_5, \quad \cdots \cdots, \quad p^{(n)}(a) = n! A_n.$$

(Here $p^{(k)}$ is the k-th derivative.)

If $p(x)$ is a polynomial of degree n and if a is any number, then

$$p(x) = p(a) + p'(a)(x - a) + \frac{1}{2!} p''(a)(x - a)^2$$

$$+ \frac{1}{3!} p'''(a)(x - a)^3 + \cdots + \frac{1}{n!} p^{(n)}(a)(x - a)^n.$$

■ **EXAMPLE 1** Express $p(x) = x^3 - x^2 + 1$ in powers of $x - \frac{1}{2}$.

Solution

$$p(x) = x^3 - x^2 + 1, \qquad p'(x) = 3x^2 - 2x, \qquad p''(x) = 6x - 2, \qquad p'''(x) = 6,$$

$$p(\tfrac{1}{2}) = \tfrac{7}{8}, \qquad\qquad p'(\tfrac{1}{2}) = -\tfrac{1}{4}, \qquad\qquad p''(\tfrac{1}{2}) = 1, \qquad\qquad p'''(\tfrac{1}{2}) = 6.$$

Therefore

$$p(x) = \frac{7}{8} - \frac{1}{4}\left(x - \frac{1}{2}\right) + \frac{1}{2!}\left(x - \frac{1}{2}\right)^2 + \frac{6}{3!}\left(x - \frac{1}{2}\right)^3$$

$$= \frac{7}{8} - \frac{1}{4}\left(x - \frac{1}{2}\right) + \frac{1}{2}\left(x - \frac{1}{2}\right)^2 + \left(x - \frac{1}{2}\right)^3. \qquad ■$$

The next example illustrates the computational advantages gained by expanding polynomials in powers of $x - a$.

■ **EXAMPLE 2** Let $p(x) = x^3 - x^2 + 1$. Compute $p(0.50028)$ to 5 places.

Solution Use Example 1:

$$p(0.50028) = p(\tfrac{1}{2} + 0.00028) = \tfrac{7}{8} - \tfrac{1}{4}(0.00028) + \tfrac{1}{2}(0.00028)^2 + (0.00028)^3.$$

The last two terms on the right are smaller than 10^{-7}. Therefore to 5 places,

$$p(0.50028) \approx \tfrac{7}{8} - \tfrac{1}{4}(0.00028) = 0.87500 - 0.00007 = 0.87493. \qquad ■$$

Summation Notation Because we shall write polynomials frequently, it is convenient to use summation notation:

$$\sum_{i=0}^{n} A_i x^i = A_0 + A_1 x + A_2 x^2 + \cdots + A_n x^n.$$

The formula for an n-th degree polynomial in powers of $x - a$ can be abbreviated:

$$p(x) = \sum_{i=0}^{n} \frac{p^{(i)}(a)}{i!} (x - a)^i.$$

Here $p^{(i)}(a)$ denotes the i-th derivative of $p(x)$ evaluated at $x = a$, with the special convention $p^{(0)}(a) = p(a)$. (Also recall the convention $0! = 1$.)

Polynomial Approximations   Let us return to the problem: given a function $f(x)$ and a number a, find a polynomial $p(x)$ of degree n that approximates $f(x)$ for values of x near a. In view of the way we found $p_2(x)$, it now seems reasonable to construct a polynomial $p_n(x)$ so that

$$p_n(a) = f(a), \quad p'_n(a) = f'(a), \quad p''_n(a) = f''(a), \quad \cdots \quad , \quad p_n^{(n)}(a) = f^{(n)}(a).$$

Thus $p_n(x)$ mimics $f(x)$ and its first n derivatives at $x = a$.

Let us find $p_n(x)$ explicitly. We write

$$p_n(x) = A_0 + A_1(x - a) + A_2(x - a)^2 + \cdots + A_n(x - a)^n$$

and choose the coefficients A_k appropriately. We know that $A_k = p_n^{(k)}(a)/k!$. Since we want $p_n^{(k)}(a) = f^{(k)}(a)$, we must choose $A_k = f^{(k)}(a)/k!$.

The n-th **degree Taylor polynomial** of $f(x)$ at $x = a$ is

$$p_n(x) = f(a) + f'(a)(x - a) + \frac{1}{2!} f''(a)(x - a)^2 + \cdots + \frac{1}{n!} f^{(n)}(a)(x - a)^n$$

$$= f(a) + \sum_{k=1}^{n} \frac{f^{(k)}(a)}{k!} (x - a)^k.$$

Remark   When $f(x)$ is itself a polynomial of degree n, then $p_n(x) = f(x)$ is precisely the expression for $f(x)$ in powers of $x - a$. Furthermore, in this case,

$$p_n(x) = p_{n+1}(x) = p_{n+2}(x) = \cdots.$$

(Why?) Thus for an n-th degree polynomial $f(x)$, the n-th degree and all higher Taylor polynomials *equal* $f(x)$.

Here are the first three Taylor polynomials explicitly:

$$p_1(x) = f(a) + f'(a)(x - a),$$
$$p_2(x) = f(a) + f'(a)(x - a) + \tfrac{1}{2}f''(a)(x - a)^2,$$
$$p_3(x) = f(a) + f'(a)(x - a) + \tfrac{1}{2}f''(a)(x - a)^2 + \tfrac{1}{6}f'''(a)(x - a)^3.$$

The first two are old friends from the previous section. In general, each Taylor polynomial is derived from the preceding one by addition of a single term:

$$p_{n+1}(x) = p_n(x) + \frac{1}{(n+1)!} f^{(n+1)}(a)(x - a)^{n+1}.$$

We anticipate that $p_n(x)$ is a good approximation to $f(x)$; the error is $f(x) - p_n(x)$. We try to reduce this error by adding an additional term $f^{(n+1)}(a)(x - a)^{n+1}/(n + 1)!$ to $p_n(x)$, thereby obtaining $p_{n+1}(x)$, an even better approximation (we hope).

■ **EXAMPLE 3**   Find the n-th degree Taylor polynomial of

(a) $f(x) = e^x$,   at $x = 0$ (b) $f(x) = \ln x$,   at $x = 1$.

Solution (a) Compute successive derivatives of $f(x) = e^x$ and evaluate them at $x = 0$:

$$f(x) = e^x, \quad f'(x) = e^x, \quad f''(x) = e^x, \cdots;$$
$$f(0) = 1, \quad f'(0) = 1, \quad f''(0) = 1, \cdots.$$

Hence, according to the recipe,

$$p_n(x) = f(0) + f'(0)x + \frac{1}{2!}f''(0)x^2 + \cdots + \frac{1}{n!}f^{(n)}(0)x^n$$

$$= 1 + x + \frac{x^2}{2!} + \frac{x^3}{3!} + \cdots + \frac{x^n}{n!} = \sum_{k=0}^{n} \frac{x^k}{k!}.$$

(b) Compute successive derivatives of $f(x) = \ln x$ and evaluate them at $x = 1$:

$$f' = \frac{1}{x}, \quad f'' = -\frac{1}{x^2}, \quad f''' = \frac{2!}{x^3}, \quad f^{(4)} = -\frac{3!}{x^4}, \quad \cdots, \quad f^k = (-1)^{k-1}\frac{(k-1)!}{x^k},$$

hence $\quad f(1) = 0, \quad f'(1) = 1, \quad f''(1) = -1, \quad f'''(1) = 2!,$

$$f^{(4)}(1) = -3!, \cdots, f^{(k)}(1) = (-1)^{k-1}(k-1)!.$$

Therefore $\quad \dfrac{f^{(k)}(1)}{k!} = \dfrac{(-1)^{k-1}(k-1)!}{k!} = \dfrac{(-1)^{k-1}}{k}$

and $\quad p_n(x) = \displaystyle\sum_{k=1}^{n}(-1)^{k-1}\frac{(x-1)^k}{k}$

$$= (x-1) - \frac{(x-1)^2}{2} + \frac{(x-1)^3}{3} - + \cdots + (-1)^{n-1}\frac{(x-1)^n}{n}. \qquad \blacksquare$$

Remark Table 1 gives some evidence on the accuracy of these approximations for e^x.

Table 1

x	e^x	$p_2(x)$	$p_3(x)$	$p_4(x)$
0.1	1.10517	1.10500	1.10517	1.10517
0.2	1.22140	1.22000	1.22133	1.22140
0.3	1.34986	1.34500	1.34950	1.34984
0.4	1.49182	1.48000	1.49067	1.49173
0.5	1.64872	1.62500	1.64583	1.64844
0.75	2.11700	2.03125	2.10156	2.11475
1.0	2.71828	2.50000	2.66667	2.70833

■ **EXAMPLE 4** Compute the Taylor polynomials at $x = 0$ for

(a) $f(x) = \sin x$,   (b) $f(x) = \cos x$.

Solution (a)   Compute derivatives: $f(x) = \sin x$,

$$f'(x) = \cos x, \quad f''(x) = -\sin x, \quad f'''(x) = -\cos x, \quad f^{(4)}(x) = \sin x, \cdots,$$

repeating in cycles of four. At $x = 0$, the values are

$$0, \quad 1, \quad 0, \quad -1, \quad 0, \quad 1, \quad 0, \quad -1, \quad 0, \cdots.$$

Hence the n-th degree Taylor polynomial of $\sin x$ is

$$p_n(x) = x - \frac{x^3}{3!} + \frac{x^5}{5!} - \frac{x^7}{7!} + \frac{x^9}{9!} - \cdots,$$

where the last term is $\pm x^n/n!$ if n is odd and $\pm x^{n-1}/(n-1)!$ if n is even. For example,

$$p_3(x) = p_4(x) = x - \frac{x^3}{3!}, \qquad p_5(x) = p_6(x) = x - \frac{x^3}{3!} + \frac{x^5}{5!},$$

$$p_7(x) = p_8(x) = x - \frac{x^3}{3!} + \frac{x^5}{5!} - \frac{x^7}{7!}.$$

In general, $p_{2m-1}(x) = p_{2m}(x)$, and the sign of the last term is plus if m is odd, minus if m is even:

$$p_{2m-1}(x) = p_{2m}(x) = x - \frac{x^3}{3!} + \frac{x^5}{5!} - \cdots + (-1)^{m-1} \frac{x^{2m-1}}{(2m-1)!}$$

$$= \sum_{i=1}^{m} (-1)^{i-1} \frac{x^{2i-1}}{(2i-1)!}.$$

(b)   Since $\cos x$ is the derivative of $\sin x$, we can read off its derivatives at $x = 0$ from those of $\sin x$ found in (a). They are

$$1, \quad 0, \quad -1, \quad 0, \quad 1, \quad 0, \quad -1, \quad 0, \quad \cdots,$$

repeating in cycles of 4. Using these values, we obtain

$$p_{2m}(x) = p_{2m+1}(x) = \sum_{i=0}^{m} (-1)^i \frac{x^{2i}}{(2i)!} = 1 - \frac{x^2}{2!} + \frac{x^4}{4!} - + \cdots + (-1)^m \frac{x^{2m}}{(2m)!}. \qquad \blacksquare$$

Remark   Table 2 gives further evidence on the accuracy of Taylor polynomials for approximation.

Table 2

x	$\sin x$	$p_1(x) = p_2(x)$	$p_3(x) = p_4(x)$	$p_5(x) = p_6(x)$
0.1	0.09983	0.10000	0.09983	0.09983
0.2	0.19867	0.20000	0.19867	0.19867
0.3	0.29552	0.30000	0.29550	0.29552
0.4	0.38942	0.40000	0.38933	0.38942
0.5	0.47943	0.50000	0.47917	0.47943
1.0	0.84147	1.00000	0.83333	0.84167

Express in powers of $x - a$

1   $x^2 + 5x + 2$ $a = 1$
3   $2x^3 + 5x^2 + 13x + 10$ -1
5   $2x^4 + 5x^3 + 4x + 16$ -2
7   $5x^5 + 4x^4 - 3x^3 - 2x^2 + x + 1$ -1

2   $x^3 - 3x^2 + 4x$ $a = 2$
4   $3x^3 - 2x^2 - 2x + 1$ 1
6   $x^4 - 5x^2 + x + 2$ 2
8   $x^5 + 2x^4 + 3x^2 + 4x + 5$ -2.

Compute $p(a)$ to 5 places

9   $x^3 - 3x^2 + 2x + 1$ $a = 1.004$
11   $4x^4 - 3x^2 + 10x + 12$ -0.995

10   $x^4 + x^3 + x^2 + x + 1$ 1.999
12   $x^5 + 2x^2 - 6x - 5$ -3.0001.

Compute the 5-th degree Taylor polynomial of

13   xe^{-x} at $x = 0$
15   $(\sin x)/(1 + x^3)$ at $x = 0$
17   $\sin x$ at $x = \frac{1}{4}\pi$
19   $1/x^3$ at $x = -1$

14   $x^2 \cos x$ at $x = 0$
16   $x^3/(1 + x^2)$ at $x = 0$
18   $\cos x$ at $x = \frac{1}{6}\pi$
20   e^{x^2} at $x = 0$.

4. TAYLOR'S FORMULA

We introduced the Taylor polynomial $p_n(x)$ in the last section with the hope that $p_n(x)$ would be a good approximation to a given function $f(x)$. Our optimism will be justified if the error

$$r_n(x) = f(x) - p_n(x)$$

is very small compared to $x - a$. The next two statements assure us that it is. The first gives an explicit formula for the error, the second an estimate of its size.

In this subject the error is usually called the **remainder**. That is why we wrote $r_n(x)$ above rather than our usual $e_n(x)$.

Taylor's Formula with Remainder   Suppose $f(x)$ has continuous derivatives up to and including $f^{(n+1)}(x)$ near $x = a$. Write

$$f(x) = p_n(x) + r_n(x),$$

where $p_n(x)$ is the n-th degree Taylor polynomial at $x = a$ and $r_n(x)$ is the remainder (or error). Then

$$r_n(x) = \frac{1}{n!} \int_a^x (x - t)^n f^{(n+1)}(t)\, dt.$$

Usually the integral expressing $r_n(x)$ cannot be computed exactly. Nevertheless, the integral can be estimated. One important estimate depends on a bound for the $(n + 1)$-th derivative, $f^{(n+1)}(x)$.

> **Estimate of Remainder**   Let
> $$f(x) = p_n(x) + r_n(x),$$
> where $p_n(x)$ is the n-th Taylor polynomial of $f(x)$ at $x = a$. Suppose that
> $$|f^{(n+1)}(x)| \le M$$
> in some interval including a, say $b \le a \le c$. Then
> $$|r_n(x)| \le \frac{M}{(n+1)!} |x - a|^{n+1} \qquad \text{for} \quad b \le x \le c.$$

Let us assume Taylor's Formula temporarily. (It will be derived at the end of this section.) The remainder estimate then follows easily:

$$|r_n(x)| = \frac{1}{n!} \left| \int_a^x (x-t)^n f^{(n+1)}(t) \, dt \right| \le \frac{1}{n!} \left| \int_a^x |(x-t)^n f^{(n+1)}(t)| \, dt \right|$$

$$\le \frac{M}{n!} \left| \int_a^x |(x-t)^n| \, dt \right| = \frac{M}{(n+1)!} |x-a|^{n+1}.$$

Note   We have used inequality (5), p. 247. The absolute value signs outside of the integral take care of the possibility that $x < a$.

■ **EXAMPLE 1**   Suppose that the function e^x is approximated by its n-th degree Taylor polynomial at $x = 0$. Estimate the remainder assuming $|x| \le B$.

Solution   The $(n+1)$-th derivative is $f^{(n+1)}(x) = e^x$. If $x \ge 0$, the largest value of $f^{(n+1)}(x)$ between 0 and B is $e^B < 3^B$. By the remainder estimate with $M = 3^B$,

$$|r_n(x)| < \frac{3^B}{(n+1)!} x^{n+1} \qquad \text{for} \quad 0 \le x \le B.$$

If $x \le 0$, the largest value of $f^{(n+1)}(t)$ between 0 and $-B$ is $e^0 = 1$. By the remainder estimate with $M = 1$,

$$|r_n(x)| \le \frac{|x|^{n+1}}{(n+1)!} \qquad \text{for} \quad x \le 0. \qquad \blacksquare$$

How could we compute a 5-place table of e^x for a certain range of x? One method is to approximate e^x by a Taylor polynomial $p_n(x)$ with n large enough that $|r_n(x)| < 5 \times 10^{-6}$ on the given range. (For the sake of economy we should use the smallest n that does the trick.) This is a practical method because computing many values of a polynomial is relatively easy, especially with a computer. (We remark that there are other ways of approaching this problem, for instance by interpolation. See Sections 7 and 8.)

■ **EXAMPLE 2**   Use the conclusion of Example 1 to estimate the smallest n such that the remainder in Taylor's Formula for e^x at $x = 0$ satisfies $|r_n(x)| < 5 \times 10^{-6}$ for

(a) $|x| < 0.5$ (b) $|x| < 1.$

Solution By Example 1, $|r_n(x)| \le \dfrac{3^B}{(n+1)!} B^{n+1}$ for $|x| \le B$.

The problem is to choose n as small as possible so that

$$\frac{3^B}{(n+1)!} B^{n+1} < 5 \times 10^{-6}.$$

(a) Here $B = \frac{1}{2}$, so we want

$$\frac{3^{1/2}}{(n+1)!} \left(\frac{1}{2}\right)^{n+1} < 5 \times 10^{-6}, \qquad \text{that is,} \quad 2^{n+1}(n+1)! > \frac{\sqrt{3}}{5 \times 10^{-6}}.$$

Now

$$\frac{\sqrt{3}}{5} \times 10^6 < \frac{2}{5} \times 10^6 = 4 \times 10^5,$$

so it suffices to choose n such that

$$2^{n+1}(n+1)! \ge 4 \times 10^5.$$

By trial and error,

$$2^7 \times 7! > 6 \times 10^5 \qquad \text{but} \qquad 2^6 \times 6! < 5 \times 10^4.$$

The correct choice is $n + 1 = 7$, that is, $n = 6$.

(b) This time $B = 1$. By similar reasoning, n must satisfy

$$\frac{3}{(n+1)!} < 5 \times 10^{-6},$$

so it suffices to choose n such that

$$(n+1)! > \frac{3}{5} \times 10^6 = 6 \times 10^5.$$

Since $9! < 4 \times 10^5$ and $10! > 3 \times 10^6$, the correct choice is $n + 1 = 10$, that is, $n = 9$. ∎

Remark By calculation, $|r_5(0.5)| \approx 2.3 \times 10^{-5}$, so $n = 5$ doesn't work; 6 is the smallest possible n for part (a). For part (b), $n = 9$ is not the smallest possible. It can be shown with some difficulty that $|r_8(x)| < 4 \times 10^{-6}$ for $|x| < 1$. Since $|r_7(1)| \approx 2.8 \times 10^{-5}$, we see that $n = 8$ is the smallest possible. Can you see where we lost ground in our estimates?

■ **EXAMPLE 3** Estimate the remainders in Taylor's Formula for (a) $\sin x$ (b) $\cos x$. (Refer to Example 4, p. 457.)

Solution (a) A bound for the $(n+1)$-th derivative is easy: $|f^{(n+1)}(x)| \le 1$ because $f^{(n+1)}(x) = \pm \sin x$ or $\pm \cos x$. Hence

$$|r_n(x)| \le \frac{|x|^{n+1}}{(n+1)!}.$$

By Example 4a, $p_{2m-1}(x) = p_{2m}(x)$; it follows that

$$|r_{2m-1}(x)| = |r_{2m}(x)| \le \frac{|x|^{2m+1}}{(2m+1)!}.$$

(b)   The estimate for $|r_n(x)|$ is exactly the same as in (a). This time, by Example 4b, $p_{2m}(x) = p_{2m+1}(x)$ and it follows that

$$|r_{2m}(x)| = |r_{2m+1}(x)| = \frac{|x|^{2m+2}}{(2m+2)!}. \qquad \blacksquare$$

■   **EXAMPLE 4** Find the lowest degree Taylor polynomial for $\sin x$ at $x = 0$ such that the estimate of Example 3a implies

$$|r_n(x)| < 5 \times 10^{-6} \qquad \text{for} \quad |x| \leq \tfrac{1}{4}\pi.$$

Solution By Example 3a, $|r_{2m-1}(x)| \leq \dfrac{|x|^{2m+1}}{(2m+1)!},$

hence we want to choose m so that

$$\frac{(\tfrac{1}{4}\pi)^{2m+1}}{(2m+1)!} < 5 \times 10^{-6}.$$

By trial and error we find

$$\frac{(\tfrac{1}{4}\pi)^7}{7!} > \frac{(\tfrac{3}{4})^7}{7!} > 2 \times 10^{-5} \qquad \text{and} \qquad \frac{(\tfrac{1}{4}\pi)^9}{9!} < \frac{1}{9!} < 3 \times 10^{-6}.$$

Therefore the correct choice is $2m + 1 = 9$, that is, $m = 4$, and the corresponding Taylor polynomial is

$$p_{2m-1}(x) = p_7(x) = x - \frac{x^3}{3!} + \frac{x^5}{5!} - \frac{x^7}{7!},$$

by Example 4a, p. 457. $\qquad \blacksquare$

Remark   Since $|r_5(\tfrac{1}{4}\pi)| \approx 3.6 \times 10^{-5}$, the smallest possible n is 7. Compare the remark after Example 2.

■   **EXAMPLE 5** If $f(x) = \ln x$ is approximated by its n-th degree Taylor polynomial at $x = 1$, estimate the remainder for $|x - 1| < 0.5$. (Refer to Example 3b, p. 456.)

Solution As was shown in that example, $f^{(n+1)}(x) = \pm n!/x^{n+1}$. If $|x - 1| < 0.5$, then $x > 0.5$ and hence

$$|f^{(n+1)}(x)| < \frac{n!}{(0.5)^{n+1}} = 2^{n+1}n!.$$

Therefore $|r_n(x)| \leq \dfrac{2^{n+1}n!}{(n+1)!}|x-1|^{n+1} = \dfrac{|2(x-1)|^{n+1}}{n+1} < \dfrac{1}{n+1},$

since $|2(x-1)| < 1 \qquad \text{for} \quad |x - 1| < 0.5. \qquad \blacksquare$

Dividing out Zeros   If $f(x)$ is a polynomial* and $x = a$ is a zero of multiplicity m, then $(x - a)^m$ can be divided out of $f(x)$; the quotient is another polynomial $g(x)$:

$$f(x) = (x - a)^m g(x).$$

---

* For a review of some basic properties of polynomials, see pp. 49–50.

It is remarkable that an analogous result holds for arbitrary functions that have enough continuous derivatives.

Let us first look at the easiest case, that of a simple zero. Suppose $f(a) = 0$ and $f(x)$ is differentiable at $x = a$. Then

$$\lim_{x \to a} \frac{f(x)}{x - a} = \lim_{x \to a} \frac{f(x) - f(a)}{x - a} = f'(a).$$

We define a function $g(x)$ by

$$\begin{cases} g(x) = \dfrac{f(x)}{x - a} & \text{for} \quad x \neq a \\[2mm] g(a) = f'(a). \end{cases}$$

Then

$$f(x) = (x - a)g(x)$$

(even at $x = a$). Furthermore, $g(x)$ is continuous at $x = a$ because

$$\lim_{x \to a} g(x) = \lim_{x \to a} \frac{f(x)}{x - a} = f'(a) = g(a).$$

(Also, $g(x)$ is continuous at each x where $f(x)$ is continuous.)

Example

$$\sin x = xg(x), \qquad \text{where} \quad \begin{cases} g(x) = \dfrac{\sin x}{x} & \text{for} \quad x \neq 0 \\[2mm] g(0) = 1, \end{cases}$$

and $g(x)$ is continuous, even at $x = 0$.

Assuming that $f'(x)$ is continuous, we can find a neat formula for $g(x)$ by starting with

$$f(x) = \int_a^x f'(t)\, dt, \qquad f(a) = 0.$$

We change variables, setting $t = a + (x - a)u$, so u goes from 0 to 1 as t goes from a to x. Then $dt = (x - a)\, du$ and

$$f(x) = \int_a^x f'(t)\, dt = \int_0^1 f'[a + (x - a)u](x - a)\, du = (x - a)\int_0^1 f'[a + (x - a)u]\, du.$$

Thus

$$f(x) = (x - a)g(x), \qquad g(x) = \int_0^1 f'[a + (x - a)u]\, du.$$

The continuity of $f'(t)$ implies the continuity of $g(x)$ (we skip the details); hence

$$g(0) = \int_0^1 f'(a + 0)\, du = \int_0^1 f'(a)\, du = f'(a).$$

Now we state the general case; the proof is left for the exercises.

Dividing out Zeros Suppose $f(x)$ has $n + 1$ continuous derivatives near $x = a$ and

$$f(a) = f'(a) = f''(a) = \cdots = f^{(n)}(a) = 0.$$

Then $f(x) = (x - a)^{n+1}g(x)$, where $g(x)$ is a continuous function near $x = a$ and

$$g(a) = \lim_{x \to a} \frac{f(x)}{(x - a)^{n+1}} = \frac{1}{(n + 1)!} f^{(n+1)}(a).$$

What is more,

$$g(x) = \frac{1}{n!} \int_0^1 (1 - u)^n f^{(n+1)}[a + (x - a)u] \, du.$$

Example $f(x) = 1 - \cos x$ at $x = 0$. Then $f(0) = f'(0) = 0$ and $f''(x) = \cos x$, so

$$1 - \cos x = x^2 g(x), \qquad g(x) = \int_0^1 (1 - u) \cos(xu) \, du,$$

$$\lim_{x \to 0} \frac{1 - \cos x}{x^2} = g(0) = \int_0^1 (1 - u) \, du = \frac{1}{2}.$$

Derivation of Taylor's Formula Recall that $p_n(x)$ is constructed so that

$$p_n(a) = f(a), \quad p_n'(a) = f'(a), \quad p_n''(a) = f''(a), \quad \cdots \quad, \quad p_n^{(n)}(a) = f^{(n)}(a).$$

Consequently the remainder, $r_n(x) = f(x) - p_n(x)$, satisfies

$$r_n(a) = r_n'(a) = r_n''(a) = \cdots = r_n^{(n)}(a) = 0.$$

The following lemma asserts that such a function can be expressed as an integral.

Lemma Let $g(x)$ have continuous derivatives near $x = a$ up to and including $g^{(n+1)}(x)$, and suppose

$$g(a) = g'(a) = g''(a) = \cdots = g^{(n)}(a) = 0.$$

Then

$$g(x) = \frac{1}{n!} \int_a^x (x - t)^n g^{(n+1)}(t) \, dt.$$

The lemma is proved by repeated integration by parts. Fix a and x, and set

$$u(t) = \frac{(x - t)^n}{n!} \qquad \text{and} \qquad v(t) = g^{(n)}(t).$$

Then

$$du = \frac{-(x - t)^{n-1}}{(n - 1)!} \, dt \qquad \text{and} \qquad dv = g^{(n+1)}(t) \, dt.$$

The integral in the lemma is

$$\frac{1}{n!}\int_a^x (x - t)^n g^{(n+1)}(t)\, dt = \int_a^x u(t)\, dv(t)$$

$$= u(t)v(t)\Big|_a^x - \int_a^x v(t)\, du(t) = 0 - 0 + \frac{1}{(n-1)!}\int_a^x (x - t)^{n-1} g^{(n)}(t)\, dt,$$

because $g^{(n)}(a) = 0$. Thus n has decreased by one. Repeat the process until the exponent of $x - t$ reaches 0:

$$\frac{1}{n!}\int_a^x (x - t)^n g^{(n+1)}(t)\, dt = \frac{1}{(n-1)!}\int_a^x (x - t)^{n-1} g^{(n)}(t)\, dt$$

$$= \frac{1}{(n-2)!}\int_a^x (x - t)^{n-2} g^{(n-1)}(t)\, dt = \cdots = \frac{1}{0!}\int_a^x g'(t)\, dt = g(x) - g(a) = g(x).$$

This completes the proof of the lemma.

To derive Taylor's Formula from the lemma, take $g(x) = r_n(x) = f(x) - p_n(x)$. Then $g(x)$ satisfies the hypotheses of the lemma and $g^{(n+1)}(x) = f^{(n+1)}(x)$ since $p_n(x)$ is a polynomial of degree n or less. Hence

$$r_n(x) = \frac{1}{n!}\int_a^x (x - t)^n f^{(n+1)}(t)\, dt.$$

EXERCISES

Find $p_n(x)$ and an upper bound for $|r_n(x)|$

1	$\sin 2x$	$a = 0$	**2**	$\sin 2x$	$a = \frac{1}{2}\pi$	**3**	xe^x	$a = 0$
4	xe^x	$a = 1$	**5**	$x^2 \ln x$	$a = 1$	**6**	$x^2 \ln x$	$a = e$
7	$x^2 e^{-x}$	$a = 0$	**8**	$x^2 e^{-x}$	$a = 1$	**9**	$x \sin x$	$a = 0$
10	$x \sin x$	$a = \frac{1}{2}\pi$	**11**	$\sin x + \cos x$	$a = 0$	**12**	$\cosh x$	$a = 0$
13	$\sinh x$	$a = 0$	**14**	$1 + e^x + e^{2x}$	$a = 0$	**15**	$1/(1 + x)$	$a = 1$
16*	$x/(1 + x^4)$	$a = 0$.						

17   Estimate the error in approximating $f(x) = \ln(1 + x)$ for $-\frac{1}{3} \le x \le \frac{1}{3}$ by its 10-th degree Taylor polynomial about $x = 0$.

18   Approximate $f(x) = 1/(1 - x)^2$ for $-\frac{1}{4} \le x \le \frac{1}{4}$ to 3 decimal places by a Taylor polynomial about $x = 0$.

19   Approximate $\sin^2 x$ by its 4-th degree Taylor polynomial about $x = 0$. Estimate the error if $|x| \le 0.1$. [*Hint* $\sin^2 x = \frac{1}{2}(1 - \cos 2x)$.]

20   Show that for $100 \le x \le 101$, the approximation $\sqrt{x} \approx 10 + \frac{1}{20}(x - 100)$ is correct to within 0.0002.

Let $r_n(x)$ be the remainder for $\sin x$ at $a = 0$. Show that

21 $|r_3(x)| < 5 \times 10^{-6}$ for $|x| < 0.22 \approx 12.6°$

22 $|r_5(x)| < 5 \times 10^{-6}$ for $|x| < 0.59 \approx 33.8°$

23 $|r_7(x)| < 5 \times 10^{-6}$ for $|x| < 1.06 \approx 60.7°$

24 $|r_9(x)| < 5 \times 10^{-6}$ for $|x| < 1.61 \approx 92.2°$.

25   Find k so that $r_2(x)$ for $\cos x$ at $a = 0$ satisfies $|r_2(x)| < 5 \times 10^{-6}$ for $|x| < k\pi$.

26   (cont.) How can you estimate quite simply $\sin(\frac{19}{40}\pi)$ to 5 places?

27 How many terms of the Taylor Polynomial for ln x at $a = 1$ are needed to compute ln 1.25 to 5 places?

28 (cont.) The same for ln 0.75.

29 Show that $p_3(x)$ yields 5-place accuracy in the approximation of sin x at $a = \frac{1}{4}\pi$ when $|x - \frac{1}{4}\pi| < 0.1 \approx 5.7°$.

30 (cont.) Show that $p_3(x)$ yields 8-place accuracy for $44° < x < 46°$. Write out $p_3(x)$.

31 Find $p_3(x)$ for $f(x) = \sqrt{1 - x}$ at $a = 0$, and estimate the error.

32 In Ex. 12, p. 200, we found the diameter D of a hole in which a needle gauge of length L has "rock" x is $D = L^2/\sqrt{L^2 - x^2}$. Justify the approximation

$$D \approx L + \frac{x^2}{2L} + \frac{3x^4}{8L^3}.$$

Find

33 $\displaystyle\lim_{x \to 0} \frac{e^x - 1 - x - \frac{1}{2}x^2}{x^3}$

34 $\displaystyle\lim_{x \to 0} \frac{\tan x - x - \frac{1}{3}x^3}{x^5}.$

35 Suppose $f(0) = g(0) = 0$, $f'(x)$ and $g'(x)$ are continuous near 0, and $g'(0) \neq 0$. Prove

$$\lim_{x \to 0} \frac{f(x)}{g(x)} = \frac{f'(0)}{g'(0)}.$$

36* (cont.) More generally, suppose $f(0) = f'(0) = \cdots = f^{(n)}(0) = 0$, $g(0) = g'(0) = \cdots = g^{(n)}(0) = 0$, $f^{(n+1)}(x)$ and $g^{(n+1)}(x)$ are continuous near 0, and $g^{(n+1)}(0) \neq 0$. Prove $g(x) \neq 0$ for $0 < |x| < \delta$ with δ sufficiently small and

$$\lim_{x \to 0} \frac{f(x)}{g(x)} = \frac{f^{(n+1)}(0)}{g^{(n+1)}(0)}.$$

37* Suppose $f^{(n+1)}(x)$ is continuous near $x = a$ and $f'(a) = f''(a) = \cdots = f^{(n)}(a) = 0$. Suppose n is odd. Prove if $f^{(n+1)}(a) > 0$, then $f(a)$ is a local min; if $f^{(n+1)}(a) < 0$, then $f(a)$ is a local max. (This generalizes the second derivative test.)

38* (cont.) Suppose n is even and $f^{(n+1)}(a) \neq 0$. What conclusion?

39* Use Taylor's Formula to derive the formula $f(x) = (x - a)^{n+1}g(x)$, where $g(x)$ is given by the last expression in the first box on p. 464.

40* (cont.) Prove $g(x)$ is continuous near $x = a$.

5. ROLLE'S THEOREM

In Chapter 3, on pp. 148–149, we introduced Rolle's Theorem and the Mean Value Theorem in order to prove some basic properties of differentiable functions. In this and the next section we shall take a second look at these theorems and their applications.

Let us recall the statement of Rolle's Theorem, in its greatest generality.

Let $f(x)$ be a continuous function on the closed interval $[a, b]$, and suppose $f'(x)$ exists for $a < x < b$. Assume $f(a) = f(b)$. Then $f'(c) = 0$ for some c between a and b.

If you read the proof in Chapter 3 carefully, you will see that this is exactly what was proved.

Sometimes we won't bother with the fussy "continuous on $[a, b]$ and differentiable on $a < x < b$," but merely say "differentiable on $a \leq x \leq b$," which assumes slightly more. You should keep in mind the example $f(x) = \sqrt{1 - x^2}$ on $[-1, 1]$ to remember the difference between differentiable at the end points and just continuous.

A number of interesting results are obtained by applying Rolle's Theorem to some cleverly chosen auxiliary function. The Mean Value Theorem itself is one case in point. Here is another.

■ **EXAMPLE 1**   Let $f(x)$ be differentiable on $[a, b]$, let $f(a) = f(b) = 0$, and let k be a constant. Prove that

$$f'(c) = kf(c)$$

for some c satisfying $a < c < b$.

Solution   Set $g(x) = e^{-kx}f(x)$. Then $g(x)$ is differentiable, $g(a) = 0$, and $g(b) = 0$, so $g(a) = g(b)$. By Rolle's Theorem, $g'(c) = 0$ for some c between a and b. But

$$g'(x) = e^{-kx}[-kf(x) + f'(x)], \qquad \text{hence} \quad f'(c) = kf(c). \qquad ■$$

Application to Polynomials   Long ago (p. 35) we claimed that the graph of a polynomial $y = f(x)$ of degree n can go up and down at most n times. From the calculus point of view this is obvious because $f'(x)$, a polynomial of degree $n - 1$, has at most $n - 1$ zeros, hence at most $n - 1$ sign changes. If these zeros are $x_1 < x_2 < \cdots < x_k$, where $k \leq n - 1$, then $f'(x)$ has constant sign on each of the intervals

$$-\infty < x < x_1, \quad x_1 < x < x_2, \quad x_2 < x < x_3, \cdots, x_{k-1} < x < x_k, \quad x_k < x < \infty.$$

Therefore $f(x)$ is strictly increasing or strictly decreasing on each of these at most n intervals.

For instance if $n = 5$, then in the most extreme case $f'(x)$ has 4 sign changes, so (assuming $f(x)$ has positive leading coefficient) $f(x)$ goes up–down–up–down–up, 5 ups and downs in all. See Fig. 1.

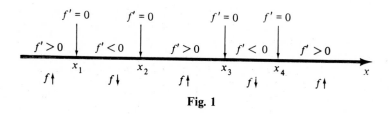

Fig. 1

Factored Polynomials   A polynomial $f(x)$ of degree n has at most n real zeros.* (We shall not discuss complex zeros here.) If $x = r_1$ is a zero, then $f(x)$ is divisible by $x - r_1$, that is, $f(x) = (x - r_1)g(x)$, where $g(x)$ is a polynomial of degree $n - 1$. If $x = r_2$ is another zero, then $g(x)$ is divisible by $x - r_2$. Hence $f(x) = (x - r_1)(x - r_2)h(x)$, where $h(x)$ is a polynomial of degree $n - 2$. Thus we can factor

---

* Compare the Factor Theorem, p. 50, and p. 462.

out successive zeros. In the important case that $f(x)$ has n zeros (the maximum possible), then

$$f(x) = a(x - r_1)(x - r_2)\cdots(x - r_n) \qquad (a \neq 0).$$

If the r_j are distinct we say that $f(x)$ has n **simple zeros**. Otherwise some zeros are **simple zeros**, and others are **multiple** (repeated) zeros. Collecting like factors, we write

$$f(x) = a(x - r_1)^{m_1}(x - r_2)^{m_2}\cdots(x - r_s)^{m_s}.$$

Such a polynomial of degree n is called a **polynomial in factored form**, or a polynomial with n real zeros. Here $r_1, \cdots, r_s$ are the distinct zeros of $f(x)$ and $m_1, \cdots, m_s$ their **multiplicities**.* The simple zeros are those r_j for which $m_j = 1$, and the multiple zeros are those r_j for which $m_j \geq 2$. Since $\deg f(x) = n$, we have

$$m_1 + m_2 + \cdots + m_s = n.$$

Let us study the derivative of a polynomial $f(x)$ in factored form. First we look at the nature of $f'(x)$ at a zero of $f(x)$. Suppose $x = r$ is a zero of multiplicity m of $f(x)$. Then

$$f(x) = (x - r)^m g(x), \qquad g(r) \neq 0.$$

Consequently

$$f'(x) = m(x - r)^{m-1}g(x) + (x - r)^m g'(x) = (x - r)^{m-1}h(x),$$

where
$$h(x) = mg(x) + (x - r)g'(x).$$

Clearly $h(r) = mg(r) \neq 0$. Hence if $m = 1$, then $f'(r) = h(r) \neq 0$; if $m \geq 2$, then $x = r$ is a zero of $f'(x)$ of multiplicity $m - 1$.

Let $x = r$ be a zero of multiplicity m of a polynomial $f(x)$. If $m = 1$, then $f'(r) \neq 0$. If $m \geq 2$, then $x = r$ is a zero of $f'(x)$ of multiplicity $m - 1$.

This result is important in itself. But our main purpose is using it and Rolle's Theorem to derive the basic result about derivatives of factored polynomials.

Factored Polynomials Let
$$f(x) = a(x - r_1)^{m_1}\cdots(x - r_s)^{m_s},$$

where $r_1 < r_2 < \cdots < r_s$, be a polynomial of degree n with n zeros. Then $f'(x)$ is a polynomial of degree $n - 1$ with $n - 1$ zeros. More precisely, $f'(x)$ has one simple zero between every two successive r_j and a zero of multiplicity $m_j - 1$ at r_j if $m_j \geq 2$.

In particular, if $f(x)$ has all simple zeros, then $f'(x)$ has all simple zeros, and the zeros of $f'(x)$ alternate with the zeros of $f(x)$.

---

* The multiplicity of a zero is also called its **order**.

The proof is a matter of counting. By Rolle's Theorem, $f'(x)$ has a zero between every two successive zeros of $f(x)$. Thus $f'(x)$ has at least one zero between r_1 and r_2, at least one zero between r_2 and r_3, ..., and at least one zero between r_{s-1} and r_s. These total to at least $s - 1$ zeros *between* the r's. Also $f'(x)$ has a zero of multiplicity $m_1 - 1$ at r_1, a zero of multiplicity $m_2 - 1$ at r_2, ..., and a zero of multiplicity $m_s - 1$ at r_s. These total to exactly

$$(m_1 - 1) + (m_2 - 1) + \cdots + (m_s - 1) = n - s$$

zeros *at* the r's. The grand total is *at least*

$$(s - 1) + (n - s) = n - 1$$

zeros of $f'(x)$. But deg $f'(x) = n - 1$, so $f'(x)$ has *at most* $n - 1$ zeros. Therefore it has *precisely* $n - 1$ zeros. Furthermore, there can be just one simple zero in each interval $r_1 < x < r_2, r_2 < x < r_3$, etc., or the grand total would exceed $n - 1$. This completes the proof.

■ **EXAMPLE 2** Let $\quad f(x) = (x^2 - 4)^5 x^3 (x + 1)^2 (x - 3)(x - 5)^4$.
Locate the zeros of $f'(x)$.

Solution   Rewrite $f(x)$ in the form

$$f(x) = (x + 2)^5 (x + 1)^2 x^3 (x - 2)^5 (x - 3)(x - 5)^4,$$

exhibiting the successive zeros and their multiplicities (Fig. 2).

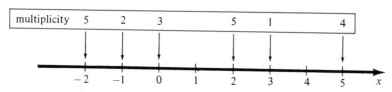

Fig. 2   Zeros of $f(x) = (x^2 - 4)^5 x^3 (x + 1)^2 (x - 3)(x - 5)^4$

By inspection, we get a picture of where the zeros of $f'(x)$ are located. However, we do not get the exact location of every zero. For example, there is one simple zero of $f'(x)$ between 2 and 3, but we cannot pinpoint it further by the methods of this section. The result is shown in Fig. 3. Note that $f(x)$ has degree 20 and 20 zeros, counting multiplicities, while $f'(x)$ has degree 19 and 19 zeros.

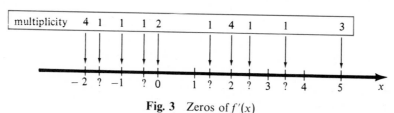

Fig. 3   Zeros of $f'(x)$ ■

Generalized Rolle's Theorem   The following result has applications in approximation theory. We shall see one in Section 8.

> **Generalized Rolle's Theorem**   Let $f(x)$ be an n-times differentiable function on an interval $[a, b]$ and suppose
> $$f(x_0) = 0, \quad f(x_1) = 0, \quad \cdots \quad, \quad f(x_n) = 0,$$
> where $x_0, x_1, \cdots, x_n$ are $n + 1$ distinct points of the interval. Then $f^{(n)}(c) = 0$ for some point c of the interval $a < c < b$.

The proof is by repeated application of Rolle's Theorem, first to $f(x)$, then to $f'(x)$, then to $f''(x)$, etc., to $f^{(n-1)}(x)$.

We may assume

$$a \leq x_0 < x_1 < x_2 < \cdots < x_{n-1} < x_n \leq b.$$

Since $f(x_0) = f(x_1)$, there is a point x_0' such that

$$x_0 < x_0' < x_1 \quad \text{and} \quad f'(x_0') = 0.$$

Similarly, there is a point x_1' such that

$$x_1 < x_1' < x_2 \quad \text{and} \quad f'(x_1') = 0.$$

Continuing, we find points

$$a < x_0' < x_1' < \cdots < x_{n-1}' < b$$

such that

$$f'(x_0') = f'(x_1') = \cdots = f'(x_{n-1}') = 0.$$

The same reasoning applied to $f'(x)$ yields points such that

$$a < x_0'' < x_1'' < \cdots < x_{n-2}'' < b$$

and

$$f''(x_0'') = f''(x_1'') = \cdots = f''(x_{n-2}'') = 0.$$

Continuing, we find $n - 2$ zeros of $f'''(x)$, $n - 3$ zeros of $f^{(4)}(x)$, etc., and finally $n - (n - 1) = 1$ zero of $f^{(n)}(x)$. That does it.

EXERCISES

Interpret Example 1 for

1   $\cos x$ on $[-\tfrac{1}{2}\pi, \tfrac{1}{2}\pi]$

2   $\sqrt{1 - x^2}$ on $[-1, 1]$.

3   Show that $f(x) = 10x^9 - 16x^7 + 6x^2 - 1$ has a zero on $0 < x < 1$. [*Hint*   Integrate.]

4   (cont.)   Suppose $\dfrac{a_0}{n+1} + \dfrac{a_1}{n} + \cdots + \dfrac{a_{n-1}}{2} + a_n = 0$.   Show that
$f(x) = a_0 x^n + a_1 x^{n-1} + \cdots + a_n$ has a zero on $0 < x < 1$.

5   Suppose $f(x)$ is differentiable on $[a, b]$ and $f(a) = f(b) = 0$. Prove there exists c such that $a < c < b$ and $f(c) + cf'(c) = 0$.

6   (cont.) What does Ex. 5 imply if $f(x) = \sin x$?

7   (cont.) What does Ex. 5 imply if $f(x) = x^2 - 1 + \sin \pi x$?

8   Suppose $0 < a < b$, $f(x)$ is differentiable on $[a, b]$, and $f(a) = f(b) = 0$. Prove there exists c such that $a < c < b$ and $f(c) - cf'(c) = 0$.

9   (cont.) Apply Ex. 8 to $\sin x$.

10   (cont.) Apply Ex. 8 to $x^4 - 11x^2 + 30$.

11   Let $f(x)$ and $g(x)$ be differentiable on $[a, b]$ and $f(a) = f(b) = 0$. Suppose $f'g - fg'$ has no zeros in $[a, b]$. Prove g has a zero in $a < x < b$.

12   (cont.) Find an example of Ex. 11.

13   Let $f(x)$ be differentiable on $[a, b]$. Prove there exists c such that

$$a < c < b \qquad \text{and} \qquad f'(c) = \frac{f(b) - f(c)}{c - a}.$$

14   Apply Ex. 13 to $f(x) = \sin^2 x$ on $[0, \tfrac{1}{2}\pi]$.

15   Let $f(x)$ be a polynomial with n distinct zeros, and let $k \neq 0$ be a constant. Prove that $g(x) = f(x) + kf'(x)$ has at least $n - 1$ distinct zeros.

16* (cont.) Assume $f(x)$ has degree n and n simple zeros. Prove that $g(x)$ has n simple zeros. [*Hint* By considering $g(x)/f(x)$, show that $g(x)$ has a zero either in $-\infty < x < r_1$ or in $r_n < x < \infty$.]

17   Suppose the polynomial $f(x)$ has n zeros in $[a, b]$, counting multiplicities. For each $k = 1, 2, \cdots, n$, show that $f^{(k)}(x)$ has $n - k$ zeros in $[a, b]$.

18   (cont.) Show that $\dfrac{d^n}{dx^n}[(x^2 - 1)^n]$ has n simple zeros in $-1 < x < 1$.

19   Let $a < c < b$, let f''' exist on $[a, b]$, and assume $f(a) = f(c) = f(b) = f'(c) = 0$. Prove f''' has a zero in $a < x < b$.

20   Let $f(x)$ be a polynomial. Assume $f(a) = f(b) = 0$ and $f(x) \neq 0$ for all x on $a < x < b$. Prove $f'(x)$ has an odd number of zeros, counting multiplicities, on $a < x < b$. [*Hint* Logarithmic differentiation.]

21   Suppose a polynomial $f(x)$ has a zero of multiplicity $n + 1$ at $x = a$. Express $f(x)$ in powers of $x - a$, and write the corresponding expression for $f^{(n+1)}(x)$.

22   (cont.) Apply the Dividing out Zeros formula, p. 464, to prove

$$\int_0^1 (1 - u)^n u^k \, du = \frac{n! \, k!}{(n + k + 1)!}.$$

23* Use the Generalized Rolle's Theorem for a new proof of a result on convex functions (Ex. 21, p. 152): if $f(a) = f(b) = 0$ and $f'' > 0$ on $a < x < b$, then $f < 0$ on $a < x < b$. [*Hint* Consider $g(x) = f(x) - k(x - a)(b - x)$ for various k.]

24* (cont.) Suppose $f(0) = f(1) = 0$ and $|f''(x)| \leq 1$ on $[0, 1]$. Prove $|f(x)| \leq \tfrac{1}{8}$ on $[0, 1]$.

6. MEAN VALUE THEOREMS AND LHOSPITAL'S RULE

Let us recall the Mean Value Theorem (MVT) of differential calculus. (As we shall see later, there is also a mean value theorem of integral calculus.)

Let $f(x)$ be a continuous function on the closed interval $[a, b]$, and suppose $f'(x)$ exists for $a < x < b$. Then

$$\frac{f(b) - f(a)}{b - a} = f'(c)$$

for some c between a and b.

This was proved in Chapter 3, p. 149. Note carefully where the Mean Value Theorem differs from Rolle's Theorem. In the latter, we add the hypothesis $f(a) = f(b)$ and draw the conclusion that $f'(c) = 0$ for some c between a and b.

Geometrically, the Mean Value Theorem says that somewhere between $x = a$ and $x = b$, the tangent is parallel to the chord (Fig. 1a). Rolle's Theorem is the special case where the chord is horizontal (Fig. 1b). Or you might say that the MVT is a tilted version of Rolle's Theorem.

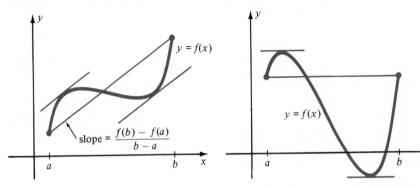

(a) Mean Value Theorem: Somewhere the tangent is parallel to the chord.

(b) Rolle's Theorem: If $f(a) = f(b)$, then somewhere the tangent is horizontal.

Fig. 1

In most applications, the MVT is expressed in the form

$$f(b) - f(a) = f'(c)(b - a) \qquad (a < c < b)$$

and is used to estimate the difference between the values of $f(x)$ at two points.

Examples   1. $|\sin b - \sin a| \leq |b - a|$.

Set $f(x) = \sin x$. Then $|f'(x)| = |\cos x| \leq 1$, so

$$|\sin b - \sin a| = |\cos c|\,|b - a| \leq |b - a|.$$

2. $|\sqrt{102} - \sqrt{101}| < 0.05$.

Set $f(x) = \sqrt{x}$. Then

$$|\sqrt{102} - \sqrt{101}| = |f'(c)|\,|102 - 101| = \frac{1}{2\sqrt{c}} < \frac{1}{2\sqrt{100}} = 0.05.$$

3. If $f'(x) > 0$ for $a \leq x \leq b$, then $f(b) > f(a)$.

$$f(b) - f(a) = f'(c)(b - a) = (\text{positive})(\text{positive}) > 0.$$

Application to Second Derivatives   The derivative of $f(x)$ is defined by the formula

$$\lim_{h \to 0} \frac{f(c + h) - f(c)}{h} = f'(c).$$

With the Mean Value Theorem we can establish an analogous formula for the second derivative.

Let $f(x)$ have a continuous second derivative on the interval $a < x < b$. If c is any point of this interval, then

$$\lim_{h \to 0} \frac{f(c+h) - 2f(c) + f(c-h)}{h^2} = f''(c).$$

Proof The idea is to show that the big fraction after the limit sign equals $f''(w)$ for some w near c.

Note that the fraction is unchanged if h is replaced by $-h$. If $h < 0$, we do so; hence we can assume $h > 0$. We hold h fixed for the moment and write

$$f(c+h) - 2f(c) + f(c-h) = [f(c+h) - f(c)] + [f(c) - f(c-h)]$$
$$= g(c+h) - g(c),$$

where $$g(x) = f(x) - f(x-h).$$

By the MVT, $$g(c+h) - g(c) = hg'(z),$$

where $c < z < c + h$. Now $$g'(z) = f'(z) - f'(z-h),$$

and by the MVT again, $$f'(z) - f'(z-h) = hf''(w),$$

where $z - h < w < z$. Note that $c - h < w < c + h$.

Putting the pieces together, we have

$$\frac{f(c+h) - 2f(c) + f(c-h)}{h^2} = \frac{g(c+h) - g(c)}{h^2} = \frac{hg'(z)}{h^2}$$

$$= \frac{f'(z) - f'(z-h)}{h} = \frac{hf''(w)}{h} = f''(w).$$

Now we let $h \longrightarrow 0$. Since w is squeezed between $c - h$ and $c + h$, we have $w \longrightarrow c$. But $f''(x)$ is *continuous*; hence $f''(w) \longrightarrow f''(c)$. Therefore

$$\lim_{h \to 0} \frac{f(c+h) - 2f(c) + f(c-h)}{h^2} = f''(c).$$

■ **EXAMPLE 1** Find   (a) $\displaystyle\lim_{h \to 0} \frac{e^h + e^{-h} - 2}{h^2}$ (b) $\displaystyle\lim_{\theta \to 0} \frac{1 - \cos\theta}{\theta^2}$.

Solution (a) $\displaystyle\lim_{h \to 0} \frac{e^h + e^{-h} - 2}{h^2} = \lim_{h \to 0} \frac{e^h - 2e^0 + e^{-h}}{h^2} = \left.\frac{d^2}{dx^2} e^x\right|_{x=0} = \left.e^x\right|_{x=0} = 1.$

(b) $\displaystyle\frac{1 - \cos\theta}{\theta^2} = -\frac{1}{2}\frac{2\cos\theta - 2}{\theta^2} = -\frac{1}{2}\frac{\cos\theta - 2\cos 0 + \cos(-\theta)}{\theta^2},$

and $\displaystyle\lim_{\theta \to 0} \frac{\cos\theta - 2\cos 0 + \cos(-\theta)}{\theta^2} = \left.\frac{d^2}{d\theta^2}\cos\theta\right|_0 = -1.$

Hence $$\lim_{\theta \to 0} \frac{1 - \cos\theta}{\theta^2} = \frac{1}{2}.$$ ■

Generalized MVT   The MVT compares the change in $f(x)$ to the change in the function $g(x) = x$:

$$\frac{f(b) - f(a)}{b - a} = \frac{f(b) - f(a)}{g(b) - g(a)}.$$

The following generalized MVT compares the change in $f(x)$ to the change in a more general function $g(x)$. The only restriction is that like x, the function $g(x)$ must be increasing (or decreasing).

Generalized Mean Value Theorem   Let $f(x)$ and $g(x)$ be differentiable functions on $[a, b]$ and suppose $g'(x) > 0$ (or $g'(x) < 0$) for all x, so that $g(x)$ is a strictly increasing (decreasing) function. Then

$$\frac{f(b) - f(a)}{g(b) - g(a)} = \frac{f'(c)}{g'(c)}$$

for some c between a and b.

It is tempting to give a phony proof. Apply the MVT to $f(x)$ and $g(x)$ separately, then divide:

$$f(b) - f(a) = g'(c)(b - a), \qquad g(b) - g(a) = g'(c)(b - a)$$

Therefore, $$\frac{f(b) - f(a)}{g(b) - g(a)} = \frac{f'(c)(b - a)}{g'(c)(b - a)} = \frac{f'(c)}{g'(c)}.$$

Very nice, but WRONG! The c that works for $f(x)$ and the c that works for $g(x)$ are usually not the same. Try $f(x) = x^3$, $g(x) = x^2$, $[a, b] = [0, 1]$ if you are not convinced.

For a correct proof, set

$$h(x) = [g(b) - g(a)][f(x) - f(a)] - [f(b) - f(a)][g(x) - g(a)].$$

Then $h(x)$ is differentiable and $h(a) = h(b) = 0$. By Rolle's Theorem, $h'(c) = 0$ for some c such that $a < c < b$. But

$$h'(x) = [g(b) - g(a)]f'(x) - [f(b) - f(a)]g'(x).$$

Therefore $$[g(b) - g(a)]f'(c) = [f(b) - f(a)]g'(c).$$

Since $g'(c) \neq 0$ and $g(b) - g(a) \neq 0$ by hypothesis, we can divide to obtain the desired formula.

Lhospital's Rule   We sometimes have to find limits of the form

$$\lim_{x \to a} \frac{f(x)}{g(x)}$$

where $f(a) = g(a) = 0$, what we might loosely call "limits of the form $0/0$."

Examples $\displaystyle\lim_{x \to 0} \frac{\tan x}{e^x - 1}$, $\displaystyle\lim_{x \to \pi} \frac{x - \pi}{\sin x}$, $\displaystyle\lim_{x \to \infty} \frac{\frac{1}{2}\pi - \arctan x}{x^{-1}}$.

Lhospital's Rule helps evaluate such limits. We shall look at several forms of the rule, and also limits of the form ∞/∞. (Incidentally, Lhospital* is pronounced *lō′-pee-tal.*)

Lhospital's Rule I   Let $f(x)$ and $g(x)$ be differentiable functions near $x = a$ such that $f(a) = g(a) = 0$. Suppose that $g(x) \neq 0$ and $g'(x) \neq 0$ for $x \neq a$. If

$$\lim_{x \to a} \frac{f'(x)}{g'(x)} = L,$$

where L is a real number or $\pm\infty$, then also

$$\lim_{x \to a} \frac{f(x)}{g(x)} = L.$$

Briefly,
$$\lim_{x \to a} \frac{f(x)}{g(x)} = \lim_{x \to a} \frac{f'(x)}{g'(x)}.$$

This rule is an easy consequence of the Generalized MVT. If $x \neq a$, then there exists c between a and x such that

$$\frac{f(x)}{g(x)} = \frac{f(x) - f(a)}{g(x) - g(a)} = \frac{f'(c)}{g'(c)}.$$

If $x \longrightarrow a$, then $c \longrightarrow a$, so

$$\frac{f(x)}{g(x)} = \frac{f'(c)}{g'(c)} \longrightarrow L.$$

■ **EXAMPLE 2**   Find   (a) $\displaystyle \lim_{x \to 0} \frac{\tan x}{e^x - 1}$ (b) $\displaystyle \lim_{x \to 0} \frac{1 - \cos x}{x^2}$.

Solution   (a) $\tan 0 = e^0 - 1 = 0$, so Lhospital's Rule applies:

$$\lim_{x \to 0} \frac{\tan x}{e^x - 1} = \lim_{x \to 0} \frac{(\tan x)'}{(e^x - 1)'} = \lim_{x \to 0} \frac{\sec^2 x}{e^x} = \frac{\sec^2 0}{e^0} = 1.$$

(b) $1 - \cos 0 = 0^2 = 0$, so Lhospital's Rule applies:

$$\lim_{x \to 0} \frac{1 - \cos x}{x^2} = \lim_{x \to 0} \frac{(1 - \cos x)'}{(x^2)'} = \lim_{x \to 0} \frac{\sin x}{2x}.$$

This is another limit of the form 0/0, so Lhospital's Rule applies a second time:

$$\lim_{x \to 0} \frac{1 - \cos x}{x^2} = \lim_{x \to 0} \frac{\sin x}{2x} = \lim_{x \to 0} \frac{\cos x}{2} = \frac{1}{2}. \qquad ■$$

---

* The spelling of Lhospital varies, and in fact our spelling of the name is not a common one. Yet according to an article by R. P. Boas in the American Mathematical Monthly, 1969, it is the spelling the Marquis De Lhospital used himself.

Lhospital's Rule II   Let $f(x)$ and $g(x)$ be differentiable functions near $x = a$ such that

$$f(x) \longrightarrow \pm \infty \qquad \text{and} \qquad g(x) \longrightarrow \infty$$

as $x \longrightarrow a$. Suppose that $g(x) \neq 0$ and $g'(x) \neq 0$ for $x \neq a$. If

$$\lim_{x \to a} \frac{f'(x)}{g'(x)} = L,$$

where L is finite or $\pm \infty$, then also

$$\lim_{x \to a} \frac{f(x)}{g(x)} = L.$$

If the first limit is a one-sided limit (say $x \longrightarrow a+$), then so is the second.

The proof of this rule is tricky. Let us just sketch the idea for the case when L is finite.

Let $a < x < z$. Then

$$\frac{f(z) - f(x)}{g(z) - g(x)} = \frac{f'(c)}{g'(c)}$$

for some $x < c < z$. Choose z very close to a. Then c is very close to a, hence $f'(c)/g'(c)$ is close to L. Therefore,

$$\frac{f(z) - f(x)}{g(z) - g(x)} \approx L,$$

and this is true for any x such that $a < x < z$.

Now hold z fixed and let $x \longrightarrow a$. By hypothesis, $f(x)$ and $g(x)$ grow overwhelmingly large. Hence

$$\frac{f(z) - f(x)}{g(z) - g(x)} = \frac{f(x)}{g(x)} \left(\frac{1 - f(z)/f(x)}{1 - g(z)/g(x)} \right) \approx \frac{f(x)}{g(x)}.$$

Therefore $f(x)/g(x)$ is very close to L when x is close to a.

■ **EXAMPLE 3**   Find $\lim_{x \to 0+} x \ln x$.

Solution   Write $x \ln x = \dfrac{\ln x}{1/x}$,

which is of the form $f(x)/g(x)$, where

$$f(x) = \ln x \longrightarrow -\infty \qquad \text{and} \qquad g(x) = \frac{1}{x} \longrightarrow \infty$$

as $x \longrightarrow 0+$. By Lhospital's Rule II,

$$\lim_{x \to 0+} x \ln x = \lim_{x \to 0+} \frac{\ln x}{1/x} = \lim_{x \to 0+} \frac{1/x}{-1/x^2} = \lim_{x \to 0+} (-x) = 0. \qquad ■$$

Remark   Compare to the discussion on p. 317, bottom, to p. 318.

Lhospital's Rule III   Let $f(x)$ and $g(x)$ be differentiable functions for x large such that either

$$\begin{vmatrix} f(x) \longrightarrow 0 \\ g(x) \longrightarrow 0 \end{vmatrix} \quad \text{or} \quad \begin{vmatrix} f(x) \longrightarrow \pm\infty \\ g(x) \longrightarrow \infty \end{vmatrix}$$

as $x \longrightarrow \infty$. Suppose that $g(x) \neq 0$ and $g'(x) \neq 0$. If

$$\lim_{x \to \infty} \frac{f'(x)}{g'(x)} = L,$$

where L is finite or $\pm\infty$, then also

$$\lim_{x \to \infty} \frac{f(x)}{g(x)} = L.$$

We shall omit the rather complicated proof of this rule.

■ **EXAMPLE 4**   Find   (a) $\lim_{x \to \infty} x \operatorname{arc\,cot} x$ (b) $\lim_{x \to \infty} \dfrac{(\ln x)^2}{x}$.

Solution   (a)   Write $x \operatorname{arc\,cot} x = \dfrac{\operatorname{arc\,cot} x}{1/x}$,

which is of the form $f(x)/g(x)$, where

$$f(x) = \operatorname{arc\,cot} x \longrightarrow 0 \quad \text{and} \quad g(x) = \frac{1}{x} \longrightarrow 0$$

as $x \longrightarrow \infty$. By Lhospital's Rule,

$$\lim_{x \to \infty} x \operatorname{arc\,cot} x = \lim_{x \to \infty} \frac{\operatorname{arc\,cot} x}{1/x} = \lim_{x \to \infty} \frac{-1/(1+x^2)}{-1/x^2} = \lim_{x \to \infty} \frac{x^2}{1+x^2} = 1.$$

(b)   This is a form ∞/∞, so Lhospital applies:

$$\lim_{x \to \infty} \frac{(\ln x)^2}{x} = \lim_{x \to \infty} \frac{2(\ln x)/x}{1} = 2 \lim_{x \to \infty} \frac{\ln x}{x}.$$

Once again:

$$\lim_{x \to \infty} \frac{(\ln x)^2}{x} = 2 \lim_{x \to \infty} \frac{\ln x}{x} = 2 \lim_{x \to \infty} \frac{1/x}{1} = 2 \lim_{x \to \infty} \frac{1}{x} = 0. \qquad ■$$

Remark   Compare to the discussion on the middle of p. 317.

Many limits of the form $\lim f(x)^{g(x)}$ can be handled by Lhospital's Rule. The trick is to take logs first, then use the continuity of $\ln x$ and its inverse function e^x.

■ **EXAMPLE 5**   Find $\lim_{x \to 0+} x^{x^2}$.

Solution
$$\ln x^{x^2} = x^2 \ln x.$$

By Lhospital's Rule (or what we know about the rate of growth of $\ln x$),

$$\lim_{x \to 0+} \ln x^{x^2} = \lim_{x \to 0+} x^2 \ln x = 0.$$

Hence

$$\lim_{x \to 0+} x^{x^2} = e^0 = 1. \qquad \blacksquare$$

The Lhospital Habit   Try to resist the bad habit of automatically applying Lhospital's Rule to every limit problem you face. Often there is a simpler way, for instance, use of limits already known or algebraic simplifications.

■ **EXAMPLE 6** Find   (a) $\displaystyle \lim_{x \to \infty} \frac{\sqrt{2 + 5x^2}}{x}$ (b) $\displaystyle \lim_{x \to 0} \frac{\sin^2 x}{x^2}$

Solution   (a) This is a form ∞/∞, so by Lhospital,

$$\lim_{x \to \infty} \frac{\sqrt{2 + 5x^2}}{x} = \lim_{x \to \infty} \frac{5x/\sqrt{2 + 5x^2}}{1}$$

$$= \lim_{x \to \infty} \frac{5x}{\sqrt{2 + 5x^2}} = \frac{5}{\displaystyle \lim_{x \to \infty} \frac{\sqrt{2 + 5x^2}}{x}} = \frac{5}{?},$$

back to where we started. Ugh!

Alternative solution for (a)

$$\frac{\sqrt{2 + 5x^2}}{x} = \sqrt{\frac{2}{x^2} + 5} \longrightarrow \sqrt{0 + 5} = \sqrt{5}.$$

(b) This is a $0/0$ form, so by Lhospital,

$$\lim_{x \to 0} \frac{\sin^2 x}{x^2} = \lim_{x \to 0} \frac{2 \sin x \cos x}{2x} = \lim_{x \to 0} \frac{\cos^2 x - \sin^2 x}{1}$$

$$= \lim_{x \to 0} \cos^2 x - \lim_{x \to 0} \sin^2 x = 1 - 0 = 1.$$

This solution is correct, but long-winded.

Alternative solution for (b)   As we have seen dozens of times, $\sin x/x \longrightarrow 1$ as $x \longrightarrow 0$. Therefore

$$\frac{\sin^2 x}{x^2} = \left(\frac{\sin x}{x} \right)^2 \longrightarrow 1^2 = 1. \qquad \blacksquare$$

Be alert to the possibility of adapting known limits to new situations. For example, from $(\sin x)/x \longrightarrow 1$ as $x \longrightarrow 0$ follow (without Lhospital)

$$\frac{x}{\tan x} = \frac{x \cos x}{\sin x} = \frac{\cos x}{(\sin x)/x} \longrightarrow \frac{1}{1} = 1,$$

$$\frac{\sin 3x}{2x} = \frac{\sin 3x}{3x} \cdot \frac{3x}{2x} \longrightarrow 1 \cdot \frac{3}{2} = \frac{3}{2}, \qquad \frac{\sin(x^2)}{x} = \frac{\sin(x^2)}{x^2} \cdot x \longrightarrow 1 \cdot 0 = 0.$$

MVT of Integral Calculus The MVT of differential calculus starts with a function on an interval $[a, b]$ and gives some information about its derivative at an intermediate point. The MVT of integral calculus starts with a definite integral on $[a, b]$ and gives some information about the integrand (or a factor of it) at an intermediate point.

Mean Value Theorem of Integral Calculus Let $f(x)$ and $g(x)$ be continuous functions on the interval $[a, b]$, with $g(x) \geq 0$. Then

$$\int_a^b f(x)g(x)\, dx = f(c) \int_a^b g(x)\, dx$$

for some c on the interval.

Proof We may assume

$$\int_a^b g(x)\, dx > 0.$$

For if not, then $g(x) = 0$ and both sides of the desired relation equal 0; any c will do then.

Since $f(x)$ is continuous on the closed interval $[a, b]$ it takes on its max and min (see p. 146):

$$f(c_0) \leq f(x) \leq f(c_1),$$

where c_0 and c_1 lie in the interval. Since $g(x) \geq 0$, we have

$$f(c_0)g(x) \leq f(x)g(x) \leq f(c_1)g(x).$$

Therefore $\qquad f(c_0) \int_a^b g(x)\, dx \leq \int_a^b f(x)g(x)\, dx \leq f(c_1) \int_a^b g(x)\, dx,$

that is, $\qquad f(c_0) \leq \dfrac{\int_a^b f(x)g(x)\, dx}{\int_a^b g(x)\, dx} \leq f(c_1).$

Thus the quotient in the middle falls between two values of $f(x)$. By the Intermediate Value Theorem (p. 346), it is equal to $f(c)$ for some c in $[a, b]$. This completes the proof.

Examples   1. $\displaystyle\int_0^\pi x \sin x\, dx = c \int_0^\pi \sin x\, dx = 2c, \qquad$ where $\quad 0 \leq c \leq \pi.$

The other way: $\displaystyle\int_0^\pi x \sin x\, dx = (\sin c) \int_0^\pi x\, dx = \tfrac{1}{2}\pi^2 \sin c, \qquad$ where $\quad 0 \leq x \leq \pi.$

2.   The special case $g(x) = 1$ is worth noting. Then

$$\int_a^b f(x)\, dx = f(c) \int_a^b dx = (b-a)f(c).$$

We conclude that the average value of $f(x)$ on $[a, b]$ is equal to one of its values.

> Let $f(x)$ be a continuous function on the interval $[a, b]$. Then
> $$\frac{1}{b-a}\int_a^b f(x)\,dx = f(c)$$
> for some c on the interval.

EXERCISES

Prove

1    $\ln 51 - \ln 50 < 0.02$

2    $\sqrt[3]{1001} < 10.00034$

3    $\arctan 6 - \arctan 5 < 0.04$

4    $\tan \frac{2}{9}\pi > 0.82$

5    $\arcsin \frac{3}{5} < 37.5°$

6    $\arcsin \frac{2}{5} > 23°.$

Find

7    $\displaystyle\lim_{x\to 0}\frac{x}{e^x - 1}$

8    $\displaystyle\lim_{x\to 1}\frac{\ln x}{1 - x}$

9    $\displaystyle\lim_{x\to 0}\frac{x - \sin x}{x^3}$

10    $\displaystyle\lim_{x\to 0}\frac{\cos x - 1 + \frac{1}{2}x^2}{x^4}$

11    $\displaystyle\lim_{x\to 0}\frac{\cosh x - 1}{x^2}$

12    $\displaystyle\lim_{x\to 0}\frac{\tan x^2}{\sin^2 x}$

13    $\displaystyle\lim_{x\to \frac{1}{2}\pi^-}\left(x - \tfrac{1}{2}\pi\right)\tan x$

14    $\displaystyle\lim_{x\to \pi^+}\left(x - \pi\right)\csc x$

15    $\displaystyle\lim_{x\to 0+} x^{\tan x}$

16    $\displaystyle\lim_{x\to 0+} x^{\ln(1+x)}$

17    $\displaystyle\lim_{x\to 0+}\left[\ln(1+x)\right]^x$

18    $\displaystyle\lim_{x\to \infty}\left(1 + \frac{1}{x^2}\right)^x$

19    $\displaystyle\lim_{x\to \infty}\frac{(\ln x)^3}{x}$

20    $\displaystyle\lim_{x\to \infty}\frac{e^x}{x^{10}}$

21    $\displaystyle\lim_{x\to \infty} x\left(\tfrac{1}{2}\pi - \arctan x\right)$

22    $\displaystyle\lim_{x\to \infty} x^2\left(\operatorname{arc\,cot} x\right)^3$

23    $\displaystyle\lim_{x\to \infty}\left(\sqrt{x^2 + 1} - x\right)$

24    $\displaystyle\lim_{x\to \infty}\frac{\sqrt{x^2 + 1}}{x}$

[*Hint for 22*   Don't overdo the Lhospital habit.]

25    $\displaystyle\lim_{x\to 0}\left(\frac{1}{x} - \frac{1}{\sin x}\right)$

26    $\displaystyle\lim_{x\to \infty} x^3\left(1 - e^{-1/x^2}\right)$

27    $\displaystyle\lim_{x\to \infty} x \arctan x$

28    $\displaystyle\lim_{x\to 0}\left(\frac{1}{x} - \frac{1}{\sin x}\right)\Big/(e^x - 1).$

29   Suppose $f'(x)$ is continuous for $|x - a| < \delta$. Use the MVT to find $\lim_{h\to 0}\left[f(a + h) - f(a - h)\right]/2h.$

30   Suppose $f(x)$ is continuous for $|x - a| < \delta$ and differentiable near $x = a$ except possibly at $x = a$. Suppose $\lim_{x\to a} f'(x) = L$. Prove $f(x)$ is differentiable at $x = a$ and $f'(a) = L.$

31   Suppose $f(x)$ is differentiable for $a < x < b$ and $|f'(x)| \le M$. Prove $f(x)$ is uniformly continuous. (See p. 255.)

32   Suppose $f(x)$ is differentiable for $x \ge a$ and $f'(x) \ge m > 0$. Prove $f(x) \longrightarrow \infty$ as $x \longrightarrow \infty.$

33   Let $f(x) = a_0 x^n + a_1 x^{n-1} + \cdots + a_n$, where $a_0 > 0$. Find $\lim_{x\to \infty} f(x)^{1/x}.$

34   Let $f(x), g(x), h(x)$ be continuous on $[a, b]$ and differentiable for $a < x < b$. Prove there exists c such that $a < c < b$ and

$$\begin{vmatrix} f(a) & g(a) & h(a) \\ f(b) & g(b) & h(b) \\ f'(c) & g'(c) & h'(c) \end{vmatrix} = 0.$$

35 Suppose $f(x)$ is differentiable for $x \geq a$, $f(x) \longrightarrow L$ and $f'(x) \longrightarrow M$ as $x \longrightarrow \infty$, where L is finite. Prove $M = 0$.

36* Suppose $f'(x) \longrightarrow \infty$ as $x \longrightarrow \infty$. Prove $f(x)/x \longrightarrow \infty$.

37 Suppose $f'(x)/e^x$ is bounded for $x \geq a$. Prove that $f(x)/e^x$ is also bounded.

38* (cont.) Suppose $f'(x)/e^x \longrightarrow 0$ as $x \longrightarrow \infty$. Prove $f(x)/e^x \longrightarrow 0$.

39 Suppose $g(x) + g'(x) \longrightarrow 0$ as $x \longrightarrow \infty$. Prove $g(x) \longrightarrow 0$. [*Hint* Ex. 38.]

40* Suppose $f(x) \longrightarrow L$ and $f''(x) \longrightarrow 0$ as $x \longrightarrow \infty$, where L is finite. Prove $f'(x) \longrightarrow 0$.

41* Suppose $f(x)$ is continuous on $[a, b]$, $f'(x)$ exists and satisfies $|f'(x)| \leq M$ on $a < x < b$, and $f(a) = f(b) = 0$. Prove

$$\left| \int_a^b f(x)\,dx \right| \leq \tfrac{1}{4} M |b - a|^2.$$

(Compare Ex. 22, p. 260, where f' is also assumed integrable.) [*Hint* Consider the intervals $[a, c]$ and $[c, b]$, where c is the midpoint of $[a, b]$, and use the MVT.]

42* Another proof of Taylor's Formula. Suppose $f'(x), f''(x), \cdots, f^{(n)}(x)$ are continuous for $a \leq x \leq b$ and $f^{(n+1)}(x)$ exists for $a < x < b$. (Or if $b < a$, $b \leq x \leq a$, etc.) Set

$$g(x) = f(b) - f(x) - f'(x)(b - x) - \frac{1}{2} f''(x)(b - x)^2$$

$$- \cdots - \frac{1}{n!} f^{(n)}(x)(b - x)^n - \frac{A}{(n + 1)!} (b - x)^{n+1}.$$

Then $g(b) = 0$. Choose the constant A so $g(a) = 0$. Apply Rolle's Theorem to $g(x)$. Conclusion?

7. INTERPOLATION

An important scientific technique is fitting a function to data. In this and the next section, we develop methods for doing so using polynomials.

Suppose in an accurate experiment, a quantity x is measured at 11 time readings. For example,

t	0	0.1	0.2	$\cdots$	0.8	0.9	1.0
x	2.783	3.142	4.003	$\cdots$	2.001	1.833	1.801

In order to find a mathematical relation between these readings, we seek a function that fits the data, that is, a function $x(t)$ defined for all values of t in the interval $0 \leq t \leq 1$ and satisfying the conditions $x(0) = 2.783$, $x(0.1) = 3.142$, $\cdots$, $x(1.0) = 1.801$. In other words, we want a function whose graph passes through 11 given points. Finding such a function is called **interpolation**.

Linear Interpolation The simple case of linear interpolation is familiar. For example, from a 5-place table,

$$\log 3.1920 \approx 0.50406 \quad \text{and} \quad \log 3.1930 \approx 0.50420.$$

To estimate $\log 3.1927$, add $\frac{7}{10}$ of the difference $(0.50420 - 0.50406)$ to 0.50406:

$$\log 3.1927 \approx 0.50416.$$

This is equivalent to finding the linear function $f(x)$ that fits the two points (3.1920, 0.50406) and (3.1930, 0.50420), and writing

$$\log 3.1927 \approx f(3.1927).$$

As is easily computed,

$$f(x) = 0.50406 + (0.14)(x - 3.1920).$$

Hence

$$\log 3.1927 \approx 0.50406 + (0.14)(3.1927 - 3.1920) = 0.50406 + (0.14)(0.0007) \approx 0.50416.$$

In this illustration, the data consisted of two points only, so we could fit a linear polynomial (straight line graph). When the data consist of more than two points, we shall seek a higher degree polynomial that fits.

Polynomial Interpolation   Here is the basic problem. Given n points

$$(x_1, y_1), \quad (x_2, y_2), \cdots, (x_n, y_n), \quad \text{where} \quad x_1 < x_2 < \cdots < x_n,$$

find a polynomial $y = p(x)$ of least degree whose graph passes through the given points, i.e.,

$$p(x_1) = y_1, \quad p(x_2) = y_2, \quad \cdots \quad , p(x_n) = y_n.$$

Let us start with three points. Suppose we wish to interpolate

$$(x_1, y_1), \quad (x_2, y_2), \quad (x_3, y_3), \quad x_1 < x_2 < x_3,$$

by a polynomial of least degree. A first degree polynomial generally won't do since its graph is a straight line. So we try a quadratic polynomial

$$y = A + Bx + Cx^2$$

and hope to choose its three coefficients so that its graph passes through the three given points.

■ **EXAMPLE 1**   Fit a quadratic to the points (0, 1), (1, 2), (2, 4).

Solution   Set $y = p(x) = A + Bx + Cx^2$.
Substitute $(x, y) = (0, 1)$, (1, 2), and (2, 4):

$$\begin{cases} A & = 1 \\ A + B + C = 2 \\ A + 2B + 4C = 4. \end{cases}$$

The solution of this linear system is $A = 1$, $B = \frac{1}{2}$, $C = \frac{1}{2}$, hence the answer is

$$p(x) = 1 + \tfrac{1}{2}x + \tfrac{1}{2}x^2. \qquad ■$$

This example illustrates the **method of undetermined coefficients**. The next example demonstrates the method for four data points.

■ **EXAMPLE 2**   Fit a cubic to the four points $(-1, \frac{1}{2})$, (0, 1), (1, 2), (2, 4).

Solution   Set $p(x) = A + Bx + Cx^2 + Dx^3$.

To fit the four points, the coefficients must satisfy

$$\begin{cases} A - B + C - D = \frac{1}{2} \\ A \qquad\qquad\qquad = 1 \\ A + B + C + D = 2 \\ A + 2B + 4C + 8D = 4. \end{cases}$$

After some labor one finds the solution:

$$A = 1, \qquad B = \tfrac{2}{3}, \qquad C = \tfrac{1}{4}, \qquad D = \tfrac{1}{12}.$$

Therefore
$$p(x) = 1 + \tfrac{2}{3}x + \tfrac{1}{4}x^2 + \tfrac{1}{12}x^3. \qquad\blacksquare$$

Polynomial interpolation is important in approximation problems. For example, we may need many values of a function $f(x)$ that is difficult to compute. Then it is convenient to have a polynomial (easy to compute) that approximates $f(x)$. A basic method of finding one is fitting a polynomial to several points on the graph of $y = f(x)$.

■ **EXAMPLE 3**   Fit a quadratic to $y = 2^x$ that is exact at $x = 0, 1, 2$.

Solution   We want the quadratic $p(x)$ that interpolates

$$(0, 2^0) = (0, 1), \qquad (1, 2^1) = (1, 2), \qquad (2, 2^2) = (2, 4).$$

By Example 1, $\qquad p(x) = 1 + \tfrac{1}{2}x + \tfrac{1}{2}x^2.$ $\qquad\qquad\blacksquare$

Remark   The approximation of 2^x by $p(x)$ is quite good on the interval $0 \le x \le 2$. Let $\varepsilon(x)$ denote the error:

$$\varepsilon(x) = 2^x - (1 + \tfrac{1}{2}x + \tfrac{1}{2}x^2).$$

Table 1 shows values of 2^x, $p(x)$, and $\varepsilon(x)$ to 3 places.

Table 1

x	0.0	0.2	0.4	0.6	0.8	1.0	1.2	1.4	1.6	1.8	2.0
2^x	1.000	1.149	1.320	1.516	1.741	2.000	2.297	2.639	3.031	3.482	4.000
$p(x)$	1.000	1.120	1.280	1.480	1.720	2.000	2.320	2.680	3.080	3.520	4.000
$\varepsilon(x)$	0.000	0.029	0.040	0.036	0.021	0.000	−0.023	−0.041	−0.049	−0.038	0.000

Newton Interpolation   The method of undetermined coefficients involves a lot of computation when there are more than a few points. We shall examine two more satisfactory methods, Newton interpolation and, in the next section, Lagrange interpolation.

Newton interpolation applies when the x_j are equally spaced, as often happens with experimental data. To simplify the notation, we shall consider first the special case

$$x_1 = 1, \quad x_2 = 2, \quad \cdots, \quad x_n = n.$$

By shifting and changing the scale, we can easily pass to the general case

$$x_1 = a + h, \quad x_2 = a + 2h, \quad \cdots, \quad x_n = a + nh.$$

Newton interpolation uses a set of standard polynomials, made to fit the problem:

$$p_0(x) = 1, \quad p_1(x) = (x - 1), \quad p_2(x) = (x - 1)(x - 2),$$

$$\cdots , p_n(x) = (x - 1)(x - 2) \cdots (x - n).$$

Thus $p_n(x)$ is a polynomial of degree n and leading coefficient 1, having zeros at $x = 1, 2, \cdots, n$.

Let us apply these polynomials to the interpolation problem. Suppose we want to fit a polynomial of degree $n - 1$ to the data

$$(1, y_1), \quad (2, y_2), \quad \cdots , (n, y_n).$$

We try

$$p(x) = a_0 p_0(x) + a_1 p_1(x) + \cdots + a_{n-1} p_{n-1}(x)$$

$$= a_0 + a_1(x - 1) + a_2(x - 1)(x - 2)$$

$$+ \cdots + a_{n-1}(x - 1)(x - 2) \cdots (x - n + 1),$$

hoping to find suitable coefficients a_j. We set x equal to $1, 2, \cdots, n$ successively:

$$\begin{cases} y_1 = a_0 \\ y_2 = a_0 + a_1 \\ y_3 = a_0 + 2a_1 + 2a_2 \\ y_4 = a_0 + 3a_1 + 3 \cdot 2a_2 + 3 \cdot 2 \cdot 1a_3 \\ \cdots \cdots \cdots \cdots \cdots \cdots \cdots \cdots \\ y_n = a_0 + (n - 1)a_1 + \cdots + (n - 1)! \, a_{n-1}. \end{cases}$$

The value of a_0 is obvious from the first equation. Having a_0, we find a_1 from the second equation. Having a_0 and a_1, we find a_2 from the third equation, and so on. Each successive equation determines one more a_j until finally we find a_{n-1} from the last equation.

■ **EXAMPLE 4** Fit a cubic to the points $(1, 4), \quad (2, -1), \quad (3, 0), \quad (4, 1)$.

Solution Try a cubic of the form

$$p(x) = a_0 + a_1 p_1(x) + a_2 p_2(x) + a_3 p_3(x)$$

$$= a_0 + a_1(x - 1) + a_2(x - 1)(x - 2) + a_3(x - 1)(x - 2)(x - 3).$$

Now set $x = 1, 2, 3, 4$ successively. This produces a linear system for the unknown coefficients a_j:

$$\begin{cases} 4 = a_0 \\ -1 = a_0 + a_1 \\ 0 = a_0 + 2a_1 + 2a_2 \\ 1 = a_0 + 3a_1 + 6a_2 + 6a_3. \end{cases}$$

Solve successively:

$$a_0 = 4, \qquad a_1 = -1 - 4 = -5, \qquad 2a_2 = -4 + 10 = 6, \qquad a_2 = 3,$$

$$6a_3 = 1 - 4 + 15 - 18 = -6, \qquad a_3 = -1.$$

The answer is

$$p(x) = 4 - 5p_1(x) + 3p_2(x) - p_3(x)$$

$$= 4 - 5(x - 1) + 3(x - 1)(x - 2) - (x - 1)(x - 2)(x - 3). \qquad \blacksquare$$

The Difference Operator   A more systematic method for finding the coefficients $a_0, a_1, \cdots, a_{n-1}$ involves the **difference operator** Δ. For any function $f(x)$, we define

$$\Delta f(x) = f(x + 1) - f(x).$$

For example,

$$\Delta 1 = 0, \qquad \Delta x = (x + 1) - x = 1, \qquad \Delta x^2 = (x + 1)^2 - x^2 = 2x + 1.$$

More generally,

$$\Delta x^n = (x + 1)^n - x^n = (x^n + nx^{n-1} + \cdots) - x^n = nx^{n-1} + \cdots.$$

Therefore the operator Δ transforms a polynomial of degree n into a polynomial of degree $n - 1$, like the derivative operator d/dx does.

The standard polynomials behave very nicely under the operation Δ. We compute $\Delta p_n(x)$:

$$p_n(x) = (x - 1)(x - 2) \cdots (x - n),$$

$$\Delta p_n(x) = p_n(x + 1) - p_n(x) = x(x - 1) \cdots (x - n + 2)(x - n + 1)$$

$$- (x - 1)(x - 2) \cdots (x - n + 1)(x - n)$$

$$= [x - (x - n)](x - 1)(x - 2) \cdots (x - n + 1)$$

$$= n(x - 1)(x - 2) \cdots (x - n + 1) = np_{n-1}(x).$$

$$\Delta p_n(x) = np_{n-1}(x).$$

Remark   Notice the analogy to the differentiation formula $\dfrac{d}{dx} x^n = nx^{n-1}$.

The operator Δ can be applied several times in succession. We denote the operation of applying Δ twice by Δ^2, three times by Δ^3, etc. Bear in mind that Δ^3 is not the cube of anything, but merely the operation of applying Δ three times, as d^3/dx^3 is the operation of applying d/dx three times.

Examples $\Delta^2 x^3 = \Delta(\Delta x^3) = \Delta[(x + 1)^3 - x^3] = \Delta(3x^2 + 3x + 1)$

$$= [3(x + 1)^2 + 3(x + 1) + 1] - [3x^2 + 3x + 1] = 6x + 6;$$

$$\Delta^3 p_4(x) = \Delta^2[\Delta p_4(x)] = \Delta^2[4p_3(x)] = \Delta[\Delta 4p_3(x)] = \Delta[4 \cdot 3p_2(x)] = 4 \cdot 3 \cdot 2p_1(x);$$

$$\Delta^n x^{n-1} = 0.$$

Now suppose $p(x)$ is a polynomial expressed in the form

$$p(x) = a_0 + a_1 p_1(x) + \cdots + a_{n-1} p_{n-1}(x).$$

To compute the coefficients, apply Δ successively $n - 1$ times:

$$\Delta p(x) = a_1 p_0(x) + 2a_2 p_1(x) + \cdots + (n-1)a_{n-1} p_{n-2}(x)$$

$$\Delta^2 p(x) = 2a_2 p_0(x) + 6a_3 p_1(x) + \cdots + (n-1)(n-2)a_{n-1} p_{n-3}(x)$$

$$\cdots \cdots \cdots \cdots \cdots \cdots \cdots \cdots \cdots \cdots \cdots \cdots$$

$$\Delta^{n-1} p(x) = (n-1)! \, a_{n-1} p_0(x).$$

Substitute $x = 1$ into all n of these equations. Since $p_0(1) = 1$ but $p_1(1) = p_2(1) = \cdots = p_{n-1}(1) = 0$, only the first term on the right side of each equation survives:

$$\begin{cases} p(1) = a_0 \\ \Delta p(1) = a_1 \\ \Delta^2 p(1) = 2a_2 \\ \Delta^3 p(1) = 6a_3 \\ \cdots \cdots \cdots \cdots \\ \Delta^{n-1} p(1) = (n-1)! \, a_{n-1}. \end{cases}$$

To summarize:

Newton's Interpolation Formula   If $p(x)$ is a polynomial of degree $n - 1$, then

$$p(x) = p(1) + \Delta p(1) p_1(x) + \frac{1}{2!} \Delta^2 p(1) p_2(x)$$

$$+ \frac{1}{3!} \Delta^3 p(1) p_3(x) + \cdots + \frac{1}{(n-1)!} \Delta^{(n-1)} p(1) p_{n-1}(x).$$

Difference Tables   Newton's Interpolation Formula is important because it gives a complete solution to the problem of fitting a polynomial to data. It is also *practical* because there is a convenient way of computing the required differences

$$\Delta p(1), \quad \Delta^2 p(1), \quad \cdots \quad , \Delta^{(n-1)} p(1).$$

Given

$$p(1) = y_1, \quad p(2) = y_2, \quad \cdots \quad , p(n) = y_n,$$

we construct a **difference table**. In the first row we enter $y_1, y_2, \cdots, y_n$. In the second row, we enter the differences

$$y_2 - y_1, \quad y_3 - y_2, \quad \cdots \quad , y_n - y_{n-1}.$$

In the third row we enter the differences of these differences, that is, the second differences, and so on. Then the left-hand entries of successive rows form the sequence

$$p(1), \quad \Delta p(1), \quad \Delta^2 p(1), \quad \cdots \quad , \Delta^{(n-1)} p(1).$$

■ **EXAMPLE 5** Fit a polynomial of degree 4 to

$$(1, 4), \quad (2, -1), \quad (3, 0), \quad (4, 1), \quad (5, 44).$$

Solution Compute the difference table starting with 4, −1, 0, 1, 44 in the first row:

$$
\begin{array}{ccccccccc}
4 & & -1 & & 0 & & 1 & & 44 \\
 & -5 & & 1 & & 1 & & 43 & \\
 & & 6 & & 0 & & 42 & & \\
 & & & -6 & & 42 & & & \\
 & & & & 48 & & & &
\end{array}
$$

The sequence of first entries is

$$4 \ -5 \ \ 6 \ -6 \ \ 48.$$

Therefore, by Newton's Interpolation Formula,

$$p(x) = 4 - 5p_1(x) + \frac{6}{2!}p_2(x) - \frac{6}{3!}p_3(x) + \frac{48}{4!}p_4(x)$$

$$= 4 - 5p_1(x) + 3p_2(x) - p_3(x) + 2p_4(x)$$

$$= 4 - 5(x-1) + 3(x-1)(x-2)$$

$$- (x-1)(x-2)(x-3) + 2(x-1)(x-2)(x-3)(x-4). \quad ■$$

Remark Compare Examples 4 and 5. The data in Example 4 are part of the data of Example 5 where one additional point is given. The answer in Example 5 equals the answer in Example 4 plus a multiple of $p_4(x)$.

This illustrates an important feature of Newton interpolation: further data points can be added without changing the result of previous computations.

The General Case Suppose the data points are

$$(x_1, y_1), \quad (x_2, y_2), \quad \cdots, \quad (x_n, y_n),$$

where the x_j are equally spaced, that is, points in an arithmetic progression. Let h be the common increment; then the sequence of x_j's is

$$x_1, \quad x_2 = x_1 + h, \quad x_3 = x_1 + 2h, \quad \cdots, \quad x_n = x_1 + (n-1)h.$$

The corresponding standard polynomials are

$$P_0(x) = 1, \quad P_1(x) = \frac{1}{h}(x - x_1), \quad P_2(x) = \frac{1}{h^2}(x - x_1)(x - x_2), \cdots,$$

$$P_n(x) = \frac{1}{h^n}(x - x_1)(x - x_2) \cdots (x - x_n).$$

Thus $P_n(x) = \frac{1}{h^n}(x - x_1)(x - x_1 - h)(x - x_1 - 2h) \cdots [x - x_1 - (n-1)h].$

The corresponding difference operator $\Delta = \Delta_h$ is defined by

$$\Delta f(x) = f(x + h) - f(x).$$

The basic formula—verified directly as before—is

$$\Delta P_n(x) = nP_{n-1}(x).$$

To fit a polynomial $P(x)$ of degree at most $n-1$ to $(x_1, y_1), \cdots, (x_n, y_n)$, we write

$$P(x) = a_0 P_0(x) + a_1 P_1(x) + \cdots + a_{n-1} P_{n-1}(x).$$

Exactly as before, $a_j = \dfrac{1}{j!} \Delta^j P(x_1).$

The number $\Delta^j P(x_1)$ is the first entry in the j-th row of the difference table whose 0-th row is $y_1, y_2, \cdots, y_n$.

■ **EXAMPLE 6** Fit a polynomial of degree 4 to

$$(0, 1), \quad (2, 1), \quad (4, 3), \quad (6, 3), \quad (8, 3).$$

Solution   The x-coordinates 0, 2, 4, 6, 8 are equally spaced with common increment $h = 2$. The corresponding standard polynomials are

$$P_0(x) = 1, \quad P_1(x) = \tfrac{1}{2}x, \quad P_2(x) = \tfrac{1}{4}x(x-2),$$
$$P_3(x) = \tfrac{1}{8}x(x-2)(x-4), \quad P_4(x) = \tfrac{1}{16}x(x-2)(x-4)(x-6).$$

The difference table for the y-coordinates is

$$
\begin{array}{ccccccccc}
1 & & 1 & & 3 & & 3 & & 3 \\
& 0 & & 2 & & 0 & & 0 & \\
& & 2 & & -2 & & 0 & & \\
& & & -4 & & 2 & & & \\
& & & & 6 & & & &
\end{array}
$$

Therefore the required polynomial is

$$P(x) = 1 \cdot P_0(x) + \frac{2}{2!} P_2(x) - \frac{4}{3!} P_3(x) + \frac{6}{4!} P_4(x)$$

$$= 1 + \frac{1}{4} x(x-2) - \frac{1}{12} x(x-2)(x-4) + \frac{1}{64} x(x-2)(x-4)(x-6). \quad ■$$

Warning   When you use this method, don't forget the denominators: first the powers of h, second the factorials.

EXERCISES

Find the polynomial of least degree that fits the data

1 $(-2, 6), \quad (1, 0), \quad (2, 2)$
2 $(1, -3), \quad (2, 3), \quad (4, 27)$
3 $(0, 1), \quad (1, 3), \quad (3, 13)$
4 $(-3, 15), \quad (-1, 5), \quad (0, 3)$
5 $(3, 5), \quad (4, 7), \quad (7, 13)$
6 $(-2, -11), \quad (-1, -8), \quad (1, -2)$
7 $(-2, -5), \quad (-1, 1), \quad (0, 1), \quad (2, 7)$
8 $(-1, -6), \quad (-\tfrac{1}{2}, -1), \quad (\tfrac{1}{2}, 0), \quad (1, 2)$
9 $(-1, 0), \quad (0, 1), \quad (1, 0), \quad (2, 0), \quad (4, 45)$
10 $(-3, 25), \quad (-2, 0), \quad (-1, -3), \quad (1, -3), \quad (2, 0).$

Fit the data by Newton interpolation

11 $(1, 1), \quad (2, -1), \quad (3, 0)$
12 $(1, 4), \quad (2, 3), \quad (3, 0)$
13 $(1, 4), \quad (2, 3), \quad (3, -1)$
14 $(1, -1), \quad (2, -1), \quad (3, 1)$

15 (1, 1), (2, 2), (3, 3), (4, 10)   **16** (1, 0), (2, 0), (3, −1), (4, −1)
17 (1, 1), (2, 2), (3, 5), (4, 16), (5, 65)
18 (1, 1), (2, 1), (3, 2), (4, 4), (5, 8)
19 (1, 1), (3, 1), (5, 2)   **20** (0, −1), (1, 1), (2, 3)
21 (−2, 1), (−1, 0), (0, 1), (1, 0)   **22** (−3, 1), (0, −1), (3, 0), (6, 1)
23 (−2, 1), (−1, 0), (0, 1), (1, 0), (2, 1)
24 (−4, 1), (0, 2), (4, 3), (8, 0).

25 Fit a quadratic to sin x that is exact at $x = 0, \frac{1}{2}\pi, \pi$.
26 (cont.) Compute to 4 places the actual error if the quadratic is used to approximate sin $\frac{1}{4}\pi$, sin $\frac{3}{4}\pi$.
27 Fit a fourth degree polynomial to sin x that is exact at $x = -\pi, -\frac{1}{2}\pi, 0, \frac{1}{2}\pi, \pi$.
28 (cont.) Estimate the error if this polynomial is used to approximate sin $\frac{1}{4}\pi$, sin $\frac{3}{4}\pi$.
29 Use the interpolation polynomials in Exs. 25 and 27 to approximate $\int_0^\pi \sin x \, dx$. Which gives the closer approximation?
30 Fit a fourth degree polynomial to $1/(1 + x^2)$ at $x = -2, -1, 0, 1, 2$.

8. LAGRANGE INTERPOLATION

Newton's Interpolation Formula provides a neat solution to the interpolation problem, provided the points x_j are equally spaced. But what if they are not? Then all we have to fall back on is the method of undetermined coefficients—bad news. This method requires solution of a linear system of n equations. Even if we knew that the system is consistent (which is proved by linear algebra), computing the solution when n is more than 3 or 4 can be a hard grind.

Fortunately, there is a way around undetermined coefficients. The method, called Lagrange Interpolation, uses an important principle: solve the problem first with the *simplest possible data*, then superpose the resulting elementary solutions. In applied linear mathematics, this is called the principle of superposition.

The general interpolation problem is to fit a polynomial to any n points

$$(x_1, y_1), \quad (x_2, y_2), \quad \cdots \quad , (x_n, y_n),$$

provided only that $x_1 < x_2 < \cdots < x_n$. In Lagrange Interpolation, we solve the problem first for the points

$$(x_1, 1), \quad (x_2, 0), \quad (x_3, 0), \quad \cdots \quad , (x_n, 0).$$

This is what we interpret as "simplest possible data". We want a polynomial $p_1(x)$ of degree at most $n - 1$ such that

$$p_1(x_2) = p_1(x_3) = \cdots = p_1(x_n) = 0 \quad \text{and} \quad p_1(x_1) = 1.$$

Obviously,

$$q_1(x) = (x - x_2)(x - x_3) \cdots (x - x_n)$$

satisfies the first conditions, and $q_1(x_1) \neq 0$. We simply divide by the number $q_1(x_1)$, obtaining a polynomial equal to 1 at x_1. Therefore

$$p_1(x) = \frac{q_1(x)}{q_1(x_1)} = \frac{(x - x_2)(x - x_3) \cdots (x - x_n)}{(x_1 - x_2)(x_1 - x_3) \cdots (x_1 - x_n)}$$

is the answer. For example, the polynomial

$$\frac{(x - 4)(x - 5)(x - 8)}{(2 - 4)(2 - 5)(2 - 8)}$$

takes the value 0 at $x = 4, 5, 8$ and the value 1 at $x = 2$.

Similarly, for each $j = 1, \cdots, n$, we set

$$p_j(x) = \frac{(x - x_1) \cdots (x - x_{j-1})(x - x_{j+1}) \cdots (x - x_n)}{(x_j - x_1) \cdots (x_j - x_{j-1})(x_j - x_{j+1}) \cdots (x_j - x_n)}.$$

(There is no factor $x - x_j$ in the numerator and no factor $x_j - x_j$ in the denominator.)
Then $p_j(x)$ has degree $n - 1$ and

$$p_j(x_j) = 1, \quad \text{but} \quad p_j(x_i) = 0 \quad \text{if} \quad i \neq j.$$

Thus $p_j(x)$ solves the interpolation problem for the simple data

$$y_1 = 0, \cdots, y_{j-1} = 0, \quad y_j = 1, \quad y_{j+1} = 0, \cdots, y_n = 0.$$

Now we superpose these elementary solutions to solve the problem for the general data

$$(x_1, y_1), \quad (x_2, y_2), \quad \cdots, (x_n, y_n).$$

We set $$p(x) = y_1 p_1(x) + y_2 p_2(x) + \cdots + y_n p_n(x).$$

This is it! Check:

$$p(x_1) = y_1 \cdot 1 + y_2 \cdot 0 + \cdots + y_n \cdot 0 = y_1,$$

$$p(x_2) = y_1 \cdot 0 + y_2 \cdot 1 + \cdots + y_n \cdot 0 = y_2,$$

$$\cdot \quad \cdot \quad \cdot \quad \cdot \quad \cdot \quad \cdot \quad \cdot \quad \cdot \quad \cdot \quad \cdot \quad \cdot \quad \cdot \quad \cdot$$

$$p(x_n) = y_1 \cdot 0 + y_2 \cdot 0 + \cdots + y_n \cdot 1 = y_n.$$

Since $p(x)$ is the sum of $(n - 1)$-th degree polynomials, $p(x)$ has degree at most $n - 1$ (some terms may drop out). The formula for $p(x)$ is the **Lagrange interpolation formula**.

Lagrange Interpolation Formula   The polynomial of degree at most $n - 1$ that fits n points

$$(x_1, y_1), \quad (x_2, y_2), \quad \cdots, (x_n, y_n),$$

with $x_1 < x_2 < \cdots < x_n$, is

$$p(x) = y_1 p_1(x) + y_2 p_2(x) + \cdots + y_n p_n(x),$$

where for $j = 1, \cdots, n$,

$$p_j(x) = \frac{(x - x_1)(x - x_2) \cdots (x - x_{j-1})(x - x_{j+1}) \cdots (x - x_n)}{(x_j - x_1)(x_j - x_2) \cdots (x_j - x_{j-1})(x_j - x_{j+1}) \cdots (x_j - x_n)}.$$

■ **EXAMPLE 1**   Fit a cubic to $(1, 4), \quad (2, -1), \quad (3, 0), \quad (4, 1)$.

Solution Compute the polynomials $p_j(x)$ for $j = 1, 2, 3, 4$:

$$q_1(x) = (x-2)(x-3)(x-4), \quad q_1(1) = -6, \quad p_1(x) = -\tfrac{1}{6}q_1(x),$$
$$q_2(x) = (x-1)(x-3)(x-4), \quad q_2(2) = 2, \quad p_2(x) = \tfrac{1}{2}q_2(x),$$
$$q_3(x) = (x-1)(x-2)(x-4), \quad q_3(3) = -2, \quad p_3(x) = -\tfrac{1}{2}q_3(x),$$
$$q_4(x) = (x-1)(x-2)(x-3), \quad q_4(4) = 6, \quad p_4(x) = \tfrac{1}{6}q_4(x).$$

By the Lagrange Interpolation Formula, the answer is

$$p(x) = 4p_1(x) - p_2(x) + p_4(x) = -\tfrac{2}{3}(x-2)(x-3)(x-4) - \tfrac{1}{2}(x-1)(x-3)(x-4)$$
$$+ \tfrac{1}{6}(x-1)(x-2)(x-3). \qquad \blacksquare$$

Remark We did not need $p_3(x)$ because $y_3 = 0$. Actually it would have been wise at the outset to note that $p(x) = (x-3)r(x)$, where

$$r(1) = \frac{p(1)}{1-3} = -2, \quad r(2) = \frac{p(2)}{2-3} = 1, \quad r(4) = \frac{p(4)}{4-3} = 1.$$

This reduces the problem to fitting a quadratic to $(1, -2)$, $(2, 1)$, $(4, 1)$. Hindsight! Also an excuse for testing the method on unequally spaced points, as we do next.

■ **EXAMPLE 2** Fit a quadratic $r(x)$ to $(1, -2)$, $(2, 1)$, $(4, 1)$.

Solution
$$q_1(x) = (x-2)(x-4), \quad q_1(1) = 3, \quad p_1(x) = \tfrac{1}{3}q_1(x),$$
$$q_2(x) = (x-1)(x-4), \quad q_2(2) = -2, \quad p_2(x) = -\tfrac{1}{2}q_2(x),$$
$$q_3(x) = (x-1)(x-2), \quad q_3(4) = 6, \quad p_3(x) = \tfrac{1}{6}q_3(x).$$

By Lagrange Interpolation,

$$r(x) = -2p_1(x) + p_2(x) + p_3(x)$$
$$= -\tfrac{2}{3}(x-2)(x-4) - \tfrac{1}{2}(x-1)(x-4) + \tfrac{1}{6}(x-1)(x-2). \qquad \blacksquare$$

Remark The answer multiplied by $x - 3$ is the answer to Example 1, as predicted.

Uniqueness Lagrange Interpolation shows that there always exists a polynomial of degree $n - 1$ whose graph passes through any n points

$$(x_1, y_1), \quad (x_2, y_2), \cdots, (x_n, y_n)$$

provided $x_1, x_2, \cdots, x_n$ are distinct. But is there only one? If we use undetermined coefficients or Newton Interpolation, do we find others? The answer is that there is exactly one polynomial that fits the points. No matter how you find it, you've got it.

Existence and Uniqueness of Interpolation Polynomials
Given n points
$$(x_1, y_1), \quad (x_2, y_2), \cdots, (x_n, y_n),$$
where $x_1 < x_2 < \cdots < x_n$, there exists a unique polynomial $p(x)$ of degree at most $n - 1$ that passes through these points.

In the particular case $n = 2$, this amounts to saying there is one and only one straight line (linear function) through two points.

To prove uniqueness in general, suppose there were two polynomials, $p(x)$ and $q(x)$, that interpolate the given points. Set $r(x) = p(x) - q(x)$. Then

$$r(x_j) = p(x_j) - q(x_j) = y_j - y_j = 0.$$

for $j = 1, 2, \cdots, n$. Consequently $r(x)$ is a polynomial of degree at most $n - 1$ with n distinct zeros. This implies $r(x) = 0$, because $r(x)$ must be divisible by $(x - x_1)(x - x_2) \cdots (x - x_n)$, a polynomial of too high degree to divide $r(x)$ unless $r(x) = 0$. Therefore $p(x) = q(x)$.

Error in Interpolation An important application of interpolation is the approximation of functions. The idea is to approximate a given function on an interval $[a, b]$ by an interpolating polynomial that agrees with the function at n values of x in $[a, b]$. The following estimate gives a bound for the worst possible error in this type of approximation.

Error in Interpolation Let $f(x)$ be an n-times differentiable function on $[a, b]$ and suppose

$$|f^{(n)}(x)| \le M \qquad \text{for} \quad a \le x \le b.$$

Let $p(x)$ be a polynomial of degree at most $n - 1$ such that

$$p(x_1) = f(x_1), \quad p(x_2) = f(x_2), \quad \cdots \quad , \quad p(x_n) = f(x_n),$$

where $x_1 < x_2 < \cdots < x_n$ are n given points in $[a, b]$. Then

$$|f(x) - p(x)| \le \frac{M}{n!} |(x - x_1) \cdots (x - x_n)|$$

for all x on the interval.

Proof Set $g(x) = f(x) - p(x)$. We must prove the inequality

$$|g(x)| \le \frac{M}{n!} |(x - x_1) \cdots (x - x_n)|, \qquad a \le x \le b.$$

We have some information about $g(x)$. First,

$$g(x_1) = g(x_2) = \cdots = g(x_n) = 0.$$

Second, $g^{(n)}(x) = f^{(n)}(x)$ because $\deg p(x) \le n - 1$. Therefore $|g^{(n)}(x)| \le M$.

The desired inequality is certainly true if $x = x_1, x_2, \cdots, x_n$. Suppose $x = x_0$, different from $x_1, x_2, \cdots, x_n$. Choose k so that

$$g(x_0) = k(x_0 - x_1)(x_0 - x_2) \cdots (x_0 - x_n),$$

and set

$$h(x) = g(x) - k(x - x_1)(x - x_2) \cdots (x - x_n).$$

Then $h(x)$ has $n + 1$ distinct zeros in the interval $[a, b]$, at $x_1, x_2, \cdots , x_n$ and at x_0. By the Generalized Rolle's Theorem (p. 470), $h^{(n)}(c) = 0$ for some c in the interval. Therefore

$$0 = h^{(n)}(c) = g^{(n)}(c) - (n!)k, \qquad k = \frac{g^{(n)}(c)}{n!}, \qquad |k| = \left| \frac{g^{(n)}(c)}{n!} \right| \leq \frac{M}{n!}.$$

It follows that
$$|g(x_0)| \leq \frac{M}{n!} |(x_0 - x_1) \cdots (x_0 - x_n)|.$$

But x_0 was arbitrary on the interval. This completes the proof.

■ **EXAMPLE 3** The polynomial $1 + \frac{1}{2}x + \frac{1}{2}x^2$ interpolates 2^x at $x = 0, 1, 2$. Show that

$$\left| 2^x - 1 - \tfrac{1}{2}x - \tfrac{1}{2}x^2 \right| < 0.09$$

for $0 \leq x \leq 2$.

Solution   Set $f(x) = 2^x$. Then $f'''(x) = (\ln 2)^3 2^x$, so for $0 \leq x \leq 2$,
$$|f'''(x)| \leq (\ln 2)^3 2^2.$$

Therefore $\left| 2^x - 1 - \tfrac{1}{2}x - \tfrac{1}{2}x^2 \right| \leq \dfrac{(4)(\ln 2)^3}{3!} |x(x-1)(x-2)|.$

To complete the estimate we need the maximum value of
$$P(x) = x(x-1)(x-2) = x^3 - 3x^2 + 2x$$
on the interval $[0, 2]$. Since $P(0) = P(2) = 0$, we rule out the end points and set $P'(x) = 0$:

$$3x^2 - 6x + 2 = 0, \qquad \text{hence,} \quad x = \frac{3 \pm \sqrt{3}}{3}.$$

A computation yields
$$P[\tfrac{1}{3}(3 \pm \sqrt{3})] = \mp \tfrac{2}{9}\sqrt{3}, \qquad \text{so} \quad |P(x)| \leq \tfrac{2}{9}\sqrt{3}.$$
Therefore $\left| 2^x - 1 - \tfrac{1}{2}x - \tfrac{1}{2}x^2 \right| \leq \tfrac{4}{27}\sqrt{3}\,(\ln 2)^3 < 0.086.$ ■

Remark   A careful calculation based on maximizing $\left| 2^x - 1 - \tfrac{1}{2}x - \tfrac{1}{2}x^2 \right|$ on $0 \leq x \leq 2$ gives the more accurate bound 0.05. This agrees well with the table on p. 483. Also see Exs. 29 and 30, p. 511.

EXERCISES

Use Lagrange interpolation to fit a polynomial to the data
1 $(-3, 4)$, $(-2, 0)$, $(0, -2)$, $(1, 1)$
2 $(-2, -2)$, $(-1, 2)$, $(0, 3)$, $(1, 0)$, $(2, 1)$
3 $(0, 8)$, $(2, 4)$, $(4, 2)$, $(6, 1)$
4 $(-3, -1)$, $(-1, 0)$, $(1, 0)$, $(3, 1)$
5 $(-2, -5)$, $(0, 2)$, $(2, 0)$, $(4, -2)$, $(6, 3)$
6 $(-1, 0)$, $(1, 0)$, $(2, 2)$, $(5, 0)$, $(8, -1)$, $(10, 0)$.

Use Lagrange interpolation to fit the function at the given points

7 $y = \sin \frac{1}{2}\pi x$, $x = 0, 1, 2, 3, 4$

8 $y = e^{-x} \cos \frac{1}{2}\pi x$, $x = 0, 1, 2, 3$

9 $y = (1 + \sin \pi x)(\cos \frac{2}{3}\pi x)$, $x = \frac{1}{2}, \frac{3}{4}, 1$

10 $y = x^2(x + 1)^2$, $x = -1, 0, 1, 2$.

11   By inspection prove the identity

$$x^3 = \frac{(x - 2)(x - 3)(x - 4)}{(1 - 2)(1 - 3)(1 - 4)} + 8 \frac{(x - 1)(x - 3)(x - 4)}{(2 - 1)(2 - 3)(2 - 4)} + 27 \frac{(x - 1)(x - 2)(x - 4)}{(3 - 1)(3 - 2)(3 - 4)}$$

$$+ 64 \frac{(x - 1)(x - 2)(x - 3)}{(4 - 1)(4 - 2)(4 - 3)}.$$

12   Suppose $x_1, x_2, \cdots, x_n$ are equally spaced and $p(x)$ is a quadratic polynomial that fits $(x_1, y_1), (x_2, y_2), \cdots, (x_n, y_n)$. Give the third row of the difference table whose zero-th row is $y_1, y_2, \cdots, y_n$.

13   Suppose $S = \{x_1, x_2, \cdots, x_n\}$ is a set of n distinct real numbers such that $S = \{-x_1, -x_2, \cdots, -x_n\}$. Suppose $f(x)$ is an even function and $p(x)$ is a polynomial of degree at most $n - 1$ that fits $f(x)$ at $x = x_1, x_2, \cdots, x_n$. Prove that $p(x)$ is also even.

14*   Suppose $f(x)$ has 4 derivatives on $[-3, 3]$ and $|f^{(4)}(x)| \le M$. Show that

$$\left| f(0) - \frac{1}{16}[-f(-3) + 9f(-1) + 9f(1) - f(3)] \right| \le \frac{3}{8}M.$$

15   The polynomial $(4/\pi^2)x(\pi - x)$ fits $\sin x$ at $x = 0, \frac{1}{2}\pi, \pi$. Estimate the error on $[0, \pi]$.

16*   The polynomial $(8/3\pi^3)x(\pi^2 - x^2)$ fits $\sin x$ at $x = -\pi, -\frac{1}{2}\pi, 0, \frac{1}{2}\pi, \pi$. Estimate the error on $[-\pi, \pi]$.

17*   Suppose $f(a) = f(b) = 0$ and $|f''(x)| \le M$ on $[a, b]$. Prove

$$\left| \int_a^b f(x)\, dx \right| \le \frac{M}{12}(b - a)^3.$$

18   (cont.) Prove the error estimate in the Trapezoidal Rule, p. 234.

9. APPROXIMATE INTEGRATION

In Chapter 5 we learned two methods of approximate integration, rectangular approximation and the more accurate Trapezoidal Rule:

$$\int_a^b f(x)\, dx \approx \frac{1}{2} h[f_0 + 2f_1 + 2f_2 + \cdots + 2f_{n-1} + f_n],$$

where $h = (b - a)/n$ and $f_j = f(a + jh)$. We also stated the error estimate

$$|\text{error}| \le \frac{M(b - a)}{12} h^2, \qquad M \ge |f''(x)|.$$

Simpson's Rule   When $n = 1$, the Trapezoidal Rule says

$$\int_a^b f(x)\, dx \approx \frac{b - a}{2} [f(a) + f(b)] = \int_a^b p_1(x)\, dx,$$

where $p_1(x)$ is the linear function that interpolates $f(x)$ at the two end points $(a, f(a))$ and $(b, f(b))$. For greater accuracy, we shall approximate $f(x)$ by a quadratic $p(x)$

that interpolates the function at the two end points and the midpoint:

$$(a, f(a)), \qquad \left(\frac{a+b}{2}, f\left(\frac{a+b}{2}\right)\right), \qquad (b, f(b)).$$

Then we shall approximate the integral of $f(x)$ by the integral of $p(x)$.
Let us use the notation

$$h = \frac{b-a}{2}, \qquad x_0 = a, \qquad x_1 = a + h, \qquad x_2 = a + 2h = b.$$

Given a quadratic polynomial $p(x)$, we need a formula for

$$\int_a^b p(x)\, dx$$

in terms of $p_0 = p(x_0)$, $p_1 = p(x_1)$, and $p_2 = p(x_2)$.

If $p(x)$ is a quadratic polynomial, then

$$\int_a^b p(x)\, dx = \tfrac{1}{3}h[p_0 + 4p_1 + p_2].$$

Proof   Write $p(x) = A + B(x - x_1) + C(x - x_1)^2$.

Then by symmetry, the left side is

$$\int_a^b p(x)\, dx = \int_{x_1 - h}^{x_1 + h} [A + C(x - x_1)^2]\, dx = 2Ah + \tfrac{2}{3}Ch^3.$$

The right side is

$$\tfrac{1}{3}h(p_0 + 4p_1 + p_2) = \tfrac{1}{3}h[(A - Bh + Ch^2) + 4A + (A + Bh + Ch^2)]$$
$$= \tfrac{1}{3}h(6A + 2Ch^2) = 2Ah + \tfrac{2}{3}Ch^3.$$

The two sides agree.

Remark   Both sides are zero, hence agree if $p(x) = D(x - x_1)^3$. We conclude that the formula is correct not only for quadratics, but also for cubics, an important bonus.

Given a continuous function $f(x)$ on $[a, b]$, we interpolate $f(x)$ by a quadratic $p(x)$ at $x = x_0$, x_1, and x_2. Then $p_0 = f_0$, $p_1 = f_1$, and $p_2 = f_2$. Since $p(x)$ is an approximation to $f(x)$, its integral is an approximation to the integral of $f(x)$. Therefore

$$\int_a^b f(x)\, dx \approx \int_a^b p(x)\, dx = \tfrac{1}{3}h[p_0 + 4p_1 + p_2] = \tfrac{1}{3}h[f_0 + 4f_1 + f_2].$$

Simpson's Three-Point Rule $\displaystyle\int_a^b f(x)\, dx \approx \tfrac{1}{3}h[f_0 + 4f_1 + f_2]$,

where $h = \tfrac{1}{2}(b - a)$ and

$$f_0 = f(a), \qquad f_1 = f(a + h), \qquad f_2 = f(a + 2h) = f(b).$$

Remark   As noted above, Simpson's Three-Point Rule is exact if $f(x)$ is a polynomial of degree three or less.

■ **EXAMPLE 1**   Use Simpson's Three-Point Rule to estimate

(a) $\displaystyle\int_0^\pi \sin x \, dx$ (b) $\displaystyle\int_0^1 e^x \, dx$ to 4 places. What is the error?

Solution   (a)   Here $h = \frac{1}{2}\pi$, so

$$\int_0^\pi \sin x \, dx \approx \tfrac{1}{3}(\tfrac{1}{2}\pi)(\sin 0 + 4 \sin \tfrac{1}{2}\pi + \sin \pi) = \tfrac{2}{3}\pi \approx 2.0944.$$

The exact value is $\displaystyle\int_0^\pi \sin x \, dx = -\cos x \,\Big|_0^\pi = 2.$

To 4 places the error is 0.0944.

(b)   In this case $h = \frac{1}{2}$, and

$$\int_0^1 e^x \, dx \approx \tfrac{1}{3}(\tfrac{1}{2})(1 + 4e^{1/2} + e) \approx 1.7189.$$

Exact value: $\displaystyle\int_0^1 e^x \, dx = e^x \,\Big|_0^1 = e - 1 \approx 1.7183.$

To 4 places the error is 0.0006.   ■

General Simpson's Rule   Does the Three-Point Rule always provide a good approximation to

$$\int_a^b f(x) \, dx \,?$$

That depends on the graph of $f(x)$. See Fig. 1.

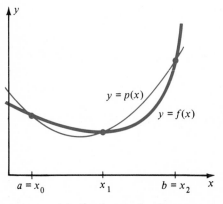

(a) Good approximation

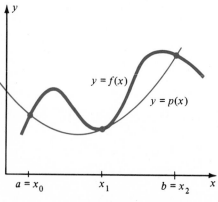

(b) Poor approximation

Fig. 1

The same interpolating quadratic $p(x)$ appears in Fig. 1a and Fig. 1b since the three points that determine $p(x)$ are the same. Yet the approximation

$$\int_a^b f(x)\,dx \approx \int_a^b p(x)\,dx$$

is good in Fig. 1a and poor in Fig. 1b.

In the first case, $p(x)$ approximates $f(x)$ closely. Furthermore,

$$\int_{x_0}^{x_1} p(x)\,dx < \int_{x_0}^{x_1} f(x)\,dx, \qquad \text{whereas} \qquad \int_{x_1}^{x_2} p(x)\,dx > \int_{x_1}^{x_2} f(x)\,dx,$$

so the errors tend to cancel.

In Fig. 1b, however, $p(x)$ is a poor approximation to $f(x)$. To make matters worse, the errors do not cancel, they accumulate. The trouble is that the points x_0, x_1, x_2 are too widely spaced. It is much better to apply the Three-Point Rule *twice*, once from x_0 to x_1 and once from x_1 to x_2 (Fig. 2).

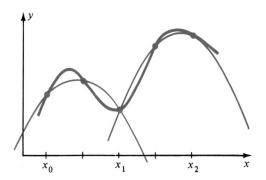

Fig. 2 Three-point rule applied twice; on $[x_0, x_1]$ and on $[x_1, x_2]$

In general, to improve the approximation we divide $[a, b]$ into an even number of equal pieces and apply Simpson's Three-Point Rule serially, taking three division points at a time.

Let us divide $[a, b]$ into $2n$ subintervals, each of length $h = (b - a)/2n$. We write

$$\int_a^b f(x)\,dx = \int_{x_0}^{x_2} + \int_{x_2}^{x_4} + \int_{x_4}^{x_6} + \cdots + \int_{x_{2n-2}}^{x_{2n}},$$

and apply three-point approximation to each integral on the right side:

$$\int_{x_0}^{x_2} f(x)\,dx \approx \frac{1}{3} h[f_0 + 4f_1 + f_2], \qquad \int_{x_2}^{x_4} f(x)\,dx \approx \frac{1}{3} h[f_2 + 4f_3 + f_4],$$

$$\int_{x_4}^{x_6} f(x)\,dx \approx \frac{1}{3} h[f_4 + 4f_5 + f_6], \ldots, \qquad \int_{x_{2n-1}}^{x_{2n}} f(x)\,dx \approx \frac{1}{3} h[f_{2n-2} + 4f_{2n-1} + f_{2n}].$$

Adding these estimates, we obtain Simpson's Rule.

Simpson's Rule

$$\int_a^b f(x)\,dx \approx \frac{1}{3} h[f_0 + 4f_1 + 2f_2 + 4f_3 + 2f_4 + 4f_5 + \cdots + 2f_{2n-2} + 4f_{2n-1} + f_{2n}],$$

where $h = (b - a)/2n$ and $f_0, f_1, \cdots, f_{2n}$ are the values of f at the successive points of division of $a \le x \le b$ into $2n$ equal parts. The rule is exact whenever $f(x)$ is a polynomial of degree 3 or less.

Note the sequence of coefficients:

$$1,\ 4,\ 2,\ 4,\ 2,\ 4,\ \cdots,\ 4,\ 2,\ 4,\ 1.$$

The number of subintervals of $a \le x \le b$ is $2n$ (even). The number of points of division, counting the end points, is $2n + 1$ (odd). These odd numbers are used to describe various versions of Simpson's Rule. Thus a 7-point Simpson approximation refers to the division shown here:

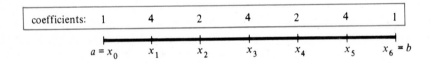

In this case $n = 3$, $2n = 6$, $2n + 1 = 7$, and the rule states

$$\int_a^b f(x)\,dx \approx \tfrac{1}{3}h[f_0 + 4f_1 + 2f_2 + 4f_3 + 2f_4 + 4f_5 + f_6].$$

■ **EXAMPLE 2** Use Simpson's Rule with 5 and with 11 points to estimate

$$\int_{-1}^1 x^4\,dx$$

to 4 places. Compare to the exact value.

Solution The exact value is

$$\int_{-1}^1 x^4\,dx = \tfrac{1}{5}x^5 \big|_{-1}^1 = \tfrac{2}{5} = 0.4000.$$

If $2n + 1 = 5$, then $n = 2$ and $h = (b - a)/2n = \tfrac{1}{2}$. The approximation is

$$\int_{-1}^1 x^4\,dx \approx \tfrac{1}{3}h[f_0 + 4f_1 + 2f_2 + 4f_3 + f_4]$$

$$= \tfrac{1}{6}[1 + 4(-\tfrac{1}{2})^4 + 0 + 4(\tfrac{1}{2})^4 + 1]$$

$$= \tfrac{1}{6} \cdot \tfrac{5}{2} = \tfrac{5}{12} \approx 0.4167.$$

Similarly, if $2n + 1 = 11$, then $n = 5$, $h = \frac{2}{10}$, and

$$\int_{-1}^{1} x^4 \, dx \approx \frac{1}{15}[1 + 4(-\frac{8}{10})^4 + 2(-\frac{6}{10})^4 + 4(-\frac{4}{10})^4 + 2(-\frac{2}{10})^4$$
$$+ 0 + 2(\frac{2}{10})^4 + 4(\frac{4}{10})^4 + 2(\frac{6}{10})^4 + 4(\frac{8}{10})^4 + 1]$$
$$= \frac{2}{15} \times 10^{-4}(10^4 + 4 \times 8^4 + 2 \times 6^4 + 4 \times 4^4 + 2 \times 2^4)$$
$$= \frac{2}{15} \times 10^{-4} \times 30032 \approx 0.4004.$$

The 5-point error is about 0.017 and the 11-point error is 0.0004.   ∎

It pays to exploit symmetry when applying Simpson's Rule. You can generally improve accuracy with no additional computation. Compare the following example to the last one.

∎ **EXAMPLE 3**   Use Simpson's Rule with 5 points to estimate $\displaystyle\int_{-1}^{1} x^4 \, dx$ to 4 places. Exploit symmetry.

Solution   The integrand is an even function, hence

$$\int_{-1}^{1} x^4 \, dx = 2\int_{0}^{1} x^4 \, dx.$$

Now use Simpson with 5 points on $[0, 1]$, which is practically as good as using Simpson with 9 points on $[-1, 1]$. The result is

$$\int_{-1}^{1} x^4 \, dx \approx \frac{2}{12}[0 + 4(\frac{1}{4})^4 + 2(\frac{1}{2})^4 + 4(\frac{3}{4})^4 + 1] \approx 0.4010.$$

The error is only 0.001, much smaller than the error 0.017 in Example 2 for the 5-point Simpson approximation.   ∎

A Further Example   The exact value of $\displaystyle\int_{0}^{2} e^{-x^2} \, dx$ is not known, but to 11 places it is

$$0.88208\ 10350\ 6.$$

In the following table we show the results of several trapezoidal and Simpson approximations to this integral.

Table 1

Trapezoidal Approximations				(2n + 1)-Point Simpson Approximations			
n	$\int_{0}^{2} e^{-x^2} \, dx$	n	$\int_{0}^{2} e^{-x^2} \, dx$	$2n + 1$	$\int_{0}^{2} e^{-x^2} \, dx$	$2n + 1$	$\int_{0}^{2} e^{-x^2} \, dx$
2	0.877037	8	0.881704	3	0.829944	9	0.882066
4	0.880618	10	0.881837	5	0.881812	11	0.882073
6	0.881415	20	0.882020	7	0.882031	21	0.882081

Table 1 shows that Simpson's Rule is much more precise than the Trapezoidal Rule. For instance, with 5 points of subdivision, the errors are approximately 0.001463 (Trapezoidal) versus 0.000269 (Simpson); with 11 points of subdivision, 0.000244 (Trapezoidal) versus 0.000008 (Simpson).

An Application to Volume   A **prismoid** is a region in space bounded by two parallel planes and one or more surfaces joining the planes (Fig. 3). There is a nice formula for the volume V of a prismoid whose cross-sectional area $f(x)$ is a *cubic* function of the height x, measured along an axis perpendicular to the base planes:

$$V = \frac{h}{6}(A_0 + 4M + A_1),$$

where h is the distance between base planes, A_0 and A_1 are the areas of the bases, and M is the cross-sectional area halfway between the bases. This is sometimes called the **Prismoidal Formula**. The proof is a direct application of Simpson's Rule. See Ex. 45. Note that "cubic" includes linear and quadratic polynomials, the usual cases in applications of the Prismoidal Formula.

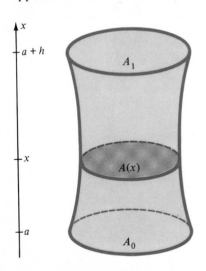

Fig. 3   Prismoid: $A(x) =$ a cubic in x.

■ **EXAMPLE 4** Find the volume of a sphere of radius a by the Prismoidal Formula.

Solution   The cross-sectional area is a quadratic function of the height, hence the formula applies with

$$h = 2a, \qquad A_0 = A_1 = 0, \qquad M = \pi a^2.$$

Conclusion: $$V = \frac{2a}{6}(0 + 4\pi a^2 + 0) = \frac{4}{3}\pi a^3.$$ ■

Error in Simpson's Rule   Simpson's Rule is exact for cubic polynomials. Now $f(x)$ is a cubic if and only if $f^{(4)}(x) = 0$. Hence $f^{(4)}(x)$ measures how far $f(x)$ differs from being a cubic, so it is reasonable that $|f^{(4)}(x)|$ should appear in error estimates for Simpson's Rule. The following estimate for the Three-Point Rule is proved in numerical analysis. (See Exs. 53 and 54.)

Error in Simpson's Rule   Suppose $f(x)$ has a continuous fourth derivative on $[a, b]$, and $|f^{(4)}(x)| \leq M$. Then

$$\int_a^b f(x)\,dx = \frac{h}{3}\,(f_0 + 4f_1 + f_2) + \varepsilon, \qquad \text{where} \quad |\varepsilon| \leq \frac{Mh^5}{90}.$$

To obtain an error estimate for the General Simpson's Rule, we divide the interval into $2n$ parts, and apply the preceding estimate n times taking $h = (b - a)/2n$. This yields

$$|\text{error}| \leq \frac{Mnh^5}{90} = \frac{M}{90 \times 2^5}\,\frac{(b - a)^5}{n^4}.$$

Thus the maximum error is proportional to $1/n^4$. That is why a large n yields high accuracy.

■ **EXAMPLE 5**   Estimate ln 2 by Simpson's Rule with $h = 0.1$. Find a bound for the error.

Solution   Write $\ln 2 = \displaystyle\int_1^2 \frac{dx}{x} = \int_{1.0}^{1.2} + \int_{1.2}^{1.4} + \int_{1.4}^{1.6} + \int_{1.6}^{1.8} + \int_{1.8}^{2.0},$

and apply the three-point Simpson's Rule to each integral on the right. In each case the error is at most

$$\frac{M(0.1)^5}{90},$$

where M is a bound for the fourth derivative of $f(x) = 1/x$. Since

$$f^{(4)}(x) = \frac{2 \cdot 3 \cdot 4}{x^5} \qquad \text{and} \qquad 1 \leq x \leq 2,$$

take $M = 2 \cdot 3 \cdot 4 = 24$. Thus each of the 5 errors is at most

$$\frac{24(0.1)^5}{90},$$

and the combined error is at most 5 times this much:

$$|\text{error}| < 5 \times \frac{24(0.1)^5}{90} < 1.4 \times 10^{-5}.$$

So much for the accuracy. The estimate is

$$\ln 2 \approx \frac{0.1}{3}\left(\frac{1}{1} + \frac{4}{1.1} + \frac{2}{1.2} + \frac{4}{1.3} + \frac{2}{1.4} + \frac{4}{1.5} + \frac{2}{1.6} + \frac{4}{1.7} + \frac{2}{1.8} + \frac{4}{1.9} + \frac{1}{2}\right)$$

$$\approx 0.693150.$$

To 6 places, ln 2 ≈ 0.693147, so the *actual* error is smaller than 4×10^{-6}.   ■

Gaussian Quadrature In addition to the Trapezoidal Rule and Simpson's Rule, there are many other rules for approximate integration based on interpolation. We shall look at the easiest case of a series of rules that go under the name **Gaussian Quadrature.**

For simplicity of notation, we shall work only on the interval $[-1, 1]$. Any other interval can be transformed into this one by shifting and changing scale, so there is no real loss of generality.

Simpson's Rule is based on quadratic interpolation through the points

$$(-1, f(-1)), \qquad (0, f(0)), \qquad (1, f(1)),$$

and is exact for cubics. The Trapezoidal Rule is based on linear interpolation through the points

$$(-1, f(-1)) \qquad \text{and} \qquad (1, f(1)),$$

and is exact for linear functions. We now ask: Is it possible that *linear* interpolation through *two other points* could produce a rule that is exact for quadratics at least?

Two other points would be

$$(\alpha, f(\alpha)) \qquad \text{and} \qquad (\beta, f(\beta)),$$

where $-1 \leq \alpha < \beta \leq 1$. The linear function through these points is

$$g(x) = f(\alpha) + \frac{f(\beta) - f(\alpha)}{\beta - \alpha}(x - \alpha) = \frac{\beta f(\alpha) - \alpha f(\beta)}{\beta - \alpha} + \frac{f(\beta) - f(\alpha)}{\beta - \alpha} x,$$

and it follows easily that

$$\int_{-1}^{1} g(x)\, dx = 2\, \frac{\beta f(\alpha) - \alpha f(\beta)}{\beta - \alpha}.$$

Therefore we obtain the approximation

$$\int_{-1}^{1} f(x)\, dx \approx 2\, \frac{\beta f(\alpha) - \alpha f(\beta)}{\beta - \alpha},$$

which is exact for linear functions.

Can we make this approximation exact for quadratics as well? It will be enough to make it exact for $f(x) = x^2$. Now

$$\int_{-1}^{1} x^2\, dx = \frac{2}{3} \qquad \text{and} \qquad 2\, \frac{\beta \alpha^2 - \alpha \beta^2}{\beta - \alpha} = -2\alpha\beta,$$

so the condition for equality is $-2\alpha\beta = \frac{2}{3}$, that is, $\alpha = -1/3\beta$. Since $-1 \leq \alpha < \beta \leq 1$, we may choose β to be any number satisfying $\frac{1}{3} \leq \beta \leq 1$. Once β is chosen, α is determined (and satisfies $-1 \leq \alpha \leq -\frac{1}{3}$).

To summarize the discussion: if $\frac{1}{3} \leq \beta \leq 1$ and $\alpha = -1/3\beta$, then the approximation

$$\int_{-1}^{1} f(x)\, dx \approx 2\, \frac{\beta f(\alpha) - \alpha f(\beta)}{\beta - \alpha}$$

is exact for quadratic functions.

We still have some freedom in the choice of β. So why not shoot the works and see if we can make the approximation exact for *cubics*? It will suffice to make it exact for $f(x) = x^3$. Now

$$\int_{-1}^{1} x^3 \, dx = 0 \qquad \text{and} \qquad 2\frac{\beta\alpha^3 - \alpha\beta^3}{\beta - \alpha} = -2\alpha\beta(\alpha + \beta),$$

so the condition for equality is $\alpha + \beta = 0$, that is, $\alpha = -\beta$. But $\alpha\beta = -\frac{1}{3}$, hence $\alpha^2 = \beta^2 = \frac{1}{3}$. Therefore we choose $\beta = \frac{1}{3}\sqrt{3}$ and $\alpha = -\frac{1}{3}\sqrt{3}$, so

$$2\frac{\beta f(\alpha) - \alpha f(\beta)}{\beta - \alpha} = 2\frac{\frac{1}{3}\sqrt{3}\,f(-\frac{1}{3}\sqrt{3}) + \frac{1}{3}\sqrt{3}\,f(\frac{1}{3}\sqrt{3})}{\frac{2}{3}\sqrt{3}} = f(-\frac{1}{3}\sqrt{3}) + f(\frac{1}{3}\sqrt{3}).$$

> **Gaussian Quadrature**   If $f(x)$ is continuous on $[-1, 1]$, then
>
> $$\int_{-1}^{1} f(x) \, dx \approx f(-\frac{1}{3}\sqrt{3}) + f(\frac{1}{3}\sqrt{3}).$$
>
> The approximation is exact if $f(x)$ is a cubic polynomial.

■ **EXAMPLE 6**   Estimate $\qquad \displaystyle\int_{-1}^{1} \frac{dx}{1 + x^2}$

by Gaussian Quadrature, and compute the error.

Solution   Set $f(x) = 1/(1 + x^2)$, an even function. The estimate is

$$\int_{-1}^{1} \frac{dx}{1 + x^2} \approx f(\frac{1}{3}\sqrt{3}) + f(-\frac{1}{3}\sqrt{3}) = 2f(\frac{1}{3}\sqrt{3}) = \frac{2}{1 + (\frac{1}{3}\sqrt{3})^2} = \frac{2}{1 + \frac{1}{3}} = \frac{3}{2} = 1.5.$$

The exact value is $2 \arctan 1 = 2 \times \frac{1}{4}\pi = \frac{1}{2}\pi$; the error is $\frac{1}{2}\pi - 1.5 \approx 0.07$.   ■

Like Simpson's Rule, Gaussian Quadrature is exact for cubics; hence the error must depend on the size of $f^{(4)}(x)$. The following error estimate is proved in numerical analysis. (See Exs. 55 and 56.)

> **Error in Gaussian Quadrature**   Suppose $f(x)$ has a continuous fourth derivative on $[-1, 1]$ and $|f^{(4)}(x)| \leq M$. Then
>
> $$\int_{-1}^{1} f(x) \, dx = f(-\frac{1}{3}\sqrt{3}) + f(\frac{1}{3}\sqrt{3}) + \varepsilon, \qquad \text{where} \qquad |\varepsilon| \leq \frac{M}{135}.$$

■ **EXAMPLE 7**   Check the error estimate for $f(x) = x^4$.

Solution $\displaystyle\int_{-1}^{1} x^4 \, dx = \frac{2}{5}$ and $\left(-\frac{1}{3}\sqrt{3}\right)^4 + \left(\frac{1}{3}\sqrt{3}\right)^4 = \frac{2}{9},$

so the actual error is $\frac{2}{5} - \frac{2}{9} = \frac{8}{45}$. In this case, $f^{(4)}(x) = 4! = 24$, so the theoretical maximum error is

$$\frac{M}{135} = \frac{24}{135} = \frac{8}{45},$$

exactly equal to the error.   ■

Remark Generally the actual error is smaller than $M/135$. If $f(x) = e^x$, for example, the error is about 40% of $M/135$. See Ex. 40.

EXERCISES

Estimate to 4 places by Simpson's Three-Point Rule, and give $|\text{error}|$

1 $\displaystyle\int_0^{\pi/2} \sin x \, dx$ **2** $\displaystyle\int_{\pi/2}^{\pi} \sin x \, dx$ **3** $\displaystyle\int_{-\pi/4}^{\pi/4} \cos x \, dx$

4 $\displaystyle\int_{-1}^{0} e^x \, dx$ **5** $\displaystyle\int_{1}^{2} e^x \, dx$ **6** $\displaystyle\int_0^1 \frac{dx}{1+x^2}$

7 $\displaystyle\int_1^2 \frac{dx}{x^2}$ **8** $\displaystyle\int_{-1}^{1} x^5 \, dx.$

[In Exs. 9–26, remember that Simpson's Rule with n implies $2n$ subintervals.]

Estimate to 5 places by Simpson's Rule with $n = 2$ and $n = 4$

9 $\displaystyle\int_0^1 e^x \, dx$ **10** $\displaystyle\int_0^{\pi} \sin x \, dx$ **11** $\displaystyle\int_0^1 x^9 \, dx$

12 $\displaystyle\int_0^2 \sqrt{1+x^3} \, dx$ **13** $\displaystyle\int_1^3 \sqrt[4]{1+x^2} \, dx$ **14** $\displaystyle\int_1^2 \frac{x^2 \, dx}{1+x^4}$

15 $\displaystyle\int_{-1}^{1} \sin x^2 \, dx$ **16** $\displaystyle\int_{-\pi/2}^{\pi/2} \frac{\sin x}{x} \, dx$ **17** $\displaystyle\int_1^4 e^{1/x} \, dx$

18 $\displaystyle\int_{\pi}^{2\pi} \frac{\cos x}{x} \, dx$ **19** $\displaystyle\int_1^3 \sqrt{1+e^x} \, dx$ **20** $\displaystyle\int_1^3 e^{\sqrt{x}} \, dx.$

Estimate to 7 places with $n = 10$

21 $\displaystyle\int_0^1 e^x \, dx$ **22** $\displaystyle\int_0^{\pi} \sin x \, dx.$

23 Compute exactly the Simpson approximation to $(1/\pi)\int_0^{4\pi} \sqrt{4 + \sin x} \, dx$ with $n = 2$.
24 (cont.) Exploit symmetry in the same problem.
25 Using error estimates, show that the error in Ex. 21 is at most 10^{-7}.
26 Show that the error estimate in Example 5 can be improved to about 4.7×10^{-6} by using the factor $1/x^5$ carefully, not just bounding it by 1.

Estimate to 4 places by Gaussian Quadrature

27 $\displaystyle\int_{-1}^{1} e^x \, dx$ **28** $\displaystyle\int_{-1}^{1} \sqrt{x^2+1} \, dx$ **29** $\displaystyle\int_{-1}^{1} x^6 \, dx$ **30** $\displaystyle\int_{-1}^{1} \frac{x^2}{1+x^4} \, dx.$

31 Show that the Gaussian Quatrature formula for the interval $[c - h, c + h]$ is

$$\int_{c-h}^{c+h} f(x) \, dx \approx h[f(c + \tfrac{1}{3}h\sqrt{3}) + f(c - \tfrac{1}{3}h\sqrt{3})].$$

32 (cont.) Find the corresponding error estimate.

Divide each interval into 4 equal parts and apply the Gaussian Quadrature of Ex. 31 to each part, then sum the results. Carry your work to 6 places

33 $\displaystyle\int_0^1 e^x \, dx$ **34** $\displaystyle\int_0^{\pi} \sin x \, dx$ **35** $\displaystyle\int_1^5 \sqrt[3]{1+x} \, dx$

36 $\displaystyle\int_{-\pi/2}^{\pi/2} \frac{\sin x}{x}\,dx$ **37** $\displaystyle\int_{1}^{4} e^{1/x}\,dx$ **38** $\displaystyle\int_{1}^{3} e^{\sqrt{x}}\,dx.$

39 The integral in Ex. 29 equals $\frac{2}{7} \approx 0.286$. The answer to the exercise is 0.0741. Why is it so bad? Does use of Ex. 31 and symmetry help?

40 Compute the ratio of the actual error in Ex. 27 to the error estimate $M/135$.

Show that the Prismoidal Formula applies to the solid of revolution and compute the volume

41 the region bounded by $x = y^2$ and $x = a$ revolved about the x-axis

42 the region bounded by the x-axis, $y = x^{3/2}$, and $x = a$ revolved about the x-axis

43 the ellipse $\dfrac{x^2}{a^2} + \dfrac{y^2}{b^2} = 1$ revolved about the x-axis

44 the region bounded by the hyperbola $-\dfrac{x^2}{a^2} + \dfrac{y^2}{b^2} = 1$, $x = -c$, and $x = c$, revolved about the x-axis.

45 Prove the Prismoidal Formula.

46 Use the error estimate to find the least n for which it is safe to estimate

$$\frac{1}{4}\pi \approx \int_{0}^{1} \frac{dx}{1 + x^2}$$

to 5 places by Simpson's Rule.

47* Suppose $f(x) + f(b - x) = 1$ for $0 \le x \le b$. Show that Simpson's Rule with any n is exact for the integral of $f(x)$ on $[0, b]$.

48* (cont) Apply this idea to evaluate

$$\int_{0}^{\pi/2} \frac{\sin x}{\sin(x + \frac{1}{4}\pi)}\,dx \quad \text{and} \quad \int_{0}^{\pi/2} \frac{(x - \frac{1}{4}\pi)^2 \sin x}{\sin x + \cos x}\,dx.$$

The remaining exercises concern proofs of error estimates and are necessarily harder.

49 Suppose $f''(x)$ is continuous on $[a, b]$ and $f(a) = f(b) = 0$. Prove

$$\int_{a}^{b} f(x)\,dx = -\frac{1}{2}\int_{a}^{b} (x - a)(b - x)f''(x)\,dx.$$

50 (cont.) Suppose also $|f''(x)| \le M$. Prove $\left|\int_{a}^{b} f(x)\,dx\right| \le \frac{1}{12}M(b - a)^2$.

51* Sharpen the reasoning used to prove Error in Interpolation (p. 492) to prove the following: if $a < c < b$, $f(a) = f(c) = f(b) = f'(c) = 0$, and $|f^{(4)}(x)| \le M$ on $[a, b]$, then

$$|f(x)| \le \tfrac{1}{24}M\,|(x - a)(x - c)^2(x - b)|.$$

52* (cont.) Prove the error estimate in Simpson's Rule (p. 501).

53* Suppose $f(a) = f'(a) = f(b) = f'(b) = 0$, that $f^{(4)}(x)$ is continuous on $[a, b]$, and $|f^{(4)}(x)| \le M$. Use the method indicated in Exs. 51 and 52 to prove

$$\left|\int_{a}^{b} f(x)\,dx\right| \le \frac{M}{720}(b - a)^5.$$

54* The midpoint approximation is

$$\int_{a}^{b} f(x)\,dx \approx (b - a)f\left[\tfrac{1}{2}(a + b)\right].$$

Suppose $|f''(x)| \le M$ on $[a, b]$. Prove that $\qquad |error| \le \dfrac{M}{24}(b - a)^3.$

[*Hint* Set $c = \frac{1}{2}(a + b)$. Subtract a suitable linear function from $f(x)$ so the result $g(x)$ has $g(c) = g'(c) = 0$.]

55* Let $0 < c \leq 1$, $3c^2 \neq 1$. Suppose $f(\frac{1}{3}\sqrt{3}) = f(-\frac{1}{3}\sqrt{3}) = 0$. Show that there is a cubic $p(x)$ such that if $g(x) = f(x) - p(x)$, then

$$\int_{-1}^{1} f(x)\,dx = \int_{-1}^{1} g(x)\,dx, \qquad g(\tfrac{1}{3}\sqrt{3}) = g(-\tfrac{1}{3}\sqrt{3}) = 0,$$

and $g(c) = g(-c) = 0$.

56* (cont.) Suppose also $|f^{(4)}(x)| \leq M$ on $[-1, 1]$. Obtain the error in Gaussian Quadrature (p. 503). [*Hint* See Ex. 53.]

10. ROOT APPROXIMATION AND HILL CLIMBING

One of the bread-and-butter problems of mathematics is solving equations, or at least approximating solutions to high accuracy. Suppose we want to solve an equation

$$f(x) = 0,$$

where $f(x)$ is continuous on an interval $[a, b]$, and somehow we know there is at least one solution in the interval. How do we go about finding a solution to say 5 decimal places?

One possibility is to use the Intermediate Value Theorem: if $f(c) < 0$ and $f(d) > 0$, then $f(x) = 0$ somewhere in the interval $[c, d]$. Hence if we can find a tiny interval of this type, we shall have a solution nearly pinpointed. These remarks suggest a brute-force way of squeezing down on a solution in a sequence of steps, each of which adds one additional decimal place accuracy to the preceding estimate.

For instance, suppose $f(4) < 0$ and $f(5) > 0$. Then $f(x)$ has a zero between 4 and 5. We compute

$$f(4.1), \quad f(4.2), \quad f(4.3), \cdots$$

until we find the first positive one. This will require from one to nine computations of $f(x)$. Suppose we find

$$f(4.1) < 0, \quad f(4.2) < 0, \cdots, f(4.7) < 0, \quad \text{but} \quad f(4.8) > 0.$$

Then there is a zero in $[4.7, 4.8]$. We repeat the process relative to this interval, computing

$$f(4.71), \quad f(4.72), \quad f(4.73), \cdots$$

until the first positive value, again as many as nine evaluations. This time we might find

$$f(4.74) < 0 \quad \text{and} \quad f(4.75) > 0.$$

Then there is a zero in $[4.74, 4.75]$. If we stop here, we choose either $x \approx 4.74$ or $x \approx 4.75$ depending on which of $|f(4.74)|$ or $|f(4.75)|$ is smaller.

Continuing this way, we can grind out one additional decimal place accuracy at a time, but at the cost of computing as many as nine values of $f(x)$ in each step. The method works, but has a Neanderthal quality.

■ **EXAMPLE 1** Estimate to 4 places the solution of $\cos x = x$.

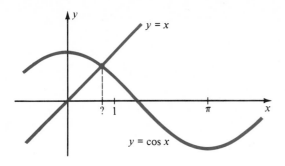

Fig. 1 Intersection of $y = \cos x$ and $y = x$.

Solution A rough sketch (Fig. 1) shows that the graphs $y = x$ and $y = \cos x$ intersect for a value of x between 0 and 1. We set $f(x) = x - \cos x$. Then

$$f(0) = -1 < 0 \qquad \text{and} \qquad f(1) = 1 - \cos 1 > 0,$$

so we are ready to start. By tables or a calculator,

$$f(0.7) \approx -0.065 < 0 \qquad \text{and} \qquad f(0.8) \approx 0.103 > 0.$$

At the next step,

$$f(0.73) \approx -0.015 < 0 \qquad \text{and} \qquad f(0.74) \approx 0.002 > 0.$$

Twice more:

$$f(0.739) \approx -0.00014 < 0 \qquad \text{and} \qquad f(0.740) \approx 0.00153 > 0$$

$$f(0.7390) \approx -0.00014 < 0 \qquad \text{and} \qquad f(0.7391) \approx 0.00002 > 0.$$

We conclude that $x \approx 0.7391$ is accurate to 4 places.   ■

Remark 1   The result of the second step is $f(0.73) \approx -0.015$ and $f(0.74) \approx 0.002$. Since $0.002 < |-0.015|$, it seems more likely at this stage that the zero of $f(x)$ in the interval $[0.73, 0.74]$ is closer to 0.74 than to 0.73. Therefore our next tests should be $f(0.739), f(0.738), \cdots$ rather than $f(0.731), f(0.732), \cdots$.

In general, it is good computing practice to start at the point where $|f(x)|$ is least and work toward the other point. Instead of $\frac{1}{2} \times 9 = 4\frac{1}{2}$ tests, you can expect about $\frac{1}{2} \times 5 = 2\frac{1}{2}$ tests on the average to find a sign change.

Remark 2   The method in Example 1 is a modification of the method called **Bisection**, in which the interval is split into two, not ten, parts. If $f(a) > 0$ and $f(b) < 0$, you compute $f(x)$ at the midpoint $c = \frac{1}{2}(a + b)$. If $f(c) < 0$, you next check the midpoint of $[c, b]$; if $f(c) > 0$, you next check the midpoint of $[a, c]$, etc. This method does not fit as well with our usual representation of numbers by decimals as does the method used above, which might be called **10-section**.

The Secant Method   Our next method is based on linear interpolation. Suppose that $f(a) < 0$ and $f(b) > 0$ and that $f(x)$ increases on $[a, b]$. The linear function through $(a, f(a))$ and $(b, f(b))$ is an approximation to $f(x)$ on $[a, b]$, so its zero should be a reasonable approximation to the zero of $f(x)$. See Fig. 2.

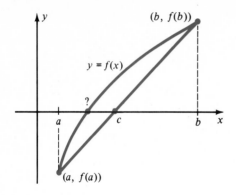

Fig. 2 Secant method: The line through $(a, f(a))$ and $(b, f(b))$ determines c, an approximation to a zero of $f(x)$. As drawn, $f(c) > 0$, so the next approximation is based on $(a, f(a))$ and $(c, f(c))$.

Call this zero c. To compute c, note that the interpolating linear function is

$$y = f(a) + \frac{f(b) - f(a)}{b - a}(x - a).$$

Set $y = 0$ and $x = c$, and solve for c:

$$0 = f(a) + \frac{f(b) - f(a)}{b - a}(c - a), \qquad c - a = -f(a)\frac{b - a}{f(b) - f(a)},$$

$$c = a - f(a)\frac{b - a}{f(b) - f(a)} = \frac{a[f(b) - f(a)] - f(a)(b - a)}{f(b) - f(a)},$$

$$\boxed{c = \frac{af(b) - bf(a)}{f(b) - f(a)}.}$$

The computation of c does not involve any *new* values of $f(x)$; presumably you have computed $f(a)$ and $f(b)$ earlier to test their signs. This is a labor saving feature of the method.

Having c, compute $f(c)$. If $f(c) < 0$, repeat the process on the interval $[c, b]$; if $f(c) > 0$, repeat the process on $[a, c]$. Continue until you reach a c such that $f(c) \approx 0$ to the desired degree of accuracy.

Remark For historical reasons, this method is known as the **Method of False Position**. The name **Secant Method** generally denotes a slightly modified process; however, we shall use this latter name here.

■ **EXAMPLE 2** Find an x by the Secant Method such that $\cos x \approx x$ to 4-place accuracy.

Solution Set $f(x) = x - \cos x$. Then $f(0) = -1$, $f(1) \approx 0.4597$, so apply the Secant Method to $[a, b] = [0, 1]$:

$$c = \frac{0 - f(0)}{f(1) - f(0)} = \frac{1}{0.4597 + 1} \approx 0.6851.$$

Since $$f(c) \approx -0.0893 < 0,$$

the solution is in the interval $[c, 1]$.

The new c is

$$c_1 = \frac{cf(1) - f(c)}{f(1) - f(c)} \approx \frac{(0.6851)(0.4597) + 0.0893}{0.4597 + 0.0893} \approx 0.7363.$$

Since
$$f(c_1) \approx -0.0047 < 0,$$

the solution is in $[c_1, 1]$. Next

$$c_2 = \frac{c_1 f(1) - f(c_1)}{f(1) - f(c_1)} \approx 0.7390.$$

Since
$$f(c_2) \approx -0.0001 < 0,$$

the solution is in $[c_2, 1]$. Finally,

$$c_3 = \frac{c_2 f(1) - f(c_2)}{f(1) - f(c_2)} \approx 0.7391, \qquad f(c_3) \approx 0.0000,$$

so $x \approx 0.7391$ is an answer. ∎

Remark In working Example 2 by the Secant Method, we evaluated $f(x)$ at 6 points: 0, 1, c, c_1, c_2, c_3. In Example 1, to get the same accuracy by the 10-section method, we needed evaluations at

$$0, \quad 1, \quad 0.1, \quad 0.2, \cdots, 0.8, \quad 0.71, \quad 0.72, \quad 0.73, \quad 0.74,$$
$$0.731, \quad 0.732, \cdots, 0.739, \quad 0.7391,$$

a total of 24 points. (However, if we follow the suggestion of Remark 1, p. 507, then 11 evaluations suffice.)

We end our discussion here. More sophisticated methods of root finding based on interpolation exist, but they are generally less satisfactory than the methods we shall discuss in the next section. Also, error estimates for interpolation-based methods are difficult to obtain.

Hill Climbing Another bread-and-butter problem of calculus, frequently called **optimization**, is to maximize a function $f(x)$ on an interval $a \leq x \leq b$. In practice, the First Derivative Test may not help because $f'(x)$ may be complicated and the equation $f'(x) = 0$ hard to solve. Such situations require numerical techniques. We shall describe a method called **hill climbing**, which is something like the 10-section method for root approximation.

The idea is to start at a value x_0 where the function is fairly large, then to move by small steps "uphill" until you pass a peak. To be precise, take a guess x_0 and compute $f(x_0)$. Then move a fixed amount h to the left and right, computing $f(x_0 - h)$ and $f(x_0 + h)$. If (say) $f(x_0 + h) > f(x_0)$, move again to the right. If $f(x_0 + 2h) > f(x_0 + h)$, move once again to the right, etc. Keep going to the right until the first time you go down. For instance, suppose

$$f(x_0) < f(x_0 + h) < f(x_0 + 2h) < \cdots < f(x_0 + nh)$$
but
$$f[x_0 + (n + 1)h] < f(x_0 + nh).$$

Then obviously the graph has a peak near $x_0 + nh$.

Now start at $x_1 = x_0 + nh$ and repeat the whole process, but with $\frac{1}{10}h$ instead of h, etc. In this way you sneak up on the point x where $f(x)$ has a max. The method only requires $f(x)$ to be continuous, not even differentiable.

■ **EXAMPLE 3** Set $f(x) = 2 \sin x - x^2$. Estimate to 2 places the point $x = c$ where $f(c)$ is a max, and estimate $f(c)$.

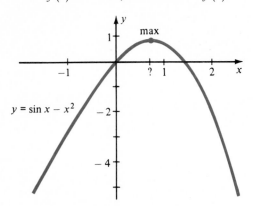

Fig. 3

Solution A rough graph (Fig. 3) shows $f(x)$ has a single local max, around $x = 1$. So start at $x_0 = 1$ with $h = 0.1$:

$$f(0.9) \approx 0.7567, \qquad f(1) \approx 0.6829, \qquad f(1.1) \approx 0.5724.$$

The curve climbs to the left. Keep going to the left in steps of 0.1 until the first time you go down:

$$f(0.8) \approx 0.7947, \qquad f(0.7) \approx 0.7984, \qquad \text{but} \qquad f(0.6) \approx 0.7693.$$

Now take $x_1 = 0.70$ and $h = 0.01$:

$$f(0.69) \approx 0.7970, \qquad f(0.71) \approx 0.7996.$$

This time you climb to the right:

$$f(0.72) \approx 0.8004, \qquad f(0.73) \approx 0.8008, \qquad f(0.74) \approx 0.8010, \qquad f(0.75) \approx 0.8008.$$

The winner is $x \approx 0.74$, and $f(0.74) \approx 0.8008$. ■

Remark The First Derivative Test leads to the equation $\cos x - x = 0$, for which the only solution is $x \approx 0.7391$ as we know from the previous examples.

Hill climbing obviously has its ups and downs. You might end up at a local max instead of the actual max. Or, you might miss the max altogether if your first h is too large. Each example requires some common sense, a good feel for the function involved, and perhaps a rough graph. On the positive side, however, the method is easy to program on a computer.

EXERCISES

Estimate the root (or roots) to 2 places by 10-section

1 $x^3 = 5$ **2** $x^3 = 60$ **3** $x^5 = 100$
4 $x^5 = -20$ **5** $x^4 = 7$ **6** $x^{10} = 35.5$
7 $x^3 + x^2 - x + 1 = 0$ **8** $3x^3 = 7x + 3$
9 $e^x = x + 2$ **10** $4 \cos x = 2 - x$.

Use the Secant Method to find an x near each root of the equation for which the equation is correct to 5 decimal places.

11	$x^3 = 29$	**12**	$x^3 = -17$	**13**	$x^3 = \frac{1}{2}x + 5$
14	$x^{2.3} = 10$	**15**	$\frac{1}{4}x^3 + 1 = x^2$	**16**	$\frac{1}{2}x^3 = x^2 + x + 1$
17	$x \ln x = 1$	**18**	$x = e^{x/5}$		
19	$x + \tan x = 1 \quad (0 < x < 2\pi)$	**20**	$x + \ln x = 0$		
21	$x^{2.5} = x + 3$	**22**	$\sec x = \sqrt{4 - x^2}$		
23	$\sin^2 x = \cos x \quad (0 < x < \frac{1}{2}\pi)$	**24**	$x^{3/2} = \dfrac{1}{1 + x^2}.$		

By hill climbing (descending) estimate to 3 places the required extremum of $f(x)$

25 $\max[(\ln x)/e^x]$ **26** $\max(x/2^x)$

27 $\max[x(x^2 - 1)(x^2 - 4)], \quad 0 < x < 1$ **28** $\min[x(x^2 - 1)(x^2 - 4)], \quad 1 < x < 2$

29 $\max(2^x - 1 - \frac{1}{2}x - \frac{1}{2}x^2), \quad 0 < x < 1$ **30** $\min(2^x - 1 - \frac{1}{2}x - \frac{1}{2}x^2), \quad 1 < x < 2$

31 $\min\left[\sin x - \dfrac{4}{\pi^2} x(\pi - x)\right], \quad 0 < x < \pi$

32 $\min\left[\sin x - \dfrac{8}{3\pi^3} x(\pi^2 - x^2)\right], \quad 0 < x < \pi$

33 $\max\left[\sin x - \dfrac{8}{3\pi^3} x(\pi^2 - x^2)\right], \quad 0 < x < \pi$

34 $\max \dfrac{1}{x^5(e^{1/x} - 1)}, \quad 0 < x.$

35 Solve $x^x = 5$ by the Secant Method, starting on $[2, 3]$. Stop when your answer is stable to 4 places. How many steps?

36 (cont.) You will notice that the method is slow because the right end point 3 never changes, so the successive intervals all have length ≈ 1. To avoid this, try alternating secant with bisection. Now how many steps are required for the same accuracy? Count each secant as one step and each bisection as one step.

11. ITERATION AND NEWTON'S METHOD

Choose any real number x_0. Set $x_1 = \cos x_0$; then $x_2 = \cos x_1$, then $x_3 = \cos x_2$, etc. You will notice that the successive x_n tend to stabilize as n grows.

For instance starting with $x_0 = 1$, a 4-place computation yields successive values of x_n as shown in Table 1.

Table 1 $x_{n+1} = \cos x_n, \quad x_0 = 1$

n	x_n	n	x_n	n	x_n	n	x_n
1	.5403	7	.7221	13	.7375	19	.7389
2	.8576	8	.7504	14	.7402	20	.7392
3	.6543	9	.7314	15	.7383	21	.7390
4	.7935	10	.7442	16	.7396	22	.7391
5	.7014	11	.7356	17	.7387	23	.7391
6	.7639	12	.7414	18	.7393		

Conclusion: To 4-place accuracy,

$$\cos(0.7391) \approx 0.7391.$$

If we start afresh with $x_0 = 0.739100$ and work to 6-place accuracy, we find

$$x_9 = x_{10} = 0.739085,$$

hence $\cos(0.739085) \approx 0.739085$ to 6-place accuracy.

The equation $x = \cos x$ is an example of an equation of the form $x = \phi(x)$. Such equations are common in practice and arise, as we shall see, in connection with root finding.

Under certain conditions, an equation $x = \phi(x)$ can be solved (approximately) by **iteration**. We start with an initial value x_0 and define successively

$$x_1 = \phi(x_0), \qquad x_2 = \phi(x_1), \qquad x_3 = \phi(x_2), \cdots, x_{n+1} = \phi(x_n).$$

We hope these values stabilize, and do so rapidly; then the process yields an approximate solution of $x = \phi(x)$. For $x_n \approx x_{n+1}$ is the same as $x_n \approx \phi(x_n)$; hence x_n is an approximate solution.

The process does not always work. For instance, if $\phi(x) = 2 \cos x$, then $x = 1.02986653$ is a solution of $\phi(x) = x$ to 8 places. However, set $x_0 = 1$ and iterate to 5-place accuracy:

$$x_1 = 1.08060, \qquad x_2 = 0.94160, \qquad x_3 = 1.17699, \qquad x_4 = 0.76741,$$

$$x_5 = 1.43942, \qquad x_6 = 0.26200, \qquad x_7 = 1.93175, \qquad x_8 = -0.70633.$$

The message is clear: even though we started within 0.03 of the answer, the process is unstable.

We can salvage this iteration by modifying it slightly. Instead of defining x_{n+1} as $2 \cos x_n$, let's try the *average* of x_n and $2 \cos x_n$. Therefore, we define a new function.

$$\phi(x) = \frac{x + 2 \cos x}{2}.$$

If we can solve $x = \phi(x)$, then $x = 2 \cos x$, so we shall have solved the original equation. Let us again start with $x_0 = 1$ and iterate (Table 2).

Table 2 $\phi(x) = \frac{1}{2}(x + 2 \cos x)$, $x_0 = 1$

n	x_n	n	x_n	n	x_n
1	1.040302	6	1.029806	11	1.029867
2	1.026111	7	1.029888	12	1.029866
3	1.031204	8	1.029859	13	1.0298667
4	1.029388	9	1.029869	14	1.0298665
5	1.030037	10	1.029866	15	1.0298665

We conclude that $x = 1.0298665$ approximates the solution of $x = 2 \cos x$ to 6 places.

Explanation   This method has its ups and downs; sometimes it works, sometimes it fails. We need a way of testing $\phi(x)$ in advance to be sure the method will work. The key is the size of $\phi'(x)$ near the solution. If $|\phi'(x)| < 1$, the method works; otherwise, it generally fails.

  Examples   If $\phi(x) = \cos x$, then $|\phi'(x)| = |-\sin x| < 1$ near $x \approx 0.7391$. Iteration works.

  If $\phi(x) = 2 \cos x$, then $|\phi'(x)| = |-\sin x| > 1$ near $x \approx 1.0299$. Iteration fails.

  If $\phi(x) = \frac{1}{2}(x + 2 \cos x)$, then $|\phi'(x)| = |\frac{1}{2} - \sin x| < 1$ near $x \approx 1.0299$. Iteration works.

  Let us state a precise result.

Iteration   Let $\phi(x)$ be a differentiable function which maps an interval $[a, b]$ into itself, that is, $a \leq \phi(x) \leq b$ whenever $a \leq x \leq b$. Let

$$|\phi'(x)| \leq M < 1$$

on the interval, where M is a constant. Finally, assume there exists a point c in the interval such that $c = \phi(c)$. Choose x_0 on the interval arbitrarily and define

$$x_1 = \phi(x_0), \quad x_2 = \phi(x_1), \quad x_3 = \phi(x_2), \quad \cdots \cdot$$

Then x_n becomes as close as we please to c as n increases, so x_n approximates c to any required degree of accuracy.

  To prove this assertion, let x be any point of the interval. By the MVT,

$$\phi(x) - \phi(c) = (x - c)\phi'(z),$$

where z is between x and c. Therefore

$$|\phi(x) - \phi(c)| = |\phi'(z)|\,|x - c| \leq M|x - c|.$$

But $\phi(c) = c$, so this implies

$$|\phi(x) - c| \leq M|x - c|.$$

Now we lift ourselves up by our bootstraps, starting with $x = x_0$:

$$|x_1 - c| = |\phi(x_0) - c| \leq M|x_0 - c|.$$

We continue with $x = x_1, x_2$, etc.:

$$|x_2 - c| = |\phi(x_1) - c| \leq M|x_1 - c| \leq M^2|x_0 - c|,$$
$$|x_3 - c| \leq |\phi(x_2) - c| \leq M|x_2 - c| \leq M^3|x_0 - c|,$$
$$\cdot \quad \cdot \quad \cdot \quad \cdot \quad \cdot \quad \cdot \quad \cdot \quad \cdot \quad \cdot \quad \cdot \quad \cdot \quad \cdot \quad \cdot$$
$$|x_n - c| \leq |\phi(x_{n-1}) - c| \leq M|x_{n-1} - c| \leq M^n|x_0 - c|.$$

Since $0 \leq M < 1$, it is clear that $M^n \longrightarrow 0$ as $n \longrightarrow \infty$. Therefore $x_n \longrightarrow c$.

Remark 1   According to the final inequality, $|x_n - c| \leq M^n|x_0 - c|$, the powers of M control how fast x_n approaches c. If $M \leq 0.1$, then each x_n is less than one-tenth the distance from c of x_{n-1}, so its accuracy is at least one more place. But if $M = 0.9$, then it takes about 22 more x's to be sure of one more decimal place accuracy, because $(0.9)^{21} > 0.1$ and $(0.9)^{22} < 0.1$.

Remark 2   Since $$|\phi(x) - c| \le M|x - c|,$$

the distance from $\phi(x)$ to c is less than the distance from x to c by a factor of M or less. This is fine, but we would much prefer an inequality like

$$|\phi(x) - c| \le K|x - c|^2.$$

Then the distance from $\phi(x)$ to c is at most proportional to the *square* of the distance from x to c. No matter what the constant K is, once x is close to c, then $\phi(x)$ is *very* close to c.

Remark 3   It is known that the assumption $c = \phi(c)$ for some c is not necessary; the other assumptions guarantee the existence of a unique solution.

Roots and Fixed Points   A point c such that $\phi(c) = c$ is called a **fixed point** of the function $\phi(x)$. It is a value of x where the graphs of $y = x$ and $y = \phi(x)$ intersect (Fig. 1). It is also a root of the equation $\phi(x) - x = 0$.

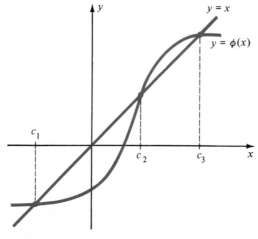

Fig. 1   Fixed points of $\phi(x)$: points c where $\phi(c) = c$

The idea of fixed points is important in solving equations. For given an equation $f(x) = 0$, we can set $\phi(x) = x + f(x)g(x)$, where $g(x)$ is any convenient function. Then each root of the equation $f(x) = 0$ is a fixed point of the function $\phi(x)$. Conversely, if $\phi(c) = c$ and $g(c) \ne 0$, then c is a root of $f(x) = 0$.

We shall now look at several powerful methods for estimating roots of $f(x) = 0$. In each case we associate with $f(x)$ a function $\phi(x)$ whose fixed points are roots of $f(x) = 0$, and whose iterates estimate the fixed points very rapidly.

Newton–Raphson Method   The **Newton–Raphson method** (also called **Newton's method**) is based on linear approximation. Suppose $x = c$ is a zero of $f(x)$. We guess an x, take the tangent to the graph $y = f(x)$ at $(x, f(x))$, and let $\phi(x)$ be its x-intercept (Fig. 2). It appears that $\phi(x)$ is a lot closer to c than x is.

The equation of the tangent is

$$\frac{Y - f(x)}{X - x} = f'(x).$$

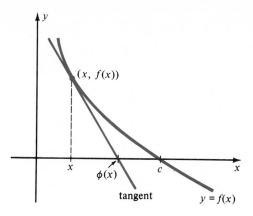

Fig. 2 Newton–Raphson method: $\phi(x)$ is the x-intercept of the tangent to $y = f(x)$ at $(x, f(x))$.

For the x-intercept, we set $Y = 0$ and solve for X:

$$\frac{-f(x)}{X - x} = f'(x), \qquad X - x = -\frac{f(x)}{f'(x)}, \qquad X = x - \frac{f(x)}{f'(x)}.$$

Therefore, given $f(x)$, we *define* $\qquad \phi(x) = x - \dfrac{f(x)}{f'(x)}$,

assuming $f'(x) \neq 0$. Obviously, if $f(c) = 0$, then $\phi(c) = c$. Conversely if $\phi(c) = c$, then $f(c) = 0$, assuming $f'(c) \neq 0$. Thus the root problem for $f(x) = 0$ is transformed into the fixed point problem for $\phi(x)$.

Still assuming that c is a root of $f(x) = 0$, we want to approximate c by iteration. We know that the smaller $|\phi'(x)|$ is near $x = c$, the faster the iterates x_0, $x_1 = \phi(x_0)$, $x_2 = \phi(x_1)$, etc. will tend to stabilize at $x = c$. So let us compute the derivative:

$$\phi'(x) = \frac{d}{dx}\left[x - \frac{f(x)}{f'(x)}\right] = 1 - \frac{[f'(x)]^2 - f(x)f''(x)}{[f'(x)]^2},$$

hence $\qquad\qquad\qquad \phi'(x) = \dfrac{f(x)f''(x)}{[f'(x)]^2}.$

Consequently $\phi'(c) = 0$; great! That means $|\phi'(x)|$ is very small in the neighborhood of $x = c$. Therefore, if x_0 is chosen sufficiently close to c in the first place, the iterates x_n will tend to $x = c$ rapidly (Fig. 3).

Let us summarize this discussion.

Newton–Raphson Method Suppose $f(c) = 0$, $f'(c) \neq 0$, and $f''(x)$ is continuous near $x = c$. Set

$$\phi(x) = x - \frac{f(x)}{f'(x)}.$$

If x_0 is sufficiently close to c, then the iterates

$$x_1 = \phi(x_0), \quad x_2 = \phi(x_1), \cdots, x_n = \phi(x_{n-1}), \cdots$$

approach as close as we please to c. (Also $\phi'(c) = 0$, so the convergence is rapid.)

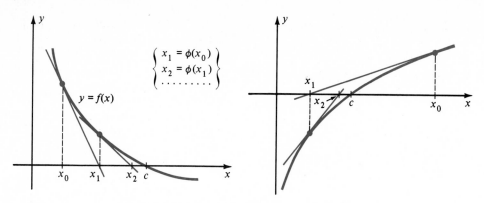

Fig. 3   Typical cases in which $x_n \longrightarrow c$ rapidly

■ **EXAMPLE 1**   Estimate to 6 places the roots of $x^5 - x + 1 = 0$.

Solution   A graph of $y = f(x) = x^5 - x + 1$ will show how many roots there are and their approximate locations. The derivative

$$f'(x) = 5x^4 - 1$$

is 0 at $x = \pm \sqrt[4]{\frac{1}{5}} \approx \pm 0.67$ and changes sign at these two points. This gives useful information about the graph:

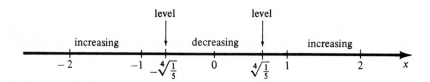

Clearly $f(x)$ has a local max at $-(\frac{1}{5})^{1/4}$ and a local min at $(\frac{1}{5})^{1/4}$. Their values are

$$f_{\max} = f[-(\tfrac{1}{5})^{1/4}] \approx 1.53 \qquad \text{and} \qquad f_{\min} = f[(\tfrac{1}{5})^{1/4}] \approx 0.47.$$

This is enough for a rough sketch (Fig. 4).

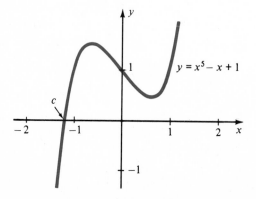

$y = x^5 - x + 1$

Fig. 4

The graph shows that there is exactly one root, and it is negative. By some trial and error, $f(-2) = -31$, and $f(-1) = 1$, hence the root is between -2 and -1, probably close to -1.

Now the work starts. Set

$$\phi(x) = x - \frac{f(x)}{f'(x)} = x - \frac{x^5 - x + 1}{5x^4 - 1} = \frac{4x^5 - 1}{5x^4 - 1}$$

and $x_0 = -1$. Then

$$x_1 = \phi(-1) = -\tfrac{5}{4} = -1.25, \qquad x_2 = \phi(-1.25) \approx -1.178459,$$

$$x_3 = \phi(x_2) \approx -1.167537, \qquad x_4 = \phi(x_3) \approx -1.167304,$$

$$x_5 = \phi(x_4) \approx -1.167304.$$

The process has stabilized, so -1.167304 is the one root of $f(x) = 0$, to 6 places. (Compare Example 1, p. 347.) ∎

■ **EXAMPLE 2** Estimate the root of $\cos x = x$ to 6 places.

Solution Set $f(x) = x - \cos x$ and

$$\phi(x) = x - \frac{f(x)}{f'(x)} = x - \frac{x - \cos x}{1 + \sin x} = \frac{x \sin x + \cos x}{1 + \sin x}.$$

Start with $x_0 = 1$. Iteration yields

$$x_1 = \phi(x_0) \approx 0.750364, \qquad x_2 = \phi(x_1) \approx 0.739113$$

$$x_3 = \phi(x_2) \approx 0.739085, \qquad x_4 = \phi(x_3) \approx 0.739085.$$

Therefore 0.739085 is a root of $\cos x = x$ to 6 places. ∎

Efficiency of Newton's Method Suppose $f(c) = 0$ and $f'(c) \neq 0$, and set

$$\phi(x) = x - \frac{f(x)}{f'(x)}.$$

As pointed out in Remark 2, p. 514, it would be nice if $\phi(x)$ satisfied an inequality of the type

$$|\phi(x) - c| \leq K|x - c|^2.$$

For that would imply that the iterates of ϕ approach c rapidly. Such an inequality does hold, and it can be derived using Lhospital's Rule.

We have already found $\qquad \phi'(x) = \dfrac{f(x)f''(x)}{[f'(x)]^2}.$

Now by Lhospital's Rule,

$$\lim_{x \to c} \frac{\phi(x) - c}{(x - c)^2} = \lim_{x \to c} \frac{\phi'(x)}{2(x - c)} = \lim_{x \to c} \frac{f(x)f''(x)}{2(x - c)[f'(x)]^2}$$

$$= \frac{f''(c)}{2[f'(c)]^2} \lim_{x \to c} \frac{f(x)}{x - c} = \frac{f''(c)}{2[f'(c)]^2} f'(c)$$

$$\lim_{x \to c} \frac{\phi(x) - c}{(x - c)^2} = \frac{f''(c)}{2f'(c)}.$$

It follows that if K is any constant larger than $|\frac{1}{2}f''(c)/f'(c)|$, then

$$|\phi(x) - c| < K|x - c|^2$$

for x sufficiently close to c.

This result is fine for theoretical purposes, but totally impractical for applications. It only guarantees the inequality for "x sufficiently close to c." That's too vague to be of any use. Furthermore, in practice we do not know c, so we cannot compute the constant $K > f''(c)/2f'(c)$ needed for error estimates.

Using more sophisticated techniques, we can prove a practical result.

Let $f(x)$ have continuous first and second derivatives on the interval $[a, b]$, satisfying

$$|f'(x)| \geq m > 0 \qquad \text{and} \qquad |f''(x)| \leq M,$$

where m and M are constants. Suppose $f(c) = 0$ for some c on the interval. Set

$$\phi(x) = x - \frac{f(x)}{f'(x)}.$$

Then

$$|\phi(x) - c| \leq \frac{M}{2m}|x - c|^2.$$

Proof

$$|\phi(x) - c| = \left| x - \frac{f(x)}{f'(x)} - c \right| = \frac{1}{|f'(x)|}|-f(x) - f'(x)(c - x)|.$$

Since $f(c) = 0$, we can write

$$|\phi(x) - c| = \frac{1}{|f'(x)|}|f(c) - f(x) - f'(x)(c - x)| = \frac{1}{|f'(x)|}|r_1(c)|,$$

where $r_1(c)$ is the remainder in the approximation of $f(c)$ by its first degree Taylor polynomial. (Note that the roles of x and c are reversed.) By Taylor's Formula with $n = 1$,

$$|r_1(c)| = \left| \int_x^c (c - t)f''(t) \, dt \right| \leq M \left| \int_x^c |c - t| \, dt \right| = \frac{1}{2}M|x - c|^2.$$

Therefore

$$|\phi(x) - c| = \frac{1}{|f'(x)|}|r_1(c)|$$

$$\leq \frac{M}{2|f'(x)|}|x - c|^2 \leq \frac{M}{2m}|x - c|^2.$$

■ **EXAMPLE 3** Suppose Newton's method is used to estimate $\sqrt{2}$, starting with $x_0 = 1$. How far should the process be continued to guarantee 20-place accuracy.

Solution Set $f(x) = x^2 - 2$ and

$$\phi(x) = x - \frac{f(x)}{f'(x)},$$

$$= x - \frac{x^2 - 2}{2x} = \frac{x^2 + 2}{2x}.$$

Since $f(1) = -1$ and $f(2) = 2$, we should seek our solution on the interval $[1, 2]$. There $|f'(x)| = |2x| \geq 2 > 0$ and $|f''(x)| = |2| = 2$, so we choose $m = 2$ and $M = 2$. Consequently, by the estimate in the last box,

$$|\phi(x) - \sqrt{2}| \leq \tfrac{1}{2}|x - \sqrt{2}|^2.$$

Since $\sqrt{2} < 1.5$,

$$|x_0 - \sqrt{2}| = |1 - \sqrt{2}| < |1.5 - 1| = \frac{1}{2}.$$

Now pull the bootstrap:

$$|x_1 - \sqrt{2}| \leq \frac{1}{2}|x_0 - \sqrt{2}|^2 < \frac{1}{2}\left(\frac{1}{2}\right)^2 = \frac{1}{2^3},$$

$$|x_2 - \sqrt{2}| \leq \frac{1}{2}|x_1 - \sqrt{2}|^2 < \frac{1}{2}\left(\frac{1}{2^3}\right)^2 = \frac{1}{2^7},$$

$$|x_3 - \sqrt{2}| < \frac{1}{2}\left(\frac{1}{2^7}\right)^2 = \frac{1}{2^{15}},$$

$$|x_4 - \sqrt{2}| < \frac{1}{2}\left(\frac{1}{2^{15}}\right)^2 = \frac{1}{2^{31}},$$

and in general,

$$|x_n - \sqrt{2}| < \frac{1}{2^N}, \qquad \text{where} \quad N = 2^{n+1} - 1.$$

We want $|x_n - \sqrt{2}| < 5 \times 10^{-21}$. This will be so provided that

$$\frac{1}{2^N} < 5 \times 10^{-21}, \qquad \text{where} \quad N = 2^{n+1} - 1,$$

that is,

$$2^N > \tfrac{1}{5} \times 10^{21}, \qquad 2^{N-1} > 10^{20}.$$

But $2^{67} \approx 1.5 \times 10^{20}$ (by logs or a calculator), so we choose $N \geq 68$. The least n for which $2^{n+1} - 1 \geq 68$ is $n = 6$. Thus x_6 provides 20-place accuracy. ■

Remark 1 Note that $\phi(x) = \dfrac{x^2 + 2}{2x} = \dfrac{1}{2}\left(x + \dfrac{2}{x}\right)$, so $\phi(x)$ is the average of x and $2/x$. This is reasonable since $\sqrt{2}$ is that number c for which $c = 2/c$.

Remark 2   The first six x_n are

$$x_0 = 1, \qquad x_1 = \frac{3}{2}, \qquad x_2 = \frac{17}{12}, \qquad x_3 = \frac{577}{408}, \qquad x_4 = \frac{665857}{470832}, \qquad x_5 = \frac{88\ 67310\ 88897}{62\ 70135\ 66048}.$$

We find by a computer that

$$\sqrt{2} \approx 1.41421\ 35623\ 73095\ 04880\ 169,$$

$$x_4 \approx 1.41421\ 35623\ 747,$$

$$x_5 \approx 1.41421\ 35623\ 73095\ 04880\ 1690.$$

The theory predicts that $|x_4 - \sqrt{2}| < 2^{-31} < 5 \times 10^{-10}$, hence that x_4 is accurate to at least 9 places. But error estimates usually lose something; actually x_4 is accurate to 11 places. Similarly the theory predicts that $|x_5 - \sqrt{2}| < 2^{-63} < 1.1 \times 10^{-19}$, assuring only 18-place accuracy for x_5; actually x_5 is accurate to at least 23 places. (See Ex. 42.)

Third Order Methods   The Newton method of solving $f(x) = 0$ is a second order method, which means that $\phi(x)$ satisfies

$$|\phi(x) - c| \leq K|x - c|^2$$

near a zero $x = c$ of $f(x)$. By a **third order method** we mean iteration with a function $\psi(x)$, defined in terms of $f(x)$ and its derivatives, such that

$$|\psi(x) - c| \leq K|x - c|^3.$$

We shall describe two third order methods. Each can be considered a refinement of Newton's method. Their derivations will be omitted except for some hints in the exercises.

We assume that $f(x)$ has a zero at $x = c$ and use the notation

$$u(x) = \frac{f(x)}{f'(x)}, \qquad \phi(x) = x - u(x), \qquad v(x) = \frac{f''(x)}{f'(x)}.$$

ψ-Method $\psi(x) = \phi(x) - \dfrac{1}{2} u(x)^2 v(x),$

$\psi(x) - c \approx K(x - c)^3,$ where $K = \left(\dfrac{1}{2} v^2 - \dfrac{1}{6} \dfrac{f'''}{f'} \right)\Big|_{x=c}.$

Halley's Method $\theta(x) = x - \dfrac{u(x)}{1 - \frac{1}{2}u(x)v(x)},$

$\theta(x) - c \approx H(x - c)^3,$ where $H = \left(\dfrac{1}{4} v^2 - \dfrac{1}{6} \dfrac{f'''}{f'} \right)\Big|_{x=c}.$

Presumably $u(x)$ will be very small for x near c. Therefore the second term in the ψ-method can be considered as a small correction to $\phi(x)$. (Without that second term we would be back to Newton's method.) In Halley's method,

$$\theta(x) \approx x - \frac{u(x)}{1 - 0} = x - u(x) = \phi(x)$$

for x near c. Therefore Halley's is also a modification of Newton's method.

■ **EXAMPLE 4** Estimate the root of $\cos x = x$ by the ψ-method; start with $x_0 = 1$, and compute x_2.

Solution Set $f(x) = x - \cos x$. Then

$$u = \frac{f}{f'} = \frac{x - \cos x}{1 + \sin x}, \qquad \phi = \frac{x \sin x + \cos x}{1 + \sin x}, \qquad v = \frac{\cos x}{1 + \sin x},$$

$$\psi(x) = \phi - \frac{1}{2} u^2 v$$

$$= \frac{x \sin x + \cos x}{1 + \sin x} - \frac{1}{2} \frac{(x - \cos x)^2 \cos x}{(1 + \sin x)^3}.$$

Now iterate, starting with $x_0 = 1$:

$$x_1 = \psi(x_0) \approx 0.74122\,15391,$$

$$x_2 = \psi(x_1) \approx 0.73908\,51352,$$

which is accurate to 8 places (to 10 places, the root is $0.73908\,51332$). ■

■ **EXAMPLE 5** Approximate $\sqrt{2}$ by Halley's method; start with $x_0 = 1$ and compute x_2.

Solution Set $f(x) = x^2 - 2$. Then

$$u = \frac{f}{f'} = \frac{x^2 - 2}{2x}, \qquad v = \frac{1}{x},$$

$$\theta(x) = x - \frac{u}{1 - \frac{1}{2} uv}$$

$$= x - \frac{\dfrac{x^2 - 2}{2x}}{1 - \dfrac{1}{2} \left(\dfrac{x^2 - 2}{2x} \right) \left(\dfrac{1}{x} \right)}$$

$$= x - \frac{2x(x^2 - 2)}{4x^2 - (x^2 - 2)}$$

$$= x - \frac{2x^3 - 4x}{3x^2 + 2} = \frac{x^3 + 6x}{3x^2 + 2}.$$

If $x_0 = 1$, then

$$x_1 = \theta(x_0) = \frac{7}{5} = 1.4,$$

$$x_2 = \theta(x_1) = \frac{1393}{985} \approx 1.41421\,3198,$$

accurate to 6 places (to 8 places, $\sqrt{2} \approx 1.41421\,356$). ■

EXERCISES

Use iteration to estimate the solution of $\phi(x) = x$; work to 4-place accuracy

1	$\phi(x) = e^{-x/2}$ $x_0 = 1$	**2**	$\phi(x) = e^{-x}$ $x_0 = 1$
3	$\phi(x) = \sqrt{3 + x}$ $x_0 = 3$	**4**	$\phi(x) = \sqrt{3 + x}$ $x_0 = 100$
5	$\phi(x) = 2 \arctan x$ $x > 0$ $x_0 = 2$	**6**	$\phi(x) = \pi + \arctan x$ $x_0 = 0$
7	$\phi(x) = \sqrt{\frac{1}{2}(1 + x)}$ $x_0 = 0$	**8**	$\phi(x) = \sqrt{\frac{1}{2}(1 + x)}$ $x_0 = 1000$
9	$\phi(x) = x(\frac{5}{4} - x^2)$ $x_0 = 1$	**10**	$\phi(x) = x(\frac{5}{4} - x^2)$ $x_0 = 1.25$.

11 Set $\phi(x) = 0.6316x + 0.7368 \cos x$ and $x_0 = 1$. Find x_3 and x_4 to 8 places.

12 (cont.) $\phi(x) = x$ is equivalent to what simple problem previously considered?

Use the Newton–Raphson method to estimate all roots to 5 places (10 places)

13	$x^2 = 10$	**14**	$x^3 = 10$	**15**	$5x^4 = 1$
16	$x \ln x = 1$	**17**	$x^3 - x + 1 = 0$	**18**	$e^{-5x} = x$
19	$x^3 + x^2 - 4 = 0$	**20**	$x^3 + x + 5 = 0$	**21**	$e^x + x = 0$
22	$x^3 - 2x^2 + x + 3 = 0$	**23**	$5x^2 + 4x - 3 = 0$	**24**	$x^3 - 3x + 1 = 0$
25	$x^3 - 2x^2 - x + 3 = 0$	**26**	$x^3 - 6x^2 + 9x - 1 = 0$	**27**	$e^x = 2 \cos x$ $x > -\pi$
28	$e^x = 3x$	**29**	$x = 2 \sin x$	**30**	$x = 2 \cos 2x$
31	$x^5 - 5x + 1 = 0$	**32**	$x^3 + 1.5x^2 - 5.75x + 3.37 = 0$		

33 $10 \cos x - 8 + x^2 = 0$ **34** $x^4 - 10^4x + 1 = 0$ **35** $e^{-x^2} = \dfrac{1}{1 + 4x^2}$

36* $x^4 - 1.73x^2 + 0.46x + 1.275 = 0$.

37 Find to 5 places the largest c such that $y = cx$ is tangent to $y = \cos x$.

38 Find to 5 places the largest c such that $\cosh x \geq cx$ for all x.

39 Find to 5 places the minimum vertical distance between the graphs $y = 1.5e^x$ and $y = x^3$.

40 Estimate to 3 place accuracy the solutions of the system $\{x^2 + y^4 = 100,\ e^x - e^y = 1\}$.

41 Show in Example 3, p. 518, that $|x_6 - \sqrt{2}| < 10^{-43}$.

42* (cont.) Show that $x_1 > x_2 > x_3 > \cdots > \sqrt{2}$. Conclude that $|x_5 - \sqrt{2}| < 5 \times 10^{-23}$.

43 Let $x_0 > 0$ and $\phi(x) = \sqrt{\frac{1}{2}(1 + x)}$. Show that $x_n \longrightarrow 1$.

44* Let $\phi(x) = x(a - x^2)$, where $1 < a < 2$, and set $b = \sqrt{a - 1}$. Verify that $0 < b < 1$ and $\phi(b) = b$. Show that there is a $K < 1$ such that $|\phi(x) - b| < K|x - b|$ whenever $b \leq x \leq 1$. If $b \leq x_0 \leq 1$, conclude that $x_n \longrightarrow b$.

45 (cont.) For what particular a does $x_n \longrightarrow b$ the fastest?

46* Set $\phi(x) = e^{-ex}$ and choose $x_0 = 1$. Show that $\phi(e^{-1}) = e^{-1}$. Convince yourself by calculation that $x_n \longrightarrow e^{-1}$, although painfully slowly. How can you speed up the process?

47 Estimate the positive root of $x^2 = 2$ by the ψ-method. Take $x_0 = 1$ and find x_1, x_2, and x_3 to 5 (10) places.

48 Estimate the root of $x^3 = 10$ by Halley's method. Take $x_0 = 1$ and find x_1, x_2, x_3, and x_4 to 5 (10) places.

49* Show that $\psi'(x) = u^2 \left(\dfrac{3}{2} v^2 - \dfrac{1}{2} \dfrac{f'''}{f'} \right)$; then use Lhospital's Rule to find

$$\lim_{c \to 0} [\psi(x) - c]/(x - c)^3.$$

50* (cont.) Show that $\theta(x) = \psi(x) - \dfrac{u^3 v^2}{4(1 - \frac{1}{2}uv)}$; then find $\lim_{c \to 0} [\theta(x) - c]/(x - c)^3$.

12. MISCELLANEOUS EXERCISES

1 Find the 5-th degree Taylor polynomial of $\cos x$ at $x = \frac{1}{3}\pi$.

2 Find the 15-th degree Taylor polynomial of e^{-x^3} at $x = 0$.

3 Examine the graphs $y = \cos x$ and $y = \sqrt{1 - x^2}$ near $(0, 1)$. Is the cosine curve inside or outside of the circle?

4 Show that

$$\tan \theta \approx \frac{1}{\frac{1}{2}\pi - \theta} \quad \text{for } \theta \approx \frac{1}{2}\pi.$$

5 Let $p(x)$ be the 4-th degree Taylor polynomial of e^x at $x = 0$. Which gives the best approximation to e: $p(1)$, $[p(0.5)]^2$, $[p(0.1)]^{10}$?

6 Let $p_2(x)$ be the second degree Taylor polynomial of $\sqrt{x}$ at $x = 100$. Show that

$$|\sqrt{x} - p_2(x)| < 5 \times 10^{-6} \quad \text{if} \quad 100 \le x < 102.$$

7 How good is the estimate $\quad \sin x \approx \frac{1}{3}x[10(\frac{1}{20}x^2 + 1)^{-1} - 7] \quad$ for x small?

8* Show that

$$\sin x \approx \frac{1}{2}(e^x - e^{-x}) - \frac{1}{3}x^3[1 + \frac{1}{840}(x^4 + \frac{1}{7920}x^8)],$$

with error less than 5×10^{-9} for $|x| < \frac{1}{2}\pi$.

9 Assume $f(x)$ is continuously differentiable near $x = c$. Find

$$\lim_{x \to 0} \frac{f(c + 2x) - f(c + x)}{x}.$$

10 Find $\quad \lim_{x \to \infty} [1 + a(b^{1/x} - 1)]^x$.

11 Suppose $f(0) = f'(0) = f''(0) = \cdots = f^{(n)}(0)$ and $f^{(n+1)}(x) \ge 0$ for all $x \ge 0$. Prove that $f(x) \ge 0$ for all $x > 0$ and that if $f(c) = 0$ for some $c > 0$, then $f(x) = 0$ for all $0 \le x \le c$.

12* The ends of a long straight metal rod of length L are clamped. When the rod is slightly warmed, its length increases to $L + \varepsilon$, so it bows into a (we shall assume) circular arc. Show that the center of the arc rises $H \approx \frac{1}{4}\sqrt{6L\varepsilon}$. Estimate H in feet when $L = 1$ mile and $\varepsilon = 1$ inch.

Estimate to 4 places

13 $\displaystyle\int_1^3 \frac{e^x}{x}\, dx$

14 $\displaystyle\int_2^5 \frac{dx}{\ln x}$

15 $\displaystyle\int_{0.6}^{1.4} \ln \tan x \, dx$

16 $\displaystyle\int_0^{\pi/2} \sqrt{1 + \cos^2 x}\, dx$

17 $\displaystyle\int_0^3 x \sin^2(\frac{1}{4}\pi x)\, dx$

18 $\displaystyle\int_1^4 (\ln x)^2\, dx.$

Estimate to 5 (10) places the roots of

19 $e^{-x/2} = x + 3$

20 $2 \arctan x = x$

21 $(2\pi + x) \tan x = 1 \quad 0 < x < \frac{1}{2}\pi$

22 $0.4x = \tan x \quad \frac{1}{2}\pi < x < \frac{3}{2}\pi.$

23   For what choice of the constant c, $0 < c < 1$, does the weighted average

$$\phi(x) = cx + (1 - c)\frac{a}{x^2}$$

iterate as rapidly as possible toward $\sqrt[3]{a}$?

24*   Let $0 < x < y < 0.4°$. Show that

$$\log \sin y - \log \sin x \approx \log y - \log x,$$

with error $< 5 \times 10^{-6}$. (Here $\log = \log_{10}$.)

Convergence ∎∎

1. SEQUENCES AND LIMITS

A sequence is a collection of numbers (elements) in a definite order.

Examples

$$2,\ 4,\ 6,\ 8,\ 10,\ 12 \qquad\qquad 1,\ -\frac{1}{2},\ \frac{1}{4},\ -\frac{1}{8},\ \cdots,\ \frac{1}{256}$$

$$1,\ 3,\ 5,\ 7,\ 9,\ \cdots \qquad\qquad \frac{6}{5},\ \frac{7}{6},\ \frac{8}{7},\ \frac{9}{8},\ \frac{10}{9},\ \cdots$$

The first two examples are finite sequences; they stop. The other two examples are infinite sequences; they go indefinitely and have no last element. We shall study infinite sequences in this chapter. The word *sequence* will always mean infinite sequences.

Notation   Standard notation for a sequence $a_1, a_2, a_3, \cdots$ is

$$\{a_n\} \qquad \text{or} \qquad \{a_n\}_{n=1}^{\infty} \qquad \text{or} \qquad \{a_n\}_1^{\infty}.$$

The first element of a sequence need not be a_1; it may be a_0, or a_3, or a_{10}, etc. We write

$$\{a_n\}_0^{\infty} \quad \text{for} \quad a_0, a_1, a_2, \cdots \qquad \text{and} \qquad \{a_n\}_{10}^{\infty} \quad \text{for} \quad a_{10}, a_{11}, a_{12}, \cdots.$$

Other enumerations are possible too. For example,

$$\{a_{2n}\}_1^{\infty} = a_2,\ a_4,\ a_6,\ \cdots, \qquad \{b_{4n+1}\}_0^{\infty} = b_1,\ b_5,\ b_9,\ \cdots.$$

Many sequences we encounter have some mathematical rule of formation, generally a formula for the *n*-th element.

Examples

(1) $a_n = \dfrac{1}{3^n}, \quad n \geq 1$: $\dfrac{1}{3},\ \dfrac{1}{3^2},\ \dfrac{1}{3^3},\ \dfrac{1}{3^4}, \cdots$

(2) $a_n = \dfrac{4n+1}{2^{n+1}}, \quad n \geq 0$: $\dfrac{1}{2},\ \dfrac{5}{2^2},\ \dfrac{9}{2^3},\ \dfrac{13}{2^4}, \cdots$

(3) $\quad a_n = \dfrac{(-1)^n n}{10n + 3}$, $\quad n \ge 2$: $\qquad \dfrac{2}{23}, \dfrac{-3}{33}, \dfrac{4}{43}, \dfrac{-5}{53}, \cdots$

(4) $\quad a_n = \left(1 + \dfrac{1}{n}\right)^n$, $\quad n \ge 1$: $\qquad (1 + 1), \left(1 + \dfrac{1}{2}\right)^2, \left(1 + \dfrac{1}{3}\right)^3, \left(1 + \dfrac{1}{4}\right)^4, \cdots$

(5) $\quad a_n = \dfrac{1 \cdot 3 \cdot 5 \cdots (2n - 1)}{n!}$, $\quad n \ge 1$: $\qquad 1, \dfrac{1 \cdot 3}{1 \cdot 2}, \dfrac{1 \cdot 3 \cdot 5}{1 \cdot 2 \cdot 3}, \dfrac{1 \cdot 3 \cdot 5 \cdot 7}{1 \cdot 2 \cdot 3 \cdot 4}, \cdots$

Not every sequence has a simple formula, or even any formula at all, for example $\{d_n\}$, where d_n is the n-th digit in the decimal representation of π, or $\{p_n\}$, where p_n is a patient's blood pressure n minutes after an operation.

A **subsequence** of $\{a_n\}_1^\infty$ is a sequence $\{a_{n_j}\}_{j=1}^\infty$, where $\{n_j\}_1^\infty$ is a strictly increasing sequence of positive integers. For example,

$$a_1, a_3, a_5, a_7, \cdots, \qquad a_4, a_5, a_6, a_7, \cdots, \qquad a_1, a_4, a_9, a_{16}, \cdots$$

are subsequences of $\{a_n\}_1^\infty$.

Limits of Sequences   Consider the sequence $\{a_n\}$, where $a_n = (n + 1)/n$:

$$2, \frac{3}{2}, \frac{4}{3}, \frac{5}{4} \quad \cdots \quad , \frac{n + 1}{n}, \quad \cdots \quad .$$

These numbers seem to "approach" the number 1. But what does that mean exactly?

By "a_n approaches 1" we mean that the numbers get close to 1. How close? To within 0.01? That is not good enough; the sequence

$$1.005, \quad 1.005, \quad 1.005, \cdots$$

gets within 0.01 of the number 1, yet this sequence does not "approach" 1. To within 0.001? Again not good enough; the sequence

$$0.9999, \quad 0.9999, \quad 0.9999, \cdots$$

gets within 0.001, yet does not "approach" 1. No matter what accuracy we specify, examples like these show that it is not good enough. The trick is to require that the numbers a_n approximate 1 to within *every possible degree* of accuracy.

Let us try to say that "in math". We specify a "degree of accuracy" by a positive number ε (usually very small). Given ε, we want a_n eventually to come within ε of 1; written "in math", we want $|1 - a_n| < \varepsilon$. Now we should clarify the vague word "eventually". This should mean "from a certain point on in the sequence". Now a "point in the sequence" is indicated by a subscript, so we want $|1 - a_n| < \varepsilon$ starting with a certain subscript N.

With these ideas in mind we formulate the general definition:

Limit of a Sequence   A sequence $\{a_n\}$ has **limit** L if for each positive ε, there exists an integer N such that

$$|L - a_n| < \varepsilon \qquad \text{for all} \quad n \ge N.$$

The definition requires all terms beyond some point to be within ε of L. In other words, the inequality $|L - a_n| < \varepsilon$ must hold with only a finite number of exceptions. In geometric terms, given any interval centered at L, no matter how small, the elements of the sequence eventually get into that interval *and stay there* (Fig. 1).

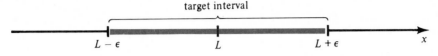

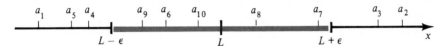

Fig. 1   Eventually all a_n fall into the target interval $L - \varepsilon < L < L + \varepsilon$. (As shown, "eventually" means starting with a_6.)

A sequence need not have a limit.

Examples $1, 2, 3, 4, \cdots, \quad 0, \tfrac{1}{2}, 0, \tfrac{3}{4}, 0, \tfrac{7}{8}, \cdots.$

The first sequence just marches off the map; it does not get close to any number. The second does get close both to 0 and to 1, but does not eventually stay close to either.

If $\{a_n\}$ has a limit L, we write

$$\lim_{n \to \infty} a_n = L, \quad \text{or simply} \quad \lim a_n = L.$$

We also write

$$a_n \longrightarrow L \quad \text{as} \quad n \longrightarrow \infty, \quad \text{or simply} \quad a_n \longrightarrow L.$$

If $a_n \longrightarrow L$, then $\{a_n\}$ **converges** to L. A sequence that has a limit is **convergent**, and a sequence without a limit is **divergent**.

If a sequence does have a limit, then it has only one limit since its elements cannot simultaneously be arbitrarily close to two different numbers.

Uniqueness of the Limit   If $a_n \longrightarrow L_1$ and $a_n \longrightarrow L_2$, then $L_1 = L_2$.

We leave a formal proof as an exercise. Because of this theorem, we may speak of *the* limit of a convergent sequence.

Proving that $a_n \longrightarrow L$ is a kind of game; I challenge you with an ε, and you must find some appropriate N. It is not necessary to find the *smallest* N such that $|L - a_n| < \varepsilon$ for all $n \geq N$; that may be hard. But for each ε, you must find *some* N.

■ **EXAMPLE 1**   If $a_n = 1/n$, prove that $a_n \longrightarrow 0$.

Solution   Let $\varepsilon > 0$ be given. We must find an N such that

$$|0 - a_n| = |a_n| < \varepsilon$$

for all $n \geq N$. We simply choose N larger than $1/\varepsilon$. (For example, if $\varepsilon = 0.001$, we take any integer $N > 1/\varepsilon = 1000$.) Then $1/N < \varepsilon$, hence

$$|a_n| = \frac{1}{n} \leq \frac{1}{N} < \varepsilon \qquad \text{for all} \quad n \geq N.$$

This completes the proof.   ■

Remark   The proof is valid for *each* choice of ε. If you produce an N that works only for $\varepsilon = 0.1$ or $\varepsilon = 0.001$, that is not sufficient. In general the smaller ε is, the larger N must be. That makes sense: the closer you want to approximate the limit, the farther out in the sequence you must go.

■ **EXAMPLE 2**   If $a_n = \dfrac{n^2}{2n^2 + 1}$, prove that $a_n \longrightarrow \frac{1}{2}$.

Solution   Write out a few terms:

$$\frac{1}{3}, \frac{4}{9}, \frac{9}{19}, \frac{16}{33}, \frac{25}{51}, \cdots.$$

These numbers do appear to approach $\frac{1}{2}$. To prove this is so, let $\varepsilon > 0$ be given. You must find an N such that $\left|\frac{1}{2} - a_n\right| < \varepsilon$ for all $n \geq N$. Now

$$\left|\frac{1}{2} - a_n\right| = \left|\frac{1}{2} - \frac{n^2}{2n^2 + 1}\right| = \frac{1}{2(2n^2 + 1)} < \frac{1}{4n^2}.$$

Choose N so large that

$$\frac{1}{4N^2} < \varepsilon, \qquad \text{that is,} \quad N^2 > \frac{1}{4\varepsilon}, \quad N > \sqrt{\frac{1}{4\varepsilon}}.$$

(For example, if $\varepsilon = 0.001$, then $N > \sqrt{1/0.004} \approx 15.8$, so $N > 15$ does it.) Then for all $n \geq N$,

$$\left|\frac{1}{2} - a_n\right| < \frac{1}{4n^2} \leq \frac{1}{4N^2} < \varepsilon.$$

This completes the proof.   ■

The idea of the limit of a sequence is fundamental to the real number system. For what does it mean really to express a number as an infinite decimal, for instance,

$$\pi = 3.141592654 \cdots ?$$

It means that π is the limit of the sequence $\{a_n\}$, where

$$a_1 = 3.1, \qquad a_2 = 3.14, \qquad a_3 = 3.141, \qquad a_4 = 3.1415, \qquad \text{etc.}$$

Note that $|\pi - a_n|$ is the error when π is approximated by n places of its decimal expansion. Hence

$$|\pi - a_n| < 10^{-n}.$$

Before continuing, we wish to dispel two old-wives' tales concerning limits of sequences:

Tale 1   If $a_n \longrightarrow L$, then the numbers a_n get closer and closer to L without ever reaching L. Not necessarily! Examples:

$$3, \ 3, \ 3, \ 3, \cdots, \qquad 0, \ \frac{1}{2}, \ 0, \ \frac{1}{3}, \ 0, \ \frac{1}{4}, \cdots.$$

The first sequence converges to 3, the second to 0. Each "reaches it limit" infinitely often.

Tale 2   If $a_n \longrightarrow L$, the numbers increase toward L or decrease toward L. Not necessarily! Example:

$$1, \ -\frac{1}{2}, \ \frac{1}{3}, \ -\frac{1}{4}, + \ - \cdots.$$

This sequence converges to 0 but jumps around. In fact, take the sequence

$$1, \ \frac{1}{2}, \ \frac{1}{3}, \ \frac{1}{4}, \cdots$$

and sprinkle in minus signs at random. The resulting irregular sequence still converges to zero.

An even more striking example is this one:

$$\frac{1}{10}, \ \frac{1}{10^2}, \ \frac{2}{10^2}, \ \frac{1}{10^3}, \ \frac{2}{10^3}, \ \frac{3}{10^3}, \ \frac{1}{10^4}, \ \frac{2}{10^4}, \ \frac{3}{10^4}, \ \frac{4}{10^4}, \cdots.$$

This sequence converges to zero, yet it has longer and longer strings of terms that move *away* from zero!

Two Principles   The ε, N definition is important in making the concept of limit precise, but tedious to apply. In practice, we use certain properties of limits that allow us to derive the convergence of many sequences from that of known convergent sequences. A very simple property of this type concerns subsequences of convergent sequences.

Convergence of Subsequences   If $a_n \longrightarrow L$, then any subsequence of $\{a_n\}$ also converges to L.

The proof is not hard; we leave it for an exercise. As an application of this principle, we easily deduce that the following sequences $\{b_n\}$ all converge to zero because each is a subsequence of $\{1/n\}$, which we know converges to zero:

(1) $1, \frac{1}{3}, \frac{1}{5}, \frac{1}{7}, \cdots$ $b_n = \dfrac{1}{2n-1}$ (2) $\frac{1}{2}, \frac{1}{4}, \frac{1}{8}, \frac{1}{16}, \cdots$ $b_n = \dfrac{1}{2^n}$

(3) $1, \frac{1}{2^2}, \frac{1}{3^2}, \frac{1}{4^2}, \cdots$ $b_n = \dfrac{1}{n^2}$ (4) $1, \frac{1}{2^3}, \frac{1}{3^3}, \frac{1}{4^3}, \cdots$ $b_n = \dfrac{1}{n^3}.$

The next principle says that a sequence trapped between two sequences that converge to the same limit is itself squeezed to the same limit.

> **Trapped Sequences** Suppose $\{a_n\}$ and $\{c_n\}$ are convergent sequences with the same limit L, and suppose $\{b_n\}$ is a sequence such that $a_n \le b_n \le c_n$ for $n \ge 1$. Then $\{b_n\}$ also converges to L.

Proof Let $\varepsilon > 0$ be given. There exist integers N_1 and N_2 such that

$$|L - a_n| < \varepsilon \quad \text{for all} \quad n \ge N_1, \qquad |L - c_n| < \varepsilon \quad \text{for all} \quad n \ge N_2.$$

If $N = \max\{N_1, N_2\}$, the larger of N_1 and N_2, then both inequalities hold for all $n \ge N$. In particular,

$$L - \varepsilon < a_n \quad \text{and} \quad c_n < L + \varepsilon.$$

Since $a_n \le b_n \le c_n$, $L - \varepsilon < a_n \le b_n \le c_n < L + \varepsilon.$

Hence $|L - b_n| < \varepsilon \quad \text{for all} \quad n \ge N.$

This proves that $b_n \longrightarrow L$.

The special case $a_n = 0$ of this theorem is particularly useful.

> **Corollary** Suppose $0 \le b_n \le c_n$, and $c_n \longrightarrow 0$. Then $b_n \longrightarrow 0$.

As applications of the corollary, we deduce immediately that the following sequences $\{b_n\}$ all converge to zero:

$$\{1/n^{3/2}\}, \qquad \{1/n^2\}, \qquad \{(\sin^2 n)/n\}.$$

In each case, $0 < b_n \le 1/n$. Since $1/n \longrightarrow 0$, the corollary shows that $b_n \longrightarrow 0$.

Arithmetic of Convergent Sequences Given a sequence $\{a_n\}$ we can form a new sequence $\{|a_n|\}$ by taking absolute values and another new sequence $\{ca_n\}$ by multiplying each term by a number c. Given two sequences $\{a_n\}$ and $\{b_n\}$ we can form new sequences $\{a_n + b_n\}$, $\{a_n b_n\}$, $\{a_n/b_n\}$ by adding, multiplying, or dividing termwise. Here are the basic rules concerning convergence.

> **Rules for Limits** Suppose $a_n \longrightarrow A$ and $b_n \longrightarrow B$. Then
>
> (1) $|a_n| \longrightarrow |A|$ $\qquad\qquad$ (2) $ca_n \longrightarrow cA$
>
> (3) $a_n \pm b_n \longrightarrow A \pm B$ $\qquad$ (4) $a_n b_n \longrightarrow AB$
>
> (5) $\dfrac{a_n}{b_n} \longrightarrow \dfrac{A}{B}$ $\quad (b_n \ne 0, \quad B \ne 0).$

We shall omit the proof, which is close in spirit to that of the corresponding result on limits of functions.

■ **EXAMPLE 3** Find $\displaystyle\lim_{n \to \infty} \left(2 + \frac{1}{n}\right)\left(3 - \frac{4}{n^2}\right).$

Solution We know that $1/n \longrightarrow 0$ and $1/n^2 \longrightarrow 0$. By Rule (2), with $c = 4$, we have $4/n^2 \longrightarrow 0$. By Rule (3),

$$2 + \frac{1}{n} \longrightarrow 2 \quad \text{and} \quad 3 - \frac{4}{n^2} \longrightarrow 3.$$

Finally, by Rule (4),

$$\left(2 + \frac{1}{n}\right)\left(3 - \frac{4}{n^2}\right) \longrightarrow 2 \cdot 3 = 6.$$ ∎

■ **EXAMPLE 4** Find $\lim\limits_{n \to \infty} \dfrac{n^2 - 2n - 5}{n^3}$.

Solution Write

$$a_n = \frac{n^2 - 2n - 5}{n^3} = \frac{1}{n} - 2\left(\frac{1}{n^2}\right) - 5\left(\frac{1}{n^3}\right).$$

By the various rules, we have

$$\lim a_n = 0 - 2 \cdot 0 - 5 \cdot 0 = 0.$$ ∎

Remark In practice, we skip many of the steps illustrated in the solution of Example 2. For example, to find

$$\lim \frac{3n^2 + n - 7}{4n^2 + 3n + 6},$$

we divide all terms by n^2, then use the various rules, combining some obvious steps:

$$\lim \frac{3n^2 + n - 7}{4n^2 + 3n + 6} = \lim \frac{3 + \dfrac{1}{n} - \dfrac{7}{n^2}}{4 + \dfrac{3}{n} + \dfrac{6}{n^2}} = \frac{3 + 0 - 0}{4 + 0 + 0} = \frac{3}{4}.$$

■ **EXAMPLE 5** Let $0 < a < 1$. Prove $a^n \longrightarrow 0$.

Solution $1/a > 1$, hence $1/a = 1 + p$ with $p > 0$. By the Binomial Theorem,

$$\frac{1}{a^n} = (1 + p)^n = 1 + pn + (\text{positive terms}) > pn.$$

Therefore $$0 < a^n < \frac{1}{pn} = \frac{1}{p} \cdot \frac{1}{n}.$$

Since $1/n \longrightarrow 0$ and $1/p$ is fixed, $a^n \longrightarrow 0$. ∎

EXERCISES

If the sequence is denoted by $\{a_n\}_1^\infty$, find a formula for a_n

1 $\dfrac{1}{3}, \dfrac{1}{7}, \dfrac{1}{11}, \dfrac{1}{15}, \cdots$

2 $\dfrac{2}{25}, \dfrac{3}{36}, \dfrac{4}{49}, \dfrac{5}{64}, \cdots$

3 $-\dfrac{3 \cdot 5}{2 \cdot 3}, \dfrac{5 \cdot 7}{2 \cdot 3 \cdot 3}, -\dfrac{7 \cdot 9}{2 \cdot 3 \cdot 3 \cdot 3}, \dfrac{9 \cdot 11}{2 \cdot 3 \cdot 3 \cdot 3 \cdot 3}, \cdots$

4 $1, \ 1 + 2, \ 1 + 2 + 3, \ 1 + 2 + 3 + 4, \ \cdots$

5 $f'(1), \ f''(1), \ f'''(1), \ f^{(4)}(1), \ \cdots \qquad f(x) = \dfrac{1}{\sqrt{x}}$

6 $f'(0), \ f''(0), \ f'''(0), \ f^{(4)}(0), \ \cdots \qquad f(x) = e^{x/2}.$

Find N so that for $n \geq N$,

7 $\left| \dfrac{3n+1}{4n+5} - \dfrac{3}{4} \right| < \dfrac{1}{1000}$

8 $\dfrac{2^n}{(2n)!} < 10^{-5}.$

Using the definition of limit, show that

9 $\displaystyle\lim_{n \to \infty} \dfrac{1}{\sqrt{n}} = 0$

10 $\displaystyle\lim_{n \to \infty} \dfrac{n^2}{(n+1)(n+2)} = 1.$

Find $\lim a_n$ if $a_n =$

11 $\dfrac{n+1}{2n+3}$

12 $\dfrac{4n}{n^2+1}$

13 $\dfrac{n^2+2n-8}{n^3+7n+9}$

14 $\dfrac{n^2+5}{3n^2+12n+5}$

15 $\dfrac{n}{\sqrt{n^2+1}}$

16 $\dfrac{2n}{\sqrt{n^3+6}}$

17 $\dfrac{\sqrt{n}}{3+2\sqrt{n}}$

18 $\dfrac{1}{n^2+\sin n}$

19 $\left(1+\dfrac{2}{n}\right)^3$

20 $\left(1+\dfrac{3n}{n^2+5}\right)^4$

21 $\dfrac{10^n}{n!}$

22 $\dfrac{(n+1)(n+2)(n+3)}{(n+4)(n+5)(n+6)}$

23 $\left[\dfrac{3n}{n+7}\right]\left[1-\left(\dfrac{1}{2}\right)^n\right]$

24 $\dfrac{7^n+1}{8^n}$

25 $\dfrac{\sin n}{n}$

26 $\dfrac{1}{1+\log n}$

27 $\dfrac{n(2-e^{-n})}{3n+1}$

28 $\dfrac{e^n-1}{e^n+1}.$

29 If $\{a_n\}$ converges and $a_n \geq 0$ for $n \geq 1$, show that $\lim a_n \geq 0$.
30 (cont.) If $\{a_n\}$ and $\{b_n\}$ converge and if $a_n \geq b_n$ for $n \geq 1$, show that $\lim a_n \geq \lim b_n$.
31 If $|a_n| \longrightarrow |a|$, does $a_n \longrightarrow a$?
32 If $\{a_n\}$ and $\{b_n\}$ diverge, does $\{a_n + b_n\}$ diverge?
33 If $\{a_n\}$ and $\{b_n\}$ diverge, does $\{a_n b_n\}$ diverge?
34 If $\lim a_n = 0$, prove that $\lim a_n^2 = 0$.
35 Prove that inserting or deleting a finite number of terms does not affect the convergence of a sequence.
36 If $a_n \longrightarrow L$ and $b_n \longrightarrow L$, show that the sequence

$$a_1, \ b_1, \ a_2, \ b_2, \ a_3, \ b_3, \ \cdots$$

also converges to L.

37 Under what circumstances can a sequence of *integers* have a limit?
38 Show by examples that a sequence may diverge yet contain subsequences which converge.
39 Suppose $a_n \longrightarrow 0$ and $|b_n| \leq B$, a positive constant. Show that $a_n b_n \longrightarrow 0$.
40 (cont.) Prove that

$$\dfrac{3+e^{-n}}{n(2+\sin^4 \pi n)} \longrightarrow 0.$$

41 Prove that if $a_n \longrightarrow L$, then any subsequence of $\{a_n\}$ also converges to L.

42 Prove that if $a_n \longrightarrow L_1$ and $a_n \longrightarrow L_2$, then $L_1 = L_2$. [*Hint* Write $L_1 - L_2 = (L_1 - a_n) - (L_2 - a_n)$.]

2. PROPERTIES OF LIMITS

Continuous Functions of Limits There is a useful principle relating limits of sequences with continuous functions. It allows us to deduce things like

$$\sin a_n \longrightarrow \sin L \qquad \text{and} \qquad e^{a_n} \longrightarrow e^L$$

from $a_n \longrightarrow L$.

> Let $a_n \longrightarrow L$, and suppose $f(x)$ is continuous at $x = L$. Then
> $$f(a_n) \longrightarrow f(L) \qquad \text{as} \qquad n \longrightarrow \infty.$$

Proof We combine the two facts (1) when x is near L, then $f(x)$ is near $f(L)$; and (2) when n is large, then a_n is near L. Let $\varepsilon > 0$. Then there exists $\delta > 0$ such that

$$|f(x) - f(L)| < \varepsilon \qquad \text{whenever} \qquad |x - L| < \delta.$$

Now choose N so $\qquad |a_n - L| < \delta \qquad$ whenever $\quad n \geq N$.

Then $\qquad\qquad |f(a_n) - f(L)| < \varepsilon \qquad$ whenever $\quad n \geq N$.

This completes the proof.

■ **EXAMPLE 1** Let $a > 0$. Prove $\sqrt[n]{a} \longrightarrow 1$.

Solution The idea is to use

$$\sqrt[n]{a} = a^{1/n} = \exp(\ln a^{1/n}) = \exp[(\ln a)/n] = e^{(\ln) a/n}.$$

Now $(\ln a)/n \longrightarrow 0$ as $n \longrightarrow \infty$, hence

$$\sqrt[n]{a} = e^{(\ln a)/n} \longrightarrow e^0 = 1$$

since $f(x) = e^x$ is continuous. ■

Existence of Limits Each time we have proved that a sequence converges, we have actually found its limit. As we shall see, sometimes it is hard or impossible to find the exact limit. However, in some applications the limit itself is not needed, only the knowledge of whether or not a given sequence converges.

In such situations we need "intrinsic" criteria for convergence. These are tests for convergence that use the nature of the sequence itself and do not require knowledge of the limit. We shall state two intrinsic criteria, one for monotone sequences and one for Cauchy sequences. Both are equivalent to a deep property of the real number system called **completeness**. We shall leave the theoretical discussion to more advanced courses.

Monotone Sequences A sequence $\{a_n\}$ is called **increasing** if

$$a_1 \leq a_2 \leq a_3 \leq \cdots.$$

Similarly, $\{a_n\}$ is **decreasing** if

$$a_1 \geq a_2 \geq a_3 \geq \cdots.$$

A sequence that is either increasing or decreasing is called **monotone** (moving in one direction).

A sequence $\{a_n\}$ is **bounded above** if there is a number B such that $a_n \leq B$ for each element a_n. A sequence is **bounded below** if there is a number A such that $a_n \geq A$ for each a_n.

Convergence of Monotone Sequences An increasing sequence that is bounded above converges; a decreasing sequence that is bounded below converges.

If $a_1 \leq a_2 \leq a_3 \leq \cdots \leq B$, then $\lim a_n \leq B$.

If $a_1 \geq a_2 \geq a_3 \geq \cdots \geq A$, then $\lim a_n \geq A$.

Thus an increasing sequence is either bounded above and crowds in upon a limit, or is unbounded above and marches off the map to ∞. An analogous statement holds for decreasing sequences.

■ **EXAMPLE 2** Prove $\{a_n\}$ converges, where

$$a_n = \left(1 - \frac{1}{2^2}\right)\left(1 - \frac{1}{3^2}\right)\left(1 - \frac{1}{4^2}\right) \cdots \left(1 - \frac{1}{n^2}\right).$$

Solution Clearly $a_n > 0$ and

$$a_{n+1} = a_n\left(1 - \frac{1}{(n+1)^2}\right) < a_n,$$

so

$$a_1 > a_2 > a_3 > \cdots > 0.$$

In other words, $\{a_n\}$ is a decreasing sequence that is bounded below; hence $\{a_n\}$ converges. ■

■ **EXAMPLE 3** Define a sequence $\{a_n\}$ by

$$a_n = 1 + \frac{1}{1!} + \frac{1}{2!} + \cdots + \frac{1}{n!}.$$

Prove that $\{a_n\}$ converges.

Solution Since $a_{n+1} = a_n + 1/(n+1)! > a_n$, the sequence is increasing. To prove convergence, we shall show the sequence is bounded above. We use an important technique: replacing the terms making up a_n by slightly larger terms that can be summed easily. Now the one thing we can sum for sure is a geometric progression. This suggests the comparison

$$n! = 1 \cdot 2 \cdot 3 \cdots n > 1 \cdot 2 \cdot 2 \cdot 2 \cdots = 2^{n-1}, \qquad \frac{1}{n!} < \frac{1}{2^{n-1}}.$$

It follows that

$$a_n = 1 + 1 + \frac{1}{2} + \frac{1}{6} + \frac{1}{24} + \cdots + \frac{1}{n!} < 1 + \left(1 + \frac{1}{2} + \frac{1}{2^2} + \frac{1}{2^3} + \cdots + \frac{1}{2^{n-1}}\right)$$

$$= 1 + \frac{1 - 1/2^n}{1 - 1/2} = 1 + 2\left(1 - \frac{1}{2^n}\right) < 3.$$

Therefore $\{a_n\}$ increases and is bounded above by 3, so $\{a_n\}$ converges. We can say (from this proof) that $\lim a_n \leq 3$, no more.   ■

Remark In elementary calculus it is shown that the sequence in Example 3 actually converges to $e \approx 2.718281828$.

It is also shown that $b_n \longrightarrow e$, where $b_n = (1 + 1/n)^n$. The sequence $\{b_n\}$ is increasing and bounded above, but that is rather tricky to prove. See Exs. 29–32.

■ **EXAMPLE 4** Define a sequence $\{a_n\}$ by

$$a_1 = 2, \qquad a_{n+1} = \frac{1}{2}\left(a_n + \frac{2}{a_n}\right).$$

Prove $\{a_n\}$ converges and find its limit.

Solution Compute a few elements:

$$a_1 = 2, \qquad a_2 = \frac{1}{2}\left(2 + \frac{2}{2}\right) = \frac{3}{2} = 1.5,$$

$$a_3 = \frac{1}{2}\left(\frac{3}{2} + \frac{2}{\frac{3}{2}}\right) = \frac{17}{12} \approx 1.4167, \qquad a_4 = \frac{1}{2}\left(\frac{17}{12} + \frac{2}{\frac{17}{12}}\right) = \frac{577}{408} \approx 1.4142.$$

The sequence appears to be decreasing. Since its elements are bounded below by 0, we are optimistic about the chances of convergence. Probably the sequence converges to a positive limit L, somewhere around 1.4.

Assuming the sequence does converge, what is its limit? We exploit the defining relation

$$a_{n+1} = \frac{1}{2}\left(a_n + \frac{2}{a_n}\right).$$

If $a_n \longrightarrow L$, then the right-hand side approaches $\frac{1}{2}(L + 2/L)$. The left side approaches L because $\{a_{n+1}\}$ is a subsequence of $\{a_n\}$. Therefore, the limit L must satisfy

$$L = \frac{1}{2}\left(L + \frac{2}{L}\right), \qquad 2L^2 = L^2 + 2, \qquad L^2 = 2.$$

Since $L \geq 0$, the only possibility is $L = \sqrt{2} \approx 1.414213562$.

This reasoning is correct *provided* $\{a_n\}$ converges. To prove it does, we need only show that $\{a_n\}$ is decreasing. We would like to argue as follows:

$$a_{n+1} = \frac{1}{2}\left(a_n + \frac{2}{a_n}\right) \leq \frac{1}{2}(a_n + a_n) = a_n,$$

hence $a_{n+1} \leq a_n$.

This argument is valid provided $2/a_n \leq a_n$, that is, provided $2 \leq a_n^2$. So the final step is to prove this inequality for $n \geq 2$. Now

$$a_n^2 = \left[\frac{1}{2} \left(a_{n-1} + \frac{2}{a_{n-1}} \right) \right]^2 = \frac{1}{4} \left(a_{n-1} - \frac{2}{a_{n-1}} \right)^2 + 2 \geq 2.$$

This ties up the last loose end.   ∎

Remark   Sometimes a sequences $\{a_n\}$ jumps around a lot at first, then eventually settles down to a nice behavior. This doesn't matter, because convergence depends on what happens in the long run, not at first.

In general, convergence tests are unaffected by a finite number of exceptions. This is an important practical point, and should be remembered. For instance, a sequence that is bounded above and *ultimately* increasing converges.

Cauchy Sequences   If $a_n \longrightarrow L$, then the numbers a_n get close to L, hence close to each other. More precisely, suppose $\varepsilon > 0$ is given. There exists an N such that if $m \geq N$ and $n \geq N$, then both $|a_m - L| < \frac{1}{2}\varepsilon$ and $|a_n - L| < \frac{1}{2}\varepsilon$. Consequently

$$|a_m - a_n| = |(a_m - L) - (a_n - L)| \leq |a_m - L| + |a_n - L| < \tfrac{1}{2}\varepsilon + \tfrac{1}{2}\varepsilon = \varepsilon.$$

Cauchy Sequences   A sequence $\{a_n\}$ is called a **Cauchy sequence** if for each $\varepsilon > 0$, there is a positive integer N such that

$$|a_m - a_n| < \varepsilon \qquad \text{for all} \quad m \geq N \quad \text{and all} \quad n \geq N.$$

(A Cauchy sequence is sometimes called a **fundamental sequence**.)

(Cauchy is pronounced koh'-shee.) As an example of a Cauchy sequence, take $\{a_n\}$ where $a_n = 1/n$. Then

$$|a_m - a_n| = \left| \frac{1}{m} - \frac{1}{n} \right| < \frac{1}{m} + \frac{1}{n}.$$

Therefore $|a_m - a_n| < \varepsilon$ if

$$\frac{1}{m} < \frac{\varepsilon}{2} \quad \text{and} \quad \frac{1}{n} < \frac{\varepsilon}{2}, \qquad \text{that is,} \quad \text{if} \ m > \frac{2}{\varepsilon} \ \text{and} \ n > \frac{2}{\varepsilon}.$$

For example, if $\varepsilon = 0.001$, then $|a_m - a_n| < \varepsilon$ if $m > 2000$ and $n > 2000$. For instance, $|a_{2177} - a_{14508}| < 0.001$.

Not only is every convergent sequence a Cauchy sequence, but conversely every Cauchy sequence is a convergent sequence.

Cauchy Criterion   A sequence converges if and only if it is a Cauchy sequence.

We proved the "only if" part of this assertion: each convergent sequence is a Cauchy sequence. The converse, the "if" part, is deep and beyond the scope of this course: if a sequence is a Cauchy sequence, then it converges.

The Cauchy Criterion is *intrinsic*; it depends on the sequence itself and on nothing else. Often it enables us to prove the convergence of a sequence without knowing its limit.

In certain situations we need only compare *successive* terms of the sequence. Two important consequences of the Cauchy Criterion guarantee that if successive terms are always close together, then the sequence converges.

First Comparison Test   Suppose $0 < c < 1$ and $0 < B$, and suppose $\{a_n\}$ is a sequence for which

$$|a_n - a_{n+1}| \leq Bc^n \qquad \text{for all } n.$$

Then $\{a_n\}$ converges.

Proof   Suppose $m < n$. We use the triangle inequality and the sum of a geometric progression to estimate $|a_m - a_n|$:

$$|a_m - a_n| = |(a_m - a_{m+1}) + (a_{m+1} - a_{m+2}) + \cdots + (a_{n-1} - a_n)|$$

$$\leq |a_m - a_{m+1}| + |a_{m+1} - a_{m+2}| + \cdots + |a_{n-1} - a_n|$$

$$\leq Bc^m + Bc^{m+1} + Bc^{m+2} + \cdots + Bc^{n-1}$$

$$= Bc^m(1 + c + c^2 + \cdots + c^{n-m-1}) = Bc^m \frac{1 - c^{n-m}}{1 - c} < \frac{Bc^m}{1 - c}.$$

Since $c^m \longrightarrow 0$ as $m \longrightarrow \infty$, this is quite good enough to show that $\{a_n\}$ is a Cauchy sequence. In fact, if $\varepsilon > 0$, we can choose N so that $Bc^m/(1 - c) < \varepsilon$ for all $m \geq N$. Then for all $m \geq N$ and $n \geq N$, we have $|a_m - a_n| < \varepsilon$, so $\{a_n\}$ is a Cauchy sequence, hence convergent by the Cauchy Criterion.

■ **EXAMPLE 5**   Given an infinite decimal $0.d_1 d_2 d_3 \cdots$, where $0 \leq d_i \leq 9$, define $a_n = 0.d_1 d_2 \cdots d_n$. Prove that $\{a_n\}$ converges.

Solution $|a_n - a_{n+1}| = \dfrac{d_{n+1}}{10^{n+1}} \leq \dfrac{9}{10^{n+1}} = \dfrac{9}{10}\left(\dfrac{1}{10}\right)^n.$

The Comparison Test applies with $B = \frac{9}{10}$ and $c = \frac{1}{10}$.   ■

Second Comparison Test   Suppose $0 < c < 1$, and suppose $\{a_n\}$ is a sequence for which

$$|a_{n+1} - a_{n+2}| \leq c|a_n - a_{n+1}| \qquad \text{for all } n.$$

Then $\{a_n\}$ converges.

Proof   We work recursively back to GO:

$$|a_n - a_{n+1}| \leq c|a_{n-1} - a_n| \leq c^2|a_{n-2} - a_{n-1}|$$

$$\leq \cdots \leq c^{n-1}|a_1 - a_2| = Bc^n,$$

where $B = |a_1 - a_2|/c$. Now the previous test applies.

Remark Note carefully the distinction between these tests. In the first you must show that the difference $|a_n - a_{n+1}|$ decreases geometrically. In the second you must show that each difference is at most a fixed proportion of the previous difference.

■ **EXAMPLE 6** Set $a_1 = 1$ and $a_{n+1} = 1/(1 + a_n)$. Prove that $\{a_n\}$ converges and find its limit.

Solution The first four terms are $1, \frac{1}{2}, \frac{2}{3}, \frac{3}{5}, \frac{5}{8}, \frac{8}{13}$. They seem to increase and decrease alternately; apparently 1 is the largest, $\frac{1}{2}$ the smallest. We can easily prove this last guess by induction. For if $\frac{1}{2} \leq a_n \leq 1$, then $\frac{3}{2} \leq 1 + a_n \leq 2$, so $\frac{2}{3} \geq a_{n+1} = 1/(1 + a_n) \geq \frac{1}{2}$. Hence certainly $\frac{1}{2} \leq a_{n+1} \leq 1$.

Next, we compare successive terms:

$$\left| a_{n+1} - a_{n+2} \right| = \left| \frac{1}{1 + a_n} - \frac{1}{1 + a_{n+1}} \right| = \left| \frac{a_{n+1} - a_n}{(1 + a_n)(1 + a_{n+1})} \right|.$$

But $(1 + a_n)(1 + a_{n+1}) \geq (1 + \frac{1}{2})^2 = \frac{9}{4}$, hence

$$\left| a_{n+1} - a_{n+2} \right| \leq \tfrac{4}{9} \left| a_n - a_{n+1} \right| \qquad \text{for all } n.$$

Therefore $\{a_n\}$ converges by the Second Comparison Test.

Let $a_n \longrightarrow L$. Then $L > 0$, and from $a_{n+1} = 1/(1 + a_n)$ follows $L = 1/(1 + L)$. Hence

$$L^2 + L - 1 = 0, \qquad L = \frac{\sqrt{5} - 1}{2} \approx 0.61803399.$$

Note that $a_{15} \approx 0.61803445$, so the sequence converges rapidly.   ■

EXERCISES

Find $\lim a_n$, where $a_n =$

1 $\quad e^{-1/n}$ **2** $\quad \cos \dfrac{2\pi}{\sqrt{n}}$ **3** $\quad \arctan\left(\dfrac{n}{n + 3}\right)$ **4** $\quad \left(\dfrac{n^2 + 5n + 1}{4n^2 + 3}\right)^{1/2}$

Prove convergence of the sequence $\{a_n\}$ by showing it is monotone and bounded

5 $\quad a_n = \dfrac{1 \cdot 4 \cdot 7 \cdots (3n - 2)}{2 \cdot 5 \cdot 8 \cdots (3n - 1)}$ **6** $\quad a_n = \left(1 - \dfrac{1}{4}\right)\left(1 - \dfrac{1}{9}\right) \cdots \left(1 - \dfrac{1}{n^2}\right)$

7 $\quad a_n = 1 + \dfrac{1}{(2!)^2} + \dfrac{1}{(3!)^2} + \cdots + \dfrac{1}{(n!)^2}$ **8*** $\quad a_n = \dfrac{1}{n + 1} + \dfrac{1}{n + 2} + \cdots + \dfrac{1}{2n}$

9 $\quad a_n = c^n, \quad 0 < c < 1$ **10** $\quad a_0 = 1, \quad a_1 = \sqrt{5}, \cdots, a_{n+1} = \sqrt{5a_n}$

11 $\quad a_0 = 1, \quad a_1 = \frac{3}{2}, \cdots, a_{n+1} = 1 + \frac{1}{2}a_n$

12 $\quad a_0 = \frac{1}{2}\pi, \quad a_1 = \sin \frac{1}{2}\pi, \cdots, \quad a_{n+1} = \sin a_n.$

Use the method of Example 4 to evaluate the limit

13 $\quad$ in Ex. 10   **14** $\quad$ in Ex. 12.

15 $\quad$ Define $\{x_n\}$ by $x_0 = 0$, $x_1 = 1$, and $x_{n+2} = \frac{1}{2}(x_n + x_{n+1})$. Prove $\{x_n\}$ is convergent.

16 $\quad$ (cont.) Find $\lim x_n$.

17 $\quad$ Define $a_1 = \sqrt{2}$ and $a_{n+1} = \sqrt{2 + a_n}$ for $n = 1, 2, \cdots$. Prove that $\{a_n\}$ is convergent.

18 (cont.) Find $\lim a_n$.

19 Suppose $x > 0$. Define $a_1 = \sqrt{x}$ and $a_{n+1} = \sqrt{x + a_n}$ for $n = 1, 2, \cdots$. Prove that $\{a_n\}$ is bounded.

20 (cont.) Prove $\{a_n\}$ is increasing, hence has a limit. Find the limit.

21 Suppose $0 < a < 1$ and $p > 0$. Find $\lim n^p a^n$. [*Hint* $(\ln n)/n \longrightarrow 0$.]

22* Let $a_1 = \sqrt{12}$ and $a_{n+1} = \sqrt{12 - a_n}$ for $n = 1, 2, \cdots$. Analyze $\{a_n\}$.

23 Suppose $0 < x < y$. Define two sequences $\{a_n\}$ and $\{b_n\}$ by $a_0 = x$, $b_0 = y$,

$$a_{n+1} = \frac{2}{(1/a_n) + (1/b_n)}, \qquad b_{n+1} = \frac{a_n + b_n}{2}.$$

Prove $a_0 < a_1 < a_2 < \cdots < b_2 < b_1 < b_0$.

24 (cont.) Find $\lim a_n$ and $\lim b_n$.

25 Suppose $0 < x < 2$. Define $\{b_n\}$ by $b_0 = 1 - x$ and $b_{n+1} = b_n^2$. Show that $b_n \longrightarrow 0$.

26 (cont.) Define $\{a_n\}$ by $a_0 = 1$, $a_{n+1} = a_n(1 + b_n)$. Prove that $a_n \longrightarrow 1/x$. [*Hint* Show that $a_n = (1 - b_n)/x$.] (This provides an algorithm for division on a calculator that has only $+$, $-$, and $\times$.)

27 Suppose $0 < x < 2$. Define $\{b_n\}$ by $b_0 = 1 - x$ and $b_{n+1} = \frac{1}{4}b_n^2(3 + b_n)$. Prove $b_n \longrightarrow 0$.

28 (cont.) Define $\{a_n\}$ by $a_0 = x$ and $a_{n+1} = a_n(1 + \frac{1}{2}b_n)$. Prove $a_n \longrightarrow \sqrt{x}$. [*Hint* Show that $a_n = \sqrt{x(1 - b_n)}$.] (This provides an algorithm for $\sqrt{}$ in terms of $+$, $-$, and $\times$.)

29 Let $0 < x < y$. Prove

$$(n + 1)x^n < \frac{y^{n+1} - x^{n+1}}{y - x} < (n + 1)y^n,$$

hence $\quad x^n[(n + 1)y - nx] < y^{n+1} \quad$ and $\quad y^n[(n + 1)x - ny] < x^{n+1}$.

30* (cont.) Set $a_n = (1 + 1/n)^n$. Prove $\{a_n\}$ is increasing and $a_n < 4$.

31* (cont.) Set $b_n = (1 + 1/n)^{n+1}$. Prove $\{b_n\}$ is decreasing.

32 (cont.) Conclude that $\{a_n\}$ and $\{b_n\}$ converge to the same limit.

Use the First Comparison Test to prove the convergence of $\{a_n\}$, where

33 $\quad a_n = 1 - \dfrac{1}{1!} + \dfrac{1}{2!} - \dfrac{1}{3!} + \cdots + \dfrac{(-1)^n}{n!}$ **34** $\quad a_n = \dfrac{1}{2} - \dfrac{2}{2^2} + \dfrac{3}{2^3} + \cdots + (-1)^{n-1}\dfrac{n}{2^n}$.

35 Suppose $0 < x$. Set $b_0 = \frac{1}{2}(x + 1/x)$ and $b_{n+1} = \sqrt{\frac{1}{2}(1 + b_n)}$. Show that $b_n \longrightarrow 1$.

36* (cont.) Set $a_0 = \frac{1}{2}(x - 1/x)$ and $a_{n+1} = a_n/b_{n+1}$. Prove that $a_n \longrightarrow \ln x$. (This provides an algorithm for computing logs on a calculator with $+$, $-$, $\times$, and $\sqrt{}$.) [*Hint* Express b_n and a_n in terms of x.]

3. INFINITE SERIES

One of the most important topics in mathematical analysis, both in theory and applications, is infinite series. The basic problem is how to add up a sum with infinitely many terms. At first that seems impossible; life is too short. However, suppose we look at the sum

$$1 + \frac{1}{2} + \frac{1}{4} + \cdots + \frac{1}{2^n} + \cdots$$

and start adding up terms. We find $1, \frac{3}{2}, \frac{7}{4}, \frac{15}{8}, \frac{31}{16}, \cdots$, numbers getting closer and closer to 2. The message is clear: these finite sums have limit 2. Therefore in some sense, the infinite sum equals 2.

If we try to add up terms of the sum

$$1 + 1 + 1 + \cdots,$$

we find $1, 2, 3, 4, \cdots$, numbers becoming larger and larger. The situation is hopeless; there is no reasonable total.

Let us now consider in some detail two important infinite sums.

Geometric Series   A **geometric series** is an infinite sum in which the ratio of any two consecutive terms is always the same:

$$a + ar + ar^2 + \cdots + ar^n + \cdots \qquad (a \neq 0, \quad r \neq 0).$$

Let s_n denote the sum of all terms up to ar^n,

$$s_n = a + ar + ar^2 + \cdots + ar^n.$$

If $r = 1$, then $s_n = a + a + \cdots + a = (n + 1)a$, so $s_n \longrightarrow \pm\infty$. If $r \neq 1$, there is a simple formula for s_n:

$$s_n = a(1 + r + r^2 + \cdots + r^n) = a\left(\frac{1 - r^{n+1}}{1 - r}\right).$$

(To check, multiply both sides by $1 - r$.) If $|r| < 1$, then $r^{n+1} \longrightarrow 0$ as n increases. Hence

$$\lim s_n = \frac{a}{1 - r}, \qquad |r| < 1,$$

so a logical choice for the "sum" of the series is $a/(1 - r)$. But if $|r| > 1$, then r^{n+1} grows beyond all bound, and the situation is hopeless. If $r = -1$, then s_n is alternately a and 0. There is no reasonable sum in this case either.

> An infinite geometric series
>
> $$a + ar + ar^2 + \cdots + ar^n + \cdots$$
>
> has the sum $a/(1 - r)$ if $|r| < 1$, but no sum if $|r| \geq 1$.

Harmonic Series   The series

$$1 + \frac{1}{2} + \frac{1}{3} + \cdots + \frac{1}{n} + \cdots$$

is known as the **harmonic series**. Although it is not at all obvious, the sums $s_n = 1 + \frac{1}{2} + \frac{1}{3} + \cdots + 1/n$ increase beyond all bound, so the series has no sum. To see why, we observe that

$$s_1 = 1 > \frac{1}{2}, \qquad s_2 = s_1 + \frac{1}{2} > \frac{1}{2} + \frac{1}{2} = \frac{2}{2},$$

$$s_4 = s_2 + \left(\frac{1}{3} + \frac{1}{4}\right) > s_2 + \left(\frac{1}{4} + \frac{1}{4}\right) > \frac{2}{2} + \frac{1}{2} = \frac{3}{2},$$

$$s_8 = s_4 + \left(\frac{1}{5} + \frac{1}{6} + \frac{1}{7} + \frac{1}{8}\right) > s_4 + \left(\frac{1}{8} + \frac{1}{8} + \frac{1}{8} + \frac{1}{8}\right) > \frac{3}{2} + \frac{1}{2} = \frac{4}{2}.$$

Similarly, $s_{16} > 5/2$, $s_{32} > 6/2$, $\cdots$, $s_{2^n} > (n + 1)/2$. Thus the sequence of sums s_n increases, and our estimates show s_n eventually passes any given positive number. (This happens very slowly it is true; around 2^{15} terms are needed before s_n exceeds 10 and around 2^{29} terms before it exceeds 20.)

Remark   Both the geometric series for $0 < r < 1$ and the harmonic series have positive terms that decrease toward zero, yet one series has a sum and the other does not. This indicates the subtlety we must expect in our further study of infinite series.

Convergence and Divergence   It is time to formulate our ideas more precisely.

An **infinite series** is a formal sum

$$\sum a_n = a_1 + a_2 + a_3 + \cdots.$$

Associated with each infinite series is its sequence $\{s_n\}$ of **partial sums** defined by

$$s_1 = a_1, \qquad s_2 = a_1 + a_2, \cdots, s_n = a_1 + a_2 + \cdots + a_n.$$

A series **converges** to the number S, or has **sum** S, if $\lim s_n = S$. A series **diverges**, or has no sum, if $\lim s_n$ does not exist. A series that converges is called **convergent**; a series that diverges is called **divergent**. If a series converges to S, we shall write

$$a_1 + a_2 + a_3 + \cdots = S, \qquad \text{or equivalently,} \qquad \sum_{n=1}^{\infty} a_n = S.$$

We have a precise definition for $\lim s_n = S$. Let us rephrase the definition of convergence of series accordingly:

The infinite series $a_1 + a_2 + a_3 + \cdots$ converges to S if for each $\varepsilon > 0$, there is a positive integer N such that

$$|(a_1 + a_2 + \cdots + a_n) - S| < \varepsilon$$

whenever $n \geq N$.

Thus, no matter how small ε, you will get within ε of S by adding up enough terms. For each ε, the N tells how many terms are "enough." Naturally the smaller ε is, the larger N will be.

■ **EXAMPLE 1**   Consider the series $3 + \frac{3}{2} + \frac{3}{4} + \frac{3}{8} + \cdots$.
  (a) Show that its sum is 6.   Find N so that
  (b) $|s_n - 6| < 10^{-4}$ for $n \geq N$ (c) $|s_n - 6| < 10^{-8}$ for $n \geq N$.

Solution   (a)   This is a geometric series with first term $a = 3$ and common ratio $r = \frac{1}{2}$. Since $|r| < 1$, the series converges to

$$\frac{a}{1-r} = \frac{3}{1-\frac{1}{2}} = 6.$$

(b)   The n-th partial sum is

$$s_n = 3\left(1 + \frac{1}{2} + \frac{1}{4} + \cdots + \frac{1}{2^{n-1}}\right) = 3\,\frac{1-(\frac{1}{2})^n}{1-\frac{1}{2}} = 6\left[1 - \left(\frac{1}{2}\right)^n\right].$$

Hence
$$|s_n - 6| = 6(\tfrac{1}{2})^n.$$

Therefore $|s_n - 6| < 10^{-4}$ provided

$$6(\tfrac{1}{2})^n < 10^{-4}, \qquad (\tfrac{1}{2})^n < \tfrac{1}{6} \times 10^{-4}, \qquad 2^n > 6 \times 10^4.$$

Now $2^{15} < 33000$ and $2^{16} > 65000$, hence $N = 16$ works.

(c)   As in (b) we must have $2^n > 6 \times 10^8$, that is,

$$n \log 2 > \log(6 \times 10^8) = 8 + \log 6, \qquad n > \frac{8 + \log 6}{\log 2} \approx 29.2.$$

Therefore $N = 30$ works.   ∎

Convergence Tests   When we study the convergence of an infinite series $\sum a_n$, we really study the convergence of the sequence $\{s_n\}$ of partial sums. Thus we actually have *two* sequences involved, the sequence $\{a_n\}$ that defines the series and the derived sequence $\{s_n\}$ of partial sums. The definition of convergence concentrates on $\{s_n\}$, and we can apply everything we know about the convergence of sequences to $\{s_n\}$. However, what we really want are tests for the convergence of $\sum a_n$ in terms of the sequence $\{a_n\}$.

For example, we know that adding a constant to each term of $\{s_n\}$ does not affect its convergence or divergence. Consequently inserting, deleting, or altering a finite number of terms of an infinite series only adds a constant to each s_n beyond a certain point, hence does not affect its convergence or divergence. For instance, if we delete the first 10 terms of the series $a_1 + a_2 + a_3 + \cdots$, then we decrease each partial sum s_n (for $n > 10$) by the amount s_{10}. If the original series diverges, then so does the modified series. If it converges to S, then the modified series converges to $S - s_{10}$.

Warning   In problems where we must decide whether a given infinite series converges or diverges, we may, without prior notice, ignore or change a (finite) batch of terms at the beginning. This does not affect convergence.

Recall the Cauchy Criterion for convergence of sequences:

A sequence $\{s_n\}$ converges if and only if for each $\varepsilon > 0$, there is a positive integer N such that

$$|s_m - s_n| < \varepsilon$$

whenever $m, n \geq N$.

Thus all elements of the sequence beyond a certain point must be within ε of each other. The advantage of the Cauchy Criterion is that it depends only on the elements of the sequence itself; you don't have to know the limit of a sequence in order to show convergence. That's a great help since often it is very hard to find the exact limit of a sequence; besides, you may only need to know that the sequence does indeed converge to some limit.

Let us apply the Cauchy Criterion to the partial sums of a series. We simply observe (for $m > n$) that

$$s_m - s_n = (a_1 + a_2 + \cdots + a_n + a_{n+1} + \cdots + a_m) - (a_1 + a_2 + \cdots + a_n)$$

$$= a_{n+1} + a_{n+2} + \cdots + a_m.$$

Cauchy Test   An infinite series $\sum a_n$ converges if and only if for each $\varepsilon > 0$, there is a positive integer N such that

$$\left| a_{n+1} + a_{n+2} + \cdots + a_m \right| < \varepsilon$$

whenever $m > n \geq N$.

Thus beyond a certain point in the series, any block of consecutive terms, *no matter how long*, must have a very small sum.

An important corollary of the "only if" part of the Cauchy Test is a necessary condition for convergence of an infinite series. If a series does converge, then the Cauchy Test is satisfied. In particular, it is satisfied in the special case $n = m - 1$. Then the block consists of just one term, a_m, so $|a_m| < \varepsilon$ when $m \geq N$. In other words, $a_m \longrightarrow 0$.

---

n-th Term Test   If $\sum a_n$ converges, then $\lim_{n \to \infty} a_n = 0$.

If $\lim a_n$ does not exist, or if $\lim a_n$ exists but $\lim a_n \neq 0$, then $\sum a_n$ diverges.

For emphasis we have stated the test in two equivalent forms. We remark that it can easily be proved from scratch, independently of the Cauchy Test. For, if $\sum a_n = S$, then $s_n \longrightarrow S$, hence

$$a_n = s_n - s_{n-1} \longrightarrow S - S = 0.$$

Warning   Pay attention here, because misuse of this test causes lots of errors. It is really a test for *divergence*, not a test for convergence. It says that if $\lim a_n \neq 0$, then $\sum a_n$ diverges. It says *nothing* if $\lim a_n = 0$. In that case the series may diverge—or it may converge. Keep in mind the harmonic series, where $a_n = 1/n$; then $\lim a_n = 0$ *and* $\sum a_n$ diverges.

EXERCISES

Compute the finite sum

1 $1 + \dfrac{1}{3} + \dfrac{1}{3^2} + \cdots + \dfrac{1}{3^9}$

2 $1 - \dfrac{1}{3} + \dfrac{1}{3^2} - + \cdots - \dfrac{1}{3^9}$

3 $\dfrac{1}{2} + \dfrac{1}{4} + \cdots + \dfrac{1}{256}$

4 $\left(\dfrac{2}{3}\right)^2 + \left(\dfrac{2}{3}\right)^3 + \cdots + \left(\dfrac{2}{3}\right)^6$

5 $3 + \dfrac{3^2}{x} + \dfrac{3^3}{x^2} + \cdots + \dfrac{3^{n+1}}{x^n}$

6 $1 - y^2 + y^4 - + \cdots + y^{20}$

7 $r^{1/2} + r + r^{3/2} + \cdots + r^4$

8 $(x + 1) + (x + 1)^2 + \cdots + (x + 1)^5.$

Sum the infinite series

9 $1 - \dfrac{2}{5} + \left(\dfrac{2}{5}\right)^2 - \left(\dfrac{3}{5}\right)^3 + - \cdots$

10 $\dfrac{1}{2} - \dfrac{1}{4} + \dfrac{1}{8} - \dfrac{1}{16} + - \cdots$

11 $\dfrac{1}{2^{10}} + \dfrac{1}{2^{11}} + \dfrac{1}{2^{12}} + \cdots$

12 $\dfrac{1}{3} + \dfrac{1}{27} + \dfrac{1}{243} + \cdots$

13 $\dfrac{4 + 1}{9} + \dfrac{8 + 1}{27} + \dfrac{16 + 1}{81} + \cdots$

14 $12 - 6 + 3 - \dfrac{3}{2} + \dfrac{3}{4} - \cdots$

15 $\dfrac{1}{2 + x^2} + \dfrac{1}{(2 + x^2)^2} + \dfrac{1}{(2 + x^2)^3} + \cdots$

16 $\dfrac{\cos \theta}{2} + \dfrac{\cos^2 \theta}{4} + \dfrac{\cos^3 \theta}{8} + \cdots.$

17 A certain rubber ball when dropped will bounce back to half the height from which it is released. If the ball is dropped from 3 ft and continues to bounce indefinitely, find the total distance through which it moves.

18 Trains A and B are 60 miles apart on the same track and start moving toward each other at the rate of 30 mph. At the same time, a fly starts at train A and flies to train B at 60 mph. Then it returns to train A, then to B, etc. Use a geometric series to compute the total distance it flies until the trains meet.

19 (cont.) Do Ex. 18 without geometric series.

20 A line segment of length L is drawn and its middle third is erased. Then (step 2) the middle third of each of the two remaining segments is erased. Then (step 3) the middle third of each of the four remaining segments is erased, etc. After step n, what is the total length of all the segments deleted?

Interpret the repeating decimals as geometric series and find their sums

21 $0.11111 \cdots$ **22** $0.101010 \cdots$ **23** $0.434343 \cdots$ **24** $0.185185185 \cdots.$

Show that the series diverge

25 $\frac{1}{2} + \frac{1}{4} + \frac{1}{6} + \frac{1}{8} + \cdots$

26 $1 + \frac{1}{3} + \frac{1}{5} + \frac{1}{7} + \cdots.$

27 Find n so large that $\dfrac{1}{101} + \dfrac{1}{102} + \cdots + \dfrac{1}{n} > 2.$

28 Aristotle summarized Zeno's paradoxes as follows:

I can't go from here to the wall. For to do so, I must first cover half the distance, then half the remaining distance, then again half of what still remains. This process can always be continued and can never be completed.

Explain what is going on here.

Use partial fractions to sum

29 $\dfrac{1}{1 \cdot 2} + \dfrac{1}{2 \cdot 3} + \dfrac{1}{3 \cdot 4} + \cdots$

30 $\displaystyle\sum_{n=1}^{\infty} \dfrac{2n + 1}{[n(n + 1)]^2}.$

31 Describe all convergent series of *integers*.

32 Show how the harmonic series fails the Cauchy Test for $\varepsilon = \frac{1}{4}$.

4. SERIES WITH POSITIVE TERMS

In this section we deal only with infinite series having non-negative terms. The partial sums of such a series form an increasing sequence, $s_1 \leq s_2 \leq s_3 \leq s_4 \leq \cdots$. Recall that an increasing sequence must be one of two types: either (a) the sequence is bounded above, in which case it converges; or (b) it is not bounded above, and it marches off the map to $+\infty$.

We deduce corresponding statements about series:

A series $a_1 + a_2 + a_3 + \cdots$ with $a_n \geq 0$ converges if and only if there exists a positive number M such that

$$a_1 + a_2 + \cdots + a_n \leq M \qquad \text{for all} \quad n \geq 1.$$

Using this fact, we can often establish the convergence or divergence of a given series by comparing it with a familiar series.

Comparison Test Suppose $\sum a_n$ and $\sum b_n$ are series with non-negative terms.
(1) If $\sum a_n$ converges and $b_n \leq a_n$ for all $n \geq 1$, then $\sum b_n$ also converges,
(2) If $\sum a_n$ diverges and $b_n \geq a_n$ for all $n \geq 1$, then $\sum b_n$ also diverges.

Proof Let s_n and t_n denote the partial sums of $\sum a_n$ and $\sum b_n$ respectively. Then $\{s_n\}$ and $\{t_n\}$ are increasing sequences.
(1) Since $\sum a_n$ converges, $s_n \leq \sum_1^\infty a_n = M$ for all $n \geq 1$. Since $b_k \leq a_k$ for all k, we have $t_n \leq s_n$ for all n. Hence $t_n \leq s_n \leq M$ for all $n \geq 1$, so $\sum b_n$ converges.
(2) Since $\sum a_n$ diverges, the sequence $\{s_n\}$ is unbounded. Since $b_k \geq a_k$, we have $t_n \geq s_n$. Hence $\{t_n\}$ is also unbounded, so $\sum b_n$ diverges.

Remarks It is important to apply the Comparison Test correctly. Roughly speaking, (1) says that "smaller than small is small" and (2) says that "bigger than big is big." However the phrases "smaller than big" and "bigger than small" contain little useful information.

The Comparison Test, as well as the other tests we shall derive, applies if the given condition holds from some point on, not necessarily starting at $n = 1$. For example, if $\sum a_n$ converges, and if $b_n \leq a_n$ for $n \geq 500$, then $\sum b_n$ converges. The finite sum $b_1 + b_2 + \cdots + b_{499}$ can be anything; only the ultimate behavior of a series counts toward convergence or divergence.

■ **EXAMPLE 1** Test for convergence or divergence:

(a) $\sum \dfrac{\sin^2 n}{3^n}$ (b) $\sum \dfrac{1}{\sqrt{n}}$ (c) $\sum \dfrac{n}{2n + 1}$.

Solution (a) $(\sin^2 n)/3^n \leq 1/3^n$. But $\sum 1/3^n$ converges, so the given series converges.
(b) $1/\sqrt{n} \geq 1/n$. But $\sum 1/n$ diverges, so the given series diverges.
(c) Diverges because $a_n = n/(2n + 1) \longrightarrow \frac{1}{2} \neq 0$. ■

p-Series The comparison test is useful provided you have a good supply of known series. An excellent class of series for comparisons are those of the form $\sum 1/n^p$.

> The series $\displaystyle\sum \frac{1}{n^p}$ diverges if $p \le 1$ and converges if $p > 1$.

Proof If $0 < p \le 1$, then $1/n^p \ge 1/n$ and the series diverges by comparison with the divergent series $\sum 1/n$.

If $p > 1$, we shall show that the partial sums of the series are bounded. We use an important trick: we interpret s_n as an area and compare it with a region below the curve $y = 1/x^p$. See Fig. 1.

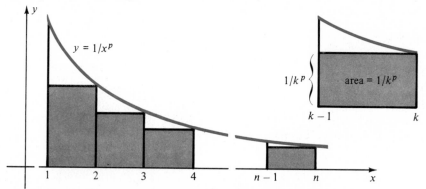

Fig. 1 The rectangular sum is less than the area under the curve.

The combined areas of the rectangles shown is less than the area under the decreasing curve between $x = 1$ and $x = n$. Therefore

$$\frac{1}{2^p} + \frac{1}{3^p} + \cdots + \frac{1}{n^p} < \int_1^n \frac{dx}{x^p} = \left.\frac{-1}{p-1}\frac{1}{x^{p-1}}\right|_1^n = \frac{1}{p-1}\left(1 - \frac{1}{n^{p-1}}\right).$$

Since $p - 1 > 0$, the right side is a positive number, a little less than $1/(p-1)$ for all values of n. Hence

$$s_n = 1 + \left(\frac{1}{2^p} + \frac{1}{3^p} + \cdots + \frac{1}{n^p}\right) < 1 + \frac{1}{p-1}$$

for $n \ge 1$. Thus the partial sums are bounded if $p > 1$, so the series converges.

Further Comparison Tests Suppose $\sum a_n$ is a given series, and $c \ne 0$. Then the two series $\sum a_n$ and $\sum ca_n$ either both converge or both diverge. For the partial sums of the series are $\{s_n\}$ and $\{cs_n\}$, sequences that converge or diverge together.

We can extend these remarks to a pair of series $\sum a_n$ and $\sum b_n$ where the ratios b_n/a_n are not constant, but restricted to a suitable range.

> Let $\sum a_n$ and $\sum b_n$ be given series with positive terms. Suppose there exist positive numbers c and d such that
>
> $$c \le \frac{b_n}{a_n} \le d$$
>
> for all sufficiently large n. Then the series both converge or both diverge.

Proof If $\sum a_n$ converges, then so does $\sum da_n$. But $b_n \leq da_n$, so $\sum b_n$ converges. If $\sum a_n$ diverges, then so does $\sum ca_n$. But $b_n \geq ca_n$, so $\sum b_n$ diverges. Done.

The conditions of the preceding test are automatically satisfied if the ratios b_n/a_n actually approach a positive limit L. Then, by the definition of limit with $\varepsilon = \frac{1}{2}L$, all ratios satisfy $\frac{1}{2}L < b_n/a_n < \frac{3}{2}L$, except perhaps for a finite number of them.

Let $\sum a_n$ and $\sum b_n$ have positive terms. If $\lim b_n/a_n = L$ exists and if $L > 0$, then either both series converge or both series diverge.

■ **EXAMPLE 2** Test for convergence or divergence:

(a) $\sum \dfrac{1}{n + \sqrt{n}}$ (b) $\sum \dfrac{4n + 1}{3n^3 - n^2 - 1}$.

Solution (a) When n is very large, n is much larger than $\sqrt{n}$. This suggests that the terms behave roughly like $1/n$, so we begin to smell divergence. Let $a_n = 1/n$ and $b_n = 1/(n + \sqrt{n})$. We could establish divergence right away if it were true that $b_n > a_n$. Unfortunately that is false. (It would be true if b_n were $1/(n - \sqrt{n})$.) Instead we examine the ratios of b_n to a_n:

$$\frac{b_n}{a_n} = \frac{n}{n + \sqrt{n}} = \frac{1}{1 + 1/\sqrt{n}} \longrightarrow \frac{1}{1 + 0} = 1, \quad \text{as} \quad n \longrightarrow \infty.$$

The ratios have a positive limit. Therefore $\sum b_n$ diverges since $\sum 1/n$ diverges.

(b) When n is very large, the terms appear to behave like $4n/3n^3 = 4/3n^2$. This suggests comparison with the convergent series $\sum 1/n^2$. Let $a_n = 1/n^2$ and $b_n = (4n + 1)/(3n^3 - n^2 - 1)$. Then

$$\frac{b_n}{a_n} = \frac{(4n + 1)n^2}{3n^3 - n^2 - 1} = \frac{4 + 1/n}{3 - 1/n - 1/n^3} \longrightarrow \frac{4}{3}.$$

The ratios have a positive limit. Therefore $\sum b_n$ converges because $\sum 1/n^2$ converges.

■

The Ratio Test In a geometric series, the ratio a_{n+1}/a_n is a constant, r. If $|r| < 1$, the series converges, basically because its terms decrease rapidly. By analogy, we should expect convergence in general if the ratios are small, not necessarily constant.

First Ratio Test Let $\sum a_n$ be a series of positive terms.

(1) The series converges if $\dfrac{a_{n+1}}{a_n} \leq r < 1$ from some point on.

(2) The series diverges if $\dfrac{a_{n+1}}{a_n} \geq 1$ from some point on.

Proof (1) Suppose $a_{n+1}/a_n \le r < 1$ starting with $n = N$. Then

$$a_{N+1} \le a_N r, \qquad a_{N+2} \le a_{N+1} r \le a_N r^2,$$

and by induction, $a_{N+k} \le a_N r^k$, that is, $a_n \le a_N r^{n-N} = (a_N r^{-N})r^n$ for all $n \ge N$. It follows that the series $\sum a_n$ converges by comparison with the convergent geometric series $\sum r^n$.

(2) From some point on, $a_{n+1} \ge a_n$. The terms increase, hence the series diverges.

Warning Note that the test for convergence requires $a_{n+1}/a_n \le r < 1$, not just $a_{n+1}/a_n < 1$. The ratios must *stay away* from 1. If $a_{n+1}/a_n < 1$ but $a_{n+1}/a_n \longrightarrow 1$, we may have divergence. For example, take $a_n = 1/n$. Then $a_{n+1}/a_n = n/(n+1) = 1 - 1/(n+1) < 1$, but $\sum 1/n$ diverges.

It often happens that the ratios a_{n+1}/a_n approach a limit. Then we can cast the Ratio Test in a different form.

Second Ratio Test Let $\sum a_n$ be a series of positive terms. Suppose

$$\lim_{n \to \infty} \frac{a_{n+1}}{a_n} = r.$$

(1) The series converges if $r < 1$.

(2) The series diverges if $r > 1$.

(3) If $r = 1$, the test is inconclusive; the series may either converge or diverge.

Proof (1) If $r < 1$, choose ε so small that $r + \varepsilon < 1$. By definition of the statement $a_{n+1}/a_n \longrightarrow r$, there is a positive integer N such that $a_{n+1}/a_n < r + \varepsilon < 1$ for all $n \ge N$. Therefore the series converges by the preceding test.

(2) Similarly, if $r > 1$, then $a_{n+1}/a_n \ge 1$ from some point on. The series diverges.

(3) If $r = 1$, this test cannot distinguish between convergent and divergent series. For example, take $a_n = 1/n^p$. The series converges for $p > 1$, diverges for $p \le 1$. But for all values of p,

$$\frac{a_{n+1}}{a_n} = \frac{n^p}{(n+1)^p} = \left(\frac{n}{n+1}\right)^p = \left(1 - \frac{1}{n+1}\right)^p \longrightarrow (1-0)^p = 1.$$

■ **EXAMPLE 3** Test for convergence or divergence

(a) $\displaystyle\sum \frac{n}{2^n}$ (b) $\displaystyle\sum \frac{10^n}{n!}$.

Solution (a) Set $a_n = n/2^n$. Then

$$\frac{a_{n+1}}{a_n} = \frac{n+1}{2^{n+1}} \bigg/ \frac{n}{2^n} = \frac{n+1}{2n} = \frac{1}{2}\left(1 + \frac{1}{n}\right) \longrightarrow \frac{1}{2}.$$

Since $\frac{1}{2} < 1$, the series converges by the ratio test.

(b) Set $a_n = 10^n/n!$. Then

$$\frac{a_{n+1}}{a_n} = \frac{10^{n+1}}{1 \cdot 2 \cdots n(n+1)} \bigg/ \frac{10^n}{1 \cdot 2 \cdots n} = \frac{10}{n+1} \longrightarrow 0.$$

Since $0 < 1$, the series converges by the ratio test.   ■

EXERCISES

Determine whether the series converges or diverges

1 $\displaystyle\sum \frac{1}{n^2 + 1}$

2 $\displaystyle\sum \frac{1}{2^n \sqrt{n}}$

3 $\displaystyle\sum \frac{n}{4n + 3}$

4 $\displaystyle\sum \frac{1}{(2n - 1)^2}$

5 $\displaystyle\sum \frac{1}{n\sqrt{n + 3}}$

6 $\displaystyle\sum \frac{1 + \sqrt[3]{n}}{n}$

7 $\displaystyle\sum \frac{n^2}{2n^4 + 7}$

8 $\displaystyle\sum \frac{1}{\ln n}$

9 $\displaystyle\sum \frac{1}{n^n}$

10 $\displaystyle\sum \frac{n}{(n + 1)(n + 3)(n + 5)}$.

11 Prove that if $\sum a_n$ and $\sum b_n$ converge, then so does $\sum (a_n + b_n)$, and find the sum.

12 If $\sum a_n$ and $\sum b_n$ diverge, show by examples that $\sum (a_n + b_n)$ may either converge or diverge.

Let $\sum a_n$ be a convergent series of positive terms

13 Prove that $\sum a_n{}^2$ converges

14 Show by examples that $\sum \sqrt{a_n}$ may either converge or diverge.

Test for convergence or divergence

15 $\displaystyle\sum \frac{1}{n^2 - 3}$

16 $\displaystyle\sum \frac{1}{\sqrt{2n^3 - n}}$

17 $\displaystyle\sum \frac{1}{4n - 1}$

18 $\displaystyle\sum \frac{5 + \sqrt{n}}{1 + n}$

19 $\displaystyle\sum \frac{n^3}{n!}$

20 $\displaystyle\sum n^3 \left(\frac{3}{4}\right)^n$

21 $\displaystyle\sum n e^{-n}$

22 $\displaystyle\sum \frac{3^n + 1}{5e^n + n}$

23 $\displaystyle\sum \frac{2^n + n}{3^n - n}$

24 $\displaystyle\sum \frac{1}{(\ln n)^n}$

25 $\displaystyle\sum \frac{n!}{1 \cdot 3 \cdot 5 \cdots (2n - 1)}$

26 $\displaystyle\sum \frac{(n!)^2}{(2n)!}$.

Find all real numbers x for which the series converges

27 $\displaystyle\sum \frac{x^{2n}}{n!}$

28 $\displaystyle\sum \frac{\sin^2 nx}{n^2}$

29 $\displaystyle\sum (3x)^{2n}$

30 $\displaystyle\sum n x^{2n}$.

31 **(Root Test)** If $a_n > 0$ and if $\sqrt[n]{a_n} \le r < 1$ for $n \ge 1$, show that $\sum a_n$ converges.

32 Let $\sum a_n$ and $\sum b_n$ be series with positive terms. Suppose $b_n/a_n \longrightarrow 0$. Find an example where $\sum b_n$ converges while $\sum a_n$ diverges. Does this contradict the text?

5. SERIES WITH POSITIVE AND NEGATIVE TERMS

Infinite series with both positive and negative terms are generally more complicated than series with terms all of the same sign. In this section, we discuss two common types of mixed series that are manageable.

Alternating Series An **alternating series** is one whose terms are alternately positive and negative. Examples:

$$1 - \frac{1}{2} + \frac{1}{3} - \frac{1}{4} + - + - \cdots,$$

$$1 - x^2 + x^4 - x^6 + - + - \cdots \qquad \text{(alternating for all } x \neq 0\text{)},$$

$$x - \frac{x^2}{4} + \frac{x^3}{9} - \frac{x^4}{16} + - + - \cdots \qquad \text{(alternating only for } x > 0\text{)}.$$

Such series have some extremely useful properties, two of which we now state.

Alternating Series Test If the terms of an alternating series decrease in absolute value to zero, then the series converges.

Remainder Estimate If such a series is broken off at the n-th term, then the remainder (in absolute value) is less than the absolute value of the $(n + 1)$-th term.

These assertions provide a very simple convergence criterion and an immediate remainder estimate for *alternating* series. Let us show geometrically that they make good sense. (A formal proof is outlined in Exs. 25 and 26.)

Suppose $\sum a_n$ is an alternating series whose terms decrease in absolute value to zero. (To be definite, assume $a_1 > 0$.) The partial sums $s_n = a_1 + a_2 + \cdots + a_n$ oscillate back and forth, as shown in Fig. 1. But since the terms decrease to zero, the oscillations become shorter and shorter. The odd partial sums decrease and the even ones increase, squeezing down on some number S. Thus, the series converges to S.

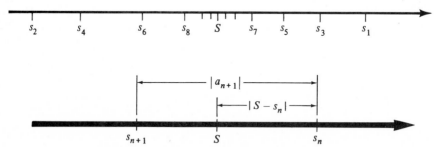

Fig. 1 The partial sums of an alternating series

If the series is broken off after n terms, the remainder is $|S - s_n|$. But from Fig. 1,

$$|S - s_n| < |s_{n+1} - s_n| = |a_{n+1}|.$$

Thus, the remainder is less than the absolute value of the $(n + 1)$-th term.

■ **EXAMPLE 1** Prove convergent

(a) $1 - \dfrac{1}{\sqrt{2}} + \dfrac{1}{\sqrt{3}} - \dfrac{1}{\sqrt{4}} + - \cdots$ \qquad (b) $\dfrac{1}{2} - \dfrac{\sqrt{2}}{3} + \dfrac{\sqrt{3}}{4} - \dfrac{\sqrt{4}}{5} + - \cdots.$

Solution (a) If $\sum a_n$ denotes the series, then $|a_n| = 1/\sqrt{n}$. The numbers $|a_n|$ clearly decrease toward 0. Hence the series converges by the alternating series test.

(b) This time $|a_n| = \sqrt{n}/(n+1)$. The series alternates, and $|a_n| \longrightarrow 0$. It will be enough to show that the sequence $\{\sqrt{n}/(n+1)\}$ decreases. To do so show that the *function* $y(x) = \sqrt{x}/(x+1)$ decreases. Compute its derivative:

$$y'(x) = \frac{(x+1)/2\sqrt{x} - \sqrt{x}}{(x+1)^2} = \frac{1-x}{2\sqrt{x}\,(x+1)^2}.$$

Clearly $y'(x) < 0$ for $x > 1$, therefore $y(x)$ decreases. In particular,

$$y(n+1) < y(n), \qquad \text{or equivalently,} \qquad |a_{n+1}| < |a_n|.$$

Hence the series converges by the alternating series test.   ∎

■ **EXAMPLE 2** Find all values of x for which the series converges

$$1 + x + \frac{x^2}{2} + \frac{x^3}{3} + \frac{x^4}{4} + \cdots.$$

Solution First take $x \geq 0$. Then all terms are positive, and

$$\frac{a_{n+1}}{a_n} = \frac{x^{n+1}/(n+1)}{x^n/n} = x\,\frac{n}{n+1} \longrightarrow x.$$

By the ratio test, the series converges for $0 \leq x < 1$ and diverges for $x > 1$. The test is inconclusive for $x = 1$. However, in that case the series is

$$1 + \tfrac{1}{2} + \tfrac{1}{3} + \tfrac{1}{4} + \cdots,$$

the harmonic series, which diverges.

Now take $x < 0$. Then the series alternates. If $x < -1$, then $|x|^n/n \to \infty$; hence the series diverges. If $-1 \leq x < 0$, then $|x|^n/n$ decreases to 0; hence the series converges by the alternating series test.

Answer The series converges precisely for $-1 \leq x < 1$.   ∎

■ **EXAMPLE 3** It is known that the series $\sum_{n=0}^{\infty} x^n/n!$ converges to e^x for all x. Use this series to estimate $1/e$ to 3-place accuracy.

Solution Set $x = -1$. Then $e^{-1} = \sum_{n=0}^{\infty} (-1)^n/n!$. The signs alternate and the terms decrease in absolute value to 0. Therefore,

$$e^{-1} = 1 - \frac{1}{1!} + \frac{1}{2!} - + \cdots + \frac{(-1)^n}{n!} + \text{remainder},$$

where

$$|\text{remainder}| < \frac{1}{(n+1)!}.$$

For 3-place accuracy, we need $|\text{remainder}| < 5 \times 10^{-4}$, so we want an n for which

$$\frac{1}{(n+1)!} < 5 \times 10^{-4}, \qquad (n+1)! > \frac{1}{5} \times 10^4 = 2000.$$

Now $6! = 720$ and $7! = 5040$. So we choose $n + 1 = 7$, that is, $n = 6$. Therefore

$$e^{-1} \approx 1 - 1 + \tfrac{1}{2} - \tfrac{1}{6} + \tfrac{1}{24} - \tfrac{1}{120} + \tfrac{1}{720} \approx 0.368. \qquad \blacksquare$$

Absolute Convergence How is it that the harmonic series $\sum 1/n$ diverges but the alternating harmonic series $\sum (-1)^{n-1}/n$ converges? Essentially the harmonic series diverges because its terms don't decrease quite fast enough, like $1/n^2$ or $1/2^n$ for example. Its partial sums consist of many small terms which have a large total. The terms of $\sum 1/2^n$, however, decrease so fast that the total of any large number of them is bounded.

The alternating harmonic series converges, not by smallness of its terms alone, but also because strategically placed minus signs cause lots of cancellation. Just look at two consecutive terms:

$$+\frac{1}{n} - \frac{1}{n+1} = \frac{1}{n(n+1)}.$$

Cancellation produces a term of a convergent series! Thus $\sum (-1)^{n-1}/n$ converges because its terms get small *and* because a delicate balance of positive and negative terms produces important cancellations.

Some series with mixed terms converge by the smallness of their terms alone; they would converge even if all the signs were $+$. We say that a series $\sum a_n$ **converges absolutely** if $\sum |a_n|$ converges. As we might expect, absolute convergence implies (is even stronger than) convergence.

> If a series $\sum a_n$ converges absolutely, then it converges.

Proof Suppose $\sum |a_n|$ converges. By the Cauchy Test, for each $\varepsilon > 0$ there is an N such that

$$|a_{n+1}| + |a_{n+2}| + \cdots + |a_m| < \varepsilon, \qquad m > n \geq N.$$

But
$$|a_{n+1} + a_{n+2} + \cdots + a_m| \leq |a_{n+1}| + \cdots + |a_m| < \varepsilon$$

by the triangle inequality. Therefore $\sum a_n$ converges by the Cauchy Test.

Remark   In studying series with mixed terms, it is a good idea to check first for absolute convergence. Just change all signs to $+$, then test for convergence of the positive series.

■ **EXAMPLE 4** Test for convergence and absolute convergence

(a) $1 + \dfrac{1}{2^2} - \dfrac{1}{3^2} + \dfrac{1}{4^2} + \dfrac{1}{5^2} - \dfrac{1}{6^2} + + - \cdots,$

(b) $1 - \dfrac{1}{\sqrt{2}} + \dfrac{1}{\sqrt{3}} \dfrac{1}{\sqrt{4}} + - \cdots.$

Solution (a) The series of absolute values is $\sum 1/n^2$, which converges. The series is absolutely convergent, hence convergent.

(b) The series of absolute values is $\sum 1/\sqrt{n}$, which diverges. Hence the given series does not converge absolutely. It does converge, nevertheless, because it satisfies the test for alternating series: the terms decrease in absolute value to zero. ■

EXERCISES

Test for convergence and absolute convergence

1 $\displaystyle\sum (-1)^n \frac{1}{\ln n}$ **2** $\displaystyle\sum (-1)^n \frac{n}{2^n}$ **3** $\displaystyle\sum (-1)^n \frac{n}{3n+1}$

4 $\displaystyle\sum (-1)^n \frac{\ln n}{n}$ **5** $\displaystyle\sum (-1)^n \frac{n^2}{(1.01)^n}$ **6** $\displaystyle\sum (-1)^n \frac{5n^2}{2n^3 - 1}$

7 $\displaystyle\sum (-1)^n \frac{1}{n + \ln n}$ **8** $\displaystyle\sum (-1)^n \sin \frac{2\pi}{n}$ **9** $\displaystyle\sum \frac{\sin n}{n^2}$

10 $\displaystyle\sum_{1}^{\infty} (-1)^n \frac{1}{|n - 100.5|}$ **11** $\displaystyle\sum \frac{\cos n\pi}{\sqrt{n+3}}$

12 $\displaystyle\sum (-1)^n \frac{1}{1 + (n - 100)^2}$ **13** $\displaystyle\sum (-1)^n \frac{1}{500 + (1/n)^2}$

14 $\displaystyle\sum (-1)^{2n-1} \frac{n-1}{(n+1)^3}$

15 $1 - \frac{1}{10} + \frac{1}{2} - \frac{1}{11} + \frac{1}{3} - \frac{1}{12} + - \cdots$ **16** $(1 - \frac{1}{10}) + (\frac{1}{2} - \frac{1}{11}) + (\frac{1}{3} - \frac{1}{12}) + \cdots$

17 $1 + \frac{1}{2} - \frac{1}{3} + \frac{1}{4} + \frac{1}{5} - \frac{1}{6} + + - \cdots$ **18** $1 + \frac{1}{2} - \frac{1}{3} - \frac{1}{4} + \frac{1}{5} + \frac{1}{6} - - + + \cdots$

19 $\frac{1}{2} - 1 + \frac{1}{4} - \frac{1}{3} + \frac{1}{6} - \frac{1}{5} + - \cdots$ **20** $(\frac{1}{2} - 1) + (\frac{1}{4} - \frac{1}{3}) + (\frac{1}{6} - \frac{1}{5}) + \cdots$.

Estimate to 4 places by the method of Example 3

21 $1/\sqrt{e}$ **22** $1/\sqrt[3]{e}$.

23 Suppose $\sum a_n^2$ and $\sum b_n^2$ both converge. Show that $\sum a_n b_n$ converges absolutely. [Hint $2xy \leq x^2 + y^2$.]

24* Suppose $\sum a_n$ converges, but not absolutely. Let $\sum b_j$ and $\sum c_k$ be the series made of the positive a_n's and negative a_n's respectively. Prove that both $\sum b_j$ and $\sum c_k$ diverge.

25* Suppose $a_1 - a_2 + a_3 - a_4 + \cdots$ is an alternating series whose terms decrease in absolute value toward 0. Suppose the first term is $a_1 > 0$. If $\{s_n\}$ denotes the sequence of partial sums, show that the subsequence $\{s_{2n}\}$ is increasing and bounded above and the subsequence $\{s_{2n-1}\}$ is decreasing and bounded below.

26* (cont.) Conclude that the two sequences converge and have the same limit S. Show that $\sum a_n = S$.

6. IMPROPER INTEGRALS

In scientific problems, one frequently meets definite integrals in which one (or both) of the limits is infinite. Here is an example.

Imagine a particle P of mass m at the origin. The **gravitational potential** at a point $x = a$ due to P is the work required to move a unit mass from $x = a$ to infinity, against the force exerted on it by P. According to Newton's Law of Gravitation, this force is km/d^2, where d is the distance between the two masses and k is a constant. The work done in moving the unit mass from $x = a$ to $x = b$ is

$$\int_a^b (\text{force})\, dx = \int_a^b \frac{km}{x^2}\, dx = km\left(-\frac{1}{x}\right)\Big|_a^b = km\left(\frac{1}{a} - \frac{1}{b}\right).$$

Let $b \longrightarrow \infty$. Then $1/b \longrightarrow 0$, hence

$$\int_a^b \frac{km}{x^2} \, dx \longrightarrow km\left(\frac{1}{a} - 0\right) = \frac{km}{a}.$$

Thus km/a is the work required to move the mass from a to ∞. It is convenient to set

$$\int_a^\infty \frac{km}{x^2} \, dx = \lim_{b \to \infty} \int_a^b \frac{km}{x^2} \, dx = \frac{km}{a}.$$

A definite integral whose upper limit is ∞, whose lower limit is $-\infty$, or both, is called an **improper integral**

Limits In order to give a precise definition of infinite integrals, we must recall the meaning of the statement $\lim_{x \to \infty} F(x) = L$.

Let $F(x)$ be defined for $x \geq a$, where a is some real number. Then

$$\lim_{x \to \infty} F(x) = L$$

if for each $\varepsilon > 0$, there is a number b such that

$$|F(x) - L| < \varepsilon \qquad \text{for all} \quad x \geq b.$$

For increasing and decreasing functions the basic fact about limits is analogous to the one for sequences. Its proof is left for a more advanced course.

Let F be an increasing function. Then $\lim_{x \to \infty} F(x)$ exists if and only if $F(x)$ is bounded above, i.e., if and only if there exists a number M such that

$$F(x) \leq M$$

for all $x \geq a$, that is, on the domain of F.

Similarly if $F(x)$ is a decreasing function, then $\lim_{x \to \infty} F(x)$ exists if and only if $F(x)$ is bounded below.

Another important fact, also similar in spirit to one for sequences, is the Cauchy Criterion:

Cauchy Criterion $\lim_{x \to \infty} F(x)$ exists if and only if for each $\varepsilon > 0$, there exists b such that

$$|F(x) - F(z)| < \varepsilon$$

whenever $x \geq b$ and $z \geq b$.

There is a similar discussion for limits of the form $\lim_{x \to -\infty} F(x)$. Obviously, if we set $G(x) = F(-x)$, then $\lim_{x \to -\infty} F(x)$ is equal to $\lim_{x \to \infty} G(x)$, so limits at $-\infty$ involve nothing new.

Definition of Improper Integrals Suppose $f(x)$ is defined for $x \geq a$ and is integrable on each interval $a \leq x \leq b$ for $b \geq a$. Set

$$F(b) = \int_a^b f(x) \, dx.$$

Now $\lim_{b \to \infty} F(b)$ may or may not exist.

Define

$$\int_a^\infty f(x) \, dx = \lim_{b \to \infty} \int_a^b f(x) \, dx,$$

provided the limit exists. If it does, the integral is said to **converge**, otherwise to **diverge**.

Similarly, define

$$\int_{-\infty}^b f(x) \, dx = \lim_{a \to -\infty} \int_a^b f(x) \, dx,$$

provided the limit exists.

Finally, define

$$\int_{-\infty}^\infty f(x) \, dx = \int_{-\infty}^0 f(x) \, dx + \int_0^\infty f(x) \, dx,$$

provided both integrals on the right converge.

Remark   An integral from $-\infty$ to ∞ may be split at any convenient finite point just as well as at 0.

An improper integral need not converge. For example, the integral

$$\int_1^\infty \frac{dx}{x} \qquad \text{diverges because} \qquad \lim_{b \to \infty} \int_1^b \frac{dx}{x} = \lim_{b \to \infty} \ln b \qquad \text{does not exist.}$$

■ **EXAMPLE 1** Evaluate   (1) $\displaystyle \int_0^\infty \frac{dx}{1 + x^2}$ (2) $\displaystyle \int_{-\infty}^3 e^x \, dx.$

Solution

(1)
$$\int_0^b \frac{dx}{1 + x^2} = \arc \tan x \Big|_0^b = \arc \tan b.$$

Let $b \longrightarrow \infty$. Then $\arc \tan b \longrightarrow \frac{1}{2}\pi$. Hence

$$\int_0^\infty \frac{dx}{1 + x^2} = \lim_{b \to \infty} \int_0^b \frac{dx}{1 + x^2} = \lim_{b \to \infty} \arc \tan b = \frac{1}{2}\pi.$$

(2)
$$\int_a^3 e^x \, dx = e^x \Big|_a^3 = e^3 - e^a.$$

Let $a \longrightarrow -\infty$. Then $e^a \longrightarrow 0$. Hence

$$\int_{-\infty}^3 e^x \, dx = \lim_{a \to -\infty} \int_a^3 e^x \, dx = \lim_{a \to -\infty} (e^3 - e^a) = e^3. \qquad ■$$

Remember that a definite integral of a positive function represents the area under a curve. We interpret the improper integral

$$\int_a^\infty f(x)\,dx, \qquad f(x) \geq 0,$$

as the area of the infinite region in Fig. 1. If the integral converges, the area is finite; if the integral diverges, the area is infinite.

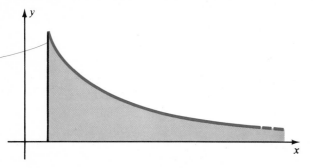

Fig. 1   Area of an infinite region

At first it may seem unbelievable that a region of infinite extent can have finite area. But it can, and here is an example. Take the region under the curve $y = 2^{-x}$ to the right of the y-axis (Fig. 2). The rectangles shown in Fig. 2 have base 1 and heights $1, \frac{1}{2}, \frac{1}{4}, \frac{1}{8}, \cdots$. Their total area is

$$1 + \frac{1}{2} + \frac{1}{4} + \frac{1}{8} + \cdots = 2.$$

Therefore, the shaded infinite region has finite area less than 2.

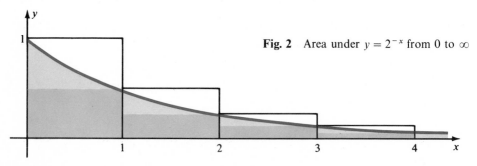

Fig. 2   Area under $y = 2^{-x}$ from 0 to ∞

■ **EXAMPLE 2**   Compute the exact area of the shaded region in Fig. 2.

Solution   The area is given by the improper integral

$$\int_0^\infty 2^{-x}\,dx = \lim_{b \to \infty} \int_0^b 2^{-x}\,dx.$$

An antiderivative of 2^{-x} is $-2^{-x}/\ln 2$ (because $2^{-x} = e^{-x \ln 2}$). Hence

$$\int_0^b 2^{-x}\,dx = \left(-\frac{1}{\ln 2}\, 2^{-x} \right)\Bigg|_0^b = \frac{1}{\ln 2}\left(1 - 2^{-b}\right),$$

$$\int_0^\infty 2^{-x}\,dx = \lim_{b\to\infty} \frac{1}{\ln 2}(1 - 2^{-b}) = \frac{1}{\ln 2} \approx 1.4427.$$ ∎

Remark The answer is reasonable. Since the darker shaded region in Fig. 2 has area

$$\frac{1}{2} + \frac{1}{4} + \frac{1}{8} + \cdots = 1,$$

the answer is between 1 and 2. A closer look shows it is slightly less than 1.5. Why?

The improper integral

$$\int_0^\infty e^{-sx} f(x)\,dx$$

arises in various applications such as electrical circuits, heat conduction, vibrating membranes, and in the solution of differential equations. It is called the **Laplace Transform** of $f(x)$.

■ **EXAMPLE 3** Evaluate $\displaystyle\int_0^\infty e^{-sx} \cos x\,dx, \quad s > 0.$

Solution From integral tables or integration by parts,

$$\int_0^b e^{-sx} \cos x\,dx = \frac{e^{-sx}}{s^2 + 1}(-s\cos x + \sin x)\Big|_0^b$$

$$= \frac{e^{-sb}}{s^2 + 1}(-s\cos b + \sin b) + \frac{s}{s^2 + 1}.$$

Now let $b \longrightarrow \infty$:

$$\int_0^\infty e^{-sx} \cos x\,dx = \lim_{b\to\infty}\int_0^b e^{-sx} \cos x\,dx = 0 + \frac{s}{s^2 + 1} = \frac{s}{s^2 + 1}.$$ ∎

Now let us try an integral where both limits are infinite.

■ **EXAMPLE 4** Evaluate $\displaystyle\int_{-\infty}^\infty \frac{dx}{3e^{2x} + e^{-2x}}.$

Solution By definition, the value of this integral is

$$\int_0^\infty \frac{dx}{3e^{2x} + e^{-2x}} + \int_{-\infty}^0 \frac{dx}{3e^{2x} + e^{-2x}},$$

provided *both* improper integrals converge. From integral tables,

$$\int_0^b \frac{dx}{3e^{2x} + e^{-2x}} = \frac{1}{2\sqrt{3}} \arctan(e^{2x}\sqrt{3})\Big|_0^b = \frac{1}{2\sqrt{3}}[\arctan(e^{2b}\sqrt{3}) - \arctan\sqrt{3}].$$

Now $\arctan(e^{2b}\sqrt{3}) \longrightarrow \frac{1}{2}\pi$ as $b \longrightarrow \infty$. Note that $\arctan\sqrt{3} = \frac{1}{3}\pi$. Hence

$$\int_0^\infty \frac{dx}{3e^{2x} + e^{-2x}} = \lim_{b\to\infty}\int_0^b \frac{dx}{3e^{2x} + e^{-2x}} = \frac{1}{2\sqrt{3}}\left(\frac{\pi}{2} - \frac{\pi}{3}\right).$$

Similarly, $$\int_{-\infty}^{0} \frac{dx}{3e^{2x} + e^{-2x}} = \lim_{a \to -\infty} \int_{a}^{0} \frac{dx}{3e^{2x} + e^{-2x}}$$

$$= \frac{1}{2\sqrt{3}} [\text{arc tan } \sqrt{3} - \text{arc tan } 0] = \frac{1}{2\sqrt{3}} \left(\frac{\pi}{3} - 0 \right).$$

Thus both improper integrals converge. The answer is the sum of their values:

$$\int_{-\infty}^{\infty} \frac{dx}{3e^{2x} + e^{-2x}} = \frac{\pi}{4\sqrt{3}}.$$ ∎

Remark   Do you prefer this snappy calculation?

$$\int_{-\infty}^{\infty} \frac{dx}{3e^{2x} + e^{-2x}} = \frac{1}{2\sqrt{3}} \text{ arc tan}(\sqrt{3}\, e^{2x}) \Big|_{-\infty}^{\infty}$$

$$= \frac{1}{2\sqrt{3}} (\text{arc tan } \infty - \text{arc tan } 0) = \frac{1}{2\sqrt{3}} \frac{\pi}{2} = \frac{\pi}{4\sqrt{3}}.$$

Warning   Try the same method on $\int_{-\infty}^{\infty} \frac{dx}{x^2}.$ It fails! Why?

The Definition of π   A good starting point for a rigorous development of the trigonometric functions is the definition

$$\text{arc tan } x = \int_{0}^{x} \frac{dt}{1 + t^2}.$$

Example 1 shows that this function arc tan x has a limit as $x \longrightarrow \infty$ and suggests the *definition* of π by the convergent integral

$$\frac{1}{2} \pi = \int_{0}^{\infty} \frac{dt}{1 + t^2}.$$

If we define π this way, can we work with it? For example, can we show, starting from scratch that arc tan $1 = \frac{1}{4}\pi$, that is,

$$\frac{1}{4} \pi = \int_{0}^{1} \frac{dt}{1 + t^2}\, ?$$

We can as follows. First, $$\frac{1}{2} \pi = \int_{0}^{\infty} \frac{dt}{1 + t^2} = \int_{0}^{1} \frac{dt}{1 + t^2} + \int_{1}^{\infty} \frac{dt}{1 + t^2}.$$

The two integrals on the right-hand side are equal. To prove this, we make the change of variable $t = 1/s$ in the second integral, but we must be careful with ∞:

$$\int_{1}^{\infty} \frac{dt}{1 + t^2} = \lim_{b \to \infty} \int_{1}^{b} \frac{dt}{1 + t^2} = \lim_{b \to \infty} \int_{1}^{1/b} \frac{-ds/s^2}{1 + (1/s)^2}$$

$$= \lim_{b \to \infty} \int_{1/b}^{1} \frac{ds}{1 + s^2} = \int_{0}^{1} \frac{ds}{1 + s^2} = \int_{0}^{1} \frac{dt}{1 + t^2}.$$

Therefore $$\frac{1}{2} \pi = 2 \int_{0}^{1} \frac{dt}{1 + t^2}, \qquad \int_{0}^{1} \frac{dt}{1 + t^2} = \frac{1}{4} \pi.$$

EXERCISES

Evaluate

1 $\displaystyle\int_2^\infty \frac{dx}{x^3}$ **2** $\displaystyle\int_5^\infty e^{-x}\,dx$ **3** $\displaystyle\int_0^\infty xe^{-x}\,dx$

4 $\displaystyle\int_{-\infty}^{-1} \frac{dx}{x^2}$ **5** $\displaystyle\int_{-\infty}^{-1} \frac{dx}{1+x^2}$ **6** $\displaystyle\int_4^\infty \frac{dx}{x\sqrt{x}}$

7 $\displaystyle\int_{-\infty}^\infty e^{-|x|}\,dx$ **8** $\displaystyle\int_{-\infty}^\infty xe^{-x^2}\,dx$ **9** $\displaystyle\int_4^\infty \frac{dx}{x\sqrt{9+x^2}}$

10 $\displaystyle\int_1^\infty \frac{dx}{x(x+3)}$ **11** $\displaystyle\int_0^\infty \frac{x\,dx}{x^4+1}$ (let $u=x^2$)   **12** $\displaystyle\int_1^\infty \frac{dx}{(x^2+1)^2}$

13 $\displaystyle\int_{-\infty}^\infty \frac{dx}{1+x^2}$ **14** $\displaystyle\int_{-\infty}^\infty e^{-|x-2|}\,dx$ **15** $\displaystyle\int_0^\infty xe^{-sx}\,dx$ $(s>0)$

16 $\displaystyle\int_0^\infty x^2 e^{-sx}\,dx$ $(s>0)$ **17** $\displaystyle\int_0^\infty x^n e^{-sx}\,dx$ $(s>0)$ **18** $\displaystyle\int_0^\infty e^{ax}e^{-sx}\,dx$ $(s>a)$

19 $\displaystyle\int_0^\infty e^{-sx}\sin x\,dx$ $(s>0)$ **20** $\displaystyle\int_0^\infty e^{-sx}\cosh x\,dx$ $(s>1)$

21 $\displaystyle\int_0^\infty xe^{ax}e^{-sx}\,dx$ $(s>0)$ **22** $\displaystyle\int_0^\infty xe^{-sx}\sin x\,dx$ $(s>0)$.

Is the area under the curve finite or infinite?

23 $y=1/x$;   from $x=5$ to $x=\infty$ **24** $y=1/x^2$;   from $x=1$ to $x=\infty$
25 $y=\sin^2 x$;   from $x=0$ to $x=\infty$ **26** $y=(1.001)^{-x}$;   from $x=0$ to $x=\infty$
27 $y=1/(3x+50)$;   from $x=2$ to $x=\infty$
28 $y=x/(x^2+10)$;   from $x=0$ to $x=\infty$.

Solve for b

29 $\displaystyle\int_0^b e^{-x}\,dx = \int_b^\infty e^{-x}\,dx$ **30** $\displaystyle\int_1^b \frac{dx}{x^2} = \int_b^\infty \frac{dx}{x^2}$.

Find b such that 99% of the area under $y=f(x)$ between $x=0$ and $x=\infty$ is contained between $x=0$ and $x=b$

31 $y=e^{-2x}$ **32** $y=\dfrac{1}{(x+5)^3}$.

Mental calculus. Without pencil and paper compute

33 $\displaystyle\int_{-\infty}^\infty x^5 e^{-x^2}\,dx$ **34** $\displaystyle\int_{-\infty}^\infty \frac{\sin 2\pi x\,dx}{x^4+7}$.

35 Suppose $a>0$ and $\displaystyle\int_0^\infty f(x)\,dx = L$. Find $\displaystyle\int_0^\infty f(ax)\,dx$.

36 Suppose $\displaystyle\int_{-\infty}^\infty f(x)\,dx = L$. Find $\displaystyle\int_{-\infty}^\infty f(x+b)\,dx$.

7. CONVERGENCE AND DIVERGENCE TESTS

Whether an improper integral converges or diverges may be a subtle matter. Let us illustrate with a useful class of improper integrals. These are analogous to p-series discussed in Section 4.

> The improper integral $\displaystyle\int_a^\infty \frac{dx}{x^p}$ $(a > 0)$
>
> diverges if $p \le 1$ and converges if $p > 1$.

Proof Suppose $p \ne 1$. Then

$$\int_a^b \frac{dx}{x^p} = -\frac{1}{p-1}\frac{1}{x^{p-1}}\Big|_a^b = \frac{1}{p-1}\left(\frac{1}{a^{p-1}} - \frac{1}{b^{p-1}}\right).$$

Now it makes a big difference whether $p - 1$ is positive or negative. For as $b \longrightarrow \infty$,

$$\frac{1}{b^{p-1}} \longrightarrow \begin{cases} 0 & \text{if } p - 1 > 0 \\ \infty & \text{if } p - 1 < 0. \end{cases}$$

Hence
$$\lim_{b \to \infty} \int_a^b \frac{dx}{x^p}$$

exists if $p > 1$, does not exist if $p < 1$. That means the given integral converges if $p > 1$, diverges if $p < 1$.

If $p = 1$, $\displaystyle\int_a^b \frac{dx}{x} = \ln b - \ln a \longrightarrow \infty$ as $b \longrightarrow \infty$;

the integral diverges.

For a graphical interpretation of these results, see Fig. 1. For $p > 0$, the curves $y = 1/x^p$ all decrease as x increases. The key is in their *rate* of decrease. If $p \le 1$, the curve decreases slowly enough that the shaded area (Fig. 1a) increases without bound as $b \longrightarrow \infty$. If $p > 1$, the curve decreases fast enough that the shaded area (Fig. 1b) is bounded by a fixed number, no matter how large b is.

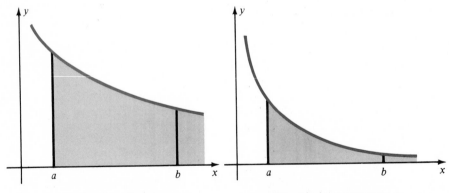

(a) $0 < p \le 1$; divergence   (b) $p > 1$; convergence

Fig. 1 $\displaystyle\int_a^\infty \frac{dx}{x^p}$

The figure suggests that for a positive function $f(x)$ the convergence or divergence of

$$\int_a^\infty f(x)\,dx$$

depends on how rapidly $f(x) \longrightarrow 0$ as $x \longrightarrow \infty$. For $f(x) = e^{-x}$, which decreases very fast, the integral converges with plenty to spare. But for $f(x) = 1/x^{1.01}$, which decreases much more slowly, the integral just about makes it.

Comparison Tests We shall obtain comparison tests for convergence and divergence of improper integrals. In spirit, the discussion will parallel that for infinite series. First, a basic principle:

> Let $f(x) \geq 0$ for $x \geq a$. Then
>
> $$\int_a^\infty f(x)\,dx \quad \text{converges} \qquad \text{if and only if} \qquad \int_a^b f(x)\,dx \leq M$$
>
> for some constant M and for all $b \geq a$.

Proof Let $\qquad F(b) = \int_a^b f(x)\,dx.$

By definition, the improper integral exists provided $\lim_{b\to\infty} F(b)$ exists. But $f(x) \geq 0$, so $F(b)$ is an increasing function of b. Therefore $\lim_{b\to\infty} F(b)$ exists if and only if $F(b)$ is bounded above, that is, if and only if

$$F(b) = \int_a^b f(x)\,dx \leq M$$

for some constant M.

Now we can state the comparison test.

> **Comparison Test for Integrals** Suppose $f(x) \geq 0$ and $g(x) \geq 0$ for $x \geq a$.
>
> (1) If $g(x) \leq f(x)$, then the convergence of $\int_a^\infty f(x)\,dx$ implies the convergence of $\int_a^\infty g(x)\,dx$.
>
> (2) If $g(x) \geq f(x)$, then the divergence of $\int_a^\infty f(x)\,dx$ implies the divergence of $\int_a^\infty g(x)\,dx$.

Proof (1) Let $\qquad F(b) = \int_a^b f(x)\,dx, \qquad G(b) = \int_a^b g(x)\,dx.$

Since $f(x) \geq 0$ and $g(x) \geq 0$, both $F(b)$ and $G(b)$ are increasing functions of b. By hypothesis, $\lim_{b\to\infty} F(b)$ exists, hence $F(b) \leq M$ for some constant M. But $g(x) \leq f(x)$, therefore $G(b) \leq F(b) \leq M$. It follows that $\lim_{b \to \infty} G(b)$ exists, that is,

$$\int_a^\infty g(x)\,dx \quad \text{converges.}$$

(2) This time $G(b) \geq F(b)$ and $F(b)$ is unbounded, so $G(b)$ is unbounded. Hence

$$\int_a^\infty g(x)\,dx \quad \text{diverges.}$$

■ **EXAMPLE 1** Show that the integrals converge

(a) $\displaystyle\int_1^\infty \frac{dx}{x^2 + \sqrt{x}}$ (b) $\displaystyle\int_0^\infty \frac{e^{-x}}{x+1}\,dx$ (c) $\displaystyle\int_3^\infty \frac{\sin^2 x}{x^3}\,dx.$

Solution $\displaystyle \frac{1}{x^2 + \sqrt{x}} < \frac{1}{x^2} \quad (1 \leq x < \infty),$

$$\frac{e^{-x}}{x+1} \leq e^{-x} \quad (0 \leq x < \infty) \qquad \text{and} \qquad \frac{\sin^2 x}{x^3} \leq \frac{1}{x^3} \quad (3 \leq x < \infty).$$

Since the integrals $\displaystyle\int_1^\infty \frac{dx}{x^2},$ $\displaystyle\int_0^\infty e^{-x}\,dx,$ $\displaystyle\int_3^\infty \frac{dx}{x^3}$

all converge, the given integrals converge by the Comparison Test.   ■

■ **EXAMPLE 2** Show that the integrals diverge

(a) $\displaystyle\int_1^\infty \frac{\sqrt{x}}{1+x}\,dx$ (b) $\displaystyle\int_2^\infty \frac{dx}{\sqrt{x} - \sqrt[3]{x}}$ (c) $\displaystyle\int_3^\infty \frac{\ln x}{x}\,dx.$

Solution $\displaystyle \frac{\sqrt{x}}{1+x} \geq \frac{1}{1+x} \quad (1 \leq x < \infty)$

$$\frac{1}{\sqrt{x} - \sqrt[3]{x}} > \frac{1}{\sqrt{x}} \quad (2 \leq x < \infty) \qquad \text{and} \qquad \frac{\ln x}{x} \geq \frac{\ln 3}{x} > \frac{1}{x}, \quad (3 \leq x < \infty).$$

Since the integrals $\displaystyle\int_1^\infty \frac{dx}{1+x},$ $\displaystyle\int_2^\infty \frac{dx}{\sqrt{x}},$ $\displaystyle\int_3^\infty \frac{dx}{x}$

all diverge, the given integrals diverge by the Comparison Test.   ■

Another Comparison Test   The following test does not require the integrand to be non-negative.

> Suppose $f(x) \geq 0$ and $g(x)$ is bounded, that is, $|g(x)| \leq M$ for some constant M. Then the convergence of
>
> $$\int_a^\infty f(x)\,dx \qquad \text{implies the convergence of} \qquad \int_a^\infty f(x)g(x)\,dx.$$

Proof   Let $\displaystyle F(b) = \int_a^b f(x)\,dx, \qquad H(b) = \int_a^b f(x)g(x)\,dx.$

We are given that $\lim_{b \to \infty} F(b)$ exists, and we must show that $\lim_{b \to \infty} H(b)$ exists. By the Cauchy Criterion, given $\varepsilon > 0$, there exists B such that

$$|F(c) - F(b)| < \frac{\varepsilon}{M} \qquad \text{whenever} \quad c > b \geq B.$$

It follows that

$$|H(c) - H(b)| = \left| \int_b^c f(x)g(x)\,dx \right| \leq \int_b^c |f(x)||g(x)|\,dx$$

$$\leq M \int_b^c f(x)\,dx = M[F(c) - F(b)] < M \cdot \frac{\varepsilon}{M} = \varepsilon$$

whenever $c > b \geq B$. Therefore by the Cauchy Criterion, $\lim_{b \to \infty} H(b)$ exists, that is,

$$\int_a^\infty f(x)g(x)\,dx \quad \text{converges.}$$

■ **EXAMPLE 3** Show that the integrals converge

(a) $\displaystyle\int_0^\infty e^{-x} \sin^3 x\,dx$ (b) $\displaystyle\int_1^\infty \frac{\ln x}{x^3}\,dx$ (c) $\displaystyle\int_{-1/2}^\infty \frac{\arctan x}{1 + x^3}\,dx.$

Solution Apply the above criterion.

(a) Since $\displaystyle\int_0^\infty e^{-x}\,dx$

converges and $|\sin^3 x| \leq 1$, the given integral converges.

(b) Write $\dfrac{\ln x}{x^3} = \dfrac{1}{x^2} \cdot \dfrac{\ln x}{x}.$

The integral $\displaystyle\int_1^\infty \frac{dx}{x^2}$

converges and $(\ln x)/x$ is bounded. (Its maximum value is $1/e$.) Hence the given integral converges.

(c) Break the integral into a sum: $\displaystyle\int_{-1/2}^\infty \frac{\arctan x}{1 + x^3}\,dx = \int_{-1/2}^1 + \int_1^\infty.$

The first integral on the right is an integral of a continuous function on a finite interval, so it exists. Therefore the problem is equivalent to showing that

$$\int_1^\infty \frac{\arctan x}{1 + x^3}\,dx$$

converges. But $|\arctan x| < \tfrac{1}{2}\pi$, and

$$\int_1^\infty \frac{dx}{1 + x^3} < \int_1^\infty \frac{dx}{x^3},$$

which converges. Hence the given integral converges.   ■

Remark   The splitting of the integral in (c) was necessary. You cannot prove the convergence of

$$\int_{-1/2}^{\infty} \frac{dx}{1 + x^3} \qquad \text{by comparison with} \qquad \int_{-1/2}^{\infty} \frac{dx}{x^3}$$

because the latter integral is undefined, due to the zero in the denominator at $x = 0$.

Absolute Convergence   An improper integral $\int_a^{\infty} f(x)\, dx$ is called **absolutely convergent** if

$$\int_a^{\infty} |f(x)|\, dx$$

is convergent. Just as with infinite series, an absolutely convergent integral is convergent. This statement will follow easily from the next comparison test.

If $\displaystyle\int_a^{\infty} f(x)\, dx$ is convergent and if $|g(x)| \le f(x)$ for $a \le x < \infty$, then

$$\int_a^{\infty} g(x)\, dx$$

is convergent and absolutely convergent.

This is proved using the Cauchy Criterion almost exactly as the last test was. We omit the details but note the critical step:

$$\left| \int_b^c g(x)\, dx \right| \le \int_b^c |g(x)|\, dx \le \int_b^c f(x)\, dx < \varepsilon.$$

The special case $f(x) = |g(x)|$ implies

If $\displaystyle\int_a^{\infty} g(x)\, dx$ is absolutely convergent, then it is convergent.

A series may converge, but not converge absolutely. Similarly, an integral may converge, but not converge absolutely. An example is

$$\int_1^{\infty} f(x)\, dx,$$

where $f(x) = (-1)^n/n$ for $n \le x < (n + 1)$. Is this cheating? Not really, but

$$\int_0^{\infty} \frac{\sin x}{x}\, dx$$

is another example, this time with a continuous integrand, as we shall see in the next section.

EXERCISES

Test for convergence

1   $\displaystyle\int_0^{\infty} \frac{dx}{x + 1}$

2   $\displaystyle\int_1^{\infty} \frac{dx}{x^2 + x}$

3   $\displaystyle\int_0^{\infty} \frac{x^2 e^{-x}}{1 + x^2}\, dx$

4 $\displaystyle\int_{2}^{\infty} e^{-x^3}\,dx$ **5** $\displaystyle\int_{0}^{\infty} \cosh x\,dx$ **6** $\displaystyle\int_{0}^{\infty} \frac{x\,dx}{\sqrt{x^2+3}}$

7 $\displaystyle\int_{-\infty}^{\infty} \frac{\sin x}{1+x^2}\,dx$ **8** $\displaystyle\int_{0}^{\infty} \sin x\,dx$ **9** $\displaystyle\int_{1}^{\infty} \frac{\cos x}{\sqrt{x}(x+4)}\,dx$

10 $\displaystyle\int_{2}^{\infty} \frac{dx}{\ln x}$ **11** $\displaystyle\int_{3}^{\infty} \frac{x^3}{x^4-1}\,dx$ **12** $\displaystyle\int_{0}^{\infty} \frac{dx}{1+x+e^x}$

13 $\displaystyle\int_{-\infty}^{0} x^2 e^x\,dx$ **14** $\displaystyle\int_{-5}^{\infty} \frac{dx}{\sqrt{(x+6)(x+7)(x+8)}}.$

15 Show that $\displaystyle\int_{2}^{\infty} \frac{dx}{x(\ln x)^p}$ converges if $p>1$, diverges if $p\le 1$. [*Hint*   Use the substitution $u=\ln x$.]

16 Show that $\displaystyle\int_{3}^{\infty} \frac{dx}{x\ln x[\ln(\ln x)]^p}$ converges if $p>1$, diverges if $p\le 1$.

17 Show that $\displaystyle\int_{1}^{\infty} \frac{\ln x}{x^p}\,dx$ converges if $p>1$, diverges if $p\le 1$. [*Hint*   Recall that $(\ln x)/x^r \longrightarrow 0$ as $r\longrightarrow\infty$ if $r>0$.]

18 Denote by R the infinite region under $y=1/x$ to the right of $x=1$. Suppose R is rotated around the x-axis, forming an infinitely long horn. Show that the volume of this horn is finite. Its surface area, however, is infinite (the surface area is certainly larger than the area of R). Here is an apparent paradox: You can fill the horn with paint, but you cannot paint it. Where is the fallacy?

Find all values of s for which the integral converges

19 $\displaystyle\int_{0}^{\infty} \frac{x\,dx}{\sqrt{1+x^s}}$ **20** $\displaystyle\int_{-1}^{\infty} (x+3)^s\,dx$ **21** $\displaystyle\int_{-\infty}^{\infty} \frac{dx}{(x^2+1)^s}$

22 $\displaystyle\int_{-\infty}^{\infty} e^{-sx}\,dx$ **23** $\displaystyle\int_{0}^{\infty} e^{-sx}e^x\,dx$ **24** $\displaystyle\int_{0}^{\infty} \frac{e^{-sx}}{1+x^2}\,dx$

25 $\displaystyle\int_{0}^{\infty} e^{-sx}e^{-x^2}\,dx$ **26** $\displaystyle\int_{1}^{\infty} \frac{x^s}{(1+x^3)^s}\,dx.$

Denote the Laplace Transform (p. 557) of $f(x)$ by

$$L(f)(s) = \int_{0}^{\infty} e^{-sx} f(x)\,dx.$$

27 Suppose $f(x)$ is continuous for $x\ge 0$ and $|f(x)|\le cx^n$ for constants c and n and all x sufficiently large. Prove $L(f)(s)$ converges for all $s>0$.

28* (cont.) Suppose $f(x)$ is differentiable for $x\ge 0$ and $f'(x)$ satisfies the conditions of Ex. 27. Prove for $s>0$ that

$$L(f')(s) = -f(0) + sL(f)(s).$$

[*Hint*   Use integration by parts.]

29 (cont.) Check this for $f(x)=\sin x$, where $L(f)(s)=1/(1+s^2)$, and $g(x)=\cos x$, where $L(g)(s)=s/(1+s^2)$.

30 (cont.) For $f(x)$ as in Ex. 27, set $g(x)=\int_0^x f(t)\,dt$. Prove

$$L(g)(s) = \frac{1}{s}L(f)(s) \qquad \text{for} \quad s>0.$$

8. RELATION BETWEEN INTEGRALS AND SERIES

We have already seen a number of similarities between infinite series and infinite integrals. In this section we discuss a useful test for convergence or divergence of a series in terms of a related integral. This is important, for usually it is easier to find the value of an integral than the sum of a series.

Consider the relation between the series

$$\frac{1}{2^2} + \frac{1}{3^2} + \cdots + \frac{1}{n^2} + \cdots$$

and the convergent integral $\displaystyle\int_1^\infty \frac{dx}{x^2}.$

(See Fig. 1.) The rectangles shown in Fig. 1 have areas $1/2^2$, $1/3^2$, $\cdots$. Obviously the sum of these areas is finite, being less than the finite area under the curve. Hence, the series converges. This illustrates a general principle:

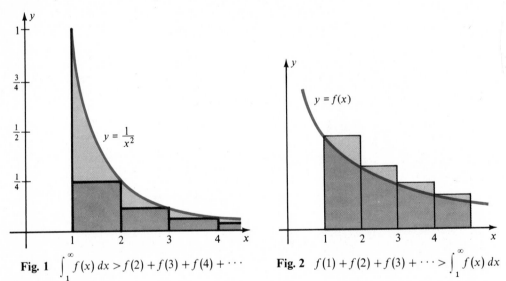

Fig. 1 $\displaystyle\int_1^\infty f(x)\,dx > f(2) + f(3) + f(4) + \cdots$

Fig. 2 $f(1) + f(2) + f(3) + \cdots > \displaystyle\int_1^\infty f(x)\,dx$

Integral Test   Suppose $f(x)$ is a positive decreasing function. Then the series

$$f(1) + f(2) + \cdots + f(n) + \cdots$$

converges if the integral $\displaystyle\int_1^\infty f(x)\,dx$

converges, and diverges if the integral diverges.

Proof The argument given above for $f(x) = 1/x^2$ holds for any positive decreasing function $f(x)$. Figure 1 indicates that

$$f(2) + f(3) + \cdots + f(n) \le \int_1^n f(x)\,dx.$$

If the infinite integral converges, then

$$s_n = f(1) + f(2) + \cdots + f(n) \leq f(1) + \int_1^n f(x)\, dx \leq f(1) + \int_1^\infty f(x)\, dx.$$

Hence the increasing partial sums are bounded; the series converges.

If the infinite integral diverges, the rectangles are drawn above the curve (Fig. 2). Their areas are $f(1), f(2), \cdots$. This time

$$s_n = f(1) + f(2) + \cdots + f(n) \geq \int_1^{n-1} f(x)\, dx.$$

But the integrals on the right are unbounded. Hence the increasing sequence $\{s_n\}$ is unbounded; the series diverges.

Remark   As usual we may ignore a finite number of terms or a finite part of the integral. The series can start at $f(k)$ and the integral can just as well be

$$\int_a^\infty f(x)\, dx, \qquad a > 0.$$

■ **EXAMPLE 1** Does the series $\qquad \dfrac{1}{2 \ln 2} + \dfrac{1}{3 \ln 3} + \dfrac{1}{4 \ln 4} + \cdots$

converge or diverge?

Solution   Let $a_n = 1/(n \ln n)$. On the one hand, $a_n < 1/n$, but that doesn't help because $\sum 1/n$ diverges. On the other hand, since $\ln n$ increases so slowly, $a_n > 1/n^2$, or even $a_n > 1/n^p$ for $p > 1$. But that again does not help because $\sum 1/n^p$ converges. In both cases, the inequalities go the wrong way.

However, $f(x) = 1/(x \ln x)$ is a positive decreasing function, so we can use the preceding test. Setting $u = \ln x$, we have

$$\int_2^\infty \frac{dx}{x \ln x} = \int_2^\infty \frac{1}{\ln x}\left(\frac{1}{x}\, dx\right) = \int_{\ln 2}^\infty \frac{1}{u}\, du.$$

This integral diverges. Hence the series diverges.   ■

Convergence of Integrals   Sometimes we can turn the tables and use the convergence of a series to establish the convergence of an integral. If the integrand changes sign regularly, we may be able to compare the integral with an alternating series.

■ **EXAMPLE 2** Prove the convergence of $\displaystyle\int_0^\infty \frac{\sin x}{x}\, dx$.

Solution   First sketch the graph of $y = (\sin x)/x$. See Fig. 3. There is no trouble at $x = 0$ because $(\sin x)/x \longrightarrow 1$ as $x \longrightarrow 0$, so the integrand is continuous, even at $x = 0$.

The figure suggests that the integral is given by an alternating series. To be precise, let

$$a_n = \left| \int_{n\pi}^{(n+1)\pi} \frac{\sin x}{x}\, dx \right| = \int_{n\pi}^{(n+1)\pi} \frac{|\sin x|}{x}\, dx.$$

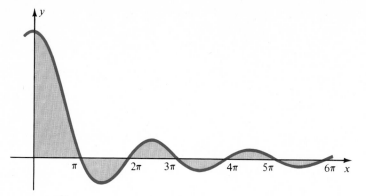

Fig. 3 $\displaystyle\int_0^\infty \frac{\sin x}{x}\,dx$

Then a_n is the area of the n-th shaded region in Fig. 3. Now $a_n \longrightarrow 0$ since

$$a_n = \int_{n\pi}^{(n+1)\pi} \frac{|\sin x|}{x}\,dx < \int_{n\pi}^{(n+1)\pi} \frac{1}{x}\,dx < \frac{\pi}{n\pi} = \frac{1}{n}.$$

Furthermore, the sequence $\{a_n\}$ decreases since

$$a_{n+1} = \int_{(n+1)\pi}^{(n+2)\pi} \frac{|\sin x|}{x}\,dx = \int_{n\pi}^{(n+1)\pi} \frac{|\sin(x+\pi)|}{x+\pi}\,dx = \int_{n\pi}^{(n+1)\pi} \frac{|\sin x|}{x+\pi}\,dx$$

$$< \int_{n\pi}^{(n+1)\pi} \frac{|\sin x|}{x}\,dx = a_n.$$

By the alternating series test, $\quad a_0 - a_1 + a_2 - a_3 + \cdots \quad$ converges. But

$$a_0 - a_1 + a_2 - \cdots + (-1)^n a_n = \int_0^{(n+1)\pi} \frac{\sin x}{x}\,dx,$$

so we have the existence of $\quad \displaystyle\lim_{n\to\infty} \int_0^{(n+1)\pi} \frac{\sin x}{x}\,dx,$

where n can take only *integer* values. This is a giant step in the right direction, but we are not quite finished yet.

If b is any positive *real* number, then there is an integer n such that $n\pi \le b < (n+1)\pi$. We write

$$\int_0^b \frac{\sin x}{x}\,dx = \int_0^{(n+1)\pi} \frac{\sin x}{x}\,dx - \int_b^{(n+1)\pi} \frac{\sin x}{x}\,dx.$$

If $b \longrightarrow \infty$, then $n \longrightarrow \infty$, so the first term on the right converges to a limit. The second term approaches 0 because

$$\left| \int_b^{(n+1)\pi} \frac{\sin x}{x}\,dx \right| \le \left| \int_{n\pi}^{(n+1)\pi} \frac{\sin x}{x}\,dx \right| = a_n \longrightarrow 0.$$

Therefore $\quad \displaystyle\int_0^\infty \frac{\sin x}{x}\,dx = \lim_{b\to\infty} \int_0^b \frac{\sin x}{x}\,dx \quad$ converges.

Alternative solution   First we get away from 0; then integrate by parts:

$$\int_0^b \frac{\sin x}{x}\, dx = K + \int_\pi^b \frac{\sin x}{x}\, dx, \qquad \text{where} \quad K = \int_0^\pi \frac{\sin x}{x}\, dx,$$

$$\int_\pi^b \frac{\sin x}{x}\, dx = -\int_\pi^b \left(\frac{1}{x}\right) d(\cos x) = -\left.\frac{\cos x}{x}\right|_\pi^b - \int_\pi^b \frac{\cos x}{x^2}\, dx.$$

As $b \longrightarrow \infty$,

$$-\left.\frac{\cos x}{x}\right|_\pi^b = \frac{\cos \pi}{\pi} - \frac{\cos b}{b} \longrightarrow \frac{\cos \pi}{\pi} = -\frac{1}{\pi},$$

and

$$\int_\pi^b \frac{\cos x}{x^2}\, dx \longrightarrow \int_\pi^\infty \frac{\cos x}{x^2}\, dx,$$

a convergent integral by the comparison $|\cos x/x^2| \leq 1/x^2$. Therefore the given integral converges:

$$\int_0^\infty \frac{\sin x}{x}\, dx = \lim_{b \to \infty} \int_0^b \frac{\sin x}{x}\, dx = \int_0^\pi \frac{\sin x}{x}\, dx - \frac{1}{\pi} + \int_\pi^\infty \frac{\cos x}{x^2}\, dx. \qquad \blacksquare$$

Remark   Clearly

$$\int_{n\pi}^{(n+1)\pi} \frac{|\sin x|}{x}\, dx \geq \frac{1}{(n+1)\pi} \int_{n\pi}^{(n+1)\pi} |\sin x|\, dx = \frac{2}{(n+1)\pi},$$

so

$$\int_0^\infty \frac{|\sin x|}{x}\, dx$$

diverges by comparison with the harmonic series. Thus the integral in Example 2 converges, but not absolutely.

EXERCISES

Use the Integral Test to test for convergence

1 $1 + \dfrac{1}{\sqrt{2}} + \dfrac{1}{\sqrt{3}} + \dfrac{1}{\sqrt{4}} + \cdots$

2 $\dfrac{1}{e} + \dfrac{2}{e^2} + \dfrac{3}{e^3} + \cdots$

3 $1 + \dfrac{1}{2^3} + \dfrac{1}{3^3} + \cdots$

4 $\dfrac{1}{1 + 1^2} + \dfrac{1}{1 + 2^2} + \dfrac{1}{1 + 3^2} + \cdots$

5 $\displaystyle\sum \frac{2n - 1}{n^2}$

6 $\displaystyle\sum \frac{1}{n(\ln n)^3}$

7 $\displaystyle\sum \frac{1 + \ln n}{n^n}$

8 $\displaystyle\sum \frac{n}{(n + 1)^3}$

9 $\displaystyle\sum (1 - \tanh n)$

10 $\displaystyle\sum (\tfrac{1}{2}\pi - \arctan n)$.

11   Show geometrically that the sum of $1 + \dfrac{1}{2^2} + \dfrac{1}{3^2} + \dfrac{1}{4^2} + \cdots$ is less than 2. See Fig. 1.

   (It is known that the exact sum is $\tfrac{1}{6}\pi^2$, a remarkable fact.)

12   Use the method of inscribing and circumscribing rectangles to show that

$$\ln(n + 1) < 1 + \frac{1}{2} + \frac{1}{3} + \cdots + \frac{1}{n} < 1 + \ln n.$$

   Is $1 + \dfrac{1}{2} + \dfrac{1}{3} + \cdots + \dfrac{1}{1000}$ more or less than 10?

13 Estimate how many terms of the series $1 + \dfrac{1}{2} + \dfrac{1}{3} + \dfrac{1}{4} + \cdots$ must be added before the sum exceeds 1000.

14* (cont.) The same for $\dfrac{1}{2 \ln 2} + \dfrac{1}{3 \ln 3} + \cdots + \dfrac{1}{n \ln n} > 10$.

15 Show geometrically that $\ln(n!) > \int_1^n \ln x \, dx$. Conclude that $n! > e(n/e)^n$.

16 (cont.) Show that $100! > 10^{157}$.

17 Prove $\displaystyle\int_0^\infty \frac{\sin x}{x} \, dx = \int_0^\infty \frac{1 - \cos x}{x^2} \, dx$.

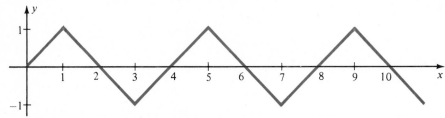

Fig. 4

18 Let $f(x)$ be the saw-toothed function (Fig. 4). Prove $\displaystyle\int_0^\infty \frac{f(x)}{x} \, dx$ converges.

19 Prove that $\displaystyle\int_1^\infty \frac{\sin x}{\sqrt{x}} \, dx$ converges. [*Hint*   Integrate by parts.]

20 (cont.) Prove that $\int_1^\infty \sin(x^2) \, dx$ converges. [*Hint*   Set $x^2 = u$.]

9. OTHER IMPROPER INTEGRALS

A definite integral $$\int_a^b f(x) \, dx, \qquad a \text{ and } b \text{ finite,}$$

is called **improper** if $f(x)$ "blows up" at one or more points in the interval $a \le x \le b$. Examples are

$$\int_0^3 \frac{dx}{x}, \qquad \int_1^2 \frac{dx}{x^2 - 4}, \qquad \int_6^{10} \frac{dx}{\ln(x - 5)}.$$

The first integrand blows up at $x = 0$, the second at $x = 2$, the third at $x = 6$. Such bad points are called **singularities** of the integrand.

We shall concentrate on integrals

$$\int_a^b f(x) \, dx,$$

where $f(x)$ has exactly one singularity. We consider first $\displaystyle\int_0^3 \frac{dx}{\sqrt{x}}$,

whose integrand has a singularity at $x = 0$. What meaning can we give to this integral? Except at $x = 0$, the integrand is well behaved. Hence if h is any positive number, no matter how small, the integral

$$\int_h^3 \frac{dx}{\sqrt{x}}$$

makes sense and is easily computed: $\displaystyle\int_h^3 \frac{dx}{\sqrt{x}} = 2\sqrt{x}\,\Big|_h^3 = 2(\sqrt{3} - \sqrt{h}).$

It is reasonable to *define*

$$\int_0^3 \frac{dx}{\sqrt{x}} = \lim_{h \to 0} \int_h^3 \frac{dx}{\sqrt{x}} = 2\sqrt{3}.$$

Next, we consider the integral $\displaystyle\int_0^3 \frac{dx}{x}.$

We try to sneak up on the integral as before. We compute

$$\int_h^3 \frac{dx}{x} = \ln 3 - \ln h,$$

then let $h \longrightarrow 0$. But $\ln h \longrightarrow -\infty$ as $h \longrightarrow 0$. Hence

$$\int_h^3 \frac{dx}{x} \longrightarrow \infty \qquad \text{as} \quad h \longrightarrow 0.$$

There is no reasonable value for this integral.

Motivated by these examples, we make the following definitions:

Suppose $f(x)$ has one singularity, at $x = a$, and that $a < b$. Define

$$\int_a^b f(x)\,dx = \lim_{h \to 0+} \int_{a+h}^b f(x)\,dx,$$

provided the limit exists. If it does, the improper integral **converges**; otherwise, it **diverges**.

Similarly, if $f(x)$ has one singularity, at $x = b$, define

$$\int_a^b f(x)\,dx = \lim_{h \to 0+} \int_a^{b-h} f(x)\,dx,$$

provided the limit exists.

Finally, if $f(x)$ has one singularity, at c where $a < c < b$, define

$$\int_a^b f(x)\,dx = \int_a^c f(x)\,dx + \int_c^b f(x)\,dx,$$

provided both improper integrals on the right converge.

Tests for Convergence Just as for other types of improper integrals, the convergence or divergence of

$$\int_a^b f(x)\,dx,$$

where $f(x)$ has one singularity, is often established by comparison with known integrals.

Comparison Test   Suppose that $f(x) \geq 0$ and $g(x) \geq 0$ on the interval $[a, b]$ and that each function has one singularity at $x = a$.

(1)   If $g(x) \leq f(x)$ for $a < x \leq b$, then the convergence of

$$\int_a^b f(x)\, dx \quad \text{implies the convergence of} \quad \int_a^b g(x)\, dx.$$

(2)   If $g(x) \geq f(x)$ for $a < x \leq b$, then the divergence of

$$\int_a^b f(x)\, dx \quad \text{implies the divergence of} \quad \int_a^b g(x)\, dx,$$

Similar statements hold if both functions have one singularity at $x = b$.

We omit the proof because it is nearly identical to the proof of the comparison test given in Section 7.

Now that we have a comparison test, we need some integrals to compare with.

The improper integral $\displaystyle\int_0^b \frac{dx}{x^p}$ converges if $p < 1$ and diverges if $p \geq 1$.

More generally, the integrals

$$\int_a^b \frac{dx}{(x-a)^p}, \qquad \int_a^b \frac{dx}{(b-x)^p}$$

converge if $p < 1$ and diverge if $p \geq 1$.

Proof   By definition, $\displaystyle\int_0^b \frac{dx}{x^p} = \lim_{h \to 0+} \int_h^b \frac{dx}{x^p},$

provided the limit exists. The case $p = 1$ was just discussed; the integral diverges. Now assume $p \neq 1$:

$$\int_h^b \frac{dx}{x^p} = -\frac{1}{p-1} \frac{1}{x^{p-1}} \Bigg|_h^b = \frac{1}{p-1}\left(\frac{1}{h^{p-1}} - \frac{1}{b^{p-1}} \right).$$

But, as $h \longrightarrow 0+$, $\dfrac{1}{h^{p-1}} \longrightarrow \begin{cases} 0 & \text{if } p - 1 < 0, \\ \infty & \text{if } p - 1 > 0. \end{cases}$

Hence the limit exists only if $p < 1$. In that case

$$\int_0^b \frac{dx}{x^p} = \lim_{h \to 0+} \int_h^b \frac{dx}{x^p} = -\frac{1}{(p-1)b^{p-1}}.$$

The assertions about $\displaystyle\int_a^b \frac{dx}{(x-a)^p}$ and $\displaystyle\int_a^b \frac{dx}{(b-x)^p}$

follow by appropriate changes of variable.

Remark   If $p \geq 1$, the curve $y = 1/x^p$ increases so fast as $x \longrightarrow 0$ that the area of the shaded region (Fig. 1a) tends to infinity. If $p < 1$, the curve rises so slowly that the area of the shaded region (Fig. 1b) is bounded.

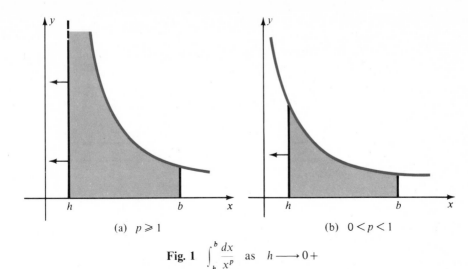

(a) $p \geqslant 1$ (b) $0 < p < 1$

Fig. 1 $\displaystyle\int_{h}^{b} \frac{dx}{x^p}$ as $h \longrightarrow 0+$

Caution   Do not confuse these results with those of Section 7 concerning

$$\int_{1}^{\infty} \frac{dx}{x^p}.$$

In fact,

$$\int_{0}^{1} \frac{dx}{x^p} \quad \begin{cases} \text{converges} & \text{if } p < 1, \\ \text{diverges} & \text{if } p \geq 1, \end{cases} \qquad \int_{1}^{\infty} \frac{dx}{x^p} \quad \begin{cases} \text{diverges} & \text{if } p \leq 1, \\ \text{converges} & \text{if } p > 1. \end{cases}$$

■ **EXAMPLE 1** Show that the integrals diverge

(a) $\displaystyle\int_{0}^{3} \frac{e^{-x}}{\sqrt{x}}\, dx$ (b) $\displaystyle\int_{7}^{10} \frac{\ln x}{\sqrt[3]{x-7}}\, dx$ (c) $\displaystyle\int_{1}^{2} \frac{dx}{\sqrt{4-x^2}}.$

Solution   Observe that

$$\frac{e^{-x}}{\sqrt{x}} \leq \frac{1}{\sqrt{x}} \quad (0 < x \leq 3), \qquad \frac{\ln x}{\sqrt[3]{x-7}} \leq \frac{\ln 10}{\sqrt[3]{x-7}} \quad (7 < x \leq 10),$$

$$\frac{1}{\sqrt{4-x^2}} = \frac{1}{\sqrt{2+x}} \frac{1}{\sqrt{2-x}} < \frac{1}{\sqrt{2-x}} \quad (1 \leq x < 2).$$

Since the integrals $\displaystyle\int_{0}^{3} \frac{1}{\sqrt{x}}\, dx,$ $\displaystyle\int_{7}^{10} \frac{dx}{\sqrt[3]{x-7}},$ $\displaystyle\int_{1}^{2} \frac{dx}{\sqrt{2-x}}$

all converge, the given integrals converge by comparison.   ■

■ **EXAMPLE 2** Show that the integrals diverge

(a) $\displaystyle\int_{0}^{\pi/3} \frac{\cos x}{x}\, dx$ (b) $\displaystyle\int_{-1}^{2} \frac{1+x^2}{(1+x)^{3/2}}\, dx$ (c) $\displaystyle\int_{2}^{3} \frac{dx}{(x^2-9)^2}.$

Solution   Observe that

$$\frac{\cos x}{x} \geq \frac{1}{2} \cdot \frac{1}{x} \quad (0 < x \leq \tfrac{1}{3}\pi), \qquad \frac{1 + x^2}{(1 + x)^{3/2}} > \frac{1}{(1 + x)^{3/2}} \quad (-1 < x \leq 2),$$

$$\frac{1}{(x^2 - 9)^2} = \frac{1}{[(x + 3)(x - 3)]^2} > \frac{1}{36} \cdot \frac{1}{(x - 3)^2} \quad (2 \leq x < 3).$$

Since the integrals $\displaystyle\int_0^{\pi/3} \frac{dx}{x}, \quad \int_{-1}^{2} \frac{dx}{(1 + x)^{3/2}}, \quad \int_{2}^{3} \frac{dx}{(x - 3)^2}$

all diverge, the given integrals diverge by comparison.   ∎

Another Test   Let us state (without proof) the following convergence criterion analogous to one proved in Section 7.

Suppose that $f(x) \geq 0$, and that $f(x)$ has a singularity at $x = a$ or at $x = b$. Suppose $g(x)$ is a bounded function. Then the convergence of

$$\int_a^b f(x)\,dx \qquad \text{implies the convergence of} \qquad \int_a^b f(x)g(x)\,dx.$$

Now any continuous function on a closed interval is bounded; hence $g(x)$ may be any continuous function on $[a, b]$.

Examples   The following integrals converge:

1. $\displaystyle\int_0^{6\pi} \frac{\cos x}{\sqrt{x}}\,dx; \qquad f(x) = \frac{1}{\sqrt{x}}, \qquad g(x) = \cos x,$

2. $\displaystyle\int_{-1}^{2} \frac{x^3 e^x}{\sqrt{4 - x^2}}\,dx; \qquad f(x) = \frac{1}{\sqrt{2 - x}}, \qquad g(x) = \frac{x^3 e^x}{\sqrt{2 + x}}.$

EXERCISES

Test for convergence

1. $\displaystyle\int_0^1 \frac{dx}{\sqrt[3]{x}}$

2. $\displaystyle\int_0^{\pi/4} \cot x\,dx$

3. $\displaystyle\int_0^2 \frac{e^x}{x}\,dx$

4. $\displaystyle\int_0^1 \frac{dx}{x^3 - 1}$

5. $\displaystyle\int_0^1 \frac{dx}{(1 - x^2)^2}$

6. $\displaystyle\int_1^3 \frac{\sin x}{\sqrt{3 - x}}\,dx$

7. $\displaystyle\int_3^5 \frac{dx}{\sqrt{x^2 - 9}}$

8. $\displaystyle\int_0^5 \frac{\sin^3 2x}{\sqrt{x}}\,dx$

9. $\displaystyle\int_0^1 \ln x\,dx$

10. $\displaystyle\int_0^3 \frac{dx}{\sqrt{x(3 + x)}}$

11. $\displaystyle\int_0^{1/2} \frac{dx}{x \ln x}$

12. $\displaystyle\int_2^4 \frac{dx}{\sqrt{-x^2 + 6x - 8}}$

13. $\displaystyle\int_1^2 \frac{dx}{\sqrt[3]{x^3 - 4x^2 + 4x}}$

14. $\displaystyle\int_0^{\pi/2} \sec x\,dx$

15. $\displaystyle\int_0^1 \sqrt{\frac{1 + x}{1 - x}}\,dx$

16. $\displaystyle\int_0^{\pi/2} \frac{\cos x}{\sqrt[3]{x}}\,dx$

17. $\displaystyle\int_0^1 \frac{dx}{e^x - 1}$

18. $\displaystyle\int_0^{\pi} \frac{dx}{\sqrt{\sin x}}$

19 $\displaystyle\int_0^\infty \frac{dx}{\sqrt{x+x^2}}$ **20** $\displaystyle\int_2^\infty \frac{dx}{x^2 - 2x}$ **21** $\displaystyle\int_{-\infty}^\infty \frac{e^{-x^2}}{x+5}\,dx$

22 $\displaystyle\int_{-\infty}^\infty \frac{dx}{\sqrt[3]{x^5+x}}$ **23** $\displaystyle\int_0^{\pi/2} \tan x\,dx$ **24** $\displaystyle\int_0^1 \frac{e^{-1/x}\,dx}{x^5}$

25 $\displaystyle\int_{-1}^1 \frac{x}{|x|^{3/2}}\,dx$ **26*** $\displaystyle\int_\pi^\infty \frac{dx}{x^2\sqrt{|\sin x|}}.$

27   Let $0 < a < b$. Prove convergent: $\displaystyle\int_0^\infty \frac{\cos ax - \cos bx}{x}\,dx$. [*Hint*   Separate the difficulties at 0 and ∞.]

28*   Let $0 < a < b$. Prove convergent: $\displaystyle\int_0^\infty \frac{\arctan bx - \arctan ax}{x}\,dx.$

Evaluate

29 $\displaystyle\int_0^1 x\ln(1+x)\,dx$ **30** $\displaystyle\int_0^1 x\ln(1-x)\,dx$ **31** $\displaystyle\int_0^{\pi/2} \ln\tan x$

32 $\displaystyle\int_0^\infty \frac{dx}{(1+x)\sqrt{x}}$ **33** $\displaystyle\int_0^\infty \operatorname{sech} x\,dx$ **34** $\displaystyle\int_0^\infty x^3 e^{-x^2}\,dx$

35 $\displaystyle\int_0^1 (\ln x)^n\,dx \quad (n \geq 0)$ **36*** $\displaystyle\int_0^{\pi/2} (\log\sin x)(\sin x)\,dx.$

37   Without evaluating, prove $\displaystyle\int_0^\infty \frac{dx}{x^3+1} = \int_0^\infty \frac{x\,dx}{x^3+1}.$

38   (cont.) Evaluate $\displaystyle\int_0^\infty \frac{dx}{x^3+1}$. [*Hint*   Add the two integrals in Ex. 37.]

10. MISCELLANEOUS EXERCISES

1   Let $\displaystyle a_n = \frac{(1+1)(1+4)(1+9)\cdots(1+n^2)}{(2+1)(2+4)(2+9)\cdots(2+n^2)}$.   Prove that $\{a_n\}$ converges.

2   (cont.) Prove that $\sum a_n$ diverges.

Test for convergence

3 $\displaystyle\frac{1}{5} + \frac{1}{15} + \frac{1}{25} + \cdots$ **4** $\displaystyle\frac{1+2}{1+3} + \frac{1+2+4}{1+3+9} + \frac{1+2+4+8}{1+3+9+27} + \cdots$

5 $\displaystyle\sum \frac{n(n+1)}{(n+2)(n+3)(n+4)}$ **6** $\displaystyle\sum \frac{n^3}{2^n}$

7 $\displaystyle\sum_{n=1}^\infty \frac{n}{4+n^3}$ **8** $\displaystyle\sum_{n=2}^\infty \frac{\ln n}{1+n^2}$ **9** $\displaystyle\sum_{n=1}^\infty \frac{n^2 e^{-n}}{1+n^2}$ **10** $\displaystyle\sum_{n=2}^\infty \frac{1}{\sqrt{n}\ln n}.$

For which x does the series converge?

11 $\displaystyle\sum_1^\infty \frac{1}{(2x+1)^n}$ **12** $\displaystyle\sum \frac{(2x)^n}{\sqrt{n}}.$

13   If a series of positive terms $\sum a_n$ converges, show that $\sum a_{2n}$ also converges.

14   Show by example that the statement in Ex. 13 is false without the assumption $a_n > 0$.

15   Suppose $\{a_n\}$ converges. Show that

$$a_1, a_2, a_2, a_3, a_3, a_3, a_4, a_4, a_4, a_4, \cdots$$

also converges.

16   Without using integration, prove that $\displaystyle\sum_1^n \frac{1}{\sqrt{k}} > 2\sqrt{n+1} - 2.$

[*Hint* $2\sqrt{k} < \sqrt{k} + \sqrt{k+1}.$]

17   Let $1 \le x < 2$. Define two sequences recursively as follows: $a_0 = x$ and $c_0 = 0$. If $a_n^2 < 2$, then $a_{n+1} = a_n^2$ and $c_{n+1} = c_n$. If $a_n^2 \ge 2$, then $a_{n+1} = \frac{1}{2}a_n^2$ and $c_{n+1} = c_n + (\frac{1}{2})^{n+1}$. Prove that

$$c_n + \frac{\log_2 a_n}{2^n} = \log_2 x.$$

18   (cont.) Prove that $1 \le a_n < 2$, that $\{c_n\}$ is increasing, and that $\lim_{n\to\infty} c_n = \log_2 x$. (This provides an algorithm for logs on a computer that only adds, squares, and divides by 2.)

19   Suppose $\displaystyle\int_0^\infty \frac{f(x)}{x}\,dx$ converges. Find $\displaystyle\int_0^\infty \frac{f(ax)}{x}\,dx$ for $a > 0$.

20*   (cont.) Suppose $f(x)$ is continuous for $x \ge 0$ and for each $\varepsilon > 0$, the integral

$$\int_\varepsilon^\infty \frac{f(x)}{x}\,dx$$

converges. Let $0 < a < b$. Prove

$$\int_0^\infty \frac{f(ax) - f(bx)}{x}\,dx = f(0)\ln\frac{b}{a}.$$

The **exponential integral** for $x < 0$ is $\mathrm{Ei}(x) = \displaystyle\int_{-\infty}^x \frac{e^t}{t}\,dt.$ Set

$$E(x) = -\,\mathrm{Ei}(-x) = \int_x^\infty \frac{e^{-t}}{t}\,dt \qquad \text{for} \quad x > 0.$$

21   Clearly $E(x)$ is very large for x near 0. How large? Prove the **asymptotic expansions**

$$E(x) = -e^{-x}\ln x + \int_x^\infty e^{-t}\ln t\,dt = -e^{-x}(\ln x + x\ln x - x) + \int_x^\infty te^{-t}(\ln t - 1)\,dt.$$

22   (cont.) Clearly $E(x)$ is very small for x large. How small? Prove

$$E(x) = e^{-x}\left(\frac{1}{x} - \frac{1}{x^2} + \frac{2!}{x^3} - \cdots + (-1)^{n-1}\frac{(n-1)!}{x^n}\right) + (-1)^n n! \int_x^\infty \frac{e^{-t}}{t^{n+1}}\,dt.$$

23   Let $K_n = \int_0^\infty x^n e^{-x^2}\,dx$. Prove $K_{n+2} = \frac{1}{2}(n+1)K_n$.   Conclude that

$K_{2n+1} = \frac{1}{2}(n!)$ and $K_{2n} = \dfrac{(2n)!}{2^{2n}n!}K_0$.   (It is known that $K_0 = \frac{1}{2}\sqrt{\pi}$. See p. 873.)

24*   Prove $K_n^2 < K_{n-1}K_{n+1}$. [*Hint*   Show that $K_{n+1} + 2tK_n + t^2 K_{n-1} > 0$.]

25   Set $F(x) = \displaystyle\int_0^\infty \frac{\arctan(xt)}{1+t^2}\,dt.$

Find $F(x) + F(1/x)$ for $x > 0$. [*Hint*   Use the change of variable $t = 1/u$.]

26* Let $\{a_n\}$ be a convergent sequence with limit 0. Define the averages

$$b_n = \frac{a_1 + \cdots + a_n}{n}.$$

Prove $\lim b_n = 0$. [*Hint*   Choose N so that $|a_n| < \frac{1}{2}\varepsilon$ for $n > N$ and write

$$b_{N+j} = \frac{a_1 + a_2 + \cdots + a_N}{N+j} + \frac{a_{N+1} + \cdots + a_{N+j}}{N+j}.]$$

27   (cont.) If $\lim a_n = L$, prove that $\lim b_n = L$. [*Hint*   Write $a_n = L + c_n$.]
28   Let $\sum a_n$ and $\sum b_n$ be series with positive terms. Suppose $\sum b_n$ converges and $a_{n+1}/a_n \le b_{n+1}/b_n$ for all n. Prove $\sum a_n$ converges.

An **infinite product** is an expression of the form

$$\prod_{n=1}^{\infty} (1 + a_n) = (1 + a_1)(1 + a_2)(1 + a_3) \cdots .$$

Its sequence of **partial products** is $\{p_n\}$, where $p_n = (1 + a_1)(1 + a_2) \cdots (1 + a_n)$. In the following exercises we shall assume $a_n \ge 0$.

29   Show that $p_n \ge 1 + a_1 + \cdots + a_n$.
30   (cont.) Suppose $\{p_n\}$ converges. Prove that $\sum a_n$ converges.
31   (cont.) Show that $\ln p_n \le a_1 + \cdots + a_n = s_n$.
32   (cont.) Suppose $\sum a_n$ converges. Prove that $\{p_n\}$ converges.
33   A certain model of a freezing ice cube involves solving the integral equation

$$\int_0^x tf(t)\, dt = \tfrac{1}{2}(1 + \beta)f(x).$$

for $f(x)$. We claim that $f(x) = x^p$ is a solution. What condition does this impose on β?

34   Define $\{a_n\}$ by $a_0 = 1$ and $a_{n+1} = \dfrac{a_n + 1}{a_n + 2}$. Prove $\{a_n\}$ is convergent and find its limit.

35   Given $1 + \dfrac{1}{2^2} + \dfrac{1}{3^2} + \dfrac{1}{4^2} + \cdots = \dfrac{\pi^2}{6}$,   show that $1 + \dfrac{1}{3^2} + \dfrac{1}{5^2} + \dfrac{1}{7^2} + \cdots = \dfrac{\pi^2}{8}$.

36   (cont.) Find $1 + \dfrac{1}{5^2} + \dfrac{1}{7^2} + \dfrac{1}{11^2} + \dfrac{1}{13^2} + \dfrac{1}{17^2} + \dfrac{1}{19^2} + \cdots .$

Power Series 12

1. BASIC PROPERTIES

In this chapter we study power series

$$a_0 + a_1(x - c) + a_2(x - c)^2 + \cdots + a_n(x - c)^n + \cdots$$

and their applications. Many of our examples will have $c = 0$, but the discussion will apply as well if $c \neq 0$.

Power series serve a number of important purposes in both theoretical and applied mathematics. First, they are particularly suitable for computation and are indispensable in many numerical problems. Second, they provide an alternative way of expressing many familiar functions and so aid in our understanding of these functions. Finally, with power series we can define functions that are hard or impossible to specify otherwise. Certainly nobody objects to defining a function by a polynomial,

$$f(x) = a_0 + a_1 x + a_2 x^2 + \cdots + a_n x^n.$$

Then why not define a function by a power series,

$$f(x) = a_0 + a_1 x + a_2 x^2 + \cdots + a_n x^n + \cdots,$$

provided, of course, that the series converges?

Convergence and Divergence Given a power series, we first ask: Does it converge? If so, where? Now for each fixed x, a power series is an infinite series of constants. So whatever we know about infinite series applies. In particular, convergence and divergence are defined in terms of the sequence of partial sums.

Convergence of Power Series A power series $\sum_{k=0}^{\infty} a_k(x - c)^k$ **converges** at a point x if the sequence of partial sums

$$s_n(x) = \sum_{k=0}^{n} a_k(x - c)^k$$

converges. The power series **diverges** at x if the sequence $\{s_n(x)\}$ diverges.

If $\lim_{n \to \infty} s_n(x) = F(x)$ for each x in a set **D**, we write

$$F(x) = a_0 + a_1(x - c) + a_2(x - c)^2 + \cdots, \qquad \text{for} \quad x \text{ in } \mathbf{D}.$$

For example, take the geometric series $1 + x + x^2 + x^3 + \cdots$.

Its n-th partial sum is $s_n(x) = 1 + x + x^2 + \cdots + x^n = \begin{cases} \dfrac{1 - x^{n+1}}{1 - x}, & x \neq 1, \\ n + 1, & x = 1. \end{cases}$

If $|x| < 1$, then $x^{n+1} \longrightarrow 0$, hence $s_n(x) \longrightarrow 1/(1 - x)$. If $|x| \geq 1$, the sequence $\{s_n(x)\}$ diverges. Therefore

$$1 + x + x^2 + x^3 + \cdots = \frac{1}{1 - x} \quad \text{if} \quad -1 < x < 1,$$

and the series diverges for all other values of x.

The geometric series is especially nice because there is a neat formula for its partial sums. This not only helps the discussion of convergence and divergence, but also leads to a neat formula for the sum of the series where it converges. Generally there is no nice formula for the partial sums of a given power series. Still we can use the techniques of the last chapter to investigate convergence and divergence. As an example, consider the power series

$$1 + x + \frac{x^2}{2^2} + \frac{x^3}{3^3} + \cdots + \frac{x^n}{n^n} + \cdots .$$

For any fixed x, eventually $n > 2|x|$. Then $\left| \dfrac{x^n}{n^n} \right| < \dfrac{1}{2^n};$

hence the series converges (absolutely) by comparison with the geometric series $\sum 1/2^n$. There is no obvious formula for the sum; we consider the power series as *defining* a new function $f(x)$ whose domain is all real x.

One further example:

$$1 + x + 2^2 x^2 + 3^3 x^3 + \cdots + n^n x^n + \cdots .$$

Obviously, this power series converges at $x = 0$. However, it diverges everywhere else; if $x \neq 0$, then

$$|n^n x^n| = |nx|^n \longrightarrow \infty \quad \text{as} \quad n \longrightarrow \infty .$$

Interval of Convergence   What is the domain of convergence of a power series? The preceding examples show that there are at least three possibilities: the domain may consist of a single point, a finite interval, or the entire real axis. In fact, these are the *only* possibilities.

Interval of Convergence   Given a power series $\sum a_n(x - c)^n$, precisely one of the following three cases holds:

(1)   The series converges only for $x = c$.

(2)   The series converges for all values of x.

(3)   There is a positive number R such that the series converges for each x satisfying $|x - c| < R$ and diverges for each x satisfying $|x - c| > R$.

The proof will be given in Section 7.

Case (1) is an extreme case. It occurs when the coefficients a_n increase so rapidly that the power series can converge only if all terms after a_0 vanish. An example is $\sum n^n x^n$. Power series of this type are of no earthly good to anybody.

Case (2) occurs when the coefficients become small very rapidly. An example is $\sum x^n/n^n$, where for any x the general term x^n/n^n eventually tends to zero quickly because the coefficients $a_n = 1/n^n$ become small so fast. Power series of this type are the nicest kind since they never cause any problems with convergence.

Case (3) lies between. The coefficients do not increase so rapidly that the series never converges (except for $x = c$), nor do they decrease so rapidly that the series always converges. A typical example is the geometric series $\sum x^n$, where each $a_n = 1$. This series converges for $|x| < 1$ and diverges for $|x| > 1$, hence $R = 1$.

In Case (3) the set of all points x for which the series $\sum a_n(x - c)^n$ converges is called its **interval of convergence** (or **domain of convergence**). This set consists of the interval $c - R < x < c + R$ and possibly one or both of its end points. The number R is called the **radius of convergence** of the power series (Fig. 1).

By convention, $R = 0$ in Case (1), convergence for $x = c$ only; and $R = \infty$ in Case (2), convergence for all x. The interval of convergence in Case (1) is the single point c; in Case (2) it is the entire x-axis.

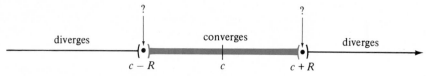

Fig. 1   Interval of convergence

■ **EXAMPLE 1**   Find the sum, radius of convergence, and interval of convergence of the power series

(a) $1 + 4x + 4^2 x^2 + 4^3 x^3 + \cdots,$

(b) $1 + \dfrac{1}{2}(x + 3) + \dfrac{1}{2^2}(x + 3)^2 + \dfrac{1}{2^3}(x + 3)^3 + \cdots,$

(c) $1 - x^2 + x^4 - x^6 + \cdots.$

Solution   Each of the three series is of the form

$$1 + y + y^2 + y^3 + \cdots = \frac{1}{1 - y}, \qquad \text{which converges for } |y| < 1.$$

(a)   Here $y = 4x$. The sum is $1/(1 - 4x)$ and the series converges if and only if $|4x| < 1$, that is, $|x| < \frac{1}{4}$. Hence $R = \frac{1}{4}$ and the interval of convergence is $-\frac{1}{4} < x < \frac{1}{4}$.

(b) $y = \frac{1}{2}(x + 3)$. The sum is $1/[1 - \frac{1}{2}(x + 3)] = -2/(1 + x)$ and the series converges if and only if $|\frac{1}{2}(x + 3)| < 1$, that is, $|x + 3| < 2$. Hence $R = 2$ and the interval of convergence is $-5 < x < -1$.

(c) $y = -x^2$. The sum is $1/(1 + x^2)$ and the series converges if and only if $|-x^2| < 1$, that is, $|x| < 1$. Hence $R = 1$ and the interval of convergence is $-1 < x < 1$.   ■

■ **EXAMPLE 2** Find the interval of convergence

(a) $x + \dfrac{x^2}{2} + \dfrac{x^3}{3} + \cdots + \dfrac{x^n}{n} + \cdots,$

(b) $\dfrac{x}{10} + \dfrac{x^2}{10^4} + \dfrac{x^3}{10^9} + \cdots + \dfrac{x^n}{10^{n^2}} + \cdots.$

Solution (a) Since $|x^n/n| < |x|^n$, the series converges for $|x| < 1$ by comparison with the geometric series. Hence its interval of convergence includes the interval $|x| < 1$. However, the series diverges at $x = 1$ (harmonic series) and converges at $x = -1$ (alternating harmonic series). Therefore its interval of convergence is $-1 \leq x < 1$.

(b) For any fixed x, choose a positive integer p such that $|x| < 10^p$. Then if $n > p$,

$$\left| \frac{x^n}{10^{n^2}} \right| < \frac{10^{np}}{10^{n^2}} = \frac{1}{10^{n(n-p)}} \leq \frac{1}{10^n}.$$

Hence the series converges by comparison with the geometric series $\sum 1/10^n$. The interval of convergence is the entire x-axis. ■

Remark The geometric series diverges at both end points of its interval of convergence. The series in Example 2a converges at one end point, $x = -1$, of its interval of convergence (alternating harmonic series) and diverges (harmonic series) at the other end point, $x = 1$. There exist power series that converge at both end points. See Ex. 32.

Ratio Test Often the radius of convergence of a given power series can be found by the following ratio test, a consequence of the ratio test for series of constants.

> **Ratio Test** Suppose the power series
>
> $$a_0 + a_1(x - c) + a_2(x - c)^2 + \cdots + a_n(x - c)^n + \cdots$$
>
> has non-zero coefficients. If
>
> $$\left| \frac{a_n}{a_{n+1}} \right| \longrightarrow R \quad \text{as} \quad n \longrightarrow \infty,$$
>
> where R is 0, positive, or ∞, then R is the radius of convergence.

Proof For simplicity of notation assume $c = 0$. Suppose $0 < R < \infty$. If $|x| < R$, then

$$\frac{|a_{n+1} x^{n+1}|}{|a_n x^n|} = \left| \frac{a_{n+1}}{a_n} \right| |x| \longrightarrow \frac{|x|}{R} < 1.$$

Hence the series $\sum a_n x^n$ converges (absolutely) by the ratio test for a series of constants (p. 548). If $|x| > R$, then

$$\frac{|a_{n+1} x^{n+1}|}{|a_n x^n|} \longrightarrow \frac{|x|}{R} > 1.$$

Hence for n sufficiently large, $\{|a_n x^n|\}$ is an increasing sequence. Therefore $\sum a_n x^n$ diverges because its terms do not approach 0. Thus the series converges for $|x| < R$ and diverges for $|x| > R$. In other words, its radius of convergence is R.

Suppose $|a_n/a_{n+1}| \longrightarrow R = 0$. If $x \neq 0$, then

$$\frac{|a_{n+1}x^{n+1}|}{|a_n x^n|} = \left|\frac{a_{n+1}}{a_n}\right| |x| \longrightarrow \infty.$$

Hence the series diverges for all $x \neq 0$; its radius of convergence is 0. Finally, suppose $|a_n/a_{n+1}| \longrightarrow R = \infty$. Then for any x,

$$\frac{|a_{n+1}x^{n+1}|}{|a_n x^n|} = \left|\frac{a_{n+1}}{a_n}\right| |x| \longrightarrow 0.$$

Hence the series converges for all x; its radius of convergence is ∞. This completes the proof in all cases.

■ **EXAMPLE 3** Find the radius of convergence

(a) $1 + \dfrac{x}{4} + \dfrac{x^2}{7} + \dfrac{x^3}{10} + \cdots + \dfrac{x^n}{3n+1} + \cdots,$

(b) $(x-1) - 4(x-1)^2 + 9(x-1)^3 - \cdots + (-1)^{n-1}n^2(x-1)^n + \cdots,$

(c) $1 + \dfrac{x}{2+1} + \dfrac{x^2}{2^2+2} + \dfrac{x^3}{2^3+3} + \cdots + \dfrac{x^n}{2^n+n} + \cdots,$

(d) $\dfrac{x^3}{1 \cdot 3 \cdot 5} + \dfrac{x^4}{1 \cdot 3 \cdot 5 \cdot 7} + \cdots + \dfrac{x^n}{1 \cdot 3 \cdot 5 \cdots (2n-1)} + \cdots,$

(e) $\dfrac{x^3}{\sqrt{3}} + \dfrac{x^6}{\sqrt{6}} + \dfrac{x^9}{\sqrt{9}} + \cdots + \dfrac{x^{3n}}{\sqrt{3n}} + \cdots.$

Solution In each case apply the Ratio Test.

(a) Here $a_n = 1/(3n+1)$ and

$$\left|\frac{a_n}{a_{n+1}}\right| = \frac{1}{3n+1} \Big/ \frac{1}{3n+4} = \frac{3n+4}{3n+1}.$$

Hence $\left|\dfrac{a_n}{a_{n+1}}\right| \longrightarrow 1$ as $n \longrightarrow \infty;$ $R = 1.$

(b) $a_n = (-1)^{n-1}n^2$. As $n \longrightarrow \infty$,

$$\left|\frac{a_n}{a_{n+1}}\right| = \frac{n^2}{(n+1)^2} = \left(\frac{n}{n+1}\right)^2 \longrightarrow 1; \qquad R = 1.$$

(c) $a_n = 1/(2^n + n)$. As $n \longrightarrow \infty$,

$$\left|\frac{a_n}{a_{n+1}}\right| = \frac{1}{2^n+n} \Big/ \frac{1}{2^{n+1}+n+1} = \frac{2^{n+1}+n+1}{2^n+n}$$

$$= \frac{2+(n+1)\cdot 2^{-n}}{1+n\cdot 2^{-n}} \longrightarrow \frac{2+0}{1+0} = 2; \qquad R = 2.$$

(d) $a_n = 1/[1 \cdot 3 \cdot 5 \cdots (2n - 1)]$. As $n \longrightarrow \infty$,

$$\left| \frac{a_n}{a_{n+1}} \right| = \frac{1 \cdot 3 \cdot 5 \cdots (2n - 1)(2n + 1)}{1 \cdot 3 \cdot 5 \cdots (2n - 1)} = 2n + 1 \longrightarrow \infty; \qquad R = \infty.$$

Notice that these ratios make sense only for $n \geq 3$ because $a_0 = a_1 = a_2 = 0$ in the given series. However, it is perfectly OK to apply the Ratio Test from some point on since we may ignore a finite number of terms when studying convergence and divergence.

(e)   The Ratio Test does not apply directly because "two-thirds" of the coefficients in this power series are zero. Nevertheless, the series may be written

$$\frac{y}{\sqrt{3}} + \frac{y^2}{\sqrt{6}} + \frac{y^3}{\sqrt{9}} + \cdots + \frac{y^n}{\sqrt{3n}} + \cdots,$$

where $y = x^3$. The Ratio Test does apply to the series in this form:

$$\frac{\sqrt{3(n + 1)}}{\sqrt{3n}} = \sqrt{\frac{n + 1}{n}} \longrightarrow 1.$$

Hence the y-series converges for $|y| < 1$ and diverges for $|y| > 1$. Therefore, the original series converges for $|x^3| < 1$ and diverges for $|x^3| > 1$, that is, for $|x| < 1$ and $|x| > 1$, respectively. Hence $R = 1$. ∎

EXERCISES

Find the radius of convergence

1   $1 + x + 2x^2 + 3x^3 + \cdots$

2   $x + \dfrac{x^2}{4} + \dfrac{x^3}{9} + \dfrac{x^4}{16} + \cdots$

3   $x - \dfrac{x^3}{3} + \dfrac{x^5}{5} - \dfrac{x^7}{7} + \cdots$

4   $1 + \dfrac{x}{1 \cdot 2} + \dfrac{x^2}{2 \cdot 2^2} + \dfrac{x^3}{3 \cdot 2^3} + \cdots$

5   $\dfrac{x}{3 + 2} + \dfrac{x^2}{3^2 + 2^2} + \dfrac{x^3}{3^3 + 2^3} + \cdots$

6   $\dfrac{x - 1}{1 \cdot 2 \cdot 3} + \dfrac{(x - 1)^2}{2 \cdot 3 \cdot 4} + \dfrac{(x - 1)^3}{3 \cdot 4 \cdot 5} + \cdots$

7   $x + \sqrt{2}\,x^2 + \sqrt{3}\,x^3 + \sqrt{4}\,x^4 + \cdots$

8   $\dfrac{1}{2} - \dfrac{x}{2^2} + \dfrac{x^2}{2^4} - \dfrac{x^3}{2^8} + \dfrac{x^4}{2^{16}} - + \cdots$

9   $(e^2 - 2)x^2 + (e^3 - 3)x^3 + (e^4 - 4)x^4 + \cdots$

10   $\dfrac{2x}{1^3} + \dfrac{3x^2}{2^3} + \dfrac{4x^3}{3^3} + \dfrac{5x^4}{4^3} + \cdots$

11   $\dfrac{x^2}{(\ln 2)^2} + \dfrac{x^3}{(\ln 3)^3} + \dfrac{x^4}{(\ln 4)^4} + \cdots$

12   $1 + x + 2!x^2 + 3!x^3 + \cdots$

13   $\dfrac{x^2}{2 + \ln 2} + \dfrac{x^3}{3 + \ln 3} + \dfrac{x^4}{4 + \ln 4} + \cdots$

14   $\dfrac{ab}{1 \cdot c}x + \dfrac{a(a + 1)b(b + 1)}{2!c(c + 1)}x^2 + \dfrac{a(a + 1)(a + 2)b(b + 1)(b + 2)}{3!c(c + 1)(c + 2)}x^3 + \cdots \quad (c > 0)$

15   $1 + x^2 + x^{10} + x^{12} + x^{20} + x^{22} + x^{30} + x^{32} + \cdots$

16   $1 + \dfrac{x^4}{2^4} + \dfrac{x^9}{2^9} + \dfrac{x^{16}}{2^{16}} + \cdots$

17   $4 \cdot 5x^4 + 8 \cdot 9x^8 + 16 \cdot 13x^{12} + 32 \cdot 17x^{16} + 64 \cdot 21x^{20} + \cdots$

18   $(1 + 2)x + (1 + 2 + 4)x^2 + (1 + 2 + 4 + 8)x^3 + \cdots$

19 $\dfrac{1 \cdot 4}{2 \cdot 5} x^2 + \dfrac{1 \cdot 4 \cdot 7}{2 \cdot 5 \cdot 8} x^3 + \dfrac{1 \cdot 4 \cdot 7 \cdot 10}{2 \cdot 5 \cdot 8 \cdot 11} x^4 + \cdots$

20 $\dfrac{(2!)^3}{6!} x^2 + \dfrac{(3!)^3}{9!} x^3 + \dfrac{(4!)^3}{12!} x^4 + \cdots.$

Find the domain of convergence

21 $\displaystyle\sum_{n=1}^{\infty} \dfrac{(\sin x)^n}{n^2}$ **22** $\displaystyle\sum_{n=1}^{\infty} \dfrac{(x^2 + 1)^n}{n(n+1)}$ **23** $\displaystyle\sum_{n=1}^{\infty} \dfrac{e^{nx}}{n}$ **24** $\displaystyle\sum_{n=1}^{\infty} n^2 (\cos \pi x)^n.$

Find the sum of the series and its domain of convergence

25 $1 + (x-3) + (x-3)^2 + (x-3)^3 + \cdots$

26 $\dfrac{x}{5} + \left(\dfrac{x}{5}\right)^2 + \left(\dfrac{x}{5}\right)^3 + \left(\dfrac{x}{5}\right)^4 + \cdots$ **27** $1 - e^x + e^{2x} - e^{3x} + \cdots$

28 $\cos^2 x + \cos^4 x + \cos^6 x + \cos^8 x + \cdots$

29 $\ln x + \ln(\sqrt{x}) + \ln(\sqrt[4]{x}) + \ln(\sqrt[8]{x}) + \cdots$

30 $\dfrac{1}{x} + \dfrac{1}{x^2} + \dfrac{1}{x^3} + \dfrac{1}{x^4} + \cdots.$

31   Give an example of a power series with radius of convergence π.

32   Give an example of a power series that converges at both end points of a finite interval of convergence.

33   Suppose that infinitely many coefficients of a power series are non-zero integers. Show that the radius of convergence is at most 1.

34   Suppose $\sum a_n x^n$ has radius of convergence R. If $|b_n| \le |a_n|$ for each n, what can be said about the radius of convergence of $\sum b_n x^n$?

35*   **(Root Test)** Suppose $|a_n| > 0$ for all n and $1/\sqrt[n]{|a_n|} \longrightarrow R$ as $n \longrightarrow \infty$. Prove that $\sum a_n x^n$ has radius of convergence R.

36*   (cont.) Prove that if $|a_n/a_{n+1}| \longrightarrow R$, then also $1/\sqrt[n]{|a_n|} \longrightarrow R$. Conclude that the Root Test applies whenever the Ratio Test applies. [*Hint* Use Ex. 27, p. 577.]

37   (cont.) Verify that the converse is not true. Show that the Root Test applies to the series

$$\dfrac{x}{10^2} + \dfrac{x^2}{10} + \dfrac{x^3}{10^4} + \dfrac{x^4}{10^3} + \dfrac{x^5}{10^6} + \dfrac{x^6}{10^5} + \cdots$$

but the Ratio Test fails.

38*   Give the interval of convergence of the following modifications of the harmonic series, with particular attention to the end points:

(a) $x + \dfrac{x^2}{2} - \dfrac{x^3}{3} - \dfrac{x^4}{4} + \dfrac{x^5}{5} + \dfrac{x^6}{6} - \dfrac{x^7}{7} - \dfrac{x^8}{8} + \cdots$

(b) $x + \dfrac{x^2}{2} - \dfrac{x^3}{3} + \dfrac{x^4}{4} + \dfrac{x^5}{5} - \dfrac{x^6}{6} + \dfrac{x^7}{7} + \dfrac{x^8}{8} - \dfrac{x^9}{9} + \cdots$

(c) $x + \dfrac{x^2}{2} + \dfrac{x^3}{3} - \dfrac{x^4}{4} + \dfrac{x^5}{5} + \dfrac{x^6}{6} + \dfrac{x^7}{7} - \dfrac{x^8}{8} + \cdots.$

2. TAYLOR SERIES

The function $f(x) = 1/(1-x)$ can be expressed as the sum of the power series $\sum_0^\infty x^n$ for $|x| < 1$. Is this just a quirk or can other functions be expressed as power series? If so, how do we find a power series for a given function?

A systematic approach to these questions is based on Taylor polynomials. Let us recall their definition: if $f(x)$ has n derivatives at $x = c$, we associate with $f(x)$ its n-th degree Taylor polynomial at $x = c$,

$$p_n(x) = f(c) + \frac{f'(c)}{1!}(x - c) + \frac{f''(c)}{2!}(x - c)^2 + \cdots + \frac{f^{(n)}(c)}{n!}(x - c)^n.$$

Hopefully, $p_n(x)$ approximates $f(x)$ near $x = c$, and the approximations improve as n increases. This suggests associating with $f(x)$ the infinite series whose partial sums are the Taylor polynomials $p_n(x)$ at $x = c$.

Taylor Series Suppose $f(x)$ has derivatives of all orders at $x = c$. We associate with $f(x)$ its **Taylor series** at $x = c$:

$$f(c) + \frac{f'(c)}{1!}(x - c) + \frac{f''(c)}{2!}(x - c)^2 + \cdots = \sum_{n=0}^{\infty} \frac{f^{(n)}(c)}{n!}(x - c)^n.$$

(We are using the conventions that $f^{(0)}(c) = f(c)$ and $0! = 1$.)

Terminology   The Taylor series of $f(x)$ at $x = c$ is often called the **Taylor expansion** of $f(x)$ at $x = c$. We speak of **expanding** or **representing** $f(x)$ in a Taylor series. A Taylor series at $x = 0$ is sometimes called a **Maclaurin series**.

■ **EXAMPLE 1**   Find the Taylor series at $x = 0$ of
(a) e^x (b) $\sin x$ (c) $\cos x$.

Solution (a) If $f(x) = e^x$, then $f^{(n)}(x) = e^x$ for each n. Hence $f^{(n)}(0) = 1$ for each n. Substitute these values into the formula. The desired Taylor series is

$$1 + x + \frac{x^2}{2!} + \frac{x^3}{3!} + \cdots = \sum_{n=0}^{\infty} \frac{x^n}{n!}.$$

(b) Compute derivatives:

$$f(x) = \sin x, \qquad f'(x) = \cos x, \qquad f''(x) = -\sin x,$$
$$f'''(x) = -\cos x, \qquad f^{(4)}(x) = \sin x, \cdots,$$

repeating in cycles of four. At $x = 0$, the values are

$$0, \quad 1, \quad 0, \quad -1, \quad 0, \quad 1, \quad 0, \quad -1, \cdots.$$

Hence the Taylor series is

$$x - \frac{x^3}{3!} + \frac{x^5}{5!} - \frac{x^7}{7!} + \cdots = \sum_{n=1}^{\infty} (-1)^{n-1} \frac{x^{2n-1}}{(2n-1)!}.$$

(c) Since $\cos x$ is the derivative of $\sin x$, read off its derivatives at $x = 0$ from those of $\sin x$ found in (b). The values are

$$1, \quad 0, \quad -1, \quad 0, \quad 1, \quad 0, \quad -1, \quad 0, \cdots,$$

repeating in cycles of four. Hence the Taylor series is

$$1 - \frac{x^2}{2!} + \frac{x^4}{4!} - \frac{x^6}{6!} + \cdots = \sum_{n=0}^{\infty} (-1)^n \frac{x^{2n}}{(2n)!}. \qquad \blacksquare$$

Remark The three series in Example 1 converge for all x as can be verified by the Ratio Test. We shall omit this, however, since the convergence will come out in the wash shortly when we make a closer study of these series.

■ **EXAMPLE 2** Find the Taylor series of $f(x) = \ln(1 + x)$ at (a) $x = 0$, (b) $x = 2$. In each case, determine the interval of convergence.

Solution Compute successive derivatives:

$$f'(x) = \frac{1}{1+x}, \qquad f''(x) = \frac{-1}{(1+x)^2}, \qquad f'''(x) = \frac{1}{(1+x)^3},$$

$$f^{(4)}(x) = \frac{-2 \cdot 3}{(1+x)^4}, \quad \cdots \quad , f^{(n)}(x) = \frac{(-1)^{n-1}(n-1)!}{(1+x)^n}.$$

Hence,
$$\frac{f^{(n)}(0)}{n!} = \frac{(-1)^{n-1}}{n}, \qquad \frac{f^{(n)}(2)}{n!} = \frac{(-1)^{n-1}}{n \cdot 3^n}.$$

Substitute these values into the formula with $c = 0$ in (a) and $c = 2$ in (b). The constant terms in the two series are respectively $\ln(1 + 0) = 0$ and $\ln(1 + 2) = \ln 3$. The resulting Taylor series are

(a)
$$x - \frac{x^2}{2} + \frac{x^3}{3} - \frac{x^4}{4} + \cdots = \sum_{n=1}^{\infty} (-1)^{n-1} \frac{x^n}{n},$$

(b)
$$\ln 3 + \frac{x-2}{1 \cdot 3} - \frac{(x-2)^2}{2 \cdot 3^2} + \cdots = \ln 3 + \sum_{n=1}^{\infty} \frac{(-1)^{n-1}}{n \cdot 3^n} (x-2)^n.$$

Now find the radius of convergence R by the Radio Test. In (a), the n-th coefficient is $a_n = (-1)^{n-1}/n$, so

$$\left| \frac{a_n}{a_{n+1}} \right| = \frac{n+1}{n} \longrightarrow 1.$$

Hence $R = 1$; since the series diverges at $x = -1$ and converges at $x = 1$ (compare Example 2a, p. 581), the interval of convergence is $-1 < x \leq 1$.

In (b), the n-th coefficient is $a_n = (-1)^{n-1}/n3^n$, so

$$\left| \frac{a_n}{a_{n+1}} \right| = \frac{(n+1)3^{n+1}}{n3^n} = \frac{3(n+1)}{n} \longrightarrow 3.$$

Hence $R = 3$; as above the interval of convergence is $-1 < x \leq 5$. ■

■ **EXAMPLE 3** Find the Taylor series of $f(x) = 1/\sqrt{x}$ at $x = 9$ and determine its radius of convergence.

Solution   Compute successive derivatives:

$$f(x) = \frac{1}{x^{1/2}}, \qquad f'(x) = \frac{-1}{2x^{3/2}}, \qquad f''(x) = \frac{3}{2^2 x^{5/2}},$$

$$f'''(x) = \frac{-3 \cdot 5}{2^3 x^{7/2}}, \cdots, f^{(n)}(x) = \frac{(-1)^n 1 \cdot 3 \cdot 5 \cdots (2n-1)}{2^n x^{(2n+1)/2}}.$$

The product in the numerator of $f^{(n)}(x)$ can be expressed in terms of factorials:

$$1 \cdot 3 \cdot 5 \cdots (2n-1) = \frac{1 \cdot 2 \cdot 3 \cdot 4 \cdot 5 \cdots (2n-1)(2n)}{2 \cdot 4 \cdot 6 \cdots (2n)} = \frac{(2n)!}{2^n n!}.$$

It follows that
$$f^{(n)}(9) = \frac{(-1)^n (2n)!}{(n!) 2^{2n} 3^{2n+1}} = \frac{(-1)^n (2n)!}{3(n!) 6^{2n}}.$$

(Note that this formula is correct for $n = 0$.) Therefore the desired Taylor series is

$$\sum_{n=0}^{\infty} \frac{f^{(n)}(9)}{n!} (x-9)^n = \sum_{n=0}^{\infty} \frac{(-1)^n (2n)!}{3(n!)^2 6^{2n}} (x-9)^n.$$

To find the radius of convergence, apply the Ratio-Test:

$$\left| \frac{a_n}{a_{n+1}} \right| = \frac{(2n)!}{3(n!)^2 6^{2n}} \frac{3[(n+1)!]^2 6^{2n+2}}{(2n+2)!} = \frac{36(n+1)^2}{(2n+1)(2n+2)} \longrightarrow 9.$$

Hence $R = 9$.   ■

Validity of Taylor Series

In the preceding examples, we formally wrote down the Taylor series associated with various functions; we did not prove that the functions actually equal the sums of their Taylor series. There is no guarantee that they do. In fact, there exist functions that are not the sums of their Taylor series. (See Exs. 21–24.)

Suppose $f(x)$ has derivatives of all orders at $x = c$. Write

$$f(x) = f(c) + \frac{f'(c)}{1!} (x-c) + \cdots + \frac{f^{(n)}(c)}{n!} (x-c)^n + r_n(x).$$

Then $f(x)$ is the sum of its Taylor series if and only if the remainders $r_n(x) \longrightarrow 0$ as $n \longrightarrow \infty$. Our trump card in dealing with $r_n(x)$ is Taylor's Formula; it says that

$$r_n(x) = \frac{1}{n!} \int_c^x (x-t)^n f^{(n+1)}(t) \, dt.$$

There follows a useful estimate for the remainder. Suppose that

$$|f^{(n+1)}(x)| \leq M$$

in some interval including c, say $a \leq x \leq b$. Then

$$|r_n(x)| \leq \frac{M}{(n+1)!} |x-c|^{n+1}, \qquad a \leq x \leq b.$$

Let us now prove that some familiar functions are actually equal to the sums of their Taylor series.

---

(1) $\qquad e^x = 1 + x + \dfrac{x^2}{2!} + \cdots = \displaystyle\sum_{n=0}^{\infty} \dfrac{x^n}{n!}, \qquad -\infty < x < \infty$

(2) $\qquad \sin x = x - \dfrac{x^3}{3!} + \dfrac{x^5}{5!} - \cdots = \displaystyle\sum_{n=1}^{\infty} (-1)^{n-1} \dfrac{x^{2n-1}}{(2n-1)!}, \qquad -\infty < x < \infty$

(3) $\qquad \cos x = 1 - \dfrac{x^2}{2!} + \dfrac{x^4}{4!} - \cdots = \displaystyle\sum_{n=0}^{\infty} (-1)^n \dfrac{x^{2n}}{(2n)!}, \qquad -\infty < x < \infty.$

Proof By Example 1, each series is the Taylor series associated with the given functions. In each case, we must write the function as $f(x) = p_n(x) + r_n(x)$ and prove that $r_n(x) \longrightarrow 0$ for all values of x.

(1) Let $f(x) = e^x$ and assume at first that $-B \le x \le B$. In this interval, $|f^{(n+1)}(x)| = e^x \le e^B$. Apply the estimate for the remainder, with $M = e^B$:

$$|r_n(x)| \le e^B \frac{|x|^{n+1}}{(n+1)!} = e^B \frac{B^{n+1}}{(n+1)!}, \qquad |x| \le B.$$

Since e^B is fixed, it is enough to show that $A_n \longrightarrow 0$, where

$$A_n = \frac{B^n}{n!},$$

for if so, then $\qquad |r_n(x)| \le e^B A_{n+1} \longrightarrow 0, \qquad$ so $\quad r_n(x) \longrightarrow 0$

for $|x| \le B$.

But we know that the exponential series $\sum x^n/n!$ converges for all x, in particular $\sum B^n/n!$ converges, hence $A_n = B^n/n! \longrightarrow 0$. Therefore $|r_n(x)| \longrightarrow 0$ for all $|x| \le B$. The argument is valid for each positive B, hence $r_n(x) \longrightarrow 0$ for all x.

(2) Let $f(x) = \sin x$ and assume that $-B \le x \le B$. Note that for $\sin x$,

$$p_{2m-1}(x) = p_{2m}(x), \qquad \text{hence} \quad r_{2m-1}(x) = r_{2m}(x).$$

Therefore, it will be enough to show that $r_{2m}(x) \longrightarrow 0$ as $m \longrightarrow \infty$.

Now $|f^{(2m+1)}(x)| = |\pm \sin x| \le 1$. Apply the estimate for the remainder, with $M = 1$:

$$|r_{2m}(x)| \le \frac{1}{(2m+1)!} |x|^{2m+1} \le \frac{B^{2m+1}}{(2m+1)!}, \qquad |x| \le B.$$

Thus $|r_{2m}(x)| \le A_{2m+1}$, where $\{A_n\}$ is the sequence in (1). But $\{A_{2m+1}\}$ is a subsequence of $\{A_n\}$ and $A_n \longrightarrow 0$; hence $A_{2m+1} \longrightarrow 0$. It follows that $r_{2m}(x) \longrightarrow 0$, provided $|x| \le B$. This is true for every positive B. Therefore $r_{2m}(x) \longrightarrow 0$ for all x.

(3) The proof is nearly the same as in (2). This time $|r_{2m}(x)| = |r_{2m+1}(x)| \leq A_{2m+2}$, and the conclusion follows in the same way. ∎

Another valid Taylor series expansion is

$$\frac{1}{1-x} = 1 + x + x^2 + x^3 + \cdots = \sum_{n=0}^{\infty} x^n, \qquad -1 < x < 1.$$

We have already proved the equality (p. 579), but we have not shown that the series actually is the *Taylor* series of $1/(1-x)$ at $x = 0$. We shall omit the routine verification since it will come out free of charge when we discuss uniqueness of power series in the next section.

■ **EXAMPLE 4** Find the sum of the series and its domain of convergence

(a) $1 - \dfrac{(2x)^2}{2!} + \dfrac{(2x)^4}{4!} - \dfrac{(2x)^6}{6!} + \cdots$,

(b) $\dfrac{x^4}{2!} - \dfrac{x^6}{3!} + \dfrac{x^8}{4!} - \dfrac{x^{10}}{5!} + \cdots$,

(c) $1 + \ln x + \dfrac{(\ln x)^2}{2!} + \dfrac{(\ln x)^3}{3!} + \cdots$.

Solution (a) The series has the form

$$1 - \frac{y^2}{2!} + \frac{y^4}{4!} - \frac{y^6}{6!} + \cdots$$

with $y = 2x$. This series converges to $\cos y$ for all real y. Therefore, given any real x, it is legitimate to substitute $y = 2x$. Thus the sum of the given series is $\cos 2x$, for all x.

(b) The series has the form

$$\left(1 + \frac{y}{1!} + \frac{y^2}{2!} + \frac{y^3}{3!} + \cdots\right) - 1 - y$$

with $y = -x^2$. The series in parentheses converges to e^y for all real values y, in particular for $y = -x^2$. Thus the given series converges to $e^{-x^2} - 1 + x^2$, for all x.

(c) The series has the form $\sum_0^{\infty} y^n/n!$ which converges to e^y for all real y. In particular, it converges for $y = \ln x$, where $x > 0$. The series that results from substituting $y = \ln x$ is not a power series; nevertheless it converges to $e^{\ln x} = x$ for $x > 0$. ∎

In Sections 3 and 4 we shall discuss ways of establishing the validity of various Taylor series without investigating the remainders. Such methods will be of great use since estimates of $r_n(x)$ can be tricky. For example, proving the validity of the series

$$\ln(1 + x) = x - \frac{x^2}{2} + \frac{x^3}{3} - \cdots$$

(Example 2) is difficult via remainders, but comes out easily in Section 4. (See Example 3, p. 599.) Let us accept it for the time being.

EXERCISES

Find the Taylor series of the function at $x = c$

1   e^{3x}, $c = 0$ 2 e^x, $c = 2$ 3 $\cos x$, $c = \frac{1}{4}\pi$

4   $\sin x$, $c = \frac{1}{6}\pi$ 5 $\dfrac{1}{1-x}$, $c = \frac{1}{3}$ 6 $\dfrac{1}{1-x}$, $c = 3$

7   $\ln(a + x)$, $c = 0$, $a > 0$ 8 $\dfrac{1}{a + bx}$, $c = 0$, $a, b > 0$

9   $(1 + x)e^x$, $c = 0$ 10 $e^{ax} + e^{bx}$, $c = 0$ 11 $\sqrt{x}$, $c = 4$

12   $\dfrac{1}{x^2}$, $c = 1$ 13 $5x^3 + 12x - 7$, $c = 0$ 14 $x^4 - 6x^2$, $c = -1$.

Find the sum of the series

15   $1 - \dfrac{(x-1)^2}{1!} + \dfrac{(x-1)^4}{2!} - \dfrac{(x-1)^6}{3!} + \cdots$ 16 $\dfrac{\pi^2}{2!}x^2 - \dfrac{\pi^4}{4!}x^4 + \dfrac{\pi^6}{6!}x^6 - + \cdots$.

17   Let $\sum_0^\infty a_n x^n$ be the Taylor series associated with $f(x)$ at $x = 0$. Find the Taylor series associated with $F(x) = \int_0^x f(t)\, dt$ at $x = 0$.

18   Suppose the derivatives of $f(x)$ have this property: for each interval $|x - c| \leq B$, there exist positive constants a and k such that $|f^{(n)}(x)| \leq ak^n$. Prove that the Taylor series of $f(x)$ at $x = c$ converges to $f(x)$ for all x. (The constants a and k may depend on B.)

19   Compute the Taylor series of $\cosh x$ at $x = 0$ and prove that it converges to $\cosh x$ for all x. [*Hint*   Use Ex. 18.]

20   Show that the Taylor series of $e^x \cos x$ at $x = 0$ converges to $e^x \cos x$ for all x. Do not compute the series. [*Hint*   Use Ex. 18.]

The following exercises establish the existence of a function that is not equal to the sum of its Taylor series.

21*   Define $f(x) = \begin{cases} e^{-1/x^2}, & x \neq 0, \\ 0, & x = 0. \end{cases}$

Show that $f(x)$ is differentiable at $x = 0$ and that $f'(0) = 0$.

22   (cont.) Let $x \neq 0$. Prove by induction that $f^{(n)}(x)$ is a sum of terms of the form $ae^{-1/x^2}/x^k$ for $n \geq 1$.

23   (cont.) Prove by induction that $f(x)$ has derivatives of all orders at $x = 0$ and that $f^{(n)}(0) = 0$ for $n \geq 1$.

24   (cont.) Conclude that the Taylor series of $f(x)$ at $x = 0$ converges for all x but not to $f(x)$.

3. EXPANSION OF FUNCTIONS

We have obtained Taylor series for various functions by computing coefficients from the formula $a_n = f^{(n)}(c)/n!$. But computing successive derivatives can be extremely laborious. Try the 7-th derivative of $\tan x$ or of $x^2/(1 + x^3)$ and you will soon agree. In some cases we were able to prove the validity of Taylor expansions by showing $r_n(x) \longrightarrow 0$. But that also can be very hard. To avoid these difficulties, we shall discuss techniques for obtaining new valid Taylor expansions from those already known.

Uniqueness One basic principle underlies all the techniques we shall develop:

Uniqueness of Power Series Suppose

$$f(x) = a_0 + \sum_{n=1}^{\infty} a_n(x-c)^n$$

in some interval, $|x - c| < R$. Then

$$a_n = \frac{f^{(n)}(c)}{n!}.$$

Thus if $f(x)$ is expressed as the sum of a power series, then that series must be the Taylor series of $f(x)$.

This principle says that there is only one possible power series for a given $f(x)$. Once you find a power series with sum $f(x)$ by any method, fair or foul, then you have its Taylor series.

For example, we showed in Section 1 that

$$\frac{1}{1-x} = 1 + x + x^2 + x^3 + \cdots, \qquad |x| < 1,$$

but we never actually verified that this series is the Taylor series of $1/(1-x)$ at $x = 0$. Now that follows automatically by the uniqueness principle.

The proof of uniqueness is based on a property of power series that will be discussed in Section 4 and proved in Section 7: a power series can be differentiated term-by-term infinitely often within its interval of convergence. Assuming this, the rest is easy. If $f(x) = \sum a_k(x-c)^k$, we differentiate n times, then set $x = c$:

$$f^{(n)}(x) = n!\, a_n + \frac{(n+1)!}{1!}\, a_{n+1}(x-c) + \frac{(n+2)!}{2!}\, a_{n+2}(x-c)^2 + \cdots,$$

$$f^{(n)}(c) = n!\, a_n, \qquad a_n = \frac{f^{(n)}(c)}{n!}.$$

Addition and Subtraction Our first principle is straightforward.

Suppose

$$f(x) = \sum_{n=0}^{\infty} a_n(x-c)^n \qquad \text{and} \qquad g(x) = \sum_{n=0}^{\infty} b_n(x-c)^n,$$

in an interval $|x - c| < R$. Then

$$f(x) \pm g(x) = \sum_{n=0}^{\infty} (a_n \pm b_n)(x-c)^n, \qquad |x - c| < R.$$

Thus two power series may be added or subtracted term-by-term within their common interval of convergence.

Proof Let $s_n(x)$ and $t_n(x)$ denote the partial sums of the two series. Then $s_n(x) \longrightarrow f(x)$ and $t_n(x) \longrightarrow g(x)$ provided $|x - c| < R$. By a basic property of limits,

$$s_n(x) \pm t_n(x) = \sum_{k=0}^{n} (a_k \pm b_k)(x - c)^k \longrightarrow f(x) \pm g(x).$$

In other words $\sum_{n=0}^{\infty} (a_n \pm b_n)(x - c)^n = f(x) \pm g(x), \qquad |x - c| < R.$

■ **EXAMPLE 1** Express as a Taylor series at $x = 0$

(a) $\cosh x$ (b) $\sinh x$.

Solution We know that

$$e^x = 1 + \frac{x}{1!} + \frac{x^2}{2!} + \frac{x^3}{3!} + \frac{x^4}{4!} + \cdots, \qquad -\infty < x < \infty,$$

hence $\qquad e^{-x} = 1 - \frac{x}{1!} + \frac{x^2}{2!} - \frac{x^3}{3!} + \frac{x^4}{4!} - \cdots, \qquad -\infty < x < \infty.$

Add these series term-by-term, then divide by 2:

$$\cosh x = \frac{1}{2}(e^x + e^{-x}) = 1 + \frac{x^2}{2!} + \frac{x^4}{4!} + \frac{x^6}{6!} + \cdots, \qquad -\infty < x < \infty.$$

Subtract and divide by 2:

$$\sinh x = \frac{1}{2}(e^x - e^{-x}) = x + \frac{x^3}{3!} + \frac{x^5}{5!} + \frac{x^7}{7!} + \cdots, \qquad -\infty < x < \infty. \qquad ■$$

A polynomial counts as a power series with infinite radius of convergence. Hence it may be legitimately added to or subtracted from any power series.

Examples

$$e^x + 4 + 3x + x^2 = \left(1 + \frac{x}{1!} + \frac{x^2}{2!} + \frac{x^3}{3!} + \cdots\right) + (4 + 3x + x^2)$$

$$= 5 + 4x + \frac{3}{2}x^2 + \frac{x^3}{3!} + \frac{x^4}{4!} + \cdots, \qquad -\infty < x < \infty,$$

$$\frac{1}{1 - x} - (1 + x + x^2 + 3x^3) = (1 + x + x^2 + x^3 + x^4 + \cdots) - (1 + x + x^2 + 3x^3)$$

$$= -2x^3 + x^4 + x^5 + x^6 + \cdots, \qquad -1 < x < 1.$$

Multiplication Our next technique is formal multiplication of power series. To simplify notation, let us take $c = 0$. Formal multiplication of two power series $\sum a_n x^n$ and $\sum b_n x^n$ is the operation of multiplying each term $a_j x^j$ of the first by each term $b_k x^k$ of the second and collecting terms, just as in multiplication of polynomials. Start with the lowest terms and work up:

$$(a_0 + a_1 x + a_2 x^2 + \cdots)(b_0 + b_1 x + b_2 x^2 + \cdots)$$
$$= a_0 b_0 + (a_0 b_1 + a_1 b_0)x + (a_0 b_2 + a_1 b_1 + a_2 b_0)x^2$$
$$+ \cdots + (a_0 b_n + a_1 b_{n-1} + \cdots + a_k b_{n-k} + \cdots + a_n b_0)x^n + \cdots.$$

Suppose $\qquad f(x) = \displaystyle\sum_{n=0}^{\infty} a_n x^n \qquad$ and $\qquad g(x) = \displaystyle\sum_{n=0}^{\infty} b_n x^n$

in an interval $|x| < R$. Then $\qquad f(x)g(x) = \displaystyle\sum_{n=0}^{\infty} \left(\sum_{k=0}^{n} a_k b_{n-k} \right) x^n, \qquad |x| < R.$

Thus two power series may be formally multiplied within their common interval of convergence.

The proof is difficult and best postponed to a more advanced course.

Since a polynomial counts as a power series, we can deduce by inspection such expansions as

$$x^3 \cos x = x^3 - \frac{x^5}{2!} + \frac{x^7}{4!} - \frac{x^9}{6!} + \cdots, \qquad -\infty < x < \infty,$$

$$x^4 e^x = x^4 + \frac{x^5}{1!} + \frac{x^6}{2!} + \frac{x^7}{3!} + \cdots, \qquad -\infty < x < \infty.$$

■ **EXAMPLE 2** Find the Taylor series of $\qquad \dfrac{1}{1-x} \ln \dfrac{1}{1-x} \qquad$ at $\quad x = 0.$

Solution In Example 2, p. 586, we found the Taylor series for $\ln(1 + x)$ at $x = 0$. Replacing x by $-x$, we have

$$\ln \frac{1}{1-x} = -\ln(1-x) = x + \frac{x^2}{2} + \frac{x^3}{3} + \cdots, \qquad |x| < 1.$$

Hence by formal multiplication,

$$\frac{1}{1-x} \ln \frac{1}{1-x} = (1 + x + x^2 + \cdots)\left(x + \frac{x^2}{2} + \frac{x^3}{3} + \cdots\right)$$

$$= x + \left(1 + \frac{1}{2}\right)x^2 + \left(1 + \frac{1}{2} + \frac{1}{3}\right)x^3 + \cdots$$

$$= \sum_{n=1}^{\infty} \left(1 + \frac{1}{2} + \frac{1}{3} + \cdots + \frac{1}{n}\right)x^n, \qquad |x| < 1. \qquad ■$$

■ **EXAMPLE 3** Compute the terms up to x^6 in the Taylor series of $x^2 e^x \sin 2x$ at $x = 0$.

Solution $\quad x^2 e^x \sin 2x = x^2 \left(1 + \dfrac{x}{1!} + \dfrac{x^2}{2!} + \dfrac{x^3}{3!} + \cdots\right)\left(2x - \dfrac{(2x)^3}{3!} + \dfrac{(2x)^5}{5!} - \cdots\right),$

where the series on the right are valid for all x. Since only terms involving x^6 and lower powers are required, it suffices to compute the product

$$x^2\left(1 + \frac{x}{1!} + \frac{x^2}{2!} + \frac{x^3}{3!}\right)\left(2x - \frac{(2x)^3}{3!}\right) = x^2\left(1 + x + \frac{x^2}{2} + \frac{x^3}{6}\right)\left(2x - \frac{4x^3}{3}\right)$$

$$= x^2\left[2x + 2x^2 + \left(1 - \frac{4}{3}\right)x^3 + \left(\frac{1}{3} - \frac{4}{3}\right)x^4 + \cdots\right].$$

Indeed, each higher term in e^x or in $\sin 2x$ contributes only to powers of x higher than x^6. For instance, the *lowest* power of x that the term $x^4/4!$ in e^x contributes to is

$$(x^2)\left(\frac{x^4}{4!}\right)(2x) = (\ )x^7, \qquad \text{etc.}$$

We conclude that

$$x^2 e^x \sin 2x = 2x^3 + 2x^4 - \tfrac{1}{3}x^5 - x^6 + \cdots, \qquad -\infty < x < \infty. \qquad \blacksquare$$

Substitution   This is a technique we have used already: if a power series $\sum a_n x^n$ converges for $|x| < R$ and if $|g(x)| < R$, then we may substitute $g(x)$ for x.

Examples:

$$\frac{1}{1 + x^2} = 1 + (-x^2) + (-x^2)^2 + (-x^2)^3 + \cdots$$

$$= 1 - x^2 + x^4 - x^6 + \cdots, \qquad |x| < 1,$$

$$\frac{1}{1 - 2x^3} = 1 + 2x^3 + 4x^6 + 8x^9 + \cdots, \qquad |x| < 1/\sqrt[3]{2},$$

$$e^{-x^2/2} = 1 + \left(-\frac{x^2}{2}\right) + \frac{1}{2!}\left(-\frac{x^2}{2}\right)^2 + \frac{1}{3!}\left(-\frac{x^2}{2}\right)^3 + \cdots$$

$$= 1 - \frac{x^2}{2} + \frac{x^4}{2^2 \cdot 2!} - \frac{x^6}{2^3 \cdot 3!} + \cdots, \qquad |x| < \infty.$$

There is a more sophisticated type of substitution than the type illustrated in these examples. Under certain circumstances we can actually substitute a power series into a power series!

Let $f(z) = \sum_0^\infty a_n z^n$ for $|z| < R$, and let

$$g(x) = b_1 x + b_2 x^2 + b_3 x^3 + \cdots,$$

a power series with zero constant term. Set $z = g(x)$ and write formally

$$f[g(x)] = a_0 + a_1 g(x) + a_2[g(x)]^2 + \cdots.$$

The series on the right can be converted into a power series by formally squaring, cubing, etc., and collecting terms. The resulting power series converges to $f[g(x)]$ in some interval $|x| < r$ in which $|g(x)| < R$.

A proof of the assertion is beyond the scope of this course. (Without the restriction $b_0 = 0$, infinitely many terms would contribute to each coefficient, a difficult situation to handle.)

■ **EXAMPLE 4** Compute the terms up to x^8 in the Taylor series of $f(x) = 1/(1 - \sin x^2)$ at $x = 0$.

Solution Expand $f(x)$ as a geometric series in $\sin x^2$:

$$f(x) = \frac{1}{1 - \sin x^2} = 1 + \sin x^2 + (\sin x^2)^2 + (\sin x^2)^3 + (\sin x^2)^4 + \cdots$$

From the power series for $\sin x$,

$$\sin x^2 = x^2 - \frac{x^6}{6} + \frac{x^{10}}{120} - \cdots.$$

Substitute this series into the expression for $f(x)$. Square, cube, etc. and collect powers of x up to x^8:

$$f(x) = 1 + \left(x^2 - \frac{x^6}{6} + \cdots\right) + \left(x^2 - \frac{x^6}{6} + \cdots\right)^2 + \left(x^2 - \frac{x^6}{6} + \cdots\right)^3$$

$$+ \left(x^2 - \frac{x^6}{6} + \cdots\right)^4 + \cdots$$

$$= 1 + \left(x^2 - \frac{x^6}{6} + \cdots\right) + \left(x^4 - \frac{x^8}{3} + \cdots\right) + (x^6 + \cdots) + (x^8 + \cdots) + \cdots$$

$$= 1 + x^2 + x^4 + \left(1 - \frac{1}{6}\right)x^6 + \left(1 - \frac{1}{3}\right)x^8 + \cdots$$

$$= 1 + x^2 + x^4 + \frac{5}{6}x^6 + \frac{2}{3}x^8 + \cdots.$$

This expansion is valid in some interval $|x| < r$ in which $|\sin x^2| < 1$. It can be proved valid for $|x| < \sqrt{\frac{1}{2}\pi}$, in this case. ■

Even and Odd Functions The power series at $x = 0$ of the odd function $\sin x$ involves only odd powers of x; the power series of the even function $\cos x$ involves only even powers of x. These examples illustrate a general principle:

If $f(x)$ is an odd function, $f(-x) = -f(x)$, then its Taylor series at $x = 0$ has the form

$$a_1 x + a_3 x^3 + a_5 x^5 + a_7 x^7 + \cdots.$$

If $f(x)$ is an even function, $f(-x) = f(x)$, then its Taylor series at $x = 0$ has the form

$$a_0 + a_2 x^2 + a_4 x^4 + a_6 x^6 + \cdots.$$

Thus the Taylor series of $x = 0$ of an odd (even) function contains only odd (even) powers of x.

The proof follows easily from the basic fact that the derivative of an odd function is even and the derivative of an even function is odd. See Ex. 31 and Exs. 33 and 34 for another proof.

Not all functions are even or odd. However, every function defined in an interval $|x| < R$ can be expressed as the sum of an even function and an odd function.

Let $f(x)$ be defined in an interval $|x| < R$. Then

$$f(x) = \frac{f(x) + f(-x)}{2} + \frac{f(x) - f(-x)}{2} = g(x) + h(x),$$

where $g(x)$ is an even function and $h(x)$ is an odd function. If

$$f(x) = \sum_{n=0}^{\infty} a_n x^n, \qquad \text{then} \qquad g(x) = \sum_{n=0}^{\infty} a_{2n} x^{2n} \quad \text{and} \quad h(x) = \sum_{n=0}^{\infty} a_{2n+1} x^{2n+1}.$$

Thus the Taylor series at $x = 0$ of $f(x)$ splits into two power series, one corresponding to the "even part" of $f(x)$ and the other corresponding to the "odd part" of $f(x)$.

Example:

$$f(x) = e^x = 1 + \frac{x}{1!} + \frac{x^2}{2!} + \frac{x^3}{3!} + \frac{x^4}{4!} + \cdots,$$

$$g(x) = \frac{f(x) + f(-x)}{2} = \frac{e^x + e^{-x}}{2} = \cosh x = 1 + \frac{x^2}{2!} + \frac{x^4}{4!} + \cdots,$$

$$h(x) = \frac{f(x) - f(-x)}{2} = \frac{e^x - e^{-x}}{2} = \sinh x = \frac{x}{1!} + \frac{x^3}{3!} + \frac{x^5}{5!} + \cdots.$$

■ **EXAMPLE 5** Find the sum of the series

$$x + \frac{x^3}{3} + \frac{x^5}{5} + \frac{x^7}{7} + \cdots.$$

Solution This is the "odd part" of the series

$$f(x) = -\ln(1 - x) = x + \frac{x^2}{2} + \frac{x^3}{3} + \frac{x^4}{4} + \cdots.$$

(See Example 2.) Therefore the sum of the given series is

$$\frac{f(x) - f(-x)}{2} = \frac{-\ln(1 - x) + \ln(1 + x)}{2} = \frac{1}{2} \ln \frac{1 + x}{1 - x}. \qquad ■$$

Undetermined Coefficients In this technique, we assume a power series $\sum_0^{\infty} a_n x^n$ for a given function $f(x)$, then try to determine the coefficients a_n from some property of $f(x)$.

■ **EXAMPLE 6** Compute the terms up to x^7 in the Taylor series for $\tan x$ at $x = 0$.

Solution   In theory, the problem merely requires long division of the series for $\sin x$ by the series for $\cos x$. In practice, the long division is carried out using undetermined coefficients. Set

$$\tan x = a_1 x + a_3 x^3 + a_5 x^7 + a_9 x^9 + \cdots.$$

(Only odd powers are necessary since $\tan x$ is an odd function.) Now write the identity $\tan x \cos x = \sin x$ in terms of power series:

$$\left(a_1 x + a_3 x^3 + a_5 x^5 + a_7 x^7 + \cdots\right)\left(1 - \frac{x^2}{2} + \frac{x^4}{24} - \frac{x^6}{720} + \cdots\right)$$

$$= x - \frac{x^3}{6} + \frac{x^5}{120} - \frac{x^7}{5040} + \cdots.$$

Multiply the two series on the left, then equate coefficients.

$$
\begin{array}{ll}
x: & a_1 = 1; \\
x^3: & a_3 - \tfrac{1}{2}a_1 = -\tfrac{1}{6}; \\
x^5: & a_5 - \tfrac{1}{2}a_3 + \tfrac{1}{24}a_1 = \tfrac{1}{120}; \\
x^7: & a_7 - \tfrac{1}{2}a_5 + \tfrac{1}{24}a_3 - \tfrac{1}{720}a_1 = -\tfrac{1}{5040}.
\end{array}
$$

Solve these equations successively for a_1, a_3, a_5, a_7:

$$a_1 = 1, \qquad a_3 = \tfrac{1}{3}, \qquad a_5 = \tfrac{2}{15}, \qquad a_7 = \tfrac{17}{315}.$$

Hence
$$\tan x = x + \tfrac{1}{3}x^3 + \tfrac{2}{15}x^5 + \tfrac{17}{315}x^7 + \cdots. \qquad\blacksquare$$

Remark   It can be shown that the Taylor series for $\tan x$ at $x = 0$ converges to $\tan x$ for $|x| < \tfrac{1}{2}\pi$. This is the largest interval about $x = 0$ in which the denominator $\cos x$ is non-zero. See Exs. 37–38.

EXERCISES

Find the Taylor series of the given function at $x = 0$

1 $\dfrac{1}{1 - 5x^2}$

2 $\dfrac{x}{1 - x^3}$

3 $\dfrac{1 - x}{1 + x}$

4 $\dfrac{1 + x^2}{1 + x^4}$

5 $x(\sin x + \sin 3x)$

6 $\sinh x + \cosh x$

7 $\dfrac{1 - \cos x}{x^2}$

8 $\dfrac{\sin x - x}{x^3}$

9 $\dfrac{2 - 2x + 3x^2}{1 - x}$

10 $(x^2 - 1)\cosh x$

11 $\sin^2 x$

12 $\dfrac{1}{1 + x + x^2 + x^3}.$

Compute the terms up to x^6 in the Taylor series at $x = 0$

13 $\dfrac{1}{(1 - 2x^2)(1 - 3x^2)}$

14 $e^x \sin(x^2)$

15 $\dfrac{1}{1 - x^2 e^x}$

16 $\dfrac{1}{1 + x + x^3}$

17 $\sin^3 x$

18 $\ln \cos x$

19 $e^{x^3} \cos x$

20 $x \cot x.$

Compute $f^{(7)}(0)$

21 $\dfrac{2x + 3}{1 - x^3}$

22 $x \cos \tfrac{1}{2}x$

23 $\tan x$

24 $\dfrac{x^2}{1 + x^3}.$

25   If $f(x) = \sum_0^\infty a_n x^n$, show that $\dfrac{1}{1-x} f(x) = \sum_{n=0}^\infty (a_0 + a_1 + \cdots + a_n)x^n.$

26   (cont.) Find the sum of the series $\sum_0^\infty (n+1)x^n$.

Find the sum of the series

27 $x + x^2 - x^3 + x^4 + x^5 - x^6 + + - \cdots$ **28** $\dfrac{1}{2!} + \dfrac{x}{3!} + \dfrac{x^2}{4!} + \dfrac{x^3}{5!} + \cdots$

29* $1 + \dfrac{x^4}{4!} + \dfrac{x^8}{8!} + \dfrac{x^{12}}{12!} + \cdots$ **30** $\dfrac{x^2}{2} + \dfrac{x^4}{4} + \dfrac{x^6}{6} + \dfrac{x^8}{8} + \cdots.$

31   Prove that the derivative of an odd (even) function is even (odd). Conclude that the Taylor series at $x = 0$ of an even (odd) function contains only even (odd) powers of x.

32   Prove that a function defined for $|x| < R$ can be written in *only* one way as the sum of an even function and an odd function.

33   Suppose $f(x) = \sum_0^\infty a_n x^n$ is even. Use *uniqueness* to prove that $a_{2n-1} = 0$.

34   (cont.) Give the corresponding proof for odd functions.

35   Set $f(x) = \sum_1^\infty nx^n$. Compute $(1-x)f(x)$ and use your result to find $f(x)$.

36   (cont.) Set $g(x) = \sum_1^\infty n(n+1)x^n$ and use the same technique to find $g(x)$.

37   Use $\sec x = 1/[1 - (1 - \cos x)]$ to find the terms up to x^6 in $\sec x$ at $x = 0$. What is the radius of convergence?

38   (cont.) Use the same technique to compute the Taylor series of $\tan x = \sin x/\cos x$ at $x = 0$ up to x^7. What is the radius of convergence?

39   Suppose the coefficients of $f(x) = \sum_0^\infty a_n x^n$ are **periodic** of period p. That is, p is a positive integer and $a_{n+p} = a_n$ for all n. Prove that $f(x)$ is a rational function.

40*   Suppose $p(t)$ is a polynomial and $f(x) = \sum_0^\infty p(n)x^n$. Show that $f(x)$ is a rational function. [*Hint*   See Exs. 35 and 36, and use induction on the degree of $p(t)$.]

Find the Taylor series at $x = 0$ up to x^5 of

41 $e^{\sin x}$ **42** $e^{\cos x}$.

4. FURTHER TECHNIQUES

Differentiation and Integration   These are important techniques for deriving new power series from known ones. The proof will be given in Section 7.

Let $f(x) = a_0 + a_1 x + a_2 x^2 + \cdots = \displaystyle\sum_{n=0}^\infty a_n x^n,$ $|x| < R.$

Then $f'(x) = a_1 + 2a_2 x + 3a_3 x^2 + \cdots = \displaystyle\sum_{n=1}^\infty n a_n x^{n-1},$ $|x| < R,$

and $\displaystyle\int_0^x f(t)\, dt = a_0 x + \dfrac{a_1}{2} x^2 + \dfrac{a_2}{3} x^3 + \cdots = \sum_{n=0}^\infty \dfrac{a_n}{n+1} x^{n+1},$ $|x| < R.$

Thus a power series may be differentiated or integrated term-by-term within its interval of convergence.

■ **EXAMPLE 1**   Find the Taylor series at $x = 0$ for $1/(1 - x)^2$ and $1/(1 - x)^3$.

Solution First observe that $\dfrac{1}{(1 - x)^2} = \dfrac{d}{dx}\left(\dfrac{1}{1 - x}\right)$. Then expand $1/(1 - x)$ in a Taylor series and differentiate term-by-term. For $|x| < 1$,

$$\frac{1}{(1 - x)^2} = \frac{d}{dx}(1 + x + x^2 + \cdots + x^n + \cdots)$$

$$= \frac{d}{dx}(1) + \frac{d}{dx}(x) + \frac{d}{dx}(x^2) + \cdots + \frac{d}{dx}(x^n) + \cdots$$

$$= 1 + 2x + 3x^2 + \cdots + nx^{n-1} + \cdots = \sum_{n=1}^{\infty} nx^{n-1} = \sum_{n=0}^{\infty} (n + 1)x^n.$$

Differentiate again:

$$\frac{2}{(1 - x)^3} = \frac{d}{dx}\left(\frac{1}{(1 - x)^2}\right) = \frac{d}{dx}(1 + 2x + 3x^2 + \cdots + nx^{n-1} + \cdots).$$

Hence $\dfrac{1}{(1 - x)^3} = \dfrac{1}{2}[2 + 6x + 12x^2 + \cdots + n(n - 1)x^{n-2} + \cdots]$

$$= \sum_{n=2}^{\infty} \frac{n(n - 1)}{2} x^{n-2} = \sum_{n=0}^{\infty} \frac{(n + 2)(n + 1)}{2} x^n. \qquad ■$$

■ **EXAMPLE 2**   Find the sum of the series $x + \dfrac{x^3}{3} + \dfrac{x^5}{5} + \cdots + \dfrac{x^{2n-1}}{2n - 1} + \cdots$.

Solution By the Ratio Test, the series converges for $|x| < 1$ to some function

$$f(x) = x + \frac{x^3}{3} + \frac{x^5}{5} + \cdots + \frac{x^{2n-1}}{2n - 1} + \cdots.$$

Each term is the integral of a power of x. This suggests that $f(x)$ is the integral of some simple function. Differentiate term-by-term:

$$f'(x) = 1 + x^2 + x^4 + \cdots + x^{2n-2} + \cdots = \frac{1}{1 - x^2}.$$

Therefore, $f(x)$ is an antiderivative of $1/(1 - x^2)$. Since $f(0) = 0$, it follows that

$$f(x) = \int_0^x \frac{dt}{1 - t^2} = \frac{1}{2}\ln\frac{1 + t}{1 - t}\bigg|_0^x = \frac{1}{2}\ln\frac{1 + x}{1 - x} \qquad \text{for} \quad |x| < 1. \qquad ■$$

■ **EXAMPLE 3**   Find the Taylor series at $x = 0$ of $\ln(1 + x)$.

Solution Integrate the derivative $1/(1 + x) = 1 - x + x^2 - x^3 + \cdots$ for $|x| < 1$:

$$\ln(1 + x) = \int_0^x \frac{dt}{1 + t} = \int_0^x (1 - t + t^2 - + \cdots + (-1)^n t^n + \cdots)\, dt$$

$$= \int_0^x dt - \int_0^x t\, dt + \int_0^x t^2\, dt - \cdots + (-1)^n \int_0^x t^n\, dt + \cdots.$$

Hence,
$$\ln(1+x) = x - \frac{x^2}{2} + \frac{x^3}{3} - \cdots + (-1)^n \frac{x^{n+1}}{n+1} + \cdots$$

$$= \sum_{n=1}^{\infty} (-1)^{n-1} \frac{x^n}{n}, \qquad |x| < 1.$$ ■

■ **EXAMPLE 4**   Find the Taylor series at $x = 0$ of arc tan x.

Solution   The derivative is $1/(1 + x^2) = 1 - x^2 + x^4 - \cdots$ for $|x| < 1$.

Therefore
$$\text{arc tan } x = \int_0^x \frac{dt}{1 + t^2} = \int_0^x (1 - t^2 + t^4 - t^6 + \cdots) \, dt$$

$$= x - \frac{x^3}{3} + \frac{x^5}{5} - \frac{x^7}{7} + \cdots = \sum_{n=0}^{\infty} (-1)^n \frac{x^{2n+1}}{2n+1}, \qquad |x| < 1.$$ ■

■ **EXAMPLE 5**   Find the sum of the series

$$x + 4x^2 + 9x^3 + \cdots = \sum_{n=1}^{\infty} n^2 x^n.$$

Solution   By the Ratio Test, the series converges for $|x| < 1$ to a function $f(x)$. Write

$$f(x) = x + 4x^2 + 9x^3 + \cdots + n^2 x^n + \cdots = xg(x),$$

where
$$g(x) = 1 + 2^2 x + 3^2 x^2 + \cdots + n^2 x^{n-1} + \cdots.$$

Now
$$g(x) = \frac{d}{dx}(x + 2x^2 + 3x^3 + 4x^4 + \cdots + nx^n + \cdots)$$

$$= \frac{d}{dx}[x(1 + 2x + 3x^2 + 4x^3 + \cdots + nx^{n-1} + \cdots)].$$

By Example 1,
$$1 + 2x + 3x^2 + \cdots + nx^{n-1} + \cdots = \frac{1}{(1-x)^2}.$$

Hence
$$g(x) = \frac{d}{dx}\left[\frac{x}{(1-x)^2}\right] = \frac{1+x}{(1-x)^3},$$

so
$$f(x) = xg(x) = \frac{x + x^2}{(1-x)^3}, \qquad |x| < 1.$$ ■

Check   Test the answer at $x = 0.1$ where

$$f(0.1) = \frac{1}{10} + \frac{4}{10^2} + \frac{9}{10^3} + \frac{16}{10^4} + \frac{25}{10^5} + \cdots.$$

According to the answer,

$$f(0.1) = \frac{(0.1) + (0.1)^2}{(1 - 0.1)^3} = \frac{0.11}{(0.9)^3} \approx 0.15089\ 16324.$$

It is easy to make a convincing numerical check. Start with the first term and add successive terms: 0.1, 0.14, 0.149, 0.1506, 0.15085, 0.15088 6, 0.15089 09, 0.15089 154, 0.15089 1621, 0.15089 16310.

An Application to Probability   Imagine an experiment in which the outcome is one of the integers 0, 1, 2, $\cdots$, such as counting the number of cars that pass a certain spot in a given hour, or the number of tails that precede the first head in a sequence of tosses of a coin. Suppose the probability of the outcome n is a number p_n, where $0 \le p_n \le 1$ and $\sum_0^\infty p_n = 1$. We say that $\{p_n\}$ is a **probability distribution** on the set $\{0, 1, 2, \cdots\}$.

The **expected value (mean)** of this distribution is defined to be

$$E = \sum_{n=0}^{\infty} np_n.$$

Since $\sum p_n = 1$, we can think of E as a weighted average: each integer is weighted by the probability that it will occur. The **variance** of the distribution is defined to be

$$\sigma^2 = \sum_{n=0}^{\infty} (n - E)^2 p_n.$$

It is a measure of the spread of the various outcomes about their expected value.

Since $\sum p_n = 1$ and $\sum np_n = E$, we have

$$\sigma^2 = \sum_{n=0}^{\infty} (n^2 - 2nE + E^2)p_n = \sum_{n=0}^{\infty} n^2 p_n - 2E \sum_{n=0}^{\infty} np_n + E^2 \sum_{n=0}^{\infty} p_n$$

$$= \sum_{n=0}^{\infty} n^2 p_n - 2E \cdot E + E^2 \cdot 1 = \sum_{n=0}^{\infty} n^2 p_n - E^2.$$

The quantity $\sum n^2 p_n$ is called the **second moment** of the distribution. Sometimes the mean, $\sum np_n$, is called the **first moment**.

■ **EXAMPLE 6**   In a sequence of throws of a dice, what is the expected number of the throw at which the first 4 occurs?

Solution   The probability of a 4 turning up is $\frac{1}{6}$; the probability of any other number turning up is $\frac{5}{6}$. The first 4 occurs at throw n provided that the first $n-1$ throws are not 4, but the n-th throw is 4. The probability of this event is

$$p_n = \frac{5}{6} \cdot \frac{5}{6} \cdots \frac{5}{6} \cdot \frac{1}{6} = \frac{1}{6}\left(\frac{5}{6}\right)^{n-1}, \qquad n = 1, 2, 3, \cdots.$$

(You should confirm that $\sum_1^\infty p_n = 1$.) By definition, the expected number of throws until a 4 shows up is

$$E = \sum_{n=1}^{\infty} np_n = \sum_{n=1}^{\infty} n \cdot \frac{1}{6}\left(\frac{5}{6}\right)^{n-1} = \frac{1}{6}\sum_{n=1}^{\infty} n\left(\frac{5}{6}\right)^{n-1}.$$

We can find this sum explicitly. According to Example 1,

$$\sum_{n=1}^{\infty} nx^{n-1} = \frac{1}{(1-x)^2}; \qquad \text{hence} \quad E = \frac{1}{6}\frac{1}{(1-\frac{5}{6})^2} = 6.$$

The expected number of throws is 6. ∎

■ **EXAMPLE 7** Compute the variance σ^2 of the probability distribution in Example 6.

Solution Compute the second moment:

$$\sum_{n=1}^{\infty} n^2 p_n = \frac{1}{6}\sum_{n=1}^{\infty} n^2\left(\frac{5}{6}\right)^{n-1} = \frac{1}{6}\cdot\frac{6}{5}\sum_{n=1}^{\infty} n^2\left(\frac{5}{6}\right)^{n}.$$

By Example 5

$$\sum_{n=1}^{\infty} n^2\left(\frac{5}{6}\right)^{n} = \frac{\frac{5}{6}(1+\frac{5}{6})}{(1-\frac{5}{6})^3} = 5\cdot 36\left(1+\frac{5}{6}\right).$$

Since $E = 6$, $\sigma^2 = \displaystyle\sum_{n=1}^{\infty} n^2 p_n - E^2 = \frac{1}{5}\cdot 5\cdot 36\left(1+\frac{5}{6}\right) - 36 = 36\cdot\frac{5}{6} = 30.$ ∎

Partial Fractions First let us mention a standard trick for exploiting the geometric series. To expand $1/(2-x)$, just write

$$\frac{1}{2-x} = \frac{1}{2(1-\frac{1}{2}x)} = \frac{1}{2}\left[1 + \frac{x}{2} + \left(\frac{x}{2}\right)^2 + \cdots\right] = \sum_{n=0}^{\infty}\frac{x^n}{2^{n+1}}.$$

Similarly, $$\frac{1}{x-5} = -\frac{1}{5(1-\frac{1}{5}x)} = -\sum_{n=0}^{\infty}\frac{x^n}{5^{n+1}}.$$

■ **EXAMPLE 8** Find the Taylor series at $x = 0$ of $\dfrac{1}{(x-2)(x-5)}.$

Solution By partial fractions $$\frac{1}{(x-2)(x-5)} = \frac{1}{3}\left(\frac{1}{x-5} - \frac{1}{x-2}\right).$$

Expand each fraction, using the preceding trick:

$$\frac{1}{x-2} = -\sum_{n=0}^{\infty}\frac{x^n}{2^{n+1}}, \qquad \frac{1}{x-5} = -\sum_{n=0}^{\infty}\frac{x^n}{5^{n+1}}.$$

Both series converge in the interval $|x| < 2$. For these values of x,

$$\frac{1}{(x-2)(x-5)} = \frac{1}{3}\left(\frac{1}{x-5} - \frac{1}{x-2}\right) = \frac{1}{3}\sum_{n=0}^{\infty}\left(\frac{1}{2^{n+1}} - \frac{1}{5^{n+1}}\right)x^n.$$ ∎

■ **EXAMPLE 9**   Find the Taylor series at $x = 0$ of $\dfrac{-1}{x^2 + x - 1}$.

Solution   The denominator can be factored:

$$x^2 + x - 1 = (x - a)(x - b),$$

where a and b are the roots of the equation $x^2 + x - 1 = 0$. By the quadratic formula,

$$a = \frac{-1 + \sqrt{5}}{2}, \qquad b = \frac{-1 - \sqrt{5}}{2}.$$

Notice that $ab = -1$ because the product of the roots of $x^2 + px + q = 0$ is q.
 Express the given function in partial fractions:

$$\frac{-1}{x^2 + x - 1} = \frac{-1}{(x - a)(x - b)} = \frac{1}{a - b}\left(\frac{1}{x - b} - \frac{1}{x - a}\right).$$

Now expand each fraction in power series, which is permissible if $|x| < \frac{1}{2}(\sqrt{5} - 1)$, the smaller of the numbers $|a|$ and $|b|$:

$$\frac{-1}{x^2 + x - 1} = \frac{1}{a - b}\left(\sum_{n=0}^{\infty} \frac{x^n}{a^{n+1}} - \sum_{n=0}^{\infty} \frac{x^n}{b^{n+1}}\right) = \sum_{n=0}^{\infty} \frac{1}{a - b}\left(\frac{1}{a^{n+1}} - \frac{1}{b^{n+1}}\right)x^n.$$

Note that $$a - b = \frac{-1 + \sqrt{5}}{2} - \frac{-1 - \sqrt{5}}{2} = \sqrt{5},$$

and (since $ab = -1$) $$\frac{1}{a} = -b = \frac{1 + \sqrt{5}}{2}, \qquad \frac{1}{b} = -a = \frac{1 - \sqrt{5}}{2}.$$

Hence, for $|x| < \frac{1}{2}(\sqrt{5} - 1)$,

$$\frac{-1}{x^2 + x - 1} = \sum_{n=0}^{\infty} c_n x^n, \qquad c_n = \frac{1}{\sqrt{5}}\left[\left(\frac{1 + \sqrt{5}}{2}\right)^{n+1} - \left(\frac{1 - \sqrt{5}}{2}\right)^{n+1}\right]. \qquad ■$$

Remark   It may not seem so, but the numbers c_n are actually integers! The first few are 1, 1, 2, 3, 5, 8, 13, $\cdots$; each is the sum of the previous two. See Exs. 25–28.

■ **EXAMPLE 10**   Find a power series for $\dfrac{1}{(1 - x)(1 + x^2)}$.

Solution   By partial fractions,

$$\frac{1}{(1 - x)(1 + x^2)} = \frac{1}{2}\left(\frac{1}{1 - x} + \frac{1 + x}{1 + x^2}\right) = \frac{1}{2}\left(\frac{1}{1 - x} + \frac{1}{1 + x^2} + \frac{x}{1 + x^2}\right)$$

$$= \frac{1}{2}[(1 + x + x^2 + x^3 + \cdots) + (1 - x^2 + x^4 - x^6 + - \cdots)$$

$$+ (x - x^3 + x^5 - x^7 + - \cdots)]$$

$$= 1 + x + x^4 + x^5 + x^8 + x^9 + \cdots.$$

Thus
$$\frac{1}{(1-x)(1+x^2)} = \sum_{n=0}^{\infty} (x^{4n} + x^{4n+1}) \qquad \text{for} \quad |x| < 1. \qquad \blacksquare$$

Remark This example also can be done by multiplying together the series for $(1-x)^{-1}$ and $(1+x^2)^{-1}$, also by multiplying the series for $(1-x^4)^{-1}$ by $1+x$.

EXERCISES

Verify by expressing both sides as power series

1 $\dfrac{d}{dx}(\sin x) = \cos x$

2 $\dfrac{d}{dx}(e^{kx}) = ke^{kx}$

3 $\dfrac{d^2}{dx^2}(\cosh kx) = k^2 \cosh kx$

4 $\dfrac{d^2}{dx^2}(xe^x) = (x+2)e^x$

5 $\displaystyle\int_0^x \frac{2t\,dt}{1-t^2} = -\ln(1-x^2), \quad |x| < 1$

6 $\displaystyle\int_0^x \arctan t\,dt = x \arctan x - \tfrac{1}{2}\ln(1+x^2), \quad |x| < 1$

7 $\displaystyle\int_0^x \frac{dt}{(1+t^2)^2} = \frac{1}{2}\left[\frac{x}{1+x^2} + \arctan x\right]$

8 $\displaystyle\int_0^x t^2 \sin t\,dt = 2x \sin x + (2-x^2)\cos x - 2.$

Find the sum of the power series by using differentiation and integration

9 $4 + 5x + 6x^2 + 7x^3 + \cdots$

10 $1 + 4x + 9x^2 + 16x^3 + 25x^4 + \cdots$

11 $\dfrac{x^4}{4} + \dfrac{x^8}{8} + \dfrac{x^{12}}{12} + \dfrac{x^{16}}{16} + \cdots$

12 $\dfrac{x^2}{1 \cdot 2} - \dfrac{x^3}{2 \cdot 3} + \dfrac{x^4}{3 \cdot 4} - \dfrac{x^5}{4 \cdot 5} + \cdots$

13 $x - \dfrac{x^3}{3^2} + \dfrac{x^5}{5^2} - \dfrac{x^7}{7^2} + \cdots$

14 $x + 2^3 x^2 + 3^3 x^3 + 4^3 x^4 + \cdots.$

[Express as an integral.]

Find the sum of the numerical series

15 $\dfrac{3}{2 \cdot 1!} - \dfrac{5}{2^2 \cdot 2!} + \dfrac{7}{2^3 \cdot 3!} - \dfrac{9}{2^4 \cdot 4!} + \cdots$ **16*** $\dfrac{1}{2!} + \dfrac{2}{3!} + \dfrac{3}{4!} + \dfrac{4}{5!} + \cdots$

[*Hint* Start with the power series for $e^{-x^2/2}$.]

17 Before the advent of fertility drugs, an empirical law asserted that one out of 87 births resulted in twins, $1/87^2$ in triplets, $1/87^3$ in quadruplets, etc. Assuming this law, compute the expected number of babies born per 100,000 live births.

18 A coin is tossed repeatedly until either two consecutive heads or two consecutive tails occur. What is the expected number of tosses?

19 Let's play this game. We toss a coin. If heads, I pay you $1 and the game ends. If tails, you pay me $1 and we toss again. This time, if it's heads, I pay you $2 and the game ends. If it's tails, you pay me $2 and we play again for $3, etc. Compute your expected gain or loss, $\sum_1^\infty a_n p_n$, where a_n is your payoff (gain or loss) at throw n, and p_n is the probability that the game ends at throw n.

20 A certain dime slot machine has three identical wheels, each with ten different fruits. You continue to play without gain until you hit three lemons, when the payoff is the jackpot, J. Find J so the game is fair, assuming all outcomes of the spinning wheels are equally likely.

Expand in Taylor series at $x = 0$

21 $\dfrac{x + 1}{x^2 - 4x + 3}$

22 $\dfrac{1}{(2x + 1)(3x + 4)}$

23 $\dfrac{1}{(x - 1)(x - 2)(x - 3)}$

24 $\dfrac{1}{(x - 3)(x^2 + 1)}.$

Define a sequence of integers $\{c_n\}$ by $c_0 = c_1 = 1$ and $c_{n+2} = c_{n+1} + c_n$. Thus the sequence begins $1, 1, 2, 3, 5, 8, 13, 21, 34, 55, \cdots$. These numbers are called the **Fibonacci numbers**. It is remarkable that there exists an exact formula for them.

25   Let $f(x) = \sum_{n=0}^{\infty} c_n x^n$. This power series is the **generating function** of the Fibonacci sequence. Show that $(1 - x - x^2)f(x) = 1$.

26   (cont.) From Example 9, derive an exact formula for c_n.

27   (cont.) Conclude that c_n is the closest integer to $\dfrac{1}{\sqrt{5}}\left[\dfrac{1}{2}\left(1 + \sqrt{5}\right)\right]^{n+1}.$

28   (cont.) Leonardo of Pisa (nicknamed Fibonacci, sometimes translated "son of an ass") in 1202 modeled rabbit breeding as follows: suppose it takes a pair of rabbits one month to mature and one month to produce a litter. Assume the parents and one pair from the litter survive for future generations, and dead rabbits are replaced. How many pairs of rabbits will there be after n months if there is one newborn pair initially?

29   Show that the Fibonacci sequence $\{c_n\}$ of Ex. 25 satisfies $c_{2n}^2 - 1 = c_{2n-1}c_{2n+1}.$

30*   (cont.) Evaluate $\left(\displaystyle\sum_{k=1}^{n} \arctan \dfrac{1}{c_{2k}}\right) + \arctan \dfrac{1}{c_{2n+1}}.$

31   Define a sequence $\{a_n\}$ by $a_0 = 0$, $a_1 = 1$, and $a_{n+2} = 5a_{n+1} - 6a_n$. Use the method of Ex. 25 to find the generating function $f(x) = \sum_0^{\infty} a_n x^n$.

32   (cont.) Expand $f(x)$ in power series and deduce an exact formula for a_n.

5. BINOMIAL SERIES

The Binomial Theorem asserts that for each positive integer p,

$$(1 + x)^p = 1 + px + \frac{p(p - 1)}{2!}x^3 + \frac{p(p - 1)(p - 2)}{3!}x^3 + \cdots$$

$$+ \frac{p(p - 1)(p - 2)\cdots(p - n + 1)}{n!}x^n + \cdots + \frac{p!}{p!}x^p.$$

Standard notation for the binomial coefficients is

$$\binom{p}{0} = 1, \qquad \binom{p}{n} = \frac{p!}{n!\,(p - n)!} = \frac{p(p - 1)(p - 2)\cdots(p - n + 1)}{n!}, \qquad 1 \leq n \leq p.$$

With this notation the expansion of $(1 + x)^p$ can be abbreviated:

$$(1 + x)^p = \sum_{n=0}^{p} \binom{p}{n}x^n.$$

A generalization of the Binomial Theorem is the **binomial series** for $(1 + x)^p$, where p is not necessarily a positive integer.

Binomial Series   For any number p, and $|x| < 1$,

$$(1 + x)^p = \sum_{n=0}^{\infty} \binom{p}{n} x^n,$$

where the coefficients are

$$\binom{p}{0} = 1, \qquad \binom{p}{n} = \frac{p(p - 1)(p - 2) \cdots (p - n + 1)}{n!}, \qquad n \geq 1.$$

Remark   In case p happens to be a positive integer and $n > p$, then the coefficient $\binom{p}{n}$ equals 0 because it has a factor $(p - p)$. In this case, the series breaks off after the term in x^p. The resulting formula,

$$(1 + x)^p = \sum_{n=0}^{p} \binom{p}{n} x^n,$$

is the old Binomial Theorem again. But if p is not a positive integer or zero, then each coefficient is non-zero, so the series has infinitely many terms.

The binomial series is just the Taylor series for $y(x) = (1 + x)^p$ at $x = 0$ because

$$y'(x) = p(1 + x)^{p-1},$$
$$y''(x) = p(p - 1)(1 + x)^{p-2},$$
$$\cdots \cdots \cdots \cdots \cdots \cdots \cdots \cdots \cdots \cdots \cdots$$
$$y^{(n)}(x) = p(p - 1)(p - 2) \cdots (p - n + 1)(1 + x)^{p-n}.$$

Therefore the coefficient of x^n in the Taylor series is

$$\frac{y^{(n)}(0)}{n!} = \frac{p(p - 1)(p - 2) \cdots (p - n + 1)}{n!} = \binom{p}{n}.$$

The binomial series converges for $|x| < 1$. When p is an integer this is obvious because the series terminates. When p is not an integer the Ratio Test applies:

$$\left| \frac{a_n}{a_{n+1}} \right| = \left| \binom{p}{n} \middle/ \binom{p}{n+1} \right| = \left| \frac{n + 1}{p - n} \right| \longrightarrow 1,$$

so the radius of convergence is 1. This, however, does not prove that the sum of the series *is* $(1 + x)^p$. That requires a delicate piece of analysis beyond the scope of this course.

■ **EXAMPLE 1**   Find the Taylor series for $\dfrac{1}{(1 + x)^2}$ at $x = 0$.

Solution   Use the binomial series with $p = -2$. The coefficient of x^n is

$$\binom{-2}{n} = \frac{(-2)(-3)(-4)\cdots(-2-n+1)}{n!}$$

$$= (-1)^n \frac{2\cdot3\cdot4\cdots n\cdot(n+1)}{n!} = (-1)^n(n+1).$$

Hence for $|x| < 1$,

$$\frac{1}{(1+x)^2} = \sum_{n=0}^{\infty} \binom{-2}{n}x^n = \sum_{n=0}^{\infty}(-1)^n(n+1)x^n$$

$$= 1 - 2x + 3x^2 - 4x^3 + \cdots.$$

Check $\dfrac{1}{(1+x)^2} = -\dfrac{d}{dx}\left(\dfrac{1}{1+x}\right) = -\dfrac{d}{dx}(1 - x + x^2 - x^3 + - \cdots)$

$$= -(-1 + 2x - 3x^2 + - \cdots). \qquad \blacksquare$$

The following example will be useful in Section 6.

■ **EXAMPLE 2**   Find the Taylor series for $\sqrt{1-x}$ at $x = 0$.

Solution   Use the binomial series with $p = \frac{1}{2}$ and x replaced by $-x$. The constant term in the series is

$$\binom{\frac{1}{2}}{0} = 1.$$

The term in x^n is $\dbinom{\frac{1}{2}}{n}(-x)^n = \dfrac{\left(\frac{1}{2}\right)\left(-\frac{1}{2}\right)\left(-\frac{3}{2}\right)\cdots\left(-\dfrac{2n-3}{2}\right)}{n!}(-x)^n$

$$= (-1)^{n-1}\frac{1\cdot3\cdot5\cdots(2n-3)}{2^n\cdot n!}(-1)^n x^n.$$

But $1\cdot3\cdot5\cdots(2n-3) = \dfrac{1\cdot2\cdot3\cdot4\cdots(2n-2)}{2\cdot4\cdot6\cdot8\cdots(2n-2)}$

$$= \frac{(2n-2)!}{2^{n-1}(n-1)!} = \frac{(2n)!}{(2^n\cdot n!)(2n-1)},$$

so the term in x^n is $-\dfrac{(2n)!}{(2^n\cdot n!)(2n-1)}\cdot\dfrac{1}{2^n\cdot n!}x^n = -\dfrac{(2n)!}{(2^n\cdot n!)^2(2n-1)}x^n.$

Therefore $\sqrt{1-x} = 1 - \displaystyle\sum_{n=1}^{\infty}\frac{(2n)!}{(2^n\cdot n!)^2(2n-1)}x^n$ for $|x| < 1.$ $\blacksquare$

■ **EXAMPLE 3**   Find the Taylor series for $\dfrac{1}{\sqrt{x}}$ at $x = 9$.

Solution   Write $\sqrt{x} = \sqrt{9 + (x - 9)}$. Then

$$\frac{1}{\sqrt{x}} = \frac{1}{3\sqrt{1 + \frac{1}{9}(x - 9)}} = \frac{1}{3}\left(1 + \frac{x - 9}{9}\right)^{-1/2} = \frac{1}{3}\sum_{n=0}^{\infty}\binom{-\frac{1}{2}}{n}\frac{(x - 9)^n}{9^n}.$$

The constant term is 1. For $n \geq 1$, the coefficients are

$$\binom{-\frac{1}{2}}{n} = \frac{\left(-\frac{1}{2}\right)\left(-\frac{3}{2}\right)\left(-\frac{5}{2}\right)\cdots\left(-\frac{2n-1}{2}\right)}{n!}$$

$$= (-1)^n\frac{1 \cdot 3 \cdot 5 \cdots (2n - 1)}{2^n n!} = (-1)^n\frac{(2n)!}{2^{2n}(n!)^2}.$$

(This formula is correct also for $n = 0$.) Therefore

$$\frac{1}{\sqrt{x}} = \sum_{n=0}^{\infty}\frac{(-1)^n(2n)!}{3 \cdot 2^{2n} \cdot 9^n(n!)^2}(x - 9)^n.$$

This answer checks with Example 3, p. 586.   ∎

∎ **EXAMPLE 4**   Find the Taylor series for arc sin x at $x = 0$.

Solution   Expand its derivative $1/\sqrt{1 - x^2}$ in power series and integrate term-by-term. For $|x| < 1$,

$$\text{arc sin } x = \int_0^x \frac{dt}{\sqrt{1 - t^2}} = \int_0^x \sum_{n=0}^{\infty}\binom{-\frac{1}{2}}{n}(-t^2)^n \, dt$$

$$= \sum_{n=0}^{\infty}(-1)^n\binom{-\frac{1}{2}}{n}\int_0^x t^{2n} \, dt = \sum_{n=0}^{\infty}\frac{(2n)!}{2^{2n}(n!)^2(2n + 1)}x^{2n+1}.$$ ∎

Remark   The formula can be written

$$\text{arc sin } x = x + \frac{x^3}{2 \cdot 3} + \frac{1 \cdot 3}{2 \cdot 4 \cdot 5}x^5 + \frac{1 \cdot 3 \cdot 5}{2 \cdot 4 \cdot 6 \cdot 7}x^7 + \cdots$$

because

$$(-1)^n\binom{-\frac{1}{2}}{n} = \frac{1 \cdot 3 \cdot 5 \cdots (2n - 1)}{2^n n!} = \frac{1 \cdot 3 \cdot 5 \cdots (2n - 1)}{2 \cdot 4 \cdot 6 \cdots (2n)}.$$

∎ **EXAMPLE 5**   Estimate $\sqrt[3]{1001}$ to 8 places.

Solution   Write $\sqrt[3]{1001} = [1000(1 + 10^{-3})]^{1/3} = 10(1 + 10^{-3})^{1/3}$.

Use the binomial series with $x = 10^{-3}$ and $p = \frac{1}{3}$:

$$10(1 + 10^{-3})^{1/3} = 10\left[1 + \binom{\frac{1}{3}}{1}(10^{-3}) + \binom{\frac{1}{3}}{2}(10^{-3})^2 + \cdots\right]$$

$$= 10\left[1 + \frac{1}{3} \cdot 10^{-3} - \frac{1}{9} \cdot 10^{-6} + \cdots\right].$$

The first three terms yield the estimate

$$\sqrt[3]{1001} \approx 10.00333\ 22222 \cdots.$$

The error in this estimate is precisely the remainder

$$r_3 = 10\left[\binom{\frac{1}{3}}{3}(10^{-3})^3 + \binom{\frac{1}{3}}{4}(10^{-3})^4 + \cdots\right].$$

Now $\quad \left|\binom{\frac{1}{3}}{n}\right| = \dfrac{1}{n!}\left|\left(\dfrac{1}{3}\right)\left(\dfrac{-2}{3}\right)\left(\dfrac{-5}{3}\right)\cdots\left(-\dfrac{3n-4}{3}\right)\right| = \dfrac{1}{3}\cdot\dfrac{2}{6}\cdot\dfrac{5}{9}\cdots\dfrac{3n-4}{3n}.$

This is a decreasing sequence. Therefore

$$|r_3| \leq 10\left[\left|\binom{\frac{1}{3}}{3}\right|10^{-9} + \left|\binom{\frac{1}{3}}{4}\right|10^{-12} + \cdots\right]$$

$$< 10\left|\binom{\frac{1}{3}}{3}\right|(10^{-9} + 10^{-12} + \cdots) = 10\cdot\dfrac{1}{3}\cdot\dfrac{2}{6}\cdot\dfrac{5}{9}(10^{-9} + 10^{-12} + \cdots)$$

$$= \dfrac{100}{162}(10^{-9} + 10^{-12} + \cdots) < 10^{-9}.$$

Hence the estimate is accurate to at least 8 places.   ∎

Remark   There is a table of the most used power series inside the front cover of this book.

EXERCISES

Expand in power series at $x = 0$

1 $\quad \dfrac{1}{(1+x)^3}$ 　　　　**2** $\quad \dfrac{1}{(1-x)^{1/3}}$ 　　　　**3** $\quad \dfrac{1}{(1-4x^2)^2}$

4 $\quad \dfrac{1}{(3-4x^2)^2}$ 　　　　**5** $\quad (1+2x^3)^{1/4}$ 　　　　**6** $\quad \sqrt{2-x}$

7 $\quad \dfrac{x^2}{\sqrt{1+2x}}$ 　　　　**8** $\quad \left(\dfrac{x}{1-x}\right)^{10}$ 　　　　**9** $\quad \sinh^{-1} x$

10 $\quad x \arc\sin x + \sqrt{1-x^2}.$

Expand in power series at $x = 1$

11 $\quad \sqrt{1+x}$ 　　　　　　　　　　**12** $\quad \dfrac{1}{(3+x)^2}.$

Compute the terms up to x^4 in the Taylor series at $x = 0$

13 $\quad \sqrt{1+x^2e^x}$ 　　　**14** $\quad \dfrac{1}{(1+\sin x)^2}$ 　　　**15** $\quad (\sin 2x)\sqrt{3+x}$

16 $\quad \dfrac{1}{\sqrt{1+x+x^2}}$ 　　　**17** $\quad \dfrac{\cos 2x}{(1+\frac{1}{3}x^2)^4}$ 　　　**18** $\quad \sqrt{\cos x}.$

Compute to 4-place accuracy, using the binomial series

19 $\quad \sqrt{16.1}$ 　　　**20** $\quad \sqrt[4]{82}$ 　　　**21** $\quad \dfrac{1}{(1.03)^5}$ 　　　**22** $\quad \dfrac{1}{\sqrt[3]{970}}.$

23 　　Derive the approximation $\quad \ln(1+x) \approx -1 + \sqrt{1+2x}.$

24* Derive $\dfrac{2}{\pi} \displaystyle\int_0^{\pi/2} \dfrac{d\theta}{1 - x^2 \sin^2 \theta} = \dfrac{1}{\sqrt{1 - x^2}}$, $|x| < 1$.

25 Show that $1/\sqrt{1 - 2xz + z^2} = \sum_0^\infty P_n(x)z^n$, where $P_n(x)$ is a polynomial of degree n, and compute $P_1(x)$, $P_2(x)$, and $P_3(x)$. The polynomials $P_n(x)$ are called the **Legendre polynomials**; they are important in applied mathematics, for instance, in the study of heat flow in spherical bodies.

26 (cont.) Find formulas for $P_n(0)$ and $P_n(1)$.

27 Let $p_n = 2n \sin(\pi/n)$ be the perimeter of the regular polygon of n sides inscribed in a unit circle. Show that there is a relation of the form

$$\pi = \frac{1}{2} p_n + \frac{a_1}{n^2} + \frac{a_2}{n^4} + \frac{a_3}{n^6} + \cdots .$$

28 (cont.) Show that $3\pi \approx 2p_n - \frac{1}{2}p_{2n}$. Estimate π to 4 places using $n = 3$ and $n = 6$.

29* (cont.) Obtain a similar estimate involving p_n, p_{2n}, and p_{4n}. Test it to 5 places with $n = 3$.

30 Show that $\sqrt{2} = \frac{100}{71}(1 + \varepsilon)^{1/2}$, where $\varepsilon = 0.0082$. Estimate $\sqrt{2}$ by expanding up to ε^2.

31 (cont.) Estimate $\sqrt{3}$ similarly. [*Hint* Find an integer k for which $3k^2$ is near 10^5.]

32* (cont.) The general idea for $\sqrt{2}$ is to find $a = 10^n$ and an integer b so

$$2 = \left(\frac{a}{b}\right)^2 \left(1 + \frac{2b^2 - a^2}{a^2}\right)$$

and $(2b^2 - a^2)$ is as small as possible. Find b for $a = 1000$ and use it to estimate $\sqrt{2}$.

6. NUMERICAL APPLICATIONS

Alternating Series   In Chapter 11, Section 5 we saw the great advantage of working with alternating series whose terms decrease in absolute value toward 0. For such series, remainder estimates are immediate and quite accurate: if a series is broken off, then the remainder is less than the absolute value of the first term omitted.

■ **EXAMPLE 1**   The power series at $x = 0$ for $\ln(1 + x^2)$ is broken off after n terms. Estimate the remainder.

Solution   The series is obtained by integrating from 0 to x the series for the derivative of $\ln(1 + t^2)$:

$$\frac{d}{dt}[\ln(1 + t^2)] = \frac{2t}{1 + t^2} = 2t(1 - t^2 + t^4 - t^6 + - \cdots)$$

$$= 2(t - t^3 + t^5 - t^7 + - \cdots)$$

for $|t| < 1$. It follows that

$$\ln(1 + x^2) = \int_0^x \frac{2t\, dt}{1 + t^2} = 2\left(\frac{x^2}{2} - \frac{x^4}{4} + \frac{x^6}{6} - \frac{x^8}{8} + - \cdots\right)$$

$$= x^2 - \frac{x^4}{2} + \frac{x^6}{3} - \frac{x^8}{4} + - \cdots$$

for $|x| < 1$. This series alternates; since $|x| < 1$, its terms decrease in absolute value to zero. Therefore, the remainder after n terms is less than the absolute value of the $(n + 1)$-th term:

$$|\text{remainder}| < \frac{x^{2n+2}}{n+1}$$ ∎

Occasionally we encounter an alternating series whose terms ultimately decrease in magnitude, but whose first few terms do not. If successive terms decrease starting at the k-th term, the series will still converge, and the remainder estimate is still valid—beyond the k-th term. The front end of the series, up to the $(k-1)$-th term, is a finite sum; it causes no trouble. The important part is the tail end, that is, the series starting with the k-th term. It is this series that we test for convergence.

∎ **EXAMPLE 2** The power series for e^{-x} at $x = 0$ is broken off after n terms. Estimate the remainder for positive values of x.

Solution The power series $\quad e^{-x} = 1 - x + \dfrac{x^2}{2!} - \dfrac{x^3}{3!} + - \cdots$

alternates for positive x. If $0 < x \leq 1$, the terms decrease to 0, and the above remainder estimate for alternating series applies. If $x > 1$, however, the first few terms may not decrease. (Take $x = 6$ for example:

$$e^{-6} = 1 - 6 + \frac{6^2}{2!} - \frac{6^3}{3!} + - \cdots = 1 - 6 + 18 - 36 + - \cdots.$$

Nevertheless, for any fixed x, the terms do decrease ultimately. To see why, note that the ratio of successive terms is

$$\frac{x^{n+1}}{(n+1)!} \bigg/ \frac{x^n}{n!} = \frac{x}{n+1}.$$

For x fixed, $\qquad \dfrac{x}{n+1} < 1$

as soon as $n + 1 > x$; from then on the terms decrease. Furthermore

$$\frac{x}{n+1} < \frac{1}{2}$$

as soon as $n + 1 > 2x$. From then on each term is less than one-half the preceding term. Hence, the terms decrease to 0.

Therefore if the series is broken off at the n-th term, the remainder estimate for an alternating series applies:

$$|\text{remainder}| < \frac{x^{n+1}}{(n+1)!} \qquad \text{for} \quad n + 1 > x.$$ ∎

Applications to Definite Integrals Power series are useful for approximating definite integrals that cannot be computed exactly.

∎ **EXAMPLE 3** Estimate $\displaystyle\int_0^x e^{-t^2} \, dt$ to 6 places for $|x| \leq \frac{1}{2}$.

Solution   A numerical integration formula such as Simpson's Rule would work, but a power series method is simpler. Expand the integrand in a power series:

$$e^{-t^2} = 1 + (-t^2) + \frac{(-t^2)^2}{2!} + \frac{(-t^2)^3}{3!} + \cdots = 1 - t^2 + \frac{t^4}{2!} - \frac{t^6}{3!} + \cdots.$$

Since this series converges for all x, it can be integrated term-by-term:

$$\int_0^x e^{-t^2}\, dt = \int_0^x \left(1 - t^2 + \frac{t^4}{2!} - \frac{t^6}{3!} + \cdots \right) dt$$

$$= x - \frac{x^3}{3} + \frac{x^5}{5 \cdot 2!} - \frac{x^7}{7 \cdot 3!} + \frac{x^9}{9 \cdot 4!} - \cdots.$$

Because of the large denominators, this series converges rapidly if x is fairly small.

For $|x| \le \frac{1}{2}$, the sixth term is at most $\left(\frac{1}{2} \right)^{11} \frac{1}{11 \cdot 5!} < 4 \times 10^{-7}$.

Since the series alternates, the remainder after five terms is less than 4×10^{-7}. Therefore five terms provide 6-place accuracy. Consequently if $|x| \le \frac{1}{2}$, then

$$\int_0^x e^{-t^2}\, dt = x - \frac{x^3}{3} + \frac{x^5}{5 \cdot 2!} - \frac{x^7}{7 \cdot 3!} + \frac{x^9}{9 \cdot 4!} + \varepsilon, \qquad \text{where} \quad |\varepsilon| < 4 \times 10^{-7}. \qquad ∎$$

Remark   The series converges for all value of x, but for large x it converges slowly. For example, it would be ridiculous to compute $\int_0^{10} e^{-t^2}\, dt$ by this method, since more than 100 terms at the beginning are greater than 1.

Elliptic Integrals   A fairly simple problem can lead to a difficult definite integral. An example is the problem of computing the arc length of an ellipse. Suppose an ellipse is given in the parametric form

$$x = a \cos \theta, \qquad y = b \sin \theta,$$

where $b > a$. Then, as will be shown in Chapter 14, its arc length is

$$L = \int_0^{2\pi} \sqrt{\left(\frac{dx}{d\theta} \right)^2 + \left(\frac{dy}{d\theta} \right)^2}\, d\theta.$$

The expression under the radical is

$$(-a \sin \theta)^2 + (b \cos \theta)^2 = b^2(\sin^2 \theta + \cos^2 \theta) - (b^2 - a^2) \sin^2 \theta = b^2(1 - k^2 \sin^2 \theta),$$

where $k^2 = (b^2 - a^2)/b^2 < 1$. Hence

$$L = \int_0^{2\pi} b\sqrt{1 - k^2 \sin^2 \theta}\, d\theta = 4b \int_0^{\pi/2} \sqrt{1 - k^2 \sin^2 \theta}\, d\theta.$$

The definite integral in the last expression, called an **elliptic integral**, is impossible to compute exactly. (More precisely, it is called a "complete elliptic integral of the second kind.")

■ **EXAMPLE 4** Express $\displaystyle\int_0^{\pi/2} \sqrt{1 - k^2 \sin^2 t}\, dt$ as a power series in k.

Solution By Example 2, p. 607, for $|x| < 1$,

$$\sqrt{1-x} = (1-x)^{1/2} = 1 - \sum_{n=1}^{\infty} \binom{\frac{1}{2}}{n} x^n, \quad \text{where} \quad \binom{\frac{1}{2}}{n} = \frac{(2n)!}{(2^n n!)^2 (2n-1)}.$$

If $k^2 < 1$, then $k^2 \sin^2 t \le k^2 < 1$, so we can substitute $x = k^2 \sin^2 t$:

$$\sqrt{1 - k^2 \sin^2 t} = 1 - \sum_{n=1}^{\infty} \binom{\frac{1}{2}}{n} k^{2n} \sin^{2n} t.$$

Now integrate term-by-term:

$$\int_0^{\pi/2} \sqrt{1 - k^2 \sin^2 t}\, dt = \frac{\pi}{2} - \sum_{n=1}^{\infty} \binom{\frac{1}{2}}{n} k^{2n} \int_0^{\pi/2} \sin^{2n} t\, dt.$$

From a table of integrals (also see Ex. 24, p. 610) $\displaystyle\int_0^{\pi/2} \sin^{2n} x\, dx = \frac{(2n)!}{(2^n n!)^2}\frac{\pi}{2}.$

Therefore

$$\binom{\frac{1}{2}}{n} \int_0^{\pi/2} \sin^{2n} t\, dt = \left[\frac{(2n)!}{(2^n \cdot n!)^2 (2n-1)}\right]\left[\frac{(2n)!}{(2^n \cdot n!)^2} \cdot \frac{\pi}{2}\right] = \left[\frac{(2n)!}{(2^n \cdot n!)^2}\right]^2 \frac{1}{(2n-1)} \cdot \frac{\pi}{2}.$$

Substitute this expression to obtain the answer: for $k^2 < 1$,

$$\int_0^{\pi/2} \sqrt{1 - k^2 \sin^2 t}\, dt = \frac{\pi}{2} - \frac{\pi}{2} \sum_{n=1}^{\infty} \frac{[(2n)!]^2}{(2^n n!)^4 (2n-1)} k^{2n}. \quad ■$$

Period of a Pendulum A pendulum consists of a bob at the end of a uniform string of length L. When displaced from the vertical by an angle $\alpha < \frac{1}{2}\pi$ and released from rest it will oscillate with a period $T = T(\alpha)$. It can be shown* that

$$T = 4\sqrt{\frac{L}{2g}} \int_0^{\alpha} \frac{d\theta}{\sqrt{\cos\theta - \cos\alpha}}.$$

* Briefly (Fig. 1)

kinetic energy + potential energy = const.,

$\frac{1}{2}v^2 + Lg(1 - \cos\theta) = 0 + Lg(1 + \cos\alpha),$

$\frac{1}{2}v^2 = Lg(\cos\theta - \cos\alpha).$

Fig. 1

But $\quad v = ds/dt = L\, d\theta/dt,$

so $\quad \dfrac{dt}{d\theta} = \sqrt{\dfrac{L}{2g}}\, \dfrac{1}{\sqrt{\cos\theta - \cos\alpha}}, \quad$ etc.

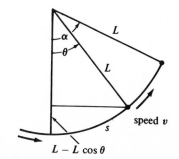

We shall see that this is a type of elliptic integral and expand it in a Taylor series. We change its form in two steps. First we write

$$\sqrt{\cos\theta - \cos\alpha} = \sqrt{(1 - 2\sin^2\tfrac{1}{2}\theta) - (1 - 2\sin^2\tfrac{1}{2}\alpha)} = \sqrt{2}\sqrt{\sin^2\tfrac{1}{2}\alpha - \sin^2\tfrac{1}{2}\theta}.$$

Next, we set $k = \sin\tfrac{1}{2}\alpha$ and change the variable from θ to ϕ via

$$\sin\tfrac{1}{2}\theta = k\sin\phi = \sin\tfrac{1}{2}\alpha \sin\phi.$$

Then ϕ varies from 0 to $\tfrac{1}{2}\pi$ as θ varies from 0 to α, and

$$\sqrt{\cos\theta - \cos\alpha} = \sqrt{2}\sqrt{k^2 - k^2\sin^2\phi} = k\sqrt{2}\cos\phi.$$

Furthermore, from $\sin\tfrac{1}{2}\theta = k\sin\phi$ follows

$$\tfrac{1}{2}\cos\tfrac{1}{2}\theta\, d\theta = k\cos\phi\, d\phi,$$

$$d\theta = \frac{2k\cos\phi}{\cos\tfrac{1}{2}\theta}\, d\phi = \frac{2k\cos\phi\, d\phi}{\sqrt{1 - \sin^2\tfrac{1}{2}\theta}} = \frac{2k\cos\phi\, d\phi}{\sqrt{1 - k^2\sin^2\phi}}.$$

Collecting all this information, we express the integral for T in terms of ϕ:

$$T = 4\sqrt{\frac{L}{2g}}\int_0^{\pi/2} \frac{2k\cos\phi\, d\phi}{(k\sqrt{2}\cos\phi)\sqrt{1 - k^2\sin^2\phi}} = 4\sqrt{\frac{L}{g}}\int_0^{\pi/2} \frac{d\phi}{\sqrt{1 - k^2\sin^2\phi}}.$$

The final integral is called a *complete* **elliptic integral** of the *first kind*. Since $k^2\sin^2\phi = \sin^2\tfrac{1}{2}\alpha \sin^2\phi < 1$, we can expand the integrand in a binomial series and integrate term-by-term:

$$T = 4\sqrt{\frac{L}{g}}\left[\int_0^{\pi/2} d\phi + \sum_{n=1}^{\infty}\binom{-\frac{1}{2}}{n}(-k^2)^n\int_0^{\pi/2}\sin^{2n}\phi\, d\phi\right].$$

But (see pp. 608 and 613)

$$\binom{-\frac{1}{2}}{n}\int_0^{\pi/2}\sin^{2n}\phi\, d\phi = \left[\frac{(-1)^n(2n)!}{2^{2n}(n!)^2}\right]\left[\frac{(2n)!}{2^{2n}(n!)^2}\frac{\pi}{2}\right].$$

Therefore we obtain the expansion

$$T = 2\pi\sqrt{\frac{L}{g}}\left(1 + \sum_{n=1}^{\infty}\left[\frac{(2n)!}{2^{2n}(n!)^2}\right]^2 k^{2n}\right)$$

$$= 2\pi\sqrt{\frac{L}{g}}\left[1 + \left(\frac{1}{2}\right)^2 k^2 + \left(\frac{1\cdot3}{2\cdot4}\right)^2 k^4 + \left(\frac{1\cdot3\cdot5}{2\cdot4\cdot6}\right)^2 k^6 + \cdots\right],$$

where $k = \sin\tfrac{1}{2}\alpha$. For α small, $k \approx \tfrac{1}{2}\alpha$ and

$$T \approx 2\pi\sqrt{\frac{L}{g}}\left(1 + \frac{\alpha^2}{16}\right).$$

Differential Equations   Suppose we must solve an initial value problem like

$$\frac{d^2y}{dx^2} = xy, \qquad y(0) = a, \quad y'(0) = b.$$

One method is to *assume* that the solution $y = y(x)$ can be expanded in a Taylor series at $x = 0$, substitute the series into the differential equation, then solve for the coefficients.

■ **EXAMPLE 5** Solve by Taylor's series

$$\frac{dy}{dx} = xy + 1, \qquad y(0) = a.$$

Solution Assume $y = a + \displaystyle\sum_{n=1}^{\infty} a_n x^n$, so $\dfrac{dy}{dx} = \displaystyle\sum_{n=1}^{\infty} n a_n x^{n-1}$.

The initial condition $y(0) = a$ is satisfied, and the differential equation becomes

$$\sum_{n=1}^{\infty} n a_n x^{n-1} = 1 + ax + \sum_{n=1}^{\infty} a_n x^{n+1}.$$

We want to equate coefficients. To keep the indices straight we rewrite the equation as

$$\sum_{n=1}^{\infty} n a_n x^{n-1} = 1 + ax + \sum_{n=3}^{\infty} a_{n-2} x^{n-1}.$$

Now we equate coefficients:

$$a_1 = 1, \qquad 2a_2 = a, \qquad 3a_3 = a_1, \qquad 4a_4 = a_2, \cdots, n a_n = a_{n-2}, \cdots.$$

Clearly the odd indices and the even indices go their separate ways:

$$a_1 = 1, \qquad a_3 = \frac{a_1}{3} = \frac{1}{1 \cdot 3}, \qquad a_5 = \frac{a_3}{5} = \frac{1}{1 \cdot 3 \cdot 5}, \cdots,$$

$$a_{2n-1} = \frac{1}{1 \cdot 3 \cdot 5 \cdots (2n-1)} = \frac{2^n n!}{(2n)!},$$

$$a_2 = \frac{a}{2}, \qquad a_4 = \frac{a_2}{4} = \frac{a}{2 \cdot 4}, \qquad \cdots, \qquad a_{2n} = \frac{a}{2 \cdot 4 \cdots 2n} = \frac{a}{2^n n!}.$$

Hence $y = a \displaystyle\sum_{n=0}^{\infty} \frac{x^{2n}}{2^n n!} + \sum_{n=1}^{\infty} \frac{2^n n!}{(2n)!} x^{2n-1}.$ ■

Remark The first sum is $e^{x^2/2}$. The second sum can be expressed in terms of exponential functions and an integral. See Ex. 32.

■ **EXAMPLE 6** Solve $\dfrac{d^2 y}{dx^2} = xy, \qquad y(0) = a, \quad y'(0) = b.$

Solution Assume $y = a + bx + \displaystyle\sum_{n=2}^{\infty} a_n x^n$, so $\dfrac{d^2 y}{dx^2} = \displaystyle\sum_{n=2}^{\infty} n(n-1) a_n x^{n-2}.$

The initial conditions are satisfied, and the differential equation becomes

$$\sum_{n=2}^{\infty} n(n-1)a_n x^{n-2} = ax + bx^2 + \sum_{n=2}^{\infty} a_n x^{n+1},$$

that is,

$$\sum_{n=2}^{\infty} n(n-1)a_n x^{n-2} = ax + bx^2 + \sum_{n=5}^{\infty} a_{n-3} x^{n-2}.$$

Equate coefficients:

$$2 \cdot 1 \cdot a_2 = 0, \qquad 3 \cdot 2 \cdot a_3 = a, \qquad 4 \cdot 3 \cdot a_4 = b,$$

$$5 \cdot 4 \cdot a_5 = a_2, \qquad 6 \cdot 5 \cdot a_6 = a_3, \quad \cdots, \qquad n(n-1)a_n = a_{n-3}, \quad \cdots.$$

Solve: $a_2 = 0, \qquad a_5 = 0, \cdots, a_{3n-1} = 0,$

$$a_3 = \frac{a}{2 \cdot 3}, \qquad a_6 = \frac{a_3}{5 \cdot 6} = \frac{a}{2 \cdot 3 \cdot 5 \cdot 6} = \frac{1 \cdot 4}{6!} a,$$

$$a_9 = \frac{a_6}{8 \cdot 9} = \frac{1 \cdot 4 \cdot 7}{9!} a, \cdots,$$

$$a_4 = \frac{b}{3 \cdot 4} = \frac{2}{4!} b, \qquad a_7 = \frac{a_4}{6 \cdot 7} = \frac{2 \cdot 5}{7!} b, \cdots.$$

Therefore

$$y = a\left(1 + \frac{1}{3!} x^3 + \frac{1 \cdot 4}{6!} x^6 + \frac{1 \cdot 4 \cdot 7}{9!} x^9 + \cdots\right)$$

$$+ b\left(x + \frac{2}{4!} x^4 + \frac{2 \cdot 5}{7!} x^7 + \frac{2 \cdot 5 \cdot 8}{10!} x^{10} + \cdots\right). \qquad \blacksquare$$

Remark   The function y is sometimes written $A(a, b; x)$ and called an **Airy function**.

Convergence to e   We know that the sequence $(1 + 1/n)^n$ converges to e, but rather slowly. Using power series we shall determine just how slowly and also construct related sequences that converge to e rapidly.

The key is the power series for the function $y(x) = (1 + x)^{1/x}$ for $0 < |x| < 1$, $y(0) = e$. It is hard to compute this series directly, so we use the following important technique: we show that $y(x)$ is the solution of a certain initial value problem, then find a power series solution.

By logarithmic differentiation,

$$\ln y = \frac{\ln(1 + x)}{x}, \qquad \frac{y'}{y} = z,$$

where

$$z = \frac{x/(1 + x) - \ln(1 + x)}{x^2}.$$

Thus $y(x)$ is the solution of the initial value problem

$$\frac{dy}{dx} = zy, \qquad y(0) = e.$$

To find a power series solution, we first expand z:

$$z = \frac{(x - x^2 + x^3 - x^4 + \cdots) - (x - \frac{1}{2}x^2 + \frac{1}{3}x^3 - \frac{1}{4}x^4 + \cdots)}{x^2}$$

$$= -\frac{1}{2} + \frac{2}{3}x - \frac{3}{4}x^2 + \frac{4}{5}x^3 - \cdots.$$

Next we assume a solution in the convenient form

$$y = e(1 + a_1 x + a_2 x^2 + a_3 x^3 + \cdots).$$

The differential equation becomes

$$a_1 + 2a_2 x + 3a_3 x^2 + 4a_4 x^3 + \cdots$$
$$= (-\frac{1}{2} + \frac{2}{3}x - \frac{3}{4}x^2 + \frac{4}{5}x^3 - \cdots)(1 + a_1 x + a_2 x^2 + a_3 x^3 + \cdots).$$

We multiply on the right and equate coefficients:

$$a_1 = -\frac{1}{2}, \qquad 2a_2 = \frac{2}{3} - \frac{1}{2}a_1 = \frac{11}{12}, \qquad a_2 = \frac{11}{24},$$

$$3a_3 = -\frac{3}{4} + \frac{2}{3}a_1 - \frac{1}{2}a_2 = -\frac{21}{16}, \qquad a_3 = -\frac{7}{16},$$

$$4a_4 = \frac{4}{5} - \frac{3}{4}a_1 + \frac{2}{3}a_2 - \frac{1}{2}a_3 = \frac{2447}{1440}, \qquad a_4 = \frac{2447}{5760}.$$

That is enough. We conclude that

$$(1 + x)^{1/x} = e(1 - \frac{1}{2}x + \frac{11}{24}x^2 - \frac{7}{16}x^3 + \frac{2447}{5760}x^4 + \cdots).$$

Hence for large n,

$$\left| \left(1 + \frac{1}{n}\right)^n - e \right| \approx \frac{e}{2n}.$$

This shows clearly why it takes very large values of n to make $(1 + 1/n)^n$ anywhere near e.

To do better, we multiply the power series for $(1 + x)^{1/x}$ by a polynomial of the form $1 + b_1 x + b_2 x^2 + \cdots$, chosen to knock out terms in x, x^2, etc. For instance,

$$(1 + x)^{1/x}(1 + \frac{1}{2}x) = e(1 + \frac{5}{24}x^2 - \frac{5}{24}x^3 + \cdots).$$

Now the left-hand side approximates e with a second order error term, $\frac{5}{24}ex^2$, which is much smaller than the previous first order error term, $-\frac{1}{2}ex$. For $x = 1/n$ we have

$$\left| \left(1 + \frac{1}{n}\right)^n \left(1 + \frac{1}{2n}\right) - e \right| \approx \frac{5e}{24n^2} < \frac{0.6}{n^2}.$$

A numerical comparison to $e \approx 2.718282$ is striking:

n	10	50	100	200
$(1 + 1/n)^n$	2.594	2.6916	2.70481	2.711517
$(1 + 1/n)^n(1 + 1/2n)$	2.723	2.7185	2.71834	2.718296

EXERCISES

1   Compute $e^{-1/5}$ to 5-place accuracy, using power series.

2   For what x is $\cos x \approx 1 - \dfrac{x^2}{2} + \dfrac{x^4}{24}$ accurate to 5 places?

3   How many terms of the power series for $\ln x$ at $x = 10$ are needed to compute $\ln(10.5)$ with 5-place accuracy? (Assume the value of $\ln 10$ is known.)

4   How many terms of the binomial series for $\sqrt{1 + x}$ will yield 5-place accuracy for $0 < x \leq 0.1$?

Compute to 5-place accuracy

5 $\displaystyle\int_0^{0.1} e^{x^3}\, dx$ **6** $\displaystyle\int_0^{0.2} e^{-x^2}\, dx$

7 $\displaystyle\int_0^{1/4} \sqrt{1 + x^3}\, dx$ **8** $\displaystyle\int_{3.00}^{3.01} \dfrac{e^x}{1 + x}\, dx$ [Expand at $x = 3$.]

9   Compute to 4-place accuracy the arc length of an ellipse with semi-axes 40 and 41.

10   Estimate the value of x for which

$$\int_0^x \frac{t}{1 + t^4}\, dt = 0.1.$$

[*Hint*   Approximate the integral by the first significant term of its power series.]

11   (cont.) Refine your estimate of x to 4-place accuracy by taking the first two significant terms of the power series. Use Newton's Method to solve approximately the resulting equation.

12   The quantity $\omega = 2\pi/T$ is the **angular frequency** of the pendulum. Expand ω in terms of α up to terms in α^2.

13   (cont.) Expand T in terms of α up to terms in α^4.

14   (cont.) Expand ω in terms of α up to terms in α^4.

Solve with the initial condition(s) $y(0) = a$ (and $y'(0) = b$ if second order)

15 $\dfrac{dy}{dx} = x + y$ **16** $\dfrac{dy}{dx} + 3y = e^x$ **17** $\dfrac{dy}{dx} = x^2 + y$

18 $\dfrac{d^2y}{dx^2} + 4y = 0$ **19** $\dfrac{d^2y}{dx^2} + x\dfrac{dy}{dx} - y = 0$ **20** $x\dfrac{d^2y}{dx^2} = y$

21 $x^2\dfrac{d^2y}{dx^2} + (x^2 + x)\dfrac{dy}{dx} = y$ **22** $x\dfrac{d^2y}{dx^2} + \dfrac{dy}{dx} + xy = 0.$

Find the power series solution up to the x^4 term

23 $\dfrac{dy}{dx} = 1 - x^2 - y^2, \quad y(0) = 2$ **24** $\dfrac{dy}{dx} = \dfrac{x}{x + y + 1}, \quad y(0) = 0$

25 $\dfrac{dy}{dx} = 1 - y + x^3y^2, \quad y(0) = -1$ **26** $\dfrac{d^2y}{dx^2} + y = \dfrac{e^x}{1 - x}, \quad y(0) = y'(0) = 1.$

27   Compute $f(x) = (1 + x)^{1/x}(1 + \frac{1}{2}x - \frac{5}{24}x^2)$ up to x^4.

28   (cont.) Compute $g(x) = (1 + x)^{1/x}(1 + \frac{1}{2}x - \frac{5}{24}x^2 + \frac{5}{48}x^3)$ up to x^4.

29   (cont.) Compute $h(x) = (1 + x)^{1/x}\left(\dfrac{12 + 11x}{12 + 5x}\right)$ up to x^3.

30   (cont.) Tabulate $f(1/n)$, $g(1/n)$, $h(1/n)$ for $n = 10, 20, 50, 100$.

31*   We know that $x^x \longrightarrow 1-$ as $x \longrightarrow 0+$. How fast? That is, estimate $1 - x^x$ for x small positive.

32* Prove $\displaystyle\sum_{n=1}^{\infty} \frac{2^n n!}{(2n)!} x^{2n-1} = e^{x^2/2} \int_0^x e^{-t^2/2}\, dt.$ [*Hint* See Example 5.]

7. SEQUENCES AND SERIES OF FUNCTIONS

Interval of Convergence Let us prove the theorem stated in Section 1: a power series converges either at a single point, on an interval, or on the whole line. For simplicity of notation, we shall take all power series at $x = 0$. First, an important preliminary result:

> **Lemma** If a power series $\sum a_n x^n$ converges at $x = c$, where $c \neq 0$, then it converges absolutely in the interval $-|c| < x < |c|$.
>
> If the series diverges at $x = b$, then it diverges at each x for which $|x| > |b|$.

Proof If $\sum a_n c^n$ converges, its terms approach 0, that is, $a_n c^n \longrightarrow 0$ as $n \longrightarrow \infty$. It follows that the terms are bounded, that is, there exists a positive number M such that $|a_n c^n| \leq M$ for all n. Hence for any x,

$$|a_n x^n| = \left| a_n c^n \left(\frac{x}{c}\right)^n \right| \leq M \left| \frac{x}{c} \right|^n.$$

In particular, if $-|c| < |x| < |c|$, then $|x/c| < 1$, so $\sum a_n x^n$ converges absolutely by comparison with the convergent geometric series $M \sum |x/c|^n$.

If $\sum a_n b^n$ diverges and $|x| > |b|$, then $\sum a_n x^n$ must diverge. For otherwise it converges, and then so does $\sum a_n b^n$ by the first statement in the lemma (with c replaced by x and x by b). End of proof.

Now to the proof of the theorem on p. 579. Let $\sum a_n x^n$ be any power series and let **D** be its domain of convergence, the set of all points x for which the given series converges. There are two cases:

Case 1 **D** is unbounded. Then there are numbers c with $|c|$ arbitrarily large and $\sum a_n c^n$ convergent. By Lemma 1, for each such c, the set **D** contains the whole interval $|x| < |c|$. Since c can be taken arbitrarily large, this means that **D** contains all real numbers.

Case 2 **D** is bounded. We draw upon the basic completeness property of the real number system; it asserts that **D** has a *unique least upper bound*. This means there is a *smallest* number R such that $x \leq R$ for all x in **D**. Clearly $R \geq 0$ since 0 is in **D**. We claim R is the radius of convergence.

On the one hand, suppose $|c| < R$. If $\sum a_n c^n$ diverges, then by Lemma 1, $\sum a_n x^n$ diverges whenever $|x| > |c|$. This means that $|c|$ is an upper bound for **D**, smaller than R, the *least* upper bound. Impossible. Therefore $\sum a_n c^n$ converges whenever $|c| < R$.

On the other hand, suppose $|c| > R$. If $\sum a_n c^n$ converges, then by Lemma 1, $\sum a_n x^n$ converges for all x such that $|x| < |c|$, in particular for any x satisfying

$R < x < |c|$, so R is *not* an upper bound for **D**. Impossible. Therefore $\sum a_n c^n$ diverges whenever $|c| > R$.

Suppose $R > 0$. Then it follows from the preceding two paragraphs that **D** consists precisely of the interval $-R < x < R$, possibly including one or both of its end points.

Finally, if $R = 0$, then the series diverges for all x with $|x| > 0$. Hence it converges only at the single point $x = 0$. This completes the proof.

A power series converges absolutely at each point of its interval of convergence except perhaps at the end points (if any).

Proof If x is any interior point of the interval of convergence, there exists another interior point $c > 0$ with $c > |x|$. According to Lemma 1, $\sum a_n x^n$ converges absolutely.

Sequences of Functions   Our next goal is to prove that power series can be differentiated and integrated term-by-term. These properties are actually special cases of something much more general, and it is worth while to look at the general situation.

When $f(x) = \sum_0^\infty a_n x^n$, then $f(x)$ is the limit of the sequence of partial sums:

$$f(x) = \lim_{n \to \infty} s_n(x), \qquad s_n(x) = \sum_{k=0}^{n} a_k x^k$$

for each x in the interval of convergence. Thus *a function can be the limit of a sequence of functions.* Another example:

$$e^x = \lim_{n \to \infty} \left(1 + \frac{x}{n}\right)^n,$$

valid for all x.

Now we shall study the general situation in which a function is the limit of a sequence of functions. What we are after are conditions under which the derivative of the limit is the limit of the derivatives or the integral of the limit is the limit of the integrals. For this we need an important new concept called uniform convergence.

Uniform Convergence

Uniform Convergence   Let $\{u_n(x)\}$ be a sequence of functions, all with the same domain **D**. The sequence **converges uniformly** on **D** to $u(x)$ if given any $\varepsilon > 0$, there exists an N such that

$$|u_n(x) - u(x)| < \varepsilon$$

for all $n \geq N$ and all x in **D**.

The words "all x in **D**" are the key to this concept. We can control the degree of approximation of $u_n(x)$ to $u(x)$ *independently of x*. The next three results show the usefulness of uniform convergence.

Continuity of the Limit Let $\{u_n(x)\}$ be a sequence of continuous functions on **D**, and let $u_n(x) \longrightarrow u(x)$ uniformly on **D**. Then $u(x)$ is continuous on **D**.

Proof Let $\varepsilon > 0$. Then there is an N such that $\left|u_N(x) - u(x)\right| < \frac{1}{3}\varepsilon$ for all x in **D**. Take any point c in **D**. Since $u_N(x)$ is continuous at c, there exists $\delta > 0$ such that $\left|u_N(x) - u_N(c)\right| < \frac{1}{3}\varepsilon$ for all x in **D** such that $|x - c| < \delta$. If x is such a point, then

$$\left|u(x) - u(c)\right| = \left|u(x) - u_N(x) + u_N(x) - u_N(c) + u_N(c) - u(c)\right|$$
$$\leq \left|u(x) - u_N(x)\right| + \left|u_N(x) - u_N(c)\right| + \left|u_N(c) - u(c)\right|$$
$$< \tfrac{1}{3}\varepsilon + \tfrac{1}{3}\varepsilon + \tfrac{1}{3}\varepsilon = \varepsilon.$$

This proves the continuity of $u(x)$ at c.

Integral of the Limit Let $\{u_n(x)\}$ be a sequence of continuous functions on a closed interval $a \leq x \leq b$, and let $u_n(x) \longrightarrow u(x)$ uniformly on this interval. Then

$$\int_a^b u(x)\, dx = \lim_{n \to \infty} \int_a^b u_n(x)\, dx.$$

Proof By the preceding result, $u(x)$ is continuous, hence integrable. Let $\varepsilon > 0$. Then there exists an N such that $\left|u_n(x) - u(x)\right| < \varepsilon/(b - a)$ for all $n \geq N$ and all x in the interval $[a, b]$. If $n \geq N$, it follows that

$$\left|\int_a^b u_n(x)\, dx - \int_a^b u(x)\, dx\right| = \left|\int_a^b [u_n(x) - u(x)]\, dx\right|$$
$$\leq \int_a^b \left|u_n(x) - u(x)\right|\, dx < \int_a^b \frac{\varepsilon}{b - a}\, dx = \varepsilon.$$

Therefore

$$\int_a^b u_n(x)\, dx \longrightarrow \int_a^b u(x)\, dx.$$

Derivative of the Limit Let $\{u_n(x)\}$ be a sequence of continuously differentiable functions on an interval $a \leq x \leq b$, let $u_n(x) \longrightarrow u(x)$ for each x on the interval, and let $u_n'(x) \longrightarrow v(x)$ uniformly on the interval. Then $u(x)$ is differentiable and $u'(x) = v(x)$.

Proof   If $a \leq x \leq b$, then

$$u_n(x) - u_n(a) = \int_a^x u_n'(t)\, dt \longrightarrow \int_a^x v(t)\, dt,$$

by the previous result. But $u_n(x) \longrightarrow u(x)$ and $u_n(a) \longrightarrow u(a)$; hence

$$u(x) - u(a) = \int_a^x v(t)\, dt.$$

By the Fundamental Theorem of Calculus, the right-hand side is differentiable and its derivative is $v(x)$. Hence so is the left-hand side and $u'(x) = v(x)$.

Remark   The last two results can be interpreted in terms of interchanging operations. They say, under suitable hypotheses, that

$$\int_a^b \lim_{n \to \infty} = \lim_{n \to \infty} \int_a^b \qquad \text{and} \qquad \frac{d}{dx} \lim_{n \to \infty} = \lim_{n \to \infty} \frac{d}{dx}.$$

Note the essential role played by uniform convergence.

Infinite Series   It is a routine matter to translate, via partial sums, statements about limits of sequences into statements about infinite series. We shall simply state the analogues for series of the three previous results.

Series of Functions

(a)   Let $\{u_n(x)\}$ be a sequence of continuous functions on $a \leq x \leq b$, and let

$$u(x) = \sum_{n=1}^{\infty} u_n(x) \qquad \text{uniformly on} \quad a \leq x \leq b.$$

Then $u(x)$ is continuous on $a \leq x \leq b$ and

$$\int_a^b u(x)\, dx = \sum_{n=1}^{\infty} \int_a^b u_n(x)\, dx.$$

(b)   Let $\{u_n(x)\}$ be a sequence of continuously differentiable functions on $a \leq x \leq b$, let

$$u(x) = \sum_{n=1}^{\infty} u_n(x) \qquad \text{for each } x \text{ on} \quad a \leq x \leq b,$$

and let $v(x) = \displaystyle\sum_{n=1}^{\infty} u_n'(x) \qquad$ uniformly on $a \leq x \leq b$.

Then $u(x)$ is differentiable on $a \leq x \leq b$ and

$$\frac{du(x)}{dx} = v(x) = \sum_{n=1}^{\infty} \frac{du_n(x)}{dx}.$$

The M-test In applying these results, the first step is always proving the uniform convergence of some series. The following **Weierstrass M-test** for uniform convergence is often adequate.

M**-test** Suppose $\{u_n(x)\}$ is a sequence of functions on the domain **D** and $\{M_n\}$ is a sequence of constants such that

(a) $\sum M_n$ converges

(b) $|u_n(x)| \le M_n$ for all x in **D** and all n.

Then $\sum u_n(x)$ converges uniformly on **D**.

Proof Let $\varepsilon > 0$. By the Cauchy Test (p. 543) there is an N such that

$$M_{n+1} + M_{n+2} + \cdots + M_m < \tfrac{1}{2}\varepsilon$$

whenever $N \le n < m$. Therefore

$$\left| \sum_{k=n+1}^{m} u_k(x) \right| \le \sum_{k=n+1}^{m} |u_k(x)| \le \sum_{k=n+1}^{m} M_k < \tfrac{1}{2}\varepsilon$$

for all x in **D** and for all m and n such that $m > n \ge N$.

By the Cauchy Test again, for each x in **D**, the series $\sum u_n(x)$ converges to a number $u(x)$. Because of this convergence we may let $m \longrightarrow \infty$ in the last displayed inequality:

$$\left| \sum_{k=n+1}^{\infty} u_k(x) \right| \le \tfrac{1}{2}\varepsilon < \varepsilon, \qquad \text{that is,} \quad \left| u(x) - \sum_{k=1}^{n} u_k(x) \right| < \varepsilon$$

for all $n \ge N$ and all x in **D**. Therefore $u(x) = \sum u_n(x)$ uniformly on **D**.

Power Series We are almost ready to prove that a power series $\sum a_n(x - c)^n$ can be differentiated and integrated term-by-term. We shall simplify the notation by assuming $c = 0$; it is a routine matter to pass from the case $c = 0$ to a general c. First we use the M-test to establish the necessary uniform convergence.

Lemma 2 Suppose $\sum a_n x^n$ has radius of convergence $R > 0$. Let $0 < R_1 < R$. Then

$$\sum a_n x^n \qquad \text{and} \qquad \sum n a_n x^{n-1}$$

converge uniformly on $|x| \le R_1$.

Proof By Lemma 1, $\sum |a_n R_1{}^n|$ converges. If $|x| \le R_1$, then

$$|a_n x^n| \le M_n = |a_n R_1{}^n|,$$

hence $\sum a_n x^n$ converges uniformly by the M-test.

The second assertion looks and is a little harder to prove. That extra factor n might make $na_n x^{n-1}$ grow out of control. Not so, because we can compensate for it by increasing R_1 slightly and using the natural tendency of geometric growth to overwhelm arithmetic growth.

Precisely, we choose, R_2 so $R_1 < R_2 < R$. Then $\sum |a_n R_2{}^n|$ converges by Lemma 1. This implies $\sum |a_n R_2{}^{n-1}|$ converges. Now let $|x| \le R_1$. Then

$$|na_n x^{n-1}| \le M_n = |na_n R_1{}^{n-1}| = \left[n \left(\frac{R_1}{R_2} \right)^{n-1} \right] |a_n R_2{}^{n-1}|.$$

But $0 < R_1/R_2 < 1$ so $n(R_1/R_2)^{n-1} \longrightarrow 0$ (by Lhospital's Rule for instance). Therefore $\sum M_n$ converges, so $\sum na_n x^{n-1}$ converges uniformly by the M-test.

Term-by-Term Operations on Power Series   Suppose $f(x) = \sum a_n x^n$ has radius of convergence $R > 0$. Then

(a) $f(x)$ is continuously differentiable on $|x| < R$, and

$$f'(x) = \sum na_n x^{n-1};$$

(b) $\displaystyle\int_0^x f(t)\, dt = \sum \frac{a_n}{n+1} x^{n+1} \qquad \text{for} \quad |x| < R.$

Proof If $|x| < R$, fix R_1 so $|x| < R_1 < R$. By Lemma 2, $\sum a_n x^n$ and $\sum na_n x^{n-1}$ converge uniformly on $|x| \le R_1$, so

$$f'(x) = \sum na_n x^{n-1} \qquad \text{and} \qquad \int_0^x f(t)\, dt = \sum \frac{a_n}{n+1} x^{n+1}$$

by the results on series of functions, p. 622.

Remark 1   (a) can be applied to f', then to f'', etc. Thus f can be differentiated repeatedly, and

$$\frac{d^k f}{dx^k} = \sum_{n=k}^{\infty} n(n-1) \cdots (n-k+1) x^{n-k} \qquad \text{for} \quad |x| < R.$$

Remark 2   We have carefully avoided the end points of the interval of convergence. If $f(x) = \sum a_n x^n$ has radius of convergence R, we do not know what happens at $x = R$ and $x = -R$. But *suppose* for instance that $\sum a_n R^n$ converges. Then it can be proved that $\sum a_n x^n$ converges *uniformly* on $-R < x \le R$, and hence that $f(x)$ is continuous and can be integrated on $0 \le x \le R$. This result is fairly deep, and its proof is beyond our scope. It is needed to prove statements like

$$\ln 2 = \int_0^1 \frac{dx}{1 + x} = 1 - \frac{1}{2} + \frac{1}{3} - \frac{1}{4} + \cdots,$$

$$\frac{\pi}{4} = \arctan 1 = \int_0^1 \frac{dx}{1 + x^2} = 1 - \frac{1}{3} + \frac{1}{5} - \frac{1}{7} + \cdots.$$

EXERCISES

1 Let $f_n(x) =$

$$nx \quad \text{for} \quad 0 \le x \le \frac{1}{n}, \qquad 2 - nx \quad \text{for} \quad \frac{1}{n} \le x \le \frac{2}{n}, \qquad 0 \quad \text{for} \quad \frac{2}{n} \le x.$$

Prove that $f_n(x) \longrightarrow 0$ on $0 \le x < \infty$, but not uniformly.

2 Let $f_n(x) = x^n$. Show that $\{f_n(x)\}$ converges uniformly on $0 \le x \le \frac{9}{10}$, but not uniformly on $0 \le x \le 1$.

3 Prove that $xe^{-nx} \longrightarrow 0$ uniformly on $0 \le x < \infty$. [*Hint* Find max xe^{-nx}.]

4 Determine whether or not $x^2 e^{-nx} \longrightarrow 0$ uniformly on $0 \le x < \infty$.

5 Prove that $\sum_1^\infty (\sin nx)/n^2$ is continuous for $-\infty < x < \infty$.

6 Prove that $\sum_1^\infty 1/(1 + x^n)$ is continuous for $x > 1$.

7 Prove that $f(x) = \sum_1^\infty e^{-nx} \sin nx$ is continuous for $x > 0$.

8 (cont.) Justify the formula $\int_1^2 f(x)\, dx = \sum_{n=1}^\infty \int_1^2 e^{-nx} \sin nx\, dx$.

9 Justify $\dfrac{d}{dx} \displaystyle\sum_{n=1}^\infty \dfrac{\sin nx}{n^3} = \displaystyle\sum_{n=1}^\infty \dfrac{\cos nx}{n^2}$.

10 Find an example where $f_n(x) \longrightarrow f(x)$ uniformly on $a \le x \le b$ and $\lim_{n\to\infty} f_n'(x) \ne f'(x)$.

11 If $\sum a_n$ is absolutely convergent, prove $\sum a_n \sin nx$ is uniformly convergent.

12* Suppose that in the "Derivative of the Limit" box on p. 621 we replace the assumption that $u_n(x) \longrightarrow u(x)$ for *all* $a \le x \le b$ by the assumption that $u_n(c) \longrightarrow u(c)$ for *some* c such $a \le c \le b$. Prove that $u_n(x) \longrightarrow u(x)$ for all $a \le x \le b$ follows anyhow.

Expand in a power series at $x = 0$

13 $\displaystyle\int_0^1 e^{xt^2}\, dt$

14 $J_0(x) = \dfrac{1}{\pi} \displaystyle\int_0^\pi \cos(x \cos \theta)\, d\theta.$

The next four exercises show how one can start "from scratch" with power series and derive the main properties of the exponential function. Set $E(x) = \sum_0^\infty x^n/n!$.

15 Show that $E(x)$ is continuously differentiable for all x and $E'(x) = E(x)$. Also $E(0) = 1$.

16 Prove $E(x)E(-x) = 1$, hence $E(x) > 0$ for all x. [*Hint* Differentiate.]

17 Prove $E(c + x)E(-x) = E(c)$.

18 Prove $E(c + x) = E(c)E(x)$.

The next 10 exercises show how properties of the trigonometric functions can be derived "from scratch" using power series. Define $S(x) = \sum_1^\infty (-1)^{n-1} x^{2n-1}/(2n - 1)!$ and $C(x) = \sum_0^\infty (-1)^n x^{2n}/(2n)!$.

19 Prove that $S(x)$ and $C(x)$ are continuously differentiable for all x, that $S'(x) = C(x)$, and that $C'(x) = -S(x)$.

20 Prove that $S(0) = 0$, $C(0) = 1$, $S(x)$ is odd, and $C(x)$ is even.

21 Prove that $S^2(x) + C^2(x) = 1$. Conclude that $|S(x)| \leq 1$ and $|C(x)| \leq 1$. [*Hint* Use Ex. 19.]

22 Prove $S(a + x) = S(a)C(x) + C(a)S(x)$. [*Hint* Expand $S(a + x)$.]

23 Deduce from Ex. 22 that $C(a + x) = C(a)C(x) - S(a)S(x)$.

24 Use $C(x) = 1 - \frac{1}{2}x^2 + r_2(x)$ to prove $C(x) > 0$ for $0 \leq x \leq 1$ and $C(2) < 0$. From these results the first positive zero of $C(x)$ is a number $\frac{1}{2}\pi$ such that $2 < \pi < 4$.

25 Prove $S(\frac{1}{2}\pi) = 1$.

26 Prove $S(\pi) = 0$ and $C(\pi) = -1$.

27 Prove $S(x + \pi) = -S(x)$ and $C(x + \pi) = -C(x)$.

28 Finally, prove $S(x)$ and $C(x)$ are periodic of period 2π.

8. MISCELLANEOUS EXERCISES

Expand in a Taylor series at $x = 0$

1 $x \cos x - \sin x$

2 $[\ln(1 - x)]/x$

3 $\displaystyle\int_0^x \frac{\sin t}{t}\, dt$

4 $\displaystyle\int_0^x \frac{1 - \cos t}{t^2}\, dt$

5 $\displaystyle\int_0^x \frac{t}{1 + t^4}\, dt$

6 $\displaystyle\int_0^x \sin t^2\, dt$.

Sum the series

7 $\dfrac{1}{4} + \dfrac{3}{16} + \dfrac{5}{64} + \dfrac{7}{256} + \cdots$

8 $\dfrac{1}{4} + \dfrac{5}{64} + \dfrac{9}{1024} + \dfrac{13}{16384} + \cdots$

9 $x + \dfrac{x^5}{5!} + \dfrac{x^9}{9!} + \dfrac{x^{13}}{13!} + \cdots$

10 $2^2 x^2 + 4^2 x^4 + 6^2 x^6 + 8^2 x^8 + \cdots$

11 $1 + \dfrac{2}{3} + \dfrac{3}{9} + \dfrac{4}{27} + \cdots$

12 $1 + \dfrac{1}{300} + \dfrac{1}{50,000} + \dfrac{1}{7,000,000} + \cdots$.

13 How many terms of $\frac{1}{4}\pi = 1 - \frac{1}{3} + \frac{1}{5} - \frac{1}{7} + \cdots$ are needed to estimate π to 4 decimal places?

14 (cont.) Use the formula $\frac{1}{4}\pi = 4 \arctan \frac{1}{5} - \arctan \frac{1}{239}$ and the series for $\arctan x$ to estimate π to 4 places.

15 Solve $(x^2 - 1)y'' = 6y$, $y(0) = a$, $y'(0) = b$.

16 Solve $(1 - x^2)y'' - xy' + 49y = 0$, $y(0) = 0$, $y'(0) = 1$.

Evaluate

17 $\dfrac{d^{20}}{dx^{20}} (x^6 \cos 2x)\bigg|_{x=0}$

18 $\dfrac{d^5}{dx^5} (e^x \sin x)\bigg|_{x=0}$.

19 Is this game fair? We roll a die. If a 6 turns up, I pay you \$5 and the game ends; otherwise you pay me \$1 and we play again for \$10 or \$2, then for \$15 or \$3, etc. until a 6 appears.

20 Let a_n be the number of ways that a sum of n cents can be made using only pennies, nickels, and dimes. Prove that

$$1 + \sum_{n=1}^{\infty} a_n x^n = \frac{1}{(1 - x)(1 - x^5)(1 - x^{10})}.$$

21 Let p_n be the probability that in n tosses of a coin, two consecutive heads do not appear. Show that $p_1 = 1$, $p_2 = \frac{3}{4}$, and $p_{n+2} = \frac{1}{2}p_{n+1} + \frac{1}{4}p_n$. [*Hint* Any such sequence of tosses must start with either T or HT.]

22 (cont.) Find the generating function $\sum_1^\infty p_n x^n$ and derive an exact formula for p_n.

23 Explain why the approximation $\dfrac{\sin x}{x} \approx \sqrt{1 - \frac{1}{3}x^2}$ is accurate for $|x|$ small.

24 Show for each integer $p \geq 1$ that $\sum_1^\infty n^p/n!$ is an integer times e.

Space Geometry and Vectors **13**

1. RECTANGULAR COORDINATES

In this chapter we shall present the analytic geometry of three-dimensional euclidean space, denoted $\mathbf{R}^3$. This subject is traditionally called **solid analytic geometry**; our treatment will emphasize vector algebra. Three-dimensional geometry is important and interesting in its own right, and it is the key to our future study of geometric applications of calculus and of the calculus of functions of two or more variables.

Space Coordinates Plane analytic geometry begins with the introduction of two perpendicular coordinate axes in the Euclidean plane $\mathbf{R}^2$. One is directed (oriented) and called the x-axis. Then the other, called the y-axis, is directed so that the pair in the order x, y is a *right-handed* system, that is, so a positive (counter-clockwise) rotation through $90°$ from the positive x-axis brings us to the positive y-axis (Fig. 1a).

Having this in mind, let us pass to space. Solid analytic geometry begins with the introduction of three mutually perpendicular coordinate axes in $\mathbf{R}^3$. They must pass through a fixed point **0** called the **origin**. Any two of these axes are chosen, directed, and labeled the x-axis and the y-axis respectively (Fig. 1b). The third axis is labeled the z-axis, and its direction is determined by the **right-hand rule** (Fig. 1c):

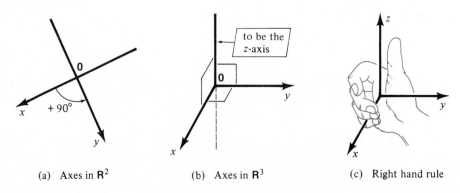

(a) Axes in $\mathbf{R}^2$ (b) Axes in $\mathbf{R}^3$ (c) Right hand rule

Fig. 1 Placing coordinate axes

when a right hand curls around the z-axis with its fingers going from the positive x-axis to the positive y-axis, its thumb points in the positive z-direction.

Since many people have trouble at first "seeing" space figures from plane drawings, a few more words and figures are in order. In Fig. 2a, drawn in perspective, we are facing the far wall of a rectangular room, and the origin of the coordinate system is the far left corner of the floor. The x-axis is chosen running toward us at the intersection of the floor and left side wall. The y-axis runs along the back of the floor to the right. Thus the z-axis runs straight up or straight down; by the right-hand rule it is *up*.

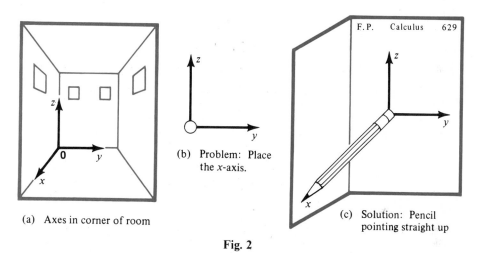

(a) Axes in corner of room

(b) Problem: Place the x-axis.

(c) Solution: Pencil pointing straight up

Fig. 2

In Fig. 2b, the right-hand page of the open book is flat on a desk. The y, z-axes are drawn on the page in the usual position. The problem is to place a pointed pencil representing the x-axis, eraser end at **0**. The solution, in agreement with the right-hand rule, is shown in Fig. 2c.

The three axes determine three **coordinate planes**. For instance, the plane of the y- and z-axes is called the y, z-**plane**, etc. (Fig. 3a).

Now take any point **p** in space. Through **p** pass planes parallel to the three coordinate planes. Their intersections with the coordinate axes determine three numbers x, y, z, called the **coordinates** of **p**. See Fig. 3b. Conversely, each triple (x, y, z) of real numbers determines a unique point **p** in space. We shall write

$$\mathbf{p} = (x, y, z).$$

A point (x, y, z) is located by marking its projection $(x, y, 0)$ in the x, y-plane and going up or down the corresponding amount z. (From the habit of living in the x, y-plane for so long, we think of the z-direction as "up".) See Fig. 3c for examples.

The portion of space where x, y, and z are positive is called the first **octant**. (No one numbers the other seven octants.) In our figures so far, we have projected into the y, z-plane, that is, we have taken the y, z-plane in the plane of the page. Then the angle at which we draw the x-axis is arbitrary. We try to choose it so our drawing is as uncluttered as possible. We can just as well project into one of the

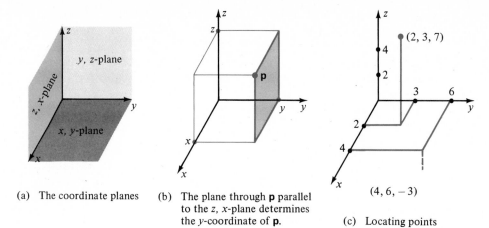

(a) The coordinate planes   (b) The plane through **p** parallel
to the *z, x*-plane determines
the *y*-coordinate of **p**.

(c) Locating points

Fig. 3

other two coordinate planes, or into some other plane altogether. But no matter
how we set up our axes, points are located in basically the same way (Fig. 4).
A little care is needed to make sure the drawing is right-handed.

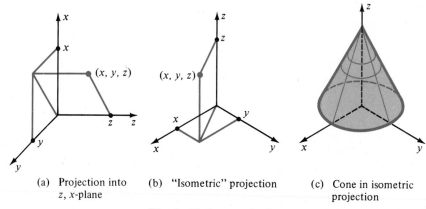

(a) Projection into   (b) "Isometric" projection   (c) Cone in isometric
z, x-plane   projection

Fig. 4 Various projections

EXERCISES

In Fig. 2a, draw the coordinate axes from the given origin along edges of the room

1 origin rear lower left, *y*-axis forward
2 origin rear lower right, *x*-axis to left
3 origin rear upper right, *z*-axis drawn
4 origin rear upper left, *y*-axis down
5 origin front lower left, *x*-axis to right
6 origin front upper left, *y*-axis down.

Locate the points in Fig. 5

7 (1, 2, 3), (1, 3, 4)   **8** (2, 4, 3), (2, −3, 3)
9 (1, −2, 1), (2, −3, −1)   **10** (1, −3, −2), (3, 2, −2).

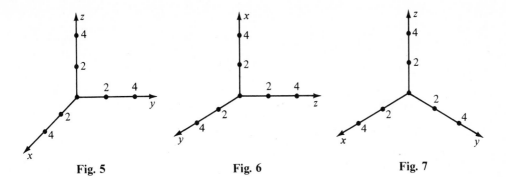

Fig. 5   **Fig. 6**   **Fig. 7**

Locate the points in Fig. 6

11 $(3, 4, -1)$, $(-3, -3, 1)$ **12** $(0, 0, -3)$, $(-2, -2, 3)$
13 $(3, -2, 2)$ $(-2, 4, 4)$ **14** $(0, -3, 2)$, $(3, 3, 3)$.

Locate the points in Fig. 7

14 $(1, 2, 3)$, $(0, 1, 4)$ **16** $(3, 4, 2)$, $(2, -3, 3)$
17 $(3, -2, 3)$, $(4, 4, 4)$ **18** $(2, 4, -1)$, $(-3, 2, 4)$.

2. VECTOR ALGEBRA

We now introduce the concept of a vector. Vectors are useful for handling problems in space because (1) equations in vector form are independent of coice of coordinate axes, hence are well suited to describe physical situations, (2) each vector equation replaces three ordinary equations, and (3) several frequently occurring procedures can be summarized neatly in vector form.

Let the origin **0** be fixed once and for all. A **vector** in space is a directed line segment that begins at **0**; it is completely determined by its terminal point. We shall denote vectors by bold-faced letters **x**, **v**, **F**, **r**, etc. (In written work use $\underline{x}$ or $\bar{x}$.) A point (x, y, z) in space is often identified with the vector **x** from the origin to the point. The **zero vector** (origin) will be written **0** $= (0, 0, 0)$. For this vector only, direction is undefined. When we write **x** $= (x, y, z)$, the numbers x, y, and z are called the **components**, or **coordinates** of **x**.

Notation   We shall frequently call the three axes the x_1-axis, the x_2-axis, and the x_3-axis instead of the x-, y-, and z-axes. This has the big advantage that as soon as we give a name to a vector, we automatically have names for its components. Thus

$$\mathbf{x} = (x_1, x_2, x_3), \qquad \mathbf{v} = (v_1, v_2, v_3), \qquad \mathbf{a} = (a_1, a_2, a_3), \qquad \text{etc.}$$

A vector is determined by two quantities, *length* (or *magnitude*) and *direction*. Many physical quantities are vectors: force, velocity, acceleration, electric field intensity, etc.

Remember that the origin **0** is fixed, and that each vector starts at **0**. We often draw vectors starting at other points, but in computations they all originate at **0**. For example, if a force **F** is applied at a point **x**, we may draw Fig. 1a because it is suggestive. But the correct figure is Fig. 1b. One must specify both the force vector **F** (magnitude and direction) and its point of application **x**.

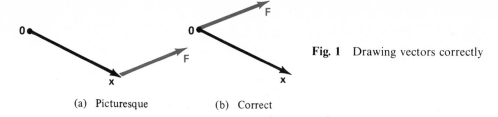

Fig. 1   Drawing vectors correctly

(a) Picturesque   (b) Correct

Addition of Vectors   The **sum u + v** of two vectors is defined by the parallelo-gram law (Fig. 2). The points **0, u, v, u + v** are the vertices of a parallelogram, with **u + v** opposite to **0**.

Vectors are added numerically by adding their components:

$$(u_1, u_2, u_3) + (v_1, v_2, v_3) = (u_1 + v_1, u_2 + v_2, u_3 + v_3).$$

For example.

$$(-1, 3, 2) + (1, 1, 4) = (0, 4, 6), \qquad (0, 0, 1) + (-1, 0, 1) = (-1, 0, 2).$$

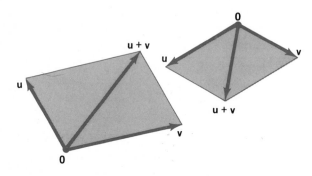

Fig. 2   Parallelogram law for vector addition

Let us prove that the sum of vectors, defined *geometrically* by the parallelogram law, can be computed *algebraically* by adding corresponding components. We pass planes P, Q, R through **u**, **v**, and **w** = **u** + **v** parallel to the x_3, x_1-plane (Fig. 3).

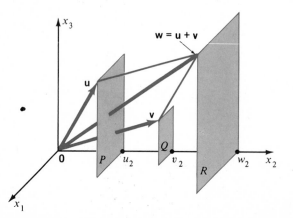

Fig. 3   Proof that vector addition is componentwise

They meet the x_2-axis at u_2, v_2, and w_2. Because $\overline{\mathbf{vw}}$ and $\overline{\mathbf{Ou}}$ are parallel, the directed distance from Q to R equals the directed distance from the x_3, x_1-plane to P. Hence $w_2 - v_2 = u_2$, that is, $w_2 = u_2 + v_2$. Similarly, $w_1 = u_1 + v_1$ and $w_3 = u_3 + v_3$.

Multiplication by Scalars Let $\mathbf{v}$ be a vector and let a be a number (scalar). We define the product $a\mathbf{v}$ to be the vector whose length is $|a|$ times the length of $\mathbf{v}$ and which points in the same direction as $\mathbf{v}$ if $a > 0$, in the opposite direction if $a < 0$. If $a = 0$, then $a\mathbf{v} = \mathbf{0}$.

There is a simple physical idea behind this definition. If a particle moving in a certain direction doubles its speed, its velocity vector is doubled; if a horse pulling a cart in a certain direction triples its effort, the force vector triples. Figure 4 illustrates multiples of a vector.

Scalar multiples are computed in components by the following rule.

$$a(v_1, v_2, v_3) = (av_1, av_2, av_3).$$

This rule is proved by similar triangles (Fig. 5). The triangle $\mathbf{0}v_2\mathbf{v}$ is similar to $\mathbf{0}w_2\mathbf{w}$, hence $w_2 = av_2$, etc.

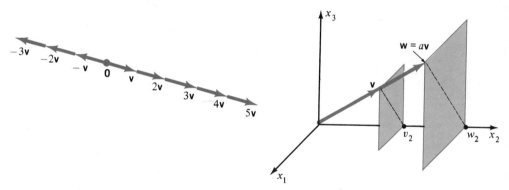

Fig. 4 Scalar multiples

Fig. 5 Proof that scalar multiplication
is componentwise

The difference $\mathbf{v} - \mathbf{w}$ of two vectors is defined by

$$\mathbf{v} - \mathbf{w} = \mathbf{v} + (-\mathbf{w}).$$

See Fig. 6. (The vector $-\mathbf{w}$ has the same length as $\mathbf{w}$ but points in the opposite direction.)

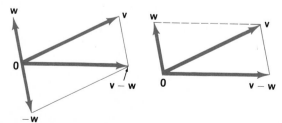

(a)   (b)

Fig. 6 Difference of vectors

The segment from the tip of **w** to the tip of **v** (the dashed line in Fig. 6b) has the same length and direction as **v** − **w**. Hence if two points are represented by vectors **v** and **w**, the distance between them is the length of **v** − **w**.

Rules of Vector Algebra   The basic rules of vector algebra follow directly from the coordinate formulas for addition and multiplication by a scalar.

Rules of Vector Algebra

$$\mathbf{v} + \mathbf{0} = \mathbf{0} + \mathbf{v} = \mathbf{v} \qquad\qquad \mathbf{v} + (-\mathbf{v}) = (-\mathbf{v}) + \mathbf{v} = \mathbf{0}$$

$$\mathbf{u} + \mathbf{v} = \mathbf{v} + \mathbf{u} \qquad\qquad \mathbf{u} + (\mathbf{v} + \mathbf{w}) = (\mathbf{u} + \mathbf{v}) + \mathbf{w}$$

$$0\mathbf{v} = \mathbf{0} \qquad 1\mathbf{v} = \mathbf{v} \qquad\qquad a(b\mathbf{v}) = (ab)\mathbf{v}$$

$$(a + b)\mathbf{v} = a\mathbf{v} + b\mathbf{v} \qquad\qquad a(\mathbf{v} + \mathbf{w}) = a\mathbf{v} + a\mathbf{w}$$

The Midpoint Formula   As an example of the convenience of vector algebra, let us find the midpoint **m** of a segment $\overline{\mathbf{uv}}$. It is clear in Fig. 7 that **m** is the midpoint of the segment from **0** to **u** + **v** (because the diagonals of a parallelogram bisect each other), hence $\mathbf{m} = \frac{1}{2}(\mathbf{u} + \mathbf{v})$.

Midpoint Formula   If **u** and **v** are any two points in $\mathbf{R}^3$, then their midpoint is

$$\mathbf{m} = \tfrac{1}{2}(\mathbf{u} + \mathbf{v}) = (\tfrac{1}{2}(u_1 + v_1), \tfrac{1}{2}(u_2 + v_2), \tfrac{1}{2}(u_3 + v_3)).$$

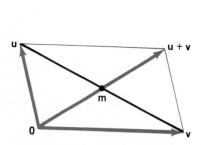

Fig. 7   The midpoint of $\overline{\mathbf{uv}}$ is

$$\mathbf{m} = \tfrac{1}{2}(\mathbf{u} + \mathbf{v}).$$

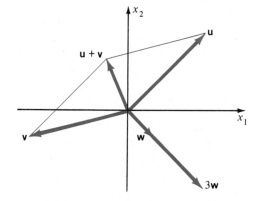

Fig. 8   Vector operations in the plane

Vectors in the Plane   The Euclidean plane $\mathbf{R}^2$ can be considered as part of $\mathbf{R}^3$, for example as the x_1, x_2-plane. By itself, this plane consists of all vectors $\mathbf{u} = (u_1, u_2)$. But as part of $\mathbf{R}^3$, it consists of all vectors $\mathbf{u} = (u_1, u_2, 0)$.

Vector addition and multiplication by scalars are defined in the plane (Fig. 8) just as in space. In coordinates,

$$(u_1, u_2) + (v_1, v_2) = (u_1 + u_2, v_1 + v_2), \qquad a(u_1, u_2) = (au_1, au_2).$$

EXERCISES

Compute

1	$(1, 2, -3) + (4, 0, 7)$	**2**	$(-1, -1, 0) + (3, 5, 2)$
3	$(4, 0, 7) - (1, 2, -3)$	**4**	$(2, 1, 1) - (3, -1, -2)$
5	$(1, 2, 3) - 6(0, 3, -1)$	**6**	$4[(1, -2, -7) - (1, 1, 1)]$
7	$3(1, 4, 2) - 2(2, 1, 1)$	**8**	$4(1, -1, 2) - 3(1, -1, 2).$
9	$3(1, 1, 0) - 2(0, 1, 1) + (1, 0, 1)$	**10**	$-5[3(1, 1, 1) - (2, 1, 4)] + 4(-2, -2, 1).$

Prove

11	$\mathbf{u} + \mathbf{v} = \mathbf{v} + \mathbf{u}$	**12**	$\mathbf{u} + (\mathbf{v} + \mathbf{w}) = (\mathbf{u} + \mathbf{v}) + \mathbf{w}$
13	$(a + b)\mathbf{u} = a\mathbf{u} + b\mathbf{u}$	**14**	$a(\mathbf{v} + \mathbf{w}) = a\mathbf{v} + a\mathbf{w}$
15	$(ab)\mathbf{u} = a(b\mathbf{u}) = b(a\mathbf{u}).$		

16   Show that the segments joining the midpoints of opposite sides of a (skew) quadrilateral bisect each other.

17   (cont.) Find the point of intersection of these segments when the vertices of the quadrilateral in order are

$$\mathbf{v}_1 = \mathbf{0}, \qquad \mathbf{v}_2 = (1, 0, 0), \qquad \mathbf{v}_3 = (0, 1, 0), \qquad \mathbf{v}_4 = (0, 0, 1).$$

18   Find the intersection of the medians of the triangle with vertices $\mathbf{a}, \mathbf{b}, \mathbf{c}$.

19   (cont.) In a tetrahedron, prove that the four lines joining each vertex to the centroid (intersection of the medians) of the opposite face are concurrent.

20*   Space billiards—no gravity. An astronaut cues a ball toward the corner of a rectangular room, with velocity $\mathbf{v}$. The ball misses the corner, but rebounds off of each of the three adjacent walls. Find its returning velocity vector.

3. LENGTH AND INNER PRODUCT

The **length** of a vector $\mathbf{v}$, denoted $|\mathbf{v}|$, is the distance of its terminal point from $\mathbf{0}$. From Fig. 1a, we may regard $|\mathbf{v}|$ as the length of a diagonal of a rectangular solid. From the figure:

> **Length Formula** $|\mathbf{v}| = \sqrt{v_1^2 + v_2^2 + v_3^2}.$

The following properties of length are clear geometrically:

> **Properties of Length** $|\mathbf{0}| = 0, \qquad |\mathbf{v}| > 0 \quad \text{if} \quad \mathbf{v} \neq \mathbf{0},$
>
> $$|a\mathbf{v}| = |a| \cdot |\mathbf{v}|,$$
>
> triangle inequality: $|\mathbf{v} + \mathbf{w}| \leq |\mathbf{v}| + |\mathbf{w}|.$

The distance between two points can be expressed in terms of vector length. In fact, by Fig. 1b, the distance between the points $\mathbf{v}$ and $\mathbf{w}$ equals the length of the vector $\mathbf{v} - \mathbf{w}$.

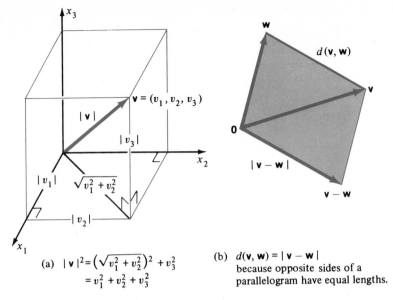

(a) $|\mathbf{v}|^2 = \left(\sqrt{v_1^2 + v_2^2}\right)^2 + v_3^2$
$= v_1^2 + v_2^2 + v_3^2$

(b) $d(\mathbf{v}, \mathbf{w}) = |\mathbf{v} - \mathbf{w}|$
because opposite sides of a
parallelogram have equal lengths.

Fig. 1 Length and distance

Distance Formula The distance between two points $\mathbf{v} = (v_1, v_2, v_3)$ and $\mathbf{w} = (w_1, w_2, w_3)$ is

$$d(\mathbf{v}, \mathbf{w}) = |\mathbf{v} - \mathbf{w}| = \sqrt{(v_1 - w_1)^2 + (v_2 - w_2)^2 + (v_3 - w_3)^2}.$$

Examples (1) The distance from $(1, 2, 4)$ to $(3, 0, -1)$ is

$$|(1, 2, 4) - (3, 0, -1)| = \sqrt{(1-3)^2 + (2-0)^2 + (4+1)^2} = \sqrt{4 + 4 + 25} = \sqrt{33}.$$

(2) The set of all points (v_1, v_2, v_3) satisfying

$$v_1{}^2 + v_2{}^2 + v_3{}^2 = 1$$

is the sphere with center $\mathbf{0}$ and radius 1.

(3) A vector equation for the sphere with center $\mathbf{x}_0$ and radius r is

$$|\mathbf{x} - \mathbf{x}_0| = r.$$

Inner Product Another important vector operation is the **inner product** of two vectors, also called the **dot product**.

Inner Product Let $\mathbf{v}$ and $\mathbf{w}$ be vectors and θ the angle between them. Their **inner product** is

$$\mathbf{v} \cdot \mathbf{w} = |\mathbf{v}| \cdot |\mathbf{w}| \cos \theta$$

(If $\mathbf{v} = \mathbf{0}$ or $\mathbf{w} = \mathbf{0}$, we define $\mathbf{v} \cdot \mathbf{w} = 0$.)

See Fig. 2a. The angle θ between the vectors can be measured either from **v** to **w** or from **w** to **v**. We shall always take $0 \leq \theta \leq \pi$. We see from Fig. 2b that $|\mathbf{w}| \cos \theta$ is the (signed) projection of **w** on **v**, hence $\mathbf{v} \cdot \mathbf{w}$ is $|\mathbf{v}|$ times the projection of **w** on **v**.

Warning The inner product of two vectors is *not* a vector; it is a *scalar* (number).

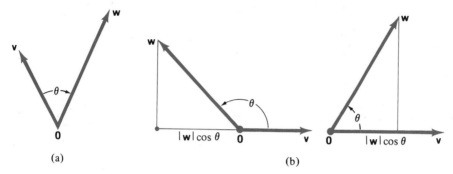

(a)   (b)

Fig. 2 Inner product

To work with inner products effectively, we need a formula for computing $\mathbf{v} \cdot \mathbf{w}$ in terms of the components of **v** and **w**. Fortunately there is a remarkably simple rule:

Formula for Inner Products If $\mathbf{v} = (v_1, v_2, v_3)$ and $\mathbf{w} = (w_1, w_2, w_3)$, then

$$\mathbf{v} \cdot \mathbf{w} = v_1 w_1 + v_2 w_2 + v_3 w_3.$$

Examples $(5, -1, 3) \cdot (2, 7, 2) = 5 \cdot 2 + (-1) \cdot 7 + 3 \cdot 2 = 9,$

$(3, 4, 0) \cdot (1, -5, 6) = 3 \cdot 1 + 4(-5) + 0 \cdot 6 = -17.$

Proof of the formula By Fig. 3 and the Law of Cosines

$$|\mathbf{v} - \mathbf{w}|^2 = |\mathbf{v}|^2 + |\mathbf{w}|^2 - 2|\mathbf{v}||\mathbf{w}| \cos \theta = |\mathbf{v}|^2 + |\mathbf{w}|^2 - 2\mathbf{v} \cdot \mathbf{w}.$$

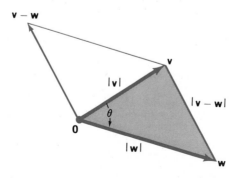

Fig. 3 $|\mathbf{v} - \mathbf{w}|^2 = |\mathbf{v}|^2 + |\mathbf{w}|^2 - 2|\mathbf{v}| \cdot |\mathbf{w}| \cos \theta$

Hence

$$2\mathbf{v} \cdot \mathbf{w} = |\mathbf{v}|^2 + |\mathbf{w}|^2 - |\mathbf{v} - \mathbf{w}|^2$$
$$= |(v_1, v_2, v_3)|^2 + |(w_1, w_2, w_3)|^2 - |(v_1 - w_1, v_2 - w_2, v_3 - w_3)|^2$$
$$= (v_1^2 + v_2^2 + v_3^2) + (w_1^2 + w_2^2 + w_3^2)$$
$$- (v_1 - w_1)^2 - (v_2 - w_2)^2 - (v_3 - w_3)^2$$
$$= 2(v_1 w_1 + v_2 w_2 + v_3 w_3).$$

The main algebraic properties of the inner product follow easily from the formula:

Properties of the Inner Product $\mathbf{v} \cdot \mathbf{w} = \mathbf{w} \cdot \mathbf{v}$

$(a\mathbf{v}) \cdot \mathbf{w} = \mathbf{v} \cdot (a\mathbf{w}) = a(\mathbf{v} \cdot \mathbf{w})$ $(\mathbf{u} + \mathbf{v}) \cdot \mathbf{w} = \mathbf{u} \cdot \mathbf{w} + \mathbf{v} \cdot \mathbf{w}.$

The proofs are left as exercises.

Perpendicular Vectors Two non-zero vectors $\mathbf{v}$ and $\mathbf{w}$ are perpendicular when the angle between them is $\theta = \frac{1}{2}\pi$. Then $\mathbf{v} \cdot \mathbf{w} = |\mathbf{v}| \cdot |\mathbf{w}| \cos \theta = 0$, and conversely. We shall consider the zero vector $\mathbf{0}$ as perpendicular to all vectors.

Perpendicular Vectors   Vectors $\mathbf{v}$ and $\mathbf{w}$ are perpendicular if and only if

$$\mathbf{v} \cdot \mathbf{w} = 0.$$

For example, $(1, 2, 3) \cdot (-1, -1, 1) = -1 - 2 + 3 = 0$, so $(1, 2, 3)$ and $(-1, -1, 1)$ are perpendicular.

Terminology   When speaking of vectors, **orthogonal** is a frequent synonym for perpendicular. Thus $(1, 2, 3)$ and $(-1, -1, 1)$ are *orthogonal* vectors.

Relations Involving Inner Products There are several useful relations between length, inner product, and the angle between vectors. First, length can be expressed in terms of the inner product:

$$|\mathbf{v}|^2 = \mathbf{v} \cdot \mathbf{v} = v_1^2 + v_2^2 + v_3^2.$$

Conversely, the inner product can be expressed in terms of length by means of a short calculation:

$$|\mathbf{v} + \mathbf{w}|^2 = (\mathbf{v} + \mathbf{w}) \cdot (\mathbf{v} + \mathbf{w}) = |\mathbf{v}|^2 + 2\mathbf{v} \cdot \mathbf{w} + |\mathbf{w}|^2,$$
$$|\mathbf{v} - \mathbf{w}|^2 = (\mathbf{v} - \mathbf{w}) \cdot (\mathbf{v} - \mathbf{w}) = |\mathbf{v}|^2 - 2\mathbf{v} \cdot \mathbf{w} + |\mathbf{w}|^2,$$

therefore $$|\mathbf{v} + \mathbf{w}|^2 - |\mathbf{v} - \mathbf{w}|^2 = 4\mathbf{v} \cdot \mathbf{w}.$$

The angle θ between $\mathbf{v}$ and $\mathbf{w}$ can be found from

$$\mathbf{v} \cdot \mathbf{w} = |\mathbf{v}||\mathbf{w}| \cos \theta$$

by solving for $\cos \theta$.

> **Relations** Let $\mathbf{v}$ and $\mathbf{w}$ be two vectors and let θ be the angle between them. Then
>
> $$|\mathbf{v}|^2 = \mathbf{v} \cdot \mathbf{v} = v_1^2 + v_2^2 + v_3^2, \qquad \mathbf{v} \cdot \mathbf{w} = \tfrac{1}{4}(|\mathbf{v} + \mathbf{w}|^2 - |\mathbf{v} - \mathbf{w}|^2),$$
>
> $$\cos \theta = \frac{\mathbf{v} \cdot \mathbf{w}}{|\mathbf{v}||\mathbf{w}|} \qquad (\mathbf{v} \neq \mathbf{0}, \quad \mathbf{w} \neq \mathbf{0}).$$

■ **EXAMPLE 1** Find the angle between $\mathbf{v} = (1, 2, 1)$ and $\mathbf{w} = (3, -1, 1)$.

Solution
$$\mathbf{v} \cdot \mathbf{w} = 3 - 2 + 1 = 2,$$
$$|\mathbf{v}|^2 = 1 + 4 + 1 = 6, \qquad |\mathbf{w}|^2 = 9 + 1 + 1 = 11.$$

Hence
$$\cos \theta = \frac{\mathbf{v} \cdot \mathbf{w}}{|\mathbf{v}||\mathbf{w}|} = \frac{2}{\sqrt{6}\sqrt{11}}, \qquad \theta = \arccos\left(\frac{2}{\sqrt{66}}\right). \qquad ■$$

■ **EXAMPLE 2** The point $(1, 1, 2)$ is joined to the points $(1, -1, -1)$ and $(3, 0, 4)$ by lines L_1 and L_2. What is the angle θ between these lines?

Solution The vector
$$\mathbf{v} = (1, -1, -1) - (1, 1, 2) = (0, -2, -3)$$
is parallel to L_1 (but starts at $\mathbf{0}$). Likewise
$$\mathbf{w} = (3, 0, 4) - (1, 1, 2) = (2, -1, 2)$$
is parallel to L_2. Hence
$$\cos \theta = \frac{\mathbf{v} \cdot \mathbf{w}}{|\mathbf{v}||\mathbf{w}|} = \frac{0 + 2 - 6}{\sqrt{0 + 4 + 9}\sqrt{4 + 1 + 4}} = \frac{-4}{\sqrt{13}\sqrt{9}},$$
$$\theta = \arccos\left(\frac{-4}{3\sqrt{13}}\right). \qquad ■$$

Note When we find $\cos \theta < 0$ for an angle between two lines, then θ is an angle in the second quadrant, hence θ is not the smaller angle between the lines, but its supplement. The basic fact here is that $\cos(\pi - \theta) = -\cos \theta$.

Direction Cosines The three unit-length vectors along the positive coordinate axes (Fig. 4) are denoted by
$$\mathbf{i} = (1, 0, 0), \qquad \mathbf{j} = (0, 1, 0), \qquad \mathbf{k} = (0, 0, 1).$$
They provide a useful way of expressing general vectors. For if $\mathbf{v}$ is any vector, then
$$\mathbf{v} = (v_1, v_2, v_3) = v_1(1, 0, 0) + v_2(0, 1, 0) + v_2(0, 0, 1) = v_1\mathbf{i} + v_2\mathbf{j} + v_3\mathbf{k}.$$
Thus $\mathbf{v}$ is the sum of three vectors $v_1\mathbf{i}$, $v_2\mathbf{j}$, $v_3\mathbf{k}$ that lie along the three coordinate axes. The components v_1, v_2, v_3 can be interpreted as dot products:
$$\mathbf{v} \cdot \mathbf{i} = (v_1, v_2, v_3) \cdot (1, 0, 0) = v_1.$$
Similarly, $v_2 = \mathbf{v} \cdot \mathbf{j}$ and $v_3 = \mathbf{v} \cdot \mathbf{k}$.

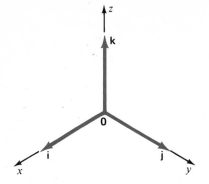

Fig. 4 The basic unit vectors

Each vector $\mathbf{v} = (v_1, v_2, v_3)$ can be expressed as

$$\mathbf{v} = v_1\mathbf{i} + v_2\mathbf{j} + v_3\mathbf{k},$$

where $\quad v_1 = \mathbf{v} \cdot \mathbf{i}, \qquad v_2 = \mathbf{v_2} \cdot \mathbf{j}, \qquad v_3 = \mathbf{v_3} \cdot \mathbf{k}.$

Now suppose $\mathbf{u}$ is any **unit vector**, that is, a vector of length one (Fig. 5a). Let α be the angle from $\mathbf{i}$ to $\mathbf{u}$. Define β and γ similarly. Then $\mathbf{u} \cdot \mathbf{i} = \cos \alpha$, $\mathbf{u} \cdot \mathbf{j} = \cos \beta$, and $\mathbf{u} \cdot \mathbf{k} = \cos \gamma$. Hence

$$\mathbf{u} = (\cos \alpha)\mathbf{i} + (\cos \beta)\mathbf{j} + (\cos \gamma)\mathbf{k} = (\cos \alpha, \cos \beta, \cos \gamma).$$

Since $|\mathbf{u}| = 1,$ $\qquad\qquad \cos^2 \alpha + \cos^2 \beta + \cos^2 \gamma = 1.$

Unit vectors are direction indicators. Any non-zero vector $\mathbf{v}$ is a positive multiple of a unit vector $\mathbf{u}$ in the same direction as $\mathbf{v}$. In fact $\mathbf{v} = |\mathbf{v}|\mathbf{u}$, so

$$\mathbf{u} = \frac{1}{|\mathbf{v}|}\,\mathbf{v} \qquad (\mathbf{v} \neq \mathbf{0}).$$

Direction Cosines Each non-zero vector $\mathbf{v}$ can be expressed as

$$\mathbf{v} = |\mathbf{v}|\mathbf{u}, \quad \mathbf{u} \text{ a unit vector}, \qquad \text{or as} \quad \mathbf{v} = |\mathbf{v}|(\cos \alpha, \cos \beta, \cos \gamma).$$

The numbers $\cos \alpha$, $\cos \beta$, $\cos \gamma$ are called the **direction cosines** of $\mathbf{v}$. They satisfy

$$\cos^2 \alpha + \cos^2 \beta + \cos^2 \gamma = 1.$$

The Plane In $\mathbf{R}^2$ the formulas of this section specialize as follows: if $\mathbf{v} = (v_1, v_2)$ and $\mathbf{w} = (w_1, w_2)$, then

$$|\mathbf{v}|^2 = \mathbf{v} \cdot \mathbf{v} = v_1^2 + v_2^2, \qquad \mathbf{v} \cdot \mathbf{w} = v_1 w_1 + v_2 w_2 = |\mathbf{v}||\mathbf{w}| \cos \theta.$$

Non-zero vectors $\mathbf{v}$ and $\mathbf{w}$ are perpendicular if $\mathbf{v} \cdot \mathbf{w} = 0$, that is, if

$$v_1 w_1 + v_2 w_2 = 0, \qquad \frac{w_2}{w_1} = \frac{-1}{v_2/v_1}$$

(provided the divisions are permissible). This is just the familiar condition that

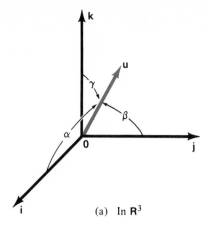

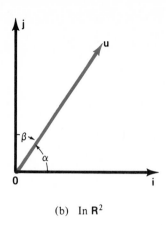

(a) In $\mathbf{R}^3$ (b) In $\mathbf{R}^2$

Fig. 5   Direction cosines

the slopes of perpendicular lines are negative reciprocals of each other. (The slope of a line in the direction of $\mathbf{v} = (v_1, v_2)$ is v_2/v_1.)

The basic unit vectors in $\mathbf{R}^2$ are

$$\mathbf{i} = (1, 0) \quad \text{and} \quad \mathbf{j} = (0, 1).$$

If $\mathbf{u}$ is any unit vector (Fig. 5b), then

$$\mathbf{u} = (\cos \alpha)\mathbf{i} + (\cos \beta)\mathbf{j},$$

and
$$\cos^2 \alpha + \cos^2 \beta = |\mathbf{u}|^2 = 1.$$

By the figure, $\alpha + \beta = \tfrac{1}{2}\pi$, hence $\cos \beta = \sin \alpha$, so

$$\mathbf{u} = (\cos \alpha, \sin \alpha) = (\cos \alpha)\mathbf{i} + (\sin \alpha)\mathbf{j}.$$

The assumption $|\mathbf{u}| = 1$ simply means $\cos^2 \alpha + \sin^2 \alpha = 1$.

EXERCISES

Compute

1 $(8, 2, 1) \cdot (3, 0, 5)$
3 $(1, 0, 2) \cdot [(1, 4, 1) + (2, 0, -3)]$
5 $|3\mathbf{i} - \mathbf{j} + \mathbf{k}|$
7 $|\tfrac{1}{3}\sqrt{3}(-1, 1, 1)|$

2 $(-1, -1, -1) \cdot (1, 2, 3)$
4 $|(2, -4, 7)|$
6 $|(-1, -1, 0) - (3, 5, 2)|$
8 $[3\mathbf{j} - (1, 1, 2)] \cdot (4\mathbf{j} - \mathbf{k})$.

Find the angle between the vectors

9 $(4, 3, 0)$, $(-3, 0, 4)$
11 $(6, 1, 5)$, $(-2, -3, 3)$
13 $(1, 1, -1)$, $(2, 0, 4)$

10 $(1, 2, 2)$, $(-2, 1, -2)$
12 $(-5, 6, 1)$, $(2, 3, -8)$
14 $(2, 2, 2)$, $(-2, 2, -2)$.

Compute the distance between the points

15 $(0, 1, 2)$, $(5, -3, 1)$
17 $(7, 0, 0)$, $(2, 3, 4)$

16 $(1, 1, 1)$, $(1, -1, 2)$
18 $(8, 5, -1)$, $(7, 9, 3)$.

Find the direction cosines

19 $(1, 0, 1)$ **20** $(-1, -1, -1)$ **21** $(2, 1, -3)$ **22** $(4, -7, -4)$.

23   Find two non-collinear vectors perpendicular to $(1, -1, 2)$.

24   Find the angle between the line joining $(0, 0, 0)$ to $(1, 1, 1)$ and the line joining $(1, 0, 0)$ to $(0, 1, 0)$.

Prove

25 $\mathbf{v} \cdot \mathbf{w} = \mathbf{w} \cdot \mathbf{v}$ **26** $(a\mathbf{v}) \cdot \mathbf{w} = \mathbf{v} \cdot (a\mathbf{w}) = a(\mathbf{v} \cdot \mathbf{w})$

27 $(\mathbf{u} + \mathbf{v}) \cdot \mathbf{w} = \mathbf{u} \cdot \mathbf{w} + \mathbf{v} \cdot \mathbf{w}$.

28*   Let $\mathbf{u}$ be a unit vector. Show that the formula $\mathbf{v} = (\mathbf{v} \cdot \mathbf{u})\mathbf{u} + [\mathbf{v} - (\mathbf{v} \cdot \mathbf{u})\mathbf{u}]$ expresses $\mathbf{v}$ as the sum of two vectors, one parallel to $\mathbf{u}$, the other perpendicular to $\mathbf{u}$, and is the only such expression.

29   Prove the **Cauchy–Schwarz inequality**:

$$|\mathbf{v} \cdot \mathbf{w}| \leq |\mathbf{v}| \cdot |\mathbf{w}|.$$

30   (cont.) Now prove the triangle inequality: $|\mathbf{v} + \mathbf{w}| \leq |\mathbf{v}| + |\mathbf{w}|$.
[*Hint* $|\mathbf{v} + \mathbf{w}|^2 = |(\mathbf{v} + \mathbf{w}) \cdot (\mathbf{v} + \mathbf{w})| = |(\mathbf{v} + \mathbf{w}) \cdot \mathbf{v} + (\mathbf{v} + \mathbf{w}) \cdot \mathbf{w}|$

$\leq |\mathbf{v} + \mathbf{w}| \cdot |\mathbf{v}| + |\mathbf{v} + \mathbf{w}| \cdot |\mathbf{w}|.$]

4. LINES AND PLANES

In this section we shall learn how to describe lines and planes in $\mathbf{R}^3$ by equations. Let us begin with lines. A line in space can be given geometrically in three ways: as the line through two points, as the intersection of two planes, or as the line through a point in a specified direction. Let us start with the third way.

Parametric Form of a Line   A direction in space is described by a non-zero vector $\mathbf{a}$ in that direction. Let L be the line (Fig. 1a) through $\mathbf{0}$ in the direction of $\mathbf{a}$. Then, as Fig. 1a shows, each point $\mathbf{x}$ of L is a multiple $t\mathbf{a}$ of $\mathbf{a}$, where $t > 0$ if $\mathbf{x}$ is on the same side of $\mathbf{0}$ as the terminal point of $\mathbf{a}$ and $t < 0$ if $\mathbf{x}$ is on the opposite side of $\mathbf{0}$. Therefore L is the set of *all* multiples $\mathbf{x} = t\mathbf{a}$ of $\mathbf{a}$, including $\mathbf{0} = 0 \cdot \mathbf{a}$.

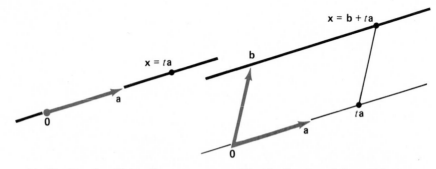

(a)   The line of multiples of $\mathbf{a}$ (b)   The line through $\mathbf{b}$ parallel to $\mathbf{a}$

Fig. 1   Line in parametric form

Now suppose we are given $\mathbf{a} \neq \mathbf{0}$ and a point $\mathbf{b}$. We want the line through $\mathbf{b}$ in the direction determined by $\mathbf{a}$, that is, the line through $\mathbf{b}$ parallel to $\mathbf{a}$.

Take any point $\mathbf{x}$ of this line. The line through $\mathbf{x}$ parallel to the vector $\mathbf{b}$

meets the line through **a**, determining a parallelogram (Fig. 1b) with vertices **0, b, x**, and *t***a**. By the parallelogram rule for vector addition, $\mathbf{x} = t\mathbf{a} + \mathbf{b}$.

Conversely, each point of the form $\mathbf{x} = t\mathbf{a} + \mathbf{b}$ is on the line through **b** parallel to **a**, again by the parallelogram rule for vector addition.

Parametric Vector Equation   The line through **b** parallel to $\mathbf{a} \neq \mathbf{0}$ consists of all points

$$\mathbf{x} = t\mathbf{a} + \mathbf{b},$$

where the **parameter** t varies over all real numbers.

The vector equation $\mathbf{x} = t\mathbf{a} + \mathbf{b}$ can be expressed as a system of three scalar equations by writing out its components:

Parametric Scalar Equations   The line through (b_1, b_2, b_3) parallel to $(a_1, a_2, a_3) \neq (0, 0, 0)$ consists of all points (x_1, x_2, x_3) such that

$$\begin{cases} x_1 = a_1 t + b_1 \\ x_2 = a_2 t + b_2 \\ x_3 = a_3 t + b_3 \end{cases} \quad \text{where} \quad -\infty < t < \infty.$$

■ **EXAMPLE 1**   Find all points (x, y, z) on the line through $(-2, 1, 2)$ parallel to $(2, -1, 3)$. Are $(0, 0, 4)$ and $(-6, 3, -4)$ on this line?

Solution $(x, y, z) = t(2, -1, 3) + (-2, 1, 2) = (2t - 2, -t + 1, 3t + 2)$.
In scalar form, the line consists of all (x, y, z) such that

$$x = 2t - 2, \qquad y = -t + 1, \qquad z = 3t + 2, \qquad -\infty < t < \infty.$$

Clearly $x = 0$ only if $2t - 2 = 0$, that is, $t = 1$. But then $y = 0$ and $z = 5 \neq 4$, so $(0, 0, 4)$ is *not* on the line. However, $x = -6$ implies $2t - 2 = -6$, $t = -2$; then $y = 3$ and $z = -4$. Hence $(-6, 3, -4)$ *is* a point of the line.   ■

Suppose we want the line through two distinct points **a** and **b**. Its direction is determined by the vector $\mathbf{b} - \mathbf{a}$. Hence the line we want is the line that passes through **a** (or **b**) and is parallel to $\mathbf{b} - \mathbf{a}$. This line we know is

$$\mathbf{x} = t(\mathbf{b} - \mathbf{a}) + \mathbf{a} = (1 - t)\mathbf{a} + t\mathbf{b}, \qquad -\infty < t < \infty.$$

Line through Two Points   Given $\mathbf{a} \neq \mathbf{b}$, the line through **a** and **b** consists of all points

$$\mathbf{x} = (1 - t)\mathbf{a} + t\mathbf{b} \qquad -\infty < t < \infty.$$

In scalar form,

$$x_1 = (1 - t)a_1 + tb_1, \qquad x_2 = (1 - t)a_2 + tb_2, \qquad x_3 = (1 - t)a_3 + tb_3.$$

■ **EXAMPLE 2**   Find the line through $(3, -1, 2)$ and $(4, 1, 1)$. Where does it meet the $x_1, x_2 -$ plane?

Solution $\mathbf{x} = (1 - t)(3, -1, 2) + t(4, 1, 1) = (t + 3, 2t - 1, -t + 2)$.

The line meets the x_1, x_2-plane where $x_3 = 0$, that is, where $-t + 2 = 0, t = 2$. Then $x_1 = 5$ and $x_2 = 3$, so $\mathbf{x} = (5, 3, 0)$.   ∎

Division of a Segment   Suppose we want the point on a segment $\overline{\mathbf{ab}}$ that is $\frac{2}{5}$ of the way from $\mathbf{a}$ to $\mathbf{b}$. By vectors this is easy; we just add $\frac{2}{5}(\mathbf{b} - \mathbf{a})$ to $\mathbf{a}$. The result (Fig. 2a) is $\mathbf{a} + \frac{2}{5}(\mathbf{b} - \mathbf{a}) = \frac{3}{5}\mathbf{a} + \frac{2}{5}\mathbf{b}$.

In general, if $0 \le t \le 1$, the point (Fig. 2b) that is t of the way from $\mathbf{a}$ to $\mathbf{b}$ is

$$\mathbf{x} = \mathbf{a} + t(\mathbf{b} - \mathbf{a}) = (1 - t)\mathbf{a} + t\mathbf{b}.$$

Do you see why this formula agrees with the parametric vector equation of a line (at least for $0 \le t \le 1$)? The special case $t = \frac{1}{2}$ yields the **midpoint formula**

$$\mathbf{m} = \tfrac{1}{2}\mathbf{a} + \tfrac{1}{2}\mathbf{b}$$

for the midpoint of $\overline{\mathbf{ab}}$.

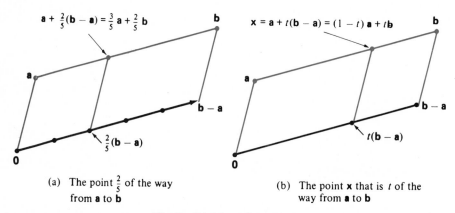

(a)   The point $\frac{2}{5}$ of the way   (b)   The point $\mathbf{x}$ that is t of the
        from $\mathbf{a}$ to $\mathbf{b}$ way from $\mathbf{a}$ to $\mathbf{b}$

Fig. 2   Division of segments

Points on the segment $\overline{\mathbf{ab}}$ can be thought of as **weighted averages** of $\mathbf{a}$ and $\mathbf{b}$. For instance the point $\frac{5}{8}$ of the way from $\mathbf{a}$ to $\mathbf{b}$ is $\frac{3}{8}\mathbf{a} + \frac{5}{8}\mathbf{b}$. This is a weighted average with weights $\frac{3}{8}$ assigned to $\mathbf{a}$ and $\frac{5}{8}$ to $\mathbf{b}$. (The point is closer to $\mathbf{b}$, so $\mathbf{b}$ gets the larger weight.) Its coordinates are weighted averages of the coordinates of $\mathbf{a}$ and $\mathbf{b}$:

$$\tfrac{3}{8}\mathbf{a} + \tfrac{5}{8}\mathbf{b} = (\tfrac{3}{8}a_1 + \tfrac{5}{8}b_1, \tfrac{3}{8}a_2 + \tfrac{5}{8}b_2, \tfrac{3}{8}a_3 + \tfrac{5}{8}b_3).$$

Remark   Bear in mind that the sum of the weights is 1. This applies also to weighted averages of more than two points.

Equation of a Plane   A plane can be described in several ways, for example as the plane through three non-collinear points, as the plane through a line and a point not on the line, or as the plane through two intersecting (or parallel) lines.

We begin our study of planes with yet another possibility: the plane P through a point $\mathbf{c}$ and perpendicular to the direction determined by a vector $\mathbf{a} \ne \mathbf{0}$. See Fig. 3. Now a point $\mathbf{x}$ lies on P if and only if the segment $\overline{\mathbf{xc}}$ is perpendicular to $\mathbf{a}$, that is,

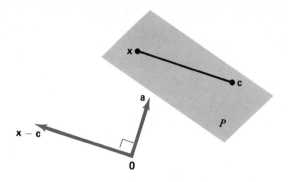

Fig. 3   The plane through **c** perpendicular to **a**

the vector $\mathbf{x} - \mathbf{c}$ is perpendicular to **a**. But two vectors are perpendicular if and only if their inner product is 0.

The plane that passes through a point **c** and is perpendicular to a vector $\mathbf{a} \neq \mathbf{0}$ consists of all points **x** such that

$$(\mathbf{x} - \mathbf{c}) \cdot \mathbf{a} = 0, \qquad \text{that is,} \quad \mathbf{x} \cdot \mathbf{a} = \mathbf{c} \cdot \mathbf{a}.$$

The corresponding scalar equation is

$$a_1(x_1 - c_1) + a_2(x_2 - c_2) + a_3(x_3 - c_3) = 0,$$

that is,

$$a_1 x_1 + a_2 x_2 + a_3 x_3 = b,$$

where $\quad b = a_1 c_1 + a_2 c_2 + a_3 c_3.$

Conversely, each equation

$$a_1 x_1 + a_2 x_2 + a_3 x_3 = b,$$

where $(a_1, a_2, a_3) \neq (0, 0, 0)$, is the equation of a plane. For set $\mathbf{a} = (a_1, a_2, a_3)$. Then the equation can be written

$$\mathbf{a} \cdot \mathbf{x} = b.$$

We need one definite solution of this equation, one vector **c** such that $\mathbf{a} \cdot \mathbf{c} = b$. Not hard to find! For instance, if $a_3 \neq 0$, then $\mathbf{c} = (0, 0, b/a_3)$ fits the bill. More systematically, $\mathbf{c} = (b/|\mathbf{a}|^2)\mathbf{a}$ does the trick in all cases because

$$\mathbf{a} \cdot \mathbf{c} = \mathbf{a} \cdot (b/|\mathbf{a}|^2)\mathbf{a} = (b/|\mathbf{a}|^2)\mathbf{a} \cdot \mathbf{a} = b$$

since $|\mathbf{a}|^2 = \mathbf{a} \cdot \mathbf{a}$.

Thus there is a vector **c** such that $\mathbf{a} \cdot \mathbf{c} = b$, so the equation $\mathbf{a} \cdot \mathbf{x} = b$ can be written

$$\mathbf{a} \cdot \mathbf{x} = \mathbf{a} \cdot \mathbf{c}, \qquad \text{that is,} \quad \mathbf{a} \cdot (\mathbf{x} - \mathbf{c}) = 0.$$

This is exactly the condition that **x** is on the plane through **c** perpendicular to **a**.

Example $\qquad\qquad\qquad 3x_1 - x_2 - 4x_3 = 2.$

Obviously, $\mathbf{c} = (0, -2, 0)$ is a point on this locus, so the locus can be written as

$$(3, -1, -4) \cdot [(x_1, x_2, x_3) - (0, -2, 0)] = 0.$$

This equation describes the plane through $(0, -2, 0)$ orthogonal to $(3, -1, -4)$.

Normal Form We have seen that a given plane P can be described by an equation of the form

$$\mathbf{a} \cdot \mathbf{x} = b, \qquad \mathbf{a} \neq \mathbf{0}.$$

If the vector $\mathbf{a}$ is a *unit* vector (length one), we say that the equation is in **normal form**.

Suppose $\mathbf{a} \cdot \mathbf{x} = b$ is not in normal form. Then there is an equivalent equation for P that is. For the equation

$$(c\mathbf{a}) \cdot \mathbf{x} = cb$$

describes P just as well. In particular, there are two choices of c that make $c\mathbf{a}$ a unit vector,

$$c = \frac{1}{|\mathbf{a}|} \quad \text{and} \quad c = \frac{-1}{|\mathbf{a}|}.$$

We choose either one of these to obtain a normal form. Thus we set

$$\mathbf{n} = \pm \frac{1}{|\mathbf{a}|} \mathbf{a} \quad \text{and} \quad p = \pm \frac{1}{|\mathbf{a}|} b$$

and obtain

$$\mathbf{n} \cdot \mathbf{x} = p, \qquad |\mathbf{n}| = 1.$$

Basically, all we do is replace the vector $\mathbf{a}$, which is perpendicular (normal) to the plane P, by a unit vector $\mathbf{n}$ in the direction of $\mathbf{a}$, hence equally perpendicular to P.

Let us show that the constant p in the normal form has a neat geometric interpretation (Fig. 4).

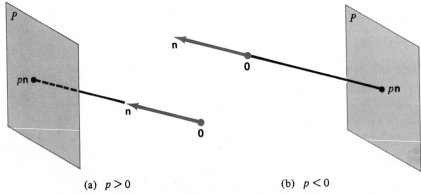

(a) $p > 0$ **(b)** $p < 0$

Fig. 4   Normal form of a plane: $\mathbf{n} \cdot \mathbf{x} = p$ where $|\mathbf{n}| = 1$

Since $\mathbf{n}$ is perpendicular to P, the line $\mathbf{x} = t\mathbf{n}$ of the vector $\mathbf{n}$ pierces P at the point of P closest to $\mathbf{0}$. This must be the point $p\mathbf{n}$, because $p\mathbf{n}$ obviously lies on the line and also lies on the plane since

$$\mathbf{n} \cdot (p\mathbf{n}) = p(\mathbf{n} \cdot \mathbf{n}) = p.$$

Therefore the distance from **0** to the plane P is $|p\mathbf{n}| = |p|$. The number p itself is the *signed* distance from **0** to P. If **n** points toward P, then $p > 0$; if **n** points away from P, then $p < 0$.

Normal Form　Each plane P has two normal forms

$$\mathbf{n} \cdot \mathbf{x} = p, \qquad |\mathbf{n}| = 1.$$

(There are two choices of the unit vector **n**.) The constant p is the signed distance from **0** to P.

■ **EXAMPLE 3**　Give a normal form for the plane

$$2x_1 - x_2 + 2x_3 = -15.$$

Find the distance from the origin to the plane.

Solution　Set $\mathbf{a} = (2, -1, 2)$ so the given equation in vector form is $\mathbf{a} \cdot \mathbf{x} = -15$. Now

$$|\mathbf{a}|^2 = 2^2 + (-1)^2 + 2^2 = 9, \qquad \text{so} \quad |\mathbf{a}| = 3.$$

Set
$$\mathbf{n} = (1/|\mathbf{a}|)\mathbf{a} = \tfrac{1}{3}\mathbf{a} = (\tfrac{2}{3}, -\tfrac{1}{3}, \tfrac{2}{3}).$$

Then **n** is a unit vector. From the given equation $\mathbf{a} \cdot \mathbf{x} = -15$ we have $(\tfrac{1}{3}\mathbf{a}) \cdot \mathbf{x} = \tfrac{1}{3}(-15)$, that is,

$$\mathbf{n} \cdot \mathbf{x} = -5, \qquad \tfrac{2}{3}x_1 - \tfrac{1}{3}x_2 + \tfrac{2}{3}x_3 = -5.$$

Either of these is a normal form. Here $p = -5$, and the distance from **0** to the plane is $|p| = 5$. ■

Distance from a Point to a Plane　Suppose we are given a plane P and a point **c** in space. We want the distance D from **c** to P. See Fig. 5a.

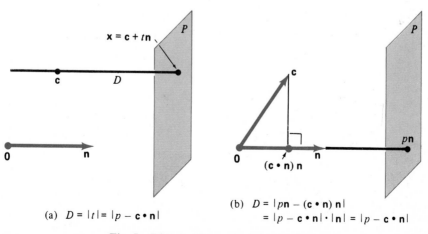

(a) $D = |t| = |p - \mathbf{c} \cdot \mathbf{n}|$

(b) $D = |p\mathbf{n} - (\mathbf{c} \cdot \mathbf{n})\,\mathbf{n}|$
$\quad = |p - \mathbf{c} \cdot \mathbf{n}| \cdot |\mathbf{n}| = |p - \mathbf{c} \cdot \mathbf{n}|$

Fig. 5　Distance from a point to a plane

First we express the plane in normal form $\mathbf{x} \cdot \mathbf{n} = p$. Then we drop a perpendicular from $\mathbf{c}$ onto P. It is parallel to $\mathbf{n}$, hence lies on the line $\mathbf{x} = \mathbf{c} + t\mathbf{n}$. Its foot is the point $\mathbf{x}$ on this line that satisfies the equation of the plane:

$$(\mathbf{c} + t\mathbf{n}) \cdot \mathbf{n} = p, \qquad \text{that is,} \quad \mathbf{c} \cdot \mathbf{n} + t = p.$$

Thus $t = p - \mathbf{c} \cdot \mathbf{n}$ and $\mathbf{x} = \mathbf{c} + (p - \mathbf{c} \cdot \mathbf{n})\mathbf{n}$. The required distance is

$$D = |\mathbf{x} - \mathbf{c}| = |(p - \mathbf{c} \cdot \mathbf{n})\mathbf{n}| = |p - \mathbf{c} \cdot \mathbf{n}|$$

since $|\mathbf{n}| = 1$. In Fig. 5b we show another way to find the same formula.

Distance Rule   The distance from a point $\mathbf{c}$ to a plane P in normal form $\mathbf{n} \cdot \mathbf{x} = p$ is

$$D = |p - \mathbf{c} \cdot \mathbf{n}|.$$

■ **EXAMPLE 4**   Find the distance from $(2, -1, -4)$ to the plane

$$3x - 6y + 2z = 28.$$

Solution   Since $3^2 + (-6)^2 + 2^2 = 49 = 7^2$, a normal form is

$$\tfrac{3}{7}x - \tfrac{6}{7}y + \tfrac{2}{7}z = \tfrac{28}{7} = 4.$$

Thus $\mathbf{n} = (\tfrac{3}{7}, -\tfrac{6}{7}, \tfrac{2}{7})$ and $p = 4$, so by the Distance Rule the required distance is

$$D = |4 - (2, -1, -4) \cdot (\tfrac{3}{7}, -\tfrac{6}{7}, \tfrac{2}{7})| = |4 - \tfrac{4}{7}| = \tfrac{24}{7}. \qquad ■$$

We shall continue our study of planes after a digression to develop some necessary tools.

EXERCISES

Find whether or not $\mathbf{c}$ is on $\overline{\mathbf{ab}}$, where $\mathbf{a}, \mathbf{b}, \mathbf{c}$ are

1   $(0, 1, 0)$, $(1, 0, 1)$, $(2, 2, 2)$ 2 $(2, 1, -1)$, $(5, -2, 1)$, $(1, 2, -1)$
3   $(3, 3, -2)$, $(-2, 3, -2)$ $(2, 3, -2)$ 4 $(2, 2, -2)$, $(4, -6, 1)$, $(3, -2, -1)$.

Find an angle between $\overline{\mathbf{ab}}$ and $\overline{\mathbf{cd}}$, where $\mathbf{a}, \mathbf{b}; \mathbf{c}, \mathbf{d}$ are

5   $(3, 3, 1)$, $(-2, 1, -1)$; $(1, 1, 1)$, $(2, 1, -2)$
6   $(-1, -1, -3)$, $(1, 1, 1)$; $(2, 2, 2)$, $(0, 1, 1)$.

7   Find the point $\tfrac{1}{3}$ of the way from $(1, 1, 1)$ to $(0, 0, 0)$.
8   Find the point $\tfrac{2}{7}$ of the way from $(1, 0, 1)$ to $(-1, -1, -1)$.

Find the intersections of the line $\overline{\mathbf{ab}}$ with the three coordinate planes, where $\mathbf{a}$ and $\mathbf{b}$ are

9    $(-2, 3, -4)$, $(-1, -2, 5)$ 10 $(-3, 4, 4)$, $(-4, 4, -3)$
11   $(-1, 1, -1)$, $(2, -2, -1)$ 12 $(1, 1, 1)$, $(-3, -3, -3)$.

13   Find the distance from $\mathbf{0}$ to the line through $(1, 1, 1)$ and $(1, 0, 1)$.
14   Find the distance from $(1, 1, 1)$ to the line through $(1, 2, 3)$ and $(-3, -2, -1)$.
15   If t is time, when and where does the parametric line $\mathbf{x} = t(4, 4, 1) + (3, 2, -5)$ hit the x, y-plane?
16   Describe the locus $\mathbf{x}(t) = t^2(4, 4, 1) + (3, 2, -5)$ as $-\infty < t < \infty$.

17   Give a parametric vector equation for the line through **a** parallel to $\overline{\mathbf{bc}}$.
18   Give parametric equations for the line through $(1, 2, -2)$ perpendicular to the plane $x + y + z = 0$.

Express in normal form

19 $x_1 - 2x_2 + 2x_3 = 1$ **20** $2x_1 + 6x_2 - 3x_3 = 14$ **21** $-8x_1 + x_2 - 4x_3 = 27$
22 $3x_1 - 2x_2 - 6x_3 = 4$ **23** $x_1 + x_2 + x_3 = 3$ **24** $x_1 - x_2 + x_3 = -12$.

Find the distance from the plane to the point

25 $x + y + z = 2$, $(1, 1, 1)$ **26** $2x - y - 2z = 4$, $(0, 0, 1)$
27 $-3x - y + 4z = 8$, $(0, 0, 2)$ **28** $-3x + 12y + 4z = 13$, $(1, 0, -1)$.

29   Find the angle θ between the line $\mathbf{x} = t\mathbf{a} + \mathbf{b}$ and the plane in normal form $\mathbf{n} \cdot \mathbf{x} = p$.
30   Find the distance between the parallel planes $4x - y - 3z = 1$ and $4x - y - 3z = 6$.

Let $\mathbf{x} = t\mathbf{a} + \mathbf{b}$ be a line and $\mathbf{n} \cdot \mathbf{x} = p$ be a plane in normal form

31   Prove that the line and plane are parallel if and only if $\mathbf{a} \cdot \mathbf{n} = 0$.
32   (cont.) Prove that the line is on the plane if and only if $\mathbf{a} \cdot \mathbf{n} = 0$ and $\mathbf{b} \cdot \mathbf{n} = p$.
33   (cont.) Suppose $\mathbf{a} \cdot \mathbf{n} \neq 0$. Prove that the point of intersection of the line and the plane is $\mathbf{z} = [(p - \mathbf{n} \cdot \mathbf{b})/(\mathbf{a} \cdot \mathbf{n})]\mathbf{a} + \mathbf{b}$.
34   Let $\mathbf{m} \cdot \mathbf{x} = p$ and $\mathbf{n} \cdot \mathbf{x} = q$ be two non-parallel planes in normal form. Let θ be one of their (dihedral) angles of intersection. Find $\cos \theta$.

Let $\mathbf{x} = t\mathbf{u} + \mathbf{b}$ be a parametric line, where $\mathbf{u}$ is a unit vector, and let $\mathbf{c}$ be a point

35*   Find the point on the line closest to $\mathbf{c}$.
36*   Find the distance D from $\mathbf{c}$ to the line (in terms of $\mathbf{u}$, $\mathbf{b}$, and $\mathbf{c}$).

5. LINEAR SYSTEMS

Introduction   Suppose we are given three planes

$$\mathbf{a}_1 \cdot \mathbf{x} = d_1, \qquad \mathbf{a}_2 \cdot \mathbf{x} = d_2, \qquad \mathbf{a}_3 \cdot \mathbf{x} = d_3.$$

How can we find their intersection? In coordinates, the problem is to solve the system of three linear equations

$$\begin{cases} a_1 x + b_1 y + c_1 z = d_1 \\ a_2 x + b_2 y + c_2 z = d_2 \\ a_3 x + b_3 y + c_3 z = d_3 \end{cases}$$

for x, y, z, where $a_1, \cdots, d_3$ are given constants. Generally, there is a single common point (Fig. 1a). However, if the planes are parallel or if one is parallel to the intersection of the other two, then there is no common point (Fig. 1b). In this case, the corresponding system of equations is called **inconsistent**. For example, the system

$$\begin{cases} x + y + z = 1 \\ x + y + z = 2 \\ 3x - 2y + 4z = 7 \end{cases}$$

is obviously inconsistent; the first two equations cannot both be satisfied. Geometrically, the first two planes are parallel. [From their equations, we see that both planes are perpendicular to the vector $(1, 1, 1)$.]

Three planes have more than one common point if they pass through a common line (Figs. 1c, 1d) or if all of them coincide. In this case, the corresponding system of equations is called **underdetermined**. For example the system

$$\begin{cases} x - 2y + 3z = 5 \\ 8x + 7y + \ z = 2 \\ 2x - 4y + 6z = 10 \end{cases}$$

is underdetermined; the third equation is twice the first. Geometrically, the first and third planes coincide. The system represents two distinct planes that have a line in common.

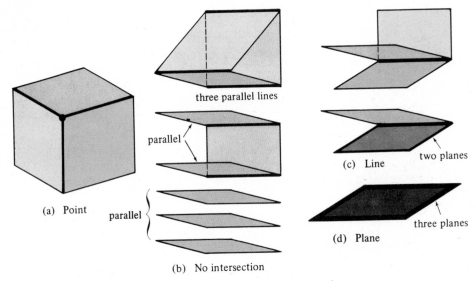

three parallel lines

parallel

(a) Point   parallel

(b) No intersection

(c) Line

two planes

(d) Plane

three planes

Fig. 1   Possible intersections of three planes

Elimination   Geometric reasoning shows that the set of solutions of a linear system of three equations in three unknowns is either (a) empty (b) a single point, (c) a line, (d) a plane, or (e) the whole space $\mathbf{R}^3$. Let us review briefly two methods of solving linear systems. Probably you have seen them in previous courses. The first, elimination, is practically self-explanatory.

■ **EXAMPLE 1**   Solve the system

$$\begin{cases} 2x - \ y + \ z = 4 \\ \qquad\quad 3y + 2z = -1 \\ \qquad\qquad\quad -z = 3. \end{cases}$$

Solution   By the third equation, $z = -3$. Substitute this into the first two equations. The result is a new system of two equations for x and y:

$$\begin{cases} 2x - y = 4 - (-3) = 7 \\ \quad\ 3y = -1 - 2(-3) = 5. \end{cases}$$

By the second equation, $y = \frac{5}{3}$. Substitute this into the first equation; the result is a single equation for x:

$$2x = 7 + \tfrac{5}{3} = \tfrac{26}{3}, \qquad x = \tfrac{13}{3}.$$

The solution is $\quad \mathbf{x} = (\tfrac{13}{3}, \tfrac{5}{3}, -3).$ ■

This example was very easy because we could solve for the unknowns one at a time. To solve a more general system, we reduce it to a system of this type by eliminating the unknowns one by one.

■ **EXAMPLE 2** Solve the system

$$\begin{cases} 2x - y + z = 4 \\ 2x + 2y + 3z = 3 \\ 6x - 9y - 2z = 17. \end{cases}$$

Solution Eliminate x from the second and third equations as follows: subtract the first equation from the second, and subtract 3 times the first equation from the third. The result is an equivalent system of three equations (the first the same as before):

$$\begin{cases} 2x - y + z = 4 \\ \quad 3y + 2z = -1 \\ \quad -6y - 5z = 5. \end{cases}$$

Now eliminate y from the third equation. Add twice the second equation to the third, but keep the first two equations:

$$\begin{cases} 2x - y + z = 4 \\ \quad 3y + 2z = -1 \\ \quad -z = 3. \end{cases}$$

This is the system in Example 1, so $\mathbf{x} = (\tfrac{13}{3}, \tfrac{5}{3}, -3).$ ■

Practical hint When you apply the method of elimination, you do not *have* to eliminate first x and then y. Eliminate any two of the unknowns in an order that makes the computation easiest.

■ **EXAMPLE 3** Solve

$$\begin{cases} 3x + y - 2z = 4 \\ -5x \quad + 2z = 5 \\ -7x - y + 3z = -2. \end{cases}$$

Solution Since y is missing from the second equation, add the first to the third; then y is eliminated from two equations:

$$\begin{cases} 3x + y - 2z = 4 \\ -5x \quad + 2z = 5 \\ -4x \quad + z = 2. \end{cases}$$

Now add -2 times the third to the second; this eliminates z:

$$\begin{cases} 3x + y - 2z = 4 \\ 3x \qquad\quad = 1 \\ -4x \quad + z = 2. \end{cases}$$

By the second equation, $x = \frac{1}{3}$. By the third equation, $z = 2 + 4x = 2 + 4(\frac{1}{3}) = \frac{10}{3}$. Finally,

$$y = 4 - 3x + 2z = 4 - 1 + \frac{20}{3} = \frac{29}{3},$$

so

$$\mathbf{x} = (\tfrac{1}{3}, \tfrac{29}{3}, \tfrac{10}{3}).$$

■

■ **EXAMPLE 4**   Solve

(a) $\begin{cases} x + y + z = 1 \\ x - 2y + 2z = 4 \\ 2x - y + 3z = 5 \end{cases}$ (b) $\begin{cases} x + y + z = 1 \\ 2x + 2y + 2z = 2 \\ 3x + 3y + 3z = 3. \end{cases}$

Solution   (a)   Eliminate x from the second and third equations:

$$\begin{cases} x + y + z = 1 \\ -3y + z = 3 \\ -3y + z = 3. \end{cases}$$

The last two equations both say the same thing. Therefore the system is equivalent to the system

$$\begin{cases} x + y + z = 1 \\ -3y + z = 3, \end{cases}$$

and that is as far as the elimination method will go. Geometrically, this reduced system represents the intersection of *two* planes. These planes are not parallel since their normal vectors, $(1, 1, 1)$ and $(0, -3, 1)$, point in different directions. Hence it is a safe bet that the solutions form a line. To get a parametric solution, set $y = t$. Then $z = 3 + 3y = 3 + 3t$, and

$$x = 1 - y - z = 1 - t - (3 + 3t) = -2 - 4t.$$

The most general solution is

$$(x, y, z) = (-2 - 4t, t, 3 + 3t),$$

where $-\infty < t < \infty$. The set of solutions is the line

$$\mathbf{x} = t(-4, 1, 3) + (-2, 0, 3).$$

As a geometric check, note that this line is in the direction of the vector $(-4, 1, 3)$. Since the line supposedly lies in *both* planes, it should be perpendicular to both normals, that is, $(-4, 1, 3)$ should be perpendicular to both $(1, 1, 1)$ and $(0, -3, 1)$. It is perpendicular:

$$(-4, 1, 3) \cdot (1, 1, 1) = -4 + 1 + 3 = 0, \qquad (-4, 1, 3) \cdot (0, -3, 1) = 0 - 3 + 3 = 0.$$

(b)   This system is obviously equivalent to the single equation $x + y + z = 1$. Therefore the set of solutions (x, y, z) is a plane in space.   ∎

Review of Determinants   We recall from elementary algebra the definition of **determinants** of orders two and three:

$$\begin{vmatrix} a_1 & b_1 \\ a_2 & b_2 \end{vmatrix} = a_1 b_2 - a_2 b_1,$$

$$\begin{vmatrix} a_1 & b_1 & c_1 \\ a_2 & b_2 & c_2 \\ a_3 & b_3 & c_3 \end{vmatrix} = a_1 b_2 c_3 + a_2 b_3 c_1 + a_3 b_1 c_2 - a_1 b_3 c_2 - a_2 b_1 c_3 - a_3 b_2 c_1.$$

From the defining formulas: (1) if two rows (columns) are equal, the determinant is zero; (2) if two rows (columns) are transposed, the determinant changes sign; (3) if a multiple of one row (column) is added to another row (column), the determinant is unchanged; and (4) if all the terms in one row (column) are multiplied by a scalar, the determinant is multiplied by the same scalar.

Also the defining formulas imply various expansions by minors of a row (column), for instance

$$\begin{vmatrix} a_1 & b_1 & c_1 \\ a_2 & b_2 & c_2 \\ a_3 & b_3 & c_3 \end{vmatrix} = a_1 \begin{vmatrix} b_2 & c_2 \\ b_3 & c_3 \end{vmatrix} - b_1 \begin{vmatrix} a_2 & c_2 \\ a_3 & c_3 \end{vmatrix} + c_1 \begin{vmatrix} a_2 & b_2 \\ a_3 & b_3 \end{vmatrix}$$

is the expansion by minors of the first row. Here, for instance,

$$\begin{vmatrix} a_2 & c_2 \\ a_3 & c_3 \end{vmatrix}$$

is the **minor** of b_1. It is the 2×2 determinant remaining after the row and the column containing b_1 are crossed off.

A system of equations

$$\begin{cases} a_1 x + b_1 y + c_1 z = d_1 \\ a_2 x + b_2 y + c_2 z = d_2 \\ a_3 x + b_3 y + c_3 z = d_3 \end{cases}$$

is both **consistent** (not inconsistent) and **determined** (not underdetermined) if and only if the **system determinant** $D \neq 0$, where

$$D = \begin{vmatrix} a_1 & b_1 & c_1 \\ a_2 & b_2 & c_2 \\ a_3 & b_3 & c_3 \end{vmatrix}.$$

When this is so, the system has a unique solution, given explicitly by **Cramer's Rule**:

$$x = \frac{1}{D} \begin{vmatrix} d_1 & b_1 & c_1 \\ d_2 & b_2 & c_2 \\ d_3 & b_3 & c_3 \end{vmatrix}, \qquad y = \frac{1}{D} \begin{vmatrix} a_1 & d_1 & c_1 \\ a_2 & d_2 & c_2 \\ a_3 & d_3 & c_3 \end{vmatrix}, \qquad z = \frac{1}{D} \begin{vmatrix} a_1 & b_1 & d_1 \\ a_2 & b_2 & d_2 \\ a_3 & b_3 & d_3 \end{vmatrix}.$$

Cramer's Rule is our second method for solving linear systems. It is derived in linear algebra courses.

Intersection of Planes

■ **EXAMPLE 5**   Find the intersection of the three planes

$$2x - y + z = 1, \qquad 2x - 3y - 2z = 4, \qquad 6x + 2y + 9z = -5.$$

Solution   We shall try Cramer's Rule. The system determinant, expanded by the first row, is

$$D = \begin{vmatrix} 2 & -1 & 1 \\ 2 & -3 & -2 \\ 6 & 2 & 9 \end{vmatrix} = 2 \begin{vmatrix} -3 & -2 \\ 2 & 9 \end{vmatrix} + \begin{vmatrix} 2 & -2 \\ 6 & 9 \end{vmatrix} + \begin{vmatrix} 2 & -3 \\ 6 & 2 \end{vmatrix}$$

$$= (2)(-23) + 30 + 22 = 6 \neq 0.$$

Hence

$$x = \frac{1}{6} \begin{vmatrix} 1 & -1 & 1 \\ 4 & -3 & -2 \\ -5 & 2 & 9 \end{vmatrix} = -\frac{2}{3}.$$

Similarly $y = -2$ and $z = \frac{1}{3}$. The planes intersect in one point: $(-\frac{2}{3}, -2, \frac{1}{3})$.   ■

Plane through Three Points   Suppose $\mathbf{p}_1, \mathbf{p}_2,$ and $\mathbf{p}_3$ are three non-collinear points in space. How can we find the unique plane that they determine? Suppose $\mathbf{p}_i = (x_i, y_i, z_i)$ for $i = 1, 2, 3$. We take the plane in the form

$$ax + by + cz = d.$$

The constants a, b, c, d must be found to satisfy the system

$$ax_i + by_i + cz_i = d, \qquad i = 1, 2, 3.$$

This seems to be a system of three equations in *four* unknowns! Actually not, because if $\lambda \neq 0$, then (a, b, c, d) is a solution if and only if $(\lambda a, \lambda b, \lambda c, \lambda d)$ is a solution. Thus by our choice of the scale factor λ we can reduce to the case where one of the unknowns is 1; then there are three equations in three unknowns.

Simplest is to try first $d = 1$. If the resulting system has a unique solution (a, b, c), done. If, however, the original system forces $d = 0$, then this approach leads to an inconsistent system. When that happens, you set $d = 0$ and find a, b, c anyhow.

■ **EXAMPLE 6**   Find the plane through

$$(1, -1, -3), \qquad (2, 2, 4), \qquad \text{and} \qquad (2, 1, 1).$$

Solution   We assume the plane can be written in the form $ax + by + cz = 1$. Then (a, b, c) must satisfy the system

$$\begin{cases} a - b - 3c = 1 \\ 2a + 2b + 4c = 1 \\ 2a + b + c = 1. \end{cases}$$

Subtract the second from the third, then twice the first from the second:

$$\begin{cases} a - b - 3c = 1 \\ 4b + 10c = -1 \\ -b - 3c = 0. \end{cases}$$

From the resulting second and third equations, $b = -\frac{3}{2}$, $c = \frac{1}{2}$. From the first equation, $a = 1$. The equation of the plane is

$$x - \tfrac{3}{2}y + \tfrac{1}{2}z = 1, \qquad \text{that is,} \quad 2x - 3y + z = 2.$$

Alternative solution Another possibility is to find a normal vector to the plane. Clearly the vectors

$$\mathbf{u} = (2, 2, 4) - (1, -1, -3) = (1, 3, 7), \qquad \mathbf{v} = (2, 1, 1) - (1, -1, -3) = (1, 2, 4)$$

are parallel to the plane. Any normal vector $\mathbf{a}$ must satisfy $\mathbf{u} \cdot \mathbf{a} = 0$ and $\mathbf{v} \cdot \mathbf{a} = 0$. If $\mathbf{a} = (a, b, c)$, we must have

$$\begin{cases} a + 3b + 7c = 0 \\ a + 2b + 4c = 0. \end{cases}$$

Subtract:
$$\begin{cases} a + 3b + 7c = 0 \\ -b - 3c = 0. \end{cases}$$

It follows that $b = -3c$ and $a = 2c$. Thus the system has a whole line of solutions $\mathbf{a} = (2c, -3c, c) = c(2, -3, 1)$, consisting of vectors perpendicular to the plane. Choose for instance $c = 1$, that is, $\mathbf{a} = (2, -3, 1)$. The plane has the form $\mathbf{a} \cdot \mathbf{x} = k$,

$$2x - 3y + z = k.$$

Finally, to determine k, simply plug in one of the three points known to be on the plane, for instance $(1, -1, -3)$: $2 + 3 - 3 = k$, $k = 2$. The result:

$$2x - 3y + z = 2. \qquad \blacksquare$$

Remark 1 We shall pursue this second method further in Section 7.

Remark 2 We have ignored the question of exactly when the three points are non-collinear. This also will be considered in Section 7.

EXERCISES

Solve by elimination

1 $\quad \begin{cases} x + 2y = 1 \\ 3y = 2 \end{cases}$

2 $\quad \begin{cases} 2x = 3 \\ -x + y = 0 \end{cases}$

3 $\quad \begin{cases} x + 2y = 1 \\ x + 3y = 2 \end{cases}$

4 $\quad \begin{cases} x + y = a \\ x - y = b \end{cases}$

5 $\quad \begin{cases} 2x - 3y = -1 \\ 3x + 5y = 2 \end{cases}$

6 $\quad \begin{cases} 2x - 3y = -1 \\ -3x + 5y = 2 \end{cases}$

7 $\quad \begin{cases} x + y + z = 0 \\ 2y - 3z = -1 \\ 3y + 5z = 2 \end{cases}$

8 $\quad \begin{cases} 2x - y - z = 1 \\ 2y - 3z = -1 \\ -3y + 5z = 2 \end{cases}$

9 $\quad \begin{cases} x + y - z = 0 \\ x - y + z = 0 \\ -x + y + z = 0 \end{cases}$

10 $\begin{cases} 2x + y + 3z = 1 \\ -x + 4y + 2z = 0 \\ 3x + y + z = -1 \end{cases}$ 11 $\begin{cases} 2x - y - 3z = 1 \\ -x - 4y - 2z = 1 \\ 3x - y - z = 1 \end{cases}$ 12 $\begin{cases} 4x + 2y - z = 0 \\ x + 3y + 2z = 0 \\ x + y + 3z = 4. \end{cases}$

Show that the system is inconsistent and interpret geometrically

13 $\begin{cases} x - y = 1 \\ -x + y = 1 \end{cases}$ 14 $\begin{cases} x = 2 \\ x = 3 \end{cases}$

15 $\begin{cases} x + y + 2z = 1 \\ 3x + 5y + 7z = 2 \\ -x - y - 2z = 0 \end{cases}$ 16 $\begin{cases} x + y + 2z = 1 \\ -x + 2y + z = 3 \\ y + z = 1. \end{cases}$

Find all solutions of each underdetermined system

17 $\begin{cases} 2x - 3y = 1 \\ -4x + 6y = -2 \end{cases}$ 18 $\begin{cases} x + y = 0 \\ x + y = 0 \end{cases}$

19 $\begin{cases} 2x - y + z = 1 \\ 3x + y + z = 0 \\ 7x - y + 3z = 2 \end{cases}$ 20 $\begin{cases} 11x + 10y + 9z = 5 \\ x + 2y + 3z = 1 \\ 3x + 2y + z = 1. \end{cases}$

Solve by Cramer's Rule

21 $\begin{cases} 2x + y + 2z = 1 \\ -x + 4y + 2z = 0 \\ 3x + y + z = -1 \end{cases}$ 22 $\begin{cases} 2x - y - 3z = 1 \\ -x + 4y - 2z = 1 \\ 3x - y - z = 1 \end{cases}$

23 $\begin{cases} 3x - y - z = 6 \\ -x - 4y + 2z = 0 \\ 2x - y - 3z = -1 \end{cases}$ 24 $\begin{cases} x + y + 3z = 1 \\ 3x + y + z = -1 \\ -x + 4y + 2z = 0. \end{cases}$

25 Show that $\begin{vmatrix} x & y & 1 \\ a_1 & b_1 & 1 \\ a_2 & b_2 & 1 \end{vmatrix} = 0$ is an equation for the line $\mathbf{R}^2$ through the two distinct

points (a_1, b_1) and (a_2, b_2).

Evaluate

26 $\begin{vmatrix} a^2 & ab & ac \\ ba & b^2 & bc \\ ca & cb & c^2 \end{vmatrix}$ 27 $\begin{vmatrix} 1 & 1 & 1 \\ a & b & c \\ a^2 & b^2 & c^2 \end{vmatrix}.$

28 (cont.) Suppose $a \neq b$, $b \neq c$, $c \neq a$. Show that the system

$$\begin{cases} x + y + z = d_1 \\ ax + by + cz = d_2 \\ a^2x + b^2y + c^2z = d_3 \end{cases}$$

has a unique solution.

Find the intersection of the three planes

29 $\begin{cases} 2x - y + 3z = 6 \\ -x - 4y - z = 2 \\ 3x + 2y + z = 2 \end{cases}$ 30 $\begin{cases} 4x + y + z = 10 \\ 2x + 3y + z = 8 \\ -x + 2y + 3z = 3 \end{cases}$

31 $\begin{cases} x + 2y + 3z = 2 \\ -x + 8y + 7z = -2 \\ 2x - y + z = 4 \end{cases}$ **32** $\begin{cases} -x - y + 2z = 4 \\ 2x + y + 6z = -1 \\ 3x + 2y + 4z = -3. \end{cases}$

Find the plane through the three points

33 $(1, 0, 1)$, $(2, 2, -2)$, $(-3, 3, 2)$
35 $(1, 2, 2)$, $(-1, 1, 7)$, $(3, 5, 1)$

34 $(1, -1, 1)$, $(4, -1, -2)$, $(-3, 2, 8)$
36 $(2, 1, -1)$, $(1, 2, 3)$, $(-4, 1, 8)$.

6. CROSS PRODUCT

Geometric Definition Given a pair of vectors **v** and **w**, we define a new vector **v** × **w**.

Cross Product (geometric definition) The **cross product** of **v** and **w**, written

$$\mathbf{v} \times \mathbf{w},$$

is the *vector* perpendicular to **v** and **w** whose direction is determined by the right-hand rule from the pair **v**, **w**, and whose magnitude is the area of the parallelogram based on **v** and **w**. See Fig. 1.

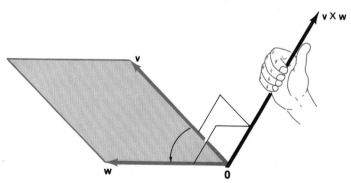

Fig. 1 Cross product, geometric definition

We can note some immediate properties of the cross product. First, if the vectors **v** and **w** are parallel, then the parallelogram collapses so **v** × **w** = **0**. In particular **v** × **v** = **0**. Next, if **v** and **w** are interchanged, then the thumb reverses direction, hence **w** × **v** = −**v** × **w**. Finally, for pairs of the basic unit vectors

$$\mathbf{i} = (1, 0, 0), \qquad \mathbf{j} = (0, 1, 0), \qquad \mathbf{k} = (0, 0, 1),$$

cross products are obvious, for instance **i** × **j** = **k**. Let us summarize:

$$\mathbf{v} \times \mathbf{v} = \mathbf{0}, \qquad \mathbf{w} \times \mathbf{v} = -\mathbf{v} \times \mathbf{w},$$

$$\mathbf{i} \times \mathbf{j} = \mathbf{k}, \qquad \mathbf{j} \times \mathbf{k} = \mathbf{i}, \qquad \mathbf{k} \times \mathbf{i} = \mathbf{j}.$$

Analytic Definition   Motivated by the geometry above, we first *define* analytically all possible cross products of pairs of basic unit vectors:

$$\mathbf{i} \times \mathbf{i} = \mathbf{j} \times \mathbf{j} = \mathbf{k} \times \mathbf{k} = \mathbf{0},$$

$$\mathbf{i} \times \mathbf{j} = \mathbf{k}, \qquad \mathbf{j} \times \mathbf{k} = \mathbf{i}, \qquad \mathbf{k} \times \mathbf{i} = \mathbf{j},$$

$$\mathbf{j} \times \mathbf{i} = -\mathbf{k}, \qquad \mathbf{k} \times \mathbf{j} = -\mathbf{i}, \qquad \mathbf{i} \times \mathbf{k} = -\mathbf{j}.$$

We use this multiplication table and the distributive law to extend the definition to any pair of vectors $\mathbf{v} = v_1\mathbf{i} + v_2\mathbf{j} + v_3\mathbf{k}$ and $\mathbf{w} = w_1\mathbf{i} + w_2\mathbf{j} + w_3\mathbf{k}$:

$$\mathbf{v} \times \mathbf{w} = (v_1\mathbf{i} + v_2\mathbf{j} + v_3\mathbf{k}) \times (w_1\mathbf{i} + w_2\mathbf{j} + w_3\mathbf{k})$$

$$= v_1 w_2 \mathbf{i} \times \mathbf{j} + v_2 w_1 \mathbf{j} \times \mathbf{i} + v_1 w_3 \mathbf{i} \times \mathbf{k} + v_3 w_1 \mathbf{k} \times \mathbf{i} + v_2 w_3 \mathbf{j} \times \mathbf{k} + v_3 w_2 \mathbf{k} \times \mathbf{j}$$

$$= (v_2 w_3 - v_3 w_2)\mathbf{i} + (v_3 w_1 - v_1 w_3)\mathbf{j} + (v_1 w_2 - v_2 w_1)\mathbf{k}.$$

Cross Product (analytic definition)   Let

$$\mathbf{v} = (v_1, v_2, v_3) \qquad \text{and} \qquad \mathbf{w} = (w_1, w_2, w_3).$$

Then $\mathbf{v} \times \mathbf{w} = (v_2 w_3 - v_3 w_2, \; v_3 w_1 - v_1 w_3, \; v_1 w_2 - v_2 w_1)$

$$= \left(\begin{vmatrix} v_2 & v_3 \\ w_2 & w_3 \end{vmatrix}, \; \begin{vmatrix} v_3 & v_1 \\ w_3 & w_1 \end{vmatrix}, \; \begin{vmatrix} v_1 & v_2 \\ w_1 & w_2 \end{vmatrix} \right).$$

Examples

$$(4, 3, -1) \times (-2, 2, 1) = \left(\begin{vmatrix} 3 & -1 \\ 2 & 1 \end{vmatrix}, \; \begin{vmatrix} -1 & 4 \\ 1 & -2 \end{vmatrix}, \; \begin{vmatrix} 4 & 3 \\ -2 & 2 \end{vmatrix} \right)$$

$$= (3 + 2, \; 2 - 4, \; 8 + 6) = (5, -2, 14).$$

$$(1, 0, 1) \times (0, 1, 1) = \left(\begin{vmatrix} 0 & 1 \\ 1 & 1 \end{vmatrix}, \; \begin{vmatrix} 1 & 1 \\ 1 & 0 \end{vmatrix}, \; \begin{vmatrix} 1 & 0 \\ 0 & 1 \end{vmatrix} \right) = (-1, -1, 1).$$

A device for remembering the cross product is a symbolic determinant—expanded by the first row:

$$(v_1, v_2, v_3) \times (w_1, w_2, w_3) = \begin{vmatrix} \mathbf{i} & \mathbf{j} & \mathbf{k} \\ v_1 & v_2 & v_3 \\ w_1 & w_2 & w_3 \end{vmatrix}$$

$$= \begin{vmatrix} v_2 & v_3 \\ w_2 & w_3 \end{vmatrix} \mathbf{i} - \begin{vmatrix} v_1 & v_3 \\ w_1 & w_3 \end{vmatrix} \mathbf{j} + \begin{vmatrix} v_1 & v_2 \\ w_1 & w_2 \end{vmatrix} \mathbf{k}.$$

From this determinant form, we see again that the cross product is anti-commutative: $\mathbf{w} \times \mathbf{v} = -\mathbf{v} \times \mathbf{w}$. For interchanging $\mathbf{v}$ and $\mathbf{w}$ switches two rows of the determinant, hence reverses its sign.

Scalar Triple Product   We have two definitions of the cross product. We shall connect them using the following formula; it expresses as a determinant the inner product of a vector with the cross product of two other vectors:

$$\mathbf{u} \cdot (\mathbf{v} \times \mathbf{w}) = \begin{vmatrix} u_1 & u_2 & u_3 \\ v_1 & v_2 & v_3 \\ w_1 & w_2 & w_3 \end{vmatrix}.$$

To prove the formula, we simply expand the determinant by its first row:

$$\begin{vmatrix} u_1 & u_2 & u_3 \\ v_1 & v_2 & v_3 \\ w_1 & w_2 & w_3 \end{vmatrix} = u_1 \begin{vmatrix} v_2 & v_3 \\ w_2 & w_3 \end{vmatrix} - u_2 \begin{vmatrix} v_1 & v_3 \\ w_1 & w_3 \end{vmatrix} + u_3 \begin{vmatrix} v_1 & v_2 \\ w_1 & w_2 \end{vmatrix}$$

$$= (u_1, u_2, u_3) \cdot \left(\begin{vmatrix} v_2 & v_3 \\ w_2 & w_3 \end{vmatrix}, \begin{vmatrix} v_3 & v_1 \\ w_3 & w_1 \end{vmatrix}, \begin{vmatrix} v_1 & v_2 \\ w_1 & w_2 \end{vmatrix} \right) = \mathbf{u} \cdot (\mathbf{v} \times \mathbf{w}).$$

Examples

(1) $\mathbf{u} = (u_1, u_2, u_3)$, $\mathbf{v} = (0, v_2, v_3)$, $\mathbf{w} = (0, 0, w_3)$,

$$\mathbf{u} \cdot (\mathbf{v} \times \mathbf{w}) = \begin{vmatrix} u_1 & u_2 & u_3 \\ 0 & v_2 & v_3 \\ 0 & 0 & w_3 \end{vmatrix} = u_1 v_2 w_3.$$

(2) $\mathbf{u} = (2, 3, 4)$, $\mathbf{v} = (1, 1, -2)$, $\mathbf{w} = (5, 3, 1)$,

$$\mathbf{u} \cdot (\mathbf{v} \times \mathbf{w}) = \begin{vmatrix} 2 & 3 & 4 \\ 1 & 1 & -2 \\ 5 & 3 & 1 \end{vmatrix} = 2 \begin{vmatrix} 1 & -2 \\ 3 & 1 \end{vmatrix} - 3 \begin{vmatrix} 1 & -2 \\ 5 & 1 \end{vmatrix} + 4 \begin{vmatrix} 1 & 1 \\ 5 & 3 \end{vmatrix}$$

$$= 2 \cdot 7 - 3 \cdot 11 + 4(-2) = -27.$$

The quantity $\mathbf{u} \cdot (\mathbf{v} \times \mathbf{w})$ occurs frequently in applications; it is called the **scalar triple product** and written $[\mathbf{u}, \mathbf{v}, \mathbf{w}]$.

Scalar Triple Product

$$[\mathbf{u}, \mathbf{v}, \mathbf{w}] = \mathbf{u} \cdot (\mathbf{v} \times \mathbf{w}) = \begin{vmatrix} u_1 & u_2 & u_3 \\ v_1 & v_2 & v_3 \\ w_1 & w_2 & w_3 \end{vmatrix}.$$

Because the scalar triple product has an expression as a determinant, properties of determinants imply properties of the scalar triple product. We list some important ones:

Properties of the Scalar Triple Product

(1) If any two of the vectors $\mathbf{u}$, $\mathbf{v}$, $\mathbf{w}$ are equal, then $[\mathbf{u}, \mathbf{v}, \mathbf{w}] = 0$.

(2) If two of its arguments are interchanged, then $[\mathbf{u}, \mathbf{v}, \mathbf{w}]$ changes sign, for instance $[\mathbf{w}, \mathbf{v}, \mathbf{u}] = -[\mathbf{u}, \mathbf{v}, \mathbf{w}]$.

(3) $[\mathbf{u}, \mathbf{v}, \mathbf{w}]$ is a homogeneous linear function of each of its arguments, for example

$$[\mathbf{u}, \mathbf{v}, \mathbf{w}_1 + \mathbf{w}_2] = [\mathbf{u}, \mathbf{v}, \mathbf{w}_1] + [\mathbf{u}, \mathbf{v}, \mathbf{w}_2], \qquad [\mathbf{u}, \mathbf{v}, c\mathbf{w}] = c[\mathbf{u}, \mathbf{v}, \mathbf{w}].$$

Warning   Remember, the cross product of two vectors is a *vector*. The inner product of two vectors is a *scalar* (number). The scalar triple product of three vectors is a *scalar* (because it is an inner product of two vectors).

Equivalence of the Geometric and Analytic Definitions   Let us show that the analytic definition of the cross product satisfies the three conditions of the geometric definition:

(a) $\mathbf{v} \times \mathbf{w}$ is orthogonal to $\mathbf{v}$ and to $\mathbf{w}$.

(b) $\mathbf{v}, \mathbf{w}, \mathbf{v} \times \mathbf{w}$ is a right-handed system (provided $\mathbf{v}$ and $\mathbf{w}$ are not parallel).

(c) $|\mathbf{v} \times \mathbf{w}|$ is the area of the parallelogram based on $\mathbf{v}$ and $\mathbf{w}$ (that is, the parallelogram with vertices $\mathbf{0}, \mathbf{v}, \mathbf{w}, \mathbf{v} + \mathbf{w}$).

Geometric condition (a) is a consequence of a property of the scalar triple product:

$$\mathbf{v} \cdot (\mathbf{v} \times \mathbf{w}) = [\mathbf{v}, \mathbf{v}, \mathbf{w}] = 0, \qquad \mathbf{w} \cdot (\mathbf{v} \times \mathbf{w}) = [\mathbf{w}, \mathbf{v}, \mathbf{w}] = 0,$$

hence $\mathbf{v}$ and $\mathbf{w}$ are both orthogonal to $\mathbf{v} \times \mathbf{w}$.

Geometric condition (c) requires a formula for the length of $\mathbf{v} \times \mathbf{w}$, where $\mathbf{v} \times \mathbf{w}$ is taken in its analytic sense:

$$|\mathbf{v} \times \mathbf{w}|^2 = |\mathbf{v}|^2 |\mathbf{w}|^2 - (\mathbf{v} \cdot \mathbf{w})^2.$$

By direct computation:

$$|\mathbf{v} \times \mathbf{w}|^2 = (v_2 w_3 - v_3 w_2)^2 + (v_3 w_1 - v_1 w_3)^2 + (v_1 w_2 - v_2 w_1)^2$$

$$= \underbrace{\sum v_2{}^2 w_3{}^2}_{6 \text{ terms}} - 2 \underbrace{\sum v_2 v_3 w_2 w_3}_{3 \text{ terms}},$$

and

$$|\mathbf{v}|^2 |\mathbf{w}|^2 - (\mathbf{v} \cdot \mathbf{w})^2 = (v_1{}^2 + v_2{}^2 + v_3{}^2)(w_1{}^2 + w_2{}^2 + w_3{}^2)$$
$$- (v_1 w_1 + v_2 w_2 + v_3 w_3)^2.$$

$$= \left(\underbrace{\sum v_1{}^2 w_1{}^2}_{3 \text{ terms}} + \underbrace{\sum v_2{}^2 w_3{}^2}_{6 \text{ terms}} \right)$$

$$- \left(\underbrace{\sum v_1{}^2 w_1{}^2}_{3 \text{ terms}} + 2 \underbrace{\sum v_2 v_3 w_2 w_3}_{3 \text{ terms}} \right).$$

All terms like $v_1{}^2 w_1{}^2$ cancel; the formula follows.

Now let θ be the angle between $\mathbf{v}$ and $\mathbf{w}$. Then $\mathbf{v} \cdot \mathbf{w} = |\mathbf{v}| \cdot |\mathbf{w}| \cos \theta$, hence

$$|\mathbf{v} \times \mathbf{w}|^2 = |\mathbf{v}|^2 |\mathbf{w}|^2 - (\mathbf{v} \cdot \mathbf{w})^2 = |\mathbf{v}|^2 |\mathbf{w}|^2 (1 - \cos^2 \theta) = |\mathbf{v}|^2 |\mathbf{w}|^2 \sin^2 \theta,$$

$$|\mathbf{v} \times \mathbf{w}| = |\mathbf{v}| \cdot |\mathbf{w}| \sin \theta.$$

This last expression is precisely the required area (Fig. 2).

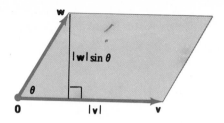

Fig. 2 $|\mathbf{v} \times \mathbf{w}| = |\mathbf{v}| \cdot |\mathbf{w}| \sin \theta = \text{area}$

It remains to verify geometric condition (b). To do so, we need some analytic way of deciding whether a given triple $\mathbf{u}, \mathbf{v}, \mathbf{w}$ is a right-handed system or not.

Observe that $\mathbf{u}, \mathbf{v}, \mathbf{w}$ and $\mathbf{v}, \mathbf{u}, \mathbf{w}$ have opposite *orientations*, that is, one is a right-handed system and the other is left-handed. By analogy, the determinants $[\mathbf{u}, \mathbf{v}, \mathbf{w}]$ and $[\mathbf{v}, \mathbf{u}, \mathbf{w}]$ have opposite signs. This suggests that the sign of $[\mathbf{u}, \mathbf{v}, \mathbf{w}]$ corresponds to the orientation of $\mathbf{u}, \mathbf{v}, \mathbf{w}$. Since $\mathbf{i}, \mathbf{j}, \mathbf{k}$ is right-handed and $[\mathbf{i}, \mathbf{j}, \mathbf{k}] = 1$, we suspect that $\mathbf{u}, \mathbf{v}, \mathbf{w}$ is right-handed if $[\mathbf{u}, \mathbf{v}, \mathbf{w}] > 0$. This indeed is the case, but instead of proving it, we shall simply take the determinant criterion as the definition of right-handedness.

In view of this definition, we must prove that $[\mathbf{v}, \mathbf{w}, \mathbf{v} \times \mathbf{w}] > 0$. But

$$[\mathbf{v}, \mathbf{w}, \mathbf{v} \times \mathbf{w}] = -[\mathbf{v}, \mathbf{v} \times \mathbf{w}, \mathbf{w}] = [\mathbf{v} \times \mathbf{w}, \mathbf{v}, \mathbf{w}]$$
$$= (\mathbf{v} \times \mathbf{w}) \cdot (\mathbf{v} \times \mathbf{w}) = |\mathbf{v} \times \mathbf{w}|^2 > 0.$$

This completes our proof that the vector $\mathbf{v} \times \mathbf{w}$, defined analytically, satisfies the three properties that define $\mathbf{v} \times \mathbf{w}$ geometrically. Hence the definitions are equivalent.

Algebraic Properties   We summarize the main algebraic properties of the cross product. They follow readily from our discussion.

> $$\mathbf{v} \times \mathbf{v} = \mathbf{0}, \qquad \mathbf{w} \times \mathbf{v} = -\mathbf{v} \times \mathbf{w},$$
> $$(a\mathbf{u} + b\mathbf{v}) \times \mathbf{w} = a(\mathbf{u} \times \mathbf{w}) + b(\mathbf{v} \times \mathbf{w}),$$
> $$\mathbf{u} \times (a\mathbf{v} + b\mathbf{w}) = a(\mathbf{u} \times \mathbf{v}) + b(\mathbf{u} \times \mathbf{w}),$$
> $$\mathbf{u} \cdot (\mathbf{v} \times \mathbf{w}) = \mathbf{v} \cdot (\mathbf{w} \times \mathbf{u}) = \mathbf{w} \cdot (\mathbf{u} \times \mathbf{v}) = [\mathbf{u}, \mathbf{v}, \mathbf{w}],$$
> $$\mathbf{v} \times \mathbf{w} = \mathbf{0} \quad \text{if and only if } \mathbf{v} \text{ and } \mathbf{w} \text{ are parallel.}$$

Remark   The associative law and the commutative law are not true in general for the cross product. For instance

$$\mathbf{i} \times \mathbf{j} = \mathbf{k} \quad \text{and} \quad \mathbf{j} \times \mathbf{i} = -\mathbf{k}, \quad \text{so} \quad \mathbf{i} \times \mathbf{j} \neq \mathbf{j} \times \mathbf{i};$$
$$\mathbf{i} \times (\mathbf{j} \times \mathbf{j}) = \mathbf{i} \times \mathbf{0} = \mathbf{0} \quad \text{and} \quad (\mathbf{i} \times \mathbf{j}) \times \mathbf{j} = \mathbf{k} \times \mathbf{j} = -\mathbf{i},$$

so

$$\mathbf{i} \times (\mathbf{j} \times \mathbf{j}) \neq (\mathbf{i} \times \mathbf{j}) \times \mathbf{j}.$$

Torque   The original motivation for the cross product of vectors came from physics. Consider this situation.

Suppose a rigid body is free to turn about the origin. A force $\mathbf{F}$ acts at a point $\mathbf{x}$ of the body. As a result the body wants to rotate about an axis through $\mathbf{0}$ perpendicular to the plane of $\mathbf{x}$ and $\mathbf{F}$ (unless $\mathbf{x}$ and $\mathbf{F}$ are collinear; then there is no turning). See Fig. 3a. As usual, the force vector **F** is *drawn* at its point of application **x**. But analytically it starts at **0**. See Fig. 3b. The positive axis of rotation is determined by the right-hand rule as applied to the pair **x**, **F** in that order: **x** first, **F** second.

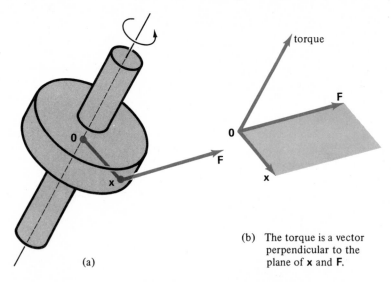

(a)

(b) The torque is a vector perpendicular to the plane of **x** and **F**.

Fig. 3   The torque due to a force **F** applied at **x**

In physics, one speaks of the **torque** (about the origin) resulting from the force **F** applied at **x**. Roughly speaking, torque is a measure of the tendency of a body to rotate under the action of forces. (Torque will be defined precisely in a moment.)

By experiment, if **F** is tripled in magnitude, the torque is tripled; if **x** is moved out twice as far along the same line and the same **F** is applied there, the torque is doubled. Hence the torque is proportional to the length of **x** and to the length of **F**. Therefore (Fig. 3b) the torque is proportional to the area of the parallelogram determined by **x** and **F**.

Let us resolve **F** into a component **G** parallel to **x** and a component **H** perpendicular to **x**. See Fig. 4a. The component **G**, being on line with **x**, produces no

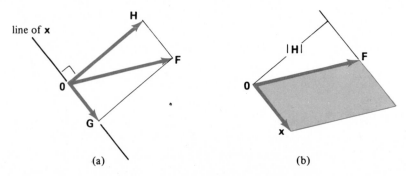

(a)

(b)

Fig. 4 |torque| = area of parallelogram

tendency to turn; only **H** produces torque. By the argument above the amount of torque is $|\mathbf{x}| \cdot |\mathbf{H}|$, the length of the lever arm times the length of **H**. But this product is the area of the parallelogram (Fig. 4b) determined by **x** and **F**.

Therefore the torque about the origin is completely described by the vector **x** × **F**. The length of **x** × **F** is the magnitude of the torque. The direction of **x** × **F** is the positive axis of rotation; with your right thumb along **x** × **F**, your fingers curl in the direction of turning. In physics, torque about the origin is *defined* to be the vector **x** × **F**.

EXERCISES

Compute the cross product

1   $(-2, 2, 1) \times (4, 3, -1)$
2   $(1, 0, 1) \times (1, 1, 0)$
3   $(1, 2, 3) \times (3, 2, 1)$
4   $(3, 1, -1) \times (3, -1, -1)$
5   $(-2, -2, -2) \times (1, 1, 0)$
6   $(-1, 2, 2) \times (3, -1, 2)$.
7   $(\mathbf{i} + \mathbf{j}) \times (\mathbf{i} + \mathbf{j} + \mathbf{k})$
8   $(\cos\theta\,\mathbf{i} + \sin\theta\,\mathbf{j}) \times (-\sin\theta\,\mathbf{i} + \cos\theta\,\mathbf{j})$
9   $(2\mathbf{i} + \mathbf{j} + 3\mathbf{k}) \times (2\mathbf{i} + 2\mathbf{j} - \mathbf{k})$
10   $(\mathbf{i} + 2\mathbf{j} + 3\mathbf{k}) \times (4\mathbf{i} + 5\mathbf{j} + 6\mathbf{k})$.

11   Find a unit vector perpendicular to both $(-3, 0, 1)$ and $(2, -1, -1)$.
12   Find an equation for the line through $(6, 1, -2)$ and perpendicular to the plane through **0**, $(-1, 1, 2)$, and $(2, 3, 4)$.
13   Three vertices of a parallelogram are $(0, 0, 0)$, $(1, 1, 1)$, and $(2, 3, 5)$. Compute its area.
14   Given **a**, find all vectors **x** such that $\mathbf{a} \times \mathbf{x} = \mathbf{x}$.

A force **F** is applied at point **x**. Find its torque about the origin

15   $\mathbf{F} = (-1, 1, 1), \quad \mathbf{x} = (10, 0, 0)$
16   $\mathbf{F} = (3, 0, 0), \quad \mathbf{x} = (0, 0, 1)$
17   $\mathbf{F} = (-1, 1, 1), \quad \mathbf{x} = (2, 2, -1)$
18   $\mathbf{F} = (2, -1, 5), \quad \mathbf{x} = (-7, 1, 0)$.

Compute the scalar triple product

19   $[\mathbf{i}, \mathbf{i} + \mathbf{j}, \mathbf{i} + \mathbf{j} + \mathbf{k}]$
20   $[3\mathbf{i} - \mathbf{j}, 2\mathbf{j} + \mathbf{k}, \mathbf{i} - 4\mathbf{j} - 5\mathbf{k}]$
21   $(4, 1, 1) \cdot (3, 6, 0) \times (2, 5, 4)$
22   $(1, 1, 1) \cdot (1, 2, 3) \times (1, 4, 9)$.

23   Find all vectors **v** for which $[\mathbf{i}, \mathbf{j}, \mathbf{v}] = 3$.
24   Let **u**, **v**, **w** be mutually orthogonal unit vectors. Find $[\mathbf{u}, \mathbf{v}, \mathbf{w}]$.

Prove

25   $\mathbf{u} \cdot (\mathbf{v} \times \mathbf{w}) = \mathbf{v} \cdot (\mathbf{w} \times \mathbf{u})$
26   $(\mathbf{u} + \mathbf{v}) \times \mathbf{w} = \mathbf{u} \times \mathbf{w} + \mathbf{v} \times \mathbf{w}$
27   $(a\mathbf{v}) \times \mathbf{w} = a(\mathbf{v} \times \mathbf{w})$
28   $\mathbf{v} \times (b\mathbf{w}) = b(\mathbf{v} \times \mathbf{w})$.

29   Show that $\begin{vmatrix} \mathbf{a} \cdot \mathbf{u} & \mathbf{a} \cdot \mathbf{v} \\ \mathbf{b} \cdot \mathbf{u} & \mathbf{b} \cdot \mathbf{v} \end{vmatrix} = (\mathbf{a} \times \mathbf{b}) \cdot (\mathbf{u} \times \mathbf{v})$ provided **a** and **b** are chosen from **i**, **j**, **k**.
30   (cont.) Now prove the formula in general.
31   Suppose **a** and **b** are vectors such that $\mathbf{a} \cdot \mathbf{v} = \mathbf{b} \cdot \mathbf{v}$ for *all* vectors **v**. Prove that $\mathbf{a} = \mathbf{b}$.
32* (cont.) Use this and the result of Ex. 30 to prove the identity

$$\mathbf{u} \times (\mathbf{v} \times \mathbf{w}) = (\mathbf{u} \cdot \mathbf{w})\mathbf{v} - (\mathbf{u} \cdot \mathbf{v})\mathbf{w}.$$

33   Use the result of Ex. 30 for a proof of $|\mathbf{v} \times \mathbf{w}|^2 = |\mathbf{v}|^2 |\mathbf{w}|^2 - (\mathbf{v} \cdot \mathbf{w})^2$ different from the one given in the text.
34   Prove that $\mathbf{v} = (\mathbf{a} \times \mathbf{b}) \times (\mathbf{a} \times \mathbf{c})$ is parallel to **a**.

7. APPLICATIONS OF THE CROSS PRODUCT

Volume Two non-collinear (non-parallel) vectors **u** and **v** determine a parallelogram, whose area is $|\mathbf{u} \times \mathbf{v}|$. Similarly, three non-coplanar vectors determine a parallelepiped, (Fig. 1a) whose volume is given by a formula involving a scalar triple product.

> The volume of the parallelepiped determined by three non-coplanar vectors **u**, **v**, **w** is
>
> $$V = |[\mathbf{u}, \mathbf{v}, \mathbf{w}]| = |\mathbf{u} \cdot (\mathbf{v} \times \mathbf{w})|.$$

Proof The volume is

$$V = (\text{area of base})(\text{height}).$$

For the base, take the parallelogram determined by **u** and **v**. Its area is $|\mathbf{u} \times \mathbf{v}|$; furthermore, the vector **u** × **v** is perpendicular to the base.

Suppose first that **w** lies on the same side of the base as **u** × **v** lies. By definition, the height of the parallelpiped is the projection (Fig. 1b) of **w** onto **u** × **v**:

$$(\text{height}) = |\mathbf{w}| \cos \theta.$$

Therefore
$$V = |\mathbf{u} \times \mathbf{v}||\mathbf{w}| \cos \theta = (\mathbf{u} \times \mathbf{v}) \cdot \mathbf{w}$$
$$= [\mathbf{u}, \mathbf{v}, \mathbf{w}] = |[\mathbf{u}, \mathbf{v}, \mathbf{w}]|.$$

In case **w** and **u** × **v** lis on opposite sides of the base, then

$$V = -[\mathbf{u}, \mathbf{v}, \mathbf{w}] = |[\mathbf{u}, \mathbf{v}, \mathbf{w}]|.$$

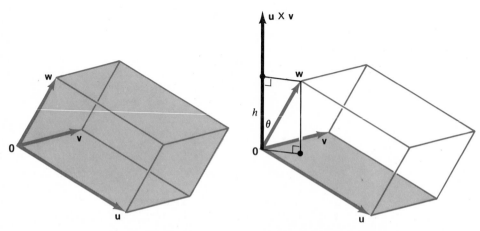

(a) The parallelepiped determined by three non-coplanar vectors

(b) Its volume equals its base area times its height.

Fig. 1 Volume of a parallelepiped

As a corollary, we derive a test for three vectors to lie on a plane through **0**. This happens if and only if the parallelepiped they determine collapses, that is, has volume 0. But its volume is $|[\mathbf{u}, \mathbf{v}, \mathbf{w}]|$, so we have the following test.

> Three vectors, **u**, **v**, and **w** lie on a plane through **0** if and only if
> $$[\mathbf{u}, \mathbf{v}, \mathbf{w}] = 0.$$

■ **EXAMPLE 1** Show that $\mathbf{w} = (1, -3, 3)$ lies in the plane of $\mathbf{u} = (5, 6, 1)$ and $\mathbf{v} = (2, 3, 0)$.

Solution Compute $[\mathbf{u}, \mathbf{v}, \mathbf{w}]$ by minors of the second row:

$$[\mathbf{u}, \mathbf{v}, \mathbf{w}] = \begin{vmatrix} 5 & 6 & 1 \\ 2 & 3 & 0 \\ 1 & -3 & 3 \end{vmatrix} = -2 \begin{vmatrix} 6 & 1 \\ -3 & 3 \end{vmatrix} + 3 \begin{vmatrix} 5 & 1 \\ 1 & 3 \end{vmatrix} = (-2)(21) + (3)(14) = 0.$$

Hence **u**, **v**, and **w** are coplanar. Obviously **u** and **v** are not parallel, so they determine a plane through **0**, and **w** must lie in this plane. ■

Intersection of Two Planes Given two planes $\mathbf{x} \cdot \mathbf{m} = p$ and $\mathbf{x} \cdot \mathbf{n} = q$, how can we find their line of intersection? We must assume the planes are not parallel, that is, their normal vectors **m** and **n** are not parallel. Then $\mathbf{a} = \mathbf{m} \times \mathbf{n}$ is perpendicular to both **m** and **n**, so **a** is parallel to the line of intersection (Fig. 2).

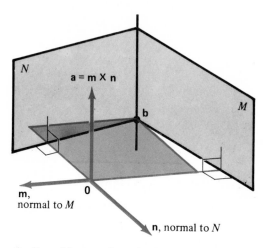

Fig. 2 To find **b** on the line of intersection of M and N and on the plane of their normals

If we can find a single point **b** on both planes, then the desired line is

$$\mathbf{x} = t\mathbf{a} + \mathbf{b}.$$

We shall look for such a vector **b** in the plane of **m** and **n**, that is, a vector of the form $\mathbf{b} = u\mathbf{m} + v\mathbf{n}$.

Since **b** lies on both planes, **b** must satisfy both $\mathbf{b} \cdot \mathbf{m} = p$ and $\mathbf{b} \cdot \mathbf{n} = q$. These conditions lead to a pair of linear equations for u and v:

$$\begin{cases} \mathbf{m} \cdot \mathbf{m}\, u + \mathbf{m} \cdot \mathbf{n}\, v = p \\ \mathbf{m} \cdot \mathbf{n}\, u + \mathbf{n} \cdot \mathbf{n}\, v = q. \end{cases}$$

The system determinant is

$$\begin{vmatrix} \mathbf{m} \cdot \mathbf{m} & \mathbf{m} \cdot \mathbf{n} \\ \mathbf{m} \cdot \mathbf{n} & \mathbf{n} \cdot \mathbf{n} \end{vmatrix} = |\mathbf{m}|^2 |\mathbf{n}|^2 - (\mathbf{m} \cdot \mathbf{n})^2 = |\mathbf{m} \times \mathbf{n}|^2 > 0.$$

Therefore the system has a unique solution.

■ **EXAMPLE 2**   Find the line of intersection of the planes $x + y + z = -1$ and $2x + y - z = 3$.

Solution 1   The equations of these planes are $\mathbf{x} \cdot \mathbf{m} = p$ and $\mathbf{x} \cdot \mathbf{n} = q$, where

$$\mathbf{m} = (1, 1, 1), \qquad \mathbf{n} = (2, 1, -1), \qquad p = -1, \qquad q = 3.$$

Set

$$\mathbf{a} = \mathbf{m} \times \mathbf{n} = (-2, 3, -1).$$

This vector is perpendicular to **m** and **n**, so it is parallel to the line of intersection.

To find a point on the line of intersection, set $\mathbf{b} = u\mathbf{m} + v\mathbf{n}$ and choose u and v so that $\mathbf{b} \cdot \mathbf{m} = -1$ and $\mathbf{b} \cdot \mathbf{n} = 3$. Now

$$|\mathbf{m}|^2 = 3, \qquad \mathbf{m} \cdot \mathbf{n} = 2, \qquad |\mathbf{n}|^2 = 6,$$

so the equations $\mathbf{b} \cdot \mathbf{m} = -1$ and $\mathbf{b} \cdot \mathbf{n} = 3$ become

$$\begin{cases} 3u + 2v = -1 \\ 2u + 6v = 3. \end{cases}$$

The solution is $u = -\frac{6}{7}$, $v = \frac{11}{14}$; therefore

$$\mathbf{b} = -\tfrac{6}{7}(1, 1, 1) + \tfrac{11}{14}(2, 1, -1) = \left(\tfrac{10}{14}, -\tfrac{1}{14}, -\tfrac{23}{14}\right).$$

The required line is

$$\mathbf{x} = t\mathbf{a} + \mathbf{b} = t(-2, 3, -1) + \left(\tfrac{10}{14}, -\tfrac{1}{14}, -\tfrac{23}{14}\right)$$
$$= \tfrac{1}{14}(-28t + 10, 42t - 1, -14t - 23).$$

Solution 2   Treat the given planes as a linear system:

$$\begin{cases} x + y + z = -1 \\ 2x + y - z = 3. \end{cases}$$

Keep the first equation; add -2 times the first to the second to eliminate x:

$$\begin{cases} x + y + z = -1 \\ -y - 3z = 5. \end{cases}$$

Set $z = -s$, then solve for y and x:

$$y = -3z - 5 = 3s - 5,$$
$$x = -1 - y - z = -1 - (3s - 5) + s = -2s + 4.$$

Hence $\mathbf{x} = (-2s + 4, \ 3s - 5, \ -s)$.

This is another parametric form of the line of intersection. Note that this line is parallel to $(-2, 3, -1) = \mathbf{m} \times \mathbf{n}$ and passes through $(4, -5, 0)$, a point on both planes. You should check that the change of parameter $s = t + \frac{23}{14}$ brings this second equation to the parametric form in the first solution.   ∎

Homogeneous Equations   Suppose we are given three planes through the origin,

$$\mathbf{x} \cdot \mathbf{u} = 0, \qquad \mathbf{x} \cdot \mathbf{v} = 0, \qquad \mathbf{x} \cdot \mathbf{w} = 0.$$

In general, the planes will have only the point $\mathbf{0}$ in common. However, it may happen that they have a line in common, or even coincide. This occurs precisely when their three normal vectors $\mathbf{u}, \mathbf{v}, \mathbf{w}$ lie in the same plane. The situation can be described algebraically:

A system of three linear homogeneous equations

$$\mathbf{x} \cdot \mathbf{u} = 0, \qquad \mathbf{x} \cdot \mathbf{v} = 0, \qquad \mathbf{x} \cdot \mathbf{w} = 0$$

has a solution $\mathbf{x} \neq \mathbf{0}$ (a non-trivial solution) if and only if

$$[\mathbf{u}, \mathbf{v}, \mathbf{w}] = \mathbf{u} \cdot (\mathbf{v} \times \mathbf{w}) = 0.$$

If $\mathbf{x}$ is any solution, then $t\mathbf{x}$ is a solution for each t.

It is useful to restate this result in terms of determinants and linear equations.

A homogeneous system of linear equations

$$\begin{cases} u_1 x_1 + u_2 x_2 + u_3 x_3 = 0 \\ v_1 x_1 + v_2 x_2 + v_3 x_3 = 0 \\ w_1 x_1 + w_2 x_2 + w_3 x_3 = 0 \end{cases}$$

has a solution $(x_1, x_2, x_3) \neq (0, 0, 0)$ if and only if

$$\begin{vmatrix} u_1 & u_2 & u_3 \\ v_1 & v_2 & v_3 \\ w_1 & w_2 & w_3 \end{vmatrix} = 0.$$

If (x_1, x_2, x_3) is any solution, then (tx_1, tx_2, tx_3) is also a solution for each t.

Remark   Recall that Cramer's Rule (Section 4) guarantees a *unique* solution if $[\mathbf{u}, \mathbf{v}, \mathbf{w}] \neq 0$. Since $(0, 0, 0)$ is obviously a solution to the homogeneous system, it is the *only* solution when $[\mathbf{u}, \mathbf{v}, \mathbf{w}] \neq 0$. This proves again that for a homogeneous system to have a non-trivial solution, its determinant must be zero.

∎ **EXAMPLE 3**   Find a non-trivial solution of

$$\mathbf{x} \cdot \mathbf{u} = \mathbf{x} \cdot \mathbf{v} = \mathbf{x} \cdot \mathbf{w} = 0,$$

where $\mathbf{u} = (1, -2, 2)$, $\mathbf{v} = (3, 1, -2)$,   and $\mathbf{w} = (5, -3, 2)$.

Solution   First

$$[\mathbf{u}, \mathbf{v}, \mathbf{w}] = \mathbf{u} \cdot (\mathbf{v} \times \mathbf{w}) = \begin{vmatrix} 1 & -2 & 2 \\ 3 & 1 & -2 \\ 5 & -3 & 2 \end{vmatrix} = 0.$$

Therefore the vectors $\mathbf{u}, \mathbf{v}, \mathbf{w}$ are coplanar, so the corresponding planes have a line L in common. (Note that $\mathbf{w} = 2\mathbf{u} + \mathbf{v}$, direct evidence of the coplanarity.) Certainly L contains the point $\mathbf{0}$, which is common to all three planes. Furthermore, it contains $\mathbf{u} \times \mathbf{v}$, since this vector starts at $\mathbf{0}$ and is parallel to L. Therefore L is the set of all multiples $t(\mathbf{u} \times \mathbf{v})$, provided $\mathbf{u} \times \mathbf{v} \neq \mathbf{0}$. (A similar statement holds for $\mathbf{v} \times \mathbf{w}$ and $\mathbf{w} \times \mathbf{u}$.) Now

$$\mathbf{u} \times \mathbf{v} = \begin{vmatrix} \mathbf{i} & \mathbf{j} & \mathbf{k} \\ 1 & -2 & 2 \\ 3 & 1 & -2 \end{vmatrix} = (2, 8, 7) \neq \mathbf{0}$$

so each point $\mathbf{x} = t(2, 8, 7)$ is a solution.   ∎

Remark   Note that

$$\mathbf{v} \times \mathbf{w} = (-4, -16, -14) = -2(\mathbf{u} \times \mathbf{v}) \quad \text{and} \quad \mathbf{w} \times \mathbf{u} = (-2, -8, -7) = -(\mathbf{u} \times \mathbf{v}).$$

Hence the vectors $\mathbf{v} \times \mathbf{w}$ and $\mathbf{w} \times \mathbf{u}$ also lead to the same set of solutions.

Skew Lines   Let $\mathbf{x} = s\mathbf{u} + \mathbf{x}_0$ and $\mathbf{x} = t\mathbf{v} + \mathbf{y}_0$ be two lines in $\mathbf{R}^3$ that do not intersect and are not parallel, i.e., two skew lines. We ask how far apart they are (Fig. 3a).

The vector $\mathbf{u} \times \mathbf{v}$ is perpendicular to both lines, so $\mathbf{n} = (\mathbf{u} \times \mathbf{v})/|\mathbf{u} \times \mathbf{v}|$ is a unit vector perpendicular to both lines. From Fig. 3b we see that the required distance is the length of the projection of $\mathbf{x}_0 - \mathbf{y}_0$ on $\mathbf{n}$, that is, $|(\mathbf{x}_0 - \mathbf{y}_0) \cdot \mathbf{n}|$.

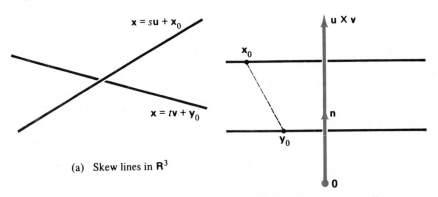

(a)   Skew lines in $\mathbf{R}^3$

(b)   As seen from a direction that makes the skew lines appear to be parallel

Fig. 3

■ **EXAMPLE 4**   Find the distance between the lines

$$\mathbf{x} = (s - 1, s, 2s + 2) \quad \text{and} \quad \mathbf{x} = (-t + 1, -t + 1, -t + 1).$$

Solution   The two lines are $\mathbf{x} = s\mathbf{u} + \mathbf{x}_0$ and $\mathbf{x} = t\mathbf{v} + \mathbf{y}_0$, where

$$\mathbf{u} = (1, 1, 2), \qquad \mathbf{v} = (-1, -1, -1), \qquad \mathbf{x}_0 = (-1, 0, 2), \qquad \mathbf{y}_0 = (1, 1, 1).$$

Therefore $\mathbf{u} \times \mathbf{v} = (1, -1, 0)$ and

$$\mathbf{n} = \frac{\mathbf{u} \times \mathbf{v}}{|\mathbf{u} \times \mathbf{v}|} = \frac{1}{2}\sqrt{2}(1, -1, 0).$$

Finally, the distance between the lines is

$$|(\mathbf{x}_0 - \mathbf{y}_0) \cdot \mathbf{n}| = |(-2, -1, 1) \cdot \tfrac{1}{2}\sqrt{2}(1, -1, 0)| = |-\tfrac{1}{2}\sqrt{2}| = \tfrac{1}{2}\sqrt{2}. \qquad \blacksquare$$

Parametric Form of a Plane   Suppose the *vectors* $\mathbf{a}$ and $\mathbf{b}$ are not parallel, that is, the three *points* $\mathbf{0}, \mathbf{a}, \mathbf{b}$ are non-collinear. Then $\mathbf{a}$ and $\mathbf{b}$ lie on a unique plane P. See Fig. 4a. All scalar multiples $s\mathbf{a}$ and $t\mathbf{b}$ of $\mathbf{a}$ and of $\mathbf{b}$ also lie on P; by the parallelogram law, so do all sums $s\mathbf{a} + t\mathbf{b}$.

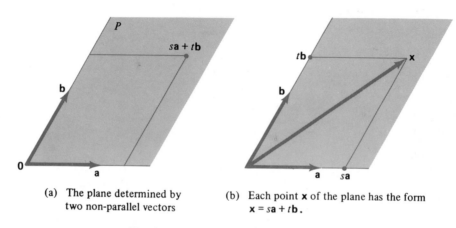

(a)   The plane determined by   (b)   Each point $\mathbf{x}$ of the plane has the form
two non-parallel vectors $\mathbf{x} = s\mathbf{a} + t\mathbf{b}$.

Fig. 4   The plane spanned by $\mathbf{a}$ and $\mathbf{b}$

Conversely each vector $\mathbf{x}$ in the plane P can be expressed in the form $s\mathbf{a} + t\mathbf{b}$. For we can construct a parallelogram with $\mathbf{x}$ as a diagonal as shown in Fig. 4b. This displays $\mathbf{x}$ as a sum of two vectors, a multiple of $\mathbf{a}$ and a multiple of $\mathbf{b}$. We say that $\mathbf{a}$ and $\mathbf{b}$ **span** the plane P.

Two non-parallel vectors $\mathbf{a}$ and $\mathbf{b}$ span the plane consisting of all vectors

$$\mathbf{x} = s\mathbf{a} + t\mathbf{b},$$

where $-\infty < s < \infty$ and $-\infty < t < \infty$.

Remark   In the subject called Linear Algebra, vectors $\mathbf{a}$ and $\mathbf{b}$ that are non-parallel are called **linearly independent**. It means in effect that neither is a scalar multiple of the other. Alternatively, if $s\mathbf{a} + t\mathbf{b} = \mathbf{0}$, then $s = t = 0$.

Next, suppose two non-parallel vectors $\mathbf{a}$ and $\mathbf{b}$ span a plane Q, and suppose $\mathbf{c}$ is a point not on Q. We seek the plane P through $\mathbf{c}$ parallel to Q. Clearly P consists of all $\mathbf{x} = \mathbf{z} + \mathbf{c}$, where $\mathbf{z}$ belongs to Q. See Fig. 5. But we know all such $\mathbf{z}$, namely, $\mathbf{z} = s\mathbf{a} + t\mathbf{b}$.

Given a point $\mathbf{c}$ and two non-parallel vectors $\mathbf{a}$ and $\mathbf{b}$, the plane through $\mathbf{c}$ parallel to the plane spanned by $\mathbf{a}$ and $\mathbf{b}$ consists of all points

$$\mathbf{x} = s\mathbf{a} + t\mathbf{b} + \mathbf{c},$$

where $-\infty < s < \infty$ and $-\infty < t < \infty$.

The variables s and t are called **parameters**, and a plane presented in this fashion is said to be in **parametric form**.

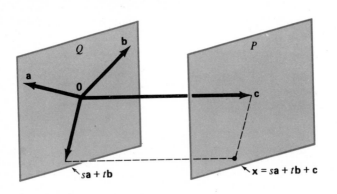

Fig. 5 Parametric form of a plane

Example $\mathbf{a} = (1, 0, 1)$, $\mathbf{b} = (1, 1, -1)$, $\mathbf{c} = (-1, 1, 2)$.

Clearly neither $\mathbf{a}$ nor $\mathbf{b}$ is a multiple of the other, so they are non-parallel. The plane through $\mathbf{c}$ parallel to the plane spanned by $\mathbf{a}$ and $\mathbf{b}$ consists of all $\mathbf{x}$ such that

$$\mathbf{x} = s\mathbf{a} + t\mathbf{b} + \mathbf{c} = s(1, 0, 1) + t(1, 1, -1) + (-1, 1, 2).$$

In coordinates: $x = s + t - 1$, $y = t + 1$, $z = s - t + 2$.

Given a plane in parametric form, $\mathbf{x} = s\mathbf{a} + t\mathbf{b} + \mathbf{c}$, how do we find a non-parametric equation for the plane, that is, an equation of the form $\mathbf{n} \cdot \mathbf{x} = p$ or $ax + by + cz = d$? First we find a vector $\mathbf{n}$ normal to the plane. That is easy: we take $\mathbf{n} = \mathbf{a} \times \mathbf{b}$, which is orthogonal to both $\mathbf{a}$ and $\mathbf{b}$, hence to the plane. Because $\mathbf{a}$ and $\mathbf{b}$ are non-parallel, the vector $\mathbf{n}$ is guaranteed to be non-zero. Then we note that

$$\mathbf{n} \cdot \mathbf{x} = s\,\mathbf{n} \cdot \mathbf{a} + t\,\mathbf{n} \cdot \mathbf{b} + \mathbf{n} \cdot \mathbf{c} = \mathbf{n} \cdot \mathbf{c}.$$

Thus $\mathbf{n} \cdot \mathbf{x} = \mathbf{n} \cdot \mathbf{c}$ is a non-parametric equation of the plane, where $\mathbf{n} = \mathbf{a} \times \mathbf{b}$.

Example   In the previous example

$$\mathbf{n} = \mathbf{a} \times \mathbf{b} = (1, 0, 1) \times (1, 1, -1) = (-1, 2, 1)$$

and $$\mathbf{n} \cdot \mathbf{c} = (-1, 2, 1) \cdot (-1, 1, 2) = 5,$$

so an equation of the plane is $\mathbf{n} \cdot \mathbf{x} = 5$, that is,

$$-x + 2y + z = 5.$$

Plane through Three Points   Let us take a second look at a problem discussed earlier (p. 654): to find the plane P through three non-collinear points $\mathbf{a}$, $\mathbf{b}$, $\mathbf{c}$.

The vectors $\mathbf{b} - \mathbf{a}$ and $\mathbf{c} - \mathbf{a}$ are parallel to P, and not parallel to each other; their cross product

$$\mathbf{n} = (\mathbf{b} - \mathbf{a}) \times (\mathbf{c} - \mathbf{a})$$

is non-zero and orthogonal to P. Hence

$$\mathbf{n} \cdot \mathbf{x} = \mathbf{n} \cdot \mathbf{a}$$

is an equation of the plane P.

Clearly

$$\mathbf{x} = s(\mathbf{b} - \mathbf{a}) + t(\mathbf{c} - \mathbf{a}) + \mathbf{a}$$

is a parametric equation of P. This equation is equivalent to

$$\mathbf{x} = (1 - s - t)\mathbf{a} + s\mathbf{b} + t\mathbf{c},$$

which we can write in the following symmetric form.

> The plane through three non-collinear points $\mathbf{a}$, $\mathbf{b}$, $\mathbf{c}$ consists of all points
>
> $$\mathbf{x} = r\mathbf{a} + s\mathbf{b} + t\mathbf{c},$$
>
> where r, s, t take on all real values subject to $r + s + t = 1$.

Equilibrium   Suppose forces $\mathbf{F}_1, \cdots, \mathbf{F}_n$ are applied at points $\mathbf{x}_1, \cdots, \mathbf{x}_n$ of a rigid body. Now a rigid body is in equilibrium when both the sum of the forces vanishes and the sum of the turning moments (torques) of the forces about $\mathbf{0}$ vanishes. Thus the conditions for equilibrium are the two vector equations:

$$\mathbf{F}_1 + \mathbf{F}_2 + \cdots + \mathbf{F}_n = \mathbf{0},$$
$$\mathbf{x}_1 \times \mathbf{F}_1 + \mathbf{x}_2 \times \mathbf{F}_2 + \cdots + \mathbf{x}_n \times \mathbf{F}_n = \mathbf{0}.$$

EXERCISES

Find the volume of the parallelepiped determined by

1 $(1, 1, 0)$ $(0, 1, 1)$ $(1, 1, 0)$

2 $(4, -1, 0)$ $(3, 0, 2)$ $(1, 1, 1)$.

Find the line of intersection (parametric form) of the planes

3 $x + 2y + 3z = 0$ $y - z = 1$

4 $x - y + z = 0$ $x + y + z = 3$

5 $x + 2y + z = 3$ $2x - y + z = 4$

6 $x + y = 1$ $y + z = -1$.

Find all solutions

7 $\begin{cases} -2x + 6y = 0 \\ 3x - 9y = 0 \end{cases}$

8 $\begin{cases} -12x + 4y = 0 \\ 3x - y = 0 \end{cases}$

9 $\begin{cases} 5x + 4y + 3z = 0 \\ -x + 2y + z = 0 \\ 3x + y + z = 0 \end{cases}$

10 $\begin{cases} -2x + 2y + 4z = 0 \\ 3x + 8y + 5z = 0 \\ -3x - y + 2z = 0 \end{cases}$

11 $\begin{cases} 3x - 4y + 2z = 0 \\ 5x + 6y = 0 \\ x + 5y - z = 0 \end{cases}$

12 $\begin{cases} 3x + 3y + 2z = 0 \\ 7x + 5y + 12z = 0 \\ x + 2y - 3z = 0 \end{cases}$

13 $\begin{cases} 4x + 3y + 5z = 0 \\ -4x - 3y - 5z = 0 \\ 12x + 9y + 15z = 0 \end{cases}$

14 $\begin{cases} 6x - 9y + 12z = 0 \\ 2x - 3y + 4z = 0 \\ -10x + 15y - 20z = 0. \end{cases}$

Find a non-parametric equation for the parametric plane

15 $\mathbf{x} = (1, s, t)$

16 $\mathbf{x} = (s, s + t, t - 1)$

17 $\mathbf{x} = (s + 2, s + t + 1, s - t)$

18 $\mathbf{x} = (3s, 2s - t, 2t + 1)$

19 $\mathbf{x} = (-s + t + 1, s - t + 2, -2s - t + 3)$

20 $\mathbf{x} = (s - 3t + 2, 2s + t + 1, 2s - t + 3)$.

Find the distance between the lines

21 $\mathbf{x} = s(1, 1, 1) + (0, 0, -1)$ and $\mathbf{x} = t(-1, 2, 2) + (2, 3, 4)$

22 $\mathbf{x} = s(-1, 1, 0) + (1, 0, 0)$ and $\mathbf{x} = t(-1, -1, 0) + (1, 1, 1)$

23 $\overline{\mathbf{ab}}$ and $\overline{\mathbf{cd}}$, where

$\mathbf{a} = (0, 0, 1),$ $\mathbf{b} = (1, 2, 3),$ $\mathbf{c} = (1, 1, 0),$ $\mathbf{d} = (-1, -1, -1)$

24 $\overline{\mathbf{ac}}$ and $\overline{\mathbf{bd}}$ with the same $\mathbf{a}, \mathbf{b}, \mathbf{c}, \mathbf{d}$.

Find the nearest points on the two lines

25   the lines of Ex. 21

26   the lines of Ex. 22.

27   Find a parametric form for the plane through $(1, 1, 0)$, $(1, 2, 1)$, and $(-1, -1, -1)$.

28   Find a non-parametric equation for the same plane.

29   A seesaw with unequal arms of lengths a and b is in horizontal equilibrium. Find the relations between weights A and B at the ends and the upward force C at the fulcrum.

30   Unit vertical forces act downward at the points $\mathbf{p}_1, \cdots, \mathbf{p}_n$ of the horizontal x, y-plane. Suppose a single force $\mathbf{F}$ acts at another point $\mathbf{p}$ of the plane so that the rigid system is in equilibrium. Find $\mathbf{F}$ and $\mathbf{p}$.

31   A force $\mathbf{F}$ is applied at a point $\mathbf{x}$. Its **torque about a point** $\mathbf{p}$ is $(\mathbf{x} - \mathbf{p}) \times \mathbf{F}$. Suppose $\mathbf{F}_1, \cdots, \mathbf{F}_n$ are applied at points $\mathbf{x}_1, \cdots, \mathbf{x}_n$ of a rigid body and the body is in equilibrium. Show that the sum of the torques about $\mathbf{p}$ equals zero. (Here $\mathbf{p}$ is any point of space, not just $\mathbf{0}$.)

32   A **couple** consists of a pair of opposite forces $\mathbf{F}$ and $-\mathbf{F}$ applied at two different points $\mathbf{p}$ and $\mathbf{q}$. Show that the total torque is unchanged if $\mathbf{p}$ and $\mathbf{q}$ are displaced the same amount, i.e., replaced by $\mathbf{p} + \mathbf{c}$ and $\mathbf{q} + \mathbf{c}$.

33   Prove $[\mathbf{a}, \mathbf{b}, \mathbf{c}]^2 \le |\mathbf{a}|^2 |\mathbf{b}|^2 |\mathbf{c}|^2$.

34   Prove that the points $\mathbf{a}, \mathbf{b}, \mathbf{c}$ are collinear if and only if $\mathbf{a} \times \mathbf{b} + \mathbf{b} \times \mathbf{c} + \mathbf{c} \times \mathbf{a} = \mathbf{0}$.

35   Suppose $\mathbf{a}, \mathbf{b}, \mathbf{c}$ are non-collinear points. Describe the set of points $\mathbf{x} = r\mathbf{a} + s\mathbf{b} + t\mathbf{c}$ where $r \ge 0, \ s \ge 0, \ t \ge 0$, and $r + s + t = 1$.

36   (cont.) Suppose in addition that $\mathbf{0}$ does not lie on the plane of $\mathbf{a}, \mathbf{b}$, and $\mathbf{c}$, and replace the last condition by $r + s + t \le 1$. Now what is the set?

37*   Vectors $\mathbf{a}, \mathbf{b}, \mathbf{c}$ are called **linearly independent** if the only solution of $r\mathbf{a} + s\mathbf{b} + t\mathbf{c} = \mathbf{0}$ is $r = s = t = 0$. Show that $\mathbf{a}, \mathbf{b}, \mathbf{c}$ are linearly independent if and only if $[\mathbf{a}, \mathbf{b}, \mathbf{c}] \ne 0$.

38*   Suppose the points $\mathbf{a}, \mathbf{b}, \mathbf{c}$ are non-collinear. Prove that

$$[\mathbf{x} - \mathbf{a}, \mathbf{x} - \mathbf{b}, \mathbf{x} - \mathbf{c}] = 0$$

is an equation of the plane through $\mathbf{a}, \mathbf{b}$, and $\mathbf{c}$.

8. MISCELLANEOUS EXERCISES

1 Let $\mathbf{a}$ and $\mathbf{b}$ be points of $\mathbf{R}^3$ and $c > 0$. Show that $(\mathbf{x} - \mathbf{a}) \cdot (\mathbf{x} - \mathbf{b}) = c^2$ is the equation of a sphere. Find its center and radius.

2 The same for $\sum_1^n |\mathbf{x} - \mathbf{a}_i|^2 = k^2$ provided k is sufficiently large.

3 Show that the line through the midpoints of two sides of a triangle is parallel to the third side.

4 Let $\mathbf{a}, \mathbf{b}, \mathbf{c}, \mathbf{d}$ be the vertices in order of a skew quadrilateral. Show that the midpoints of its sides are the vertices of a parallelogram.

5 (cont.) What is the area of this parallelogram?

6 Suppose $\mathbf{u}, \mathbf{v}, \mathbf{w}$ are mutually orthogonal and $\mathbf{x} = a\mathbf{u} + b\mathbf{v} + c\mathbf{w}$. Find $|\mathbf{x}|$.

7 Prove that the diagonals of a rhombus are perpendicular.

8 The 9 coordinates of the vertices of a certain triangle are all integers. Show that the area of the triangle is at least $\frac{1}{2}$.

9 Solve $\begin{cases} x + y + 2z = 6 \\ x + 3y + 4z = 11 \\ 2x - y + 2z = 4. \end{cases}$

10 Suppose $\mathbf{a}$ and $\mathbf{b}$ are not parallel. Interpret $[\mathbf{x} - \mathbf{c}, \mathbf{a}, \mathbf{b}] = 0$ geometrically.

11 Interpret geometrically the conditions

$$|\mathbf{a}|^2 = |\mathbf{b}|^2 = 2\mathbf{a} \cdot \mathbf{b}.$$

12 Let $\mathbf{u} = (\cos \alpha_1, \cos \alpha_2, \cos \alpha_3)$ and $\mathbf{v} = (\cos \beta_1, \cos \beta_2, \cos \beta_3)$ be two unit vectors. Show that the angle θ between them satisfies

$$\cos \theta = \cos \alpha_1 \cos \beta_1 + \cos \alpha_2 \cos \beta_2 + \cos \alpha_3 \cos \beta_3.$$

Interpret the formula when $\alpha_3 = \beta_3 = \frac{1}{2}\pi$.

13 Let $\mathbf{a}_1, \mathbf{a}_2, \cdots, \mathbf{a}_n$ be the vertices of a regular polygon in the plane. What is the vector $(\sum_1^n \mathbf{a}_i)/n$?

14 Let L be the line of intersection of two non-parallel planes $\mathbf{a} \cdot \mathbf{x} = c$ and $\mathbf{b} \cdot \mathbf{x} = d$. Show that the most general plane containing L is

$$(s\mathbf{a} + t\mathbf{b}) \cdot \mathbf{x} = sc + td,$$

where $s^2 + t^2 = 1$.

15 Suppose $a > 0$, $b > 0$, $c > 0$. Find the area of the triangle with vertices $(a, 0, 0)$, $(0, b, 0)$, $(0, 0, c)$.

16 Prove the **Jacobi identity**

$$\mathbf{u} \times (\mathbf{v} \times \mathbf{w}) + \mathbf{v} \times (\mathbf{w} \times \mathbf{u}) + \mathbf{w} \times (\mathbf{u} \times \mathbf{v}) = \mathbf{0}.$$

[*Hint* Use Ex. 32, p. 663.]

17 Let $\mathbf{a}_1, \cdots, \mathbf{a}_r$ be points of $\mathbf{R}^3$. Consider all planes $\mathbf{n} \cdot \mathbf{x} = p$ in normal form such that

$$(\mathbf{a}_1 \cdot \mathbf{n} - p) + \cdots + (\mathbf{a}_r \cdot \mathbf{n} - p) = 0.$$

Show that these planes pass through a common point.

18 Let $\mathbf{x} \cdot \mathbf{m} = p$ and $\mathbf{x} \cdot \mathbf{n} = q$ be non-parallel planes in normal form. Show that $\mathbf{x} \cdot \mathbf{m} - p = \pm(\mathbf{x} \cdot \mathbf{n} - q)$ are the two planes through their intersection that bisect their dihedral angles.

19* Let $\mathbf{v}_1$, $\mathbf{v}_2$, $\mathbf{v}_3$ be the vertices of a triangle and a_1, a_2, a_3 the lengths of the opposite sides. Prove that the **incenter** (intersection of angle bisectors = center of inscribed circle) is

$$\mathbf{p} = \frac{a_1\mathbf{v}_1 + a_2\mathbf{v}_2 + a_3\mathbf{v}_3}{a_1 + a_2 + a_3}.$$

20* Prove that the three altitudes of a triangle **abc** intersect in a common point (the **orthocenter**). [*Hint*   Take **0** at the intersection of the altitudes from **a** and **b**. Then prove **c** is orthogonal to **b** − **a**, etc.]

21 Show that $\mathbf{a} = (1, 0, 0)$, $\mathbf{b} = \left(-\frac{1}{2}, \frac{1}{2}\sqrt{3}, 0\right)$, $\mathbf{c} = \left(-\frac{1}{2}, -\frac{1}{2}\sqrt{3}, 0\right)$, $\mathbf{d} = \left(0, 0, \sqrt{2}\right)$ are the vertices of a regular tetrahedron.

22 (cont.) Find the dihedral angle θ between any two faces of a regular tetrahedron.

Vector Functions and Curves   **14**

1. DIFFERENTIATION

In this chapter we study functions whose values are vectors. For example, the position **x** of a moving particle at time t, or the gravitational force **F** on an orbiting satellite at time t are vector functions. To indicate that **x** is a function of time, we write

$$\mathbf{x} = \mathbf{x}(t);$$

in components, $\qquad\qquad \mathbf{x}(t) = (x(t),\ y(t),\ z(t)).$

Thus a vector function is a single expression for three ordinary (scalar) functions

$$x = x(t), \qquad y = y(t), \qquad z = z(t).$$

We often think of a vector function as describing a curve in space, the trajectory of a moving particle. For example, if **a** and **b** are fixed vectors, then the function

$$\mathbf{x}(t) = t\mathbf{a} + \mathbf{b}$$

describes a line traversed by a moving point which is at **b** when $t = 0$, at $\mathbf{a} + \mathbf{b}$ when $t = 1$, at $2\mathbf{a} + \mathbf{b}$ when $t = 2$, etc. Note that $\mathbf{y}(t) = 2t\mathbf{a} + \mathbf{b}$ describes the same line but traversed at twice the speed, and $\mathbf{z}(t) = t^3\mathbf{a} + \mathbf{b}$ is again the same line, but traversed by an accelerating particle. We shall deal with speed and acceleration shortly.

Limits In order to do calculus in space, we need derivatives of vector functions. We want to define the derivative of a vector function as the *limit* of a difference quotient. Obviously, we better know first what we mean by the limit of a vector function.

Limit   Let $\mathbf{x}(t)$ be defined for t near a. Then

$$\lim_{t \to a} \mathbf{x}(t) = \mathbf{c}$$

means $\qquad |\mathbf{x}(t) - \mathbf{c}| \longrightarrow 0 \qquad$ as $\ t \longrightarrow a.$

Thus, the point $\mathbf{x}(t)$ approaches the point **c** as $t \longrightarrow a$ provided the distance between $\mathbf{x}(t)$ and **c** approaches 0 as $t \longrightarrow a$. But the distance $|\mathbf{x}(t) - \mathbf{c}|$ is a *real-valued* function, so we are back on familiar ground.

The definition of limit, as it stands, has nothing to do with coordinates. However, when we actually come to computing a limit, we generally use an alternative definition in terms of coordinates:

> Let $\mathbf{x}(t) = (x_1(t), x_2(t), x_3(t))$ and $\mathbf{c} = (c_1, c_2, c_3)$. Then
> $$\lim_{t \to a} \mathbf{x}(t) = \mathbf{c}$$
> if and only if
> $$\lim_{t \to a} x_j(t) = c_j \qquad \text{for} \quad j = 1, 2, 3.$$

The proof follows from the relation

$$|\mathbf{x}(t) - \mathbf{c}|^2 = \sum_{j=1}^{3} |x_j(t) - c_j|^2,$$

which implies $|\mathbf{x}(t) - \mathbf{c}| \geq |x_j(t) - c_j|$ for each j. Therefore, if

$$|\mathbf{x}(t) - \mathbf{c}| \longrightarrow 0, \qquad \text{then} \qquad |x_j(t) - c_j| \longrightarrow 0$$

for each j as $t \longrightarrow a$. Conversely, if $|x_j(t) - c_j| \longrightarrow 0$ for each j, then

$$|\mathbf{x}(t) - \mathbf{c}|^2 = \sum_{1}^{3} |x_j(t) - c_j|^2 \longrightarrow 0,$$

hence
$$|\mathbf{x}(t) - \mathbf{c}| \longrightarrow 0 \qquad \text{as} \quad t \longrightarrow a.$$

The Derivative   Now we are ready for derivatives. Think of $\mathbf{x} = \mathbf{x}(t)$ as tracing a path in space (Fig. 1). For h small, the difference vector

$$\mathbf{x}(t + h) - \mathbf{x}(t)$$

represents the chord from $\mathbf{x}(t)$ to $\mathbf{x}(t + h)$. Hence the difference quotient

$$\frac{\mathbf{x}(t + h) - \mathbf{x}(t)}{h}$$

represents this (short) chord divided by the small number h. The limit as $h \longrightarrow 0$ is called the **derivative** of the vector function:

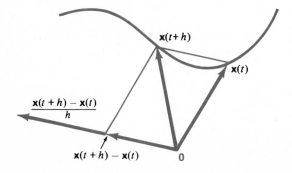

Fig. 1   Definition of $\dfrac{d\mathbf{x}}{dt}$

$$\dot{\mathbf{x}}(t) = \frac{d\mathbf{x}}{dt} = \lim_{h \to 0} \frac{\mathbf{x}(t+h) - \mathbf{x}(t)}{h}.$$

Because $\dot{\mathbf{x}}$ is a vector in the limiting position of the secant, we say that $\dot{\mathbf{x}}$ is **tangent** to the curve.

Examples   (1) $\mathbf{x}(t) = t\mathbf{a} + \mathbf{b}$

$$\frac{d\mathbf{x}}{dt} = \lim_{h \to 0} \frac{\mathbf{x}(t+h) - \mathbf{x}(t)}{h}$$

$$= \lim_{h \to 0} \frac{[(t+h)\mathbf{a} + \mathbf{b}] - [t\mathbf{a} + \mathbf{b}]}{h} = \lim_{h \to 0} \frac{h\mathbf{a}}{h} = \mathbf{a}.$$

Hence $\dot{x}(t)$ is the constant vector $\mathbf{a}$. This is reasonable because $\mathbf{x}(t)$ represents a line parallel to $\mathbf{a}$.

(2) $\mathbf{x}(t) = 2t\mathbf{a} + \mathbf{b}$ $\dot{\mathbf{x}}(t) = 2\mathbf{a}.$

(3) $\mathbf{x}(t) = t^3\mathbf{a} + \mathbf{b}$ $\dot{\mathbf{x}}(t) = 3t^2\mathbf{a}.$

In these three examples, all representing the same line, the derivatives are all parallel to $\mathbf{a}$. Their lengths, however, vary because the line is traced at different speeds.

Usually, we compute $\dot{\mathbf{x}}(t)$ in coordinates. Let $\mathbf{x}(t) = (x(t), y(t), z(t))$. From the definition of $\dot{\mathbf{x}}(t)$ as the limit of difference quotients it follows, for example, that the second coordinate of $\dot{\mathbf{x}}(t)$ is

$$\lim_{h \to 0} \frac{y(t+h) - y(t)}{h} = \frac{dy}{dt}.$$

The same holds for the other coordinates of $\dot{\mathbf{x}}(t)$.

> The derivative of a vector function $\mathbf{x}(t) = (x(t), y(t), z(t))$
>
> is the vector function $\dfrac{d\mathbf{x}}{dt} = \left(\dfrac{dx}{dt}, \dfrac{dy}{dt}, \dfrac{dz}{dt}\right).$

Example $\mathbf{x}(t) = t^3\mathbf{a} + \mathbf{b} = (a_1 t^3 + b_1, a_2 t^3 + b_2, a_3 t^3 + b_3).$

Then $\dfrac{d\mathbf{x}}{dt} = (3a_1 t^2, 3a_2 t^2, 3a_3 t^2) = 3t^2(a_1, a_2, a_3) = 3t^2\mathbf{a}.$

Velocity and Speed   Even though these two words are used interchangeably in everyday life, there is an important distinction between them in mathematics. Velocity is a *vector* function; speed is a *scalar* (real-valued) function.

If $\mathbf{x} = \mathbf{x}(t)$ is the position of a particle at time t, its **velocity** is

$$\mathbf{v} = \mathbf{v}(t) = \dot{\mathbf{x}}(t) = \frac{d\mathbf{x}}{dt};$$

its **speed** is $|\mathbf{v}(t)| = |\dot{\mathbf{x}}(t)|.$

Example $\mathbf{x}(t) = (t, t^2, t^3)$.   Then $\mathbf{v}(t) = \dot{\mathbf{x}}(t) = (1, 2t, 3t^2)$,

$|\mathbf{v}(t)|^2 = 1 + (2t)^2 + (3t^2)^2 = 1 + 4t^2 + 9t^4$,   speed $= |\mathbf{v}(t)| = \sqrt{1 + 4t^2 + 9t^4}$.

Like any other vector, the velocity vector $\mathbf{v}(t)$ starts at the origin. However there is no law against imagining $\mathbf{v}(t)$ attached at each point of $\mathbf{x}(t)$. Then $\mathbf{v}(t)$ is tangential to the curve; its direction at each point is the direction of the motion and its length is the speed. Double the speed and you double the velocity vector. (It still points in the same direction, but is twice as long.)

Zero Speed   Clearly, $\mathbf{x}(t)$ has zero speed if and only if it has zero velocity. When this is so for all t on an interval, then

$$\dot{\mathbf{x}}(t) = (\dot{x}_1(t), \dot{x}_2(t), \dot{x}_3(t)) = (0, 0, 0).$$

Hence $\dot{x}_i(t) = 0$, and so $x_i(t) = c_i$ (constant) for $i = 1, 2, 3$. Therefore

$$\mathbf{x}(t) = (c_1, c_2, c_3) = \mathbf{c}.$$

Physically, this simply says that an object with zero speed (or velocity) is standing still.

Differentiation Formulas   The following formulas are essential for differentiating vector functions.

$$\frac{d}{dt}[\mathbf{x}(t) + \mathbf{y}(t)] = \dot{\mathbf{x}}(t) + \dot{\mathbf{y}}(t), \qquad \frac{d}{dt}[g(t)\mathbf{x}(t)] = \dot{g}(t)\mathbf{x}(t) + g(t)\dot{\mathbf{x}}(t),$$

$$\frac{d}{dt}[\mathbf{x}(t) \cdot \mathbf{y}(t)] = \dot{\mathbf{x}}(t) \cdot \mathbf{y}(t) + \mathbf{x}(t) \cdot \dot{\mathbf{y}}(t). \qquad \frac{d}{dt}(\mathbf{v} \times \mathbf{w}) = \dot{\mathbf{v}} \times \mathbf{w} + \mathbf{v} \times \dot{\mathbf{w}}.$$

$$\frac{d}{dt}\mathbf{x}[u(t)] = \frac{d\mathbf{x}}{du}\frac{du}{dt} \qquad \text{(Chain Rule)}.$$

Note the similarity to ordinary differentiation formulas. In particular, the second, third, and fourth formulas, each involving a kind of product, all resemble the Product Rule for derivatives.

To prove the second formula, for example, note that the first coordinate of $g\mathbf{x}$ is gx_1, and

$$(gx_1)^{\cdot} = \dot{g}x_1 + g\dot{x}_1,$$

which is the first component of $\dot{g}\mathbf{x} + g\dot{\mathbf{x}}$, etc. The other formulas can be verified similarly. See the exercises.

Examples

(1) $\dfrac{d}{dt}(t^3, t^4, t^5) = \dfrac{d}{dt}[t^3(1, t, t^2)] = \left(\dfrac{d}{dt}t^3\right)(1, t, t^2) + t^3 \dfrac{d}{dt}(1, t, t^2)$

$$= 3t^2(1, t, t^2) + t^3(0, 1, 2t) = (3t^2, 4t^3, 5t^4).$$

(2) $(1, t, -t^2) \times (1, t, t^2) = (2t^3, -2t^2, 0)$, $\dfrac{d}{dt}(2t^3, -2t^2, 0) = (6t^2, -4t, 0)$.

On the other hand,

$$\left[\frac{d}{dt}(1, t, -t^2)\right] \times (1, t, t^2) + (1, t, -t^2) \times \frac{d}{dt}(1, t, t^2)$$
$$= (0, 1, -2t) \times (1, t, t^2) + (1, t, -t^2) \times (0, 1, 2t)$$
$$= (3t^2, -2t, -1) + (3t^2, -2t, 1) = (6t^2, -4t, 0).$$

■ **EXAMPLE 1**   Let $\mathbf{x} = \mathbf{x}(t)$ be the path of a point moving on the sphere $|\mathbf{x}| = a$. Show that the velocity $\mathbf{v}(t)$ at each instant is perpendicular to $\mathbf{x}(t)$.

Solution $\mathbf{x}(t) \cdot \mathbf{x}(t) = |\mathbf{x}(t)|^2 = a^2,$ hence $\dfrac{d}{dt}[\mathbf{x}(t) \cdot \mathbf{x}(t)] = \dfrac{d}{dt}a^2 = 0.$

But $\dfrac{d}{dt}[\mathbf{x}(t) \cdot \mathbf{x}(t)] = \dot{\mathbf{x}}(t) \cdot \mathbf{x}(t) + \mathbf{x}(t) \cdot \dot{\mathbf{x}}(t) = 2\mathbf{x}(t) \cdot \dot{\mathbf{x}}(t).$

Therefore $\mathbf{x}(t) \cdot \dot{\mathbf{x}}(t) = 0$, that is, $\mathbf{x}(t) \cdot \mathbf{v}(t) = 0$.   ■

We shall use the result of Example 1 several times in the following sections. Let us restate it in slightly different terms:

> If $\mathbf{x}(t)$ has constant length, then $\mathbf{x}(t)$ is perpendicular to its derivative $\dot{\mathbf{x}}(t)$.

This statement makes good sense geometrically. For the curve $\mathbf{x}(t)$ lies on the surface of a sphere $|\mathbf{x}| = a$. The vector $\dot{\mathbf{x}}(t)$, imagined attached at each point of the curve, is tangent to the curve, hence is tangent to the sphere. Therefore $\dot{\mathbf{x}}(t)$ is perpendicular to the radius vector, which is $\mathbf{x}(t)$. See Fig. 2.

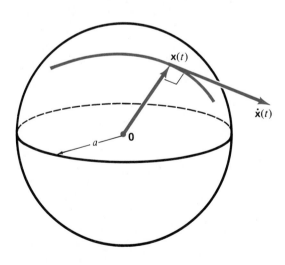

Fig. 2   The tangent to a curve on the sphere is tangent to the sphere, hence perpendicular to the radius vector.

EXERCISES

Differentiate $\mathbf{x}(t) =$

1 (e^t, e^{2t}, e^{3t})

2 (t^4, t^5, t^6)

3 $(t + 1, 3t - 1, 4t)$

4 $(t^2, 0, t^3)$

5 $(t, \cos t, \sin t)$

6 $(t^2, \arctan t, e^{-t}).$

Find the speed at t of $\mathbf{x}(t) =$

7 $(t^2, t^3 + t^4, 1)$ 8 $(2t - 1, 3t + 1, -2t + 1)$ 9 $(A \cos \omega t, A \sin \omega t, B)$
10 $(A \cos \omega t, A \sin \omega t, Bt)$ 11 $(1, t^4, t^6)$ 12 $(t, \cosh t, 3)$.

Prove

13 $\dfrac{d}{dt} [\mathbf{x}(t) + \mathbf{y}(t)] = \dot{\mathbf{x}}(t) + \dot{\mathbf{y}}(t)$

14 $\dfrac{d}{dt} [\mathbf{x}(t) \cdot \mathbf{y}(t)] = \dot{\mathbf{x}} \cdot \mathbf{y} + \mathbf{x} \cdot \dot{\mathbf{y}}$

15 $\dfrac{d}{dt} (\mathbf{x} \times \mathbf{y}) = \dot{\mathbf{x}} \times \mathbf{y} + \mathbf{x} \times \dot{\mathbf{y}}$

16 $\dfrac{d}{dt} \mathbf{x}[u(t)] = \dfrac{d\mathbf{x}}{du}\bigg|_{u(t)} \cdot \dfrac{du}{dt}$.

17 Suppose $\dot{\mathbf{x}}(t) = \mathbf{a}$. Find $\mathbf{x}(t)$.   18 Suppose $\dot{\mathbf{x}}(t) = t\mathbf{a} + \mathbf{b}$. Find $\mathbf{x}(t)$.
19 Suppose $\dot{\mathbf{x}}(t) = k\mathbf{x}(t)$. Find $\mathbf{x}(t)$.   20 Suppose $\ddot{\mathbf{x}}(t) + \mathbf{x}(t) = \mathbf{0}$. Find $\mathbf{x}(t)$.
21 Suppose that $\mathbf{x} = \mathbf{x}(t)$ is a moving point such that $\dot{\mathbf{x}}(t)$ is always perpendicular to $\mathbf{x}(t)$. Show that $\mathbf{x}(t)$ moves on a sphere with center at $\mathbf{0}$. [*Hint* Differentiate $|\mathbf{x}|^2$.]
22 Suppose $\mathbf{x}(t) \neq \mathbf{0}$. Show $\dfrac{d}{dt} |\mathbf{x}(t)| = \dfrac{1}{|\mathbf{x}|} \mathbf{x} \cdot \dot{\mathbf{x}}$.
23 A particle oscillates on the line segment **ab**. It starts at **a**, then moves halfway to **b**, then halfway back toward **a**, then half again as far toward **b**, etc. Find its position after n steps.
24 (cont.) Find its limit.
25 Prove $\dfrac{d}{dt} [\mathbf{u}(t), \mathbf{v}(t), \mathbf{w}(t)] = [\dot{\mathbf{u}}, \mathbf{v}, \mathbf{w}] + [\mathbf{u}, \dot{\mathbf{v}}, \mathbf{w}] + [\mathbf{u}, \mathbf{v}, \dot{\mathbf{w}}]$.

26 (cont.) Find $\dfrac{d}{dt} [\mathbf{u}(t), \dot{\mathbf{u}}(t), \ddot{\mathbf{u}}(t)]$.

2. ARC LENGTH

Let $\mathbf{x} = \mathbf{x}(t)$ be a space curve defined for $a \leq t \leq b$. We want to assign a length to the curve. Actually, we shall do a bit more: we shall define a function $s = s(t)$ that gives the length of the arc from $\mathbf{x}(a)$ to $\mathbf{x}(t)$ in general.

We shall give two definitions for $s(t)$. The first is motivated by the physical idea of speed. The second, at the end of this section, is based on the length of polygons. Both definitions involve integration in a natural way.

Arc Length via Velocity Let $\mathbf{x} = \mathbf{x}(t)$ be a space curve defined for $a \leq t \leq b$ by a smooth (continuously differentiable) vector function. At each point of the curve, the velocity vector $\dot{\mathbf{x}}(t)$ is tangential (Fig. 1). We feel intuitively that its length $|\dot{\mathbf{x}}(t)|$ represents speed, the rate at which the length of the curve increases with respect to time. Therefore we *define*

$$\frac{ds}{dt} = |\mathbf{v}(t)| = |\dot{\mathbf{x}}(t)|, \qquad s(a) = 0.$$

Thus the arc length function $s(t)$ is the solution of this initial value problem. Now

$$|\dot{\mathbf{x}}(t)|^2 = \left| \left(\frac{dx}{dt}, \frac{dy}{dt}, \frac{dz}{dt} \right) \right|^2 = \left(\frac{dx}{dt} \right)^2 + \left(\frac{dy}{dt} \right)^2 + \left(\frac{dz}{dt} \right)^2,$$

hence

$$\frac{ds}{dt} = \sqrt{\left(\frac{dx}{dt} \right)^2 + \left(\frac{dy}{dt} \right)^2 + \left(\frac{dz}{dt} \right)^2} = \sqrt{\dot{x}(t)^2 + \dot{y}(t)^2 + \dot{z}(t)^2}.$$

To obtain $s(t)$ itself we integrate starting at $t = a$ so that $s(a) = 0$:

Arc Length   Let $\mathbf{x} = \mathbf{x}(t)$ describe a space curve. Then the **arc length** from $\mathbf{x}(a)$ to $\mathbf{x}(t)$ is

$$s(t) = \int_a^t \sqrt{\dot{x}(u)^2 + \dot{y}(u)^2 + \dot{z}(u)^2}\, du.$$

Also,
$$\frac{ds}{dt} = |\dot{\mathbf{x}}(t)| = \sqrt{\dot{x}^2 + \dot{y}^2 + \dot{z}^2}\,.$$

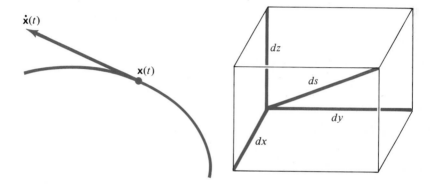

Fig. 1 $\dfrac{ds}{dt} = |\dot{\mathbf{x}}(t)|$ **Fig. 2**   Geometric interpretation of ds

Remark   The arc length formula has a direct geometric interpretation (Fig. 2). The tiny bit of arc length ds corresponds to three "displacements" dx, dy, and dz along the coordinate axis. By the Distance Formula,

$$(ds)^2 = (dx)^2 + (dy)^2 + (dz)^2.$$

Divide formally by $(dt)^2$ and take square roots. The result is

$$\frac{ds}{dt} = \sqrt{\left(\frac{dx}{dt}\right)^2 + \left(\frac{dy}{dt}\right)^2 + \left(\frac{dz}{dt}\right)^2}\,.$$

Let us write the arc length formula for the special case of a curve in the x, y-plane:

Plane Curves   Let $\mathbf{x} = (x(t),\, y(t))$ be a plane curve for $a \le t \le b$. Then its length is

$$L = \int_a^b \sqrt{\dot{x}(t)^2 + \dot{y}(t)^2}\, dt.$$

If the curve is the graph of a function $y = f(x)$ for $a \le x \le b$, then its length is

$$L = \int_a^b \sqrt{1 + [f'(x)]^2}\, dx.$$

The second formula is a special case of the first one. Just set $x = t$, $y = f(t)$, where $a \le t \le b$. Then $\dot{x} = 1$ and $\dot{y} = \dot{f}$, so

$$\frac{ds}{dt} = \sqrt{\dot{x}^2 + \dot{y}^2} = \sqrt{1 + \dot{f}^2} = \sqrt{1 + (f')^2}.$$

The formula for L follows.

■ **EXAMPLE 1**   Find the length

(a) $\mathbf{x}(t) = (t, t^2)$, $0 \le t \le 1$ (b) $y = \sin x$, $0 \le x \le \pi$.

Solution   Use the formula for arc length and an integral table:

(a) $L = \displaystyle\int_0^1 \sqrt{\dot{x}^2 + \dot{y}^2}\, dt = \int_0^1 \sqrt{1 + (2t)^2}\, dt = 2\int_0^1 \sqrt{\tfrac{1}{4} + t^2}\, dt$

$$= 2 \cdot \tfrac{1}{2}[t\sqrt{\tfrac{1}{4} + t^2} + \tfrac{1}{4}\ln(t + \sqrt{\tfrac{1}{4} + t^2})]\,\Big|_0^1 = \tfrac{1}{2}\sqrt{5} + \tfrac{1}{4}\ln(2 + \sqrt{5}) \approx 1.479.$$

The curve is a parabola because $y = t^2 = x^2$.

(b) $L = \displaystyle\int_0^\pi \sqrt{1 + \left(\frac{dy}{dx}\right)^2}\, dx = \int_0^\pi \sqrt{1 + \cos^2 x}\, dx \approx 3.820.$

The exact integral (an elliptic integral) is impossible to evaluate; the approximation is by Simpson's Rule.   ■

■ **EXAMPLE 2**   Find the length of the curve

$$\mathbf{x}(t) = (t \cos t, t \sin t, 2t) \qquad \text{for}\quad 0 \le t \le 4\pi.$$

Solution

$$\dot{x}(t)^2 + \dot{y}(t)^2 + \dot{z}(t)^2 = (\cos t - t'\sin t)^2 + (\sin t + t \cos t)^2 + 2^2 = 5 + t^2.$$

Hence, by the arc length formula,

$$L = \int_0^{4\pi} \sqrt{\dot{x}^2 + \dot{y}^2 + \dot{z}^2}\, dt = \int_0^{4\pi} \sqrt{5 + t^2}\, dt = \tfrac{1}{2}[t\sqrt{5 + t^2} + 5\ln(t + \sqrt{5 + t^2})]\,\Big|_0^{4\pi}$$

$$= \tfrac{1}{2}[4\pi a + 5\ln(4\pi + a) - \tfrac{5}{2}\ln 5],$$

where $a = (5 + 16\pi^2)^{1/2}$. Approximately, $L \approx 86.3$.   ■

The curve in Example 2 is a spiral (Fig. 3). As t increases, z increases at a steady rate, while the projection of $\mathbf{x}(t)$ on the x, y-plane traces the spiral $(t \cos t, t \sin t) = t(\cos t, \sin t)$. Actually, the curve lies on a right circular cone because

$$x^2 + y^2 = (t \cos t)^2 + (t \sin t)^2 = t^2 = \tfrac{1}{4}z^2,$$

an equation describing a cone, as will be shown in Section 9 of the next chapter.

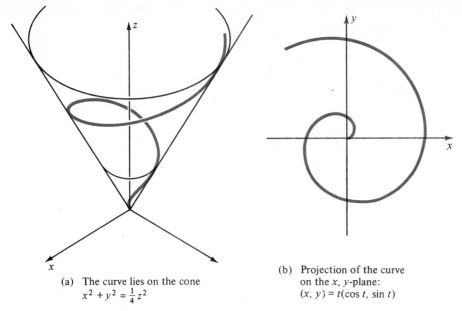

(a) The curve lies on the cone
$$x^2 + y^2 = \tfrac{1}{4}z^2$$

(b) Projection of the curve
on the x, y-plane:
$(x, y) = t(\cos t, \sin t)$

Fig. 3   Graph of $\mathbf{x}(t) = (t \cos t,\ t \sin t,\ 2t)$

Length is Independent of the Parameter   Suppose that a curve has two different parameterizations. For instance, it may be given by $\mathbf{x} = \mathbf{x}(t)$, where $a \leq t \leq b$, and also by $\mathbf{x} = \mathbf{x}(\tau)$, where $\alpha \leq \tau \leq \beta$. How do we know that the arc length formula yields the same length in each case? We suppose that either parameterization can be obtained from the other by a smooth change of variable. For example, let us take $t = t(\tau)$ for the change of variable and assume that $a = t(\alpha)$, $b = t(\beta)$, and $dt/d\tau > 0$. The t-length and the τ-length of the curve are

$$L_t = \int_a^b \left| \frac{d\mathbf{x}}{dt} \right| dt \qquad \text{and} \qquad L_\tau = \int_\alpha^\beta \left| \frac{d\mathbf{x}}{d\tau} \right| d\tau.$$

By the Chain Rule, $d\mathbf{x}/d\tau = (d\mathbf{x}\, dt)(dt/d\tau)$. The formula for change of variable in a definite integral implies

$$L_\tau = \int_\alpha^\beta \left| \frac{d\mathbf{x}}{dt} \right| \frac{dt}{d\tau}\, d\tau = \int_a^b \left| \frac{d\mathbf{x}}{dt} \right| dt = L_t\,.$$

Thus the formula yields the same length in each case. This proves that the length of a curve is a geometric quantity, independent of the analytic representation of the curve.

Arc Length as the Parameter   The arc length of a space curve is a built-in property, independent of how the curve is parameterized. Therefore a natural parameter for a curve is its own arc length.

  Example $\mathbf{x}(t) = (a \cos 2\pi t,\ a \sin 2\pi t)$.

This describes a circle of radius a traced counterclockwise at constant speed. The complete circle (of length $2\pi a$) is traced for $0 \leq t \leq 1$. Take $s = 0$ at $t = 0$ and $s > 0$ for $t > 0$. Then $s = 2\pi a t$, so $2\pi t = s/a$ and the motion can be described by

$$\mathbf{x} = \left(a \cos \frac{s}{a},\, a \sin \frac{s}{a} \right).$$

This is a formula for $\mathbf{x}$ as a function of s. In other words, it is a parameterization of the circle with the arc length itself as the parameter.

   In principle, a similar parameterization is possible for every reasonable space curve. Given $\mathbf{x} = \mathbf{x}(t)$, we fix a point $\mathbf{x}_0 = \mathbf{x}(a)$ and measure s from $\mathbf{x}_0$, positive in the direction of increasing t, negative in the opposite direction (Fig. 4).

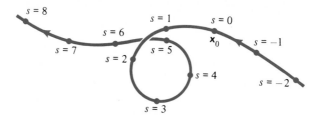

Fig. 4   Arc length as the parameter

We assume $\dot{\mathbf{x}}(t)$ is never $\mathbf{0}$. Then

$$s(t) = \int_a^t |\dot{\mathbf{x}}(u)|\,du$$

is a *strictly increasing* function of t because

$$\frac{ds}{dt} = |\dot{\mathbf{x}}(t)| > 0.$$

It follows that $s = s(t)$ has an inverse function $t = t(s)$. Therefore we may write $\mathbf{x} = \mathbf{x}(t)$ in the form

$$\mathbf{x} = \mathbf{x}[t(s)],$$

showing $\mathbf{x}$ as a function of s. This is the desired parameterization of the curve in terms of its own arc length.

Remark 1   (bad news)   For most curves it is difficult or impossible to carry out these computations. Usually the integral defining $s(t)$ is hard to evaluate because of the square root in the integrand,

$$|\mathbf{x}(t)| = \sqrt{\dot{x}^2 + \dot{y}^2 + \dot{z}^2},$$

and even if $s = s(t)$ can be computed explicitly, it is hard to find its inverse function.
   The above example of a circle is one of the few cases in which we can find $s(t)$ explicitly. There the speed is *constant*, so s is just a constant multiple of t.

Remark 2   (good news)   We seldom actually need $s = s(t)$ itself or $\mathbf{x}$ expressed in terms of s explicitly. The *idea* of using arc length as the parameter is important for understanding curves, but for calculations we can usually manage with ds/dt (and d^2s/dt^2), which we know.

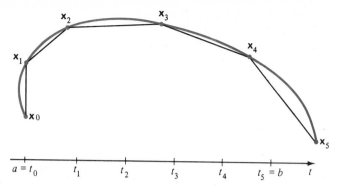

Fig. 5   Inscribed polygonal arc in a space arc

Arc Length as a Limit   Let $\mathbf{x} = \mathbf{x}(t)$ be a space curve, where $a \leq t \leq b$. One way to approximate its arc length is to inscribe a polygonal arc in the given curve and take the length of the polygonal arc as an estimate of the arc length.

To this end, we partition $[a, b]$ by

$$a = t_0 < t_1 < \cdots < t_n = b.$$

The corresponding points on the curve we denote by $\mathbf{x}_0, \mathbf{x}_1, \cdots, \mathbf{x}_n$. We connect these points to form a polygonal arc (Fig. 5) consisting of the n straight segments

$$\overline{\mathbf{x}_{i-1}\mathbf{x}_i}, \qquad i = 1, \cdots, n.$$

Thus $\mathbf{x}_i = (x_i, y_i, z_i)$, where $x_i = x(t_i)$, $y_i = y(t_i)$, and $z_i = z(t_i)$. The length of the i-th segment is

$$\left| \mathbf{x}_i - \mathbf{x}_{i-1} \right| = \sqrt{(x_i - x_{i-1})^2 + (y_i - y_{i-1})^2 + (z_i - z_{i-1})^2},$$

and the length of the polygonal arc—it depends on the partition—is

$$L = L(\text{part.}) = \sum_{i=1}^{n} \left| \mathbf{x}_i - \mathbf{x}_{i-1} \right|.$$

Now we seek the limit of L over finer and finer partitions, that is, partitions such that

$$\max_{i=1, \cdots, n} \{t_{i+1} - t_i\} \longrightarrow 0.$$

We shall use the Mean Value Theorem to approximate $\left| \mathbf{x}_i - \mathbf{x}_{i-1} \right|$ and use the result to find $\lim L$.

By one application of the MVT,

$$x_i - x_{i-1} = x(t_i) - x(t_{i-1}) = \dot{x}(\xi_i)(t_i - t_{i-1}),$$

where

$$t_{i-1} < \xi_i < t_i.$$

Similarly,

$$y_i - y_{i-1} = \dot{y}(\eta_i)(t_i - t_{i-1}) \qquad \text{and} \qquad z_i - z_{i-1} = \dot{z}(\zeta_i)(t_i - t_{i-1}),$$

where $\qquad\qquad\qquad t_{i-1} < \eta_i < t_i \qquad$ and $\qquad t_{i-1} < \zeta_i < t_i.$

Hence $\qquad\qquad\qquad |\mathbf{x}_i - \mathbf{x}_{i-1}| = (\sqrt{\dot{x}(\xi_i)^2 + \dot{y}(\eta_i)^2 + \dot{z}(\zeta_i)^2}\,)(t_i - t_{i-1}),$

so $\qquad\qquad L(\text{part.}) = \sum_{i=1}^{n} (\sqrt{\dot{x}(\xi_i)^2 + \dot{y}(\eta_i)^2 + \dot{z}(\zeta_i)^2}\,)(t_i - t_{i-1}).$

This sum smells very much like a definite integral. In order to produce one, let us assume that $\mathbf{x}(t)$ is *continuously differentiable*, that is, each of its coordinate functions $x(t)$, $y(t)$, $z(t)$ is a continuously differentiable function of t on $[a, b]$. Then it can be shown by a technical argument involving uniform continuity that

$$\left| \sqrt{\dot{x}(\xi_i)^2 + \dot{y}(\eta_i)^2 + \dot{z}(\zeta_i)^2} - \sqrt{\dot{x}(t_i)^2 + \dot{y}(t_i)^2 + \dot{z}(t_i)^2} \right|$$

can be made as small as we please for all i provided we choose the partition sufficiently fine. It follows that

$$L \approx \sum_{i=1}^{n} (\sqrt{\dot{x}(t_i)^2 + \dot{y}(t_i)^2 + \dot{z}(t_i)^2}\,)(t_i - t_{i-1}) \longrightarrow \int_a^b \sqrt{\dot{x}^2 + \dot{y}^2 + \dot{z}^2}\, dt = \int_a^b |\dot{\mathbf{x}}(t)|\, dt.$$

Thus polygonal approximation leads, in the limit, to the same formula for arc length that we arrived at earlier by a physical argument. Note that the method works for piecewise smooth curves, that is, curves that are smooth (continuously differentiable) except for a finite number of exceptional points (corners).

EXERCISES

Find the length of $\mathbf{x}(t) =$

1. $(a_1 t + b_1, a_2 t + b_2, a_3 t + b_3),\quad 0 \le t \le 1$
2. $(t^2, t^3),\quad 0 \le t \le 2 \qquad$ 3. $(t^4, t^5),\quad 0 \le t \le 1 \qquad$ 4. $(t^2, t^3, t^2),\quad 0 \le t \le 1.$

Set up each arc length as an integral, but do not evaluate

5. $y = x^3,\quad 0 \le x \le b$ $\qquad\qquad\qquad\qquad$ 6. $y = e^x,\quad a \le x \le b$
7. $\mathbf{x}(t) = (t^m, t^n, t^r),\quad a \le t \le b$
8. $\mathbf{x}(t) = (\cos t, \sin t, \cos t + \sin t),\quad 0 \le t \le 2\pi.$

Find the arc length

9. $y = \ln x,\quad 1 \le x \le 2$ $\qquad\qquad\qquad\qquad$ 10. $y = 2 \sec x,\quad -\tfrac{1}{4}\pi \le x \le \tfrac{1}{4}\pi$
11. $\mathbf{x}(t) = (t, t^2, \tfrac{4}{3}t^{3/2}),\quad 0 \le t \le b$
12. $\mathbf{x}(t) = (t, \sqrt{2}\cos t, \tfrac{1}{2}t - \tfrac{1}{4}\sin 2t),\quad 0 \le t \le \pi.$

13. Show that the curve $\mathbf{x} = (\sin^2 t, \sin t \cos t, \cos t)$ lies on the unit sphere, and verify the relation $\mathbf{x} \cdot \dot{\mathbf{x}} = 0$.
14. (cont.) Express its length for $a \le t \le b$ as an integral.
15. For the curve of Ex. 11, express t in terms of s; take $s = 0$ at $t = 0$.
16. For which functions $z(t)$ is the curve $\mathbf{x}(t) = (\cos t, \sin t, z(t))$ a plane curve?
17. Let $\mathbf{x}(t)$ be a curve in the x, y-plane joining $(a, 0)$ to $(b, 0)$, where $a < b$. Use the formula for arc length to prove the length $L \ge b - a$.
18. (cont.) Suppose instead that $\mathbf{x}(t)$ joins (a, A) to (b, B). Show that

$$L \ge \sqrt{(b - a)^2 + (B - A)^2}.$$

(Hence the shortest curve joining two points is a line segment!) [*Hint* Rotate coordinates.]

3. PLANE CURVES

In this section, we shall examine some special features of plane curves and look at some well-known examples.

Derivatives Sometimes a plane parametric curve $\mathbf{x} = (x(t), y(t))$ may be considered as the graph of a function $y = f(x)$. Suppose, for instance, that $\dot{x}(t) > 0$ for $t_0 < t < t_1$. Then $x = x(t)$ is strictly increasing on this interval, hence has an inverse function $t = t(x)$. The substitution $y = y(t) = y[t(x)]$ expresses y as a function of x. The given curve is the graph of $y = y(x)$ on the interval $x(t_0) \leq x \leq x(t_1)$.

How do you compute the derivatives

$$\frac{dy}{dx} \quad \text{and} \quad \frac{d^2y}{dx^2}$$

when the function $y(x)$ is presented parametrically? The key is the Chain Rule. First

$$\dot{y} = \frac{dy}{dt} = \frac{dy}{dx}\frac{dx}{dt} = \frac{dy}{dx}\dot{x}, \quad \text{so} \quad \frac{dy}{dx} = \frac{\dot{y}}{\dot{x}}.$$

Next, the same reasoning applied to dy/dx yields

$$\frac{d}{dx}\left(\frac{dy}{dx}\right) = \frac{(dy/dx)\dot{}}{\dot{x}} = \frac{(\dot{y}/\dot{x})\dot{}}{\dot{x}} = \frac{\ddot{y}\dot{x} - \ddot{x}\dot{y}}{(\dot{x})^3}.$$

Let $\mathbf{x} = \mathbf{x}(t) = (x(t), y(t))$ for $t_0 < t < t$, and assume $\dot{x}(t)$ is never 0. Then the curve is the graph of a function $y = y(x)$ and

$$\frac{dy}{dx} = \frac{\dot{y}}{\dot{x}} \quad \text{and} \quad \frac{d^2y}{dx^2} = \frac{\ddot{y}\dot{x} - \ddot{x}\dot{y}}{(\dot{x})^3}.$$

■ **EXAMPLE 1** Show that the equations $x(t) = t^2 + 1$, $y(t) = t^4 + 1$ define y as a function of x for $t > 0$ and compute

$$\frac{dy}{dx} \quad \text{and} \quad \frac{d^2y}{dx^2}.$$

Solution Since $\dot{x}(t) = 2t > 0$ for $t > 0$, the equations determine y as a function of x.

$$\frac{dy}{dx} = \frac{\dot{y}}{\dot{x}} = \frac{4t^3}{2t} = 2t^2, \quad \frac{d^2y}{dx^2} = \frac{\ddot{y}\dot{x} - \ddot{x}\dot{y}}{(\dot{x})^3} = \frac{(12t^2)(2t) - (2)(4t^3)}{(2t)^3} = \frac{16t^3}{8t^3} = 2.$$

Check Eliminate t:

$$t^2 = x - 1, \quad y = (t^2)^2 + 1 = (x - 1)^2 + 1.$$

Hence

$$\frac{dy}{dx} = 2(x - 1) = 2t^2, \quad \frac{d^2y}{dx^2} = 2. \qquad ■$$

Area Let $\mathbf{x} = \mathbf{x}(t) = (x(t), y(t))$ be a parametric plane curve such that $\dot{x}(t) > 0$ for $t_0 \leq t \leq t_1$. As we have seen, the curve may be considered as the graph of a

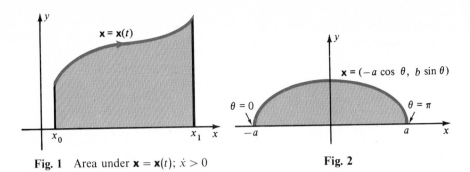

Fig. 1   Area under $\mathbf{x} = \mathbf{x}(t)$; $\dot{x} > 0$ **Fig. 2**

function $y = f(x)$. We want the area under this graph (Fig. 1) expressed as an integral involving $x(t)$, $y(t)$, etc.

We use the Change of Variable formula for definite integrals. The area is

$$A = \int_{x_0}^{x_1} y[t(x)]\, dx = \int_{t_0}^{t_1} y(t)\, \frac{dx}{dt}\, dt = \int_{t_0}^{t_1} y\dot{x}\, dt.$$

If $\dot{x}(t) > 0$ for $t_0 \leq t \leq t_1$, then the area under $\mathbf{x} = \mathbf{x}(t) = (x(t), y(t))$ is

$$A = \int_{t_0}^{t_1} y\dot{x}\, dt.$$

■ **EXAMPLE 2**   Find the area under the ellipse $\dfrac{x^2}{a^2} + \dfrac{y^2}{b^2} = 1$, $\ y \geq 0$.

Solution   The ellipse can be parameterized by $x = a \cos \theta$, $y = b \sin \theta$. However $dx/d\theta < 0$ as θ runs from 0 to π. To have $dx/d\theta > 0$, we could let θ decrease from π to 0, or, we can use the equivalent parameterization, $x = -a \cos \theta$, $y = b \sin \theta$. See Fig. 2. By the second method we obtain

$$A = \int_0^\pi y\, \frac{dx}{d\theta}\, d\theta = \int_0^\pi (b \sin \theta)(a \sin \theta)\, d\theta = ab \int_0^\pi \sin^2 \theta\, d\theta = \frac{1}{2}\, \pi ab. \qquad ■$$

For the area surrounded by a *closed* curve, there is a formula worth knowing at this point, although we postpone its proof until later (Chapter 18). A closed curve is given by a periodic vector function of period p:

$$\mathbf{x}(t + p) = \mathbf{x}(t).$$

Thus $\mathbf{x}(p) = \mathbf{x}(0)$, so a loop is formed as t runs from 0 to p. We assume the curve is **simple**, that is, never crosses itself. Precisely,

$$\mathbf{x}(t_1) \neq \mathbf{x}(t_2) \qquad \text{for} \ \ 0 \leq t_1 < t_2 < p.$$

We also assume the curve is traversed in the counterclockwise direction, that is, if you stand on the curve facing in the direction of increasing t, then the region enclosed by the curve is on your immediate left side. (This isn't very precise, but we only intend application to simple examples.) See Fig. 3.

Closed Curves   Let $\mathbf{x} = \mathbf{x}(t)$ parameterize a simple closed curve with period p, traversed counterclockwise. Then the area enclosed by the curve is

$$A = \tfrac{1}{2} \int_0^p (x\dot{y} - y\dot{x})\, dt.$$

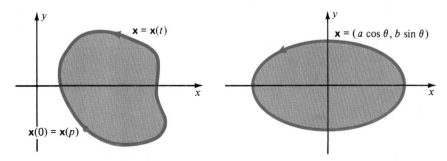

Fig. 3   Simple closed curve: **Fig. 4**
     $\mathbf{x}(t)$ periodic of period p

■ **EXAMPLE 3**   Find the area enclosed by the ellipse $\dfrac{x^2}{a^2} + \dfrac{y^2}{b^2} = 1$.

Solution   (Fig. 4)   The curve is parameterized by the vector function

$$\mathbf{x}(\theta) = (a \cos \theta, b \sin \theta),$$

periodic of period 2π. Clearly the ellipse does not cross itself, and is traversed counterclockwise, hence

$$
\begin{aligned}
A &= \frac{1}{2} \int_0^{2\pi} \left(x \frac{dy}{d\theta} - y \frac{dx}{d\theta} \right) d\theta \\
&= \frac{1}{2} \int_0^{2\pi} [(a \cos \theta)(b \cos \theta) - (b \sin \theta)(-b \sin \theta)]\, d\theta \\
&= \frac{1}{2} \int_0^{2\pi} ab\, d\theta = \pi ab.
\end{aligned}
$$
    ■

Parameterization of the Circle   The unit circle has the parameterization $\mathbf{x} = (\cos \theta, \sin \theta)$, where θ is the central angle. Another parameterization in terms of rational rather than trigonometric functions is sometimes useful.

Consider a variable line through $(-1, 0)$ with slope t. See Fig. 5. Its equation is

$$y = t(1 + x).$$

The line meets the circle in two points; one is $(-1, 0)$. To find the other we eliminate y from the system

$$x^2 + y^2 = 1, \qquad y = t(1 + x)$$

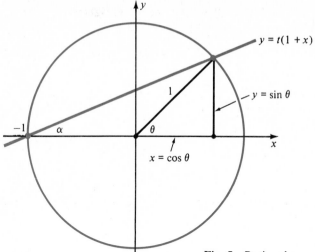

Fig. 5   Rational parameterization of the unit circle

and solve for x: $x^2 + t^2(1 + x)^2 = 1,$ $x^2 + t^2(x^2 + 2x + 1) = 1,$

$$(t^2 + 1)x^2 + 2t^2x + (t^2 - 1) = 0,$$

$$x = \frac{-t^2 \pm \sqrt{t^4 - (t^2 + 1)(t^2 - 1)}}{t^2 + 1} = \frac{-t^2 \pm \sqrt{1}}{t^2 + 1} = \frac{-t^2 \pm 1}{t^2 + 1}.$$

The minus sign leads to $x = -1$, the known solution, so we choose $+$:

$$x = \frac{1 - t^2}{1 + t^2}, \qquad y = t(1 + x) = \frac{2t}{1 + t^2}.$$

This is the desired rational parameterization of the circle. Note incidentally that the parameter t can be expressed rationally in terms of x and y:

$$y = t(1 + x), \qquad t = \frac{y}{1 + x}.$$

For $-\infty < t < \infty$, we get every point of the circle with one exception, the point $(-1, 0)$.

Remark   In a certain sense of algebraic justice, the point $(-1, 0)$ corresponds to $t = \pm\infty$. For if the slope is $\pm\infty$, the variable line is vertical, tangent to the circle at $(-1, 0)$, so it meets the circle *twice* at $(-1, 0)$.

Rational Parameterization of the Circle   The unit circle $x^2 + y^2 = 1$ is parameterized by

$$x = \frac{1 - t^2}{1 + t^2}, \qquad y = \frac{2t}{1 + t^2}.$$

Also,

$$t = \frac{y}{1 + x}.$$

Have a second look at Fig. 5. The slope t equals $\tan \alpha$. But the inscribed angle α is half the corresponding central angle θ, so $t = \tan \frac{1}{2}\theta$. Also $x = \cos \theta$ and $y = \sin \theta$. We may substitute

$$t = \tan \tfrac{1}{2}\theta, \qquad x = \cos \theta, \qquad y = \sin \theta$$

in the formulas above. Result:

Half-Angle Formulas

$$\cos \theta = \frac{1 - \tan^2 \frac{1}{2}\theta}{1 + \tan^2 \frac{1}{2}\theta}, \qquad \sin \theta = \frac{2 \tan \frac{1}{2}\theta}{1 + \tan^2 \frac{1}{2}\theta}, \qquad \tan \tfrac{1}{2}\theta = \frac{\sin \theta}{1 + \cos \theta}.$$

The Hyperbola   The hyperbola $\dfrac{x^2}{a^2} - \dfrac{y^2}{b^2} = 1$ can be parameterized by hyperbolic functions:

$$x = a \cosh t, \qquad y = b \sinh t.$$

For a geometric interpretation of the parameter t, we compute the shaded area A in Fig. 6.

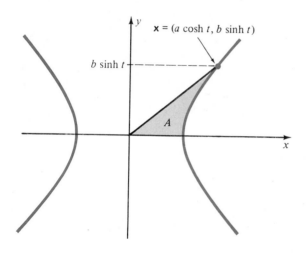

$\mathbf{x} = (a \cosh t, b \sinh t)$

$b \sinh t$

A

Fig. 6 Hyperbola; to interpret t geometrically

This area is the difference between the area bounded by the curve and the y-axis (for $0 \le y \le b \sinh t$) and the area of the triangle. Thus

$$A = \int_0^t \dot y x \, dt - \frac{1}{2} (a \cosh t)(b \sinh t) = \int_0^t ab \cosh^2 t \, dt - \frac{ab}{4} \sinh 2t$$

$$= ab \left(\frac{1}{4} \sinh 2t + \frac{1}{2} \right) \Big|_0^t - \frac{ab}{4} \sinh 2t = \frac{1}{2} abt.$$

Hence

$$t = \frac{2A}{ab}.$$

This gives us an expression for t in terms of geometric quantities.

The Cycloid The cycloid is the curve traced by an outermost point on a bicycle tire.

■ **EXAMPLE 4** A circle of radius a rolls along the x-axis in the upper half-plane. The point on the circle initially at the origin traces a **cycloid**.

(a) Parameterize the cycloid by the central angle θ in Fig. 7a.

(b) Find the length L of one arch of the cycloid.

(c) Find the area A under one arch of the cycloid.

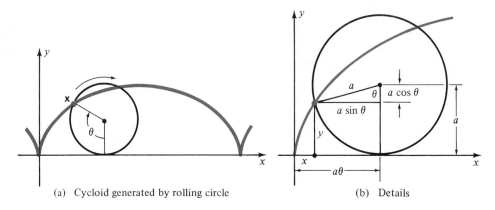

(a) Cycloid generated by rolling circle   (b) Details

Fig. 7 The cycloid

Solution (a) By carefully marking various lengths (Fig. 7b) we can read the relations

$$x + a \sin \theta = a\theta, \qquad y + a \cos \theta = a.$$

(Note that the corresponding circular arc has length $a\theta$, which equals the distance from 0 to the point of contact because the circle *rolls*.) Hence

$$\mathbf{x} = \mathbf{x}(\theta) = a(\theta - \sin \theta, \ 1 - \cos \theta).$$

(b) $\dfrac{d\mathbf{x}}{d\theta} = (\dot{x}(\theta), \dot{y}(\theta)) = a(1 - \cos \theta, \sin \theta),$

$$\dot{x}(\theta)^2 + \dot{y}(\theta)^2 = [a(1 - \cos \theta)]^2 + (a \sin \theta)^2 = a^2(2 - 2 \cos \theta) = 4a^2 \sin^2 \tfrac{1}{2}\theta.$$

Therefore $L = \displaystyle\int_0^{2\pi} \sqrt{\dot{x}^2 + \dot{y}^2} \, d\theta = \int_0^{2\pi} 2a \sin \tfrac{1}{2}\theta \, d\theta = -4a \cos \tfrac{1}{2}\theta \, \Big|_0^{2\pi} = 8a.$

(c) $A = \displaystyle\int_0^{2\pi} y \dfrac{dx}{d\theta} \, d\theta = \int_0^{2\pi} [a(1 - \cos \theta)][a(1 - \cos \theta)] \, d\theta$

$$= a^2 \int_0^{2\pi} (1 - \cos \theta)^2 \, d\theta = a^2 \int_0^{2\pi} (1 - 2 \cos \theta + \cos^2 \theta) \, d\theta$$

$$= a^2 \int_0^{2\pi} (1 + \cos^2 \theta) \, d\theta = a^2(2\pi + \pi) = 3\pi a^2. \qquad ■$$

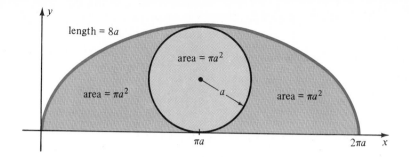

Fig. 8 Geometry of the cycloid. Note that $2\pi a \approx 6.28a < 8a$.

Remark Figure 8 illustrates the results of Example 3.

The Tractrix Suppose a point-weight attached to a string of length a is placed at $(0, a)$ on the rough horizontal x, y-plane, and the end of the string is placed at $(0, 0)$. See Fig. 9a. Then the end is moved slowly along the x-axis. The weight traces a curve called the **tractrix**. Because of friction, the string is always tangent to the curve.

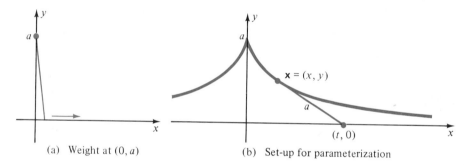

(a) Weight at $(0, a)$ (b) Set-up for parameterization

Fig. 9 The tractrix

■ **EXAMPLE 5**   Parameterize the tractrix by t, the x-intercept of the tangent from (x, y).

Solution   See Fig. 9b. The length of the segment joining (x, y) and $(t, 0)$ is a, hence

$$(t - x)^2 + y^2 = a^2.$$

The slope of the segment is $dy/dx = \dot{y}/\dot{x}$, hence $\dfrac{\dot{y}}{\dot{x}} = \dfrac{-y}{t - x}.$

Thus we have the system $(t - x)^2 + y^2 = a^2,$ $(t - x)\dot{y} = -y\dot{x},$

with the initial data $t = 0,$ $x = 0,$ $y = a.$

Our aim is to express x and y as functions of t. To simplify things a bit, we set $u = t - x$. Then $\dot{u} = 1 - \dot{x}$, so our system becomes

$$u^2 + y^2 = a^2, \qquad u\dot{y} = -y\dot{x} = y(\dot{u} - 1),$$

with the initial data $u(0) = 0,$ $y(0) = a.$

There are too many quantities to juggle, so let us eliminate y and $\dot{y}$. To do so we need one more relation, which comes by differentiating the first equation:

$$u\dot{u} + y\dot{y} = 0.$$

We multiply this relation by u, then use the equations of our system to simplify the result:

$$u^2\dot{u} + y(u\dot{y}) = 0, \qquad u^2\dot{u} + y^2(\dot{u} - 1) = 0,$$

$$u^2\dot{u} + (a^2 - u^2)(\dot{u} - 1) = 0, \qquad a^2\dot{u} - a^2 - u^2 = 0.$$

This last equation can be expressed as

$$\frac{du}{a^2 - u^2} = \frac{1}{a^2}\, dt.$$

We integrate, taking into account the initial data $u(0) = 0$:

$$\frac{1}{a}\tanh^{-1}\frac{u}{a} = \frac{1}{a^2}\, t, \qquad u = a\tanh\frac{t}{a}.$$

Also
$$y^2 = a^2 - u^2 = a^2 - a^2\tanh^2\frac{t}{a} = a^2\,\mathrm{sech}^2\frac{t}{a}.$$

Since $u = t - x$, the final result is

$$x = t - a\tanh\frac{t}{a}, \qquad y = a\,\mathrm{sech}\frac{t}{a}. \qquad\blacksquare$$

Remark The tractrix is an example of a **pursuit curve**. Imagine a fox chasing a rabbit. When $t = 0$ the rabbit is at $(0, 0)$ and the fox is at $(0, a)$. The rabbit runs along the positive x-axis and the fox pursues it so that he is always pointed directly at the rabbit. If the gap between them remains constant, then the fox's path is a tractrix. A similar argument might apply to a submarine tracking a target ship.

EXERCISES

1   Show that $\mathbf{x}(t) = (t^2 - 1,\ t^3 - t)$ parameterizes $y^2 = x^3 + x^2$.
2   (cont.) Sketch the curve and identify t geometrically.
3   Show that the **witch of Agnesi** $y = a^3/(a^2 + x^2)$ is parameterized by $\mathbf{x}(\theta) = a(\cot\theta,\ \sin^2\theta)$.
4   (cont.) Sketch the curve. Try to identify the geometric angle θ.
5   Show that the **serpentine** $(a^2 + x^2)y = abx$ is parameterized by $\mathbf{x}(\theta) = (a\cot\theta,\ b\sin\theta\cos\theta)$.
6   (cont.) Sketch the curve.
7   Show that the **folium** of Descartes $x^3 + y^3 = 3axy$ is parameterized by $\mathbf{x}(t) = 3a(t/(1 + t^3),\ t^2/(1 + t^3))$.
8* (cont.) Sketch the curve.
9   Show that the **cissoid** of Diocles $y^2(a - x) = x^3$ can be parameterized by $\mathbf{x}(t) = a(t^2/(1 + t^2),\ t^3/(1 + t^2))$.
10 Sketch the curve.
11 As a wheel of radius a rolls along the x-axis, a point *inside* the wheel at distance b from the center, $b < a$, traces a **curtate cycloid** (Fig. 10). Apply the method of Example 4 to parameterize the curve.

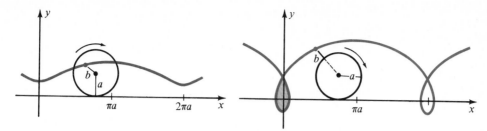

Fig. 10   Curtate cycloid   **Fig. 11**   Prolate cycloid

12   (cont.) Find the area under one arch of the curtate cycloid.

13   (cont.) Suppose $b > a$, so the point lies on a flange outside the wheel. (Think of the outer rim of a railroad wheel.) Then the curve is a **prolate cycloid** (Fig. 11). Parameterize this curve.

14*   (cont.) Find the shaded area in Fig. 11.

15   A (spool of) thread is wound clockwise around the unit circle so its outer end is at $(1, 0)$. Now it is unwound, always kept taut. The end traces a curve called the **involute** of the circle. Parameterize the curve, using the central angle θ in Fig. 12 as the parameter.

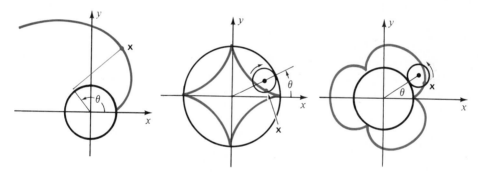

Fig. 12   Involute of circle   **Fig. 13**   Hypocycloid   **Fig. 14**   Epicycloid

16   (cont.) Parameterize in terms of the arc length s, measured from $(1, 0)$.

17   Find the arc length functions $s = s(t)$, with $s(0) = 0$, for the tractrix of Example 5.

18   (cont.) Show that $y = ae^{-s/a}$. This relation, so simple, must have a shortcut proof. Does it?

The following exercises deal with hypocycloids. A circle of radius b rolls on the inside of a fixed circle of radius $a > b$. The locus of a point on its boundary is called a **hypocycloid**. We take $|\mathbf{x}| = a$ for the fixed circle and $(a, 0)$ for the initial position of the moving point (Fig. 13).

19   Parameterize the curve by the angle θ.

20   Describe the curve if $a = 2b$.

21   Suppose $a = 3b$. The curve is a hypocycloid of three cusps, or **deltoid**. Give the parametric equation, and find the area enclosed by the curve.

22   (cont.) Sketch the curve and find its length.

23   Suppose $a = 4b$, so the curve is a hypocycloid of 4 cusps (Fig. 13). Give the parametric equation in as simple a form as possible.

24   (cont.) Find the length of the curve.
25   (cont.) Find the area enclosed by the curve.
26*   Fix a and let $b = a/n$. Find the length L_n of the hypocycloid of n cusps inscribed in a circle of radius a.
27*   (cont.) Show that $\{L_n\}$ is an increasing sequence and find $\lim L_n$.
28*   (cont.) Find the area A_n enclosed by the curve and find $\lim A_n$.

Suppose a circle of radius b rolls on the *outside* of a fixed circle of radius a. A point on its rim traces an **epicycloid**. (These curves are important in the design of gear teeth.) Let the fixed circle be $|\mathbf{x}| = a$ and take the point initially at $(a, 0)$. See Fig. 14.
29   Parameterize the curve by the angle θ.
30*   Fix a and let $b = a/n$. Find the area A_n of the resulting epicycloid of n cusps, and find $\lim A_n$.
31   Let $a = 2b$. Find the length of the corresponding **nephroid** (epicycloid of two cusps).
32*   Fix a and let $b = a/n$. Find the length L_n of the resulting epicycloid of n cusps, and find $\lim L_n$.

4. TANGENT, NORMAL, AND CURVATURE

Let $\mathbf{x} = \mathbf{x}(t)$ be a space curve. In this section we shall usually operate under the assumption that $\dot{\mathbf{x}}(t)$ is never $\mathbf{0}$. As we know (p. 683), this implies that we can parameterize the curve in terms of arc length if we wish.

Let us assume $\mathbf{v} = \dot{\mathbf{x}} \neq \mathbf{0}$. Then at each point, the curve has a non-zero tangent vector $\mathbf{v}$, and

$$\mathbf{t} = \frac{\mathbf{v}}{|\mathbf{v}|}$$

is a unit vector in the direction of this tangent. The vector $\mathbf{t}$ is the **unit tangent vector** of the curve $\mathbf{x}(t)$.

■ **EXAMPLE 1**   Find the unit tangent vector to the curve $\mathbf{x}(t) = (t, t^2, t^3)$ at the point $\mathbf{x}(1) = (1, 1, 1)$.

Solution
$$\mathbf{v}(1) = \dot{\mathbf{x}}(1) = (1, 2t, 3t^2)\Big|_{t=1} = (1, 2, 3),$$

$$|\mathbf{v}(1)| = \sqrt{1 + 4 + 9} = \sqrt{14}, \qquad \mathbf{t}(1) = \frac{\mathbf{v}(1)}{|\mathbf{v}(1)|} = \frac{1}{\sqrt{14}}(1, 2, 3). \qquad ■$$

Let us find an alternative formula for $\mathbf{t}$ in terms of arc length. Since $\mathbf{v} \neq \mathbf{0}$, we can parameterize $\mathbf{x}(t)$ in terms of arc length. Then, by the Chain Rule,

$$\mathbf{v} = \frac{d\mathbf{x}}{dt} = \frac{d\mathbf{x}}{ds}\frac{ds}{dt} = \frac{d\mathbf{x}}{ds}|\mathbf{v}|.$$

Therefore
$$\mathbf{t} = \frac{\mathbf{v}}{|\mathbf{v}|} = \frac{d\mathbf{x}}{ds}.$$

Unit Tangent Vector   If $\mathbf{x} = \mathbf{x}(t)$ is a space curve with $\mathbf{v}(t) \neq \mathbf{0}$, then its unit tangent vector is

$$\mathbf{t} = \frac{\mathbf{v}}{|\mathbf{v}|} = \frac{d\mathbf{x}}{ds},$$

where s is arc length.

Cusps   It is worth exploring what can happen at a point where $\mathbf{v}(t) = \mathbf{0}$. For example, consider the plane curve $\mathbf{x}(t) = (t^3, t^2)$. Then

$$\mathbf{v} = \dot{\mathbf{x}} = (3t^2, 2t),$$

so $\mathbf{v}(t) \neq \mathbf{0}$ if $t \neq 0$, but $\mathbf{v}(0) = \mathbf{0}$. Plotting the curve (Fig. 1), we see there is a sharp point at $\mathbf{0}$, called a **cusp**. At the cusp, the curve changes direction abruptly. If $t < 0$ and is very small, the tangent points nearly in the direction of the negative y-axis. But if $t > 0$ and is very small, the tangent points nearly in the direction of the positive y-axis. Thus the curve does an almost instantaneous about-face. A particle moving on the curve slows up as it comes toward $\mathbf{0}$, stops instantaneously at $\mathbf{0}$, then speeds up as it leaves $\mathbf{0}$ in the opposite direction.

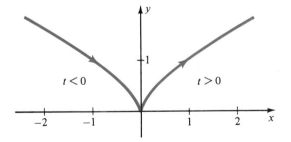

Fig. 1   Graph of $\mathbf{x} = (t^3, t^2)$. Note that $y = x^{2/3}$.

For the remainder of this section we shall assume that $\mathbf{v} \neq \mathbf{0}$, so that $\mathbf{t}$ is defined and unpleasant cusps are ruled out.

Curvature   The unit tangent vector indicates the direction of a space curve at each of its points. As we move along the curve, the direction generally changes, rapidly if the curve bends sharply, less rapidly if the curve is fairly straight. We define curvature to be the rate of change of direction. We measure this rate relative to arc length, so that curvature does not depend on how the curve was originally parameterized.

Curvature   If $\mathbf{x} = \mathbf{x}(s)$ is a space curve, where s denotes arc length, its **curvature** is

$$k = \left| \frac{d\mathbf{t}}{ds} \right|.$$

By definition, curvature is a non-negative real number.

■ **EXAMPLE 2**   Find all curves with curvature identically zero.

Solution   A natural guess is all straight lines. Let us prove this is so. We are given $k = 0$. Therefore

$$\left|\frac{d\mathbf{t}}{ds}\right| = 0; \quad \text{hence} \quad \frac{d\mathbf{t}}{ds} = 0.$$

It follows that $\mathbf{t} = \mathbf{a}$, a constant vector. Consequently

$$\frac{d\mathbf{x}}{ds} = \mathbf{t} = \mathbf{a} = \frac{d}{ds}(s\mathbf{a}), \quad \text{so} \quad \mathbf{x}(s) = s\mathbf{a} + \mathbf{b},$$

where $\mathbf{b}$ is constant. This is the parametric vector equation of a straight line through $\mathbf{b}$ parallel to $\mathbf{a}$.   ■

Computation of Curvature   We shall derive several formulas for computing curvature. They all follow from one basic formula.

If $\mathbf{x} = \mathbf{x}(t)$ is a space curve, then $k = \dfrac{|\dot{\mathbf{x}} \times \ddot{\mathbf{x}}|}{|\dot{\mathbf{x}}|^3}.$

Proof   By Chain Rule,

$$\dot{\mathbf{x}} = \frac{ds}{dt}\frac{d\mathbf{x}}{ds} = \frac{ds}{dt}\mathbf{t}, \quad \ddot{\mathbf{x}} = \frac{d^2s}{dt^2}\mathbf{t} + \frac{ds}{dt}\frac{d\mathbf{t}}{dt} = \frac{d^2s}{dt^2}\mathbf{t} + \left(\frac{ds}{dt}\right)^2\frac{d\mathbf{t}}{ds}.$$

We compute $\dot{\mathbf{x}} \times \ddot{\mathbf{x}}$, using $\mathbf{t} \times \mathbf{t} = \mathbf{0}$: $\dot{\mathbf{x}} \times \ddot{\mathbf{x}} = \left(\dfrac{ds}{dt}\right)^3 \mathbf{t} \times \dfrac{d\mathbf{t}}{ds}.$

Now $\mathbf{t}$, being a unit vector, is perpendicular to $d\mathbf{t}/ds$, hence

$$\left|\mathbf{t} \times \frac{d\mathbf{t}}{ds}\right| = |\mathbf{t}| \cdot \left|\frac{d\mathbf{t}}{ds}\right| = 1 \cdot k = k.$$

Therefore, $\left|\dot{\mathbf{x}} \times \ddot{\mathbf{x}}\right| = \left(\dfrac{ds}{dt}\right)^3 k = |\dot{\mathbf{x}}|^3 \, k,$

which is equivalent to the stated formula.

Now follow three formulas for curvature. The first two apply to curves given in parametric form, the third to the graph of a function.

Curvature Formulas   If $\mathbf{x} = \mathbf{x}(t)$ is a space curve, then

$$k = \frac{[|\dot{\mathbf{x}}|^2 \, |\ddot{\mathbf{x}}|^2 - (\dot{\mathbf{x}} \cdot \ddot{\mathbf{x}})^2]^{1/2}}{|\dot{\mathbf{x}}|^3}.$$

If $\mathbf{x} = (x(t), y(t))$ is a plane curve, then $k = \dfrac{|\dot{x}\ddot{y} - \dot{y}\ddot{x}|}{(\dot{x}^2 + \dot{y}^2)^{3/2}}.$

If a plane curve is the graph of a function $y = f(x)$, then $k = \dfrac{|f''(x)|}{[1 + f'(x)^2]^{3/2}}.$

The first assertion is a restatement of the basic formula because of the identity

$$|\mathbf{u} \times \mathbf{v}|^2 = |\mathbf{u}|^2 \, |\mathbf{v}|^2 - (\mathbf{u} \cdot \mathbf{v})^2,$$

proved on p. 660.

If $\mathbf{x} = (x(t), y(t))$ is a plane curve, then

$$\dot{\mathbf{x}} = (\dot{x}, \dot{y}) \qquad \text{and} \qquad \ddot{\mathbf{x}} = (\ddot{x}, \ddot{y}),$$

hence $|\dot{\mathbf{x}}|^2 \, |\ddot{\mathbf{x}}|^2 - (\dot{\mathbf{x}} \cdot \ddot{\mathbf{x}})^2 = (\dot{x}^2 + \dot{y}^2)(\ddot{x}^2 + \ddot{y}^2) - (\dot{x}\ddot{x} + \dot{y}\ddot{y})^2 = (\dot{x}\ddot{y} - \dot{y}\ddot{x})^2,$

so the second formula follows.

Finally, if the plane curve is the graph of $y = f(x)$, apply the second formula with $t = x$ and $\mathbf{x} = (t, f(t)) = (x, f(x))$. Then $\dot{x} = 1$, $\ddot{x} = 0$, $\dot{y} = f'(x)$, and $\ddot{y} = f''(x)$, so the third formula follows by direct substitution.

■ **EXAMPLE 3**   Find the curvature of a circle of radius a.

Solution   Let the equation of the circle be $x^2 + y^2 = a^2$. Thus

$$y = \pm\sqrt{a^2 - x^2}.$$

(This equation describes either the upper or lower half of the circle depending on whether the positive or negative square root is chosen.) Differentiate:

$$y' = \frac{-x}{\pm\sqrt{a^2 - x^2}} = -\frac{x}{y}.$$

Differentiate again:

$$y'' = -\frac{y - xy'}{y^2} = -\frac{y - x(-x/y)}{y^2} = -\frac{x^2 + y^2}{y^3} = -\frac{a^2}{y^3}.$$

Now

$$1 + y'^2 = 1 + \left(-\frac{x}{y}\right)^2 = \frac{y^2 + x^2}{y^2} = \frac{a^2}{y^2}.$$

Hence by the formula for curvature,

$$k = \frac{|-a^2/y^3|}{(a^2/y^2)^{3/2}} = \frac{a^2}{a^3} = \frac{1}{a}.$$

Alternative solution   Write $\mathbf{x}(t) = (a \cos t, a \sin t)$.

(This describes the circle by its central angle t.) Then

$$\dot{\mathbf{x}} = (-a \sin t, a \cos t), \qquad |\dot{\mathbf{x}}| = \frac{ds}{dt} = \sqrt{(-a \sin t)^2 + (a \cos t)^2} = a.$$

Hence

$$\mathbf{t} = \frac{\dot{\mathbf{x}}}{|\dot{\mathbf{x}}|} = \frac{1}{a}(-a \sin t, a \cos t) = (-\sin t, \cos t).$$

Differentiate with respect to t: On the one hand,

$$\frac{d\mathbf{t}}{dt} = (-\cos t, -\sin t), \qquad \left|\frac{d\mathbf{t}}{dt}\right| = 1.$$

On the other hand,

$$\frac{d\mathbf{t}}{dt} = \frac{ds}{dt}\frac{d\mathbf{t}}{ds} = a\frac{d\mathbf{t}}{ds}, \qquad \left|\frac{d\mathbf{t}}{dt}\right| = a\left|\frac{d\mathbf{t}}{ds}\right| = ak.$$

Therefore $\qquad\qquad\qquad\qquad ak = 1, \qquad a = \frac{1}{k}.$ ∎

Remark The curvature of a circle is the reciprocal of its radius. This is reasonable on two counts. First, the curvature is the same at all points of a circle. Second, it is small for large circles, since the larger the circle the more slowly its direction changes per unit of arc length.

The Unit Normal The vector $d\mathbf{t}/ds$ has length k, the curvature. Suppose $k \neq 0$; then

$$\frac{d\mathbf{t}}{ds} = k\mathbf{n},$$

where $\mathbf{n}$ is a unit vector in the direction of $d\mathbf{t}/ds$. Since $\mathbf{t}$ is a unit vector, $\mathbf{t}$ and $d\mathbf{t}/ds$ are perpendicular, that is, $\mathbf{t}$ and $\mathbf{n}$ are perpendicular (Fig. 2). The vector $\mathbf{n}$ is called the **unit normal** to the curve.

Let $\mathbf{x} = \mathbf{x}(s)$ be a space curve with curvature $k(s) \neq 0$. Then

$$\frac{d\mathbf{x}}{ds} = \mathbf{t}, \qquad \frac{d\mathbf{t}}{ds} = k\mathbf{n}, \qquad |\mathbf{t}| = |\mathbf{n}| = 1, \qquad \mathbf{t} \cdot \mathbf{n} = 0.$$

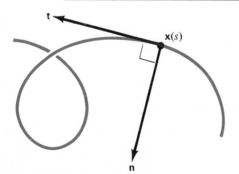

Fig. 2 Unit tangent and normal

The further study of space curves, not pursued here, begins with an analysis of $d\mathbf{n}/ds$. That leads to another quantity, torsion, which measures how fast the plane of $\mathbf{t}$ and $\mathbf{n}$ is turning around the tangent line.

■ **EXAMPLE 4** Compute $\mathbf{t}, \mathbf{n}$, and k for the helix (circular spiral)

$$\mathbf{x}(t) = (a\cos t, a\sin t, bt),$$

where $a > 0$ and $b > 0$.

Solution The projection of $\mathbf{x}(t)$ on the x, y-plane is $(a\cos t, a\sin t, 0)$. As a particle describes the curve $\mathbf{x}(t)$, its projection (Fig. 3a) describes a circle of radius a. The third component of $\mathbf{x}(t)$ is bt; the particle moves upward at a steady rate. Thus, the curve is a spiral, circular and rising steadily (Fig. 3b).

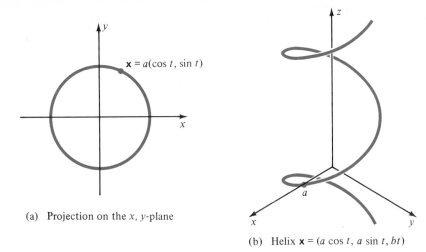

(a) Projection on the x, y-plane

(b) Helix $\mathbf{x} = (a \cos t, a \sin t, bt)$

Fig. 3

To find $\mathbf{t}$, differentiate $\mathbf{x} = (a \cos t, a \sin t, bt)$:

$$\dot{\mathbf{x}} = (-a \sin t, a \cos t, b), \qquad |\dot{\mathbf{x}}| = \sqrt{a^2 + b^2} = c.$$

Hence, $$\mathbf{t} = \frac{\dot{\mathbf{x}}}{|\dot{\mathbf{x}}|} = \frac{1}{c}(-a \sin t, a \cos t, b) \quad \text{and} \quad \frac{ds}{dt} = c.$$

To find k and $\mathbf{n}$, use the relation $k\mathbf{n} = \dfrac{d\mathbf{t}}{ds}$:

$$k\mathbf{n} = \frac{d\mathbf{t}}{ds} = \frac{d\mathbf{t}}{dt}\frac{dt}{ds} = \frac{d\mathbf{t}}{dt} \bigg/ \frac{ds}{dt} = \frac{1}{c}\frac{d\mathbf{t}}{dt} = \frac{a}{c^2}(-\cos t, -\sin t, 0).$$

The left-hand side is $k\mathbf{n}$, where $k \geq 0$ and $\mathbf{n}$ is a unit vector. The right-hand side is also a positive constant times a unit vector. It follows that

$$k = \frac{a}{c^2}, \qquad \mathbf{n} = (-\cos t, -\sin t, 0).$$

Answer $\mathbf{t} = \dfrac{1}{\sqrt{a^2 + b^2}}(-a \sin t, a \cos t, b), \quad \mathbf{n} = (-\cos t, -\sin t, 0),$

$$k = \frac{a}{a^2 + b^2}. \qquad\qquad\qquad \blacksquare$$

Remark   If $b = 0$, the spiral degenerates into a circle of radius a and the curvature k reduces to $1/a$, which agrees with Example 3.

Ovals   An **oval** is a simple closed plane curve with $k > 0$ at each point. If an oval has length L, then $\mathbf{x} = \mathbf{x}(s)$ is periodic of period L.

Let C be an oval taken counterclockwise (Fig. 4). Let α be the angle between the positive x-axis and the unit tangent $\mathbf{t}$. Then α increases from 0 to 2π as s increases from 0 to L. If we like, we can parameterize the curve in terms of α instead of s.

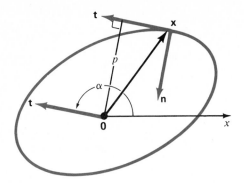

Fig. 4   Geometry of an oval

Clearly
$$\mathbf{t} = (\cos \alpha, \sin \alpha).$$

It follows that
$$k\mathbf{n} = \frac{d\mathbf{t}}{ds} = \frac{d\alpha}{ds}\frac{d\mathbf{t}}{d\alpha} = \frac{d\alpha}{ds}(-\sin \alpha, \cos \alpha).$$

Since $d\alpha/ds > 0$ and $(-\sin \alpha, \cos \alpha)$ is a unit vector, we have
$$k = \frac{d\alpha}{ds}, \qquad \mathbf{n} = (-\sin \alpha, \cos \alpha).$$

The normal $\mathbf{n}$ is directed inward because $\mathbf{t}$ is turning counterclockwise.

The component of the position vector $\mathbf{x}$ in the normal direction is important in the study of ovals; it is called the **support function** of the oval and written $p = p(\alpha)$:
$$p = -\mathbf{x} \cdot \mathbf{n} = x \sin \alpha - y \cos \alpha.$$

Note that p is the distance from $\mathbf{0}$ to the **supporting line** $x \sin \alpha - y \cos \alpha = p$. This line is tangent to the oval at $\mathbf{x}$ and separates the plane into two half-planes, one containing the oval and the other completely apart from it.

For a circle of radius a, we have $p = a$ and $k = 1/a$, hence $p = 1/k$. For more general ovals, p and k vary, but there is still a relation between them:
$$p + \frac{d^2 p}{d\alpha^2} = \frac{1}{k}.$$

We postpone the proof to the exercises. For the moment, let us assume this formula and show how it implies some nice properties of ovals.

■ **EXAMPLE 5**   Suppose an oval has length L and area A. Prove that

(a) $L = \displaystyle\int_0^{2\pi} p \, d\alpha$ (b) $A = \dfrac{1}{2}\displaystyle\int_0^{L} p \, ds.$

Solution   (a)   From the relation quoted above,
$$\int_0^{2\pi} p \, d\alpha + \int_0^{2\pi} \frac{d^2 p}{d\alpha^2} \, d\alpha = \int_0^{2\pi} \frac{d\alpha}{k}.$$

By inspection, the second integral is 0. That is because $p(\alpha)$ is periodic with period 2π, hence so is its derivative $dp/d\alpha$. Therefore $(dp/d\alpha)(2\pi) - (dp/d\alpha)(0) = 0$.

For the integral on the right, we recall that

$$k = \frac{d\alpha}{ds}, \qquad \text{hence} \quad \frac{1}{k} = \frac{ds}{d\alpha}.$$

Therefore
$$\int_0^{2\pi} p \, d\alpha = \int_0^{2\pi} \frac{ds}{d\alpha} \, d\alpha = s(2\pi) - s(0) = L.$$

(b)   We use the formula

$$A = \frac{1}{2} \int_0^L \left(x \frac{dy}{ds} - y \frac{dx}{ds} \right) ds = \frac{1}{2} \int_0^L (x, y) \cdot \left(\frac{dy}{ds}, -\frac{dx}{ds} \right) ds.$$

But
$$\mathbf{t} = \frac{d\mathbf{x}}{ds} = \left(\frac{dx}{ds}, \frac{dy}{ds} \right), \qquad \text{hence} \quad \mathbf{n} = \left(-\frac{dy}{ds}, \frac{dx}{ds} \right).$$

It follows that
$$A = \tfrac{1}{2} \int_0^L \mathbf{x} \cdot (-\mathbf{n}) \, ds = \tfrac{1}{2} \int_0^L p \, ds. \qquad \blacksquare$$

EXERCISES

Find the unit tangent to the curve

1   $y = x^2$
2   $xy = 1$
3   $\mathbf{x} = (t^2, t^3), t \neq 0$
4   $\mathbf{x} = (t \cos t, t \sin t)$
5   $\mathbf{x} = t\mathbf{a} + \mathbf{b}, \quad \mathbf{a} \neq \mathbf{0}$
6   $\mathbf{x} = (t \cos t, t \sin t, 2t)$
7   $\mathbf{x} = \tfrac{1}{2}t^2 \mathbf{a} + t\mathbf{b} + \mathbf{c}$
8   $\mathbf{x} = (t^2, t^3, t^4)$

Find the curvature

9   $y = x^2$
10   $xy = 1$
11   $\mathbf{x} = (t^3, t^2), \quad t \neq 0$
12   $\mathbf{x} = (t, t^2, t^3)$
13   $\mathbf{x} = t\mathbf{a} + t^3 \mathbf{b}$
14   $\mathbf{x} = (t \cos t, t \sin t)$
15   $\mathbf{x} = (\cos t, \sin t, \sin 2t)$
16   $\mathbf{x} = (e^{-t}, t, e^t)$.

Find the maximum curvature of

17   $y = \sin x$
18   $y = \ln x$.
19   A point moves along $y = e^x$ at the rate of r cm/sec. How fast is the tangent turning when the point is at (a, e^a) ?
20   Compute the maximum and minimum curvature of an ellipse with semi-major axis a and semi-minor axis b. Check the case $a = b$.

Plot carefully near $t = 0$:

21   $\mathbf{x} = (t^2, t^5)$
22   $\mathbf{x} = (t^2, 2t^4 + t^5)$.

23   Let $\mathbf{x} = \mathbf{x}(s)$ be a plane curve. Show that $d\mathbf{n}/ds = k\mathbf{t}$. [*Hint*   Differentiate $\mathbf{t} \cdot \mathbf{n} = 0$ and $\mathbf{n} \cdot \mathbf{n} = 1$.]

24*   Let $\mathbf{x} = \mathbf{x}(s)$ be a plane curve such that $\mathbf{x}(0) = \mathbf{0}$, $\mathbf{t}(0) = (1, 0)$, $\mathbf{n}(0) = (0, 1)$. Let $k = k(s) = a + bs + \cdots$ be the Taylor expansion of $k(s)$ at $s = 0$, with $a > 0$. Show that

$$x(s) = s - \tfrac{1}{6}a^2 s^3 + \cdots, \qquad y(s) = \tfrac{1}{2}as^2 + \tfrac{1}{6}bs^3 + \cdots.$$

25   Let $\mathbf{x}(s)$ be a space curve on the unit sphere, that is, $|\mathbf{x}| = 1$. Show that $k(s) \geq 1$. [*Hint* Differentiate twice.]

The next 11 exercises concern plane ovals. See pp. 701–2 for the notation.

26   Prove $\dfrac{d\mathbf{t}}{d\alpha} = \mathbf{n}, \quad \dfrac{d\mathbf{n}}{d\alpha} = -\mathbf{t}$

27   Express $\mathbf{x} \cdot \mathbf{t}$ in terms of p and α.   **28***   Express $\mathbf{x}$ in terms of p and α.

29   Find $\int_0^L k \, ds$.

30   Prove $\dfrac{1}{k} = p + \dfrac{d^2 p}{d\alpha^2}$.   [*Hint*   Use Ex. 27.]

31*   Prove $A = \frac{1}{2} \int_0^{2\pi} \left[p^2 - \left(\dfrac{dp}{d\alpha} \right)^2 \right] d\alpha$.   [*Hint*   Compute $\dfrac{d}{d\alpha} \left(p \dfrac{dp}{d\alpha} \right)$ and integrate by parts.]

32*   Find the support function for the ellipse $x^2/a^2 + y^2/b^2 = 1$. [*Hint*   Parameterize.]

33*   Let $\mathbf{x} = \mathbf{x}(\alpha)$ be an oval and $a > 0$. Define the **parallel oval** $\mathbf{x}_1 = \mathbf{x}_1(\alpha)$ by $\mathbf{x}_1 = \mathbf{x} - a\mathbf{n}$. Show that $\mathbf{t}_1 = \mathbf{t}$.

34*   (cont.) Find p_1.

35*   (cont.) Prove $L_1 = L + 2\pi a$.

36*   (cont.) Prove $A_1 = A + La + \pi a^2$.

5. VELOCITY AND ACCELERATION

Suppose $\mathbf{x} = \mathbf{x}(t)$ is the path of a moving particle. The **velocity** of the particle is the vector

$$\mathbf{v}(t) = \dot{\mathbf{x}}(t).$$

This velocity vector is tangential to the curve; its length is the **speed** of the particle. The **acceleration** of the particle is the vector

$$\mathbf{a}(t) = \dot{\mathbf{v}}(t) = \ddot{\mathbf{x}}(t).$$

Its length and direction are generally not as apparent as those of the velocity vector.

Example $\mathbf{x}(t) = (b \cos \omega t, \, b \sin \omega t)$,

where $b > 0$ and $\omega > 0$. Since $|\mathbf{x}| = b$, the particle moves on a circle of radius b. Its

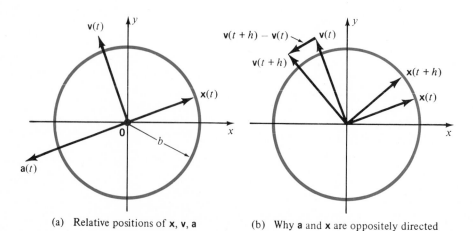

(a) Relative positions of $\mathbf{x}$, $\mathbf{v}$, $\mathbf{a}$ (b) Why $\mathbf{a}$ and $\mathbf{x}$ are oppositely directed

Fig. 1   Uniform circular motion: $\mathbf{x}(t) = b(\cos \omega t, \, \sin \omega t)$

velocity and acceleration are

$$\mathbf{v}(t) = \dot{\mathbf{x}}(t) = b\omega(-\sin \omega t, \cos \omega t),$$

$$\mathbf{a}(t) = \dot{\mathbf{v}}(t) = b\omega^2(-\cos \omega t, -\sin \omega t) = -b\omega^2\mathbf{x}.$$

The speed is $|\mathbf{v}| = b\omega$, a constant, so the motion is uniform circular motion. The velocity vector $\mathbf{v}(t)$ is perpendicular to the position vector since $\mathbf{x}(t) \cdot \mathbf{v}(t) = 0$. This is not surprising since each tangent to a circle is perpendicular to the corresponding radius. The acceleration is $\mathbf{a}(t) = -b\omega^2\mathbf{x}(t)$, so the acceleration vector $\mathbf{a}(t)$ is directed opposite to the position vector $\mathbf{x}(t)$. See Fig. 1a. Why should that be?

The acceleration $\mathbf{a}(t)$ measures the rate of change of the velocity. Observe the velocity vectors at t and at an instant later, $t + h$. See Fig. 1b. The difference $\mathbf{v}(t + h) - \mathbf{v}(t)$ is nearly parallel to $\mathbf{x}(t)$, but oppositely directed. Therefore the instantaneous rate of change of the velocity is in a direction opposite to that of $\mathbf{x}(t)$.

Newton's Law of Motion   This famous principle states that

$$\text{force} = \text{mass} \times \text{acceleration}.$$

But force and acceleration are vectors, both having magnitude and direction. Thus Newton's Law is a vector equation:

$$\mathbf{F} = m\ddot{\mathbf{x}}.$$

It is equivalent to three scalar equations for the components:

$$F_1 = m\ddot{x}_1, \qquad F_2 = m\ddot{x}_2, \qquad F_3 = m\ddot{x}_3.$$

■ **EXAMPLE 1**   A particle of mass m is subject to zero force. Find its trajectory.

Solution   By Newton's Law, $m\ddot{\mathbf{x}} = \mathbf{0}$, $\ddot{\mathbf{x}} = \mathbf{0}$

Since $\ddot{\mathbf{x}} = \dot{\mathbf{v}}$, $\dfrac{d\mathbf{v}}{dt} = \mathbf{0}$.

Integrate once; $\mathbf{v}$ is constant: $\mathbf{v} = \mathbf{v}_0$, $\dfrac{d\mathbf{x}}{dt} = \mathbf{v}_0$.

Integrate again: $\mathbf{x} = t\mathbf{v}_0 + \mathbf{x}_0$.

Therefore the trajectory is a straight line, traversed at constant speed.   ■

■ **EXAMPLE 2**   A shell is fired at an angle α with the ground and initial speed v_0. What is its path? Neglect air resistance.

Solution   Draw a figure, taking the axes as indicated (Fig. 2). Let $\mathbf{v}_0$ be the initial velocity vector, so $\mathbf{v}_0 = v_0(\cos \alpha, \sin \alpha)$. Let m denote the mass of the shell. The force of gravity at each point is constant,

$$\mathbf{F} = (0, -mg).$$

The equation of motion is $m\mathbf{a} = \mathbf{F}$,   that is, $\dfrac{d^2\mathbf{x}}{dt^2} = (0, -g)$.

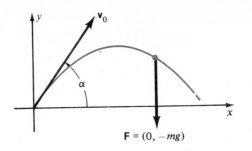

Fig. 2

Integrate:
$$\frac{d\mathbf{x}}{dt} = (0, -gt) + \mathbf{v}_0.$$

Integrate again, noting that $\mathbf{x}_0 = \mathbf{0}$ by the choice of axes:

$$\mathbf{x} = (0, -\tfrac{1}{2}gt^2) + t\mathbf{v}_0.$$

Hence $(x(t), y(t)) = (0, -\tfrac{1}{2}gt^2) + tv_0(\cos \alpha, \sin \alpha) = (v_0 t \cos \alpha, v_0 t \sin \alpha - \tfrac{1}{2}gt^2).$

To describe the path, eliminate t: $x = v_0 t \cos \alpha,$ $t = \dfrac{x}{v_0 \cos \alpha},$

$$y = v_0 t \sin \alpha - \frac{1}{2} gt^2 = x \tan \alpha - \frac{g}{2v_0{}^2 \cos^2 \alpha} x^2 = -ax^2 + bx,$$

where $a = g/2v_0{}^2 \cos^2 \alpha$ and $b = \tan \alpha$. The graph of this quadratic is a parabola. ∎

■ **EXAMPLE 3**   In Example 2, what is the maximum range (ground distance) for fixed v_0?

Solution   The shell hits ground when $y = 0$:

$$(v_0 \sin \alpha - \tfrac{1}{2}gt)t = 0.$$

This equation has two roots. The root $t = 0$ indicates the initial point. We want the other root, $t = 2v_0(\sin \alpha)/g$. The range is the value of x at this time:

$$x = (v_0 \cos \alpha)\left(\frac{2v_0 \sin \alpha}{g}\right) = \frac{v_0{}^2}{g} \sin 2\alpha.$$

Clearly x is maximum when $\sin 2\alpha = 1$, or $\alpha = \tfrac{1}{4}\pi$.

Therefore the maximum range is $v_0{}^2/g$, and it is achieved by firing at $45°$.   ∎

Components of Acceleration   If a particle moves on a curve, it is useful to express its velocity and acceleration in terms of $\mathbf{t}, \mathbf{n}$, and k since these are quantities built into the curve (independent of parameterization). We already know

$$\mathbf{v} = \frac{ds}{dt}\mathbf{t},$$

which says that the motion is directed along the tangent with speed ds/dt.

For further information, differentiate **v** with respect to time, using the Chain Rule carefully:

$$\mathbf{a} = \frac{d\mathbf{v}}{dt} = \frac{d^2s}{dt^2}\mathbf{t} + \frac{ds}{dt}\frac{d\mathbf{t}}{dt}.$$

But $\dfrac{d\mathbf{t}}{dt} = \dfrac{ds}{dt}\dfrac{d\mathbf{t}}{ds} = \dfrac{ds}{dt}k\mathbf{n},$ where k is the curvature. Therefore

---

Tangential and Normal Components of Acceleration

$$\mathbf{a} = \frac{d^2s}{dt^2}\mathbf{t} + k\left(\frac{ds}{dt}\right)^2\mathbf{n}.$$

This is an important equation in mechanics. It resolves the acceleration into two components, one tangential to the direction of motion, the other normal (perpendicular) to the direction of motion. The normal component $k\dot{s}^2\mathbf{n}$ is called the **centripetal acceleration**.

Normal and tangential components of acceleration have a natural interpretation. Remember that acceleration is the rate of change of the velocity vector **v**. Now a vector can change for two reasons: (a) its length changes, (b) its direction changes. Since $|\mathbf{v}| = \dot{s}$, a change in $|\mathbf{v}|$ is indicated by the second derivative $\ddot{s}$. Hence the tangential component of **a** corresponds to the changing *length* of $|\mathbf{v}|$, that is, the changing speed. A change in the direction of **v** is measured by the curvature k. Hence the normal component of **a** corresponds to the changing *direction* of **v**.

■ **EXAMPLE 4**   A particle moves counterclockwise on a circle of radius b. Resolve its acceleration into tangential and normal components.

Solution   Place the circle in the x, y-plane with center at **0**. Let $\theta = \theta(t)$ denote the central angle at time t. Then the path is given by

$$\mathbf{x}(t) = b(\cos\theta, \sin\theta).$$

Differentiate: $\mathbf{v} = \dot{\mathbf{x}} = b\dot{\theta}(-\sin\theta, \cos\theta).$

On the other hand, $\mathbf{v} = |\mathbf{v}|\mathbf{t}.$

Since $b\dot{\theta} > 0$ and $(-\sin\theta, \cos\theta)$ is a unit vector, it follows that

$$|\mathbf{v}| = b\dot{\theta}, \qquad \mathbf{t} = (-\sin\theta, \cos\theta).$$

The **angular speed** $\dot{\theta}$ is usually denoted by $\omega = \omega(t)$. Therefore

$$\mathbf{v} = b\omega(-\sin\theta, \cos\theta) = b\omega\mathbf{t}.$$

To find the acceleration, differentiate again:

$$\mathbf{a} = \dot{\mathbf{v}} = b\dot{\omega}\mathbf{t} + b\omega\dot{\mathbf{t}} = b\dot{\omega}\mathbf{t} + b\omega^2(-\cos\theta, -\sin\theta).$$

Since $(-\cos\theta, -\sin\theta)$ is perpendicular to **t**, it must equal $\pm\mathbf{n}$. The correct sign is plus because the normal component of **a** is a *positive* multiple of **n** and $b\omega^2 > 0$. Therefore

$$\mathbf{a} = b\dot{\omega}\mathbf{t} + b\omega^2\mathbf{n}.$$

Remark   When the motion is uniform (ω constant), then $\mathbf{a} = b\omega^2\mathbf{n}$, so the acceleration is all centripetal, perpendicular to the direction of motion. This agrees with the example on p. 704.

■ **EXAMPLE 5**   The position of a moving particle is given by $\mathbf{x}(t) = (5t + 1, t^2, 3t^2)$. Resolve its acceleration into tangential and normal components.

Solution   Write the acceleration vector as $\mathbf{a} = \mathbf{a}_t + \mathbf{a}_n$,

where $\mathbf{a}_t$ and $\mathbf{a}_n$ are its tangential and normal components. It will be enough to find $\mathbf{a}_t$ since $\mathbf{a}_n = \mathbf{a} - \mathbf{a}_t = \ddot{\mathbf{x}} - \mathbf{a}_t$. Now

$$\mathbf{a}_t = \frac{d^2s}{dt^2}\,\mathbf{t}, \qquad \text{where} \quad \mathbf{t} = \frac{\dot{\mathbf{x}}}{|\dot{\mathbf{x}}|} \quad \text{and} \quad \frac{ds}{dt} = |\dot{\mathbf{x}}|.$$

From $\dot{\mathbf{x}} = (5, 2t, 6t)$, we have

$$\frac{ds}{dt} = |\dot{\mathbf{x}}| = (25 + 40t^2)^{1/2}, \qquad \frac{d^2s}{dt^2} = 40t(25 + 40t^2)^{-1/2}.$$

Hence $\mathbf{a}_t = \dfrac{d^2s}{dt^2}\dfrac{1}{|\dot{\mathbf{x}}|}\dot{\mathbf{x}} = \dfrac{40t}{25 + 40t^2}\,(5, 2t, 6t) = \dfrac{8t}{5 + 8t^2}\,(5, 2t, 6t).$

It follows that $\mathbf{a}_n = \ddot{\mathbf{x}} - \mathbf{a}_t = (0, 2, 6) - \mathbf{a}_t = \dfrac{10}{5 + 8t^2}\,(-4t, 1, 3).$ ■

Angular Velocity   Suppose that a rigid body rotates about an axis through **0**. See Fig. 3a. The central angle is $\theta = \theta(t)$, so $\omega = \dot{\theta}$ is its **angular speed**, the rate of rotation in radians per second. Its **angular velocity** is defined to be the vector $\boldsymbol{\omega}$ having magnitude $\dot{\theta}$ and pointing along the (positive) axis of rotation according to the right-hand rule (Fig. 3b).

Once the angular velocity vector $\boldsymbol{\omega}$ is known, it is easy to find the velocity $\mathbf{v}$ of any point $\mathbf{x}$ in the rigid body. See Fig. 3c.

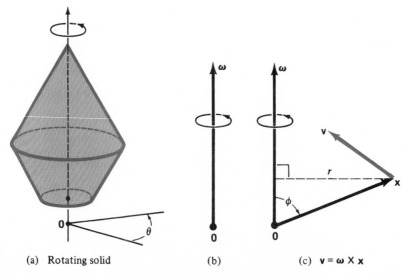

(a) Rotating solid   (b)   (c) $\mathbf{v} = \boldsymbol{\omega} \times \mathbf{x}$

Fig. 3   Angular velocity

Since the point **x** is rotating about the axis of ω, its velocity vector **v** is perpendicular to the plane of ω and **x**. By the right-hand rule, **v** points in the direction of ω × **x**. The speed $|\mathbf{v}|$ is the product of the angular speed $\omega = |\omega|$ and the distance r of **x** from the axis of rotation. But $r = |\mathbf{x}| \sin \phi$, hence

$$|\mathbf{v}| = |\omega| \, |\mathbf{x}| \sin \phi = |\omega \times \mathbf{x}|.$$

Therefore:

> Suppose a rigid body rotates with angular velocity ω about an axis through **0**. The velocity of a point **x** in the body is
>
> $$\mathbf{v} = \omega \times \mathbf{x}.$$

EXERCISES

1 A hill makes angle β with the ground (Fig. 4). A shell is fired with initial speed v_0 from the base of the hill at angle α with the ground. Show that the x-component of the position where the shell strikes the hill is $x = (2v_0{}^2/g)(\sin \alpha \cos \alpha - \tan \beta \cos^2 \alpha)$.
2 (cont.) Find the maximum of x as a function of α and for what α it occurs.

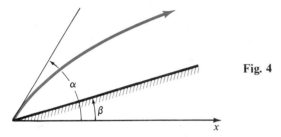

Fig. 4

Find the tangential and normal components of the acceleration vector.

3 $\mathbf{x} = (t, t^2)$ 4 $\mathbf{x} = (t^2, t^3), \quad t > 0$ 5 $\mathbf{x} = (t, \sin t)$
6 $\mathbf{x} = (e^t, t)$ 7 $\mathbf{x} = (\cos^2 t, \sin^2 t)$ 8 $\mathbf{x} = (2 \cos t, 3 \sin t)$
9 $\mathbf{x} = (b \cos \omega t, b \sin \omega t, ct), \quad \omega$ constant
10 $\mathbf{x} = (t, t^2, t^3)$ 11 $\mathbf{x} = (\sin t, \cos t, \sin t)$ 12 $(t, \quad t^2)$.

13 A particle moves with constant speed 1 on the surface of the unit sphere $|\mathbf{x}| = 1$. Show that the normal component of the acceleration has magnitude at least 1.
14 A particle moves on the surface $z = x^2 + y^2$ with constant speed 1. At a certain instant t_0 it passes through **0**. Show that the tangential component of **a** is **0** and the normal component is $(\ddot{x}(t_0), \ddot{y}(t_0), 2)$. Show also with $\dot{x}\ddot{x} + \dot{y}\ddot{y} = 0$ at t_0.

6. CURVES IN POLAR COORDINATES

We shall study plane curves whose polar coordinates are given as functions of time:

$$r = r(t), \qquad \theta = \theta(t).$$

Area   Certain problems require the area swept out by the segment joining **0** to a moving point on a curve (Fig. 1a). Suppose the curve is given in parametric polar coordinates

$$r = r(t), \qquad \theta = \theta(t), \qquad a \le t \le b.$$

In a small interval of time the segment sweeps out a thin triangle of base $r\, d\theta$ and height r (ignoring negligible errors (Fig. 1b). Hence

$$dA = \frac{1}{2} r^2\, d\theta = \frac{1}{2} r^2 \frac{d\theta}{dt}\, dt,$$

$$A = \frac{1}{2} \int_a^b r^2 \frac{d\theta}{dt}\, dt.$$

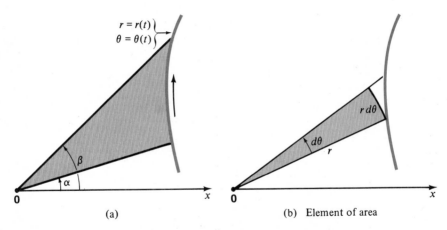

(a)

(b) Element of area

Fig. 1   Area swept out by the radius vector

If the curve is given by $r = r(\theta)$ for $\alpha \le \theta \le \beta$, choose $t = \theta$. Then the formula specializes to

$$A = \frac{1}{2} \int_\alpha^\beta [r(\theta)]^2\, d\theta.$$

(The rough argument given here can be made rigorous using approximating sums for integrals, but we shall not go into the details.)

■ **EXAMPLE 1**   Compute the area of the four-petal rose $r = a \cos 2\theta$.

Solution   Figure 2 shows the graph, emphasizing the part where $0 \le \theta \le \frac{1}{4}\pi$. Because of symmetry it suffices to compute the area of half of one petal. Thus

$$A = 8 \int_0^{\pi/4} \tfrac{1}{2}(a \cos 2\theta)^2\, d\theta = 4a^2 \int_0^{\pi/4} \cos^2 2\theta\, d\theta = 2a^2 \int_0^{\pi/4} (1 + \cos 4\theta)\, d\theta = \tfrac{1}{2}\pi a^2.$$

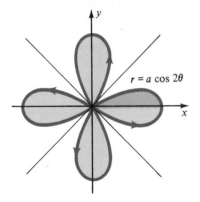

$r = a \cos 2\theta$

Fig. 2   Four-petal rose curve

The Natural Frame   When dealing with curves given in parametric polar form, we use a pair of unit vectors that do for polar coordinates what **i** and **j** do for rectangular coordinates. Note that at each point (x, y), the vector **i** points in the direction of increasing x and **j** points in the direction of increasing y. Now we define at each point $\{r, \theta\}$, where $r \neq 0$, a unit vector **u** in the direction of increasing r and a unit vector **w** in the direction of increasing θ. See Fig. 3. From the figure we find that

$$\mathbf{u} = (\cos \theta, \sin \theta), \qquad \mathbf{w} = (-\sin \theta, \cos \theta).$$

Like **i** and **j**, clearly **u** and **w** are perpendicular. But unlike **i** and **j**, the vectors **u** and **w** associated with $\mathbf{x} = \{r, \theta\}$ vary with **x**. Actually, they depend on θ alone. For future reference we note their derivatives: $d\mathbf{u}/d\theta = \mathbf{w}$ and $d\mathbf{w}/d\theta = -\mathbf{u}$.

The unit vectors $\mathbf{u} = (\cos \theta, \sin \theta)$ and $\mathbf{w} = (-\sin \theta, \cos \theta)$

indicate, respectively, the directions of increasing r and increasing θ at each point $\{r, \theta\}$ where $r \neq 0$. Their derivatives satisfy the relations

$$\frac{d\mathbf{u}}{d\theta} = \mathbf{w} \quad \text{and} \quad \frac{d\mathbf{w}}{d\theta} = -\mathbf{u}.$$

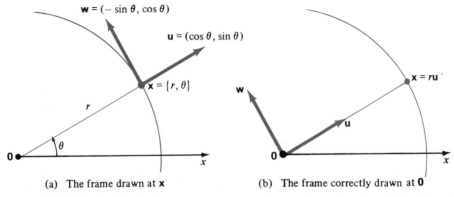

(a)   The frame drawn at **x**   (b)   The frame correctly drawn at **0**

Fig. 3   The natural frame (pair of orthogonal unit vectors) attached to $\mathbf{x} = \{r, \theta\}$

In applications, **u** and **w** provide at each point a natural coordinate system (frame) in which computations are often simpler than in rectangular coordinates. For example, a curse given by $r = r(t)$ and $\theta = \theta(t)$ has a convenient vector form (Fig. 3b):

$$\mathbf{x} = (r \cos \theta, r \sin \theta) = r\mathbf{u}.$$

Velocity and Acceleration   Suppose the path of a particle is described in polar form, $r = r(t)$, $\theta = \theta(t)$. It is natural to express its velocity and acceleration in terms of the perpendicular unit vectors **u** and **w**. (You can think of this pair of vectors as a frame moving along the path, providing at each point a convenient coordinate system.)

We shall need the time derivatives $\dot{\mathbf{u}}$ and $\dot{\mathbf{w}}$. We already have the derivatives $d\mathbf{u}/d\theta = \mathbf{w}$ and $d\mathbf{w}/d\theta = -\mathbf{u}$. From this information we find the derivatives with respect to t by applying the Chain Rule:

$$\dot{\mathbf{u}} = \frac{d\mathbf{u}}{dt} = \frac{d\theta}{dt} \frac{d\mathbf{u}}{d\theta} = \dot{\theta}\, \mathbf{w}. \qquad \dot{\mathbf{w}} = \frac{d\mathbf{w}}{dt} = \frac{d\theta}{dt} \frac{d\mathbf{w}}{d\theta} = -\dot{\theta}\, \mathbf{u}.$$

To find the velocity and acceleration vectors, we first write the given curve in vector form, $\mathbf{x} = r\mathbf{u}$. Then we differentiate twice using the Product Rule carefully:

$$\mathbf{v} = \dot{\mathbf{x}} = \dot{r}\mathbf{u} + r\dot{\mathbf{u}} = \dot{r}\mathbf{u} + r\dot{\theta}\,\mathbf{w}$$

and $\qquad \mathbf{a} = \dot{\mathbf{v}} = (\ddot{r}\mathbf{u} + \dot{r}\dot{\mathbf{u}}) + (\dot{r}\dot{\theta}\,\mathbf{w} + r\ddot{\theta}\mathbf{w} + r\dot{\theta}\,\dot{\mathbf{w}})$

$$= (\ddot{r}\mathbf{u} + \dot{r}\dot{\theta}\,\mathbf{w}) + (\dot{r}\dot{\theta}\,\mathbf{w} - r\dot{\theta}\,^2\mathbf{u}) = (\ddot{r} - r\dot{\theta}\,^2)\mathbf{u} + (r\ddot{\theta} + 2\dot{r}\dot{\theta}\,)\mathbf{w}.$$

Velocity and Acceleration in Polar Coordinates
If the motion of a particle is given by $r = r(t)$, $\theta = \theta(t)$, then

$$\mathbf{v} = \dot{r}\mathbf{u} + r\dot{\theta}\,\mathbf{w}, \qquad \mathbf{a} = (\ddot{r} - r\dot{\theta}\,^2)\mathbf{u} + (r\ddot{\theta} + 2\dot{r}\dot{\theta}\,)\mathbf{w}.$$

Example   The spiral $r = t$, $\theta = t$. Then $\dot{r} = 1$, $\ddot{r} = 0$, $\dot{\theta} = 1$, $\ddot{\theta} = 0$; hence

$$\mathbf{v}(t) = \mathbf{u} + t\mathbf{w}, \qquad |\mathbf{v}| = \sqrt{1 + t^2}, \qquad \mathbf{a} = -t\mathbf{u} + 2\mathbf{w}.$$

Central Force   Suppose a particle moves under the influence of a **central force**

$$\mathbf{F} = f(t)\mathbf{u}.$$

At each instant, the force is directed toward or away from the origin. Since $m\mathbf{a} = \mathbf{F}$, the component of **a** in the direction of **w** is zero:

$$r\ddot{\theta} + 2\dot{r}\dot{\theta}\, = 0, \qquad \text{that is,} \quad \tfrac{1}{2}r^2\ddot{\theta} + r\dot{r}\dot{\theta}\, = 0.$$

This is the same as $\qquad \dfrac{d}{dt}(\tfrac{1}{2}r^2\, \dot{\theta}) = 0, \qquad$ that is, $\quad \tfrac{1}{2}r^2\dot{\theta}\, = \text{constant}.$

But $$\tfrac{1}{2}r^2\dot{\theta}\, = \frac{dA}{dt},$$

the rate at which central area is swept out by the curve. It follows that the same area is swept out in equal time anywhere along the path. This is **Kepler's Second Planetary Law.**

Slope   A curve is presented in parametric polar form $r = r(t)$, $\theta = \theta(t)$. We seek its slope.

From $x = r \cos \theta$ and $y = y \sin \theta$, we have

$$\dot{x} = \dot{r} \cos \theta - r\dot{\theta} \sin \theta, \qquad \dot{y} = \dot{r} \sin \theta + r\dot{\theta} \cos \theta.$$

If $\dot{x} \neq 0$, then by the Chain Rule, $dy/dx = \dot{y}/\dot{x}$, hence

$$\frac{dy}{dx} = \frac{\dot{r} \sin \theta + r\dot{\theta} \cos \theta}{\dot{r} \cos \theta - r\dot{\theta} \sin \theta}.$$

In the special case $r = r(\theta)$, the formula becomes

$$\frac{dy}{dx} = \frac{(dr/d\theta) \sin \theta + r \cos \theta}{(dr/d\theta) \cos \theta - r \sin \theta}.$$

Example   The spiral $r = \theta$: $\dfrac{dy}{dx} = \dfrac{\sin \theta + \theta \cos \theta}{\cos \theta - \theta \sin \theta} = \dfrac{\tan \theta + \theta}{1 - \theta \tan \theta}$.

The slope is $\tan \alpha$, where α is the angle between the tangent to the curve and the x-axis. Another useful angle is the angle ψ between the tangent to the curve and the radius vector (Fig. 4). From the geometry we easily see that $\tan \psi = r d\theta/dr$, hence

$$\tan \psi = \frac{r\dot{\theta}}{\dot{r}} = \frac{r}{dr/d\theta}.$$

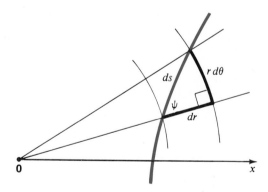

Fig. 4   Local geometry of a polar curve

Arc Length   We shall derive a formula for the arc length of a curve in parametric polar form $r = r(t)$, $\theta = \theta(t)$, where $a \leq t \leq b$. We use the formula $ds/dt = |\mathbf{v}| = |\dot{\mathbf{x}}|$ and the formula $\mathbf{v} = \dot{r}\mathbf{u} + r\dot{\theta}\,\mathbf{w}$ found earlier. Since $\mathbf{u}$ and $\mathbf{w}$ are orthogonal unit vectors,

$$|\mathbf{v}|^2 = \dot{r}^2 + (r\dot{\theta})^2, \qquad \text{hence} \qquad \left(\frac{ds}{dt}\right)^2 = \dot{r}^2 + r^2\dot{\theta}^2.$$

Arc Length   The length of a curve $r = r(t)$, $\theta = \theta(t)$ for $a \le t \le b$ is

$$L = \int_a^b \sqrt{\dot{r} + r^2 \dot{\theta}'^2}$$

The length of a curve $r = r(\theta)$ for $\alpha \le \theta \le \beta$ is

$$L = \int_\alpha^\beta \sqrt{r^2 + \left(\frac{dr}{d\theta}\right)^2}\, d\theta.$$

The second formula follows from the first by setting $\theta = t$, $r = r(t)$.

Figure 4 provides an aid to memory. The "right triangle" has sides dr, $r\, d\theta$, and ds, so the Pythagorean Theorem suggests

$$(ds)^2 = (dr)^2 + r^2(d\theta)^2.$$

■ **EXAMPLE 2**   Find the length of the spiral $r = \theta^2$ for $0 \le \theta \le 2\pi$.

Solution

$$L = \int_0^{2\pi} \sqrt{r^2 + \left(\frac{dr}{d\theta}\right)^2}\, d\theta = \int_0^{2\pi} \sqrt{\theta^4 + (2\theta)^2}\, d\theta$$

$$= \int_0^{2\pi} \theta\sqrt{\theta^2 + 4}\, d\theta = \frac{1}{3}(\theta^2 + 4)^{3/2}\Big|_0^{2\pi} = \frac{8}{3}(\pi^2 + 1)^{3/2} - \frac{8}{3} \approx 92.90. \quad ■$$

EXERCISES

Find the area enclosed by

1	$r = a \sin \theta$	2	$r = a \cos 3\theta$ (rose)
3	$r = a \cos(2n + 1)\theta$ (rose)	4	$r = a \cos 2n\theta$ (rose)
5	$r = a \cos^2 2n\theta$	6	$r = a(1 - \cos \theta)$ (cardioid)
7	$r^2 = a^2 \cos 2\theta$ (lemniscate)		
8	the closed loop of the strophoid $r = a \cos 2\theta \sec \theta$.		

9   Find the area outside the circle $r = 1$ and inside the rose $r = 2 \cos 2\theta$.
10*   Find the area between the two loops of the limaçon $r = b + a \cos \theta$, $0 < b < a$.
11   Find the area common to $r = a \cos \theta$ and $r = a \sin \theta$.
12   Find the area swept out by the segment from $\mathbf{0}$ to the spiral $r = \theta$ for $0 \le \theta \le 2\pi$.

Compute $\mathbf{v}$ and $\mathbf{a}$ for

13	$r = t, \quad \theta = 2t$	14	$r = t, \quad \theta = t^2$
15	$r = \cos t \quad \theta = t$	16	$r = \sin 2t, \quad \theta = t$.

17   Compute $\mathbf{v}$ and $\mathbf{a}$ for $r = e^t$, $\theta = t$.
18   (cont.) Find the angle between $\mathbf{v}$ and $\mathbf{a}$.

Find the slope at $\{r, \theta\}$ and $\tan \psi$, where ψ is the angle between the tangent and the radius vector

19	$r = \theta$	20	$r = \theta^2$	21	$r = a \cos 3\theta$	22	$r = ae^\theta$.

Set up, but do not evaluate, an integral for the arc length

23	$r = a\theta \quad 0 \le \theta \le 2\pi$	24	$r = a \cos 2\theta$
25	$r = a \cos 3\theta$	26	$r = a(1 - \cos \theta)$.

The remaining exercises sketch Kepler's First and Third Laws of Planetary Motion. Assume a particle of unit mass is moving in a central force field given by the inverse square law:

$$\mathbf{F} = -\frac{1}{r^2}\,\mathbf{u}.$$

27   Show that the equations of motion are $r^2\theta^{\cdot} = J$, $\ddot{r} - \dfrac{J^2}{r^3} = -\dfrac{1}{r^2}$, where J is a constant.

28   Show that $\dot{r}^2 + \dfrac{J^2}{r^2} = \dfrac{2}{r} + C$, where C is a constant. This equation is essentially the Law of Conservation of Energy. [*Hint*   Multiply the second equation in Ex. 27 by $\dot{r}$ and integrate.]

29   Set $p = \dfrac{1}{r}$. Show that $\dfrac{\dot{p}^2}{p^4} + J^2 p^2 = 2p + C$.

30   Imagine $\theta = \theta(t)$ solved for t as a function of θ and this substituted into $p = p(t)$. Thus p may be considered as a function of θ. Show that

$$J^2\left(\frac{dp}{d\theta}\right)^2 = \frac{\dot{p}^2}{p^4}, \qquad \text{and conclude that} \quad J^2\left[\left(\frac{dp}{d\theta}\right)^2 + p^2\right] = 2p + C.$$

31   Show that $\dfrac{d^2 p}{d\theta^2} + p = \dfrac{1}{J^2}$. [*Hint*   Differentiate the previous relation.]

32   Show that $p = A\cos\theta + B\sin\theta + 1/J^2$, where A and B are constants, is a solution of the preceding differential equation.

33   Show that by a suitable choice of the x-axis, the solution may be written

$$\frac{1}{r} = \frac{1}{J^2}\,(1 - e\cos\theta), \qquad \text{where } e \text{ is a constant, the } \textbf{eccentricity} \text{ of the orbit, } e \geq 0.$$

34   Suppose $e = 0$. Show that the orbit is a circle with center at $\mathbf{0}$, and that the speed is constant.

35   Suppose $e = 1$. Show the orbit is a parabola with focus at the origin, directrix $x = -J^2$, and opening in the positive x-direction.

36   Suppose $e > 1$. Show that the orbit is a branch of a hyperbola with one focus at the origin.

37   Suppose $0 < e < 1$. Show that the orbit is the ellipse $\dfrac{(x-c)^2}{a^2} + \dfrac{y^2}{b^2} = 1$, where

$a = \dfrac{J^2}{1-e^2}$, $b = \dfrac{J^2}{\sqrt{1-e^2}}$, and $c = ae$. [By Ex. 34–36 each closed orbit is an ellipse (or circle), **Kepler's First Law.**]

38   (cont.) Show that $a^2 = b^2 + c^2$. Conclude that the foci of the ellipse are $(0, 0)$ and $(2c, 0)$.

39   (cont.) Let T denote the **period** of the orbit, the time necessary for a complete revolution. Show that $\dfrac{J}{2}T = \pi ab$. [*Hint*   Use Kepler's Second Law].

40   Conclude that $T^2 = 4\pi^2 a^3$. This is **Kepler's Third Law:** the square of the period of a planetary orbit is proportional to the cube of its semimajor axis.

7. MISCELLANEOUS EXERCISES

1   Find the length of the catenary $y = a\cosh(x/a)$ for $-b \leq x \leq b$.
2   Let $\mathbf{x}(t) = (x(t),\ y(t),\ z(t))$ for $a \leq t \leq b$ be a space curve with length L. Suppose

$\mathbf{x}_1(t) = (x(t), y(t))$, its projection on the x, y-plane, has length L_1. Prove that

$$L_1 \leq L \quad \text{and} \quad L \leq L_1 + \int_a^b |\dot{z}| \, dt.$$

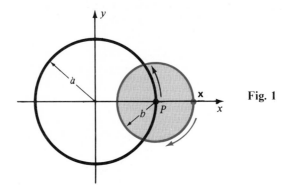

Fig. 1

3 The point P in Fig. 1 moves steadily around the circle of radius a. Also P is the center of a circle of radius $b < a$ that moves with it. Point $\mathbf{x}$ on the boundary of the small circle rotates steadily *backward*, completing one revolution in one revolution of P. Find the locus of $\mathbf{x}$, assuming it starts at $\mathbf{x}_0$. (Ptolemy proposed this eccentric circle as the orbit of the planet $\mathbf{x}$ around the Earth $\mathbf{0}$. He called it an **epicycle-deferent**.)

4 A point moves with constant unit speed along a curve C. Show that at each point of C, the curvature is the same as the length of the acceleration vector.

5 Find the quadratic $y = a + bx + cx^2$ that passes through $(0, 1)$ and agrees with the curve $y = e^x$ in slope and curvature at $(0, 1)$.

6 Find the maximum curvature of $y = x^2$.

7 The ellipse with foci $\mathbf{p}$ and $\mathbf{q}$ and length sum $2a$ is defined by $|\mathbf{x} - \mathbf{p}| + |\mathbf{x} - \mathbf{q}| = 2a$. Prove the reflection property of the ellipse by differentiating this relation with respect to arc length. [*Hint* $|\mathbf{x} - \mathbf{p}|^2 = (\mathbf{x} - \mathbf{p}) \cdot (\mathbf{x} - \mathbf{p})$.]

8 (cont.) Do the same for the parabola.

9 Describe the curve $\mathbf{x} = (a \sec t, b \tan t)$

10 Find the area of the portion of the lateral surface of the cylinder $x^2 + y^2 = 1$ between the planes $z = 0$ and $y + z = 1$.

Let $\mathbf{x} = \mathbf{x}(t)$ for $a \leq t \leq b$. Define

$$\int_a^b \mathbf{x}(t) \, dt = \left(\int_a^b x(t) \, dt, \int_a^b y(t) \, dt, \int_a^b z(t) \, dt \right).$$

Prove

11 $\displaystyle\int_a^b c\mathbf{x} \, dt = c \int_a^b \mathbf{x} \, dt$

12 $\displaystyle\int_a^b (\mathbf{x} + \mathbf{y}) \, dt = \int_a^b \mathbf{x} \, dt + \int_a^b \mathbf{y} \, dt$

13 $\displaystyle\int_a^b \mathbf{c} \cdot \mathbf{x} \, dt = \mathbf{c} \cdot \int_a^b \mathbf{x} \, dt$

14 $\displaystyle\int_a^b \mathbf{c} \times \mathbf{x} \, dt = \mathbf{c} \times \int_a^b \mathbf{x} \, dt$

15 $\displaystyle\left| \mathbf{c} \cdot \int_a^b \mathbf{x} \, dt \right| \leq |\mathbf{c}| \int_a^b |\mathbf{x}| \, dt$

16 $\displaystyle\left| \int_a^b \mathbf{x} \, dt \right| \leq \int_a^b |\mathbf{x}| \, dt.$

[*Hint* Use Ex. 15.]

17 Find all curves $r = r(\theta)$ such that the angle ψ between the tangent and the radius vector is constant.

18 Describe the locus $\mathbf{x} = (a \tan(t + \alpha), b \tan(t + \beta))$, where a, b, α, β are constants and $\alpha - \beta$ is not a multiple of π.

19 Show that the curve $\mathbf{x} = (\sin t \cos t, \sin^2 t, \cos t)$ lies on a sphere.

20 (cont.) Find its curvature.

Functions of Several Variables 15

1. FUNCTIONS AND GRAPHS

Up to now we have been concerned with functions such as $y = f(x)$, where y depends on one real variable x. In all sorts of situations, however, a quantity may depend on several real variables. Here are two examples:

(1)   The speed v of sound in an ideal gas is

$$v = \sqrt{\gamma \frac{p}{d}},$$

where d is the density of the gas, p is the pressure, and γ is a constant characteristic of the gas. Then v depends on (is a function of) the two variables p and d. We may write

$$v = f(p, d) \qquad \text{or} \qquad v = v(p, d).$$

(2)   The area of a triangle with sides x, y, z is

$$A = \sqrt{s(s - x)(s - y)(s - z)}$$

where s is the semiperimeter $\frac{1}{2}(x + y + z)$. Then A depends on the three variables x, y, and z. We may write

$$A = f(x, y, z) \qquad \text{or} \qquad A = A(x, y, z).$$

Note that x, y, z are not three arbitrary numbers but must satisfy the inequalities $x > 0$, $y > 0$, $z > 0$ and $z < x + y$, $x < y + z$, $y < z + x$.

In (1), the quantity v is a function of p and d, defined for a certain set of pairs (p, d), which we can think of as a subset of the p, d-plane. In (2), the area A is a function of x, y, z defined for a certain set of points (x, y, z) in space.

In general, a real-valued function f of two variables is an assignment of a real number to each point of a subset **D** of the plane $\mathbf{R}^2$. A real-valued function of three variables is an assignment of a real number to each point of a subset **D** of space $\mathbf{R}^3$. The set **D** is the **domain** of f.

Suppose f is a function of two variables and (x, y) is a point of its domain. We denote the real number assigned to (x, y) by $f(x, y)$. We often use vector notation, writing $\mathbf{x} = (x, y)$ and $f(\mathbf{x}) = f(x, y)$. There is similar notation for functions of three variables.

718

A common notation is

$$f \colon \mathbf{D} \longrightarrow \mathbf{R},$$

suggesting that f carries, or maps, the set **D** into the set **R** of real numbers.

We want to extend the concepts of one-variable calculus to functions of several variables, concepts such as continuity, derivative, and integral. Now in the one-variable situation, for these concepts to be meaningful, the domain of a function has to be a reasonably nice set, generally an interval or the union of several intervals. The same is so in the plane and in three-space. Generally the functions we shall study in the plane have as their domains plane regions bounded by a few simple arcs of curves. In space, the domains are usually bounded by portions of familiar surfaces such as planes and spheres. Part of the boundary may be excluded from the domain. Let us look at some examples.

Polynomials such as

$$3x^5 - 2xy^2 + x^3y^3 - 4y^7 \quad \text{and} \quad (x - y)^{10} - 10x^5y^5$$

are defined on the whole plane. Rational functions like

$$\frac{x^2 + y^2}{2xy} \quad \text{and} \quad \frac{x + y}{x - y}$$

are defined wherever their denominators are non-zero. The first is defined on the whole x, y-plane except for the x-axis and the y-axis; its domain consists of the four quadrants without their boundaries (Fig. 1a). The second is defined everywhere except on the line $x = y$; its domain consists of two half-planes without their common boundary (Fig. 1b).

The function

$$\sqrt{x + y}$$

is defined wherever $x + y \ge 0$; its domain consists of a half-plane including its boundary (Fig. 1c).

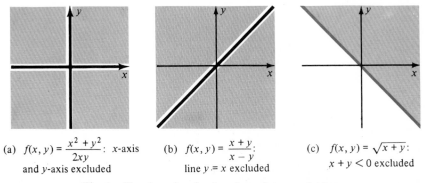

(a) $f(x, y) = \dfrac{x^2 + y^2}{2xy}$: x-axis and y-axis excluded   (b) $f(x, y) = \dfrac{x + y}{x - y}$: line $y = x$ excluded   (c) $f(x, y) = \sqrt{x + y}$: $x + y < 0$ excluded

Fig. 1   The domain of a function of two variables

Convergence   We need the notion of convergence of points in space. Henceforth, to avoid repetition, we shall use the word *space* to mean either two-space (plane) $\mathbf{R}^2$ or three-space $\mathbf{R}^3$.

The definition of convergence in space looks just like the definition on the line with the "nearness" of points **x** and **y** measured by $|\mathbf{x} - \mathbf{y}|$, the distance between the points. We shall write

$$\mathbf{x} \longrightarrow \mathbf{a}$$

provided
$$|\mathbf{x} - \mathbf{a}| \longrightarrow 0.$$

We know from the last chapter that $\quad \mathbf{x} \longrightarrow \mathbf{a} \quad$ if and only if

$$x_1 \longrightarrow a_1, \qquad x_2 \longrightarrow a_2, \qquad x_3 \longrightarrow a_3 .$$

Continuous Functions For a function of several variables to be useful, it must have some reasonable properties. The most basic of such properties is continuity. Here is the formal definition, a direct generalization of the definition of continuity for a function of one variable.

Continuity Let f be a real-valued function defined on **D**, a subset of $\mathbf{R}^2$ or $\mathbf{R}^3$. Let **a** be a point of **D**. We say f is **continuous** at **a** if $f(\mathbf{x}) \longrightarrow f(\mathbf{a})$ as $\mathbf{x} \longrightarrow \mathbf{a}$. Precisely, for each $\varepsilon > 0$ there exists $\delta > 0$ such that $|f(\mathbf{x}) - f(\mathbf{a})| < \varepsilon$ whenever $\mathbf{x} \in \mathbf{D}$ and $|\mathbf{x} - \mathbf{a}| < \delta$.

We say f is **continuous on D** if f is continuous at each point of **D**.

As for functions of one variable, this definition requires that a continuous function be "predictable"; you should be able to predict the value of the function at **a** from its values near **a**.

The elementary properties of continuous functions of one variable carry over easily. In particular, *sums, products, and quotients* (with non-zero denominator) *of continuous functions are continuous.*

Obviously the functions defined by $f(x, y) = x$ and $g(x, y) = y$ are continuous. By forming products and sums we conclude that *each polynomial is a continuous function on* $\mathbf{R}^2$ (on $\mathbf{R}^3$). From this we deduce that *each rational function is continuous wherever its denominator is not zero.* (Recall that a rational function is a quotient of polynomials.)

If $f(x)$ is a continuous function of *one* variable, then it is almost obvious that $F(x, y) = f(x)$ is continuous as a function of *two* variables. Similarly, if $g(y)$ is continuous in y, then $G(x, y) = g(y)$ is continuous in x and y. Therefore the product

$$h(x, y) = f(x)g(y)$$

is continuous. For example $h(x, y) = x^3 \ln y$ is continuous with domain all (x, y) such that $y > 0$.

Composite Functions Suppose we want to prove that $f(x, y) = y^x$ is continuous on the domain $x > 0$, $y > 0$. We could write

$$f(x, y) = e^{x \ln y} = e^{g(x, y)}.$$

Thus $f(x, y)$ is the composite of the continuous functions $h(t) = e^t$ and $t = x \ln y$. It seems reasonable that $f(x, y)$ is continuous also.

Here is a more complicated type of example. Suppose we somehow manage to prove that

$$K(x, y, z) = \int_0^x (y^3 + t^4) \sin(zt^2) \, dt$$

is continuous in x, y, z-space. We want to conclude that $K(u^v, v^u, uv)$ is continuous on the domain $u > 0$, $v > 0$. What we need is a theorem stating (roughly) that if $K(x, y, z)$ is continuous, and if x, y, z are continuous functions of u and v, then K is continuous as a function of u and v.

Composite Functions   Let $K(x, y, z)$ be continuous on a domain **D** in x, y, z-space. Let f, g, h be continuous on a domain **E** of the u, v-plane, and suppose that $(f(u, v), g(u, v), h(u, v))$ is a point of **D** whenever (u, v) is a point of **E**. Then the composite function

$$k(u, v) = K[f(u, v), g(u, v), h(u, v)]$$

is continuous on **E**.

Proof   Let $(u, v) \longrightarrow (u_0, v_0)$. Then $f(u, v) \longrightarrow f(u_0, v_0)$, $g(u, v) \longrightarrow g(u_0, v_0)$, and $h(u, v) \longrightarrow h(u_0, v_0)$. Hence

$$(f(u, v), g(u, v), h(u, v)) \longrightarrow (f(u_0, v_0), g(u_0, v_0), h(u_0, v_0)).$$

But K is continuous, so

$$k(u, v) = K[f(u, v), g(u, v), h(u, v)] \longrightarrow K[f(u_0, v_0), g(u_0, v_0), h(u_0, v_0)] = k(u_0, v_0),$$

therefore k is continuous.

Note   The theorem above is stated for a function of three variables, where each variable is replaced by a function of two variables. Clearly, there is nothing special about three and two, and the result may be modified as needed.

Graphs   Given a function of one variable $y = f(x)$, its graph is the set of points $(x, f(x))$ in the plane, where x is in the domain of f. Similarly, given a function of two variables $z = f(x, y)$, its **graph** is the set of points $(x, y, f(x, y))$, where (x, y) is in the domain of f. See Fig. 2. For a function of one variable, the graph is a curve in $\mathbf{R}^2$; for a function of two variables, the graph is a surface in $\mathbf{R}^3$. In either case, we picture the graph as lying above (or below) the domain.

Since a surface can be difficult to visualize, we use various techniques for picturing the graph of a function of two variables. One technique is to slice the surface by various planes and examine the cross sections.

■ **EXAMPLE 1**   Graph the function $z = x^2 + y^2$.

Solution   Each cross section by a plane $z = c$ parallel to the x, y-plane is a circle. Therefore the graph is a surface of revolution. It intersects the y, z-plane in the curve $z = y^2$, a parabola (Fig. 3a). This is enough information for a sketch (Fig. 3b). The surface is called a paraboloid of revolution.   ■

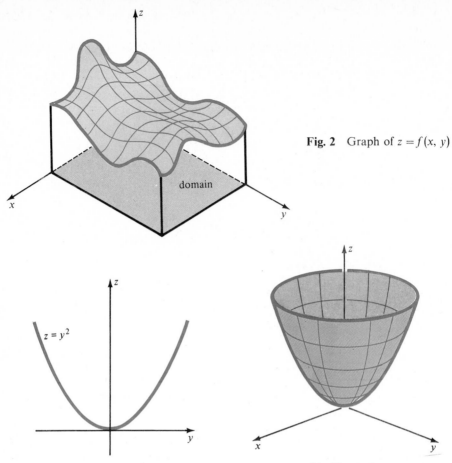

Fig. 2 Graph of $z = f(x, y)$

$z = y^2$

(a) Cross-section by the y, z-plane

(b) The graph

Fig. 3 Graph of $z = x^2 + y^2$

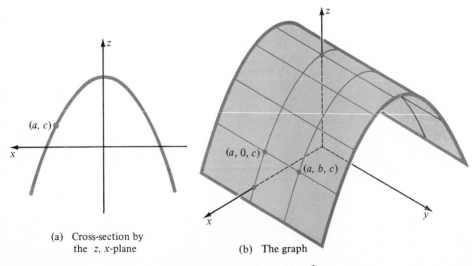

(a, c)

$(a, 0, c)$

(a, b, c)

(a) Cross-section by
the z, x-plane

(b) The graph

Fig. 4 Graph of $z = 1 - x^2$

■ **EXAMPLE 2** Graph the function $z = f(x, y) = 1 - x^2$.

Solution The function $f(x, y)$ is independent of y. Its graph is a cylinder with generators parallel to the y-axis. To see this, first graph the parabola $z = 1 - x^2$ in the z, x-plane (Fig. 4a). If (a, c) is any point on this parabola and b is any value of y whatsoever, then (a, b, c) is on the graph of $z = f(x, y)$. Therefore the graph is a parabolic cylinder with generators parallel to the y-axis (Fig. 4b). ■

Level Curves A systematic way of visualizing a surface $z = f(x, y)$ is drawing its contour map. We slice the surface by planes $z = c$ at various levels (Fig. 5).

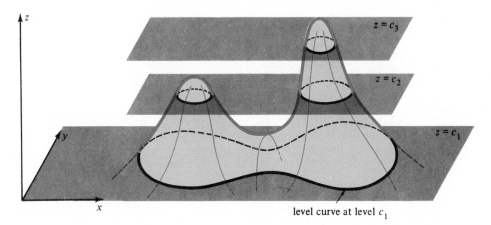

Fig. 5 Level curves

Each plane $z = c$ intersects the surface in a curve. The projection of this curve onto the x, y-plane is the **level curve** or **contour line** at level c. It is the locus of $f(x, y) = c$ and indicates where the surface has "height" c. Taken together the level curves form the **contour map** of the surface. Where level curves are close together the surface is steep; where they are far apart it is relatively flat. Figure 6 shows the contour map of the surface in Fig. 5.

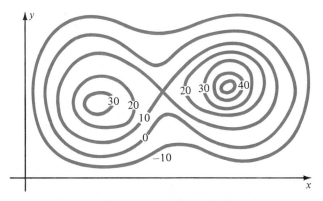

Fig. 6 Contour map of a function

A contour map is the next best thing to a good drawing of a surface. Consider the surface $z = xy$ for example. A drawing is difficult (though not impossible). Still, a reasonable idea of the graph is given by a contour map, which is easy to draw; each level curve $xy = c$ is a hyperbola $(c \neq 0)$ or degenerate hyperbola $(c = 0)$. See Fig. 7.

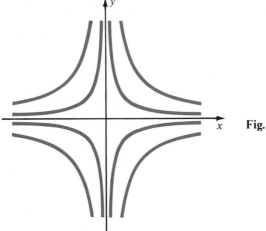

Fig. 7   Contour map of $z = xy$.

Surfaces   If $f(x, y, z)$ is a function of three variables, the relation

$$f(x, y, z) = k \qquad (k \text{ constant})$$

generally determines a surface in $\mathbf{R}^3$. For example, if

$$f(x, y, z) = x + 2y + 3z,$$

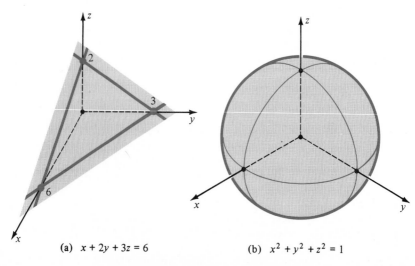

(a) $x + 2y + 3z = 6$ (b) $x^2 + y^2 + z^2 = 1$

Fig. 8   Surfaces of the form $f(x, y, z) = k$

then $f(x, y, z) = 6$ determines a plane (Fig. 8a). If

$$f(x, y, z) = x^2 + y^2 + z^2,$$

then $f(x, y, z) = 1$ determines the unit sphere (Fig. 8b).

Level Surfaces   We cannot graph a function of three variables

$$w = f(x, y, z)$$

(the graph would be four-dimensional). We can, however, learn a good deal about the function by plotting in three-space the **level surfaces**

$$f(x, y, z) = k.$$

For example, the level surfaces of $w = x^2 + y^2 + z^2$ are the family of all spheres with center **0** (and the point **0** itself—a degenerate level surface).

EXERCISES

Express as a function of two variables

1   the volume V of a cone of radius r and height h
2   the total area A of a tin can, including top and bottom, radius r, height h
3   the distance D from $(2, 4, 3)$ to a point on the plane $x + y + z = 1$, variables x and y
4   the maximum M of $x^p e^{-qx}$ for $x \geq 0$, where $p > 0$ and $q > 0$.

Express as a function of three variables

5   the gravitational attraction F between masses m_1 and m_2 at distance d apart
6   the value V of a deposit P after n years, interest at per annum rate $r\%$ compounded quarterly
7   the side c of a triangle, given the sides a and b and their included angle θ
8   the distance d from **0** to the plane $ax + by + cz = 1$, variables a, b, c.

Give the domain of the function

9   $4x - 11y + 2$ 10 $(x - 3y)^2 e^{xy}$ 11 $\dfrac{1 + x^2 y^3}{x^2 - y^2}$

12   $\ln(y - 2x)$ 13 $\sqrt{x^2 - y}$ 14 $\sqrt{4x^2 + 9y^2 - 36}$

15   $x \sec y$ 16 $\tan(x - y)$ 17 $\ln(x + 2y + 3z - 4)$

18   $\dfrac{1}{xyz}$ 19 $\arcsin(x^2 + y^2 + z^2)$ 20 $\dfrac{x + 2y}{x^4 - (y + 3z)^4}$.

Make a contour map of the function

21   $f(x, y) = x - 3y$ 22 $f(x, y) = x^2 - y^2$ 23 $f(x, y) = x^2 + 4y^2$

24   $f(x, y) = |x| + |y|$ 25 $f(x, y) = \ln(y - x^2)$ 26 $f(x, y) = \dfrac{1}{x + y}$.

Sketch the graph

27   $z = 1 - 2x$ 28 $z = x^2$ 29 $z = \sqrt{y}$

30   $z = x + y^2$ 31 $z = x + \frac{1}{2}y$ 32 $z = 1 - x^2 - y^2$.

33   The gravitational potential at (x, y, z) due to a point mass at the origin is $k/\sqrt{x^2 + y^2 + z^2}$. What are the equipotential surfaces?

34 The atmospheric pressure p at sea level around the center of an anticyclone is given by $p = a^2 - (x - bt)^2 - y^2$. Plot the isobars (level curves of p). Show that the weather system is moving with constant velocity.

Suppose $\mathbf{x}_n \longrightarrow \mathbf{a}$ and $\mathbf{y}_n \longrightarrow \mathbf{b}$. Prove

35 $\mathbf{x}_n + \mathbf{y}_n \longrightarrow \mathbf{a} + \mathbf{b}$ **36** $\mathbf{x}_n \cdot \mathbf{y}_n \longrightarrow \mathbf{a} \cdot \mathbf{b}$.

37 If $f(x, y)$ and $g(x, y)$ are continuous, prove that their sum is continuous.
38 If $f(x, y)$ and $g(x, y)$ are continuous, prove that $h(x, y) = f(x, y) g(x, y)$ is continuous.

39 Can $f(x, y) = \dfrac{\sin(x^2 + y^2)}{x^2 + y^2}$ be defined at $(0, 0)$ so as to be continuous?

40 Answer the same question for $f(x, y) = \dfrac{xy}{x^2 + y^2}$.

2. PARTIAL DERIVATIVES

Let $z = f(x, y)$ be a function of two variables and let $\mathbf{a} = (a, b)$ be a point of its domain, not on the boundary. Suppose we set $y = b$ and allow only x to vary. Then $f(x, b)$ is a function of the single variable x, defined at least in some open interval including a. We define

$$\frac{\partial z}{\partial x}(a, b) = \frac{d}{dx} f(x, b) \bigg|_{x=a}.$$

This is called the **partial derivative** (or simply **partial**) of z with respect to x. (The ∂ is a curly d.) It measures the rate of change of z with respect to x while y is held constant.

Similarly, we define the partial derivative of z with respect to y:

$$\frac{\partial z}{\partial y}(a, b) = \frac{d}{dy} f(a, y) \bigg|_{y=b}.$$

In like manner, given a function $w = f(x, y, z)$ of three variables, we define the three partial derivatives $\partial w/\partial x$, $\partial w/\partial y$, and $\partial w/\partial z$. For instance,

$$\frac{\partial w}{\partial y}(a, b, c) = \frac{d}{dy} f(a, y, c) \bigg|_{y=b}.$$

Each of the partials is the derivative of w with respect to the variable in question, taken while all other variables are held fixed.

■ **EXAMPLE 1**   Let $z = f(x, y) = xy^2$. Find

$$\frac{\partial z}{\partial x}(1, 3), \qquad \frac{\partial z}{\partial y}(-4, 2), \qquad \frac{\partial z}{\partial x} \quad \text{and} \quad \frac{\partial z}{\partial y} \quad \text{in general.}$$

Solution   Set $y = 3$. Then $z = 9x$ and

$$\frac{\partial z}{\partial x}(1, 3) = \frac{d}{dx}(9x) \bigg|_{x=1} = 9.$$

Likewise, set $x = -4$. Then $z = -4y^2$ and

$$\frac{\partial z}{\partial y}(-4, 2) = \frac{d}{dy}(-4y^2)\Big|_{y=2} = -16.$$

To compute $\partial z/\partial x$ in general, just differentiate as usual, treating y as a constant:

$$\frac{\partial z}{\partial x} = \frac{\partial(xy^2)}{\partial x} = y^2 \frac{d}{dx}(x) = y^2.$$

To compute $\partial z/\partial y$, differentiate, treating x as a constant:

$$\frac{\partial z}{\partial y} = \frac{\partial(xy^2)}{\partial y} = x \frac{d}{dy}(y^2) = 2xy. \qquad\blacksquare$$

We consider two further examples of partial derivatives.

(1)   The gas law for a fixed mass of n moles of an ideal gas is $P = nRTN$, where R is the universal gas constant. Thus P is a function of the two variables T and V:

$$\frac{\partial P}{\partial T} = nR\,\frac{1}{V}, \qquad \frac{\partial P}{\partial V} = -nR\,\frac{T}{V^2}.$$

(2)   The area A of a parallelogram of base b, slant height s, and angle α is $A = sb \sin \alpha$, a function of b, s, and α. The partial derivatives are

$$\frac{\partial A}{\partial s} = b \sin \alpha, \qquad \frac{\partial A}{\partial b} = s \sin \alpha, \qquad \frac{\partial A}{\partial \alpha} = sb \cos \alpha.$$

Notation   There are several different notations for partial derivatives in common use. Become familiar with them; they come up again and again in applications. Suppose $w = f(x, y, z)$.   Common notations for $\partial w/\partial x$ are

$$f_x, \quad f_x(x, y, z), \quad w_x, \quad w_x(x, y, z), \quad D_x f.$$

For example, if

$$w = f(x, y, z) = x^3 y^2 \sin z,$$

then

$$f_x = 3x^2 y^2 \sin z, \quad w_y = 2x^3 y \sin z, \quad D_z f = x^3 y^2 \cos z.$$

Geometric Interpretation   The graph of $z = f(x, y)$ is a surface in three dimensions. A plane $x = a$ cuts the graph in a plane curve $x = a$, $z = f(a, y)$. See Fig. 1a. The projection of this curve straight back onto the y, z-plane is the graph of the function $z = f(a, y)$. See Fig. 1b. The partial derivative

$$\frac{\partial f}{\partial y}(a, y)$$

is the slope of this graph.

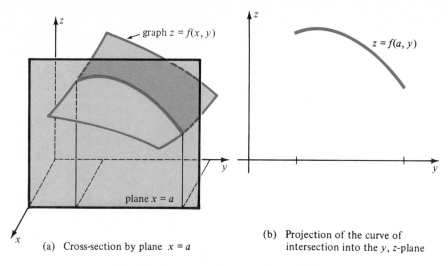

(a) Cross-section by plane $x = a$

(b) Projection of the curve of intersection into the y, z-plane

Fig. 1   Geometric interpretation of $\partial f/\partial y$

For example, suppose the graph of the function $z = x^2 + y^2$ is sliced by the plane $x = a$. The resulting curve is the parabola $x = a$, $z = y^2 + a^2$. Its projection onto the y, z-plane is the parabola $z = y^2 + a^2$, with slope

$$2y = \frac{\partial z}{\partial y}.$$

The Chain Rule   The Chain Rule for functions of one variable states that the composite of differentiable functions is itself differentiable and gives a formula for the derivative of a composite function: if $y = f(x)$ where $x = x(t)$, then

$$\frac{dy}{dt} = \frac{dy}{dx}\frac{dx}{dt}.$$

The Chain Rule for functions of several variables also has two parts, an assertion that a composite function built out of differentiable functions is itself differentiable, and a practical formula for computing partial derivatives of a composite function. For the time being, we shall just state the formula and try to get a feel for how it works. We shall postpone a precise statement of the Chain Rule and its proof until Section 11.

Suppose $z = f(x, y)$, where $x = x(t)$ and $y = y(t)$. Then z is indirectly a function of t. The Chain Rule asserts that

$$\frac{dz}{dt} = \frac{\partial z}{\partial x}\frac{dx}{dt} + \frac{\partial z}{\partial y}\frac{dy}{dt}.$$

It helps to interpret this formula geometrically. We write $z = f(x, y)$ as $z = f(\mathbf{x})$ and the composite function $z(t) = f[x(t), y(t)]$ as $z(t) = f[\mathbf{x}(t)]$. We think of t as time and $\mathbf{x}(t)$ as the path of a moving particle in the plane. Then the composite function $f[\mathbf{x}(t)]$ assigns a number z at each instant of the motion. The Chain Rule is a formula for dz/dt. For instance if $f(\mathbf{x})$ is the temperature at position

x, then the Chain Rule tells how fast the temperature is changing as the particle moves along the curve **x**(t).

Chain Rule   Let $z = f(x, y)$, where $x = x(t)$ and $y = y(t)$. Then

$$\frac{dz}{dt} = \frac{\partial z}{\partial x}\frac{dx}{dt} + \frac{\partial z}{\partial y}\frac{dy}{dt},$$

where $\dfrac{\partial z}{\partial x}$ and $\dfrac{\partial z}{\partial y}$ are evaluated at $(x(t), y(t))$. Briefly, $\dot{z} = f_x \dot{x} + f_y \dot{y}$.

In terms of vectors, if $z = f[\mathbf{x}(t)]$, then $\dfrac{dz}{dt} = f_x[\mathbf{x}(t)]\dot{x}(t) + f_y[\mathbf{x}(t)]\dot{y}(t)$.

Similar rules hold for functions of more than two variables. For instance, if $w = f(x, y, z)$, where $x = x(t)$, $y = y(t)$, $z = z(t)$, then

$$\frac{dw}{dt} = \frac{\partial w}{\partial x}\frac{dx}{dt} + \frac{\partial w}{\partial y}\frac{dy}{dt} + \frac{\partial w}{\partial z}\frac{dz}{dt}.$$

■ **EXAMPLE 2**   Let $w = f(x, y, z) = xy^2z^3$, where $x = \cos t$, $y = e^t$, and $z = \ln(t + 2)$.   Compute $\dfrac{dw}{dt}$ at $t = 0$.

Solution   There is a direct but tedious way to do the problem. Write

$$w = (\cos t)e^{2t}[\ln(t + 2)]^3.$$

Differentiate, then set $t = 0$. That's quite a job!
    Use of the Chain Rule splits the computation into small easy pieces. First,

$$\dot{w}(0) = \frac{\partial w}{\partial x}\dot{x}(0) + \frac{\partial w}{\partial y}\dot{y}(0) + \frac{\partial w}{\partial z}\dot{z}(0),$$

where the partials are evaluated at $\mathbf{a} = (x(0),\ y(0),\ z(0)) = (1,\ 1,\ \ln 2)$. From $w = xy^2z^3$ follows

$$\left.\frac{\partial w}{\partial x}\right|_{\mathbf{a}} = y^2z^3\Big|_{\mathbf{a}} = (\ln 2)^3, \qquad \left.\frac{\partial w}{\partial y}\right|_{\mathbf{a}} = 2xyz^3\Big|_{\mathbf{a}} = 2(\ln 2)^3,$$

$$\left.\frac{\partial w}{\partial z}\right|_{\mathbf{a}} = 3xy^2z^2\Big|_{\mathbf{a}} = 3(\ln 2)^2.$$

Therefore $\dot{w}(0) = (\ln 2)^3\dot{x}(0) + 2(\ln 2)^3\dot{y}(0) + 3(\ln 2)^2\dot{z}(0)$.

But $\dot{x}(0) = -\sin t\Big|_0 = 0, \qquad \dot{y}(0) = e^t\Big|_0 = 1, \qquad \dot{z}(0) = \dfrac{1}{t + 2}\Big|_0 = \dfrac{1}{2},$

so $\dot{w}(0) = 2(\ln 2)^3 + \frac{3}{2}(\ln 2)^2.$ ■

Another Version of the Chain Rule   Suppose $z = f(x, y)$, where this time x and y are functions of *two* variables, $x = x(s, t)$ and $y = y(s, t)$. Then indirectly,

z is a function of the variables s and t. There is a chain rule for computing $\partial z/\partial s$ and $\partial z/\partial t$:

Chain Rule If $z = f(x, y)$ is a function of two variables x and y, where $x = x(s, t)$ and $y = y(s, t)$, then

$$\frac{\partial z}{\partial s} = \frac{\partial z}{\partial x}\frac{\partial x}{\partial s} + \frac{\partial z}{\partial y}\frac{\partial y}{\partial s} \quad\text{and}\quad \frac{\partial z}{\partial t} = \frac{\partial z}{\partial x}\frac{\partial x}{\partial t} + \frac{\partial z}{\partial y}\frac{\partial y}{\partial t},$$

where $\dfrac{\partial z}{\partial x}$ and $\dfrac{\partial z}{\partial y}$ are evaluated at $(x(s, t), y(s, t))$.

This Chain Rule is a consequence of the previous one. For instance, to compute $\partial z/\partial s$, hold t fixed, making $x(s, t)$ and $y(s, t)$ effectively functions of the one variable s. Then apply the previous Chain Rule.

■ **EXAMPLE 3**   Let $w = x^2 y$, where $x = s^2 + t^2$ and $y = \cos st$. Compute $\dfrac{\partial w}{\partial s}$.

Solution $\dfrac{\partial w}{\partial s} = \dfrac{\partial w}{\partial x}\dfrac{\partial x}{\partial s} + \dfrac{\partial w}{\partial y}\dfrac{\partial y}{\partial s} = (2xy)(2s) + (x^2)(-t \sin st)$

$$= 2(s^2 + t^2)(\cos st)(2s) + (s^2 + t^2)^2(-t \sin st)$$

$$= (s^2 + t^2)[4s \cos st - t(s^2 + t^2) \sin st]. \qquad\blacksquare$$

The next example is important in physical applications.

■ **EXAMPLE 4**   If $w = f(x, y)$, where $x = r \cos \theta$ and $y = r \sin \theta$, show that

$$\left(\frac{\partial w}{\partial x}\right)^2 + \left(\frac{\partial w}{\partial y}\right)^2 = \left(\frac{\partial w}{\partial r}\right)^2 + \frac{1}{r^2}\left(\frac{\partial w}{\partial \theta}\right)^2.$$

Solution   Use the Chain Rule to compute $\partial w/\partial r$ and $\partial w/\partial \theta$:

$$\frac{\partial w}{\partial r} = \frac{\partial w}{\partial x}\frac{\partial x}{\partial r} + \frac{\partial w}{\partial y}\frac{\partial y}{\partial r} = \frac{\partial w}{\partial x}\cos \theta + \frac{\partial w}{\partial y}\sin \theta;$$

$$\frac{\partial w}{\partial \theta} = \frac{\partial w}{\partial x}\frac{\partial x}{\partial \theta} + \frac{\partial w}{\partial y}\frac{\partial y}{\partial \theta} = \frac{\partial w}{\partial x}(-r \sin \theta) + \frac{\partial w}{\partial y}r \cos \theta = r\left(-\frac{\partial w}{\partial x}\sin \theta + \frac{\partial w}{\partial y}\cos \theta\right).$$

From these formulas follow

$$\left(\frac{\partial w}{\partial r}\right)^2 = \left(\frac{\partial w}{\partial x}\right)^2\cos^2 \theta + 2\frac{\partial w}{\partial x}\frac{\partial w}{\partial y}\sin \theta \cos \theta + \left(\frac{\partial w}{\partial y}\right)^2\sin^2 \theta,$$

$$\frac{1}{r^2}\left(\frac{\partial w}{\partial \theta}\right)^2 = \left(\frac{\partial w}{\partial x}\right)^2\sin^2 \theta - 2\frac{\partial w}{\partial x}\frac{\partial w}{\partial y}\sin \theta \cos \theta + \left(\frac{\partial w}{\partial y}\right)^2\cos^2 \theta.$$

Add:

$$\left(\frac{\partial w}{\partial r}\right)^2 + \frac{1}{r^2}\left(\frac{\partial w}{\partial \theta}\right)^2 = \left[\left(\frac{\partial w}{\partial x}\right)^2 + \left(\frac{\partial w}{\partial y}\right)^2\right](\cos^2 \theta + \sin^2 \theta) = \left(\frac{\partial w}{\partial x}\right)^2 + \left(\frac{\partial w}{\partial y}\right)^2. \qquad\blacksquare$$

EXERCISES

Find $\partial z/\partial x$ and $\partial z/\partial y$ for $z =$

1	$x + 2y$	2	$3x + 4y$	3	$3xy$
4	$x^2 y$	5	$2x^2/(y + 1)$	6	$x^3 y^2 - 2xy^4$
7	$x \sin y$	8	$y^2 \cos x$	9	$\tan 2x + \cot 3y$
10	$x \tan y + y \tan x$	11	$\sin 2xy$	12	$\cos(2x + y)$

13	$\dfrac{x}{x^2 + y^2}$	14	$\dfrac{1}{x + 2y + 5}$	15	$\ln(x^2 + 3y)$
16	$\sqrt{x^2 + 3y}$	17	$e^{2x} \sin(x - y)$	18	$(1 + 2x^2 - 3y^3)^5$

$$19 \quad \sqrt{\frac{x - y}{x + y}} \qquad\qquad 20 \quad \frac{(x + y)(2x + y)}{(x - y)(2x - y)}.$$

[*Hint* Logarithmic differentiation.]

21 Let $z = x^2 y$. Find $\partial z/\partial x$ for $y = 2$, and $\partial z/\partial y$ for $x = -1$.

22 Let $z = y^2/x$. Find z_x for $y = 3$.

23 Let $w = xy^2 z^3$. Find w_x for $y = 2$ and $z = 2$, find w_y for $x = 1$ and $z = 0$, and find w_z for $x = y$.

24 Let $w = xy - xz - yz$. Find $\dfrac{\partial w}{\partial x} + \dfrac{\partial w}{\partial y} + \dfrac{\partial w}{\partial z}$.

Show that

25 $z = (3x - y)^2$ satisfies $\partial z/\partial x + 3 \, \partial z/\partial y = 0$

26 $z = f(x) + y^2$ satisfies $\partial z/\partial y = 2y$

27 $z = x^2 - y^2$ satisfies $(\partial z/\partial x)^2 - (\partial z/\partial y)^2 = 4z$

28 $z = x^6 - x^5 y + 7x^3 y^3$ satisfies $x(\partial z/\partial x) + y \, (\partial z/\partial y) = 6z$

29 $w = \dfrac{xyz}{x^4 + y^4 + z^4}$ satisfies $xw_x + yw_y + zw_z = -w$

30 $w = e^{nr} \cos n\theta$ satisfies $(w_r{}^2 + w_\theta{}^2) \cos^2 n\theta = n^2 w^2$.

Find dz/dt by the Chain Rule

31 $z = e^{xy}$; $\quad x = 3t + 1$, $y = t^2$ $\qquad$ **32** $z = x/y$; $\quad x = t + 1$, $y = t - 1$

33 $z = x^2 \cos y - x$; $\quad x = t^2$, $y = 1/t$ $\qquad$ **34** $z = x/y$; $\quad x = \cos t$, $y = 1 + t^2$.

Find dw/dt by the Chain Rule

35 $w = xyz$; $\quad x = t^2$, $y = t^3$, $z = t^4$

36 $w = e^x \cos(y + z)$; $\quad x = 1/t$, $y = t^2$, $z = -t$

37 $w = e^{-x} y^2 \sin z$; $\quad x = t$, $y = 2t$, $z = 4t$

38 $w = (e^{-x} \sec z)/y^2$; $\quad x = t^2$, $y = 1 + t$, $z = t^3$.

Find $\partial z/\partial s$ and $\partial z/\partial t$ by the Chain Rule

39 $z = x^3/y^2$; $\quad x = s^2 - t$, $y = 2st$

40 $z = (x + y^2)^4$; $\quad x = se^t$, $y = se^{-t}$

41 $z = \sqrt{1 + x^2 + y^2}$; $\quad x = st^2$, $y = 1 + st$

42 $z = e^{x^2 y}$; $\quad x = \dfrac{s}{\sqrt{1 + t^2}}$, $y = st$.

43 The radius r and height h of a conical tank increase at rates $\dot{r} = 0.3$ in./hr and $\dot{h} = 0.5$ in./hr. Find the rate of increase $\dot{V}$ of the volume when $r = 6$ ft and $h = 30$ ft.

44 (cont.) Also find the rate of increase $\dot{S}$ at the same instant of the total area (base plus lateral).

45 A metal bar of length 120 cm lies on the x-axis from $x = 0$ to $x = 120$. At each point x

of the bar, the temperature at time t is $u(x, t) = 100e^{-kt} \sin(\pi x/120)$, where $k > 0$. Show that the bar is cooling at all points except its ends, and that at each instant the rate of cooling at the midpoint is twice the rate at the points $\frac{1}{6}$ of the way across.

46   (cont.) Show that $u(x, t)$ satisfies $\partial u/\partial t = -ku$. Now find *all* solutions $u(x, t)$ of this partial differential equation.

47   Given $F(x, y)$, show that

$$\frac{\partial}{\partial u} F(u + v, u - v) + \frac{\partial}{\partial v} F(u + v, u - v) = 2 \frac{\partial}{\partial x} F(u + v, u - v)$$

Verify the formula for $F = y \sin(xy)$.

48   Given $f(x)$ show that

$$b \frac{\partial}{\partial u} f(au + bv) = a \frac{\partial}{\partial v} f(au + bv).$$

Verify the formula for $f(x) = x^3$.

49   Let $z = \arctan(y/x)$. If $x = r \cos \theta$ and $y = r \sin \theta$, compute $\partial z/\partial r$ and $\partial z/\partial \theta$ by the Chain Rule. Check your answers by expressing z in terms of r and θ first, then differentiating.

50*   Prove that $\quad \dfrac{d}{dx} \displaystyle\int_{g(x)}^{h(x)} F(t) \, dt = F[h(x)]h'(x) - F[g(x)]g'(x).$

3. GRADIENTS AND DIRECTIONAL DERIVATIVES

Given a function f of several variables, we associate to each point of its domain a vector called the gradient of f.

Gradient Suppose $f(x, y)$ is defined on a domain **D** in the x, y-plane. Let **a** be a point of **D**, not on the boundary. The **gradient** of f at **a** is the vector

$$\operatorname{grad} f(\mathbf{a}) = (f_x, f_y) \Big|_{\mathbf{a}}.$$

Similarly, if $f(x, y, z)$ is defined on a domain **D** in space, and **a** is a point inside **D**, then the **gradient** of f at **a** is the vector

$$\operatorname{grad} f(\mathbf{a}) = (f_x, f_y, f_z) \Big|_{\mathbf{a}}.$$

The **gradient field** of f is the assignment of the vector $\operatorname{grad} f(\mathbf{a})$ to each point **a** inside **D**.

Examples (1) $f(x, y) = x^2 + y^3$, $\operatorname{grad} f = (2x, 3y^2)$.

(2) $f(x, y, z) = |\mathbf{x}|^2 = x^2 + y^2 + z^2$, $\operatorname{grad} f = (2x, 2y, 2z) = 2\mathbf{x}$.

(3) $f(x, y, z) = xyz$, $\operatorname{grad} f = (yz, zx, xy)$.

We can visualize the gradient field of f by drawing the vector $\operatorname{grad} f(\mathbf{a})$ at each point **a**. For example, the field $\operatorname{grad} \frac{1}{4}(x^2 + y^2)$ is shown in Fig. 1.

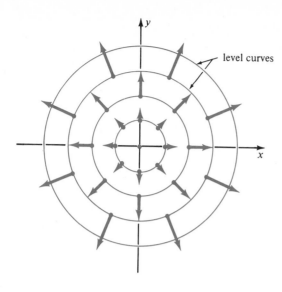

Fig. 1 Gradient field of $f(x, y) = \frac{1}{4}(x^2 + y^2)$

Notation   A common notation for grad f is ∇f. The upside-down delta is called "nabla".

Relation to the Chain Rule   The Chain Rule can be expressed in a concise vector notation using the gradient. Suppose $w = f(x, y, z) = f(\mathbf{x})$ and $\mathbf{x} = \mathbf{x}(t)$. According to the Chain Rule,

$$\dot{w} = f_x \dot{x} + f_y \dot{y} + f_z \dot{z}.$$

Notice that the right-hand side is the inner product of two vectors, $(f_x, f_y, f_z) =$ grad f and $(\dot{x}, \dot{y}, \dot{z}) = \dot{\mathbf{x}}$.

> If $w = f(\mathbf{x})$ and $\mathbf{x} = \mathbf{x}(t)$, then
>
> $$\dot{w} = (\text{grad } f) \cdot \dot{\mathbf{x}}.$$

Gradients and Level Curves   Recall that the level curves of a function $f(x, y)$ are the curves $f(x, y) = c$ in the x, y-plane. There is an important relation between the level curves and the gradient field: at each point of a level curve, the gradient vector is perpendicular to the curve.

> The field   grad f is orthogonal (perpendicular) to the level curves of f.

Before giving a proof, let us note that Fig. 1 illustrates this statement. There $f(x, y) = \frac{1}{4}(x^2 + y^2)$, so the level curves are the circles $x^2 + y^2 = a^2$, while the gradient field is grad $f = (x, y) = \frac{1}{2}\mathbf{x}$. At each $\mathbf{x}$, the vector grad $f(\mathbf{x})$ points along the radius, hence is orthogonal to the circle.

Proof   Suppose a particle moves along a level curve $f(x, y) = c$. Let its position at time t be $\mathbf{x} = \mathbf{x}(t)$. Then the composite function $f[\mathbf{x}(t)] = c$ is constant, so its time derivative is 0. By the Chain Rule

$$\dot{f} = [\text{grad} f(\mathbf{x})] \cdot \dot{\mathbf{x}}(t) = 0,$$

hence grad $f(\mathbf{x})$ is perpendicular to $\dot{\mathbf{x}}$, a vector tangent to the level curve.

The Tangent Line   Here is a practical application of the orthogonality of gradients and level curves. Suppose we want an equation for the tangent line to a level curve $f(x, y) = c$ at a point $\mathbf{a}$. This tangent line passes through $\mathbf{a}$ and is perpendicular to grad $f(\mathbf{a})$, hence an equation is

$$[\text{grad} f(\mathbf{a})] \cdot (\mathbf{x} - \mathbf{a}) = 0.$$

Tangent Line   Let $\mathbf{a} = (a, b)$ be a point of the curve $f(x, y) = c$. Then the tangent to the curve at $\mathbf{a}$ is

$$[\text{grad} f(\mathbf{a})] \cdot (\mathbf{x} - \mathbf{a}) = 0.$$

In coordinates,

$$\frac{\partial f}{\partial x} \cdot (x - a) + \frac{\partial f}{\partial y} \cdot (y - b) = 0,$$

where the partials $\partial f/\partial x$ and $\partial f/\partial y$ are evaluated at (a, b).

■ **EXAMPLE 1**   Find the tangent to the cubic $x^2 = y^3$ at $(-1, 1)$.

Solution   Set $f(x, y) = x^2 - y^3$, so the cubic is the level curve $f = 0$. Then $\mathbf{a} = (-1, 1)$ and

$$\text{grad} f(\mathbf{a}) = (2x, -3y^2)\Big|_{(-1, 1)} = (-2, -3).$$

Hence the tangent is

$$(-2, -3) \cdot [(x, y) - (-1, 1)] = 0, \qquad (-2, -3) \cdot (x + 1, y - 1) = 0,$$
$$-2(x + 1) - 3(y - 1) = 0, \qquad 2x + 3y = 1. \qquad ■$$

Directional Derivatives   The partial derivatives of a function give the rates of change of the function in directions parallel to the axes. Now we ask for the rate of change in an arbitrary direction indicated by a unit vector $\mathbf{u}$.

Let $f(x, y, z)$ be a function and $\mathbf{a}$ a point of its domain. Imagine a particle moving along a straight line with constant velocity $\mathbf{u}$ and passing through $\mathbf{a}$ when $t = 0$.

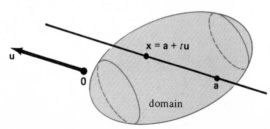

Fig. 2   Set-up for directional derivatives

See Fig. 2. To each point $\mathbf{a} + t\mathbf{u}$ of its path is assigned the number

$$w(t) = f(\mathbf{a} + t\mathbf{u}).$$

We define the directional derivative of f at $\mathbf{a}$ in the direction $\mathbf{u}$ to be $\dot{w}(0)$.

> The **directional derivative** of $f(x, y, z)$ at a point $\mathbf{a}$ in the direction $\mathbf{u}$ is
>
> $$D_{\mathbf{u}} f(\mathbf{a}) = \frac{d}{dt} f(\mathbf{a} + t\mathbf{u}) \Big|_{t=0}.$$

For example, suppose $f(x, y, z)$ is the steady temperature at each point (x, y, z) of a fluid. Suppose a particle moves with unit speed through a point $\mathbf{a}$ in the direction $\mathbf{u}$. Then $D_{\mathbf{u}} f(\mathbf{a})$ measures the time rate of change of the particle's temperature.

To compute the directional derivative, let

$$\mathbf{x}(t) = \mathbf{a} + t\mathbf{u},$$

and $w(t) = f[\mathbf{x}(t)]$. By definition, $D_{\mathbf{u}} f(\mathbf{a}) = \dot{w}(0)$. But by the Chain Rule,

$$\dot{w}(0) = [\operatorname{grad} f(\mathbf{a})] \cdot \dot{\mathbf{x}}(0) = [\operatorname{grad} f(\mathbf{a})] \cdot \mathbf{u}.$$

Since $\mathbf{u}$ is a unit vector, $(\operatorname{grad} f) \cdot \mathbf{u}$ is the projection of the vector $\operatorname{grad} f$ on $\mathbf{u}$.

> The derivative of f in the direction $\mathbf{u}$ is the projection of $\operatorname{grad} f$ on $\mathbf{u}$:
>
> $$D_{\mathbf{u}} f(\mathbf{a}) = [\operatorname{grad} f(\mathbf{a})] \cdot \mathbf{u}.$$

In particular if $\mathbf{u} = \mathbf{i} = (1, 0, 0)$, then

$$D_{\mathbf{i}} f(\mathbf{a}) = (\operatorname{grad} f) \cdot (1, 0, 0) = \left(\frac{\partial f}{\partial x}, \frac{\partial f}{\partial y}, \frac{\partial f}{\partial z} \right) \cdot (1, 0, 0) = \frac{\partial f}{\partial x} \Big|_{\mathbf{a}}.$$

A similar situation holds for $\mathbf{j} = (0, 1, 0)$ and $\mathbf{k} = (0, 0, 1)$.

> The directional derivatives of $f(x, y, z)$ in the directions $\mathbf{i}$, $\mathbf{j}$, and $\mathbf{k}$ are the partial derivatives:
>
> $$D_{\mathbf{i}} f = \frac{\partial f}{\partial x}, \qquad D_{\mathbf{j}} f = \frac{\partial f}{\partial y}, \qquad D_{\mathbf{k}} f = \frac{\partial f}{\partial z}.$$

A completely analogous discussion holds for functions $f(x, y)$ of two variables.

■ **EXAMPLE 2**   Compute the directional derivatives of $f(x, y, z) = xy^2 z^3$ at $(3, 2, 1)$, in the direction of the vectors

(a) $(-2, -1, 0)$,   (b) $(5, 4, 1)$

Solution $D_{\mathbf{u}} f(\mathbf{a}) = [\operatorname{grad} f(\mathbf{a})] \cdot \mathbf{u},$

where $\mathbf{u}$ is a unit vector in the desired direction and $\mathbf{a} = (3, 2, 1)$. Now

$$\operatorname{grad} f(\mathbf{a}) = (y^2 z^3, 2xyz^3, 3xy^2 z^2) \Big|_{(3, 2, 1)} = (4, 12, 36),$$

hence $\qquad\qquad D_{\mathbf{u}} f(\mathbf{x}) = (4, 12, 36) \cdot \mathbf{u}.$

(a) $\mathbf{u} = \dfrac{1}{\sqrt{5}}(-2, -1, 0),\qquad D_{\mathbf{u}} f(\mathbf{x}) = (4, 12, 36) \cdot \dfrac{1}{\sqrt{5}}(-2, -1, 0) = \dfrac{-20}{\sqrt{5}}.$

(b) $\mathbf{u} = \dfrac{1}{\sqrt{42}}(5, 4, 1),\qquad D_{\mathbf{u}} f(\mathbf{x}) = (4, 12, 36) \cdot \dfrac{1}{\sqrt{42}}(5, 4, 1) = \dfrac{104}{\sqrt{42}}.$ ∎

Directions of Most Rapid Increase and Decrease   In which direction is a function $f(\mathbf{x})$ increasing fastest? decreasing fastest? In other words, given a point **a** in space, for which unit vector **u** is $D_{\mathbf{u}} f(\mathbf{a})$ largest? smallest? The answers follow immediately from the formula

$$D_{\mathbf{u}} f(\mathbf{a}) = [\operatorname{grad} f(\mathbf{a})] \cdot \mathbf{u} = |\operatorname{grad} f(\mathbf{a})| \cos \theta,$$

where θ is the angle between $\operatorname{grad} f(\mathbf{a})$ and **u**. Therefore the largest value of $D_{\mathbf{u}} f(\mathbf{a})$ is $|\operatorname{grad} f(\mathbf{a})|$, taken where $\cos \theta = 1$, that is, for $\theta = 0$. The smallest value is $-|\operatorname{grad} f(\mathbf{a})|$, taken in exactly the opposite direction, that is, for $\theta = \pi$.

> The direction of most rapid increase of $f(x, y, z)$ at a point **a** is the direction of the gradient. The derivative in that direction is $|\operatorname{grad} f(\mathbf{a})|$.
>
> The direction of most rapid decrease is opposite to the direction of the gradient. The derivative in that direction is $-|\operatorname{grad} f(\mathbf{a})|$.

∎ **EXAMPLE 3**   Find the direction of most rapid increase of $f(x, y, z) = x^2 + yz$ at $(1, 1, 1)$ and give the rate of increase in this direction.

Solution $\qquad\qquad \operatorname{grad} f \Big|_{(1, 1, 1)} = (2x, z, y) \Big|_{(1, 1, 1)} = (2, 1, 1).$

The most rapid increase is

$$D_{\mathbf{u}} f = |\operatorname{grad} f| = \sqrt{2^2 + 1^2 + 1^2} = \sqrt{6},$$

where **u** is the direction of most rapid increase, the direction of $\operatorname{grad} f$:

$$\mathbf{u} = \frac{\operatorname{grad} f}{|\operatorname{grad} f|} = \frac{1}{\sqrt{6}}(2, 1, 1). \qquad\qquad ∎$$

Here is an application of the direction of most rapid decrease. Suppose on a contour map (Fig. 3) you want to plot the path of water flowing downhill from a spring at P. A physical principle says that gravity will cause the water to flow in such a way that its potential energy decreases as rapidly as possible. But the potential energy of a water particle equals its mass (constant) times its height. Therefore the particle "chooses" the direction of steepest descent (most rapid change of altitude). If the hill is represented by the surface $z = f(x, y)$, then water will flow in the direction of $-\operatorname{grad} f$, that is, *perpendicular to the level curves.*

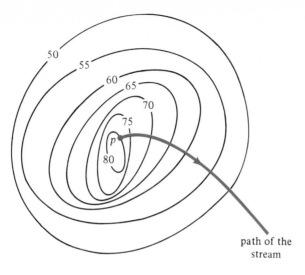

Fig. 3 Contour map; path of quickest descent from spring at P

path of the
stream

The General Directional Derivative   It is customary to define $D_{\mathbf{v}} f$ for *any* vector $\mathbf{v}$, not just unit vectors. Formally, the definition is the same as before:

General Directional Derivative $D_{\mathbf{v}} f(\mathbf{a}) = \dfrac{d}{dt} f(\mathbf{a} + t\mathbf{v}) \Big|_{t=0}.$

We shall refer to this derivative as "$D_{\mathbf{v}} f(\mathbf{a})$", and reserve the name "directional derivative" for $D_{\mathbf{u}} f(\mathbf{a})$, where $\mathbf{u}$ is a unit vector.

As before, by the Chain Rule

$$D_{\mathbf{v}} f(\mathbf{a}) = [\operatorname{grad} f(\mathbf{a})] \cdot \mathbf{v}.$$

Suppose $\mathbf{v} = k\mathbf{u}$, where $\mathbf{u}$ is a unit vector and $k = |\mathbf{v}|$. Then

$$D_{\mathbf{v}} f(\mathbf{a}) = D_{k\mathbf{u}} f(\mathbf{a}) = [\operatorname{grad} f(\mathbf{a})] \cdot (k\mathbf{u})$$

$$= k[\operatorname{grad} f(\mathbf{a})] \cdot \mathbf{u} = k\, D_{\mathbf{u}} f(\mathbf{a}).$$

Therefore $D_{\mathbf{v}} f(\mathbf{x}) = |\mathbf{v}| D_{\mathbf{u}} f(\mathbf{x}).$

Hence, for a fixed vector $\mathbf{v}$ the derivative $D_{\mathbf{v}} f(\mathbf{x})$ is nothing terribly new: it is simply proportional to the directional derivative $D_{\mathbf{u}} f(\mathbf{x})$ in the direction $\mathbf{u}$ or $\mathbf{v}$.

Example $f(\mathbf{x}) = xyz,$ $\mathbf{a} = (2, 1, -1),$ $\mathbf{v} = (1, 1, 2).$

$$D_{\mathbf{v}} f(\mathbf{a}) = \left[\operatorname{grad} f(\mathbf{x}) \Big|_{(2,\,1,\,-1)} \right] \cdot (1, 1, 2)$$

$$= \left[(yz, zx, xy) \Big|_{(2,\,1,\,-1)} \right] \cdot (1, 1, 2) = (-1, -2, 2) \cdot (1, 1, 2) = 1.$$

Let us look again at the definition $D_{\mathbf{v}} f(\mathbf{a}) = \dfrac{d}{dt} f(\mathbf{a} + t\mathbf{v}) \Big|_{t=0}$

It gives special importance to the *line*

$$\mathbf{x} = \mathbf{x}(t) = \mathbf{a} + t\mathbf{v},$$

which passes through $\mathbf{a}$ with velocity $\mathbf{v}$.

Consider instead *any curve* $\mathbf{x} = \mathbf{x}(t)$ that passes through $\mathbf{x} = \mathbf{a}$ at time $t = 0$ with velocity $\mathbf{v}$:

$$\mathbf{x}(0) = \mathbf{a} \qquad \text{and} \qquad \frac{d\mathbf{x}}{dt}\bigg|_{t=0} = \mathbf{v}.$$

Now differentiate the composite function $f[\mathbf{x}(t)]$ at $t = 0$:

$$\frac{d}{dt} f[\mathbf{x}(t)]\bigg|_{t=0} = \left[\operatorname{grad} f\bigg|_{\mathbf{x}(0)}\right] \cdot \frac{d\mathbf{x}}{dt}\bigg|_{t=0} = [\operatorname{grad} f(\mathbf{a})] \cdot \mathbf{v} = D_{\mathbf{v}} f(\mathbf{a}).$$

The result is completely independent of what curve $\mathbf{x}(t)$ you take; it depends only on $\mathbf{x}(0) = \mathbf{a}$ and $\dot{\mathbf{x}}(0) = \mathbf{v}$.

Let $\mathbf{x}(t)$ be any differentiable curve such that $\mathbf{x}(0) = \mathbf{a}$ and $\dot{\mathbf{x}}(0) = \mathbf{v}$. Then

$$\frac{d}{dt} f[\mathbf{x}(t)]\bigg|_{t=0} = D_{\mathbf{v}} f(\mathbf{a}).$$

Example $f(x, y) = x^2 y$, $\quad \mathbf{a} = (1, 0)$, $\quad \mathbf{v} = (0, 1)$, $\quad \mathbf{x}(t) = (\cos t, \sin t)$.

On the one hand,

$$D_{\mathbf{v}} f(\mathbf{a}) = [\operatorname{grad} f(\mathbf{a})] \cdot \mathbf{v} = \left[(2xy, x^2)\bigg|_{(1, 0)}\right] \cdot (0, 1) = (0, 1) \cdot (0, 1) = 1.$$

On the other hand, the curve $\mathbf{x}(t)$ satisfies

$$\mathbf{x}(0) = (1, 0) = \mathbf{a}, \qquad \dot{\mathbf{x}}(0) = (0, 1) = \mathbf{v}.$$

Now check the derivative of $f[\mathbf{x}(t)] = \cos^2 t \sin t$:

$$\frac{d}{dt} f[\mathbf{x}(t)]\bigg|_{t=0} = (-2 \cos t \sin^2 t + \cos^3 t)\bigg|_{t=0} = \cos^3 0 = 1.$$

EXERCISES

Compute $\operatorname{grad} f$

1 $f = x^2 y + 3xy^3$

2 $f = y^2 e^{xy}$

3 $f = \dfrac{ax + by}{cx + dy}$

4 $f = \sqrt{x^2 + 3y^2}$

5 $f = (x^2 + y^2)e^z$

6 $f = \log(3x - y - 4z)$

7 $f = \sqrt{1 + x^2 y^2 z^4}$

8 $f = x^2 \cos(yz)$.

9 Let $f(x, y, z) = x^2 + y^2 + z^2 - 2xy + 3yz + 6zx$. Find all points where $\operatorname{grad} f$ is parallel to the x, y-plane.

10 Let $z = 1/(x^2 + y^2 + 10)$. Show that $\operatorname{grad} z$ points toward $\mathbf{0}$ at all points $\mathbf{x} \neq \mathbf{0}$.

11 Suppose $z = f(r, \theta)$ is given in terms of polar coordinates. Show that
    $$\operatorname{grad} z = f_r \mathbf{u} + r^{-1} f_\theta \mathbf{w}, \qquad \text{where} \quad \mathbf{u} = (\cos \theta, \sin \theta) \quad \text{and} \quad \mathbf{w} = (-\sin \theta, \cos \theta).$$

12 (cont.) Find $\text{grad}(r^{-2}\cos 2\theta)$.

13 (cont.) Use Ex. 11 to do Ex. 10.

14 (cont.) Let $f(x, y) = \arctan(y/x)$. Compute $\text{grad } f$ in both rectangular and polar co-ordinates. Show that your results agree.

15 Suppose $f(a, b) = g(a, b)$, and $\text{grad } f(a, b) \cdot \text{grad } g(a, b) = 0$. Assuming the two gradients are non-zero, translate into a statement about level curves.

16 (cont.) What can you say if the condition on the gradients is replaced by $\text{grad } f(a, b) = \lambda \text{ grad } g(a, b)$, where λ is a constant?

Find the tangent line to

17 $x^2 - 3xy + y^2 = -1$ at $(1, 2)$

18 $x + x^3 y^4 - y = 0$ at $(0, 0)$

19 $y + \sin xy = 1$ at $(0, 1)$

20 $x\sqrt{3x + y} = 6$ at $(2, 3)$.

Compute the directional derivatives at $\mathbf{a}$ in the directions of $\mathbf{v}_1$ and $\mathbf{v}_2$

21 $f(x, y) = e^x \cos y$; $\mathbf{a} = \mathbf{0}$, $\mathbf{v}_1 = (1, 0)$, $\mathbf{v}_2 = (0, 1)$

22 $f(x, y) = \ln(x + 2y)$; $\mathbf{a} = (0, 1)$, $\mathbf{v}_1 = (1, 1)$, $\mathbf{v}_2 = (3, 4)$

23 $f(x, y, z) = xyz$; $\mathbf{a} = (1, -1, 2)$, $\mathbf{v}_1 = (1, 1, 0)$, $\mathbf{v}_2 = (1, 0, 1)$

24 $f(x, y, z) = \sqrt{x + 3y + 5z}$; $\mathbf{a} = (1, 1, 1)$, $\mathbf{v}_1 = (1, 1, -1)$, $\mathbf{v}_2 = (1, 2, 3)$.

Find the largest directional derivative of $f(x, y, z)$ at $\mathbf{a}$:

25 $f(x, y, z) = x^3 + y^2 + z$; $\mathbf{a} = \mathbf{0}$

26 $f(x, y, z) = x^2 - y^2 + 4z^2$; $\mathbf{a} = (-1, -1, 1)$.

Find all unit vectors $\mathbf{u}$ for which $D_{\mathbf{u}} f(\mathbf{a}) = 0$:

27 $f(x, y) = (2 + xy^2)^5$; $\mathbf{a} = (0, 1)$

28 $f(x, y, z) = x^2 + xy + yz$; $\mathbf{a} = (-1, 1, 1)$.

29 Given $f(x, y)$ and a point $\mathbf{a}$, show that the value of $[D_{\mathbf{u}} f(\mathbf{a})]^2 + [D_{\mathbf{v}} f(\mathbf{a})]^2$ is constant for all pairs of perpendicular unit vectors $\mathbf{u}$ and $\mathbf{v}$. What is the constant value?

30 (cont.) Given k, show that there exist perpendicular unit vectors $\mathbf{u}$ and $\mathbf{v}$ such that $D_{\mathbf{v}} f(\mathbf{a}) = k D_{\mathbf{u}} f(\mathbf{a})$.

Compute the general directional derivative $D_{\mathbf{v}} f(\mathbf{a})$

31 $f(x, y) = 1/(2x + 5y)$; $\mathbf{a} = (2, 1)$, $\mathbf{v} = (-3, 4)$

32 $f(x, y, z) = xye^{yz}$; $\mathbf{a} = (1, 1, 1)$, $\mathbf{v} = (2, 2, 5)$.

33 Through each point $\mathbf{a} \neq \mathbf{0}$ of the plane pass a level curve of xy and a level curve of $x^2 - y^2$. Find their angle of intersection.

34 Show that the curves of steepest ascent for the function $f(x, y) = \frac{1}{2}x^2 + y^2$ are the parabolas $y = kx^2$. Draw a contour map of f to see if this looks reasonable.

Prove

35 $D_{\mathbf{v}+\mathbf{w}} f(\mathbf{a}) = D_{\mathbf{v}} f(\mathbf{a}) + D_{\mathbf{w}} f(\mathbf{a})$

36 $D_{\mathbf{v}}(fg)(\mathbf{a}) = f(\mathbf{a}) D_{\mathbf{v}} g(\mathbf{a}) + g(\mathbf{a}) D_{\mathbf{v}} f(\mathbf{a})$.

4. SURFACES

We shall study surfaces given as the locus of equations

$$F(x, y, z) = k \qquad (k \text{ constant});$$

briefly, $F(\mathbf{x}) = k$. We shall always assume that the locus is not the empty set, as in the case of the locus $F(\mathbf{x}) = |\mathbf{x}|^2 + 1 = 0$. We shall also make a technical assumption to rule out corners, edges, and other types of singularities, namely that

$$\text{grad } F(\mathbf{a}) \neq \mathbf{0} \quad \text{whenever} \quad F(\mathbf{a}) = k.$$

Examples (1) $F(x, y, z) = ax + by + cz$, where $(a, b, c) \neq \mathbf{0}$.

The surface $F(x, y, z) = k$ is the plane

$$ax + by + cz = k,$$

and
$$\text{grad } F(x, y, z) = (a, b, c) \neq (0, 0, 0).$$

In vector notation:

$$F(\mathbf{x}) = \mathbf{a} \cdot \mathbf{x}, \qquad \text{grad } F(\mathbf{x}) = \mathbf{a} \neq \mathbf{0},$$

and $F(\mathbf{x}) = k$ is equivalent to $\mathbf{a} \cdot \mathbf{x} = k$.

(2) $F(x, y, z) = x^2 + y^2 + z^2$, $k = a^2$, $a > 0$.

The surface $F(x, y, z) = a^2$ is the sphere

$$x^2 + y^2 + z^2 = a^2$$

with center $\mathbf{0}$ and radius a. In vector notation

$$F(\mathbf{x}) = |\mathbf{x}|^2, \qquad \text{grad } F(\mathbf{x}) = (2x, 2y, 2z) = 2\mathbf{x},$$

and $F(\mathbf{x}) = a^2$ is equivalent to $|\mathbf{x}| = a$. For each $\mathbf{x}$ on the sphere, grad $F(\mathbf{x}) = 2\mathbf{x} \neq \mathbf{0}$ since $|\mathbf{x}| = a > 0$.

(3) $F(x, y, z) = z - x^2 - y^2$.

The surface $F(x, y, z) = 0$ is

$$z = x^2 + y^2, \qquad \text{and} \qquad \text{grad } F(\mathbf{x}) = (-2x, -2y, 1) \neq \mathbf{0}.$$

In Example 1, p. 721, we showed that this surface is a paraboloid of revolution.

The Tangent Plane Given a surface $F(x, y, z) = k$ and one of its points $\mathbf{a}$, we want to define the plane tangent to the surface at $\mathbf{a}$ and find an equation for this plane. On p. 734 we found the tangent *line* to the level *curve* $f(x, y) = c$. Now we want the tangent *plane* to the level *surface* $F(x, y, z) = k$. It is not surprising, therefore, that the discussion and resulting equation will be similar.

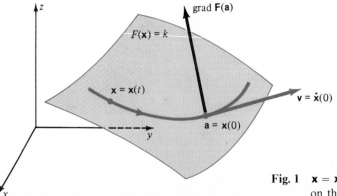

Fig. 1 $\mathbf{x} = \mathbf{x}(t)$ is an arbitrary curve on the surface $F(\mathbf{x}) = k$.

Let us begin by proving a fundamental geometric fact: for any curve lying on the surface $F(\mathbf{x}) = k$, its velocity vector at $\mathbf{a}$ is perpendicular to grad $F(\mathbf{a})$. See Fig. 1.

This fact is remarkable because grad $F(\mathbf{a})$ is a fixed vector, whereas the assertion holds for any curve whatsoever lying on the surface and passing through $\mathbf{a}$.

Let $\mathbf{a}$ be a point of the surface $F(\mathbf{x}) = k$. Then the velocity vector $\mathbf{v}$ at $\mathbf{a}$ of each curve on the surface through $\mathbf{a}$ satisfies

$$\mathbf{v} \cdot \text{grad } F(\mathbf{a}) = 0.$$

Proof Let $\mathbf{x} = \mathbf{x}(t)$ be any differentiable curve lying on $F(\mathbf{x}) = k$ and passing through $\mathbf{a}$ at $t = 0$. Thus

(1) $F[\mathbf{x}(t)] = k$,   (2) $\mathbf{x}(0) = \mathbf{a}$.

The composite function in (1) is constant, hence its derivative with respect to t is zero. Therefore, by the vector form of the Chain Rule,

$$\dot{\mathbf{x}}(t) \cdot \text{grad } F[\mathbf{x}(t)] = 0.$$

We set $t = 0$ in this relation. Since $\mathbf{x}(0) = \mathbf{a}$ and $\dot{\mathbf{x}}(0) = \mathbf{v}$, we obtain

$$\mathbf{v} \cdot \text{grad } F(\mathbf{a}) = 0,$$

the desired equation.

We now define the tangent plane at $\mathbf{a}$ as the plane through $\mathbf{a}$ perpendicular to grad $F(\mathbf{a})$. This definition is reasonable geometrically. The plane passes through $\mathbf{a}$ and is parallel to the velocity vector at $\mathbf{a}$ of any curve through $\mathbf{a}$ that lies on the surface. To obtain an equation of this plane, let $\mathbf{x}$ be any point on it. Then $\mathbf{x} - \mathbf{a}$ is perpendicular to grad $F(\mathbf{a})$, hence

$$(\mathbf{x} - \mathbf{a}) \cdot \text{grad } F(\mathbf{a}) = 0.$$

Tangent Plane   Let $\mathbf{a} = (a, b, c)$ be a point of the surface $F(\mathbf{x}) = k$. The **tangent plane** to the surface at $\mathbf{x} = \mathbf{a}$ is the plane

$$[\text{grad}.F(\mathbf{a})] \cdot (\mathbf{x} - \mathbf{a}) = 0.$$

In coordinate notation, this equation is

$$\frac{\partial F}{\partial x} \cdot (x - a) + \frac{\partial F}{\partial y} \cdot (y - b) + \frac{\partial F}{\partial z} \cdot (z - c) = 0,$$

where the partial derivatives are evaluated at (a, b, c).

Note the similarity to the corresponding statement about the tangent line to a level curve (p. 734).

■ **EXAMPLE 1**   Find the tangent plane to

(a) $x + 2y + 3z = 6$ at $(1, 1, 1)$ (b) $x^2 + y^2 + z^2 = 14$ at $(3, 2, 1)$.

Solution (a)   Set $F(x, y, z) = x + 2y + 3z$. Then the given surface (plane) is

$F(\mathbf{x}) = 6$, and the point $\mathbf{a} = (1, 1, 1)$ is on the surface since $F(\mathbf{a}) = 6$. We have

$$\mathrm{grad}\, F(\mathbf{a}) = \mathrm{grad}\, F(\mathbf{x}) \Big|_{\mathbf{a}} = (1, 2, 3) \Big|_{\mathbf{a}} = (1, 2, 3).$$

Hence the tangent plane at $\mathbf{a}$ is

$$(1, 2, 3) \cdot (\mathbf{x} - \mathbf{a}) = 0, \qquad (1, 2, 3) \cdot (x - 1, y - 1, z - 1) = 0,$$

$$x + 2y + 3z = 1 + 2 + 3 = 6.$$

Not surprising! The tangent plane to a plane at any of its points is the plane itself.

(b)   Set $F(\mathbf{x}) = x^2 + y^2 + z^2 = |\mathbf{x}|^2$. Then the given surface (sphere) is $F(\mathbf{x}) = 14$, and the point $\mathbf{a} = (3, 2, 1)$ is on the surface since $F(\mathbf{a}) = 14$. We have

$$\mathrm{grad}\, F(\mathbf{a}) = \mathrm{grad}\, F(\mathbf{x}) \Big|_{\mathbf{a}} = 2(x, y, z) \Big|_{\mathbf{a}} = 2\mathbf{a}.$$

Hence the tangent plane at $\mathbf{a}$ is

$$2\mathbf{a} \cdot (\mathbf{x} - \mathbf{a}) = 0 \qquad \mathbf{a} \cdot \mathbf{x} = \mathbf{a} \cdot \mathbf{a} = 14, \qquad 3x + 2y + z = 14. \qquad \blacksquare$$

The Unit Normal   The vector $\mathrm{grad}\, F(\mathbf{a})$ is normal (perpendicular) to the tangent plane at $\mathbf{x} = \mathbf{a}$. Therefore a *unit* vector normal to the tangent plane is $\mathrm{grad}\, F(\mathbf{a})$ divided by its length.

Unit Normal   Let $\mathbf{a}$ be a point of the surface $F(\mathbf{x}) = k$. The **unit normal** to the surface at $\mathbf{x} = \mathbf{a}$ is the vector

$$\mathbf{n} = \frac{1}{|\mathrm{grad}\, F(\mathbf{a})|} \,\mathrm{grad}\, F(\mathbf{a}).$$

■ **EXAMPLE 2**   Find the unit normal to

$$xyz = 3 \qquad \text{at} \quad (-1, -3, 1).$$

Solution   Set $F(\mathbf{x}) = xyz$ and $\mathbf{a} = (-1, -3, 1)$. Then $F(\mathbf{a}) = 3$ and

$$\mathrm{grad}\, F(\mathbf{a}) = \mathrm{grad}\, F(\mathbf{x}) \Big|_{\mathbf{a}} = (yz, zx, xy) \Big|_{\mathbf{a}} = (-3, -1, 3).$$

Since $|\mathrm{grad}\, F(\mathbf{a})|^2 = 19$, we have

$$\mathbf{n} = \frac{1}{|\mathrm{grad}\, F(\mathbf{a})|} \,\mathrm{grad}\, F(\mathbf{a}) = \frac{1}{\sqrt{19}} (-3, -1, 3). \qquad \blacksquare$$

Remark   "The" unit normal is not quite fair, because really $\mathbf{n}$ is only determined up to sign. If we define the surface by $-F(\mathbf{x}) = -k$, then $\mathbf{n}$ is changed to $-\mathbf{n}$. Nonetheless, the inaccurate terminology "*the* unit normal" is common, and either $+\mathbf{n}$ or $-\mathbf{n}$ can be chosen.

Graph of a Function   Let us consider the graph of

$$z = f(x, y).$$

Here $f(x, y)$ is a function of two variables, defined over some plane domain $\mathbf{D}$ and assumed to have continuous partial derivatives (Fig. 2). The graph is a surface; to fit it into our previous pattern, set

$$F(x, y, z) = z - f(x, y).$$

Then the graph $z = f(x, y)$ is the locus of $F(x, y, z) = 0$. Now

$$\text{grad } F(\mathbf{x}) = (-f_x, -f_y, 1).$$

Since the third component is 1, this gradient is never $\mathbf{0}$, so our basic assumption is satisfied free of charge.

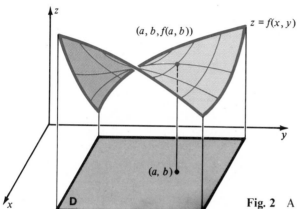

Fig. 2   A surface defined as the graph of a function $z = f(x, y)$

If (a, b) is a point of $\mathbf{D}$, then $(a, b, f(a, b))$ is the corresponding point on the surface. The gradient there is

$$\text{grad } F \bigg|_{(a, b, f(a, b))} = (-f_x(a, b), -f_y(a, b), 1),$$

and the tangent plane is

$$(\text{grad } F) \cdot (x - a, y - b, z - f(a, b)) = 0,$$

which simplifies to $z = f(a, b) + f_x(a, b)(x - a) + f_y(a, b)(y - b).$

The unit normal is $\mathbf{n} = \text{grad } F / |\text{grad } F|.$

(Note that this one of the two possible choices has a positive z-coordinate.) Clearly $|\text{grad } F|^2 = 1 + f_x^2 + f_y^2.$

Tangent Plane and Unit Normal to a Graph   Let (a, b) be a (non-boundary) point of the domain of $f(x, y)$. Then the tangent plane and unit normal to the graph $z = f(x, y)$ at $(a, b, f(a, b))$ are

$$z = f(a, b) + f_x(a, b)(x - a) + f_y(a, b)(y - b)$$

and $$\mathbf{n} = \frac{1}{\sqrt{1 + f_x^2(a, b) + f_y^2(a, b)}} (-f_x(a, b), -f_y(a, b), 1).$$

■ **EXAMPLE 3**   Find the tangent plane and unit normal to $z = x^2 y$ for $(x, y) = (3, 2)$.

Solution   Set $f(x, y) = x^2 y$. Then $f(3, 2) = 18$, and

$$\left. (-f_x, -f_y, 1) \right|_{(3,2)} = \left. (-2xy, -x^2, 1) \right|_{(3,2)} = (-12, -9, 1).$$

The tangent plane is

$$z = 18 + f_x \cdot (x - 3) + f_y \cdot (y - 2) = 18 + 12(x - 3) + 9(y - 2),$$

that is,

$$12x + 9y - z = 36.$$

The unit normal is

$$\mathbf{n} = \frac{1}{|(-12, -9, 1)|} (-12, -9, 1) = \frac{1}{\sqrt{226}} (-12, -9, 1). \qquad ■$$

Level Surfaces   Earlier we showed that the gradient field of a function of two variables is orthogonal (perpendicular) to the level curves. Now we observe that the same holds for a function of three variables. Recall that the **level surfaces** of $F(x, y, z)$ are the surfaces

$$F(x, y, z) = k.$$

The **gradient field** of F is the assignment of the vector grad $F(\mathbf{x})$ to each point $\mathbf{x}$ of the domain of F.

> The **gradient field** of $F(x, y, z)$ is orthogonal to the level surfaces of F.

This statement means that the tangent plane at each point of $F(x, y, z) = k$ is orthogonal to grad $F(\mathbf{x})$ at that point. But this is practically the definition of the tangent plane.

As an application we can compute the angle at which a line intersects a surface $F(x, y, z) = k$. We define that angle as the complement of the angle between the line and the normal to the surface at the point of intersection. See Fig. 3. If θ is the angle between the line and the normal then $\cos \theta = \mathbf{u} \cdot \mathbf{n}$, where $\mathbf{u}$ is a unit vector parallel to the line and $\mathbf{n}$ is the unit normal at the point of intersection.

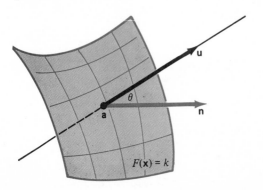

Fig. 3   The line and the surface intersect at angle $\frac{1}{2}\pi - \theta$.

■ **EXAMPLE 4**   Find the angle at which the line $\mathbf{x}(t) = (t,\ t,\ t)$ intersects the surface $x^2 + y^2 + 2z^2 = 1$ in the first octant.

Solution   A point of intersection corresponds to a value of t satisfying

$$t^2 + t^2 + 2t^2 = 1, \qquad 4t^2 = 1, \qquad t = \pm\tfrac{1}{2}.$$

There are two points of intersection. The one in the first octant is $\mathbf{a} = (\tfrac{1}{2},\ \tfrac{1}{2},\ \tfrac{1}{2})$.
   Write the surface as $F(x,\ y,\ z) = x^2 + y^2 + 2z^2 = 1$. Then

$$\operatorname{grad} F(\mathbf{a}) = (2x,\ 2y,\ 4z)\Big|_{\mathbf{a}} = (1,\ 1,\ 2),$$

and the unit normal at $\mathbf{a}$ is

$$\mathbf{n} = \frac{1}{|(1,\ 1,\ 2)|}(1,\ 1,\ 2) = \frac{1}{\sqrt{6}}(1,\ 1,\ 2).$$

Obviously a vector parallel to the given line is $(1,\ 1,\ 1)$. A unit vector in the same direction is

$$\mathbf{u} = \frac{1}{|(1,\ 1,\ 1)|}(1,\ 1,\ 1) = \frac{1}{\sqrt{3}}(1,\ 1,\ 1).$$

Therefore, if θ is the angle between the line and the normal to the surface, then

$$\cos\theta = \mathbf{u}\cdot\mathbf{n} = \frac{1}{\sqrt{3}}(1,\ 1,\ 1)\cdot\frac{1}{\sqrt{6}}(1,\ 1,\ 2) = \frac{4}{\sqrt{18}} = \frac{2}{3}\sqrt{2} \approx 0.94281.$$

Hence $\theta \approx 19.47°$. The desired angle is the complement,

$$90 - \theta = \operatorname{arc\,sin}\left(\tfrac{2}{3}\sqrt{2}\right) \approx 70.53°. \qquad\blacksquare$$

EXERCISES

Find the tangent plane and the unit normal to the given surface at the given point

1   $x^3 + y^3 + z^3 = 3$, $(4, 4, -5)$
2   $x^2 + y^3 + z^4 = 18$, $(3, 2, 1)$
3   $x^2 + yz = 7$, $(1, 2, 3)$
4   $x^3 + y^3 + z^3 + 3xyz = 16$, $(1, 1, 2)$
5   $xy^2z^3 + yz^5 = 14$, $(3, 2, 1)$
6   $z^5 - xz^4 + yz^3 - 1 = 0$, $(1, 1, 1)$
7   $xy + yz + zx = 0$, $(2, 2, -1)$
8   $z\cos(xy) = y^2$, $(0, 1, 1)$
9   $e^{xy} + y = e^z + 1$, $(1, 1, 1)$
10   $x\sin y + y\sin z + xyz = 0$, $(1, 0, 0)$
11   $x^3 + xe^{yz} = 10$, $(2, 0, 1)$
12   $(x + 2y)/\cosh z = 1$, $(5, -2, 0)$.

Find the tangent plane and unit normal to the graph $z = f(x, y)$ at the point with the given (x, y)

13   $z = x^2 - y^2$, $(0, 0)$
14   $z = x^2 - y^2$, $(1, -1)$
15   $z = x^2 + 4y^2$, $(2, 1)$
16   $z = x^2 e^y$, $(-1, 2)$
17   $z = x^2y + y^3$, $(-1, 2)$
18   $z = x\cos y + y\cos x$, $(0, 0)$
19   $z = x + x^2y^3 + y$, $(0, 0)$
20   $z = x^3 + y^3$, $(1, -1)$
21   $z = \ln(1 + x^2 + 2y^2)$, $(1, 1)$
22   $z = \sqrt{1 - x^2 - y^2}$, $(\tfrac{1}{2}, \tfrac{1}{2})$.

23   Find the angle of intersection of the line through $\mathbf{0}$ and $(1, 1, 1)$ and the surface $z = x^2 + y^2$ at each of their intersections.

24   Compute the angle at which a line through the north pole $\mathbf{k}$ of the unit sphere $|\mathbf{x}| = 1$ intersects the sphere at a typical point $\mathbf{a} = (a, b, c)$.

25 Find all points on the surface $xyz = 1$ where the normal line intersects the z-axis. (The normal line is the line through the point with the direction of the unit normal.)

26 Suppose $F(\mathbf{x})$ is an **even** function, that is, $F(-\mathbf{x}) = F(\mathbf{x})$. Show that the tangent planes at $\mathbf{a}$ and $-\mathbf{a}$ to the surface $F(\mathbf{x}) = k$ are parallel.

5. PARAMETRIC SURFACES AND SURFACE OF REVOLUTION

We have seen that a curve can be parameterized by a vector function $\mathbf{x}(t)$ of one real variable. Let us discuss parameterization of a surface.

Physically speaking, a curve is a one-dimensional set of points; you can move on a curve with one degree of freedom (forward or backward). In slightly different terms, the points of a curve can be specified by one real number (parameter), for instance, by the directed distance along the curve from a fixed point. A surface is a two-dimensional set of points; you can move on a surface with two degrees of freedom. Thus it requires *two* real numbers to specify a point, for instance, latitude and longitude on a sphere.

This discussion leads us to define a surface by a vector function

$$\mathbf{x} = \mathbf{x}(u, v)$$

of two real variables. In coordinates,

$$\mathbf{x} = (x, y, z) = (x(u, v), y(u, v), z(u, v)).$$

Here (u, v) varies over a domain **D** in the u, v-plane. Each point of **D** is identified by a pair of real numbers (u, v); to this pair the function assigns a point $\mathbf{x}(u, v)$ on the surface (Fig. 1). In other words, $\mathbf{x}(u, v)$ maps the plane domain **D** onto the surface. We assume $\mathbf{x}(u, v)$ is continuously differentiable, that is, the six partials

$$\frac{\partial x}{\partial u}, \quad \frac{\partial x}{\partial v}, \quad \frac{\partial y}{\partial u}, \quad \frac{\partial y}{\partial v}, \quad \frac{\partial z}{\partial u}, \quad \frac{\partial z}{\partial v}$$

exist and are continuous.

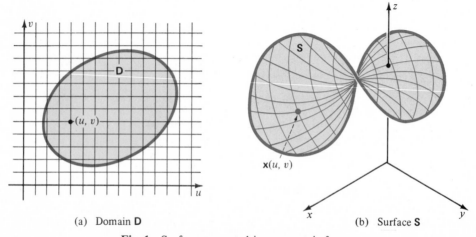

(a) Domain **D**   (b) Surface **S**

Fig. 1 Surface presented in parametric form

Examples (1) $\mathbf{x} = (u + v, u - v, u)$, $-\infty < u < \infty$, $-\infty < v < \infty$,
parameterizes the plane $x + y = 2z$.

(2) $\mathbf{x} = (u \cos v, u \sin v, u)$
parameterizes the surface $x^2 + y^2 = z^2$, a right circular cone, as we shall see shortly.

Look at one more example, $\mathbf{x} = (u + v, u + v, u + v)$.
There is something wrong here because this is just another version of the *line* $\mathbf{x} = (t, t, t)$. To assure that a surface is genuinely two-dimensional we shall assume that

$$\left(\frac{\partial x}{\partial u}, \frac{\partial y}{\partial u}, \frac{\partial z}{\partial u}\right) \times \left(\frac{\partial x}{\partial v}, \frac{\partial y}{\partial v}, \frac{\partial z}{\partial v}\right) \neq \mathbf{0}.$$

In briefer notation this condition is $\dfrac{\partial \mathbf{x}}{\partial u} \times \dfrac{\partial \mathbf{x}}{\partial v} \neq \mathbf{0}$, or shorter yet, $\mathbf{x}_u \times \mathbf{x}_v \neq \mathbf{0}$.

This assumption plays a similar role to that of grad $F(\mathbf{x}) \neq \mathbf{0}$ for surfaces $F(\mathbf{x}) = k$. It guarantees the existence of a well-defined tangent plane, as we are about to show.

Tangent Plane and Unit Normal   Let us find the tangent plane to a parametric surface $\mathbf{S}: \mathbf{x} = \mathbf{x}(u, v)$ at the point $\mathbf{c} = \mathbf{x}(a, b)$. We shall show that the velocity vectors at $\mathbf{c}$ of all curves on $\mathbf{S}$ that pass through $\mathbf{c}$ fill out a plane. The parallel plane through $\mathbf{c}$ is the tangent plane.

To start, let us look at two special curves on the surface:

$$\mathbf{x}(u) = [x(u, b), y(u, b), z(u, b)], \qquad \mathbf{x}(v) = [x(a, v), y(a, v), z(a, v)].$$

Both pass through $\mathbf{c}$, the first for $u = a$, the second for $v = b$. Their velocity vectors at $\mathbf{c}$ are

$$\left(\frac{\partial x}{\partial u}, \frac{\partial y}{\partial u}, \frac{\partial z}{\partial u}\right)\bigg|_{(a, b)} = \mathbf{x}_u(a, b), \qquad \left(\frac{\partial x}{\partial v}, \frac{\partial y}{\partial v}, \frac{\partial z}{\partial v}\right)\bigg|_{(a, b)} = \mathbf{x}_v(a, b).$$

Our basic assumption $\mathbf{x}_u \times \mathbf{x}_v \neq \mathbf{0}$ guarantees that $\mathbf{x}_u$ and $\mathbf{x}_v$ are non-parallel. Hence they span a plane $\mathbf{P}$ consisting of all vectors

$$h\mathbf{x}_u + k\mathbf{x}_v.$$

Let us show that the velocity vectors at $\mathbf{c}$ of all curves on $\mathbf{S}$ that pass through $\mathbf{c}$ fill out the plane $\mathbf{P}$. Any such curve is of the form

$$\mathbf{x}(t) = (x(u, v), y(u, v), z(u, v)),$$

where $u = u(t)$, $v = v(t)$, and $u(0) = a$, $v(0) = b$.
Its velocity vector at $\mathbf{c}$ is $\dot{\mathbf{x}}(0)$. By the Chain Rule,

$$\dot{\mathbf{x}}(0) = \left(\frac{\partial x}{\partial u} \dot{u}(0) + \frac{\partial x}{\partial v} \dot{v}(0), \frac{\partial y}{\partial u} \dot{u}(0) + \frac{\partial y}{\partial v} \dot{v}(0), \frac{\partial z}{\partial u} \dot{u}(0) + \frac{\partial z}{\partial v} \dot{v}(0)\right)$$

$$= \dot{u}(0)\left(\frac{\partial x}{\partial u}, \frac{\partial y}{\partial u}, \frac{\partial z}{\partial u}\right) + \dot{v}(0)\left(\frac{\partial x}{\partial v}, \frac{\partial y}{\partial v}, \frac{\partial z}{\partial v}\right) = \dot{u}(0)\mathbf{x}_u + \dot{v}(0)\mathbf{x}_v.$$

Hence $\dot{\mathbf{x}}(0)$ is in the plane $\mathbf{P}$.

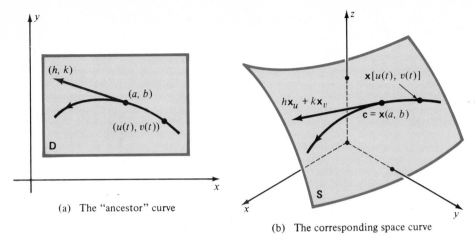

(a) The "ancestor" curve

(b) The corresponding space curve

Fig. 2   Curve on a parametric surface

Conversely, suppose

$$h\mathbf{x}_u + k\mathbf{x}_v$$

is an arbitrary vector in **P**. There is a curve $(u(t), v(t))$ in the u, v-plane such that

$$u(0) = a, \quad v(0) = b \qquad \text{and} \qquad \dot{u}(0) = h, \quad \dot{v}(0) = k.$$

There corresponds (Fig. 2) a curve $\mathbf{x}(t) = \mathbf{x}[u(t), v(t)]$ on **S** passing through **c**, whose velocity vector at **c** is

$$\dot{\mathbf{x}}(0) = \dot{u}(0)\mathbf{x}_u + \dot{v}(0)\mathbf{x}_v = h\mathbf{x}_u + k\mathbf{x}_v.$$

Thus the velocity vectors fill out the plane **P**. See Fig. 3.

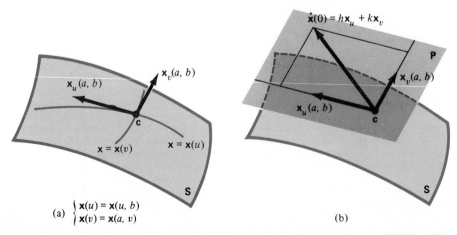

(a) $\begin{cases} \mathbf{x}(u) = \mathbf{x}(u, b) \\ \mathbf{x}(v) = \mathbf{x}(a, v) \end{cases}$

(b)

Fig. 3   The set of all velocity vectors $\dot{\mathbf{x}}(0)$ of all curves on **S** through $\mathbf{c} = \mathbf{x}(0)$ fill a plane **P**.

The vector $\mathbf{x}_u \times \mathbf{x}_v$ is perpendicular to $\mathbf{P}$, hence normal to the surface at $\mathbf{c}$. Therefore the unit normal is

$$\mathbf{n} = \frac{\mathbf{x}_u \times \mathbf{x}_v}{|\mathbf{x}_u \times \mathbf{x}_v|}.$$

Let $\mathbf{x} = \mathbf{x}(u, v)$ be a surface in parametric form and let $\mathbf{c} = \mathbf{x}(a, b)$. The tangent plane at $\mathbf{c}$, in parametric form, is

$$\mathbf{x} = \mathbf{c} + h\mathbf{x}_u(a, b) + k\mathbf{x}_v(a, b),$$

where h and k are arbitrary real numbers.

The unit normal at $\mathbf{c}$ is $\qquad \mathbf{n} = \dfrac{\mathbf{x}_u \times \mathbf{x}_v}{|\mathbf{x}_u \times \mathbf{x}_v|}.$

■ **EXAMPLE 1**   Find the unit normal and the tangent plane in both parametric and non-parametric form for the surface

$$\mathbf{x} = (u \cos v, u \sin v, v), \qquad u > 0, \quad \text{at } \mathbf{c} = (0, 1, \tfrac{1}{2}\pi).$$

Solution   The point $\mathbf{c}$ corresponds to $(u, v) = (1, \tfrac{1}{2}\pi)$. We have

$$\mathbf{x}_u(1, \tfrac{1}{2}\pi) = (\cos v, \sin v, 0)\Big|_{(1, \pi/2)} = (0, 1, 0),$$

$$\mathbf{x}_v(1, \tfrac{1}{2}\pi) = (-u \sin v, u \cos v, 1)\Big|_{(1, \pi/2)} = (-1, 0, 1).$$

Now $\mathbf{x}_u \times \mathbf{x}_v = (0, 1, 0) \times (-1, 0, 1) = (1, 0, 1)$, so

$$\mathbf{n} = \frac{(1, 0, 1)}{|(1, 0, 1)|} = \frac{1}{\sqrt{2}} (1, 0, 1).$$

The tangent plane in parametric form is

$$\mathbf{x} = \mathbf{c} + h\mathbf{x}_u(1, \tfrac{1}{2}\pi) + k\mathbf{x}_v(1, \tfrac{1}{2}\pi) = (0, 1, \tfrac{1}{2}\pi) + h(0, 1, 0) + k(-1, 0, 1),$$

$$\mathbf{x} = (-k, 1 + h, \tfrac{1}{2}\pi + k).$$

In non-parametric form the tangent plane is

$$(\mathbf{x} - \mathbf{c}) \cdot \mathbf{n} = 0, \qquad (\mathbf{x} - \mathbf{c}) \cdot (1, 0, 1) = 0,$$

$$(x, y - 1, z - \tfrac{1}{2}\pi) \cdot (1, 0, 1) = 0, \qquad x + z = \tfrac{1}{2}\pi.$$

The two forms agree, as is easily checked. The surface is a spiral ramp. For a figure, see p. 870. ■

Surfaces of Revolution   Let $f(x, z) = 0$ be the equation of a curve in the part $x \geq 0$ of the x, z-plane (Fig. 4a). If the curve is rotated about the z-axis, a surface, called a **surface of revolution**, is swept out (Fig. 4b).

To find its equation, let $\mathbf{x} = (x, y, z)$ be any point of the surface. The distance from $\mathbf{x}$ to the z-axis is $\sqrt{x^2 + y^2}$. See Fig. 4b. Hence $\mathbf{x}$ is swept out by the point

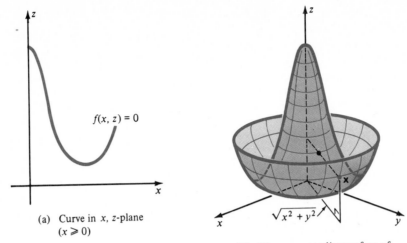

(a) Curve in x, z-plane
     $(x \geqslant 0)$

(b) The corresponding surface of
     revolution about the z-axis

Fig. 4   Surface of revolution

$(\sqrt{x^2 + y^2}, z)$ on the original curve. This point satisfies the equation of the curve so

$$f(\sqrt{x^2 + y^2}, z) = 0.$$

Thus $\mathbf{x} = (x, y, z)$ satisfies this equation, hence it is the equation of the surface.

Examples (1) A paraboloid of revolution is swept out when the curve $x^2 - z = 0$, $x \geq 0$ is rotated about the z-axis. Its equation is

$$(\sqrt{x^2 + y^2})^2 - z = 0, \qquad \text{that is,} \quad z = x^2 + y^2.$$

(2) A right circular cone is swept out when the line $x - z = 0$, $x \geq 0$ is rotated about the z-axis. Its equation is

$$\sqrt{x^2 + y^2} - z = 0, \qquad \text{or equivalently,} \quad z^2 = x^2 + y^2, \qquad z \geq 0.$$

Similar reasoning applies to surfaces of revolution about the other coordinate axes. For example if $f(x, y) = 0$, $y \geq 0$, is rotated about the x-axis, the equation of the resulting surface is

$$f(x, \sqrt{y^2 + z^2}) = 0.$$

EXERCISES

Express the parametric surface in the form $F(x, y, z) = k$; describe the surface

1    $\mathbf{x} = (u, v, u + v)$

2    $\mathbf{x} = (2u - v + 3, -u + 4v + 1, \\ \qquad\qquad -u - 2v - 6)$

3    $\mathbf{x} = (v \cos u, v \sin u, v)$

4    $\mathbf{x} = (a \cos u, a \sin u, v)$

5    $\mathbf{x} = (a \sin u, v, a \cos u)$

6    $\mathbf{x} = (u^3, u^3, v)$.

Find all points (a, b) in the u, v-domain where the surface is degenerate, that is, where $\mathbf{x}_u$ and $\mathbf{x}_v$ are parallel $(\mathbf{x}_u \times \mathbf{x}_v = \mathbf{0})$

7    $\mathbf{x} = (u + v, u - v, uv)$

8    $\mathbf{x} = (u, u^2, v^3)$

9    $\mathbf{x} = (u \cos v, u \sin v, v)$

10    $\mathbf{x} = (u + v, uv, u^2 + v^2)$

11 $\mathbf{x} = (u^2, v, u^3)$ **12** $\mathbf{x} = (1 - u^2, u - u^3, v)$.

Give the tangent plane in parametric form and the unit normal for the surface at the given parametric values (u, v)

13 $\mathbf{x} = (u - v, u + v, u + 3v)$, $(1, 0)$ **14** $\mathbf{x} = (u \sin v, v, u \cos v)$, $(2, \frac{1}{6}\pi)$
15 $\mathbf{x} = (\sin u \cos v, \sin u \sin v, 2 + \cos u)$, $(\frac{1}{2}\pi, \frac{1}{4}\pi)$
16 $\mathbf{x} = (u, v/u, 1/v)$, $(1, -2)$ **17** $\mathbf{x} = (u^2 - v^2, 2uv, u^3 + v^3)$, $(0, -2)$
18 $\mathbf{x} = (u^2 v^2, u^3 + v^3, u^2 v + uv^2)$, $(-1, -1)$.

19   Let $0 < a < b$. Revolve about the z-axis the circle in the x, z-plane with center $(b, 0)$ and radius a. Give an equation for the resulting terms.
20*   (cont.) Parameterize the surface in terms of two angles.
21   What is a simple way of parameterizing the graph $z = f(x, y)$ of a function?
22*   Suppose a curve in the x, z-plane is revolved about the z-axis. Show that the normal line at each point of the resulting surface of revolution either intersects, or is parallel to, the z-axis. (See Ex. 25, p. 746.)
23*   Let $\mathbf{x} = \mathbf{x}(s)$ be a curve on the unit sphere $|\mathbf{x}| = 1$, where s is arc length. Consider the cone $\mathbf{x} = \mathbf{x}(u, v) = v\mathbf{x}(u)$, $v > 0$. Show that this surface has a tangent plane at each of its points.
24*   (cont.) Show that the unit normal to the surface is constant along each generator, and the tangent plane is also the same at each point of a generator.
25   Let $\mathbf{x} = \mathbf{x}(s)$ be a space curve with arc length s and curvature $k = k(s) > 0$. The set of (positive) tangent lines to the curve sweep out the surface $\mathbf{x} = \mathbf{x}(u, v) = \mathbf{x}(u) + v\mathbf{t}(u)$, $v > 0$, where $\mathbf{t}(s)$ is the unit tangent at $\mathbf{x}(s)$. Find $\mathbf{x}_u$ and $\mathbf{x}_v$ and show that these vectors are never parallel.
26   (cont.) Find an equation for the tangent plane at $\mathbf{x}(u, v)$, and show that that it is independent of v.

6. QUADRIC SURFACES

A **quadric surface** is the locus of an equation $f(x, y, z) = 0$, where $f(x, y, z)$ is a quadratic polynomial. The most general quadratic polynomial is

$$f(x, y, z) = Ax^2 + By^2 + Cz^2 + Dxy + Eyz + Fzx + px + qy + rz + k.$$

We shall study quadratic surfaces only for the following special types of quadratic polynomials:

$$\text{(1)}\quad Ax^2 + By^2 + Cz^2 + k$$

$$\text{(2)}\quad Ax^2 + By^2 + rz \qquad\qquad \text{(2')}\quad Ax^2 + By^2 + k$$

$$\text{(3)}\quad Ax^2 + qy \qquad\qquad\qquad \text{(3')}\quad Ax^2 + k.$$

Later we shall observe that every quadric surface is one of these types with respect to a suitably chosen rectangular coordinate system. We assume throughout that *some* second degree terms are present; otherwise the locus of $f(x, y, z) = 0$ is a plane, which doesn't interest us here.

We begin with type (1), assuming $ABC \neq 0$ and $k \neq 0$. With a bit of juggling of constants, $f = 0$ can be written in the form

$$\pm \frac{x^2}{a^2} \pm \frac{y^2}{b^2} \pm \frac{z^2}{c^2} = 1.$$

The nature of this quadric surface depends on how many of the signs are plus and how many are minus. If all three signs are minus, the locus is empty. Otherwise it is symmetric in each coordinate plane, because if (x, y, z) is on the surface, then so are all eight points $(\pm x, \pm y, \pm z)$. Therefore the whole surface is determined by symmetry from its part in the first octant.

Ellipsoids   Consider the quadric surface

$$\frac{x^2}{a^2} + \frac{y^2}{b^2} + \frac{z^2}{c^2} = 1, \qquad a, b, c > 0.$$

Since squares are non-negative, each of its points satisfies

$$\frac{x^2}{a^2} \le 1, \qquad \frac{y^2}{b^2} \le 1, \qquad \frac{z^2}{c^2} \le 1.$$

This means the surface is confined to the box

$$-a \le x \le a, \qquad -b \le y \le b, \qquad -c \le z \le c.$$

Suppose $-c \le z_0 \le c$. The horizontal plane $z = z_0$ intersects the surface in the curve

$$\frac{x^2}{a^2} + \frac{y^2}{b^2} = 1 - \frac{z_0{}^2}{c^2}, \qquad z = z_0.$$

This curve is an ellipse. It is as large as possible when $z_0 = 0$, and it becomes smaller and smaller as $z_0 \longrightarrow c$ or $z_0 \longrightarrow -c$. Thus each such cross section by a horizontal plane is an ellipse, except at the extremes $z_0 = \pm c$, where it is a single point.

The same argument applies to plane sections parallel to the other coordinate planes. This gives us enough information for a sketch. The surface is called an **ellipsoid** (Fig. 1). In the special case $a = b = c$, it is a sphere.

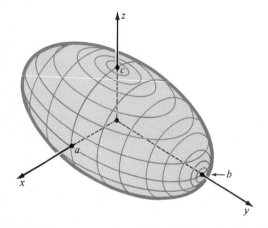

Fig. 1   Ellipsoid: $\dfrac{x^2}{a^2} + \dfrac{y^2}{b^2} + \dfrac{z^2}{c^2} = 1$

There is a nice formula for the tangent plane to an ellipsoid.

Let $\mathbf{m} = (\ell, m, n)$ be a point of the ellipsoid $\dfrac{x^2}{a^2} + \dfrac{y^2}{b^2} + \dfrac{z^2}{c^2} = 1.$

Then the tangent plane at $\mathbf{m}$ is $\dfrac{\ell x}{a^2} + \dfrac{my}{b^2} + \dfrac{nz}{c^2} = 1.$

To derive the formula, we first compute the gradient:

$$\text{grad}\left(\frac{x^2}{a^2} + \frac{y^2}{b^2} + \frac{z^2}{c^2}\right)\bigg|_{\mathbf{m}} = 2\left(\frac{x}{a^2}, \frac{y}{b^2}, \frac{z}{c^2}\right)\bigg|_{\mathbf{m}} = 2\left(\frac{\ell}{a^2}, \frac{m}{b^2}, \frac{n}{c^2}\right).$$

Therefore $\mathbf{x}$ is on the tangent plane at $\mathbf{m}$ if and only if

$$2\left(\frac{\ell}{a^2}, \frac{m}{b^2}, \frac{n}{c^2}\right) \cdot (\mathbf{x} - \mathbf{m}) = 0.$$

This condition is equivalent to

$$\frac{\ell x}{a^2} + \frac{my}{b^2} + \frac{nz}{c^2} = \frac{\ell^2}{a^2} + \frac{m^2}{b^2} + \frac{n^2}{c^2} = 1.$$

The right-hand side equals one because $\mathbf{m}$ is on the ellipsoid.

Hyperboloids of One Sheet   Consider the locus of

$$\frac{x^2}{a^2} + \frac{y^2}{b^2} - \frac{z^2}{c^2} = 1, \qquad a, b, c > 0.$$

Each horizontal cross section is an ellipse, $\dfrac{x^2}{a^2} + \dfrac{y^2}{b^2} = 1 + \dfrac{z_0{}^2}{c^2}, \qquad z = z_0,$

no matter what value z_0 is. The smallest ellipse occurs for $z_0 = 0$; as $z_0 \longrightarrow \infty$ or $z_0 \longrightarrow -\infty$, the ellipses get larger and larger.

The surface meets the y, z-plane in the hyperbola $\dfrac{y^2}{b^2} - \dfrac{z^2}{c^2} = 1,$

and it meets the z, x-plane in the hyperbola $\dfrac{x^2}{a^2} - \dfrac{z^2}{c^2} = 1.$

This information is enough to sketch the surface, called a **hyperboloid of one sheet** (Fig. 2).

The equation for the tangent plane to $\dfrac{x^2}{a^2} + \dfrac{y^2}{b^2} - \dfrac{z^2}{c^2} = 1$ at $\mathbf{m} = (\ell, m, n)$

is $\dfrac{\ell x}{a^2} + \dfrac{my}{b^2} - \dfrac{nz}{c^2} = 1.$

It can be derived exactly as for the ellipsoid.

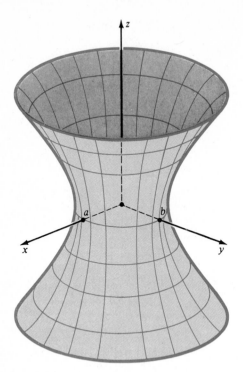

Fig. 2   Hyperboloid of one sheet:

$$\frac{x^2}{a^2} + \frac{y^2}{b^2} - \frac{z^2}{c^2} = 1$$

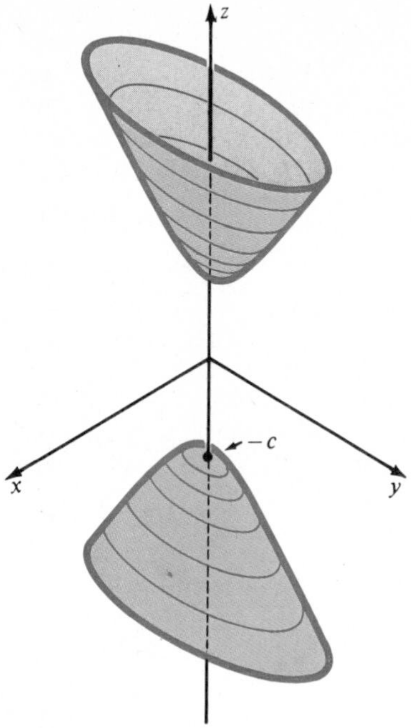

Fig. 3   Hyperboloid of two sheets:

$$-\frac{x^2}{a^2} - \frac{y^2}{b^2} + \frac{z^2}{c^2} = 1$$

Hyperboloids of Two Sheets   Consider the equation

$$-\frac{x^2}{a^2} - \frac{y^2}{b^2} + \frac{z^2}{c^2} = 1, \qquad a, b, c > 0.$$

If (x, y, z) is a point on the surface, then

$$\frac{z^2}{c^2} = 1 + \frac{x^2}{a^2} + \frac{y^2}{b^2} \geq 1,$$

hence $z^2 \geq c^2$. This means either $z \geq c$ or $z \leq -c$, that is, there are no points of the surface between the horizontal planes $z = c$ and $z = -c$.

If $z_0{}^2 > c^2$, the horizontal plane $z = z_0$ meets the surface in the curve

$$\frac{x^2}{a^2} + \frac{y^2}{b^2} = \frac{z_0{}^2}{c^2} - 1 > 0, \qquad z = z_0,$$

an ellipse. Also the surface meets the y, z-plane and the z, x-plane in the hyperbolas

$$-\frac{y^2}{b^2} + \frac{z^2}{c^2} = 1 \qquad \text{and} \qquad -\frac{x^2}{a^2} + \frac{z^2}{c^2} = 1$$

respectively. The surface breaks into two parts, and it is called a **hyperboloid of two sheets** (Fig. 3).

As before, the tangent plane at a point **m** of the surface is

$$-\frac{\ell x}{a^2} - \frac{my}{b^2} + \frac{nz}{c^2} = 1.$$

Quadratic Cones   To complete the study of type (1) on p. 751 for $ABC \neq 0$, we now consider the case $k = 0$. Then $f = 0$ can be written

$$\pm \frac{x^2}{a^2} \pm \frac{y^2}{b^2} \pm \frac{z^2}{c^2} = 0, \qquad a, b, c > 0.$$

If the signs are all the same, then $(0, 0, 0)$ is the only point on the graph; not interesting. Otherwise two signs are equal, the other opposite. Changing signs if necessary, and renaming constants, we have

$$z^2 = \frac{x^2}{a^2} + \frac{y^2}{b^2}, \qquad a, b > 0.$$

This surface has the following property. For each point $\mathbf{x}_0$ on the surface, the entire line $\mathbf{x} = t\mathbf{x}_0$ lies on the surface. Such a surface is called a **cone**, and the lines $\mathbf{x} = t\mathbf{x}_0$ are called **generators** of the cone.

To show that

$$z^2 = \frac{x^2}{a^2} + \frac{y^2}{b^2}$$

really has the cone property, we take any point $\mathbf{x}_0$ on the surface and check that

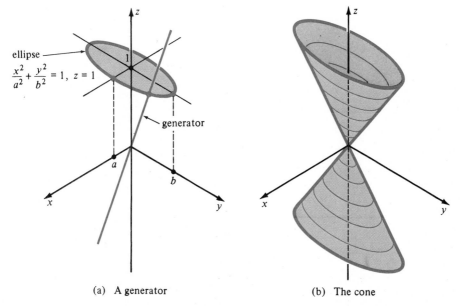

(a) A generator   (b) The cone

Fig. 4   Quadratic cone: $z^2 = \dfrac{x^2}{a^2} + \dfrac{y^2}{b^2}$

$t\mathbf{x}_0$ is also on the surface. If $\mathbf{x}_0 = (x_0, y_0, z_0)$, then $t\mathbf{x}_0 = (tx_0, ty_0, tz_0)$ and

$$(tz_0)^2 = t^2 z_0{}^2 = t^2 \left(\frac{x_0{}^2}{a^2} + \frac{y_0{}^2}{b^2} \right) = \frac{(tx_0)^2}{a^2} + \frac{(ty_0)^2}{b^2}.$$

Thus $t\mathbf{x}_0$ is on the surface.

To sketch the cone, we note that it meets the horizontal plane $z = 1$ in the ellipse

$$\frac{x^2}{a^2} + \frac{y^2}{b^2} = 1.$$

For each point on this ellipse, we draw the line through the point and $\mathbf{0}$. See Fig. 4.

The origin is a singular point on the cone (no tangent plane). If $\mathbf{m} \neq \mathbf{0}$ is any other point on the cone, then we see in the usual way that the tangent plane at $\mathbf{m}$ is

$$nz = \frac{\ell x}{a^2} + \frac{my}{b^2}.$$

Paraboloids   Next we take up type (2) on p. 751 and study the locus of $f = 0$, where

$$f(x, y, z) = Ax^2 + By^2 + rz, \qquad AB \neq 0, \qquad r \neq 0.$$

We may write the equation $f = 0$ in the form

$$z = \pm \frac{x^2}{a^2} \pm \frac{y^2}{b^2}, \qquad a, b > 0.$$

Changing z to $-z$ merely turns the surface upside-down, so there are essentially two distinct cases: $++$ and $-+$.

The first case is the surface

$$z = \frac{x^2}{a^2} + \frac{y^2}{b^2}, \qquad a, b > 0,$$

called an **elliptic paraboloid**. It lies above the x, y-plane and is symmetric in the y, z- and z, x-planes. Each horizontal cross-section

$$\frac{x^2}{a^2} + \frac{y^2}{b^2} = z_0 > 0, \qquad z = z_0$$

is an ellipse. As z_0 increases, the ellipse grows larger. The surface is called a paraboloid because it meets the y, z-plane and the z, x-plane in the parabolas $z = y^2/b^2$ and $z = x^2/a^2$ respectively (Fig. 5).

The tangent plane to the elliptic paraboloid at a point $\mathbf{m} = (\ell, m, n)$ is

$$z = 2 \left(\frac{\ell x}{a^2} + \frac{m^2}{b^2} \right) - n.$$

The derivation is left as an exercise.

The second case is the **hyperbolic paraboloid**, the locus of

$$z = -\frac{x^2}{a^2} + \frac{y^2}{b^2}, \qquad a, b > 0.$$

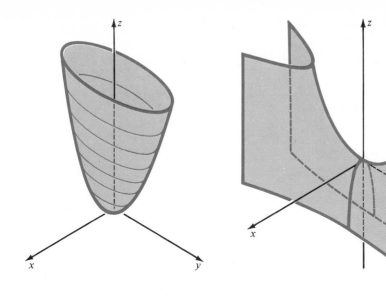

Fig. 5 Elliptic paraboloid:

$$z = \frac{x^2}{a^2} + \frac{y^2}{b^2}$$

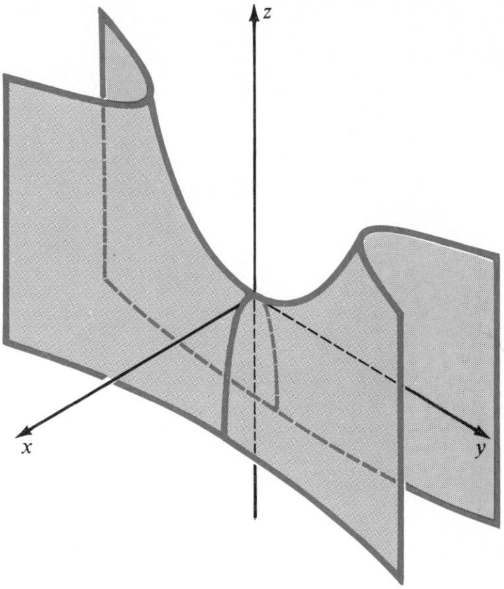

Fig. 6 Hyperbolic paraboloid:

$$z = -\frac{x^2}{a^2} + \frac{y^2}{b^2}$$

This surface is symmetric in the y, z- and z, x-planes. The horizontal planes $z = z_0 > 0$ meet it in hyperbolas whose branches open out in the y-direction. The horizontal planes $z = z_0 < 0$ meet it in hyperbolas that open out in the x-direction. The y, z-plane meets the locus in the parabola $z = y^2/b^2$, which opens upward; and the z, x-plane meets it in the parabola, $z = -x^2/a^2$, which opens downward. The best description is "saddle-shaped". See Fig. 6.

The tangent plane at **m** is

$$z = 2\left(-\frac{\ell x}{a^2} + \frac{my}{b^2}\right) - n.$$

Again, the derivation is left as an exercise.

Quadratic Cylinders In types (2′) and (3) on p. 751 the variable z is missing. Generally, when one variable is missing the locus is a cylinder. Take for example $Ax^2 + By^2 + k = 0$, where $A > 0$, $B > 0$, and $k < 0$. This can be written in the form

$$\frac{x^2}{a^2} + \frac{y^2}{b^2} = 1, \qquad a, b > 0.$$

The surface meets each horizontal plane $z = z_0$ in the same ellipse. If (x_0, y_0, z_0) is any point of the surface, the whole vertical line (x_0, y_0, z) for $-\infty < z < \infty$ lies on the surface. The surface is an **elliptic cylinder** and these vertical lines that lie on the surface are called **generators** of the cylinder (Fig. 7a). Any curve $f(x, y) = 0$ in the x, y-plane generates a cylinder in space consisting of all points (x_0, y_0, z) for which $f(x_0, y_0) = 0$. In particular, a circle generates a (right)

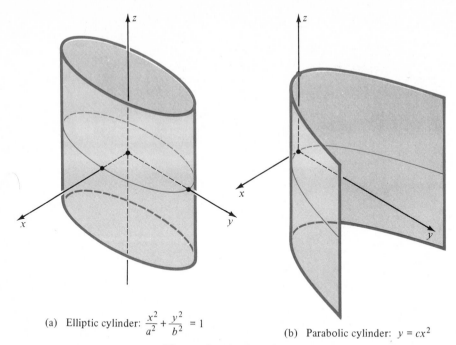

(a) Elliptic cylinder: $\dfrac{x^2}{a^2} + \dfrac{y^2}{b^2} = 1$

(b) Parabolic cylinder: $y = cx^2$

Fig. 7   Quadratic cylinders

circular cylinder, a hyperbola generates a hyperbolic cylinder, etc. Type (3) on p. 751 leads to a parabolic cylinder (Fig. 7b).

In the remaining type (3′) both y and z are missing. Depending on the signs of A and k, the locus of $Ax^2 + k = 0$ is empty or consists of one plane or two planes parallel to the y, z-plane. In general, the locus in $\mathbf{R}^3$ of $f(x) = 0$ is a set of planes parallel to the y, z-plane. For each zero x_0 of $f(x)$, the plane $x = x_0$ is included in the locus.

Reduction of Quadratic Polynomials   Given a general quadratic polynomial

$$f(x, y, z) = Ax^2 + By^2 + Cz^2 + Dxy + Eyz + Fzx + px + qy + rz + k,$$

a fairly deep theorem in linear algebra says that the mixed terms, xy, yz, zx, can be eliminated by a suitable rotation of the coordinate system. Obviously, a rotation does not affect the nature of the quadric surface $f(x, y, z) = 0$, so in studying quadrics we may assume

$$f(x, y, z) = Ax^2 + By^2 + Cz^2 + px + qy + rz + k.$$

If $A \neq 0$, the translation $x = x' - p/2A$ eliminates the term px, and similarly if $B \neq 0$ or $C \neq 0$. This reduces the study of the following types of f:

(i) $Ax^2 + By^2 + Cz^2 + k$

(ii) $Ax^2 + By^2 + rz + k$

(iii) $Ax^2 + qy + rz + k.$

These include all possibilities, provided we are willing to permute the variables. For instance, the polynomial $Cz^2 + px + qy + k$ becomes type (iii) when x and z are interchanged.

In type (ii), if $r \neq 0$, then a translation in the z-direction eliminates k. In type (iii), a rotation in the y, z-plane, taken so that $qy + rz = 0$ is the new z-axis, changes the function to the form $ax^2 + qy + k$. Again, k can be eliminated by translation if $q \neq 0$.

This discussion shows that we may study all quadric surfaces by concentrating on the five types of quadratic polynomials listed at the beginning of this section.

EXERCISES

Sketch the first octant portion

1	$\frac{1}{4}x^2 + y^2 + \frac{1}{4}z^2 = 1$	**2**	$x^2 + \frac{1}{9}y^2 + \frac{1}{4}z^2 = 1$	**3**	$x^2 + y^2 - z^2 = 1$
4	$-x^2 - y^2 + z^2 = 1$	**5**	$x^2 - y^2 + z^2 = 1$	**6**	$-x^2 + y^2 - z^2 = 1$
7	$z = x^2 + y^2$	**8**	$z = \frac{1}{4}x^2 + y^2$	**9**	$z = -x^2 + y^2$
10	$z = x^2 - y^2.$				

Identify the quadric surface

11 $z = x^2 + 2x + y^2$ **12** $x^2 + 2y^2 + 3z^2 - 2x - 8y + 6z = 0$
[*Hint*   Complete the square.]

13 $x^2 + y^2 - a^2(z-1)^2 = 0$ **14** $z = xy$ [*Hint*   Rotate $45°$ about the
z-axis.]

Sketch the paraboloids

15 $x = y^2 + z^2$ **16** $y = x^2 - z^2.$

Sketch the surface in x, y, z-space

17	$x - z = 1$	**18**	$y = x^2$	**19**	$xy = 1$
20	$-x^2 + y^2 = 1$	**21**	$x = z^2$	**22**	$y^2 + 4z^2 = 1$
23	$z = x^2 - x$	**24**	$x^2 + 4z^2 = 1$	**25**	$y^2 = z^2 + 4x^2$
26	$x^2 = y^2.$				

27   Suppose $f(x, y) = 0$, $z = 1$ is a curve on the plane $z = 1$. Find an equation for the cone obtained by taking all points on all lines through **0** and points of the curve.

28   (cont.) Test your result on $x^2 + y^2 = 1$, $z = 1$.

Find the tangent plane to the paraboloid at $\mathbf{m} = (\ell, m, n)$

29   elliptic: $z = \dfrac{x^2}{a^2} + \dfrac{y^2}{b^2}$ **30**   hyperbolic: $z = -\dfrac{x^2}{a^2} + \dfrac{y^2}{b^2}.$

31   Show that all cross sections of the ellipsoid $x^2/a^2 + y^2/b^2 + z^2/c^2 = 1$ by planes $x = k$, where $-a < x < a$, are ellipses of the same eccentricity.

32   Given an ellipsoid and a plane P, prove that there are exactly two points on the ellipsoid where the tangent plane is parallel to P.

33   Let $\mathbf{n}$ be a fixed unit vector and α a fixed acute angle. Find a vector equation for the cone with vertex **0** whose generators all make angle α with $\mathbf{n}$.

34   (cont.) Find an equation for the cone with vertex **0** whose generators make angle $45°$ with the line through **0** and $(1, 1, 0)$.

35   Show that the tangent planes at all points (u, v, k), where $k > 0$ is fixed, of the quadric $z = x^2/a^2 + y^2/b^2$ have a common point. What point?   [*Hint*   See Ex. 29.]

36   Find the intersection of the tangent plane at **0** to the hyperbolic paraboloid $x = y^2/a^2 - z^2/b^2$ with the quadric itself.

The following exercises assume some knowledge of matrices. Let A be a 3×3 non-zero symmetric matrix and $\mathbf{x} = (x, y, z)$. Then $\mathbf{x}'$ is the transpose of $\mathbf{x}$, a column vector.

37*   Set $f(\mathbf{x}) = \mathbf{x}A\mathbf{x}'$. Show that grad $f(\mathbf{x}) = 2\mathbf{x}A$.

38   (cont.) Show that the tangent plane at a point $\mathbf{m}$ of the quadric $f(\mathbf{x}) = 1$ is $\mathbf{m}A\mathbf{x}' = 1$.

39   (cont.) Let $\mathbf{b}$ be a fixed vector and set $f(\mathbf{x}) = \mathbf{x}A\mathbf{x}' + 2\mathbf{b}\mathbf{x}' + k$, the most general quadratic polynomial. Find grad $f(\mathbf{x})$.

40*   (cont.) Let $\mathbf{m}$ be a point of the quadric $f(\mathbf{x}) = 0$. Find the tangent plane at $\mathbf{m}$.

41   Let $\mathbf{n}$ be a unit vector. Show that the set of points on the ellipsoid $x^2/a^2 + y^2/b^2 + z^2/c^2 = 1$, where the tangent plane is parallel to $\mathbf{n}$ is an ellipse. (You may assume that any plane section of an ellipsoid is an ellipse.)

42   (cont.) What does this say about the shadow of an ellipsoid when the sun is directly overhead?

7. OPTIMIZATION

The word means choosing the most favorable value, for us, usually the max or min of a function of several variables. Before we plunge in, let us recall some facts about extrema of functions of one variable. Let $f(x)$ be a differentiable function. Suppose $f(x)$ has a local max or a local min at a point $x = c$ that is not a boundary point of its domain. Then $y = f(x)$ has a horizontal tangent at $x = c$. See Figs. 1a and 1b. But as Fig. 1c shows, a horizontal tangent does not guarantee a local max or min.

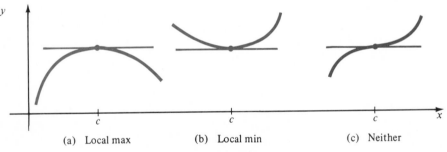

(a) Local max   (b) Local min   (c) Neither

Fig. 1   Horizontal tangent

Now let $f(x, y)$ be a continuously differentiable function on a domain $\mathbf{D}$. Suppose $f(x, y)$ has a local min at $(x, y) = (a, b)$, an interior point of $\mathbf{D}$, that is, not a boundary point. This means

$$f(x, y) \geq f(a, b)$$

for all points (x, y) of $\mathbf{D}$ sufficiently near (a, b).

Let us show that the tangent plane to the surface $z = f(x, y)$ at (a, b) is horizontal. We set $g(x) = f(x, b)$. Then $g(x)$, a function of one variable, has a local min at $x = a$. Therefore $g'(a) = 0$. But by definition, $g'(a) = f_x(a, b)$. Thus $f_x(a, b) = 0$ and similarly, $f_y(a, b) = 0$.

The equation of the tangent plane is

$$z = f(a, b) + f_x(a, b)(x - a) + f_y(a, b)(y - b).$$

Since $f_x(a, b) = f_y(a, b) = 0$, this reduces to $z = f(a, b)$, a constant. Hence the tangent plane is horizontal. Conversely, if the plane is horizontal, then $f_x(a, b) = f_y(a, b) = 0$.

Let $f(x, y)$ be a continuously differentiable function and let (a, b) be an interior point of its domain. Suppose $f(x, y)$ has either a local max or a local min at (a, b). Then the tangent plane to $z = f(x, y)$ at $(a, b, f(a, b))$ is horizontal. Equivalently,

$$\frac{\partial f}{\partial x}(a, b) = 0 \quad \text{and} \quad \frac{\partial f}{\partial y}(a, b) = 0.$$

Remark Intuitively it is almost obvious that the tangent plane is horizontal at a local max or local min. Suppose, for example, that $f(x, y) \le f(a, b)$ for (x, y) near (a, b). Then the surface $z = f(x, y)$ is on or below the horizontal plane $z = f(a, b)$ but touches it at $(x, y) = (a, b)$. See Fig. 2a.

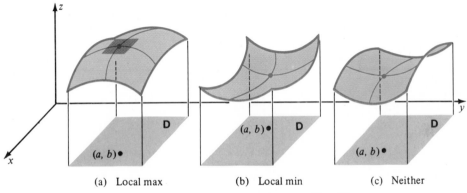

(a) Local max   (b) Local min   (c) Neither

Fig. 2 Horizontal tangent plane, that is, $\dfrac{\partial f}{\partial x} = \dfrac{\partial f}{\partial y} = 0$

The conditions $\qquad f_x(a, b) = 0, \qquad f_y(a, b) = 0$

are the analogue of the first derivative test for functions of one variable. As in the one variable case, this test has certain shortcomings that you should be aware of. First, the conditions $f_x = f_y = 0$ are *necessary* for a local max or a local min, but they are *not sufficient*. The graph $z = f(x, y)$ may have a saddle point (Fig. 2c) where both partials are 0, yet there is neither a max nor a min. Obviously a second derivative test is needed, but that is a matter for the next chapter.

Second, the test does not apply at the boundary of the domain. For example, take $f(x, y) = x^2 + y^2$ on the circular domain $x^2 + y^2 \le 1$. Obviously the maximum value is 1, taken on the boundary. But $f_x = f_y = 0$ only at $(0, 0)$.

Applications   In practice, we look for the max or min of a given function by solving the system

$$\frac{\partial f}{\partial x} = 0, \qquad \frac{\partial f}{\partial y} = 0$$

of two equations in two unknowns. This may give a number of possibilities. Then we try by hook or by crook to sort out the maxs, mins, and neithers. We should also give special attention to the boundary of the domain. However, we can often rule out the boundary, using additional information, physical properties, or if all else fails, common sense.

■ **EXAMPLE 1**   Find the extrema of

$$f(x, y) = x^2 - xy + y^2 + 3x.$$

Solution   The function is defined for all values of x and y; there is no boundary. Begin by finding all points (x, y) at which both

$$\frac{\partial f}{\partial x} = 0 \quad \text{and} \quad \frac{\partial f}{\partial y} = 0.$$

Now

$$\frac{\partial f}{\partial x} = \frac{\partial}{\partial x}(x^2 - xy + y^2 + 3x) = 2x - y + 3,$$

$$\frac{\partial f}{\partial y} = \frac{\partial}{\partial y}(x^2 - xy + y^2 + 3x) = -x + 2y,$$

so the conditions are

$$\begin{cases} 2x - y + 3 = 0 \\ -x + 2y = 0. \end{cases}$$

Solve:

$$x = -2, \quad y = -1.$$

The corresponding value of $f(x, y)$ is

$$f(-2, -1) = (-2)^2 - (-2)(-1) + (-1)^2 + 3(-2) = 4 - 2 + 1 - 6 = -3.$$

Is this a max, a min, or neither? We suspect a minimum because the values of $f(x, y)$ seem to increase as $|x|$ increases or $|y|$ increases. For a fixed $y = b$,

$$f(x, b) = x^2 + (3 - b)x + b^2 \longrightarrow \infty \quad \text{as} \quad x \longrightarrow \pm\infty.$$

Similarly, $f(a, y) = y^2 - ay + (a^2 + 3a) \longrightarrow \infty \quad \text{as} \quad y \longrightarrow \pm\infty.$

Let us prove our conjecture by a little algebra. First we move the origin to $(-2, -1)$ by setting

$$x = u - 2 \quad \text{and} \quad y = v - 1.$$

Then

$$f(x, y) = (u - 2)^2 - (u - 2)(v - 1) + (v - 1)^2 + 3(u - 2) = u^2 - uv + v^2 - 3.$$

Next, we complete the square:

$$f(x, y) = \left(u - \frac{v}{2}\right)^2 - \frac{v^2}{4} + v^2 - 3 = \left(u - \frac{v}{2}\right)^2 + \frac{3}{4}v^2 - 3 \geq -3.$$

Conclusion: the min of $f(x, y)$ is

$$f_{min} = f(-2, -1) = -3.$$

There is no other local max or min.   ■

■ **EXAMPLE 2**   Find the rectangular solid of maximum volume whose total edge length is a given constant k.

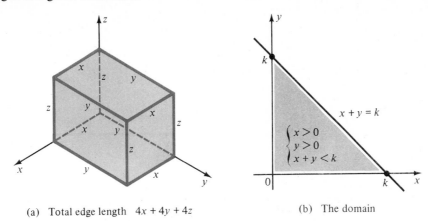

(a) Total edge length $4x + 4y + 4z$ (b) The domain

Fig. 3

Solution   As drawn in Fig. 3a, the total length of the 12 edges is $4x + 4y + 4z$. Thus $4x + 4y + 4z = 4k$, that is,

$$x + y + z + k.$$

The volume is $V = xyz = xy(k - x - y) = kxy - x^2y - xy^2,$

so write $V = V(x, y) = kxy - x^2y - xy^2.$

By the nature of the problem, $x > 0$ and $y > 0$; also $z > 0$, hence $x + y < k$. These conditions describe the domain of $V(x, y)$, the interior of a triangle (Fig. 3b). Clearly $V > 0$ on this domain and $V = 0$ on its boundary (since either $x = 0$, $y = 0$, or $z = 0$ at each point of the boundary). Hence it is plausible that V has a positive maximum on the domain. To locate it, solve

$$\frac{\partial V}{\partial x} = 0, \qquad \frac{\partial V}{\partial y} = 0,$$

that is, $\begin{cases} ky - 2xy - y^2 = 0 \\ kx - x^2 - 2xy = 0. \end{cases}$

Since $x > 0$ and $y > 0$, cancel y from the first equation and x from the second:

$$\begin{cases} 2x + y = k \\ x + 2y = k. \end{cases}$$

This pair of simultaneous linear equations has the unique solution

$$x = \tfrac{1}{3}k, \qquad y = \tfrac{1}{3}k.$$

Hence $z = k - x - y = \tfrac{1}{3}k$; the solid is a cube.   ■

Remark   We have used an important principle: suppose $f(x, y) > 0$ on a domain **D** and $f(x, y) = 0$ on the boundary of **D**. If $f_x = f_y = 0$ at only one point (a, b) of **D**, then

$f(x, y)$ has its max at (a, b). For assuming the maximum exists, it certainly does not occur on the boundary. But (a, b) is the only interior point that satisfies the necessary conditions for a max, so $f(a, b)$ is the max.

■ **EXAMPLE 3**   What is the largest possible volume, and what are the dimensions of an open rectangular aquarium constructed from 12 ft² of Plexiglas? Ignore the thickness of the plastic.

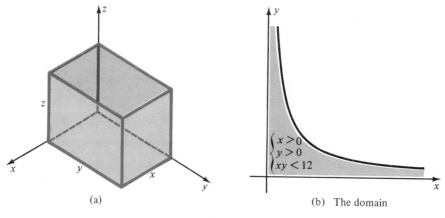

(a)   (b)   The domain

Fig. 4

Solution   See Fig. 4a. The volume is $V = xyz$. The total surface area of the bottom and four sides is

$$xy + 2yz + 2zx = 12.$$

Solve for z, then substitute into the formula for V:

$$z = \frac{12 - xy}{2(x + y)}, \qquad V = \frac{(12 - xy)xy}{2(x + y)}.$$

By the nature of the problem, $x > 0$, $y > 0$, and $z > 0$, hence $xy < 12$. Therefore the domain is the region of the first quadrant bounded by the axes and the curve $xy = 12$. See Fig. 4b. Since $V > 0$ in the interior and $V = 0$ on the boundary, there is probably a max in the interior.

The domain extends to ∞ in the x- and y-directions. What happens to V if, say, x is taken very large? Well, $0 < xy < 12$, hence $(12 - xy)xy < 12 \times 12 = 144$, so

$$0 < V = \frac{(12 - xy)xy}{2(x + y)} < \frac{144}{2(x + y)} < \frac{72}{x + y} < \frac{72}{x}.$$

Therefore V is small if x is large, and similarly, V is small if y is large.

It follows that the max of V occurs for values of x and y neither too near 0 nor too large. To find the max, compute the partial derivatives of V (only one computation is needed because of the symmetry in x and y):

$$\frac{\partial V}{\partial x} = \frac{y^2(-x^2 - 2xy + 12)}{2(x + y)^2}, \qquad \frac{\partial V}{\partial y} = \frac{x^2(-y^2 - 2xy + 12)}{2(x + y)^2}.$$

Now find all points (x, y) where both partials are zero. Such points must satisfy

$$\begin{cases} y^2(-x^2 - 2xy + 12) = 0 \\ x^2(-y^2 - 2xy + 12) = 0. \end{cases}$$

Since both x and y are positive, the factors y^2 and x^2 may safely be canceled:

$$\begin{cases} -x^2 - 2xy + 12 = 0 \\ -y^2 - 2xy + 12 = 0. \end{cases}$$

Solve these equations for x and y. Subtract: $x^2 - y^2 = 0$, hence $y = \pm x$. Since both x and y are positive, only $y = x$ applies. Now substitute $y = x$ into the first equation:

$$-x^2 - 2x^2 + 12 = 0, \qquad 3x^2 = 12, \qquad x^2 = 4, \qquad x = 2.$$

Therefore $(x, y) = (2, 2)$ is the *only* point where $V_x = V_y = 0$, so must yield V_{max}. Now when $x = 2$ ft and $y = 2$ ft, then

$$z = \frac{12 - xy}{2(x + y)} = \frac{12 - 4}{2 \cdot 4} = 1 \text{ ft}.$$

Hence

$$V_{max} = xyz \Big|_{(2, 2, 1)} = 4 \text{ ft}^3.\qquad\blacksquare$$

EXERCISES

Find the max and min of $f(x, y) =$

1   $4 - 2x^2 - y^2$
2   $x^2 + y^2 - 1$
3   $(x - 2)^2 + (y + 3)^2$
4   $(x - 1)^2 + y^2 + 3$
5   $x^2 - 2xy + 2y^2 + 4$
6   $xy - x^2 - 2y^2 + x + 2y$
7   $xye^{-x^2 - y^2}$
8   $e^{-x^2 - y^2 + y}$
9   $\sin x + \sin y + \sin(x + y)$
10 $\cos x + \cos y + \cos(x + y)$.

Find the min for $x \geq 0$, $y \geq 0$ of $f(x, y) =$

11 $\dfrac{1}{xy} + x + y^2$
12 $x^3 + y^3 - 3axy, \quad a > 0$
13 $xy + \dfrac{a}{x} + \dfrac{b}{y}, \quad a > 0, \quad b > 0$
14 $x^5 + y^3 - 15ax^3y, \quad a > 0$
15 $xy^3(2x + y - 10)$
16 $xy^2(x + y - 1)^3$.

Find the max and min of $f(x, y)$ in the given domain. Be sure to check the boundary!

17 $xy + \dfrac{1}{xy}, \quad x > 0, \quad y > 0, \quad x + y \leq 1$
18 $3(x - 1)^2 + (y - 2)^2, \quad 0 \leq x \leq 2, \quad 0 \leq y \leq 3$
19 $x^2(x^2 + y^2), \quad x^2 + y^2 \leq r^2$
20 $x^3 - y^2 + 6xy, \quad 0 \leq x \leq 1, \quad 0 \leq y \leq 1$.

21 Find the largest possible volume of a rectangular solid inscribed in the unit sphere.
22 Compute the largest possible volume of a rectangular box, edges parallel to the axes, and inscribed in the solid bounded below by the x, y-plane and above by the paraboloid $z = c - x^2/a^2 - y^2/b^2$.
23 Find the dimensions of an open-top rectangular box of minimal surface area whose volume V is given.
24 Find the dimensions of the cheapest (open-top) aquarium of given volume V whose slate base costs 3 times as much per unit area as its glass sides.

25   The base of a prism is a right triangle, its lateral sides are perpendicular to its base, and its volume is V, given. Find its smallest possible total surface area.

26   Work Example 3, p. 764, using the face areas as variables.

27   Find the closest plane to **0** among all planes tangent to the surface $xyz^2 = 1$ at a point of the first octant.

28*   A rectangle of dimension $a \times b$ is cut into 4 subrectangles by two lines. If $p > 0$, find the max and the min of $A_1{}^p + A_2{}^p + A_3{}^p + A_4{}^p$, where A_i is the area of the i-th piece.

8. FURTHER OPTIMIZATION PROBLEMS

Three Variables   Suppose $f(a, b, c)$ is a local max or local min of $f(x, y, z)$ and (a, b, c) is an interior point of its domain. We claim that

$$\frac{\partial f}{\partial x}(a, b, c) = 0, \qquad \frac{\partial f}{\partial y}(a, b, c) = 0, \qquad \frac{\partial f}{\partial z}(a, b, c) = 0.$$

The proof is practically the same as for two variables. Set $g(x) = f(x, b, c)$. Then $g(x)$ has a local max or min at $x = a$, hence $g'(a) = 0$. But $g'(a) = f_x(a, b, c)$, etc.

The practical procedure for locating extrema for functions of three (or more) variables is quite similar to that for two, except that we start by solving the system

$$f_x = 0, \qquad f_y = 0, \qquad f_z = 0$$

of three equations in three unknowns.

■ EXAMPLE 1   Find the max of $f(x, y, z) = \dfrac{x + 2y + 3z}{1 + x^2 + y^2 + z^2}.$

Solution   The function is defined on the whole of space. However, there is no sense in looking for a max far from the origin because $|f|$ is very small if $|(x, y, z)|$ is large. To see why, suppose $x^2 + y^2 + z^2 = \rho^2$. Then $|x| \le \rho$, $|y| \le \rho$, and $|z| \le \rho$, hence

$$|f| = \left| \frac{x + 2y + 3z}{1 + x^2 + y^2 + z^2} \right| \le \frac{\rho + 2\rho + 3\rho}{1 + \rho^2} = \frac{6\rho}{1 + \rho^2} < \frac{6}{\rho}.$$

Clearly $f \longrightarrow 0$ as $\rho \longrightarrow \infty$. For example $|f| < 0.1$ if $|\mathbf{x}| \ge 60$. But $f(0, 0, \pm 1) = \pm\frac{3}{2}$. Therefore, f certainly has a positive max (and a negative min) inside the sphere $|\mathbf{x}| = 60$ (actually inside the sphere $|\mathbf{x}| = 4$, as the same reasoning shows).

To find the max, solve the equations $f_x = 0, f_y = 0, f_z = 0$. This computation gets messy, so it is best to organize it. For simplicity, write

$$f = \frac{u}{v}, \qquad \text{where} \quad u = x + 2y + 3z \quad \text{and} \quad v = 1 + x^2 + y^2 + z^2.$$

Then $f_x = \dfrac{vu_x - uv_x}{v^2}, \qquad f_y = \dfrac{vu_y - uv_y}{v^2}, \qquad f_z = \dfrac{vu_z - uv_z}{v^2}.$

The system of equations to be solved becomes

$$vu_x = uv_x, \qquad vu_y = uv_y, \qquad vu_z = uv_z,$$

that is, $v = 2xu, \qquad 2v = 2yu, \qquad 3v = 2zu.$

Now $v > 0$ for (x, y, z), so $u \neq 0$. Divide each equation by u:

$$2x = y = \frac{2}{3}z = \frac{v}{u}.$$

Therefore $y = 2x$, $z = 3x$, and from here on it's smooth sailing. Substitute these expressions for y and z:

$$u = x + 2y + 3z = x + 4x + 9x = 14x,$$

$$v = 1 + x^2 + y^2 + z^2 = 1 + x^2 + 4x^2 + 9x^2 = 1 + 14x^2.$$

Now substitute these values of u and v into the equation $v = 2xu$ and solve for x:

$$1 + 14x^2 = 2x(14x) = 28x^2, \qquad 14x^2 = 1, \qquad x = \pm\tfrac{1}{14}\sqrt{14}.$$

Also $y = 2x$, and $z = 3x$, so the conditions $f_x = f_y = f_z = 0$ hold only at

$$(x, y, z) = \pm\tfrac{1}{14}\sqrt{14}\,(1, 2, 3).$$

At these points, $v = 1 + 14x^2 = 2$ and $u = 14x = \pm\sqrt{14}$, hence $f = u/v = \pm\tfrac{1}{2}\sqrt{14}$.

Therefore $f_{\max} = f\left(\tfrac{1}{14}\sqrt{14}, \tfrac{2}{14}\sqrt{14}, \tfrac{3}{14}\sqrt{14}\right) = \tfrac{1}{2}\sqrt{14}$ ∎

Remark   The solution shows also that $f_{\min} = -\tfrac{1}{2}\sqrt{14}$. Note that $\left|\tfrac{1}{14}\sqrt{14}\,(1, 2, 3)\right| = 1$, confirming the prediction that the max occurs inside the sphere $|\mathbf{x}| = 4$.

Quadratic Functions   In the next chapter we shall develop a second derivative test for extrema. The test hinges on the behavior near $(0, 0)$ of a **homogeneous quadratic polynomial** (also called **quadratic form**), that is, a quadratic polynomial with *only* second degree terms:

$$Q(x, y) = ax^2 + 2bxy + cy^2.$$

Of particular importance are those homogeneous quadratic polynomials that have an overall maximum or minimum at $(0, 0)$. Now the extrema of $Q(x, y)$ are found from solutions of the system $Q_x = 0$, $Q_y = 0$, that is,

$$\begin{cases} ax + by = 0 \\ bx + cy = 0. \end{cases}$$

Certainly $(0, 0)$ is a solution; hence $(0, 0)$ is a candidate for a max or a min. Since $Q(0, 0) = 0$, obviously Q has an overall min at $(0, 0)$ if

$$Q(x, y) > 0 \qquad \text{for} \quad (x, y) \neq (0, 0).$$

When this condition is satisfied, we say that $Q(x, y)$ is **positive definite**. It is helpful to have a criterion for positive definiteness:

Test for Positive Definiteness   Let $Q(x, y) = ax^2 + 2bxy + cy^2$.   Then $Q(x, y)$ is positive definite if and only if

$$a > 0 \qquad \text{and} \qquad \begin{vmatrix} a & b \\ b & c \end{vmatrix} > 0.$$

Proof Suppose $Q(x, y)$ is positive definite, that is

$$Q(x, y) = ax^2 + 2bxy + cy^2 > 0$$

whenever $(x, y) \neq (0, 0)$. In particular $a = Q(1, 0) > 0$. Now complete the square:

$$Q(x, y) = a\left(x + \frac{b}{a} y\right)^2 + \left(\frac{ac - b^2}{a}\right) y^2.$$

Then $Q(-b/a, 1) > 0$, hence

$$\frac{ac - b^2}{a} > 0, \qquad ac - b^2 = a\left(\frac{ac - b^2}{a}\right) > 0.$$

Therefore $a > 0$ and $ac - b^2 > 0$.

Conversely, suppose $a > 0$ and $ac - b^2 > 0$. Then certainly

$$Q(x, y) = a\left(x + \frac{b}{a} y\right)^2 + \left(\frac{ac - b^2}{a}\right) y^2 \geq 0$$

for any (x, y). Furthermore, $Q(x, y) = 0$ only if each of the squared quantities is zero:

$$x + \frac{b}{a} y = 0, \qquad y = 0,$$

hence only for $(x, y) = (0, 0)$. This completes the proof.

Examples   Positive definite

(1) $3x^2 - 2xy + y^2$ $a = 3 > 0, \quad ac - b^2 = 2 > 0$
(2) $5x^2 + 6xy + 2y^2$ $a = 5 > 0, \quad ac - b^2 = 1 > 0$

Not positive definite

(3) $x^2 + 4xy + y^2$ $a = 1 > 0, \quad ac - b^2 = -3 \leq 0$
(4) $-2x^2 + 5xy + y^2$ $a = -2 \leq 0, \quad ac - b^2 = -\frac{33}{4} \leq 0$
(5) $-2x^2 + 2xy - y^2$ $a = -2 \leq 0, \quad ac - b^2 = 1 > 0$
(6) $x^2 + 6xy + 9y^2$ $a = 1 > 0, \quad ac - b^2 = 0 \leq 0$

We define $Q(x, y)$ to be **negative definite** if $Q(x, y) < 0$ whenever $(x, y) \neq (0, 0)$. This is the same as $-Q(x, y)$ positive definite, so the conditions are

$$-a > 0, \qquad \begin{vmatrix} -a & -b \\ -b & -c \end{vmatrix} = \begin{vmatrix} a & b \\ b & c \end{vmatrix} > 0.$$

Test for Negative Definiteness   Let $Q(x, y) = ax^2 + 2bxy + cy^2$. Then $Q(x, y)$ is negative definite if and only if

$$a < 0 \qquad \text{and} \qquad \begin{vmatrix} a & b \\ b & c \end{vmatrix} > 0.$$

Application to Least Squares (Regressions)   Certain experiments produce a sequence of readings

$$(x_1, y_1), \quad (x_2, y_2), \quad \cdots \quad , \quad (x_n, y_n).$$

When plotted (Fig. 1a) the points may cluster in roughly the form of a straight line, suggesting that y is a linear function of x (which may be just what the experimenter would like to establish). Assuming that there is a linear relationship and that the points deviate from a straight line because of experimental error, round-off error, etc., a practical question arises: for which constants A and B does the straight line $y = Ax + B$ *most closely fit* the data?

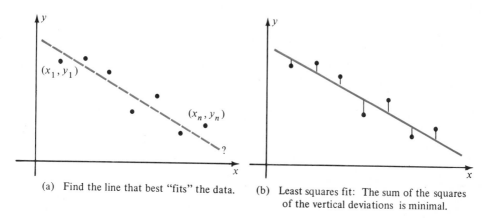

(a) Find the line that best "fits" the data.

(b) Least squares fit: The sum of the squares of the vertical deviations is minimal.

Fig. 1

The answer depends on what is meant by "fit". Probably the most popular measure of fit is by **least squares**: the line is chosen to minimize the sum of the squares of the *vertical* deviations from the line (Fig. 1b).

■ **EXAMPLE 2**   Find the line $y = Ax + B$ that is the least squares fit to the points $(0, 2)$, $(1, 3)$, $(2, 3)$.

Solution   Write $y(x) = Ax + B$, and choose A and B to minimize

$$f(A, B) = [y(0) - 2]^2 + [y(1) - 3]^2 + [y(2) - 3]^2$$

$$= (B - 2)^2 + (A + B - 3)^2 + (2A + B - 3)^2.$$

Necessary conditions for a minimum are

$$\frac{\partial f}{\partial A} = 0 \quad \text{and} \quad \frac{\partial f}{\partial B} = 0.$$

Differentiate:

$$\frac{\partial f}{\partial A} = 2(A + B - 3) + 4(2A + B - 3) = 2(5A + 3B - 9),$$

$$\frac{\partial f}{\partial B} = 2(B - 2) + 2(A + B - 3) + 2(2A + B - 3) = 2(3A + 3B - 8).$$

Set these partial derivatives equal to zero: $\begin{cases} 5A + 3B - 9 = 0 \\ 3A + 3B - 8 = 0. \end{cases}$

The system has a unique solution: $A = \frac{1}{2}$ and $B = \frac{13}{6}$.

Now $f(A, B)$ must have a minimum at $(\frac{1}{2}, \frac{13}{6})$; here is why. If either A or B is large, then $f(A, B)$ is large (by inspection). On the circle $A^2 + B^2 = (1000)^2$, for example, $f(A, B)$ is very large. The minimum of $f(A, B)$ in the region bounded by this circle occurs either on the boundary or at a point where $\partial f/\partial A = \partial f/\partial B = 0$. But the boundary is ruled out, hence the minimum occurs at $(\frac{1}{2}, \frac{13}{6})$, the only point where $\partial f/\partial A = \partial f/\partial B = 0$.

Therefore the answer (Fig. 2) is $y = \frac{1}{2}x + \frac{13}{6}$.   ∎

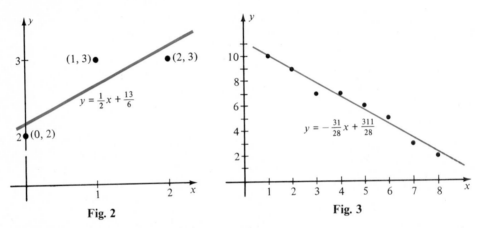

Fig. 2   Fig. 3

To solve the general problem of least squares fit, we imitate the method used in this example. Given readings

$$(x_1, y_1), \quad (x_2, y_2), \quad \cdots, \quad (x_n, y_n)$$

for n distinct values of x, we seek a linear function $y = Ax + B$ that minimizes

$$f(A, B) = \sum_{i=1}^{n} [(Ax_i + B) - y_i]^2.$$

Now $\dfrac{\partial f}{\partial A} = 2 \sum x_i(Ax_i + B - y_i) = 2\left[A\left(\sum x_i^2\right) + B\left(\sum x_i\right) - \left(\sum x_i y_i\right) \right],$

$\dfrac{\partial f}{\partial B} = 2 \sum (Ax_i + B - y_i) = 2\left[A\left(\sum x_i\right) + nB - \left(\sum y_i\right) \right].$

We set these partial derivatives equal to zero and obtain two equations for the two unknowns A and B:

$$\begin{cases} \left(\sum x_i^2\right)A + \left(\sum x_i\right)B = \sum x_i y_i \\[2mm] \left(\sum x_i\right)A + nB = \sum y_i. \end{cases}$$

All coefficients in this system of equations are computable from the data. We now show that there is a unique solution (A, B), assuming the x_i are not all equal. It can be checked (Ex. 38) that this solution indeed minimizes $f(A, B)$.

To express the solution in a concise form, we introduce some standard statistical terminology:

$$\bar{x} = \frac{1}{n} \sum x_i, \qquad \bar{y} = \frac{1}{n} \sum y_i, \qquad \sigma_x^2 = \frac{1}{n} \sum (x_i - \bar{x})^2,$$

$$s_{xy} = \frac{1}{n} \sum (x_i - \bar{x})(y_i - \bar{y}).$$

Then $\bar{x}$ is called the **mean** (average) of $x_1, \cdots, x_n$ and σ_x^2 their **variance**; $\bar{y}$ is the mean of the y_i. Also s_{xy} is called the **covariance** of the x's and y's.

It is an easy exercise to verify the relations

$$\sigma_x^2 = \frac{1}{n} \sum x_i^2 - \bar{x}^2 \qquad \text{and} \qquad s_{xy} = \frac{1}{n} \sum x_i y_i - \bar{x}\bar{y}.$$

Let us return to the linear system for A and B. We divide both equations by n. Then in terms of this new notation, the system becomes

$$\begin{cases} (\sigma_x^2 + \bar{x}^2)A + \bar{x}B = s_{xy} + \bar{x}\bar{y} \\ \bar{x}A + B = \bar{y}. \end{cases}$$

We multiply the second by $\bar{x}$, subtract, and divide by σ_x^2, which is non-zero since the x_i are not all the same:

$$\sigma_x^2 A = s_{xy}, \qquad A = \frac{s_{xy}}{\sigma_x^2}.$$

Since $B = \bar{y} - \bar{x}A$, the desired linear fit $y = Ax + B$ can be written

$$y = Ax + \bar{y} - \bar{x}A, \qquad y - \bar{y} = A(x - \bar{x}),$$

that is,

$$y - \bar{y} = \frac{s_{xy}}{\sigma_x^2}(x - \bar{x}).$$

This is the required least squares fit to the data. In statistics it is called the **regression line** and $A = s_{xy}/\sigma_x^2$ is called the **regression coefficient**.

> **Regression Line**   The best linear fit to the data
>
> $$(x_1, y_1), \quad (x_2, y_2), \cdots, (x_n, y_n)$$
>
> in the sense of least squares is the line
>
> $$y - \bar{y} = \frac{s_{xy}}{\sigma_x^2}(x - \bar{x}).$$

Remark   We should mention that σ_x is the **standard deviation** of the x's, and $\rho_{xy} = s_{xy}/\sigma_x \sigma_y$ is the **correlation coefficient**. Note that the regression line passes through $(\bar{x}, \bar{y})$. Does that seem reasonable?

■ **EXAMPLE 3**   Find the regression line for the data

x	1	2	3	4	5	6	7	8
y	10	9	7	7	6	5	3	2

Solution   Here $n = 8$ and

$$\bar{x} = \frac{1}{8} \sum x_i = \frac{1}{8}(1 + 2 + \cdots + 8) = \frac{9}{2},$$

$$\bar{y} = \frac{1}{8} \sum y_i = \frac{1}{8}(10 + 9 + 7 + \cdots + 2) = \frac{49}{8},$$

$$\sigma_x^2 = \frac{1}{8} \sum x_i^2 - \bar{x}^2 = \frac{1}{8}(1^2 + 2^2 + \cdots + 8^2) - \left(\frac{9}{2}\right)^2 = \frac{1}{8}(204) - \frac{81}{4} = \frac{21}{4},$$

$$s_{xy} = \frac{1}{8} \sum x_i y_i - \bar{x}\bar{y} = \frac{1}{8}(1 \cdot 10 + 2 \cdot 9 + \cdots + 8 \cdot 2) - \left(\frac{9}{2}\right)\left(\frac{49}{8}\right)$$

$$= \frac{1}{8}(174) - \frac{441}{16} = -\frac{93}{16}.$$

The regression coefficient is $A = \dfrac{s_{xy}}{\sigma_x^2} = -\dfrac{31}{28},$

and the least square fit is $y - \dfrac{49}{8} = -\dfrac{31}{28}\left(x - \dfrac{9}{2}\right),$

that is, $y = -\dfrac{31}{28}x + \dfrac{311}{28} \approx -1.107x + 11.12.$

See Fig. 3, p. 770.   ■

The least squares idea can be used in more complicated situations. For example, one might seek the *quadratic* polynomial $y = Ax^2 + Bx + C$ that most closely fits the data

$$(x_1, y_1), \quad (x_2, y_2), \cdots, (x_n, y_n).$$

Then A, B, and C must be found so that

$$F(A, B, C) = \sum_{i=1}^{n} [(Ax_i^2 + Bx_i + C) - y_i]^2$$

is minimized.

Still more complicated is the problem of approximating a *function*, rather than discrete data. For example, suppose $y = x^2$ is to be approximated on the interval

$0 \leq x \leq 1$ by a linear function $y = Ax + B$ in the sense of least squares. Then

$$F(A, B) = \int_0^1 [(Ax + B) - x^2]^2 \, dx$$

must be minimized.

EXERCISES

1 Find the min of $\dfrac{9}{x} + \dfrac{4}{y} + \dfrac{1}{z} + xyz$ for $x > 0, y > 0, z > 0$.

2 Find the min of $x^5 + y^5 + z^5 - 5xyz$ for $x > 0, y > 0, z > 0$.

3 Find the max of $xyz(10 - x^2 - 2y^2 - 3z^2)$ for $x > 0, y > 0, z > 0$.

4 Find the max of $e^{-x^2 - y^2 - z^2}(x + 2y + 4z)$.

5 Find the max of $\dfrac{ax + by + cz}{1 + x^2 + y^2 + z^2}$.

6 Find the max and min of $\dfrac{xyz}{(1 + x^2 + y^2 + z^2)^2}$.

7 Find the max and min of $\dfrac{xyz}{(1 + x^2 + 2y^2 + 3z^2)^2}$.

8 Find the max and min of $\cos x + \cos y + \cos z - \cos(x + y + z)$.

9 A segment of length L is cut into four pieces. What is the largest possible product of their four lengths?

10 Find the min of $Q(x, y, z) = x^2 + 3y^2 + 14z^2 - 2xy + 2yz - 6zx$. Complete squares to prove you really have the min.

Find if the quadratic form is positive definite, negative definite, or neither

11 $x^2 + 4xy + 2y^2$ **12** $9x^2 - 12xy + y^2$ **13** $2x^2 - xy + 3y^2$

14 $7x^2 + 5xy + 4y^2$ **15** $-x^2 + 3xy - 3y^2$ **16** $-5x^2 + 2xy + y^2$

17 $4xy$ **18** $x^2 + 2xy$.

19 Suppose $a > 0$ and $ac - b^2 = 0$. Show that $Q = ax^2 + 2bxy + cy^2$ is **positive semidefinite**, that is, $Q(x, y) \geq 0$ for all (x, y) and $Q(x, y) = 0$ for some $(x, y) \neq (0, 0)$. [*Hint* Complete square.]

20* Suppose $ac - b^2 < 0$. Show that $Q = ax^2 + 2bxy + cy^2$ is **indefinite**, that is, takes on both positive and negative values. [*Hint* Complete squares.]

Find the least squares straight line fit to the data

21 $(0, 0), \quad (1, 1), \quad (2, 3)$ **22** $(1, 0), \quad (2, -1), \quad (3, -4)$

23 $(0, 0), \quad (\tfrac{1}{2}, \tfrac{1}{4}), \quad (1, 1), \quad (\tfrac{3}{2}, \tfrac{9}{4}), \quad (2, 4)$ **24** $(1, 1), \quad (2, \tfrac{1}{2}), \quad (3, \tfrac{1}{3}), \quad (4, \tfrac{1}{4})$

25 $(1, 5.0), \quad (2, 5.3), \quad (3, 5.4), \quad (4, 5.6)$ **26** $(-2, 3.0), \quad (-1, 1.8), \quad (0, 1.1), \quad (1, 0.6)$

27 $(0, 0), \quad (1, -1), \quad (2, -2), \quad (3, -1), \quad (4, 0)$

28 $(-3, 0), \quad (-2, 0), \quad (-1, 0), \quad (0, 1), \quad (1, 2), \quad (2, 2), \quad (3, 2)$

29 $(-1, 6.2), \quad (0, 5.4), \quad (1, 5.5), \quad (2, 5.0)$

30 $(1, -1), \quad (2, 1), \quad (3, -1), \quad (4, 1), \cdots, (9, -1)$.

Approximate $y = f(x)$ on the interval $[0, 1]$ by a linear function $y = Ax + B$ in the sense of least squares, that is, minimize

$$F(A, B) = \int_0^1 [(Ax + B) - f(x)]^2 \, dx$$

31 $f(x) = x^2$ **32** $f(x) = x^3$ **33** $f(x) = e^x$.

34   (cont.) Prove in general that $A = 6(2\beta - \alpha)$ and $B = 2(2\alpha - 3\beta)$, where

$$\alpha = \int_0^1 f(x)\, dx \qquad \text{and} \qquad \beta = \int_0^1 xf(x)\, dx.$$

35   Find the best least squares fit to the data $(1, 7)$, $(2, 4)$, $(3, 3)$ by $y = ax + b/x$.

36   The population of a city (in thousands) was

year	1970	1972	1974	1976	1978
population	100	104	108	113	120

A demographer assumes an exponential curve $y = ae^{bt}$, with $t = 0$ at year 1970, which he obtains by fitting a straight line by least squares to $\ln y$. What growth curve does he find?

37   Verify the relations $\sigma_x^2 = n^{-1} \sum x_i^2 - \bar{x}^2$ and $s_{xy} = n^{-1} \sum x_i y_i - \bar{x}\bar{y}$ given on p. 771.

38*   Show that $f(A, B)$ on p. 770 really has its unique minimum at the solution (A, B) of the system $\partial f/\partial A = 0$, $\partial f/\partial B = 0$. [You may assume that a function of (x, y) of the form (pos. def. quad. form) + (linear) $\to \infty$ as $x^2 + y^2 \to \infty$. See Ex. 28, p. 792.]

39*   (cont.) Express the actually minimum in terms of the statistical functions σ_x^2, etc.

40*   (cont.) Prove that $f_{\min} \geq 0$. Describe the position of the given points that yields minimum 0.

9. IMPLICIT FUNCTIONS

Sometimes a function $y = f(x)$ of *one* variable is specified as a root of an equation

$$F(x, y) = 0,$$

where $F(x, y)$ is a function of *two* variables. In such a case, the equation is said to define an **implicit function** $y = f(x)$. For example, Fig. 1 shows part of the graph of

$$y^6 + y + xy - x = 0.$$

Near the origin, this equation defines y as an implicit function of x. (It is pretty hopeless to express y as an *explicit* function of x.)

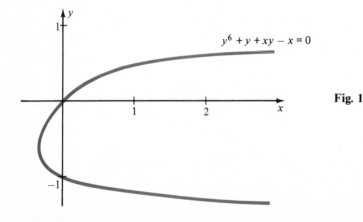

Fig. 1

Suppose we have a function $y = f(x)$ satisfying $F(x, y) = 0$. We want a formula for $dy/dx = f'(x)$ in terms of $F(x, y)$. To find it, we substitute $y = f(x)$ into $F(x, y) = 0$:

$$F[x, f(x)] = 0.$$

We differentiate with respect to x by the Chain Rule:

$$F_x[x, f(x)] + F_y[x, f(x)]f'(x) = 0.$$

Therefore

$$f'(x) = -\frac{F_x(x, f(x))}{F_y(x, f(x))}.$$

Implicit Differentiation Let $y = f(x)$ be a function defined implicitly by a relation

$$F(x, y) = 0.$$

Then

$$\frac{dy}{dx} = f'(x) = -\frac{F_x(x, y)}{F_y(x, y)}\bigg|_{y = f(x)}$$

at each point $(x, y) = (x, f(x))$ where $F_y(x, y) \neq 0$.

Remark The minus sign in the formula seems to contradict common sense. In the ordinary Chain Rule, differentials "cancel":

$$\frac{dy}{dx} = \frac{dy}{du}\frac{du}{dx} \qquad \text{implies} \qquad \frac{dy}{dx}\bigg/\frac{dy}{du} = \frac{du}{dx};$$

the dy appears to have "canceled". But here we are writing

$$\frac{\partial F}{\partial x}\bigg/\frac{\partial F}{\partial y} = -\frac{dy}{dx},$$

and ∂F "cancels" with a mysterious sign change.

The reason for the sign change is that when we write $F(x, y) = 0$, we have "taken y to the other side". The equation $y = f(x)$ is equivalent to

$$F(x, y) = y - f(x) = 0.$$

Now

$$\frac{F_x}{F_y} = \frac{-f'(x)}{1} = -f'(x).$$

There is the minus sign!

■ **EXAMPLE 1** Find $\dfrac{dy}{dx}\bigg|_{(0, 0)}$ where $y = f(x)$ satisfies $y^6 + y + xy - x = 0$.

Solution

$$F(x, y) = y^6 + y + xy - x, \qquad F_x = y - 1, \qquad F_y = 6y^5 + 1 + x.$$

Hence $\dfrac{dy}{dx} = -\dfrac{F_x}{F_y} = -\dfrac{y - 1}{6y^5 + 1 + x}$, so $\dfrac{dy}{dx}\bigg|_{(0, 0)} = -\dfrac{-1}{1} = 1.$

Alternative Solution   Differentiate the equation

$$y^6 + y + xy - x = 0,$$

treating y as a function of x:

$$6y^5 \frac{dy}{dx} + \frac{dy}{dx} + x\frac{dy}{dx} + y - 1 = 0, \qquad \frac{dy}{dx} = -\frac{y-1}{6y^5 + 1 + x}. \qquad ∎$$

Remark   The technique in the alternative solution is equivalent to use of the rule

$$\frac{dy}{dx} = -\frac{F_x}{F_y}$$

because the rule was derived by that very technique.

∎ **EXAMPLE 2**   Let $y = y(x) = \sqrt{1 - x^2}$. Express y' and y'' in terms of x and y by differentiating $x^2 + y^2 - 1 = 0$ implicitly.

Solution   Differentiate $x^2 + y^2 - 1 = 0$ with respect to x:

$$2x + 2yy' = 0, \qquad y' = -\frac{x}{y}.$$

Once again: $y'' = -\dfrac{y - xy'}{y^2} = -\dfrac{y - x\left(-\dfrac{x}{y}\right)}{y^2} = -\dfrac{y^2 + x^2}{y^3} = -\dfrac{1}{y^3}.$ ∎

Existence of Implicit Functions   An advanced theorem, called the Implicit Function Theorem, states that a relation $F(x, y) = 0$ determines a function $y = f(x)$ in the neighborhood of any point where $F_y \neq 0$. More precisely, suppose $F(x, y)$ is continuously differentiable, $F(a, b) = 0$ and $F_y(a, b) \neq 0$. Then there exists a unique function $y = f(x)$, defined and differentiable in a neighborhood of $x = a$, such that $f(a) = b$ and $F(x, f(x)) = 0$. The function $f(x)$ is differentiable and $f'(x) = -F_x[x, f(x)]/F_y[x, f(x)]$.

For example take $F(x, y) = x^2 + y^2 - 25$ and $(a, b) = (4, 3)$. Since $F_y(4, 3) = 6 \neq 0$, there is a function $f(x)$ defined near $x = 4$ for which $f(4) = 3$ and $x^2 + [f(x)]^2 - 25 = 0$. Obviously, $f(x) = \sqrt{25 - x^2}$. Note that $F(4, -3) = 0$ and $F_y(4, -3) = -6 \neq 0$, so there also exists a function $g(x)$ defined near $x = 4$ for which $g(4) = -3$ and $x^2 + [g(x)]^2 - 25 = 0$. Obviously, $g(x) = -\sqrt{25 - x^2}$. Thus $F(x, y) = 0$ defines *two* implicit functions near $x = 4$. To distinguish between them, you need their values at $x = 4$.

As another example, see Fig. 1. The relation $F(x, y) = y^6 + y + xy - x = 0$ determines two implicit functions near $x = 0$. One takes the value 0, the other the value -1 at $x = 0$.

Applications   Implicit differentiation is useful when a function must be maximized or minimized subject to certain restrictions.

∎ **EXAMPLE 3**   A cylindrical container (right circular) is required to have a given volume V. The material on the top and bottom is k times as expensive as the material on the sides. What are the proportions of the most economical container?

Solution   The cost of the container is proportional to

$$C = (\text{area of side}) + k(\text{area of top} + \text{area of bottom}).$$

Let r and h denote the radius and height of the container. Then

$$C = 2\pi rh + k(2\pi r^2).$$

The problem is to minimize C subject to the restriction

$$\pi r^2 h = V, \qquad \text{a constant.}$$

One approach is obvious: solve the last equation for h and substitute into the equation for C. Then C is an explicit function of r which can be minimized.

It is simpler, however, not to make the substitution, but to consider C as a function of r anyway (as if the substitution had been made). Differentiate:

$$\frac{dC}{dr} = 2\pi\left(r\frac{dh}{dr} + h + 2kr\right).$$

Before setting this derivative equal to zero, eliminate dh/dr. Differentiate the equation for V with respect to r:

$$2\pi rh + \pi r^2\frac{dh}{dr} = 0, \qquad \frac{dh}{dr} = -\frac{2h}{r}.$$

Substitute this value of dh/dr into the preceding equation:

$$\frac{dC}{dr} = 2\pi\left[r\left(\frac{-2h}{r}\right) + h + 2kr\right] = 2\pi(2kr - h).$$

From this simpler expression for the derivative it follows that

$$\frac{dC}{dr} = 0 \qquad \text{when} \quad h = 2kr.$$

To verify that C is minimal for $h = 2kr$, examine the sign of dC/dr. Since $\pi r^2 h$ is constant, h is large if r is small and decreases as r increases. Therefore $(2kr - h)$ increases from negative to positive as r increases. Thus dC/dr satisfies a condition for C to have a minimum at $h = 2kr$.   ∎

Remark   The special case $k = 1$ is interesting: all parts of the cylinder are equally expensive. Then the cheapest cylinder is the one with least surface area. Conclusion: of all cylinders with fixed volume, the one with least surface area is the one whose height is twice its radius.

■ **EXAMPLE 4**   Find the greatest distance between the origin and a point of the curve $C\colon x^4 + 4y^4 = 1$.

Solution   By symmetry, it is enough to consider only that part of the curve in the first quadrant. See Fig. 2.

The curve C lies outside of the ellipse $x^2 + 4y^2 = 1$. For if (x, y) is on the ellipse and $x > 0$ and $y > 0$, then $x^2 < 1$ and $y^2 < 1$. Hence

$$x^4 + 4y^4 = x^2 x^2 + 4y^2 y^2 < x^2 + 4y^2 = 1.$$

Thus $x^4 + 4y^4 < 1$, so (x, y) is inside the curve C. Therefore the ellipse lies inside C.

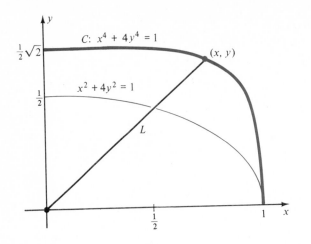

Fig. 2

Let $L(x, y)$ be the distance from $(0, 0)$ to a point (x, y) on C. Figure 2 suggests that the minimum distance is $L(0, 1) = 1/\sqrt{2}$ and that the maximum distance is a bit more than 1.

It suffices to maximize

$$L^2 = x^2 + y^2 \qquad \text{subject to} \quad x^4 + 4y^4 = 1.$$

Differentiate the second equation implicitly:

$$4x^3 + 16y^3 y' = 0, \qquad y' = -\frac{x^3}{4y^3}.$$

Therefore $\quad \dfrac{d}{dx}(L^2) = 2x + 2yy' = 2x + 2y\left(-\dfrac{x^3}{4y^3}\right) = \dfrac{2x(4y^2 - x^2)}{4y^2}.$

This derivative equals 0 in the interior of the first quadrant only for $x = 2y$. Hence the maximum distance from the origin occurs at the point (x, y) of the first quadrant for which

$$x^4 + 4y^4 = 1, \qquad x = 2y.$$

It follows that $\quad (2y)^4 + 4y^4 = 1, \qquad 20y^4 = 1, \qquad y = \dfrac{1}{\sqrt[4]{20}}, \qquad x = \dfrac{2}{\sqrt[4]{20}}.$

Therefore $\qquad (L_{\max})^2 = x^2 + y^2 = \dfrac{5}{\sqrt{20}} = \dfrac{1}{2}\sqrt{5},$

$$L_{\max} = \frac{1}{2}\sqrt{2}\sqrt[4]{5} = \frac{1}{2}\sqrt[4]{20} \approx 1.0574. \qquad\blacksquare$$

EXERCISES

Find explicitly the function $y = f(x)$ defined implicitly by $F(x, y) = 0$

1 $\quad F(x, y) = 2x - y - 6$ **2** $\quad F(x, y) = x^2 y - 2x + 3y$

3 $\quad F(x, y) = 2 - \ln(5x^2 - y)$ **4** $\quad F(x, y) = 1 + 2\,\text{arc}\sin(x - 3y + 7)$

5 $\quad F(x, y) = 4 + 2xy - y^2 \quad$ at $\quad (0, -2)$ **6** $\quad F(x, y) = \sqrt{x} + \sqrt{y} - 3, \; x \geq 0, \; y \geq 0.$

7   If $F(x, y) = 4x^4 - 5x^2y^2 + y^4$, show that $F(x, y) = 0$ defines implicitly four functions of x near $x = 1$. Compute dy/dx at $(1, 2)$ implicitly and check by direct differentiation.

8   If $x^2 + y^2 = r^2$, then $dy/dx = -x/y$. Interpret this statement geometrically.

Compute dy/dx

9	$x + y = x \sin y$	**10**	$x^2 + y^3 = xy$	**11**	$e^{xy} = 3xy^2$
12	$x^4 - y^4 = 3x^2y^3$	**13**	$e^x \sin y = e^y \cos x$	**14**	$y + \sinh y = x^3$
15	$x^4 + 3y^6 = 1$	**16**	$y = \arctan(y/x)$		

Compute d^2y/dx^2

17 $x^3 + y^2 = 1$ **18** $e^y - y = x$ **19** $x^4y^5 = 1$ **20** $xy = x + y$.

Work these maxima and minima problems using implicit differentiation

21   Find the maximum value of y on the ellipse $6x^2 + 3xy + 2y^2 = 1$.

22   Find the greatest distance from the origin to a point of the curve $\dfrac{x^4}{a^4} + \dfrac{y^4}{b^4} = 1$.

23   Find the ratio (height)/(base radius) of the right circular cone of largest volume inscribed in a given sphere.

24   Can a sphere contain a cylinder whose volume is more than 60% the volume of the sphere?

25   A line tangent to the ellipse $x^2/a^2 + y^2/b^2 = 1$ at a point in the first quadrant forms a triangle with the coordinate axes. Find the least possible area of the triangle.

26   Find the maximal volume of a right circular cone of slant height a.

27   A metal container is capsule-shaped, a right circular cylinder with hemispherical caps at both ends. The material for the caps is k times as expensive as the material for the cylindrical part. Find the dimensions of the cheapest container for a given volume V. Assume $k \geq 1$.

28   (cont.) Express the cheapest total cost in terms of k, V, and C, where C is the per unit area cost of the cylindrical part.

29   The output of a certain power plant is proportional to $\sqrt{10xy + 2x^2 + 3y^2}$, where x is the quantity of bituminous coal burned and y is the quantity of anthracite coal burned. If bituminous costs \$50/ton and anthracite \$75/ton, how should the plant spend a \$5,000,000 fuel allotment for maximal output?

30   Suppose $z(x, y)$ is defined implicitly by $F(x, y, z) = 0$. Assuming $F_z(x, y, z) \neq 0$, prove

$$\frac{\partial z}{\partial x} = -\frac{F_x}{F_z}, \qquad \frac{\partial z}{\partial y} = -\frac{F_y}{F_z},$$

where F_x, F_y, F_z are evaluated at $(x, y, z(x, y))$.

Use Ex. 30 to compute $\partial z/\partial x$ and $\partial z/\partial y$

31 $x^2z + y^2z^2 - 4xy - z^3 = 0$ **32** $e^z - e^{xy} - xz^2 = 0$.

33   Find the largest value of z on the surface (ellipsoid) $3x^2 + 2y^2 + z^2 + xz - yz = 1$. Use Ex. 30.

34*   The equations $x + 2y + 3z + 4w = 0$ and $x^2 + y^2 + z^2 + w^2 = 1$ define z and w implicitly as functions of x and y. Compute $\partial z/\partial x$ and $\partial w/\partial y$.

In each case, state whether or not there exists a differentiable function $y = y(x)$ defined near $x = 0$ such that $y(0) = 0$ and $F[x, y(x)] = 0$

35	$F(x, y) = y^2$	**36**	$F(x, y) = (y - x)^3$
37	$F(x, y) = 2y - x^4 - x^3y^5$	**38**	$F(x, y) = (x + 1)y - (y + 1)^3x^2$

39 $F(x, y) = y^2 - 6x^5 + 3x^3y - 2x^2y$
40* $F(x, y) = y^3 - x^2y + x^4$. [*Hint*   Try a power series, or parameterization.]

10. DIFFERENTIALS AND APPROXIMATION

First Order Approximation   If $f(x)$ is a differentiable function of one variable, then its first order approximation at $x = a$ is

$$f(a + t) \approx f(a) + f'(a)t,$$

highly accurate if t is sufficiently small.
   There is a similar result for functions of several variables.

First Order Approximation   Let $f(\mathbf{x})$ have continuous partials near $\mathbf{x} = \mathbf{a}$. Then

$$f(\mathbf{a} + \mathbf{v}) \approx f(\mathbf{a}) + D_\mathbf{v} f(\mathbf{a}) = f(\mathbf{a}) + [\operatorname{grad} f(\mathbf{a})] \cdot \mathbf{v}$$

for $|\mathbf{v}|$ small.

   To see why, write $\mathbf{v} = t\mathbf{u}$, where $\mathbf{u}$ is a unit vector and t is small, and set

$$g(t) = f(\mathbf{a} + \mathbf{v}) = f(\mathbf{a} + t\mathbf{u}).$$

Then since t is small,

$$g(t) \approx g(0) + g'(0)t = f(\mathbf{a}) + g'(0)t.$$

But $g'(0) = D_\mathbf{u} f(\mathbf{a}),$ (definition of directional derivative)

hence $g(t) \approx f(\mathbf{a}) + tD_\mathbf{u} f(\mathbf{a}) = f(\mathbf{a}) + D_\mathbf{v} f(\mathbf{a}),$

and the assertion follows.

■ **EXAMPLE 1**   Given $f(5, 7) = 10$ and grad $f(5, 7) = (2, 3)$, estimate $f(5.01, 6.98)$.

Solution   Set $\mathbf{a} = (5, 7)$ and $\mathbf{v} = (0.01, -0.02)$.

Then $f(5.01, 6.98) = f(\mathbf{a} + \mathbf{v})$
$$\approx f(\mathbf{a}) + [\operatorname{grad} f(\mathbf{a})] \cdot \mathbf{v} = 10 + (2, 3) \cdot (0.01, -0.02)$$
$$= 10 + 0.02 - 0.06 = 9.96. \qquad ■$$

   Later we shall make the meaning of the symbol $\approx$ in first order approximation more precise.

Differentials   In the first order approximation

$$f(\mathbf{a} + \mathbf{v}) \approx f(\mathbf{a}) + [\operatorname{grad} f(\mathbf{a})] \cdot \mathbf{v},$$

suppose we allow *both* $\mathbf{a}$ and $\mathbf{v}$ to vary. It is customary to replace $\mathbf{a}$ by $\mathbf{x}$ and $\mathbf{v}$ by a new quantity

$$d\mathbf{x} = (dx, dy, dz).$$

In this notation, $\qquad f(\mathbf{x} + d\mathbf{x}) \approx f(\mathbf{x}) + [\text{grad } f(\mathbf{x})] \cdot d\mathbf{x}.$

The second term on the right is especially important. We call it the **differential** of f, and denote it by df:

Differential $\quad df = [\text{grad } f(\mathbf{x})] \cdot d\mathbf{x} = \dfrac{\partial f}{\partial x}\, dx + \dfrac{\partial f}{\partial y}\, dy + \dfrac{\partial f}{\partial z}\, dz.$

Examples (1) $\;f(x, y, z) = xy^2z^3 \qquad df = y^2z^3\, dx + 2xyz^3\, dy + 3xy^2z^2\, dz$

(2) $\;f(x, y) = \dfrac{x}{y} \qquad\qquad df = \dfrac{1}{y}\, dx - \dfrac{x}{y^2}\, dy.$

For a function $f(x, y, z)$ of three variables, the differential df is a function of six variables x, y, z, dx, dy, dz. Technically we should use a notation such as $df = df(\mathbf{x}, d\mathbf{x})$, but that is cumbersome.

For each fixed $\mathbf{x}$ in the domain of f, the differential is a linear function of the variables dx, dy, dz, whose domains are unrestricted. For instance, if $f(x, y, z) = xy^2z^3$ as in Example (1), then

$$df = dx + 2\, dy + 3\, dz \qquad\qquad \text{at}\;\; (1, 1, 1),$$
$$df = -27\, dx - 108\, dy + 54\, dz \qquad \text{at}\;\; (2, 1, -3).$$

The differential has elementary algebraic properties, which correspond to analogous results for derivatives:

$d(f + g) = df + dg \qquad\qquad d(af) = a\, df$

$d(fg) = (df)g + f\, dg \qquad\quad d(f/g) = \dfrac{(df)g - f\, dg}{g^2}, \quad g \neq 0.$

For example,

$$d(f + g) = \frac{\partial}{\partial x}\,(f + g)\, dx + \frac{\partial}{\partial y}\,(f + g)\, dy + \frac{\partial}{\partial z}\,(f + g)\, dz$$

$$= \left(\frac{\partial f}{\partial x} + \frac{\partial g}{\partial x}\right) dx + \left(\frac{\partial f}{\partial y} + \frac{\partial g}{\partial y}\right) dy + \left(\frac{\partial f}{\partial z} + \frac{\partial g}{\partial z}\right) dz$$

$$= \left(\frac{\partial f}{\partial x}\, dx + \frac{\partial f}{\partial y}\, dy + \frac{\partial f}{\partial z}\, dz\right) + \left(\frac{\partial g}{\partial x}\, dx + \frac{\partial g}{\partial y}\, dy + \frac{\partial g}{\partial z}\, dz\right)$$

$$= df + dg.$$

Numerical Approximations The variables dx, dy, dz that appear in the differential are free to take all real values. However, in practice, we usually think of them as small changes in x, y, and z respectively. A typical application is in the approximation

$$f(\mathbf{x} + d\mathbf{x}) \approx f(\mathbf{x}) + df.$$

■ **EXAMPLE 2**   Given $f(x, y, z) = \dfrac{xy^2}{1 + z^4}$,   estimate

(a) $f(1.96, 3.02, 0.99)$ (b) $f(2.06, 3, 1.01)$.

Solution   Both parts ask for values of $f(\mathbf{x})$ near $\mathbf{x} = (2, 3, 1)$. Use the approximation $f(\mathbf{x} + d\mathbf{x}) \approx f(\mathbf{x}) + df$ for $\mathbf{x} = (2, 3, 1)$. Now $f(2, 3, 1) = 9$, and at $(2, 3, 1)$

$$df = \frac{y^2}{1 + z^4}\, dx + \frac{2xy}{1 + z^4}\, dy - \frac{4xy^2 z^3}{(1 + z^4)^2}\, dz = \tfrac{9}{2}\, dx + 6\, dy - 18\, dz.$$

(a)   Set $d\mathbf{x} = (-0.04, 0.02, -0.01)$:

$$f(1.96, 3.02, 0.99) \approx f(2, 3, 1) + df$$

$$= 9 + \tfrac{9}{2}(-0.04) + 6(0.02) - 18(-0.01) = 9.12.$$

(b)   Set $d\mathbf{x} = (0.06, 0, 0.01)$:

$$f(2.06, 3, 1.01) \approx 9 + \tfrac{9}{2}(0.06) + 6(0) - 18(0.01) = 9.09. \qquad ■$$

Substitution   The differential has a useful formal property of remaining unchanged when the variables are replaced by functions of other variables. (This property is another form of the Chain Rule.)

For instance, suppose $f = f(x, y, z)$ so

$$df = f_x\, dx + f_y\, dy + f_z\, dz.$$

Now suppose $x = x(u, v)$, $y = y(u, v)$, $z = z(u, v)$. We can look at the composite function $f[x(u, v)]$ and compute its differential, another "df":

$$df = f_u\, du + f_v\, dv.$$

Is this different? No, because by the Chain Rule

$$f_u\, du + f_v\, dv = (f_x x_u + f_y y_u + f_z z_u)\, du + (f_x x_v + f_y y_v + f_z z_v)\, dv$$

$$= f_x(x_u\, du + x_v\, dv) + f_y(y_u\, du + y_v\, dv) + f_z(z_u\, du + z_v\, dv)$$

$$= f_x\, dx + f_y\, dy + f_z\, dz.$$

This may appear as simply a consequence of sloppy notation for composite functions. Still there is something more to it.

Suppose we have a function f of independent variables u, v, but there are some intermediate variables in the way. After some computation we arrive at an expression

(1) $df = M\, du + N\, dv.$

Then we know automatically that $M = f_u$ and $N = f_v$ no matter how we obtained (1).

■ **EXAMPLE 3**   Let $z = z(x, y)$ be defined implicitly by $x^2 + y^2 + z^2 = 1$. Compute $\partial z/\partial x$ and $\partial z/\partial y$.

Solution   Since the differential of a constant function is 0,

$$x\, dx + y\, dy + z\, dz = 0.$$

Hence $dz = -\dfrac{x}{z}\, dx - \dfrac{y}{z}\, dy.$

This is a relation of the form $dz = M\,dx + N\,dy$, from which it follows that $\partial z/\partial x = M$ and $\partial z/\partial y = N$. Hence

$$\frac{\partial z}{\partial x} = -\frac{x}{z} = -\frac{x}{\sqrt{1 - x^2 - y^2}}, \qquad \frac{\partial z}{\partial y} = -\frac{y}{z} = -\frac{y}{\sqrt{1 - x^2 - y^2}}.$$

We can check these answers directly from $z = \pm\sqrt{1 - x^2 - y^2}$. ∎

The technique of Example 3 applies in general. Suppose $z = z(x, y)$ is defined implicitly by $F(x, y, z) = 0$. Then

$$F_x\,dx + F_y\,dy + F_z\,dz = 0, \qquad \text{which implies} \qquad dz = -\frac{F_x}{F_z}\,dx - \frac{F_y}{F_z}\,dy$$

whenever $F_z \neq 0$ at $(x, y, z(x, y))$. Therefore

$$\partial z/\partial x = -F_x/F_z \qquad \text{and} \qquad \partial z/\partial y = -F_y/F_z.$$

Implicit Differentiation   Let $z = z(x, y)$ be defined implicitly by $F(x, y, z) = 0$. Then

$$\frac{\partial z}{\partial x} = -\frac{F_x}{F_z}, \qquad \frac{\partial z}{\partial y} = -\frac{F_y}{F_z}$$

at each point $(x, y, z(x, y))$ where $F_z \neq 0$.

Error in First Order Approximation   We shall now establish a basic property of functions of several variables. For simplicity, we shall confine our proof to two variables, but the same method goes over to three or more.

Suppose $f(\mathbf{x})$ is continuously differentiable near $\mathbf{x} = \mathbf{a}$. Then

$$f(\mathbf{a} + \mathbf{v}) = f(\mathbf{a}) + [\operatorname{grad} f(\mathbf{a})] \cdot \mathbf{v} + e(\mathbf{v}),$$

where the error $e(\mathbf{v})$ satisfies $\dfrac{e(\mathbf{v})}{|\mathbf{v}|} \longrightarrow 0 \qquad$ as $\quad \mathbf{v} \longrightarrow \mathbf{0}$.

The main point is that $e(\mathbf{v})$ is not only small when $\mathbf{v}$ is small, but it is small relative to the size of $\mathbf{v}$. When $\mathbf{v}$ is small, $e(\mathbf{v})$ is very small, and when $\mathbf{v}$ is very small, then $e(\mathbf{v})$ is very very small; the *ratio* $e(\mathbf{v})/|\mathbf{v}|$ can be made as small as we please by taking $\mathbf{v}$ sufficiently small.

Let us restate the result in coordinate notation.

Suppose $f(x, y)$ is continuously differentiable near (a, b). Then

$$f(a + u, b + v) = f(a, b) + f_x(a, b)u + f_y(a, b)v + e(u, v),$$

where $\dfrac{e(u, v)}{\sqrt{u^2 + v^2}} \longrightarrow 0 \qquad$ as $\quad (u, v) \longrightarrow (0, 0)$.

Proof Recall that "continuously differentiable" means that $f_x(x, y)$ and $f_y(x, y)$ exist and are continuous for all (x, y) sufficiently near (a, b).

The proof depends on the old trick of subtracting and adding a term. Write

$$f(a + u, b + v) - f(a, b) = [f(a + u, b + v) - f(a, b + v)] + [f(a, b + v) - f(a, b)].$$

Apply the Mean Value Theorem twice:

$$\begin{cases} f(a + u, b + v) - f(a, b + v) = uf_x(a + \theta u, b + v) \\ \quad\quad f(a, b + v) - f(a, b) = vf_y(a, b + \lambda v). \end{cases}$$

where $0 < \theta < 1$ and $0 < \lambda < 1$. Therefore

$$f(a + u, b + v) = f(a, b) + uf_x(a + \theta u, b + v) + vf_y(a, b + \lambda v).$$

Now define $e(u, v)$ by

$$f(a + u, b + v) = f(a, b) + uf_x(a, b) + vf_y(a, b) + e(u, v)$$

Then

$$e(u, v) = ug(u, v) + vh(u, v),$$

where

$$\begin{cases} g(u, v) = f_x(a + \theta u, b + v) - f_x(a, b) \\ h(u, v) = f_y(a, b + \lambda v) - f_y(a, b). \end{cases}$$

By the triangle inequality,

$$\left| \frac{e(u, v)}{\sqrt{u^2 + v^2}} \right| = \left| \frac{u}{\sqrt{u^2 + v^2}} g(u, v) + \frac{v}{\sqrt{u^2 + v^2}} h(u, v) \right|$$

$$\leq \frac{|u|}{\sqrt{u^2 + v^2}} |g(u, v)| + \frac{|v|}{\sqrt{u^2 + v^2}} |h(u, v)| \leq |g(u, v)| + |h(u, v)|$$

since $|u| \leq \sqrt{u^2 + v^2}$ and $|v| \leq \sqrt{u^2 + v^2}$.

Now suppose $(u, v) \longrightarrow (0, 0)$. Then

$$(a + \theta u, b + v) \longrightarrow (a, b) \quad \text{and} \quad (a, b + \lambda v) \longrightarrow (a, b).$$

Therefore

$$g(u, v) \longrightarrow 0 \quad \text{and} \quad h(u, v) \longrightarrow 0$$

because $f_x(x, y)$ and $f_y(x, y)$ are continuous at (a, b). It follows that

$$\frac{e(u, v)}{\sqrt{u^2 + v^2}} \longrightarrow 0 \quad \text{as} \quad (u, v) \longrightarrow (0, 0).$$

This completes the proof.

We shall have more to say about this result in the next section.

EXERCISES

Compute df at $(x, y) = (2, 1)$ and at $(1, -3)$

1   $f(x, y) = x^3 y$ 2 $f(x, y) = \sqrt{x^2 + y^2}$.

Compute df in general

3   $f(x, y) = 3x^2 y - xy^2$ 4 $f(x, y) = xe^{xy}$ 5 $f(x, y) = \ln(x^2 + 3y^2)$

6   $f(x, y) = \sin(x + 2y)$ 7 $f(x, y, z) = \dfrac{x}{y^2 z}$ 8 $f(x, y, z) = \dfrac{xyz}{1 + y^2}$

9 $f(x, y, z) = e^x \cos y + e^y \cos z$

10 $f(x, y, z) = x^2 - y^2 - 5z^2 + xy + yz + 3x.$

Prove

11 $d(fg) = (df)g + f \, dg$

12 $d(f/g) = \dfrac{(df)g - f \, dg}{g^2}, \quad g \neq 0.$

13   If $x = r \cos \theta$ and $y = r \sin \theta$, prove that $x \, dx + y \, dy = r \, dr$ and $x \, dy - y \, dx = r^2 \, d\theta.$

14   Do Ex. 13 by computing $d(r^2)$ and $d\theta$ from $r^2 = x^2 + y^2$ and $\theta = \arctan(y/x).$

Estimate using differentials

15 $5.1 \times 7.1 \times 9.9$ **16** $\sqrt{(5.99)^2 + (8.03)^2}$

17 $(2.01)^{0.98}$ **18** $e^{-0.1} \tan(.24\pi)$

19   the distance from the origin to $(3.05, 4.02, 11.96)$

20   the percentage increase in the volume V of a rectangular box caused by a 1% increase in each of its dimensions. [*Hint*   Look at dV/V.]

21   The intensity I of illumination at a point of space due to a point source of light is proportional to the power of the source and inversely proportional to the square of the distance from the source. Estimate the percentage change in I caused by 1% increases in both the power of the source and in the distance from the source. [*Hint*   Look at dI/I.]

22   The sides of a triangle are 20, 30, and 40 cm. Estimate the change in the largest angle if each side is shortened by 1 cm.

23   The range of a shell fixed at angle α with the ground is $x = (v^2 \sin 2\alpha)/g$, where v is the muzzle velocity and g is the gravitational constant. If $v = 300 \text{ m/sec}$ and $g = 10 \text{ m/sec}^2$, the shell will hit a target at distance $4500\sqrt{3} \approx 7794 \text{ m}$ if fixed at a $30°$ angle. Suppose, in an actual test, $v = 297.0$, $g = 10.1$, and $\alpha = 30.2°$. Estimate how far the shell will miss the target.

24   (cont.) Show that $\dfrac{dx}{x} = 2\dfrac{dv}{v} + (2\alpha \cot 2\alpha)\dfrac{d\alpha}{\alpha} - \dfrac{dg}{g}.$ Discuss the relative importance of a 1% error in α as compared to a 1% error in v for $\alpha = 30°$ and $45°$.

Compute dz implicitly and check explicitly

25 $x^2yz = 1$ **26** $a^2x^2 + b^2y^2 + c^2z^2 = 10.$

Compute dz implicitly

27 $z^5 + 6xz^2 + 8y^2z = 3$ **28** $z^3 = e^{-xyz}.$

Find $f(x, y)$ if

29 $df = 3(x^2 + y) \, dx + (3x - 2) \, dy, \quad f(0, 0) = 4$

30 $df = -\dfrac{y}{x^2 + y^2} \, dx + \dfrac{x}{x^2 + y^2} \, dy, \quad f(1, 1) = 0.$

31*   An investment fund keeps equal amounts of capital in three types of account, A, B, C, returning respectively 6%, 7%, and 10% annually compounded continuously. Due to a recession, these interest rates are cut by $\frac{1}{2}\%$, $\frac{1}{2}\%$, and $\frac{1}{4}\%$. If the fund has $\$10^7$ in each account before the recession, estimate how much more capital it must invest to maintain the same total earnings.

32   Suppose $f(x, y, z) = F[g(x, y, z)].$ Give a formula for df.

11. DIFFERENTIABLE FUNCTIONS

It may be said that differential calculus is the study of functions that have a linear approximation at each point of their domain. We shall extend this point of view to functions of several variables, but first let us review the one variable situation.

Suppose $f(x)$ has derivative $f'(c)$ at $x = a$. Then for x near a, the approximation $f(x) \approx f(a) + f'(a)(x - a)$ is quite accurate. The term "accurate" is not well defined; let us express what we mean geometrically. We know that $y = f(a) + f'(a)(x - a)$ is the equation of the tangent line to the graph of $y = f(x)$ at $(a, f(a))$. Geometrically, the tangent "hugs" the curve near the point of tangency. Therefore if $e(x)$ is the vertical distance between the curve and the tangent line, $e(x)$ ought to be small compared to $|x - a|$. In fact, the closer x is to a, the smaller the ratio $e(x)/|x - a|$ should be.

Let us now state these ideas more precisely. It is convenient to set $v = x - a$.

Suppose $f(x)$ is differentiable at $x = a$. Set $f(a + v) = f(a) + f'(a)v + e(v)$.

$$\text{Then} \qquad \frac{e(v)}{v} \longrightarrow 0 \qquad \text{as} \quad v \longrightarrow 0.$$

Example $f(x) = x^2$, $\quad a = 3$

$$(3 + v)^2 = 9 + 6v + v^2, \qquad \text{that is,} \quad f(3 + v) = f(3) + f'(3)v + v^2,$$

and

$$\frac{e(v)}{v} = \frac{v^2}{v} = v \longrightarrow 0 \qquad \text{as} \quad v \longrightarrow 0.$$

The relation $e(v)/v \longrightarrow 0$ actually characterizes differentiability. For suppose $f(x)$ satisfies

$$f(a + v) = f(a) + kv + e(v),$$

where k is a constant and $e(v)/v \longrightarrow 0$ as $v \longrightarrow 0$. Then

$$\frac{f(a + v) - f(v)}{v} = k + \frac{e(v)}{v} \longrightarrow k \qquad \text{as} \quad v \longrightarrow 0.$$

This proves the following:

Suppose $f(x)$ is defined near $x = a$ and satisfies

$$f(a + v) = f(a) + kv + e(v),$$

where k is a constant and $e(v)/v \longrightarrow 0$ as $v \longrightarrow 0$. Then $f(x)$ is differentiable at $x = a$ and $f'(a) = k$.

The relation $e(v)/v \longrightarrow 0$ means that the linear function $f(a) + kv$ closely approximates $f(a + v)$, and as $v \longrightarrow 0$: the error $e(v) = f(a + v) - [f(a) + kv]$ approaches 0 *faster* than v approaches 0.

For a function of *one* variable, the relation between derivative and linear approximation is clear: if the derivative exists, the linear approximation exists; if the linear approximation exists, the derivative exists.

For a function of *several* variables, the situation is not at all like this. It can happen that $\partial f/\partial x$ and $\partial f/\partial y$ both exist at a point (a, b), but no linear approximation $c + hx + ky$ exists!

■ **EXAMPLE 1**   Let $f(0, 0) = 0$ and

$$f(x, y) = \frac{xy(x + y)}{x^2 + y^2} \qquad \text{if} \quad (x, y) \neq (0, 0).$$

Show   (a) f is continuous at $(0, 0)$,   (b) $f_x(0, 0)$ and $f_y(0, 0)$ exist,
       (c) f does not have a linear approximation at $(0, 0)$.

Solution   (a)   From $(x \pm y)^2 = x^2 + y^2 \pm 2xy \geq 0$ we have $2|xy| \leq x^2 + y^2$. Therefore if $(x, y) \neq (0, 0)$,

$$|f(x, y)| = \frac{|xy|}{x^2 + y^2}|x + y| \leq \frac{1}{2}|x + y|,$$

so $f(x, y) \longrightarrow 0 = f(0, 0)$ as $(x, y) \longrightarrow (0, 0)$. This says that f is continuous at $(0, 0)$.

(b) $\dfrac{f(x, 0) - f(0, 0)}{x} = 0 \longrightarrow 0 \qquad \text{as} \quad x \longrightarrow 0.$

Hence $f_x(0, 0)$ exists, and $f_x(0, 0) = 0$. Similarly, $f_y(0, 0) = 0$.

(c)   Since $f(x, y) = 0$ on the x-axis and on the y-axis, the only possible linear approximation is 0. If 0 is truly a linear approximation to $f(x, y)$, then as (x, y) approaches $(0, 0)$ the quantity $f(x, y) - 0$ must approach 0 *faster* than (x, y) approaches $(0, 0)$. But along the line $x = y$,

$$f(x, x) - 0 = f(x, x) = \frac{x^2(x + x)}{x^2 + x^2} = x,$$

which obviously approaches 0 at the same rate as (x, x) approaches $(0, 0)$. Hence, $f(x, y)$ does *not* have a linear approximation at $(0, 0)$.   ■

Differentiable Functions of Two Variables   Example 1 shows that the existence of both partials of $f(x, y)$ at a point does not guarantee that $f(x, y)$ is a reasonably behaved function. Now we shall examine the most useful functions of two variables, those that have good linear approximations.

A function f is **differentiable** at (a, b) if there exists a linear function $hx + ky$ such that

$$f(a + x, b + y) = f(a, b) + hx + ky + e(x, y),$$

where $\dfrac{e(x, y)}{\sqrt{x^2 + y^2}} \longrightarrow 0 \qquad \text{as} \quad (x, y) \longrightarrow (0, 0).$

First we shall show that differentiability implies both continuity and the existence of partials.

Theorem   If f is differentiable at (a, b), then f is continuous at (a, b).

Proof   Let $f(a + x, b + y) = f(a, b) + hx + ky + e(x, y)$ as in the definition. If $(x, y) \longrightarrow (0, 0)$, then $x \longrightarrow 0$, $y \longrightarrow 0$, and $e(x, y) \longrightarrow 0$. Consequently $f(a + x, b + y) \longrightarrow f(a, b)$. This is continuity.

Theorem   Let f be differentiable at an *interior* point (a, b) of its domain, and let

$$f(a + x, b + y) = f(a, b) + hx + ky + e(x, y),$$

where $e(x, y)/|(x, y)| \longrightarrow 0$ as $(x, y) \longrightarrow (0, 0)$. Then the first partials of f exist at (a, b) and

$$\frac{\partial f}{\partial x}(a, b) = h, \qquad \frac{\partial f}{\partial y}(a, b) = k.$$

Proof   Here (a, b) is an **interior** point of the domain of f means that $(a + x, b + y)$ is in the domain of f whenever x and y are both small. Now we have

$$\frac{\partial f}{\partial x}(a, b) = \lim_{x \to 0} \frac{f(a + x, b) - f(a, b)}{x}$$

$$= \lim_{x \to 0} \frac{hx + e(x, 0)}{x} = h + \lim_{x \to 0} \frac{e(x, 0)}{x} = h.$$

Similarly, $f_y(a, b) = k$.

A Test for Differentiability   The existence of partials $\partial f/\partial x$ and $\partial f/\partial y$ does not necessarily mean that f is differentiable. However, if the partials are *continuous*, then f is indeed differentiable. This assertion is equivalent to the statement proved on p. 784. It provides a practical test for differentiability.

Theorem   Suppose f has partials $\partial f/\partial x$ and $\partial f/\partial y$ at each point of its domain, and (a, b) is an interior point of the domain. Assume $\partial f/\partial x$ and $\partial f/\partial y$ are continuous at (a, b). Then $f(x, y)$ is differentiable at (a, b).

Remark   We actually assumed more on p. 783: that the partials are continuous on the whole domain. But the proof only uses their continuity at (a, b).

The Chain Rule   Now we are in a position to give a precise statement and proof of the Chain Rule. This is important because the Chain Rule is the backbone of several-variable differential calculus. For simplicity we shall limit the discussion to two variables.

Chain Rule Let $f(x, y)$ be differentiable at $(x, y) = (a, b)$. Suppose $x = x(t)$ and $y = y(t)$ are differentiable at $t = c$ and $(x(c), y(c)) = (a, b)$. Then the function

$$z(t) = f[x(t), y(t)]$$

is differentiable at $t = c$ and

$$\dot{z}(c) = f_x(a, b)\dot{x}(c) + f_y(a, b)\dot{y}(c).$$

Proof Since $f(x, y)$ is differentiable at (a, b),

$$f(a + u, b + v) = f(a, b) + hu + kv + e(u, v),$$

where $h = f_x(a, b),$ $k = f_y(a, h),$ and $\dfrac{e(u, v)}{\sqrt{u^2 + v^2}} \longrightarrow 0$

as $(u, v) \longrightarrow (0, 0)$. Substitute

$$u = x(c + t) - a,   v = y(c + t) - b$$

into this relation:

$$f[x(c + t), y(c + t)] = f(a, b) + h \cdot [x(c + t) - a] + k \cdot [y(c + t) - b] + e(u, v).$$

We form the difference quotient

$$\frac{f[x(c + t), y(c + t)] - f(a, b)}{t} = h\,\frac{x(c + t) - a}{t} + k\,\frac{y(c + t) - b}{t} + \frac{e(u, v)}{t},$$

and see what happens as $t \longrightarrow 0$. Certainly $u \longrightarrow 0$, $v \longrightarrow 0$,

$$\frac{x(c + t) - a}{t} \longrightarrow \dot{x}(c),   \text{and}   \frac{y(c + t) - b}{t} \longrightarrow \dot{y}(c),$$

since $a = x(c)$ and $b = y(c)$. Therefore

$$\lim_{t \to 0} \frac{z(c + t) - z(c)}{t} = \lim_{t \to 0} \frac{f[x(c + t), y(c + t)] - f(a, b)}{t} = h\dot{x}(c) + k\dot{y}(c) + \lim_{t \to 0} \frac{e(u, v)}{t}.$$

The proof will be complete when we show that $e(u, v)/t \longrightarrow 0$ as $t \longrightarrow 0$. We may write

$$e(u, v) = \sqrt{u^2 + v^2}\, E(u, v),$$

where $E(u, v) \longrightarrow 0$ as $(u, v) \longrightarrow (0, 0)$.

$$\left| \frac{\sqrt{u^2 + v^2}}{t} \right| = \sqrt{\left(\frac{x(c + t) - a}{t} \right)^2 + \left(\frac{y(c + t) - b}{t} \right)^2} \longrightarrow \sqrt{\dot{x}(c)^2 + \dot{y}(c)^2}.$$

Therefore $\sqrt{u^2 + v^2}/t$ is bounded. But $E(u, v) \longrightarrow 0$, so

$$\frac{e(u, v)}{t} = \frac{\sqrt{u^2 + v^2}}{t}\, E(u, v) \longrightarrow 0$$

as $t \longrightarrow 0$. This completes the proof.

Maxima and Minima   One of the main concerns of calculus is maximum and minimum values of functions. Recall one of the basic facts about continuous functions of one variable:

If f is continuous on a closed interval $[a, b]$, then there exist points x_0 and x_1 in the interval such that

$$f(x_0) \leq f(x) \leq f(x_1)$$

for all $x \in [a, b]$.

The result says that

$$f(x_0) = \min\{f(x)\,|\,a \leq x \leq b\}, \qquad f(x_1) = \max\{f(x)\,|\,a \leq x \leq b\}.$$

If the interval is not closed, then f need not have a maximum or a minimum. For example, $f(x) = x$ has neither a maximum nor a minimum on the *open* interval $a < x < b$. The same holds for any continuous increasing or decreasing function.

Furthermore, the result is not true on a domain which is unbounded, that is, contains points arbitrarily far from the origin. For example, on the domain $0 \leq x < \infty$ the function $f(x) = e^{-x}$ has a maximum but no minimum; on $0 < x < \infty$ it has neither.

The correct generalization of the preceding theorem requires a domain that is both closed and bounded.

Definition   A domain **D** in space is **bounded** if there is a number B such that $|\mathbf{x}| \leq B$ for all **x** in **D**. In other words **D** is obtained in some sphere centered at the origin.

It is not obvious what is the proper definition of "closed" in space, to generalize a closed interval on the line in some sense. The following turns out to be satisfactory.

Definition   A domain **D** in space is **closed** if whenever the points $\mathbf{x}_n$ are in **D** and $\mathbf{x}_n \longrightarrow \mathbf{x}$, then **x** is in **D**.

Intuitively, closed means "includes its boundary points".

Now we can state the basic existence property of maxima and minima.

Existence of Maxima and Minima   Let $f(\mathbf{x})$ be a continuous function on a bounded, closed domain **D**. Then there exist points $\mathbf{x}_0$ and $\mathbf{x}_1$ in **D** such that

$$f(\mathbf{x}_0) \leq f(\mathbf{x}) \leq f(\mathbf{x}_1) \qquad \text{for all} \quad \mathbf{x} \text{ in } \mathbf{D}.$$

This result justifies all of our work on max and min. We leave its proof to more advanced courses.

EXERCISES

Show that $f(x, y)$ is differentiable by finding a linear function $hx + ky$ and a function $e(x, y)$ that satisfy the definition of differentiability. [Use the estimates $|x| \le |\mathbf{x}|$, $|y| \le |\mathbf{x}|$ as needed to prove $e(\mathbf{x})/|\mathbf{x}| \longrightarrow 0$.]

1 $3x - 7y + 4$ at (a, b) **2** $x^2 + y^2$ at $(-2, 1)$ **3** xy^2 at $(0, 0)$

4 xy^2 at $(3, 2)$ **5** y/x at $(1, 1)$ **6** $1/xy$ at $(-1, 2)$.

7 Prove that the sum of two differentiable functions is differentiable.

8 Prove that the product of two differentiable functions is differentiable.

Given $f(x, y) = 0$, determine (a) whether $f(x, y)$ is continuous at $(0, 0)$, (b) whether $f_x(0, 0)$ and $f_y(0, 0)$ exist and their values if they do, (c) whether $f(x, y)$ is differentiable at $(0, 0)$

9 $f(x, y) = xy/(x^2 + y^2)$ if $(x, y) \ne (0, 0)$

10 $f(x, y) = x^3/(x^2 + y^2)$ if $(x, y) \ne (0, 0)$

11 $f(x, y) = (x^6 + y^6)/(x^4 + y^4)$ if $(x, y) \ne (0, 0)$

12 $f(x, y) = (x^5 + y^6)/(x^4 + y^4)$ if $(x, y) \ne (0, 0)$.

13 Prove that $f(x, y) = x^2 - 6xy + 10y^2$ has a positive minimum value p on the circle $x^2 + y^2 = 1$.

14 If $f(x, y) = ax^2 + bxy + cy^2$ has a positive minimum p on $x^2 + y^2 = 1$, prove that $f(x, y) > 0$ for all (x, y) except $(0, 0)$. [*Hint* Find the minimum of $f(x, y)$ on $x^2 + y^2 = r^2$.]

12. MISCELLANEOUS EXERCISES

A function $w = f(x, y, z)$ is **homogeneous of degree n** if $f(tx, ty, tz) = t^n f(x, y, z)$ for all $t > 0$. The condition of homogeneity can be written vectorially:

$$f(t\mathbf{x}) = t^n f(\mathbf{x}).$$

Show that the function is homogeneous: What degree?

1 $x^2 + yz$ **2** $x - y + 2z$ **3** $x^3 + y^3 + z^3 - 3xyz$

4 $x^2 e^{-y/z}$ **5** $\dfrac{xyz}{x^4 + y^4 + z^4}$ **6** $\dfrac{1}{x + y}$.

7 Suppose f and g are homogeneous of degree m and n respectively. Show that fg is homogeneous of degree mn.

8 Let $f(x, y, z)$ be homogeneous of degree n. Show that f_x is homogeneous of degree $n - 1$. (Exception: $n = 0$ and f constant.)

9 Let $f(x, y, z)$ be homogeneous of degree n. Prove **Euler's Relation:** $xf_x + yf_y + zf_z = nf$. [*Hint* Differentiate $f(tx, ty, tz) = t^n f(x, y, z)$ with respect to t, using the Chain Rule; then set $t = 1$.]

10* (cont.) (Converse of Euler's Relation) Let $f(\mathbf{x})$ be differentiable for $\mathbf{x} \ne \mathbf{0}$, and suppose $\mathbf{x} \cdot \operatorname{grad} f = nf$. Prove f is homogeneous of degree n. [*Hint* Show that $\partial[t^{-n} f(t\mathbf{x})]/\partial t = 0$.]

11 Let $\mathbf{a} = (a, b, c)$ be a point not on the paraboloid $z = x^2 + y^2$, that is, such that $c \ne a^2 + b^2$. Show that the set of all points $\mathbf{x}$ on the paraboloid such that the tangent plane at $\mathbf{x}$ passes through $\mathbf{a}$ lies on a plane.

12 (cont.) Assume $c < a^2 + b^2$ so $\mathbf{a}$ is *outside* of the paraboloid. Show that the intersection of the plane found in Ex. 11 and the paraboloid is also the intersection of the paraboloid with a right circular cylinder.

13 Let $\mathbf{a}$ be a point of the hyperbolic paraboloid $z = x^2 - y^2$. Show that the tangent plane at $\mathbf{a}$ intersects the surface in two straight lines. [*Hint* Eliminate z from the equations of the quadric and the tangent plane.]

14   (cont.) Show the same thing for the hyperboloid of one sheet $z^2 = x^2 + y^2 - 1$.

15   Find the ratio (height)/(radius) of a right circular cone of maximal volume with given lateral surface area.

16   Show that the plane tangent to the surface $xyz = k$ at any point forms with the coordinate planes a tetrahedron of fixed volume.

17   Find the polynomial $p(x) = x^2 + ax + b$ that minimizes $\int_{-1}^{1} p(x)^2\, dx$.

18   (cont.) The same for $p(x) = x^3 + ax^2 + bx + c$.

19   Let $z = z(x, y)$ be the larger solution of $z^2 + 2xz + y = 0$, where $x^2 - y > 0$. Find dz in terms of x, y, dx, dy only.

20*   Show that the function $z = z(x, y)$ defined implicitly by

$$\left|\begin{array}{l} z\phi'(t) = [y - \phi(t)]^2 \\ (x + t)\phi'(t) = y - \phi(t) \end{array}\right.$$

(eliminate t) satisfies $z = \dfrac{\partial z}{\partial x}\dfrac{\partial z}{\partial y}$.   [*Hint*   Use differentials and eliminate dt.]

21   Find the general solution of $\dfrac{\partial z}{\partial x} = \dfrac{\partial z}{\partial y}$.   [*Hint*   Set $u = x + y$ and $v = x - y$.]

22*   Given $f(x, y, z)$ and $\mathbf{a} = (a, b, c)$ such that $f(\mathbf{a}) = 0, f_x(\mathbf{a}) \neq 0, f_y(\mathbf{a}) \neq 0, f_z(\mathbf{a}) \neq 0$. Then $f(x, y, z) = 0$ can be solved for either x, y, or z as a function of the other two. Thus $x = P(y, z)$, where $a = P(b, c)$ and $f[P(y, z), y, z] = 0$. Let $y = Q(z, x)$ and $z = R(x, y)$ be the other two implicit functions. Find the product $P_y Q_z R_x$.

23   Let $y = y(x)$ be a curve in the x, y-plane with $y'' \neq 0$. Then $(u, v) = (dy/dx, x\, dy/dx - y)$ is a curve in the u, v-plane. Express its slope simply in terms of x and y.

24*   Find the most general differentiable function $f(\mathbf{x})$ with domain $\mathbf{R}^3$ such that $f(\mathbf{u} + \mathbf{v}) = f(\mathbf{u}) + f(\mathbf{v})$ for all $\mathbf{u}$ and $\mathbf{v}$. [*Hint*   Prove grad $f(\mathbf{x}) = \mathbf{c}$.]

25   The line $x = -1$ is parameterized by $\mathbf{x} = (-1, \ln u)$. Similarly $x = 0$ and $x = 1$ are parameterized by $\mathbf{x} = (0, \frac{1}{2}\ln w)$ and $\mathbf{x} = (1, \ln v)$. Now a ruler is placed connecting any point $(-1, \ln u)$ of $x = -1$ with any point $(1, \ln v)$ of $x = 1$. The ruler crosses the y-axis at $(0, \frac{1}{2}\ln w)$. Express w as a function of u and v. (Write in the u, v, and w scales; the result is an example of an **alignment chart**, a useful graphical device for quick estimates of functions of two or more variables.)

26   (cont.) Construct an alignment chart for $w = \sqrt{u^2 + 2v^2}$.

27   Let $Q(x, y) = ax^2 + 2bxy + cy^2$ be an indefinite quadratic form, that is, $ac - b^2 < 0$. (See Ex. 20, p. 773.) Show that Q can be expressed as the product of two linear forms. [*Hint*   Solve $at^2 + 2bt + c = 0$.]

28*   Let $f(x, y) = ax^2 + 2bxy + cy^2 + px + qy + r$, where $ax^2 + 2bxy + cy^2$ is positive definite. Prove that $f(x, y) \longrightarrow \infty$ as $|\mathbf{x}| \longrightarrow \infty$. [*Hint*   First prove that $ax^2 + 2bxy + cy^2 \geq m|\mathbf{x}|^2$, where $m > 0$.]

The analogue of **Newton's method** for solving a system $f(x, y) = 0$, $g(x, y) = 0$ is an iterative scheme that replaces an approximation $\mathbf{x}$ to a solution by $\phi(\mathbf{x})$, a closer approximation. The formula is

$$\phi(\mathbf{x}) = \mathbf{x} - \frac{1}{\begin{vmatrix} f_x & f_y \\ g_x & g_y \end{vmatrix}}\left(\begin{vmatrix} f & f_y \\ g & g_y \end{vmatrix}, \begin{vmatrix} f_x & f \\ g_x & g \end{vmatrix}\right),$$

all functions evaluated at $\mathbf{x}$. (The denominator must be non-zero near the solution.) Given an initial "guess" $\mathbf{x}_0$, we construct $\mathbf{x}_1 = \phi(\mathbf{x}_0)$, $\mathbf{x}_2 = \phi(\mathbf{x}_1)$, $\mathbf{x}_3 = \phi(\mathbf{x}_2), \cdots$. Under reasonable conditions the sequence $\{\mathbf{x}_n\}$ converges to a solution of the system $f(\mathbf{x}) = 0$, $g(\mathbf{x}) = 0$.

Test this: in each case compute $\mathbf{x}_1$ and $\mathbf{x}_2$

29 $f(x, y) = x + y^2$, $g(x, y) = y + x^2$, $\mathbf{x}_0 = (\frac{1}{2}, 0)$
30 $f(x, y) = x + y^2$, $g(x, y) = x + y$, $\mathbf{x}_0 = (1, \frac{1}{4})$
31 $f(x, y) = x^2 + y^2 - 1$, $g(x, y) = x + 4y - 2$, $\mathbf{x}_0 = (1, 1)$.

32*   Use the definition of differentiability, with the error ignored, to derive Newton's method.
33   Consider the cone $x^2 + y^2 = z^2 \tan^2 \alpha$ and the cylinder $(z - a \csc \alpha)^2 + y^2 = a^2$. where $a > 0$ and $0 < \alpha < \frac{1}{2}\pi$. Find the two points of tangency of the two surfaces.
34*   (cont.) Show that the intersection of the two surfaces consists of a pair of ellipses.

Higher Partials and Applications 16

1. MIXED PARTIALS

A function of two variables $f(x, y)$ has two first partial derivatives,

$$f_x(x, y) \quad \text{and} \quad f_y(x, y),$$

each itself a function of two variables. Each in turn has two first partial derivatives; these four new functions are the second derivatives of $f(x, y)$. Figure 1 shows their evolution:

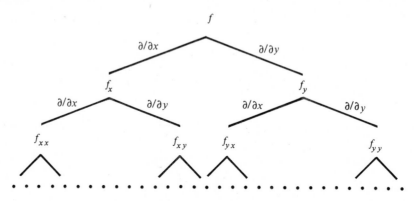

Fig. 1 Family tree of partial derivatives

The **pure** second partials f_{xx} and f_{yy} represent nothing really new. Each is found by holding one variable constant and differentiating twice with respect to the other variable.

Alternative notation:

$$f_{xx} = \frac{\partial^2 f}{\partial x^2}, \qquad f_{yy} = \frac{\partial^2 f}{\partial y^2}.$$

For example, if $f(x, y) = x^3 y^4 + \cos 5y$, then

$$f_x = 3x^2 y^4, \qquad f_y = 4x^3 y^3 - 5 \sin 5y,$$
$$f_{xx} = 6xy^4, \qquad f_{yy} = 12x^3 y^2 - 25 \cos 5y.$$

The **mixed** second partials

$$f_{xy} = \frac{\partial}{\partial y}\left(\frac{\partial f}{\partial x}\right) = \frac{\partial^2 f}{\partial y\, \partial x} \quad \text{and} \quad f_{yx} = \frac{\partial}{\partial x}\left(\frac{\partial f}{\partial y}\right) = \frac{\partial^2 f}{\partial x\, \partial y}$$

are new. The mixed partial f_{xy} measures the rate of change in the y-direction of the rate of change of f in the x-direction. The other mixed partial f_{yx} measures the rate of change in the x-direction of the rate of change of f in the y-direction. It is not easy to see how, if at all, the two mixed partials are related to each other.

Let us compute the mixed partials of the function $f(x, y) = x^3y^4 + \cos 5y$:

$$f_x = 3x^2y^4, \qquad\qquad f_{xy} = 3 \cdot 4x^2y^3,$$
$$f_y = 4x^3y^3 - 5\sin 5y, \qquad f_{yx} = 4 \cdot 3x^2y^3 + 0.$$

The mixed partials are equal! This is not an accident but a special case of a general phenomenon, true for functions normally encountered in applications.

Equality of Mixed Partials Let $f(x, y)$ be defined on a domain **D**. If f_{xy} and f_{yx} exist at each point of **D** and are continuous at (a, b), then

$$\frac{\partial^2 f}{\partial x\, \partial y} = \frac{\partial^2 f}{\partial y\, \partial x} \qquad \text{at} \quad (a, b).$$

We postpone the proof until Section 6.

Higher Partials The next discussion assumes continuity of all the partials involved.

A function $z = f(x, y)$ has two distinct first partials,

$$\frac{\partial z}{\partial x} \quad \text{and} \quad \frac{\partial z}{\partial y},$$

and three distinct second partials,

$$\frac{\partial^2 z}{\partial x^2}, \quad \frac{\partial^2 z}{\partial x\, \partial y}, \quad \frac{\partial^2 z}{\partial y^2}.$$

Because the mixed second partials are equal, so are certain mixed third partials. For example,

$$\frac{\partial^2}{\partial x^2}\left(\frac{\partial z}{\partial y}\right) = \frac{\partial}{\partial y}\left(\frac{\partial^2 z}{\partial x^2}\right)$$

since

$$\frac{\partial^2}{\partial x^2}\left(\frac{\partial z}{\partial y}\right) = \frac{\partial}{\partial x}\left[\frac{\partial}{\partial x}\left(\frac{\partial z}{\partial y}\right)\right] = \frac{\partial}{\partial x}\left[\frac{\partial}{\partial y}\left(\frac{\partial z}{\partial x}\right)\right] = \frac{\partial}{\partial y}\left[\frac{\partial}{\partial x}\left(\frac{\partial z}{\partial x}\right)\right] = \frac{\partial}{\partial y}\left(\frac{\partial^2 z}{\partial x^2}\right).$$

Thus there are precisely four distinct third partials:

$$\frac{\partial^2 z}{\partial x^3}, \quad \frac{\partial^3 z}{\partial x^2\, \partial y}, \quad \frac{\partial^3 z}{\partial x\, \partial y^2}, \quad \frac{\partial^3 z}{\partial y^3}.$$

In general there are $n + 1$ distinct partials of order n:

$$\frac{\partial^n z}{\partial x^k \, \partial y^{n-k}} \qquad k = 0, 1, 2, \cdots, n.$$

Examples   In the following examples we shall find all functions whose partials satisfy given relations. We assume all necessary partials exist and, for simplicity, that the functions are defined on the whole plane. All the solutions are based on one fact: if

$$\frac{\partial z}{\partial x} = 0 \quad \text{and} \quad \frac{\partial z}{\partial y} = 0,$$

then z is a constant. For $\partial z / \partial x = 0$ means $z(x, y)$ is constant in x, that is, $z = z(y)$, a function of y alone. But $\partial z / \partial y = 0$ implies this function of y alone is a constant function.

■ **EXAMPLE 1**   Find all functions $z = f(x, y)$ that satisfy the system of partial differential equations $\dfrac{\partial^2 z}{\partial x^2} = 0, \quad \dfrac{\partial^2 z}{\partial x \, \partial y} = 0, \quad \dfrac{\partial^2 z}{\partial y^2} = 0.$

Solution   First, look at the partials of $\partial z / \partial x$:

$$\frac{\partial}{\partial x}\left(\frac{\partial z}{\partial x}\right) = \frac{\partial^2 z}{\partial x^2} = 0 \quad \text{and} \quad \frac{\partial}{\partial y}\left(\frac{\partial z}{\partial x}\right) = \frac{\partial^2 z}{\partial y \, \partial x} = \frac{\partial^2 z}{\partial x \, \partial y} = 0.$$

It follows that $\partial z / \partial x = A$, a constant. Consequently

$$\frac{\partial}{\partial x}\,[z(x, y) - Ax] = \frac{\partial z}{\partial x} - A = 0,$$

so $z - Ax$ is a function of y alone,

$$z = Ax + g(y).$$

But $\partial^2 z / \partial y^2 = 0$, hence $g''(y) = 0$ so $g(y) = By + C$. Therefore

$$z(x, y) = Ax + By + C,$$

a linear polynomial.   ■

Remark   We spelled out in detail the passage from $\partial z / \partial x = A$ to $z = Ax + g(y)$. After this we shall just refer to such a step as "integrating (with respect to x)".

■ **EXAMPLE 2**   Find all functions $z = f(x, y)$ whose third partials are all 0.

Solution   The second partials of $\partial z / \partial x$ are all 0. By the last example,

$$\frac{\partial z}{\partial x} = Ax + By + C.$$

Integrate: $\qquad z = \dfrac{1}{2} Ax^2 + Bxy + Cx + g(y).$

But $\qquad\qquad 0 = \dfrac{\partial^3 z}{\partial y^3} = \dfrac{d^3 g}{dy^3}$

so $g(y)$ is a quadratic polynomial in y. Therefore z is a quadratic in x and y:

$$z(x, y) = ax^2 + bxy + cy^2 + dx + ey + f. \qquad \blacksquare$$

■ **EXAMPLE 3**   Find all functions $z = f(x, y)$ that satisfy $\dfrac{\partial^2 z}{\partial x^2} = 0$.

Solution   Write the condition in the form

$$\frac{\partial}{\partial x}\left(\frac{\partial z}{\partial x}\right) = 0, \quad \text{then integrate:} \quad \frac{\partial z}{\partial x} = g(y).$$

Integrate again: $\qquad\qquad\qquad z = g(y)x + h(y),$

where $g(y)$ and $h(y)$ are arbitrary functions of y alone.

Check $\qquad \dfrac{\partial^2 z}{\partial x^2} = \dfrac{\partial}{\partial x}\left[\dfrac{\partial}{\partial x}\big(g(y)x + h(y)\big)\right] = \dfrac{\partial}{\partial x} g(y) = 0. \qquad \blacksquare$

■ **EXAMPLE 4**   Find all functions $z = f(x, y)$ that satisfy $\dfrac{\partial^2 z}{\partial x\,\partial y} = 0$.

Solution   Write the condition in the form

$$\frac{\partial}{\partial y}\left(\frac{\partial z}{\partial x}\right) = 0, \quad \text{then integrate:} \quad \frac{\partial z}{\partial x} = p(x).$$

Integrate again: $\qquad\qquad z = g(x) + h(y),$

where $g(x)$ is an antiderivative of $p(x)$. Note that $g(x)$ is an arbitrary function of x since $p(x)$ is.

Check $\quad \dfrac{\partial^2}{\partial x\,\partial y}[g(x) + h(y)] = \dfrac{\partial}{\partial x}\left[\dfrac{\partial}{\partial y}\big(g(x) + h(y)\big)\right] = \dfrac{\partial}{\partial x}[h'(y)] = 0. \qquad \blacksquare$

■ **EXAMPLE 5**   Find all functions $z = f(x, y)$ that satisfy the system of partial differential equations $\dfrac{\partial z}{\partial x} = y, \quad \dfrac{\partial z}{\partial y} = 1.$

Solution   Integrate the first equation:

$$z = xy + g(y).$$

Substitute this into the second equation:

$$\frac{\partial}{\partial y}[xy + g(y)] = 1, \qquad x + g'(y) = 1, \qquad g'(y) = 1 - x.$$

This is impossible since the left-hand side is a function of y alone. Therefore the problem has no solution. $\qquad\qquad\qquad\qquad\qquad\qquad\qquad\qquad\qquad\qquad\qquad \blacksquare$

Remark   The example illustrates an important point. A system of partial differential equations may have no solution at all! Could we have foreseen this catastrophe for the system above? Yes; for suppose there *were* a function $f(x, y)$ satisfying

$$\frac{\partial f}{\partial x} = y \quad \text{and} \quad \frac{\partial f}{\partial y} = 1.$$

Then
$$\frac{\partial^2 f}{\partial y\, \partial x} = \frac{\partial}{\partial y}(y) = 1, \qquad \frac{\partial^2 f}{\partial x\, \partial y} = \frac{\partial}{\partial x}(1) = 0,$$

so the mixed partials would be unequal, a contradiction.

If the system of equations $\quad \dfrac{\partial z}{\partial x} = p(x, y), \qquad \dfrac{\partial z}{\partial y} = q(x, y)$

has a solution, then $\quad \dfrac{\partial p}{\partial y} = \dfrac{\partial q}{\partial x}.$

Indeed,
$$\frac{\partial p}{\partial y} = \frac{\partial}{\partial y}\left(\frac{\partial z}{\partial x}\right) = \frac{\partial}{\partial x}\left(\frac{\partial z}{\partial y}\right) = \frac{\partial q}{\partial x}.$$

More Variables   All that has been said applies to functions of three or more variables. For example, suppose $w = f(x, y, z)$. Then w has three first partials:

$$\frac{\partial w}{\partial x}, \qquad \frac{\partial w}{\partial y}, \qquad \frac{\partial w}{\partial z}.$$

The nine possible second partials may be written in matrix form:

$$\begin{bmatrix} \dfrac{\partial^2 w}{\partial x^2} & \dfrac{\partial^2 w}{\partial x\, \partial y} & \dfrac{\partial^2 w}{\partial x\, \partial z} \\[2ex] \dfrac{\partial^2 w}{\partial y\, \partial x} & \dfrac{\partial^2 w}{\partial y^2} & \dfrac{\partial^2 w}{\partial y\, \partial z} \\[2ex] \dfrac{\partial^2 w}{\partial z\, \partial x} & \dfrac{\partial^2 w}{\partial z\, \partial y} & \dfrac{\partial^2 w}{\partial z^2} \end{bmatrix}.$$

This matrix is symmetric since the mixed second partials are equal in pairs:

$$\frac{\partial^2 w}{\partial y\, \partial x} = \frac{\partial^2 w}{\partial x\, \partial y}, \qquad \frac{\partial^2 w}{\partial x\, \partial z} = \frac{\partial^2 w}{\partial z\, \partial x}, \qquad \frac{\partial^2 w}{\partial z\, \partial y} = \frac{\partial^2 w}{\partial y\, \partial z}.$$

EXERCISES

Compute $\dfrac{\partial^2 f}{\partial x^2}, \quad \dfrac{\partial^2 f}{\partial x\, \partial y}, \quad$ and $\quad \dfrac{\partial^2 f}{\partial y^2}$

1   $\sin(x - 3y)$
3   $x^2 \arcsin y$
5   $ax^2 + 2bxy + cy^2 + dx + ey$

2   xy^6
4   $e^{2x} \cosh y$
6   $\ln(1 + x - 2y).$

Verify that $\quad \dfrac{\partial^2 f}{\partial x\, \partial y} = \dfrac{\partial^2 f}{\partial y\, \partial x}$

7   x/y^2
9   $x^m y^n$
11 $\dfrac{x + y}{x - y}$

8   $x + x^3 y + y^4$
10 $g(x)h(y)$
12 $(x - y)(x - 2y)(x - 3y).$

Compute $\dfrac{\partial^3 f}{\partial x^2\, \partial y}$ and $\dfrac{\partial^3 f}{\partial x\, \partial y^2}$

13 $x^3 y^4$	**14** $x^4 y^5 - xy^2$	**15** $\cos(xy)$
16 $\sin(x^2 y)$	**17** $e^{2y} \sin(x + y)$	**18** x^y.

Find all functions $f(x, y)$ satisfying

19 $\dfrac{\partial^3 f}{\partial x^2\, \partial y} = 0$

[*Hint* Use Examples 3 and 4.]

20 $\dfrac{\partial^3 f}{\partial x^2\, \partial y} = 0$ and $\dfrac{\partial^3 f}{\partial x\, \partial y^2} = 0$

21 all fourth partial derivatives equal 0

22 $\dfrac{\partial^4 f}{\partial x^2\, \partial y^2} = 0$

23 $\dfrac{\partial f}{\partial x} = a$, $\dfrac{\partial f}{\partial y} = b$

24 $\dfrac{\partial f}{\partial x} = y$, $\dfrac{\partial f}{\partial y} = x$

25 $\dfrac{\partial f}{\partial x} = y^2$, $\dfrac{\partial f}{\partial y} = x^2$

26 $\dfrac{\partial^2 f}{\partial x^2} = 2y^3$, $\dfrac{\partial f}{\partial y} = 3x^2 y^2$.

Write the matrix of 9 second partials

27 $xy + yz + zx$ **28** $x^m y^n z^p$ **29** $\sin(x + 2y + 3z)$ **30** $x^2 e^{yz}$.

31 How many distinct third partials does $f(x, y, z)$ have?
32 (cont.) Find a function for which they really are distinct.
33 How many distinct second partials does $f(x, y, z, w)$ have? How many distinct third partials?
34 Show that each function of the form $f(x, y) = g(x + y) + h(x - y)$ satisfies the partial differential equation $\dfrac{\partial^2 f}{\partial x^2} - \dfrac{\partial^2 f}{\partial y^2} = 0$.

35* Show that each solution of the partial differential equation $\dfrac{\partial f}{\partial x} + 2\dfrac{\partial f}{\partial y} = 0$ is of the form

$f(x, y) = g(y - 2x)$. [*Hint* Set $x = u$ and $y = 2u + v$, then solve a partial differential equation for $h(u, v) = f(u, 2u + v)$.]

36* (cont.) Apply the same technique to solve $\dfrac{\partial^2 f}{\partial x^2} + 4\dfrac{\partial^2 f}{\partial x\, \partial y} + 4\dfrac{\partial^2 f}{\partial y^2} = 0$.

[*Hint* Use Example 3.]

37 A uniform metal bar (surrounded by insulation) lies on the x-axis from $x = 0$ to $x = L$. Its temperature $u(x, t)$ at $(x, 0)$ at time t satisfies the **heat equation** $\dfrac{\partial^2 u}{\partial x^2} = \dfrac{1}{k}\dfrac{\partial u}{\partial t}$. Here k is a physical constant. Suppose the ends of the bar are kept at temperature 0. Show that the functions $u_n(x, t) = c_n e^{-kn^2\pi^2 t/L^2} \sin \dfrac{n\pi x}{L}$ are solutions.

38 A taut guitar or piano string is fixed at $(0, 0)$ and $(L, 0)$ and plucked or struck when $t = 0$ so as to vibrate in the x, y-plane. Its displacement $y(x, t)$ from the x-axis (if small) satisfies the **wave equation** $\dfrac{\partial^2 y}{\partial x^2} = \dfrac{1}{c^2}\dfrac{\partial^2 y}{\partial t^2}$, where c^2 depends on physical properties of the string. Verify that the functions

$$y_n(x, t) = \left[\sin\left(\frac{n\pi x}{L}\right)\right]\left[a_n \cos\left(\frac{n\pi ct}{L}\right) + b_n \sin\left(\frac{n\pi ct}{L}\right)\right]$$

are solutions.

39   If the bar in Ex. 37 is moving with constant speed v, then $k\dfrac{\partial^2 u}{\partial x^2} = v\dfrac{\partial u}{\partial x} + \dfrac{\partial u}{\partial t}$.

Show that for each integer n there is an a such that $u(x, t) = e^{-at + vx/2k}\sin(n\pi x/L)$ is a solution.

40   The current $I(x, t)$ along a transmission line satisfies

$$LC\frac{\partial^2 I}{\partial t^2} + RC\frac{\partial I}{\partial t} = \frac{\partial^2 I}{\partial x^2}, \qquad R, L, C \text{ constant.}$$

Find the system of ordinary differential equations that $Y(x)$ and $Z(x)$ must satisfy so that $I = Y(x)\cos nt + Z(x)\sin nt$ is a solution.

41   Show that $V(x, y) = e^{-ax}\cos a(y - y_0)$ satisfies Laplace's equation

$$\partial^2 V/\partial x^2 + \partial^2 V/\partial y^2 = 0.$$

42   Find all quadratic forms in x, y, z that satisfy Laplace's equation

$$\partial^2 V/\partial x^2 + \partial^2 V/\partial y^2 + \partial^2 V/\partial z^2 = 0.$$

43   Find conditions under which

$$z(x, y, t) = \left(\sin\frac{m\pi x}{a}\right)\left(\sin\frac{n\pi y}{b}\right)(A\cos pt + B\sin pt)$$

satisfies the **wave equation** in two dimensions:

$$\frac{\partial^2 z}{\partial x^2} + \frac{\partial^2 z}{\partial y^2} = \frac{1}{c^2}\frac{\partial^2 z}{\partial t^2}.$$

44   The temperature at distance r from an instantaneous line source of heat is

$$u(r, t) = \frac{a}{t}e^{-r^2/4kt}.$$

Show that
$$\frac{\partial^2 u}{\partial r^2} + \frac{1}{r}\frac{\partial u}{\partial r} = \frac{1}{k}\frac{\partial u}{\partial t}.$$

2. TAYLOR APPROXIMATION

Let us recall some facts about Taylor approximation of functions of one variable. If $y = f(x)$, then

$$f(x) = f(a) + f'(a)(x - a) + r_1(x),$$

and
$$f(x) = f(a) + f'(a)(x - a) + \frac{1}{2}f''(a)(x - a)^2 + r_2(x),$$

where
$$|r_1(x)| \le \frac{M_2}{2!}(x - a)^2, \qquad |r_2(x)| \le \frac{M_3}{3!}|x - a|^3,$$

and where M_2 and M_3 are bounds for $|f''(x)|$ and $|f'''(x)|$ respectively. The Taylor polynomial

$$p_1(x) = f(a) + f'(a)(x - a)$$

is constructed so that $p_1(a) = f(a)$ and $p_1'(a) = f'(a)$. The Taylor polynomial

$$p_2(x) = f(a) + f'(a)(x - a) + \frac{1}{2}f''(a)(x - a)^2$$

is constructed so that $p_2(a) = f(a)$, $p_2'(a) = f'(a)$, $p_2''(a) = f''(a)$.

In a similar way, one can construct linear and quadratic polynomials in two variables approximating a given function of two variables.

Taylor Polynomials   Let $f(x, y)$ have continuous first and second partials on a domain **D**. The **first degree** and **second degree Taylor polynomials** of f at (a, b) are

$$p_1(x, y) = f(a, b) + f_x \cdot (x - a) + f_y \cdot (y - b),$$

$$p_2(x, y) = p_1(x, y) + \tfrac{1}{2}[f_{xx} \cdot (x - a)^2 + 2f_{xy} \cdot (x - a)(y - b) + f_{yy} \cdot (y - b)^2],$$

where all the partials are evaluated at (a, b).

It is easy to check that $p_1(a, b) = f(a, b)$ and that the first partials of p_1 agree with those of f at (a, b). Similarly, $p_2(a, b) = f(a, b)$ and all first and second partials of p_2 agree with the corresponding partials of f at (a, b).

Now we ask how closely these Taylor polynomials approximate $f(x, y)$ for (x, y) near (a, b). In other words, we want estimates for the errors in the approximations $f(x, y) \approx p_1(x, y)$ and $f(x, y) \approx p_2(x, y)$. The first of these approximations is familiar. For

$$z = f(a, b) + f_x \cdot (x - a) + f_y \cdot (y - b)$$

is the equation of the tangent plane to the graph of $z = f(x, y)$ at the point $(a, b, f(a, b))$. So we are approximating a surface by its tangent plane.

Let us state some error estimates subject to the mild restriction that f is defined on a convex domain. A domain **D** in the plane or space is called **convex** if it contains the whole segment joining any two of its points. For example, a domain bounded by a triangle, square, or ellipse is convex.

Error in Taylor Approximation   Let $f(x, y)$ have a convex domain **D** and let **a** be a point of **D**.

(1)   Suppose f has continuous first and second derivatives on **D** and the second derivatives satisfy

$$|f_{xx}| \leq M_2, \qquad |f_{xy}| \leq M_2, \qquad |f_{yy}| \leq M_2, \qquad \text{for all} \quad \mathbf{x} \text{ in } \mathbf{D}.$$

Let p_1 be the first degree Taylor polynomial of f at **a**. Then

$$f(\mathbf{x}) = p_1(\mathbf{x}) + r_1(\mathbf{x}),$$

where $|r_1(\mathbf{x})| \leq M_2 |\mathbf{x} - \mathbf{a}|^2.$

Error in Taylor Approximation (cont.)

(2)   Suppose f has continuous first, second, and third derivatives on **D** and the third derivatives satisfy

$$|f_{xxx}| \le M_3, \qquad |f_{xxy}| \le M_3, \qquad |f_{xyy}| \le M_3, \qquad |f_{yyy}| \le M_3.$$

Let p_2 be the second degree Taylor polynomial of f at **a**. Then

$$f(\mathbf{x}) = p_2(\mathbf{x}) + r_2(\mathbf{x}),$$

where $|r_2(\mathbf{x})| \le \dfrac{\sqrt{2}}{3} M_3 |\mathbf{x} - \mathbf{a}|^3.$

We shall postpone the proof until Section 6.

Remark   There are Taylor polynomials of higher degree and corresponding error estimates. The notation for these polynomials is complicated, and since we shall not need them, we leave their study to an advanced calculus course.

■ **EXAMPLE 1**   Compute the Taylor polynomials $p_1(x, y)$ and $p_2(x, y)$ of the function $f(x, y) = \sqrt{x^2 + y^2}$ at $(3, 4)$.

Solution

$$\frac{\partial f}{\partial x} = \frac{x}{\sqrt{x^2 + y^2}}, \qquad \frac{\partial f}{\partial y} = \frac{y}{\sqrt{x^2 + y^2}},$$

$$\frac{\partial^2 f}{\partial x^2} = \frac{y^2}{(x^2 + y^2)^{3/2}}, \qquad \frac{\partial^2 f}{\partial x\, \partial y} = \frac{-xy}{(x^2 + y^2)^{3/2}}, \qquad \frac{\partial^2 f}{\partial y^2} = \frac{x^2}{(x^2 + y^2)^{3/2}}.$$

At $(3, 4)$,

$$\frac{\partial f}{\partial x} = \frac{3}{5}, \qquad \frac{\partial f}{\partial y} = \frac{4}{5}, \qquad \frac{\partial^2 f}{\partial x^2} = \frac{16}{125}, \qquad \frac{\partial^2 f}{\partial x\, \partial y} = -\frac{12}{125}, \qquad \frac{\partial^2 f}{\partial y^2} = \frac{9}{125}.$$

Therefore

$$p_1(x, y) = 5 + \tfrac{3}{5}(x - 3) + \tfrac{4}{5}(y - 4),$$

$$p_2(x, y) = p_1(x, y) + \tfrac{1}{2}[\tfrac{16}{125}(x - 3)^2 - \tfrac{24}{125}(x - 3)(y - 4) + \tfrac{9}{125}(y - 4)^2]. \qquad ■$$

■ **EXAMPLE 2**   Estimate $\sqrt{(3.10)^2 + (4.02)^2}$

   (a) by $p_1(x, y)$,   (b) by $p_2(x, y)$,

the Taylor approximations of Example 1.

Solution   Set $f(x, y) = \sqrt{x^2 + y^2}$. Near $(3, 4)$,

$$f(x, y) \approx \tfrac{1}{5}[25 + 3(x - 3) + 4(y - 4)],$$

$$f(3.1, 4.02) \approx \tfrac{1}{5}[25 + 3(0.1) + 4(0.02)] = 5.076.$$

$$f(x, y) \approx p_1(x, y) + \tfrac{1}{250}[16(x - 3)^2 - 24(x - 3)(y - 4) + 9(y - 4)^2],$$

$$f(3.1, 4.02) \approx p_1(3.1, 4.02) + \tfrac{1}{250}[16(0.1)^2 - 24(0.1)(0.02) + 9(0.02)^2]$$

$$= 5.076 + \frac{0.1156}{250} = 5.076 + 0.0004624 = 5.0764624.$$

(Actual value to 7 places: 5.0764555.)

■ **EXAMPLE 3**   If $|x| < 0.1$ and $|y| < 0.1$, prove that

$$|e^x \sin(x + y) - (x + y)| < 0.05.$$

Solution   Set $f(x, y) = e^x \sin(x + y)$. Then $f(0, 0) = 0$ and

$$f_x(0, 0) = e^x \sin(x + y) + e^x \cos(x + y)\Big|_{(0,0)} = 1$$

and

$$f_y(0, 0) = e^x \cos(x + y)\Big|_{(0,0)} = 1,$$

hence at $(0, 0)$, $p_1(x, y) = x + y.$

Therefore the problem is to show that $|r_1(\mathbf{x})| < 0.05$ for points $\mathbf{x} = (x, y)$ with $|x| < 0.1$ and $|y| < 0.1$. Such points satisfy $|\mathbf{x}|^2 = (0.1)^2 + (0.1)^2$, so restrict the domain of f to the disk $|\mathbf{x}|^2 < 0.02$.

According to the error estimate,

$$|r_1(\mathbf{x})| \leq M_2|\mathbf{x}|^2 = (0.02)M_2,$$

where M_2 is a bound for $|f_{xx}|, |f_{xy}|, |f_{yy}|$. To find a suitable value for M_2, compute the second partials:

$$f_{xx} = 2e^x \cos(x + y), \quad f_{xy} = e^x[\cos(x + y) - \sin(x + y)], \quad f_{yy} = -e^x \sin(x + y).$$

Since $|\sin(x + y)| \leq 1$ and $|\cos(x + y)| \leq 1$,

$$|f_{xx}| \leq 2e^x, \quad |f_{xy}| \leq 2e^x, \quad |f_{yy}| \leq e^x.$$

Furthermore, $|x| < 0.1$, so

$$e^x < e^{0.1} \approx 1.1052 < 1.11.$$

Therefore $M_2 = 2(1.11) = 2.22$ is a suitable bound. It follows that

$$|r_1(\mathbf{x})| < (0.02)(2.22) = 0.0444 < 0.05.$$

EXERCISES

Compute the Taylor polynomials $p_1(x, y)$ and $p_2(x, y)$

1 x^2y^2 at $(1, 1)$ 2 x^4y^3 at $(2, -1)$ 3 $\sin(xy)$ at $(0, 0)$
4 e^{xy} at $(0, 0)$ 5 x^y at $(1, 0)$ 6 x^y at $(1, 1)$
7 $\cos(x + y)$ at $(0, \pi/2)$ 8 $1 + xy$ at $(1, 1)$ 9 $\ln(x + 2y)$ at $(\tfrac{1}{2}, \tfrac{1}{4})$
10 x^2e^y at $(1, 0)$.

Estimate using the second degree Taylor polynomial; carry your work to 5 significant figures

11 $(1.1)^{1.2}$ 12 $[(1.2)^2 + 7.2]^{1/3}$

13 $f(1.01, 2.01)$,   where $f(x, y) = x^3y^2 - 2xy^4 + y^5$

14 $f(2.01, 0.98)$,   where $f(x, y) = \dfrac{1}{2^7}x^7y^{10}$.

15 If $p_1(x)$, $p_2(x)$ and $q_1(x)$, $q_2(x)$ are the first and second degree Taylor polynomials of $f(x)$ and $g(x)$ at a and b respectively, find the first and second degree Taylor polynomials of $h(x, y) = f(x)g(y)$ at (a, b).

16 Let $p_1(x, y)$ be the first degree Taylor polynomial of $f(x, y)$ at $(a, b) = \mathbf{a}$. Show that $p_1(\mathbf{x}) = f(\mathbf{a}) + D_{\mathbf{x} - \mathbf{a}} f(\mathbf{a})$.

Prove the inequality, given $|x| < 0.1$ and $|y| < 0.1$

17 $\left|\sqrt{1 + x + 2y} - (1 + \tfrac{1}{2}x + y)\right| < 0.04$ **18** $\left|e^x \sin(x + y) - (1 + x)(x + y)\right| < 0.01$
(even < 0.005 with more careful estimates).

19 Given a function $f(x, y, z)$ of three variables, define its first and second order Taylor polynomials $p_1(x, y, z)$ and $p_2(x, y, z)$ at (a, b, c). Check that all of their first order and second order partial derivatives agree respectively with those of $f(x, y, z)$ at (a, b, c).

20 (cont.) Assuming grad $f(a, b, c) \neq 0$, identify the graph of $p_1(x, y, z) = f(a, b, c)$.

21 Compute $p_2(x, y)$ at $(0, 0)$ for $f(x, y) = \dfrac{1}{1 - x - 2y + 3xy}$. Verify that you get the same result by expanding $\dfrac{1}{1 - z}$ in power series, where $z = x + 2y - 3xy$, then collecting all terms of degree 2 or less.

22 (cont.) Use a similar technique for $f(x) = \sqrt{1 + 3x - 4y + x^2}$.

3. STABILITY

In this section we shall develop second derivative tests for maxima and minima. Stability is a modern term for the behavior of a function near a critical point. See the remark on p. 808.

We shall assume here and in the rest of this chapter that all functions have continuous first, second, and third partial derivatives. With this assumption, the Taylor approximations of the previous section apply.

Let us begin with a brief review of the one-variable case. We consider a function $g(t)$ and a point c where $g'(c) = 0$. Suppose $g''(c) > 0$. We want to conclude that $g(c)$ is a local minimum of g. For this purpose, an excellent tool is the second degree Taylor approximation of g at c:

$$\begin{cases} g(t) = g(c) + \tfrac{1}{2}g''(c)(t - c) + r_2(t), \\ |r_2(t)| \leq k|t - c|^3, \qquad k > 0. \end{cases}$$

It follows that

$$g(t) - g(c) = \tfrac{1}{2}g''(c)(t - c)^2 + r_2(t) \geq \tfrac{1}{2}g''(c)(t - c)^2 - k|t - c|^3$$
$$= (t - c)^2[\tfrac{1}{2}g''(c) - k|t - c|].$$

Since $g''(c) > 0$, the quantity on the right is positive if $0 < |t - c| < \tfrac{1}{2}g''(c)/k$. Thus there is a positive number $\delta = \tfrac{1}{2}g''(c)/k$ such that $g(t) - g(c) > 0$ when $0 < |t - c| < \delta$. In other words, $g(c)$ is smaller than any other value of g in an interval of radius δ and center c. Hence $g(c)$ is a local minimum of g.

Second Derivative Test   Let us try to generalize these ideas to the two variable case. Given $f(x, y)$, suppose (a, b) is an interior point of the domain of f where $f_x(a, b) = f_y(a, b) = 0$. We would like a condition on the second derivatives of f, analogous to $g''(c) > 0$, guaranteeing that $f(a, b)$ is a relative minimum of f. Now the condition $g''(c) > 0$ can be interpreted as meaning that

$$g''(c)(t - c)^2$$

is a positive definite quadratic form in the one variable $t - c$. It is natural, therefore, to examine the quadratic part of the second degree Taylor polynomial of $f(x, y)$ at (a, b). If it is positive definite, we suspect that $f(a, b)$ is a minimum of f.

Second Derivative Test for a Minimum   Let (a, b) be an interior point of the domain of $f(x, y)$, that is, not a boundary point. Suppose $f_x(a, b) = 0$ and $f_y(a, b) = 0$, and in addition

$$f_{xx}(a, b) > 0, \qquad f_{xx}f_{yy} - f_{xy}^2 \Big|_{(a, b)} > 0.$$

Then $f(a, b)$ is a strong local minimum of $f(x, y)$. Precisely, there exists $\delta > 0$ such that

$$f(x, y) > f(a, b) \qquad \text{whenever} \quad 0 < |(x, y) - (a, b)| < \delta.$$

This test is a consequence of the second order Taylor approximation

$$f(a + x, b + y) = f(a, b) + \tfrac{1}{2}[Ax^2 + 2Bxy + Cy^2] + r_2(x, y),$$

where $A = f_{xx}(a, b), \qquad B = f_{xy}(a, b), \qquad C = f_{yy}(a, b).$

The idea is that the second order term is positive and dominates the remainder for (x, y) small, hence

$$f(a + x, b + y) > f(a, b)$$

for (x, y) small. There are some technical difficulties in completing this argument, so we postpone the proof until Section 6.

Remark   Recall the test for positive definiteness (p. 767): $Ax^2 + 2Bxy + Cy^2$ is positive definite if and only if $A > 0$ and $AC - B^2 > 0$.

EXAMPLE 1   Find the point on the paraboloid $z = \dfrac{x^2}{4} + \dfrac{y^2}{9}$ closest to $\mathbf{i} = (1, 0, 0)$.

Solution   Let $\mathbf{x} = (x, y, z)$ be a point on the surface. Then

$$|\mathbf{x} - \mathbf{i}|^2 = (x - 1)^2 + y^2 + z^2$$

$$= (x - 1)^2 + y^2 + \left(\frac{x^2}{4} + \frac{y^2}{9}\right)^2 = f(x, y).$$

The function $f(x, y)$ must be minimized, so solve $f_x = f_y = 0$. Now

$$f_x = 2(x - 1) + x\left(\frac{x^2}{4} + \frac{y^2}{9}\right), \qquad f_y = 2y + \frac{4y}{9}\left(\frac{x^2}{4} + \frac{y^2}{9}\right) = 2y\left[1 + \frac{2}{9}\left(\frac{x^2}{4} + \frac{y^2}{9}\right)\right].$$

The condition $f_y = 0$ is satisfied if either

$$y = 0 \qquad \text{or} \qquad 1 + \frac{2}{9}\left(\frac{x^2}{4} + \frac{y^2}{9}\right) = 0.$$

The latter is impossible. Therefore $f_y = 0$ implies $y = 0$. But if $y = 0$, the condition $f_x = 0$ means

$$2(x - 1) + \tfrac{1}{4}x^3 = 0, \qquad \text{that is,} \quad x^3 + 8x - 8 = 0.$$

A rough sketch shows this cubic has only one real root, and the root is near $x = 1$. By Newton's method iterated twice, $x \approx 0.9068$.

Thus $f(x, y)$ has a possible minimum only at the unique point $(a, 0)$, where $a^3 + 8a - 8 = 0$. Test the second derivatives at $(a, 0)$:

$$f_{xx} = 2 + \tfrac{3}{4}a^2 > 0, \qquad f_{xy} = 0, \qquad f_{yy} = 2 + \tfrac{1}{9}a^2 > 0, \qquad f_{xx}f_{yy} - f_{xy}{}^2 > 0.$$

Therefore the minimum does occur at $(a, 0)$, so the closest point is $(a, 0, \tfrac{1}{4}a^2)$, where $a^3 + 8a - 8 = 0$. ∎

The second derivative test for a minimum, applied to $-f(x, y)$, implies a test for a maximum.

Second Derivative Test for a Maximum   Let (a, b) be an interior point of the domain of $f(x, y)$. Suppose $f_x(a, b) = 0$ and $f_y(a, b) = 0$, and the quadratic form

$$Ax^2 + 2Bxy + Cy^2$$

is negative definite $(A < 0,\ AC - B^2 > 0)$, where

$$A = f_{xx}(a, b), \qquad B = f_{xy}(a, b), \qquad C = f_{yy}(a, b).$$

Then $f(a, b)$ is a strong local maximum of $f(x, y)$. Precisely, there exists $\delta > 0$ such that

$$f(x, y) > f(a, b) \qquad \text{whenever} \quad |(x, y) - (a, b)| < \delta.$$

Remark   Recall the conditions (p. 768) for the quadratic form to be negative definite: $A < 0,\ AC - B^2 > 0$.

■ **EXAMPLE 2**   Of all triangles with fixed perimeter $2S$, which has the largest area?

Solution   Let the sides be x, y, z. By Heron's formula, the area A satisfies

$$A^2 = S(S - x)(S - y)(S - z).$$

Since $2S = x + y + z$, you can eliminate z:

$$S - z = x + y - S, \qquad A^2 = S(S - x)(S - y)(x + y - S).$$

To maximize A, it suffices to maximize

$$f(x, y) = (S - x)(S - y)(x + y - S)$$

for $x < S$, $y < S$, $x + y > S$. Now

$$f_x = (S - y)[-(x + y - S) + (S - x)] = (S - y)(2S - 2x - y),$$
$$f_y = (S - x)[-(x + y - S) + (S - y)] = (S - x)(2S - x - 2y).$$

Hence $f_x = 0$ and $f_y = 0$ imply

$$2S - 2x - y = 0 \quad \text{and} \quad 2S - x - 2y = 0.$$

The only solution is $x = \frac{2}{3}S$, $y = \frac{2}{3}S$. For these values of x and y,

$$f_{xx} = -2(S - y) = -\tfrac{2}{3}S < 0, \qquad f_{yy} = -2(S - x) = -\tfrac{2}{3}S,$$
$$f_{xy} = -(2S - 2x - y) - (S - y) = -3S + 2x + 2y = -\tfrac{1}{3}S,$$
$$f_{xx}f_{yy} - f_{xy}{}^2 = \tfrac{4}{9}S^2 - \tfrac{1}{9}S^2 = \tfrac{1}{3}S^2 > 0.$$

Therefore $f(x, y)$ has its only local maximum at $(\frac{2}{3}S, \frac{2}{3}S)$. Since $f(x, y) = 0$ on the boundary, this point yields an absolute maximum. Now $z = 2S - x - y = 2S - \frac{2}{3}S - \frac{2}{3} = \frac{2}{3}S$. The triangle of largest area is equilateral. ∎

Saddle Points Let us review our method. We find an interior point (a, b) of the domain of $f(x, y)$ where $f_x = 0$ and $f_y = 0$. Then we compute the **(Hessian)** matrix

$$H = \begin{bmatrix} f_{xx} & f_{xy} \\ f_{yx} & f_{yy} \end{bmatrix}_{(a, b)}.$$

There are three possibilities for det H:

Case 1 det $H > 0$. Then

$$f_{xx}f_{yy} > f_{xy}{}^2 \geq 0$$

so $f_{xx}(a, b) \neq 0$. Either $f_{xx} > 0$ and $f(a, b)$ is a strong local min, or $f_{xx} < 0$ and $f(a, b)$ is a strong local max.

Case 2 det $H = 0$. No conclusion can be drawn. Look at these examples; in both cases $(a, b) = (0, 0)$ and all first and second partials equal 0 at $(0, 0)$:

(i) $f(x, y) = x^4 y^4$, local min;

(ii) $f(x, y) = x^3 y^3$, neither local max nor min.

Case 3 det $H < 0$. In this case we are *sure* there is neither a local max nor a local min. Here is the exact statement:

Saddle Point Test Let $f_x(a, b) = 0$ and $f_y(a, b) = 0$ at an interior point (a, b) of the domain of $f(x, y)$. Suppose

$$\left. f_{xx}f_{yy} - f_{xy}{}^2 \right|_{(a, b)} < 0.$$

Then $f(a, b)$ is neither a relative max nor a relative min of $f(x, y)$. Precisely, there exists points (x, y) arbitrarily close to (a, b) where $f(x, y) < f(a, b)$ and there exist points (x, y) arbitrarily close to (a, b) where $f(x, y) > f(a, b)$.

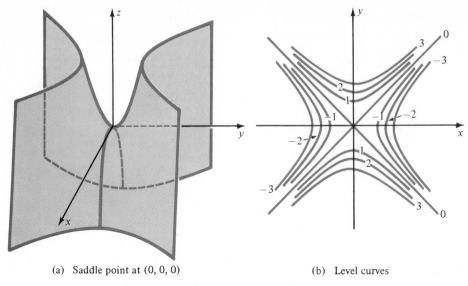

(a) Saddle point at $(0, 0, 0)$ (b) Level curves

Fig. 1   Example of saddle point: $z = -x^2 + y^2$

In this situation $f(x, y)$ is said to have a **saddle point** at (a, b); the surface $z = f(x, y)$ is shaped like a saddle or a mountain pass near $(a, b, f(a, b))$. The tangent plane is horizontal; the surface rises in some directions, falls in others, so that it crosses its tangent plane. See Fig. 1 for an example. We shall postpone a proof of the saddle point test until Section 6. Let us summarize our results:

Summary   Let (a, b) be an interior point of the domain of $f(x, y)$ and suppose
$$f_x(a, b) = f_y(a, b) = 0.$$
(1)   If $f_{xx} > 0$ and $f_{xx} f_{yy} - f_{xy}^2 > 0$, then $f(a, b)$ is a strong local min.
(2)   If $f_{xx} < 0$ and $f_{xx} f_{yy} - f_{xy}^2 > 0$, then $f(a, b)$ is a strong local max.
(3)   If $f_{xx} f_{yy} - f_{xy}^2 < 0$, then $f(a, b)$ is a saddle point, neither a local max nor min.
(4)   If $f_{xx} f_{yy} - f_{xy}^2 = 0$, no conclusion can be drawn from this information alone.

Remark   Imagine a particle constrained to move on the surface $z = f(x, y)$ and subject to the downward force of gravity. Suppose the particle is at rest (in equilibrium) at a point $(a, b, f(a, b))$, that is, its height z is stationary: $f_x(a, b) = 0$ and $f_y(a, b) = 0$.

Now suppose the particle is displaced slightly—given a little shove. What happens? If $f(a, b)$ is a strong local max of z, then the particle tumbles downward—its equilibrium was **unstable**. If $f(a, b)$ is a strong local min of z, then the particle returns to its equilibrium point—its equilibrium was **stable**. The saddle point case is smack in between. For some displacement directions the particle tumbles downward, and for others it returns to equilibrium.

As a concrete example of such a surface, imagine an inflated inner tube (or a doughnut) hanging from a string. Then at the top is a max of z, at the bottom a min of z. There are two saddle points of z, one at the top and one at the bottom of the "hole".

■ **EXAMPLE 3**   Find the points on the ellipsoid $\frac{1}{9}x^2 + y^2 + \frac{1}{4}z^2 = 1$ nearest to and farthest from the origin.

Solution   The square of the distance from (x, y, z) to $(0, 0, 0)$ is

$$D^2 = f(x, y) = x^2 + y^2 + z^2 = x^2 + y^2 + 4(1 - \tfrac{1}{9}x^2 - y^2) = 4 + \tfrac{5}{9}x^2 - 3y^2.$$

Since

$$\tfrac{1}{9}x^2 + y^2 \le \tfrac{1}{9}x^2 + y^2 + \tfrac{1}{4}z^2 = 1,$$

the domain of $f(x, y)$ is the elliptical region

$$\tfrac{1}{9}x^2 + y^2 \le 1.$$

The first partials of $f(x, y)$ are

$$f_x = \tfrac{10}{9}x \qquad \text{and} \qquad f_y = -6y,$$

which are simultaneously 0 only for $(x, y) = (0, 0)$. However, at $(0, 0)$

$$f_{xx} f_{yy} - f_{xy}{}^2 = (\tfrac{10}{9})(-6) - 0 < 0.$$

Therefore $f(x, y)$ has neither a maximum nor a minimum at $(0, 0)$. There seems to be no possible maximum or minimum.

We have forgotten the boundary! The continuous function $f(x, y)$ has both a maximum and a minimum in its *bounded closed* domain, and since they do not occur inside the domain, they must occur on the boundary curve (Fig. 2), an ellipse. On this ellipse, $z = 0$ so $f(x, y) = x^2 + y^2$. By inspection, $f_{\min} = f(0, \pm 1) = 1$ and $f_{\max} = f(\pm 3, 0) = 9$.

Therefore the points of the original ellipsoid nearest to the origin are $(0, \pm 1, 0)$ and the points farthest from the origin are $(\pm 3, 0, 0)$. [The points $(0, 0, \pm 2)$ are saddle points for the distance function.]

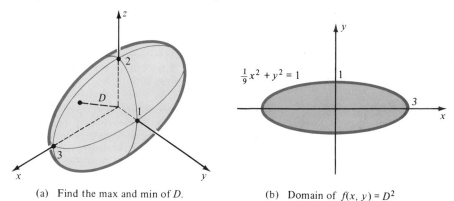

(a)   Find the max and min of D.   (b)   Domain of $f(x, y) = D^2$

Fig. 2   ■

EXERCISES

Does the function $f(x, y)$ have a local max, local min, or saddle point at the origin? (Use the first and second derivative tests; if all else fails, sketch a few level curves near the origin.)

1	$x^2 + 3y^2$	**2**	xy	**3**	$x^2 - 4y^2$
4	$x^2 + 2xy + y^2$	**5**	$-x^2 + 2xy - y^2$	**6**	$x^4 + y^2$

7	x^4	**8**	$-x^2y^2$	**9**	$x^3 + y^3$
10	$x^2 + y^5$	**11**	$x^2y^2 - 3x^4y^4$	**12**	$x^6 + y^6 - 6x^2y^2.$

Find all local maxs, local mins, and saddle points

13	$x^4 + y^4 - 2(x - y)^2$	**14**	$2x^2y - y^2 + 4x$
15	$(x^2 + y^2)^2 - 4(x^2 + 2y^2) + 1$	**16**	$x^3 + y^3 - 3axy, \quad a > 0$
17	$x^3 - 4x^2 - xy - y^2$	**18**	$e^{4y}(x^2 + x + y)$
19	$(y - x)(x^2 + y^2 - 1)$	**20***	$[(x + 1)^2 + y^2 - 1][(x - 1)^2 + y^2 - 1]$
21	$(y - x^2)(2 - x - y)$	**22**	$(y - x^2)(x^2 + y^2 - 1).$

23 Suppose $f(x)$ has a local max at $x = a$ and a local min at $x = b$, and $g(y)$ has a local max at $y = c$ and a local min at $y = d$. What can you say about $h(x, y) = f(x) + g(y)$ at (a, c), (a, d), (b, c), and (b, d) ?

24 Explain geometrically the remark about saddle points at the end of the solution of Example 3.

25 A manufacturer produces two lines of a product, at a cost of $2 per unit for the regular model and $3 per unit for the special model. If he fixes the price at x dollars and y dollars respectively, the demand for the regular model is $y - x$ and the demand for the special model is $14 + x - 2y$ in thousand units per week. What prices maximize his profits?

26 Suppose **a** does not lie on a closed surface $F(\mathbf{x}) = 0$ and **b** is a point of the surface that maximizes or minimizes the distance $|F(\mathbf{x}) - \mathbf{a}|$. Show that $\mathbf{a} - \mathbf{b}$ is normal to the surface at **b**. [*Hint* Parameterize.]

4. CONSTRAINED OPTIMIZATION

Here are several problems that have a common feature.

(a) Of all rectangles with perimeter one, which has the shortest diagonal? That is, minimize $(x^2 + y^2)^{1/2}$ subject to $2x + 2y = 1$.

(b) Of all right triangles with perimeter one, which has largest area? That is, maximize $\frac{1}{2}xy$ subject to $x + y + (x^2 + y^2)^{1/2} = 1$.

(c) Find the largest value of $x + 2y + 3z$ for points (x, y, z) on the unit sphere. That is, maximize $x + 2y + 3z$ subject to $x^2 + y^2 + z^2 = 1$.

(d) Of all rectangular boxes with fixed surface area, which has greatest volume? That is, maximize xyz subject to $xy + yz + zx = c$.

Each of these problems asks for the maximum (or minimum) of a function of several variables, where the variables must satisfy a certain relation (constraint).

Such problems may be analyzed geometrically. Suppose you are asked to maximize a function $f(x, y)$, subject to a constraint $g(x, y) = c$. On the same graph plot $g(x, y) = c$ and several level curves of $f(x, y)$, noting the direction of increase of the level (Fig. 1a). To find the largest value of $f(x, y)$ on the curve $g(x, y) = c$, find the highest level curve that intersects $g = c$. If there is a highest one and the intersection does not take place at an end point, this level curve and the graph $g = c$ are tangent.

Suppose $f(x, y) = M$ is a level curve tangent to $g(x, y) = c$ at a point (x, y). See Fig. 1b. Since the two graphs are tangent at (x, y), their normals at (x, y) are parallel. But the vectors

$$\text{grad } f(x, y) \qquad \text{and} \qquad \text{grad } g(x, y)$$

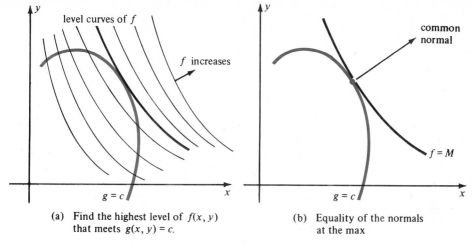

(a) Find the highest level of $f(x, y)$
     that meets $g(x, y) = c$.

(b) Equality of the normals
     at the max

Fig. 1   Maximize $f(x, y)$ subject to $g(x, y) = c$.

point in the respective normal directions, hence one is a multiple of the other:

$$\text{grad } f(x, y) = \lambda \text{ grad } g(x, y)$$

for some number λ. (The argument presupposes that grad $g \neq \mathbf{0}$ at the point in question.)

This geometric argument yields a practical rule for locating points on $g(x, y) = c$ where $f(x, y)$ may have a maximum or minimum. [However, note that where the condition of tangency is satisfied, there may be a maximum, a minimum, or neither (Fig. 2).]

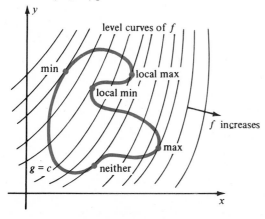

Fig. 2   Points of tangency

Lagrange Multiplier Rule   To maximize or minimize a function $f(x, y)$ subject to a constraint $g(x, y) = c$, solve the system of equations

$$(f_x, f_y) = \lambda(g_x, g_y), \qquad g(x, y) = c$$

in the three unknowns x, y, λ. Each resulting point (x, y) is a candidate.
The number λ is called a **Lagrange multiplier**, or simply **multiplier**.

This rule requires three simultaneous equations

$$\begin{cases} f_x(x, y) = \lambda g_x(x, y) \\ f_y(x, y) = \lambda g_y(x, y) \\ g(x, y) = c, \end{cases}$$

to be solved for three unknowns x, y, λ.

■ **EXAMPLE 1**   Find the largest and smallest values of $f(x, y) = x + 2y$ on the circle $x^2 + y^2 = 1$.

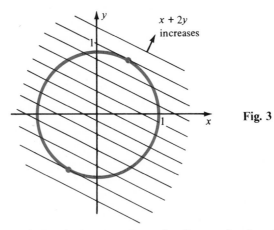

$x + 2y$ increases

Fig. 3

Solution   Draw a figure (Fig. 3). As seen from the figure, f takes its maximum at a point in the first quadrant, and its minimum at a point in the third quadrant. Apply the method of Lagrange multipliers with

$$f(x, y) = x + 2y, \qquad g(x, y) = x^2 + y^2,$$

$$\operatorname{grad} f = (1, 2), \qquad \operatorname{grad} g = (2x, 2y).$$

The conditions $(f_x, f_y) = \lambda(g_x, g_y), \qquad g(x, y) = 1$

become $(1, 2) = \lambda(2x, 2y), \qquad x^2 + y^2 = 1.$

Thus

$$x = \frac{1}{2\lambda}, \qquad y = \frac{1}{\lambda}, \qquad \left(\frac{1}{2\lambda}\right)^2 + \left(\frac{1}{\lambda}\right)^2 = 1.$$

By the third equation, $\lambda^2 = \frac{5}{4}, \qquad \lambda = \pm\frac{1}{2}\sqrt{5}$.

The value $\lambda = \frac{1}{2}\sqrt{5}$ yields

$$x = \frac{1}{\sqrt{5}}, \qquad y = \frac{2}{\sqrt{5}}, \qquad f(x, y) = \frac{5}{\sqrt{5}} = \sqrt{5};$$

the value $\lambda = -\frac{1}{2}\sqrt{5}$ yields

$$x = -\frac{1}{\sqrt{5}}, \qquad y = -\frac{2}{\sqrt{5}}, \qquad f(x, y) = -\frac{5}{\sqrt{5}} = -\sqrt{5}.$$

The largest value is $\sqrt{5}$, the smallest, $-\sqrt{5}$. ∎

■ **EXAMPLE 2** Find the largest and smallest values of xy on the segment $x + 2y = 2$, $x \geq 0$, $y \geq 0$.

Solution Draw a graph (Fig. 4). Evidently the smallest value of xy is 0, taken at either end point. To find the largest value, use the multiplier technique with

$$f(x, y) = xy, \qquad g(x, y) = x + 2y.$$

The relevant system of equations is

$$(y, x) = \lambda(1, 2), \qquad x + 2y = 2.$$

Consequently $x = 2\lambda$, $y = \lambda$, and

$$2\lambda + 2\lambda = 2, \qquad \lambda = \tfrac{1}{2}.$$

Therefore $\qquad\qquad (x, y) = (1, \tfrac{1}{2}), \qquad f_{\max} = f(1, \tfrac{1}{2}) = \tfrac{1}{2}.$

As already noted, $f_{\min} = f(2, 0) = f(0, 1) = 0$. ∎

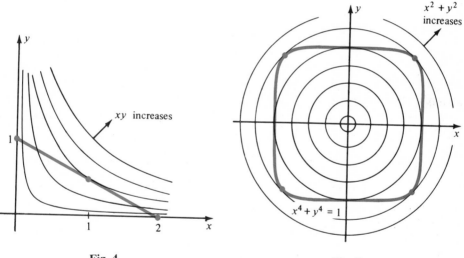

Fig. 4 　　　　　　　　　　　　　　**Fig. 5**

■ **EXAMPLE 3** Find the largest and smallest values of $x^2 + y^2$, subject to $x^4 + y^4 = 1$.

Solution Graph the curve $x^4 + y^4 = 1$ and the level curves of $f(x, y) = x^2 + y^2$. See Fig. 5. By drawing $x^4 + y^4 = 1$ *accurately*, you see that the graph is quite flat where it crosses the axes and most sharply curved where it crosses the 45° lines $y = \pm x$. It is closest to the origin ($x^2 + y^2$ is least) at $(\pm 1, 0)$ and $(0, \pm 1)$, and farthest where $y = \pm x$.

The analysis confirms this. Use the multiplier technique with

$$f(x, y) = x^2 + y^2, \qquad g(x, y) = x^4 + y^4.$$

The relevant equations are

$$(2x, 2y) = \lambda(4x^3, 4y^3), \qquad x^4 + y^4 = 1.$$

Obvious solutions are

$$x = 0, \quad y = \pm 1, \quad \lambda = \tfrac{1}{2}; \qquad y = 0, \quad x = \pm 1, \quad \lambda = \tfrac{1}{2}.$$

Thus the points $(0, \pm 1)$ and $(\pm 1, 0)$ are candidates for the maximum or minimum. At each of these points $f(x, y) = 1$.

Suppose both $x \neq 0$ and $y \neq 0$. From

$$2x = 4\lambda x^3, \quad 2y = 4\lambda y^3 \qquad \text{follows} \quad x^2 = y^2 = \frac{1}{2\lambda}.$$

Hence $\lambda = 1/(2x^2) > 0$. From $x^4 + y^4 = 1$ follow

$$\left(\frac{1}{2\lambda}\right)^2 + \left(\frac{1}{2\lambda}\right)^2 = 1, \qquad \lambda^2 = \frac{1}{2}, \qquad \lambda = \frac{1}{\sqrt{2}}, \qquad x^2 = y^2 = \frac{1}{2\lambda} = \frac{\sqrt{2}}{2} = \frac{1}{\sqrt{2}}.$$

Consequently, the four points $\left(\pm \dfrac{1}{\sqrt[4]{2}}, \pm \dfrac{1}{\sqrt[4]{2}}\right)$

are candidates for the maximum or minimum. At each of these points $f(x, y) = x^2 + y^2 = 2/\sqrt{2} = \sqrt{2}$. Therefore the largest value is $\sqrt{2}$, the smallest, 1.   ∎

■ **EXAMPLE 4**   Find the proportions of a right circular cone with fixed lateral area and maximal volume.

Solution   Denote the radius, height, fixed lateral area, and volume by r, h, A, and V. Then

$$V = \tfrac{1}{3}\pi r^2 h \qquad \text{and} \qquad A = \tfrac{1}{2}(2\pi r)\sqrt{r^2 + h^2}$$

because $\sqrt{r^2 + h^2}$ is the slant height. Since A is fixed, so is A^2/π^2. Therefore the problem is equivalent to maximizing

$$f(r, h) = r^2 h \qquad \text{subject to} \quad g(r, h) = r^2(r^2 + h^2) = c.$$

The Lagrange multiplier equations are

$$2rh = \lambda(4r^3 + 2rh^2), \qquad r^2 = 2\lambda r^2 h,$$

that is,

$$h = (2r^2 + h^2)\lambda, \qquad 2\lambda h = 1.$$

To eliminate λ, multiply the first equation by $2h$:

$$2h^2 = (2r^2 + h^2)(2\lambda h) = 2r^2 + h^2, \qquad h^2 = 2r^2, \qquad h = r\sqrt{2}.$$

Let us show that these proportions yield maximal volume. By the nature of the problem $r > 0$. Also $h > 0$ and the relation

$$\pi r \sqrt{r^2 + h^2} = A = \text{constant}$$

implies that $r \longrightarrow \sqrt{A/\pi}$ as $h \longrightarrow 0$ and conversely.

As a function of r, the volume V is defined for $0 < r < \sqrt{A/\pi}$. Furthermore,

$h < \sqrt{r^2 + h^2}$, so as $r \longrightarrow 0 +$,

$$V = \tfrac{1}{3}\pi r^2 h < \tfrac{1}{3}\pi r^2 \sqrt{r^2 + h^2} = \tfrac{1}{3}rA \longrightarrow 0.$$

As $r \longrightarrow \sqrt{A/\pi}$ from below, $h \longrightarrow 0+$ and

$$V = \tfrac{1}{3}\pi r^2 h \longrightarrow \tfrac{1}{3}\pi(A/\pi) \cdot 0 = 0.$$

Thus V is a positive function of r on the interval $0 < r < \sqrt{A/\pi}$ and V approaches 0 toward the end points of the interval. Therefore V certainly has a maximum in the interval, but not necessarily a minimum. Since $h = r\sqrt{2}$ represents the only candidate for either a max or a min, this condition must yield the maximal volume. ∎

Second Derivative Test As you might expect, there is a second derivative test for constrained extrema problems. The following test follows from the one discussed in Section 3; we leave its proof as an exercise.

Second Derivative Test Given $f(x, y)$ and $g(x, y)$, let (a, b, λ) be a solution of

$$\operatorname{grad} f(a, b) = \lambda \operatorname{grad} g(a, b), \qquad g(a, b) = c.$$

Set $\qquad A = f_{xx} - \lambda g_{xx}, \qquad B = f_{xy} - \lambda g_{xy}, \qquad C = f_{yy} - \lambda g_{yy},$

all evaluated at (a, b), and let

$$Q(x, y) = Ax^2 + 2Bxy + Cy^2$$

be the corresponding quadratic form.

(a) If $Q(x, y)$ is positive definite, then there exists $\delta > 0$ such that

$$f(x, y) > f(a, b)$$

whenever $\quad g(x, y) = c \quad$ and $\quad 0 < |(x, y) - (a, b)| < \delta.$

(b) If $Q(x, y)$ is negative definite, then there exists $\delta > 0$ such that

$$f(x, y) < f(a, b)$$

whenever $\quad g(x, y) = c \quad$ and $\quad 0 < |(x, y) - (a, b)| < \delta.$

The test is inconclusive in all other cases.

Let us try it on Example 3. There

$$f(x, y) = x^2 + y^2, \qquad g(x, y) = x^4 + y^4, \qquad \lambda = \frac{1}{\sqrt{2}}, \qquad a^2 = b^2 = \frac{1}{\sqrt{2}}.$$

At one of the four possible points (a, b),

$$A = 2 - 12\lambda a^2 = -4, \qquad B = 0, \qquad C = 2 - 12\lambda b^2 = -4.$$

Therefore $A < 0$, $AC - B^2 > 0$, so $Q(x, y)$ is negative definite. Under these conditions, the test guarantees as local maximum, which checks with Example 3.

EXERCISES

Find the max and min of the function subject to the constraint, and state where they occur

1   xy on $x^2 + y^2 = 1$ 2 xy on $\frac{1}{4}x^2 + \frac{1}{9}y^2 = 1$
3   $x + y$ on $\frac{1}{4}x^2 + \frac{1}{9}y^2 = 1$ 4 xy on $x^2 + xy + 4y^2 = 1$
5   xy^2 on $x^2 + y^2 = 1$ 6 $x + y^2$ on $x^2 + y^2 = 1$
7   $x - y$ on the branch $x > 0$ of $\frac{1}{9}x^2 - \frac{1}{4}y^2 = 1$
8   $x - y$ on $\frac{1}{9}x^2 - \frac{1}{4}y^2 = 1$
9   $x - y$ on the branch $x > 0$ of $\frac{1}{4}x^2 - \frac{1}{9}y^2 = 1$
10   $x - y$ on the branch $x > 0$ of $x^2 - y^2 = 1$
11* $8y - (x - 3)^2$ on $y^2 = x$ 12* $y - (x - 1)^2$ on $4y^2 - 3x = 0$
13   x^3y on $\sqrt{x} + \sqrt{y} = 1$ 14 y/x on $(x - 3)^2 + (y - 3)^2 = 6$
15   $x + y$ on $y^2 = x^2 + x^3$, $x \le 0$ 16 x^2y^3 on $\frac{1}{2}x + \frac{1}{6}y = 1$, $-\frac{1}{4} \le x \le 2$
17   $y - 2x$ on $x^3 = y^2$, $y \ge 0$ 18 x^2y^3 on $x^2 + y^2 = 15$.

19   Find the dimensions of the right circular cylinder of fixed total surface area A, including top and bottom, with maximum volume.
20   Find the dimensions of the right circular cylinder of fixed lateral area A with maximum volume.
21   Find the maximal area of a right triangle of perimeter 1.
22   Find the dimensions of the right circular cone of fixed total surface area A, including the base, with maximum volume.
23   Maximize $x + y - \sqrt{x^2 + y^2}$, where x and y are the legs of a right triangle of area 1.
24   Maximize $x + y - \sqrt{x^2 + y^2}$, where x and y are the legs of a right triangle of perimeter $3 + 2\sqrt{2}$.
25   Maximize $(x^2 + y^2)(2x + y)$ on $x^2 + y^2 \le 9$.
26   Let $f(x, y) = x^{3/2} + 2y^{3/2}$. Maximize $|\operatorname{grad} f|$ on $x^2 + (y - 1)^2 \le 1$.
27   Given $\mathbf{a}_1, \cdots, \mathbf{a}_n$ in the plane, find the $\mathbf{x}$ on the circle $|\mathbf{x}| = 1$ for which $\sum_1^n |\mathbf{x} - \mathbf{a}_j|^2$ is least.
28   Derive the second derivative test for constrained maxima and minima from the second derivative test for unconstrained maxima and minima.
29   Let $0 < p < q$ and $b > 0$. Find the max and min of $x^p + y^p$ on $x^q + y^q = b^q$ for $x \ge 0$ and $y \ge 0$.
30*   (cont.) Let $0 < p < q$ and $x \ge 0$, $y \ge 0$. Prove

$$\frac{1}{2^{1/p - 1/q}} \left(\frac{x^q + y^q}{2} \right)^{1/q} \le \left(\frac{x^p + y^p}{2} \right)^{1/p} \le \left(\frac{x^q + y^q}{2} \right)^{1/q}.$$

[Hint   Set $x^q + y^q = b^q$.]

5. FURTHER CONSTRAINT PROBLEMS

The constraint problem in three variables is to maximize (or minimize) a function $f(x, y, z)$ subject to a constraint $g(x, y, z) = c$. Geometrically, this problem is completely analogous to the two variable problem treated in the preceding section. We try to find the level surface of $f(x, y, z)$ of highest level that intersects the surface $g(x, y, z) = c$. See Fig. 1a. Each point of tangency is a candidate for a maximum or a minimum. At such a point, the normals to the two surfaces are parallel (Fig. 1b). But the vectors

$$\operatorname{grad} f(\mathbf{x}) \qquad \text{and} \qquad \operatorname{grad} g(\mathbf{x})$$

point in the respective normal directions, hence one is a multiple of the other:

$$\operatorname{grad} f(\mathbf{x}) = \lambda \operatorname{grad} g(\mathbf{x})$$

for some number λ, provided $\operatorname{grad} g(\mathbf{x}) \neq \mathbf{0}$. This observation leads to a practical method for locating possible maxima and minima (proof omitted).

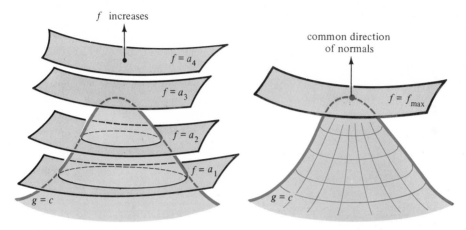

(a) To find the highest level surface   (b) The highest level $f = f_{\max}$ and $g = c$ are
     of f that meets $g = c$.   are tangent at their common contact point.

Fig. 1   Lagrange multiplier rule

Lagrange Multiplier Rule To maximize (minimize) a function $f(x, y, z)$ subject to a constraint $g(x, y, z) = 0$, solve the system of equations

$$(f_x, f_y, f_z) = \lambda(g_x, g_y, g_z), \qquad g(x, y, z) = 0$$

in the four unknowns x, y, z, λ. Each resulting point (x, y, z) is a candidate.

In applications, the usual precautions concerning the boundary must be observed.

■ **EXAMPLE 1**   Find the points on the ellipsoid $\frac{1}{9}x^2 + y^2 + \frac{1}{4}z^2 = 1$ nearest to and farthest from the origin.

Solution   The problem calls for the extrema of

$$f(x, y, z) = x^2 + y^2 + z^2 \quad \text{subject to} \quad g(x, y, z) = \tfrac{1}{9}x^2 + y^2 + \tfrac{1}{4}z^2 = 1.$$

Now $\operatorname{grad} f = (2x, 2y, 2z)$ and $\operatorname{grad} g = (\tfrac{2}{9}x, 2y, \tfrac{1}{2}z)$,

so according to the Lagrange Multiplier Rule, we must solve the equations

$$(2x, 2y, 2z) = \lambda(\tfrac{2}{9}x, 2y, \tfrac{1}{2}z), \qquad \tfrac{1}{9}x^2 + y^2 + \tfrac{1}{4}z^2 = 1,$$

that is, $\quad 2x = \tfrac{2}{9}\lambda x, \quad 2y = 2\lambda y, \quad 2z = \tfrac{1}{2}\lambda z, \quad \tfrac{1}{9}x^2 + y^2 + \tfrac{1}{4}z^2 = 1.$

At least one of x, y, z is non-zero. If $x \neq 0$, then the first equation implies $\lambda = 9$; hence the second and third equations imply $y = z = 0$. Now by the last equation, $x^2 = 9$, $x = \pm 3$. Thus the points $(\pm 3, 0, 0)$ are candidates for extrema.

If $y \neq 0$, then the second equation implies $\lambda = 1$, which leads in the same way to candidates $(0, \pm 1, 0)$. If $z \neq 0$, then $\lambda = 4$ and the candidates are $(0, 0, \pm 2)$. Since

$$f(\pm 3, 0, 0) = 9, \qquad f(0, \pm 1, 0) = 1, \qquad f(0, 0, \pm 2) = 4,$$

the max distance is $\sqrt{9} = 3$, the min is 1.   ∎

Remark   Compare this procedure with the previous solution of the same problem, Example 3, p. 809, and you will see the advantage of the multiplier method.

■ **EXAMPLE 2**   Find the volume of the largest rectangular solid with sides parallel to the coordinate axes that can be inscribed in the ellipsoid $x^2 + \frac{1}{9}y^2 + \frac{1}{4}z^2 = 1$.

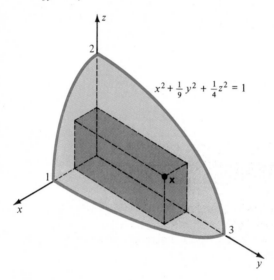

$$x^2 + \frac{1}{9}y^2 + \frac{1}{4}z^2 = 1$$

Fig. 2

Solution   As Fig. 2 shows, one-eighth of the volume is xyz, where $x > 0, y > 0, z > 0$. Hence it suffices to maximize

$$f(x, y, z) = xyz \qquad \text{subject to} \quad g(x, y, z) = x^2 + \tfrac{1}{9}y^2 + \tfrac{1}{4}z^2 = 1.$$

Apply the Lagrange Multiplier Rule by setting grad $f = \lambda$ grad g and $g = 1$:

$$(yz, zx, xy) = \lambda(2x, \tfrac{2}{9}y, \tfrac{1}{2}z), \qquad x^2 + \tfrac{1}{9}y^2 + \tfrac{1}{4}z^2 = 1,$$

that is, $yz = 2\lambda x, \qquad zx = \tfrac{2}{9}\lambda y, \qquad xy = \tfrac{1}{2}\lambda z, \qquad x^2 + \tfrac{1}{9}y^2 + \tfrac{1}{4}z^2 = 1.$

To solve these equations, multiply the first two and cancel xy:

$$z^2 = \tfrac{4}{9}\lambda^2. \qquad \text{Likewise} \qquad x^2 = \tfrac{1}{9}\lambda^2, \quad y^2 = \lambda^2.$$

(This is valid because if x or y is zero, then the volume is 0; not for us.) Substitute into the fourth equation:

$$\tfrac{1}{9}\lambda^2 + \tfrac{1}{9}\lambda^2 + \tfrac{1}{9}\lambda^2 = 1, \qquad \lambda^2 = 3, \qquad \text{hence} \quad x^2 = \tfrac{1}{3}, \quad y^2 = 3, \quad z^2 = \tfrac{4}{3}.$$

Therefore $f(x, y, z)^2 = x^2 y^2 z^2 = \tfrac{4}{3}, \qquad f_{max} = \tfrac{2}{3}\sqrt{3}.$

The maximal volume is 8 times this: $V_{max} = \tfrac{16}{3}\sqrt{3}$.   ∎

Remark   Another way of solving for x, y, z, λ is to solve the first three equations for λ and equate the results:

$$\lambda = \frac{yz}{2x} = \frac{9zx}{2y} = \frac{2xy}{2}.$$

It follows easily that $y^2 = 9x^2$ and $z^2 = 4x^2$. Substituting these values into the fourth equation, we find $3x^2 = 1$, $x^2 = \frac{1}{3}$, etc.

Two Constraints   Suppose the problem is to maximize (minimize) $f(x, y, z)$, where (x, y, z) is subject to *two* constraints, $g(x, y, z) = a$ and $h(x, y, z) = b$. Each constraint defines a surface, and these two surfaces in general have a curve of intersection (Fig. 3a). A candidate for a maximum or minimum of $f(\mathbf{x})$ is a point $\mathbf{x}$ where a level surface of f is tangent to this curve of intersection (Fig. 3b). The vector $\operatorname{grad} f(\mathbf{x})$ is normal to the level surface at $\mathbf{x}$, hence normal to the curve. But the vectors $\operatorname{grad} g(\mathbf{x})$ and $\operatorname{grad} h(\mathbf{x})$ *determine* the normal plane to the curve at $\mathbf{x}$. Hence for some constants λ and μ,

$$\operatorname{grad} f(\mathbf{x}) = \lambda \operatorname{grad} g(\mathbf{x}) + \mu \operatorname{grad} h(\mathbf{x}).$$

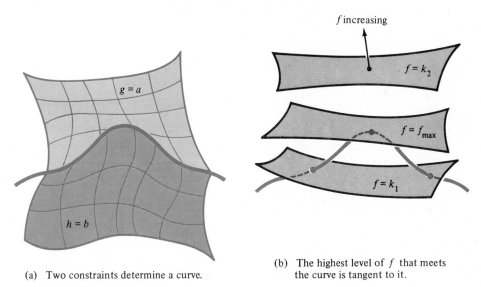

(a) Two constraints determine a curve.

(b) The highest level of f that meets the curve is tangent to it.

Fig. 3   Lagrange multiplier rule for two constraints

Remark   The existence of such **multipliers** λ and μ presupposes that $\operatorname{grad} g \neq \mathbf{0}$, $\operatorname{grad} h \neq \mathbf{0}$, and that neither is a multiple of the other.

Lagrange Multiplier Rule   To maximize (minimize) a function $f(x, y, z)$ subject to two constraints $g(x, y, z) = a$ and $h(x, y, z) = b$, solve the system of five equations

$$\begin{cases} (f_x, f_y, f_z) = \lambda(g_x, g_y, g_z) + \mu(h_x, h_y, h_z) \\ g(x, y, z) = a, \qquad h(x, y, z) = b \end{cases}$$

in five unknowns x, y, z, λ, μ. Each resulting point (x, y, z) is a candidate.

■ EXAMPLE 3 Find the maximum and minimum of $f(x, y, z) = x + 2y + z$ on the ellipse $x^2 + y^2 = 1$, $y + z = 1$.

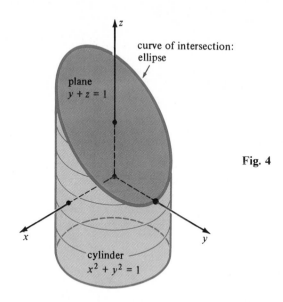

curve of intersection:
ellipse

plane
$y + z = 1$

Fig. 4

x

y

cylinder
$x^2 + y^2 = 1$

Solution The ellipse is the intersection of the cylinder $x^2 + y^2 = 1$ and the plane $y + z = 1$. See Fig. 4. Maximizing $f(x, y, z)$ on the ellipse is equivalent to maximizing $f(x, y, z)$ subject to the constraints $g(x, y, z) = 1$ and $h(x, y, z) = 1$, where

$$g(x, y, z) = x^2 + y^2 \qquad \text{and} \qquad h(x, y, z) = y + z.$$

According to the Lagrange Multiplier Rule, the equations to be solved are

$$\begin{cases} (1, 2, 1) = \lambda(2x, 2y, 0) + \mu(0, 1, 1) \\ x^2 + y^2 = 1, \qquad y + z = 1. \end{cases}$$

From the first equation,

$$1 = 2\lambda x, \qquad 2 = 2\lambda y + \mu, \qquad 1 = \mu; \qquad \text{hence} \qquad x = \frac{1}{2\lambda}, \qquad y = \frac{1}{2\lambda}.$$

Therefore $\qquad \left(\dfrac{1}{2\lambda}\right)^2 + \left(\dfrac{1}{2\lambda}\right)^2 = 1, \qquad \lambda^2 = \dfrac{1}{2}, \qquad \lambda = \pm\dfrac{\sqrt{2}}{2}.$

The solution $\lambda = \frac{1}{2}\sqrt{2}$ yields $\mathbf{x}_1 = (\frac{1}{2}\sqrt{2}, \frac{1}{2}\sqrt{2}, 1 - \frac{1}{2}\sqrt{2})$.

The solution $\lambda = -\frac{1}{2}\sqrt{2}$ yields $\mathbf{x}_2 = (-\frac{1}{2}\sqrt{2}, -\frac{1}{2}\sqrt{2}, 1 + \frac{1}{2}\sqrt{2})$.

These are the only candidates for maxima or minima. But $f(\mathbf{x}_1) = 1 + \sqrt{2}$ and $f(\mathbf{x}_2) = 1 - \sqrt{2}$. Therefore

$$f_{\max} = f(\mathbf{x}_1) = 1 + \sqrt{2} \qquad \text{and} \qquad f_{\min} = f(\mathbf{x}_2) = 1 - \sqrt{2}. \qquad ■$$

EXERCISES

Find the max and the min, and the points where they are taken

1 $2x + y - 5z$ on $x^2 + y^2 + z^2 = 1$ **2** $x + y + z$ on $\dfrac{x^2}{a^2} + \dfrac{y^2}{b^2} + \dfrac{z^2}{c^2} = 1$

3 $x + y + z$ on $\dfrac{1}{x} + \dfrac{4}{y} + \dfrac{9}{z} = 1$, $\quad x > 0, \; y > 0, \; z > 0$

4 xy^2z^3 on $x + y + z = 3$, $\quad x \geq 0, \; y \geq 0, \; z \geq 0$

5 xyz on $x^2 + y^2 + 3z^2 = 5$

6 $e^x yz$ on $4x + 2y + z = 12$, $\quad x \geq 0, \; y \geq 0, \; z \geq 0$

7 xyz on $\dfrac{x}{a} + \dfrac{y}{b} + \dfrac{z}{c} = 1$, $\quad x \geq 0, \; y \geq 0, \; z \geq 0$, where $a, b, c > 0$

8 $x^4 + y^4 + z^4$ on $x^2 + y^2 + z^2 = 1$.

9   Show that the maximum of $ax + by + cz$ subject to $x^2 + y^2 + z^2 = r^2$ is $r\sqrt{a^2 + b^2 + c^2}$. Use Lagrange multipliers. Assume $(a, b, c) \neq \mathbf{0}$.

10   (cont.) Obtain the same conclusion by means of vectors and the Cauchy–Schwarz inequality.

Use the result of Ex. 9 to maximize

11 $\dfrac{2x - 2y + z}{1 + x^2 + y^2 + z^2}$ **12** $\dfrac{ax + by + cz}{e^{x^2 + y^2 + z^2}}$.

13   Find the rectangular solid of fixed volume with minimum surface area.

14   Find the rectangular solid of fixed total edge length with maximum surface area.

15   A silo is built in the form of a right circular cylinder topped by a right circular cone. What dimensions will give a given volume with the least possible surface area?

16   Given $\mathbf{a}_1, \cdots, \mathbf{a}_n$ in space, find the point $\mathbf{x}$ of the plane $\mathbf{x} \cdot \mathbf{c} = p$ such that $\sum_1^n |\mathbf{x} - \mathbf{a}_j|^2$ is least. Assume $|\mathbf{c}| = 1$, and express your answer in terms of $\mathbf{a} = n^{-1} \sum \mathbf{a}_j$.

17   Maximize $x^2 y^2 z^2$ on $x^2 + y^2 + z^2 = r^2$.

18   (cont.) Prove the inequality $\sqrt[3]{xyz} \leq \dfrac{x + y + z}{3}$ $\quad$ for $\quad x > 0, \; y > 0, \; z > 0$.

19   Prove the inequalities $-\tfrac{1}{2}(x^2 + y^2 + z^2) \leq xy + yz + zx \leq x^2 + y^2 + z^2$.

20   Let $\mathbf{a}_1, \cdots, \mathbf{a}_n$ be points of space. Set $\mathbf{a} = n^{-1} \sum_1^n \mathbf{a}_j$ and assume $\mathbf{a} \neq \mathbf{0}$. Find the points $\mathbf{x}$ of the unit sphere $|\mathbf{x}| = 1$ that yield the max and min of $\sum |\mathbf{x} - \mathbf{a}_i|^2$.

21   Use Heron's formula and Lagrange multipliers to maximize the area of a triangle with fixed semi-perimeter s.

22   Find the maximal area in Fig. 5 if the perimeter is $4 + 2\sqrt{3}$.

23   Prove that $\cot x - 1/x$ is strictly decreasing for $0 < x < \pi$.

24*   (cont.) Let x, y, z be the angles of a triangle. Find the maximum of

$$\frac{\sin x \, \sin y \, \sin z}{xyz}$$

and where this maximum is taken.

Find the point of the first octant nearest the origin and on the surface

25 $xyz = 1$ **26** $xyz^2 = 1$ **27** $x^2 + yz = 1$

28 $x^3 + 3yz = 3$ **29** $x(y^2 + z^2) = 1$ **30** $xy^2z^3 = 6\sqrt{3}$.

31   Let $0 < p < q$. Find the maximum and minimum of $x^p + y^p + z^p$ on the surface
$x^q + y^q + z^q = b^q$ where $x \geq 0$, $y \geq 0$, and $z \geq 0$.

32   (cont.) Let $0 < p < q$ and $x \geq 0$, $y \geq 0$, $z \geq 0$. Show that

$$\left(\frac{x^p + y^p + z^p}{3}\right)^{1/p} \leq \left(\frac{x^q + y^q + z^q}{3}\right)^{1/q}$$

33   Find the max and min of $\sqrt{x^2 + y^2 + z^2}$ on $x^3 + y^2 z = 62$, $x \geq 0$, $y \geq 0$, $z \geq 0$,
and where they occur.

34   Find the max of xyz on $x^2 + y^3 + z^4 = \frac{13}{3}$, $x \geq 0$, $y \geq 0$, $z \geq 0$, and where it
occurs.

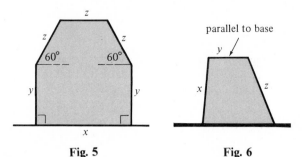

Fig. 5   **Fig. 6**

35*   Find the maximal area in Fig. 6, provided $x + y + z = 6k$. [*Hint* Introduce the
altitude and treat this as a problem in four variables with one constraint.]

36*   Suppose $p > 1$, $q > 1$, $\dfrac{1}{p} + \dfrac{1}{q} = 1$, $a > 0$, $b > 0$, $c > 0$, and $a^q + b^q + c^q = d^q$.
Find the max of $ax + by + cz$ on $x^p + y^p + z^p = r^p$, $x \geq 0$, $y \geq 0$, $z \geq 0$.

Find the closest point to the origin

37   on the line $x + y + z = 1$, $x + 2y + 3z = 1$

38   on the ellipse $x + y + z = 0$, $x^2 + \frac{1}{4}y^2 + \frac{1}{4}z^2 = 2$.

39   Find the maximum and minimum volumes of a rectangular solid whose total edge
length is 24 ft and whose surface area is 22 ft^2.

40   Find the max of x on the circle $x^2 + y^2 + z^2 = 1$, $x + 2y + 3z = 0$.

41   Find the min of $2x^2 + y^2 + z^2$ on the line $2x - y + z = 1$, $x - 2y - z = 2$.

42*   Find the max and min of $x + y + z$ on the curve $xyz = 1$, $x^2 + y^2 + z^2 = 5$,
$x > 0$, $y > 0$, $z > 0$.

43   A rectangular solid has volume 30. The sum of the area of its base and the square of its
height is 19. Find the max and min of the perimeter of its base.

44   Find the maximal area of a right triangle of perimeter 1 by treating its three edge
lengths as variables subject to 2 constraints.

6. ADDITIONAL TOPICS

In this section we complete some unfinished business of the previous sections.

Equality of Mixed Partials   We must prove that $f_{xy}(a, b) = f_{yx}(a, b)$ under the
hypotheses that f_{xy} and f_{yx} exist at all points near (a, b) and are continuous at
(a, b).

We consider the mixed second difference

(1) $\qquad \Delta = [f(a + h, b + k) - f(a + h, b)] - [f(a, b + k) - f(a, b)]$

and its alternative form

(2) $\qquad \Delta = [f(a + h, b + k) - f(a, b + k)] - [f(a + h, b) - f(a, b)].$

We shall apply the Mean Value Theorem (twice) to (1). To do this, we set

$$g(x) = f(x, b + k) - f(x, b).$$

Then $\qquad \Delta = g(a + h) - g(a) = hg'(x_1),$

where x_1 is between a and $a + h$. Next,

$$g'(x_1) = f_x(x_1, b + k) - f_x(x_1, b) = kf_{xy}(x_1, y_1),$$

where y_1 is between b and $b + k$. Hence

(3) $\qquad\qquad\qquad \Delta = hkf_{xy}(x_1, y_1).$

We apply similar reasoning to the second expression for Δ to obtain

(4) $\qquad\qquad\qquad \Delta = hkf_{yx}(x_2, y_2),$

where x_2 is between a and $a + h$ and y_2 is between b and $b + k$.
    Now we take $h = k \neq 0$. By (3) and (4),

(5) $\qquad\qquad\qquad f_{xy}(x_1, y_1) = f_{yx}(x_2, y_2).$

Let $h \longrightarrow 0$. Then $(x_1, y_1) \longrightarrow (a, b)$ and $(x_2, y_2) \longrightarrow (a, b)$. Since f_{xy} and f_{yx} are continuous at (a, b), we deduce from (5) that

$$f_{xy}(a, b) = f_{yx}(a, b).$$

Error in Taylor Approximation   We shall now prove the boxed statements on pp. 801–2. The idea is to interpret the problem in such a way that we can use the error estimates on p. 800 for a function of one variable.
    Assume at first that $\mathbf{a} = \mathbf{0}$; this will simplify the notation considerably. Now fix a point $\mathbf{x}$ in $\mathbf{D}$. By convexity, $\mathbf{D}$ contains the entire line segment connecting $\mathbf{0}$ and $\mathbf{x}$, that is, $\mathbf{D}$ contains all points $t\mathbf{x}$ for $0 \leq t \leq 1$.
    Set $g(t) = f(t\mathbf{x})$. Then $g(t)$ is a function of one variable defined for $0 \leq t \leq 1$ and

$$g(0) = f(\mathbf{0}), \qquad g(1) = f(\mathbf{x}).$$

Let us compute the first and second degree Taylor polynomials of g at $t = 0$. By the Chain Rule,

$$g'(t) = f_x(t\mathbf{x})x + f_y(t\mathbf{x})y, \qquad g'(0) = f_x(\mathbf{0})x + f_y(\mathbf{0})y,$$

$$g''(t) = f_{xx}(t\mathbf{x})x^2 + 2f_{xy}(t\mathbf{x})xy + f_{yy}(t\mathbf{x})y^2,$$

$$g''(0) = f_{xx}(\mathbf{0})x^2 + 2f_{xy}(\mathbf{0})xy + f_{yy}(\mathbf{0})y^2,$$

$$g'''(t) = f_{xxx}(t\mathbf{x})x^3 + 3f_{xxy}(t\mathbf{x})x^2y + 3f_{xyy}(t\mathbf{x})xy^2 + f_{yyy}(t\mathbf{x})y^3.$$

From these calculations we see that

$$p_1(\mathbf{x}) = g(0) + g'(0), \qquad p_2(\mathbf{x}) = g(0) + g'(0) + \tfrac{1}{2}g''(0).$$

Thus $p_1(\mathbf{x})$ and $p_2(\mathbf{x})$ are the first and second degree Taylor polynomials of $g(t)$ evaluated at $t = 1$. Therefore

$$|r_1(\mathbf{x})| = |f(\mathbf{x}) - p_1(\mathbf{x})| \leq \frac{\max|g''(t)|}{2!},$$

$$|r_2(\mathbf{x})| = |f(\mathbf{x}) - p_2(\mathbf{x})| \leq \frac{\max|g'''(t)|}{3!}.$$

It remains to estimate $|g''(t)|$ and $|g'''(t)|$. We have

$$|g''(t)| \leq M_2|x|^2 + 2M_2|x||y| + M_2|y|^2 = M_2(|x| + |y|)^2,$$

$$|g'''(t)| \leq M_3|x|^3 + 3M_3|x|^2|y| + 3M_3|x||y|^2 + M_3|y|^3 = M_3(|x| + |y|)^3.$$

Now we modify these estimates slightly as follows. From $(|x| - |y|)^2 \geq 0$ we have $2|x||y| \leq |x|^2 + |y|^2$, hence

$$(|x| + |y|)^2 = |x|^2 + 2|x||y| + |y|^2 \leq 2(|x|^2 + |y|^2) = 2|\mathbf{x}|^2.$$

Take the $\tfrac{3}{2}$ power: $\qquad (|x| + |y|)^3 \leq 2^{3/2}|\mathbf{x}|^3.$

Therefore, $\qquad |g''(t)| \leq 2M_2|\mathbf{x}|^2, \qquad |g'''(t)| \leq 2\sqrt{2}\,M_3|\mathbf{x}|^3.$

Combining results we obtain

$$|r_1(\mathbf{x})| \leq \frac{2M_2|\mathbf{x}|^2}{2!}, \qquad |r_2(\mathbf{x})| \leq \frac{2\sqrt{2}\,M_3|\mathbf{x}|^3}{3!}.$$

This completes the proof assuming $\mathbf{a} = \mathbf{0}$. In the general case, we define $g(t) = f[\mathbf{a} + t(\mathbf{x} - \mathbf{a})]$. The proof proceeds as before, except that (x, y) is replaced by $(x - a, y - b)$ and the partials of f are all evaluated at $\mathbf{a}$.

The Second Derivative Test The second derivative test (for a minimum) was stated on p. 805, and its proof postponed until now. We require a preliminary lemma about positive definite quadratic forms.

Lemma Let $Ax^2 + 2Bxy + Cy^2$ be a positive definite quadratic form. Then there exists a constant $k > 0$ such that

$$Ax^2 + 2Bxy + Cy^2 > k(x^2 + y^2) \qquad \text{for all} \quad (x, y) \neq (0, 0).$$

Proof We are given $A > 0$ and $AC - B^2 > 0$. We choose $k > 0$ so small that

$$A - k > 0 \qquad \text{and} \qquad (A - k)(C - k) - B^2 > 0.$$

[Reason that $g(A, C) = AC$ is continuous so

$$g(A - k, C - k) \longrightarrow g(A, C) = AC > B^2 \text{ as } k \longrightarrow 0+.]$$

These inequalities imply that

$$(A - k)x^2 + 2Bxy + (C - k)y^2$$

is positive definite, that is, takes positive values for all $(x, y) \neq (0, 0)$. This means

$$Ax^2 + 2Bxy + Cy^2 > k(x^2 + y^2)$$

whenever $(x, y) \neq (0, 0)$.

Now we can prove the boxed statement on p. 805. To make the notation simple, let us take $(a, b) = (0, 0)$. Then the second degree Taylor approximation of f at $(0, 0)$ is

$$f(x, y) = f(0, 0) + \tfrac{1}{2}[Ax^2 + 2Bxy + Cy^2] + r(x, y), \qquad |r(x, y)| < h|\mathbf{x}|^3,$$

where $A = f_{xx}(0, 0)$, etc. and $h > 0$ is a constant. Since $Ax^2 + 2Bxy + Cy^2$ is positive definite, the lemma provides a constant $k > 0$ such that

$$Ax^2 + 2Bxy + Cy^2 \geq k|\mathbf{x}|^2.$$

Consequently $f(x, y) - f(0, 0) \geq \tfrac{1}{2}k|\mathbf{x}|^2 - h|\mathbf{x}|^3 = |\mathbf{x}|^2(\tfrac{1}{2}k - h|\mathbf{x}|)$.

This implies $f(x, y) - f(0, 0) > 0$, that is $f(x, y) > f(0, 0)$, provided $0 < |\mathbf{x}| < \tfrac{1}{2}k/h$. Finally, for δ we may choose any number such that $0 < \delta \leq \tfrac{1}{2}k/h$ and so small that the disk $|\mathbf{x}| < \delta$ is contained in the domain of $f(x, y)$.

Remark   The essential point in the proof is that for $|\mathbf{x}|$ small, the positive definite quadratic form $Ax^2 + 2Bxy + Cy^2$ is much larger than the remainder $r(x, y)$ because r is of third order: $|r(x, y)| \leq h|\mathbf{x}|^3$.

The Saddle Point Test   Before we prove the saddle point test (p. 807), we need another lemma about quadratic forms.

Lemma   Let $Q(x, y) = Ax^2 + 2Bxy + Cy^2$, and suppose $AC - B^2 < 0$. Then there are points (x_1, y_1) and (x_2, y_2) such that

$$Q(x_1, y_1) < 0 \qquad \text{and} \qquad Q(x_2, y_2) > 0.$$

Proof   We assume $A \neq 0$ and complete the square:

$$Q(x, y) = Ax^2 + 2Bxy + Cy^2$$

$$= A\left(x + \frac{B}{A}y\right)^2 + \frac{AC - B^2}{A}y^2 = A\left(x + \frac{B}{A}y\right)^2 + C_1 y^2,$$

where
$$AC_1 = A\left(\frac{AC - B^2}{A}\right) = AC - B^2 < 0.$$

Clearly
$$Q(1, 0) = A \qquad \text{and} \qquad Q(-B/A, 1) = C_1$$

have opposite signs.

The same argument applies if $A = 0$ but $C \neq 0$. If both $A = 0$ and $C = 0$, then $B \neq 0$ since $-B^2 = AC - B^2 < 0$. In that case $Q(x, y) = 2Bxy$ and $Q(1, 1)$ and $Q(1, -1)$ have opposite signs.

Now we can prove the boxed statement on p. 807. We take $(a, b) = (0, 0)$ for simplicity and consider the Taylor approximation

$$f(x, y) = f(0, 0) + \tfrac{1}{2}Q(x, y) + r(x, y), \qquad |r(x, y)| < h|\mathbf{x}|^3,$$

where $Q(x, y) = Ax^2 + 2Bxy + Cy^2$. Since

$$AC - B^2 < 0,$$

the lemma gives us points (x_1, y_1) and (x_2, y_2) such that $Q(x_1, y_1) < 0$ and $Q(x_2, y_2) > 0$.

Consider $(x, y) = (tx_1, ty_1)$, where $t > 0$ and small. Then

$$\tfrac{1}{2}Q(tx_1, ty_1) = \tfrac{1}{2}Q(x_1, y_1)t^2 = kt^2, \qquad k < 0,$$

$$|r(tx_1, ty_1)| < h(x_1{}^2 + y_1{}^2)^{3/2}t^3 = h_1 t^3,$$

hence $\qquad f(tx_1, ty_1) - f(0, 0) \le kt^2 + h_1 t^3 = t^2(k + h_1 t).$

Now for t sufficiently small, $t^2(k + h_1 t) < 0$, hence

$$f(tx_1, ty_1) < f(0, 0)$$

for points (tx_1, ty_1) as close to $(0, 0)$ as we please.

Similarly $f(tx_2, ty_2) > f(0, 0)$ for points (tx_2, ty_2) as close to $(0, 0)$ as we please. Therefore $f(0, 0)$ is neither a max nor a min of f.

Functions of Three Variables A straightforward extension of the second derivative test applies to functions of three (or more) variables. We shall state it only for a local minimum. The corresponding result for a maximum is obtained by replacing f by $-f$.

Second Derivative Test Let $\mathbf{a}$ be an interior point of the domain of $f(x, y, z)$ such that $f_x(\mathbf{a}) = f_y(\mathbf{a}) = f_z(\mathbf{a}) = 0$. Suppose at $\mathbf{a}$ that

$$f_{xx} > 0, \qquad \begin{vmatrix} f_{xx} & f_{xy} \\ f_{yx} & f_{yy} \end{vmatrix} > 0, \qquad \begin{vmatrix} f_{xx} & f_{xy} & f_{xz} \\ f_{yx} & f_{yy} & f_{yz} \\ f_{zx} & f_{zy} & f_{zz} \end{vmatrix} > 0.$$

Then $f(\mathbf{a})$ is a strong local minimum of $f(\mathbf{x})$. Precisely, there exists $\delta > 0$ such that

$$f(\mathbf{x}) > f(\mathbf{a}) \qquad \text{whenever} \quad 0 < |\mathbf{x} - \mathbf{a}| < \delta.$$

For a strong local maximum the corresponding conditions are

$$f_{xx} < 0, \qquad \begin{vmatrix} f_{xx} & f_{xy} \\ f_{yx} & f_{yy} \end{vmatrix} > 0, \qquad \begin{vmatrix} f_{xx} & f_{xy} & f_{xz} \\ f_{yx} & f_{yy} & f_{yz} \\ f_{zx} & f_{zy} & f_{zz} \end{vmatrix} < 0.$$

The proof is similar to the proof for two variables except that some more advanced linear algebra is involved, so we shall omit it. Note that the conditions for a max of f are simply the conditions for a min of $-f$.

■ **EXAMPLE 1** Find all local minima of

$$f(x, y, z) = x^4 + y^4 + z^4 - 108x + 4y - 4z.$$

Solution $\operatorname{grad} f = (4x^3 - 108, 4y^3 + 4, 4z^3 - 4)$, so $\operatorname{grad} f = 0$ only for $(x, y, z) = (3, -1, 1)$. All the mixed second partials equal zero. The pure ones are $f_{xx} = 12x^2, f_{yy} = 12y^2, f_{zz} = 12z^2$, so

$$f_{xx}(3, -1, 1) = 108, \qquad f_{yy}(3, -1, 1) = 12, \qquad f_{zz}(3, -1, 1) = 12.$$

Therefore the second derivative test is satisfied: at $(3, -1, 1)$,

$$f_{xx} = 108 > 0, \qquad \begin{vmatrix} f_{xx} & f_{xy} \\ f_{yx} & f_{yy} \end{vmatrix} = \begin{vmatrix} 108 & 0 \\ 0 & 12 \end{vmatrix} = 108 \cdot 12 > 0,$$

$$\begin{vmatrix} f_{xx} & f_{xy} & f_{xz} \\ f_{yx} & f_{yy} & f_{yz} \\ f_{zx} & f_{zy} & f_{zz} \end{vmatrix} = \begin{vmatrix} 108 & 0 & 0 \\ 0 & 12 & 0 \\ 0 & 0 & 12 \end{vmatrix} = 108 \cdot 12 \cdot 12 > 0.$$

Therefore the only local minimum of $f(x, y, z)$ is

$$f(3, -1, 1) = 81 + 1 + 1 - 324 - 4 - 4 = -249.$$

[It is relatively easy to show that $f(\mathbf{x}) \longrightarrow \infty$ as $|\mathbf{x}| \longrightarrow \infty$, hence we have an absolute minimum.] ■

There is a corresponding second derivative test for constrained minima which we shall just mention. Suppose $f(x, y, z)$ and $g(x, y, z)$ are given, and

$$\operatorname{grad} f(\mathbf{a}) = \lambda \operatorname{grad} g(\mathbf{a}), \qquad g(\mathbf{a}) = c.$$

Suppose these inequalities hold at $\mathbf{a}$:

$$f_{xx} - \lambda g_{xx} > 0, \qquad \begin{vmatrix} f_{xx} - \lambda g_{xx} & f_{xy} - \lambda g_{xy} \\ f_{yx} - \lambda g_{yx} & f_{yy} - \lambda g_{yy} \end{vmatrix} > 0,$$

$$\begin{vmatrix} f_{xx} - \lambda g_{xx} & f_{xy} - \lambda g_{xy} & f_{xz} - \lambda g_{xz} \\ f_{yx} - \lambda g_{yx} & f_{yy} - \lambda g_{yy} & f_{yz} - \lambda g_{yz} \\ f_{zx} - \lambda g_{zx} & f_{zy} - \lambda g_{zy} & f_{zz} - \lambda g_{zz} \end{vmatrix} > 0.$$

Then there exists $\delta > 0$ such that $f(\mathbf{x}) > f(\mathbf{a})$ whenever $g(\mathbf{x}) = c$ and $0 < |\mathbf{x} - \mathbf{a}| < \delta$.

EXERCISES

The second derivative test is inconclusive at $(0, 0, 0)$ for the following functions. Determine nonetheless if the function has a local max, local min, or neither at the origin

1	$x^2 + y^2 + z^4$	2	$x^2 + y^2 z^2$	3	$x^2 + y^2$
4	$x^4 + y^2 - z^6$	5	$x^2 + y^4 + z^6$	6	$x^3 y^3 z^3$
7	$x^4 + y^3 z^3$	8	$x^4 y^4 - z^5$.	9	$x^4 + y^2 z^2$
10	$x^3 + y^3 + z^3$	11	$x^4 y^6 z^3$	12	$x^2 y^2 z^2$.

Use the first and second derivative tests to find all local maxs and mins.

13 $-2x^2 - y^2 - 3z^2 + 2xy - 2xz$ **14** $x^2 + 2y^2 + z^2 + 2xy - 4yz$
15 $2x^2 + y^2 + 2z^2 + 2xy + 2yz + 2zx$ **16** $x^2 + 3xy + y^2 - z^2 - x - 2y + z + 3.$
$+ x - 3z$

Does the function have a local max, local min, or neither at $(0, 0, 0)$?

17 $x^2 + y^2 + z^2 + xy + yz + zx$ **18** $x^2 + 4y^2 + 9z^2 - xy - 2yz$
19 $-x^2 - 2y^2 - z^2 + yz$ **20** $x^2 + y^2 + 2z^2 - 10yz$
21 $x^2 - y^2 + 3z^2 + 12xy$ **22** $3x^2 + y^2 + 4z^2 - xy - yz - zx.$

The next four exercises show that f_{xy} and f_{yx} are not necessarily equal if they fail to be continuous

23   Set $f(0, 0) = 0$ and

$$f(x, y) = \frac{(x - y)(x^3 + y^3)}{x^2 + y^2} \qquad \text{for} \quad (x, y) \neq (0, 0).$$

Show that $f(x, y)$ is continuous everywhere.
24   (cont.) Compute f_x and f_y at $(x, y) \neq (0, 0)$.
25   (cont.) Compute $f_x(0, 0)$ and $f_y(0, 0)$. Conclude that f_x and f_y are continuous everywhere.
26   (cont.) Compute $f_{xy}(0, 0)$ and $f_{yx}(0, 0)$.

Suppose $f(x, y)$ has continuous second partials

27   Assume $f(x, y) = f(y, x)$. Prove $f_{xx}(c, c) = f_{yy}(c, c)$
28   Assume $f(x, y) = -f(y, x)$. Prove $f_{xv}(c, c) = 0$.
29   The form $x^2 + 2xy + 2y^2$ is positive definite so there is a $k > 0$ such that $x^2 + 2xy + 2y^2 \geq k(x^2 + y^2)$ for all (x, y). Find the largest such k.
30*   (cont.) Solve this problem in general for a positive definite $f(\mathbf{x}) = ax^2 + 2bxy + cy^2$.
31   Suppose $f(x, y)$ is a function of two variables and $x = x(t)$ and $y = y(t)$ are functions of time. Form the composite function $g(t) = f[x(t), y(t)]$. By the Chain Rule, $\dot{g}(t) = f_x[x(t), y(t)]\dot{x}(t) + f_y[x(t), y(t)]\dot{y}(t)$. Show that

$$\ddot{g} = f_{xx}\dot{x}^2 + 2f_{xy}\dot{x}\dot{y} + f_{yy}\dot{y}^2 + f_x\ddot{x} + f_y\ddot{y}.$$

32   (cont.) Suppose also that $f_x(0, 0) = f_y(0, 0) = 0$, and that only curves $\mathbf{x}(t) = (x(t), y(t))$ are allowed which pass through $(0, 0)$ with speed 1 at $t = 0$. Suppose for each such curve $\ddot{g}(0) > 0$. Show that $\dot{g}(0) = 0$ and $f_{xx}(0, 0) > 0$ and $f_{yy}(0, 0) > 0$.
33*   (cont.) Show also that $f_{xx}(0, 0)f_{yy}(0, 0) - f_{xy}(0, 0)^2 > 0$.
34   Suppose $g(x, y) = Ax^2 + 2Bxy + Cy^2$ and $AC - B^2 = 0$. Conclude that $g(x, y) = \pm(ax + by)^2$.

7. MISCELLANEOUS EXERCISES

Find the second degree Taylor polynomial

1 $\arctan \dfrac{x - y}{1 + xy}$ at $(0, 0)$ **2** $\dfrac{x}{1 - x + y}$ at $(3, 1)$.
3   Given a function $f(x, y)$, let $u = x + y$ and $v = x - y$. Show that

$$\frac{\partial^2 f}{\partial x^2} - \frac{\partial^2 f}{\partial y^2} = 4\frac{\partial^2 f}{\partial u \, \partial v}.$$

4 (cont.) Find all functions that satisfy the partial differential equation $\dfrac{\partial^2 f}{\partial x^2} - \dfrac{\partial^2 f}{\partial y^2} = 0$.

5 The period of a pendulum is approximately $T = 2\pi\sqrt{L/g}$ sec, where L is its length in meters and g is the acceleration of gravity in m/sec^2. Suppose *relative* errors of h and k are made in measuring L and g. Show that the relative error in T, up to second order, is

$$\tfrac{1}{2}(h - k) - \tfrac{1}{8}[h^2 + 2hk - 3k^2].$$

6 Find all functions $f(x, y, z)$ that satisfy $\dfrac{\partial^3 f}{\partial x\, \partial y\, \partial z} = 0$.

A function is called **harmonic** if it satisfies **Laplaces' equation**

$$\frac{\partial^2 f}{\partial x^2} + \frac{\partial^2 f}{\partial y^2} + \frac{\partial^2 f}{\partial z^2} = 0.$$

Verify that the function is harmonic

7 $\arctan(y/x)$ **8** $x \sin x \cosh y - y \cos x \sinh y$

9 $x^5 - 10x^3 y^2 + 5xy^4$ **10** $e^{13x} \sin 5y \cos 12z$.

11 $\ln(x^2 + y^2)$ **12** $1/\sqrt{x^2 + y^2 + z^2}$.

13 Show that Laplace's equation in two variables, $\partial^2 f/\partial x^2 + \partial^2 f/\partial y^2 = 0$, expressed in polar coordinates, becomes

$$\frac{\partial^2 f}{\partial r^2} + \frac{1}{r}\frac{\partial f}{\partial r} + \frac{1}{r^2}\frac{\partial^2 f}{\partial \theta^2} = 0.$$

14 (cont.) Show that the functions $g_k(x, y) = r^k \cos k\theta$ and $h_k(x, y) = r^k \sin k\theta$ are harmonic for all integers k.

15 There is a function $z = f(x, y)$ such that $z^5 - xz^4 + yz^3 - 1 = 0$ and $f(1, 1) = 1$. Find $\partial^2 f/\partial x\, \partial y$ at $(1, 1)$.

16 Compute the maximum of $xy^2 z^3$ on the sphere $x^2 + y^2 + z^2 = 1$.

17 Find the greatest distance from the ellipse $4x^2 - 2xy + y^2 = 1$ to the x-axis.

18 Use the first and second derivative tests to locate the point where $f(x, y, z) = x^3 + x^2 + 4y^2 + 9z^2 + 4xy - 12zx$ has a local min.

19 Find the maximum of $x^3 y^2(12 - x - y)$ in the first quadrant.

20 Solve Ex. 19 by maximizing $x^3 y^2 z$ subject to a suitable condition on x, y, and z.

21 Find the largest possible area of a triangle of perimeter 1, if one of its angles is α.

22 Assume $a, b, c > 0$. Find the volume of the largest rectangular solid (with sides parallel to the coordinate planes) inscribed in the ellipsoid

$$\frac{x^2}{a^2} + \frac{y^2}{b^2} + \frac{z^2}{c^2} = 1.$$

23 Find the min of $\dfrac{a}{x} + \dfrac{b}{y} + \dfrac{c}{z} + xyz$ for $x > 0$, $y > 0$, $z > 0$. Assume $a > 0$, $b > 0$, $c > 0$.

24* Find the largest A and the smallest B so that

$$A(x^4 + y^4 + z^4)^3 \le (x^6 + y^6 + z^6)^2 \le B(x^4 + y^4 + z^4)^3 \qquad \text{for all} \quad (x, y, z).$$

25* Consider an ammonia synthesis reaction

$$N_2 + 3H_2 \;\rightleftharpoons\; 2NH_3$$

at fixed pressure P and fixed temperature. Then $P = x + y + z$, where x, y, z are the partial pressures of N_2, H_2, and NH_3 respectively. In this situation it is known that $z^2/xy^3 = 3k^2$, where k is a constant. Also, the concentration of each of the three ingredients is proportional to its partial pressure. The problem is to maximize z, so as much ammonia as possible is produced.

26   A production process uses input materials $X_1, \cdots, X_n$, and the output is a product Y. Let p_i be the (fixed) price per unit amount of X_i and x_i the amount of X_i used. Then the cost of producing Y is $C = \sum p_i x_i$. The amount of Y produced is $y = f(x_1, \cdots, x_n)$. We assume the **production function** f is homogeneous of degree 1. (In economics, this property of f is called "constant returns to scale".) Suppose $(x_1, \cdots, x_n)$ minimizes C subject to the constraint $f(x_1, \cdots, x_n) = k$, and all $x_i > 0$. Prove the **law of marginal productivity**: for each i,

$$\frac{p_i}{\partial f/\partial x_i} = \frac{C}{k} \qquad \text{at} \quad (x_1, \cdots, x_n).$$

(In a competitive market, C/k will be the unit price of Y.) [*Hint* Use Euler's relation, Ex. 9, p. 791.]

Double Integrals **17**

1. INTRODUCTION

The Volume Problem   In single variable calculus we discussed the problem of finding the area under a curve $y = f(x)$ and above an interval $[a, b]$. Its solution led to the definite integral (also called the **simple integral**). Now we take up the problem of finding the *volume* under a surface $z = f(x, y)$ and above a domain **D** of the x, y-plane (Fig. 1). This leads to a new kind of integral, called the **double integral**.

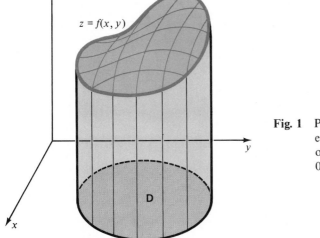

$z = f(x, y)$

Fig. 1   Problem: To define and
evaluate the volume
of the space region
$0 \le z \le f(x, y)$, (x, y) in **D**.

We attack the volume problem in basically the same way as we did the area problem. Assuming the volume of the region in Fig. 1 exists, we make reasonable approximations to it, then try to sneak up on the exact value by taking a limit.

First, we partition the domain **D** into many small domains $\mathbf{D}_1, \mathbf{D}_2, \cdots, \mathbf{D}_n$. See Fig. 2a. In each $\mathbf{D}_i$, we choose a point (x_i, y_i). Then we approximate that part of the region above $\mathbf{D}_i$ by a thin solid of height $f(x_i, y_i)$. Its volume is

$$(\text{height})(\text{area of base}) = f(x_i, y_i)|\mathbf{D}_i|.$$

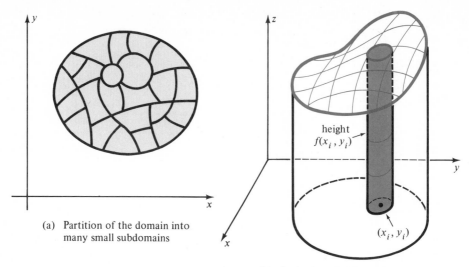

(a) Partition of the domain into
many small subdomains

(b) Long thin cylinder approximates part
of the volume under the surface

Fig. 2   Approximating the volume

See Fig. 2b. The sum of these volumes,

$$\sum_{i=1}^{n} f(x_i, y_i)|\mathbf{D}_i|,$$

ought to be close to the desired volume provided the $\mathbf{D}_i$'s are small enough. Therefore, we look for a limit as the $\mathbf{D}_i$'s shrink smaller and smaller. If the limit exists, we define it to be the volume and denote it by

$$\iint_{\mathbf{D}} f(x, y)\, dx\, dy.$$

So far everything looks practically the same as in a discussion of simple integrals. However, in two dimensions certain technical difficulties arise that do not exist at all in one dimension. For one thing, plane domains **D** can be much more complicated than intervals $[a, b]$. Consequently partitions of **D** into smaller domains can be nasty. Also there is the question of which functions are integrable.

These matters certainly require attention, but we shall not attempt a full treatment of double integrals. That is a hard subject, best left to courses in advanced calculus or the theory of real functions. We hope to give enough of the discussion to convey the flavor of the subject. We shall try to justify intuitively the various properties of double integrals, and concentrate our energy on learning how to evaluate and apply double integrals.

Domains   The simple integral integrates a function $f(x)$ over a *closed interval* $[a, b]$. But all closed intervals on the line look alike; if you've seen one, you've seen them all. In contrast, plane domains can be exceedingly complicated. We avoid difficulties by limiting attention to a restricted type of domain, which we shall call a domain of integration. Such domains are general enough for all practical purposes. They are bounded (stay in a finite part of the plane—don't go off to infinity in any direction) and have reasonable boundaries.

A **domain of integration** in the plane is a bounded domain **D** that includes all points of its boundary. Its boundary must consist of a finite number of graphs of convex or concave functions.

Here "graph of a convex function" means either

$$y = f(x), \quad a \leq x \leq b, \quad f''(x) \geq 0 \qquad \text{or} \qquad x = g(y), \quad c \leq y \leq d, \quad g''(y) \geq 0,$$

with a similar interpretation for "concave". Figure 3 shows some examples of domains of integration.

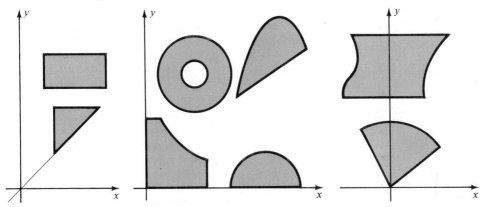

Fig. 3 Typical domains of integration

Partitions We shall be partitioning a domain of integration into smaller domains. Exactly what does it mean to say a domain is small? Certainly its area should be small, but that is not enough. For instance a rectangle with dimensions 10^6 and 10^{-12} has smaller area, yet is very long. We want small to mean that any two points of the domain are close together.

As a measure of smallness, we define the **radius** of a domain **D**, written rad(**D**), as the radius of the smallest circle that includes **D**. See Fig. 4. Clearly

$$|\mathbf{D}| \leq \pi[\text{rad}(\mathbf{D})]^2,$$

so $|\mathbf{D}|$ is small if rad(**D**) is small.

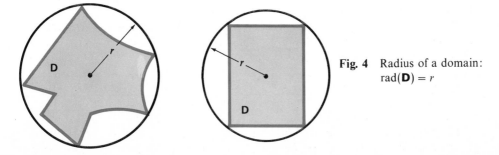

Fig. 4 Radius of a domain: rad(**D**) = r

Now given a domain of integration **D**, let Π be a partition of **D** into subdomains $\mathbf{D}_1, \mathbf{D}_2, \cdots, \mathbf{D}_n$. We define the **mesh** of Π:

$$\text{mesh}(\Pi) = \max\{\text{rad}(\mathbf{D}_1), \text{rad}(\mathbf{D}_2), \cdots, \text{rad}(\mathbf{D}_n)\}.$$

Just as for partitions of an interval, the smaller the mesh, the finer the partition.

The Double Integral   We are ready to define the double integral of a bounded function $f(x, y)$ on a domain of integration **D**. Let Π be a partition of **D** into subdomains $\mathbf{D}_1, \mathbf{D}_2, \cdots, \mathbf{D}_n$, and let $\overline{\mathbf{x}}$ denote a choice of one point (x_i, y_i) in each subdomain $\mathbf{D}_i$. The corresponding **approximating sum** is

$$S(f, \Pi, \overline{\mathbf{x}}) = \sum_{i=1}^{n} f(x_i, y_i)|\mathbf{D}_i|.$$

The function $f(x, y)$ will be called integrable on **D** if these approximating sums approach a limit as $\text{mesh}(\Pi) \longrightarrow 0$.

Definition   A bounded function $f(x, y)$ on a domain of integration **D** is **integrable** on **D** if there is a number L such that

$$S(f, \Pi, \overline{\mathbf{x}}) \longrightarrow L \quad \text{as} \quad \text{mesh}(\Pi) \longrightarrow 0.$$

Precisely, if $\varepsilon > 0$, there must exist $\delta > 0$ such that

$$|S(f, \Pi, \overline{\mathbf{x}}) - L| < \varepsilon \qquad \text{whenever} \quad \text{mesh}(\Pi) < \delta.$$

Then L is called the **double integral** of $f(x, y)$ on **D**, and we write

$$L = \iint_{\mathbf{D}} f(x, y)\, dx\, dy = \lim_{\text{mesh}(\Pi) \to 0} S(f, \Pi, \overline{\mathbf{x}}).$$

This theoretical definition is fine, but immediately two practical questions arise: (1) For what functions, if any, does the integral exist? (2) If the integral does exist, how do we evaluate it?

The answer to the second question is the business of Sections 2 and 3. A complete answer to the first question is too technical, but the following basic assertion covers most situations that arise in practice.

Integrability of Continuous Functions   If $f(x, y)$ is continuous on a domain of integration **D**, then $f(x, y)$ is integrable on **D**, that is, the double integral

$$\iint_{\mathbf{D}} f(x, y)\, dx\, dy$$

exists.

We omit the proof. It is similar to but harder than the proof of the corresponding theorem for functions of one variable. It depends on a property of continuous functions called uniform continuity.

Remark The theorem solves (at least theoretically) the volume problem for continuous functions $f(x, y) > 0$. However, the theorem does not require that f be positive. If f takes both positive and negative values, the double integral represents an *algebraic volume* rather than a geometric volume. The volume between the surface $z = f(x, y)$ and the x, y-plane counts positively where $f > 0$ and negatively where $f < 0$.

Properties of the Double Integral Several basic properties of double integrals follow from the corresponding properties of approximating sums, just as they do for simple integrals. For instance, the formulas

$$\sum_{i=1}^{n} kf(x_i, y_i)|\mathbf{D}_i| = k \sum_{i=1}^{n} f(x_i, y_i)|\mathbf{D}_i|,$$

$$\sum_{i=1}^{n} [f(x_i, y_i) + g(x_i, y_i)]|\mathbf{D}_i| = \sum_{i=1}^{n} f(x_i, y_i)|\mathbf{D}_i| + \sum_{i=1}^{n} g(x_i, y_i)|\mathbf{D}_i|$$

imply the relations

(1) $$\iint_{\mathbf{D}} kf(x, y)\, dx\, dy = k \iint_{\mathbf{D}} f(x, y)\, dx\, dy$$

(2) $$\iint_{\mathbf{D}} [f(x, y) + g(x, y)]\, dx\, dy = \iint_{\mathbf{D}} f(x, y)\, dx\, dy + \iint_{\mathbf{D}} g(x, y)\, dx\, dy.$$

Next, suppose $f(x, y) \le g(x, y)$ on $\mathbf{D}$. The inequality carries over to approximating sums:

$$\sum_{1}^{n} f(x_i, y_i)|\mathbf{D}_i| \le \sum_{1}^{n} g(x_i, y_i)|\mathbf{D}_i|.$$

The relation carries over to the limit.

If $f(x, y) \le g(x, y)$ on $\mathbf{D}$, then

(3) $$\iint_{\mathbf{D}} f(x, y)\, dx\, dy \le \iint_{\mathbf{D}} g(x, y)\, dx\, dy.$$

This inequality has all of the usual consequences. We next mention two inequalities that are most useful in making estimates:

(4) $$\left| \iint_{\mathbf{D}} f(x, y)\, dx\, dy \right| \le \iint_{\mathbf{D}} |f(x, y)|\, dx\, dy$$

(5) If $|f(x, y)| \le M$, then

$$\left| \iint_{\mathbf{D}} f(x, y)g(x, y)\, dx\, dy \right| \le M \iint_{\mathbf{D}} |g(x, y)|\, dx\, dy.$$

In the next two sections, we shall evaluate double integrals. Meanwhile, we observe that one double integral is obvious, that of a constant function.

(6)
$$\iint_{\mathbf{D}} k \, dx \, dy = k|\mathbf{D}|.$$

This is clear because all the approximating sums have the same value:

$$\sum_{1}^{n} f(x_i, y_i)|\mathbf{D}_i| = \sum_{1}^{n} k|\mathbf{D}_i| = k \sum_{1}^{n} |\mathbf{D}_i| = k|\mathbf{D}|.$$

2. RECTANGULAR DOMAINS

How do we actually evaluate double integrals? In most cases, direct application of the definition is practically hopeless. (It is tough enough for simple integrals on intervals.) Fortunately, there is a way of reducing the problem to the evaluation of simple integrals. This method makes the double integral a practical, as well as theoretical, tool.

Let us avoid difficulties with domains by dealing throughout this section only with rectangular domains (Fig. 1) of the form

$$\mathbf{D} = \Big\{ (x, y), \quad \text{where} \quad a \le x \le b \text{ and } c \le y \le d \Big\}.$$

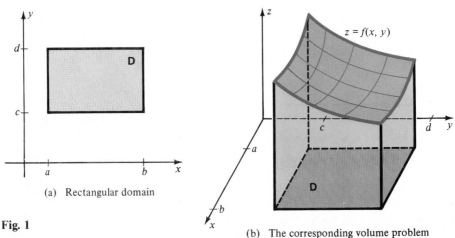

(a) Rectangular domain

Fig. 1

(b) The corresponding volume problem

The simplest way to partition a rectangular domain is into subrectangles. Take arbitrary partitions

$$a = x_0 < x_1 < x_2 < \cdots < x_m = b \quad \text{and} \quad c = y_0 < y_1 < y_2 < \cdots < y_n = d$$

of the intervals $[a, b]$ and $[c, d]$ and put together the corresponding parallel rulings of the plane (Fig. 2). The rectangle $\mathbf{D}$ is partitioned into mn subrectangles $\mathbf{D}_{11}, \cdots, \mathbf{D}_{mn}$, where

$$\mathbf{D}_{ij} = \Big\{ (x, y), \quad \text{where} \quad x_{i-1} \le x \le x_i \text{ and } y_{j-1} \le y \le y_j \Big\},$$

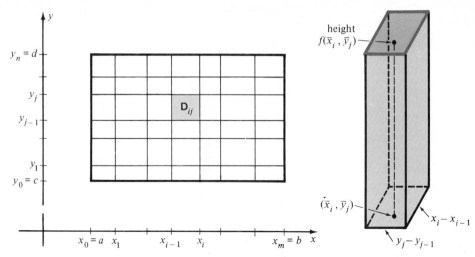

Fig. 2 Partition of **D** into subintervals $\mathbf{D}_{ij}$ **Fig. 3** Typical term in the approximating sum: volume $= f(\bar{x}_i, \bar{y}_j)(x_i - x_{i-1})(y_j - y_{j-1})$

and
$$|\mathbf{D}_{ij}| = (x_i - x_{i-1})(y_j - y_{j-1}).$$

Clearly rad($\mathbf{D}_{ij}$) equals half the diagonal of $\mathbf{D}_{ij}$. It follows that rad($\mathbf{D}_{ij}$) is small if $x_i - x_{i-1}$ and $y_j - y_{j-1}$ are both small.

Now choose any points $\bar{x}_i$ and $\bar{y}_j$ satisfying

$$x_{i-1} \le \bar{x}_i \le x_i \qquad \text{and} \qquad y_{j-1} \le \bar{y}_j \le y_j.$$

Then $(\bar{x}_i, \bar{y}_j)$ is a point of $\mathbf{D}_{ij}$. Form the corresponding approximating sum:

$$S_{mn} = \sum_{i=1}^{m} \sum_{j=1}^{n} f(\bar{x}_i, \bar{y}_j)(x_i - x_{i-1})(y_j - y_{j-1}).$$

Each term in this sum is the (algebraic) volume of a thin rectangular solid (Fig. 3).

The terms in the approximating sum can be added together in any order. Think of the mn terms as arranged in a rectangle with n rows and m columns. First add the i-th column (i fixed, and j running), then sum the column totals:

$$S_{mn} = \sum_{i=1}^{m} \left(\sum_{j=1}^{n} f(\bar{x}_i, \bar{y}_j)(y_j - y_{j-1}) \right)(x_i - x_{i-1}).$$

Now here is the crucial step. Look closely at the inner sum

$$\sum_{j=1}^{n} f(\bar{x}_i, \bar{y}_j)(y_j - y_{j-1}).$$

It looks familiar. In fact, since $\bar{x}_i$ is constant in each term, this inner sum is an approximating sum for the integral $\displaystyle\int_c^d f(\bar{x}_i, y)\, dy.$

Note that one variable is held fixed and the other is integrated out. Thus

$$\sum_{j=1}^n f(\bar{x}_i, \bar{y}_j)(y_j - y_{j-1}) \approx \int_c^d f(\bar{x}_i, y)\, dy.$$

Substitute in the approximating sum: $\displaystyle S_{mn} \approx \sum_{i=1}^m \left(\int_c^d f(\bar{x}_i, y)\, dy \right)(x_i - x_{i-1}).$

Theory (omitted) says this is a good approximation if all $x_i - x_{i-1}$ and $y_j - y_{j-1}$ are small.

Again the expression looks like an approximating sum for a simple integral. To see this clearly, set

$$g(x) = \int_c^d f(x, y)\, dy.$$

Then $\displaystyle S_{mn} \approx \sum_{i=1}^m g(\bar{x}_i)(x_i - x_{i-1}) \approx \int_a^b g(x)\, dx.$

Putting everything together, we have $\displaystyle S_{mn} \approx \int_a^b \left(\int_c^d f(x, y)\, dy \right) dx.$

But S_{mn} is an approximating sum for $\displaystyle\iint_D f(x, y)\, dx\, dy,$

so there is strong reason to believe that $\displaystyle\iint_D f(x, y)\, dx\, dy = \int_a^b \left(\int_c^d f(x, y)\, dy \right) dx.$

The right-hand expression is called an **iterated integral** or **repeated integral**. The value of the inner integral depends on x. In other words the inner integral is a function of x, so it makes sense to integrate it with respect to x.

The above formula was derived in a non-rigorous way. However, the theory of double integration confirms that it is correct.

Iteration Formula   Let $f(x, y)$ be continuous on the domain

$$\mathbf{D} = \{(x, y), \quad \text{where} \quad a \le x \le b \text{ and } c \le y \le d\}.$$

Then
$$g(x) = \int_c^d f(x, y)\, dy$$

is continuous on $[a, b]$ and $\displaystyle\iint_D f(x, y)\, dx\, dy = \int_a^b \left(\int_c^d f(x, y)\, dy \right) dx.$

Similarly $\displaystyle\iint_D f(x, y)\, dx\, dy = \int_c^d \left(\int_a^b f(x, y)\, dx \right) dy.$

This result is a real breakthrough, for it reduces the calculation of double integrals to the calculation of simple integrals, something we are pretty good at by now. True, a *rectangular* domain is rather special, but we shall overcome that restriction in the next section.

Volume by Slicing   Before working some examples, let us take another intuitive look at the iteration formulas. Again we compute the volume

$$V = \iint_{\mathbf{D}} f(x, y) \, dx \, dy,$$

for $f(x, y) > 0$ and **D** a rectangular domain $a \le x \le b, c \le y \le d$. But this time we use a slicing technique.

We fix a value of y and slice the solid by the corresponding plane parallel to the x, z-plane (Fig. 4a).

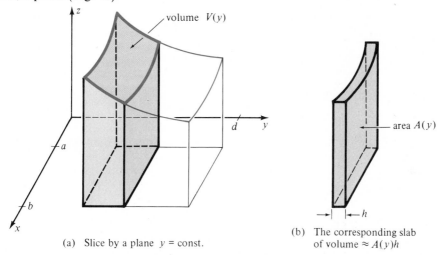

(a)   Slice by a plane y = const.

(b)   The corresponding slab of volume $\approx A(y)h$

Fig. 4   Volume by slicing

Let $A(y)$ be the area of the cross section, and let $V(y)$ be the volume to left of the slice. Then $V(c) = 0$ and $V(d) = V$. To compute $V(y)$ for general values of y, we need the basic relation

$$\frac{dV}{dy} = A(y).$$

We justify this relation intuitively. The derivative dV/dy is the limit as $h \longrightarrow 0$ of

$$\frac{V(y + h) - V(y)}{h}.$$

For h very small, the numerator is the volume of a slab of width h and cross-sectional area approximately $A(y)$. Hence the quotient is approximately $A(y)$, so as $h \longrightarrow 0$, the limit of the quotient, dV/dy, is $A(y)$.

Now we find V by integrating dV/dy:

$$V = \int_{c}^{d} \frac{dV}{dy} \, dy = \int_{c}^{d} A(y) \, dy.$$

But $A(y)$, the area of the cross section (Fig. 5), can be expressed as a simple integral. Indeed, $A(y)$ is the area under the curve $z = f(x, y)$ from $x = a$ to $x = b$. Thus

$$A(y) = \int_a^b f(x, y) \, dx,$$

where y is treated as a constant in computing the integral. Hence once again the iteration formula appears:

$$V = \iint_D f(x, y) \, dx \, dy = \int_c^d \left(\int_a^b f(x, y) \, dx \right) dy.$$

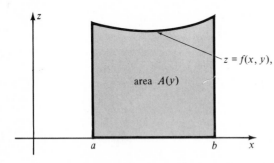

Fig. 5   Area of the cross-section

Examples   The first three examples will have integrands of the special form

$$f(x, y) = g(x)h(y).$$

It is worth seeing what the iteration formula boils down to in this case:

$$\iint_D f(x, y) \, dx \, dy = \iint_D g(x)h(y) \, dx \, dy = \int_a^b \left(\int_c^d g(x)h(y) \, dy \right) dx.$$

But $g(x)$ is constant in the inner integration, so

$$\int_c^d g(x)h(y) \, dy = g(x) \int_c^d h(y) \, dy, \qquad \iint_D f(x, y) \, dx \, dy = \int_a^b \left(g(x) \int_c^d h(y) \, dy \right) dx.$$

But $\int_c^d h(y) \, dy$ is a constant, so

$$\iint_D f(x, y) \, dx \, dy = \left(\int_c^d h(y) \, dy \right) \left(\int_a^b g(x) \, dx \right).$$

$$\boxed{\; \iint_D g(x)h(y) \, dx \, dy = \left(\int_a^b g(x) \, dx \right) \left(\int_c^d h(y) \, dy \right), \;}$$

where $D = \{ (x, y), \text{ where } a \le x \le b \text{ and } c \le y \le d \}.$

■ **EXAMPLE 1**   Find $\iint x^2 y^3 \, dx \, dy$ over $0 \le x \le 1, \; 1 \le y \le 3.$

Solution

$$\iint\limits_{D} x^2 y^3 \, dx \, dy = \left(\int_0^1 x^2 \, dx \right)\left(\int_1^3 y^3 \, dy \right)$$

$$= \left(\frac{1}{3} x^3 \, \Big|_0^1 \right)\left(\frac{1}{4} y^4 \, \Big|_1^3 \right) = \frac{1}{3}\left(\frac{81}{4} - \frac{1}{4} \right) = \frac{20}{3}. \qquad \blacksquare$$

■ **EXAMPLE 2**   Find $\iint e^x \cos y \, dx \, dy$ over $0 \le x \le 1, \ \frac{1}{2}\pi \le y \le \pi$.

Solution

$$\iint\limits_{D} e^x \cos y \, dx \, dy = \left(\int_0^1 e^x \, dx \right)\left(\int_{\pi/2}^{\pi} \cos y \, dy \right)$$

$$= \left(e^x \, \Big|_0^1 \right)\left(\sin y \, \Big|_{\pi/2}^{\pi} \right) = (e - 1)(-1) = 1 - e. \qquad \blacksquare$$

■ **EXAMPLE 3**   Find $\iint e^{x-y} \, dx \, dy$ over $0 \le x \le 1, \ -2 \le y \le -1$.

Solution

$$\iint\limits_{D} e^{x-y} \, dx \, dy = \iint\limits_{D} e^x e^{-y} \, dx \, dy = \left(\int_0^1 e^x \, dx \right)\left(\int_{-2}^{-1} e^{-y} \, dy \right)$$

$$= \left(e^x \, \Big|_0^1 \right)\left(-e^{-y} \, \Big|_{-2}^{-1} \right) = (e - 1)(e^2 - e) = e(e - 1)^2. \qquad \blacksquare$$

■ **EXAMPLE 4**   Find $V = \iint (x^2 y - 3xy^2) \, dx \, dy$

over $1 \le x \le 2, \ -1 \le y \le 1$.

Solution   Use the linear property of the double integral:

$$\iint\limits_{D} (x^2 y - 3xy^2) \, dx \, dy = \iint\limits_{D} x^2 y \, dx \, dy - 3 \iint\limits_{D} xy^2 \, dx \, dy.$$

Evaluate these two integrals separately:

$$\iint\limits_{D} x^2 y \, dx \, dy = \left(\int_1^2 x^2 \, dx \right)\left(\int_{-1}^1 y \, dy \right) = 0;$$

$$\iint\limits_{D} xy^2 \, dx \, dy = \left(\int_1^2 x \, dx \right)\left(\int_{-1}^1 y^2 \, dy \right) = \tfrac{3}{2} \cdot \tfrac{2}{3} = 1.$$

Therefore $V = 0 - 3 \cdot 1 = -3.$ ■

   Now we look at some examples where the integrand is not of the form $g(x)h(y)$ or a sum of such functions. These examples require the iteration formulas in their general form.

■ **EXAMPLE 5**   Find $\displaystyle\iint \frac{dx\,dy}{x+y}$ over $0 \leq x \leq 1, \quad 0 \leq y \leq 2.$

Solution

$$\iint_{D} \frac{dx\,dy}{x+y} = \int_0^1 \left(\int_1^2 \frac{dy}{x+y} \right) dx.$$

For fixed x,

$$\int_1^2 \frac{dy}{x+y} = \ln(x+y) \Big|_{y=1}^{y=2} = \ln(2+x) - \ln(1+x),$$

hence

$$\iint_{D} \frac{dx\,dy}{x+y} = \int_0^1 \left[\ln(2+x) - \ln(1+x) \right] dx.$$

But

$$\int \ln u \, du = u \ln u - u + C,$$

hence

$$\iint_{D} \frac{dx\,dy}{x+y} = \int_0^1 \left[\ln(2+x) - \ln(1+x) \right] dx$$

$$= \left[(2+x)\ln(2+x) - (2+x) - (1+x)\ln(1+x) + (1+x) \right] \Big|_0^1$$

$$= 3 \ln 3 - 2 \ln 2 - 2 \ln 2 = 3 \ln 3 - 4 \ln 2 = \ln(27/16). \qquad ■$$

Remark   Note carefully the expression

$$\ln(x+y) \Big|_{y=1}^{y=2}.$$

We wrote $y = 1$ and $y = 2$ because there were two variables. If we had written only 1 and 2, it would not have been clear how to evaluate $\ln(x+y)$.

An important feature of the iteration formulas is that the iteration may be done in either order. Sometimes the computation is difficult in one order but relatively easy in the opposite order.

■ **EXAMPLE 6**   Find $\displaystyle\iint y \cos(xy) \, dx \, dy$ over $0 \leq x \leq 1, \quad 0 \leq y \leq \pi.$

Solution   Here is one setup:

$$\iint_{D} = \int_0^1 \left(\int_0^\pi y \cos(xy) \, dy \right) dx.$$

The inner integral,

$$\int_0^\pi y \cos(xy) \, dy,$$

while not impossible to integrate (by parts), is tricky. Another procedure is iteration in the opposite order:

$$\iint_{D} = \int_0^\pi \left(\int_0^1 y \cos(xy) \, dx \right) dy.$$

Since y is constant in the inner integration, this can be rewritten as

$$\iint_D = \int_0^\pi y \left(\int_0^1 \cos(xy) \, dx \right) dy.$$

Now

$$\int_0^1 \cos(xy) \, dx = \frac{1}{y} \sin(xy) \Big|_{x=0}^{x=1} = \frac{\sin y}{y}.$$

Hence

$$\iint_D = \int_0^\pi y \frac{\sin y}{y} \, dy = \int_0^\pi \sin y \, dy = 2.$$

Another way of writing the solution:

$$\iint_D y \cos(xy) \, dx \, dy = \int_0^\pi y \, dy \int_0^1 \cos(xy) \, dx$$

$$= \int_0^\pi y \left(\frac{\sin(xy)}{y} \Big|_0^1 \right) dy = \int_0^\pi \sin y \, dy = 2. \qquad \blacksquare$$

EXERCISES

Compute $\displaystyle\int_0^1 f(x, y) \, dx$ and $\displaystyle\int_1^3 f(x, y) \, dy$

1 $f(x, y) = xy^3$ 2 $f(x, y) = e^{xy}$
3 $f(x, y) = (2x - y - 1)^4$ 4 $f(x, y) = \sqrt{x + y + 2}$
5 $f(x, y) = \dfrac{1 + y^2}{1 + x^2}$ 6 $f(x, y) = x^3 y^2 + 5x - 7y.$

Evaluate

7   $\displaystyle\iint (3x - 1) \, dx \, dy$ $-1 \le x \le 2$ $0 \le y \le 5$

8   $\displaystyle\iint e^y \, dx \, dy$ $-1 \le x \le 1$ $0 \le y \le \ln 2$

9   $\displaystyle\iint x^2 y^2 \, dx \, dy$ $-1 \le x \le 1$ $-1 \le y \le 1$

10 $\displaystyle\iint x^3 y^3 \, dx \, dy$ $-1 \le x \le 1$ $-1 \le y \le 1$

11 $\displaystyle\iint (x^5 - y^5) \, dx \, dy$ $0 \le x \le 1$ $0 \le y \le 1$

12 $\displaystyle\iint (e^{x^2} - e^{y^2}) \, dx \, dy$ $0 \le x \le 1$ $0 \le y \le 1$

13 $\displaystyle\iint (1 + x - 2y)^3 \, dx \, dy$ $0 \le x \le 1$ $1 \le y \le 2$

14 $\displaystyle\iint \sqrt{x + y + 2} \, dx \, dy$ $0 \le x \le 1$ $1 \le y \le 3$

15 $\displaystyle\iint \frac{dx \, dy}{(1 + x + y)^2}$ $0 \le x \le 1$ $0 \le y \le 1$

16 $\displaystyle\iint \sin(x + y) \, dx \, dy$ $0 \le x \le \frac{1}{2}\pi$ $0 \le y \le \frac{1}{2}\pi$

17 $\iint (1 - 2x) \sin(y^2)\, dx\, dy$ $\qquad 0 \leq x \leq 1 \qquad 0 \leq y \leq 1$

18 $\iint \dfrac{x^2}{y^3}\, dx\, dy$ $\qquad\qquad 1 \leq x \leq 2 \qquad 1 \leq y \leq 4$

19 $\iint \dfrac{x}{1 + y^2}\, dx\, dy$ $\qquad\quad 0 \leq x \leq 2 \qquad 0 \leq y \leq 1$

20 $\iint xy \ln x\, dx\, dy$ $\qquad\qquad 1 \leq x \leq 4 \qquad -1 \leq y \leq 2$

21 $\iint x \ln(xy)\, dx\, dy$ $\qquad\quad 2 \leq x \leq 3 \qquad 1 \leq y \leq 2$

22 $\iint e^{x+y} \cos 2x\, dx\, dy$ $\qquad 0 \leq x \leq \pi \qquad 1 \leq y \leq 2.$

Find the volume under the surface and above the portion of the x, y-plane indicated

23 $z = 2 - (x^2 + y^2)$ $\qquad -1 \leq x \leq 1 \qquad -1 \leq y \leq 1$
24 $z = 1 - xy$ $\qquad\qquad\quad\ 0 \leq x \leq 1 \qquad 0 \leq y \leq 1$
25 $z = x^2 + 4y^2$ $\qquad\qquad\ 0 \leq x \leq 2 \qquad 0 \leq y \leq 1$
26 $z = \sin x \sin y$ $\qquad\qquad 0 \leq x \leq \pi \qquad 0 \leq y \leq \pi$
27 $z = x^2y + y^2x$ $\qquad\qquad 1 \leq x \leq 2 \qquad 2 \leq y \leq 3$
28 $z = (1 + x^3)y^2$ $\qquad\quad -1 \leq x \leq 1 \qquad -1 \leq y \leq 1.$

29 Suppose $f(x, -y) = -f(x, y)$. Prove that $\iint f(x, y)\, dx\, dy = 0$ on each rectangle of the form $a \leq x \leq b$, $-c \leq y \leq c$. Verify for $f(x, y) = x^2y^3$.

30 Suppose $f(-x, -y) = -f(x, y)$. Prove that $\iint f(x, y)\, dx\, dy = 0$ on each rectangle of the form $-a \leq x \leq a$, $-b \leq y \leq b$. Verify for $f(x, y) = (3x - 2y)^5$.

31 Find the constant A that best approximates $f(x, y)$ on the domain $0 \leq x \leq 1, 0 \leq y \leq 1$ in the least squares sense. In other words, minimize

$$\iint [f(x, y) - A]^2\, dx\, dy.$$

32 (cont.) Show that the coefficients of the least squares linear approximation $A + Bx + Cy$ to $f(x, y)$ on the domain $0 \leq x \leq 1, 0 \leq y \leq 1$ satisfy

$$A + \tfrac{1}{2}B + \tfrac{1}{2}C = \iint f\, dx\, dy, \quad \tfrac{1}{2}A + \tfrac{1}{3}B + \tfrac{1}{4}C = \iint xf\, dx\, dy, \quad \tfrac{1}{2}A + \tfrac{1}{4}B + \tfrac{1}{3}C = \iint yf\, dx\, dy.$$

3. DOMAIN BETWEEN GRAPHS

Let us venture away from the security of rectangles and consider double integrals over some harder domains. We shall discuss an important class of domains, harder than rectangles but not too hard. They are "4-sided" domains, bounded on two sides by smooth curves and on the other sides by parallel lines. To be precise, they are described either by

$$g(x) \leq y \leq h(x), \quad a \leq x \leq b, \quad \text{or by} \quad g(y) \leq x \leq h(y), \quad c \leq y \leq d,$$

where $g(x)$ and $h(x)$ are smooth functions. See Fig. 1.

Sometimes we describe the domains in Fig. 1a by saying that for each fixed x from a to b, the variable y runs from $g(x)$ to $h(x)$.

Let us examine the volume problem for such a domain. Suppose $f(x, y) \geq 0$ is a continuous function on the domain

$$\mathbf{D} = \big\{ (x, y), \quad \text{where} \quad g(y) \leq x \leq h(y) \quad \text{and} \quad c \leq y \leq d \big\}.$$

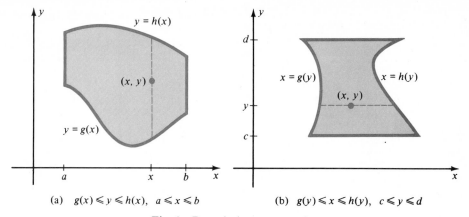

(a) $g(x) \leqslant y \leqslant h(x), \quad a \leqslant x \leqslant b$ (b) $g(y) \leqslant x \leqslant h(y), \quad c \leqslant y \leqslant d$

Fig. 1 Domain between graphs

The solid under the graph consists of all points (x, y, z) in space such that (x, y) is in **D** and

$$0 \leq z \leq f(x, y).$$

See Fig. 2. Its volume is

$$V = \iint_{\mathbf{D}} f(x, y) \, dx \, dy.$$

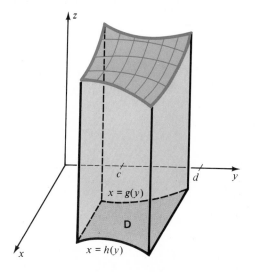

Fig. 2 Volume problem for the solid
$$0 \leq z \leq f(x, y), \quad (x, y) \in \mathbf{D}$$

We compute V by slicing as in the preceding section. We slice the solid by a plane $y = $ constant and let $V(y)$ denote the volume to the left of the plane (Fig. 3a). The area of the cross-section (Fig. 3b) is

$$A(y) = \int_{g(y)}^{h(y)} f(x, y) \, dx.$$

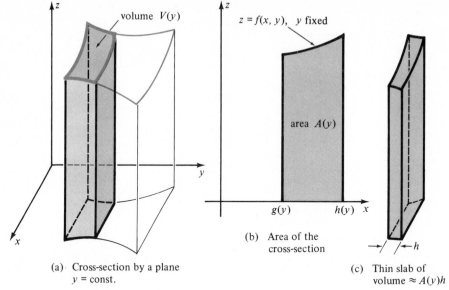

(a) Cross-section by a plane
$y = $ const.

(b) Area of the
cross-section

(c) Thin slab of
volume $\approx A(y)h$

Fig. 3 Solution by sectioning

Again we argue intuitively that

$$\frac{dV}{dy} = A(y).$$

For the derivative is the limit as $h \longrightarrow 0$ of

$$\frac{V(y+h) - V(y)}{h}.$$

For h very small, the numerator is the volume of a thin slab (Fig. 3c) of width h and cross-sectional area approximately $A(y)$. Hence the quotient is approximately $A(y)$.

It follows that $V = \displaystyle\int_c^d \frac{dV}{dy}\,dy = \int_c^d A(y)\,dy = \int_c^d \left(\int_{g(y)}^{h(y)} f(x, y)\,dx \right) dy.$

We are led to the following iteration rules:

Iteration Formulas Let $f(x, y)$ be continuous on **D**.

(1) If $\mathbf{D} = \big\{ (x, y), \text{ where } g(y) \leq x \leq h(y) \text{ and } c \leq y \leq d \big\}$, then

$$\iint_{\mathbf{D}} f(x, y)\,dx\,dy = \int_c^d \left(\int_{g(y)}^{h(y)} f(x, y)\,dx \right) dy.$$

(2) If $\mathbf{D} = \big\{ (x, y), \text{ where } g(x) \leq y \leq h(x) \text{ and } a \leq x \leq b \big\}$, then

$$\iint_{\mathbf{D}} f(x, y)\,dx\,dy = \int_a^b \left(\int_{g(x)}^{h(x)} f(x, y)\,dy \right) dx.$$

Remark In the theory of double integrals, these formulas are proved via approximating sums.

■ **EXAMPLE 1** Find the volume under the surface $z = x + y$ and over the domain of the x, y-plane bounded by the y-axis, the parabola $x = y^2$, and the lines $y = 1$ and $y = 2$.

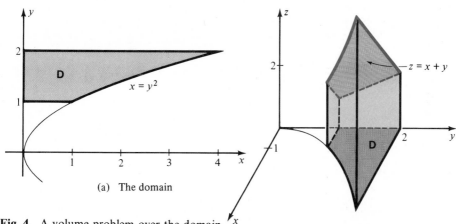

(a) The domain

Fig. 4 A volume problem over the domain

$$\mathbf{D} = \left\{ (x, y) \ \middle| \ 0 \le x \le y^2, \ 1 \le y \le 2 \right\}$$

(b) The solid

Solution First draw the domain (Fig. 4a). It doesn't hurt to draw the solid also (Fig. 4b). The domain is the region between the graphs $x = 0$ and $x = y^2$ for $1 \le y \le 2$. Therefore

$$V = \iint_{\mathbf{D}} (x + y) \, dx \, dy = \int_1^2 \left(\int_0^{y^2} (x + y) \, dx \right) dy.$$

(This is the crucial step. Study the set-up carefully and be sure you understand it.)

Now $\displaystyle \int_0^{y^2} (x + y) \, dx = \tfrac{1}{2}x^2 + xy \ \bigg|_{x=0}^{x=y^2} = \tfrac{1}{2}y^4 + y^3.$

so $$V = \int_1^2 \left(\tfrac{1}{2}y^4 + y^3 \right) dy = \tfrac{1}{10}y^5 + \tfrac{1}{4}y^4 \ \bigg|_1^2$$

$$= \tfrac{1}{10}(32 - 1) + \tfrac{1}{4}(16 - 1) = \tfrac{31}{10} + \tfrac{15}{4} = \tfrac{137}{20}. \qquad ■$$

■ **EXAMPLE 2** Find the volume under the surface $z = e^{-(x+y)}$, and over the domain of the x, y-plane bounded by the x-axis, the line $y = x$, and the lines $x = \tfrac{1}{2}$ and $x = 1$.

Solution First draw the domain and the solid (Fig. 5). By iteration,

$$V = \iint_{\mathbf{D}} e^{-(x+y)} \, dx \, dy = \int_{1/2}^1 \left(\int_0^x e^{-(x+y)} \, dy \right) dx.$$

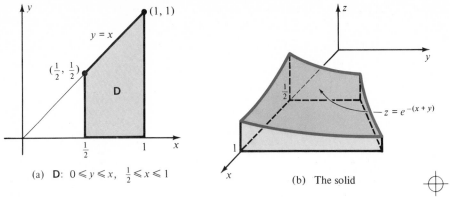

(a) **D**: $0 \leqslant y \leqslant x$, $\frac{1}{2} \leqslant x \leqslant 1$ (b) The solid

Fig. 5

Now

$$\int_0^x e^{-(x+y)} \, dx \, dy = \int_0^x e^{-x} e^{-y} \, dy = e^{-x} \int_0^x e^{-y} \, dy = -e^{-x}(e^{-y}) \Big|_{y=0}^{y=x} = e^{-x} - e^{-2x},$$

so $V = \int_{1/2}^1 (e^{-x} - e^{-2x}) \, dx = (\tfrac{1}{2} e^{-2x} - e^{-x}) \Big|_{1/2}^1$

$$= \tfrac{1}{2} e^{-2} - \tfrac{3}{2} e^{-1} + e^{-1/2} = \tfrac{1}{2} e^{-2}(1 - 3e + 2e^{3/2}). \qquad \blacksquare$$

■ **EXAMPLE 3**   Find the volume under the surface $z = 1 - x^2 - y^2$, lying over the square with vertices $(\pm 1, 0)$ and $(0, \pm 1)$.

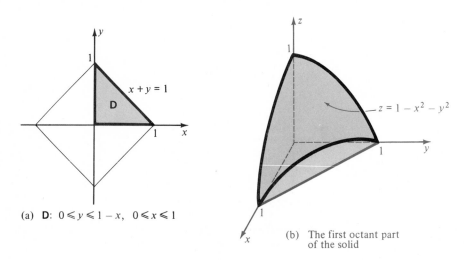

(a) **D**: $0 \leqslant y \leqslant 1 - x$, $0 \leqslant x \leqslant 1$ (b) The first octant part of the solid

Fig. 6

Solution   First draw the square (Fig. 6a). Observe that by symmetry, it suffices to find the volume over the triangular portion in the first quadrant, and then quadruple it. The corresponding quarter of the solid is shown in Fig. 6b. By the iteration formula,

$$V = 4 \iint_D (1 - x^2 - y^2)\, dx\, dy$$

$$= 4 \int_0^1 \left(\int_0^{1-x} (1 - x^2 - y^2)\, dy \right) dx = 4 \int_0^1 \left[(y - x^2 y - \tfrac{1}{3}y^3) \Big|_{y=0}^{y=1-x} \right] dx$$

$$= 4 \int_0^1 [(1 - x) - x^2(1 - x) - \tfrac{1}{3}(1 - x)^3]\, dx$$

$$= 4 \int_0^1 [1 - x - x^2 + x^3 - \tfrac{1}{3}(1 - x)^3]\, dx$$

$$= 4[1 - \tfrac{1}{2} - \tfrac{1}{3} + \tfrac{1}{4} - \tfrac{1}{12}] = 4 \cdot \tfrac{4}{12} = \tfrac{4}{3}.$$

Alternative Solution (using symmetry) The surface $z = 1 - x^2 - y^2$ is unchanged under rotation around the z-axis, so the x and y axes can be rotated. Rotate them 45°. The resulting domain is shown in Fig. 7.

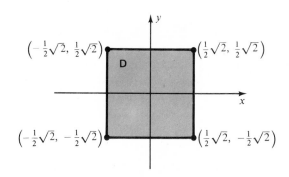

Fig. 7   The domain after a 45° rotation

Accordingly,

$$V = \iint_D (1 - x^2 - y^2)\, dx\, dy = \iint_D dx\, dy - \iint_D x^2\, dx\, dy - \iint_D y^2\, dx\, dy.$$

By symmetry,

$$\iint_D y^2\, dx\, dy = \iint_D x^2\, dx\, dy,$$

hence

$$V = 2 - 2 \iint_D x^2\, dx\, dy = 2 - 2 \left(\int_{-\sqrt{2}/2}^{\sqrt{2}/2} x^2\, dx \right) \left(\int_{-\sqrt{2}/2}^{\sqrt{2}/2} dy \right)$$

$$= 2 - 2 \left(\tfrac{1}{3}x^3 \Big|_{-\sqrt{2}/2}^{\sqrt{2}/2} \right)(2\sqrt{2}) = 2 - 2(\tfrac{1}{6}\sqrt{2})(\sqrt{2}) = \tfrac{4}{3}. \qquad ■$$

■ **EXAMPLE 4**   Find the volume under the plane $z = 1 + x + y$, and over the domain bounded by the curves $x = \tfrac{1}{2}$, $x = 1$, $y = x^2$, $y = 2x^2$.

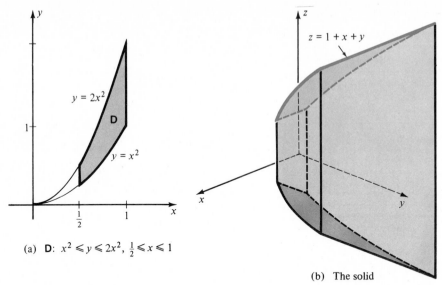

(a) **D**: $x^2 \leqslant y \leqslant 2x^2$, $\frac{1}{2} \leqslant x \leqslant 1$

(b) The solid

Fig. 8

Solution   The domain and the solid are sketched in Fig. 8. Since the domain is between the graphs of two functions, the iteration formula applies. First integrate on y:

$$V = \iint_{\mathbf{D}} (1 + x + y) \, dx \, dy = \int_{1/2}^{1} \left(\int_{x^2}^{2x^2} (1 + x + y) \, dy \right) dx.$$

The inside integral equals

$$(y + xy + \tfrac{1}{2} y^2) \Big|_{y=x^2}^{y=2x^2} = (2x^2 + 2x^3 + 2x^4) - (x^2 + x^3 + \tfrac{1}{2} x^4) = x^2 + x^3 + \tfrac{3}{2} x^4$$

so

$$V = \int_{1/2}^{1} (x^2 + x^3 + \tfrac{3}{2} x^4) \, dx = (\tfrac{1}{3} x^3 + \tfrac{1}{4} x^4 + \tfrac{3}{10} x^5) \Big|_{1/2}^{1}$$

$$= (\tfrac{1}{3} + \tfrac{1}{4} + \tfrac{3}{10}) - \tfrac{1}{8}(\tfrac{1}{3} + \tfrac{1}{8} + \tfrac{3}{40}) = \tfrac{49}{60}. \qquad \blacksquare$$

■ **EXAMPLE 5**   Find $\displaystyle\iint xy \, dx \, dy$

over the domain **D** bounded by $y = x$ and $y = x^2$.

Solution   The first problem is to describe **D** in a way that shows the limits of integration. The line $y = x$ and the parabola $y = x^2$ intersect at $(0, 0)$ and at $(1, 1)$. This information and a drawing (Fig. 9a) suggest that **D** is the domain between $y = x^2$ and $y = x$ for $0 \leq x \leq 1$. Therefore

$$\iint_{\mathbf{D}} xy \, dx \, dy = \int_{0}^{1} \left(\int_{x^2}^{x} xy \, dy \right) dx = \int_{0}^{1} \left(\tfrac{1}{2} xy^2 \Big|_{y=x^2}^{y=x} \right) dx$$

$$= \int_{0}^{1} \tfrac{1}{2}(x^3 - x^5) \, dx = \tfrac{1}{2}(\tfrac{1}{4} - \tfrac{1}{6}) = \tfrac{1}{24}.$$

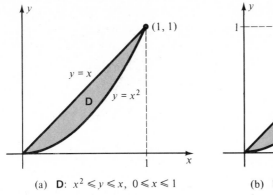 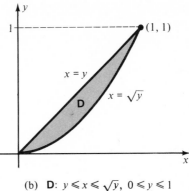

(a) **D**: $x^2 \leqslant y \leqslant x, \; 0 \leqslant x \leqslant 1$ (b) **D**: $y \leqslant x \leqslant \sqrt{y}, \; 0 \leqslant y \leqslant 1$

Fig. 9

Alternative Solution The domain **D** may be thought of as bounded by the curves $x = y$ (below) and $x = \sqrt{y}$ (above), where $0 \leq y \leq 1$. See Fig. 9b. For each y, the range of x is $y \leq x \leq \sqrt{y}$. Therefore the set-up for the iteration is

$$\iint\limits_{\mathbf{D}} xy \, dx \, dy = \int_0^1 \left(\int_y^{\sqrt{y}} xy \, dx \right) dy = \int_0^1 \left(\tfrac{1}{2} x^2 y \, \Big|_{x=y}^{x=\sqrt{y}} \right) dy$$

$$= \int_0^1 \tfrac{1}{2}(y^2 - y^3) \, dy = \tfrac{1}{2}(\tfrac{1}{3} - \tfrac{1}{4}) = \tfrac{1}{24}. \qquad \blacksquare$$

EXERCISES †

Compute the volume under the surface $z = f(x, y)$ and over the indicated domain of the x, y-plane

1 $z = 1$	**5** $z = x^2$
2 $z = y$	**6** $z = xy$
3 $z = x$	**7** $z = \sqrt{xy}$
4 $z = y^2 \sin \pi x$	**8** $z = 1/(1 + x^2)$

Fig. 10   Fig. 11

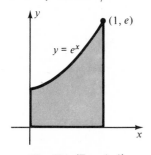

Fig. 10   (Exs. 1–4)

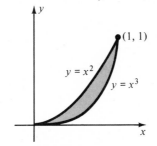

Fig. 11   (Exs. 5–8)

9 $z = x/y^2$	**13** $z = ye^x$
10 $z = x + 2y$	**14** $z = x^2 y$
11 $z = 1/x^3 y^3$	**15** $z = y^3$
12 $z = x^2 e^{xy}$	**16** $z = x\sqrt{x^2 + y^4}.$

Fig. 12   Fig. 13

† The table of *definite* integrals inside the front cover will prove useful for this and subsequent exercise sets.

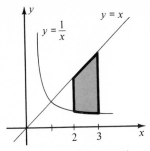

Fig. 12 (Exs. 9–12)

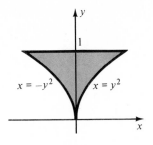

Fig. 13 (Exs. 13–16)

Express $\iint f(x, y)\, dx\, dy$ as one iterated integral (not a sum) over the domain indicated, showing suitable limits of integration. Do it in two ways if possible

17. triangle, vertices $(0, 0)$, $(0, 2)$, $(3, 3)$

18 triangle, vertices $(-2, 5)$, $(-1, 1)$, $(5, 5)$

19 parallelogram, vertices $(-4, 1)$, $(0, 1)$, $(-3, 3)$, $(1, 3)$

20 trapezoid, vertices $(0, 2)$, $(1, 4)$, $(4, 4)$, $(8, 2)$

21 the domain bounded by the x-axis and the uper half of the circle with center $(5, 0)$ and radius 5.

22 the domain bounded by the lines $x = 3$ and $y = 1$ and the hyperbola $xy = 1$

23 the domain in the first quandrant bounded by the ellipse $x^2 + \frac{1}{9}y^2 = 1$ and the line $y = 3 - 3x$

24 the domain bounded by the parabola $y = x^2$ and the line $5x + y = 14$.

Compute the double integral over the domain bounded by the given curves

25 $\iint xe^{xy}\, dx\, dy$ $x = 1$ $x = 3$ $xy = 1$ $xy = 2$

26 $\iint x^2 y\, dx\, dy$ $y = 0$ $x = 0$ $x = (y - 1)^2$

27 $\iint (x^3 + y^3)\, dx\, dy$ $x^2 + y^2 = 1$

28 $\iint (x + y)^2\, dx\, dy$ $x + y = 0$ $y = x^2 + x$

29 $\iint (1 + xy)\, dx\, dy$ $y = 0$ $y = x$ $y = 1 - x$

30 $\iint e^{x^2}\, dx\, dy$ $x = 1$ $x = 2$ $y = x$ $y = x^3$

31 $\iint x^2 y\, dx\, dy$ $y = x^4$ $y = x^2$

32 $\iint x^4 y^2\, dx\, dy$ $|x + y| = 1$ $|x - y| = 1$

33 $\iint (1 + x)\, dx\, dy$ $x + y = 0$ $x^2 + y = 1$

34 $\iint x\, dx\, dy$ $y = x$ $x = 3$ $y = x^2 - x$.

Justify the formulas

35 $\displaystyle\int_0^1 \left(\int_0^x f(x, y)\, dy \right) dx = \int_0^1 \left(\int_y^1 f(x, y)\, dx \right) dy$

36 $\displaystyle\int_{-a}^{a}\left(\int_{0}^{\sqrt{a^2-x^2}}f(x,y)\,dy\right)dx = \int_{0}^{a}\left(\int_{-\sqrt{a^2-y^2}}^{\sqrt{a^2-y^2}}f(x,y)\,dx\right)dy.$

Interchange the order of integration

37 $\displaystyle\int_{4}^{9}\left(\int_{\sqrt{x}}^{3}f(x,y)\,dy\right)dx$ **38** $\displaystyle\int_{0}^{2}\left(\int_{0}^{3\sqrt{1-y^2/4}}f(x,y)\,dx\right)dy.$

4. ARBITRARY DOMAINS

The domains studied in Section 3 are fundamental in double integration. They are simple enough that the iteration formulas hold; you can compute double integrals on them. But at the same time these domains are complicated enough that each domain of integration (p. 833) can be partitioned into a finite number of them.

Each domain of integration **D** can be partitioned into a finite number of domains of integration

$$\mathbf{D}_1, \quad \mathbf{D}_2, \cdots, \mathbf{D}_n,$$

each a domain between the graphs of two functions.

Some examples are shown in Fig. 1.

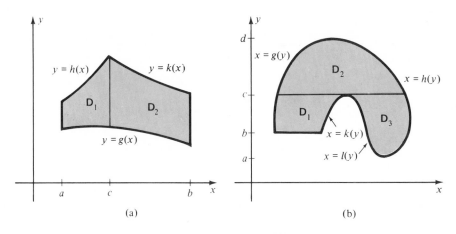

Fig. 1 Arbitrary domains

This is the key to integration over arbitrary domains. Split the domain, preferably using horizontal or vertical line segments, into subdomains, each between the graphs of two functions. Then apply iteration on each subdomain and add the results. In Fig. 1a the domain splits into two nice domains **D**₁ and **D**₂. The integration set-up is

$$\iint_{\mathbf{D}} f(x, y)\, dx\, dy = \iint_{\mathbf{D}_1} f(x, y)\, dx\, dy + \iint_{\mathbf{D}_2} f(x, y)\, dx\, dy$$

$$= \int_a^c \left(\int_{g(x)}^{h(x)} f(x, y)\, dy \right) dx + \int_c^b \left(\int_{g(x)}^{k(x)} f(x, y)\, dy \right) dx.$$

The set-up for Fig. 1b is

$$\iint_{\mathbf{D}} f(x, y)\, dx\, dy = \left(\iint_{\mathbf{D}_1} + \iint_{\mathbf{D}_2} + \iint_{\mathbf{D}_3} \right) f(x, y)\, dx\, dy$$

$$= \int_b^c \left(\int_{g(y)}^{k(y)} f(x, y)\, dx \right) dy + \int_c^d \left(\int_{g(y)}^{h(y)} f(x, y)\, dx \right) dy + \int_a^c \left(\int_{\ell(y)}^{h(y)} f(x, y)\, dx \right) dy.$$

■ **EXAMPLE 1**   Set up $\displaystyle \iint f(x, y)\, dx\, dy$ as an iterated integral over the domain in Fig. 2a.

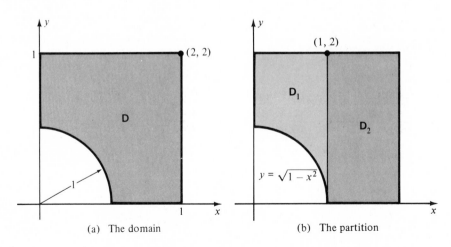

(a) The domain   (b) The partition

Fig. 2

Solution   Partition **D** by a vertical segment through $(1, 0)$. See Fig. 2b. Domain **D**$_1$ is the domain between $y = \sqrt{1 - x^2}$ and $y = 2$ for $0 \le x \le 1$, and **D**$_2$ is the (rectangular) domain between $y = 0$ and $y = 2$ for $1 \le x \le 2$. Therefore

$$\iint_{\mathbf{D}} f(x, y)\, dx\, dy = \left(\iint_{\mathbf{D}_1} + \iint_{\mathbf{D}_2} \right) f(x, y)\, dx\, dy$$

$$= \int_0^1 \left(\int_{\sqrt{1 - x^2}}^2 f(x, y)\, dy \right) dx + \int_1^2 \left(\int_0^2 f(x, y)\, dy \right) dx.$$

Alternative Solution   Partition with a horizontal segment:

$$\iint_{\mathbf{D}} f(x, y)\, dx\, dy = \int_0^1 \left(\int_{\sqrt{1 - y^2}}^2 f(x, y)\, dx \right) dy + \int_1^2 \left(\int_0^2 f(x, y)\, dx \right) dy. \qquad ■$$

Domains and Inequalities   A domain in the plane (or space) is frequently specified by a system of inequalities. A single inequality $f(x, y) \geq 0$ determines a domain whose boundary is $f(x, y) = 0$. To find the domain described by several such inequalities, draw the domain each describes, then form their intersection.

■ **EXAMPLE 2** Sketch $\mathbf{D} = \left\{ (x, y), \text{ where } x + y \leq 0 \text{ and } y \geq x^2 + 2x \right\}$ and set up a double integral on $\mathbf{D}$.

Solution   The first inequality determines the domain below (and on) the line $x + y = 0$. See Fig. 3a. The second inequality determines the domain above (and on) the parabola $y = x^2 + 2x$. See Fig. 3b. The line and parabola intersect at $(0, 0)$ and $(-3, 3)$. The domain $\mathbf{D}$, satisfying both inequalities, is shown in Fig. 3c. Clearly $\mathbf{D}$ is a domain between two functions of x, hence

$$\iint\limits_{\mathbf{D}} f(x, y) \, dx \, dy = \int_{-3}^{0} \left(\int_{x^2 + 2x}^{-x} f(x, y) \, dy \right) dx. \qquad ■$$

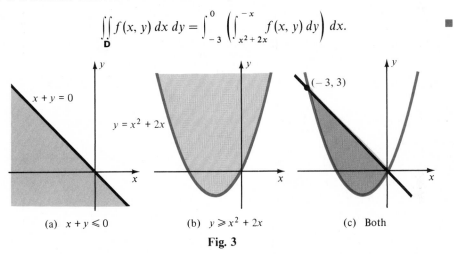

(a) $x + y \leq 0$ (b) $y \geq x^2 + 2x$ (c) Both

Fig. 3

■ **EXAMPLE 3**   Sketch the domain $\mathbf{D}$ specified by $0 \leq x \leq y \leq b$. Equate the two corresponding iterated integrals.

Solution   The region is described by the three inequalities

$$x \geq 0, \qquad y \geq x, \qquad y \leq b.$$

Draw the corresponding domains and take their intersection, a triangle (Fig. 4).

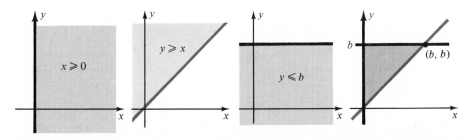

Fig. 4

The domain can be considered in two ways as the domain between two graphs:

$$\mathbf{D} = \left\{ (x, y), \quad \text{where} \quad x \leq y \leq b \text{ and } 0 \leq x \leq b \right\},$$

$$\mathbf{D} = \left\{ (x, y), \quad \text{where} \quad 0 \leq x \leq y \text{ and } 0 \leq y \leq b \right\}.$$

Correspondingly, $\quad \displaystyle\iint_{\mathbf{D}} f(x, y)\, dx\, dy = \int_0^b \left(\int_x^b f(x, y)\, dy \right) dx$

and $\quad \displaystyle\iint_{\mathbf{D}} f(x, y)\, dx\, dy = \int_0^b \left(\int_0^y f(x, y)\, dx \right) dy.$

Therefore $\quad \displaystyle\int_0^b \left(\int_0^y f(x, y)\, dx \right) dy = \int_0^b \left(\int_x^b f(x, y)\, dy \right) dx.$ ∎

Remark A special case is interesting. Suppose $f(x, y) = g(x)$, a function of x alone. Then the right-hand side is

$$\int_0^b \left(\int_x^b g(x)\, dy \right) dx = \int_0^b g(x) \left(\int_x^b dy \right) dx = \int_0^b (b - x)g(x)\, dx.$$

Therefore $\quad \displaystyle\int_0^b \left(\int_0^y g(x)\, dx \right) dy = \int_0^b (b - x)g(x)\, dx.$

This formula expresses the repeated integral of a function of one variable as a simple integral.

Domains are sometimes described by inequalities involving absolute values. To draw such a domain, find its boundary, and be sure to take into account all possible signs.

■ **EXAMPLE 4** Sketch the domain:

(a) $|x| + |y| \leq 1$ (b) $|y| \leq x^2$ and $|x| \leq 1$.

Solution (a) The boundary satisfies $|x| + |y| = 1$. In each quadrant this has a different expression, for instance, in the fourth quadrant it becomes $x - y = 1$. Thus four lines bound the region:

$$x + y = 1, \quad -x + y = 1, \quad -x - y = 1, \quad x - y = 1.$$

These lines cut the plane into 9 pieces (Fig. 5a). Which is the right one? Obviously the central square because $(0, 0)$ satisfies $|x| + |y| \leq 1$; alternatively because $|x| + |y| \leq 1$ certainly describes a bounded region, and the square is the only one that is bounded.

(b) The boundary consists of $|y| = x^2$ (that is, two parabolas $y = x^2$ and $-y = x^2$) and $|x| = 1$ (that is, two vertical lines $x = -1$, $x = 1$). This is enough for a sketch (Fig. 5b). ∎

Remark There are several accepted ways of writing iterated integrals. The following string of equalities will give the idea:

$$\int_1^3 \left(\int_x^{x^2} f(x, y)\, dy \right) dx = \int_1^3 dx \int_x^{x^2} f(x, y)\, dy = \int_1^3 \int_x^{x^2} f(x, y)\, dy\, dx.$$

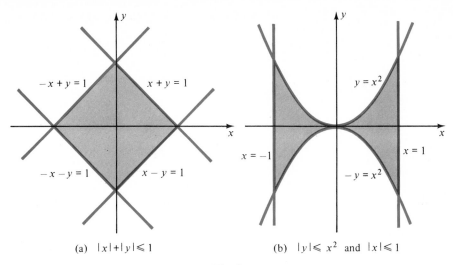

(a) $|x| + |y| \leq 1$ (b) $|y| \leq x^2$ and $|x| \leq 1$

Fig. 5

Element of Area If **D** is a domain of integration, then

$$\iint\limits_{\mathbf{D}} 1 \, dx \, dy = \text{area}(\mathbf{D}).$$

For this reason the symbol $dx \, dy$ is sometimes called the **element of area** in rectangular coordinates. We may think of dx and dy as tiny displacements parallel to the coordinate axes. Thus in a fine partition of **D** by lines parallel to the axes, the typical subdomain is a rectangle of area $dx \, dy$. See Fig. 6. The double integral adds up these little bits of area to give the total area of **D**. Since we picture dx and dy as lengths, $dx \, dy$ then has the dimension of area, that is, length squared.

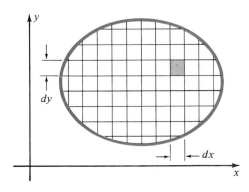

Fig. 6 Element of area $dx \, dy$

EXERCISES

Sketch the domain

1 $x^2 + y^2 \leq 1$ $y + x^2 \geq 0$
3 $x^2 + y^2 \geq 1$ $(x - 2)^2 + y^2 \leq 9$
5 $1 \leq x \leq y \leq 4$

2 $x^2 + y^2 \leq 1$ $-x^2 \leq y \leq x^2$
4 $x \geq 3$ $y \leq -5$ $y - x \geq -10$
6 $\frac{1}{2} \leq y \leq \sqrt{1 - x^2}$

7 $(x + y)^2 \le 1$ $(x - y)^2 \le 1$ **8** $x + y \le 0$ $xy \le 1$ $(x - y)^2 \le 1$.

Express the double integral of $f(x, y)$ over the specified domain as a sum of one or more iterated integrals in which y is the first variable integrated

9 $x^2 + y^2 \le 1$ $x^2 + (y - 1)^2 \le 1$ **10** $y \ge (x + 1)^2$ $y + 2x \le 3$

11 $x \ge 0$ $0 \le y \le \pi$ $x \le \sin y$ **12** $x \ge 0$ $x^2 - y^2 \ge 1$ $x^2 + y^2 \le 9$

13 the triangle with vertices $(0, 0)$, $(-1, 4)$, $(2, 3)$

14 the parallelogram with vertices $(0, 0)$, $(1, 5)$, $(6, 7)$, $(5, 2)$.

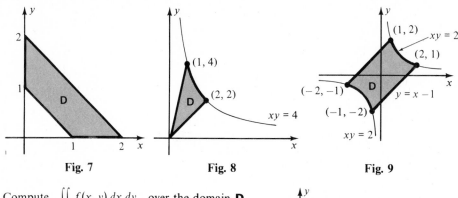

Fig. 7   Fig. 8   Fig. 9

Compute $\displaystyle\iint f(x, y)\, dx\, dy$ over the domain **D**

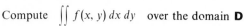
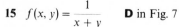

15 $f(x, y) = \dfrac{1}{x + y}$ **D** in Fig. 7

16 $f(x, y) = xy$ **D** in Fig. 7

17 $f(x, y) = xy$ **D** in Fig. 8

18 $f(x, y) = \sqrt{xy^3}$ **D** in Fig. 8

19 $f(x, y) = x^2$ **D** in Fig. 9

20 $f(x, y) = xy$ **D** in Fig. 9

21 $f(x, y) = 2x + y$ **D** in Fig. 10

22 $f(x, y) = x^2 y^2$ **D** in Fig. 10

23 $f(x, y) = e^y$ **D** bounded by $y = \pm x$, $y = \frac{1}{2}x + 3$

24 $f(x, y) = 1 + xy^2$ **D** bounded by $x = -y^2$ and the segments from $(2, 0)$ to $(-1, \pm1)$

25 $f(x, y) = x$ **D** determined by $x^2 + y^2 \le 1$, $y \le 2x + 1$

26 $f(x, y) = y/x^2$ **D** the quadrilateral with vertices $(1, 1)$, $(2, 0)$, $(4, 0)$, $(7, 3)$

27 $f(x, y) = x$ **D** is determined by $x^2 + y^2 \le 1$, $x + y \ge 0$

28 $f(x, y) = x$ **D** bounded by $y = 0$, $y = 2x$, and $5y - 3x = 21$.

Express as an iterated integral in which x is the first variable integrated

29 $\displaystyle\int_0^1 \left(\int_0^x f(x, y)\, dy \right) dx + \int_1^e \left(\int_{\ln x}^1 f(x, y)\, dy \right) dx$

30 $\displaystyle\int_1^4 \left(\int_1^{\sqrt{x}} f(x, y)\, dy \right) dx + \int_4^5 \left(\int_1^{6-x} f(x, y)\, dy \right) dx$.

5. POLAR COORDINATES

Sometimes it is convenient to set up a double integral in the polar coordinates r, θ rather than the rectangular coordinates x, y. Certainly polar coordinates are more suitable for domains of the type shown in Fig. 1.

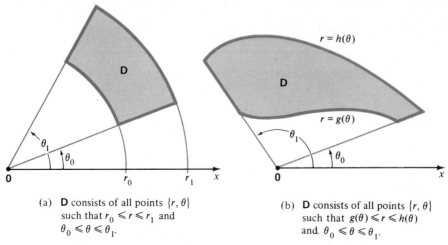

(a) **D** consists of all points $\{r, \theta\}$
such that $r_0 \leqslant r \leqslant r_1$ and
$\theta_0 \leqslant \theta \leqslant \theta_1$.

(b) **D** consists of all points $\{r, \theta\}$
such that $g(\theta) \leqslant r \leqslant h(\theta)$
and, $\theta_0 \leqslant \theta \leqslant \theta_1$.

Fig. 1 Domains for which polar coordinates are prefered

Given such a domain **D**, the problem is to express

$$\iint_{\mathbf{D}} f(x, y)\, dx\, dy$$

in terms of r and θ. Since $x = r \cos \theta$ and $y = r \sin \theta$,

$$f(x, y) = f(r \cos \theta, r \sin \theta),$$

so there is no trouble with the integrand. But what do we do with $dx\, dy$, just replace it with $dr\, d\theta$? No, this won't do because $dx\, dy$ represents an area (with dimension *length squared*) whereas $dr\, d\theta$ has dimension *length* since the angle θ is dimensionless.

If that argument doesn't convince you, here is another. If we could use $dr\, d\theta$ in place of $dx\, dy$, we would have

$$\iint_{\mathbf{D}} 1 \cdot dr\, d\theta = |\mathbf{D}|.$$

Try it. Suppose **D** is the circle $0 \leq r \leq a$, $0 \leq \theta \leq 2\pi$. Its area is πa^2, but

$$\iint_{\mathbf{D}} dr\, d\theta = \int_0^{2\pi} \left(\int_0^a dr \right) d\theta = 2\pi a.$$

The answer is wrong; $dr\, d\theta$ will not do. Notice that $2\pi a$ is the *length* of the circle. We told you so!

Element of Area   The "element of area" $dA = dx\, dy$ in rectangular coordinates is the area of the rectangle swept out by an increase dx in x and an increase dy in y. See Fig. 2a. Now suppose in polar coordinates r increases from r to $r + dr$ and θ from θ to $\theta + d\theta$. Then a small region (Fig. 2b) is swept out. It is almost a rectangle of sides dr and $r\, d\theta$ (the arc of a circle of radius r with central angle $d\theta$). A natural guess, therefore, is that the corresponding polar "element of area" is $dA = r\, dr\, d\theta$.

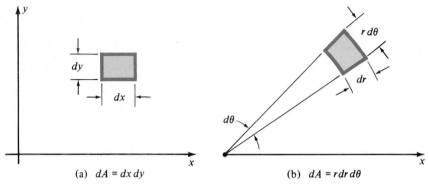

(a) $dA = dx\,dy$ (b) $dA = r\,dr\,d\theta$

Fig. 2   Element of area

If our guess is right, we should have

$$\iint\limits_{\mathbf{D}} r\,dr\,d\theta = |\mathbf{D}|.$$

Let us confirm this formula in the case that $\mathbf{D}$ is the circle $0 \le r \le a$, $0 \le \theta \le 2\pi$. The formula yields

$$\iint\limits_{\mathbf{D}} r\,dr\,d\theta = \left(\int_0^{2\pi} d\theta\right)\left(\int_0^a r\,dr\right) = 2\pi \cdot \tfrac{1}{2}a^2 = \pi a^2,$$

which is the correct area.

We shall assume, without formal proof, that $r\,dr\,d\theta$ is the correct element of area in polar coordinates.

Element of Area $dA = dx\,dy = r\,dr\,d\theta.$

Integrals   Now we can write general double integrals in polar coordinates.

$$\iint\limits_{\mathbf{D}} f(x, y)\,dx\,dy = \iint f(r\cos\theta,\, r\sin\theta)\,r\,dr\,d\theta.$$

■ **EXAMPLE 1**   Find

(a) $\displaystyle\iint\limits_{r \le 1} x^2\,dx\,dy$ (b) $\displaystyle\iint\limits_{\substack{r \le a \\ 0 \le \theta \le \pi/2}} y\,dx\,dy.$

Solution   (a)   The domain is a full circle, so $0 \le \theta \le 2\pi$:

$$\iint\limits_{r \le 1} x^2\,dx\,dy = \iint\limits_{r \le 1} (r\cos\theta)^2(r\,dr\,d\theta) = \iint\limits_{r \le 1} r^3\cos^2\theta\,dr\,d\theta$$

$$= \left(\int_0^{2\pi}\cos^2\theta\,d\theta\right)\left(\int_0^1 r^3\,dr\right) = \tfrac{1}{4}\pi.$$

(b)   The domain is a quarter circle of radius a:

$$\iint y\,dx\,dy = \iint r\sin\theta\,r\,dr\,d\theta = \left(\int_0^{\pi/2}\sin\theta\,d\theta\right)\left(\int_0^a r^2\,dr\right) = \tfrac{1}{3}a^3. \qquad \blacksquare$$

■ **EXAMPLE 2**   Find the volume under the cone $z = 3\sqrt{x^2 + y^2}$ and over the circle $x^2 + y^2 \le a^2$.

Solution   Use polar coordinates. The integrand is $z = 3r$ and the domain is $r \le a$. Hence

$$V = \iint z\,dx\,dy = \iint 3r\,r\,dr\,d\theta = 3\iint r^2\,dr\,d\theta$$

$$= 3\left(\int_0^{2\pi}d\theta\right)\left(\int_0^a r^2\,dr\right) = 3(2\pi)(\tfrac{1}{3}a^3) = 2\pi a^3. \qquad \blacksquare$$

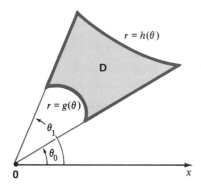

Fig. 3   Domain between polar graphs

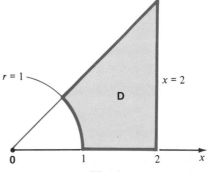

Fig. 4

Now we look at some examples where the domain is like that of Fig. 3. Assume the integrand is already expressed as a function of r and θ. Then the set-up is

$$\iint_D F(r,\theta)\,r\,dr\,d\theta = \int_{\theta_0}^{\theta_1}\left(\int_{g(\theta)}^{h(\theta)} r\,F(r,\theta)\,dr\right)d\theta.$$

■ **EXAMPLE 3**   Find $\displaystyle\iint_D \cos^2\theta\,dx\,dy$ over the domain **D** of Fig. 4.

Solution   The domain is bounded by $\theta = 0$, $\theta = \tfrac{1}{4}\pi$, $r = 1$, and $x = 2$. Since $x = r\cos\theta$, the line $x = 2$ has the polar equation $r\cos\theta = 2$, that is, $r = 2\sec\theta$. Hence

$$\mathbf{D} = \big\{\{r,\theta\},\quad\text{where}\quad 1 \le r \le 2\sec\theta \text{ and } 0 \le \theta \le \tfrac{1}{4}\pi\big\},$$

$$\iint_D \cos^2\theta\,dx\,dy = \iint_D \cos^2\theta\,r\,dr\,d\theta$$

$$= \int_0^{\pi/4}\cos^2\theta\,d\theta\left(\int_1^{2\sec\theta} r\,dr\right) = \int_0^{\pi/4}\cos^2\theta\,(2\sec^2\theta - \tfrac{1}{2})\,d\theta$$

$$= \int_0^{\pi/4}(2 - \tfrac{1}{2}\cos^2\theta)\,d\theta = \tfrac{1}{2}\pi - (\tfrac{1}{16}\pi + \tfrac{1}{8}) = \tfrac{7}{16}\pi - \tfrac{1}{8}. \qquad \blacksquare$$

■ **EXAMPLE 4**   Compute $\displaystyle\iint_{D} xy\,dx\,dy,$

where **D** is the domain bounded by $r = \sin 2\theta$ for $0 \le \theta \le \tfrac{1}{2}\pi$.

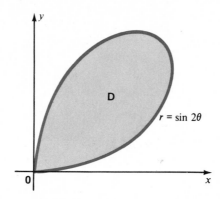

Fig. 5   Domain $\mathbf{D} = \Big\{ 0 \le r \le \sin 2\theta, \quad 0 \le \theta \le \tfrac{1}{2}\pi \Big\}$

Solution   The domain is bounded by one petal of a 4-petal rose curve (Fig. 5). We have

$$\mathbf{D} = \Big\{ \{r, \theta\}, \quad \text{where} \quad 0 \le r \le \sin 2\theta \text{ and } 0 \le \theta \le \tfrac{1}{2}\pi \Big\},$$

$$\iint_{D} xy\,dx\,dy = \iint_{D} (r \cos \theta)(r \sin \theta)(r\,dr\,d\theta)$$

$$= \iint_{D} r^3 \cdot \tfrac{1}{2} \sin 2\theta\,dr\,d\theta = \tfrac{1}{2} \int_{0}^{\pi/2} (\sin 2\theta)\Big(\int_{0}^{\sin 2\theta} r^3\,dr \Big)\,d\theta$$

$$= \tfrac{1}{2} \int_{0}^{\pi/2} (\sin 2\theta)(\tfrac{1}{4} \sin^4 2\theta)\,d\theta = \tfrac{1}{8} \int_{0}^{\pi/2} \sin^5 2\theta\,d\theta = \tfrac{1}{16} \int_{0}^{\pi} \sin^5 \alpha\,d\alpha$$

$$= \tfrac{1}{8} \int_{0}^{\pi/2} \sin^5 \alpha\,d\alpha = \tfrac{1}{8} \cdot \frac{2 \cdot 4}{1 \cdot 3 \cdot 5} = \tfrac{1}{15} \quad \text{(by tables).} \qquad ■$$

■ **EXAMPLE 5**   Two solid right circular cylinders of radius 1 intersect at a right angle on center. Find their common volume.

Solution   Choose coordinates (Fig. 6a) so the solid cylinders are

$$x^2 + y^2 \le 1 \qquad \text{and} \qquad x^2 + z^2 \le 1.$$

Their intersection is the solid **S** consisting of all points (x, y, z) that satisfy both inequalities. From the first inequality, **S** lies above the circle $r \le 1$ in the x, y-plane. From the second, $z^2 \le 1 - x^2$ so that **S** lies between the graphs

$$z = +\sqrt{1 - x^2} \qquad \text{and} \qquad z = -\sqrt{1 - x^2}.$$

Set up a double integral in polar coordinates for the volume V. By symmetry V equals 8 times the volume of the part in the first octant. Hence use the domain

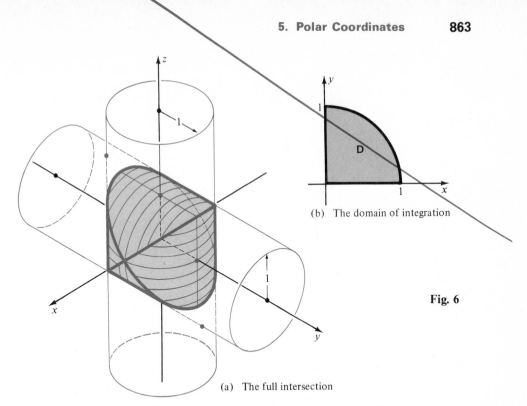

(b) The domain of integration

Fig. 6

(a) The full intersection

D: $0 \le r \le 1, \quad 0 \le \theta \le \tfrac{1}{2}\pi$:

$$V = 8 \iint\limits_{D} \sqrt{1 - x^2} \, dx \, dy = 8 \iint\limits_{D} \sqrt{1 - r^2 \cos^2 \theta} \; r \, dr \, d\theta$$

$$= 8 \int_0^{\pi/2} \left[\int_0^1 r\sqrt{1 - r^2 \cos^2 \theta} \; dr \right] d\theta$$

$$= 8 \int_0^{\pi/2} \left[\left(\frac{-1}{3 \cos^2 \theta} \right)(1 - r^2 \cos^2 \theta)^{3/2} \Big|_{r=0}^{r=1} \right] d\theta = \frac{8}{3} \int_0^{\pi/2} \frac{1 - \sin^3 \theta}{\cos^2 \theta} \, d\theta.$$

Now

$$\frac{1 - \sin^3 \theta}{\cos^2 \theta} = \frac{1 - \sin^3 \theta}{1 - \sin^2 \theta} = \frac{1 + \sin \theta + \sin^2 \theta}{1 + \sin \theta} = \sin \theta + \frac{1}{1 + \sin \theta}.$$

From tables, $$\int \frac{d\theta}{1 + \sin \theta} = -\tan\left(\frac{\pi}{4} - \frac{\theta}{2} \right) + C,$$

hence $$V = \frac{8}{3} \left[-\cos \theta - \tan\left(\frac{\pi}{4} - \frac{\theta}{2} \right) \right] \Big|_0^{\pi/2} = \frac{8}{3}(1 + 1) = \frac{16}{3}.$$ ∎

■ **EXAMPLE 6**   A cylindrical hole of radius a is bored through a sphere of radius $2a$. The surface of the hole passes through the center of the sphere. How much material is removed.

Solution   Choose coordinates (Fig. 7a) so **0** is the center of the sphere, the axis of the cylinder is parallel to the z-axis, and the cylinder intersects the x, y-plane in a circle

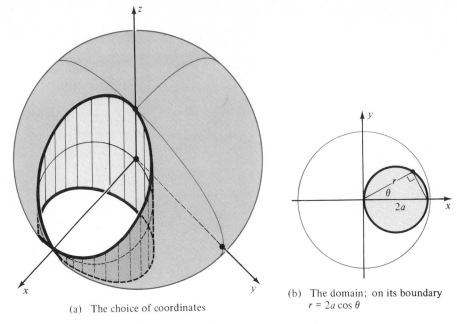

(a)   The choice of coordinates

(b)   The domain; on its boundary
$r = 2a \cos \theta$

Fig. 7   Cylindrical hole through sphere

D centered on the positive x-axis. By Fig. 7b, the base **D** of the cylinder is described by

$$-\tfrac{1}{2}\pi \leq \theta \leq \tfrac{1}{2}\pi, \qquad 0 \leq r \leq 2a \cos \theta.$$

The upper surface of the hole is

$$z = \sqrt{(2a)^2 - x^2 - y^2} = \sqrt{4a^2 - r^2}$$

and similarly the lower surface is $z = -\sqrt{4a^2 - r^2}$. Therefore the volume of the hole is

$$V = \iint\limits_{\mathbf{D}} 2\sqrt{4a^2 - r^2}\, r\, dr\, d\theta = 2 \int_{-\pi/2}^{\pi/2} d\theta \int_0^{2a\cos\theta} r\sqrt{4a^2 - r^2}\, dr$$

$$= 4 \int_0^{\pi/2} d\theta \int_0^{2a\cos\theta} r\sqrt{4a^2 - r^2}\, dr = \frac{4}{3} \int_0^{\pi/2} \left. -(4a^2 - r^2)^{3/2} \right|_0^{2a\cos\theta} d\theta$$

$$= \frac{4}{3} \int_0^{\pi/2} [(4a^2)^{3/2} - (4a^2 \sin^2\theta)^{3/2}]\, d\theta = \frac{32a^3}{3} \int_0^{\pi/2} (1 - \sin^3\theta)\, d\theta$$

$$= \frac{32a^3}{3} \left(\frac{\pi}{2} - \frac{2}{3} \right).$$

(See definite integral 5 inside the front cover.)   ■

Remark   An instructive mistake is possible in this example. By symmetry we equated the integral from $-\tfrac{1}{2}\pi$ to $\tfrac{1}{2}\pi$ to twice the integral from 0 to $\tfrac{1}{2}\pi$. Suppose we had not done this.

Then we would have reached

$$V = \frac{2}{3} \int_{-\pi/2}^{\pi/2} [(4a^2)^{3/2} - (4a^2 \sin^2 \theta)^{3/2}] \, d\theta$$

$$= \frac{16a^3}{3} \int_{-\pi/2}^{\pi/2} (1 - \sin^3 \theta) \, d\theta = \frac{16a^3}{3} \int_{-\pi/2}^{\pi/2} d\theta = \frac{16\pi a^3}{3},$$

since $\sin^3 \theta$ is an odd function. This is certainly a different answer! Where is the goof?

Well, it is a subtle application of the phony argument

$$3 = \sqrt{3^2} = \sqrt{(-3)^2} = [(-3)^2]^{1/2} = -3.$$

The point is that $\sin \theta$ is *negative* in the fourth quadrant, so for $-\frac{1}{2}\pi < \theta < 0$,

$$(\sin^2 \theta)^{3/2} = -\sin^3 \theta, \quad \text{not} \quad \sin^3 \theta.$$

Therefore, the correct argument is

$$V = \frac{16a^3}{3} \int_{-\pi/2}^{\pi/2} [1 - (\sin^2 \theta)^{3/2}] \, d\theta = \frac{16a^3}{3} \left[\int_{-\pi/2}^{0} (1 + \sin^3 \theta) \, d\theta + \int_{0}^{\pi/2} (1 - \sin^3 \theta) \, d\theta \right].$$

Avoid this blunder!

Change of Variables   The polar coordinate transformation

$$x = r \cos \theta, \quad y = r \sin \theta$$

can be interpreted as a correspondence between domains. To a domain **D** in the x, y-plane corresponds a domain **E** in the r, θ-plane. For instance, the domain (Fig. 8a)

$$\mathbf{D} = \left\{ \{r, \theta\}, \quad \text{where} \quad g(\theta) \leq r \leq h(\theta) \text{ and } \theta_0 \leq \theta \leq \theta_1 \right\}$$

in the x, y-plane corresponds to the domain (Fig. 8b)

$$\mathbf{E} = \left\{ (r, \theta), \quad \text{where} \quad g(\theta) \leq r \leq h(\theta) \text{ and } \theta_0 \leq \theta \leq \theta_1 \right\}$$

in the r, θ-plane.

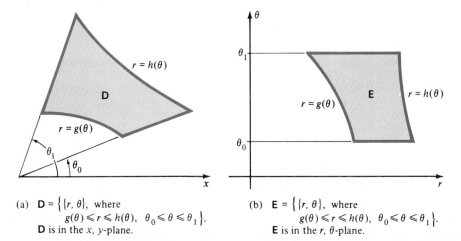

(a) $\mathbf{D} = \{\{r, \theta\},$ where
    $g(\theta) \leq r \leq h(\theta), \; \theta_0 \leq \theta \leq \theta_1\}.$
 $\mathbf{D}$ is in the x, y-plane.

(b) $\mathbf{E} = \{\{r, \theta\},$ where
    $g(\theta) \leq r \leq h(\theta), \; \theta_0 \leq \theta \leq \theta_1\}.$
 $\mathbf{E}$ is in the r, θ-plane.

Fig. 8 Correspondence between domains

The formula for integrating in polar coordinates can be expressed as

$$\iint_D f(x, y)\, dx\, dy = \iint_E f(r \cos \theta, r \sin \theta)\, r\, dr\, d\theta.$$

Thus you may substitute $x = r \cos \theta$ and $y = r \sin \theta$ in the integral on the left, but then you must replace $dx\, dy$ by $r\, dr\, d\theta$ and **D** by the corresponding domain **E** in the r, θ-plane. (The latter corresponds to changing the limits in a simple integral.)

The formula is a special case of a very general formula for changing variables in double integrals. Its proof is beyond the scope of this course, but we shall state the result. First we need a definition.

Jacobian   Suppose

$$\begin{cases} x = x(u, v) \\ y = y(u, v) \end{cases}$$

is a pair of differentiable functions of two variables. Their **Jacobian** is the determinant

$$\frac{\partial(x, y)}{\partial(u, v)} = \begin{vmatrix} x_u & x_v \\ y_u & y_v \end{vmatrix} = x_u y_v - x_v y_u.$$

Example $x = r \cos \theta, \quad y = r \sin \theta$:

$$\frac{\partial(x, y)}{\partial(r, \theta)} = \begin{vmatrix} x_r & x_\theta \\ y_r & y_\theta \end{vmatrix} = \begin{vmatrix} \cos \theta & -r \sin \theta \\ \sin \theta & r \cos \theta \end{vmatrix} = r \cos^2 \theta + r \sin^2 \theta = r.$$

Now we can state the rule for changing variables.

Change of Variables   Suppose

$$\begin{cases} x = x(u, v) \\ y = y(u, v) \end{cases}$$

maps a domain **E** of the u, v-plane in a one-to-one manner onto a domain **D** of the x, y-plane, and suppose

$$\frac{\partial(x, y)}{\partial(u, v)} > 0$$

on **E**. Then

$$\iint_D f(x, y)\, dx\, dy = \iint_E f[x(u, v), y(u, v)] \frac{\partial(x, y)}{\partial(u, v)}\, du\, dv.$$

The formula is usually valid even if the Jacobian equals zero at some points of the boundary of **E**.

■ **EXAMPLE 7**   Find the area enclosed by the ellipse

$$\frac{x^2}{a^2} + \frac{y^2}{b^2} = 1, \qquad a > 0, \quad b > 0.$$

Solution   Consider the transformation

$$x = au, \qquad y = bv.$$

It takes the circle **E**: $u^2 + v^2 \le 1$ onto the domain **D** enclosed by the ellipse. The Jacobian is

$$\frac{\partial(x, y)}{\partial(u, v)} = \begin{vmatrix} a & 0 \\ 0 & b \end{vmatrix} = ab > 0.$$

Therefore $\displaystyle |\mathbf{D}| = \iint_{\mathbf{D}} dx\, dy = \iint_{\mathbf{E}} ab\, du\, dv = ab \iint_{\mathbf{E}} du\, dv = ab\,|\mathbf{E}| = \pi ab.$ ■

EXERCISES

Evaluate $\displaystyle \iint f(x, y)\, dx\, dy$ over the disk $x^2 + y^2 \le a^2$

1 $f = xy^2$

2 $f = e^{x^2 + y^2}$

3 $f = \ln(a^2 + x^2 + y^2)$

4 $f = \dfrac{1}{a + \sqrt{x^2 + y^2}}$

5 $f = x^4$

6 $f = x^2 y^2$

7 $f = \dfrac{1}{a^2 + x^2 + y^2}$

8 $f = \sin(xy).$

Use polar coordinates to compute the volume

9 $0 \le z \le r^2$ $1 \le r \le 2$ $0 \le \theta \le \frac{1}{2}\pi$
10 $0 \le z \le x$ $0 \le r \le 1$ $-\frac{1}{2}\pi \le \theta \le \frac{1}{2}\pi$
11 $0 \le z \le xy$ $1 \le r \le 2$ $\frac{1}{4}\pi \le \theta \le \frac{1}{2}\pi$
12 $0 \le z \le r^6$ $0 \le r \le 1$ $0 \le \theta \le \frac{1}{2}\pi$
13   hemisphere of radius a
14   region bounded by the two paraboloids $z = x^2 + y^2$ and $z = 4 - 3(x^2 + y^2)$
15   lens-shaped region common to the sphere of radius 1 centered at $(0, 0, 0)$ and the sphere of radius 1 centered at $(0, 0, 1)$
16   the material removed when a drill of radius b bores on center through a sphere of radius a, where $b < a$
17   region under the cone $z = 5r$ and over one petal of the rose $r = \sin 3\theta$
18   The region under the paraboloid $z = x^2 + y^2$ and over the circle $x^2 + y^2 = 2x$
19 $0 \le z \le x^4 y^4$ $x^2 + y^2 \le 1$
20 $0 \le z \le r^3$ $0 \le r \le \theta, \ \ 0 \le \theta \le \pi$
21 $0 \le z \le x^2 y^4$ $x^2 + y^2 \le 1$
22*   the region common to three right circular cylinders of radius 1 intersecting at mutual right angles on center.

Compute the Jacobian $\dfrac{\partial(x, y)}{\partial(u, v)}$

23 $x = Au + Bv$ $y = Cu + Dv$
24 $x = u + v$ $y = uv$
25 $x = u^2 + v^2$ $y = u - v$
26 $x = ue^v$ $y = u^2 - v^2.$

27 Show that $x = u - v$, $y = v$ takes the domain

$$\mathbf{E} = \left\{ (u, v) \ \middle| \ a \le u \le b, \ \ 0 \le v \le u \right\}$$

onto

$$\mathbf{D} = \left\{ (x, y) \ \middle| \ x \ge 0, \ \ y \ge 0, \ \ a \le x + y \le b \right\}$$

and compute the Jacobian. (See Fig. 7, p. 858.)

28 (cont.) Assume $0 < a < b$ and compute $\iint xy \, dx \, dy$ over **D** by changing variables.

29 (cont.) Express $\iint f(x + y) \, dx \, dy$ over **D** as a simple integral.

30 (cont.) Evaluate $\iint e^{(x+y)^2} \, dx \, dy$ over **D**.

31 Compute the Jacobian of $x = u/v$, $y = v$.

32 (cont.) Suppose $0 < a < b$ and let

$$\mathbf{D} = \left\{ (x, y) \ \middle| \ ax \le by, \ \ ay \le bx, \ \ xy \le ab \right\}.$$

To what domain in the u, v-plane does **D** correspond? (See Fig. 8, p. 858.)

33 (cont.) Use the change of variables in Ex. 31 to compute $\iint \sqrt{xy^3} \, dx \, dy$ over **D**.

34 Use the transformation of Example 7 and polar coordinates to evaluate $\iint x^2 y \, dx \, dy$ over the domain $x^2/a^2 + y^2/b^2 \le 1$, $y \ge 0$.

35 Let $0 < a < b$ and consider the domain

$$\mathbf{D} = \left\{ (x, y) \ \middle| \ 0 \le y - x \le b - a, \ \ xy \le ab \right\}.$$

To what domain in the u, v-plane does **D** correspond under the transformation $x = \frac{1}{2}(u - v)$, $y = \frac{1}{2}(u + v)$? (See the part of Fig. 9, p. 858, where $y \ge x$.)

36 (cont.) Use the transformation to compute $\iint (y - x) \, dx \, dy$ over **D**.

6. APPLICATIONS

Leibniz Rule   In this section we take up several important applications of double integrals. The first concerns differentiation under the integral sign. Consider a function defined by a definite integral,

$$F(t) = \int_a^b f(x, t) \, dx.$$

When the variable x is "integrated out", there remains a function of t. Problem: find the derivative $F'(t)$. The answer is called the Leibniz Rule, or the rule for differentiating under the integral sign.

Leibniz Rule   Suppose $f(x, t)$ and the partial derivative $f_t(x, t)$ are continuous on a rectangle

$$a \le x \le b, \qquad c \le t \le d.$$

Then
$$\frac{d}{dt} \int_a^b f(x, t) \, dx = \int_a^b f_t(x, t) \, dx \qquad \text{for} \quad c \le t \le d.$$

Proof   Fix t and let **D** be the rectangle

$$\mathbf{D} = \left\{ (x, s), \ \text{ where } \ a \le x \le b \ \text{ and } \ c \le s \le t \right\}$$

in the x, s-plane. Let $G(t) = \iint_{\mathbf{D}} f_s(x, s) \, dx \, ds.$

The idea is to iterate the double integral both ways, then to compute $G'(t)$. On the one hand, $G(t) = \int_a^b \left(\int_c^t f_s(x, s)\, ds \right) dx = \int_a^b [f(x, t) - f(x, c)]\, dx,$

hence $\dfrac{d}{dt} G(t) = \dfrac{d}{dt} \int_a^b f(x, t)\, dx - \dfrac{d}{dt} \int_a^b f(x, c)\, dx = \dfrac{d}{dt} \int_a^b f(x, t)\, dx.$

On the other hand, $G(t) = \int_c^t \left(\int_a^b f_s(x, s)\, dx \right) ds.$

By the Fundamental Theorem of Calculus,

$$\frac{d}{dt} G(t) = \frac{d}{dt} \int_c^t \left(\int_a^b f_s(x, s)\, dx \right) ds = \int_a^b f_t(x, t)\, dx.$$

The Leibniz Rule follows upon equating these expressions for $G'(t)$.

■ **EXAMPLE 1**   Find $\dfrac{d}{dt} \int_0^\pi \dfrac{\sin tx}{x}\, dx$ at $t = \tfrac{1}{2}$.

Solution $\dfrac{d}{dt} \int_0^\pi \dfrac{\sin tx}{x}\, dx = \int_0^\pi \dfrac{\partial}{\partial t} \left(\dfrac{\sin tx}{x} \right) dx = \int_0^\pi \cos tx\, dx = \dfrac{\sin \pi t}{t}.$ ◨

When $t = \tfrac{1}{2}$, the value is $(\sin \tfrac{1}{2}\pi)/\tfrac{1}{2} = 2$.

Remark   It is known that $F(t) = \int [(\sin tx)/x]\, dx$ cannot be expressed in terms of (a finite number of) the usual functions of calculus; you won't find it in a table of integrals, except as an infinite series. Nevertheless, $F(t)$ is a perfectly good differentiable function. But to compute its derivative, you need the Leibniz Rule.

Surface Area   We are given a piece of surface in $\mathbf{R}^3$. What is its area, i.e., how much paint is needed to cover one side? Suppose the surface is given in parametric form:

$$\mathbf{x} = \mathbf{x}(u, v),$$

where (u, v) varies over a domain **D** of the u, v-plane and, as usual,

$$\mathbf{x}_u \times \mathbf{x}_v \neq \mathbf{0}.$$

Recall that the unit normal to the surface is $\mathbf{n} = \dfrac{\mathbf{x}_u \times \mathbf{x}_v}{|\mathbf{x}_u \times \mathbf{x}_v|}.$

A small rectangle in **D** with sides du and dv maps to a small region on the surface. According to the formula $d\mathbf{x} = \mathbf{x}_u\, du + \mathbf{x}_v\, dv$, this region is closely approximated by the parallelogram (Fig. 1) in the tangent plane with sides $\mathbf{x}_u\, du$ and $\mathbf{x}_v\, dv$, whose area is

$$dA = |(\mathbf{x}_u\, du) \times (\mathbf{x}_v\, dv)| = |\mathbf{x}_u \times \mathbf{x}_v|\, du\, dv.$$

Recall that $\mathbf{x}_u \times \mathbf{x}_v = \begin{vmatrix} \mathbf{i} & \mathbf{j} & \mathbf{k} \\ x_u & y_u & z_u \\ x_v & y_v & z_v \end{vmatrix} = \begin{vmatrix} y_u & z_u \\ y_v & z_v \end{vmatrix} \mathbf{i} + \begin{vmatrix} z_u & x_u \\ z_v & x_v \end{vmatrix} \mathbf{j} + \begin{vmatrix} x_u & y_u \\ x_v & y_v \end{vmatrix} \mathbf{k}$

$$= \frac{\partial(y, z)}{\partial(u, v)} \mathbf{i} + \frac{\partial(z, x)}{\partial(u, v)} \mathbf{j} + \frac{\partial(x, y)}{\partial(u, v)} \mathbf{k}.$$

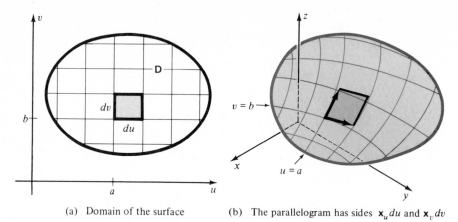

(a)   Domain of the surface   (b)   The parallelogram has sides $\mathbf{x}_u\,du$ and $\mathbf{x}_v\,dv$

Fig. 1   Approximation to area by a small "parallelogram"

Consequently $\left|\mathbf{x}_u \times \mathbf{x}_v\right|^2 = \left(\dfrac{\partial(y, z)}{\partial(u, v)}\right)^2 + \left(\dfrac{\partial(z, x)}{\partial(u, v)}\right)^2 + \left(\dfrac{\partial(x, y)}{\partial(u, v)}\right)^2.$

Substitute this into $dA = \left|\mathbf{x}_u \times \mathbf{x}_v\right| du\,dv$; the result leads to the following definition:

Surface Area   Let $\mathbf{x} = \mathbf{x}(u, v)$ define a parametric surface with domain **D**. Its **surface area** is

$$A = \iint\limits_{\mathbf{D}} \left|\mathbf{x}_u \times \mathbf{x}_v\right| du\,dv = \iint\limits_{\mathbf{D}} \sqrt{\left(\dfrac{\partial(y, z)}{\partial(u, v)}\right)^2 + \left(\dfrac{\partial(z, x)}{\partial(u, v)}\right)^2 + \left(\dfrac{\partial(x, y)}{\partial(u, v)}\right)^2}\; du\,dv.$$

EXAMPLE 2   Find the area of the spiral ramp $\mathbf{x} = (u \cos v, u \sin v, bv)$ corresponding to the rectangle **D**: $0 \le u \le a$, $0 \le v \le c$.

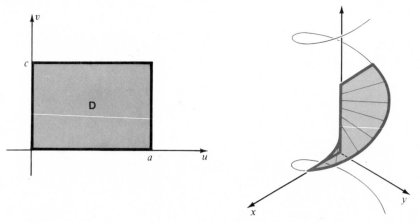

Solution   Although not necessary, it is nice to sketch the surface (Fig. 2). Since

$$\mathbf{x}_u = (\cos v, \sin v, 0) \qquad \text{and} \qquad \mathbf{x}_v = (-u \sin v, u \cos v, b),$$

the element of area is

$$dA = \sqrt{\begin{vmatrix} \sin v & 0 \\ u \cos v & b \end{vmatrix}^2 + \begin{vmatrix} 0 & \cos v \\ b & -u \sin v \end{vmatrix}^2 + \begin{vmatrix} \cos v & \sin v \\ -u \sin v & u \cos v \end{vmatrix}^2} \; du\, dv$$

$$= \sqrt{b^2 \sin^2 v + b^2 \cos^2 v + u^2}\; du\, dv = \sqrt{b^2 + u^2}\; du\, dv.$$

As (u, v) ranges over the rectangle, the point $\mathbf{x}(u, v)$ runs over the spiral ramp. Hence

$$A = \iint_D \sqrt{b^2 + u^2}\; du\, dv = \left(\int_0^a \sqrt{b^2 + u^2}\; du \right)\left(\int_0^c dv \right)$$

$$= \frac{c}{2}\left[a\sqrt{a^2 + b^2} + b^2 \ln\left(\frac{a + \sqrt{a^2 + b^2}}{b} \right) \right]. \qquad \blacksquare$$

Area of a Graph Suppose a surface is given as the graph of a function $z = f(x, y)$, where (x, y) varies over a domain **D** in the x, y-plane. This is a special case of a parametric surface, where the parameters are x and y and the surface is defined by

$$\mathbf{x} = (x, y, f(x, y)).$$

Now $dA = |\mathbf{x}_x \times \mathbf{x}_y|\, dx\, dy$. But

$$\frac{\partial \mathbf{x}}{\partial x} = (1, 0, f_x) \qquad \text{and} \qquad \frac{\partial \mathbf{x}}{\partial y} = (0, 1, f_y),$$

so

$$\frac{\partial \mathbf{x}}{\partial x} \times \frac{\partial \mathbf{x}}{\partial y} = (1, 0, f_x) \times (0, 1, f_y) = (-f_x, -f_y, 1).$$

Hence the resulting formula for the element of area is $\;dA = \sqrt{1 + f_x^2 + f_y^2}\; dx\, dy.$

> The area of the graph of a function $z = f(x, y)$ with domain **D** is
>
> $$A = \iint_D \sqrt{1 + f_x^2 + f_y^2}\; dx\, dy.$$

The formula for dA has a geometric interpretation (Fig. 3). The unit normal to the surface is

$$\mathbf{n} = \frac{1}{\sqrt{1 + f_x^2 + f_y^2}}\, (-f_x, -f_y, 1).$$

Its third component (direction cosine) is

$$\cos \gamma = \frac{1}{\sqrt{1 + f_x^2 + f_y^2}},$$

where γ is the angle between the normal and the z-axis. Thus

$$(\cos \gamma)\, dA = dx\, dy,$$

which means that the small piece of surface of area dA projects onto a small portion of the x, y-plane of area $dx\, dy$.

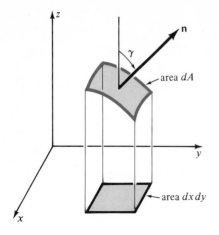

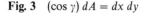

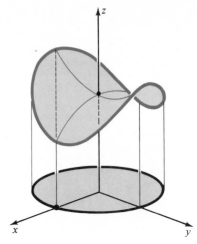

Fig. 3 $(\cos \gamma)\, dA = dx\, dy$ **Fig. 4** $z = 2 - x^2 + y^2$

■ **EXAMPLE 3**   Find the area of the graph of $z = 2 - x^2 + y^2$ over the circle $x^2 + y^2 \le a^2$.

Solution   The saddle-shaped surface is part of a hyperbolic paraboloid (Fig. 4). We have

$$A = \iint\limits_{\mathbf{D}} \sqrt{1 + z_x{}^2 + z_y{}^2}\, dx\, dy = \iint\limits_{\mathbf{D}} \sqrt{1 + (-2x)^2 + (2y)^2}\, dx\, dy$$

$$= \iint\limits_{\mathbf{D}} \sqrt{1 + 4x^2 + 4y^2}\, dx\, dy.$$

It pays to use polar coordinates. Then $x^2 + y^2 = r^2$ and $dx\, dy = r\, dr\, d\theta$, so

$$A = \iint\limits_{\mathbf{D}} \sqrt{1 + 4r^2}\, r\, dr\, d\theta = \int_0^{2\pi} d\theta \int_0^a r\sqrt{1 + 4r^2}\, dr$$

$$= (2\pi) \cdot \tfrac{1}{12}[(1 + 4a^2)^{3/2} - 1] = \tfrac{1}{6}\pi[(1 + 4a^2)^{3/2} - 1].$$ ■

A Probability Integral   The improper integral

$$I = \int_{-\infty}^{\infty} e^{-x^2}\, dx$$

is important in probability. Its exact value can be found by using polar coordinates and a clever trick. Write

$$I = \int_{-\infty}^{\infty} e^{-y^2}\, dy.$$

Then $$I^2 = \left(\int_{-\infty}^{\infty} e^{-x^2}\, dx\right)\!\left(\int_{-\infty}^{\infty} e^{-y^2}\, dy\right) = \iint\limits_{\mathbf{R}^2} e^{-(x^2 + y^2)}\, dx\, dy.$$

Now switch from rectangular to polar coordinates. Then $x^2 + y^2 = r^2$ and

$dx\,dy = r\,dr\,d\theta$. Hence

$$I^2 = \int_0^{2\pi} \left(\int_0^{\infty} e^{-r^2} r\,dr \right) d\theta = 2\pi \left(-\tfrac{1}{2} e^{-r^2} \right) \Big|_0^{\infty} = 2\pi(\tfrac{1}{2}) = \pi.$$

Since I is positive, $$\int_{-\infty}^{\infty} e^{-x^2}\,dx = \sqrt{\pi}.$$

The function $$\phi(x) = \frac{1}{\sqrt{2\pi}} e^{-x^2/2}$$

is known as the density function of the **normal distribution**. Its graph is the familiar bell-shaped curve (Fig. 5) and encloses area 1 (by a simple modification of the definite integral above).

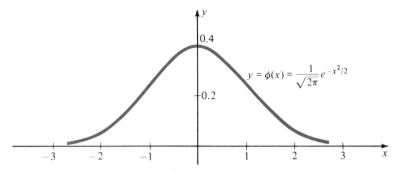

Fig. 5   Normal distribution

Remark 1   Expressing an antiderivative of e^{-x^2} in terms of elementary functions is known to be impossible. Thus it is quite remarkable that the integral of this function over the whole x-axis can be evaluated.

Remark 2   The derivation above is not quite complete because we have not discussed improper double integrals, a subject we leave for a later course. A basic result of this subject is that for non-negative integrands, any device leading to an answer yields the correct answer. You may take this for granted in the exercises.

EXERCISES

Evaluate $F(t)$ for $t > 0$ and then compute $F'(t)$. Now compute $F'(t)$ by the Leibniz Rule and compare your answers

1   $F(t) = \int_0^1 e^{-tx}\,dx$

2   $F(t) = \int_0^1 \left(\arctan \frac{x}{t} \right) dx, \quad t > \dfrac{2}{\pi}$

3   $F(t) = \int_0^1 (t + x)^n\,dx$

4   $F(t) = \int_0^1 x^t\,dx, \quad t > -1.$

Parameterize the surface and set up an integral for its surface area; evaluate the area if you can

5   triangle with vertices **a**, **b**, **c**

6   (cont.) triangle with vertices $(a, 0, 0)$, $(0, b, 0)$, $(0, 0, c)$

7   lateral surface of a right circular cylinder of radius a and height h

8   lateral surface of a right circular cone of radius a and lateral height L

9   right circular torus obtained by revolving a circle of radius a about an axis in its plane at distance A from its center (where $a < A$)

10   sphere $|\mathbf{x}| = a$ [*Hint* $\mathbf{x} = a(\sin u \cos v, \sin u \sin v, \cos u).$]

11*   ellipsoid $\dfrac{x^2}{a^2} + \dfrac{y^2}{b^2} + \dfrac{z^2}{c^2} = 1$ [*Hint* $\mathbf{x} = (a \sin u \cos v, b \sin u \sin v, c \cos u).$]

12*   (cont.) Reduce the double integral to a simple integral in the special case $a = b$.

Set up an integral for the area of the given graph, and evaluate it if you can

13 $z = ax + by$ (x, y) in a domain **D**

14 $z = x^2 + y^2$ $x^2 + y^2 \leq 1$

15 $z = \sqrt{1 - x^2 - y^2}$ $x^2 + y^2 \leq 1$.

16 $bz = xy$ $x^2 + y^2 \leq a^2$ where $a > 0$ and $b > 0$.

17   Find the area of the top piece of the spherical surface that is inside the hole in Example 6, p. 863.

18.   (cont.) Find the lateral surface area of the hole.

19*   A curve in the part $x > 0$ of the z, x-plane is given parametrically: $z = z(s)$, $x = x(s)$, where s is arc length and $a \leq s \leq b$. The curve is rotated about the z-axis, generating a surface. Express the area of this surface as a simple integral.

20*   Suppose a graph in $\mathbf{R}^3$ is given in the form $z = g(r, \theta)$, where (r, θ) varies over a domain **E** in the r, θ-plane. Express the area of the graph as an integral over **E**.

Evaluate

21 $\displaystyle\lim_{a \to 0+} \iint_{a \leq r \leq 1} \ln r \, dx \, dy$

22 $\displaystyle\lim_{a \to 0+} \iint_{a \leq r \leq 1} (1 - r)^p \, dx \, dy$ $p > -1$

23 $\displaystyle\lim_{a \to \infty} \iint_{r \leq a} \frac{dx \, dy}{1 + x^2 + y^2}$

24 $\displaystyle\lim_{a \to \infty} \iint_{r \leq a} \frac{dx \, dy}{(1 + x^2 + y^2)^p}$ $p > 1$

25 $\displaystyle\lim_{a \to 0+} \iint_{a \leq r \leq 1} \frac{dx \, dy}{r^{2p}}$ $p < 1$

26 $\displaystyle\lim_{a \to \infty} \iint_{1 \leq r \leq a} \frac{dx \, dy}{r^{2p}}$ $p > 1$

27 $\displaystyle\int_0^\infty e^{-ax^2} \, dx$ $a > 0$

28* $\displaystyle\int_0^\infty x^2 e^{-x^2} \, dx$.

29   Set $f_n(t) = \displaystyle\int_0^\infty x^{2n+1} e^{-tx^2} \, dx$ for $t > 0$

and $n \geq 0$. Assume the Leibniz Rule applies, and use it to derive a relation between $f_n(t)$ and $f_{n+1}(t)$.

30   (cont.) Evaluate $f_0(t)$ and then $f_n(t)$ in general.

7. PHYSICAL APPLICATIONS

Mass and Density   Suppose a sheet of non-homogeneous material covers a plane domain **D**. See Fig. 1a. At each point (x, y), let $\rho(x, y)$ denote the **density** of the material, i.e., the mass per unit area. (Dimensionally, planar density is mass divided by length squared. Common units are gm/cm^2 and lb/ft^2.) The mass of a small rectangular portion of the sheet (Fig. 1b) is

$$dM \approx \rho(x, y) \, dx \, dy.$$

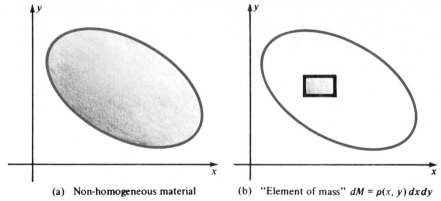

(a) Non-homogeneous material   (b) "Element of mass" $dM = \rho(x, y)\,dx\,dy$

Fig. 1   Mass and density

Therefore the total mass of the sheet is

$$M = \iint_D \rho(x, y)\, dx\, dy.$$

■ **EXAMPLE 1**   The density (lb/ft^2) at each point of a one-foot square of plastic is the product of the four distances of the point from the sides of the square. Find the total mass.

Solution   Take the square in the position $0 \leq x \leq 1$, $0 \leq y \leq 1$. Then

$$\rho(x, y) = x(1 - x)y(1 - y),$$

$$M = \iint \rho(x, y)\, dx\, dy = \left(\int_0^1 x(1 - x)\, dx\right)\left(\int_0^1 y(1 - y)\, dy\right) = \tfrac{1}{6} \cdot \tfrac{1}{6} = \tfrac{1}{36} \text{ lb.} \qquad ■$$

Moment and Center of Gravity   Suppose gravity (perpendicular to the plane of the figure) acts on the sheet of Fig. 1a. The sheet is to be suspended by a single point so that it balances parallel to the floor. This point of balance is the **center of gravity** of the sheet and is denoted $\bar{\mathbf{x}} = (\bar{x}, \bar{y})$.

To motivate the formulas that follow, let us recall what happens when, instead of a sheet, we have a finite system of point masses glued to the weightless plane, mass M_1 at $\mathbf{x}_1$, mass M_2 at $\mathbf{x}_2, \cdots$, mass M_n at $\mathbf{x}_n$. Then the center of gravity $\bar{\mathbf{x}}$ of the system is the weighted average of the $\mathbf{x}_i$, weighted with the M_i:

$$\bar{\mathbf{x}} = \frac{M_1\mathbf{x}_1 + M_2\mathbf{x}_2 + \cdots + M_n\mathbf{x}_n}{M_1 + M_2 + \cdots + M_n} = \frac{1}{\sum M_i} \sum M_i \mathbf{x}_i.$$

We can write

$$\bar{\mathbf{x}} = (\bar{x}, \bar{y}) = \frac{1}{M}\,\mathbf{m} = \frac{1}{M}\,(m_x, m_y),$$

where $M = \sum M_i$, the total mass, and the moments of the system of masses are

$$m_x = \sum M_i x_i, \qquad m_y = \sum M_i y_i.$$

Note that $\bar{x} = m_x/M$ is the weighted average of the x-coordinates of the individual masses, and similarly for $\bar{y}$.

Now suppose mass is distributed continuously with density $\rho(x, y)$ over a plane sheet **D**. How do we define the center of gravity? We argue very roughly as follows. We partition **D** into a large number of tiny pieces and pretend each piece is a point mass. The mass of the i-th piece **D**$_i$ is approximately $M_i = \rho(x_i, y_i)|\mathbf{D}_i|$, where (x_i, y_i) is a point in **D**$_i$. For this finite system, the moments are

$$\sum x_i M_i = \sum x_i \rho(x_i, y_i)|\mathbf{D}_i|, \qquad \sum y_i M_i = \sum y_i \rho(x_i, y_i)|\mathbf{D}_i|.$$

The center of gravity $\dfrac{1}{\sum M_i}\left(\sum x_i M_i, \sum y_i M_i\right)$

approximates that of the sheet. The approximation should improve as the partitions become finer and finer. But when that happens, $\sum x_i M_i$, $\sum y_i M_i$, and $\sum M_i$ approach double integrals. This suggests definitions, first for the **moment** of the sheet:

$$\boxed{\mathbf{m} = (m_x, m_y) = \left(\iint_D x\,\rho(x, y)\,dx\,dy, \iint_D y\,\rho(x, y)\,dx\,dy\right) = \iint_D \rho\,\mathbf{x}\,dx\,dy.}$$

Then for the center of gravity of the sheet:

$$\boxed{\bar{\mathbf{x}} = (\bar{x}, \bar{y}) = \frac{1}{M}\mathbf{m} = \frac{1}{M}(m_x, m_y),}$$

where $$M = \iint_D \rho(x, y)\,dx\,dy$$

is the total mass of the sheet.

We can still think of $\bar{x}$ as the weighted average of x over the sheet:

$$\bar{x} = \iint_D x\,\rho(x, y)\,dx\,dy \Big/ \iint_D \rho(x, y)\,dx\,dy,$$

and similarly for $\bar{y}$.

After two examples, we shall prove that the sheet really balances if suspended from $\bar{\mathbf{x}}$.

■ **EXAMPLE 2**   Find the center of gravity of a homogeneous rectangular sheet.

Solution   "Homogeneous" means the density ρ is constant. Take the sheet in the position $0 \le x \le a$ and $0 \le y \le b$. The mass is $M = \rho ab$, and the moment is

$$\mathbf{m} = \iint \rho\,\mathbf{x}\,dx\,dy = \rho \iint \mathbf{x}\,dx\,dy = \rho\left(\iint x\,dx\,dy, \iint y\,dx\,dy\right)$$

$$= \rho\left(\int_0^a x\,dx \int_0^b dy, \int_0^a dx \int_0^b y\,dy\right) = \rho(\tfrac{1}{2}a^2 b, \tfrac{1}{2}ab^2).$$

Therefore the center of gravity is

$$\overline{\mathbf{x}} = \frac{1}{M}\,\mathbf{m} = \frac{1}{\rho ab}\,\rho(\tfrac{1}{2}a^2b, \tfrac{1}{2}ab^2) = \tfrac{1}{2}(a, b).$$

This is the midpoint (intersection of the diagonals) of the rectangle. (Of course the rectangle balances on its midpoint; no one needs calculus for this, but it is reassuring that the analytic method gives the right answer.) ∎

Remark We sometimes speak of the "center of gravity of a domain **D**", without reference to a density. Then it is understood that $\rho = 1$, so the center of gravity is a purely geometric quantity associated with **D**. It is also known as the **centroid** of **D**.

■ **EXAMPLE 3** The triangular sheet $x \geq 0$, $y \geq 0$, $x + y \leq 1$ has density $\rho = xy$. Find its center of gravity.

Solution
$$M = \iint_{\mathbf{D}} xy\,dx\,dy = \int_0^1 y\,dy \int_0^{1-y} x\,dx = \tfrac{1}{2}\int_0^1 y(1-y)^2\,dy$$

$$= \tfrac{1}{2}\int_0^1 (y - 2y^2 + y^3)\,dy = \tfrac{1}{2}(\tfrac{1}{2} - \tfrac{2}{3} + \tfrac{1}{4}) = \tfrac{1}{24}.$$

By symmetry, $m_x = m_y$, and

$$m_x = \iint_{\mathbf{D}} x \cdot xy\,dx\,dy = \int_0^1 y\,dy \int_0^{1-y} x^2\,dx = \tfrac{1}{3}\int_0^1 y(1-y)^3\,dy$$

$$= \tfrac{1}{3}\int_0^1 (1-u)\,u^3\,du = \tfrac{1}{3}\int_0^1 (u^3 - u^4)\,du = \tfrac{1}{3}(\tfrac{1}{4} - \tfrac{1}{5}) = \tfrac{1}{60}.$$

Therefore $\bar{x} = \bar{y} = \tfrac{1}{60}/\tfrac{1}{24} = \tfrac{2}{5}$ and $\overline{\mathbf{x}} = (\tfrac{2}{5}, \tfrac{2}{5})$. ∎

Proof That the Sheet Balances Suppose a knife-edge passes through $\overline{\mathbf{x}}$ and the sheet balances (Fig. 2). Divide the sheet into many small rectangles. The turning moments of these pieces about the knife-edge must add up to zero.

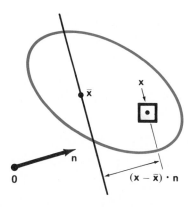

Fig. 2 Proof of the center of gravity formula for a non-homogeneous sheet

Let **n** be a unit vector in the plane of the rectangle perpendicular to the knife edge. A small rectangle with sides dx and dy located at **x** has (signed) distance $(\mathbf{x} - \overline{\mathbf{x}}) \cdot \mathbf{n}$ from the knife edge and has mass $\rho\,dx\,dy$. Hence its turning moment is

$$(\mathbf{x} - \bar{\mathbf{x}}) \cdot \mathbf{n} \, \rho \, dx \, dy.$$

The sum of all such turning moments must be zero:

$$\iint\limits_D (\mathbf{x} - \bar{\mathbf{x}}) \cdot \mathbf{n} \, \rho \, dx \, dy = 0.$$

Since $\bar{\mathbf{x}}$ and $\mathbf{n}$ are constant, this relation may be written

$$\mathbf{n} \cdot \iint\limits_D \mathbf{x} \, \rho \, dx \, dy = (\mathbf{n} \cdot \bar{\mathbf{x}}) \iint\limits_D \rho \, dx \, dy,$$

or $$\mathbf{n} \cdot \mathbf{m} = M\mathbf{n} \cdot \bar{\mathbf{x}}.$$

This equation of balance is true for each choice of the knife-edge (each choice of the unit vector $\mathbf{n}$). Hence

$$\mathbf{n} \cdot (\mathbf{m} - M\bar{\mathbf{x}}) = 0$$

for each unit vector $\mathbf{n}$. This means the component of $\mathbf{m} - M\bar{\mathbf{x}}$ in each direction is zero. Therefore

$$\mathbf{m} - M\bar{\mathbf{x}} = \mathbf{0}, \qquad \mathbf{m} = M\bar{\mathbf{x}}.$$

Mass and Center of Gravity in Polar Coordinates   Suppose a non-homogeneous sheet covers a domain **D** described by polar coordinates. The element of area is $dA = r \, dr \, d\theta$, so the element of mass is

$$dM = \rho \, r \, dr \, d\theta,$$

where the density $\rho = \rho(r, \theta)$ is expressed as a function of r and θ. Therefore the mass of the sheet is

$$M = \iint\limits_D \rho \, r \, dr \, d\theta.$$

To find its moment, express $\mathbf{x}$ in polar coordinates:

$$\mathbf{x} = (x, y) = (r \cos \theta, r \sin \theta).$$

Then

$$\mathbf{m} = \iint\limits_D \mathbf{x} \, dM = \iint\limits_D \mathbf{x} \, \rho \, r \, dr \, d\theta = \iint\limits_D \rho \, r^2(\cos \theta, \sin \theta) \, dr \, d\theta$$

$$= \left(\iint\limits_D \rho \, r^2 \cos \theta \, dr \, d\theta, \iint\limits_D \rho \, r^2 \sin \theta \, dr \, d\theta \right).$$

Once M and $\mathbf{m}$ are computed, we know $\bar{\mathbf{x}} = \mathbf{m}/M$.

■ **EXAMPLE 4** Find the center of gravity of a uniform semi-circular sheet of radius a.

Solution Take the center at **0** and the diameter along the x-axis. Then the domain **D** covered by the sheet is $0 \leq r \leq a$, $0 \leq \theta \leq \pi$. Since ρ is constant, $M = \frac{1}{2}\pi a^2 \rho$ and

$$\mathbf{m} = \rho \left(\iint\limits_{\mathbf{D}} r^2 \cos\theta \, dr \, d\theta, \quad \iint\limits_{\mathbf{D}} r^2 \sin\theta \, dr \, d\theta \right)$$

$$= \rho \left(\int_0^\pi \cos\theta \, d\theta \int_0^a r^2 \, dr, \quad \int_0^\pi \sin\theta \, d\theta \int_0^a r^2 \, dr \right) = \rho(0, \tfrac{2}{3}a^3).$$

Therefore $$\bar{\mathbf{x}} = \frac{1}{M}\mathbf{m} = \frac{2}{\pi a^2 \rho} \rho(0, \tfrac{2}{3}a^3) = \left(0, \ \frac{4a}{3\pi}\right).$$

See Fig. 3.   ■

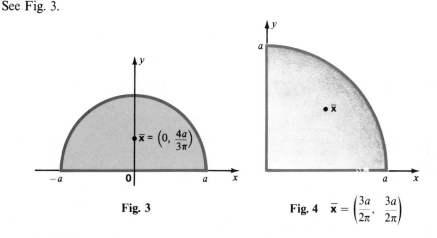

Fig. 3   Fig. 4 $\bar{\mathbf{x}} = \left(\dfrac{3a}{2\pi}, \dfrac{3a}{2\pi}\right)$

In the figure: $\bar{\mathbf{x}} = \left(0, \dfrac{4a}{3\pi}\right)$

■ **EXAMPLE 5** Find the center of gravity of a sheet in the shape of a quarter circle, whose density is proportional to the distance from the center of the circle.

Solution Place the sheet as in Fig. 4. Then $\rho = kr$, where k is a constant. Obviously k will cancel in the end, so we may take $k = 1$, that is, $\rho = r$. The mass of the sheet is

$$M = \iint \rho \, r \, dr \, d\theta = \iint r^2 \, dr \, d\theta = \int_0^a r^2 \, dr \int_0^{\pi/2} d\theta = \frac{\pi a^3}{6},$$

and its moment is $$\mathbf{m} = \iint \mathbf{x} \, \rho \, r \, dr \, d\theta = \iint (r\cos\theta, r\sin\theta) \, r^2 \, dr \, d\theta$$

$$= \int_0^a r^3 \, dr \int_0^{\pi/2} (\cos\theta, \sin\theta) \, d\theta = \frac{a^4}{4}(1, 1).$$

Therefore $$(\bar{\mathbf{x}}) = \frac{1}{M}\mathbf{m} = \frac{6}{\pi a^3} \cdot \frac{a^4}{4}(1, 1) = \left(\frac{3a}{2\pi}, \frac{3a}{2\pi}\right).$$

Could you have predicted that $\bar{\mathbf{x}}$ lies on the line $y = x$ and $|\bar{\mathbf{x}}| > \frac{1}{2}a$?   ■

The First Pappus Theorem   There is a useful connection between the centers of gravity and volumes of revolution.

First Pappus Theorem   Suppose a region **D** in the x, y-plane, to the right of the y-axis, is revolved about the y-axis. Then the volume of the resulting solid is

$$V = 2\pi \bar{x} A,$$

where A is the area of the plane region **D** and $\bar{x}$ is the x-coordinate of its center of gravity (Fig. 5).

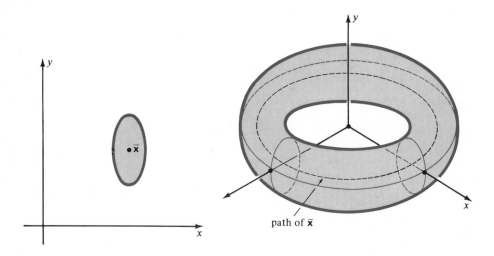

Fig. 5   First Pappus Theorem

In words, the volume is the area times the length of the circle traced by the center of gravity. Proof: a small portion $dx\,dy$ at **x** revolves into a thin ring of volume

$$dV = 2\pi x\,dx\,dy.$$

Hence
$$V = \iint\limits_{\mathbf{D}} 2\pi x\,dx\,dy = 2\pi m_x.$$

But $m_x = \bar{x}A$, hence $V = 2\pi \bar{x} A$.

Wires   A non-homogeneous wire is described by its position, a space curve $\mathbf{x} = \mathbf{x}(s)$, where $a \le s \le b$, and its density $\delta = \delta(s)$. (Here s denotes arc length.)

Its mass is $M = \displaystyle\int_a^b \delta(s)\,ds$ and its moment is $\mathbf{m} = \displaystyle\int_a^b \mathbf{x}(s)\,\delta(s)\,ds,$

so its center of gravity is $\bar{\mathbf{x}} = \dfrac{1}{M}\,\mathbf{m}.$

If the wire is uniform, then $\delta(s)$ is a constant. In this case, the center of gravity is independent of δ, hence it is a property of the curve $\mathbf{x} = \mathbf{x}(s)$ alone; you can take $\delta = 1$ and replace M by L, the length.

■ **EXAMPLE 6**   Find the center of gravity of the uniform semi-circle $r = a$, $y \geq 0$.

Solution   The length is $L = \pi a$. The moment is

$$\mathbf{m} = \int \mathbf{x} \, ds = \int_0^\pi (a \cos \theta, a \sin \theta) \, a \, d\theta = a^2 \int_0^\pi (\cos \theta, \sin \theta) \, d\theta = a^2(0, 2).$$

Hence
$$\bar{\mathbf{x}} = \frac{1}{L} \mathbf{m} = \frac{1}{\pi a} a^2(0, 2) = \frac{a}{\pi}(0, 2) = \left(0, \frac{2}{\pi} a\right).$$ ■

Suppose a plane curve is revolved about an axis in its plane, generating a surface of revolution. There is a useful relation between the center of gravity of the curve and the area of the surface.

Second Pappus Theorem   Suppose a curve in the x, y-plane to the right of the y-axis is revolved about the y-axis. Then the area of the resulting surface is

$$A = 2\pi \bar{x} L,$$

where L is the length of the curve and $\bar{x}$ is the x-coordinate of the center of gravity (Fig. 6).

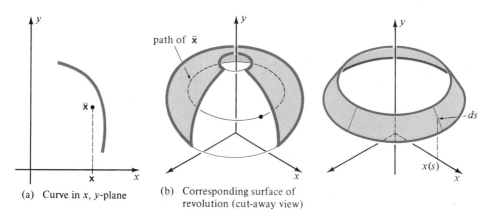

(a)   Curve in x, y-plane

(b)   Corresponding surface of revolution (cut-away view)

path of $\bar{\mathbf{x}}$

$-ds$

$x(s)$

Fig. 6   Second Pappus Theorem   **Fig. 7**   Element of rotated area

In words, the area is the length of the curve times the length of the circle traced by the center of gravity. Proof: a short segment of length ds of the curve at the point $\mathbf{x}(s)$ revolves into the frustum of a cone with lateral area $dA = 2\pi x \, ds$. See Fig. 7. Hence

$$A = \int_a^b 2\pi x \, ds = 2\pi \int_a^b x \, ds = 2\pi m_x = 2\pi \bar{x} L.$$

Moment of Inertia Suppose a plane sheet over domain **D** rotates about **0** with angular speed ω rad/sec (Fig. 8). An element of mass dM at distance r from **0** has speed $r\omega$, so its element of kinetic energy is

$$dE = \tfrac{1}{2}(dM)(\text{speed})^2 = \tfrac{1}{2}(\rho \, dx \, dy)(r\omega)^2$$
$$= \tfrac{1}{2}\omega^2 r^2 \rho \, dx \, dy = \tfrac{1}{2}\omega^2(x^2 + y^2)\rho \, dx \, dy.$$

Therefore the total kinetic energy of the rotating sheet is

$$E = \frac{\omega^2}{2} \iint_{\mathbf{D}} (x^2 + y^2)\,\rho \, dx \, dy = \frac{1}{2}\, I\omega^2,$$

where I is called the moment of inertia of the sheet (with respect to the origin).

$$\boxed{\;\textbf{Moment of Inertia} \qquad I = \iint_{\mathbf{D}} (x^2 + y^2)\rho(x, y)\,dx\,dy.\;}$$

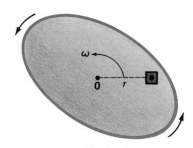

Fig. 8 Rotating sheet, angular speed ω rad/sec

■ **EXAMPLE 7** Find the moment of inertia of a uniform square of side a about its center.

Solution Place the square in the position $-\tfrac{1}{2}a \le x \le \tfrac{1}{2}a$, $-\tfrac{1}{2}a \le y \le \tfrac{1}{2}a$. The density ρ is constant, and

$$I = \iint_{\mathbf{D}} (x^2 + y^2)\,\rho \, dx \, dy = \rho \iint_{\mathbf{D}} x^2 \, dx \, dy + \rho \iint_{\mathbf{D}} y^2 \, dx \, dy$$

$$= \rho\left(\int_{-a/2}^{a/2} x^2 \, dx \int_{-a/2}^{a/2} dy + \int_{-a/2}^{a/2} dx \int_{-a/2}^{a/2} y^2 \, dy \right)$$

$$= \rho[\tfrac{2}{3}(\tfrac{1}{2}a)^3 a + a(\tfrac{2}{3})(\tfrac{1}{2}a)^3] = \tfrac{1}{6}\rho a^4.$$

Since the mass of the square is $M = \rho a^2$, the answer can be written $I = \tfrac{1}{6}Ma^2$. ■

EXERCISES

Find the mass and the center of gravity for the density ρ on the indicated region

1 $\rho = (1 + x)(1 + y)$ · $0 \le x \le 1$ $0 \le y \le 1$
2 $\rho = xy$ $1 \le x \le 2$ $1 \le y \le 3$
3 $\rho = 2 - x$ $0 \le x \le 1$ $-1 \le y \le 1$

4 $\rho = 1 + x$ $\qquad\qquad$ $0 \le x \le 3$ $\qquad$ $0 \le y \le 2$
5 $\rho = (1 - x)(1 - y) + 1$ $\quad$ $0 \le x \le 1$ $\qquad$ $0 \le y \le 1$
6 $\rho = \sin x$ $\qquad\qquad\quad$ $0 \le x \le \pi$ $\qquad$ $0 \le y \le 1$
7 $\rho = 1 + x^2 + y^2$ $\qquad\;$ $-1 \le x \le 1$ $\qquad$ $1 \le y \le 4$
8 $\rho = 2 + x^2 y^2$ $\qquad\quad\;$ $-1 \le x \le 1$ $\qquad$ $0 \le y \le 1$
9 $\rho = 1$ $\qquad\qquad\qquad\;$ $0 \le y \le 1 - x^2$
10 $\rho = y$ $\qquad\qquad\qquad$ $0 \le y \le 1 - x^2$
11 $\rho = 1$ $\qquad\qquad\qquad$ $0 \le x \le 1$ $\qquad$ $0 \le y \le x^2$
12 $\rho = x$ $\qquad\qquad\qquad$ $0 \le x \le 1$ $\qquad$ $0 \le y \le x^2$

13 $\rho = 1$ $\qquad\qquad\qquad$ $\dfrac{x^2}{a^2} + \dfrac{y^2}{b^2} \le 1$ $\quad$ $y \ge 0$

14 $\rho = 2 + \dfrac{x}{a}$ $\qquad\qquad$ $\dfrac{x^2}{a^2} + \dfrac{y^2}{b^2} \le 1$

15 $\rho = 1$ $\qquad\qquad\qquad$ $r \le a$ $\quad$ $0 \le \theta \le \tfrac{1}{2}\pi$

16 $\rho = 1 - \dfrac{r^2}{a^2}$ $\qquad\quad$ $r \le a$ $\quad$ $0 \le \theta \le \tfrac{1}{2}\pi$

17 $\rho = 1$ $\qquad\qquad\qquad$ $r \le a$ $\quad$ $0 \le \theta \le \alpha.$

18 (cont.) Let $\alpha \longrightarrow 0$ in Ex. 17. Find the limiting position of $\bar{\mathbf{x}}$.

Find the mass and the center of gravity for the density ρ on the indicated wire

19 $\rho = 1$ $\qquad$ $r = a$ $\quad$ $0 \le \theta \le \tfrac{1}{2}\pi$ $\qquad$ **20** $\rho = \sin \theta$ $\quad$ $r = a$ $\quad$ $0 \le \theta \le \pi$
21 $\rho = k\theta$ $\qquad$ $r = a$ $\quad$ $0 \le \theta \le \pi$ $\qquad$ **22** $\rho = 1 + k\theta$ $\quad$ $r = a$ $\quad$ $0 \le \theta \le \pi$
23 $\rho = 1$ $\qquad$ triangle: vertices $(0, 0)$, $(4, 0)$, $(0, 3)$
24 $\rho = 1$ $\qquad$ triangle: vertices $(0, 0)$, $(1, 0)$, $(0, 1)$.

25 Verify the First Pappus Theorem for a semi-circle revolved about its diameter.
26 Use the First Pappus Theorem to find the volume of a right circular torus.
27 Verify the First Pappus Theorem for a right triangle revolved about a leg.
28 Use the Second Pappus Theorem to find the surface area of a right circular torus.
29 Use the Second Pappus Theorem to obtain another solution of Example 6.
30 Use the Second Pappus Theorem to obtain another solution of Ex. 23.

Find the moment of inertia with respect to the origin; give the answer in the form $I = M \cdot (?)$

31 $\rho = 1$ $\qquad$ circle $r \le a$
32 $\rho = r^n$ $\qquad$ circle $r \le a$
33 $\rho = 1$ $\qquad$ circle $(x - a)^2 + y^2 \le a^2$
34 $\rho = 1$ $\qquad$ rectangle $|x| \le a$ $\;$ $|y| \le b$
35 $\rho = 1 + x$ $\quad$ triangle vertices $(0, 0)$, $(1, 0)$, $(0, 1)$
36 $\rho = xy$ $\qquad$ square $0 \le x \le b$ $\;$ $0 \le y \le b.$

8. APPROXIMATE INTEGRATION

In this section we discuss an extension of Simpson's Rule to approximation of double integrals. For simplicity, we shall allow only rectangular domains.

Let us recall Simpson's Rule. To approximate the integral

$$\int_a^b f(x)\, dx,$$

we divide the interval $a \leq x \leq b$ into $2m$ equal parts of length h:

$$a = x_0 < x_1 < x_2 < \cdots < x_{2m} = b, \qquad h = \frac{b-a}{2m},$$

and use the formula $\qquad \displaystyle\int_a^b f(x)\,dx \approx \frac{h}{3} \sum_{i=0}^{2m} B_i f(x_i),$

where the coefficients B_i are $1, 4, 2, 4, 2, 4, 2, \cdots, 2, 4, 1$.

We extend Simpson's Rule to double integrals in the following way. To approximate

$$\iint_{\mathbf{D}} f(x, y)\,dx\,dy,$$

where $\mathbf{D}$ denotes the rectangle $a \leq x \leq b$ and $c \leq y \leq d$, we divide the x-interval into $2m$ parts as before and also divide the y-interval into $2n$ equal parts of length k:

$$c = y_0 < y_1 < y_2 < \cdots < y_{2n} = d, \qquad k = \frac{d-c}{2n}.$$

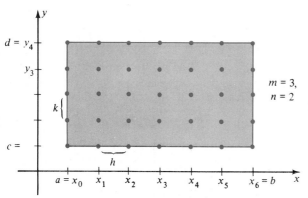

$m = 3,$
$n = 2$

Fig. 1   Division of rectangular domain

We obtain $(2m + 1)(2n + 1)$ points of the rectangle (Fig. 1). The rule is

$$\iint_{\mathbf{D}} f(x, y)\,dx\,dy \approx \frac{hk}{9} \sum_{i=0}^{2m} \sum_{j=0}^{2n} A_{ij} f(x_i, y_j),$$

where the coefficients A_{ij} are certain products of the coefficients in the ordinary Simpson's Rule. Precisely,

$$A_{ij} = B_i C_j,$$

where $B_0, B_1, \cdots, B_{2m}$ are the coefficients in the ordinary Simpson's Rule

$$\int_a^b p(x)\,dx \approx \frac{h}{3} \sum_{i=0}^{2m} B_i p(x_i),$$

and C_0, C_1, $\cdots$, C_{2n} are the coefficients in the ordinary Simpson's Rule

$$\int_c^d q(y)\,dy \approx \frac{k}{3}\sum_{j=0}^{2n} C_j q(y_j).$$

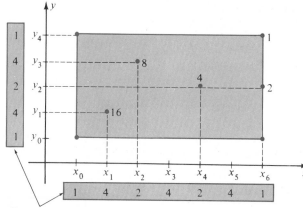

Fig. 2 Coefficients for double Simpson's Rule

ordinary Simpson's
Rule coefficients

In Fig. 2, several of these products are formed. Since B_i and C_j take values 1, 2, and 4, the coefficients A_{ij} take values 1, 2, 4, 8, and 16. The A_{ij} can be written in a matrix corresponding to the points (x_i, y_j) as in Fig. 1. For example, if $m = 3$ and $n = 2$, the matrix is

$$\begin{bmatrix} 1 & 4 & 2 & 4 & 2 & 4 & 1 \\ 4 & 16 & 8 & 16 & 8 & 16 & 4 \\ 2 & 8 & 4 & 8 & 4 & 8 & 2 \\ 4 & 16 & 8 & 16 & 8 & 16 & 4 \\ 1 & 4 & 2 & 4 & 2 & 4 & 1 \end{bmatrix}.$$

■ **EXAMPLE 1**   Estimate $I = \displaystyle\iint_D (x + y)^3\,dx\,dy$

by Simpson's Rule with $m = n = 1$, where **D** is the rectangle $0 \le x \le 1$, $0 \le y \le 1$. Compare the result with the exact answer.

Solution   Here $h = k = \frac{1}{2}$. The coefficient matrix is $[A_{ij}] = \begin{bmatrix} 1 & 4 & 1 \\ 4 & 16 & 4 \\ 1 & 4 & 1 \end{bmatrix}.$

Write the value $(x_i + y_j)^3$ in a matrix:

$$[(x_i + y_j)^3] = \begin{bmatrix} (1+0)^3 & (1+\tfrac{1}{2})^3 & (1+1)^3 \\ (\tfrac{1}{2}+0)^3 & (\tfrac{1}{2}+\tfrac{1}{2})^3 & (\tfrac{1}{2}+1)^3 \\ (0+0)^3 & (0+\tfrac{1}{2})^3 & (0+1)^3 \end{bmatrix} = \begin{bmatrix} 1 & \tfrac{27}{8} & 8 \\ \tfrac{1}{8} & 1 & \tfrac{27}{8} \\ 0 & \tfrac{1}{8} & 1 \end{bmatrix}.$$

Now estimate the integral by $I \approx \dfrac{hk}{9} \displaystyle\sum_{i=0}^{2} \sum_{j=0}^{2} A_{ij}(x_i + y_j)^3.$

To evaluate the sum, multiply corresponding terms of the two matrices and add the nine products:

$$I \approx \tfrac{1}{9}(\tfrac{1}{2})(\tfrac{1}{2})[1 \cdot 1 + 4 \cdot \tfrac{27}{8} + 1 \cdot 8 + 4 \cdot \tfrac{1}{8} + 16 \cdot 1 + 4 \cdot \tfrac{27}{8} + 1 \cdot 0 + 4 \cdot \tfrac{1}{8} + 1 \cdot 1]$$

$$= \tfrac{1}{36}[1 + \tfrac{27}{2} + 8 + \tfrac{1}{2} + 16 + \tfrac{27}{2} + \tfrac{1}{2} + 1] = \tfrac{54}{36} = \tfrac{3}{2}.$$

The exact value is

$$I = \iint (x + y)^3 \, dx \, dy = \int_0^1 \left(\int_0^1 (x + y)^3 \, dx \right) dy$$

$$= \tfrac{1}{4} \int_0^1 [(y + 1)^4 - y^4] \, dy = \tfrac{1}{20}[(y + 1)^5 - y^5] \Big|_0^1 = \tfrac{30}{20} = \tfrac{3}{2},$$

so the estimate is exact in this case.   ■

Remark 1   Because Simpson's Rule is exact for cubics, the double integral rule is exact for cubics in two variables. (See Exs. 19–20.)

Remark 2   The matrix of values $[f(x_i, y_i)]$ is arranged to conform to the layout of points (x_i, y_j) in the plane (Fig. 1).

The next example is an integral that cannot be evaluated exactly, only approximated.

■ **EXAMPLE 2**   Estimate $I = \displaystyle\iint_{\mathbf{D}} \sin(xy) \, dx \, dy,$

using $m = n = 1$, where **D** is the square $0 \le x \le \tfrac{1}{2}\pi, \quad 0 \le y \le \tfrac{1}{2}\pi.$

Solution   Here $h = k = \tfrac{1}{4}\pi$, and the coefficient matrix is $[A_{ij}] = \begin{bmatrix} 1 & 4 & 1 \\ 4 & 16 & 4 \\ 1 & 4 & 1 \end{bmatrix}$

The matrix of values of $\sin xy$ is $[\sin x_i y_j] = \begin{bmatrix} \sin 0 & \sin \tfrac{1}{8}\pi^2 & \sin \tfrac{1}{4}\pi^2 \\ \sin 0 & \sin \tfrac{1}{16}\pi^2 & \sin \tfrac{1}{8}\pi^2 \\ \sin 0 & \sin 0 & \sin 0 \end{bmatrix}$

Therefore $I = \displaystyle\iint_{\mathbf{D}} \sin xy \, dx \, dy \approx \dfrac{\pi^2}{144} \left(16 \sin \dfrac{\pi^2}{16} + 8 \sin \dfrac{\pi^2}{8} + \sin \dfrac{\pi^2}{4} \right) \approx 1.195.$ ■

Error Estimate   The error estimate for Simpson's Rule in two variables is analogous to that in one variable:

$$|error| \le \dfrac{(b - a)(d - c)}{180} [h^4 M + k^4 N],$$

where $\qquad \left| \dfrac{\partial^4 f}{\partial x^4} \right| \le M \quad$ and $\quad \left| \dfrac{\partial^4 f}{\partial y^4} \right| \le N.$

We omit the proof.

■ **EXAMPLE 3**   Estimate the error in Example 2.

Solution $\dfrac{\partial^4}{\partial x^4}(\sin xy) = y^4 \sin xy, \qquad \dfrac{\partial^4}{\partial y^4}(\sin xy) = x^4 \sin xy.$

But $|\sin xy| \le 1$. Hence in the square $0 \le x \le \frac{1}{2}\pi, \;\; 0 \le y \le \frac{1}{2}\pi$, the inequalities

$$\left| \frac{\partial^4 f}{\partial x^4} \right| = |y^4 \sin xy| \le \left(\frac{\pi}{2} \right)^4, \qquad \left| \frac{\partial^4 f}{\partial y^4} \right| = |x^4 \sin xy| \le \left(\frac{\pi}{2} \right)^4$$

hold. Apply the error estimate with $m = n = 1$, $h = k = \frac{1}{4}\pi$, and $M = N = (\frac{1}{2}\pi)^4$:

$$|\text{error}| \le \frac{1}{180}\left(\frac{\pi}{2} \right)^2 \left[2\left(\frac{\pi}{4} \right)^4 \left(\frac{\pi}{2} \right)^4 \right] = \frac{1}{45} \cdot \frac{\pi^{10}}{2^{15}} < 0.064.$$ ■

EXERCISES

Estimate to 4 significant figures; take $m = n = 1$

1   $\displaystyle\iint \cos(xy)\, dx\, dy$
$\;0 \le x \le \frac{1}{4}\pi$
$\;0 \le y \le \frac{1}{4}\pi$

2   $\displaystyle\iint \frac{dx\, dy}{1 + x + y}$
$\;0 \le x \le 1$
$\;0 \le y \le 1$

3   $\displaystyle\iint \frac{dx\, dy}{1 + x^2 + y^2}$
$\;0 \le x \le 1$
$\;0 \le y \le 1$

4   $\displaystyle\iint e^{x^2/y}\, dx\, dy$
$\;0 \le x \le 1$
$\;1 \le y \le 2$

5   $\displaystyle\iint e^{-x^2 - y^2}\, dx\, dy$
$\;0 \le x \le 2$
$\;0 \le y \le 2$

6   $\displaystyle\iint e^{-x^2 y^2}\, dx\, dy$
$\;0 \le x \le 1$
$\;0 \le y \le 1$

7   $\displaystyle\iint x^4 y^3\, dx\, dy$
$\;0 \le x \le 1$
$\;0 \le y \le 1$

8   $\displaystyle\iint \tan(xy)\, dx\, dy.$
$\;0 \le x \le \frac{1}{4}\pi$
$\;0 \le y \le 1$

Estimate to 5 significant figures; take $m = n = 2$

9   $\displaystyle\iint \sin(xy)\, dx\, dy$
$\;0 \le x \le \frac{1}{2}\pi$
$\;0 \le y \le \frac{1}{2}\pi$

10   $\displaystyle\iint \cos(xy)\, dx\, dy$
$\;0 \le x \le \frac{1}{2}\pi$
$\;0 \le y \le \frac{1}{2}\pi$

11   $\displaystyle\iint \frac{xy^2}{x + y}\, dx\, dy$
$\;0 \le x \le 1$
$\;1 \le y \le 2$

12   $\displaystyle\iint e^{-x^2 - y^2}\, dx\, dy$
$\;0 \le x \le 2$
$\;0 \le y \le 2$

13   $\displaystyle\iint \sin(xy)\, dx\, dy$
$\;0 \le x \le 1$
$\;0 \le y \le \pi$

14   $\displaystyle\iint \frac{dx\, dy}{1 + x^3 + y^4}.$
$\;0 \le x \le 1$
$\;0 \le y \le 2$

Give an upper bound for the error in

15   Ex. 1   16   Ex. 2   17   Ex. 9   18   Ex. 13.

19   Suppose $f(x, y) = p(x)q(y)$. Show that the double integral Simpson's Rule estimate is just the product of the Simpson's Rule estimate for $\int p(x)\, dx$ by that for $\int q(y)\, dy$.

20   (cont.) Conclude that the rule is exact for polynomials involving only $x^3 y^3$, $x^3 y^2$, $x^2 y^3$, $x^3 y$, $x^2 y^2$, xy^3, and lower degree terms.

21   The analogue of the Trapezoidal Rule is

$$\iint\limits_{\substack{0 \le x \le 1 \\ 0 \le y \le 1}} f(x, y)\, dx\, dy \approx \frac{1}{4}[f(0, 0) + f(0, 1) + f(1, 1) + f(1, 0)].$$

Show that this rule is exact for polynomials $f(x, y) = A + Bx + Cy + Dxy$.

22   (cont.) Find the corresponding rule for a rectangle $a \le x \le b$, $c \le x \le d$, divided into rectangles of size h by k with $h = (b - a)/m$ and $k = (d - c)/n$.

23   (cont.) Test the resulting rule on $\iint\limits_{\substack{0 \le x \le 1 \\ 0 \le y \le 1}} x^4 y^4\, dx\, dy$ with $m = n = 4$.

24*   Let **I** denote the unit square $0 \le x, y \le 1$. Suppose $f(x, y) = 0$ at its four vertices. Prove that

$$\iint\limits_{\mathbf{I}} f(x, y)\, dx\, dy = -\tfrac{1}{2} \iint\limits_{\mathbf{I}} y(1 - y) f_{yy}(x, y)\, dx\, dy - \tfrac{1}{4} \int_0^1 x(1 - x)[f_{xx}(x, 1) + f_{xx}(x, 0)]\, dx.$$

25   (cont.) Suppose also that $|f_{xx}| \le M$ and $|f_{yy}| \le N$ on **I**. Prove that

$$\left| \iint\limits_{\mathbf{I}} f(x, y)\, dx\, dy \right| \le \tfrac{1}{12} M + \tfrac{1}{12} N.$$

26   (cont.) Conclude that for any function on the square $0 \le x \le 1$, $0 \le y \le 1$, the error in the trapezoidal estimate (Ex. 21) is at most $\frac{1}{12}(M + N)$. [*Hint*   Use the result of Ex. 21 and interpolation.]

27   (cont.) Suppose $f(x, y)$ has domain $a \le x \le b$, $c \le y \le d$ and satisfies $|f_{xx}| \le M$, $|f_{yy}| \le N$. Show that the error in the trapezoidal approximation (Ex. 22) with $m = n = 1$ is at most $\dfrac{hk}{12}(h^2 M + k^2 N)$,   where $h = b - a$ and $k = d - c$.

28   (cont.) Deduce the corresponding error estimate for arbitrary m and n in Ex. 22.

29   (cont.) What does this result give for Ex. 23?

30   (cont.) Suppose the trapezoidal approximation is used to estimate the integral of Ex. 9, but with any m and n. Give an upper bound for the error.

9. MISCELLANEOUS EXERCISES

1   Let C denote the volume of the cone with base $x^2/a^2 + y^2/b^2 = 1$ and apex $(0, 0, c)$. Let P denote the volume of the inverted paraboloid $0 \le z \le c(1 - x^2/a^2 - y^2/b^2)$. Assume a, b, c all positive. Find the (Archimedes') relation between P and C.

2   Find the volume of the solid $0 \le z \le cy/b$, $x^2/a^2 + y^2/b^2 \le 1$, where $a > 0$, $b > 0$, $c > 0$.

3   Integrate xy over the domain $0 \le x \le a$, $0 \le y \le a$, $x^2 + y^2 \ge a^2$.

4   Evaluate $\iint dx\, dy$ over $a \le x \le y \le b$.

5   An exponential horn loudspeaker is bounded by the 6 surfaces $x = e^{az}$, $x = -e^{az}$, $y = e^{az}$, $y = -e^{az}$, $z = 0$, $z = b$. Find its volume.

6   Evaluate $\iint \dfrac{dx\, dy}{r^4}$ over $1 \le r \le 2$, $r^2 \ge 2x$.

7   A certain solid lies between the planes $z = a$ and $z = b$, where $a < b$. Let $A(z)$ be the cross-sectional area of the solid at height z. Give a formula for the volume of the solid (**Cavalieri's principle**).

8   (cont.) Let $a > 0$, $b > 0$, $c > 0$. Join each point $(x, y, 0)$ in the elliptic domain $x^2/a^2 + y^2/b^2 \le 1$, $z = 0$ by a segment to $(0, y, c)$. These segments then sweep out a tentlike solid. Find its volume.

9   Let $\mathbf{x} = \mathbf{x}(s)$ be a curve of length L on the unit sphere $|\mathbf{x}| = 1$. The segment $\overline{0\mathbf{x}(s)}$ sweep out a (conical) surface. Find its area.

10*   To find the volume of the portion of a cone cut off by a plane. Let $0 < a < b$ and $c > 0$. Find the volume of the solid $cy/b \le z \le c(1 - r/a)$, where $r^2 = x^2 + y^2$ as usual.

11*   Find the area and the moment m_x for **D** in Fig. 1; $\rho = 1$. [*Hint* You may use Ex. 17, p. 883.]

12   (cont.) A sphere of radius a is inscribed in one mappe of a right circular cone of apex angle 2α. Find the volume of the portion of the cone between the sphere and the apex.

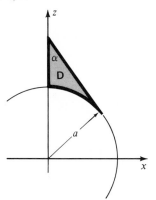

Fig. 1   Fig. 2

13   Find the moment m_x for **D** in Fig. 2; $\rho = 1$. [*Hint* Use Example 4, p. 879.]

14   (cont.) Find the volume of the hole in a donut, **D** rotated about the z-axis.

15   Find the mass and center of gravity of the spiral wire $r = \theta$, $0 \le \theta \le 2\pi$, where the density is $\rho = \sqrt{1 + \theta^2}$.

16*   Suppose $f(a) = 0$. Express $2\int_a^b \left(\int_a^y f(x)f(y)[1 - f'(x)]\, dx \right) dy$

in terms of $\int_a^b f(x)\, dx$ and $\int_a^b f(x)^3\, dx$.

17   (cont.) Suppose $0 = f(a) \le f(x)$ and $f'(x) \le 1$ for $a \le x \le b$. Prove

$$\int_a^b [f(x)]^3\, dx \le \left(\int_a^b f(x)\, dx \right)^2.$$

18   Evaluate $\displaystyle\int_0^1 \frac{dx}{(1 + x^2)^2}$ by differentiating $\displaystyle\int_0^1 \frac{dx}{x^2 + t^2} = \frac{1}{t} \arctan \frac{1}{t}$.

Suppose $f(x)$ and $g(x)$ are continuous for $x \ge 0$. Their **convolution** is the function $h = f * g$ defined by $\displaystyle h(x) = \int_0^x f(t)\, g(x - t)\, dt.$

19   Prove $g * f = f * g$.

20   (cont.) Prove $f * (g * h) = (f * g) * h$.

21*   (cont.) Recall that the **Laplace transform** of $f(x)$ is the function

$$L(f)(s) = \int_0^\infty e^{-sx} f(x)\, dx$$

where convergent. Assuming everything in sight converges and changing the order of integration is valid, prove $L(f * g)(s) = L(f)(s) \cdot L(g)(s)$.

22   Criticize:

$$I = \iint\limits_{\substack{0 \le x \le 1 \\ 0 \le y \le 1}} \frac{x^2 - y^2}{(x^2 + y^2)^2} \, dx \, dy = \int_0^1 \left(\int_0^1 \frac{x^2 - y^2}{(x^2 + y^2)^2} \, dx \right) dy = \int_0^1 \left(\frac{-x}{x^2 + y^2} \right) \Big|_{x=0}^{x=1} dy$$

$$= \int_0^1 \frac{-1}{1 + y^2} \, dy = -\frac{1}{4} \pi.$$

In a snow pile, let $p = p(\mathbf{x})$ denote the pressure and $\rho = \rho(\mathbf{x})$ the density. Assume that p and ρ depend only on the depth z of the snow above $\mathbf{x}$. Thus in a vertical column of unit cross-sectional area, we have

$$p(z) = \int_0^z \rho(u) \, du \qquad \text{at depth} \quad z.$$

Assume also that the density at any point depends only on the pressure there, so $\rho = f(p)$. If the snow is not too deep (so it doesn't pack into ice), a reasonable assumption is $\rho = \rho_0 + kp$, where $\rho_0 > 0$ and $k > 0$.

23   Prove $p(z) = (\rho_0/k)(e^{kz} - 1)$ and $\rho(z) = \rho_0 e^{kz}$.

24   (cont.) Suppose the snow pile has the shape of a right circular cone of radius a and height h. Find the weight of the snow pile.

25   (cont.) Suppose the snow pile has the shape of a hemisphere of radius a. Find the weight of the snow pile.

26   (cont.) Suppose the snow pile has the shape of an inverted paraboloid of revolution of height h and base radius a. Find its weight.

27   Suppose $f(x)$ is increasing on $a \le x \le b$. Prove

$$(b + a) \int_a^b f(x) \, dx \le 2 \int_a^b x f(x) \, dx.$$

[*Hint*   Consider $(x - y)[f(x) - f(y)]$.]

28* Suppose $f(x)$ is increasing on $a \le x \le b$ and $f(x) > 0$. Prove

$$\frac{\int_a^b x[f(x)]^2 \, dx}{\int_a^b [f(x)]^2 \, dx} \ge \frac{\int_a^b x f(x) \, dx}{\int_a^b f(x) \, dx}.$$

29   Let a line segment in $\mathbf{R}^2$ have length L, and let L_1 and L_2 be the lengths of its projections on the axes. Prove $L^2 = L_1^2 + L_2^2$.

30   (cont.) Find the analogous relation for a plane region in $\mathbf{R}^3$ of area A.

Multiple Integrals 18

1. TRIPLE INTEGRALS

Triple integrals arise from problems of the following type. Suppose we have a bounded domain **D** in space filled by a non-homogeneous solid. At each point **x** the density of the solid is $\delta(\mathbf{x})$ gm/cm^3. What is the total mass?

Our previous experience suggests an approach to the problem. We decompose **D** into many small subdomains **D**$_i$ and choose a point **x**$_i$ in **D**$_i$. Then the mass in **D**$_i$ is approximately $\delta(\mathbf{x}_i)|\mathbf{D}_i|$, where $|\mathbf{D}_i|$ is the volume. The total mass is approximately

$$\sum \delta(\mathbf{x}_i)|\mathbf{D}_i|.$$

Finally, we take the limit of these approximating sums as the subdivisions of **D** become finer and finer. The limit, if it exists, is the triple integral

$$\iiint\limits_{\mathbf{D}} \delta(\mathbf{x})\, dx\, dy\, dz.$$

All of this reminds us of double integrals. In fact the theory of triple integrals is similar to the theory of double integrals and presents no really new difficulties, so we shall omit most of it. The main theoretical fact is that the triple integral exists if $\delta(\mathbf{x})$ is continuous and the boundary of **D** is not too complicated. The main practical fact is that the integral can be evaluated by iteration.

Iteration A convenient domain for triple integration is the part of a cylinder bounded between two surfaces, each the graph of a function. Precisely, suppose two surfaces $z = g(x, y)$ and $z = h(x, y)$ are defined over a domain **S** in the x, y-plane, and that $g(x, y) < h(x, y)$. See Fig. 1. These surfaces can be considered as the top and bottom of a domain **D** in the cylinder over **S**. Thus **D** consists of all points (x, y, z) where (x, y) is in **S** and

$$g(x, y) \le z \le h(x, y).$$

In this situation the iteration formula is

$$\iiint\limits_{\mathbf{D}} \delta(\mathbf{x})\, dx\, dy\, dz = \iint\limits_{\mathbf{S}} \left(\int_{z=g(x,\, y)}^{z=h(x,\, y)} \delta(x, y, z)\, dz \right) dx\, dy.$$

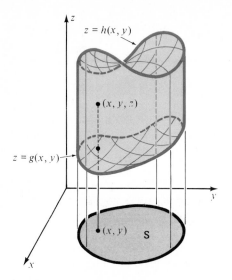

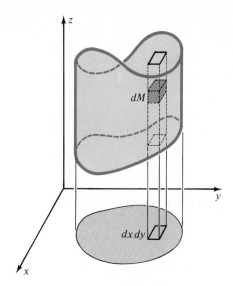

Fig. 1   Domain in $\mathbf{R}^3$ between two
graphs of functions

Fig. 2   Intuitive reason for the iteration
formula

Some prefer the notation

$$\iint_{\mathbf{S}} dx\, dy \int_{g(x,\,y)}^{h(x,\,y)} \delta(x,\, y,\, z)\, dz.$$

The reason for the iteration formula is illustrated in Fig. 2. First, the elements of mass $dM = \delta\, dx\, dy\, dz$ in one column are added up by an integral in the vertical direction (x and y fixed, z variable). The result is

$$\left(\int_{g(x,\,y)}^{h(x,\,y)} \delta(x,\, y,\, z)\, dz \right) dx\, dy.$$

Then the masses of these individual columns are totaled by a double integral over $\mathbf{S}$.

■ **EXAMPLE 1**   Find $\iiint (x^2 + y) z\, dx\, dy\, dz$,   taken over the block

$1 \le x \le 2$, $0 \le y \le 1$, $3 \le z \le 5$.

Solution   The upper and lower boundaries are the planes $z = 5$ and $z = 3$. Therefore

$$\iiint (x^2 + y) z\, dx\, dy\, dz = \iint_{\mathbf{S}} \left(\int_{3}^{5} (x^2 + y) z\, dz \right) dx\, dy,$$

where $\mathbf{S}$ is the rectangle $1 \le x \le 2$ and $0 \le y \le 1$. Now x and y are constant in the inner integral, so

$$\int_{3}^{5} (x^2 + y) z\, dz = (x^2 + y) \int_{3}^{5} z\, dz = 8(x^2 + y).$$

Hence·

$$\iiint (x^2 + y)\, z\, dx\, dy\, dz = \iint_{\mathbf{S}} 8(x^2 + y)\, dx\, dy$$

$$= \int_1^2 \left(\int_0^1 8(x^2 + y)\, dy \right) dx = 8\int_1^2 (x^2 + \tfrac{1}{2})\, dx = 8(\tfrac{7}{3} + \tfrac{1}{2}) = \tfrac{68}{3}. \qquad \blacksquare$$

Remark   The solution can be set up in the form

$$\iiint (x^2 + y)z\, dx\, dy\, dz = \int_1^2 \left[\int_0^1 \left(\int_3^5 (x^2 + y)\, z\, dz \right) dy \right] dx.$$

■ **EXAMPLE 2**   Compute $\iiint x^3 y^2 z\, dx\, dy\, dz$ over the domain **D** bounded by $x = 1,\ x = 2;\ \ y = 0,\ y = x^2;\ $ and $\ z = 0,\ z = 1/x.$

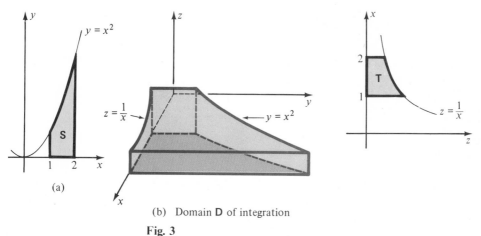

(a)

(b) Domain **D** of integration

Fig. 3

Solution   The domain **D** is the portion between the surfaces $z = 0$ and $z = 1/x$ of a solid cylinder parallel to the z-axis. The cylinder has base **S** in the x, y-plane, where **S** is shown in Fig. 3a. The solid **D** itself is sketched in Fig. 3b. (A rough sketch showing the general shape is satisfactory.) The iteration is

$$\iiint_{\mathbf{D}} x^3 y^2 z\, dx\, dy\, dz = \iint_{\mathbf{S}} \left(\int_0^{1/x} x^3 y^2 z\, dz \right) dx\, dy = \iint_{\mathbf{S}} x^3 y^2 \left(\tfrac{1}{2} z^2 \Big|_0^{1/x} \right) dx\, dy$$

$$= \tfrac{1}{2} \iint_{\mathbf{S}} x y^2\, dx\, dy = \tfrac{1}{2} \int_1^2 x \left(\int_0^{x^2} y^2\, dy \right) dx$$

$$= \tfrac{1}{6} \int_1^2 x^7\, dx = \tfrac{1}{48}(2^8 - 1) = \tfrac{255}{48} = \tfrac{85}{16}.$$

Alternative Solution   The domain may be considered as the portion between the surfaces $y = 0$ and $y = x^2$ of a solid cylinder parallel to the y-axis. The cylinder has base **T** in the z, x-plane (Fig. 4). From this view point, the first integration is with respect to y; the iteration is

$$\iiint\limits_{D} x^3 y^2 z \, dx \, dy \, dz = \iint\limits_{T} x^3 z \left(\int_{0}^{x^2} y^2 \, dy \right) dx \, dz = \iint\limits_{T} \tfrac{1}{3} x^9 z \, dx \, dz$$

$$= \tfrac{1}{3} \int_{1}^{2} x^9 \left(\int_{0}^{1/x} z \, dz \right) dx = \tfrac{1}{6} \int_{1}^{2} x^7 \, dx = \tfrac{1}{48}(2^8 - 1) = \tfrac{85}{16}. \quad\blacksquare$$

Remark It is bad technique to consider the region as a solid cylinder parallel to the x-axis because the projection of the solid into the y, z-plane breaks into four parts. Therefore, the solid **D** itself must be decomposed into four parts, and the triple integral correspondingly expressed as a sum of four triple integrals (Fig. 5). The resulting computation is much longer than that in either of the previous solutions.

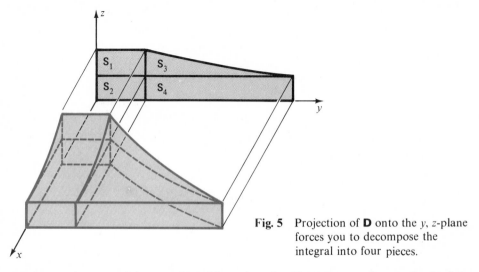

Fig. 5 Projection of **D** onto the y, z-plane forces you to decompose the integral into four pieces.

In general, try to pick an order of iteration that decomposes the required triple integral into as few summands as possible, at best only one. The typical summand has the form

$$\int_{a}^{b} \left[\int_{k(x)}^{h(x)} \left(\int_{g(x, y)}^{f(x, y)} \delta(x, y, z) \, dz \right) dy \right] dx.$$

(Possibly the variables are in some other order.) Once the integral

$$\int_{g(x, y)}^{f(x, y)} \delta(x, y, z) \, dz$$

is evaluated, the result is a function of x and y alone; z does not appear. Likewise, once the integral

$$\int_{k(x)}^{h(x)} \left(\int_{g(x, y)}^{f(x, y)} \delta(x, y, z) \, dz \right) dy$$

is evaluated, the result is a function of x alone; y does not appear.

Remember there are six possible orders of iteration for triple integrals. If you encounter an integrand you cannot find in tables, try a different order of iteration.

Domains and Inequalities   If a domain **D** is specified by inequalities, it may be possible to arrange the inequalities so that limits of integration can be set up automatically. For example, suppose the inequalities can be arranged in this form:

$$a \leq x \leq b, \qquad h(x) \leq y \leq k(x), \qquad g(x, y) \leq z \leq f(x, y).$$

Then
$$\iiint_{D} \delta(x, y, z)\, dx\, dy\, dz = \int_{a}^{b} \left[\int_{h(x)}^{k(x)} \left(\int_{g(x, y)}^{f(x, y)} \delta(x, y, z)\, dz \right) dy \right] dx.$$

Tetrahedral domains can be expressed by such inequalities, and they occur frequently enough that it is useful to practice setting up integrals over them.

■ **EXAMPLE 3**   A tetrahedron **T** has vertices at $(0, 0, 0)$, $(a, 0, 0)$, $(0, b, 0)$, $(0, 0, c)$,   where $a, b, c > 0$. Set up $\iiint_{T} \delta(x, y, z)\, dx\, dy\, dz$ as an iterated integral.

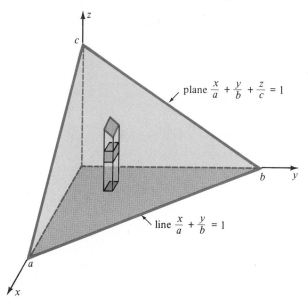

plane $\dfrac{x}{a} + \dfrac{y}{b} + \dfrac{z}{c} = 1$

Fig. 6   Tetrahedral domain

line $\dfrac{x}{a} + \dfrac{y}{b} = 1$

Solution   The slanted surface (Fig. 6) has equation

$$\frac{x}{a} + \frac{y}{b} + \frac{z}{c} = 1.$$

The domain is defined by the inequalities

$$0 \leq x, \qquad 0 \leq y, \qquad 0 \leq z, \qquad \frac{x}{a} + \frac{y}{b} + \frac{z}{c} \leq 1.$$

Any order of iteration is satisfactory; for instance, choose the order of integration

$$\int \left[\int \left(\int \delta(x, y, z)\, dz \right) dy \right] dx.$$

To find the limits of integration, we must replace the system of inequalities defining

D by an equivalent system of the form

$$a \leq x \leq b, \qquad h(x) \leq y \leq k(x), \qquad g(x, y) \leq z \leq f(x, y).$$

Obviously the original inequalities imply $0 \leq x \leq a$. Once we choose such an x, then

$$0 \leq y \leq b\left(1 - \frac{x}{a} - \frac{z}{c}\right) \leq b\left(1 - \frac{x}{a}\right),$$

since $z \geq 0$. Once we choose x and y then

$$0 \leq z \leq c\left(1 - \frac{x}{a} - \frac{y}{b}\right).$$

Thus we obtain the equivalent system of inequalities:

$$0 \leq x \leq a, \qquad 0 \leq y \leq b\left(1 - \frac{x}{a}\right), \qquad 0 \leq z \leq c\left(1 - \frac{x}{a} - \frac{y}{b}\right).$$

The corresponding iteration is

$$\iiint_{\mathbf{T}} \delta(x, y, z) \, dx \, dy \, dz = \int_0^a \left[\int_0^{b\left(1 - \frac{x}{a}\right)} \left(\int_0^{c\left(1 - \frac{x}{a} - \frac{y}{b}\right)} \delta(x, y, z) \, dz \right) dy \right] dx. \qquad \blacksquare$$

■ **EXAMPLE 4**   Set up an evaluation of the triple integral

$$I = \iiint_{\mathbf{D}} f(x, y, z) \, dx \, dy \, dz,$$

where the domain of integration **D** is specified by the inequalities

$$0 \leq x \leq 2, \qquad 0 \leq y \leq 2, \qquad 0 \leq z \leq 2, \qquad x + y \leq 3, \qquad y + z \leq 3.$$

Solution   The first order of business in solving such a problem is drawing **D**. We start with the cube $0 \leq x \leq 2$, $0 \leq y \leq 2$, $0 \leq z \leq 2$. The plane $y + z = 3$, shown in Fig. 7a, cuts the cube into two pieces. Clearly $y + z \leq 3$ is the lower one, so we chop off the prism $y + z \geq 3$. The result is the solid in Fig. 7b. The plane $x + y = 3$,

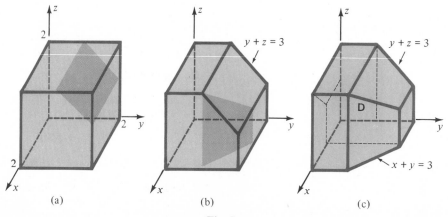

(a)   (b)   (c)

Fig. 7

shown in Fig. 7b, cuts the solid into two pieces, and we want the rear one, the one that includes the origin. We chop off the front piece and are left with **D** itself, shown in Fig. 7c.

Clearly we shall have to split the integral into pieces. If for instance we project **D** into the z, x-plane, then the resulting plane domain splits naturally into three subdomains (shown faintly in Fig. 7c). Better is to project **D** into the x, y-plane or into the y, z-plane; in either case the resulting plane domain splits only into two pieces. Again refer to Fig. 7c.

We shall choose the projection into the y, z-plane. The shadow of **D** is a plane domain $\mathbf{S} = \mathbf{S}_1 + \mathbf{S}_2$, where

$$\mathbf{S}_1 = \left\{0 \le y \le 1, \quad 0 \le z \le 2\right\} \quad \text{and} \quad \mathbf{S}_2 = \left\{1 \le y \le 2, \quad 0 \le z \le 3 - y\right\}.$$

The part $\mathbf{D}_1$ of **D** that projects onto $\mathbf{S}_1$ is specified by $0 \le x \le 2$. The part $\mathbf{D}_2$ that projects onto $\mathbf{S}_2$ is specified by $0 \le x \le 3 - y$. Accordingly,

$$I = \iiint_{\mathbf{D}_1} f\, dx\, dy\, dz + \iiint_{\mathbf{D}_2} f\, dx\, dy\, dz = \iint_{\mathbf{S}_1} \left(\int_0^2 f\, dx\right) dy\, dz + \iint_{\mathbf{S}_2} \left(\int_0^{3-y} f\, dx\right) dy\, dz$$

$$= \int_0^1 \left[\int_0^2 \left(\int_0^2 f\, dx\right) dz\right] dy + \int_1^2 \left[\int_0^{3-y} \left(\int_0^{3-y} f\, dx\right) dz\right] dy. \quad \blacksquare$$

EXERCISES

Evaluate the triple integral over the indicated domain

1 $\iiint xy^2 z\, dx\, dy\, dz$

$0 \le x \le 1 \quad 0 \le y \le 1 \quad 0 \le z \le 2$

2 $\iiint \dfrac{xy}{z}\, dx\, dy\, dz$

$0 \le x \le 1 \quad 1 \le y \le 2 \quad 1 \le z \le 3$

3 $\iiint \dfrac{x}{y+z}\, dx\, dy\, dz$

$-1 \le x \le 2 \quad 0 \le y \le 1 \quad 1 \le z \le 3$

4 $\iiint xy^2 \sin(xyz)\, dx\, dy\, dz$

$0 \le x \le 1 \quad 0 \le y \le \pi \quad 0 \le z \le 1$

5 $\iiint (x-y)(y-z)(z-x)\, dx\, dy\, dz$

$0 \le x \le 1 \le y \le 2 \le z \le 3$

6 $\iiint (x+y)(y+z)(z+x)\, dx\, dy\, dz$

$0 \le x \le 1 \quad 0 \le y \le 1 \quad 0 \le z \le 1$

7 $\iiint 120(x+y+z)^3\, dx\, dy\, dz$

$0 \le x \le a \quad 0 \le y \le b \quad 0 \le z \le c$

8 $\iiint \dfrac{x+y}{y+z}\, dx\, dy\, dz$

$1 \le x \le 3 \quad 1 \le y \le 2 \quad 0 \le z \le 1$

9 $\iiint z\, dx\, dy\, dz$

$0 \le x \quad 0 \le y \quad x + y \le 1$
$0 \le z \le 1 - x^2$

10 $\iiint y\, dx\, dy\, dz$

$0 \le x \quad 0 \le y \quad x + y \le 1$
$0 \le z \le x^2 + 2y^2$

11 $\iiint \dfrac{xz}{(1+y)^2}\, dx\, dy\, dz$

$0 \le x \quad 0 \le z \le 1 - y^2 \quad y \ge x^2$

12 $\iiint (3x^2 - z^2)\, dx\, dy\, dz$

$y \le 1 \quad -y \le x \le y \quad -y^2 \le z \le y^2$

13 $\iiint z^3\, dx\, dy\, dz$

pyramid with apex $(0, 0, 1)$, base the square with vertices $(\pm 1, \pm 1, 0)$

14 the same as Ex. 13, except the square base has vertices $(\pm 1, 0, 0)$, $(0, \pm 1, 0)$

15 $\iiint xyz \, dx \, dy \, dz$

tetrahedron with vertices
$(0, 0, 0),\quad (1, 0, 0),\quad (0, 1, 0),\quad (0, 0, 1)$

16 $\iiint y \, dx \, dy \, dz$

tetrahedron with vertices $(0, 0, 0),$
$(0, 0, 1),\quad (1, 1, 0),\quad (-1, 1, 0)$

17 $\iiint x \, dx \, dy \, dz$

tetrahedron with vertices
$(0, 0, 0),\quad (0, 0, 1),\quad (0, 1, 0),\quad (1, 1, 1)$

18 $\iiint (y + z) \, dx \, dy \, dz$

tetrahedron with vertices
$(1, 0, 0),\quad (0, 0, 2),\quad (1, 0, 1),\quad (1, 1, 1)$

19* $\iiint x \, dx \, dy \, dz$

tetrahedron with vertices $(1, 0, 0),$
$(-1, 1, 0),\quad (1, 1, 1),\quad (2, 2, 0)$

20* $\iiint x^2 \, dx \, dy \, dz$

tetrahedron with vertices $(0, 0, 0),$
$(1, 1, 0),\quad (2, -2, 0),\quad (3, 0, 2)$

21 $\iiint (x + y + z)^2 \, dx \, dy \, dz$

$0 \le x \le 1 \quad 0 \le y \le 1 \quad 0 \le z \le 1$
$x + y + z \le 2$

22 $\iiint xy \, dx \, dy \, dz$

$0 \le x \le 2 \quad 0 \le y \le 2 \quad 0 \le z$
$x + y + 3z \le 3$

23 $\iiint (x + 2y + 3z) \, dx \, dy \, dz$

$0 \le x \le 2y \le 3z \le 6$

24 $\iiint z^2 \, dx \, dy \, dz$

$(x - 1)^2 + y^2 \le 4 \quad (x + 1)^2 + y^2 \le 4$
$0 \le z \le y.$

25   A solid cube has side a. Its density at each point is k times the product of the 6 distances of the point to the faces of the cube, where k is constant. Find the mass.

26   Charge is distributed over the tetrahedron with vertices **0**, **i**, **j**, **k**. The charge density at each point is a constant k times the product of the 4 distances from the point to the faces of the tetrahedron. Find the total charge.

27   Express $\displaystyle\int_0^a \left[\int_0^z \left(\int_0^y g(x) \, dx \right) dy \right] dz$ as a simple integral.

28*   Find a formula for $\iiint x^p y^q z^r (1 - x - y - z)^s \, dx \, dy \, dz$ taken over $0 \le x,\ 0 \le y,$
$0 \le z,\quad x + y + z \le 1,$ where p, q, r, s are non-negative integers. You may take as known $\displaystyle\int_0^1 x^m (1 - x)^n \, dx = \frac{m! \, n!}{(m + n + 1)!}.$

29   Take four vertices of a unit cube, no two adjacent. Find the volume of the tetrahedron with these points as vertices.

30   (cont.) Now take the tetrahedron whose vertices are the remaining four vertices of the cube. The two tetrahedra intersect in a certain polyhedron. Describe it and find its volume.

And a few more integrals for those who like this sort of thing

31 $\displaystyle\iiint \frac{z}{(x + y)^2} \, dx \, dy \, dz$

$x \ge 0 \quad y \ge 0 \quad 1 \le x + y \le 2$
$y^2 \le z \le x^2$

32 $\iiint z \, dx \, dy \, dz$

$0 \le x \le 2 \quad 0 \le y \le 2 \quad 0 \le z \le xy$
$y \le (x - 2)^2 \quad x \le (y - 2)^2$

33 $\iiint x^2 y^2 z^2 \, dx \, dy \, dz$

regular octahedron with vertices
$(\pm 1, 0, 0),\quad (0, \pm 1, 0),\quad (0, 0, \pm 1)$

34 $\iiint (x^2 + 2xy) \, dx \, dy \, dz$

$|2xy| \le z \le 1 - x^2 - y^2.$

2. CYLINDRICAL COORDINATES

Cylindrical coordinates are designed to fit situations with rotational (axial) symmetry about an axis, usually taken to be the z-axis.

The **cylindrical coordinates** of a point $\mathbf{x} = (x, y, z)$ are $\{r, \theta, z\}$, where $\{r, \theta\}$ are the polar coordinates of (x, y) and z is the third rectangular coordinate (Fig. 1a). Each surface $r = $ constant is a right circular cylinder, hence the name, cylindrical coordinates (Fig. 1b).

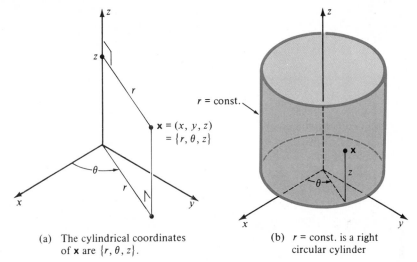

(a) The cylindrical coordinates
of **x** are $\{r, \theta, z\}$.

(b) $r = $ const. is a right
circular cylinder

Fig. 1 Cylindrical coordinates

Through each point **x** (not on the z-axis) pass three surfaces, $r = $ constant, $\theta = $ constant, $z = $ constant (Fig. 2). Each is orthogonal (perpendicular) to the other two at their common intersection **x**.

The relations between the rectangular coordinates (x, y, z) and the cylindrical coordinates $\{r, \theta, z\}$ of a point are

$$
\begin{cases} x = r \cos \theta \\ y = r \sin \theta \\ z = z \end{cases}
\qquad
\begin{cases} r^2 = x^2 + y^2 \\ \cos \theta = x/r, \qquad \sin \theta = y/r \\ z = z. \end{cases}
$$

The origin in the plane is given in polar coordinates by $r = 0$; the angle θ is undefined. Similarly, a point on the z-axis is given in cylindrical coordinates by $r = 0$, $z = $ constant; θ is undefined.

■ **EXAMPLE 1** Graph the surfaces   (a) $z = 2r$,   (b) $z = r^2$.

Solution Both are surfaces of revolution about the z-axis, as is any surface $z = f(r)$. Since z depends only on r, not on θ, the height of the surface is constant above each circle $r = c$ in the x, y-plane. Thus the level curves are circles in the x, y-plane centered at the origin.

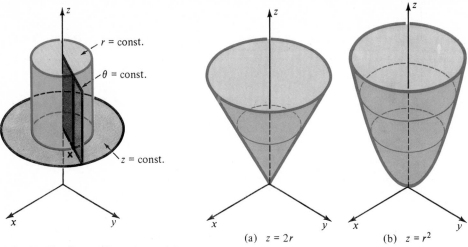

Fig. 2   The (mutually orthogonal) level surfaces of each cylindrical coordinate

(a) $z = 2r$ (b) $z = r^2$

Fig. 3

In (a), the surface meets the first quadrant of the y, z-plane in the line $z = 2y$. (Note that in the first quadrant of the y, z-plane, $x = 0$ and $y \geq 0$. Since $r^2 = x^2 + y^2 = y^2$, it follows that $r = y$.) Rotated about the z-axis, this line spans a cone with apex at **0**. See Fig. 3a.

In (b), the surface meets the y, z-plane in the parabola $z = y^2$. Rotated about the z-axis, this parabola generates a paraboloid of revolution (Fig. 3b).   ∎

Integrals   If a solid has axial symmetry, it is often convenient to place the z-axis on the axis of symmetry, and use cylindrical coordinates $\{r, \theta, z\}$ for the computation of integrals.

In polar coordinates $\{r, \theta\}$, the element of area is $r \, dr \, d\theta$. Correspondingly, the element of volume in cylindrical coordinates $\{r, \theta, z\}$ is

> **Element of Volume** $dV = r \, dr \, d\theta \, dz.$

Let us justify this formula intuitively. We start at a point $\mathbf{x} = \{r, \theta, z\}$ and give small displacements $dr, d\theta, dz$ to its cylindrical coordinates. According to Fig. 4 the displacement of $\mathbf{x}$ in the r-direction has length dr, that in the θ-direction has length $r \, d\theta$, and that in the z-direction has length dz. These three displacements are mutually orthogonal, so they span a "rectangular" box of volume $(dr)(r \, d\theta)(dz)$, hence the formula $dV = r \, dr \, d\theta \, dz$.

∎ **EXAMPLE 2**   Evaluate $\displaystyle\iiint z\sqrt{x^2 + y^2} \, dx \, dy \, dz$ taken over the first octant portion of the solid cone with apex $(0, 0, 2)$ and base $x^2 + y^2 \leq 1$.

Solution   The axial symmetry of the cone (Fig. 5) plus the expression $\sqrt{x^2 + y^2} = r$ in the integrand make this problem a natural for cylindrical coordinates. The

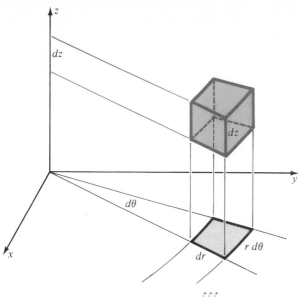

Fig. 4   Intuitive proof of
$dV = (dr)(r\,d\theta)(dz)$

integral becomes $\qquad I = \iiint (rz)\, r\, dr\, d\theta\, dz.$

The surface of the cone must be of the form $z = f(r)$. Since $f(r)$ is obviously linear, and $f(0) = 2$ and $f(1) = 0$, the surface is $z = 2 - 2r$. Therefore the solid domain of integration is described by

$$0 \le \theta \le \tfrac{1}{2}\pi, \qquad 0 \le r \le 1, \qquad 0 \le z \le 2 - 2r.$$

Hence $\quad I = \left(\int_0^{\pi/2} d\theta\right)\left(\int_0^1 r^2\, dr \int_0^{2-2r} z\, dz\right) = \dfrac{\pi}{2}\int_0^1 \dfrac{1}{2} r^2 (2 - 2r)^2\, dr$

$$= \pi \int_0^1 r^2 (1 - r)^2\, dr = \pi \int_0^1 (r^2 - 2r^3 + r^4)\, dr = \dfrac{\pi}{30}. \qquad \blacksquare$$

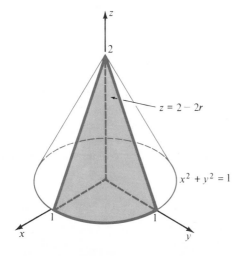

$z = 2 - 2r$

$x^2 + y^2 = 1$

Fig. 5

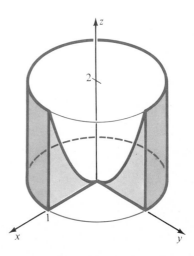

Fig. 6

■ **EXAMPLE 3**   A region **D** in space is generated by revolving the plane region bounded by $z = 2x^2$, the x-axis, and $x = 1$ about the z-axis. Mass is distributed in **D** so that the density at each point is proportional to the distance of the point from the plane $z = -1$, and to the square of the distance of the point from the z-axis. Compute the total mass.

Solution   The density is

$$\delta = k(x^2 + y^2)(z + 1) = kr^2(z + 1),$$

where k is a constant. A cut-away view of the solid is shown in Fig. 6. In cylindrical coordinates, the solid is described by the inequalities

$$0 \le \theta \le 2\pi, \qquad 0 \le r \le 1, \qquad 0 \le z \le 2r^2.$$

For as Fig. 6 shows, $0 \le r \le 1$. And fixing a value of r in this range determines the surface of a vertical cylinder on which z runs from the level $z = 0$ to the level $z = 2r^2$.

The total mass of the solid is

$$\iiint \delta(x, y, z)\, dx\, dy\, dz = \iiint kr^2(z + 1)\, r\, dr\, d\theta\, dz$$

$$= k \left(\int_0^{2\pi} d\theta \right) \int_0^1 r^3 \left[\int_0^{2r^2} (z + 1)\, dz \right] dr$$

$$= 2\pi k \int_0^1 r^3 [\tfrac{1}{2}(2r^2)^2 + (2r^2)]\, dr$$

$$= 4\pi k \int_0^1 (r^7 + r^5)\, dr = \tfrac{7}{6}\pi k. \qquad ■$$

The Natural Frame   It is convenient to fit a frame of three mutually perpendicular vectors to cylindrical coordinates just as the frame **i**, **j**, **k** fits rectangular coordinates. At each point $\{r,\ \theta,\ z\}$ of space attach three mutually perpendicular unit vectors **u**, **w**, **k** chosen so

$$\left.\begin{array}{c} \mathbf{u} \\ \mathbf{w} \\ \mathbf{k} \end{array}\right\} \quad \text{points in the direction of increasing} \quad \left\{\begin{array}{c} r \\ \theta. \\ z \end{array}\right.$$

Thus (Fig. 7)

$$\boxed{\begin{array}{l} \mathbf{u} = \dfrac{1}{r}(x, y, 0) = (\cos\theta, \sin\theta, 0), \\[2mm] \mathbf{w} = \dfrac{1}{r}(-y, x, 0) = (-\sin\theta, \cos\theta, 0), \\[2mm] \mathbf{k} = (0, 0, 1). \end{array}}$$

Note also the relation $\mathbf{x} = r\mathbf{u} + z\mathbf{k}.$

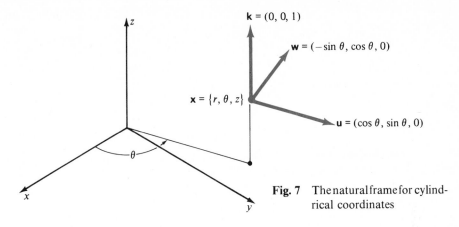

Fig. 7 The natural frame for cylindrical coordinates

The vectors **u**, **w**, **k** form a right-hand system:

$$\mathbf{u} \times \mathbf{w} = \mathbf{k}. \qquad \mathbf{w} \times \mathbf{k} = \mathbf{u}, \qquad \mathbf{k} \times \mathbf{u} = \mathbf{w}.$$

Note that **u** and **w** depend on θ alone, while **k** is a constant vector, our old friend from the trio **i**, **j**, **k**. Note also that

$$\frac{\partial \mathbf{u}}{\partial \theta} = \mathbf{w}, \qquad \frac{\partial \mathbf{w}}{\partial \theta} = -\mathbf{u}.$$

In situations with axial symmetry, it is frequently better to express vectors in terms of **u**, **w**, **k** rather than **i**, **j**, **k**.

Let us express $d\mathbf{x}$ in terms of dr, $d\theta$, and dz. Intuitively (Fig. 8), if r, θ, z are given small increments dr, $d\theta$, dz, then the displacement of **x** in the **u**-direction is $dr\mathbf{u}$, in the **w**-direction is $r\, d\theta\, \mathbf{w}$, and in the **k**-direction is $dz\, \mathbf{k}$. Accordingly

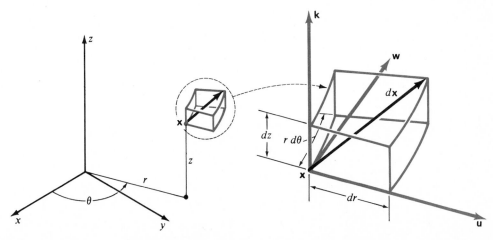

Fig. 8 "Proof" that $d\mathbf{x} = dr\, \mathbf{u} + r\, d\theta\, \mathbf{w} + dz\, \mathbf{k}$

$$dx = dr\,\mathbf{u} + r\,d\theta\,\mathbf{w} + dz\,\mathbf{k}.$$

This formula has a direct analytic derivation. We have

$$\begin{cases} x = r\cos\theta \\ y = r\sin\theta \\ z = z \end{cases} \qquad \begin{cases} dx = dr\cos\theta - r\,d\theta\sin\theta \\ dy = dr\sin\theta + r\,d\theta\cos\theta \\ dz = dz. \end{cases}$$

Therefore

$$\begin{aligned} dx &= (dx, dy, dz) = (dr\cos\theta - r\,d\theta\sin\theta, \, dr\sin\theta + r\,d\theta\cos\theta, \, dz) \\ &= dr\,(\cos\theta, \sin\theta, 0) + r\,d\theta|(-\sin\theta, \cos\theta, 0) + dz\,(0,0,1) \\ &= dr\,\mathbf{u} + r\,d\theta\,\mathbf{w} + dz\,\mathbf{k}. \end{aligned}$$

EXERCISES

Give an equation for the surface in cylindrical coordinates

1 $\dfrac{x}{a} + \dfrac{y}{b} + \dfrac{z}{c} = 1$

2 sphere, center **0**, radius a

3 cylinder parallel to z-axis, base the circle in the x, y-plane with center $(0, a)$ and radius a

4 hyperboloid $z = 2xy$.

Use cylindrical coordinates to evaluate the integral over the indicated domain

5 $\iiint xyz\,dx\,dy\,dz$

$x \geq 0 \quad y \geq 0 \quad 0 \leq z \leq b$
$x^2 + y^2 \leq a^2$

6 $\iiint (x^2 + y^2 + z^2)\,dx\,dy\,dz$

$x^2 + y^2 \leq a^2 \quad |z| \leq b$

7 $\iiint yz\,dx\,dy\,dz$

$0 \leq z \leq y \quad x^2 + y^2 \leq a^2$

8 $\iiint z^2\,dx\,dy\,dz$

$x^2 + y^2 + z^2 \leq a^2 \quad x^2 + y^2 \leq b^2$
$(0 < b < a)$

9 $\iiint e^z\,dx\,dy\,dz$

$x^2 + y^2 \leq z \leq 2(x^2 + y^2) \leq 2$

10 $\iiint z^4\,dx\,dy\,dz$

$-(x^2 + y^2) \leq z \leq 0 \quad x \geq 0$
$y \leq 0 \quad x^2 + y^2 \leq a^2$

11 $\iiint z\,dx\,dy\,dz$

$1 \leq x^2 + y^2 + z^2$
$1 \leq x^2 + y^2 + (z-2)^2 \quad x^2 + y^2 \leq 1$
$0 \leq z \leq 2$

12 $\iiint e^z\,dx\,dy\,dz$

$b \leq z \leq 2b + \sqrt{a^2 - x^2 - y^2}$
$x^2 + y^2 \leq a^2 \quad (b > 0)$

13 $\iiint z\,dx\,dy\,dz$

$(x - a)^2 + y^2 \leq a^2 \quad 2x \leq z \leq 3x$

14 $\iiint z\,dx\,dy\,dz$

$0 \leq r \leq \cos 2\theta \quad -\tfrac{1}{4}\pi \leq \theta \leq \tfrac{1}{4}\pi$
$0 \leq z \leq 1 - r^2$

15 $\iiint (y - 5z)\,dx\,dy\,dz$

$x^2 + z^2 \leq 4 \quad 0 \leq y \leq 1$

16 $\iiint (x^3 + y^3)\,dx\,dy\,dz$

$0 \leq y \leq 2z \quad z^2 + x^2 \leq a^2$

17 $\displaystyle\iiint xy\, dx\, dy\, dz$

$y^2 + z^2 \le a^2 \quad y \ge 0$
$z \le x \le z + b$

18 $\displaystyle\iiint (y - x)\, dx\, dy\, dz$

$0 \le ax \le y^2 + z^2 \le a^2 \quad y \le 0$

19 $\displaystyle\iiint z^2\, dx\, dy\, dz$

$\dfrac{x^2}{4} + \dfrac{y^2}{9} \le 1 \quad 0 \le z \le 2$

20* $\displaystyle\iiint z\, dx\, dy\, dz$

$\dfrac{x^2}{4} + \dfrac{y^2}{9} \le 1 \quad x \ge 0 \quad y \ge 0$
$x + y + z \le \sqrt{13} \quad z \ge 0.$

A space curve is given in the parametric form

$$r = r(t), \qquad \theta = \theta(t), \qquad z = z(t), \qquad a \le t \le b$$

21 Express its velocity in terms of the natural frame **u**, **w**, **k**
22 Express its acceleration in terms of the natural frame
23 Use Ex. 21 to express its arc length in terms of r, θ, z and their time derivatives
24 Find the length of the spiral $r = A$, $\theta = Bt$, $z = Ct$, $a \le t \le b$.

A surface is given in the parametric form

$$r = r(u, v), \qquad \theta = \theta(u, v), \qquad z = z(u, v),$$

where (u, v) varies over a domain **D**

25 Express $\partial\mathbf{x}/\partial u$ and $\partial\mathbf{x}/\partial v$ in terms of the natural frame **u**, **w**, **k**
26 Express $\partial^2\mathbf{x}/\partial u\, \partial v$ in terms of the natural frame
27 Use Ex. 25 to express the surface area in terms of r, θ, z and their derivatives.

Use Ex. 27 to find the area

28 lateral surface of a right circular cone of radius a and height h
29 hemisphere of radius a
30 paraboloid $bz = x^2 + y^2$, $x^2 + y^2 \le a^2$.

31 Given a function f on a domain in $\mathbf{R}^3$, express df in terms of dr, $d\theta$, and dz.
32 (cont.) Express grad f in terms of the natural frame **u**, **w**, **k**.
      [*Hint* $df = (\text{grad } f) \cdot d\mathbf{x}$.]

3. SPHERICAL COORDINATES

Spherical coordinates are designed to fit situations with central symmetry. The **spherical coordinates** $[\rho, \phi, \theta]$ of a point **x** are its distance $\rho = |\mathbf{x}|$ from the origin, its elevation angle ϕ, and its azimuth angle θ. (Often θ is called the longitude and ϕ the co-latitude.). Note that θ is not determined on the z-axis, so points of this axis are usually avoided. In general θ is determined up to a multiple of 2π, and $0 < \phi < \pi$. See Fig. 1a.

Relations between the rectangular coordinates (x, y, z) of a point and its spherical coordinates may be read from Fig. 1b. They are

$$\begin{cases} x = \rho \sin\phi \cos\theta \\ y = \rho \sin\phi \sin\theta \\ z = \rho \cos\phi \end{cases} \qquad \begin{cases} \rho^2 = x^2 + y^2 + z^2 \\ \cos\phi = z/\rho \\ \tan\theta = y/x. \end{cases}$$

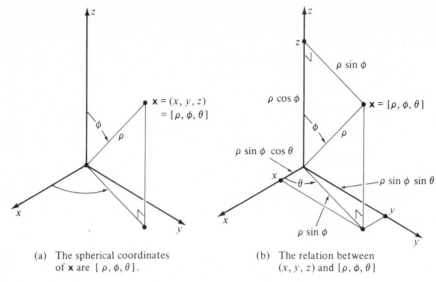

(a) The spherical coordinates
     of **x** are $[\rho, \phi, \theta]$.

(b) The relation between
     (x, y, z) and $[\rho, \phi, \theta]$

Fig. 1   Spherical coordinates

The level surfaces

$$\left.\begin{array}{l} \rho = \text{constant} \\ \phi = \text{constant} \\ \theta = \text{constant} \end{array}\right\} \quad \text{are} \quad \left\{\begin{array}{l} \text{concentric spheres about } \mathbf{0} \\ \text{right circular cones, apex } \mathbf{0} \\ \text{planes through the } z\text{-axis.} \end{array}\right.$$

At each point **x** the three level surfaces intersect orthogonally (Fig. 2).

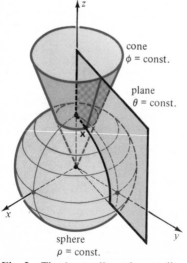

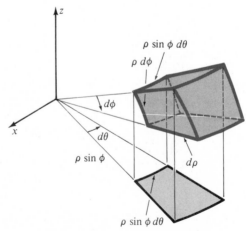

Fig. 2   The (mutually orthogonal)
            level surfaces of each
            spherical coordinate

Fig. 3   Intuitive proof of
            $dV = (d\rho)(\rho \, d\phi)(\rho \sin \phi \, d\theta)$

Integrals If a solid has central symmetry, it is often convenient to place the origin at the center of symmetry and use spherical coordinates $[\rho, \phi, \theta]$ for the computation of integrals.

Let us find a formula for the element of volume dV in terms of ρ, ϕ, and θ. We start at a point **x** and give small increments $d\rho$, $d\phi$, $d\theta$ to its spherical coordinates. According to Fig. 3 the displacement of **x** in the ρ-direction has length $d\rho$, that in the ϕ-direction has length $\rho \, d\phi$, and that in the θ-direction has length $\rho \sin \phi \, d\theta$. The three displacements are mutually orthogonal, so they span a "rectangular" box of volume $(d\rho)(\rho \, d\phi)(\rho \sin \phi \, d\theta)$, hence the formula

> **Element of Volume** $dV = \rho^2 \sin \phi \, d\rho \, d\phi \, d\theta.$

■ **EXAMPLE 1** Use spherical coordinates to find the volume of a sphere of radius a.

Solution

$$V = \iiint dV = \iiint \rho^2 \sin \phi \, d\rho \, d\phi \, d\theta$$

$$= \left(\int_0^a \rho^2 \, d\rho \right) \left(\int_0^\pi \sin \phi \, d\phi \right) \left(\int_0^{2\pi} d\theta \right) = (\tfrac{1}{3}a^3)(2)(2\pi) = \tfrac{4}{3}\pi a^3. \qquad ■$$

■ **EXAMPLE 2** Find the volume of the portion of the unit sphere that lies in the right circular cone having its apex at the origin and making angle α with the positive z-axis.

Solution The cone is specified by $0 \le \phi \le \alpha$, so the portion of the sphere is determined by $0 \le \theta \le 2\pi, 0 \le \phi \le \alpha$, and $0 \le \rho \le 1$. See Fig. 4. Hence the volume is

$$V = \left(\int_0^{2\pi} d\theta \right) \left(\int_0^\alpha \sin \phi \, d\phi \right) \left(\int_0^1 \rho^2 \, d\rho \right) = (2\pi)(1 - \cos \alpha)\left(\frac{1}{3}\right) = \frac{2\pi}{3}(1 - \cos \alpha). \qquad ■$$

Remark As a check, let $\alpha \longrightarrow \pi$. Then the volume should approach the volume of a sphere of radius 1. Does it?

■ **EXAMPLE 3** A solid fills the region between concentric spheres of radii a and b, where $0 < a < b$. The density at each point is inversely proportional to its distance from the center. Find the total mass.

Solution The solid is specified by $a \le \rho \le b$; the density is $\delta = k/\rho$. Hence

$$M = \iiint \delta(\mathbf{x}) \, dV = \iiint \frac{k}{\rho} \rho^2 \sin \phi \, d\rho \, d\phi \, d\theta = k \left(\int_0^{2\pi} d\theta \right) \left(\int_0^\pi \sin \phi \, d\phi \right) \left(\int_a^b \rho \, d\rho \right)$$

$$= (2\pi k)(2)\left(\frac{b^2 - a^2}{2} \right) = 2\pi k(b^2 - a^2). \qquad ■$$

Remark As $a \longrightarrow 0$, the solid tends to the whole sphere, with infinite density at the center. But $M \longrightarrow 2\pi k b^2$, which is finite.

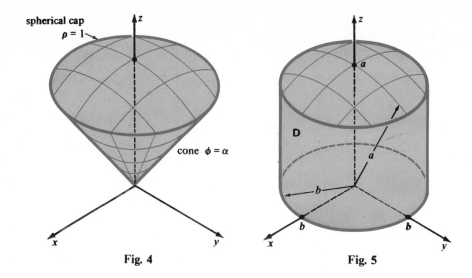

Fig. 4   **Fig. 5**

The following example can be solved either by cylindrical or spherical coordinates. Indeed, it is hard to decide at first which to use. Therefore we shall work the problem both ways.

■ **EXAMPLE 4**   Find $I = \iiint_{\mathbf{D}} z \, dx \, dy \, dz$,   where **D** is described by

$$0 \le z, \qquad x^2 + y^2 \le b^2, \qquad x^2 + y^2 + z^2 \le a^2,$$

and $0 \le b < a$.

Solution   The domain is like a cylindrical can with a spherical cap at one end (Fig. 5). Let us first set up the integral in cylindrical coordinates. Since $x^2 + y^2 = r^2$, the sphere is given by $r^2 + z^2 = a^2$. Therefore the domain **D** is described by

$$0 \le \theta \le 2\pi, \qquad 0 \le r \le b, \qquad 0 \le z \le \sqrt{a^2 - r^2}.$$

Consequently $I = \iiint_{\mathbf{D}} z \, dx \, dy \, dz = \iiint_{\mathbf{D}} z \, r \, dr \, d\theta \, dz$

$$= \int_0^{2\pi} d\theta \int_0^b r \, dr \int_0^{\sqrt{a^2 - r^2}} z \, dz$$

$$= (2\pi) \int_0^b \tfrac{1}{2} r (\sqrt{a^2 - r^2})^2 \, dr = \pi \int_0^b (a^2 r - r^3) \, dr$$

$$= \pi(\tfrac{1}{2} a^2 b^2 - \tfrac{1}{4} b^4) = \tfrac{1}{4} \pi b^2 (2a^2 - b^2).$$

Now let us set up the solution in spherical coordinates. The domain **D** naturally splits into two subdomains (Fig. 6a). The first, $\mathbf{D}_1$, an ice cream cone, is described by

$$0 \le \theta \le 2\pi, \qquad 0 \le \phi \le \arcsin b/a, \qquad 0 \le \rho \le a.$$

The limit $\arcsin b/a$ on the apex angle ϕ is seen from Fig. 6b. The second

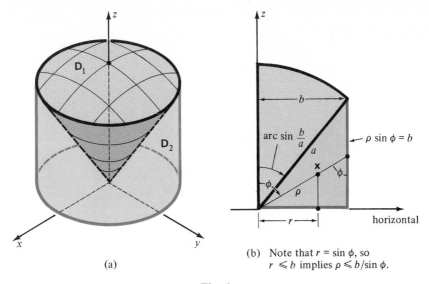

(b) Note that $r = \sin \phi$, so $r \le b$ implies $\rho \le b/\sin \phi$.

(a)

Fig. 6

subdomain $\mathbf{D}_2$, a cylindrical can with a cone removed, is described by

$$0 \le \theta \le 2\pi, \qquad \text{arc sin } b/a \le \phi \le \tfrac{1}{2}\pi, \qquad 0 \le \rho \le b/\sin \phi.$$

Consequently

$$I = \iiint_{\mathbf{D}} z \, dx \, dy \, dz = \iiint_{\mathbf{D}} (\rho \cos \phi) \, \rho^2 \sin \phi \, d\rho \, d\phi \, d\theta$$

$$= \iiint_{\mathbf{D}_1} \rho^3 \sin \phi \cos \phi \, d\rho \, d\phi \, d\theta + \iiint_{\mathbf{D}_2} \rho^3 \sin \phi \cos \phi \, d\rho \, d\bar\phi \, d\theta.$$

Now $\quad \displaystyle\iiint_{\mathbf{D}_1} = \int_0^{2\pi} d\theta \int_0^{\text{arc sin } b/a} \sin \phi \cos \phi \, d\phi \int_0^a \rho^3 \, d\rho$

$$= (2\pi)(\tfrac{1}{4}a^4)(\tfrac{1}{2} \sin^2 \phi) \, \Big|_0^{\text{arc sin } b/a} = \tfrac{1}{4}\pi a^4 (b/a)^2 = \tfrac{1}{4}\pi a^2 b^2.$$

Next, $\quad \displaystyle\iiint_{\mathbf{D}_2} = \int_0^{2\pi} d\theta \int_{\text{arc sin } b/a}^{\pi/2} \sin \phi \cos \phi \, d\phi \int_0^{b/\sin \phi} \rho^3 \, d\rho$

$$= (2\pi)(\tfrac{1}{4}b^4) \int_{\text{arc sin } b/a}^{\pi/2} \frac{\cos \phi}{\sin^3 \phi} \, d\phi = (\tfrac{1}{2}\pi b^4)[-\tfrac{1}{2}(\sin \phi)^{-2}] \, \Big|_{\text{arc sin } b/a}^{\pi/2}$$

$$= (\tfrac{1}{4}\pi b^4)[(a/b)^2 - 1] = \tfrac{1}{4}\pi b^2 (a^2 - b^2).$$

Therefore $\qquad I = \tfrac{1}{4}\pi a^2 b^2 + \tfrac{1}{4}\pi b^2 (a^2 - b^2) = \tfrac{1}{4}\pi b^2 (2a^2 - b^2).$ ∎

Spherical Area Suppose a domain $\mathbf{S}$ lies on the surface of the sphere $\rho = a$. We should be able to use spherical coordinates to find its area $|\mathbf{S}|$. What we need is a formula for the element of spherical area dA. Now ρ is constant. Small displacements $d\phi$ and $d\theta$ result in a small rectangular region (Fig. 7a) whose sides are $a\,d\phi$ and $a\sin\phi\,d\theta$. Therefore

Element of Spherical Area $dA = a^2 \sin\phi\,d\phi\,d\theta.$

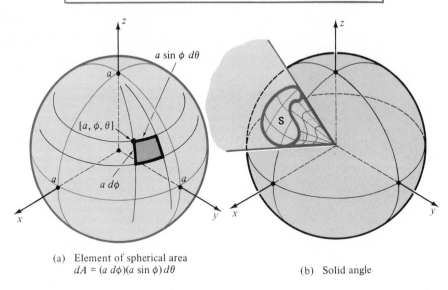

(a)   Element of spherical area
$dA = (a\,d\phi)(a\sin\phi)\,d\theta$

(b)   Solid angle

Fig. 7   Spherical area and solid angle

■ **EXAMPLE 5**   Find the area of the polar cap, all points of co-latitude α or less on the unit sphere.

Solution   Refer back to Fig. 4, p. 908. The region is defined on the sphere $\rho = 1$ by $0 \le \phi \le \alpha$, hence

$$A = \int_0^{2\pi} \left(\int_0^{\alpha} \sin\phi\,d\phi \right) d\theta = 2\pi(1 - \cos\alpha). \qquad ■$$

Remark   Suppose **S** is a region on the *unit* sphere. The totality of infinite rays starting at **0** and passing through points of **S** is a cone which is called a **solid angle** (Fig. 7b). A solid angle is measured by the area of the base region **S**. The unit for solid angles is the **steradian** (sr). The solid angle determined by the whole sphere equals 4π sr. The solid angle determined by the first octant equals $\frac{1}{2}\pi$ sr. The solid angle determined by the polar cap in Example 5 equals $2\pi(1 - \cos\alpha)$ sr.

The Natural Frame   As for cylindrical coordinates, there is a natural frame of unit vectors suited to spherical coordinates. At each point $[\rho, \phi, \theta]$, we select unit vectors $\boldsymbol{\lambda}, \boldsymbol{\mu}, \boldsymbol{\nu}$:

$$\left.\begin{array}{c} \boldsymbol{\lambda} \\ \boldsymbol{\mu} \\ \boldsymbol{\nu} \end{array}\right\} \quad \text{points in the direction of increasing} \quad \left\{\begin{array}{c} \rho \\ \phi. \\ \theta \end{array}\right.$$

Points on the z-axis must be excluded because θ is not defined there. See Fig. 8.

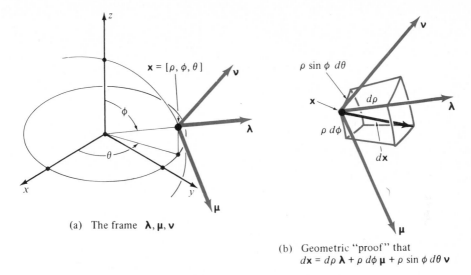

(a) The frame $\boldsymbol{\lambda}, \boldsymbol{\mu}, \boldsymbol{\nu}$

(b) Geometric "proof" that
$d\mathbf{x} = d\rho\,\boldsymbol{\lambda} + \rho\,d\phi\,\boldsymbol{\mu} + \rho\,\sin\phi\,d\theta\,\boldsymbol{\nu}$

Fig. 8   The natural frame for spherical coordinates

To express $\boldsymbol{\lambda}, \boldsymbol{\mu}, \boldsymbol{\nu}$ in terms of ρ, ϕ, θ, we use a short cut:

$$\mathbf{x} = \rho(\sin\phi\cos\theta, \sin\phi\sin\theta, \cos\phi),$$

$$d\mathbf{x} = (\sin\phi\cos\theta, \sin\phi\sin\theta, \cos\phi)\,d\rho + \rho(\cos\phi\cos\theta, \cos\phi\sin\theta, -\sin\phi)\,d\phi$$

$$+ \rho(-\sin\phi\sin\theta, \sin\phi\cos\theta, 0)\,d\theta$$

$$= \boldsymbol{\lambda}\,d\rho + \boldsymbol{\mu}\,\rho\,d\phi + \boldsymbol{\nu}\,\rho\,\sin\phi\,d\theta.$$

Conclusion:

$$\begin{cases} \boldsymbol{\lambda} = (\sin\phi\cos\theta, \sin\phi\sin\theta, \cos\phi) \\ \boldsymbol{\mu} = (\cos\phi\cos\theta, \cos\phi\sin\theta, -\sin\phi) \\ \boldsymbol{\nu} = (-\sin\theta, \cos\theta, 0), \end{cases}$$
$$d\mathbf{x} = d\rho\,\boldsymbol{\lambda} + \rho\,d\phi\,\boldsymbol{\mu} + \rho\,\sin\phi\,d\theta\,\boldsymbol{\nu}.$$

As is easily verified, the vectors $\boldsymbol{\lambda}, \boldsymbol{\mu}, \boldsymbol{\nu}$, computed in this way, are mutually orthogonal unit vectors. Furthermore, they point in the directions of increasing ρ, ϕ, θ respectively: if only ρ increases, then $d\phi = d\theta = 0$, hence $d\mathbf{x} = \boldsymbol{\lambda}\,d\rho$. Similarly, if only ϕ increases, then $d\mathbf{x} = \boldsymbol{\mu}\,\rho\,d\phi$, and if only θ increases, then $d\mathbf{x} = \boldsymbol{\nu}\,\rho\,\sin\phi\,d\theta$.

We are now in a position to prove analytically the formula for dV, which was derived intuitively at the beginning of this section. We have proved that $d\mathbf{x}$ is the sum of three mutually orthogonal displacements $d\rho\,\boldsymbol{\lambda}$, $\rho\,d\phi\,\boldsymbol{\mu}$, and $\rho\,\sin\phi\,d\theta\,\boldsymbol{\nu}$. These vectors span a box whose volume dV is a scalar triple product:

$$dV = [d\rho\,\boldsymbol{\lambda}, \ \rho\,d\phi\,\boldsymbol{\mu}, \ \rho\,\sin\phi\,d\theta\,\boldsymbol{\nu}]$$

$$= (d\rho)(\rho\,d\phi)(\rho\,\sin\phi\,d\theta)[\boldsymbol{\lambda}, \boldsymbol{\mu}, \boldsymbol{\nu}] = \rho^2\,\sin\phi\,d\rho\,d\phi\,d\theta.$$

The scalar triple product $[\boldsymbol{\lambda}, \boldsymbol{\mu}, \boldsymbol{\nu}]$ equals 1 because $\boldsymbol{\lambda}, \boldsymbol{\mu}, \boldsymbol{\nu}$ are mutually perpendicular unit vectors and form a right-handed system. (You should check that $\boldsymbol{\mu} \times \boldsymbol{\nu} = \boldsymbol{\lambda}$.]

Change of Variables   We now mention very briefly the general rule for changing variables in triple integrals. It includes the cylindrical and spherical coordinate changes as special cases. This discussion is a continuation of that in the previous chapter, p. 865.

First we define the **Jacobian** of a change of variables.

$$\text{If} \quad \begin{cases} x = x(u, v, w) \\ y = y(u, v, w), \\ z = z(u, v, w) \end{cases} \quad \text{then} \quad \frac{\partial(x, y, z)}{\partial(u, v, w)} = \begin{vmatrix} x_u & x_v & x_w \\ y_u & y_v & y_w \\ z_u & z_v & z_w \end{vmatrix}$$

Change of Variables   Suppose

$$\begin{cases} x = x(u, v, w) \\ y = y(u, v, w) \\ z = z(u, v, w) \end{cases}$$

is a one–one transformation of a domain **E** in **u**-space onto a domain **D** in **x**-space. Suppose the functions $x(u, v, w)$, $\cdots$ are continuously differentiable and that

$$\frac{\partial(x, y, z)}{\partial(u, v, w)} > 0$$

at all points of **E**. Then

$$\iiint_D f(x, y, z)\, dx\, dy\, dz = \iiint_E f[x(u, v, w), y(u, v, w), z(u, v, w)]\, \frac{\partial(x, y, z)}{\partial(u, v, w)}\, du\, dv\, dw.$$

The main thing to remember is that

$$dx\, dy\, dz \quad \text{is replaced by} \quad \frac{\partial(x, y, z)}{\partial(u, v, w)}\, du\, dv\, dw.$$

For spherical coordinates $u = \rho$, $v = \phi$, $w = \theta$ and

$$\begin{cases} x = \rho \sin \phi \cos \theta \\ y = \rho \sin \phi \sin \theta \\ z = \rho \cos \phi, \end{cases}$$

$$\frac{\partial(x, y, z)}{\partial(\rho, \phi, \theta)} = \begin{vmatrix} \sin \phi \cos \theta & o \cos \phi \cos \theta & -\rho \sin \phi \sin \theta \\ \sin \phi \sin \theta & \rho \cos \phi \sin \theta & \rho \sin \phi \cos \theta \\ \cos \phi & -\rho \sin \phi & 0 \end{vmatrix}$$

$$= \rho^2 \sin \phi \begin{vmatrix} \sin \phi \cos \theta & \cos \phi \cos \theta & -\sin \theta \\ \sin \phi \sin \theta & \cos \phi \sin \theta & \cos \theta \\ \cos \phi & -\sin \phi & 0 \end{vmatrix} = \rho^2 \sin \phi.$$

(The determinant can be expanded easily by minors of the third row.) Therefore $dx\, dy\, dz$ is replaced by $\rho^2 \sin\phi\, d\rho\, d\phi\, d\theta$.

■ **EXAMPLE 6**   Find the volume enclosed by the ellipsoid

$$\frac{x^2}{a^2} + \frac{y^2}{b^2} + \frac{z^2}{c^2} = 1, \qquad a, b, c > 0.$$

Solution   We want

$$|\mathbf{D}| = \iiint_{\mathbf{D}} dx\, dy\, dz, \qquad \text{where} \quad \mathbf{D} = \left\{(x, y, z), \quad \text{where} \quad \frac{x^2}{a^2} + \frac{y^2}{b^2} + \frac{z^2}{c^2} \le 1\right\}.$$

Let $\mathbf{E} = \left\{(u, v, w), \quad \text{where} \quad u^2 + v^2 + w^2 \le 1\right\}$ be the unit sphere in $\mathbf{u}$-space and define the transformation

$$x = au, \qquad y = bv, \qquad z = cw,$$

which takes $\mathbf{E}$ onto $\mathbf{D}$ in a one–one manner. Also

$$\frac{\partial(x, y, z)}{\partial(u, v, w)} = \begin{vmatrix} a & 0 & 0 \\ 0 & b & 0 \\ 0 & 0 & c \end{vmatrix} = abc > 0.$$

Therefore

$$|\mathbf{D}| = \iiint_{\mathbf{D}} dx\, dy\, dz = \iiint_{\mathbf{E}} abc\, du\, dv\, dw = abc \iiint_{\mathbf{E}} du\, dv\, dw$$

$$= abc\,|\mathbf{E}| = \tfrac{4}{3}\pi abc. \qquad\qquad ■$$

EXERCISES

Give an equation for the surface in spherical coordinates

1   sphere, center $(0, 0, a)$, radius a
2   the cylinder of all points at distance a from the z-axis
3   paraboloid $z = x^2 + y^2$
4   hyperbolic paraboloid $z = x^2 - y^2$
5   right circular cylinder, axis through $(a, 0, 0)$ and parallel to the z-axis, radius a.
6   (cont.) Find the intersection of this cylinder with the sphere of radius $2a$ and center $\mathbf{0}$. Give your answer in the form of two relations between the spherical coordinates.

Use spherical coordinates to evaluate the integral over the indicated domain

7   $\displaystyle\iiint z\, dx\, dy\, dz$
$\rho \le a \quad x \ge 0 \quad y \ge 0 \quad z \ge 0$

8   $\displaystyle\iiint x^2\, dx\, dy\, dz$
$\rho \le a \quad x \ge 0 \quad y \ge 0$

9   $\displaystyle\iiint \rho^n\, dx\, dy\, dz \quad n \ge 0$
$\rho \le a$

10   $\displaystyle\iiint (a - \rho)^n\, dx\, dy\, dz \quad n \ge 0$
$\rho \le a$

11   $\displaystyle\iiint z\, dx\, dy\, dz$
$1 \le z \quad \rho \le \tfrac{2}{3}\sqrt{3}$

12   $\displaystyle\iiint \rho^{-2}\, dx\, dy\, dz$
$a \le \rho \le b \qquad (0 < a < b)$

13 $\iiint\limits_{\rho \leq a} x^4 \, dx \, dy \, dz$

14 $\iiint\limits_{\rho \leq a} x^2 y^2 \, dx \, dy \, dz$

15 $\iiint\limits_{\rho \leq a} (x^2 + y^2)^2 \, dx \, dy \, dz$

16 $\iiint\limits_{\rho \leq a} (x^2 + z^2)^2 \, dx \, dy \, dz$

17 $\iiint\limits_{\rho \leq a \;\; \frac{1}{4}\pi \leq \phi \leq \frac{3}{4}\pi} z^2 \, dx \, dy \, dz$

18 $\iiint\limits_{\rho \leq a \;\; \frac{5}{12}\pi \leq \theta \leq \frac{7}{12}\pi} x^2 \, dx \, dy \, dz$

19 $\lim\limits_{\varepsilon \to 0^+} \iiint\limits_{\varepsilon \leq \rho \leq 1} (\ln \rho) \, dx \, dy \, dz$

20 $\lim\limits_{\varepsilon \to 0^+} \iiint\limits_{\varepsilon \leq \rho \leq 1} \dfrac{dx \, dy \, dz}{\rho^{5/2}}$

21 $\iiint\limits_{\rho \geq 1 \;\; x^2 + y^2 \leq 1 \;\; 0 \leq z \leq 1} z \, dx \, dy \, dz$

22 $\iiint\limits_{\substack{\rho \geq 1 \;\; 0 \leq x \leq 1 \;\; 0 \leq y \leq 1 \\ 0 \leq z \leq 1.}} \rho^2 \, dx \, dy \, dz$

23   Given a function f on a domain in $\mathbf{R}^3$, express grad f in terms of the natural frame $\boldsymbol{\lambda}$, $\boldsymbol{\mu}$, $\mathbf{v}$. [*Hint* $df = (\text{grad } f) \cdot d\mathbf{x}$.]

24   Derive the three formulas

$$d\boldsymbol{\lambda} = d\phi \, \boldsymbol{\mu} + \sin \phi \, d\theta \, \mathbf{v}, \quad d\boldsymbol{\mu} = -d\phi \, \boldsymbol{\lambda} + \cos \phi \, d\theta \, \mathbf{v}, \quad d\mathbf{v} = -\sin \phi \, d\theta \, \boldsymbol{\lambda} - \cos \phi \, d\theta \, \boldsymbol{\mu}.$$

A space curve is given in the parametric form

$$\rho = \rho(t), \quad \phi = \phi(t), \quad \theta = \theta(t), \quad a \leq t \leq b.$$

25   Express its velocity $\mathbf{v}$ in terms of the natural frame $\boldsymbol{\lambda}$, $\boldsymbol{\mu}$, $\mathbf{v}$.

26   Do the same for its acceleration $\mathbf{a}$. [*Hint*   Use Ex. 24.]

27   Use Ex. 25 to express its arc length in terms of ρ, ϕ, θ and their time derivatives.

28   (cont.) A **rhumb line** on a sphere of radius a is a curve that intersects each meridian at the same angle α. (Follow a constant compass setting.) Find the length of a rhumb line from the equator to the north pole.

29   (cont.) Set up a definite integral for the length of the conical spiral $\phi = \alpha$, $\rho = \theta$, $a \leq \theta \leq b$. Here α is a constant.

30   (cont.) Set up a definite integral for the length of the upper part of the curve of intersection in Ex. 6.

A surface is given in the parametric form

$$\rho = \rho(u, v), \quad \phi = \phi(u, v), \quad \theta = \theta(u, v),$$

where (u, v) varies over a domain $\mathbf{D}$

31   Express $\partial \mathbf{x}/\partial u$ and $\partial \mathbf{x}/\partial v$ in terms of the natural frame $\boldsymbol{\lambda}$, $\boldsymbol{\mu}$, $\mathbf{v}$

32   Express $\partial^2 \mathbf{x}/\partial u \, \partial v$ in terms of the natural frame [*Hint*   Use the result of Ex. 24.]

33   Use Ex. 31 to express the surface area in terms of ρ, ϕ, θ and their derivatives

34   Give a simple expression for the element of area on the cone $\phi = \alpha$, a constant

35   (cont.) Find the area of the region

$$0 \leq \rho \leq a + b \sin n\theta$$

on the cone, $a > b > 0$.

36   Use the result of Ex. 33 to set up the area of the cylindrical surface

$$x^2 + y^2 = a^2, \; 0 \leq z \leq h.$$

Use one or more changes of variables to evaluate

37 $\iiint xyz\, dx\, dy\, dz$

$\dfrac{x^2}{a^2} + \dfrac{y^2}{b^2} + \dfrac{z^2}{c^2} \le 1$

$x \ge 0 \quad y \ge 0 \quad z \ge 0$

38 $\iiint \dfrac{dx\, dy\, dz}{(x + y + z)^2}$

$a \le x + y + z \le b \qquad (0 < a < b)$

$x \ge 0 \quad y \ge 0 \quad z \ge 0$

39 $\iiint z^3\, dx\, dy\, dz$

$\dfrac{x^2}{a^2} + \dfrac{y^2}{b^2} \le 1$

$0 \le \dfrac{z}{c} \le 1 - \dfrac{x^2}{a^2} - \dfrac{y^2}{b^2}$

40* $\iiint (ax + by + cz)^{2n}\, dx\, dy\,\ dz.$

$x^2 + y^2 + z^2 \le k^2$

41 The tetrahedron **T** has vertices **a**, **b**, **c**, **d**. Set up $I = \iiint f(\mathbf{x})\, dx\, dy\, dz$ over **T**, using the change of variables

$$\mathbf{x} = (1 - u - v - w)\mathbf{a} + u\mathbf{b} + v\mathbf{c} + w\mathbf{d}.$$

42 (cont.) Evaluate $\iiint x^2\, dx\, dy\, dz$ over the tetrahedron with vertices $(0, 0, 0)$, $(1, 1, 0)$, $(2, -2, 0)$, $(3, 0, 2)$.

4. CENTER OF GRAVITY

We have studied the center of gravity for non-homogeneous plane sheets and for wires in the plane. Now we extend this concept to solids.

Suppose a solid **D** has density $\delta(\mathbf{x})$ at each point **x**. Its element of mass is $dM = \delta(\mathbf{x})\, dV$, and its (total) mass is

$$M = \iiint_{\mathbf{D}} dM = \iiint_{\mathbf{D}} \delta(\mathbf{x})\, dV = \iiint_{\mathbf{D}} \delta(\mathbf{x})\, dx\, dy\, dz.$$

The **moment** of **D** is

$$\mathbf{m} = (m_x, m_y, m_z) = \iiint_{\mathbf{D}} \mathbf{x}\, dM = \iiint_{\mathbf{D}} \delta(\mathbf{x})\, \mathbf{x}\, dx\, dy\, dz.$$

The **center of gravity** of **D** is $\quad \bar{\mathbf{x}} = (\bar{x}, \bar{y}, \bar{z}) = \dfrac{1}{M}\, \mathbf{m}.$

Thus for instance, $\qquad \bar{x} = \iiint_{\mathbf{D}} x\, dM \bigg/ \iiint_{\mathbf{D}} dM,$

so $\bar{x}$ is the weighted average of x over **D**. Similar statements apply to $\bar{y}$ and $\bar{z}$. The center of gravity may be considered as a weighted average of the points of the solid. Recall in this connection that the center of gravity of a system of point-masses $M_1, \cdots, M_n$ located at $\mathbf{x}_1, \cdots, \mathbf{x}_n$ is

$$\bar{\mathbf{x}} = \dfrac{1}{M}\, (M_1\mathbf{x}_1 + M_2\mathbf{x}_2 + \cdots + M_n\mathbf{x}_n),$$

where $M = M_1 + \cdots + M_n$.

Symmetry If a solid is symmetric in a coordinate plane, then the center of gravity lies on that coordinate plane. For example, suppose **D** is **symmetric** in the x, y-plane. This means that whenever a point (x, y, z) is in the solid, then $(x, y, -z)$ is in the solid, *and* $\delta(x, y, z) = \delta(x, y, -z)$. The contribution to m_z at (x, y, z) is

$$\delta(x, y, z) \, z \, dx \, dy \, dz;$$

it is cancelled by the contribution

$$\delta(x, y, -z)(-z) \, dx \, dy \, dz = -\delta(x, y, z) \, z \, dx \, dy \, dz$$

at $(x, y, -z)$. Hence $m_z = 0$ and $\bar{z} = 0$.

Similarly, if **D** is symmetric in a coordinate axis, then the center of gravity lies on that axis. Finally, if **D** is symmetric in the origin, then $\bar{\mathbf{x}} = \mathbf{0}$.

Similar conclusions apply to symmetries in abitrary planes, lines, or points. Another physically intuitive fact is that the center of gravity of a solid **D** depends only on the solid, not on how the rectangular coordinate system is chosen.

Remark Suppose **D** is a domain in space. When we refer to the **center of gravity** (or **centroid**) of **D** without mentioning a density function, then it is understood that **D** is a uniform solid with $\delta = 1$. For such a solid, $M = V$ and $\bar{\mathbf{x}}$ depends only on the shape of the solid.

Examples To compute the center of gravity of a solid, exploit any symmetry it has by choosing an appropriate coordinate system and expressing the element of volume dV in that system.

■ **EXAMPLE 1** Find the center of gravity of a hemisphere of radius a.

Solution Here $\delta = 1$ so

$$M = V = \tfrac{1}{2}(\tfrac{4}{3}\pi a^3) = \tfrac{2}{3}\pi a^3.$$

To exploit symmetry, choose spherical coordinates. The hemisphere is symmetric in the z-axis, hence $\bar{\mathbf{x}}$ lies on the z-axis, that is, $\bar{x} = \bar{y} = 0$. Therefore we need only compute $\bar{z}$:

$$m_z = \iiint_{\mathbf{D}} z \, dV = \iiint_{\mathbf{D}} \rho \cos \phi \, \rho^2 \sin \phi \, d\rho \, d\phi \, d\theta$$

$$= \int_0^{2\pi} d\theta \int_0^{\pi/2} \cos \phi \sin \phi \, d\phi \int_0^a \rho^3 \, d\rho = (2\pi)(\tfrac{1}{2})(\tfrac{1}{4}a^4) = \tfrac{1}{4}\pi a^4.$$

Therefore

$$\bar{z} = \frac{m_z}{M} = \frac{\tfrac{1}{4}\pi a^4}{\tfrac{2}{3}\pi a^3} = \frac{3}{8} a.$$

It follows that the center of gravity lies on the axis of the hemisphere, $\tfrac{3}{8}$ of the distance from the center to the pole (Fig. 1).   ■

■ **EXAMPLE 2** Find the center of gravity of a right circular cone of radius a and height h.

Solution $\delta = 1$, so

$$M = V = \tfrac{1}{3}\pi a^2 h.$$

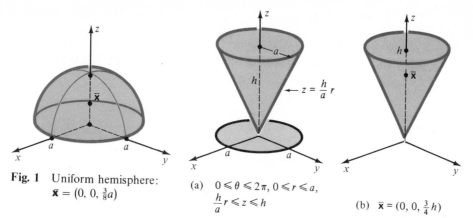

Fig. 1 Uniform hemisphere:
$\bar{\mathbf{x}} = (0, 0, \frac{3}{8}a)$

(a) $0 \leqslant \theta \leqslant 2\pi, 0 \leqslant r \leqslant a,$
$\frac{h}{a} r \leqslant z \leqslant h$

(b) $\bar{\mathbf{x}} = (0, 0, \frac{3}{4}h)$

Fig. 2 Center of gravity of uniform right circular cone

Use cylindrical coordinates with the axes placed as in Fig. 2a. By symmetry, $\bar{x} = \bar{y} = 0$, so we need only compute $\bar{z}$:

$$m_z = \iiint_D z\, dV = \iiint_D z\, r\, dr\, d\theta\, dz = \int_0^{2\pi} d\theta \int_0^a r\, dr \int_{hr/a}^h z\, dz$$

$$= (2\pi)\left(\frac{h^2}{2}\right) \int_0^a r\left(1 - \frac{r^2}{a^2}\right) dr = \pi h^2\left(\frac{a^2}{2} - \frac{a^4}{4a^2}\right) = \tfrac{1}{4}\pi a^2 h^2.$$

Therefore
$$\bar{z} = \frac{m_z}{M} = \frac{\tfrac{1}{4}\pi a^2 h^2}{\tfrac{1}{3}\pi a^2 h} = \frac{3}{4} h.$$

It follows that the center of gravity is on the cone's axis, $\frac{1}{4}$ of the distance from its base to its apex (Fig. 2b). ∎

■ **EXAMPLE 3** The solid $\quad 0 \leq x \leq 1, \quad 0 \leq y \leq 2, \quad 0 \leq z \leq 3 \quad$ meters has density xyz kg/m³. Find its center of gravity.

Solution

$$M = \iiint xyz\, dx\, dy\, dz = \int_0^1 x\, dx \int_0^2 y\, dy \int_0^3 z\, dz = \frac{1}{2} \cdot \frac{4}{2} \cdot \frac{9}{2} = \frac{9}{2} \text{ kg.}$$

$$\mathbf{m} = \iiint (x, y, z)\, xyz\, dx\, dy\, dz$$

$$= \left(\iiint x^2 yz\, dx\, dy\, dz, \iiint xy^2 z\, dx\, dy\, dz, \iiint xyz^2\, dx\, dy\, dz\right)$$

$$= \left(\int_0^1 x^2\, dx \int_0^2 y\, dy \int_0^3 z\, dz, \quad \int_0^1 x\, dx \int_0^2 y^2\, dy \int_0^3 z\, dz, \quad \int_0^1 x\, dx \int_0^2 y\, dy \int_0^3 z^2\, dz\right)$$

$$= \left(\frac{1}{3} \cdot \frac{4}{2} \cdot \frac{9}{2}, \quad \frac{1}{2} \cdot \frac{8}{3} \cdot \frac{9}{2}, \quad \frac{1}{2} \cdot \frac{4}{2} \cdot \frac{27}{3}\right) = \frac{4 \cdot 9}{3 \cdot 2 \cdot 2}(1, 2, 3) = 3(1, 2, 3) \text{ kg-m.}$$

Hence $$\bar{\mathbf{x}} = \frac{\mathbf{m}}{M} = \tfrac{2}{9}\mathbf{m} = \tfrac{2}{9}(3)(1, 2, 3) = (\tfrac{2}{3}, \tfrac{4}{3}, 2)\ \text{m}.$$ ∎

Addition Law   Sometimes it is convenient to decompose a solid **D** into two or more pieces. Then the center of gravity of **D** can be expressed in terms of the c.g.'s of the pieces.

Addition Law   Suppose a solid **D** of mass M and center of gravity $\bar{\mathbf{x}}$ is made up of two pieces $\mathbf{D}_0$ and $\mathbf{D}_1$, of masses M_0 and M_1, and centers of gravity $\bar{\mathbf{x}}_0$ and $\bar{\mathbf{x}}_1$. Then

$$M = M_0 + M_1, \qquad \bar{\mathbf{x}} = \frac{1}{M}(M_0\bar{\mathbf{x}}_0 + M_1\bar{\mathbf{x}}_1).$$

The first formula is obvious. The second is just a decomposition of the moment integral.

$$M\bar{\mathbf{x}} = \iiint\limits_{\mathbf{D}} \delta(\mathbf{x})\,\mathbf{x}\,dV = \iiint\limits_{\mathbf{D}_0} + \iiint\limits_{\mathbf{D}_1} = M_0\bar{\mathbf{x}}_0 + M_1\bar{\mathbf{x}}_1$$

Remark   A similar principle applies to plane sheets and wires.

■ **EXAMPLE 4**   A solid consists of a cylindrical can of radius a and height h capped by a hemisphere at one end (Fig. 3a). Locate its center of gravity.

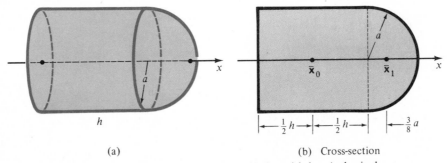

(a)   (b) Cross-section

Fig. 3   Center of gravity of uniform cylinder with hemispherical cap

Solution   Since no density is given, we may assume that $\delta = 1$ and work with volumes instead of masses. We choose the x-axis along the axis of symmetry of the solid, with 0 at its base. The cylinder has volume

$$V_0 = \pi a^2 h \qquad \text{and centroid} \quad \bar{x}_0 = \tfrac{1}{2}h.$$

By Example 1, the hemisphere has volume

$$V_1 = \tfrac{2}{3}\pi a^3 \qquad \text{and centroid} \quad \bar{x}_1 = h + \tfrac{3}{8}a.$$

See Fig. 3b. Therefore, the volume of the whole solid is

$$V = V_0 + V_1 = \pi a^2 h + \tfrac{2}{3}\pi a^3 = \tfrac{1}{3}\pi a^2(3h + 2a),$$

and its center of gravity is at $\bar{x}$ on the axis, where

$$\bar{x} = \frac{1}{V}(V_0 \bar{x}_0 + V_1 \bar{x}_1) = \frac{1}{V}[(\pi a^2 h)(\tfrac{1}{2}h) + (\tfrac{2}{3}\pi a^3)(h + \tfrac{3}{8}a)]$$

$$= \frac{1}{V}(\pi a^2)[\tfrac{1}{2}h^2 + (\tfrac{2}{3}a)(h + \tfrac{3}{8}a)]$$

$$= \frac{1}{V}(\tfrac{1}{12}\pi a^2)(6h^2 + 8ah + 3a^2) = \frac{6h^2 + 8ah + 3a^2}{4(3h + 2a)}.$$

As a check, note that $\bar{x} \longrightarrow \tfrac{3}{8}a$ as $h \longrightarrow 0$, and $\bar{x} \longrightarrow \tfrac{1}{2}h$ as $a \longrightarrow 0$.   ■

EXERCISES

Find the center of gravity of

1   the first octant portion of the uniform sphere $\rho \leq a$
2   the hemisphere $\rho \leq a, \quad z \geq 0, \quad$ density $\delta = a - \rho$
3   the uniform spherical cone $\rho \leq a, \quad 0 \leq \phi \leq \alpha$
4   the uniform hemispherical shell $a \leq \rho \leq b, \quad z \geq 0$
5   the uniform solid $0 < a \leq r \leq b, \quad 0 \leq z \leq r$
6   the uniform solid $x^2 + y^2 \leq z \leq 1$
7   the uniform spherical cap (surface) $\rho = a, \quad 0 \leq \phi \leq \alpha$
8   the uniform solid spherical cap $\rho \leq a, \quad a - h \leq z$
9   the uniform sheet $x^2 + y^2 \leq 4a^2, \quad (x - a)^2 + y^2 \geq a^2$
10   the uniform solid $x^2 + y^2 + z^2 \leq 4a^2, \quad (x - a)^2 + y^2 + z^2 \geq a^2$
11   the lateral surface of the cone $\phi = \alpha, \quad 0 \leq \rho \leq a$
12   the uniform solid $r \leq a, \quad 0 \leq az \leq r^2$
13   the uniform spherical triangle $\rho = a, \quad x \geq 0, \quad y \geq 0, \quad z \geq 0$
14   the uniform wedge $\rho \leq a, \quad -\alpha \leq \theta \leq \alpha$
15   the uniform lune $\rho = a, \quad -\alpha \leq \theta \leq \alpha$
16   the sphere $\rho \leq a$ with density $\delta = a + z$
17   the uniform frustum of a right circular cone of height h and base radii $a < b$
18* the uniform cone with apex $(0, 0, c)$ and base a domain **D** in the x, y-plane
19   the octant of a uniform ellipsoid $x \geq 0, \quad y \geq 0, \quad z \geq 0, \quad x^2/a^2 + y^2/b^2 + z^2/c^2 \leq 1$
20* the uniform tetrahedron with vertices **a**, **b**, **c**, **d**. [*Hint*   Use Ex. 41, p. 915.]

5. MOMENTS OF INERTIA

Let **D** be a solid with density $\delta(\mathbf{x})$ and let α be any straight line (axis) in space. The **moment of inertia** of **D** about α is

$$I_\alpha = \iiint\limits_{\mathbf{D}} p(\mathbf{x})^2\, \delta(\mathbf{x})\, dV = \iiint\limits_{\mathbf{D}} p(\mathbf{x})^2\, dM,$$

where $p(\mathbf{x})$ is the distance from **x** to the axis α. See Fig. 1a.

The quantity I_α is used in computing kinetic energies of rotating bodies. Suppose **D** rotates about α with angular speed ω. A point **x** in **D** moves with speed $p(\mathbf{x})\omega$. Since kinetic energy equals one-half the mass times the speed squared, an element of mass dM at **x** has kinetic energy $dK = \tfrac{1}{2}[p(\mathbf{x})\omega]^2\, dM$. Hence the total kinetic energy

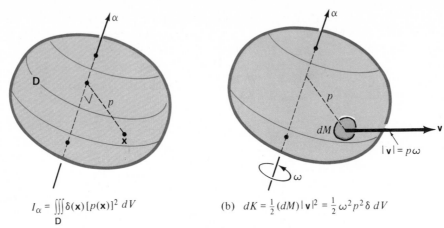

$$I_\alpha = \iiint\limits_D \delta(\mathbf{x})\,[p(\mathbf{x})]^2\,dV$$

(b) $dK = \tfrac{1}{2}(dM)|\mathbf{v}|^2 = \tfrac{1}{2}\omega^2 p^2 \delta\, dV$

Fig. 1   Moment of inertia and kinetic energy

of **D** is

$$K = \tfrac{1}{2}\omega^2 \iiint\limits_D p(\mathbf{x})^2\,dM = \tfrac{1}{2}I_\alpha\,\omega^2.$$

We denote the moments of inertia of **D** with respect to the x-axis, y-axis, and z-axis, respectively, by I_x, I_y, and I_z. For these quantities, $p(\mathbf{x})^2$ is particularly simple. For instance, if $\mathbf{x} = (x,\ y,\ z)$ and $p(\mathbf{x})$ is the distance to the x-axis, then $p(\mathbf{x})^2 = y^2 + z^2$. Similar formulas hold relative to the y-axis and the z-axis.

$$I_x = \iiint\limits_D (y^2 + z^2)\,\delta(x,y,z)\,dV, \qquad I_y = \iiint\limits_D (z^2 + x^2)\,\delta(x,y,z)\,dV,$$

$$I_z = \iiint\limits_D (x^2 + y^2)\,\delta(x,y,z)\,dV.$$

■ **EXAMPLE 1**   Compute the moments of inertia I_x, I_y, I_z of a uniform sphere of center **0**, radius a, and mass M.

Solution   If δ is the constant density, then $M = \tfrac{4}{3}\pi a^3\delta$. By symmetry $I_x = I_y = I_z$. It seems most natural to use spherical coordinates to compute I_z:

$$I_z = \delta \iiint (x^2 + y^2)\,dV = \delta \iiint (\rho^2 \sin^2\phi\cos^2\theta + \rho^2 \sin^2\phi\sin^2\theta)\,dV$$

$$= \delta \iiint (\rho^2 \sin^2\phi)\,\rho^2 \sin\phi\,d\rho\,d\phi\,d\theta$$

$$= \delta \int_0^{2\pi} d\theta \int_0^{\pi} \sin^3\phi\,d\phi \int_0^a \rho^4\,d\rho = \delta(2\pi)(\tfrac{4}{3})(\tfrac{1}{5}a^5) = \tfrac{2}{5}(\tfrac{4}{3}\pi a^3\delta)a^2 = \tfrac{2}{5}Ma^2. \quad ■$$

Units   If M is measured in kilograms and a in meters, then I_z is measured in kg-m^2. In the formula for kinetic energy $K = \tfrac{1}{2}I_\alpha\,\omega^2$, if I_α is in kg-m^2 and ω in rad/sec, then K is in joules = newton-meters. A newton, the metric unit of force, equals one kg-m/sec^2.

Parallel Axes Theorem Suppose α is an axis through the center of gravity of a solid **D**, and β is an axis parallel to α. There is a formula that expresses the moment of inertia I_β in terms of I_α.

Parallel Axes Theorem If α is an axis through the center of gravity of a solid **D**, and β is an axis parallel to α, then

$$I_\beta = I_\alpha + Md^2,$$

where M is the mass of **D** and d is the distance between the axes.

Proof Choose the coordinate system with $\bar{\mathbf{x}} = \mathbf{0}$, and with α the z-axis and β the line $x = d$, $y = 0$. See Fig. 2. For any point $\mathbf{x} = (x, y, z)$ of **D**, let $p(\mathbf{x})$ be the distance from $\mathbf{x}$ to the axis α and $q(\mathbf{x})$ the distance from $\mathbf{x}$ to the axis β. Then

$$p(\mathbf{x})^2 = x^2 + y^2,$$

$$q(\mathbf{x})^2 = (x - d)^2 + y^2 = x^2 + y^2 - 2dx + d^2 = p(\mathbf{x})^2 - 2dx + d^2.$$

Therefore

$$I_\beta = \iiint_{\mathbf{D}} q(\mathbf{x})^2 \, dM = \iiint_{\mathbf{D}} [p(\mathbf{x})^2 - 2\,dx + d^2] \, dM = I_\alpha - 2d \iiint_{\mathbf{D}} x \, dM + d^2 \iiint_{\mathbf{D}} dM.$$

The second integral is the moment m_x, and the third integral is M. But $m_x = 0$ because the center of gravity is at **0**. Hence

$$I_\beta = I_\alpha + Md^2.$$

■ **EXAMPLE 2** Find the moment of inertia of a uniform sphere of radius a and mass M about an axis tangent to the sphere.

Solution From Example 1, the moment of inertia about any axis through the center (c.g.) is $\frac{2}{5}Ma^2$. The distance from a tangent axis β to the center is a, so the Parallel Axis Theorem implies

$$I_\beta = \tfrac{2}{5}Ma^2 + Ma^2 = \tfrac{7}{5}Ma^2. \qquad\qquad ■$$

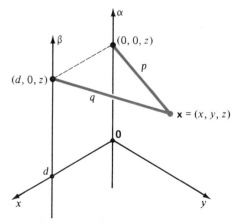

Fig. 2 Parallel Axes Theorem

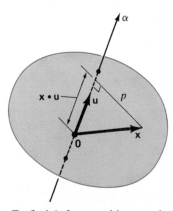

Fig. 3 To find I_z for an arbitrary axis α

Products of Inertia   Suppose α is an axis through the origin. It seems that there should be a formula for I_α in terms of I_x, I_y, and I_z. Let us try to discover one.

We represent the axis α by a unit vector $\mathbf{u} = (u, v, w)$. See Fig. 3. Then $\mathbf{x} \cdot \mathbf{u}$ is the projection of $\mathbf{x}$ on α, hence

$$|\mathbf{x}|^2 = (\mathbf{x} \cdot \mathbf{u})^2 + p(\mathbf{x})^2,$$

$$p(\mathbf{x})^2 = |\mathbf{x}|^2 - (\mathbf{x} \cdot \mathbf{u})^2 = (x^2 + y^2 + z^2) - (ux + vy + wz)^2$$

$$= (1 - u^2)x^2 + (1 - v^2)y^2 + (1 - w^2)z^2 - 2(vwyz + wuzx + uvxy).$$

Since $|\mathbf{u}|^2 = 1$,

$$1 - u^2 = v^2 + w^2, \qquad 1 - v^2 = u^2 + w^2, \qquad 1 - w^2 = u^2 + v^2.$$

After some rearrangement we obtain

$$p(\mathbf{x})^2 = (y^2 + z^2)u^2 + (z^2 + x^2)v^2 + (x^2 + y^2)w^2 - 2vwyz - 2wuzx - 2uvxy.$$

We multiply by dM and integrate:

$$I_\alpha = \iiint_D p(\mathbf{x})^2 \, dM$$

$$= u^2 \iiint_D (y^2 + z^2) \, dM + v^2 \iiint_D (z^2 + x^2) \, dM + w^2 \iiint_D (x^2 + y^2) \, dM$$

$$- 2vw \iiint_D yz \, dM - 2wu \iiint_D zx \, dM - 2uv \iiint_D xy \, dM$$

$$= u^2 I_x + v^2 I_y + w^2 I_z + \text{(three mixed terms)}.$$

This is not quite the type of formula we expected because of the mixed terms. We are forced to introduce three new quantities:

Products of Inertia

$$I_{yz} = -\iiint_D yz \, dM, \qquad I_{zx} = -\iiint_D zx \, dM, \qquad I_{xy} = -\iiint_D xy \, dM.$$

In terms of these new quantities and the moments of inertia, we obtain a formula for I_α:

Let α be an axis through the origin determined by the unit vector $\mathbf{u} = (u, v, w)$. Then

$$I_\alpha = I_x u^2 + I_y v^2 + I_z w^2 + 2I_{yz} vw + 2I_{zx} wu + 2I_{xy} uv.$$

Remark   It is customary to define $I_{yx} = I_{xy}$, etc.

■ **EXAMPLE 3**   A uniform solid of mass M fills the first octant of the sphere $|\mathbf{x}| \le a$. Find its products of inertia.

Solution   If the density is δ, the mass is

$$M = \tfrac{1}{8}(\tfrac{4}{3}\pi a^3)\delta = \tfrac{1}{6}\pi a^3 \delta.$$

By symmetry $I_{yz} = I_{zx} = I_{xy}$. It is easiest to compute I_{xy}—by spherical coordinates since the z-axis plays a special role in spherical coordinates:

$$I_{xy} = -\delta \iiint xy\, dx\, dy\, dz = -\delta \iiint (\rho^2 \sin^2 \phi \cos \theta \sin \theta)\, \rho^2 \sin \phi\, d\rho\, d\phi\, d\theta$$

$$= -\delta \int_0^{\pi/2} \cos \theta \sin \theta\, d\theta \int_0^{\pi/2} \sin^3 \phi\, d\phi \int_0^a \rho^4\, d\rho = -\delta \left(\frac{1}{2}\right)\left(\frac{2}{3}\right)\left(\frac{a^5}{5}\right)$$

$$= -\frac{2}{5}\left(\frac{1}{6}a^3\delta\right)a^2 = -\frac{2}{5}\left(\frac{M}{\pi}\right)a^2 = -\frac{2}{5\pi}Ma^2.$$ ∎

■ **EXAMPLE 4**   The solid in Example 3 rotates with angular speed ω about the axis α through **0** and $(1, 1, 1)$. Find its kinetic energy.

Solution $K = \tfrac{1}{2}I_\alpha \omega^2$, so we must find I_α. The axis α is determined by the unit vector

$$\mathbf{u} = \tfrac{1}{3}\sqrt{3}\,(1,\ 1,\ 1) = (\tfrac{1}{3}\sqrt{3},\ \tfrac{1}{3}\sqrt{3},\ \tfrac{1}{3}\sqrt{3}\,).$$

From the formula for I_α on the previous page, with $u = v = w = \tfrac{1}{3}\sqrt{3}$,

$$I_\alpha = \tfrac{1}{3}I_x + \tfrac{1}{3}I_y + \tfrac{1}{3}I_z + \tfrac{2}{3}I_{yz} + \tfrac{2}{3}I_{zx} + \tfrac{2}{3}I_{xy}.$$

By symmetry, $I_x = I_y = I_z$ and $I_{yz} = I_{zx} = I_{xy}$. Therefore, the formula boils down to

$$I_\alpha = I_x + 2I_{yz}.$$

But I_{yz} is known from Example 3, and I_x equals $\tfrac{1}{8}$ of the corresponding moment of inertia of a sphere of mass $8M$. Hence $I_x = \tfrac{2}{5}Ma^2$ by Example 1. Therefore

$$I_\alpha = \frac{2}{5}Ma^2 - \frac{4}{5\pi}Ma^2 = \frac{2}{5}\left(1 - \frac{2}{\pi}\right)Ma^2.$$

Finally, $$K = \frac{1}{2}I_\alpha \omega^2 = \frac{1}{5}\left(1 - \frac{2}{\pi}\right)Ma^2\omega^2.$$ ∎

EXERCISES

Find the moments of inertia; give your answer in the form $I_x = M \cdot (\ \ )$, etc.

1   uniform solid $|x| \le a,\ \ |y| \le b,\ \ |z| \le c$
2   uniform hemisphere $\rho \le a,\ \ z \ge 0$
3   uniform cylinder $r \le a,\ \ |z| \le h$
4   uniform cone $hr \le az \le ah$
5   uniform cylinder $(x - a)^2 + y^2 \le a^2,\ \ |z| \le h$
6*   uniform solid torus $(r - A)^2 + z^2 \le a^2,\ \ 0 < a < A$
7   uniform double cone $a|z| \le h(a - r)$
8   sphere $\rho \le a$,   density $\delta = 1/\rho$
9   uniform solid paraboloid $0 \le z \le h\left(1 - \dfrac{r^2}{a^2}\right)$
10   uniform solid $\rho \ge a,\ \ |x| \le a,\ \ |y| \le a,\ \ |z| \le a$
11   uniform spherical shell $\rho = a$

12   uniform lateral surface of cone $r \leq a$, $hr = az$
13   uniform toroidal shell $(r - A)^2 + z^2 = a^2$, $0 < a < A$ [*Hint* Use the parameteriza-
     tion $x = (A + a \cos \phi) \cos \theta$, $y = (A + a \cos \phi) \sin \theta$, $z = a \sin \phi$, $0 \leq \phi \leq 2\pi$,
     $0 \leq \theta \leq 2\pi$.]
14   uniform ellipsoid $x^2/a^2 + y^2/b^2 + z^2/c^2 \leq 1$.

15   In Example 1, use symmetry to prove, without integrating, that $I_z = \frac{2}{3}\delta \iiint \rho^2 \, dV$.
     Then evaluate the integral
16*   A domain **D** in the x, y-plane has density $\delta = 1$, area A, and moment of inertia
     $I = Ad^2$ (with respect to the z-axis). Find I_z for the cone with base **D**, apex $(0, 0, c)$
     and density 1.

Find the products of inertia for the

17   uniform box $0 \leq x \leq a$, $0 \leq y \leq b$, $0 \leq z \leq c$
18   uniform tetrahedron, vertices $(0, 0, 0)$, $(a, 0, 0)$, $(0, b, 0)$ $(0, 0, c)$
19   uniform quarter cylinder $r \leq a$, $0 \leq \theta \leq \frac{1}{2}\pi$, $0 \leq z \leq h$
20   uniform hemisphere $(x - a)^2 + y^2 + z^2 \leq a^2$, $z \geq 0$
21   cube $|x| \leq a$, $|y| \leq a$, $|z| \leq a$,   density $\delta = (x + a)(y + a)(z + a)$
22   prism $0 \leq x$, $0 \leq y$, $|z| \leq h$, $x + y \leq a$,   density $\delta = xy(z + h)$.

6. LINE INTEGRALS

The remainder of this chapter deals with certain new kinds of integrals and their relations to double and triple integrals. The discussion starts here with integrals over curves in space and in the plane. The next section studies an important relation between integrals over closed plane curves and certain double integrals. The final section deals with integrals over surfaces in space and their relation to certain triple integrals.

Line Integrals   A **vector field** is the assignment of a vector **F(x)** to each point **x** of a region **D** in space. A familiar example of a vector field is the gradient of a function.

Let **F(x)** be a vector field on **D** and suppose $\mathscr{C}$ is a directed curve in **D**. See Fig. 1. Written in components,

$$\mathbf{F(x)} = (P(\mathbf{x}), Q(\mathbf{x}), R(\mathbf{x})).$$

We define a new kind of integral denoted by

$$\int_{\mathscr{C}} \mathbf{F} \cdot d\mathbf{x} \qquad \text{or} \qquad \int_{\mathscr{C}} P \, dx + Q \, dy + R \, dz.$$

Line Integral   Let $\mathbf{F} = \mathbf{F(x)}$ be a continuous vector field on **D** and let $\mathscr{C}$ be a directed curve in **D**. Suppose $\mathscr{C}$ is described parametrically by $\mathbf{x} = \mathbf{x}(t)$, where $a \leq t \leq b$. Define the **line integral**

$$\int_{\mathscr{C}} \mathbf{F} \cdot d\mathbf{x} = \int_a^b \mathbf{F}[\mathbf{x}(t)] \cdot \mathbf{x}(t) \, dt.$$

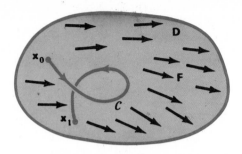

Fig. 1   A vector field **F** and directed curve $\mathscr{C}$ in domain **D**

It follows from the change of variable rule for simple integrals that the line integral depends on $\mathscr{C}$, not on how $\mathscr{C}$ is parameterized. Precisely, if τ is another parameter for the curve, running from α to β as t runs from a to b, then

$$\int_a^b \mathbf{F} \cdot \frac{d\mathbf{x}}{dt} \, dt = \int_\alpha^\beta \mathbf{F} \cdot \frac{d\mathbf{x}}{dt} \frac{dt}{d\tau} \, d\tau = \int_\alpha^\beta \mathbf{F} \cdot \frac{d\mathbf{x}}{d\tau} \, d\tau.$$

The definition is similar for a plane vector field $\mathbf{F}(\mathbf{x}) = (P(\mathbf{x}), Q(\mathbf{x}))$ and a plane curve $\mathscr{C}$. In this case, the line integral can be written

$$\int_{\mathscr{C}} P \, dx + Q \, dy.$$

■ **EXAMPLE 1**   Let $\mathbf{F}(\mathbf{x}) = (x, y, x + y + z)$.   Compute the line integral

$$\int_{\mathscr{C}} \mathbf{F} \cdot d\mathbf{x}$$

over the path $\mathscr{C}$ given by

(a) $\mathbf{x}(t) = (t, t, t), \quad 0 \le t \le 1$ (b) $\mathbf{x}(t) = (t, t^2, t^3), \quad 0 \le t \le 1.$

Solution   (a) $\mathbf{x}(t) = (t, t, t)$ and $\dot{\mathbf{x}}(t) = (1, 1, 1)$. Along the curve, $\mathbf{F}(\mathbf{x}) = \mathbf{F}[\mathbf{x}(t)] = (t, t, 3t)$. Therefore

$$\int_{\mathscr{C}} \mathbf{F} \cdot d\mathbf{x} = \int_0^1 (t, t, 3t) \cdot (1, 1, 1) \, dt = \int_0^1 5t \, dt = \tfrac{5}{2}.$$

(b) $\mathbf{x}(t) = (t, t^2, t^3)$ and $\dot{\mathbf{x}}(t) = (1, 2t, 3t^2)$. Also $\mathbf{F}[\mathbf{x}(t)] = (t, t^2, t + t^2 + t^3)$. Therefore

$$\int_{\mathscr{C}} \mathbf{F} \cdot d\mathbf{x} = \int_0^1 (t, t^2, t + t^2 + t^3) \cdot (1, 2t, 3t^2) \, dt = \int_0^1 [t + 2t^3 + 3(t^3 + t^4 + t^5)] \, dt$$

$$= \int_0^1 (t + 5t^3 + 3t^4 + 3t^5) = \tfrac{1}{2} + \tfrac{5}{4} + \tfrac{3}{5} + \tfrac{1}{2} = \tfrac{57}{20}.$$ ■

For an interpretation of line integrals, suppose $\mathbf{x} = \mathbf{x}(s)$ is a parameterization of $\mathscr{C}$ with arc length as the parameter, and $a \le s \le b$. Then $d\mathbf{x}/ds = \mathbf{t}$, the unit tangent. Therefore the definition of the line integral in this case is

$$\int_{\mathscr{C}} \mathbf{F} \cdot d\mathbf{x} = \int_a^b \mathbf{F}[\mathbf{x}(s)] \cdot \mathbf{t}(s) \, ds.$$

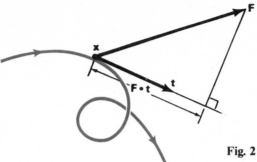

Fig. 2   Interpretation of $\mathbf{F} \cdot d\mathbf{x} = (\mathbf{F} \cdot \mathbf{t})\, ds$
as the element of work

The quantity $\mathbf{F} \cdot \mathbf{t}$ is the tangential component of $\mathbf{F}$. See Fig. 2. Hence

$$\int_{\mathscr{C}} \mathbf{F} \cdot d\mathbf{x} = \int_a^b [\text{tangential component of } \mathbf{F}(s)]\, ds.$$

This formula gives an immediate physical interpretation of the line integral as *work*. Recall that work equals force times distance. When a particle moves on a curve in a force field $\mathbf{F}(\mathbf{x})$, it is the tangential component of the force that does work. Thus the element of work is

$$dW = (\mathbf{F} \cdot \mathbf{t})\, ds.$$

The line integral adds up these small bits of work and yields the total work done by the field in moving the particle along its path.

Independence of Path   In Example 1, the same vector field is integrated over two different curves. Both curves have the same end points, $\mathbf{x}(0) = (0, 0, 0)$ and $\mathbf{x}(1) = (1, 1, 1)$, yet the line integrals are unequal. Therefore, in general, a line integral depends on the curve, not just on its end points.

In an important special case, however, the line integral does *not* depend on the curve $\mathscr{C}$, but only on its initial and terminal points:

Suppose the vector field $\mathbf{F}(\mathbf{x})$ is the gradient of some function $f(\mathbf{x})$:

$$\mathbf{F} = \operatorname{grad} f.$$

Then for each curve $\mathscr{C}$ going from $\mathbf{x}_0$ to $\mathbf{x}_1$,

$$\int_{\mathscr{C}} \mathbf{F} \cdot d\mathbf{x} = f(\mathbf{x}_1) - f(\mathbf{x}_0).$$

Therefore the line integral depends only on the end points, not on what path is taken between them.

In particular, if $\mathscr{C}$ is a *closed* path, then

$$\int_{\mathscr{C}} \mathbf{F} \cdot d\mathbf{x} = 0.$$

Proof By the Chain Rule (read backwards),

$$\mathbf{F} \cdot \dot{\mathbf{x}} = [\operatorname{grad} f(\mathbf{x})] \cdot \dot{\mathbf{x}} = \frac{d}{dt} f[\mathbf{x}(t)].$$

Therefore

$$\int_{\mathscr{C}} \mathbf{F} \cdot d\mathbf{x} = \int_a^b \mathbf{F} \cdot \dot{\mathbf{x}} \, dt = \int_a^b \frac{d}{dt} f[\mathbf{x}(t)] \, dt = f[\mathbf{x}(t)] \Big|_a^b = f(\mathbf{x}_1) - f(\mathbf{x}_0).$$

If $\mathscr{C}$ is closed, then $\mathbf{x}_1 = \mathbf{x}_0$ so $f(\mathbf{x}_1) - f(\mathbf{x}_0) = 0.$

Conservation of Energy Suppose $\mathbf{F} = \mathbf{F}(\mathbf{x})$ is a force field and a particle of mass m moves under the influence of this force along a path $\mathscr{C}$ from $\mathbf{x}_0$ to $\mathbf{x}_1$. Say the path is parameterized by time: $\mathbf{x} = \mathbf{x}(t)$, where $t_0 \le t \le t_1$.

The work done by $\mathbf{F}$ in moving the particle is

$$W = \int_{\mathscr{C}} \mathbf{F} \cdot d\mathbf{x} = \int_{t_0}^{t_1} \mathbf{F}[\mathbf{x}(t)] \cdot \dot{\mathbf{x}} \, dt.$$

Now the motion of the particle is determined by Newton's law

$$\mathbf{F} = m\ddot{\mathbf{x}}.$$

It follows that

$$\mathbf{F} \cdot \dot{\mathbf{x}} = m\dot{\mathbf{x}} \cdot \ddot{\mathbf{x}} = \frac{1}{2} m \frac{d}{dt} (\dot{\mathbf{x}} \cdot \dot{\mathbf{x}}) = \frac{1}{2} m \frac{d}{dt} |\mathbf{v}|^2,$$

where $\mathbf{v} = \dot{\mathbf{x}}$, the velocity. Therefore

$$W = \int_{t_0}^{t_1} \frac{1}{2} m \frac{d}{dt} |\mathbf{v}|^2 = \frac{1}{2} m |\mathbf{v}_1|^2 - \frac{1}{2} m |\mathbf{v}_0|^2.$$

The quantity $K = \frac{1}{2} m |\mathbf{v}|^2$ is the kinetic energy of the particle. The result of this example is the Law of Conservation of Energy: work done equals change in kinetic energy.

Inverse Square Law If $\mathbf{F} = \operatorname{grad} f$ is a force, the net work done by this force in moving a particle from $\mathbf{x}_0$ to $\mathbf{x}_1$ is $f(\mathbf{x}_1) - f(\mathbf{x}_0)$, independent of the path. The function f, which is unique up to an additive constant, is called the **potential** of the force.

An important example is that of a central force subject to the inverse square law, for instance, the electric force $\mathbf{E}$ on a unit charge at $\mathbf{x}$ due to a unit charge of the same sign at the origin. The magnitude of the vector $\mathbf{E}$ is inversely proportional to $|\mathbf{x}|^2$. Its direction is the same as that of $\mathbf{x}$. See Fig. 3.

The unit vector in the direction of $\mathbf{x}$ is

$$\frac{\mathbf{x}}{|\mathbf{x}|} = \frac{\mathbf{x}}{\rho}, \qquad \rho = |\mathbf{x}|.$$

Therefore, expressed in suitable units,

$$\mathbf{E} = \frac{1}{\rho^2} \frac{\mathbf{x}}{\rho} = \frac{\mathbf{x}}{\rho^3}.$$

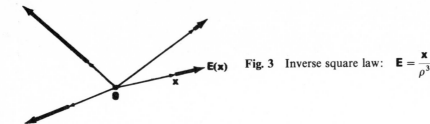

Fig. 3   Inverse square law: $\mathbf{E} = \dfrac{\mathbf{x}}{\rho^3}$

The force field $\mathbf{E}$ is defined at all points of space except the origin. We shall prove that $\mathbf{E}$ is the gradient of a function, in fact that

$$\mathbf{E} = \operatorname{grad} f, \quad \text{where} \quad f(x, y, z) = -\frac{1}{\rho} = \frac{-1}{\sqrt{x^2 + y^2 + z^2}}.$$

Let us compute the gradient of $f = -1/\rho$:

$$\frac{\partial f}{\partial x} = \frac{x}{(x^2 + y^2 + z^2)^{3/2}} = \frac{x}{\rho^3}.$$

Similarly,
$$\frac{\partial f}{\partial y} = \frac{y}{\rho^3} \quad \text{and} \quad \frac{\partial f}{\partial z} = \frac{z}{\rho^3}.$$

Therefore
$$\operatorname{grad} f = \left(\frac{x}{\rho^3}, \frac{y}{\rho^3}, \frac{z}{\rho^3} \right) = \frac{1}{\rho^3}(x, y, z) = \frac{\mathbf{x}}{\rho^3} = \mathbf{E}.$$

It follows that

$$\int_{\mathbf{x}_0}^{\mathbf{x}_1} \mathbf{E} \cdot d\mathbf{x} = f(\mathbf{x}_1) - f(\mathbf{x}_0) = \frac{1}{|\mathbf{x}_0|} - \frac{1}{|\mathbf{x}_1|}.$$

The right-hand side is the **potential difference** or **voltage**. It represents the work done by the electric force when a unit charge moves from $\mathbf{x}_0$ to $\mathbf{x}_1$ *along any path*.

If $\mathbf{x}_1$ is far out, then $1/|\mathbf{x}_1|$ is small, so

$$\int_{\mathbf{x}_0}^{\mathbf{x}_1} \mathbf{E} \cdot d\mathbf{x} \approx \frac{1}{|\mathbf{x}_0|}.$$

As $\mathbf{x}_1$ moves farther out, the approximation improves:

$$\int_{\mathbf{x}_0}^{\mathbf{x}_1} \mathbf{E} \cdot d\mathbf{x} \longrightarrow \frac{1}{|\mathbf{x}_0|} \quad \text{as} \quad |\mathbf{x}_1| \longrightarrow \infty,$$

that is,
$$\int_{\mathbf{x}_0}^{\infty} \mathbf{E} \cdot d\mathbf{x} = \frac{1}{|\mathbf{x}_0|} = \frac{1}{\rho_0}.$$

The final result, $1/|\mathbf{x}_0|$, is called the **potential** at $\mathbf{x}_0$. It is the work done by the force in moving a unit charge from $\mathbf{x}_0$ to infinity, along any path.

Conservation of Momentum   Recall that the integral (simple or multiple) of a vector-valued function is defined componentwise. For instance, if

$$\mathbf{u}(t) = (u(t), v(t), w(t)),$$

then
$$\int_a^b \mathbf{u}(t)\, dt = \left(\int_a^b u(t)\, dt, \int_a^b v(t)\, dt, \int_a^b w(t)\, dt \right).$$

The following useful formula can be easily proved componentwise:

$$\int_a^b \frac{d\mathbf{w}}{dt}\, dt = \mathbf{w}(b) - \mathbf{w}(a).$$

For example,
$$\int_0^2 (2t, 3t^2, 4t^3)\, dt = \int_0^2 \frac{d}{dt}(t^2, t^3, t^4)\, dt = (t^2, t^3, t^4)\Big|_0^2 = (4, 8, 16).$$

Now suppose a particle of mass m moves on a path $\mathscr{C}$ under the influence of the force field $\mathbf{F} = \mathbf{F}(\mathbf{x})$. According to Sir Isaac,

$$\mathbf{F} = m\ddot{\mathbf{x}} = m\frac{d}{dt}\,\dot{\mathbf{x}} = m\frac{d}{dt}\,\mathbf{v}.$$

Say the path is described by $\mathbf{x} = \mathbf{x}(t)$, where $t_0 \le t \le t_1$. Then

$$\int_{t_0}^{t_1} \mathbf{F}[\mathbf{x}(t)]\, dt = \int_{t_0}^{t_1} m\frac{d}{dt}\,\mathbf{v}\, dt = m\mathbf{v}_1 - m\mathbf{v}_0.$$

The quantity $m\mathbf{v}$ is the **momentum** of the particle. The quantity

$$\int_{t_0}^{t_1} \mathbf{F}\, dt$$

is the **impulse** of the force during the time interval $[t_0, t_1]$. The equation

$$\int_{t_0}^{t_1} \mathbf{F}\, dt = m\mathbf{v}\,\Big|_{t_0}^{t_1}$$

is the Law of Conservation of Momentum: impulse equals change in momentum.

The **angular momentum** of the particle with respect to the origin $\mathbf{0}$ is defined as

$$m\mathbf{x} \times \mathbf{v} = m\mathbf{x} \times \dot{\mathbf{x}}.$$

Now
$$\frac{d}{dt}(m\mathbf{x} \times \mathbf{v}) = \frac{d}{dt}(m\mathbf{x} \times \dot{\mathbf{x}}) = m\dot{\mathbf{x}} \times \dot{\mathbf{x}} + m\mathbf{x} \times \ddot{\mathbf{x}} = m\mathbf{x} \times \ddot{\mathbf{x}}$$

since $\dot{\mathbf{x}} \times \dot{\mathbf{x}} = \mathbf{0}$. But $m\ddot{\mathbf{x}} = \mathbf{F}$, hence

$$m\mathbf{x} \times \ddot{\mathbf{x}} = \mathbf{x} \times m\ddot{\mathbf{x}} = \mathbf{x} \times \mathbf{F},$$

which is the torque of $\mathbf{F}$ at $\mathbf{x}$. Therefore

$$\frac{d}{dt}(m\mathbf{x} \times \mathbf{v}) = \mathbf{x} \times \mathbf{F}.$$

Integrate:
$$\int_{t_0}^{t_1} \mathbf{x} \times \mathbf{F}\, dt = m\mathbf{x} \times \mathbf{v}\,\Big|_{t_0}^{t_1}.$$

This result, called the Law of Conservation of Angular Momentum, asserts that the time integral of the torque equals the change in angular momentum.

EXERCISES

Evaluate the line integral over the indicated path

1 $\displaystyle\int y\, dx - x\, dy$

$\mathbf{x} = (\cos t, \sin t) \quad 0 \le t \le \tfrac{1}{2}\pi$

2 $\displaystyle\int x\, dx + y\, dy$

$\mathbf{x} = (t^3, t^2), \quad 0 \le t \le 1$

3 $\displaystyle\int z\, dx + x\, dy + y\, dz$

straight path from
$(0, 0, 0)$ to (a, b, c)

4 $\displaystyle\int z\, dx + x\, dy + y\, dz$

straight path from
$(1, 1, 2)$ to $(-1, 0, 4)$

5 $\displaystyle\int xy\, dx + (1 + 2y)\, dy$

straight path from
$(0, 1)$ to $(0, -1)$

6 $\displaystyle\int xy\, dx + (1 + 2y)\, dy$

semi-circle $r = 1$
$\tfrac{1}{2}\pi \le \theta \le \tfrac{3}{2}\pi$

7 $\displaystyle\int -dy + x\, dz$

$\mathbf{x} = (t^2, t^3, t^4) \quad 1 \le t \le 2$

8 $\displaystyle\int z\, dx + x\, dz$

$\mathbf{x} = (\sin t, \cos t, t^2)$
$0 \le t \le \tfrac{1}{2}\pi$

9 $\displaystyle\int yz\, dx + 2x\, dy + xy\, dz$

$\mathbf{x} = (t^2, t^3, t^{-4})$
$a \le t \le b \quad (0 < a)$

10 $\displaystyle\int \sin y \cos z\, dx + x \cos y \cos z\, dy$
$\qquad\qquad\qquad\qquad\qquad - x \sin y \sin z\, dz$

$\mathbf{x} = (\cos t, \sin t, \cos t)$
$0 \le t \le 2\pi.$

11 Let $\mathbf{F} = (3x^2 y^2 z, 2x^3 yz, x^3 y^2)$. Show that $\displaystyle\int_{(0,0,0)}^{(1,1,1)} \mathbf{F} \cdot d\mathbf{x}$ is independent of the path, and evaluate it.

12 Let $\mathbf{F} = (x^2 + yz, y^2 + zx, z^2 + xy)$. Show that $\displaystyle\int_{(0,0,0)}^{(a,b,c)} \mathbf{F} \cdot d\mathbf{x}$ is independent of the path, and evaluate it.

13 Let θ denote the polar angle in the plane. Show that $\operatorname{grad} \theta = \left(\dfrac{-y}{x^2 + y^2}, \dfrac{x}{x^2 + y^2}\right).$

14 (cont.) Find $\displaystyle\int \frac{-y\, dx + x\, dy}{x^2 + y^2}$ over the circle $|\mathbf{x}| = a$.

15 Find the work done by the central force field $\mathbf{F} = -\dfrac{1}{|\mathbf{x}|^3}\,\mathbf{x}$ in moving a particle from $(1, 8, 4)$ to $(2, 1, 2)$ along a straight path.

16 Find the work done by the uniform gravitational field $\mathbf{F} = (0, 0, -g)$ in moving a particle from $(0, 0, 1)$ to $(1, 1, 0)$ along a straight path.

17 Let $\mathbf{F} = \mathbf{x}/|\mathbf{x}|^5$ and suppose $\mathbf{a} \ne \mathbf{0}$. Show that $\displaystyle\int_{\mathbf{a}}^{\infty} \mathbf{F} \cdot d\mathbf{x}$, taken along any path from $\mathbf{a}$ not passing through $\mathbf{0}$ and going out indefinitely, depends only on $a = |\mathbf{a}|$. Evaluate the integral.

18 (cont.) Do the same for $\mathbf{F} = \mathbf{x}/|\mathbf{x}|^n$, for any $n > 2$.

19 (cont.) Show that $\mathbf{F} = \mathbf{x}/|\mathbf{x}|^2$ is a gradient.

20 (cont.) What value should be assigned to $\displaystyle\int_{\mathbf{a}}^{\infty} \mathbf{F} \cdot d\mathbf{x}$, where $\mathbf{F} = \mathbf{x}/|\mathbf{x}|^2$, taken along a path from $\mathbf{a}$ not passing through $\mathbf{0}$ and going out indefinitely?

Evaluate

21 $\displaystyle\int_0^1 (1 + t, 1 + 2t, 1 + 3t)\, dt$ **22** $\displaystyle\int_0^{2\pi} (\cos t, \sin t, 1)\, dt$

23 $\displaystyle\int_{-1}^1 (t^3, t^4, t^5)\, dt$ **24** $\displaystyle\int_1^4 \left(\frac{1}{t}, \frac{1}{t^2}, \frac{1}{t^3}\right) dt.$

25 The force $\mathbf{F}(t) = (1 - t, 1 - t^2, 1 - t^3)$ acts from $t = 0$ to $t = 2$. Find its impulse.
26 The force $\mathbf{F}(t) = (e^t, e^{2t}, e^{3t})$ acts from $t = -1$ to $t = 0$. Find its impulse.
27 An electron of mass m in a uniform magnetic field follows the spiral path

$$\mathbf{x}(t) = (a \cos t, a \sin t, bt).$$

Find its angular momentum with respect to $\mathbf{0}$.
28 A particle of unit mass moves on the unit sphere $|\mathbf{x}| = 1$ with unit speed. Show that its angular momentum with respect to $\mathbf{0}$ is a unit vector.

7. GREEN'S THEOREM

In this section we discuss some properties of line integrals in the plane, that is, integrals of the form

$$\int_{\mathscr{C}} P\, dx + Q\, dy$$

where $P = P(x, y)$, $Q = Q(x, y)$, and $\mathscr{C}$ is a directed plane curve.

Before coming to the main business of this section, let us make some practical remarks about the evaluation of such integrals. By definition, we parameterize $\mathscr{C}$ by $x = x(t)$, $y = y(t)$ and then evaluate

$$\int P[x(t), y(t)] \frac{dx}{dt}\, dt + Q[x(t), y(t)] \frac{dy}{dt}\, dt.$$

Now suppose $\mathscr{C}$ happens to be part of the graph of a function $y = f(x)$, say for x running from a to b. Then it is most natural to parameterize $\mathscr{C}$ by x itself. Accordingly we write $x = x$, $y = f(x)$. For example, suppose $\mathscr{C}$ is a horizontal segment. Then we write $x = x$ and $y = c$, so $dy/dx = 0$ and

$$\int_{\mathscr{C}} P\, dx + Q\, dy = \int_{\mathscr{C}} P(x, c)\, dx.$$

The $Q\, dy$ part of the integral drops out. We describe this situation by saying that on a horizontal segment $dy = 0$. Similarly, on a vertical segment $dx = 0$ and the $P\, dx$ part of the integral drops out.

Green's Theorem   There is a useful connection between certain line integrals and double integrals. Suppose **D** is a domain in the x, y-plane bounded by one or more closed curves, each composed of arcs with continuously turning tangent vectors. We assign the counterclockwise direction to each of the boundary curves (Fig. 1). If we walk around any boundary curve in this direction, the region **D** is always on our left.

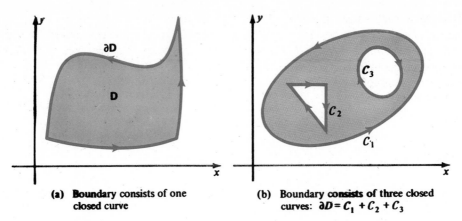

(a) **Boundary** consists of one
     closed curve

(b) **Boundary consists of three closed**
     curves: $\partial D = C_1 + C_2 + C_3$

Fig. 1   Orientation of the boundary $\partial \mathbf{D}$ of a plane domain $\mathbf{D}$

The symbol $\partial \mathbf{D}$ will denote the whole directed boundary of $\mathbf{D}$. We are interested in the line integral

$$\int_{\partial \mathbf{D}} P \, dx + Q \, dy.$$

If $\partial \mathbf{D}$ consists of the directed closed paths, $\mathscr{C}_1, \mathscr{C}_2, \cdots, \mathscr{C}_n$, we write

$$\partial \mathbf{D} = \mathscr{C}_1 + \mathscr{C}_2 + \cdots + \mathscr{C}_n$$

and define

$$\int_{\partial \mathbf{b}} P \, dx + Q \, dy = \int_{\mathscr{C}_1} (P \, dx + Q \, dy) + \cdots + \int_{\mathscr{C}_n} (P \, dx + Q \, dy).$$

Notation   There is a time-honored tradition of writing

$$\oint_{\mathscr{C}} P \, dx + Q \, dy$$

for a line integral over a *closed* path.

An important theorem says that the *line* integral over the boundary $\partial \mathbf{D}$ is equal to a certain *double* integral over the region $\mathbf{D}$.

Green's Theorem   Suppose $P(x, y)$ and $Q(x, y)$ are continuously differentiable functions on a plane domain $\mathbf{D}$. Then

$$\oint_{\partial \mathbf{D}} P \, dx + Q \, dy = \iint_{\mathbf{D}} \left(\frac{\partial Q}{\partial x} - \frac{\partial P}{\partial y} \right) dx \, dy.$$

The theorem may be viewed as two independent formulas,

$$\oint_{\partial \mathbf{D}} P \, dx = -\iint_{\mathbf{D}} \frac{\partial P}{\partial y} \, dx \, dy \quad \text{and} \quad \oint_{\partial \mathbf{D}} Q \, dy = \iint_{\mathbf{D}} \frac{\partial Q}{\partial x} \, dx \, dy.$$

It is fairly easy to see that the second follows from the first by interchanging x and y. (This changes the sense of turning in the plane, which accounts for the sign change.) We shall prove the first formula, not in the most general case, but when **D** can be decomposed by segments into subdomains, each of which is the domain between the graphs of two functions of x.

If **D** is so decomposed,

$$\mathbf{D} = \mathbf{D}_1 + \cdots + \mathbf{D}_n;$$

then

$$\oint_{\partial\mathbf{D}} P\,dx = \oint_{\partial\mathbf{D}_1} P\,dx + \cdots + \oint_{\partial\mathbf{D}_n} P\,dx$$

because the contributions over the common division segments cancel in pairs (Fig. 2a).

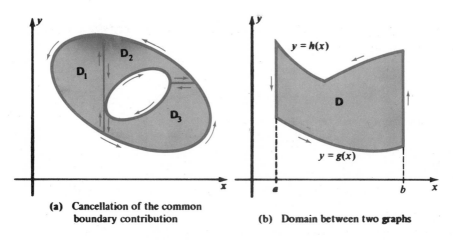

(a) Cancellation of the common boundary contribution

(b) Domain between two graphs

Fig. 2   Proof of Green's Theorem

We shall prove

$$\oint_{\partial\mathbf{D}} P\,dx = -\iint_{\mathbf{D}} \frac{\partial P}{\partial y}\,dx\,dy$$

for the domain **D** of Fig. 2b. Its boundary $\partial\mathbf{D}$ consists of four pieces; accordingly the line integral decomposes into four summands. On the two vertical sides x is constant, hence $dx = 0$, no contribution. Therefore

$$\oint_{\partial\mathbf{D}} P\,dx = \int_b^a P[x, h(x)]\,dx + \int_a^b P[x, g(x)]\,dx = \int_a^b \left\{ -P[x, h(x)] + P[x, g(x)]\right\}\,dx.$$

$$= -\int_a^b \left(\int_{g(x)}^{h(x)} \frac{\partial P}{\partial y}\,dy \right) dx = -\iint_{\mathbf{D}} \frac{\partial P}{\partial y}\,dx\,dy.$$

This completes the proof.

Remark   Green's Theorem is a two-dimensional generalization of the Fundamental Theorem of Calculus,

$$\int_a^b f'(x)\,dx = f(b) - f(a),$$

which equates the integral of an exact derivative $f'(x)$ over an interval $[a, b]$ to something computed at the *boundary* (the two points a and b) of the interval.

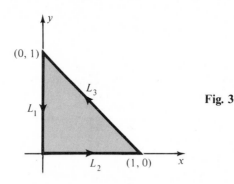

Fig. 3

■ **EXAMPLE 1**   Compute the line integral $\displaystyle\oint_{\mathscr{C}} (x^2 + y^2)\,dx + (x + 2)\,dy,$

where $\mathscr{C}$ is the boundary of the triangle **T** with vertices at $(0, 1)$, $(0, 0)$, and $(1, 0)$. Use   (a) direct computation   (b) Green's Theorem.

Solution   (a)   The integral breaks into integrals over line segments L_1, L_2, L_3 as shown in Fig. 3. Compute each separately.

On L_1, we have $x = 0$ and $dx = 0$ since x is constant. Hence

$$\int_{L_1} (x^2 + y^2)\,dx + (x + 2)\,dy = \int_1^0 2\,dy = -2.$$

On L_2, we have $y = 0$ and $dy = 0$. Hence

$$\int_{L_2} (x^2 + y^2)\,dx + (x + 2)\,dy = \int_0^1 x^2\,dx = \tfrac{1}{3}.$$

On L_3 we use x as the parameter running from 1 to 0. Then $y = 1 - x$ and $dy = -dx$. Hence

$$\int_{L_2} (x^2 + y^2)\,dx + (x + 2)\,dy = \int_1^0 [x^2 + (1 - x)^2 - (x + 2)]\,dx$$

$$= \int_1^0 (-1 - 3x + 2x^2)\,dx = \int_0^1 (1 + 3x - 2x^2)\,dx = \tfrac{11}{6}.$$

Adding the results, we have

$$\oint_{\mathscr{C}} (x^2 + y^2)\,dx + (x + 2)\,dy = -2 + \tfrac{1}{3} + \tfrac{11}{6} = \tfrac{1}{6}.$$

(b)   Let $P(x, y) = x^2 + y^2$ and $Q(x, y) = x + 2$. By Green's Theorem

$$\oint_{\mathscr{C}} P\, dx + Q\, dy = \iint_T \left(\frac{\partial Q}{\partial x} - \frac{\partial P}{\partial y} \right) dx\, dy = \iint_T (1 - 2y)\, dx\, dy$$

$$= \int_0^1 (1 - 2y)\left(\int_0^{1-y} dx \right) dy = \int_0^1 (1 - 2y)(1 - y)\, dy$$

$$= \int_0^1 (1 - 3y + 2y^2)\, dy = \tfrac{1}{6}. \qquad \blacksquare$$

Area Formula   A useful application of Green's Theorem is a formula for the area of a plane domain in terms of an integral over its boundary.

Area Formula   If **D** is a plane domain with a smooth boundary, then

$$|\mathbf{D}| = \frac{1}{2} \oint_{\partial \mathbf{D}} -y\, dx + x\, dy.$$

Proof   Apply Green's Theorem with $P = -y$ and $Q = x$. Then $Q_x - P_y = 2$, so

$$\oint_{\partial \mathbf{D}} -y\, dx + x\, dy = \iint_{\mathbf{D}} 2\, dx\, dy = 2\,|\mathbf{D}|.$$

Remark   By similar applications of Green's Theorem, we also have

$$|\mathbf{D}| = \oint_{\partial \mathbf{D}} x\, dy = -\oint_{\partial \mathbf{D}} y\, dx.$$

The boxed formula is often more convenient than either of these because it has a certain amount of symmetry.

$\blacksquare$ **EXAMPLE 2**   Find the area enclosed by the ellipse $\dfrac{x^2}{a^2} + \dfrac{y^2}{b^2} = 1$.

Solution   Let **D** denote the domain bounded by the ellipse. Parameterize the ellipse as usual by

$$x = a \cos \theta, \qquad y = b \sin \theta, \qquad 0 \le \theta \le 2\pi.$$

Then $|\mathbf{D}| = \tfrac{1}{2} \oint_{\partial \mathbf{D}} -y\, dx + x\, dy$

$$= \tfrac{1}{2} \int_0^{2\pi} -(b \sin \theta)(-a \sin \theta\, d\theta) + (a \cos \theta)(b \cos \theta\, d\theta)$$

$$= \tfrac{1}{2} \int_0^{2\pi} (ab \sin^2 \theta + ab \cos^2 \theta)\, d\theta = \tfrac{1}{2} \int_0^{2\pi} ab\, d\theta = \pi ab. \qquad \blacksquare$$

EXERCISES

Evaluate the line integral over the indicated closed curve

1 $\oint 2y\,dx + 4x\,dy$ Fig. 4   **2** $\oint x^2\,dx + xy\,dy$ Fig. 5

3 $\oint y^2 e^x\,dx + 2ye^x\,dy$ Fig. 5   **4** $\oint -y^2\,dx + x^2\,dy$ Fig. 6

5 $\oint (y - x^2)\,dx + (2x + y^2)\,dy$ Fig. 6   **6** $\oint \cos x \sin y\,dx + \sin x \cos y\,dy$ Fig. 6

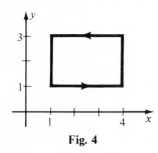
Fig. 4

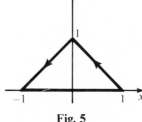
Fig. 5

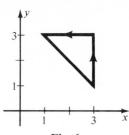
Fig. 6

7 $\oint xy\,dx + xy\,dy$ Fig. 5   **8** $\oint xy\,dx + xy\,dy$ Fig. 7

9 $\oint -y^3\,dx + x^3\,dy$ Fig. 7   **10** $\oint -y^3\,dx + x^3\,dy$ Fig. 8

11 $\oint x^2 y^2\,dx - x^3 y\,dy$ Fig. 8   **12** $\oint \dfrac{x\,dx}{(x^2 + y^2)^3} + \dfrac{y\,dy}{(x^2 + y^2)^3}$ Fig. 9.

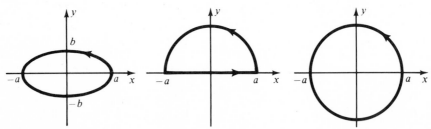

Fig. 7 Ellipse   **Fig. 8** Semicircle   **Fig. 9** Circle

13 Find $\displaystyle\oint \dfrac{-y\,dx + x\,dy}{x^2 + y^2}$ over the circle in Fig. 9.

[*Hint*   Express the integral in terms of polar coordinates. The answer is *not* 0.]

14 (cont.) Evaluate the integral over the closed curves in (a) Fig. 4, (b) Fig. 6, (c) Fig. 10.

15 (cont.) Evaluate $\displaystyle\oint \dfrac{-y\,dx + (x - 1)\,dy}{(x - 1)^2 + y^2} + \dfrac{y\,dx - (x + 1)\,dy}{(x + 1)^2 + y^2}$ over the contour in Fig. 11.

16 (cont.) Evaluate the same integral over the contour in Fig. 12.

17 Evaluate $\displaystyle\oint \dfrac{(-3x^2 y + y^3)\,dx + (x^3 - 3xy^2)\,dy}{(x^2 + y^2)^3}$ over the rectangle in Fig. 4.

18 (cont.) Evaluate the same integral over the contour in Fig. 10.

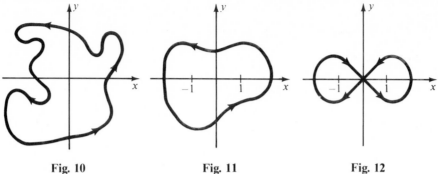

Fig. 10   Fig. 11   Fig. 12

19   Prove **Green's formula** under suitable hypotheses:

$$\oint_{\partial \mathbf{D}} (-uv_y \, dx + uv_x \, dy) - \oint_{\partial \mathbf{D}} (-vu_y \, dx + vu_x \, dy) = \iint_{\mathbf{D}} [u(v_{xx} + v_{yy}) - v(u_{xx} + u_{yy})] \, dx \, dy.$$

20   (cont.) Test this formula when **D** is the unit disk, $u = 1$, and $v = \ln r$. Explain the result.

21   Suppose P and Q are continuously differentiable on a rectangle $a \le x \le b$, $c \le y \le d$. For each point (x, y) of the rectangle define

$$F(x, y) = \int_a^x P(u, c) \, du + \int_c^y Q(x, v) \, dv \qquad \text{and} \qquad G(x, y) = \int_a^x P(u, y) \, dy + \int_c^y Q(a, v) \, dv.$$

Find $\partial F / \partial y$ and $\partial G / \partial x$.

22   (cont.) Assume also that $\partial P / \partial y = \partial Q / \partial x$. Prove that $F = G$ and deduce that $(P, Q) = \text{grad } F$. [*Hint*   Use Green's Theorem.]

23   Express the area formula $A = \frac{1}{2} \oint - y \, dx + x \, dy$ in terms of polar coordinates.

24   (cont.) Apply your answer to find the total area enclosed by the 5-petal rose curve $r = a \cos 5\theta$.

8. SURFACE INTEGRALS

In this section we shall study integrals of the form

$$\iint_{\mathbf{S}} A \, dy \, dz + B \, dz \, dx + C \, dx \, dy,$$

where **S** is an oriented surface in $\mathbf{R}^3$. By oriented surface we mean a two-sided surface† with one side designated as the top. We choose the unit normal vector **n** to point out of the top of the surface.

Such **surface integrals** arise in problems of the following type. Suppose fluid is flowing in a region of space with velocity field $\mathbf{A} = \mathbf{A}(\mathbf{x})$. Given an oriented surface **S** fixed somewhere in space, the problem is to find how fast the fluid flows through **S**, through the bottom and out the top. In other words, to compute how much fluid crosses **S** per unit time. (The rate of flow across **S** is called the **flux** of **A** through **S**.)

---

† There do exist one-sided surfaces such as the Möbius band, but why go looking for trouble?

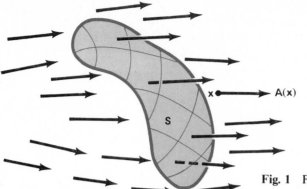

Fig. 1 Find the flux through surface **S** of velocity field $\mathbf{A} = \mathbf{A}(\mathbf{x})$.

To find the flux Φ, we decompose **S** into small "elements" of surface, compute the flux through each, and sum the results (integrate). We consider an element of surface (Fig. 2a) and two associated quantities, its area $d\sigma$ and its unit normal **n**, pointing out of the *top* of the surface. The fluid velocity **A** at the surface element decomposes into a tangential component and a normal component. The tangential component doesn't cross the surface; its contribution to the flux is 0. The normal component (Fig. 2b), which flows straight through the surface, equals $\mathbf{A} \cdot \mathbf{n}$; its contribution to the flux is

$$d\Phi = (\mathbf{A} \cdot \mathbf{n})\, d\sigma = \mathbf{A} \cdot (\mathbf{n}\, d\sigma).$$

Therefore
$$\Phi = \iint_{\mathbf{S}} (\mathbf{A} \cdot \mathbf{n})\, d\sigma,$$

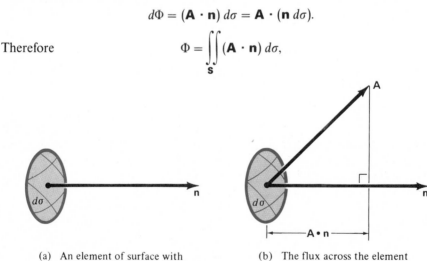

(a) An element of surface with area $d\sigma$ and unit normal **n**

(b) The flux across the element of surface is $d\Phi = (\mathbf{A} \cdot \mathbf{n}) d\sigma$.

Fig. 2 Element of flux

To evaluate the integral, we suppose **S** is parameterized by $\mathbf{x} = \mathbf{x}(u, v)$, where (u, v) runs over a domain **D** in the u, v-plane (Fig. 3). We recall the formulas (pp. 749 and 870) for the unit normal and the element of area:

$$\mathbf{n} = \frac{\mathbf{x}_u \times \mathbf{x}_v}{|\mathbf{x}_u \times \mathbf{x}_v|}, \qquad d\sigma = |\mathbf{x}_u \times \mathbf{x}_v|\, du\, dv.$$

It follows that $\mathbf{n}\, d\sigma = \mathbf{x}_u \times \mathbf{x}_v\, du\, dv.$

This quantity is important in the study of surface integrals and is given a special name:

> **Element of Vectorial Area** $d\boldsymbol{\sigma} = \mathbf{n}\, d\sigma = \mathbf{x}_u \times \mathbf{x}_v\, du\, dv.$

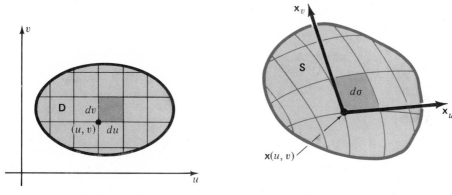

Fig. 3   Parameterization of **S** by $\mathbf{x} = \mathbf{x}(u, v)$, where (u, v) is in **D**

In terms of $d\boldsymbol{\sigma}$, the element of flux is

$$d\Phi = \mathbf{A} \cdot d\boldsymbol{\sigma} = \mathbf{A} \cdot (\mathbf{x}_u \times \mathbf{x}_v)\, du\, dv = [\mathbf{A}, \mathbf{x}_u, \mathbf{x}_v]\, du\, dv$$

$$= \begin{vmatrix} A & B & C \\ x_u & y_u & z_u \\ x_v & y_v & z_v \end{vmatrix} du\, dv = \left[A\,\frac{\partial(y, z)}{\partial(u, v)} + B\,\frac{\partial(z, x)}{\partial(u, v)} + C\,\frac{\partial(x, y)}{\partial(u, v)} \right] du\, dv.$$

This completes our set-up for the formal definition of surface integrals.

> **Surface Integral**   Let $\mathbf{A} = \mathbf{A}(\mathbf{x})$ be a continuous vector field in a region of $\mathbf{R}^3$ and let **S** be an oriented surface in that region, given parametrically by $\mathbf{x} = \mathbf{x}(u, v)$, where (u, v) varies over a domain **D**. Assume $\mathbf{x}(u, v)$ is continuously differentiable. Then
>
> $$\iint_S A\, dy\, dz + B\, dz\, dx + C\, dx\, dy = \iint_S \mathbf{A} \cdot d\boldsymbol{\sigma}$$
>
> $$= \iint_D \left[A\,\frac{\partial(y, z)}{\partial(u, v)} + B\,\frac{\partial(z, x)}{\partial(u, v)} + C\,\frac{\partial(x, y)}{\partial(u, v)} \right] du\, dv.$$
>
> It is understood that $A = A[\mathbf{x}(u, v)]$, etc. in the last integral.

The two forms of the integrand suggest writing

$$d\boldsymbol{\sigma} = \left(\frac{\partial(y, z)}{\partial(u, v)},\ \frac{\partial(z, x)}{\partial(u, v)},\ \frac{\partial(x, y)}{\partial(u, v)} \right) du\, dv = (dy\, dz,\ dz\, dx,\ dx\, dy).$$

This last expression, $\qquad d\boldsymbol{\sigma} = (dy\,dz,\ dz\,dx,\ dx\,dy),$

is important because it gives the vectorial area element in a form independent of the parameterization.

Remark 1   The physical interpretation of the surface integral as flux makes it clear that the integral depends only on the vector field **A** and the oriented surface **S**, not on how the surface is parameterized. An analytic proof of this fact is possible using the change of variables rule. See the exercises.

Remark 2   If an oriented surface cannot be parameterized (because it is closed or has handles, etc.), cut it into pieces and parameterize each piece; then sum the corresponding integrals to obtain the surface integral over the complete surface.

Remark 3   Choosing a parameterization $\mathbf{x} = \mathbf{x}(u, v)$ for a surface **S** fixes a normal vector

$$\mathbf{n} = \mathbf{x}_u \times \mathbf{x}_v = \left(\frac{\partial(y, z)}{\partial(u, v)},\ \frac{\partial(z, x)}{\partial(u, v)},\ \frac{\partial(x, y)}{\partial(u, v)}\right).$$

This normal either agrees with the normal determined by the orientation of **S** or is the negative of it. In the latter case, the surface integral will give $-\Phi$, not Φ.

EXAMPLE 1   Let **T** be the triangle with vertices $(1, 0, 0)$, $(0, 1, 0)$, $(0, 0, 1)$ and normal pointing away from **0**. Let $\mathbf{A}(\mathbf{x}) = (y, z, 0)$. Find

$$\iint_{\mathbf{T}} \mathbf{A} \cdot d\boldsymbol{\sigma} = \iint_{\mathbf{T}} y\,dy\,dz + z\,dz\,dx.$$

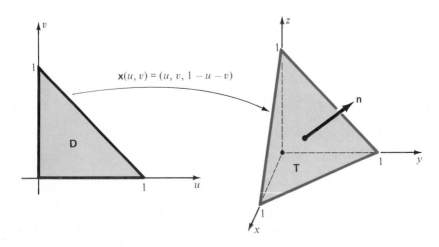

Solution   The triangle is given by $x \geq 0$, $y \geq 0$, $z \geq 0$ and $x + y + z = 1$. Parameterize it by

$$x = u, \qquad y = v, \qquad z = 1 - u - v,$$

where $u \geq 0$, $v \geq 0$, $u + v \leq 1$. See Fig. 4. Then

$$\frac{\partial(y, z)}{\partial(u, v)} = \begin{vmatrix} y_u & y_v \\ z_u & z_v \end{vmatrix} = \begin{vmatrix} 0 & 1 \\ -1 & -1 \end{vmatrix} = 1.$$

Similarly, $\qquad \dfrac{\partial(z, x)}{\partial(u, v)} = \begin{vmatrix} -1 & -1 \\ 1 & 0 \end{vmatrix} = 1, \qquad \dfrac{\partial(x, y)}{\partial(u, v)} = \begin{vmatrix} 1 & 0 \\ 0 & 1 \end{vmatrix} = 1.$

Hence $\qquad\qquad\qquad\qquad d\boldsymbol{\sigma} = (1, 1, 1)\, du\, dv,$

$$\mathbf{A} \cdot d\boldsymbol{\sigma} = (y, z, 0) \cdot (1, 1, 1)\, du\, dv = (y + z)\, du\, dv = (1 - u)\, du\, dv.$$

Therefore

$$\iint_{\mathbf{T}} \mathbf{A} \cdot d\boldsymbol{\sigma} = \iint_{\mathbf{D}} (1 - u)\, du\, dv = \int_0^1 dv \int_0^{1-v} (1 - u)\, du$$

$$= \int_0^1 \left[(1 - v) - \tfrac{1}{2}(1 - v)^2 \right] dv = \int_0^1 \left(\tfrac{1}{2} - \tfrac{1}{2}v^2 \right) dv = \tfrac{1}{2} - \tfrac{1}{6} = \tfrac{1}{3}.$$

Note that the parameterization chosen does indeed yield the normal directed away from **0**, because all three Jacobians are positive. ∎

■ **EXAMPLE 2** Find $\qquad \Phi = \iint_{\mathbf{H}} dx\, dy + 2\, dy\, dz + 3\, dz\, dx,$

where **H** is the hemisphere $\rho = a$, $z \geq 0$, taken with its normal *inward*.

Solution It is easiest to treat the three terms separately. For the $dx\, dy$ term, parameterize by projection onto the circle **D**: $x^2 + y^2 \leq a^2$. That is, let x and y be the parameters, with (x, y) running over **D** and parameterize **H** by

$$\mathbf{x}(x, y) = (x, y, \sqrt{a^2 - x^2 - y^2}).$$

This parameterization determines the outward (upward) normal **n**, because the z-component of **n** is

$$\frac{\partial(x, y)}{\partial(x, y)} = \begin{vmatrix} 1 & 0 \\ 0 & 1 \end{vmatrix} = 1 > 0.$$

Since this is *opposite* to the given orientation,

$$\iint_{\mathbf{H}} dx\, dy = -\iint_{\mathbf{D}} dx\, dy = -|\mathbf{D}| = -\pi a^2.$$

For the $dy\, dz$ term, break the hemisphere into two quarter spheres, $x \geq 0$ and $x \leq 0$. Parameterize each by projection onto the semi-circle **E** in the y, z-plane: $y^2 + z^2 \leq a^2$, $z \geq 0$. For the quarter sphere $x \geq 0$, this parameterization defines the outward normal; for the quarter sphere $x \leq 0$, it defines the inward normal. (The key fact is that the x-component of **n** is 1, proved by the same reasoning as above.) Therefore

$$\iint_{\mathbf{H}} dy\, dz = -\iint_{\mathbf{E}} dy\, dz + \iint_{\mathbf{E}} dy\, dz = 0.$$

Likewise $\qquad \iint_{\mathbf{H}} dz\, dx = 0,$ so $\Phi = \iint_{\mathbf{H}} dx\, dy = -\pi a^2.$

Alternative Solution   Parameterize **H** by spherical coordinates ϕ, θ:

$$x = a \sin \phi \cos \theta, \qquad y = a \sin \phi \sin \theta, \qquad z = a \cos \phi,$$

where $0 \le \theta \le 2\pi$, $0 \le \phi \le \tfrac{1}{2}\pi$. Then

$$\frac{\partial(y, z)}{\partial(\phi, \theta)} = \begin{vmatrix} a \cos \phi \sin \theta & a \sin \phi \cos \theta \\ -a \sin \phi & 0 \end{vmatrix} = a^2 \sin^2 \phi \cos \theta,$$

$$\frac{\partial(z, x)}{\partial(\phi, \theta)} = \begin{vmatrix} -a \sin \phi & 0 \\ a \cos \phi \cos \theta & -a \sin \phi \sin \theta \end{vmatrix} = a^2 \sin^2 \phi \sin \theta,$$

$$\frac{\partial(x, y)}{\partial(\phi, \theta)} = \begin{vmatrix} a \cos \phi \cos \theta & -a \sin \phi \sin \theta \\ a \cos \phi \sin \theta & a \sin \phi \cos \theta \end{vmatrix} = a^2 \sin \phi \cos \phi.$$

Therefore

$$d\boldsymbol{\sigma} = (dy\, dz,\, dz\, dx,\, dx\, dy). = (a^2 \sin^2 \phi \cos \theta,\, a^2 \sin^2 \phi \sin \theta,\, a^2 \sin \phi \cos \phi)\, d\phi\, d\theta.$$

In the first octant $(0 < \theta < \tfrac{1}{2}\pi,\ 0 < \phi < \tfrac{1}{2}\pi)$, all components are positive, so the parameterization yields the *outward* normal, the wrong one, and we must change sign. We have

$$\Phi = \iint_{\mathbf{H}} 2\, dy\, dz + 3\, dz\, dx + dx\, dy$$

$$= -\iint_{\mathbf{D}} (2a^2 \sin^2 \phi \cos \theta + 3a^2 \sin^2 \phi \sin \theta + a^2 \sin \phi \cos \phi)\, d\phi\, d\theta,$$

where **D** is the rectangle $0 \le \theta \le 2\pi$, $0 \le \phi \le \tfrac{1}{2}\pi$ in the ϕ, θ-plane. Clearly

$$\iint_{\mathbf{D}} \sin^2 \phi \cos \theta\, d\phi\, d\theta = \int_0^{2\pi} \cos \theta\, d\theta \int_0^{\pi/2} \sin^2 \phi\, d\phi = 0$$

and similarly,

$$\iint_{\mathbf{D}} \sin^2 \phi \sin \theta\, d\phi\, d\theta = 0,$$

hence

$$\Phi = -\iint_{\mathbf{D}} a^2 \sin \phi \cos \phi\, d\phi\, d\theta = -2\pi a^2 \int_0^{\pi/2} \sin \phi \cos \phi\, d\phi$$

$$= -2\pi a^2 (\tfrac{1}{2}) = -\pi a^2. \qquad \blacksquare$$

Divergence   We now pave the way towards an important theorem that relates surface integrals over closed surfaces to certain triple integrals. It is a generalization of Green's Theorem for line integrals. First we need a definition.

Divergence   Let $\mathbf{A} = \mathbf{A}(\mathbf{x}) = (A, B, C)$ be a differentiable vector field on a domain in $\mathbf{R}^3$. The **divergence** of $\mathbf{A}$ is the real-valued function

$$\operatorname{div} \mathbf{A}(\mathbf{x}) = \frac{\partial A}{\partial x} + \frac{\partial B}{\partial y} + \frac{\partial C}{\partial z}.$$

Examples $\qquad$ div $\mathbf{x} = \operatorname{div}(x, y, z) = 1 + 1 + 1 = 3,$

$$\operatorname{div}(x^3,\, xyz,\, xz^2) = \frac{\partial}{\partial x}(x^3) + \frac{\partial}{\partial y}(xyz) + \frac{\partial}{\partial z}(xz^2) = 3x^2 + xz + 2xz = 3(x^2 + xz).$$

There is an important relation satisfied by the divergence operator:

$$\operatorname{div}(f\mathbf{A}) = (\operatorname{grad} f)\,\mathbf{A} + f \operatorname{div} \mathbf{A}.$$

Its proof is left for Ex. 13.

The Divergence Theorem Now we are ready for the principal result of this section, a formula which has numerous applications in mathematics and in the physical sciences.

Divergence Theorem Let $\mathbf{A} = \mathbf{A}(\mathbf{x})$ be a continuously differentiable vector field on a domain $\mathbf{D}$ in $\mathbf{R}^3$. Orient the boundary $\partial\mathbf{D}$ by the outward normal. Then

$$\iint_{\partial\mathbf{D}} \mathbf{A} \cdot d\boldsymbol{\sigma} = \iiint_{\mathbf{D}} (\operatorname{div} \mathbf{A})\, dV.$$

In coordinate notation,

$$\iint_{\partial\mathbf{D}} A\, dy\, dz + B\, dz\, dx + C\, dx\, dy = \iiint_{\mathbf{D}} \left(\frac{\partial A}{\partial x} + \frac{\partial B}{\partial y} + \frac{\partial C}{\partial z} \right) dx\, dy\, dz.$$

As with Green's Theorem, we shall give only a partial proof. We shall prove

$$\iint_{\partial\mathbf{D}} C\, dx\, dy = \iiint_{\mathbf{D}} \frac{\partial C}{\partial z}\, dx\, dy\, dz$$

when $\mathbf{D}$ is the domain between two graphs of functions of (x, y):

$$\mathbf{D} = \big\{ (x, y, z), \quad \text{where} \quad (x, y) \text{ is in } \mathbf{E} \quad \text{and} \quad g(x, y) \leq z \leq h(x, y) \big\}.$$

Here $\mathbf{E}$ is a domain in the x, y-plane (Fig. 5a).

The boundary of $\mathbf{D}$ splits naturally into top, bottom, and lateral side (Fig. 5b):

$$\partial\mathbf{D} = \mathbf{T} + \mathbf{B} + \mathbf{S}.$$

We observe first that $\qquad \displaystyle\iint_{\mathbf{S}} C\, dx\, dy = 0.$

For, by definition, $\qquad \displaystyle\iint_{\mathbf{S}} C\, dx\, dy = \iint_{\mathbf{S}} (0, 0, C) \cdot \mathbf{n}\, d\sigma.$

But on the *vertical* side $\mathbf{S}$, the normal vector $\mathbf{n}$ is parallel to the x, y-plane, so its z-component is 0. Therefore, $(0, 0, C) \cdot \mathbf{n} = 0$.

Next we parameterize both the top $\mathbf{T}$ and the bottom $\mathbf{B}$ by projection onto $\mathbf{E}$. This parameterization gives the outward normal on $\mathbf{T}$ and the inward normal on $\mathbf{B}$.

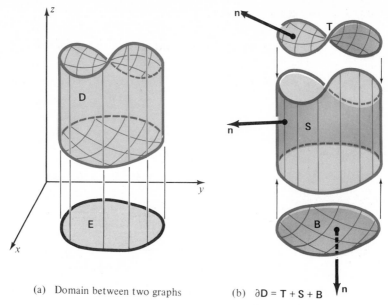

(a) Domain between two graphs   (b) ∂**D** = **T** + **S** + **B**

Fig. 5 Proof of the Divergence Theorem

Therefore

$$\iint_{\partial \mathbf{D}} C \, dx \, dy = \left(\iint_{\mathbf{T}} + \iint_{\mathbf{B}} + \iint_{\mathbf{S}} \right) C \, dx \, dy = \iint_{\mathbf{T}} C \, dx \, dy + \iint_{\mathbf{B}} C \, dx \, dy$$

$$= \iint_{\mathbf{E}} C[x, y, h(x, y)] \, dx \, dy - \iint_{\mathbf{E}} C[x, y, g(x, y)] \, dx \, dy.$$

On the other hand, we iterate the triple integral:

$$\iiint_{\mathbf{D}} \frac{\partial C}{\partial z} \, dx \, dy \, dz = \iint_{\mathbf{E}} \left(\int_{g(x, y)}^{h(x, y)} \frac{\partial C}{\partial z} \, dz \right) dx \, dy$$

$$= \iint_{\mathbf{E}} \left(C(x, y, z) \Big|_{z = g(x, y)}^{z = h(x, y)} \right) dx \, dy = \iint_{\mathbf{E}} \left\{ C[x, y, h(x, y)] - C[x, y, g(x, y)] \right\} dx \, dy.$$

The results are equal so the proof is complete.

Remark   The Divergence Theorem is also known as Gauss's Theorem.

Volume Formula   Just as Green's Theorem implies the Area Formula, so does the Divergence Theorem imply a formula for the volume of a domain in terms of an integral over its boundary.

Volume Formula   Let **D** be a domain in **R**^{3}. Then

$$|\mathbf{D}| = \frac{1}{3} \iint_{\partial \mathbf{D}} x \, dy \, dz + y \, dz \, dx + z \, dx \, dy = \frac{1}{3} \iint_{\partial \mathbf{D}} \mathbf{x} \cdot d\boldsymbol{\sigma}.$$

Proof By the Divergence Theorem,

$$\iint_{\partial \mathbf{D}} x \, dy \, dz + y \, dz \, dx + z \, dx \, dy = \iiint_{\mathbf{D}} \operatorname{div}(x, y, z) \, dV = \iiint_{\mathbf{D}} 3 \, dV = 3 \, |\mathbf{D}|.$$

■ **EXAMPLE 3** Find the volume V of the ellipsoid

$$\frac{x^2}{a^2} + \frac{y^2}{b^2} + \frac{z^2}{c^2} \le 1.$$

Solution Parameterize the boundary of the ellipsoid by

$$x = a \sin \phi \cos \theta, \qquad y = b \sin \phi \sin \theta, \qquad z = c \cos \phi,$$

where $0 \le \theta \le 2\pi$, $0 \le \phi \le \pi$. Then

$$x \, dy \, dz + y \, dz \, dx + z \, dx \, dy$$

$$= \begin{vmatrix} x & y & z \\ x_\phi & y_\phi & z_\phi \\ x_\theta & y_\theta & z_\theta \end{vmatrix} d\phi \, d\theta = abc \begin{vmatrix} \sin \phi \cos \theta & \sin \phi \sin \theta & \cos \phi \\ \cos \phi \cos \theta & \cos \phi \sin \theta & -\sin \phi \\ -\sin \phi \sin \theta & \sin \phi \cos \theta & 0 \end{vmatrix} d\phi \, d\theta$$

$$= abc \sin \phi \, d\phi \, d\theta.$$

Therefore by the Volume Theorem,

$$V = \frac{1}{3} \iint x \, dy \, dz + y \, dz \, dx + z \, dx \, dy = \frac{1}{3} abc \iint \sin \phi \, d\phi \, d\theta$$

$$= \frac{1}{3} abc \int_0^{2\pi} d\theta \int_0^\pi \sin \phi \, d\phi = \frac{1}{3} abc (2\pi)(2) = \frac{4}{3} \pi abc. \qquad ■$$

Inverse Square Law Suppose a closed surface $\mathbf{S}$ bounds a domain that includes the origin (Fig. 6a). We want to compute the flux

$$\Phi = \iint_{\mathbf{S}} \mathbf{F} \cdot d\boldsymbol{\sigma}$$

where the force field $\mathbf{F}$ obeys the inverse square law:

$$\mathbf{F} = \frac{1}{\rho^3} \mathbf{x}, \qquad \rho = |\mathbf{x}|.$$

First suppose $\mathbf{S}$ is the sphere $\rho = a$. Then

$$\iint_{\mathbf{S}} \mathbf{F} \cdot d\boldsymbol{\sigma} = \frac{1}{a^3} \iint_{\mathbf{S}} \mathbf{x} \cdot d\boldsymbol{\sigma} = \frac{1}{a^3} (3V)$$

according to the Volume Theorem, where V is the volume of the ball $\rho \le a$, that is, $V = \frac{4}{3}\pi a^3$. Hence

$$\iint_{\rho = a} \mathbf{F} \cdot d\boldsymbol{\sigma} = 4\pi.$$

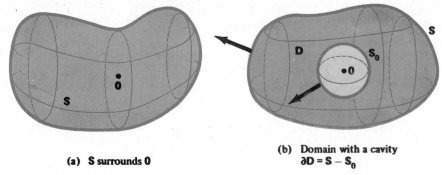

(a) **S** surrounds **0**

(b) **Domain with a cavity**
$\partial \mathbf{D} = \mathbf{S} - \mathbf{S_0}$

Fig. 6 Flux of inverse square law field through a closed surface **S**

Note that $\Phi = 4\pi$ no matter what the radius a is.

Let us pass to the general case. We shall prove a remarkable fact: $\Phi = 4\pi$ for every closed surface **S** that includes the origin in its interior. We let **D** be the domain inside of **S** and outside of a small sphere $\mathbf{S_0}$ with center **0**. See Fig. 6b. We take the *outward* normal on $\mathbf{S_0}$. Then the oriented boundary of **D** is

$$\partial \mathbf{D} = \mathbf{S} - \mathbf{S_0},$$

hence by the Divergence Theorem

$$\iint_{\mathbf{S}} \mathbf{F} \cdot d\boldsymbol{\sigma} - \iint_{\mathbf{S_0}} \mathbf{F} \cdot d\boldsymbol{\sigma} = \iint_{\partial \mathbf{D}} \mathbf{F} \cdot d\boldsymbol{\sigma} = \iiint_{\mathbf{D}} (\text{div } \mathbf{F})\, dV.$$

We must compute div **F**. Now $\rho^2 = x^2 + y^2 + z^2$, hence $\rho \rho_x = x,\ \rho \rho_y = y,\ \rho \rho_z = z$, and

$$\text{div } \mathbf{F} = \frac{\partial}{\partial x}\left(\frac{x}{\rho^3}\right) + \frac{\partial}{\partial y}\left(\frac{y}{\rho^3}\right) + \frac{\partial}{\partial z}\left(\frac{z}{\rho^3}\right) = \frac{\rho^3 - 3\rho x^2}{\rho^6} + \frac{\rho^3 - 3\rho y^2}{\rho^6} + \frac{\rho^3 - 3\rho z^2}{\rho^3}$$

$$= \frac{3\rho^3 - 3\rho(x^2 + y^2 + z^2)}{\rho^6} = \frac{3\rho^3 - 3\rho^3}{\rho^6} = 0.$$

It follows that

$$\iint_{\mathbf{S}} \mathbf{F} \cdot d\boldsymbol{\sigma} = \iint_{\mathbf{S_0}} \mathbf{F} \cdot d\boldsymbol{\sigma} = 4\pi.$$

The flux is constant, independent of the surface **S**.

Let **S** be a closed surface in $\mathbf{R}^3$ that surrounds **0**, and take its normal outward. Then

$$\iint_{\mathbf{S}} \frac{x\, dy\, dz + y\, dz\, dx + z\, dx\, dy}{(x^2 + y^2 + z^2)^{3/2}} = 4\pi.$$

Stokes's Theorem   We close with a brief mention of another important result relating integrals. It may also be considered as a generalization of Green's Theorem. We first require the definition of the curl (or rotation) of a vector field.

Curl Let $\mathbf{F} = (P, Q, R)$ be a differentiable vector field on a domain in $\mathbf{R}^3$. The **curl** of $\mathbf{F}$ is the *vector field*

$$\text{curl } \mathbf{F}(\mathbf{x}) = (R_y - Q_z, P_z - R_x, Q_x - P_y).$$

Example $\text{curl}(y^2 + z^2, z^2 + x^2, x^2 + y^2) = (2y - 2z, 2z - 2x, 2x - 2y).$

Stokes's Theorem Let $\mathbf{F} = \mathbf{F}(\mathbf{x})$ be a continuously differentiable vector field in a region of space and let $\mathbf{S}$ be an oriented surface in that region. Orient the boundary $\partial \mathbf{S}$ of $\mathbf{S}$ to be consistent with the orientation of $\mathbf{S}$. Then

$$\oint_{\partial \mathbf{S}} \mathbf{F} \cdot d\mathbf{x} = \iint_{\mathbf{S}} (\text{curl } \mathbf{F}) \cdot d\boldsymbol{\sigma}.$$

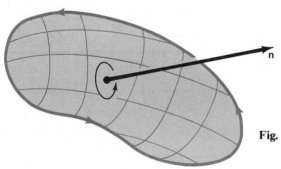

Fig. 7 Orientation of the boundary $\partial \mathbf{S}$ of a directed surface $\mathbf{S}$: right-hand rule

A proof of this is best left to a later course. See Fig. 7 for the orientation of the boundary.

EXERCISES

Evaluate the surface integral over the indicated portion of the surface; always take the normal to have a non-negative z-component

1 $\iint x \, dy \, dz + y \, dz \, dx + z \, dx \, dy$

    plane $2x - 2y + z = 3$
    $0 \le x \le 1$ $0 \le y \le 1$

2 $\iint x^2 \, dy \, dz + xy \, dz \, dx$

    plane $x + 2y + 3z = 1$
    $x \ge 0$, $y \ge 0$, $x + y \le 1$

3 $\iint xz \, dy \, dz$

    sphere $\rho = a$ $z \ge 0$

4 $\iint 3x \, dy \, dz + 2z^2 \, dx \, dy$

    paraboloid $z = x^2 + 2y^2$
    $x^2 + y^2 \le 1$

5 $\iint z \, dy \, dz + z \, dz \, dx + dx \, dy$

    cone $z = \sqrt{x^2 + y^2}$ $x^2 + y^2 \le 4$
    $x \ge 0$

6 $\iint z \, dy \, dz + z \, dz \, dx + dx \, dy$

    cone $z = 2 - \sqrt{x^2 + y^2}$
    $x^2 + y^2 \le 4$

7 $\iint 4\,dy\,dz - 3\,dz\,dx$

cylinder $z = \sqrt{1-x^2}$ $0 \le x \le 1$
$0 \le y \le 2$

8 $\iint x^2\,dy\,dz$

cylinder $z = -\sqrt{1-x^2}$
$-1 \le x \le 0$ $-1 \le y \le 1$

9 $\iint x^6\,dy\,dz + z^5\,dx\,dy$

cylinder $x^2 + y^2 = 1$ $0 \le z \le 1$
(outward normal)

10 $\iint x\,dy\,dz + y\,dz\,dx + z\,dx\,dy$

cylinder $x^2 + y^2 = a^2$ $0 \le z \le h$
(inward normal)

11 $\iint x^n\,dy\,dz$ sphere $\rho = a$

12 $\iint y^n\,dy\,dz$ sphere $\rho = a$.

13 Prove the formula $\operatorname{div}(f\mathbf{A}) = (\operatorname{grad} f)\cdot\mathbf{A} + f\operatorname{div}\mathbf{A}$.

14 Prove the following vector relation, a consequence of the Divergence Theorem:

$$\iint_{\partial \mathbf{D}} f\,d\boldsymbol{\sigma} = \iiint_{\mathbf{D}} (\operatorname{grad} f)\,dV.$$

Note that both sides are vectors. [*Hint* Dot an arbitrary constant vector into both sides.]

15 (Archimedes' Principle) A body $\mathbf{D}$ is completely submerged in a fluid of constant density δ. Show that the buoyant force on $\mathbf{D}$ due to the surrounding fluid pressure equals the weight of the fluid that $\mathbf{D}$ displaces. Recall that the pressure at depth z equals the weight of a vertical column of fluid of unit cross-sectional area over z, and use Ex. 14.

16 (cont.) Complete the proof of Archimedes' Principle by deriving the same result for a partly submerged body.

Evaluate over a closed surface

17 $\iint_{\mathbf{S}} d\boldsymbol{\sigma}$

[*Hint* Use Ex. 14.]

18* $\iint_{\mathbf{S}} \mathbf{x} \times d\boldsymbol{\sigma}$.

[*Hint* Use the hint in Ex. 14.]

The gravitational force of a mass m_1 at $\mathbf{x}_1$ on a mass m_2 at $\mathbf{x}_2$ is

$$\mathbf{F} = Gm_1 m_2 \frac{\mathbf{x}_1 - \mathbf{x}_2}{|\mathbf{x}_1 - \mathbf{x}_2|^3}.$$

The gravitational attraction of one rigid body on another is obtained from this basic inverse square law of Newton by integration.

19 Find the gravitational attraction $\mathbf{F}$ on a unit mass at $(0, 0, c)$ of the uniform disk $x^2 + y^2 \le a^2$, $z = 0$, of density δ.

20 (cont.) Find $\lim_{a\to\infty} \mathbf{F}$.

21 (cont.) Express $\mathbf{F}$ in terms of the mass M of the disk instead of δ. Find $\lim_{a\to 0} \mathbf{F}$, assuming M constant.

22 Hold a closed "ideally" transparent surface near a point source of light. Light rays pass in at some places and out at others; none remains inside. Explain that by mathematics.

23 The solid sphere $\rho \le a$ has density $\delta = \delta(\rho)$ with radial symmetry. Using symmetry only, what can you say about the gravitational force field $\mathbf{F} = \mathbf{F}(\mathbf{x})$ of the sphere on a unit mass at a typical point $\mathbf{x}$ of space. Your answer should involve $\mathbf{x}$ and $\rho = |\mathbf{x}|$.

24* (cont.) Let $\mathbf{S}$ be *any* closed surface that includes the sphere in its interior. What is the flux of $\mathbf{F}$ over $\mathbf{S}$?

25* (cont.) Now choose $\mathbf{S}$ carefully to prove the famous result of Newton, $\mathbf{F} = -GM\mathbf{x}/|\mathbf{x}|^3$

for **x** outside of the sphere. It says that the gravitational attraction of the sphere on an external point **x** is the same as that of its whole mass concentrated at its center.

26 (cont.) Find the gravitational attraction of the sphere on a point **x** *inside* the sphere. Take the case of constant density δ.

27 (cont.) Suppose that a spherical planet of radius a consists of a homogeneous fluid of density δ. Find the pressure $p = p(\rho)$ in the planet at distance ρ from its center. Then find the pressure at the center.

Prove

28 $\operatorname{div}[f(\rho)\mathbf{x}] = [\rho^3 f(\rho)]'/\rho^2, \quad \rho = |\mathbf{x}|$ 29 $\operatorname{curl}(f\mathbf{A}) = (\operatorname{grad} f) \times \mathbf{A} + f \operatorname{curl} \mathbf{A}$

30 $\operatorname{curl}(\operatorname{grad} f) = \mathbf{0}$ 31 $\operatorname{curl}[f(\rho)\mathbf{x}] = \mathbf{0}$

32 $\operatorname{div}(\operatorname{curl} \mathbf{A}) = 0$ 33 $\operatorname{curl}(\mathbf{a} \times \mathbf{x}) = 2\mathbf{a}, \quad \mathbf{a} \text{ constant}$

34 $\operatorname{div}[\operatorname{grad} f(\rho)] = [\rho f(\rho)]''/\rho$.

The **Laplacian** Δf of a function f is defined by

$$\Delta f = \frac{\partial^2 f}{\partial x^2} + \frac{\partial^2 f}{\partial y^2} + \frac{\partial^2 f}{\partial z^2}.$$

35 Express Δ in terms of grad and div.

36 Express $\operatorname{div}(f \operatorname{grad} g - g \operatorname{grad} f)$ in terms of Laplacians.

37 Compute $\Delta(\rho^n)$.

38 Suppose $\operatorname{div} \mathbf{A} = 0$ and $\operatorname{curl} \mathbf{A} = 0$. Prove $\Delta A = \Delta B = \Delta C = 0$, where $\mathbf{A} = (A, B, C)$.

39 Suppose **D** is rotating about an axis with angular velocity $\boldsymbol{\omega}$. (Possibly $\boldsymbol{\omega}$ varies with time.) Let $\mathbf{v} = \mathbf{v}(\mathbf{x})$ denote the velocity of **x** in **D**. Prove $\boldsymbol{\omega} = \frac{1}{2} \operatorname{curl} \mathbf{v}$. [*Hint* Express **v** in terms of $\boldsymbol{\omega}$ and **x** and use Ex. 33.]

40* From Ex. 30 we know that a necessary condition for a vector field **A** to be a gradient is curl $\mathbf{A} = \mathbf{0}$. Find a necessary condition (involving **A** alone) that $\mathbf{A} = g \operatorname{grad} f$, where f and g are functions and $g \neq 0$.

Set $\mathbf{F} = (y^2 z, z^2 x, x^2 y)$. Compute both sides of Stokes's Theorem and verify their equality

41 $\mathbf{S} = \{\rho = 1, \quad x \geq 0, \quad y \geq 0, \quad z \geq 0\}$, spherical octant

42 $\mathbf{S} = \{z = r, \quad 0 \leq r \leq 1\}$, cone

43 $\mathbf{S} = \{r = 1, 0 \leq z \leq 1\}$, cylinder

44 $\mathbf{S} = \{(1 - z)^2 = x^2 + y^2, \quad z \geq 0, \quad x^2 + y^2 \leq 1\}$, cone.

9. MISCELLANEOUS EXERCISES

1 Two parallel planes at distance h intersect a sphere of radius a. Find the surface area of the spherical zone between the planes.

2 Find the center of gravity of the uniform spiral $\mathbf{x}(t) = (a \cos t, a \sin t, bt), 0 \leq t \leq T$.

3 Suppose $f[\rho, \phi, \theta]$ is homogeneous of degree n with respect to *rectangular* coordinates, that is, $f(t\mathbf{x}) = t^n f(\mathbf{x})$ for all $t > 0$. Prove $f[\rho, \phi, \theta] = \rho^n g(\phi, \theta)$, where $g(\phi, \theta) = f[1, \phi, \theta]$.

4 Evaluate $\oint \frac{3x^2 y^2}{z} \, dx + \frac{2x^3 y}{z} \, dy - \frac{x^3 y^2}{z^2} \, dz$ over the curve
$\mathbf{x} = (e^{\sin t} \cos t, e^{\cos t} \sin t, 2 + \sin 7t), 0 \leq t \leq 2\pi$.

5 Evaluate $\iiint (x^n + y^n + z^n) \, dx \, dy \, dz$ over $\rho \leq a$.

6 Satellite observations determine that the moment of inertia of Earth about its axis of

rotation is $I = \frac{1}{3}Ma^2$, where M is the mass and a the radius of Earth. Assume Earth is a sphere whose density δ at any point $\mathbf{x}$ depends only on $\rho = |\mathbf{x}|$ and is linear: $\delta = ab - c\rho$, where b and c are constants. Prove that the density at the center is five times the surface density.

7 **Toroidal coordinates** ρ, ϕ, θ of $\mathbf{x}$ in $\mathbf{R}^3$ are defined by the change of variables

$$x = (A + \rho \sin \phi) \cos \theta, \qquad y = (A + \rho \sin \phi) \sin \theta, \qquad z = \rho \cos \phi.$$

Here $0 \le \rho \le a$, where $a \le A$, $0 \le \phi \le 2\pi$, $0 \le \theta \le 2\pi$. What are the level surfaces $\rho = $ const., $\phi = $ const., $\theta = $ const.? Find their mutal angles of intersection.

8 (cont.) Evaluate the Jacobian $\dfrac{\partial(x, y, z)}{\partial(\rho, \phi, \theta)}$.

9 Decribe the domain $0 \le x \le y \le 1$, $0 \le z \le y$ and find its volume.

10 The infinite wire $x = y = 0$, $-\infty < z < \infty$ has uniform density δ. Find its gravitational attraction on a unit mass at $(b, 0, 0)$.

11 (cont.) The infinite cylinder $x^2 + y^2 = a^2$, $-\infty < z < \infty$ has uniform density δ. Find its gravitational attraction on a unit mass at $(b, 0, 0)$, where $b > a$.

$$\left[Hint \quad \int_0^{2\pi} \frac{d\theta}{A + B \cos \theta} = \frac{2\pi}{\sqrt{A^2 - B^2}}. \right]$$

12 (cont.) What is the gravitational force if $0 < b < a$?

13 By a long series of steps involving reduction formulas one arrives at the formulas

$$\int_0^{\pi/2} \sin^{2n} \theta \, d\theta = \frac{(2n)!}{2^{2n}(n!)^2} \frac{\pi}{2} \qquad \text{and} \qquad \int_0^{\pi/2} \sin^{2n+1} \theta \, d\theta = \frac{2^{2n}(n!)^2}{(2n+1)!}.$$

A consequence is the simple looking formula

$$\left(\int_0^{\pi/2} \sin^{2n} \theta \, d\theta \right) \left(\int_0^{\pi/2} \sin^{2n+1} \theta \, d\theta \right) = \frac{\pi}{2(2n+1)}.$$

Use multiple integrals to give a simple proof of the latter formula.

14 If $\mathbf{v} = (v_1, v_2, v_3)$ is a vector field, its divergence is div $\mathbf{v} = \sum \partial v_i / \partial x_i$. Suppose we take another rectangular coordinate system (with the same origin). Then its coordinates are $\bar{x}_i = \sum a_{ij} x_j$, where the a_{ij} are constants. Express $\mathbf{v}$ in the new coordinate system, and show that $\overline{\text{div } \mathbf{v}} = \text{div } \mathbf{v}$, that is, that div $\mathbf{v}$ is a geometric quantity, independent of the coordinate system.

15* Find in any way the volume of the solid $x^2 + y^2 \le a^2$, $|y + z| \le b$, $|y - z| \le b$, where $0 < b \le a$. Note that the solid may be interpreted as a square hole on center through a cylinder, the sides of the hole at $45°$ to the axis of the cylinder.

16 A cylindrical hole of radius a is bored through a solid cylinder of radius $2a$; the hole is perpendicular to the solid cylinder and just touches a generator. Find in any way the volume removed.

17 From each point of the space curve $\mathbf{x} = \mathbf{x}(s)$ draw a segment of length 1 in the direction of the unit tangent. These segments sweep out a surface. Show that its area is $\frac{1}{2} \int k(s) \, ds$, where $k(s)$ is the curvature and the integral is taken over the length of the curve.

18 Find the volume of the four-dimensional sphere of radius a, $x^2 + y^2 + z^2 + w^2 \le a^2$.

19 A regular icosahedron has its center at the origin. Find the solid angle subtended by each of its faces.

20* A rigid body $\mathbf{D}$ is rotating about an axis through $\mathbf{0}$ with angular velocity $\boldsymbol{\omega}$. If $\mathbf{v} = \mathbf{v}(\mathbf{x})$ is the velocity at $\mathbf{x}$, the **angular momentum** of the rotating body is

$\mathbf{J} = \iiint\limits_{\mathbf{D}} \mathbf{x} \times \mathbf{v} \, dM$. Prove

$$\mathbf{J} = \boldsymbol{\omega} \begin{bmatrix} I_x & I_{xy} & I_{xz} \\ I_{yx} & I_y & I_{yz} \\ I_{zx} & I_{zy} & I_z \end{bmatrix},$$

the product of the row vector $\boldsymbol{\omega}$ with the inertia matrix.

21 A vector field $\mathbf{A}$ and a function f are defined on a region in space. Suppose for each $\mathbf{x}$ inside the region and for each $\varepsilon > 0$ there is a domain $\mathbf{D}$ containing $\mathbf{x}$ such that $\mathrm{rad}(\mathbf{D}) < \varepsilon$, $\mathbf{D}$ lies in the region, and

$$\iint\limits_{\partial\mathbf{D}} \mathbf{A} \cdot d\boldsymbol{\sigma} = \iiint\limits_{\mathbf{D}} f \, dV.$$

Prove $f = \mathrm{div}\ \mathbf{A}$. Assume f continuous and $\mathbf{A}$ continuously differentiable.

In the next 7 exercises we shall obtain the Laplacian $\Delta f = \mathrm{div}(\mathrm{grad}\ f)$ in spherical coordinates. (See Ex. 35, p. 949.) We need the result of Ex. 21 and the following formula:

$$\mathrm{grad}\ f = f_\rho\, \boldsymbol{\lambda} + \frac{1}{\rho}\, f_\phi\, \boldsymbol{\mu} + \frac{1}{\rho \sin \phi}\, f_\theta\, \mathbf{v} \qquad (\text{Ex. 23, p. 914}).$$

Let $\mathbf{D}$ denote the domain $\rho_0 \le \rho \le \rho_1$, $\phi_0 \le \phi \le \phi_1$, $\theta_0 \le \theta \le \theta_1$, with the outward normal on $\partial\mathbf{D}$.

22 Compute $d\boldsymbol{\sigma}$ on all six faces of $\partial\mathbf{D}$.

23 Prove $\displaystyle \iint\limits_{\partial\mathbf{D}} (A\boldsymbol{\lambda}) \cdot d\boldsymbol{\sigma} = \iiint\limits_{\mathbf{D}} \frac{1}{\rho^2} \frac{\partial}{\partial\rho} (\rho^2 A) \, dV$.

24 Obtain a similar formula for $\displaystyle \iint\limits_{\partial\mathbf{D}} (B\boldsymbol{\mu}) \cdot d\boldsymbol{\sigma}$.

25 Obtain a similar formula for $\displaystyle \iiint\limits_{\partial\mathbf{D}} (C\mathbf{v}) \cdot d\boldsymbol{\sigma}$.

26 Assemble the last three steps into one; let $\mathbf{A} = A\boldsymbol{\lambda} + B\boldsymbol{\mu} + C\mathbf{v}$ and express $\displaystyle \iint\limits_{\partial\mathbf{D}} \mathbf{A} \cdot d\boldsymbol{\sigma}$ as an integral over $\mathbf{D}$.

27 Use Ex. 21 and the result of the last exercise to express $\mathrm{div}\ \mathbf{A}$ in terms of spherical coordinates.

28 Apply the result to $\mathbf{A} = \mathrm{grad}\ f$ to obtain Δf in terms of spherical coordinates.

29 (cont.) The **Yukawa potential** is the function $f(\rho) = e^{-k\rho}/\rho$ with k constant. Show that $\Delta f = k^2 f$.

30* Use the result of Ex. 28 to show that if $f[\rho,\ \phi,\ \theta]$ satisfies $\Delta f = 0$, then its **Kelvin Transform**

$$g[\rho,\ \phi,\ \theta] = \frac{1}{\rho}\, f\left[\frac{1}{\rho},\ \phi,\ \theta\right]$$

satisfies $\Delta g = 0$.

Numerical Tables

N	0	1	2	3	4	5	6	7	8	9
1.0	000	004	009	013	017	021	025	029	033	037
1.1	041	045	049	053	057	061	064	068	072	076
1.2	079	083	086	090	093	097	100	104	107	111
1.3	114	117	121	124	127	130	134	137	140	143
1.4	146	149	152	155	158	161	164	167	170	173
1.5	176	179	182	185	188	190	193	196	199	201
1.6	204	207	210	212	215	217	220	223	225	228
1.7	230	233	236	238	241	243	246	248	250	253
1.8	255	258	260	262	265	267	270	272	274	276
1.9	279	281	283	286	288	290	292	294	297	299
2.0	301	303	305	307	310	312	314	316	318	320
2.1	322	324	326	328	330	332	334	336	338	340
2.2	342	344	346	348	350	352	354	356	358	360
2.3	362	364	365	367	369	371	373	375	377	378
2.4	380	382	384	386	387	389	391	393	394	396
2.5	398	400	401	403	405	407	408	410	412	413
2.6	415	417	418	420	422	423	425	427	428	430
2.7	431	433	435	436	438	439	441	442	444	446
2.8	447	449	450	452	453	455	456	458	459	461
2.9	462	464	465	467	468	470	471	473	474	476
3.0	477	479	480	481	483	484	486	487	489	490
3.1	491	493	494	496	497	498	500	501	502	504
3.2	505	507	508	509	511	512	513	515	516	517
3.3	519	520	521	522	524	525	526	528	529	530
3.4	531	533	534	535	537	538	539	540	542	543
3.5	544	545	547	548	549	550	551	553	554	555
3.6	556	558	559	560	561	562	563	565	566	567
3.7	568	569	571	572	573	574	575	576	577	579
3.8	580	581	582	583	584	585	587	588	589	590
3.9	591	592	593	594	595	597	598	599	600	601
4.0	602	603	604	605	606	607	609	610	611	612
4.1	613	614	615	616	617	618	619	620	621	622
4.2	623	624	625	626	627	628	629	630	631	632
4.3	633	634	635	636	637	638	639	640	641	642
4.4	643	644	645	646	647	648	649	650	651	652
4.5	653	654	655	656	657	658	659	660	661	662
4.6	663	664	665	666	667	667	668	669	670	671
4.7	672	673	674	675	676	677	678	679	679	680
4.8	681	682	683	684	685	686	687	688	688	689
4.9	690	691	692	693	694	695	695	696	697	698
5.0	699	700	701	702	702	703	704	705	706	707
5.1	708	708	709	710	711	712	713	713	714	715
5.2	716	717	718	719	719	720	721	722	723	723
5.3	724	725	726	727	728	728	729	730	731	732
5.4	732	733	734	735	736	736	737	738	739	740
N	0	1	2	3	4	5	6	7	8	9

TABLE 1   Three-Place Mantissas for Common Logarithms *(Continued)*   **A3**

N	0	1	2	3	4	5	6	7	8	9
5.5	740	741	742	743	744	744	745	746	747	747
5.6	748	749	750	751	751	752	753	754	754	755
5.7	756	757	757	758	759	760	760	761	762	763
5.8	763	764	765	766	766	767	768	769	769	770
5.9	771	772	772	773	774	775	775	776	777	777
6.0	778	779	780	780	781	782	782	783	784	785
6.1	785	786	787	787	788	789	790	790	791	792
6.2	792	793	794	794	795	796	797	797	798	799
6.3	799	800	801	801	802	803	803	804	805	806
6.4	806	807	808	808	809	810	810	811	812	812
6.5	813	814	814	815	816	816	817	818	818	819
6.6	820	820	821	822	822	823	823	824	825	825
6.7	826	827	827	828	829	829	830	831	831	832
6.8	833	833	834	834	835	836	836	837	838	838
6.9	839	839	840	841	841	842	843	843	844	844
7.0	845	846	846	847	848	848	849	849	850	851
7.1	851	852	852	853	854	854	855	856	856	857
7.2	857	858	859	859	860	860	861	862	862	863
7.3	863	864	865	865	866	866	867	867	868	869
7.4	869	870	870	871	872	872	873	873	874	874
7.5	875	876	876	877	877	878	879	879	880	880
7.6	881	881	882	883	883	884	884	885	885	886
7.7	886	887	888	888	889	889	890	890	891	892
7.8	892	893	893	894	894	895	895	896	897	897
7.9	898	898	899	899	900	900	901	901	902	903
8.0	903	904	904	905	905	906	906	907	907	908
8.1	908	909	910	910	911	911	912	912	913	913
8.2	914	914	915	915	916	916	917	918	918	919
8.3	919	920	920	921	921	922	922	923	923	924
8.4	924	925	925	926	926	927	927	928	928	929
8.5	929	930	930	931	931	932	932	933	933	934
8.6	934	935	936	936	937	937	938	938	939	939
8.7	940	940	941	941	942	942	943	943	943	944
8.8	944	945	945	946	946	947	947	948	948	949
8.9	949	950	950	951	951	952	952	953	953	954
9.0	954	955	955	956	956	957	957	958	958	959
9.1	959	960	960	960	961	961	962	962	963	963
9.2	964	964	965	965	966	966	967	967	968	968
9.3	968	969	969	970	970	971	971	972	972	973
9.4	973	974	974	975	975	975	976	976	977	977
9.5	978	978	979	979	980	980	980	981	981	982
9.6	982	983	983	984	984	985	985	985	986	986
9.7	987	987	988	988	989	989	989	990	990	991
9.8	991	992	992	993	993	993	994	994	995	995
9.9	996	996	997	997	997	998	998	999	999	1000
N	0	1	2	3	4	5	6	7	8	9

TABLE 2 Exponential Functions

x	exp(x)		exp(−x)		x	exp(x)		exp(−x)	
0.0	1.00	+0	1.00	+0	5.0	1.48	+2	6.74	−3
.1	1.11	+0	9.05	−1	5.1	1.64	+2	6.10	−3
.2	1.22	+0	8.19	−1	5.2	1.81	+2	5.52	−3
.3	1.35	+0	7.41	−1	5.3	2.00	+2	4.99	−3
.4	1.49	+0	6.70	−1	5.4	2.21	+2	4.52	−3
.5	1.65	+0	6.07	−1	5.5	2.45	+2	4.09	−3
.6	1.82	+0	5.49	−1	5.6	2.70	+2	3.70	−3
.7	2.01	+0	4.97	−1	5.7	2.99	+2	3.35	−3
.8	2.23	+0	4.49	−1	5.8	3.30	+2	3.03	−3
.9	2.46	+0	4.07	−1	5.9	3.65	+2	2.74	−3
1.0	2.72	+0	3.68	−1	6.0	4.03	+2	2.48	−3
1.1	3.00	+0	3.33	−1	6.1	4.46	+2	2.24	−3
1.2	3.32	+0	3.01	−1	6.2	4.93	+2	2.03	−3
1.3	3.67	+0	2.73	−1	6.3	5.45	+2	1.84	−3
1.4	4.06	+0	2.47	−1	6.4	6.02	+2	1.66	−3
1.5	4.48	+0	2.23	−1	6.5	6.65	+2	1.50	−3
1.6	4.95	+0	2.02	−1	6.6	7.35	+2	1.36	−3
1.7	5.47	+0	1.83	−1	6.7	8.12	+2	1.23	−3
1.8	6.05	+0	1.65	−1	6.8	8.98	+2	1.11	−3
1.9	6.69	+0	1.50	−1	6.9	9.92	+2	1.01	−3
2.0	7.39	+0	1.35	−1	7.0	1.10	+3	9.12	−4
2.1	8.17	+0	1.22	−1	7.1	1.21	+3	8.25	−4
2.2	9.03	+0	1.11	−1	7.2	1.34	+3	7.47	−4
2.3	9.97	+0	1.00	−1	7.3	1.48	+3	6.76	−4
2.4	1.10	+1	9.07	−2	7.4	1.64	+3	6.11	−4
2.5	1.22	+1	8.21	−2	7.5	1.81	+3	5.53	−4
2.6	1.35	+1	7.43	−2	7.6	2.00	+2	5.00	−4
2.7	1.49	+1	6.72	−2	7.7	2.21	+3	4.53	−4
2.8	1.64	+1	6.08	−2	7.8	2.44	+3	4.10	−4
2.9	1.82	+1	5.50	−2	7.9	2.70	+3	3.71	−4
3.0	2.01	+1	4.98	−2	8.0	2.98	+3	3.35	−4
3.1	2.22	+1	4.50	−2	8.1	3.29	+3	3.04	−4
3.2	2.45	+1	4.08	−2	8.2	3.64	+3	2.75	−4
3.3	2.71	+1	3.69	−2	8.3	4.02	+3	2.49	−4
3.4	3.00	+1	3.34	−2	8.4	4.45	+3	2.25	−4
3.5	3.31	+1	3.02	−2	8.5	4.91	+3	2.03	−4
3.6	3.66	+1	2.73	−2	8.6	5.43	+3	1.84	−4
3.7	4.04	+1	2.47	−2	8.7	6.00	+3	1.67	−4
3.8	4.47	+1	2.24	−2	8.8	6.63	+3	1.51	−4
3.9	4.94	+1	2.02	−2	8.9	7.33	+3	1.36	−4
4.0	5.46	+1	1.83	−2	9.0	8.10	+3	1.23	−4
4.1	6.03	+1	1.66	−2	9.1	8.96	+3	1.12	−4
4.2	6.67	+1	1.50	−2	9.2	9.90	+3	1.01	−4
4.3	7.37	+1	1.36	−2	9.3	1.09	+4	9.14	−5
4.4	8.15	+1	1.23	−2	9.4	1.21	+4	8.27	−5
4.5	9.00	+1	1.11	−2	9.5	1.34	+4	7.49	−5
4.6	9.95	+1	1.01	−2	9.6	1.48	+4	6.77	−5
4.7	1.10	+2	9.10	−3	9.7	1.63	+4	6.13	−5
4.8	1.22	+2	8.23	−3	9.8	1.80	+4	5.55	−5
4.9	1.34	+2	7.45	−3	9.9	1.99	+4	5.02	−5

x	exp(x)	exp(−x)	x	exp(x)	exp(−x)

TABLE 2 Exponential Functions *(Continued)* **A5**

x	exp(x)	exp(−x)	x	exp(x)	exp(−x)
10.0	2.20 +4	4.54 −5	50.0	5.18 +21	1.93 −22
11.0	5.99 +4	1.67 −5	51.0	1.41 +22	7.10 −23
12.0	1.63 +5	6.14 −6	52.0	3.83 +22	2.61 −23
13.0	4.42 +5	2.26 −6	53.0	1.04 +23	9.60 −24
14.0	1.20 +6	8.32 −7	54.0	2.83 +23	3.53 −24
15.0	3.27 +6	3.06 −7	55.0	7.69 +23	1.30 −24
16.0	8.89 +6	1.13 −7	56.0	2.09 +24	4.78 −25
17.0	2.42 +7	4.14 −8	57.0	5.69 +24	1.76 −25
18.0	6.57 +7	1.52 −8	58.0	1.55 +25	6.47 −26
19.0	1.78 +8	5.60 −9	59.0	4.20 +25	2.38 −26
20.0	4.85 +8	2.06 −9	60.0	1.14 +26	8.76 −27
21.0	1.32 +9	7.58 −10	61.0	3.10 +26	3.22 −27
22.0	3.58 +9	2.79 −10	62.0	8.44 +26	1.19 −27
23.0	9.74 +9	1.03 −10	63.0	2.29 +27	4.36 −28
24.0	2.65 +10	3.78 −11	64.0	6.24 +27	1.60 −28
25.0	7.20 +10	1.39 −11	65.0	1.69 +28	5.90 −29
26.0	1.96 +11	5.11 −12	66.0	4.61 +28	2.17 −29
27.0	5.32 +11	1.88 −12	67.0	1.25 +29	7.98 −30
28.0	1.45 +12	6.91 −13	68.0	3.40 +29	2.94 −30
29.0	3.93 +12	2.54 −13	69.0	9.25 +29	1.08 −30
30.0	1.07 +13	9.36 −14	70.0	2.52 +30	3.98 −31
31.0	2.90 +13	3.44 −14	71.0	6.84 +30	1.46 −31
32.0	7.90 +13	1.27 −14	72.0	1.86 +31	5.38 −32
33.0	2.15 +14	4.66 −15	73.0	5.05 +31	1.98 −32
34.0	5.83 +14	1.71 −15	74.0	1.37 +32	7.28 −33
35.0	1.59 +15	6.31 −16	75.0	3.73 +32	2.68 −33
36.0	4.31 +15	2.32 −16	76.0	1.01 +33	9.85 −34
37.0	1.17 +16	8.53 −17	77.0	2.76 +33	3.63 −34
38.0	3.19 +16	3.14 −17	78.0	7.50 +33	1.33 −34
39.0	8.66 +16	1.15 −17	79.0	2.04 +34	4.91 −35
40.0	2.35 +17	4.25 −18	80.0	5.54 +34	1.80 −35
41.0	6.40 +17	1.56 −18	81.0	1.51 +35	6.64 −36
42.0	1.74 +18	5.75 −19	82.0	4.09 +35	2.44 −36
43.0	4.73 +18	2.12 −19	83.0	1.11 +36	8.99 −37
44.0	1.29 +19	7.78 −20	84.0	3.03 +36	3.31 −37
45.0	3.49 +19	2.86 −20	85.0	8.22 +36	1.22 −37
46.0	9.50 +19	1.05 −20	86.0	2.24 +37	4.47 −38
47.0	2.58 +20	3.87 −21	87.0	6.08 +37	1.65 −38
48.0	7.02 +20	1.43 −21	88.0	1.65 +38	6.05 −39
49.0	1.91 +21	5.24 −22	89.0	4.49 +38	2.23 −39
x	exp(x)	exp(−x)	x	exp(x)	exp(−x)

TABLE 3 Natural Logarithms

x	ln x	x	ln x	x	ln x	x	ln x
0.0		5.0	1.609	10	2.303	60	4.09
.1	-2.303	5.1	1.629	11	2.398	61	4.11
.2	-1.609	5.2	1.649	12	2.485	62	4.13
.3	-1.204	5.3	1.668	13	2.565	63	4.14
.4	-.916	5.4	1.686	14	2.639	64	4.16
.5	-.693	5.5	1.705	15	2.708	65	4.17
.6	-.511	5.6	1.723	16	2.773	66	4.19
.7	-.357	5.7	1.740	17	2.833	67	4.20
.8	-.223	5.8	1.758	18	2.890	68	4.22
.9	-.105	5.9	1.775	19	2.944	69	4.23
1.0	0.000	6.0	1.792	20	2.996	70	4.25
1.1	.095	6.1	1.808	21	3.045	71	4.26
1.2	.182	6.2	1.825	22	3.091	72	4.28
1.3	.262	6.3	1.841	23	3.135	73	4.29
1.4	.336	6.4	1.856	24	3.178	74	4.30
1.5	.405	6.5	1.872	25	3.219	75	4.32
1.6	.470	6.6	1.887	26	3.258	76	4.33
1.7	.531	6.7	1.902	27	3.296	77	4.34
1.8	.588	6.8	1.917	28	3.332	78	4.36
1.9	.642	6.9	1.932	29	3.367	79	4.37
2.0	.693	7.0	1.946	30	3.401	80	4.38
2.1	.742	7.1	1.960	31	3.434	81	4.39
2.2	.788	7.2	1.974	32	3.466	82	4.41
2.3	.833	7.3	1.988	33	3.497	83	4.42
2.4	.875	7.4	2.001	34	3.526	84	4.43
2.5	.916	7.5	2.015	35	3.555	85	4.44
2.6	.956	7.6	2.028	36	3.584	86	4.45
2.7	.993	7.7	2.041	37	3.611	87	4.47
2.8	1.030	7.8	2.054	38	3.638	88	4.48
2.9	1.065	7.9	2.067	39	3.664	89	4.49
3.0	1.099	8.0	2.079	40	3.689	90	4.50
3.1	1.131	8.1	2.092	41	3.714	91	4.51
3.2	1.163	8.2	2.104	42	3.738	92	4.52
3.3	1.194	8.3	2.116	43	3.761	93	4.53
3.4	1.224	8.4	2.128	44	3.784	94	4.54
3.5	1.253	8.5	2.140	45	3.807	95	4.55
3.6	1.281	8.6	2.152	46	3.829	96	4.56
3.7	1.308	8.7	2.163	47	3.850	97	4.57
3.8	1.335	8.8	2.175	48	3.871	98	4.58
3.9	1.361	8.9	2.186	49	3.892	99	4.60
4.0	1.386	9.0	2.197	50	3.912	100	4.61
4.1	1.411	9.1	2.208	51	3.932	101	4.62
4.2	1.435	9.2	2.219	52	3.951	102	4.62
4.3	1.459	9.3	2.230	53	3.970	103	4.63
4.4	1.482	9.4	2.241	54	3.989	104	4.64
4.5	1.504	9.5	2.251	55	4.007	105	4.65
4.6	1.526	9.6	2.262	56	4.025	106	4.66
4.7	1.548	9.7	2.272	57	4.043	107	4.67
4.8	1.569	9.8	2.282	58	4.060	108	4.68
4.9	1.589	9.9	2.293	59	4.078	109	4.69

x	ln x	x	ln x	x	ln x	x	ln x

TABLE 4 Trigonometric Functions of Degrees   **A7**

deg	sin	csc	tan	cot	sec	cos	
0	.000		.000		1.000	1.000	90
1	.017	57.30	.017	57.29	1.000	1.000	89
2	.035	28.65	.035	28.64	1.001	.999	88
3	.052	19.11	.052	19.08	1.001	.999	87
4	.070	14.34	.070	14.30	1.002	.998	86
5	.087	11.47	.087	11.43	1.004	.996	85
6	.105	9.567	.105	9.514	1.006	.995	84
7	.122	8.206	.123	8.144	1.008	.993	83
8	.139	7.185	.141	7.115	1.010	.990	82
9	.156	6.392	.158	6.314	1.012	.988	81
10	.174	5.759	.176	5.671	1.015	.985	80
11	.191	5.241	.194	5.145	1.019	.982	79
12	.208	4.810	.213	4.705	1.022	.978	78
13	.225	4.445	.231	4.331	1.026	.974	77
14	.242	4.134	.249	4.011	1.031	.970	76
15	.259	3.864	.268	3.732	1.035	.966	75
16	.276	3.628	.287	3.487	1.040	.961	74
17	.292	3.420	.306	3.271	1.046	.956	73
18	.309	3.236	.325	3.078	1.051	.951	72
19	.326	3.072	.344	2.904	1.058	.946	71
20	.342	2.924	.364	2.747	1.064	.940	70
21	.358	2.790	.384	2.605	1.071	.934	69
22	.375	2.669	.404	2.475	1.079	.927	68
23	.391	2.559	.424	2.356	1.086	.921	67
24	.407	2.459	.445	2.246	1.095	.914	66
25	.423	2.366	.466	2.145	1.103	.906	65
26	.438	2.281	.488	2.050	1.113	.899	64
27	.454	2.203	.510	1.963	1.122	.891	63
28	.469	2.130	.532	1.881	1.133	.883	62
29	.485	2.063	.554	1.804	1.143	.875	61
30	.500	2.000	.577	1.732	1.155	.866	60
31	.515	1.942	.601	1.664	1.167	.857	59
32	.530	1.887	.625	1.600	1.179	.848	58
33	.545	1.836	.649	1.540	1.192	.839	57
34	.559	1.788	.675	1.483	1.206	.829	56
35	.574	1.743	.700	1.428	1.221	.819	55
36	.588	1.701	.727	1.376	1.236	.809	54
37	.602	1.662	.754	1.327	1.252	.799	53
38	.616	1.624	.781	1.280	1.269	.788	52
39	.629	1.589	.810	1.235	1.287	.777	51
40	.643	1.556	.839	1.192	1.305	.766	50
41	.656	1.524	.869	1.150	1.325	.755	49
42	.669	1.494	.900	1.111	1.346	.743	48
43	.682	1.466	.933	1.072	1.367	.731	47
44	.695	1.440	.966	1.036	1.390	.719	46
45	.707	1.414	1.000	1.000	1.414	.707	45
	cos	sec	cot	tan	csc	sin	deg

TABLE 5 Trigonometric Functions of Radians

rad	sin	cos	tan	rad	sin	cos	tan
0.00	0.000	1.000	0.000	1.00	.841	.540	1.557
.02	.020	1.000	.020	1.02	.852	.523	1.628
.04	.040	.999	.040	1.04	.862	.506	1.704
.06	.060	.998	.060	1.06	.872	.489	1.784
.08	.080	.997	.080	1.08	.882	.471	1.871
.10	.100	.995	.100	1.10	.891	.454	1.965
.12	.120	.993	.121	1.12	.900	.436	2.066
.14	.140	.990	.141	1.14	.909	.418	2.176
.16	.159	.987	.161	1.16	.917	.399	2.296
.18	.179	.984	.182	1.18	.925	.381	2.427
.20	.199	.980	.203	1.20	.932	.362	2.572
.22	.218	.976	.224	1.22	.939	.344	2.733
.24	.238	.971	.245	1.24	.946	.325	2.912
.26	.257	.966	.266	1.26	.952	.306	3.113
.28	.276	.961	.288	1.28	.958	.287	3.341
.30	.296	.955	.309	1.30	.964	.267	3.602
.32	.315	.949	.331	1.32	.969	.248	3.903
.34	.333	.943	.354	1.34	.973	.229	4.256
.36	.352	.936	.376	1.36	.978	.209	4.673
.38	.371	.929	.399	1.38	.982	.190	5.177
.40	.389	.921	.423	1.40	.985	.170	5.798
.42	.408	.913	.447	1.42	.989	.150	6.581
.44	.426	.905	.471	1.44	.991	.130	7.602
.46	.444	.896	.495	1.46	.994	.111	8.989
.48	.462	.887	.521	1.48	.996	.091	10.983
.50	.479	.878	.546	1.50	.997	.071	14.101
.52	.497	.868	.573	1.52	.999	.051	19.670
.54	.514	.858	.599	1.54	1.000	.031	32.461
.56	.531	.847	.627	1.56	1.000	.011	92.620
.58	.548	.836	.655	1.58	1.000	−.009	−108.649
.60	.565	.825	.684	1.60	1.000	−.029	−34.233
.62	.581	.814	.714	1.62	.999	−.049	−20.307
.64	.597	.802	.745	1.64	.998	−.069	−14.427
.66	.613	.790	.776	1.66	.996	−.089	−11.181
.68	.629	.778	.809	1.68	.994	−.109	−9.121
.70	.644	.765	.842	1.70	.992	−.129	−7.697
.72	.659	.752	.877	1.72	.989	−.149	−6.652
.74	.674	.738	.913	1.74	.986	−.168	−5.854
.76	.689	.725	.950	1.76	.982	−.188	−5.222
.78	.703	.711	.989	1.78	.978	−.208	−4.710
.80	.717	.697	1.030	1.80	.974	−.227	−4.286
.82	.731	.682	1.072	1.82	.969	−.247	−3.929
.84	.745	.667	1.116	1.84	.964	−.266	−3.624
.86	.758	.652	1.162	1.86	.958	−.285	−3.361
.88	.771	.637	1.210	1.88	.953	−.304	−3.130
.90	.783	.622	1.260	1.90	.946	−.323	−2.927
.92	.796	.606	1.313	1.92	.940	−.342	−2.746
.94	.808	.590	1.369	1.94	.933	−.361	−2.584
.96	.819	.574	1.428	1.96	.925	−.379	−2.438
.98	.830	.557	1.491	1.98	.917	−.398	−2.306
rad	sin	cos	tan	rad	sin	cos	tan

TABLE 6 Inverse Trigonometric Functions to Radians   **A9**

x	arcsin	arccos	arctan	x	arcsin	arccos	arctan
.00	.000	1.571	.000	.50	.524	1.047	.464
.01	.010	1.561	.010	.51	.535	1.036	.472
.02	.020	1.551	.020	.52	.547	1.024	.480
.03	.030	1.541	.030	.53	.559	1.012	.487
.04	.040	1.531	.040	.54	.570	1.000	.495
.05	.050	1.521	.050	.55	.582	.988	.503
.06	.060	1.511	.060	.56	.594	.976	.510
.07	.070	1.501	.070	.57	.607	.964	.518
.08	.080	1.491	.080	.58	.619	.952	.526
.09	.090	1.481	.090	.59	.631	.940	.533
.10	.100	1.471	.100	.60	.644	.927	.540
.11	.110	1.461	.110	.61	.656	.915	.548
.12	.120	1.451	.119	.62	.669	.902	.555
.13	.130	1.440	.129	.63	.682	.889	.562
.14	.140	1.430	.139	.64	.694	.876	.569
.15	.151	1.420	.149	.65	.708	.863	.576
.16	.161	1.410	.159	.66	.721	.850	.583
.17	.171	1.400	.168	.67	.734	.837	.590
.18	.181	1.390	.178	.68	.748	.823	.597
.19	.191	1.380	.188	.69	.761	.809	.604
.20	.201	1.369	.197	.70	.775	.795	.611
.21	.212	1.359	.207	.71	.789	.781	.617
.22	.222	1.349	.217	.72	.804	.767	.624
.23	.232	1.339	.226	.73	.818	.752	.631
.24	.242	1.328	.236	.74	.833	.738	.637
.25	.253	1.318	.245	.75	.848	.723	.644
.26	.263	1.308	.254	.76	.863	.707	.650
.27	.273	1.297	.264	.77	.879	.692	.656
.28	.284	1.287	.273	.78	.895	.676	.662
.29	.294	1.277	.282	.79	.911	.660	.669
.30	.305	1.266	.291	.80	.927	.644	.675
.31	.315	1.256	.301	.81	.944	.627	.681
.32	.326	1.245	.310	.82	.961	.609	.687
.33	.336	1.234	.319	.83	.979	.592	.693
.34	.347	1.224	.328	.84	.997	.574	.699
.35	.358	1.213	.337	.85	1.016	.555	.704
.36	.368	1.203	.346	.86	1.035	.536	.710
.37	.379	1.192	.354	.87	1.055	.516	.716
.38	.390	1.181	.363	.88	1.076	.495	.722
.39	.401	1.170	.372	.89	1.097	.473	.727
.40	.412	1.159	.381	.90	1.120	.451	.733
.41	.422	1.148	.389	.91	1.143	.428	.738
.42	.433	1.137	.398	.92	1.168	.403	.744
.43	.444	1.126	.406	.93	1.194	.376	.749
.44	.456	1.115	.415	.94	1.223	.348	.754
.45	.467	1.104	.423	.95	1.253	.318	.760
.46	.478	1.093	.431	.96	1.287	.284	.765
.47	.489	1.082	.439	.97	1.325	.246	.770
.48	.501	1.070	.448	.98	1.370	.200	.775
.49	.512	1.059	.456	.99	1.429	.142	.780
				1.00	1.571	0.000	.785

x	arcsin	arccos	arctan	x	arcsin	arccos	arctan

TABLE 7   Trigonometric Functions of $a = \pi \cdot x$ Radians

x	sin a	cos a	tan a	cot a	1−x
.00	.000	1.000	.000		1.00
.01	.031	1.000	.031	31.821	.99
.02	.063	.998	.063	15.895	.98
.03	.094	.996	.095	10.579	.97
.04	.125	.992	.126.	7.916	.96
.05	.156	.988	.158	6.314	.95
.06	.187	.982	.191	5.242	.94
.07	.218	.976	.224	4.474	.93
.08	.249	.969	.257	3.895	.92
.09	.279	.960	.291	3.442	.91
.10	.309	.951	.325	3.078	.90
.11	.339	.941	.360	2.778	.89
.12	.368	.930	.396	2.526	.88
.13	.397	.918	.433	2.311	.87
.14	.426	.905	.471˙	2.125	.86
.15	.454	.891	.510	1.963	.85
.16	.482	.876	.550	1.819	.84
.17	.509	.861	.591	1.691	.83
.18	.536	.844	.635	1.576	.82
.19	.562	.827	.680	1.471	.81
.20	.588	.809	.727	1.376	.80
.21	.613	.790	.776	1.289	.79
.22	.637	.771	.827	1.209	.78
.23	.661	.750	.882	1.134	.77
.24	.685	.729	.939	1.065	.76
.25	.707	.707	1.000	1.000	.75
.26	.729	.685	1.065	.939	.74
.27	.750	.661	1.134	.882	.73
.28	.771	.637	1.209	.827	.72
.29	.790	.613	1.289	.776	.71
.30	.809	.588	1.376	.727	.70
.31	.827	.562	1.471	.680	.69
.32	.844	.536	1.576	.635	.68
.33	.861	.509	1.691	.591	.67
.34	.876	.482	1.819	.550	.66
.35	.891	.454	1.963	.510	.65
.36	.905	.426	2.125	.471	.64
.37	.918	.397	2.311	.433	.63
.38	.930	.368	2.526	.396	.62
.39	.941	.339	2.778	.360	.61
.40	.951	.309	3.078	.325	.60
.41	.960	.279	3.442	.291	.59
.42	.969	.249	3.895	.257	.58
.43	.976	.218	4.474	.224	.57
.44	.982	.187	5.242	.191	.56
.45	.988	.156	6.314	.158	.55
.46	.992	.125	7.916	.126	.54
.47	.996	.094	10.579	.095	.53
.48	.998	.063	15.895	.063	.52
.49	1.000	.031	31.821	.031	.51
.50	1.000	.000		.000	.50
1−x	sin a	−cos a	−tan a	−cot a	x

Answers to Odd-Numbered Exercises

CHAPTER 1

Section 2, page 6

1. Suppose for instance that $b = 0$. We must prove $a \cdot 0 = 0$. Set $c = a \cdot 0$. Then
$c = a \cdot 0 = a(0 + 0) = a \cdot 0 + a \cdot 0 = c + c$. Hence
$0 = c + (-c) = (c + c) + (-c) = c + [c + (-c)] = c + 0 = c$.

3. $(ab)(a^{-1}b^{-1}) = (ab)(b^{-1}a^{-1}) = a[b(b^{-1}a^{-1})] = a[(bb^{-1})a^{-1}] = a(1 \cdot a^{-1}) = aa^{-1} = 1$.
Hence $a^{-1}b^{-1} = (ab)^{-1}$.

5. $(a/b)/(c/d) = (ab^{-1})/(cd^{-1}) = (ab^{-1})(cd^{-1})^{-1} = (ab^{-1})(c^{-1}d) = ab^{-1}c^{-1}d$. Next,
$(ad)/(bc) = (ad)(bc)^{-1} = (ad)(b^{-1}c^{-1}) = ab^{-1}c^{-1}d$, the same. We have used
$(d^{-1})^{-1} = d$, true because $1 = d^{-1}(d^{-1})^{-1}$ and $1 = d^{-1}d$.

7. $(a/b)(b/a) = (ab^{-1})(ba^{-1}) = a(b^{-1}b)a^{-1} = aa^{-1} = 1$, hence $(a/b)^{-1} = b/a$.

9. First, $a^2 \geq 0$ because either $a \geq 0$ or $a = -a_1$ where $a_1 \geq 0$ and $a^2 = a_1^2 \geq 0$. Similarly
$b^2 \geq 0$, $c^2 \geq 0$, $d^2 \geq 0$, so $a^2 + b^2 + c^2 + d^2 \geq 0$. But, for instance, $c \neq 0$, so $c^2 > 0$.
Hence the sum is greater than 0.

11. Either $|a| = +a$ or $|a| = -a$, hence $|a| = \pm a \leq b$.

13. Say $a > 0$ and $b < 0$. Then $a + b < a < a + (-b) = a - b$ and
$-(a + b) = -a - b < -b < a - b$, hence $|a + b| = \pm(a + b) < a - b$. But
$|a| + |b| = a - b$, so $|a + b| < |a| + |b|$.

15. $|x| = 2$ **17.** $|x - a| \leq |x - b|$ **19.** $|x - 17| < 1$ **21.** $-4 \leq x \leq 4$

23. $-0.01 \leq x \leq 0.01$ **25.** $2 \leq x \leq 4$ **27.** $-2 < x < 2$ **29.** $-1 < x < 9$

31. $x > 0$ or $x < -\frac{2}{3}$ **33.** $-3 < x < 3$ **35.** $-2.1 < x < -2$ or $-2 < x < -1.9$

37. $0 \leq x \leq 4$ **39.** $x < \frac{2}{3}$ **41.** $x < -\frac{3}{2}$ **43.** $-5 < x < 5$ **45.** $-5 < x < 3$

47. $x < -\frac{1}{4}$ or $x > 0$ **49.** $0 < x < 3$ **51.** $-1, 2$

53. The distance from 1 to 12 is 11 units, so the sum of the distances from x to 1 and
from x to 12 must be 11 or more. But $2 + 3 < 11$.

55. $|7x - 7a| = 7|x - a| < 7 \times 10^{-6} < 10^{-5}$.

57. First note that $|x - 5| < \frac{1}{10}$ implies $x < 5.1 < 6$. Then
$|xy - 35| = |x(y - 7) + 7(x - 5)| \leq |x||y - 7| + 7|x - 5| < \frac{6}{10} + \frac{7}{10} = 1.3$.

59. First, $|x + 3| = x + 3 < 7$ since $x < 3 + 10^{-6} < 4$. Next,
$|x^2 - 9| = |(x + 3)(x - 3)| = |x + 3||x - 3| < 7 \times 10^{-6} < 10^{-5}$.

Section 3, page 10

1.   **3.**

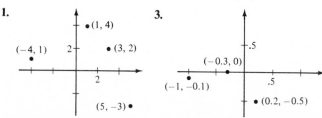

5.

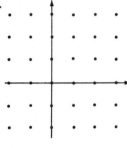

• (50, 0.6)

(150, 0.3) •

0.2

100

7.

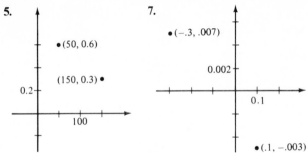

• (−.3, .007)

0.002

0.1

• (.1, −.003)

9.

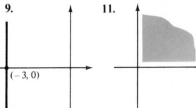

(−3, 0)

11.

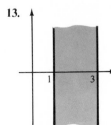

13.

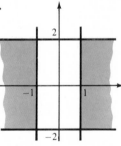

1   3

15.

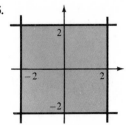

2

−2   2

−2

17.

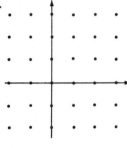

19.

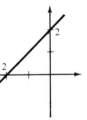

2

−1   1

−2

21.

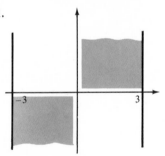

−3   3

23. $(\pm 1, \pm 1)$ **25.** $(9, 0), (0, 12)$ or $(12, 0), (0, 9)$

Section 4,   page 15

1. 5,   9,   6, $\dfrac{2}{x} + 5 = \dfrac{2 + 5x}{x}$, $2x - 1$ **3.**

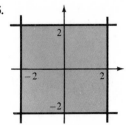

2

−2

5.

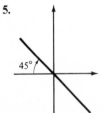

45°

7.

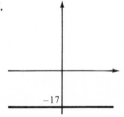

−17

9.

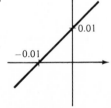

0.01

−0.01

11. 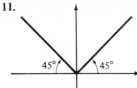   **13.** 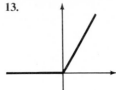   **15.**

17. all x, all y **19.** all x, all y **21.** $x \neq \frac{3}{2}, y \neq 0$ **23.** $x \neq \frac{5}{3}, y \neq \frac{1}{3}$
25. $x \geq 6, y \geq 0$ **27.** $|x| \leq \frac{2}{3}, 0 \leq y \leq 2$ **29.** $x \geq \frac{3}{2}, y \geq 0$ **31.** $|x| \leq \frac{1}{2}, 0 \leq y \leq \frac{1}{2}$
33. $x \leq 1$ or $x \geq 4, 0 \leq y$ **35.** $3x - 1, -6x - 2$ **37.** $x^2 - 2x + 1, -2x^3 + x^2$
39. No; their domains have no point in common.
41. $[f \circ g](x) = 3x - 5, [g \circ f](x) = 3x - 1$ **43.** $2x^2 + 4x + 2, -2x^2 - 1$
45. $-4x, -4x$ **47.** 9, 3   **49.** $g(x)$ **51.** x **53.** $x + 1$, 3, etc.
55. No; $f(x)$ is defined only for $x \geq \frac{5}{2}$, but $g(x) \leq 1$.
57. Yes; $f[\frac{1}{2}(x_0 + x_1)] = \frac{1}{2}a(x_0 + x_1) + b$ and
    $\frac{1}{2}[f(x_0) + f(x_1)] = \frac{1}{2}[(ax_0 + b) + (ax_1 + b)] = \frac{1}{2}a(x_0 + x_1) + b$.
59. $f[\frac{1}{2}(x_0 + x_1)] = 2/(x_0 + x_1) = 2f(x_0 + x_1)$.

Section 5,   page 20

1. 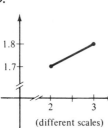   **3.** 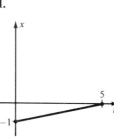   **5.** 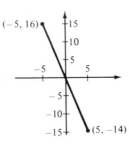   **7.**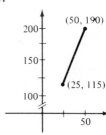

9. 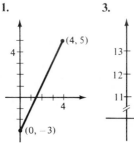   **11.** 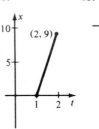   **13.** 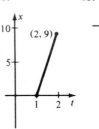   **15.**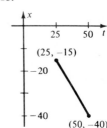

17. $\frac{4}{3}$ **19.** 0   **21.** 1   **23.** $\frac{3}{2}$ **25.** 1   **27.** $y = x + 1$ **29.** $y = 3$
31. $y = \frac{1}{2}x - 3$ **33.** $y = 2x$ **35.** $y = \frac{4}{3}x + \frac{4}{3}$ **37.** $y = x + \frac{1}{2}$ **39.** $y = -5x + 3.5$
41. 3, -7 **43.** -1, 7   **45.** x- 2, y- 3   **47.** x- $\frac{1}{2}$, y- $\frac{1}{3}$
49. By the two-point form, the line through $(c, 0)$ and $(0, d)$ is

$$\frac{y - 0}{x - c} = \frac{d - 0}{0 - c}, \quad \text{that is,} \quad \frac{y}{d} = \frac{x - c}{-c}.$$

    This simplifies to $x/c + y/d = 1$.
51. 45°   **53.** 90°

Section 6, page 26

1.

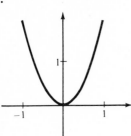

3.

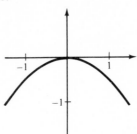

5.

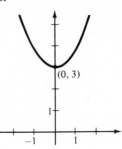

7.

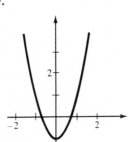

9.

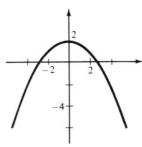

11.

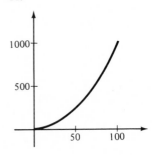

13.

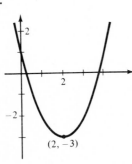

15.

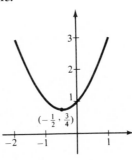

17.

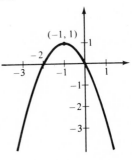

19.

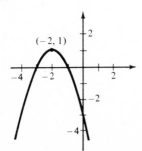

21.

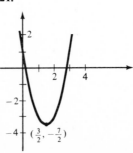

23.

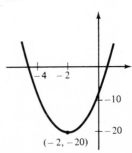

25.

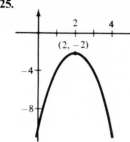

$(2, -2)$

27.

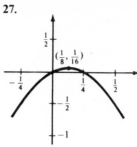

$(\frac{1}{8}, \frac{1}{16})$

29.

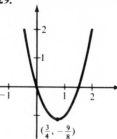

$(\frac{3}{4}, -\frac{9}{8})$

31.

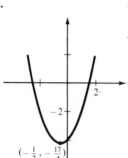

$(-\frac{1}{2}, -\frac{17}{4})$

33.

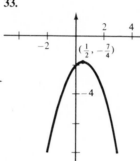

$(\frac{1}{2}, -\frac{7}{4})$

35.

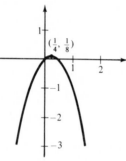

$(\frac{1}{4}, \frac{1}{8})$

37. When $x = 0$, then $y = a \cdot 0^2 + b \cdot 0 = 0$.   **39.** $b = 0$

41. $x(1 - x) = \frac{1}{4} - (x - \frac{1}{2})^2 \le \frac{1}{4}$

43. $A^2 = (\frac{1}{2}ab)^2$ where $a^2 + b^2 = 16$, hence $4A^2 = a^2(16 - a^2) = 64 - (a^2 - 8)^2 \le 64$.
Therefore $A^2 \le 16$, $A \le 4$.

Section 7, page 34

1. x^2, x^4, $1/(x^2 + 1)$

5. 1, $3/(x^2 - 9)$, $x^2/(1 - x^2)$

7. $g(-x) = \frac{1}{2}[f(-x) - f(x)] = -g(x)$

3.

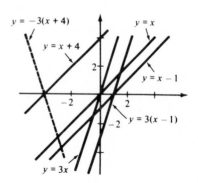

$y = -3(x + 4)$ $y = x$

$y = x + 4$

$y = x - 1$

$y = 3(x - 1)$

$y = 3x$

9.

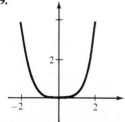

11.

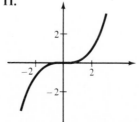

13.

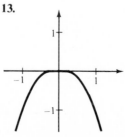

15.

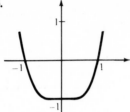

17.

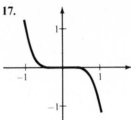

19.

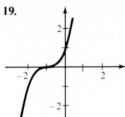

21.

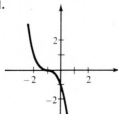

23.

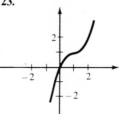

25.

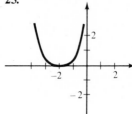

27.

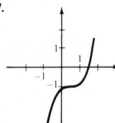

29.

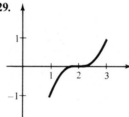

31.

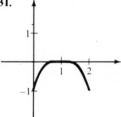

33.

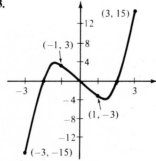

35.

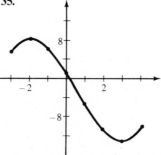

37.

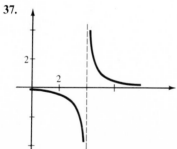

39.

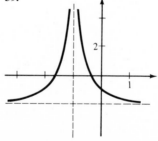

Section 8,   page 42

1.

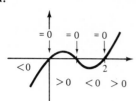

3.

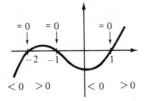

5.

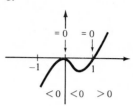

7.

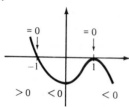

9.

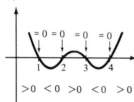

11.

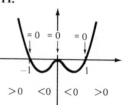

13.

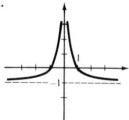

15.

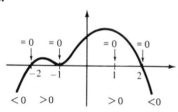

17. $3 < x < 5$ or $x > 8$ **19.** $x < -2$, or $-1 < x < 1$, or $x > 2$

21.

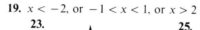

23.

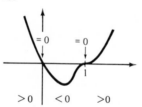

25.

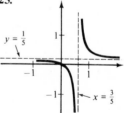

27.

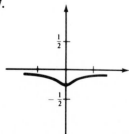

29.

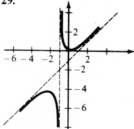

31. $r(x) \longrightarrow 0-$

33.

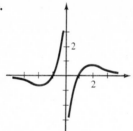

35.

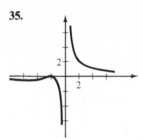

37.

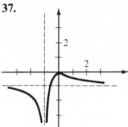

39.

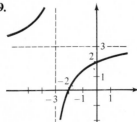

41.

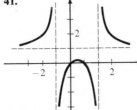

43.

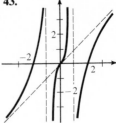

45.

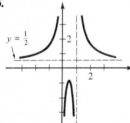

47.

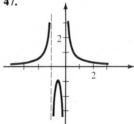

49. By long division, $f(x) = g(x)(ax + b) + h(x)$, where $a \neq 0$ and $\deg h(x) < \deg g(x)$, so that $h(x)/g(x) \longrightarrow 0$ as $x \longrightarrow \infty$. Therefore $r(x) \approx ax + b$ as $x \longrightarrow \infty$.

Section 9, page 48

1. $(x - 1)^2 + (y - 3)^2 = 36$ **3.** $(x + 4)^2 + (y - 3)^2 = 25$ **5.** $(x - 1)^2 + (y - 5)^2 = 26$

7. $(x + 5)^2 + (y - 2)^2 = 25$ **9.** $(x - \frac{3}{2})^2 + (y - 2)^2 = \frac{13}{4}$ **11.** $(x - a)^2 + (y \pm 3)^2 = 9$

13. $x^2 + y^2 - ax - by = 0$

15. No; their centers are farther apart than the sum of their radii: $(2^2 + 3^2)^{1/2} > 1 + 2$.

17. To prove: $\sqrt{x^2 + y^2} = 2\sqrt{(x - 3)^2 + y^2}$. Square and expand:
$x^2 + y^2 = 4[(x - 3)^2 + y^2]$. $3x^2 + 3y^2 - 24x + 36 = 0$, $x^2 + y^2 - 8x + 12 = 0$,
$(x - 4)^2 + y^2 = 4$. Now read backwards.

19.

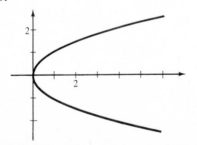

21. $y = \frac{1}{4}x^2 + 1$; parabola with focus $(0, 2)$ and directrix the x-axis

23. Their midpoint, $(\frac{1}{2}(5x + 3x + 4y), \frac{1}{2}(5y + 4x - 3y)) = (4x + 2y, 2x + y)$, lies on the line $2y = x$, and the slope of the segment joining them,

$$m = \frac{(4x - 3y) - 5y}{(3x + 4y) - 5x} = \frac{4x - 8y}{-2x + 4y} = -2,$$

is the negative reciprocal of the slope of the line $2y = x$. Hence segment and line are perpendicular.

Section 10,   page 53

1. $y = \frac{1}{4}x^2$; parabola   **3.** $f(x) = (x - 2)^2(x^3 - x + 3)$ **5.** $y = -6x - 3$
7. $y = x - 1$ **9.** $y = 3x - 6$

Section 11,   page 53

1. (a) $1, -5, 9x + 4$
  (b) $f(a + b) = 3(a + b) + 1 = 3a + 3b + 1 = (3a + 1) + (3b + 1) - 1 = f(a) + f(b) - 1$.
5. $f(x) = -x + 6$

3.

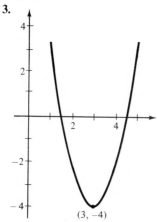

$(3, -4)$

7.

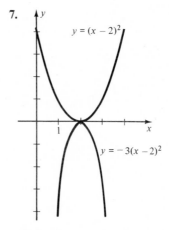

$y = (x - 2)^2$

$y = -3(x - 2)^2$

9. $x \neq -1$

11. $\dfrac{3x^2 + 1}{x(x - 4)}$, $3 + \dfrac{1}{x(x - 4)}$, etc.

13.

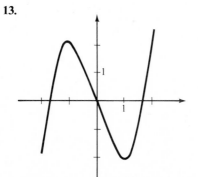

15.

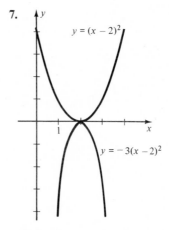

17.

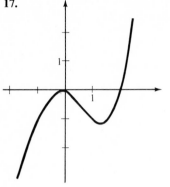

19.

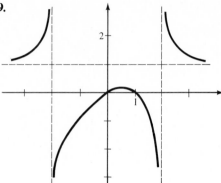

21.

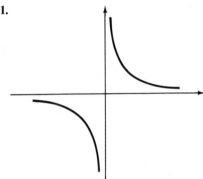

23.

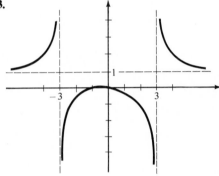

25. circle, center $(\frac{9}{8}, -\frac{1}{8})$, radius $\frac{3}{8}\sqrt{2}$

CHAPTER 2

Section 1,   page 58

1. -11.7 **3.** -11.999997 **5.** -0.9 **7.** -1.0001 **9.** 10.8 **11.** 10.99998
13. 5.41 **15.** $5 - (4 \times 10^{-10}) + 10^{-20}$ **17.** 47.1201 **19.** 47.00120001 **21.** $6a$
23. $4a + 3$

Section 2,   page 62

1. 3 **3.** 8 **5.** 1 **7.** -4 **9.** 4 **11.** -1
13. $0.04996, 0.02499, 0.00500, 0.00050$; $\lim = 0$ **15.** 3 **17.** $-\frac{1}{2}$ **19.** $-\frac{1}{3}$
21. $\frac{1}{4}$ **23.** $\frac{3}{7}$ **25.** 0 **27.** $\frac{1}{4}\sqrt{2}$

Section 3,   page 67

1. $2x$ **3.** -4 **5.** 1 **7.** 3 **9.** 8 **11.** $3y^2$ **13.** 0 **15.** $0, 27, 27$
17. 108 **19.** $48, 3a^2$ **21.** $-12, 24, 2$ **23.** $-1, -1, -1/a^2, -1/a^2$
25. $-1/a^2, -a^2$ **27.** $-1/b^2$ **29.** $6, 14, 22$ **31.** 13 **33.** 6 **35.** -32
37. $(3, 9)$ only   **39.** $(-2, -8), (2, 8)$ **41.** No   **43.** Yes, at $(0, 0)$
45. $y = x^2$ at $x = \frac{1}{2}$; $y = x^3$ at $x = 1$ and $x = 2$ **47.** $x > \frac{2}{3}$

Section 4,   page 72

1. $2x + 3$ **3.** $-4x^3 + 4x$ **5.** $1 - (1/x^2)$ **7.** $3x^2 - 3 \Big|_{x=-1} = 0$

9. $3x^2 + 1 + (2/x^2)\Big|_2 = \frac{27}{2}$ **11.** $(x + 1)(2x) + (x^2 + 1) = 3x^2 + 2x + 1$

13. $(3x + 4)(2x - 2) + 3(x^2 - 2x - 3) = 9x^2 - 4x - 17$

15. $(x^5 - 2)(3x^2 + 1) + (5x^4)(x^3 + x - 3) = 8x^7 + 6x^5 - 15x^4 - 6x^2 - 2$

17. $10x(x^2 - 1)^4$ **19.** $3(2x - 2)(x^2 - 2x + 1)^2 = 6(x - 1)^5$

21. $(x^2 + 1)^3 \cdot 2(x - 1) + 6x(x^2 + 1)^2(x - 1)^2 = 2(x - 1)(x^2 + 1)^2(4x^2 - 3x + 1)$

23. $\left(\dfrac{x^3 + 6}{x}\right)' = \left(x^2 + \dfrac{6}{x}\right)' = 2x - \dfrac{6}{x^2}$

25. $\left(\dfrac{1}{x^3}\right)' = \left(\dfrac{1}{x^2}\dfrac{1}{x}\right)' = \left(\dfrac{1}{x^2}\right)\left(\dfrac{-1}{x^2}\right) + \left(\dfrac{-2}{x^3}\right)\left(\dfrac{1}{x}\right) = \dfrac{-3}{x^4}$ **27.** $mn(mx + b)^{n-1}$

29. $(fgh)' = [f(gh)]' = f'(gh) + f(gh)' = f'gh + f(g'h + gh') = f'gh + fg'h + fgh'$

31. $(x + 2)(x + 3) + (x + 1)(x + 3) + (x + 1)(x + 2) = 3x^2 + 12x + 11$

33. $2(x - 3)(3x + 4) + (2x + 1)(3x + 4) + 3(2x + 1)(x - 3) = 18x^2 - 14x - 29$

35. $8(2x + 1)^3(x - 3)(3x + 4) + (2x + 1)^4(3x + 4) + 3(2x + 1)^4(x - 3)$
$= (2x + 1)^3(36x^2 - 44x - 101)$

37. $2x[(x^2 - 2)(x^2 - 4)(x^2 - 8) + (x^2 - 1)(x^2 - 4)(x^2 - 8) + (x^2 - 1)(x^2 - 2)(x^2 - 8)$
$+ (x^2 - 1)(x^2 - 2)(x^2 - 4)] = 2x(4x^6 - 45x^4 + 140x^2 - 120)$

39. 0 since dy/dx has a factor x^2

Section 5,   page 76

1. $\left(\dfrac{u}{v}\right)' = 0 \neq \dfrac{u'}{v'} = 1$ **3.** $\left(\dfrac{u}{v}\right)' = -\dfrac{1}{x^2} \neq \dfrac{u'}{v'} = \dfrac{1}{2x}$ **5.** $\dfrac{1}{(x + 1)^2}$ **7.** $\dfrac{14}{(x + 5)^2}$

9. $\dfrac{x^2 + 8x + 1}{(x + 4)^2}$ **11.** $\dfrac{-9}{x^{10}}$ **13.** $\dfrac{-4(2x + 1)}{(x^2 + x)^5}$ **15.** $\dfrac{4(x + 1)}{(x + 3)^3}$ **17.** $\dfrac{-12x}{(x^2 + 1)^7}$

19. $\dfrac{-6x(x^3 + x + 1)}{(2x^3 - 1)^2}$ **21.** $\dfrac{1}{2\sqrt{x + 3}}$ **23.** $\dfrac{1}{2\sqrt{x(x + 1)^3}}$ **25.** $\dfrac{1}{\sqrt{(1 - x^2)^3}}$

27. $\dfrac{3x^3 - 2a^2x}{\sqrt{x^2 - a^2}}$ **29.** $\dfrac{-16}{\sqrt{(8x - x^2)^3}}$ **31.** $\dfrac{3x^2(\sqrt{x} + 2)}{2(\sqrt{x} + 1)^4}$ **33.** $\dfrac{-(1 + 3x)}{(1 + 2x + 3x^2)^{3/2}}$

35. $\left(\dfrac{fg}{h}\right)' = \left[\dfrac{fgh}{h^2}\right]' = \dfrac{h^2(fgh)' - (fgh)(2hh')}{h^4} = \dfrac{(fgh)' - 2fgh'}{h^2}$

Section 6,   page 81

1. $6x^2(x^3 + 1)(2x^6 + 4x^3 - 1)$

3. $\left[5\left(x + \dfrac{1}{x}\right)^4 - 3\left(x + \dfrac{1}{x}\right)^2\right]\left(1 - \dfrac{1}{x^2}\right) = \dfrac{(x^2 - 1)(x^2 + 1)^2(5x^4 + 7x^2 + 5)}{x^6}$

5. $-30\left(2x - \dfrac{1}{x}\right)^9\left(2 + \dfrac{1}{x^2}\right)$ **7.** $\dfrac{-1}{(2x + 1)^{3/2}}$ **9.** $\dfrac{x + 1}{3(x^2 + 2x)^{2/3}[(x^2 + 2x)^{1/3} - 1]^{1/2}}$

11. $\dfrac{-2}{(x^2 - 1)^{1/2}[(x + 1)^{1/2} - (x - 1)^{1/2}]^2}$ **13.** $\dfrac{-1}{3x^{2/3}(2 + x^{1/3})^2}$ **15.** 1

17. $2u \cdot 2v \cdot 2x = 8x(x^2 + 1)[(x^2 + 1)^2 + 1]$ **19.** $\dfrac{1}{2\sqrt{u + 1}}$ $\left(\tfrac{2}{3}x^{-1/3}\right) = \dfrac{1}{3x^{1/3}\sqrt{x^{2/3} + 2}}$

Section 7,   page 84

1. $y = 4x - 4$ **3.** $y = -x - 2$ **5.** $y = 2x + 2$ **7.** $y = \tfrac{2}{27}x + \tfrac{1}{3}$

9. $y = 10x - 25$ **11.** $y = \pm 8x - 16$ **13.** $y = 7x - 5$, $(0, -5)$, $(\tfrac{5}{7}, 0)$

15. $y = -81x - 18$, $(0, -18)$, $(-\tfrac{2}{9}, 0)$; $y = -81x + 18$, $(0, 18)$, $(\tfrac{2}{9}, 0)$ **17.** 2

19. $(0, -9)$ **21.** $y = 1, -x^2$ **23.** $y = 27(x - 2)$, $(x - 3)^2(x + 6)$

25. $y = -3x - 2$, $9(x+1)^2/(3x+4)$

27. $t(x) = \frac{1}{2}(1+x)$, $e(x) = \sqrt{x} - \frac{1}{2}(1+x)$

x	0.90	0.98	1.02	1.10
$\sqrt{x}$	0.948683	0.989949	1.009950	1.048809
$t(x)$	0.950000	0.990000	1.010000	1.050000
$e(x)$	-0.001317	-0.000051	-0.000050	-0.001191

29. $t(x) = 8 + 12(x-2)$; $x^3 - t(x) = x^3 - 8 - 12(x-2) = (x-2)[x^2 + 2x + 4 - 12]$
$$= (x-2)(x^2 + 2x - 8) = (x-2)^2(x+4)$$
If $|x-2| < 1$, then $5 < x + 4 < 7$, hence $|x^3 - t(x)| < 7|x-2|^2$.

Section 8,   page 88

1. $2x^2 + c$ **3.** $-\frac{1}{2}x^2 + c$ **5.** $-x^2 + x + c$ **7.** $\frac{1}{3}x^3 - x^2 + c$

9. $\frac{1}{4}x^4 + \frac{1}{2}x^2 - x + c$ **11.** $\frac{1}{2}x^2 - \dfrac{1}{x} + c$ **13.** $\frac{1}{3}x^3 - \frac{5}{2}x^2 + 6x + c$

15. $-\dfrac{1}{x} + \dfrac{1}{x^3} + c$ **17.** $\dfrac{-2}{(x-3)^2} + c$ **19.** $\frac{1}{14}(2x+1)^7 + c$ **21.** $\frac{1}{4}(\frac{1}{2}x + \frac{1}{5})^8 + c$

23. $\frac{1}{3}x^3 + 4\sqrt{x} + c$ **25.** $\frac{2}{3}x^{3/2} + c$ **27.** $3x^2 - 4x^{1/4} + c$ **29.** $ax^3 + bx^2 + cx + d$

31. $y = \frac{1}{2}x^2 - 5x + \frac{13}{2}$ **33.** $y = x^3 - 2x^2 - \dfrac{1}{x} + \frac{19}{2}$ **35.** $\frac{1}{2}[f(x)]^2$ **37.** $\frac{1}{3}[F(x)]^3$

Section 9,   page 93

1. 6   **3.** $2 - 6x$ **5.** $30x^4 + 40x^3 + 12x^2$ **7.** $\dfrac{-2 + 6x^2}{(1 + x^2)^3}$ **9.** $\dfrac{4}{(x-2)^3}$

11. $\dfrac{4a(3x^2 + a)}{(x^2 - a)^3}$ **13.** $\dfrac{27}{4(1 - 3x)^{5/2}}$ **15.** $\dfrac{9}{(x^2 + 9)^{3/2}}$ **17.** $\dfrac{-2}{9x^{5/3}} + \dfrac{10}{9(1+x)^{5/3}}$

19. 16π **21.** 5/864   **23.** 18   **25.** $\dfrac{(-1)^{n-1}}{3^n} 2 \cdot 5 \cdot 8 \cdots (3n-4)x^{-(3n-1)/3}$

27. $\frac{1}{3}, 3$ **29.** $\frac{2}{3}x^3 - \frac{1}{2}x^2 + cx + d$ **31.** $12x^2(2x+1)^4(30x^2 + 12x + 1)$

33. $(8 \cdot 7 \cdot 6 \cdot 5)(9x - 15)(x - 3)^3$ **35.** $\dfrac{3 \cdot 5 \cdot 7 \cdots 15(3x - 17)}{2^{10}x^{19/2}}$ **37.** 0   **39.** $ff''/(f')^2$

Section 10,   page 101

1. $\frac{1}{2}$ **3.** $\frac{1}{5}$ **5.** 1   **7.** 0   **9.** 1   **11.** $\frac{3}{5}$

13. $\dfrac{d}{dx}\left(\dfrac{1}{x}\right)\Big|_{x=a} = \lim_{h \to 0} \dfrac{1}{h}\left(\dfrac{1}{a+h} - \dfrac{1}{a}\right) = \lim_{h \to 0}\left(-\dfrac{1}{a^2} + \dfrac{h}{a^2(a+h)}\right) = \dfrac{-1}{a^2}$

15. Use the definition of limit with $\varepsilon = \frac{1}{2} \cdot 10^{-6}$.

17. $\lim_{x \to 0+} f(x) = 1$ because $|1 - f(x)| = 0 < \varepsilon$ for $x > 0$ and $\varepsilon > 0$. Similarly
$\lim_{x \to 0-} f(x) = 0$. Now suppose $\lim_{x \to 0} f(x) = L$. Take $\varepsilon = \frac{1}{2}$. If $|x| < \delta$, then
$|L - f(x)| = |L - 1| < \frac{1}{2}$ for $x > 0$, and $|L - f(x)| = |L| < \frac{1}{2}$ for $x < 0$, impossible
because these inequalities imply $1 = |(L-1) - L| \le |L-1| + |L| < \frac{1}{2} + \frac{1}{2} = 1$.

19. If $c = 0$, there is nothing to prove. Otherwise $|f(x)| < \varepsilon/|c|$ for $|x - a| < \delta$. Then
$|cf(x)| < |c|(\varepsilon/|c|) = \varepsilon$ for $|x - a| < \delta$, hence $\lim_{x \to a}[cf(x)] = 0$.

21. Apply Theorem 1(2) to $[f(x) + g(x)] + h(x)$.

23. Use Theorem 2. Since $f(x) = x$ is continuous, repeated application of (3) shows
$x \cdot x \cdot x \cdots x = x^k$ is continuous. Then $a_k x^k$ is continuous by (1). Finally a sum of terms
$a_n x^n$, $a_{n+1}x^{n+1}$, etc. is continuous by repeated application of (2).

25. By Theorem 3

27. By Theorem 3 with $g(x) = x^2 + 3 \neq 0$ for all x.

29. By Theorem 4, and Theorem 5 with $f(x) = \sqrt{x}$ and $g(x) = 1 + x^2 > 0$.

31. If $\varepsilon > 0$, then $|\sqrt[3]{x}| < \varepsilon$ if $|x| < \delta = \varepsilon^3$. Hence $\lim \sqrt[3]{x} = 0 = \sqrt[3]{0}$.

33. Given $\varepsilon > 0$, there is a $\delta > 0$ such that $|f(x)| < \varepsilon/M$ for $0 < |x - a| < \delta$. Then $|f(x)g(x)| = |f(x)|\,|g(x)| < (\varepsilon/M)M = \varepsilon$. Hence $f(x) \longrightarrow 0$ as $x \longrightarrow a$.

35. Apply Ex. 20 with $f(x) = g(x) = h(x)$.

37. $\delta = \frac{2}{3}10^{-4}$ will do. For if $|h| < \delta$, then

$$\left| \frac{1-h}{2+h} - \frac{1}{2} \right| = \left| \frac{-3h}{2(2+h)} \right| = \frac{3|h|}{2|2+h|} < \frac{3|h|}{2} < \frac{3}{2} \cdot \frac{2}{3}10^{-4} = 10^{-4}.$$

39. By Heron's formula, $f(\varepsilon) = [(s - \frac{3}{2}\varepsilon)(s - a - \varepsilon)(s - b - \varepsilon)(s - c - \varepsilon)]^{1/2}$, a continuous function of ε. Hence $\lim_{\varepsilon \to 0} f(\varepsilon) = f(0) = A$.

Section 11, page 105

1. $6h + 4h^2 + h^3$ **3.** $\dfrac{h}{4(-2+h)}$

5. $f(x) = 4 + |x - 3|$. Since $|x - 3|$ is continuous but not differentiable at $x = 3$, the same holds for $f(x)$.

7. At all points. For $x \neq 0$, this is clear. For $x = 0$, $\dfrac{f(0+h) - f(0)}{h} = |h| \longrightarrow 0$

as $h \longrightarrow 0$. Hence $f(x)$ is differentiable and $f'(0) = 0$.

9. $\dfrac{f(a+h) - f(a)}{h} = \dfrac{[b + (a + h - a)g(a+h)] - b}{h} = g(a+h) \longrightarrow g(a)$ as $h \longrightarrow 0$.

Hence $f'(a) = g(a)$.

11. By definition of derivative, $g(x) \longrightarrow f'(c) = g(c)$ as $x \longrightarrow c$. Thus $g(x)$ is continuous at $x = c$. If $x \neq c$, then $g(x)$ is continuous, the quotient of continuous functions with non-zero denominator.

Section 12, page 105

1. $\pm\sqrt{3}$ **3.** $\dfrac{5\sqrt{3}}{2\sqrt{x}}(2 + \sqrt{3x})^4$ **5.** $2(2x + 1)(3x + 1)^2(18x^2 + 24x + 7)$

7. $x^2 y'' + xy' - 4y = x^2 \left(\dfrac{6}{x^4}\right) + x\left(\dfrac{-2}{x^3}\right) - \dfrac{4}{x^2} = \dfrac{6}{x^2} - \dfrac{2}{x^2} - \dfrac{4}{x^2} = 0$

9. The tangents are $y = 2ax - a^2$ and $y = 2(a + 1)x - (a + 1)^2$; they intersect at $x = a + \frac{1}{2}$, $y = a^2 + a$. But $a^2 + a = (a + \frac{1}{2})^2 - \frac{1}{4}$, so $y = x^2 - \frac{1}{4}$.

11. $-\frac{1}{2}$ **13.** $-\frac{1}{2}$ **15.** $2/(x + 2)^3$

17. Continuous for all x; differentiable for all x except $x = -1, -2, -3$.

19. $r(t) = -\frac{1}{2}t + \frac{1}{2}\sqrt{t^2 + 12}$, a continuous function for all t since $t^2 + 12$ is continuous and positive.

21. $h'(x) = f'[g(x)]g'(x) - g'[f(x)]f'(x)$,
hence $h'(c) = f'[g(c)]g'(c) - g'[f(c)]f'(c) = f'(c)g'(c) - g'(c)f'(c) = 0$.

CHAPTER 3

Section 1, page 113

1. convex for all x

3. convex $x > 2$, concave $x < 2$, inflection $x = 2$

5. convex $x < 0$, concave $x > 0$, inflection $x = 0$

7. convex $|x| > \frac{1}{2}\sqrt{10}$, concave $|x| < \frac{1}{2}\sqrt{10}$, inflections $x = \pm\frac{1}{2}\sqrt{10}$

9.

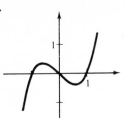

11.

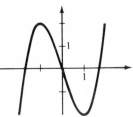

13.

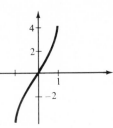

15.

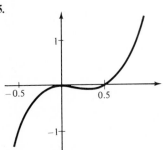

17.

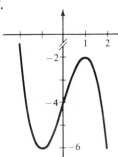

19.

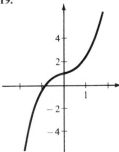

21.

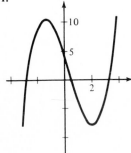

23.

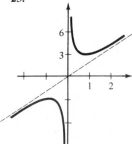

25.

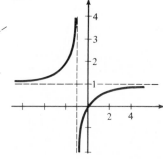

27.

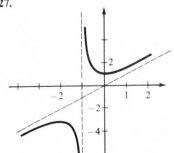

29.

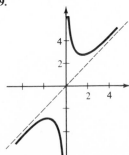

31.

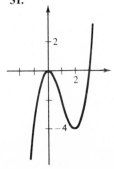

33.

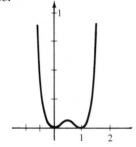

35.

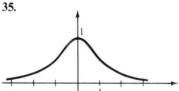

37.

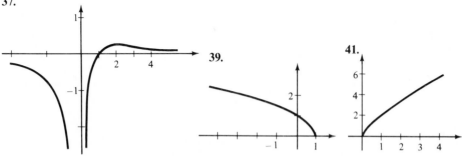

39.

41.

Section 2,   page 117

1. 900, 914.4, 916 ft/sec   **3.** 2000 m, -50 m/sec, -200 m/sec
5. $\frac{3}{4}$ sec, 16 ft/sec up or down (speed, not velocity)
7. (a) increasing for $t < 2$ or $t > 4$, decreasing for $2 < t < 4$;
   (b) increasing for $t > 3$, decreasing for $t < 3$;
   (c) 20 ft forward from 0 to 2 sec, 4 ft backward from 2 to 4 sec, and 20 ft forward
      from 4 to 6 sec; total 44 ft
9. 40 ft   **11.** 6 ft/sec^2 **13.** 480 m, 16 sec   **15.** $100 + 1200 + \frac{200}{3} = 1366\frac{2}{3}$ m
17. $y = -8x^2 + 12$ **19.** $y = -16t^2 + 64t$ **21.** $y = \frac{1}{3}t^3 - \frac{1}{2}t^2 + 3t + 5$

Section 3,   page 122

1. $-\frac{10}{27}$ **3.** -1 **5.** $\frac{5}{9}$ **7.** 24π ft/sec   **9.** $\frac{25}{3}$ ft/sec   **11.** $\frac{21}{5}\sqrt{5}$
13. 0.024 cm/sec   **15.** 620π ft^3/sec   **17.** $4\sqrt{3}$ ft/sec   **19.** $-P^2\dot{V}/k$

Section 4,   page 127

1. min $f(2) = 2$, no max   **3.** max $f(3) = 13$, no min
5. min $f(\frac{1}{6}) = 4$, no max   **7.** max $f(0)$ 1, min $f(2) = -19$
9. min $f(\frac{3}{2}) = 4$, no max   **11.** max $f(1) = f(3) = \frac{4}{3}$, min $f(\sqrt{3}) = \frac{2}{3}\sqrt{3}$
13. max $f(-3) = -\frac{11}{27}$, min $f(-1) = -1$ **15.** max $f(1) = 5$, min $f(\frac{1}{2}) = 3$
17. max $f(2) = 9$, min $f(\pm 1) = 0$ **19.** max $f(-3) = 15$, min $f(-1) = -17$
21. no max or min; $f(x) \longrightarrow -\infty$ as $x \longrightarrow 2-$ and $f(x) \longrightarrow \infty$ as $x \longrightarrow -2+$

Section 5,   page 135

1. 28 ft   **3.** 12   **5.** 75 m $\times$ 25 m   **7.** $\frac{1}{2}a \times \frac{1}{2}b$
9. At noon the ships are as close as they will ever be.

11. $(1/\sqrt[3]{2}, \pm\sqrt[6]{2})$ **13.** 1   **15.** 80 ft   **17.** radius $a\sqrt{2/3}$, height $a/\sqrt{3}$
19. $16/(4+\pi) \approx 2.24$ ft   **21.** \$2725   **23.** $1000/\sqrt{3} \approx 577.35$ mph
25. radius = height = $10\sqrt[3]{V/\pi}$ cm   **27.** $\frac{1}{2}L^2$ ft^2
29. radius $125/\pi \approx 39.8$ m, straight side 125 m
31. Rows to a point $\sqrt{3}$ mi upshore, then walks.
33. depth/width = $\sqrt{2}$ **35.** $\sqrt[4]{4\sigma g/\rho}$ at $\lambda = 2\pi\sqrt{\sigma/\rho g}$ **37.** $x = \frac{3}{7}, y = \frac{4}{7}$ **39.** $\frac{1}{4}$
41. The non-negative function

$$f(x) = \frac{\sqrt[n+1]{a_1 \cdots a_n x}}{\dfrac{1}{n+1}(a_1 + \cdots + a_n + x)}$$

satisfies $f(0) = 0$ and $\lim_{x \to \infty} f(x) = 0$. Also $f'(x) = 0$ only for one positive x, namely
$x = A_n$. Hence A_n yields a maximum, so $f(x) \leq f(A_n)$, with " = " only if $x = A_n$. In
particular, $f(a_{n+1}) \leq f(A_n)$, equality holding only if $a_{n+1} = A_n$. This implies
$G_{n+1}/A_{n+1} \leq (G_n/A_n)^{n/(n+1)}$.

Section 6,   page 143

1. max $y(0) = 0$; mins $y(-1) = y(1) = -1$
3. max $y(0) = 2$; mins $y(-1) = -3$, $y(2) = -30$
5. max $y(2) = \frac{1}{4}\sqrt{2}$; min $y(-2) = -\frac{1}{4}\sqrt{2}$ **7.** max $y(-3) = -9$; no mins
9. max $y(\frac{3}{2}) = \frac{4}{27}$; no mins   **11.** maxs $y(\pm 1) = \frac{1}{2}$; min $y(0) = 0$
13. $y[2/(n-1)] = (n-1)^{n-1}/2^{n-1}n^n$
15. $y' = 2Q(x)/\sqrt{b^2 + (a-x)^2}$, where $Q(x) = x^2 - ax + \frac{1}{2}b^2$. *Case 1:* $a^2 < 2b^2$. Then
$Q(x) > 0$ for all x; no max or min. *Case 2:* $a^2 = 2b^2$. Then $Q(x) = (x - \frac{1}{2}a)^2$; horizontal
inflection at $x = \frac{1}{2}a$, but no max or min. *Case 3:* $a^2 > 2b^2$. Then $Q(x) = (x - x_1)(x - x_2)$,
where $x_1 = \frac{1}{2}(a - \sqrt{a^2 - 2b^2})$ and $x_2 = \frac{1}{2}(a + \sqrt{a^2 - 2b^2})$; max at x_1, min at x_2.
17. If the graph has an inflection point for $x = c$, then $f''(c) = 0$ and $f''(x)$ changes sign
at $x = c$. Hence $f'(x)$ has a max or min at $x = c$ by the first derivative test.
19. $(y')' = 0$, hence $y' = \text{const} = a$. Then $(y - ax)' = 0$, hence $y - ax = \text{const} = b$,
so $y = ax + b$. But $0 = y(0) = b$, and $0 = y(1) = a$. Hence $y \equiv 0$.

Section 8,   page 152

1. On the closed interval $0 \leq x \leq 1$, the continuous function $f(x)$ has a min which is
negative since $f(0) = -1$. This is also the min for $x \geq 0$, since $f(x) > 0$ for $x \geq 1$.
3. Let $f(x) = x$ for $a \leq x < b$, and $f(b) = a$.
5. Take $f(x) = x$ on $0 < x < 1$ and $f(x) = 1/(x^2 + 1)$ on $-\infty < x < \infty$.
7. No; $f(x) = (x - 1)/x^2 \longrightarrow -\infty$ as $x \longrightarrow 0$.
9. Set $g(x) = 2x - f(x)$. Then $g(0) = 0$ and $g'(0) = 2 - f'(0) = 2 - 1 > 0$. By the theorem
on p. 147, there exists $\delta > 0$ such that $g(x) > 0$, that is, $f(x) < 2x$ for $0 < x < \delta$.
11. $f(0) = 1$, $f'(0) = f(0) = 1$. By the theorem on p. 147, there exists $\delta > 0$ such that
$f(x) > f(0)$, that is, $f(x) > 1$ for $0 < x < \delta$.
13. $(f + g)''(x) = f''(x) + g''(x) \geq 0$ if $f''(x) \geq 0$ and $g''(x) \geq 0$.
15. No; if $f(x) = x^{3/4}$, then $[f(x)]^2 = x^{3/2}$ is convex, but $f(x)$ is concave.
17. Let $h(x) = f[g(x)]$. Apply the Chain Rule twice: $h'(x) = f'[g(x)]g'(x)$. Next, $h''(x) = f''[g(x)][g'(x)]^2 + f'[g(x)]g''(x)$. Thus $h''(x) \geq 0$ since $f''(x) \geq 0$ and $g''(x) \geq 0$ by convexity
and $f'(x) \geq 0$ because $f(x)$ is increasing.

19. $f'(x)$ is increasing because its derivative is $f''(x) > 0$. Hence $f'(x) > 0$ if $x > c$; also by the Mean Value Theorem, $f(x) - f(c) = f'(z)(x - c)$, where $c < z < x$. Hence $f'(z) > 0$ so $f(x) - f(c) > 0, f(x) > f(c)$. Similarly $f(x) > f(c)$ for $x < c$.

21. Suppose $f(x) \geq 0$ for some $a < x < b$. Then $\max f = f(c) \geq 0$, where $a < c < b$. But $f''(c) < 0$, contradicting the Second Derivative Test.

23. This is the statement of the Chord Theorem applied to the interval from x to y at the interior point $tx + (1 - t)y$.

25. Exercise 18 applied to $f(x) = x^r$ yields $[\frac{1}{2}(x + y)]^r \leq \frac{1}{2}(x^r + y^r)$. Multiply by 2^r.

Section 9,   page 153

1. $\dfrac{1}{2}x^2 + \dfrac{1}{x} + \dfrac{3}{2}$ **3.** 15 sec, 675 ft   **5.** 500   **7.** $6 \times 6 \times 18$ in.^{3} **9.** $\frac{2}{27}\pi c^3\sqrt{3}$

11.   **13.** 500

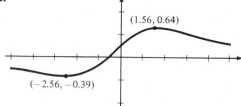

(1.56, 0.64)

(−2.56, −0.39)

15. The circumference of the circle should be $30\pi/(4 + \pi)$ in.   **17.** $y = -\frac{4}{3}x\sqrt{3} + \frac{16}{3}$

19. Let the x-axis be the border of the two regions and let the given points be $(0, a)$ and $(b, -c)$.

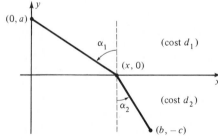

(0, a)

α_1 (cost d_1)

$(x, 0)$

x

α_2 (cost d_2)

$(b, -c)$

The most economical path must be some broken line as shown. The cost of the cable from $(0, a)$ to $(x, 0)$ is

$$(\text{cost per mile})(\text{distance}) = d_1\sqrt{x^2 + a^2} = \frac{\sqrt{x^2 + a^2}}{1/d_1},$$

and the cost from $(x, 0)$ to $(b, -c)$ is $\dfrac{\sqrt{(b - x)^2 + c^2}}{1/d_2}$.

Therefore you must minimize the total cost, $\dfrac{\sqrt{x^2 + a^2}}{1/d_1} + \dfrac{\sqrt{(b - x)^2 + c^2}}{1/d_2}$.

But this is precisely the same as Example 7, p. 134. Hence the optimal path is described by $\dfrac{\sin \alpha_1}{1/d_1} = \dfrac{\sin \alpha_2}{1/d_2}$.

21. Set $f(x) = \dfrac{(x^4 + 1)^3}{(x^3 + 1)^4}$; then $f'(x) = \dfrac{12x^2(x^4 + 1)^2(x - 1)}{(x^3 + 1)^5}$. Hence f decreases for $0 \leq x \leq 1$ and increases for $x \geq 1$, so $f_{\min} = f(1) = \frac{1}{2}$. This proves $f(x) \geq \frac{1}{2}$ for all $x \geq 0$, and the required inequality follows.

23. $v = (b/a)^{1/4}$ **25.** $(x + y)_{\min} = 4f$

CHAPTER 4

Section 1, page 159

1. $\log_{10} 2 \approx 0.30103$, $\log_{10} 3 \approx 0.47712$, $\log_{10} 10 = 1$, and $\log_{10} e \approx \log_{10} 2.7183 \approx 0.43430$.
Divide.

3. $k = (\log_{10} a)/(\log_{10} e)$

5. $\lim_{h \to 0} (2^h - 1)/h \approx 0.6931$, so $(2^h - 1)/h \approx 0.6931$ and $2^h - 1 \approx (0.6931)h$ for h small.

7. By the same reasoning, $3^h \approx 1 + (1.0986)h$, so $3^{0.1} \approx 1 + (1.0986)(0.1) \approx 1.1099$. (By calculator, $3^{0.1} \approx 1.1161$.)

9. $6xe^{3x^2}$ **11.** $4x^3 e^{x^4}$ **13.** $(x - 1)e^x/x^2$ **15.** $e^{-x}/(1 + e^{-x})^2 = e^x/(1 + e^x)^2$

17. $3x^2(1 + e^{2x} - 2xe^{2x})/(e^{2x} + 1)^4$ **19.** $-4/(e^x - e^{-x})^2$ **21.** $1, 4$ **23.** $-2, 3$

25. If $y = e^x$, then $y'' - 2y' + y = e^x - 2e^x + e^x = 0$. If $y = xe^x$, then $y'' - 2y' + y = (2e^x + xe^x) - 2(e^x + xe^x) + xe^x = 0$.

Section 2, page 164

1.

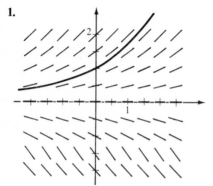

3. $y' + y = \frac{1}{2}(e^x - e^{-x}) + \frac{1}{2}(e^x + e^{-x}) = e^x$

5.

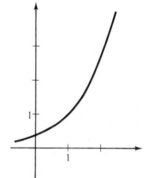

7.

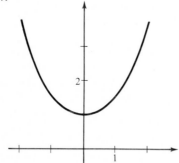

9. $y' = kae^{kx} - kbe^{-kx}$,
$y'' = k^2 ae^{kx} + k^2 be^{-kx} = k^2 y$

11. If $y = cE(x^2)$, then
$y' = 2xcE(x^2) = 2xy$.

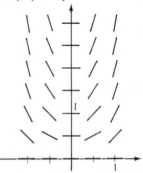

13. $E(2x) = E(x + x) = E(x)E(x) = [E(x)]^2$,
$E(3x) = E(2x + x) = E(2x)E(x) = E(x)^2 E(x) = E(x)^3, \cdots$,
$E[(n + 1)x] = E(nx + x) = E(nx)E(x) = E(x)^n E(x) = [E(x)]^{n+1}$.
This proves $[E(x)]^y = E(xy)$ when y is a positive integer. The result is obvious $(1 = 1)$ for $y = 0$. Finally, $E[x(-y)] = E(-xy) = 1/E(xy) = 1/[E(x)]^y = [E(x)]^{-y}$, so the result holds for all negative integers also.

15. $3k \cdot 10^{3x}$ where $e^k = 10$, $k \approx 2.3026$, $3k \approx 6.9078$

17. $k \cdot 5^{x-1}$ where $e^k = 5$, $k \approx 1.6094$ **19.** $4k \cdot 10^{4x-1}$ where $e^k = 10$, $4k \approx 9.2103$
21. $2k(10^{2x} - 10^{-2x})$ where $e^k = 10$, $2k \approx 4.6052$
23. $a = e^k$, $b = e^{\ell}$; $a^x b^x = e^{kx} e^{\ell x} = e^{kx+\ell x} = e^{(k+\ell)x} = (ab)^x$ since $ab = e^k e^{\ell} = e^{k+\ell}$.
25. $a = e^k$, $a^x a^y = e^{kx} e^{ky} = e^{kx+ky} = e^{k(x+y)} = a^{x+y}$.
27. $\frac{1}{2}e^{2x} + c$ **29.** $(10^x/k) + c$ where $e^k = 10$ **31.** $xe^x - e^x + c$ **33.** $y = e^x - e^{-x} + 1$

Section 3   page 170

1. 0   **3.** e^{-1} **5.** e^5 **7.** 0

9. $\left(1 + \dfrac{1}{n}\right)^n \approx e$, so $\log_{10}\left(1 + \dfrac{1}{n}\right)^n \approx \log_{10} e \approx 0.4343$. But $\log_{10}\left(1 + \dfrac{1}{n}\right)^n = n \log_{10}\left(1 + \dfrac{1}{n}\right)$

so the result follows.

11. $y(0) = 0$, $y(x) > 0$ for $x > 0$, and $\lim_{x \to \infty} y = 0$. Hence y has a max. Now $y' = 0$ only for $x = 1$, hence $y_{max} = y(1) = e^{-1}$.

13. $y(0) = 2$ and $\lim_{x \to \infty} y = -\infty$. Also $y'(0) = 1 > 0$, so $y(x) > y(0)$ for x near 0. Finally $y'(x) = 0$ only for $e^x = \frac{3}{2}$, so $y_{max} = 3(\frac{3}{2}) - (\frac{3}{2})^2 = \frac{9}{4}$.

15.

x	10	100	1000	10^6
e^x/x^x	2.2×10^{-6}	2.7×10^{-157}	2.0×10^{-2566}	3.0×10^{-5565706}

Section 4   page 174

1. $m(t) = m_0 e^{-\lambda t}$, $\lambda = \log 2/\log e^{3.64} \approx 0.1904$. Alternatively: $m(t) = m_0 2^{-t/3.64}$; about 5.77 days

3. $\log 2/\log e \approx 69.315\%$ **5.** $m(t) = (3 \times 10^6)3^{t/2}$; $\log 4/\log 3 \approx 1.26$ hr

7. $30(\frac{5}{6})^5 \approx 12.1$ in.   **9.** about 2.19 days.   **11.** $\sqrt{a/b}$

13. Set $y = a(b - x)/b(a - x)$. Then by the Chain Rule,

$$\frac{dy}{dt} = \frac{dy}{dx}\frac{dx}{dt} = \frac{a(b - a)}{b(a - x)^2} k(a - x)(b - x) = k(b - a)y,$$

hence $y = y_0 e^{k(b-a)t}$. Set $t = 0$. Then $x_0 = 0$, so $y_0 = ab/ba = 1$, $y = e^{k(b-a)t}$.

15. $u = a + (u_0 - a)e^{-kt}$. As $t \longrightarrow \infty$, $e^{-kt} \longrightarrow 0$, hence $u \longrightarrow a$.

17. $(-U)_{max} = k = -U(c)$ **19.** $v = (v_0 - g/\lambda)e^{-\lambda t} + g/\lambda$, $v \longrightarrow g/\lambda$ as $t \longrightarrow \infty$

21. $\dot{m} = -(m/u)\dot{v}$. But $\dot{m} = (dm/dv)\dot{v}$ by the Chain Rule, hence $dm/dv = -m/u$, so $m = ce^{-v/u}$. Set $t = 0$. Then $v = 0$ and $m = m_0$, so $c = m_0$, $m = m_0 e^{-v/u}$. The dot is d/dt as usual.

23. $(mv)^{\cdot} = \dot{m}v + m\dot{v} = (v - u)\dot{m} - mg$. Cancel $\dot{m}v$:

$$m\frac{dv}{dt} = -u\frac{dm}{dt} - mg, \qquad m\left(\frac{dv}{dt} + g\right) = -u\frac{dm}{dt}.$$

Set $w = v + gt$. Then $\dot{w} = \dot{v} + g$, so

$$m\frac{dw}{dt} = -u\frac{dm}{dt} = -u\frac{dm}{dw}\frac{dw}{dt}, \qquad \frac{dm}{dw} = -\frac{m}{u}, \qquad m = m_0 e^{-w/u} = m_0 e^{-(v+gt)/u}.$$

25. Let $m(t) = m_0 e^{-\lambda t}$ be the mass of ^{14}C per gram of wood killed at the time of Hammurabi. Here t is measured in years from that time, m_0 is the mass of ^{14}C per gram of living wood, and $e^{\lambda} = 2^{1/5568}$. Now $\dot{m} = -\lambda m = -\lambda m_0 e^{-\lambda t}$ is the rate of decay, proportional to the dpm. Hence

$$\frac{\dot{m}_0}{\dot{m}} = \frac{-\lambda m_0}{-\lambda m_0 e^{-\lambda t}} = e^{\lambda t}, \qquad \frac{6.68}{4.09} = e^{\lambda t}, \qquad t\log(e^{\lambda}) = \log\frac{6.68}{4.09},$$

$$t = \frac{\log(6.68/4.09)}{\log(e^{\lambda})} = \frac{\log(6.68/4.09)}{\log(2^{1/5568})} = (5568)\frac{\log(6.68/4.09)}{\log 2} \approx 3940 \text{ yr.}$$

This is the time from the date we want to 1950 A.D., so Hammurabi's reign was about 3940–1950 = 1990 B.C. (Some give his dates as 1955–1913 B.C. Of course this particular tree could have been cut earlier.)

27. Set $n(t) = n_0 e^{kt - act^2/2}$. Then $\dfrac{dn}{dt} = \dfrac{d}{dt}\left(kt - \dfrac{1}{2}act^2\right)n = (k - act)n = kn - anx$

and $n(0) = n_0$. If $t \longrightarrow \infty$, then $kt - \frac{1}{2}act^2 \longrightarrow -\infty$ so $n \longrightarrow 0$.

29. If $h = a_2/b_2$ and $p = a_1/b_1$, then

$$a_1 h - b_1 ph = 0 = h \qquad \text{and} \qquad -a_2 p + b_2 ph = 0 = p.$$

31. By the Chain Rule $\dfrac{d}{dt}\left(\dfrac{p}{a - bp}\right) = \dfrac{a}{(a - bp)^2}(ap - bp^2) = a\left(\dfrac{p}{a - bp}\right),$

hence $p/(a - bp) = ke^{at}$. From this, $p = (a - bp)ke^{at}$, $(1 + bke^{at})p = ake^{at}$, etc.

33. 2000: 5.74×10^9, 2050: 8.44×10^9, 2100: 9.49×10^9, ultimately: 9.86×10^9

35. $R = R_\infty \exp[-kH\rho_0 e^{-h/H}]$

Section 5, page 183

1. $\frac{1}{3}\pi$ $\frac{5}{6}\pi$ $-\frac{4}{3}\pi$ $\frac{13}{6}\pi$ $-\frac{5}{2}\pi$ 5π

3. $\frac{1}{4}\pi$ $-\pi$ $\frac{3}{2}\pi$ $\frac{7}{2}\pi$ $-\frac{3}{4}\pi$ $\frac{11}{4}\pi$

5. $90°$ $-120°$ $300°$ $540°$ $-480°$ $3000°$

7. $45°$ $24°$ $75°$ $5°$ $-165°$ $-12°$

9. $(\frac{1}{2}\sqrt{2}, \frac{1}{2}\sqrt{2})$ $(-\frac{1}{2}\sqrt{2}, -\frac{1}{2}\sqrt{2})$ $(-\frac{1}{2}\sqrt{2}, \frac{1}{2}\sqrt{2})$

11. $(-1, 0)$ $(-\frac{1}{2}, -\frac{1}{2}\sqrt{3})$ $(\frac{1}{2}\sqrt{3}, \frac{1}{2})$

13. $\frac{1}{4}\pi, \frac{3}{4}\pi$ **15.** $\frac{5}{6}\pi, \frac{7}{6}\pi$ **17.** $\frac{3}{2}\pi$ **19.** $\frac{1}{4}\pi, \frac{5}{4}\pi$

21. No: for functions, even $\times$ odd = odd, odd $\times$ odd = even.

23. odd **25.** even **27.** even **29.** 1 **31.** π **33.** π

35. Replace θ by $-\theta$ in $\cos(\theta + \pi) = -\cos\theta$ and $\sin(\theta + \pi) = -\sin\theta$, and use parity.

37. $\sin(\alpha \pm \pi) = -\sin\alpha$, $\cos(\alpha \pm \pi) = -\cos\alpha$

39. $\frac{1}{2}\cos\theta + \frac{1}{2}\sqrt{3}\sin\theta$ **41.** $3\sin\theta\cos^2\theta - \sin^3\theta = 3\sin\theta - 4\sin^3\theta$

43. $4\sin\theta\cos\theta(\cos^2\theta - \sin^2\theta)$

45. $\cos(\alpha + \beta) + \cos(\alpha - \beta) = (\cos\alpha\cos\beta - \sin\alpha\sin\beta) + (\cos\alpha\cos\beta + \sin\alpha\sin\beta) = 2\cos\alpha\cos\beta$

47. Use Ex. 45 six times:

$(\cos x + \cos 3x) + (\cos 5x + \cos 7x) + \cdots + (\cos 13x + \cos 15x)$

$= 2\cos x\cos 2x + 2\cos x\cos 6x + 2\cos x\cos 10x + 2\cos x\cos 14x$

$= 2\cos x[(\cos 2x + \cos 6x) + (\cos 10x + \cos 14x)]$

$= 2\cos x[2\cos 2x\cos 4x + 2\cos 2x\cos 12x]$

$= 4\cos x\cos 2x(\cos 4x + \cos 12x)$

$= 4\cos x\cos 2x(2\cos 4x\cos 8x).$

49. $\cos 2x = 1 - 2\sin^2 x$, so $\cos\theta = 1 - 2\sin^2\frac{1}{2}\theta$, $\sin^2\frac{1}{2}\theta = \frac{1}{2}(1 - \cos\theta)$, $\sin\frac{1}{2}\theta = \pm\sqrt{\frac{1}{2}(1 - \cos\theta)}$ with $+$ if $0 \le \theta \le 2\pi$ and $-$ if $-2\pi \le \theta < 0$.

51.

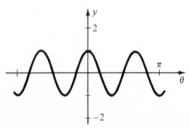

53.

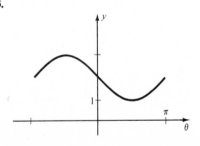

55.

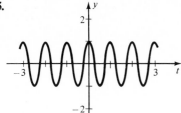

Section 6,   page 188

1. $\sin x = \dfrac{\pm 1}{\sqrt{1 + \cot^2 x}}$, $+$ in quadrants 1 and 4   **3.** $\cot^2 x = \dfrac{\cos^2 x}{1 - \cos^2 x}$

5. $\cot(\alpha - \beta) = \dfrac{1}{\tan(\alpha - \beta)} = \dfrac{1 - \tan \alpha \tan(-\beta)}{\tan \alpha + \tan(-\beta)} = \dfrac{1 + \tan \alpha \tan \beta}{\tan \alpha - \tan \beta} = \dfrac{\cot \alpha \cot \beta + 1}{\cot \beta - \cot \alpha}$.

(Last step: divide numerator and denominator by $\tan \alpha \tan \beta$.)

7. $\dfrac{\sin 2\theta}{1 + \cos 2\theta} = \dfrac{2 \sin \theta \cos \theta}{1 + (2 \cos^2 \theta - 1)} = \dfrac{\sin \theta}{\cos \theta} = \tan \theta.$ **9.** $\frac{1}{2}\pi$

11.   **13.**   **15.**

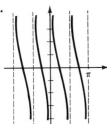 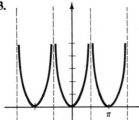 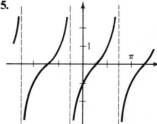

Section 7,   page 194

1. $-2x \sin x^2$ **3.** $2e^x \sin x$ **5.** $\tan x + x \sec^2 x$ **7.** $2(x \cos 2x - \sin 2x)/x^3$

9. $(\sec x)(\tan x - 1)/(1 + \tan x)^2$ **11.** $-2/(\sin x - \cos x)^2$ **13.** $-\cos x$

15. $(\cot x)' = (\cos x/\sin x)' = (-\sin^2 x - \cos^2 x)/\sin^2 x = -1/\sin^2 x = -\csc^2 x,$

$(\csc x)' = (1/\sin x)' = -(\cos x)/\sin^2 x = -\cot x \csc x$

17. $\sec^4 x - \tan^4 x = (\sec^2 x + \tan^2 x)(\sec^2 x - \tan^2 x) = \sec^2 x + \tan^2 x = 1 + 2 \tan^2 x,$

so $(\sec^4 x - \tan^4 x)' = (1 + 2 \tan^2 x)' = 4 \tan x \sec^2 x.$

19.

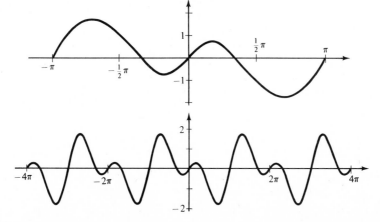

21.

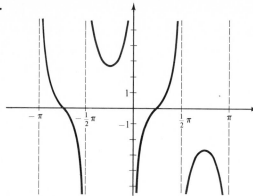

23.

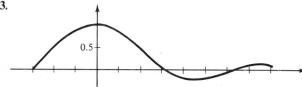

25.

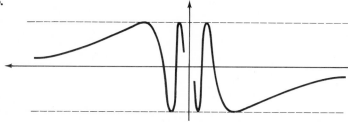

27. $(\frac{1}{3}\pi + 2n\pi, \frac{1}{2}\sqrt{3}), (-\frac{1}{3}\pi + 2n\pi, -\frac{1}{2}\sqrt{3})$

29. $|\sin x| \le 1$, so $\sin x < x$ if $x > 1$. For $0 \le x \le 1$, set $f(x) = x - \sin x$. Then $f(0) = 0$ and $f'(x) = 1 - \cos x > 0$ for $0 < x < 1$. Hence f increases, $f(x) > f(0)$ for $0 < x \le 1$, so $\sin x < x$ for all $x > 0$.

Section 8,   page 198

1. $\dot{x} = a\omega \cos \frac{1}{2}\theta$ **3.** $\frac{2}{3}(3 - \sqrt{6})\pi$

5. $6 - \sqrt{3}$ m; the angle at the moveable pulley is $120°$.   **7.** 4π in./min to the left

9. $[(a^{2/3} + b^{2/3})^3 + c^2]^{1/2}$

11. At $3:00$ the distance is *decreasing* at the rate $11\pi ab/6\sqrt{a^2 + b^2}$ cm/hr. At $8:00$ the distance is *increasing* at the rate $(11\pi ab\sqrt{3})/12(a^2 + ab + b^2)^{1/2}$ cm/hr.

13. 0.2332 deg/sec   **15.** $\frac{2}{9}a^2\sqrt{3}$

Section 9,   page 201

1. $(2x + 3)e^{-3/x}$ **3.** $(1 - x)/e^x$

5. Set $f(x) = e^x - x - 1$. Then $f'(x) = e^x - 1$, so $f'(x) < 0$ for $x < 0$, $f'(x) > 0$ for $x > 0$, and $f'(0) = 0$. Hence $f(x)$ decreases to $f(0)$, then increases, so for all x, $f(x) \ge f(0) = 0$, that is, $e^x \ge 1 + x$.

7. -1 **9.** 1

11. Use logs. For example, $\log[\exp(10^3)] = 10^3 \log e \approx 434.2945$, so $\exp(10^3) \approx 1.97 \times 10^{434}$. Similarly $\log[\exp(10^6)] \approx 434294.4819$, so $\exp(10^6) \approx 3.033 \times 10^{434294}$.

13. $e^{-1/e}$

15. $\dfrac{d}{dx}(e^{-1/x}) = \dfrac{e^{-1/x}}{x^2}$, so the assertion is true for $n = 1$. If it is true for n, then

$$\frac{d^{n+1}}{dx^{n+1}}(e^{-1/x}) = \frac{d}{dx}\left[\frac{d^n}{dx^n}(e^{-1/x})\right] = \frac{d}{dx}\left[\frac{P_n(x)e^{-1/x}}{x^{2n}}\right]$$

$$= \frac{(P_n'e^{-1/x} + P_n e^{-1/x}/x^2)x^{2n} - 2nx^{2n-1}P_n e^{-1/x}}{x^{4n}}$$

$$= \frac{[x^2 P_n' + (1 - 2nx)P_n]e^{-1/x}}{x^{2n+2}} = \frac{P_{n+1}(x)e^{-1/x}}{x^{2(n+1)}},$$

where P_{n+1} has degree at most n. The formula follows by induction.

17. As $x \longrightarrow 0+$, $(e^x - 1)/x \longrightarrow de^x/dx\,|_0 = 1$ and $e^x \longrightarrow 1$, so $f(x) \longrightarrow 1$. For $x \longrightarrow \infty$,

$$f(x) = f(x)\frac{e^{2x}}{e^{2x}} = \frac{x^2 e^x}{e^{2x}}\left(\frac{e^x}{e^x - 1}\right)^2 = \left(\frac{x^2}{e^x}\right)\left(\frac{1}{1 - e^{-x}}\right)^2 \longrightarrow 0 \cdot 1^2 = 0.$$

19. For $x < 0$, $\phi^{(n)}(x) = 0$ and for $x > 0$, $\phi^{(n)}(x)$ is given by Ex. 15. It remains only to check differentiability at $x = 0$. Let us show by induction that $\phi^{(n)}(0) = 0$, starting with $\phi^{(0)}(0) = \phi(0) = 0$. Now $\phi^{(n+1)}(0) = \lim[\phi^{(n)}(x) - \phi^{(n)}(0)]/x = \lim_{x \to 0} \phi^{(n)}(x)/x$. If $x < 0$, then $\phi^{(n)}(x)/x = 0$. If $x > 0$, then

$$\phi^{(n)}(x)/x = P_n(x)e^{-1/x}/x^{2n+1} = P_n(x)t^{2n+1}e^{-t}, \qquad t = 1/x.$$

If $x \longrightarrow 0$, then $t \longrightarrow \infty$ and $t^{2n+1}e^{-t} \longrightarrow 0$. Hence $\lim_{x \to 0} \phi^{(n)}(x)/x = 0$, so $\phi^{(n+1)}(0) = 0$, which completes the induction.

21. $f(t)e^{ct}$ is continuous for $t \geq 0$. Also $f(t)e^{ct} = t^a e^{-(b-c)t} \longrightarrow 0$ as $t \longrightarrow \infty$. Hence $f(t)e^{ct} \leq K$ for some $K > 0$. This implies $f(t) \leq Ke^{-ct}$.

23. $(1.5)^{3/2} \times 10^5 \approx 184{,}000$.　　**25.** $(e^{Rt/L}I)^{\cdot} = (E/L)e^{Rt/L}$, $I = (E/R)(1 - e^{-Rt/L})$

27. $\dfrac{1 - \tan^2 x}{1 + \tan^2 x} = \dfrac{\cos^2 x - \sin^2 x}{\cos^2 x + \sin^2 x} = \dfrac{\cos 2x}{1}$.

Alternative: $\dfrac{1 - \tan^2 x}{1 + \tan^2 x} = \dfrac{1 - \tan^2 x}{\sec^2 x} = (\cos^2 x)(1 - \tan^2 x) = \cos^2 x - \sin^2 x = \cos 2x$.

29. Set $x = \frac{1}{2}(\alpha + \beta)$ and $y = \frac{1}{2}(\alpha - \beta)$ so $x + y = \alpha$ and $x - y = \beta$. Then subtract the relations

$$\begin{cases} \sin \alpha = \sin(x + y) = \sin x \cos y + \sin y \cos x \\ \sin \beta = \sin(x - y) = \sin x \cos y - \sin y \cos x. \end{cases}$$

31. 0

33. Choose α in the first quadrant so $\tan \alpha = \sqrt{2}$ ($\alpha \approx 0.9553$). Then $y(\alpha + 2n\pi) = y(-\alpha + 2n\pi) = \frac{4}{9}\sqrt{3}$ and $y(\pi + 2n\pi) = 0$ are local maxs; $y(\pi - \alpha + 2n\pi) = y(\pi + \alpha + 2n\pi) = -\frac{4}{9}\sqrt{3}$ and $y(0) = 0$ are local mins.

35. $x_{\max} = \dfrac{v_0^2}{g}\left(\dfrac{1 - \sin \alpha}{\cos \alpha}\right)$; the maximizing angle is $\theta = \frac{1}{2}(\alpha + \frac{1}{2}\pi)$, halfway between the slope of the hill and the vertical.

37. It starts at $y = 7$ and oscillates with period $2\pi/\omega = 4\pi/\sqrt{g}$ sec between $y = 3$ and 7; also $v(t) = -2\omega \sin \omega t$ and $a(t) = -2\omega^2 \cos \omega t = -\frac{1}{2}g \cos \omega t$. This is an example of **simple harmonic motion**.

39. $\dot{y} = 2\dot{x} = 2c \sin x$, $\ddot{y} = 2c\dot{x} \cos x = 2c^2 \sin x \cos x = c^2 \sin 2x = c^2 \sin y$

41. $y_{\max} = y(\frac{11}{3}\pi) = \frac{11}{6}\pi + \frac{1}{2}\sqrt{3} \approx 6.6256$. Note that $y(4\pi) = 2\pi \approx 6.2832$ is less.

43. Set $f(t) = $ LHS. Then $f(0) = f'(0) = f''(0) = 0$ and $f'''(t) = (2 + t)e^t > 0$ for $t > 0$, etc.

CHAPTER 5

Section 2,   page 212

1. $\frac{1}{2}(b^2 - a^2)$ **3.** $\frac{1}{2}(b^2 - a^2)$ **5.** $\frac{1}{3}(b^3 - a^3)$

7. $\displaystyle\int_a^b \frac{1}{x}\, dx = \lim \sum_1^n (ar^j - ar^{j-1})(ar^j)^{-1} = \lim \sum_1^n \frac{r-1}{r} = \lim \frac{r-1}{r} n = \lim \left[\frac{1}{r} \frac{(b/a)^{1/n} - 1}{1/n} \right].$

But $\lim_{r \to 1} (1/r) = 1$ and $\displaystyle\lim_{n \to \infty} \frac{(b/a)^{1/n} - 1}{1/n} = \lim_{t \to 0} \frac{(b/a)^t - 1}{t} = \frac{d}{dt} \left(\frac{b}{a} \right)^t \Big|_{t=0}.$

9. $1 - \cos b$ **11.** $\frac{1}{4}(b^4 - a^4)$

Section 3,   page 222

1. $\frac{3}{2}$ **3.** 0   **5.** $\frac{1}{2}$ **7.** 2   **9.** $\frac{20}{3}$ **11.** $\frac{7}{2} - e^{-1}$ **13.** $\frac{1}{6}(b - a)^3$ **15.** $\frac{75}{4}$
17. 0   **19.** 1   **21.** $\sec x$ **23.** $(\cos x)\sqrt{\sin x}$ **25.** 3 cm^2/sec

Section 4,   page 228

1. 9   **3.** 72   **5.** 2   **7.** 5   **9.** $\frac{16}{15}$ **11.** 12   **13.** 2   **15.** $\frac{29}{6}$
17. The rectangle with vertices $(0, 0)$, $(b, 0)$, (b, b^2), $(0, b^2)$ has area b^3 and splits into the region between the x-axis and $y = x^2$ and the region between the y-axis and $x = \sqrt{y}$.

$$\int_0^b x^2\, dx + \int_0^{b^2} \sqrt{y}\, dy = \frac{1}{3} x^3 \Big|_0^b + \frac{2}{3} y^{3/2} \Big|_0^{b^2} = \frac{1}{3} b^3 + \frac{2}{3} b^3 = b^3.$$

19. $\frac{1}{2}\pi a^2$ **21.** 0   **23.** $\frac{81}{5}$ **25.** $\frac{1}{2}(e - e^{-1})$ **27.** \$39   **29.** 1.1 cm
31. $\frac{1}{4}n^4$; $\approx 1.97\%$

Section 5,   page 235

	$n = 4$	$n = 10$	$n = 50$	exact	6 sig. figs.
1.	0.697	0.6938	0.693172	$\ln 2$	0.693147
3.	0.606	0.6125	0.613656	$2 - 2 \ln 2$	0.613706
5.	0.458	0.4580	0.458141	$\frac{1}{2} \ln \frac{5}{2}$	0.458145
7.	9.37	9.307	9.29408	$\frac{1}{2}[4\sqrt{17} + \ln(4 + \sqrt{17})]$	9.29357
9.	0.608	0.9201	0.996173	$1 - 11e^{-10}$	0.999501
11.	0.504	0.5046	0.504785	unknown	0.504792
13.	0.167	0.1676	0.167758	$(2 \ln 2 - 1)/\ln 10$	0.167766
15.	0.350	0.3471	0.346594	$\ln \sqrt{2}$	0.346574
17.	1.18	1.148	1.14185	$\pi - 2$	1.14159
19.	0.785	0.7854	0.785398	$\frac{1}{4}\pi$	0.785398

21. 592.8 ft   **23.** exact ≈ 2.667, trap = 3, mid = 2.5
25. exact ≈ 1.0986, trap ≈ 1.1667, mid ≈ 1.0667
27. $|f''| = |-3e^{-x} \sin 2x - 4e^{-x} \cos 2x| < 7e^{-x} < \frac{7}{50}$ for $4 \le x \le 7$, hence

$$|\text{error}| < \frac{(\frac{7}{50})(100)(\frac{3}{100})^3}{12} < (\frac{7}{50})(\frac{27}{12})(10^{-4}) < 4 \times 10^{-5}$$

Section 6,   page 240

1. -2 **3.** $-\frac{1}{4}(e^2 + 3e^{-2})$ **5.** $x \tan x - F$ **7.** 102.3
9. $721/(8 \times 9^3) = 721/5832$ **11.** $\frac{1}{2}\exp(x^2)$ **13.** $-\exp(\cos x)$ **15.** $\frac{2}{3}(x^4 + x)^{3/2}$
17. Use Translation with $f(x) = \sqrt{2x + 7}$ and $c = -1$.
19. Use Reflection with $f(x) = 1/(e^x + 1)$. Note that $f(-x) = 1/(e^{-x} + 1) = e^x/(e^x + 1)$.
21. Use Translation with $f(x) = e^x \sin x$, $c = 2\pi$, $a = 0$ and $b = 2\pi$.

Section 7,   page 246

1. $x = 0$ **3.** $(-1, 0)$ **5.** $x = \frac{1}{4}\pi + n\pi,\ (-\frac{1}{4}\pi + n\pi, 0)$ **7.** $x = -1$
9. $\displaystyle\int_1^2 (x^3 - 5x)\, dx$ **11.** $\displaystyle\int_{1/2 - 5/2}^{1/2 + 5/2} \frac{2(x - \frac{1}{2})}{(x - \frac{1}{2})^2 + \frac{3}{4}}\, dx = 0$ **13.** $2\displaystyle\int_0^{\pi/2} \sin x\, dx$
15. $4\displaystyle\int_0^{\pi/2} \sin^4 x\, dx$ **17.** $-2(\sin 4)\displaystyle\int_0^{4\pi} \cos(\tfrac{1}{12}x)\, dx$ **19.** $f(x) = 0$

21. Set $x = 0$: $\frac{1}{2}[f(a) + f(a)] = b = f(a);\ \displaystyle\int_{a-h}^{a+h} f(x)\, dx = 2hf(a)$.

Section 8,   page 250

1. $1 < \exp(x^2) < e^x$ for $0 < x < 1$, hence $1 = \displaystyle\int_0^1 dx < \int_0^1 \exp(x^2)\, dx < \int_0^1 e^x\, dx = e - 1$

3. $3 < \sqrt{3 + 2x} < \sqrt{13} < 4$ for $3 < x < 5$, etc.
5. $\displaystyle\int_0^{100} e^{-x} \sin^2 x\, dx \le \int_0^{100} e^{-x}\, dx = 1 - e^{-100} < 1$
7. $25 < 5 + 4x < 49$ for $5 < x < 11$, so $5 < \sqrt{5 + 4x} < 7$ and

$$6 < \frac{48}{7} = \frac{1}{7}\int_5^{11} x\, dx < \int_5^{11} \frac{x}{\sqrt{5 + 4x}}\, dx < \frac{1}{5}\int_5^{11} x\, dx = \frac{48}{5} < 10.$$

9. $1 - k^2 < 1 - k^2 \sin^2\theta < 1$ for $0 < \theta < \frac{1}{2}\pi$, etc.
11. $\frac{1}{4}x^3 < x^3 2^{-x} < \frac{1}{2}x^3$ for $1 < x < 2$ and $\displaystyle\int_1^2 x^3\, dx = \frac{15}{4}$, etc.
13. $\sin x + \cos x = \sqrt{2}\cos(x - \frac{1}{4}\pi) < \sqrt{2}$ for $0 < x < 1$ except for $x = \frac{1}{4}\pi$, so

$$\int_0^1 \frac{\sin x + \cos x}{(1 + x)^2}\, dx < \sqrt{2}\int_0^1 \frac{dx}{(1 + x)^2} = \frac{1}{2}\sqrt{2}.$$

15. Use $\frac{2}{3} < 1/x < 1$ for $1 < x < \frac{3}{2}$ and $\frac{1}{2} < 1/x < \frac{2}{3}$ for $\frac{3}{2} < x < 2$.
17. $2 - \sqrt{2} = \displaystyle\int_1^2 \frac{dx}{x^{3/2}} < \int_1^2 \frac{dx}{x} < \int_1^2 \frac{dx}{x^{1/2}} = 2(\sqrt{2} - 1)$.
19. $\displaystyle\int_0^{100} - \int_0^{10} = \int_{10}^{100} < e^{-10}\int_{10}^{100} dx = 90e^{-10} < 0.0041$.

21. By Example 2, $\pm 2\varepsilon \displaystyle\int_a^b f(x)g(x)\, dx \le \varepsilon^2 \int_a^b g(x)^2\, dx$ for all $\varepsilon > 0$. It follows that

$\left|\displaystyle\int fg\right| \le \frac{1}{2}\varepsilon\displaystyle\int g^2$ for all $\varepsilon > 0$. This is possible only if $\displaystyle\int_a^b f(x)g(x)\, dx = 0$.

23. $\int (f + g)^2 = \int f(f + g) + \int g(f + g) \le \sqrt{(\int f^2)(\int (f + g)^2)} + \sqrt{(\int g^2)(\int (f + g)^2)}$

$= (\sqrt{\int f^2} + \sqrt{\int g^2})\sqrt{\int (f + g)^2}$.

If $\int (f + g)^2 = 0$, there is nothing to prove. Otherwise $\int f + g)^2 > 0$ and its square root can be cancelled.

Section 9, page 259

1. Let $a = x_0 < x_1 < \cdots < x_n = b$ be all of the partition points for $s_1(x)$ *and* for $s_2(x)$. Then
$s_1(x) = A_j$ and $s_2(x) = B_j$ for $x_{j-1} < x < x_j$, and $\int_a^b (s_1 + s_2)(x)\, dx =$

$$\sum_{j=1}^{n} (A_j + B_j)(x_j - x_{j-1}) = \sum A_j(x_j - x_{j-1}) + \sum B_j(x_j - x_{j-1})$$

$$= \int_a^b s_1(x)\, dx + \int_a^b s_2(x)\, dx.$$

3. $s(x) = B_j \geq 0$ on $x_{j-1} < x < x_j$, so $\int_a^b s(x)\, dx = \sum_1^n B_j(x_j - x_{j-1}) \geq 0$.

5. Choose the partition to include c so that it is $a = x_0 < x_1 < \cdots < x_m = c = y_0 < y_1 < \cdots < y_n = b$. Let $s(x) = A_i$ on (x_{i-1}, x_i) and $s(x) = B_j$ on (y_{j-1}, y_j). Then

$$\int_a^b s(x)\, dx = \sum_{i=1}^{m} A_i(x_i - x_{i-1}) + \sum_{j=1}^{n} B_j(y_j - y_{j-1}) = \int_a^c s(x)\, dx + \int_c^b s(x)\, dx.$$

7. For instance, take $\quad s(x) = \begin{cases} 0, & 0 \leq x < \frac{2}{3} \\ \frac{4}{9}, & \frac{2}{3} \leq x \leq 1 \end{cases} \quad S(x) = \begin{cases} \frac{1}{3}, & 0 \leq x \leq \frac{1}{3}\sqrt{3} \\ 1, & \frac{1}{3}\sqrt{3} < x \leq 1. \end{cases}$

Then $s(x) \leq x^2 \leq S(x)$ for $0 \leq x \leq 1$ and

$$\int_0^1 S - \int_0^1 s = (1 - \tfrac{2}{9}\sqrt{3}) - \tfrac{4}{27} = \tfrac{1}{27}(23 - 6\sqrt{3}) < 0.47.$$

(This is the best you can do with two steps only.)

9. If $\varepsilon > 0$, choose s_1 so $|f(x) - s_1(x)| < \frac{1}{3}\varepsilon/(b - a)$. Set $s(x) = s_1(x) - \frac{1}{3}\varepsilon/(b - a)$ and $S(x) = s_1(x) + \frac{1}{3}\varepsilon/(b - a)$, both step functions. We have $s(x) < f(x) < S(x)$ and

$$\int_a^b S(x)\, dx - \int_a^b s(x)\, dx = \frac{2}{3} \int_a^b \frac{\varepsilon}{b - a}\, dx = \tfrac{2}{3}\varepsilon < \varepsilon.$$

11. Let $\varepsilon > 0$. Choose step functions such that $\quad s_1 \leq f \leq S_1, \quad s_2 \leq g \leq S_2,$

$$\int_a^b S_1 - \int_a^b s_1 < \tfrac{1}{2}\varepsilon, \quad \text{and} \quad \int_a^b S_2 - \int_a^b s_2 < \tfrac{1}{2}\varepsilon.$$

Then $s_1 + s_2 \leq f + g \leq S_1 + S_2$ and

$$\int_a^b (S_1 + S_2) - \int_a^b (s_1 + s_2) = \left[\int_a^b S_1 - \int_a^b s_1 \right] + \left[\int_a^b S_2 - \int_a^b s_2 \right] < \tfrac{1}{2}\varepsilon + \tfrac{1}{2}\varepsilon = \varepsilon.$$

Hence $f + g$ is integrable. The inequality $s_1 + s_2 \leq f + g \leq S_1 + S_2$ now implies

$$\int s_1 + \int s_2 = \int (s_1 + s_2) \leq \int f + \int g \leq \int (S_1 + S_2) = \int S_1 + \int S_2.$$

On the other hand, the inequalities $s_1 \leq f \leq S_1$ and $s_2 \leq g \leq S_2$ imply

$$\int s_1 \leq \int f \leq \int S_1 \quad \text{and} \quad \int s_2 \leq \int g \leq \int S_2.$$

hence $\int s_1 + \int s_2 \leq \int (f + g) \leq \int S_1 + \int S_2$. Thus the two numbers $\int (f + g)$ and $(\int f) + (\int g)$ are squeezed into an interval of length less than ε. Since $\varepsilon > 0$ is arbitrary, they are equal: $\int (f + g) = \int f + \int g$.

13. $s(x) = 0$ is a step function and $f(x) \geq s(x)$, hence

$$\int_a^b f(x)\, dx \geq \int_a^b s(x)\, dx = 0.$$

15. Let $\varepsilon > 0$. Choose step functions s_1 and S_1 on $[a, c]$ and step functions s_2 and S_2 on $[c, b]$ such that

$$s_1(x) \le f(x) \le S_1(x) \quad \text{on } [a, c], \qquad \int_a^c S_1 - \int_a^c s_1 < \tfrac{1}{2}\varepsilon,$$

$$s_2(x) \le f(x) \le S_2(x) \quad \text{on } [c, b], \qquad \int_c^b S_2 - \int_c^b s_2 < \tfrac{1}{2}\varepsilon.$$

Define $s(x)$ on $[a, b]$ by $s(x) = s_1(x)$ for $a \le x < c$, $s(x) = s_2(x)$ for $c < x \le b$, and $s(c) = f(c)$. Define $S(x)$ similarly. Then $s(x) \le f(x) \le S(x)$ on $[a, b]$ and

$$\int_a^b S - \int_a^b s = \left(\int_a^c S_1 + \int_c^b S_2 \right) - \left(\int_a^c s_1 + \int_c^b s_2 \right) = \left(\int_a^c S_1 - \int_a^c s_1 \right) + \left(\int_c^b S_2 - \int_c^b s_2 \right)$$

$< \tfrac{1}{2}\varepsilon + \tfrac{1}{2}\varepsilon = \varepsilon.$ Therefore f is integrable on $[a, b]$.

17. $S_1 S_2 - s_1 s_2 = S_1(S_2 - s_2) + s_2(S_1 - s_1) \le M(S_2 - s_2) + M(S_1 - s_1)$,   hence

$$\int_a^b S_1 S_2 - \int_a^b s_1 s_2 = \int_a^b (S_1 S_2 - s_1 s_2) \le \int_a^b M(S_2 - s_2) + M(S_1 - s_1)$$

$$= M \int_a^b (S_2 - s_2) + M \int_a^b (S_1 - s_1) \le M\varepsilon + M\varepsilon = 2M\varepsilon.$$

19. First choose M so $f(x) \le M$ and $g(x) \le M$. Suppose $\varepsilon > 0$. By Ex. 18 there are step functions $s_1, \cdots, S_2$ such that $0 \le s_1 \le f \le S_1 \le M$, $0 \le s_2 \le g \le S_2 \le M$, and

$$\int_a^b S_1 - \int_a^b s_1 < \frac{\varepsilon}{2M}, \qquad \int_a^b S_2 - \int_a^b s_2 < \frac{\varepsilon}{2M}.$$

Then $s_1 s_2 \le fg \le S_1 S_2$ and by Ex. 17, $\displaystyle\int_a^b S_1 S_2 - \int_a^b s_1 s_2 < 2M(\varepsilon/2M) = \varepsilon$. Since $s_1 s_2$ and $S_1 S_2$ are step functions, this implies that fg is integrable.

21. Given $\varepsilon > 0$, choose integrable g and G so $g(x) \le f(x) \le G(x)$ and $\int (G - g)\, dx < \tfrac{1}{3}\varepsilon$. Choose step functions s and S so $s(x) \le g(x)$, $G(x) \le S(x)$, $\int (g - s)\, dx < \tfrac{1}{3}\varepsilon$, and $\int (S - G)\, dx < \tfrac{1}{3}\varepsilon$. Then $s(x) \le f(x) \le S(x)$ and

$$\int (S - s)\, dx = \int [(S - G) + (G - g) + (g - s)]\, dx < \tfrac{1}{3}\varepsilon + \tfrac{1}{3}\varepsilon + \tfrac{1}{3}\varepsilon < \varepsilon.$$

23. $\displaystyle\int_a^{a+p} f'(x)\, dx = f(a + p) - f(a) = 0$, $\displaystyle\int_a^{a+p} f''(x)\, dx = f'(a + p) - f'(a) = 0$

because f' is also periodic of period p. (Just differentiate $f(x + p) = f(x)$.)

Section 10,   page 260

1. $\frac{512}{15}$ **3.** 1   **5.** 0   **7.** $\frac{1}{2}$ **9.** $-\exp(-x^2)$ **11.** $\frac{15}{2}$ **13.** $M(f) = 1$

15. $f(x - c) = -f(c - x) = +f(c + x)$, hence $f(x) = f(x + 2c)$, so $f(x)$ has period $2c$. Therefore all integrals over a full period are equal.

17. $f(x) = \tfrac{1}{2}x^2$ for $x \ge 0$; $f(x) = -\tfrac{1}{2}x^2$ for $x \le 0$.   **19.** sum $\approx 2(\sqrt{2} - 1)\sqrt{n}$.

CHAPTER 6

Section 2,   page 268

1. 36   **3.** 4   **5.** $\frac{36}{25}$ **7.** $2\sqrt{2}$ **9.** $2(e^2 + 1)$ **11.** $\frac{3}{2}$ **13.** $\frac{128}{3}$

15. $\frac{3}{2}$ **17.** $\frac{1}{6}k|b - a|^3$

Section 3,   page 275

1. 117π **3.** $\frac{4}{5}\pi$ **5.** 72π **7.** 102π **9.** $\frac{1}{4}\pi(e^4 + 1)(e^4 + 3)$

11. $\frac{1}{3}\pi h(a^2 + ab + b^2)$ **13.** $\frac{1}{3}\pi(a - h)^2(2a + h)$ **15.** $\frac{1}{6}\pi h^3$ **17.** 8

Section 4, page 280

1. $84\,J$ **3.** 2500 ft-lb   **5.** 1, 4, 16 ft-lb   **7.** 4500 ft-lb
9. 2.28×10^9 J, 4.25×10^9 J   **11.** $\frac{4}{3}\pi\delta ga^3(a+b)$ **13.** $(113/1024)wh$
15. $mg(a - a\cos\phi)$

Section 5, page 284

1. $\frac{1}{6}bh^2\delta$ **3.** $\frac{1}{6}h^2(a+2b)\delta$ **5.** $\frac{16}{3}a^3\delta$ **7.** $\frac{1}{6}(3\pi-4)a^3\delta$ **9.** $\frac{1}{6}a^4\delta$ **11.** k^3F
13. $\frac{1}{2}\delta hA$ **15.** 4.79×10^7 kg
17. We have $\delta = \delta_0 e^{-h/H}$ and $p = H\delta$. Set $h = 0$ in the first relation to find $\delta_0 = 1.29$, and set $h = 0$ in the second to find $H = p_0/\delta_0$, so $1/H = \delta_0/p_0 = 1.29/(1.03 \times 10^4) \approx 1.25 \times 10^{-4}$.

Section 6, page 294

1. $(c/r)(1 - e^{-rT})$ **3.** $(b/r^2)[1 - (1 + rT)e^{-rT}]$
5. $\dfrac{c}{r}\dfrac{1 - e^{-r}}{1 + e^{-r}}(1 - e^{-2Nr})$ **7.** $\dfrac{c\pi}{\pi^2 + r^2}(1 - e^{-2Nr})$ **9.** ce^{rt}
11. $A'(t) = A_0 e^{\phi(t)}\phi' = A(t)\phi' = A(t)r(t)$, and $A(0) = A_0 e^0 = A_0$, precisely the conditions for the growth of A_0 at interest rate $r(t)$.
13. $T(x) = \sqrt{T_0^2 + \delta^2 x^2}$ **15.** $\bar{x} = \dfrac{1}{M}\displaystyle\int_a^b x\delta(x)\,dx$, where $M = \displaystyle\int_a^b \delta(x)\,dx$
17. $\approx 4.22 \times 10^4$ m/sec $\approx 94{,}400$ mile/hr   **19.** $\bar{K} = \frac{1}{4}A^2 k$ **21.** $\frac{3}{20}Ma^2\omega^2$
23. $\ddot{x} + (GM/R^3)x = 0$, $x(0) = R$, $\dot{x}(0) = 0$; $x(t) = R\cos\omega t$,   where $\omega^2 = GM/R^3$.
25. $v_{\text{term}} = F/k$, $K = \frac{1}{8}mF^2/k^2$

Section 7, page 296

1. $\frac{4}{3}$ **3.** $x = \frac{20}{11}$ **5.** $\frac{4}{3}$ **7.** $x = \frac{2}{3}(\sqrt{13} - 1)$ **9.** $Gm_0\,\delta(b-a)/ab$ **11.** $\frac{56}{5}\pi$ ft^3
13. $V = abh + \frac{1}{2}(a\tan\beta + b\tan\alpha)h^2 + \frac{1}{3}(\tan\alpha)(\tan\beta)h^3$ **15.** $F_L = \frac{1}{6}\delta gh^2(3b + h\tan\beta)$
17. $V = \displaystyle\int_{-a}^a 2\sqrt{a^2 - y^2}\,(h - y\tan\alpha)\,dy$ **19.** $W = \delta g\displaystyle\int_{-a}^a \sqrt{a^2 - y^2}\,(h - y\tan\alpha)^2\,dy$
21. $p(h_1) - p(h_2) = p_0(e^{-ah_1} - e^{-ah_2}) = p_0 e^{-ah_1}(1 - e^{a(h_1 - h_2)}) \approx p_0 ae^{-ah_1}(h_2 - h_1)$.
     Thus the pressure is approximately proportional to the depth from a fixed reference level, the same as in a fluid, so the buoyancy must be nearly the same as for a fluid.
23. $p(x) = p_0 x_0/x$ **25.** $V \approx 6.372 \times 10^6$ ft^3, floor space $\approx 4.83 \times 10^5$ ft^2.

CHAPTER 7

Section 1, page 306

1. $x = \frac{1}{3}(y + 7)$ **3.** $x = -1/y,\ y \neq 0$ **5.** $x = \dfrac{4y + 7}{-y + 2},\ y \neq 2$

7. $x = \dfrac{3y - 2}{-y + 1},\ y \neq 1$ **9.** $x = \sqrt[3]{\dfrac{3y - 2}{-y + 1}},\ y \neq 1$ **11.** $x = \frac{1}{2}(y^2 + 8),\ y \geq 0$
13. $x = \frac{1}{2}(y + \sqrt{y^2 - 4}),\ y \geq 2$ **15.** no   **17.** yes
19. If $f(x) = y$, then $f(-x) = y$, hence each y corresponds to at least two values of x except when $x = 0$.

21. $dy/dx = 3x^2 + 1 > 0$ for all x, so the inverse function $x = g(y)$ exists.

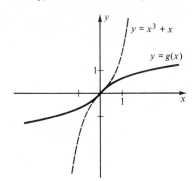

23. $\frac{1}{28}$ **25.** $\frac{25}{28}$ **27.** $-\frac{1}{12}$ **29.** $\dfrac{d}{dx} x^{4/3}\Big|_0 = \lim_{x\to 0} \dfrac{x^{4/3}}{x} = \lim_{x\to 0} x^{1/3} = 0$

31. $[(g \circ f) \circ (f^{-1} \circ g^{-1})](z) = [g \circ (f \circ f^{-1}) \circ g^{-1}](z) = (g \circ g^{-1})(z) = g[g^{-1}(z)] = z$ and
$[(f^{-1} \circ g^{-1}) \circ (g \circ f)](x) = [f^{-1} \circ (g^{-1} \circ g) \circ f](x) = (f^{-1} \circ f)(x) = f^{-1}[f(x)] = x$
for all x and z in the corresponding domains. Hence $g \circ f$ and $f^{-1} \circ g^{-1}$ are inverses
of each other, that is, $(g \circ f)^{-1} = f^{-1} \circ g^{-1}$.

33. $x = g[f(x)]$, so by the Chain Rule

$$1 = \frac{dx}{dx}\Big|_a = \frac{dg}{dy}\Big|_b \cdot \frac{df}{dx}\Big|_a = g'(b)f'(a),$$

hence $f'(a) \neq 0$.

Section 2, page 313

1. $x = -\ln y,\ y > 0$ **3.** $x = e^{1/y},\ y > 0$ **5.** $x = e^y - 5$ **7.** $a + 2$ **9.** $1/x$

11. $\frac{1}{2}$ **13.** $1/x$ **15.** $4/x$ **17.** $-1/x$ **19.** $\cot x$ **21.** $2/(1 - x^2)$

23. $1/(2x\sqrt{\ln x})$ **25.** $-1/[x(\ln x)^2]$ **27.** $1/\sqrt{x^2 + 1}$ **29.** $1/[x\sqrt{x^2 + 4}]$

31. $(\ln x)^{n-1}/x$

33. Set $e^a = x$ so $a = \ln x$. Then $e^{-a} = 1/e^a = 1/x$, so $\ln(1/x) = -a = -\ln x$.

35. $e^{b \ln x} = (e^{\ln x})^b = x^b$, hence $\ln x^b = b \ln x$.   **37.** $n < \ln 1000 \approx 6.91 < n + 1$ so $n = 6$.

39. $y_{\min} = y(1/e) = -1/e$ **41.** $y'' = -1/x^2 < 0$

43. LHS $= \ln 100 = \ln 10^2 = 2 \ln 10 =$ RHS

45. $\ln x = \displaystyle\int_1^x \frac{dt}{t} < \int_1^x dt = x - 1$ for $x > 1$.

47. $y/x > 1$, hence $[(y/x) - 1]/(y/x) < \ln(y/x) < (y/x) - 1$; use $\ln(y/x) = \ln y - \ln x$ and
divide by $y - x$.

49. $\ln x = \displaystyle\int_1^x \frac{dt}{t} < \int_1^x \frac{dt}{\sqrt{t}} = 2\sqrt{t}\,\Big|_1^x = 2(\sqrt{x} - 1)$

51. $e^x > 1$, so by Ex. 50, $x = \ln e^x < n(e^{x/n} - 1)$, hence $1 + \dfrac{x}{n} < e^{x/n}$, $\left(1 + \dfrac{x}{n}\right)^n < e^x$.

53. $e^x > 1$, so by Ex. 52, $n(1 - e^{-x/n}) < \ln e^x = x$, hence $1 - \dfrac{x}{n} < e^{-x/n}$, $\left(1 - \dfrac{x}{n}\right)^n < e^{-x}$,
$e^x < \left(1 - \dfrac{x}{n}\right)^{-n}$

55. The curve $y = 1/x$ is strictly convex for $x > 0$ and $y = \frac{1}{2} - \frac{1}{4}(x - 2)$ is its tangent line
at $(2, \frac{1}{2})$. A strictly convex curve always lies above any of its tangents. (See p. 151.) Also
$y = \frac{1}{3} - \frac{1}{9}(x - 3)$ is the tangent at $(3, \frac{1}{3})$.

57. Multiply by $2x^{3/2}$; the inequality is equivalent to $2\sqrt{x} < 1 + x$, which follows from $(\sqrt{x} - 1)^2 > 0$.

59.

x	$\ln x$	$x - 1$	$2(\sqrt{x} - 1)$	$\sqrt{x} - 1/\sqrt{x}$
1.2	0.18232	0.20000	0.19089	0.18257
1.5	0.40547	0.50000	0.44949	0.40825
2.0	0.69315	1.00000	0.82843	0.70711
e	1.00000	1.71828	1.29744	1.04219
3.0	1.09861	2.00000	1.46410	1.15470

Section 3, page 319

1. $y' = 3x^2/2y$ **3.** $y' = 6y \cdot \left(\dfrac{2x}{x^2 + 4} - \dfrac{1}{x + 7} \right)$ **5.** $y' = 2y \cdot \left(\dfrac{1}{3(2x + 3)} - x \right)$

7. $y' = y \cdot \left(1 + \dfrac{3x^2}{x^3 - 1} - \dfrac{1}{2x + 1} \right)$ **9.** $5/6x$ **11.** $\dfrac{1}{x - 3} + \dfrac{1}{x - 5}$

13. $\dfrac{1}{x + 2} + \dfrac{3}{x + 7}$ **15.** ∞ **17.** 0   **19.** ∞ **21.** 0   **23.** 0

25. $(\ln x)/x^{q/p} \longrightarrow 0$ because $q/p > 0$, hence

$$\frac{(\ln x)^p}{x^q} = \left(\frac{\ln x}{x^{q/p}} \right)^p \longrightarrow 0 \quad \text{as} \quad x \longrightarrow \infty.$$

27.

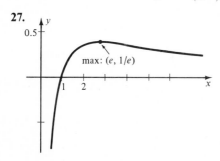

max: $(e, 1/e)$

Section 4, page 325

1. 1   **3.** 0   **5.** $x^{x-2}(x - 1 + x \ln x)$ **7.** $y' = x^{-2}y \cdot (1 - \ln x)$

9. $y' = y(\ln 3)/x$ **11.** $y' = 2xy \ln 10$ **13.** $y' = y \cdot (\ln \ln x + 1/\ln x)$

15. $y' = -y/x \ln^2 x$ **17.** $y' = 2y \cdot (\ln x)/x$

19.   **21.**   **23.**

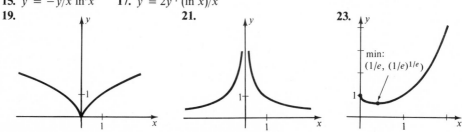

min: $(1/e, (1/e)^{1/e})$

25. $y_{\max} = y(e^{1/a}) = 1/ae$ **27.** $y_{\max} = y(x) = 1/e$

29. $\ln f(x)$ is the composite of differentiable functions, hence differentiable, so $h(x) = g(x) \ln f(x)$, the product of differentiable functions is differentiable, and finally $f(x)^{g(x)} = e^{h(x)}$ is the composite of differentiable functions, hence differentiable.

31. Take logs: the inequality is equivalent to $b \ln a > a \ln b$, that is, $(\ln a)/a > (\ln b)/b$. But $y = (\ln x)/x$ is strictly decreasing for $x > e$ since $y' = (1 - \ln x)/x^2 < 0$, so indeed $(\ln a)/a > (\ln b)/b$.

33. If $x\uparrow$, then $e^{-x}\downarrow$, $(1 - e^{-x})\uparrow$, $\ln(1 - e^{-x})\uparrow$. (The shorthand $\uparrow$ for strictly increases, etc., is useful.)

35. By the Chain Rule, $dx/dv = (dx/dt)/(dv/dt) = v/(dv/dt) = -v/(kv + g)$. The rest is algebra.

37. $m(S_1) - m(S_2) = 2.5 \log_{10}[\phi(S_2)/\phi(S_1)]$

39. $y = ka^x$, $k > 0$, $a > 0$

41. $\dfrac{E(uv)}{Ex} = \dfrac{x}{uv}\dfrac{d(uv)}{dx} = \dfrac{x}{uv}\left(v\dfrac{du}{dx} + u\dfrac{dv}{dx}\right) = \dfrac{x}{u}\dfrac{du}{dx} + \dfrac{x}{v}\dfrac{dv}{dx} = \dfrac{Eu}{Ex} + \dfrac{Ev}{Ex}$

43. $\dfrac{Ey}{Ex} = \dfrac{x}{y}\dfrac{dy}{dx} = \dfrac{x}{y}\dfrac{dy}{du}\dfrac{du}{dx} = \left(\dfrac{u}{y}\dfrac{dy}{du}\right)\left(\dfrac{x}{u}\dfrac{du}{dx}\right) = \dfrac{Ey}{Eu}\dfrac{Eu}{Ex}$

45. α **47.** kx **49.** $2\delta[b - a - a \ln(b/a)]$

Section 5, page 332

1. $\frac{1}{4}\pi$ **3.** $\frac{1}{3}\pi$ **5.** $\frac{1}{3}\pi$ **7.** $\frac{1}{2}\pi$ **9.** $\frac{4}{7}\pi$ **11.** $\frac{4}{5}$ **13.** $\frac{1}{2}\pi$

15. $x = \arcsin e^y$, $y \le 0$

17. Set $y = \arctan x$. Then $-\frac{1}{2}\pi < y < \frac{1}{2}\pi$ and $\tan y = x$, so $0 < (\frac{1}{2}\pi - y) < \pi$ and $\cot(\frac{1}{2}\pi - y) = \tan y = x$. But $z = \text{arc cot } x$ is the *unique* number such that $0 < z < \pi$ and $\cot z = x$. Hence $\frac{1}{2}\pi - y = z$.

19. Set $y = \text{arc sec } x$ and $z = \text{arc csc } x$. Then $0 \le y \le \pi$, $y \ne \frac{1}{2}\pi$ and $x = \sec y$. These imply $-\frac{1}{2}\pi \le \frac{1}{2}\pi - y \le \frac{1}{2}\pi$, $\frac{1}{2}\pi - y \ne 0$, and $\csc(\frac{1}{2}\pi - y) = \sec y = x$. But $z = \text{arc csc } x$ is the *unique* number with these properties, hence $z = \frac{1}{2}\pi - y$.

21. Set $y = \arctan(1/x)$. Then $0 < y < \frac{1}{2}\pi$ and $\cot y = 1/\tan y = 1/(1/x) = x$, hence $y = \text{arc cot } x$.

23. Set $y = \arccos x$. Then $0 \le y \le \pi$ so $\sin y \ge 0$. Hence $\sin y = +\sqrt{1 - \cos^2 y} = +\sqrt{1 - x^2}$.

25. Set $y = 2 \arccos x$. Then $0 \le y \le \pi$ and $\cos y = \cos(2 \arccos x) = 2\cos^2(\arccos x) - 1 = 2x^2 - 1$. Since $-1 \le 2x^2 - 1 \le 1$, it follows that $y = \arccos(2x^2 - 1)$.

27. Set $u = \arctan x$ and $v = \arctan y$. Then $-\frac{1}{2}\pi < u + v < \frac{1}{2}\pi$ and
$$\tan(u + v) = \frac{\tan u + \tan v}{1 - (\tan u)(\tan v)} = \frac{x + y}{1 - xy},$$ so the formula follows.

29. Set $g(x) = \arctan x$. Then $g(\frac{1}{2}) + g(\frac{1}{3}) = g\left(\dfrac{\frac{1}{2} + \frac{1}{3}}{1 - \frac{1}{2}\cdot\frac{1}{3}}\right) = g(1) = \frac{1}{4}\pi$.

31. $g(\frac{1}{5}) + g(\frac{1}{8}) = g\left(\dfrac{\frac{1}{5} + \frac{1}{8}}{1 - \frac{1}{5}\cdot\frac{1}{8}}\right) = g(\frac{1}{3})$, hence by Ex. 30,
$$2g(\tfrac{1}{5}) + g(\tfrac{1}{7}) + 2g(\tfrac{1}{8}) = g(\tfrac{1}{7}) + 2g(\tfrac{1}{3}) = \tfrac{1}{4}\pi.$$

Section 6, page 337

1. $\dfrac{1}{\sqrt{9 - x^2}}$ **3.** $\dfrac{2x}{1 + x^4}$ **5.** $\dfrac{6\arcsin 3x}{\sqrt{1 - 9x^2}}$ **7.** $\dfrac{1}{1 + x^2}$ **9.** $1/(1 + x^2)$

11. $2 \arctan 2x$ **13.** $\arcsin \frac{1}{4}x$

15. $\dfrac{d}{dx}(\text{arc cot } x) = \dfrac{d}{dx}(\tfrac{1}{2}\pi - \arctan x) = -\dfrac{d}{dx}(\arctan x) = \dfrac{-1}{1 + x^2}$

17. Use Ex. 16 and $\text{arc csc } x = \frac{1}{2}\pi - \text{arc sec } x$.

19. $\frac{1}{3}\pi$ **21.** $\frac{5}{58} \approx 0.0862$ rad/sec

23. $\dot{\theta} = \pm\frac{1}{2}\dot{x}/\sqrt{a^2 - x^2}$ with $+$ for $0 < \theta < \frac{1}{4}\pi$ and $-$ for $\frac{1}{4}\pi < \theta < \frac{1}{2}\pi$.

25. $\dot{\theta} = (1/\sqrt{a^2 - x^2} - 1/\sqrt{b^2 - x^2})\dot{x}$

27. By the solution to Ex. 25 you must show that $1/\sqrt{a^2 - x^2} > 1/\sqrt{b^2 - x^2}$. Square and clear of fractions: $b^2 - x^2 > a^2 - x^2$, that is, $b^2 > a^2$. This is true by the figure.

29. $\frac{1}{2}\pi$ **31.** $2 \arctan(2a/c)$

Section 7, page 345

1. RHS $= \frac{1}{4}(e^u - e^{-u})(e^v + e^{-v}) + \frac{1}{4}(e^u + e^{-u})(e^v - e^{-v}) = \frac{1}{2}(e^u e^v - e^{-u} e^{-v}) =$ LHS

3. Set $u = v = x$ in the text formula for $\cosh(u + v)$.

5. $\sinh 3x = \sinh(2x + x) = \sinh 2x \cosh x + \cosh 2x \sinh x = 2 \sinh x \cosh^2 x$
$+ (\cosh^2 x + \sinh^2 x) \sinh x = 3 \sinh x \cosh^2 x + \sinh^3 x = 3 \sinh x (\sinh^2 x + 1)$
$+ \sinh^3 x$, etc.

7. $5 \cosh 5x$ **9.** $x(x^2 + 1)^{-1/2} \operatorname{sech}^2(x^2 + 1)^{1/2}$

11. $e^{2x} \cosh x$ **13.** $\tanh x$ **15.** $x^2 \cosh x$ **17.** $\frac{1}{2}$ **19.** 2   **21.** 0

23. LHS $= \dfrac{d}{dx}\left(\dfrac{\sinh x}{\cosh x}\right) = \dfrac{\cosh^2 x - \sinh^2 x}{\cosh^2 x} = \dfrac{1}{\cosh^2 x} = \operatorname{sech}^2 x$

25. $y = \cosh^{-1} x$ implies $x = \cosh y = \frac{1}{2}(e^y + e^{-y})$ so $e^{2y} - 2xe^y + 1 = 0$. Solve for e^y: $e^y = x \pm \sqrt{x^2 - 1}$. Since $y > 0$ for $x > 1$, we have $e^y > 1$, so we must take $+$. Now $y = \ln(x + \sqrt{x^2 - 1})$.

27. $x = \cosh y$ implies $1 = (\sinh y)(dy/dx)$, hence

$$dy/dx = 1/\sinh y = 1/\sqrt{\cosh^2 y - 1} = 1/\sqrt{x^2 - 1}.$$

29. LHS $= \ln(\tan \theta + \sqrt{\tan^2 \theta + 1}) = \ln(\tan \theta + \sec \theta)$, and
RHS $= \ln(\sec \theta + \sqrt{\sec^2 \theta - 1}) = \ln(\sec \theta + \tan \theta)$.

31. $\ln(1 + \sqrt{2})$ **33.** $\frac{1}{2} \ln \frac{5}{3}$

35. $y' = ac \cosh cx + bc \sinh cx$, $y'' = ac^2 \sinh cx + bc^2 \cosh cx = c^2 y$

37. Let $y = \sinh x$. Then $y'' = \sinh x > 0$ for $x > 0$, so the graph is strictly convex. Since $y(0) = 0$ and $y'(0) = 1$, the tangent at $(0, 0)$ is $y = x$. By the Tangent Theorem (p. 151), $\sinh x > x$ for $x > 0$.

39. $\pi b(\cosh^{-1} b)^2 + 2\pi(b - 1) - 2\pi(\cosh^{-1} b)\sqrt{b^2 - 1}$ (Use concentric cylindrical shells.)

Section 8, page 353

1. $f(x) > 0$ and $f(-x) < 0$ for x large, hence at least one zero. $f'(x) = 3x^2 + p > 0$, hence at most one zero.

3. Set $f(x) = x^4 - 3 - e^{-x} \cos 7x$. Then $f(0) = -4 < 0$ and $f(\frac{3}{2}\pi) = (\frac{3}{2}\pi)^4 - 3 > 3^4 - 3 > 0$, hence there is a zero.

5. $[0.5, 0.6]$; also $[-1.5, -1.4]$, $[-4.8, -4.7]$, $[-7.9, -7.8]$, etc.

7. Imitate the solution of Example 2 with $g(x) = f[x + (b - a)/n] - f(x)$ on the interval $[0, (n - 1)(b - a)/n]$.

9. $g''(y) = -f''(x)/[f'(x)]^3$ evaluated at $x = g(y)$.

11. $[\ln x^p]' = (px^{p-1})/x^p = p/x = [p \ln x]'$ and $\ln x^p = p \ln x$ for $x = 1$; hence they are equal for all x.

13. $f'(x) = 1/(1 + x^4) > 0$, so $f(x)$ has an inverse

15. $f(x)$ is increasing and for $x \geq 1$,

$$f(x) = \int_0^1 \frac{dt}{1 + t^4} + \int_1^x \frac{dt}{1 + t^4} < \int_0^1 dt + \int_1^x \frac{dt}{t^4} = 1 + \frac{1}{3}\left(1 - \frac{1}{x^3}\right) < \frac{4}{3}.$$

Section 9, page 354

1. $\frac{4}{3}(x \ln x)^{1/3}(1 + \ln x)$ **3.** $\frac{1}{2}(\ln 10)10^{\sqrt{x}}/\sqrt{x}$ **5.** $(1 + x \arctan x)/\sqrt{1 + x^2}$

7. $1/(1 - x^3)$ **9.** π **11.** $\frac{5}{2}$ **13.** 5.28×10^8

15. $(\operatorname{csch} 1, -1 + \coth 1) = (2e/(e^2 - 1), 2/(e^2 - 1))$

17. $t = 2026.87 - (1.79 \times 10^{11}/N)^{1/0.99}$

19. $\frac{1}{4}\pi < \arctan x < \frac{1}{2}\pi$ for $1 < x < 100$, hence

$$\frac{\pi}{4}\int_1^{100}\frac{dx}{x^2} < \int_1^{100}\frac{\arctan x}{x^2}\,dx < \frac{\pi}{2}\int_1^{100}\frac{dx}{x^2}, \quad \text{etc.}$$

21. $z' = f'/f$ and $z'' = [ff'' - (f')^2]/f^2 < 0.$

23. LHS is the trapezoidal approximation to the RHS. It is less because $y = \ln x$ is strictly concave.

25. Set $f(x) = \frac{1}{2}\pi + \arctan x - 2\arctan 2x$. Then $f'(x) = -3/(1 + x^2)(1 + 4x^2) < 0$ so $f(x)$ is strictly decreasing. But $\lim_{x\to\infty} f(x) = \frac{1}{2}\pi + \frac{1}{2}\pi - 2(\frac{1}{2}\pi) = 0$. Therefore $f(x) > 0$ for all x.

27. $\arccos x = \arctan[(\sqrt{1 - x^2})/x]$ for $0 < x \le 1.$

29. $\arccos x = 2\arctan[(\sqrt{1 - x^2})/(1 + x)]$ for $-1 < x \le 1.$

31. LHS $= (\cosh^2 x - \sinh^2 x)(\cosh^2 x + \sinh^2 x) = \cosh^2 x + \sinh^2 x = \cosh 2x.$

33. RHS $= \dfrac{\sinh 2x + \cosh 2x - 1}{\sinh 2x + \cosh 2x + 1} = \dfrac{e^{2x} - 1}{e^{2x} + 1} = \dfrac{e^x - e^{-x}}{e^x + e^{-x}} = \tanh x$

35. $n(t) = \sqrt{\rho/\alpha}\,\tanh(\sqrt{\rho\alpha}\,t)$

37.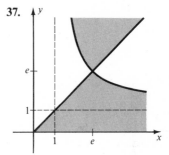

39. $\begin{bmatrix} a_{11} & a_{12} \\ a_{21} & a_{22} \end{bmatrix} = \begin{vmatrix} \varepsilon\cosh t & \varepsilon\eta\sinh t \\ \sinh t & \eta\cosh t \end{vmatrix},$

where $\varepsilon = \pm 1$ and $\eta = \pm 1$ independently.

41. Set $f(x) = x^x + y^x - (x + y)^x$. Then $f'(x) > 0$ for $x > 0$, hence $f(x) > f(0) = 0$ for $x > 0.$

CHAPTER 8*

Section 2,   page 363

1. $\frac{1}{2}\sin^2 x$ **3.** e^{5x} **5.** $-1/2(1 + x^2)^2$ **7.** $-1/\cos x$
9. $\ln(e^x + x^2 + 1)$ **11.** $\frac{1}{4}\tan^4 x$ **13.** $e^{\sqrt{x}}$ **15.** $\frac{1}{15}(3x + 1)^5$
17. $\frac{1}{3}\sin 3x$ **19.** $\frac{1}{4}\sec^4 x$ **21.** $\frac{2}{5}\sqrt{1 + 5x}$ **23.** $\frac{1}{4}[\ln(2x + 7)]^2$ **25.** $1/3(5 - 3x)$
27. $\frac{1}{3}x^3 + 2x - (1/x)$ **29.** $(ax + b)^{n+1}/a(n + 1)$ **31.** $\frac{1}{4}\ln(1 + x^4)$
33. $\ln|\ln x|$ **35.** $-\frac{15}{4}$ **37.** 1   **39.** 2   **41.** $2 - \sqrt{2}$

Section 3,   page 366

1. $\frac{2}{5}(x + 3)^{3/2}(x - 2)$ **3.** $\frac{1}{3}(x - 5)(2x + 5)^{1/2}$
5. $\ln|x - 1| - 2(x - 1)^{-1} - \frac{1}{2}(x - 1)^{-2}$ **7.** $2\sqrt{x} - 2\ln(1 + \sqrt{x})$
9. $\frac{4}{3}\ln(1 + x^{3/4})$ **11.** $\frac{1}{2}\arctan e^{2x}$ **13.** $\frac{1}{5}\arctan(5x + 2)$
15. $-\frac{1}{3}(2 + x^2)\sqrt{1 - x^2}$ **17.** $-\frac{1}{2}\ln(3 + \cos 2x)$ **19.** $3(8x - 3)(2x + 1)^{4/3}/112$
21. $\frac{1}{2}x^2 - \ln\sqrt{x^2 + 1}$ **23.** $\frac{1}{3}\ln(3x + \sqrt{9x^2 + 1})$ **25.** $\frac{49}{3}$ **27.** $\frac{11}{36}$ **29.** $\frac{1}{14}$
31. $2 - \ln\frac{9}{4}$ **33.** $\frac{16}{3} - 3\sqrt{3}$ **35.** 0

Section 4,   page 371

1. $\frac{1}{3}x^3 - 9x + \frac{82}{3}\arctan x$ **3.** $2x + 9\ln|x - 4|$ **5.** $\frac{1}{2}x^2 + 3x - 1/x + 3\ln|x|$
7. $[\arcsin(ax) - \sqrt{1 - a^2x^2}]/a$ **9.** $-\ln|\csc 2x + \cot 2x|$ **11.** $\frac{1}{5}\sin^5 x - \frac{1}{7}\sin^7 x$
13. $\frac{1}{5}\cos^5 x - \frac{1}{3}\cos^3 x$ **15.** $\frac{1}{3}\sin 3x - \frac{2}{9}\sin^3 3x + \frac{1}{15}\sin^5 3x$

---

* Constants of integration are omitted.

17. $\frac{1}{32}(\sin 4x + 8 \sin 2x + 12x)$ **19.** $\frac{1}{3}\tan^3 x - \tan x + x$ **21.** $-\frac{1}{2}\ln|\cos x^2|$

23. $(1 + \sin x)/\cos x$ **25.** $-\ln\sqrt{2}$ **27.** 1   **29.** 0   **31.** $\frac{1}{2}\arctan\frac{1}{2}(x + 1)$

33. $\arcsin\frac{1}{3}(x - 3)$

35. $2\arcsin\frac{1}{2}(x - 2) - \sqrt{4x - x^2}$

37. $\dfrac{1}{2\sqrt{3}}\ln|u + \sqrt{u^2 - \frac{1}{9}}|$, where $u = x^2 - \frac{2}{3}$ **39.** $b^{-1}\ln|ax/(b - ax)|$

Section 5,   page 376

1. $\dfrac{1}{2}\left(\dfrac{1}{x - 1} - \dfrac{1}{x + 1}\right)$ **3.** $1 + \dfrac{1}{3}\left(\dfrac{4}{x - 2} - \dfrac{1}{x + 1}\right)$

5. $\dfrac{1}{2}\left(\dfrac{-1}{x + 1} + \dfrac{4}{x + 2} - \dfrac{3}{x + 3}\right)$ **7.** $1 - \dfrac{2}{x^2 + 1} + \dfrac{1}{(x^2 + 1)^2}$

9. $\dfrac{1}{5}\left(\dfrac{2}{x - 1} + \dfrac{-2x + 3}{x^2 + 4}\right)$ **11.** $\ln|(x - 2)/(x - 1)|$

13. $\ln|x^3/(x + 1)^2|$ **15.** $3\ln|x/(x^2 + 1)^{1/2}| + 2\arctan x$

17. $\frac{1}{169}\{\ln[(x^2 + 9)^2/(x - 2)^4] - 13/(x - 2) - \frac{5}{3}\arctan(x/3)\}$

19. $\frac{1}{2}x^2 + \frac{1}{3}\ln|(x - 1)/(x^2 + x + 1)^{1/2}| + (1/\sqrt{3})\arctan[(2x + 1)/\sqrt{3}]$

21. $\frac{1}{2}x^2 - 3x + 8\ln|x + 2| - \ln|x + 1|$

23. $\frac{1}{17}[13\ln|x - 3| + 2\ln(x^2 + 2x + 2) - \arctan(x + 1)]$

25. $\ln\frac{4}{3}$ **27.** $\frac{1}{36}\pi\sqrt{3}$ **29.** $\frac{1}{3}\ln\frac{5}{4}$ **31.** $\ln\left(\dfrac{1 + \tan^2\frac{1}{2}\theta}{3 + \tan^2\frac{1}{2}\theta}\right)$

Section 6,   page 380

1. $\ln(x + \sqrt{1 + x^2})$ **3.** $-\frac{1}{3}(8 + x^2)\sqrt{4 - x^2}$ **5.** $-\frac{1}{16}x^{-1}\sqrt{16 - x^2}$

7. $-x^{-1}\sqrt{x^2 + a^2} + \ln(x + \sqrt{x^2 + a^2})$ **9.** $\frac{1}{2}\arctan x - \frac{1}{2}x(1 + x^2)^{-1}$

11. $\sinh^{-1}(x/a)$ or $\ln(x + \sqrt{x^2 + a^2})$ **13.** $\frac{1}{2}x\sqrt{a^2 + x^2} - \frac{1}{2}a^2\sinh^{-1}(x/a)$

15. $\dfrac{1}{a}\tanh^{-1}\left(\dfrac{x}{a}\right) = \dfrac{1}{2a}\ln\left(\dfrac{a + x}{a - x}\right)$ **17.** $\dfrac{dx}{xy} = \dfrac{x\,dx}{x^2 y} = \dfrac{y\,dy}{x^2 y} = \dfrac{dy}{x^2} = \dfrac{dy}{y^2 \mp a^2}$; and

$\displaystyle\int \dfrac{dy}{y^2 - a^2} = \dfrac{1}{2a}\ln\left(\dfrac{y - a}{y + a}\right)$, $\displaystyle\int \dfrac{dy}{y^2 + a^2} = \dfrac{1}{a}\arctan\dfrac{y}{a}$.   Replace y by $\sqrt{x^2 \pm a^2}$.

19. $\displaystyle\int \dfrac{y\,dx}{x^2} = \int d\left(\dfrac{-y}{x}\right) + \int \dfrac{dy}{x} = -\dfrac{y}{x} + \ln|x + y|$

21. $\dfrac{dx}{xy} = \dfrac{x\,dx}{x^2 y} = \dfrac{-y\,dy}{x^2 y} = \dfrac{-dy}{x^2} = \dfrac{dy}{y^2 - a^2}$,   hence $\displaystyle\int \dfrac{dx}{xy} = \int \dfrac{dy}{y^2 - a^2} = \dfrac{1}{2a}\ln\left|\dfrac{y - a}{y + a}\right|$.

Now replace y by $\sqrt{a^2 - x^2}$.

23. $d\left(\dfrac{y}{x}\right) = \dfrac{x\,dy - y\,dx}{x^2} = \dfrac{xy\,dy - y^2\,dx}{x^2 y} = \dfrac{-x^2\,dx - y^2\,dx}{x^2 y} = -a^2\dfrac{dx}{x^2 y}$,

hence $\displaystyle\int \dfrac{dx}{x^2 y} = \dfrac{-1}{a^2}\dfrac{y}{x}$.

Section 7,   page 384

1. $\sin x - x\cos x$ **3.** $\frac{1}{4}e^{2x}(2x - 1)$ **5.** $\frac{2}{5}x^{3/2}(3\ln x - 2)$

7. $x\arctan x - \frac{1}{2}\ln(1 + x^2)$ **9.** $\frac{1}{2}(1 + x^2)\arctan x - \frac{1}{2}x$

11. $\frac{1}{13}e^{2x}(2\sin 3x - 3\cos 3x)$ **13.** $x\sinh x - \cosh x$

15. $(a^2 x^2 \sin ax + 2ax\cos ax - 2\sin ax)/a^3$ **17.** 2   **19.** $\frac{1}{9}(24\ln 2 - 7)$

21. $10e - 5$ **23.** $\frac{1}{72}(3\pi\sqrt{3} - 5)$

25. $\int = \frac{1}{2} \int_{-1}^{1} x\, d(e^{x^2}) = \frac{1}{2} x e^{x^2} \Big|_{-1}^{1} - \frac{1}{2} \int_{-1}^{1} e^{x^2}\, dx = e - \frac{1}{2} \int_{-1}^{1} e^{x^2}\, dx$

27. $\int = \int_{0}^{2\pi} f(x)\, d(\sin x) = f(x) \sin x \Big|_{0}^{2\pi} - \int_{0}^{2\pi} f'(x) \sin x\, dx$, etc.

29. $I = \int_{0}^{2\pi} \cos x \cos 2x\, dx = \frac{1}{2} \int_{0}^{2\pi} \sin x \sin 2x\, dx = \frac{1}{4} \int_{0}^{2\pi} \cos x \cos 2x\, dx = \frac{1}{4} I$, so $I = 0$.

31. $\int_{0}^{x} \frac{\sin t}{t}\, dt = \int_{0}^{x} \frac{1}{t} d(1 - \cos t) = \frac{1 - \cos t}{t} \Big|_{0}^{x} + \int_{0}^{x} \frac{1 - \cos t}{t^2}\, dt$

$$= \frac{1 - \cos x}{x} + \int_{0}^{x} \frac{1 - \cos t}{t^2}\, dt > 0$$

because $1 - \cos t \geq 0$ and > 0 if $t \neq (2n + 1)\pi$. [To make the identity absolutely sound, replace the lower limit 0 by $\varepsilon > 0$ and let $\varepsilon \longrightarrow 0$.]

Section 8, page 387

1. $J_n = x(\ln x)^n - nJ_{n-1}$ **3.** $J_n = \frac{1}{3} x^3 (\ln x)^n = \frac{1}{3} n J_{n-1} - \frac{1}{3} n J_{n-1}$

5. $J_n = \frac{1}{2a^2(n-1)} \left[\frac{-x}{(x^2 - a^2)^{n-1}} - (2n - 3)J_{n-1} \right]$

7. $J_n = \frac{\tan^{n-1} x}{n - 1} - J_{n-2}$ **9.** $J_n = \frac{\sec^{n-2} x \tan x}{n - 1} + \frac{n - 2}{n - 1} J_{n-2}$

11. $J_n = [-\sin^{n-1} x \cos x + (n - 1)J_{n-2}]/n$

13. $(\frac{2}{3})(\frac{4}{5})(\frac{6}{7}) = \frac{16}{35}$ **15.** $\frac{1}{9} - \frac{1}{7} + \frac{1}{5} - \frac{1}{3} + 1 - \frac{1}{4}\pi$

17. $2a^4 - 8a^3 + 24a^2 - 48a + 24$, $a = \ln 2$ **19.** $J_n = \frac{e^{ax} \tan^{n-1} x}{n - 1} - \frac{a}{n - 1} J_{n-1} - J_{n-2}$

Section 9, page 390

1. $\frac{1}{29} e^{-2x}(-2 \sin 5x - 5 \cos 5x)$ **3.** $\frac{1}{64}[-4x(1 - 4x^2)^{3/2} + 2x(1 - 4x^2)^{1/2} + \arcsin 2x]$

5. $\frac{1}{25}[5x - \sqrt{10}\, \arctan(\frac{1}{2}x\sqrt{10})]$ **7.** $12(x^2 - 8) \sin \frac{1}{2}x - 2x(x^2 - 24) \cos \frac{1}{2}x$

9. $\frac{1}{10} \sqrt{10x^2 + 7} - \frac{6}{\sqrt{10}} \ln |x\sqrt{10} + \sqrt{10x^2 + 7}| + \frac{2}{\sqrt{7}} \ln \left| \frac{\sqrt{7} + \sqrt{10x^2 + 7}}{x} \right|$

11. $-\frac{1}{2}\pi$ **13.** $\frac{77}{325}(e^{3\pi} - 1)$ **15.** $\frac{3}{4} + 5 \ln \frac{7}{8}$

Section 10, page 390

1. $\frac{1}{a - b} \ln \left| \frac{x - a}{x - b} \right|$ **3.** $-\sqrt{5 + \cos 2x}$ **5.** $\frac{1}{18} \sec^6 3x - \frac{1}{12} \sec^4 3x$

7. $\frac{1}{7} \sin^7 x - \frac{1}{9} \sin^9 x$

9. $(15x^4 - 12x^2 + 8)(x^2 + 1)^{3/2}/105 = (15x^6 + 3x^4 - 4x^2 + 8)(x^2 + 1)^{1/2}/105$

11. $2(\sin \sqrt{x} - \sqrt{x} \cos \sqrt{x})$ **13.** $\arcsin(x - 1) - \sqrt{2x - x^2}$

15. $\frac{1}{6}[2x^3 \arctan x - x^2 + \ln(1 + x^2)]$

17. $\frac{-1}{a^4} \left[\frac{1}{x} + \frac{b}{a^2} \arctan \frac{bx}{a^2} \right]$ **19.** $\frac{1}{3} \ln \left(\frac{e^x + 1}{e^x + 4} \right)$,

21. $\ln \ln \ln x$ **23.** $-x^2 \cos x + 2x \sin x + 2 \cos x$ **25.** $\frac{1}{2} \ln \left| \frac{\sqrt{1 + x^2} - 1}{\sqrt{1 + x^2} + 1} \right|$

27. The integral is the area under the semi-circle $x^2 + y^2 = a^2$, $y \geq 0$, equal to $\frac{1}{2}\pi a^2$.

29. $-kt + C = \frac{1}{4a^3} \ln \frac{u - a}{u + a} - \frac{1}{2a^3} \arctan \frac{u}{a}$,

where C is a constant that depends on $u_0 = u(0)$.

31. $\int_0^b x(b-x)^n \, dx = \int_0^b x^n(b-x) \, dx = \dfrac{b^{n+2}}{(n+1)(n+2)}$

CHAPTER 9

Section 1,   page 399

1. $(8, 3)$ **3.** $(7, -3)$ **5.** $(2, -9)$ **7.** $(-6, 6)$ **9.** $(-13, -4)$ **11.** $(-14, 6)$
13. $\bar{x} = x - h$, $\bar{y} = y - k$, where $3h - 2k = 1$ **15.** $\bar{x} = x + 1$, $\bar{y} = y$
17. $\bar{x} = x - \frac{1}{6}\pi$, $\bar{y} = y + 1$ **19.** $(3, -3)$
21. It means $\mathbf{p} = (x, y)$ is on the ray from $\mathbf{p}_1$ through $\mathbf{p}_2$, beyond $\mathbf{p}_2$, so that $d(\mathbf{p}_1, \mathbf{p}) = rd(\mathbf{p}_1, \mathbf{p}_2)$, where $d(\cdot, \cdot)$ is distance. Alternatively, $\mathbf{p}_2$ is $1/r$ from $\mathbf{p}_1$ to $\mathbf{p}$.
23. circle, center $(2, 2)$, radius $\sqrt{8}$ **25.** circle, center $(-1, -3)$, radius 6
27. circle, center $(\frac{1}{2}, -1)$, radius $\frac{1}{2}\sqrt{5}$ **29.** circle, center $(\frac{3}{4}, \frac{5}{4})$, radius $\frac{1}{4}\sqrt{26}$
31. $\frac{1}{2}(-1 \pm \sqrt{17}, 1 \pm \sqrt{17})$ **33.** $(1, 2)$, point of tangency   **35.** empty
37. $\frac{1}{5}(6 \pm 4\sqrt{31}, 7 \mp 2\sqrt{31})$ **39.** $\frac{1}{8}(-21, \pm 3\sqrt{15})$ **41.** none
43. $x = -1$ **45.** $y = x - \sqrt{2}$ **47.** $y = -\sqrt{3}x + 2$
49. $x = 0$, $y = \frac{3}{4}x$ **51.** $y - 1 = m(x + 5)$, $m = \frac{2}{3}, -\frac{3}{2}$
53. distance between centers = sum of radii: $2\sqrt{5} = \frac{3}{2}\sqrt{5} + \frac{1}{2}\sqrt{5}$.
55. The second circle has radius $\frac{4}{5}$ and center $(\frac{4}{25}, \frac{3}{25})$, at distance $\frac{1}{5}$ from the center of the first circle, etc.
57. $y \pm 4 = \frac{3}{4}(x \mp 3)$ **59.** $(x - 5)^2 + (y - 5)^2 = 25$, $(x - 13)^2 + (y - 13)^2 = 169$
61. Their *only* intersection is $(\frac{11}{5}, -\frac{17}{5})$, so they are tangent.

Section 2, page 403

1. line parallel to L, halfway from $\mathbf{p}$ to L **3.** concentric circle, radius 4
5. circle, center $(\frac{25}{8}, \frac{25}{6})$, radius $\frac{25}{24}$ **7.** parabola $y = \frac{1}{2}b + 2(x - \frac{1}{2}a)^2$, where $\mathbf{a} = (a, b)$
9. $\mathbf{x} = (x, y)$; $x = (ac + b^2 u - abv)/(a^2 + b^2)$, $y = (bc + a^2 v - abu)/(a^2 + b^2)$
11. the line $ax + by = \frac{1}{2}$, where C has center (a, b)
13. a line if $a = 1$, a circle if $a \neq 1$ **15.** the parabola $y = \frac{1}{4}x^2$ and the y-axis

Section 3,   page 411

1. $(3\frac{1}{12}, 0)$, $x = 3 - \frac{1}{12}$, x-axis, $(3, 0)$ **3.** $(\frac{1}{8}, -1)$, $x = -\frac{1}{8}$, $y = -1$, $(0, -1)$
5. $(-2, \frac{5}{6})$, $y = -\frac{13}{6}$, $x = -2$, $(-2, -\frac{2}{3})$
7. $y = 3x^2$ **9.** $x - 1 = -(y - 2)^2$ **11.** $16(y + 3) = (x - 2)^2$
13. the parabola $4x = y^2$ and the x-axis
15. two parabolas, $6(x + \frac{1}{2}) = y^2$ and $2(x - \frac{3}{2}) = -y^2$
17. a half line parallel to the axis of P
19. If there is a common point, by symmetry it must lie on the y-axis. The chord $\overline{\mathbf{x}_1\mathbf{x}_2}$ has y-intercept
$$y_1 - \left(\frac{y_2 - y_1}{x_2 - x_1}\right)x_1 = \frac{y_1 x_2 - y_2 x_1}{x_2 - x_1} = \frac{1}{4p}\left(\frac{x_1^2 x_2 - x_2^2 x_1}{x_2 - x_1}\right) = \frac{-x_1 x_2}{4p}.$$
The condition for perpendicularity is $x_1 x_2 + y_1 y_2 = 0$. Hence $x_1 x_2 = -y_1 y_2 = -(x_1 x_2)^2/(4p)^2$, so $x_1 x_2 = -(4p)^2$, and $-x_1 x_2/4p = 4p$. Thus all the chords $\overline{\mathbf{x}_1\mathbf{x}_2}$ pass through $(0, 4p)$.
21. As in the solution of Ex. 19, the chord $\overline{\mathbf{x}_1\mathbf{x}_2}$ meets the y-axis in $(0, -x_1 x_2/4p)$, so $p = -x_1 x_2/4p$, hence $x_1 x_2 = -4p^2$. Let $\mathbf{x} = (x, y)$ be the midpoint, so $x = \frac{1}{2}(x_1 + x_2)$ and $y = \frac{1}{2}(y_1 + y_2)$. Then
$$x^2 = \frac{1}{4}(x_1^2 + x_2^2 + 2x_1 x_2) = \frac{1}{4}(4py_1 + 4py_2 - 8p^2) = 2py - 2p^2; \quad \text{parabola.}$$

23. The center of the circle is $\mathbf{c} = (\frac{1}{2}x_1, \frac{1}{2}(y_1 + p))$. The radius is $\frac{1}{2} d(\mathbf{x}_1, \mathbf{p})$, which equals $\frac{1}{2}(y_1 + p)$, half the distance from $\mathbf{x}_1$ to the directrix. Hence the circle is tangent to the x-axis.

25. $\mathbf{0}$, $(\pm 4, 0)$, 5, 3, $(\pm 5, 0)$, and $(0, \pm 3)$

27. $(-1, 2)$, $(-1, 1)$, and $(-1, 3)$, $\sqrt{2}$, 1, $(-1, 2 \pm \sqrt{2})$ and $(0, 2)$ and $(-2, 2)$

29. $(3, 2)$, $(3, 2 \pm \frac{1}{2}\sqrt{2})$, 1, $\frac{1}{2}\sqrt{2}$, $(3, 1)$ and $(3, 3)$ and $(3 \pm \frac{1}{2}\sqrt{2}, 2)$

31. $(x - 1)^2/81 + (y - 4)^2/4 = 1$ **33.** $(x - 5)^2/25 + y^2/16 = 1$

35. $(x - 1)^2/16 + y^2/12 = 1$

37. $y = \pm b \sqrt{1 - \dfrac{x^2}{a^2}}$, $\dfrac{dy}{dx} = \mp \dfrac{bx}{a^2 \sqrt{\cdots}}$, so $dy/dx = 0$ at $x = 0$. This accounts for the horizontal tangents. Exchange x and y for the vertical ones.

39. Let the ellipse be $x^2/a^2 + y^2/b^2 = 1$, where $a > b > 0$. Suppose (x, y) is on the ellipse and $y \neq 0$. Then $1 = x^2/a^2 + y^2/b^2 > x^2/a^2 + y^2/a^2 = (x^2 + y^2)/a^2$, hence $x^2 + y^2 < a^2$, $\sqrt{x^2 + y^2} < a$.

41. An ellipse with axes the x- and y-axes.

43. The ellipse with foci $(a, 0)$ and $(b, 0)$ and semi-major axis $\frac{1}{2}(a + b)$.

45. The ellipse with major diameter D and minor diameter of length r.

Section 4,   page 418

1. x-axis, $\mathbf{0}$, $(\pm\sqrt{13}, 0)$, $y = \pm\frac{3}{2}x$ **3.** y-axis, $\mathbf{0}$, $(0, \pm\sqrt{13})$, $y = \pm\frac{2}{3}x$

5. $y = 1$, $(-1, 1)$, $(-1 \pm \sqrt{2}, 1)$, $y = -x$, and $y = x + 2$

7. $x = -2$, $(-2, -2)$, $(-2, -2 \pm \sqrt{\frac{96}{5}})$, $y = -2 \pm \sqrt{\frac{1}{5}}(x + 2)$

9. $y = -1$, $(3, -1)$, $(3 \pm \sqrt{5}, -1)$, $y = -2x + 5$, and $y = 2x - 7$

11. $-\frac{1}{9}x^2 + \frac{1}{16}y^2 = 1$ **13.** $\frac{1}{4}x^2 - \frac{1}{16}y^2 = 1$

15. $\frac{4}{3}(x - 1)^2 - \frac{1}{3}y^2 = 1$ **17.** The asymptotes are $y - \frac{1}{2}b = \pm(x + \frac{1}{2}a)$; perpendicular.

19. Think of x as a function of y. Then on each branch

$$2\frac{x\,dx}{a^2\,dy} - 2\frac{y}{b^2} = 0, \qquad \frac{dx}{dy} = \frac{a^2\,y}{b^2\,x}.$$

At the vertex $(a, 0)$ on the right branch, $dx/dy = 0$, so the tangent is vertical with respect to the x-axis.

21. the right branch of the hyperbola with foci $(\pm a, 0)$ and absolute length difference $r - s$

23. It is the locus of the centers of all circles that are simultaneously tangent to one of the given circles externally and the other internally.

25. At the point $(\sqrt{\frac{5}{3}}, -4)$ relative to the coordinate system in which $A = (-5, 0)$ and $B = (5, 0)$

Section 5,   page 424

1. $(0, 1)$ **3.** $(\frac{1}{2}\sqrt{3}, -\frac{1}{2})$ **5.** $(-\sqrt{2}, -\sqrt{2})$ **7.** $\{\sqrt{2}, \frac{1}{4}\pi\}$ **9.** $\{\sqrt{2}, \frac{3}{4}\pi\}$

11. $\{2, -\frac{1}{6}\pi\}$ **13.** $\theta = \frac{1}{4}\pi$ **15.** $r(\cos\theta + \sin\theta) = 1$ **17.** $r = -2a\cos\theta$

19. $r = 2\cos(\theta + \frac{3}{4}\pi)$ **21.** $r(\cos\theta + 2\sin\theta) = 5$ **23.** $r^2 - 10r\cos(\theta - \frac{1}{6}\pi) + 9 = 0$

25. $\frac{2}{3}x - \frac{4}{3}y = 1$ **27.** $-\frac{7}{25}x - \frac{24}{25}y = 1$ **29.** $\frac{1}{2}x\sqrt{2} + \frac{1}{2}y\sqrt{2} = \frac{3}{2}\sqrt{2}$

31. $\bar{x}\cos\alpha + \bar{y}\sin\alpha = p - (h\cos\alpha + k\sin\alpha)$

33. $r = p\cos(\theta - \alpha) \pm \sqrt{a^2 - p^2\sin^2(\theta - \alpha)}$. If $a > p$, then $\mathbf{0}$ lies inside the circle. For *each* θ there are two distinct solutions, $r_1 > 0$ and $r_2 < 0$.

Section 6,   page 431

1.
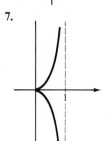
$\{2, \frac{1}{4}\pi\}$

3.
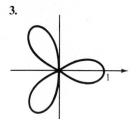
1

5.
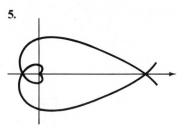

7.
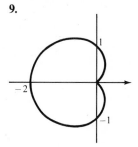
1

9.
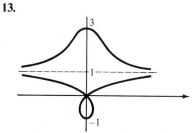
1

−2

−1

11.
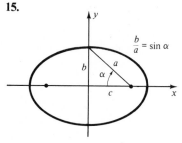
1

3

−1

13.

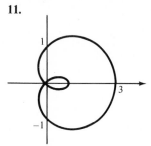
3

1

−1

15.

y

$\frac{b}{a} = \sin \alpha$

b a

α

c

x

17. $e \approx 0.206$ **19.** ellipse $r(1 - e \cos \theta) = ep$ **21.** $\sqrt{2}$

Section 7,   page 437

1. $\bar{x} = x \cos \alpha + y \sin \alpha, \bar{y} = -x \sin \alpha + y \cos \alpha; (x, y) \longrightarrow (\bar{x}, \bar{y})$ by a rotation through $-\alpha$.
3. $x_1 x_2 + y_1 y_2 = \bar{x}_1 \bar{y}_1 + \bar{x}_2 \bar{y}_2$; this means $x_1 x_2 + y_1 y_2$ is *unchanged* by a rotation.
5. The polar angle of the point with respect to the rotated axes is $\theta - \alpha$.
7. $\alpha = \frac{3}{8}\pi; -\frac{1}{2}(\sqrt{2} - 1)\bar{x}^2 + \frac{1}{2}(1 + \sqrt{2})\bar{y}^2 = 1$
9. $\alpha = \frac{3}{8}\pi; \frac{1}{2}(\sqrt{2} + 1)\bar{x}^2 - \frac{1}{2}(\sqrt{2} - 1)\bar{y}^2 = 1$ **11.** $\alpha = \frac{1}{4}\pi$; ellipse
13. $\tan 2\alpha = \frac{1}{2}$; hyperbola   **15.** $\alpha = \frac{3}{8}\pi$; ellipse   **17.** $\alpha = \frac{1}{4}\pi$; parabola
19. $\tan 2\alpha = 6$; hyperbola   **21.** Add the formulas for $\bar{a}$ and $\bar{c}$ given in the text.

Section 8,   page 445

1. $y = -x - 1$ **3.** $y = 4x - 8$ **5.** $y + 1 = \frac{1}{2}(\pm\sqrt{5} - 1)(x + 1)$
7. $y = -2x + 1, y = -6x - 3$
9. $v > au^2$ and $y + v = 2aux$, hence $y = 2aux - v < 2aux - au^2 = -a(x - u)^2 + ax^2 < ax^2$
11. $x = 1$ **13.** $\frac{1}{5}x + \frac{1}{10}y = 1$ **15.** $x = 1, 12y = 17x + 19$
17. $x = -1, 4y = x + 5$
19. The slope at (u, v) is $dy/dx = b^2u/a^2v$, so the tangent is $y - v = (b^2u/a^2v)(x - u)$, etc.
(Use $b^2u^2 - a^2v^2 = a^2b^2$.)
21. arc $\tan(\frac{4}{5}\sqrt{3})$

Section 9, page 445

1. The equation is $2(a_2 - a_1)x + 2(b_2 - b_1)y = k$, which has slope $-(a_2 - a_1)/(b_2 - b_1)$, the negative reciprocal of the slope of the line of centers.

3. The tangent and the radical axis both pass through the common point and both are perpendicular to the line of centers, so they coincide.

5. Eliminate y: $(1 + m^2)x^2 + 2(a + mb)x + c = 0$. This has a double root if and only if its discriminant vanishes.

7. Eliminate y^2: $(a^2 - b^2)x^2 = a^2(r^2 - b^2)$. This has two solutions if $r > b$, no solutions if $0 < r < b$. Similarly, eliminate x^2: $(a^2 - b^2)y^2 = b^2(a^2 - r^2)$. This has two solutions if $0 < r < a$, no solutions if $r > a$. Thus if $b < r < a$, both have solutions, and we get four possibilities for (x, y).

9. $4p(y + p) = x^2$, $p \neq 0$.

11. Let the ellipse be $x^2/a^2 + y^2/b^2 = 1$. Then
$$\frac{uv}{w^2} = \frac{(a + x)(a - x)}{y^2} = \frac{a^2 - x^2}{y^2} = \frac{a^2(y^2/b^2)}{y^2} = \frac{a^2}{b^2}.$$

13. $xy = \frac{1}{2}c^2$. Do it by rotating $\bar{x}^2 - \bar{y}^2 = c^2$ through $45°$.   15. for $-b/a < m < b/a$

17. $r^2 = 2a^2 \cos 2\theta$

19. If the firms are at $(-c, 0)$ and $(c, 0)$, the curve is the circle
$$(k^2 - 1)(x^2 + y^2) + 2c(k^2 + 1)x + c^2(k^2 - 1) = 0.$$

21. parabola

CHAPTER 10

Section 2, page 453

1. $1, x^2(-1)$ 3. $-1 + 3(x + 1), (x + 1)^2(x - 2)$ 5. $2 - 4(x - \frac{1}{2}), (x - \frac{1}{2})^2(4/x)$

7. $1 + 4(x - 1), (x - 1)^2(x^2 + 2x + 3)$ 9. $1 + 2(x + 1), (x + 1)^2[(-2x + 1)/x^2]$

11. 1 13. x 15. $1 + x, 0$ 17. $-\frac{1}{2} - \frac{1}{4}(x + 2) - \frac{1}{8}(x + 2)^2, (x + 2)^3(1/8x)$

19. $x^2, x^3[-1/(x + 1)]$ 21. $1 + 2x + 2x^2$ 23. x

25.

x	xe^x	$p_1(x)$	$p_2(x)$
0.1	0.1105	0.1000	0.1100
0.2	0.2443	0.2000	0.2400
0.3	0.4050	0.3000	0.3900
0.4	0.5967	0.4000	0.5600
0.5	0.8244	0.5000	0.7500

27. $\ln x$ and $2(x - 1)/(x + 1)$ have the same $p_2(x)$ at $x = 1$: $p_2(x) = (x - 1) - \frac{1}{2}(x - 1)^2$

x	0.5	0.8	1.2	1.5	2.0
$\ln x$	-0.6931	-0.2231	0.1823	0.4055	0.6931
$2(x - 1)/(x + 1)$	-0.6667	-0.2222	0.1818	0.4000	0.6667
$p_2(x)$	-0.6250	-0.2200	0.1800	0.3750	0.5000

Section 3, page 459

1. $8 + 7(x - 1) + (x - 1)^2$ 3. $9(x + 1) - (x + 1)^2 + 2(x + 1)^3$

5. $18(x + 2)^2 - 11(x + 2)^3 + 2(x + 2)^4$

7. $5(x + 1) - 19(x + 1)^2 + 31(x + 1)^3 - 21(x + 1)^4 + 5(x + 1)^5$ 9. 0.99600

11. 3.00052 **13.** $x - x^2 + \frac{1}{2}x^3 - \frac{1}{6}x^4 + \frac{1}{24}x^5$ **15.** $x - \frac{1}{6}x^3 - x^4 + \frac{1}{120}x^5$

17. $\frac{1}{2}\sqrt{2}\left(1 + u - \frac{1}{2}u^2 - \frac{1}{6}u^3 + \frac{1}{24}u^4 + \frac{1}{120}u^5\right)$, $u = x - \frac{1}{4}\pi$

19. $-1 - 3u - 6u^2 - 10u^3 - 15u^4 - 21u^5$, $u = x + 1$

Section 4, page 465

1. $2x - \dfrac{8}{3!}x^3 + \dfrac{2^5}{5!}x^5 - \cdots + (-1)^{n-1}\dfrac{2^{2n-1}}{(2n-1)!}x^{2n-1}$, $|r_{2n-1}(x)| \le \dfrac{2^{2n+1}}{(2n+1)!}|x|^{2n+1}$

3. $x + x^2 + \dfrac{x^3}{2!} + \dfrac{x^4}{3!} + \cdots + \dfrac{x^n}{(n-1)!}$, $|r_n(x)| \le \dfrac{(x+n+1)e^x}{(n+1)!}x^{n+1}$ if $x \ge 0$,

$|r_n(x)| \le |x|^{n+1}/n!$ if $x < 0$

5. $(x-1) + 3(x-1)^2/2! + 2(x-1)^3/3! - 2(x-1)^4/4! +$

$4(x-1)^5/5! + \cdots + (-1)^n 2(x-1)^n/n(n-1)(n-2)$;

if $n \ge 4$: $|r_n(x)| \le 2(n-2)!(x-1)^{n+1}/(n+1)!$ for $x \ge 1$,

$|r_n(x)| \le 2(n-2)!(1-x)^{n+1}/(n+1)!\,x^{n-1}$ for $0 < x \le 1$

7. $x^2 - x^3 + \frac{1}{2}x^4 - \frac{1}{6}x^5 + \cdots + (-1)^n x^n/(n-2)!$;

$|r_n(x)| \le x^{n+1}/(n-1)!$ for $x \ge 0$,

$|r_n(x)| \le [x^2 - 2(n+1)x + n(n+1)]|x|^{n+1}/e^x(n+1)!$ for $x \le 0$

9. $x^2 - x^4/3! + x^6/5! + \cdots + (-1)^{n-1}x^{2n}/(2n-1)!$;

$|r_{2n}(x)| \le [|x| + 2n + 2]|x|^{2n+2}/(2n+2)!$

11. $1 + x - \frac{1}{2}x^2 - \frac{1}{6}x^3 + x^4/4! + x^5/5! + \cdots + \sigma_n x^n/n!$, where

$\sigma_{4k} = \sigma_{4k+1} = 1$, $\sigma_{4k+2} = \sigma_{4k+3} = -1$; $|r_n(x)| \le |x|^{n+1}/(n+1)!$

13. $x + \dfrac{x^3}{3!} + \dfrac{x^5}{5!} + \cdots + \dfrac{x^{2n-1}}{(2n-1)!}$, $|r_{2n-1}(x)| \le (\cosh x)\dfrac{|x|^{2n+1}}{(2n+1)!}$

15. $\dfrac{1}{2} - \dfrac{t}{4} + \dfrac{t^2}{8} - + \cdots + (-1)^n \dfrac{t^n}{2^{n+1}}$, where $t = x - 1$,

$$|r_n(x)| \le \frac{|x-1|^{n+1}}{2^{n+1}(1+x)^{n+2}} \quad \text{for} \ -1 < x \le 1, \quad |r_n(x)| \le \frac{|x-1|^{n+1}}{2^{n+1}} \quad \text{for} \ x \ge 1$$

17. $1/(11 \times 2^{11}) < 5 \times 10^{-5}$ **19.** $p_4(x) = x^2 - \frac{1}{3}x^4$, $|\text{error}| \le \frac{1}{2}(2x)^6/6! < 5 \times 10^{-8}$

21. $(0.22)^5/5! \approx 4.3 \times 10^{-6}$ **23.** $(1.06)^9/9! \approx 4.7 \times 10^{-6}$

25. $k = 0.033$ works because $(.033\pi)^4/4! \approx 4.8 \times 10^{-6}$.

27. 7 according to our estimate applied to the interval $1 \le x \le 1.25$

29. $|r_3(x)| \le |x - \frac{1}{4}\pi|^4/4! = (0.1)^4/4! < 4.2 \times 10^{-6}$

31. $1 - \frac{1}{2}x - \frac{1}{8}x^2 - \frac{1}{16}x^3$, $|r_4(x)| \le \frac{5}{128}|x|^4$ for $x \le 0$ and $|r_4(x)| \le \frac{5}{128}|x|^4/(1-x)^{7/2}$ for $0 \le x < 1$

33. $\frac{1}{6}$

35. By "dividing out zeros," $f(x) = xF(x)$ and $g(x) = xG(x)$ where F and G are continuous, $F(0) = f'(0)$ and $G(0) = g'(0)$. Hence

$$\frac{f(x)}{g(x)} = \frac{xF(x)}{xG(x)} = \frac{F(x)}{G(x)} \longrightarrow \frac{F(0)}{G(0)} = \frac{f'(0)}{g'(0)}.$$

37. Set $n = 2m - 1$ and apply "dividing out zeros" to $f(x) - f(a)$:

$$f(x) = f(a) + (x-a)^{2m}g(x)$$

where $g(x)$ is continuous near a. Suppose for instance $f^{(n+1)}(a) > 0$. Then $g(a) = f^{(n+1)}(a)/(n+1)! > 0$ so $g(x) > 0$ for x sufficiently near a. Hence $(x-a)^{2m}g(x) > 0$, so $f(x) > f(a)$: local min, etc.

39. $p_n(x) = 0$ so $f(x) = r_n(x) = \dfrac{1}{n!}\displaystyle\int_a^x (x-t)^n f^{(n+1)}(t)\,dt$. Suppose $x \ne a$. Change variables in the integral by $t = a + (x - a)u$. The result is

$$f(x) = (x - a)^{n+1}g(x), \qquad g(x) = \frac{1}{n!}\int_0^1 (1 - u)^n f^{(n+1)}[a + (x - a)u]\, du.$$

If $x = a$, the first formula is obviously correct.

Section 5,   page 470

1. $\tan x$ takes on all real values for $-\frac{1}{2}\pi < x < \frac{1}{2}\pi$
3. Set $F(x) = x^{10} - 2x^8 + 2x^3 - x$. Then $F(0) = F(1)$ and $F' = f$. Apply Rolle's Theorem.
5. Apply Rolle's Theorem to $g(x) = xf(x)$.
7. The function $3x^2 - 1 + \sin \pi x + \pi x \cos \pi x$ has a zero on the interval $-1 < x < 1$.
9. $\tan x = x$ has a solution on each interval $n\pi < x < (n + 1)\pi$.
11. Suppose $g(x)$ has no zeros in $a < x < b$. Since $(f'g - fg')(a) = f'(a)g(a) \neq 0$, we have $g(a) \neq 0$; similarly $g(b) \neq 0$. Thus g has no zeros on $[a, b]$. Set $h(x) = f(x)/g(x)$. Then $h(a) = h(b) = 0$, and $h'(x) = [f'g - fg']/g^2$ has no zeros, contradicting Rolle's Theorem.
13. Apply Rolle's Theorem to $g(x) = (x - a)[f(b) - f(x)]$.
15. Set $h(x) = e^{x/k}f(x)$. Then $h(r_i) = f(r_i) = 0$, where $r_1 < r_2 < \cdots < r_n$ are the zeros of f. Therefore h' has at least one zero in each interval $x_{i-1} < x < x_i$. But $kh'(x) = e^{x/k}[f(x) + kf'(x)]$, etc.
17. Let $r_1 < r_2 < \cdots < r_s$ be the distinct zeros and $m_1, \cdots, m_s$ their multiplicities. Then f' has a zero of multiplicity $m_i - 1$ at r_i and a zero on each interval $r_{i-1} < x < r_i$, for a total of $\sum (m_i - 1) + (s - 1) = \sum m_i - s + s - 1 = \sum m_i - 1 = n - 1$ zeros on $[a, b]$. Apply the same result to f'; you conclude that f'' has $n - 2$ zeros or more on $[a, b]$, etc.
19. By Rolle, f' has a zero in $a < x < c$ and a zero in $c < x < b$. Thus $f'(c_1) = f'(c) = f'(c_2) = 0$, where $a < c_1 < c < c_2 < b$. By the Generalized Rolle's Theorem, $(f')'' = f'''$ has a zero in $c_1 < x < c_2$.
21. $f(x) = b_0(x - a)^{n+1} + \cdots + b_k(x - a)^{n+k+1} + \cdots,$

$$f^{(n+1)}(x) = (n + 1)!\, b_0 + \cdots + \frac{(n + k + 1)!}{k!}\, b_k(x - a)^k + \cdots.$$

23. Suppose $a < x_0 < b$ and $f(x_0) \geq 0$. Choose $k = f(x_0)/(x_0 - a)(b - x_0) \geq 0$. Then $g(a) = g(x_0) = g(b) = 0$, so there exists c such that $a < c < b$ and $g''(c) = 0$. But $g''(c) = f''(c) + 2k < 0$; contradiction.

Section 6,   page 480

1. $\ln 51 - \ln 50 = 1/c$, $50 < c < 51$, and $1/c < 1/50 = 0.02$
3. LHS $= 1/(1 + c^2)$ where $5 < c < 6$; $1 + c^2 > c^2 > 25$ so LHS $< \frac{1}{25}$.
5. $\arcsin \frac{3}{5} - \arcsin \frac{1}{2} = (\frac{3}{5} - \frac{1}{2})/\sqrt{1 - c^2}$ where $\frac{1}{2} < c < \frac{3}{5}$. Now $\sqrt{1 - c^2} > \sqrt{1 - (\frac{3}{5})^2} = \frac{4}{5}$, so $\arcsin \frac{3}{5} < \arcsin \frac{1}{2} + \frac{1}{10}/\frac{4}{5} = \arcsin \frac{1}{2} + \frac{1}{8}$. But $\frac{1}{8} = \frac{1}{8}(180/\pi)^\circ < \frac{1}{8}(180/3)^\circ = 7.5^\circ$, hence $\arcsin \frac{3}{5} < 37.5^\circ$. Note that $\arcsin \frac{3}{5} \approx 36.87^\circ$.
7. 1 9. $\frac{1}{6}$ 11. $\frac{1}{2}$ 13. -1 15. 1 17. 1 19. 0 21. 1 23. 0
25. 0 27. ∞
29. By the MVT, $[f(a + h) - f(a - h)]/2h = 2hf'(x)/2h = f'(x)$, where x is between $a - h$ and $a + h$. As $h \longrightarrow 0$, $x \longrightarrow a$, so $f'(x) \longrightarrow f'(a)$.
31. Let $\varepsilon > 0$. Choose $\delta = \varepsilon/M$. If $a < x < z < b$ and $z - x < \delta$, then $f(z) - f(x) = (z - x)f'(w)$, where $x < w < z$, hence $|f(z) - f(x)| \leq (z - x)M < \varepsilon$.
33. 1
35. For each x there is a $z = z(x)$ such that $x < z < x + 1$ and $f(x + 1) - f(x) = f'(z)$. As $x \longrightarrow \infty$, $f(x + 1) - f(x) \longrightarrow L - L = 0$, so $f'[z(x)] \longrightarrow 0$. But $z(x) \longrightarrow \infty$ so $f'[z(x)] \longrightarrow M$, hence $M = 0$.

37. Say $|f'(x)/e^x| \leq B$ for $x > a$. By the GMVT,

$$\left| \frac{f(x) - f(a)}{e^x - e^a} \right| \leq B, \quad \text{hence} \quad \left| \frac{f(x)/e^x - f(a)/e^x}{1 - e^a/e^x} \right| \leq B, \quad \left| \frac{f(x)}{e^x} \right| \leq \left| \frac{f(a)}{e^x} \right| + B \left| 1 - \frac{e^a}{e^x} \right|.$$

All terms on the RHS are bounded.

39. Apply Ex. 38 to $e^x g(x)$.

41. Let $c = \frac{1}{2}(a + b)$. If $a < x \leq c$, then $f(x) = f(x) - f(a) = (x - a)f'(z)$, $a < z < x$, hence $|f(x)| \leq M(x - a)$. Similarly $|f(x)| \leq M(b - x)$ if $c \leq x < b$. Therefore

$$\left| \int_a^b f(x)\, dx \right| \leq \int_a^c |f(x)|\, dx + \int_c^b |f(x)|\, dx \leq M \int_a^c (x - a)\, dx + M \int_c^b (b - x)\, dx, \quad \text{etc.}$$

Section 7, page 488

1. $x^2 - x$ **3.** $x^2 + x + 1$ **5.** $2x - 1$ **7.** $x^3 - x + 1$

9. $\frac{1}{4}x^4 - \frac{5}{4}x^2 + 1$ **11.** $1 - 2(x - 1) + \frac{3}{2}(x - 1)(x - 2)$

13. $4 - (x - 1) - \frac{3}{2}(x - 1)(x - 2)$ **15.** $x + (x - 1)(x - 2)(x - 3)$

17. $1 + (x - 1) + (x - 1)(x - 2) + (x - 1)(x - 2)(x - 3) + (x - 1)(x - 2)(x - 3)(x - 4)$

19. $1 + \frac{1}{8}(x - 1)(x - 3)$ **21.** $1 - (x + 2) + (x + 2)(x + 1) - \frac{2}{3}(x + 2)(x + 1)x$

23. (answer to Ex. 21) $+ \frac{1}{3}(x + 2)(x + 1)x(x - 1)$ **25.** $(4/\pi^2)x(\pi - x)$

27. $(8/3\pi^3)x(\pi^2 - x^2)$ **29.** They both give $\frac{2}{3}\pi \approx 2.0944$.

Section 8, page 493

1. $-\frac{1}{3}(x + 2)x(x - 1) + \frac{1}{3}(x + 3)(x + 2)(x - 1) + \frac{1}{12}(x + 3)(x + 2)x =$
$\frac{1}{12}(x^3 + 17x^2 + 18x - 24)$

3. $-\frac{1}{6}(x - 2)(x - 4)(x - 6) + \frac{1}{4}x(x - 4)(x - 6) - \frac{1}{8}x(x - 2)(x - 6)$
$+ \frac{1}{48}x(x - 2)(x - 4) = \frac{1}{48}(-x^3 + 18x^2 - 128x + 384)$

5. $-\frac{5}{384}x(x - 2)(x - 4)(x - 6) - \frac{1}{48}(x + 2)(x - 2)(x - 4)(x - 6)$
$+ \frac{1}{48}(x + 2)x(x - 2)(x - 6) + \frac{1}{128}(x + 2)x(x - 2)(x - 4)$
$= \frac{1}{192}(-x^4 + 40x^3 - 212x^2 + 80x + 384)$

7. $\frac{1}{3}(x^3 - 6x^2 + 8x)$ **9.** $\frac{1}{2}(8x^2 - 18x + 9)$

11. RHS interpolates the cubic x^3 at $(1, 1)$, $(2, 8)$, $(3, 27)$, and $(4, 64)$. By uniqueness it equals x^3.

13. $q(x) = p(x) - p(-x)$ is a polynomial of degree at most $n - 1$ and $q(x_i) = p(x_i) - p(-x_i) = f(x_i) - f(-x_i) = 0$. Since $q(x)$ has n distinct zeros, $q(x) = 0$ so $p(-x) = p(x)$.

15. $|f(x) - p(x)| \leq \frac{1}{6}|x(x - \frac{1}{2}\pi)(x - \pi)| < \frac{1}{4}$

17. By the error estimate applied to $p(x) = 0$, $|f(x)| \leq \frac{1}{2}M|(x - a)(x - b)|$. Therefore

$$\left| \int_a^b f(x)\, dx \right| \leq \int_a^b |f(x)|\, dx \leq \frac{M}{2} \int_a^b (x - a)(b - x)\, dx = \frac{M}{12}(b - a)^3.$$

Section 9*, page 504

1. 1.0023, 0.0023 **3.** 1.4174, 0.0032 **5.** 4.6723, 0.0015 **7.** 0.5046, 0.0046

9. 1.71832, 1.71828 **11.** 0.10869, 0.10065 **13.** 2.98209, 2.98200

15. 0.61036, 0.61989; exploiting symmetry: 0.61989, 0.62050

17. 4.87757, 4.85802 **19.** 6.03354, 6.03343 **21.** 1.7182819 **23.** 8

25. $|f^{(4)}| = |e^x| \leq e$, so $|\text{error}| \leq (10)(\frac{1}{90})e(0.05)^5 < 10^{-7}$ **27.** 2.3427

29. 0.0741 **31.** The variable change $x = c + hu$ does the trick.

33. 1.718280 **35.** 6.287173 **37.** 4.853838

39. G.Q. approximates x^6 by $(\frac{1}{3}\sqrt{3})^6 \approx 0.03704$, whereas the function grows to 1 as $x \longrightarrow 1$ or $x \longrightarrow -1$. Exploiting symmetry yields 0.2407, a lot closer to $\frac{2}{7}$.

* There may be slight round-off errors in this and the next three sections.

41. $A(x) = \pi x$, $V = \frac{1}{2}\pi a^2$ **43.** $A(x) = \pi b^2 (a^2 - x^2)/a^2$, $V = \frac{4}{3}\pi a b^2$

45. By slices, $V = \int_a^{a+h} A(x)\,dx = \dfrac{h}{6}[A(0) + 4A(a + \frac{1}{2}h) + A(a + h)] = \frac{1}{6}h(A_0 + 4M + A_1)$
since Simpson's Rule is exact for cubics.

47. By symmetry, $\int_0^b f(x)\,dx = \frac{1}{2}b$. Also, $f(x_i) + f(x_{2n-i}) = 1$ so $f_0 + 4f_1 + 2f_2 + \cdots$ reduces to $\frac{1}{2}(6n) = 3n$, hence $\frac{1}{3}h(f_0 + \cdots) = nh = n(b/2n) = \frac{1}{2}b$.

49. Integrate by parts twice on the right.

51. Choose any $x_0 \neq a$, b, c. Choose k so $g(x) = f(x) - k(x - a)(x - c)^2(x - b)$ satisfies $g(x_0) = 0$. Then $g(x)$ has 5 zeros on $[a, b]$, namely a, b, c, c, x_0, where c is a double zero. Therefore $g^{(4)}(x)$ must have a zero; $g^{(4)}(z) = 0$. Thus $f^{(4)}(z) = 24k$, $f(x_0) = \frac{1}{24}f^{(4)}(z)(x_0 - a)(x_0 - c)^2(x_0 - b)$, etc.

53. By the method used in Ex. 51 above we prove $|f(x)| \leq \frac{1}{24}M(x - a)^2(x - b)^2$. Now integrate.

55. Set $p(x) = (3x^2 - 1)(ax + b)$. Since G.Q. is exact for cubics, $\int_{-1}^1 p = 0$, so $\int_{-1}^1 f = \int_{-1}^1 g$. Choose a and b so $(3c^2 - 1)(\pm ac + b) = f(\pm c)$.

Section 10,   page 510

1. 1.71   **3.** 2.51   **5.** ± 1.63 **7.** -1.84 **9.** -1.84, 1.15   **11.** 3.0723166
13. 1.8073435   **15.** -0.9032101, 1.1939365, 3.7092729   **17.** 1.7632202
19. 0.479727, 2.2467941   **21.** 1.8862034   **23.** 0.9045568

The answers to Exs. 25–33 are given to more places than required.

25. $f(1.7632231) \approx 0.09726\ 01312\ 28$ **27.** $f(0.54391) \approx 1.418697$
29. $f(0.44094) \approx 0.039804$ **31.** $f(0.47197) \approx f(2.66962) \approx -0.056010$
33. $f(0.84222) \approx 0.082605$ **35.** $x \approx 2.1294$, 23 steps

Section 11,   page 522

1. $x_8 = x_9 = 0.7035$ **3.** $x_6 = x_7 = 2.3028$ **5.** $x_8 = x_9 = 2.3311$
7. $x_8 = x_9 = 1.0000$ **9.** $x_{16} = x_{17} = 0.4999$ **11.** $x_3 = x_4 = 1.02986\ 653$
13. $\pm 3.16227\ 76602$ **15.** $\pm 0.66874\ 03050$ **17.** $-1.32471\ 79572$
19. $1.31459\ 62123$ **21.** $-0.56714\ 32904$ **23.** $-1.27177\ 97887$, $0.47177\ 97887$
25. $-1.14789\ 90357$ **27.** $-1.45367\ 36665$, $0.53978\ 51608$
29. $\pm 1.89549\ 42670$, 0   **31.** $-1.54165\ 16841$, $0.20006\ 41026$, $1.44050\ 03973$
33. $\pm 0.72704\ 72898$, $\pm 3.90335\ 68641$ **35.** 0, $\approx 1.52861\ 47266$
37. 0.33651 [point of tangency $\approx (-2.79839, -0.94168)$]
39. 1.03159, for $x \approx -0.53984$
41. $|x_0 - \sqrt{2}| < |1.42 - 1| = 0.42$, so the estimate becomes

$$|x_6 - \sqrt{2}| < \frac{1}{2^{63}}(0.42)^{64} \approx 8.4 \times 10^{-44}.\qquad \text{In general } |x_n - \sqrt{2}| < \frac{1}{2^{2^n - 1}}|x_0 - 1|^{2^n}.$$

43. $|\phi'(x)| = \dfrac{1}{2\sqrt{2}\sqrt{1 + x}} < \dfrac{1}{2\sqrt{2}} < 1$, hence x_n approaches the fixed point $\phi(1) = 1$.

45. For $a = \frac{3}{2}$ because $\phi'(b) = a - 3b^2 = 3 - 2a$, so $\phi'(b) = 0$ only for $a = \frac{3}{2}$.

47. $\psi(x) = \dfrac{3x^4 + 12x^2 - 4}{8x^3}$, $x_1 = \frac{11}{8} = 1.375$, $x_2 \approx 1.41419\ 75019$, $x_3 \approx 1.41421\ 35624$

49. $u' = 1 - \dfrac{ff''}{(f')^2} = 1 - uv$, $\phi' = 1 - u' = uv$, hence

$$\psi' = \phi' - uu'v - \tfrac{1}{2}u^2v' = uv - u(1 - uv)v - \tfrac{1}{2}u^2v' = u^2(v^2 - \tfrac{1}{2}v').$$

But $\quad v' = \left(\dfrac{f''}{f'}\right)' = \dfrac{f'f''' - (f'')^2}{(f')^2} = \dfrac{f'''}{f'} - v^2,$

hence $\psi' = u^2[\tfrac{3}{2}v^2 - \tfrac{1}{2}(f'''/f')]$. By Lhospital's Rule $\lim_{x \to c} u(x)/(x - c) = \lim_{x \to c} u' = \lim_{x \to c} (1 - uv) = 1$, and

$$\lim_{x \to c} \frac{\psi(x) - c}{(x - c)^3} = \lim_{x \to c} \frac{\psi'(x)}{3(x - c)^2} = \frac{1}{3} \lim_{x \to c} \left(\frac{u(x)}{x - c}\right)^2 \left(\frac{3}{2}v^2 - \frac{1}{2}\frac{f'''}{f'}\right) = K = \left(\frac{1}{2}v^2 - \frac{1}{6}\frac{f'''}{f'}\right)_{x = c}.$$

Section 12,   page 523

1. $\tfrac{1}{2} - \tfrac{1}{2}\sqrt{3}\,t - \tfrac{1}{4}t^2 + \tfrac{1}{12}\sqrt{3}\,t^3 + \tfrac{1}{48}t^4 - \tfrac{1}{240}\sqrt{3}\,t^5$, where $t = x - \tfrac{1}{3}\pi$.

3. For $x \approx 0$, $\cos x \approx 1 - \tfrac{1}{2}x^2 + \tfrac{1}{24}x^4$ and $\sqrt{1 - x^2} \approx 1 - \tfrac{1}{2}x^2 - \tfrac{1}{8}x^4 < \cos x$ if $x \neq 0$, so the cosine curve is outside of the circle.

5. $p(1) < p(0.5)^2 < p(0.1)^{10} \approx 2.71827\,97 < e$.

7. Approximating $(1 + \tfrac{1}{20}x^2)^{-1}$ by its 7-th degree Taylor polynomial, we have

$$\tfrac{1}{3}x[10(1 + \tfrac{1}{20}x^2)^{-1} - 7] \approx \tfrac{1}{3}x[-7 + 10(1 - \tfrac{1}{20}x^2 + \tfrac{1}{400}x^4 - \tfrac{1}{8000}x^6)]$$

$$= x - \tfrac{1}{6}x^3 + \tfrac{1}{120}x^5 - \tfrac{1}{2400}x^7$$

$$= p_7(x) + (\tfrac{1}{2400}x^7 - \tfrac{1}{5040}x^7),$$

where $p_7(x)$ is the Taylor polynomial for $\sin x$. Hence the error is about $2.19 \times 10^{-4}x^7$ for x small.

9. By Lhospital's Rule,

$$\lim_{x \to 0} \frac{f(c + 2x) - f(c + x)}{x} = \lim_{x \to 0} [2f'(c + 2x) - f'(c + x)] = f'(c).$$

11. By the Lemma on p. 464,

$$f(x) = \frac{1}{n!} \int_0^x (x - t)^n f^{(n+1)}(t)\, dt \geq 0.$$

If $f(c) = 0$ for some $c > 0$, then clearly $f^{(n+1)}(t) = 0$ for $0 \leq t \leq c$ or else the integral would be positive. Then by the same formula, $f(x) = 0$ for $0 \leq x \leq c$.

13. 8.03871 475   **15.** 0.40910 400   **17.** 3.40757 203   **19.** $-1.18840\,99170$

21. 0.15411 28720   **23.** $c = \tfrac{2}{3}$; then $\phi'(\sqrt[3]{a}) = 0$.

CHAPTER 11

Section 1,   page 531

1. $\dfrac{1}{4n - 1}$ **3.** $(-1)^n \dfrac{(2n + 1)(2n + 3)}{2 \cdot 3^n}$ **5.** $\dfrac{(-1)^n(2n)!}{2^{2n} \cdot n!}$

7. 687   **9.** $|1/\sqrt{n} - 0| < \varepsilon$ when $n > 1/\varepsilon^2$

11. $\tfrac{1}{2}$ **13.** 0   **15.** 1   **17.** $\tfrac{1}{2}$ **19.** 1   **21.** 0   **23.** 3   **25.** 0   **27.** $\tfrac{2}{3}$

29. Suppose $a_n \longrightarrow L < 0$. Choose n so $|a_n - L| < -L$. Then $a_n - L < -L$, $a_n < 0$, a contradiction.

31. Not necessarily; try $a_n = (-1)^n$.   **33.** Not necessarily; try $a_n = b_n = (-1)^n$.

35. Given ε, always choose N larger than the index of the last inserted or deleted term.

37. Only if the sequence is eventually constant, $x_n = L$ for all $n \geq N$.

39. Given $\varepsilon > 0$, choose N so $|a_n| < \varepsilon/B$ for all $n \geq N$. Then $|a_n b_n| < (\varepsilon/B)B = \varepsilon$.

41. Let $\{b_j\}$ be a subsequence of $\{a_n\}$. Let $\varepsilon > 0$. Choose N so $|a_n - L| < \varepsilon$ for all $n \geq N$. Then $|b_j - L| < \varepsilon$ for all $j \geq N$ because $b_j = x_{n_j}$ and $n_j \geq j \geq N$.

Section 2,   page 538

1. 1   **3.** $\frac{1}{4}\pi$ **5.** $a_n > 0$ and $a_{n+1} = \dfrac{3n+1}{3n+2} a_n < a_n$

7. Clearly $\{a_n\}$ is increasing. Also $1/(n!)^2 \leq 1/n!$ and we know that $1 + 1/2! + \cdots \leq 2$, hence $a_n \leq 2$.

9. $a_n > 0$ and $a_{n+1} = ca_n < a_n$, hence $\{a_n\}$ is decreasing.

11. Assume $a_n < 2$. Then $a_{n+1} = 1 + \frac{1}{2}a_n < 1 + \frac{1}{2}(2) = 2$. Since $a_1 < 2$ it follows by induction that $a_n < 2$ for all n. This implies $a_{n+1} = 1 + \frac{1}{2}a_n > \frac{1}{2}a_n + \frac{1}{2}a_n = a_n$. Thus $a_1 < a_2 < a_3 < \cdots < 2$.

13. 5

15. $|x_{n+1} - x_{n+2}| = |x_{n+1} - \frac{1}{2}(x_n + x_{n+1})| = \frac{1}{2}|x_n - x_{n+1}|$, so the Second Comparison Test applies.

17. $a_1 < 2$. Suppose $a_n < 2$. Then $a_{n+1} < \sqrt{2+2} = 2$. Hence $a_n < 2$ for all n. Therefore $a_{n+1} = \sqrt{2 + a_n} > \sqrt{a_n + a_n} = \sqrt{2a_n} > \sqrt{a_n \cdot a_n} = a_n$. Thus $a_1 < a_2 < a_3 < \cdots < 2$; the sequence converges to a limit $L \leq 2$.

19. *Case 1:* $0 < x \leq 2$. Then $a_1 = \sqrt{x} < 2$. If $a_n < 2$, then $a_{n+1} = \sqrt{x + a_n} < \sqrt{2+2} = 2$. Therefore $a_n < 2$ for all n. *Case 2:* $2 < x$. Then $a_1 = \sqrt{x} < x$. If $a_n < x$, then $a_{n+1} = \sqrt{x + a_n} < \sqrt{x + x} = \sqrt{2x} < \sqrt{x \cdot x} = x$. Therefore $a_n < x$ for all n.

21. 0

23. Suppose $a_n < b_n$. Then b_{n+1} is the average of a_n and b_n so $a_n < b_{n+1} < b_n$. Likewise $1/a_{n+1}$ is the average of $1/a_n$ and $1/b_n$, so $1/b_n < 1/a_{n+1} < 1/a_n$, that is, $a_n < a_{n+1} < b_n$. Next, $4a_n b_n < (a_n + b_n)^2$, so $2a_n b_n/(a_n + b_n) < (a_n + b_n)/2$, that is, $a_{n+1} < b_{n+1}$.

25. By induction, $b_n = b_0^{2^n}$. Since $|b_0| = |1 - x| < 1$, the sequence $\{b_0^n\}$ converges to 0. So does the subsequence $\{b_n\}$.

27. Clearly $|b_0| < 1$. Suppose $|b_n| < |b_0|^{2^n}$. Then $|b_n| < 1$ and
$$|b_{n+1}| < \tfrac{1}{4}|b_0|^{2^{n+1}}(3 + b_n) < \tfrac{4}{4}|b_0|^{2^{n+1}} = |b_0|^{2^{n+1}}.$$
Hence $|b_n| < |b_0|^{2^n}$ for all n. It follows as in Ex. 27 that $b_n \longrightarrow 0$.

29. $(y^{n+1} - x^{n+1})/(y - x) = y^n + y^{n-1}x + y^{n-2}x^2 + \cdots + x^n$ and $x^n < y^{n-j}x^j < y^n$, etc.

31. Take $x = 1 + 1/(n+1)$ and $y = 1 + 1/n$ in the LH inequality:
$$\left(1 + \frac{1}{n+1}\right)^n \left(\frac{(n+1)^2}{n} - \frac{n(n+2)}{n+1}\right) < b_n.$$
It suffices to prove $\left(1 + \dfrac{1}{n+1}\right)^2 < \dfrac{(n+1)^2}{n} - \dfrac{n(n+2)}{n+1}$. When fractions are cleared, this boils down to
$$n^3 + 4n^2 + 4n < n^3 + 4n^2 + 4n + 1.$$

33. $|a_n - a_{n+1}| = 1/(n+1)! \leq (\frac{1}{2})^n$ for $n \geq 1$.

35. $x + 1/x = (\sqrt{x} - 1/\sqrt{x})^2 + 2 \geq 2$, hence $b_0 \geq 1$. If $b_n \geq 1$, then $b_{n+1}^2 = \frac{1}{2}(1 + b_n) \geq \frac{1}{2}(1 + 1) = 1$ so $b_{n+1} \geq 1$. By induction, $b_n \geq 1$ for all n. Next, $1 \leq b_n \leq b_n^2$, so $1 + b_n \leq 2b_n^2$, $b_{n+1}^2 = \frac{1}{2}(1 + b_n) \leq b_n^2$, $b_{n+1} \leq b_n$. Thus $\{b_n\}$ is decreasing and bounded below, so $L = \lim b_n$ exists. But $2b_{n+1}^2 = 1 + b_n$ implies $2L^2 = 1 + L$, so $L = 1$.

Section 3,   page 543

1. $\frac{3}{2}(1 - 3^{-10})$ **3.** $\frac{255}{256}$ **5.** $3(x^{n+1} - 3^{n+1})/x^n(x-3)$ if $x \neq 3$, $3(n+1)$ if $x = 3$

7. $r^{1/2}(r^4 - 1)/(r^{1/2} - 1)$ if $0 < r$ and $r \neq 1$,   8 if $r = 1$ **9.** $\frac{5}{7}$ **11.** 2^{-9}

13. $\frac{3}{2}$ **15.** $1/(1 + x^2)$ **17.** 9 ft

19. The fly travels one hour at 60 mph.   **21.** $\frac{1}{9}$ **23.** $\frac{43}{99}$

25. $\frac{1}{2} + \frac{1}{4} + \frac{1}{6} + \cdots = \frac{1}{2}(1 + \frac{1}{2} + \frac{1}{3} + \cdots) = \infty$ **27.** 1600 works:

$$\frac{1}{101} + \cdots + \frac{1}{1600} = \left(\frac{1}{101} + \cdots + \frac{1}{200}\right) + \left(\frac{1}{201} + \cdots + \frac{1}{400}\right)$$

$$+ \left(\frac{1}{401} + \cdots + \frac{1}{800}\right) + \left(\frac{1}{801} + \cdots + \frac{1}{1600}\right) > \frac{1}{2} + \cdots + \frac{1}{2} = 2.$$

(Actually, $1/101 + \cdots + 1/743 \approx 2.00121$.)

29. $\sum_{1}^{\infty} \frac{1}{n(n + 1)} = \sum_{1}^{\infty} \left(\frac{1}{n} - \frac{1}{n + 1}\right) = \left(1 - \frac{1}{2}\right) + \left(\frac{1}{2} - \frac{1}{3}\right) + \left(\frac{1}{3} - \frac{1}{4}\right) + \cdots = 1.$

31. If $\sum a_n$ converges, then $a_n \longrightarrow 0$. If also a_n is an integer, then $a_n = 0$ for all $n \geq N$. Conversely, $a_1 + a_2 + \cdots + a_{N-1} + 0 + 0 + \cdots$ obviously converges to $a_1 + \cdots + a_{N-1} = s_{N-1}$.

Section 4, page 549

1. C   **3.** D   **5.** C   **7.** C   **9.** C

11. $\sum_{j=1}^{n} (a_j + b_j) = \left(\sum_{j=1}^{n} a_j\right) + \left(\sum_{j=1}^{n} b_j\right) \longrightarrow A + B.$

13. $\sum a_n$ converges, so $a_n \longrightarrow 0$. Therefore $0 \leq a_n < 1$ for $n \geq N$, so $0 \leq a_n^2 < a_n$. Then $\sum a_n^2$ converges by comparison with $\sum a_n$.

15. C   **17.** D   **19.** C   **21.** C   **23.** C   **25.** C

27. All real x. Given x, set $a_n = x^{2n}/n!$. Then $a_{n+1}/a_n = x^2/(n + 1) \longrightarrow 0$, so the series converges by the ratio test.

29. $-\frac{1}{3} < x < \frac{1}{3}$

31. $a_n < r^n$ and $r < 1$; convergence by comparison with the geometric series

Section 5, page 553

1. C, AD   **3.** D   **5.** AC   **7.** C, AD   **9.** AC   **11.** C, AD   **13.** D

15. C, AD   **17.** D   **19.** C, AD

21. $1 - (0.5)/1! + (0.5)^2/2! - (0.5)^3/3! + (0.5)^4/4! - (0.5)^5/5! \approx 0.60651$, $e^{-0.5} \approx 0.60653$

23. The series $\sum (a_n^2 + b_n^2)$ of non-negative terms converges and $|a_n b_n| \leq \frac{1}{2}(a_n^2 + b_n^2)$, so $\sum a_n b_n$ converges absolutely by the Comparison Test.

25. First $s_{2n} = a_1 - (a_2 - a_3) - \cdots - (a_{2n-2} - a_{2n-1}) - a_{2n} < a_1$ and $s_{2n} = s_{2n-2} + (a_{2n-1} - a_{2n}) > s_{2n-2}$. Next $s_{2n-1} = (a_1 - a_2) + (a_3 - a_4) + \cdots + (a_{2n-3} - a_{2n-2}) + a_{2n-1} > 0$ and $s_{2n+1} = s_{2n-1} - (a_{2n} - a_{2n+1}) < s_{2n-1}$.

Section 6, page 559

1. $\frac{1}{8}$ **3.** 1   **5.** $\frac{1}{4}\pi$ **7.** 2   **9.** $\frac{1}{3} \ln 2$ **11.** $\frac{1}{4}\pi$ **13.** π **15.** $1/s^2$

17. $n!/s^{n+1}$ **19.** $1/(1 + s^2)$ **21.** $1/(s - a)^2$ **23.** infinite   **25.** infinite

27. infinite   **29.** $\ln 2$ **31.** $\ln 10$ **33.** 0; odd function   **35.** L/a

Section 7, page 564

1. D   **3.** C   **5.** D   **7.** C   **9.** C   **11.** D   **13.** C

15. $\displaystyle\int_{2}^{\infty} dx/x(\ln x)^p = \int_{\ln 2}^{\infty} du/u^p$, etc.

17. Suppose $p > 1$. Choose r so $p > r > 1$. Then $\dfrac{\ln x}{x^p} = \dfrac{\ln x}{x^{p-r}} \dfrac{1}{x^r} < \dfrac{1}{x^r}$

for x sufficiently large, etc. Suppose $p \leq 1$. Then $(\ln x)/x^p \geq 1/x^p$ for $x \geq e$, etc.

19. $s > 4$ **21.** $s > \frac{1}{2}$ **23.** $s > 1$ **25.** all s

27. Suppose $|f(x)| \le cx^n$ for $x \ge b$. Then

$$\int_0^\infty e^{-sx}f(x)\,dx = \int_0^b e^{-sx}f(x)\,dx + \int_b^\infty e^{-sx}f(x)\,dx$$

so we must prove the convergence of $\int_b^\infty$. But $|e^{-sx}f(x)| \le cx^n e^{-sx}$ for $b \le x$, and

$\int_b^\infty x^n e^{-sx}\,dx$ converges, so the comparison test applies.

29. $-f(0) + sL(f)(s) = -0 + s/(1+s^2) = L(\cos)(s),$
$-g(0) + sL(g)(s) = -1 + s^2/(1+s^2) = -1/(1+s^2) = L(-\sin)(s).$

Section 8,   page 569

1. D   **3.** C   **5.** D   **7.** C   **9.** C   **11.** $\displaystyle\sum_1^\infty \frac{1}{n^2} < 1 + \int_1^\infty \frac{dx}{x^2} = 2$

13. From the inequalities $\ln(n+1) < s_n < 1 + \ln n$ we conclude (1) if $n > 1.971 \times 10^{434}$, then $s_n > 1000$ and (2) if $n < 7.247 \times 10^{433}$, then $s_n \le 1000$.
For $7.247 \times 10^{433} \le n \le 1.971 \times 10^{434}$ we draw no conclusion by this method.

15. Since $\ln x$ increases,

$$\ln(n!) = \sum_{k=2}^n \ln k > \sum_{k=2}^n \int_{k-1}^k \ln x\,dx = \int_1^n \ln x\,dx = x\ln x - x\Big|_1^n = n\ln n - n + 1.$$

Hence $n! > \exp(n\ln n - n + 1) = e(n/e)^n$.

17. By parts with $u = 1 - \cos x$, $v = -1/x$:

$$\int_0^b \frac{1 - \cos x}{x^2}\,dx = -\frac{1-\cos x}{x}\Big|_0^b + \int_0^b \frac{\sin x}{x}\,dx = -\frac{1-\cos b}{b} + \int_0^b \frac{\sin x}{x}\,dx$$

Let $b \longrightarrow \infty$.

19. By parts with $u = 1/\sqrt{x}$ and $v = -\cos x$:

$$\int_1^b \frac{\sin x}{\sqrt{x}}\,dx = \cos 1 - \frac{\cos b}{\sqrt{b}} - \frac{1}{2}\int_1^b \frac{\cos x}{x^{3/2}}\,dx \longrightarrow \cos 1 - \frac{1}{2}\int_1^\infty \frac{\cos x}{x^{3/2}}\,dx.$$

Since $|(\cos x)/x^{3/2}| \le 1/x^{3/2}$, the RHS exists by a comparison test.

Section 9,   page 574

1. C   **3.** D   **5.** D   **7.** C   **9.** C   **11.** D   **13.** C   **15.** C   **17.** D
19. C   **21.** D   **23.** D   **25.** C

27. At 0, $\dfrac{\cos ax - \cos bx}{x} \approx \dfrac{(1 - \frac{1}{2}a^2x^2) - (1 - \frac{1}{2}b^2x^2)}{x} = \frac{1}{2}(b^2 - a^2)x$, so the function is

continuous at 0; no problem. Also $\displaystyle\int_1^\infty \frac{\cos ax}{x}\,dx$

exists; same proof as for the integrand $(\sin x)/x$.

29. $\frac{1}{4}$ **31.** 0   **33.** $\frac{1}{2}\pi$ **35.** $(-1)^n n!$ **37.** Substitute $x = 1/u$.

Section 10,   page 575

1. $a_n > 0$ and $\{a_n\}$ is decreasing since $a_n = a_{n-1}(1+n^2)/(2+n^2) < a_{n-1}$.

3. D   **5.** D   **7.** C   **9.** C   **11.** $x > 0$ and $x < -1$

13. $\sum_1^N a_{2n} < \sum_1^{2N} a_n < \sum_1^\infty a_n$, so the partial sums are bounded and increasing.

15. Call the new sequence $\{b_n\}$. Suppose $|a_n - L| < \varepsilon$ for $n \ge N$. Then $|b_m - L| < \varepsilon$ for $m \ge \frac{1}{2}N(N-1) = 1 + 2 + 3 + \cdots + (N-1)$.

17. In either case the quantity $c_n + (\log_2 a_n)/2^n$ doesn't change when n increases to $n + 1$. But for $n = 0$ it equals $\log_2 x$.

19. $\displaystyle\int_0^\infty \frac{f(ax)}{x}\, dx = \int_0^\infty \frac{f(t)}{t}\, dt$; substitute $t = ax$.

21. Integrate by parts with $u = e^{-t}$ and $v = \ln t$; then a second time with $u = e^{-t}$ and $v = t \ln t - t$.

23. Integrate K_{n+2} by parts with $u = x^{n+1}$, $v = -\frac{1}{2}e^{-x^2}$. This gives $K_{n+2} = \frac{1}{2}(n + 1)K_n$. Now use induction (or just multiply and telescope).

25. $\frac{1}{4}\pi^2$

27. If $a_n = L + c_n \longrightarrow L$, then $c_n \longrightarrow 0$. Hence

$$b_n = \frac{(L + c_1) + \cdots + (L + c_n)}{n} = L + \frac{c_1 + \cdots + c_n}{n} \longrightarrow L + 0 = L$$

by Ex. 26.

29. By induction, $p_1 = 1 + a_1 \geq 1 + a_1$. If $p_n \geq 1 + a_1 + \cdots + a_n$, then

$$p_{n+1} = p_n(1 + a_{n+1}) \geq (1 + a_1 + \cdots + a_n)(1 + a_{n+1})$$
$$= 1 + (a_1 + \cdots + a_{n+1}) + (\text{non-negatives}) \geq 1 + a_1 + \cdots + a_{n+1}.$$

31. We use $\ln(1 + x) \leq x$ (draw a graph!). Then $\ln p_n = \sum_1^n \ln(1 + a_j) \leq \sum_1^n a_j = s_n$.

33. $\beta > -1$, or else the integral diverges because $p + 2 > 0$ is necessary for convergence, and $1/(p + 2) = \frac{1}{2}(1 + \beta)$.

35. $\displaystyle\sum \frac{1}{n^2} = \sum \frac{1}{(2n)^2} + \sum \frac{1}{(2n + 1)^2} = \frac{1}{4}\sum \frac{1}{n^2} + \sum \frac{1}{(2n + 1)^2}$, hence

$$\sum \frac{1}{(2n + 1)^2} = \frac{3}{4}\sum \frac{1}{n^2} = \frac{3}{4}\cdot\frac{\pi^2}{6} = \frac{\pi^2}{8}.$$

CHAPTER 12

Section 1,   page 583

1. 1   **3.** 1   **5.** 3   **7.** 1   **9.** $1/e$ **11.** ∞ **13.** 1   **15.** 1   **17.** $1/\sqrt[4]{2}$
19. 1   **21.** all x **23.** $x < 0$ **25.** $1/(4 - x)$; $|x - 3| < 1$ **27.** $1/(1 + e^x)$; $x < 0$
29. $2 \ln x$; $x > 0$ **31.** $\sum (x/\pi)^n$ for instance
33. The series diverges at $x = 1$ since its terms do not $\longrightarrow 0$.
35. Let $|x| < R$. Choose r such that $|x| < r < R$. For n sufficiently large, $1/\sqrt[n]{|a_n|} > r$, hence $|a_n| < 1/r^n$, $|a_n x^n| < (|x|/r)^n$, so $\sum a_n x^n$ converges. If $|x| > R$, then $|x| > s > R$. For n large, $1/\sqrt[n]{|a_n|} < s$, hence $|a_n| > 1/s^n$, $|a_n x^n| > (|x|/s)^n$, so $\sum a_n x^n$ diverges since $|x|/s > 1$.
37. $1/a_n = 10^{n\pm 1}$, so $1/\sqrt[n]{|a_n|} = 10 \cdot 10^{\pm 1/n} \longrightarrow 10$. Hence $R = 10$. The successive ratios $\frac{1}{10}$, 10, $\frac{1}{10}$, 10, $\cdots$ do not converge, so the Ratio Test fails.

Section 2,   page 590

1. $\sum_0^\infty 3^n x^n/n!$ **3.** $\frac{1}{2}\sqrt{2} + \sum_1^\infty (-1)^n \frac{1}{2}\sqrt{2}\,[t^{2n-1}/(2n - 1)! + t^{2n}/(2n)!]$, $t = x - \frac{1}{4}\pi$
5. $\sum_0^\infty (\frac{3}{2})^{n+1}(x - \frac{1}{3})^n$ **7.** $\ln a + \sum_1^\infty (-1)^{n-1} x^n/na^n$
9. $\displaystyle 1 + \sum_1^\infty \left(\frac{1}{n!} + \frac{1}{(n - 1)!}\right)x^n = \sum_0^\infty (n + 1)x^n/n!$
11. $2 + \sum_1^\infty (-1)^{n-1} 4n(2n - 2)!\,(x - 4)^n/[(n!)^2 2^{4n}]$
13. $-7 + 12x + 5x^3$ **15.** $e^{-(x-1)^2}$ **17.** $\sum_0^\infty a_n x^{n+1}/(n + 1)$
19. $\sum_0^\infty x^{2n}/(2n)!$. Since $f^{(n+1)}(x) = \sinh x$ or $\cosh x$, we have $|f^{(n+1)}(x)| < e^x$. Apply Ex. 18 with $a = e^B$ and $k = 1$ or use the argument given in the text for the Taylor series of e^x.
21. $f'(0) = \lim_{x\to 0} [f(x) - f(0)]/x = \lim_{y\to\infty} yf\left(\frac{1}{y}\right) = \lim_{y\to\infty} ye^{-y^2} = 0.$

23. $f'(0) = 0$ by Ex. 21. Assume $f^{(n)}(0) = 0$. Then $f^{(n+1)}(0) = \lim_{x \to 0} [f^{(n)}(x) - f^{(n)}(0)]/x$. By the induction hypothesis and Ex. 22, $f^{(n+1)}(0)$ is the sum of limits of the form

$$\lim_{x \to 0} ae^{-1/x^2}/x^{k+1} = a \lim_{y \to \infty} y^{k+1}e^{-y^2} = 0.$$

Hence $f^{(n+1)}(0)$ exists and equals 0.

Section 3, page 597

1. $\sum_0^\infty 5^n x^{2n}$ **3.** $1 + 2\sum_1^\infty (-1)^n x^n$ **5.** $\sum_1^\infty (-1)^{n-1}(1 + 3^{2n-1})x^{2n}/(2n-1)!$

7. $\sum_0^\infty (-1)^n x^{2n}/(2n+2)!$ **9.** $2 + 3\sum_2^\infty x^n$

11. $\sum_1^\infty (-1)^{n-1}2^{2n-1}x^{2n}/(2n)!$

13. $1 + 5x^2 + 19x^4 + 65x^6$

15. $1 + x^2 + x^3 + \frac{3}{2}x^4 + \frac{13}{6}x^5 + \frac{73}{24}x^6$

17. $x^3 - \frac{1}{2}x^5$ **19.** $1 - \frac{1}{2}x^2 + x^3 + \frac{1}{24}x^4 - \frac{1}{2}x^5 + \frac{359}{720}x^6$ **21.** 10080   **23.** 272

25. The coefficient of x^n in $(1 + x + x^2 + \cdots)(a_0 + a_1 x + \cdots)$ is $a_0 + a_1 + \cdots + a_n$.

27. $(x + x^2 - x^3)/(1 - x^3)$ **29.** $\frac{1}{2}(\cosh x + \cos x)$

31. Suppose $f(x)$ is odd, $f(-x) = -f(x)$. Differentiate using the Chain Rule: $-f'(-x) = -f'(x)$, which says $f'(x)$ is even. Similarly the derivative of an even function is odd. If $f(x)$ is even, then $f'(x), f'''(x), f^{(5)}(x), \cdots$ are odd. Hence $f'(0) = f'''(0) = \cdots = 0$, because an odd function is 0 at $x = 0$. Therefore the odd Taylor coefficients are 0. Similarly for an even function.

33. From $f(x) = f(-x)$ we have $\sum a_n x^n = \sum (-1)^n a_n x^n$. By uniqueness, $a_n = (-1)^n a_n$ for all n, hence $a_{2n-1} = -a_{2n-1}$, $a_{2n-1} = 0$.

35. $(1 - x)f(x) = \sum_1^\infty nx^n - \sum_1^\infty nx^{n+1} = \sum_1^\infty nx^n - \sum_2^\infty (n-1)x^n = \sum_1^\infty x^n = x/(1 - x)$, hence $f(x) = x/(1 - x)^2$.

37. $\sec x = 1 + \frac{1}{2}x^2 + \frac{5}{24}x^4 + \frac{61}{720}x^6 + \cdots$, $R = \frac{1}{2}\pi$

39. $f(x) = a_0 + a_1 x + \cdots + a_{p-1}x^{p-1} + x^p f(x)$, hence $f(x) = (a_0 + \cdots + a_{p-1}x^{p-1})/(1 - x^p)$

41. $1 + x + \frac{1}{2}x^2 - \frac{1}{8}x^4 - \frac{1}{15}x^5$

Section 4, page 604

1. $\dfrac{d}{dx}\left(x - \dfrac{x^3}{3!} + \dfrac{x^5}{5!} - \cdots\right) = 1 - \dfrac{x^2}{2!} + \dfrac{x^4}{4!} - \cdots = \cos x$

3. $\dfrac{d^2}{dx^2}\left(1 + \dfrac{k^2 x^2}{2!} + \dfrac{k^4 x^4}{4!} + \cdots\right) = \dfrac{1 \cdot 2k^2}{2!} + \dfrac{3 \cdot 4k^4 x^2}{4!} + \cdots$

$$= k^2\left(1 + \dfrac{x^2}{2!} + \dfrac{x^4}{4!} + \cdots\right) = k^2 \cosh kx$$

5. $\displaystyle\int_0^x 2(t + t^3 + t^5 + \cdots)\, dt = 2\left(\dfrac{x^2}{2} + \dfrac{x^4}{4} + \dfrac{x^6}{6} + \cdots\right) = x^2 + \dfrac{(x^2)^2}{2} + \dfrac{(x^2)^3}{3} + \cdots$

$$= -\ln(1 - x^2)$$

7. Both sides equal $x - \frac{2}{3}x^3 + \frac{3}{5}x^5 - \frac{4}{7}x^7 + \cdots$

9. $(4 - 3x)/(1 - x)^2$, $|x| < 1$ **11.** $-\frac{1}{4}\ln(1 - x^4)$, $|x| < 1$

13. $\displaystyle\int_0^x \dfrac{\arctan t}{t}\, dt$ **15.** $1 - \dfrac{d}{dx}(xe^{-x^2/2})\bigg|_1 = 1$

17. The probability of a single birth is $1 - \sum_1^\infty (\frac{1}{87})^n = \frac{85}{86}$. The expected number of children born is

$$10^5\left[\dfrac{85}{86} + \sum_2^\infty n\left(\dfrac{1}{87}\right)^{n-1}\right] = 10^5\left[\dfrac{85}{86} + \sum_1^\infty n\left(\dfrac{1}{87}\right)^{n-1} - 1\right]$$

$$= 10^5\left[\dfrac{85}{86} + \left(\dfrac{87}{86}\right)^2 - 1\right] = 10^5(87^2 - 86)/86^2 \approx 101{,}176.$$

19. $\displaystyle\sum_1^\infty a_n p_n = \sum_1^\infty \left[n - \frac{n(n-1)}{2}\right]\left(\frac{1}{2}\right)^n = \frac{1}{2}\sum_1^\infty (3n - n^2)\left(\frac{1}{2}\right)^n$

$\displaystyle = \frac{3}{4}\sum_1^\infty n\left(\frac{1}{2}\right)^{n-1} - \frac{1}{2}\sum_1^\infty n^2\left(\frac{1}{2}\right)^n = \frac{3}{4}\frac{1}{(1-\frac{1}{2})^2} - \frac{1}{2}\cdot\frac{\frac{1}{2}(1+\frac{1}{2})}{(1-\frac{1}{2})^3} = 3 - 3 = 0.$

It's a fair game.

21. $\sum_0^\infty (1 - 2/3^{n+1})x^n,\ |x| < 1$ **23.** $\displaystyle\frac{1}{2}\sum_0^\infty \left(\frac{1}{2^n} - 1 - \frac{1}{3^{n+1}}\right)x^n,\ |x| < 1$

25. $(1 - x - x^2)f(x) = (1 - x - x^2)\sum_0^\infty c_n x^n$

$\qquad = c_0 + (c_1 - c_0)x + \sum_2^\infty (c_n - c_{n-1} - c_{n-2})x^n = 1$

27. $\frac{1}{5}\sqrt{5}|\frac{1}{2}(1 - \sqrt{5})|^{n+1} \le \frac{1}{5}\sqrt{5}|\frac{1}{2}(1 - \sqrt{5})| = \frac{1}{10}(5 - \sqrt{5}) < \frac{1}{2}$, hence

$c_n = \frac{1}{5}\sqrt{5}\{[\frac{1}{2}(1 + \sqrt{5})]^{n+1} - [\frac{1}{2}(1 - \sqrt{5})]^{n+1}\}$ is the closest integer to

$\frac{1}{5}\sqrt{5}[\frac{1}{2}(1 + \sqrt{5})]^{n+1}.$

29. Either use the solution formula above directly, or use induction: First it is true for $n = 1$. Next,

$c_{2n+2}^2 - 1 = (c_{2n+1} + c_{2n})^2 - 1 = c_{2n+1}^2 + 2c_{2n+1}c_{2n} + (c_{2n-1}c_{2n+1} + 1) - 1$

$\qquad = c_{2n+1}(c_{2n+1} + 2c_{2n} + c_{2n-1}) = c_{2n+1}(c_{2n+2} + c_{2n+1}) = c_{2n+1}c_{2n+3}.$

31. $f(x) = x/(1 - 2x)(1 - 3x)$

Section 5,   page 609

1. $\sum_0^\infty \frac{1}{2}(-1)^n(n + 1)(n + 2)x^n,\ |x| < 1$

3. $\sum_0^\infty (n + 1)4^n x^{2n},\ |x| < \frac{1}{2}$

5. $1 + \frac{1}{2}x^3 + \sum_2^\infty (-1)^{n-1}[3 \cdot 7 \cdot 11 \cdots (4n - 5)]x^{3n}/2^n \cdot n!,\ |x| < 2^{-1/3}$

7. $\sum_0^\infty (-1)^n(2n)!\, x^{n+2}/2^n(n!)^2,\ |x| < \frac{1}{2}$

9. $\sinh^{-1} x = \int_0^x dt/\sqrt{1 + t^2} = \sum_0^\infty (-1)^n(2n)!\, x^{2n+1}/2^{2n}(n!)^2(2n + 1),\ |x| < 1$

11. $\sqrt{2}[1 + \sum_1^\infty (-1)^{n-1}2(2n - 2)!\,(x - 1)^n/2^{3n}n!\,(n - 1)!],\ |x - 1| < 1$

13. $1 + \frac{1}{2}x^2 + \frac{1}{2}x^3 + \frac{1}{8}x^4$

15. $\sqrt{3}(2x + \frac{1}{3}x^2 - \frac{49}{36}x^3 - \frac{47}{216}x^4)$

17. $1 - \frac{10}{3}x^2 + \frac{40}{9}x^4$

19. $4(1 + 0.00625)^{1/2} \approx 4 + 0.01250 - 0.00002 \approx 4.0125$

21. $1.00000 - 0.15000 + 0.01350 - 0.00094 + 0.00006 \approx 0.8626$

23. LHS $\approx x - \frac{1}{2}x^2 + \frac{1}{3}x^3$ and RHS $\approx -1 + 1 + x - \frac{1}{2}x^2 + \frac{1}{3}x^3$

25. Expand $[1 - z(2x - z)]^{-1/2}$. Then z^n occurs in the term involving $[z(2x - z)]^n = z^n(2^n x^n + \cdots)$ and in various lower order terms. Hence the coefficient of z^n is a polynomial in x of degree n. $P_1(x) = x,\ P_2(x) = \frac{1}{2}(3x^2 - 1),\ P_3(x) = \frac{1}{2}(5x^3 - 3x).$

27. $p_n = 2n\sin\dfrac{\pi}{n} = 2n\left(\dfrac{\pi}{n} - \dfrac{\pi^3}{6n^3} + \dfrac{\pi^5}{120n^5} - \dfrac{\pi^7}{7!\,n^7} + \cdots\right)$, hence

$\pi = \dfrac{1}{2}p_n + \dfrac{\pi^3}{6n^2} - \dfrac{\pi^5}{120n^4} + \dfrac{\pi^7}{7!\,n^6} - \cdots.$

29. By Ex. 28, $3\pi = 2p_{2n} - \frac{1}{2}p_n + b_2/n^4 + b_3/n^6 + \cdots$, hence

$$3\pi = 2p_{4n} - \frac{1}{2}p_{2n} + \frac{b_2}{16n^4} + \frac{b_3}{64n^6} + \cdots.$$

Therefore

$$45\pi = 16\left(2p_{4n} - \frac{1}{2}p_{2n} + \frac{b_2}{16n^4} + \frac{b_3}{64n^6} + \cdots\right) - \left(2p_{2n} - \frac{1}{2}p_n + \frac{b_2}{n^4} + \frac{b_3}{n^6} + \cdots\right)$$

$$= 32p_{4n} - 10p_{2n} + \frac{1}{2}p_n + \frac{c_3}{n^6} + \cdots \approx 32p_{4n} - 10p_{2n} + \frac{1}{2}p_n.$$

For $n = 3$, this yields $\pi \approx \frac{1}{45} \left| 32(12\sqrt{2} - \sqrt{3}) - 60 + \frac{3}{2}\sqrt{3} \right| \approx 3.14158$.

31. $\sqrt{3} = \frac{100}{58}(1.0092)^{1/2} \approx \frac{100}{58}\left[1 + \frac{0.0092}{2} - \frac{(0.0092)^2}{8}\right]$

≈ 1.732050724 compared to $\sqrt{3} = 1.732050807\cdots$

Section 6,   page 618

1. $(\frac{1}{5})^5/5! < 4 \times 10^{-6}$, hence $e^{-1/5} \approx 1 - \frac{1}{5} + \frac{1}{2}(\frac{1}{5})^2 - \frac{1}{6}(\frac{1}{5})^3 + \frac{1}{24}(\frac{1}{5})^4 \approx 0.81874$

3. 4, including the constant

5. 0.10003   **7.** 0.25049   **9.** $k = \frac{9}{41}$, length $= (2\pi)(41)(1 - S)$,
$S \approx (\frac{1}{2})^2(\frac{9}{41})^2 + (\frac{1}{2}\cdot\frac{3}{4})^2(\frac{1}{3})(\frac{9}{41})^4 \approx 0.01216$; length ≈ 254.5

11. $\frac{1}{2}x^2 - \frac{1}{6}x^6 = 0.1$, $x \approx 0.4495$

13. $T \approx 2\pi\sqrt{\dfrac{L}{g}}\left(1 + \dfrac{1}{16}\alpha^2 + \dfrac{11}{3072}\alpha^4\right)$

15. $y = (1 + a)e^x - 1 - x$ **17.** $y = (2 + a)e^x - x^2 - 2x - 2$

19. $y = a\left(1 + \displaystyle\sum_{n=1}^{\infty} \dfrac{(-1)^{n-1}}{(2n-1)(n!)2^n}x^{2n}\right) + bx$

21. If $a = 0$, then $y = 2b(e^{-x} - 1 + x)/x$; otherwise no solution.

23. $y = 2 - 3x + 6x^2 - \frac{34}{3}x^3 + \frac{61}{3}x^4 + \cdots$

25. $y = -1 + 2x - x^2 + \frac{1}{3}x^3 + \frac{1}{6}x^4 + \cdots$

27. $e(1 - \frac{5}{48}x^3 + \frac{637}{5760}x^4)$ **29.** $e(1 - \frac{5}{288}x^3)$

31. $x \ln x \longrightarrow 0-$ as $x \longrightarrow 0+$ and

$$x^x = e^{x \ln x} = 1 + x \ln x + \tfrac{1}{2}(x \ln x)^2 + \cdots \approx 1 + x \ln x,$$

so $1 - x^x \approx -x \ln x$.

Section 7,   page 625

1. $f_n(0) = 0 \longrightarrow 0$. If $x > 0$, then $f_n(x) = 0$ for $n > 2/x$, hence $f_n(x) \longrightarrow 0$. Therefore $f_n(x) \longrightarrow 0$ for all $x \geq 0$. But $f_n(1/n) = 1$, so $|f_n(x) - 0| < 1$ for all x is impossible for any n, hence the convergence is not uniform.

3. By elementary calculus, $0 \leq xe^{-nx} \leq 1/ne$ for $0 \leq x < \infty$. Therefore $xe^{-nx} \longrightarrow 0$ uniformly on $[0, \infty)$.

5. $|(\sin nx)/n^2| \leq 1/n^2$; use the M-test, and note that $(\sin nx)/n^2$ is continuous.

7. Let $0 < a$. Then $|e^{-nx}\sin nx| \leq e^{-na}$ for $a \leq x < \infty$. By the M-test, the series converges uniformly on $[a, \infty)$. By "Continuity of the Limit", p. 621, the sum is continuous there. This is true for each $a > 0$, hence the sum is continuous for $0 < x < \infty$.

9. Both series converge uniformly by the M-test. Apply "Series of Functions (b)", p. 622.

11. $|a_n \sin nx| \leq |a_n|$. Apply the M-test with $M_n = |a_n|$.   **13.** $\sum_0^\infty x^n/n!(2n + 1)$

15. $E(x)$ converges for all x, by the Ratio Test for instance. Hence $R = \infty$ so $E(x)$ is continuously differentiable for all x and

$$E'(x) = \sum_1^\infty \frac{nx^{n-1}}{n!} = \sum_1^\infty \frac{x^{n-1}}{(n-1)!} = \sum_0^\infty \frac{x^n}{n!} = E(x).$$

Also $E(0) = 1/0! = 1$.

17. $[E(c + x)E(-x)]' = E(c + x)'E(-x) + E(c + x)[E(-x)]'$
$= E(c + x)E(-x) - E(c + x)E(-x) = 0$, so $E(c + x)E(-x)$
$= \text{constant} = E(c + 0)E(-0) = E(c)$.

19. Follow the solution of Ex. 15.

21. $(S^2 + C^2)' = 2SS' + 2CC' = 2SC - 2CS = 0$, hence $S^2 + C^2 = \text{constant} = S^2(0) + C^2(0)$
$= 1$. Therefore $|S|^2 \leq S^2 + C^2 = 1$, $|S| \leq 1$, etc.

23. Differentiate the equation in Ex. 22 w.r.t. x.

25. $1 = S^2(\frac{1}{2}\pi) + C^2(\frac{1}{2}\pi) = S^2(\frac{1}{2}\pi) + 0 = S^2(\frac{1}{2}\pi)$, hence $S(\frac{1}{2}\pi) = \pm 1$. But $S'(x) = C(x) > 0$ for $0 < x < \frac{1}{2}\pi$, so $S(x)$ strictly increases for $0 \le x \le \frac{1}{2}\pi$, that is, $S(\frac{1}{2}\pi) > S(0) = 0$, so $S(\frac{1}{2}\pi) = 1$.

27. $S(x + \pi) = S(x)C(\pi) + C(x)S(\pi) = -S(x)$, etc.

Section 8, page 626

1. $\sum_2^\infty (-1)^{n-1}(2n-2)x^{2n-1}/(2n-1)!$　　**3.** $\sum_1^\infty (-1)^{n-1}x^{2n-1}/(2n-1)!(2n-1)$

5. $\sum_0^\infty (-1)^n x^{4n+2}/(4n+2)$　　**7.** $\frac{5}{9}$　　**9.** $\frac{1}{2}(\sin x + \sinh x)$　　**11.** $\frac{9}{4}$

13. To have $4/(2n-1) < 5 \times 10^{-5}$ requires $n \ge 40{,}001$.

15. $y = 3a \sum_0^\infty \dfrac{x^{2n}}{(2n-1)(2n-3)} + b(x - x^3)$　　**17.** $-2^{14}(20!)/(14!)$

19. Yes;
$$E = \sum_1^\infty a_n p_n = \sum_1^\infty [5n - (1 + 2 + 3 + \cdots + (n-1))][\tfrac{1}{6}(\tfrac{5}{6})^{n-1}]$$
$$= \tfrac{1}{12}\sum_1^\infty (11n - n^2)(\tfrac{5}{6})^{n-1} = 0.$$

21. Clearly $p_1 = 1$ and $p_2 = 1 - \frac{1}{4} = \frac{3}{4}$. A run of $n + 2$ tosses without $\cdots HH \cdots$ can be obtained only by T followed by a run of $n + 1$ without HH, or HT followed by a run of n without HH, hence $p_{n+2} = \frac{1}{2}p_{n+1} + \frac{1}{4}p_n$.

23. $\dfrac{\sin x}{x} - \sqrt{1 - \frac{1}{3}x^2} = (1 - \frac{1}{6}x^2 + \frac{1}{120}x^4 + \cdots) - (1 - \frac{1}{6}x^2 - \frac{1}{72}x^4 + \cdots) \approx \frac{1}{45}x^4.$

CHAPTER 13

Section 1, page 630

1. x-axis up　　**3.** x-axis forward　　**5.** z-axis up

7, 9

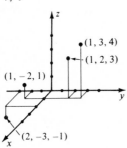

11, 13

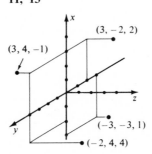

15, 17

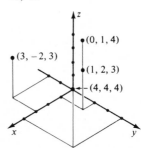

Section 2, page 635

1. $(5, 2, 4)$　　**3.** $(3, -2, 10)$　　**5.** $(1, -16, 9)$　　**7.** $(-1, 10, 4)$　　**9.** $(4, 1, -1)$

11. $u_1 + v_1 = v_1 + u_1$, etc.　　**13.** $(a + b)u_1 = au_1 + bu_1$, etc.　　**15.** $(ab)u_1 = a(bu_1)$, etc.

17. $(\frac{1}{4}, \frac{1}{4}, \frac{1}{4})$

19. Let the vertices be $\mathbf{u}$, $\mathbf{v}$, $\mathbf{w}$, $\mathbf{z}$. The centroid of $\mathbf{uvw}$ is $\frac{1}{3}(\mathbf{u} + \mathbf{v} + \mathbf{w})$. The point $\mathbf{c} = \frac{3}{4}[\frac{1}{3}(\mathbf{u} + \mathbf{v} + \mathbf{w})] + \frac{1}{4}\mathbf{z}$ is on the segment joining this centroid to $\mathbf{z}$; it is $\frac{3}{4}$ of the way from $\mathbf{z}$ to the centroid. But $\mathbf{c} = \frac{1}{4}(\mathbf{u} + \mathbf{v} + \mathbf{w} + \mathbf{z})$ is *symmetric* in the four vertices, so the same construction starting with any face and its opposite vertex leads to the same $\mathbf{c}$.

Section 3, page 641

1. 29　　**3.** -1　　**5.** $\sqrt{11}$　　**7.** 1　　**9.** $\arccos(-\frac{12}{25})$　　**11.** $\frac{1}{2}\pi$

13. $\arccos(-\frac{1}{15}\sqrt{15})$ **15.** $\sqrt{42}$ **17.** $5\sqrt{2}$ **19.** $\frac{1}{2}\sqrt{2}, 0, \frac{1}{2}\sqrt{2}$
21. $\frac{1}{7}\sqrt{14}, \frac{1}{14}\sqrt{14}, -\frac{3}{14}\sqrt{14}$ **23.** $(1, 1, 0)$ and $(0, 2, 1)$ for instance
25. $\mathbf{v} \cdot \mathbf{w} = v_1 w_1 + \cdots = w_1 v_1 + \cdots = \mathbf{w} \cdot \mathbf{v}$
27. $(\mathbf{u} + \mathbf{v}) \cdot \mathbf{w} = (u_1 + v_1, \cdots) \cdot (w_1, \cdots)$
$$= (u_1 + v_1)w_1 + \cdots = (u_1 w_1 + \cdots) + (v_1 w_1 + \cdots) = \mathbf{u} \cdot \mathbf{w} + \mathbf{v} \cdot \mathbf{w}$$
29. $|\mathbf{v} \cdot \mathbf{w}| = |\mathbf{v}| \cdot |\mathbf{w}||\cos\theta| \le |\mathbf{v}| \cdot |\mathbf{w}|$. Equality if $\mathbf{v} = \mathbf{0}$, $\mathbf{w} = \mathbf{0}$, or $\theta = 0, \pi$.

Section 4, page 648

1. not on   **3.** on   **5.** $\arccos(1/\sqrt{330})$
7. $(\frac{2}{3}, \frac{2}{3}, \frac{2}{3})$ **9.** $(0, -7, 14)$, $(-\frac{7}{5}, 0, \frac{7}{5})$, $(-\frac{14}{9}, \frac{7}{9}, 0)$
11. $(0, 0, -1)$, $(0, 0, -1)$, parallel to x, y-plane
13. $\sqrt{2}$ **15.** At time $t = 5$ and point $(23, 22, 0)$ **17.** $\mathbf{x} = t(\mathbf{b} - \mathbf{c}) + \mathbf{a}$
19. $\frac{1}{3}x_1 - \frac{2}{3}x_2 + \frac{2}{3}x_3 = \frac{1}{3}$ **21.** $-\frac{8}{9}x_1 + \frac{1}{9}x_2 - \frac{4}{9}x_3 = 3$
23. $\frac{1}{3}\sqrt{3}\,x_1 + \frac{1}{3}\sqrt{3}\,x_2 + \frac{1}{3}\sqrt{3}\,x_3 = \sqrt{3}$
25. $\frac{1}{3}\sqrt{3}$ **27.** 0 **29.** $\theta = \arcsin(\mathbf{a} \cdot \mathbf{n}/|\mathbf{a}|)$
31. The line is parallel to the plane if and only if it is perpendicular to the normal $\mathbf{n}$ to the plane if and only if $\mathbf{a} \cdot \mathbf{n} = 0$.
33. We want $\mathbf{z} = t\mathbf{a} + \mathbf{b}$ and $\mathbf{z} \cdot \mathbf{n} = p$, hence $(t\mathbf{a} + \mathbf{b}) \cdot \mathbf{n} = p$, $t = (p - \mathbf{b} \cdot \mathbf{n})/(\mathbf{a} \cdot \mathbf{n})$, so $\mathbf{z}$ is as stated.
35. $\mathbf{b} - [(\mathbf{b} - \mathbf{c}) \cdot \mathbf{u}]\mathbf{u}$

Section 5, page 655

1. $(-\frac{1}{3}, \frac{2}{3})$ **3.** $(-1, 1)$ **5.** $(\frac{1}{19}, \frac{7}{19})$ **7.** $(-\frac{8}{19}, \frac{1}{19}, \frac{7}{19})$ **9.** $(0, 0, 0)$
11. $(\frac{3}{14}, -\frac{1}{4}, -\frac{3}{28})$ **13.** two parallel lines   **15.** The first and third planes are parallel.
17. $\mathbf{x} = (t, \frac{2}{3}t - \frac{1}{3})$ **19.** $\mathbf{x} = (t, -\frac{1}{2}t - \frac{1}{2}, -\frac{5}{2}t + \frac{1}{2})$ **21.** $(-\frac{8}{15}, -\frac{13}{15}, \frac{22}{15})$
23. $\frac{1}{32}(90, 11, 67)$
25. Expanded by the first row, it is a linear equation, hence a line. If (a_i, b_i) is substituted for (x, y), a determinant with two equal rows results, value 0. Hence the line passes through the given points. Note that the coefficients of x and y are not both 0.
27. $(b - a)(c - a)(c - b)$ **29.** $(1, -1, 1)$ **31.** line $\mathbf{x} = t(-1, -1, 1) + (2, 0, 0)$
33. $x + y + z = 2$ **35.** $7x - 4y + 2z = 3$

Section 6, page 663

1. $(-5, 2, -14)$ **3.** $(-4, 8, -4)$ **5.** $(2, -2, 0)$ **7.** $\mathbf{i} - \mathbf{j}$ **9.** $-7\mathbf{i} + 8\mathbf{j} + 2\mathbf{k}$
11. $\frac{1}{11}\sqrt{11}\,(1, -1, 3)$ **13.** $\sqrt{14}$ **15.** $(0, -10, 10)$ **17.** $(3, -1, 4)$ **19.** 1
21. 87 **23.** $\mathbf{v} = v_1\mathbf{i} + v_2\mathbf{j} + 3\mathbf{k}$
25. A determinant changes sign when two rows are interchanged, hence
$$\mathbf{u} \cdot (\mathbf{v} \times \mathbf{w}) = -\mathbf{v} \cdot (\mathbf{u} \times \mathbf{w}) = +\mathbf{v} \cdot (\mathbf{w} \times \mathbf{u}).$$
27. $(av_2)w_3 - (av_3)w_2 = a(v_2 w_3 - v_3 w_2)$, etc.
29. Clearly the formula is true if $\mathbf{a} = \mathbf{b}$ because both sides equal $\mathbf{0}$. Now for example if $\mathbf{a} = \mathbf{i}$ and $\mathbf{b} = \mathbf{j}$, then
$$\text{LHS} = (\mathbf{i} \cdot \mathbf{u})(\mathbf{j} \cdot \mathbf{v}) - (\mathbf{i} \cdot \mathbf{v})(\mathbf{j} \cdot \mathbf{u}) = u_1 v_2 - u_2 v_1$$
$$= \mathbf{k} \cdot (\mathbf{u} \times \mathbf{v}) = (\mathbf{a} \times \mathbf{b}) \cdot (\mathbf{u} \times \mathbf{v}) = \text{RHS}, \quad \text{etc.}$$
31. Write $(\mathbf{a} - \mathbf{b}) \cdot \mathbf{v} = 0$ and choose
$$\mathbf{v} = \mathbf{a} - \mathbf{b}: \quad |\mathbf{a} - \mathbf{b}|^2 = (\mathbf{a} - \mathbf{b}) \cdot (\mathbf{a} - \mathbf{b}) = 0,$$
hence $\mathbf{a} - \mathbf{b} = \mathbf{0}$, $\mathbf{a} = \mathbf{b}$.
33. Take $\mathbf{a} = \mathbf{u} = \mathbf{w}$ and $\mathbf{b} = \mathbf{v}$.

Section 7,   page 671

1. 2   3. $\mathbf{x} = t(-5, 1, 1) + (-2, 1, 0)$ 5. $\mathbf{x} = t(3, 1, -5) + (\frac{8}{5}, \frac{1}{5}, 1)$
7. $\mathbf{x} = t(3, 1)$ 9. $\mathbf{x} = t(1, 4, -7)$ 11. $\mathbf{x} = t(6, -5, -19)$
13. $\mathbf{x} = s(5, 0, -4) + t(0, 5, -3)$ 15. $x = 1$ 17. $2x - y - z = 3$ 19. $x + y = 3$
21. $\sqrt{2}$ 23. $3/\sqrt{17}$ 25. $(\frac{8}{3}, \frac{8}{3}, \frac{5}{3})$ and $(\frac{8}{3}, \frac{5}{3}, \frac{8}{3})$
27. $\mathbf{x} = (1 - s - t)(1, 1, 0) + s(1, 2, 1) + t(-1, -1, -1) = (-2t + 1, s - 2t + 1, s - t)$
29. $A + B = C, \quad aA = bB$
31. $\sum (\mathbf{x}_j - \mathbf{p}) \times \mathbf{F}_j = \sum \mathbf{x}_j \times \mathbf{F}_j - \mathbf{p} \times \sum \mathbf{F}_j = 0 - 0 = 0.$
33. LHS is the square of the volume of the parallelepiped determined by $\mathbf{a}$, $\mathbf{b}$, $\mathbf{c}$, clearly at most the square of the product of the edges.
35. It is the solid triangle with vertices $\mathbf{a}$, $\mathbf{b}$, $\mathbf{c}$.
37. We show that the following three conditions are equivalent: (i) $[\mathbf{a}, \mathbf{b}, \mathbf{c}] \neq 0$, (ii) no one of the vectors $\mathbf{a}$, $\mathbf{b}$, $\mathbf{c}$ lies in the plane spanned by the other two, and (iii) $\mathbf{a}$, $\mathbf{b}$, $\mathbf{c}$ are linearly independent. Suppose (i). Then the parallelepiped determined by $\mathbf{a}$, $\mathbf{b}$, and $\mathbf{c}$ has positive volume, hence (ii). Conversely, if (ii), then the parallelepiped has positive volume, hence (i). Now assume (ii) and $r\mathbf{a} + s\mathbf{b} + t\mathbf{c} = \mathbf{0}$. If say $t \neq 0$, then $\mathbf{c} = (-r/t)\mathbf{a} + (-s/t)\mathbf{b}$, impossible. Hence $t = 0$ and similarly $r = s = 0$, so (iii) follows. Conversely, if (iii) and say $\mathbf{c} = r\mathbf{a} + s\mathbf{b}$, then $r\mathbf{a} + s\mathbf{b} + (-1)\mathbf{c} = \mathbf{0}$, impossible, hence (ii). This proves (i) if and only if (ii) if and only if (iii).

Section 8,   page 673

1. equivalent to $|\mathbf{x} - \frac{1}{2}(\mathbf{a} + \mathbf{b})|^2 = c^2 + \frac{1}{4}|\mathbf{a} + \mathbf{b}|^2$; center $\frac{1}{2}(\mathbf{a} + \mathbf{b})$, radius r, $r^2 = c^2 + \frac{1}{4}|\mathbf{a} + \mathbf{b}|^2$
3. Let $\mathbf{a}$, $\mathbf{b}$, $\mathbf{c}$ be the vertices. The midpoints of $\overline{ab}$ and $\overline{ac}$ are $\mathbf{m} = \frac{1}{2}(\mathbf{a} + \mathbf{b})$ and $\mathbf{n} = \frac{1}{2}(\mathbf{a} + \mathbf{c})$. The vector $\mathbf{n} - \mathbf{m} = \frac{1}{2}(\mathbf{c} - \mathbf{b})$ is clearly parallel to $\overline{bc}$.
5. $\frac{1}{4}|(\mathbf{c} - \mathbf{a}) \times (\mathbf{d} - \mathbf{b})| = \frac{1}{4}|\mathbf{a} \times \mathbf{b} + \mathbf{b} \times \mathbf{c} + \mathbf{c} \times \mathbf{d} + \mathbf{d} \times \mathbf{a}|$
7. Let the rhombus be generated by the vectors $\mathbf{a}$ and $\mathbf{b}$. Thus its vertices in order are $\mathbf{0}$, $\mathbf{a}$, $\mathbf{a} + \mathbf{b}$, $\mathbf{b}$, and $|\mathbf{a}| = |\mathbf{b}|$. Then $(\mathbf{a} + \mathbf{b}) \cdot (\mathbf{b} - \mathbf{a}) = |\mathbf{b}|^2 - |\mathbf{a}|^2 = 0$, which proves $\overline{0(a + b)}$ is perpendicular to $\overline{ab}$.
9. $\mathbf{x} = (4, 3, -\frac{1}{2})$
11. $\mathbf{0}$, $\mathbf{a}$, $\mathbf{b}$ are the vertices of an equilateral triangle.   13. the center of the polygon
15. $\frac{1}{2}(a^2b^2 + b^2c^2 + c^2a^2)^{1/2}$
17. The point is $r^{-1}(\mathbf{a}_1 + \cdots + \mathbf{a}_r)$.
19. Let $\mathbf{u}$ and $\mathbf{w}$ be unit vectors in the directions from $\mathbf{v}_1$ to $\mathbf{v}_2$ and from $\mathbf{v}_1$ to $\mathbf{v}_3$. Then $\mathbf{u} = (\mathbf{v}_2 - \mathbf{v}_1)/a_3$ and $\mathbf{w} = (\mathbf{v}_3 - \mathbf{v}_1)/a_2$. Thus $\mathbf{p} - \mathbf{v}_1 = [(a_1\mathbf{v}_1 + a_2\mathbf{v}_2 + a_3\mathbf{v}_3) - (a_1 + a_2 + a_3)\mathbf{v}_1]/(a_1 + a_2 + a_3) = c(\mathbf{u} + \mathbf{w})$, where $c = a_2 a_3/(a_1 + a_2 + a_3)$. Hence $\mathbf{p} - \mathbf{v}_1$ bisects the angle at vertex $\mathbf{v}_1$, etc.
21. By direct calculation, all six edges have the same length, $\sqrt{3}$.

CHAPTER 14

Section 1,   page 679

1. $(e^t, 2e^{2t}, 3e^{3t})$ 3. $(1, 3, 4)$ 5. $(1, -\sin t, \cos t)$ 7. $\sqrt{(2t)^2 + (3t^2 + 4t^3)^2}$
9. $|A\omega|$ 11. $2|t|^3\sqrt{4 + 9t^4}$ 13. $(x_1 + y_1)^{\cdot} = \dot{x}_1 + \dot{y}_1$, etc.
15. $(x_2 y_3 - x_3 y_2)^{\cdot} = (\dot{x}_2 y_3 - \dot{x}_3 y_2) + (x_2 \dot{y}_3 - x_3 \dot{y}_2)$, etc.   17. $\mathbf{x}(t) = t\mathbf{a} + \mathbf{b}$
19. $\mathbf{x}(t) = e^{kt}\mathbf{a}$ 21. $d|\mathbf{x}|^2/dt = d(\mathbf{x} \cdot \mathbf{x})/dt = 2\mathbf{x} \cdot \dot{\mathbf{x}} = 0$, hence $|\mathbf{x}|^2 = \text{const}$
23. $\mathbf{x}_n = [\frac{2}{3} + \frac{1}{3}(-\frac{1}{2})^n]\mathbf{a} + [\frac{1}{3} - \frac{1}{3}(-\frac{1}{2})^n]\mathbf{b}$
25. $[\mathbf{u}, \mathbf{v}, \mathbf{w}]^{\cdot} = [\mathbf{u} \cdot (\mathbf{v} \times \mathbf{w})]^{\cdot} = \dot{\mathbf{u}} \cdot (\mathbf{v} \times \mathbf{w}) + \mathbf{u} \cdot (\mathbf{v} \times \mathbf{w})^{\cdot}$
$= [\dot{\mathbf{u}}, \mathbf{v}, \mathbf{w}] + \mathbf{u} \cdot (\dot{\mathbf{v}} \times \mathbf{w} + \mathbf{v} \times \dot{\mathbf{w}}) = [\dot{\mathbf{u}}, \mathbf{v}, \mathbf{w}] + [\mathbf{u}, \dot{\mathbf{v}}, \mathbf{w}] + [\mathbf{u}, \mathbf{v}, \dot{\mathbf{w}}]$

Section 2,   page 686

1. $(a_1^2 + a_2^2 + a_3^2)^{1/2}$ **3.** $(1763\sqrt{41} + 2048)/9375 \approx 1.423$

5. $\displaystyle\int_0^b \sqrt{1 + 9x^4}\, dx$ **7.** $\displaystyle\int_a^b \sqrt{m^2 t^{2m-2} + n^2 t^{2n-2} + r^2 t^{2r-2}}\, dt$

9. $\sqrt{5} - \sqrt{2} + \dfrac{1}{2}\ln\left(\dfrac{3-\sqrt{5}}{6-4\sqrt{2}}\right) \approx 1.222$ **11.** $b^2 + b$

13. $|\mathbf{x}|^2 = \sin^4 t + \sin^2 t \cos^2 t + \cos^2 t = 1$,
   $\mathbf{x} \cdot \dot{\mathbf{x}} = (\sin^2 t)(2\sin t \cos t) + (\sin t \cos t)(\cos^2 t - \sin^2 t) + (\cos t)(-\sin t) = 0$.

15. $t = \frac{1}{2}(\sqrt{4s+1} - 1)$

17. $L = \displaystyle\int_{t_0}^{t_1} \sqrt{\dot{x}^2 + \dot{y}^2}\, dt \geq \int_{t_0}^{t_1} |\dot{x}|\, dt \geq \left| \int_{t_0}^{t_1} \dot{x}\, dt \right| = |x(t_1) - x(t_0)| = b - a$

Section 3,   page 694

1. $y = tx$, hence $y^2 = t^2 x^2 = (x+1)x^2 = x^3 + x^2$

3. $a^2 + x^2 = a^2 + a^2 \cot^2 \theta = a^2 \csc^2 \theta$, hence $a^3/(a^2 + x^2) = a/\csc^2 \theta = a \sin^2 \theta = y$

5. $(a^2 + x^2)y = (a^2 + a^2 \cot^2 \theta)(b \sin \theta \cos \theta)$
   $= a^2 b \csc^2 \theta \sin \theta \cos \theta = a^2 b \csc \theta \cos \theta = a^2 b \cot \theta = abx$

7. $x^3 + y^3 = 27a^3 \dfrac{t^3 + t^6}{(1+t^3)^3} = 27a^3 \dfrac{t^3}{(1+t^3)^2} = 3axy$

9. $y^2(a - x) = \left[\dfrac{a^2 t^6}{(1+t^2)^2}\right]\left[\dfrac{a}{1+t^2}\right] = \dfrac{a^3 t^6}{(1+t^2)^3} = x^3$

11. $\mathbf{x} = (a\theta - b \sin \theta, a - b \cos \theta)$. The center of the rolling circle is $(a\theta, a)$. The vector from the center to $\mathbf{x}$ has length b and positive angle $\frac{3}{2}\pi - \theta$ from the positive x-axis, so it is $b(\cos(\frac{3}{2}\pi - \theta), \sin(\frac{3}{2}\pi - \theta)) = b(-\sin \theta, -\cos \theta)$. Therefore
$$\mathbf{x} = (a\theta, a) + b(-\sin \theta, -\cos \theta), \quad \text{etc.}$$

13. Exactly the same as Ex. 11.

15. $\mathbf{x} = (\cos \theta + \theta \sin \theta, \sin \theta - \theta \cos \theta)$ **17.** $s = a \ln \cosh(t/a)$

19. $\mathbf{x} = \left((a-b)\cos \theta + b \cos\left(\dfrac{a-b}{b}\right)\theta, \ (a-b)\sin \theta - b \sin\left(\dfrac{a-b}{b}\right)\theta\right)$

21. $\mathbf{x} = (2b \cos \theta + b \cos 2\theta, 2b \sin \theta - b \sin 2\theta)$, $A = 2\pi b^2 = \frac{2}{9}\pi a^2$

23. $\mathbf{x} = (a \cos^3 \theta, a \sin^3 \theta)$

25. $\frac{3}{8}\pi a^2$ **27.** $L_n = 8a\dfrac{n-1}{n} = 8a\left(1 - \dfrac{1}{n}\right)$ increases to limit $8a$.

29. $\mathbf{x} = \left((a+b)\cos \theta - b \cos\left(\dfrac{a+b}{b}\right)\theta, \ (a+b)\sin \theta - b \sin\left(\dfrac{a+b}{b}\right)\theta\right)$

31. $L = 24b = 12a$

Section 4,   page 703

1. $(1 + 4x^2)^{-1/2}(1, 2x)$ **3.** $\pm(4 + 9t^2)^{-1/2}(2, 3t)$, + if $t > 0$, $-$ if $t < 0$

5. $|\mathbf{a}|^{-1}\mathbf{a}$ **7.** $|t\mathbf{a} + \mathbf{b}|^{-1}(t\mathbf{a} + \mathbf{b})$ **9.** $2/(1 + 4x^2)^{3/2}$

11. $6/|t|(9t^2 + 4)^{3/2}$ **13.** $6|t|\,|\mathbf{a} \times \mathbf{b}|/|\mathbf{a} + 3t^2\mathbf{b}|^3$

15. $(12 \sin^2 2t + 5)^{1/2}/(1 + 4 \cos^2 2t)^{3/2}$

17. 1   **19.** $re^a/(1 + e^{2a})^{3/2}$ **21.**

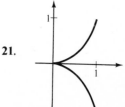

23. $(k\mathbf{n}) \cdot \mathbf{n} + \mathbf{t} \cdot d\mathbf{n}/ds = 0$ and $2\mathbf{n} \cdot d\mathbf{n}/ds = 0$, hence

$$\frac{d\mathbf{n}}{ds} \cdot \mathbf{t} = -k \qquad \text{and} \qquad \frac{d\mathbf{n}}{ds} \cdot \mathbf{n} = 0.$$

Since $\mathbf{t}$ and $\mathbf{n}$ are orthogonal unit vectors, this implies $d\mathbf{n}/ds = -k\mathbf{t}$.

25. $\mathbf{x} \cdot \mathbf{x} = 1$, hence $\mathbf{x} \cdot \mathbf{t} = 0$. Differentiate again w.r.t. s: $\mathbf{t} \cdot \mathbf{t} + \mathbf{x} \cdot (k\mathbf{n}) = 0$, that is, $k(\mathbf{x} \cdot \mathbf{n}) = -1$. But $|\mathbf{x} \cdot \mathbf{n}| \le |\mathbf{x}||\mathbf{n}| = 1$, hence $k \ge 1$.

27. Differentiate $\mathbf{x} \cdot \mathbf{n} = -p$ with respect to α:

$$\left(\mathbf{t}\frac{ds}{d\alpha} \right) \cdot \mathbf{n} + \mathbf{x} \cdot (-\mathbf{t}) = -\frac{dp}{d\alpha}, \qquad \mathbf{x} \cdot \mathbf{t} = \frac{dp}{d\alpha}.$$

29. $k\,ds = d\alpha$, so $\displaystyle\int_0^L k\,ds = \int_0^{2\pi} d\alpha = 2\pi$.

31. $\displaystyle\int_0^{2\pi} \frac{d}{d\alpha}\left(p\frac{dp}{d\alpha} \right) d\alpha = \int_0^{2\pi} \left(\frac{dp}{d\alpha} \right)^2 d\alpha + \int_0^{2\pi} p\frac{d^2p}{d\alpha^2}\,d\alpha.$

The first integral is 0 because $p\,dp/d\alpha$ is periodic of period 2π. By Ex. 30,

$$\int_0^{2\pi} p\frac{d^2p}{d\alpha^2}\,d\alpha = \int_0^{2\pi} \frac{p}{k}\,d\alpha - \int_0^{2\pi} p^2\,d\alpha.$$

But $d\alpha/k = ds$, so the result follows from Example 5b.

33. $\dfrac{ds_1}{d\alpha}\mathbf{t}_1 = \dfrac{d\mathbf{x}_1}{d\alpha} = \dfrac{d\mathbf{x}}{d\alpha} - a\dfrac{d\mathbf{n}}{d\alpha} = \dfrac{ds}{d\alpha}\mathbf{t} + a\mathbf{t} = \left(\dfrac{ds}{d\alpha} + a \right)\mathbf{t}.$

Therefore $\mathbf{t}_1 = \mathbf{t}$ since $\mathbf{t}_1$ and $\mathbf{t}$ are unit vectors.

35. From the solution of Ex. 33, $ds_1/d\alpha = ds/d\alpha + a$, hence

$$L_1 = \int_0^{2\pi} \frac{ds_1}{d\alpha}\,d\alpha = \int_0^{2\pi} \frac{ds}{d\alpha}\,d\alpha + \int_0^{2\pi} a\,d\alpha = L + 2\pi a.$$

Section 5, page 709

1. The shell strikes the hill when $y = x\tan\beta$. Substitute this into the solution $y = -ax^2 + bx$ of Example 2, cancel x, solve for x, etc.

3. $[4t/(1 + 4t^2)](1, 2t)$, $[2/(1 + 4t^2)](-2t, 1)$

5. $-\sin t \cos t\,(1 + \cos^2 t)^{-1}(1, \cos t)$, $|\sin t|(1 + \cos^2 t)^{-1}(-\cos t, 1)$

7. $2(\cos 2t)(-1, 1)$, **0**   **9. 0**, $-b\omega^2(\cos \omega t, \sin \omega t, 0)$

11. $\sin t \cos t\,(1 + \cos^2 t)^{-1}(-\cos t, \sin t, -\cos t)$, $(1 + \cos^2 t)^{-1}(-\sin t, -2\cos t, -\sin t)$

13. $\mathbf{x} \cdot \mathbf{x} = 1$, hence $\mathbf{x} \cdot \mathbf{v} = 0$, $\mathbf{x} \cdot \mathbf{a} + \mathbf{v} \cdot \mathbf{v} = 0$, $\mathbf{x} \cdot \mathbf{a} = -1$. Since $\mathbf{x}$ is a unit vector perpendicular to $\mathbf{t} = \mathbf{v}$, and since the projection $\mathbf{a} \cdot \mathbf{x}$ of $\mathbf{a}$ on $\mathbf{x}$ is -1, it follows that the normal component, the component of $\mathbf{a}$ perpendicular to $\mathbf{t}$, has length at least 1.

Section 6, page 714

1. $\frac{1}{4}\pi a^2$ **3.** $\frac{1}{4}\pi a^2$ **5.** $\frac{3}{8}\pi a^2$ **7.** a^2 **9.** $\frac{2}{3}\pi + \sqrt{3}$ **11.** $(\frac{1}{8}\pi - \frac{1}{4})a^2$

13. $\mathbf{v} = \mathbf{u} + 2t\mathbf{w} = (\cos 2t - 2t\sin 2t, \sin 2t + 2t + 2t\cos 2t)$,
$\mathbf{a} = -4t\mathbf{u} + 4\mathbf{w} = (-4t\cos 2t - 4\sin 2t, -4t\sin 2t + 4\cos 2t)$

15. $(-\sin t)\mathbf{u} + (\cos t)\mathbf{w} = (-\sin 2t, \cos 2t)$,
$-2(\cos t)\mathbf{u} - 2(\sin t)\mathbf{w} = (-2\cos 2t, -2\sin 2t)$

17. $\sqrt{2}\,e^t(\cos(t + \frac{1}{4}\pi), \sin(t + \frac{1}{4}\pi))$, $2e^t(\cos(t + \frac{1}{2}\pi), \sin(t + \frac{1}{2}\pi))$

19. $(\sin\theta + \theta\cos\theta)/(\cos\theta - \theta\sin\theta)$, θ

21. $(-3\sin 3\theta\sin\theta + \cos 3\theta\cos\theta)/(-3\sin 3\theta\cos\theta - \cos 3\theta\sin\theta)$, $-\frac{1}{3}\cot 3\theta$

23. $\displaystyle a\int_0^{2\pi} \sqrt{1 + \theta^2}\,d\theta$

25. $6a \int_0^{\pi/6} \sqrt{\cos^2 3\theta + 9\sin^2 3\theta}\, d\theta = 2a \int_0^{\pi/2} \sqrt{\cos^2\theta + 9\sin^2\theta}\, d\theta$

27. Set $\mathbf{F} = \mathbf{a}$ and equate coefficients of $\mathbf{u}$ and $\mathbf{w}$: $\ddot{r} - {}_\iota r\dot\theta^2 = -1/r^2$, $r\ddot\theta + 2\dot r\dot\theta = 0$. Then $(r^2\dot\theta)^{\cdot} = r^2\ddot\theta + 2r\dot r\dot\theta = 0$ so $r^2\dot\theta = J$, constant. Now eliminate $\dot\theta$ in the first equation.

29. $r = 1/p$ so $\dot r = -\dot p/p^2$. Substitute.   31. Differentiate and cancel $2\,dp/d\theta$.

33. The solution may be written $p = K\cos(\theta - \theta_0) + 1/J^2$. Rotate the x-axis θ_0 radians if $K \le 0$ or $\theta_0 + \pi$ radians if $K > 0$.

35. In general, $J^2 = r - er\cos\theta = r - ex$, $(J^2 + ex)^2 = r^2 = x^2 + y^2$, so the rectangular equation is $x^2(1 - e^2) - 2eJ^2 x + y^2 = J^4$. If $e = 1$, then $2J^2 x + J^4 = y^2$. Each point on the parabola is equidistant from $(0, 0)$ and the line $x = -J^2$, because $x^2 + y^2 = x^2 + 2J^2 x + J^4 = (x + J^2)^2$. Hence $(0, 0)$ is the focus and $x = -J^2$ is the directrix.

37. Suppose $e \neq 1$. Complete the square and divide by the constant term:

$$\frac{\left[x - \dfrac{eJ^2}{1 - e^2}\right]^2}{\dfrac{J^4}{(1 - e^2)^2}} + \frac{y^2}{\dfrac{J^4}{1 - e^2}} = 1.$$

Set $a^2 = J^4/(1 - e^2)^2$ and $b^2 = J^4/(1 - e^2)$ if $e < 1$, $b^2 = -J^4/(1 - e^2)$ if $e > 1$. Also set $c = eJ^2/(1 - e^2)$. If $e < 1$, then $c > 0$, so the equation is

$$\frac{(x - c)^2}{a^2} + \frac{y^2}{b^2} = 1,$$

and $c^2 = a^2 - b^2$, so this ellipse has center $(c, 0)$ and one focus at $(0, 0)$.

39. $dA/dt = \frac{1}{2}r^2\dot\theta = \frac{1}{2}J$. Integrate with respect to t from 0 to T. The area of the ellipse is πab, hence

$$\pi ab = \int_0^T \frac{dA}{dt}\, dt = \int_0^T \frac{1}{2}J\, dt = \frac{1}{2}JT.$$

Section 7, page 715

1. $2a\sinh(b/a)$ 3. ellipse $\mathbf{x} = ((a + b)\cos\theta, (a - b)\sin\theta)$ 5. $y = 1 + x + \frac{1}{2}x^2$

7. The reflection property, $\cos\alpha = \cos\beta$, where α is the angle between the tangent and $\overline{\mathbf{px}}$ and β is the angle between the tangent and $\overline{\mathbf{qx}}$.

9. hyperbola $x^2/a^2 - y^2/b^2 = 1$ 11. $\int c x_1\, dt = c \int x_1\, dt$, etc.

13. $\int (c_1 x_1 + c_2 x_2 + c_3 x_3)\, dt = c_1 \int x_1\, dt + c_2 \int x_2\, dt + c_3 \int x_3\, dt$, etc.

15. Use Ex. 11: $\left| \mathbf{c} \cdot \int \mathbf{x}\, dt \right| = \left| \int \mathbf{c} \cdot \mathbf{x}\, dt \right| \le \int |\mathbf{c} \cdot \mathbf{x}|\, dt$.

But $|\mathbf{c} \cdot \mathbf{x}(t)| \le |\mathbf{c}|\,|\mathbf{x}(t)|$, hence $\left| \mathbf{c} \cdot \int \mathbf{x}\, dt \right| \le \int |\mathbf{c}|\,|\mathbf{x}|\, dt = |\mathbf{c}| \int |\mathbf{x}|\, dt$.

17. $r = ke^{c\theta}$ 19. $|\mathbf{x}|^2 = 1$.

CHAPTER 15

Section 1, page 725

1. $V = \frac{1}{3}\pi r^2 h$ 3. $D = [(x - 2)^2 + (y - 4)^2 + (2 + x + y)^2]^{1/2}$

5. $F = Gm_1 m_2/d^2$, G constant   7. $c = [a^2 + b^2 - 2ab\cos\theta]^2$

9. the entire x, y-plane   11. the x, y-plane excluding the two lines $y = \pm x$

13. the region on and below the parabola $y = x^2$

15. the x, y-plane excluding the horizontal lines $y = \frac{1}{2}(2n - 1)\pi$, n all integers

17. the region above the plane $x + 2y + 3z = 4$

19. the unit sphere $|\mathbf{x}| = 1$ and its interior

21.

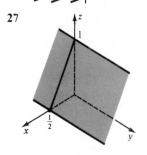

23.

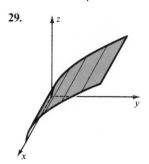

25.

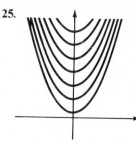
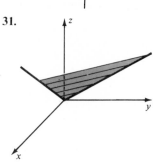

27

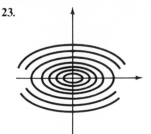

29.

31.

33. all the spheres with center **0**

35. $|(\mathbf{x}_n + \mathbf{y}_n) - (\mathbf{a} + \mathbf{b})| \le |\mathbf{x}_n - \mathbf{a}| + |\mathbf{y}_n - \mathbf{b}|$, etc.

37. $|[f(\mathbf{x}) + g(\mathbf{x})] - [f(\mathbf{a}) + g(\mathbf{a})]| \le |f(\mathbf{x}) - f(\mathbf{a})| + |g(\mathbf{x}) - g(\mathbf{a})|$, etc.

39. Yes. Set $f(0, 0) = 1$ and use $(\sin t)/t \longrightarrow 1$ as $t \longrightarrow 0$.

Section 2,   page 731

1. $1, 2$ **3.** $3y, 3x$ **5.** $4x/(y + 1), -2x^2/(y + 1)^2$

7. $\sin y, x \cos y$ **9.** $2 \sec^2 2x, -3 \csc^2 3y$ **11.** $2y \cos 2xy, 2x \cos 2xy$

13. $(y^2 - x^2)/(x^2 + y^2)^2, -2xy/(x^2 + y^2)^2$

15. $2x/(x^2 + 3y), 3/(x^2 + 3y)$ **17.** $2e^{2x} \sin(x - y) + e^{2x} \cos(x - y), -e^{2x} \cos(x - y)$

19. $\frac{1}{2}z[(x - y)^{-1} - (x + y)^{-1}], -\frac{1}{2}z[(x - y)^{-1} - (x + y)^{-1}]$

21. $4x, 1$ **23.** $32, 0, 3x^3z^2 = 3y^3z^2$ **25.** $z_x = 6(3x - y), z_y = -2(3x - y)$, etc.

27. $(z_x)^2 - (z_y)^2 = (2x)^2 - (-2y)^2 = 4(x^2 - y^2) = 4z$

29. Set $D = x^4 + y^4 + z^4$. Then $xw_x = xyz/D - 4x^4(xyz)/D^2 = w - 4x^4w/D$. By symmetry, $xw_x + yw_y + zw_z = 3w - 4(x^4 + y^4 + z^4)w/D = 3w - 4w = -w$.

31. $(9t^2 + 2t)z$ **33.** $4t^3 \cos(1/t) + t^2 \sin(1/t) - 2t$ **35.** $9t^8$

37. $4te^{-t}[(2 - t) \sin 4t + 4t \cos 4t]$

39. $(s^2 - t)^2(2s^2 + t)/2s^3t^2, -(s^2 - t)^2(2s^2 + t)/4s^2t^3$

41. $t(st^3 + st + 1)/z, s(2st^3 + st + 1)/z$ **43.** $\frac{7}{2}\pi$ ft^3/hr

45. $u_t = -100ku \le 0$, "$=0$" if and only if $x = 0$ or 120. Also,

$$u_t(60, t)/u_t(20, t) = \sin(\tfrac{1}{2}\pi)/\sin(\tfrac{1}{6}\pi) = 2.$$

47. $(\partial/\partial u)F(u + v, u - v) = (\partial F/\partial x)(\partial(u + v)/\partial u) + (\partial F/\partial y)(\partial(u - v)/\partial u)$
$= F_x(u + v, u - v) + F_y(u + v, u - v)$. Similarly $(\partial/\partial v)F(u + v, u - v) = F_x(u + v, u - v)$
$- F_y(u + v, u - v)$. Now add.

49. $0, 1$

Section 3,   page 738

1. $(2xy + 3y^3, x^2 + 9xy^2)$ **3.** $(ad - bc)(cx + dy)^{-2}(y, -x)$

5. $e^z(2x, 2y, x^2 + y^2)$ **7.** $(xyz^3/f)(yz, xz, 2xy)$

9. the plane $6x + 3y + 2z = 0$

11. RHS $= (z_x \cos\theta + z_y \sin\theta)(\cos\theta, \sin\theta) + r^{-1}(-z_x r \sin\theta + z_y r \cos\theta)(-\sin\theta, \cos\theta)$
$= (\cos^2\theta + \sin^2\theta)(z_x, z_y) = \text{grad } z$.

13. $z = 1/(r^2 + 10)$ so grad $z = f_r$, $\mathbf{u} = -[2r/(r^2 + 10)^2]\mathbf{u}$ points toward $\mathbf{0}$.

15. The level curves of f and g at (a, b) are orthogonal.   **17.** $-4x + y = -2$

19. $x + y = 1$ **21.** $1, 0$ **23.** $0, -3/\sqrt{2}$ **25.** 1 **27.** $\pm(0, 1)$

29. grad $f(\mathbf{a}) = b\mathbf{u} + c\mathbf{v}$ so $|\text{grad } f|^2 = b^2 + c^2$. But $b = (\text{grad } f) \cdot \mathbf{u} = D_\mathbf{u} f(\mathbf{a})$ and $c = (\text{grad } f) \cdot \mathbf{v} = D_\mathbf{v} f(\mathbf{a})$, hence $[D_\mathbf{u} f(\mathbf{a})]^2 + [D_\mathbf{v} f(\mathbf{a})]^2 = |\text{grad } f|^2$.

31. $-\frac{14}{81}$ **33.** $\frac{1}{2}\pi$

35. LHS $= [\text{grad } f(\mathbf{a})] \cdot (\mathbf{v} + \mathbf{w}) = [\text{grad } f(\mathbf{a})] \cdot \mathbf{v} + [\text{grad } f(\mathbf{a})] \cdot \mathbf{w} = $ RHS

Section 4,   page 745

1. $16x + 16y + 25z = 3$, $(1/\sqrt{1137})(16, 16, 25)$

3. $2x + 3y + 2z = 14$, $(1/\sqrt{17})(2, 3, 2)$

5. $4x + 13y + 46z = 84$, $(1/\sqrt{2301})(4, 13, 46)$ **7.** $x + y + 4z = 0$, $\frac{1}{6}\sqrt{2}(1, 1, 4)$

9. $ex + (e + 1)y - ez = e + 1$, $(3e^2 + 2e + 1)^{-1/2}(e, e + 1, -e)$

11. $13x + 2y = 26$, $(1/\sqrt{173})(13, 2, 0)$

13. $z = 0$, $(0, 0, 1)$ **15.** $z = 4x + 8y - 8$, $\frac{1}{9}(-4, -8, 1)$

17. $z = -4x + 13y - 20$, $(1/\sqrt{186})(4, -13, 1)$ **19.** $z = x + y$, $\frac{1}{3}\sqrt{3}(-1, -1, 1)$

21. $z = \frac{1}{2}x + y + \ln 4 - \frac{3}{2}$, $\frac{1}{3}(-1, -2, 2)$ **23.** arc sin $\frac{1}{3}\sqrt{3}$ at $\mathbf{0}$,   arc sin $\frac{1}{3}$ at $(\frac{1}{2}, \frac{1}{2}, \frac{1}{2})$

25. (a, a, a^{-2}), $(a, -a, -a^{-2})$, $a \neq 0$

Section 5,   page 750

1. $x + y - z = 0$, plane   **3.** $x^2 + y^2 - z^2 = 0$, right circular cone, axis the z-axis

5. $x^2 + z^2 = a^2$, right circular cylinder, axis the y-axis, radius a

7. none   **9.** none   **11.** all $(0, b)$

13. $\mathbf{x} = (1 + h - k, 1 + h + k, 1 + h + 3k)$, $\frac{1}{6}\sqrt{6}(1, -2, 1)$

15. $\mathbf{x} = (\frac{1}{2}\sqrt{2} - \frac{1}{2}k\sqrt{2}, \frac{1}{2}\sqrt{2} + \frac{1}{2}k\sqrt{2}, 2 - h)$, $\frac{1}{2}\sqrt{2}(1, 1, 0)$

17. $\mathbf{x} = (-4 + 4k, -4h, -8 + 12k)$, $\frac{1}{10}\sqrt{10}(-3, 0, 1)$

19. $(\sqrt{x^2 + y^2} - b)^2 + z^2 = a^2$ **21.** $\mathbf{x} = (u, v, f(u, v))$

23. Since $|\mathbf{x}(s)| = 1$, $\mathbf{x}(s)$ and the unit tangent $\mathbf{t}(s) = d\mathbf{x}/ds$ are perpendicular unit vectors. Now $\mathbf{x}_u \times \mathbf{x}_v = v\mathbf{t} \times \mathbf{x} \neq \mathbf{0}$, so there is a tangent plane.

25. $\mathbf{x}_u(u, v) = \mathbf{t}(u) + k(u)v\mathbf{n}(u)$, where $\mathbf{n}(s)$ is the unit normal to the curve at $\mathbf{x}(s)$, and $\mathbf{x}_v(u, v) = \mathbf{t}(u)$. Also $\mathbf{x}_u \times \mathbf{x}_v = k(u)v\mathbf{t}(u) \times \mathbf{n}(u) \neq \mathbf{0}$ since $k > 0$, $v > 0$, and $\mathbf{t}$ and $\mathbf{n}$ are orthogonal unit vectors.

Section 6,   page 759

1.

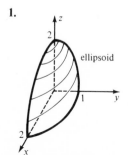

ellipsoid

3.

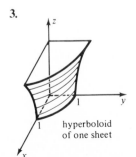

hyperboloid of one sheet

5.

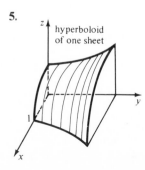

hyperboloid of one sheet

7.
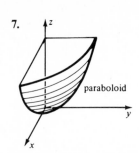

9.
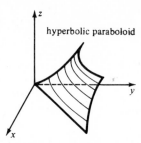

hyperbolic paraboloid

paraboloid

11. circular paraboloid, vertex $(-1, 0, -1)$, axis parallel to z-axis
13. right circular cone generated by revolving the line $y = a(z - 1)$, $x = 0$ about the z-axis

15.
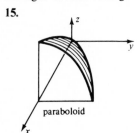

17.
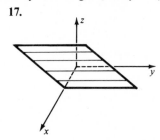

19.
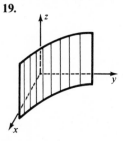

paraboloid

21.
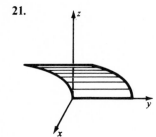

23.
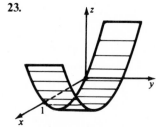

25.
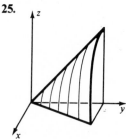

27. $f(x/z, y/z) = 0$ **29.** $z = 2\left(\dfrac{\ell x}{a^2} + \dfrac{my}{b^2}\right) - n$

31. Each cross section is $\dfrac{y^2}{b^2} + \dfrac{z^2}{c^2} = 1 - \dfrac{k^2}{a^2}$, which can be written in the form $\dfrac{y^2}{B^2} + \dfrac{z^2}{C^2} = 1$, where $B = \lambda b$, $C = \lambda c$. The eccentricity of this ellipse depends only on $B/C = b/c$, independent of λ, hence of k.

33. $(\mathbf{n} \cdot \mathbf{x})^2 = (\cos^2 \alpha)(\mathbf{x} \cdot \mathbf{x})$

35. The tangent planes $z = 2(ux/a^2 + vy/b^2) - k$ all pass through $(0, 0, -k)$.

37. The terms in $f(\mathbf{x})$ that involve x are $a_{11}x^2 + 2a_{21}xy + 2a_{31}xz$, hence $\partial f/\partial x = 2(a_{11}x + a_{21}y + a_{31}z)$, the first element of $2\mathbf{x}A$, etc.

39. $\operatorname{grad} f(\mathbf{x}) = 2(\mathbf{x}A + \mathbf{b})$

41. The tangent plane at (u, v, w) is $ux/a^2 + vy/b^2 + wz/c^2 = 1$. It is parallel to $\mathbf{n} = (\ell, m, n)$ provided its normal is orthogonal to $\mathbf{n}$.
The condition is $(u/a^2, v/b^2, w/c^2) \cdot (\ell, m, n) = 0$. Thus (u, v, w) lies on a *plane* (through $\mathbf{0}$), so the required set of points is a plane section of the ellipsoid, hence an ellipse.

Section 7, page 765

1. max 4 at $(0, 0)$, no min   **3.** no max, min 0 at $(2, -3)$ **5.** no max, min 4 at $(0, 0)$
7. max $\frac{1}{2}e^{-1}$ at $\pm(\frac{1}{2}\sqrt{2}, \frac{1}{2}\sqrt{2})$, min $-\frac{1}{2}e^{-1}$ at $\pm(\frac{1}{2}\sqrt{2}, -\frac{1}{2}\sqrt{2})$

9. max $\frac{3}{2}\sqrt{3}$ at $x = \frac{1}{3}\pi + 2\pi m$, $y = \frac{1}{3}\pi + 2\pi n$, min $-\frac{3}{2}\sqrt{3}$ at $x = -\frac{1}{3}\pi + 2\pi m$,
$y = -\frac{1}{3}\pi + 2\pi n$ 11. $\frac{5}{2}\sqrt[5]{2}$ at $(\sqrt[5]{2}, \frac{1}{2}\sqrt[5]{8})$ 13. $3\sqrt[3]{ab}$ at $(\sqrt[3]{a^2/b}, \sqrt[3]{b^2/a})$
15. -432 at $(1, 6)$ 17. no max, min $\frac{17}{4}$ at $(\frac{1}{2}, \frac{1}{2})$
19. max r^4 at $(\pm r, 0)$, min 0 at $(0, 0)$ 21. $\frac{8}{9}\sqrt{3}$; the solid is a cube.
23. base $\sqrt[3]{2V} \times \sqrt[3]{2V}$, height $\frac{1}{2}\sqrt[3]{2}V$
25. Set $\alpha = (2 + \sqrt{2})^{1/3}$. Then $A_{\min} = 3\alpha^2 V^{2/3}$, taken where the legs of the base are $x = y = \alpha V^{1/3}$ and the height is $2V^{1/3}/\alpha^2$
27. $x\sqrt[4]{2} + y\sqrt[4]{2} + z\sqrt[4]{8} = 4$ at distance $\sqrt[4]{8}$ from **0**

Section 8,   page 773

1. $4\sqrt{6}$ at $(\frac{3}{2}\sqrt{6}, \frac{2}{3}\sqrt{6}, \frac{1}{6}\sqrt{6})$ 3. $\frac{8}{3}\sqrt{3}$ at $(\sqrt{2}, 1, \sqrt{\frac{2}{3}})$
5. $\frac{1}{2}\sqrt{a^2 + b^2 + c^2}$ at $(a^2 + b^2 + c^2)^{-1/2}(a, b, c)$
7. max $\frac{1}{96}\sqrt{6}$, min $-\frac{1}{96}\sqrt{6}$ at $(\pm 1, \pm\frac{1}{2}\sqrt{2}, \pm\frac{1}{3}\sqrt{3})$, max for an odd number of pluses, min otherwise
9. $(\frac{1}{4}L)^4$ 11. neither   13. positive definite   15. negative definite   17. neither
19. $Q = a(x + by/a)^2 \geq 0$, $Q(-b/a, 1) = 0$.
21. $y = \frac{3}{2}x - \frac{1}{6}$ 23. $y = 2x - \frac{1}{2}$ 25. $y = 0.19x + 4.85$ 27. $y = -\frac{4}{5}$
29. $y = -0.35x + 5.7$ 31. $y = x - \frac{1}{6}$ 33. $y = 6(3 - e)x + 2(2e - 5) \approx 1.69x + 0.87$
35. $y = (\frac{24}{181})(2x + 51/x) \approx 0.265x + 6.762/x$
37. $\sigma_x^2 = n^{-1} \sum (x_i - \bar{x})^2 = n^{-1} \sum (x_i^2 - 2\bar{x}x_i + \bar{x}^2)$
$= n^{-1} \sum x_i^2 - 2n^{-1}\bar{x} \sum x_i + \bar{x}^2 = n^{-1} \sum x_i^2 - 2\bar{x}^2 + \bar{x}^2$, etc.
The other proof is similar.
39. $f_{\min} = n(\sigma_x^2\sigma_y^2 - s_{xy}^2)/\sigma_x^2$.

Section 9,   page 778

1. $y = 2x - 6$ 3. $y = 5x^2 - e^2$ 5. $y = x - \sqrt{x^2 + 4}$
7. $(y^2 - x^2)(y^2 - 4x^2) = 0$, hence $y = \pm x$ or $y = \pm 2x$.
At $(1, 2)$, $-F_x/F_y = -(-24)/12 = 2 = (d/dx)(2x)$.
9. $(1 - \sin y)/(x \cos y - 1)$ 11. $y(e^{xy} - 3y)/x(6y - e^{xy})$
13. $(e^y \sin x + e^x \sin y)/(e^y \cos x - e^x \cos y)$
15. $-2x^3/9y^5$ 17. $-\frac{3}{4}x(4y^2 + 3x^3)/y^3 = -\frac{3}{4}x(3 + y^2)/y^3$
19. $36y/25x^2$ 21. $4/\sqrt{26}$ 23. $\sqrt{2}$ 25. ab
27. $h = 4(k - 1)r$, where h is the height of the cylindrical part and r the radius.
29. 60,000 tons bituminous, 26,667 tons anthracite
31. $(4y - 2xz)/(x^2 + 2y^2z - 3z^2)$, $(4x - 2yz^2)/(x^2 + 2y^2z - 3z^2)$
33. $\sqrt{\frac{24}{19}}$ 35. yes: $y = 0$ 37. yes, because $(\partial F/\partial y)(0, 0) = 2 \neq 0$
39. $F(x, y) = (y - 2x^2)(y + 3x^3)$ so $y = 2x^2$ and $y = -3x^3$ are differentiable solutions.

Section 10,   page 784

1. $12\,dx + 8\,dy$, $-9\,dx + dy$ 3. $(6xy - y^2)\,dx + (3x^2 - 2xy)\,dy$
5. $(2x\,dx + 6y\,dy)/(x^2 + 3y^2)$ 7. $dx/y^2z - 2x\,dy/y^3z - x\,dz/y^2z^2$
9. $e^x(\cos y)\,dx + (-e^x \sin y + e^y \cos z)\,dy - e^y(\sin z)\,dz$
11. $d(fg) = (fg)_x\,dx + (fg)_y\,dy + (fg)_z\,dz = (f_xg + fg_x)\,dx + \cdots$
$= (f_x\,dx + f_y\,dy + f_z\,dz)g + f(g_x\,dx + g_y\,dy + g_z\,dz) = (df)g + f(dg)$
13. $x\,dx + y\,dy = (r \cos \theta)(dr \cos \theta - r \sin \theta\,d\theta) + (r \sin \theta)(dr \sin \theta + r \cos \theta\,d\theta)$
$= r\,dr(\cos^2 \theta + \sin^2 \theta) = r\,dr,$
$x\,dy - y\,dx = (r \cos \theta)(dr \sin \theta + r \cos \theta\,d\theta) - (r \sin \theta)(dr \cos \theta - r \sin \theta\,d\theta)$
$= r^2(\cos^2 \theta + \sin^2 \theta)\,d\theta = r^2\,d\theta$

15. 358.5; exact: 358.479   **17.** 1.98227; exact to 5 places: 1.98213
19. 12.98077; exact to 5 places: 12.98093
21. 1% decrease; exact to 6 places: 0.9900990% decrease   **23.** 109 m
25. $-2(z/x)\,dx - (z/y)\,dy = -(2/x^3y)\,dx - (1/x^2y^2)\,dy$
27. $-(6z^2\,dx + 16yz\,dy)/(5z^4 + 12xz + 8y^2)$ **29.** $x^3 + 3xy - 2y + 4$
31. \$1,724,138; exact to the nearest dollar: \$1,778,493

Section 11,   page 791

1. $hx + ky = 3x - 7y$, $e(x, y) = 0$
3. $hx + ky = 0$, $e(x, y) = xy^2$, $|e(x, y)|/|\mathbf{x}| \le |\mathbf{x}|^3/|\mathbf{x}| = |\mathbf{x}|^2 \longrightarrow 0$
5. $hx + ky = -x + y$, $e(x, y) = (x^2 - xy)/(1 + x)$,
   $|e(x, y)|/|\mathbf{x}|| \le 2|\mathbf{x}|^2/|\mathbf{x}||1 + x| = 2|\mathbf{x}|\,/|1 + x| \longrightarrow 0$ as $(x, y) \longrightarrow (0, 0)$.
7. Let $f_i(a + x, b + y) = f_i(a, b) + h_i x + k_i y + e_i(x, y)$ for $i = 1, 2$. Then
   $(f_1 + f_2)(a + x, b + y) = (f_1 + f_2)(a, b) + (h_1 + h_2)x + (k_1 + k_2)y + e_1(x, y) + e_2(x, y)$.
   But $[e_1(x, y) + e_2(x, y)]/|\mathbf{x}| = e_1(x, y)/|\mathbf{x}| + e_2(x, y)/|\mathbf{x}| \longrightarrow 0 + 0 = 0$ as $\mathbf{x} \longrightarrow \mathbf{0}$.
9. discontinuous, $f_x(0, 0) = f_y(0, 0) = 0$, not differentiable
11. continuous, $f_x(0, 0) = f_y(0, 0) = 0$, differentiable
13. Since f is continuous on a bounded closed set, the circle, f has a minimum there.
   But $f(x, y) = (x - 3y)^2 + y^2 > 0$ on the circle, hence the minimum is positive.

Section 12,   page 791

1. $(tx)^2 + (ty)(tz) = t^2(x^2 + yz)$; degree 2   **3.** 3   **5.** -1
7. $[fg](t\mathbf{x}) = f(t\mathbf{x})g(t\mathbf{x}) = t^m f(\mathbf{x})t^n g(\mathbf{x}) = t^{m+n}[fg](\mathbf{x})$
9. $xf_x(t\mathbf{x}) + yf_y(t\mathbf{x}) + zf_z(t\mathbf{x}) = nt^{n-1}f(\mathbf{x})$. Set $t = 1$.
11. The tangent plane at (ℓ, m, n) is $2\ell x + 2my = n + z$, and $\mathbf{a}$ is on this plane if
   $2a\ell + 2bm = c + n$. Now replace (ℓ, m, n) by (x, y, z); then $\mathbf{a}$ is on the tangent plane at
   $\mathbf{x}$ provided $x^2 + y^2 = z$ and $2ax + 2by = z + c$. The latter is the equation of a plane.
13. Take $\mathbf{a} = (a, b, a^2 - b^2)$. Then the tangent plane is $2a(x - a) - 2b(y - b) = z - (a^2 - b^2)$.
   Eliminate z from this and the equation $z = x^2 - y^2$ of the surface: $2a(x - a) - 2b(y - b) = $
   $(x^2 - a^2) - (y^2 - b^2)$, $(x - a)^2 = (y - b)^2$, $y - b = \pm(x - a)$.
   The lines are $\{\, y - b = \pm(x - a),\quad 2a(x - a) - 2b(y - b) = z - (a^2 - b^2)\,\}$.
15. $\sqrt{2}$ **17.** $x^2 - \frac{1}{3}$ **19.** $dz = -dx + (x\,dx - \frac{1}{2}\,dy)/\sqrt{x^2 - y}$
21. $z(x, y) = f(x + y)$, where $f(t)$ is any differentiable function
23. $du = y''\,dx$ and $dv = y'\,dx + xy''\,dx - y'\,dx = xy''\,dx = x\,du$, hence $dv/du = x$.
25. $w = uv$
27. If $a = 0$, then $Q = (2bx + y)y$. If $a \ne 0$, then $at^2 + 2bt + c = 0$ has two *real* roots λ and μ,
   hence $at^2 + 2bt + c = a(t - \lambda)(t - \mu)$. Replace t by x/y and multiply by y^2:

$$Q(x, y) = a(x - \lambda y)(x - \mu y) = (ax - a\lambda y)(x - \mu y).$$

29. $\mathbf{x}_1 = (0, \frac{1}{4})$, $\mathbf{x}_2 = (\frac{1}{16}, 0)$; $\mathbf{x}_n \longrightarrow (0, 0)$
31. $\mathbf{x}_1 = (\frac{4}{3}, \frac{1}{6})$, $\mathbf{x}_2 = (\frac{95}{93}, \frac{91}{372}) \approx (1.02, 0.24)$. Here $\mathbf{x}_n \longrightarrow (\frac{1}{17}(2 + 4\sqrt{13}), \frac{1}{17}(8 - \sqrt{13})) \approx$
   $(0.9660, 0.2585)$. Actually, $\mathbf{x}_3 \approx (0.9677, 0.2581)$.
33. $(\, 0, \pm a\cos\alpha, a(\cos^2\alpha)/(\sin\alpha)\,)$

CHAPTER 16

Section 1,   page 798

1. $-f$, $3f$, $-9f$ **3.** $2\arcsin y$, $2x/\sqrt{1 - y^2}$, $x^2y/(1 - y^2)^{3/2}$ **5.** $2a$, $2b$, $2c$
7. both equal $-2/y^3$ **9.** both equal $mnx^{m-1}y^{n-1}$
11. both equal $-2(x + y)/(x - y)^3$ **13.** $24xy^3$, $36x^2y^2$

15. $xy^2 \sin(xy) - 2y \cos(xy), \quad x^2y \sin(xy) - 2x \cos(xy)$

17. $-e^{2y}[2 \sin(x + y) + \cos(x + y)], \quad e^{2y}[-4 \sin(x + y) + 3 \cos(x + y)]$

19. $g(x) + xh(y) + k(y)$ **21.** any cubic polynomial in x and y

23. $ax + by + c$ **25.** none

27. $\begin{bmatrix} 0 & 1 & 1 \\ 1 & 0 & 1 \\ 1 & 1 & 0 \end{bmatrix}$ **29.** $-[\sin(x + 2y + 3z)] \begin{bmatrix} 1 & 2 & 3 \\ 2 & 4 & 6 \\ 3 & 6 & 9 \end{bmatrix}$

31. 10   **33.** 10, 20

35. By the Chain Rule, $h_u = f_x \cdot 1 + f_y \cdot 2 = 0$, hence $h(u, v) = g(v)$, that is, $f(u, 2u + v) = g(v)$. But $v = (2u + v) - 2u = y - 2x$, hence $f(x, y) = f(u, 2u + v) = g(v) = g(y - 2x)$.

37. $k \, \partial^2 u_n/\partial x^2 = -k(n\pi/L)^2 u_n = \partial u_n/\partial t$, and $u_n(0, t) = u_n(L, t) = 0$

39. Direct substitution yields $a = (n^2\pi^2k/L^2) + (v^2/4k)$.

41. $V_{xx} = a^2 V = -V_{yy}$ **43.** $\dfrac{p^2}{c^2} = \left(\dfrac{m^2}{a^2} + \dfrac{n^2}{b^2}\right)\pi^2$

Section 2,   page 803

1. $p_2(x, \ y) = 1 + 2(x - 1) + 2(y - 1) + (x - 1)^2 + 4(x - 1)(y - 1) + (y - 1)^2, \ p_1(x, \ y) = 1 + 2(x - 1) + 2(y - 1)$, the linear part of $p_2(x, y)$

3. $p_1(x, y) = 0, \quad p_2(x, y) = xy$ **5.** $p_1(x, y) = 1, \quad p_2(x, y) = 1 + (x - 1)y$

7. $p_1(x, y) = p_2(x, y) = -x - (y - \frac{1}{2}\pi)$

9. $p_1(x, y) = (x - \frac{1}{2}) + 2(y - \frac{1}{4})$,
$p_2(x, y) = p_1(x, y) - \frac{1}{2}[(x - \frac{1}{2})^2 + 4(x - \frac{1}{2})(y - \frac{1}{4}) + 4(y - \frac{1}{4})^2]$

11. 1.12000; actually $(1.1)^{1.2} \approx 1.12117$

13. 3.99930; actually $f(1.01, 2.01) \approx 3.99929$

15. $h_1(x, y) = g(b)p_1(x) + f(a)q_1(y) - f(a)g(b)$,
$h_2(x, y) = g(b)p_2(x) + f(a)q_2(y) + p_1(x)q_1(y) - f(a)g(b) - h_1(x, y)$

17. $p_1(x, \ y) = 1 + \frac{1}{2}x + y$ at $(0, \ 0)$. Also $f_{xx} = -1/4(1 + x + 2y)^{3/2}, \ f_{xy} = -1/2(\cdots)^{3/2}$, $f_{yy} = -1/(\cdots)^{3/2}$. Hence $|f_{xx}|, |f_{xy}|, |f_{yy}|$ are bounded by

$$\frac{1}{(1 + x + 2y)^{3/2}} \le \frac{1}{(1 - 0.1 - 0.2)^{3/2}} = \frac{1}{(0.7)^{3/2}} < \frac{1}{(0.64)^{3/2}} = \frac{1}{(0.8)^3} = \frac{1}{0.512} < 2.$$

Take $M_2 = 2$: $|r_1(x, y)| < M_2|(x, y)|^2 \le 2(0.02) = 0.04$.

19. $p_1(x, y, z) = f(a, b, c) + f_x \cdot (x - a) + f_y \cdot (y - b) + f_z \cdot (z - c)$,
$p_2(x, y, z) = p_1(x, y, z) + \frac{1}{2}[f_{xx} \cdot (x - a)^2 + f_{yy} \cdot (y - b)^2$
$\qquad + f_{zz} \cdot (z - c)^2 + 2f_{xy} \cdot (x - a)(y - b) + 2f_{yz} \cdot (y - b)(z - c) + 2f_{zx} \cdot (z - c)(x - a)]$,
all partials evaluated at (a, b, c)

21. $p_2(x, y) = 1 + x + 2y + x^2 + xy + 4y^2$

Section 3,   page 809

1. min   **3.** saddle   **5.** max   **7.** min   **9.** saddle   **11.** local min

13. saddle at $(0, 0)$, local mins at $(\pm\sqrt{2}, \mp\sqrt{2})$

15. local max at $(0, 0)$, local mins at $(0, \pm2)$, saddle points at $(\pm\sqrt{2}, 0)$

17. local max at $(0, 0)$, saddle point at $(\frac{5}{2}, -\frac{5}{4})$

19. local max at $(\sqrt{\frac{1}{6}}, -\sqrt{\frac{1}{6}})$, local min at $(-\sqrt{\frac{1}{6}}, \sqrt{\frac{1}{6}})$, saddle points at $\pm(\sqrt{\frac{1}{2}}, \sqrt{\frac{1}{2}})$

21. local max at $(-\frac{1}{2}, \frac{11}{8})$, saddle points at $(-2, 4)$ and $(1, 1)$

23. local max at (a, c), local min at (b, d), saddle points at (a, d) and (b, c)

25. \$3.50 and \$4.50 respectively for the regular and special

Section 4, page 816

1. $\frac{1}{2}$ at $(\pm\sqrt{\frac{1}{2}}, \pm\sqrt{\frac{1}{2}})$, $-\frac{1}{2}$ at $(\pm\sqrt{\frac{1}{2}}, \mp\sqrt{\frac{1}{2}})$

3. $\sqrt{13}$ at $\mathbf{a} = \frac{1}{13}\sqrt{13}\,(4, 9)$, $-\sqrt{13}$ at $-\mathbf{a}$

5. $\frac{2}{9}\sqrt{3}$ at $\mathbf{a} = \frac{1}{3}\sqrt{3}\,(1, \pm\sqrt{2})$. $-\frac{2}{9}\sqrt{3}$ at $-\mathbf{a}$ 7. no max, min $\sqrt{5}$ at $\frac{1}{5}\sqrt{5}\,(9, 4)$

9. no max or min   11. 15 at $(4, 2)$, -12 at $(1, -1)$

13. $9^3/16^4$ at $(\frac{9}{16}, \frac{1}{16})$, 0 at $(1, 0)$ and $(0, 1)$ 15. 0 at $(0, 0)$, $-\frac{32}{27}$ at $(-\frac{8}{9}, -\frac{8}{27})$

17. no max, $-\frac{45}{64}$ at $(\frac{9}{16}, \frac{27}{64})$ 19. $h = 2r = 2\sqrt{A/6\pi}$

21. $\frac{1}{4}(3 - 2\sqrt{2})$; the (equal) legs are $x = y = \frac{1}{2}(2 - \sqrt{2})$.

23. $2(\sqrt{2} - 1)$ 25. $27\sqrt{5}$ at $(\frac{6}{5}\sqrt{5}, \frac{3}{5}\sqrt{5})$

27. Set $\mathbf{a} = (\mathbf{a}_1 + \cdots + \mathbf{a}_n)/n$, the center of gravity of $\mathbf{a}_1, \cdots, \mathbf{a}_n$. If $\mathbf{a} = \mathbf{0}$, the sum is constant. Otherwise it is least for $\mathbf{x} = \mathbf{a}/|\mathbf{a}|$.

29. $2^{1-p/q}b^p$, b^p

Section 5, page 821

1. $\pm\sqrt{30}$ at $\pm\frac{1}{30}\sqrt{30}\,(2, 1, -5)$ 3. no max, min 36 at $(6, 12, 18)$

5. $\pm\frac{5}{9}\sqrt{5}$ at the 8 various points $(\pm\frac{1}{3}\sqrt{15}, \pm\frac{1}{3}\sqrt{15}, \pm\frac{1}{3}\sqrt{5})$

7. $\frac{1}{27}abc$ at $\frac{1}{3}(a, b, c)$, 0 on the boundary

9. From $a = 2x\lambda$, $b = 2y\lambda$, $c = 2z\lambda$ follow $x = a/2\lambda$, $y = b/2\lambda$, $z = c/2\lambda$, hence $r^2 = (a^2 + b^2 + c^2)/4\lambda^2$, $1/2\lambda = r/\sqrt{}$, $x = ar/\sqrt{}$, etc.

11. $\frac{3}{2}$ at $(\frac{2}{3}, -\frac{2}{3}, \frac{1}{3})$ 13. cube

15. The cylinder has height $\frac{1}{5}\sqrt{5}\,r$ and the cone has height $\frac{2}{5}\sqrt{5}\,r$, where r is the radius.

17. max $= \frac{1}{27}r^6$, taken at $x^2 = y^2 = z^2 = \frac{1}{3}r^2$

19. Maximize $f = xy + yz + zx$ on $x^2 + y^2 + z^2 = r^2$. The multiplier equations are $y + z = 2\lambda x$, $z + x = 2\lambda y$, $x + y = 2\lambda z$. Subtract the first two: $y - x = 2\lambda(x - y)$, etc. We conclude either $x = y = z$ or $2\lambda = -1$. If $x = y = z$, then $3x^2 = r^2$, $f = x^2 + x^2 + x^2 = r^2$. If $2\lambda = -1$, then $x + y + z = 0$. Square: $r^2 + 2f = 0$, $f = -\frac{1}{2}r^2$. Therefore $-\frac{1}{2}r^2 \leq f \leq r^2$.

21. $\frac{1}{8}s^2\sqrt{3}$ 23. Differentiate; the result is negative because $\sin x < x$ for $0 < x < \pi$.

25. $(1, 1, 1)$ 27. $(1, 0, 0)$ 29. all points $(1/\sqrt[3]{2}, y, z)$, where $y^2 + z^2 = \sqrt[3]{2}$

31. max $3^{1-p/q}b^p$, min b^p 33. $\sqrt{31}$ at $(2, 3\sqrt{2}, 3)$, $\sqrt[3]{62}$ at $(\sqrt[3]{62}, 0, 0)$

35. $3k^2\sqrt{3}$ at $x = y = z = 2k$ 37. $(\frac{5}{6}, \frac{1}{3}, -\frac{1}{6})$

39. $\frac{2}{9}(27 \pm \sqrt{3})\,ft^3$, taken at sides x, x, $6 - 2x$, where $x = \frac{1}{3}(6 \mp \sqrt{3})$ ft.

41. $\frac{3}{4}$ at $(\frac{1}{4}, -\frac{3}{4}, -\frac{1}{4})$ 43. $4\sqrt{15}$, $4\sqrt{10}$

Section 6, page 827

1. min   3. min   5. min   7. neither   9. min   11. neither

13. max 0 at $(0, 0, 0)$ 15. min $-\frac{5}{2}$ at $(-\frac{1}{2}, -1, \frac{3}{2})$ 17. min   19. max   21. neither

23. From $|x| \leq |\mathbf{x}|$ and $|y| \leq |\mathbf{x}|$ follows $|f| \leq 4|\mathbf{x}|^2$, hence $f \longrightarrow 0$ as $\mathbf{x} \longrightarrow \mathbf{0}$. This proves continuity at $\mathbf{0}$; elsewhere it is obvious.

25. $[f(x, 0) - f(0, 0)]/x = x \longrightarrow 0$, hence $f_x(0, 0) = 0$. Similarly, $f_y(0, 0) = 0$. As in Ex. 23, $|f_x| \leq 14|\mathbf{x}| \longrightarrow 0$, etc.

27. $f(x, y) = f(y, x)$ implies $f_x(x, y) = f_y(y, x)$. Now apply $\partial/\partial x$ again: $f_{xx}(x, y) = f_{yy}(y, x)$. Set $x = y = c$.

29. $\frac{1}{2}(3 - \sqrt{5})$ 31. Keep applying the Chain Rule.

33. Pick the curve $x = [B/(A^2 + B^2)^{1/2}]t$, $y = [-A/(A^2 + B^2)^{1/2}]t$, where $A = f_{xx}(0, 0)$, $B = f_{xy}(0, 0)$. Then $\ddot{g}(0) = A(AC - B^2)/(A^2 + B^2)$, where $C = f_{yy}(0, 0)$, etc.

Section 7, page 828

1. $x - y$; note that the function equals arc tan x − arc tan y

3. $f_x = f_u u_x + f_v v_x = f_u + f_v$. Similarly, $f_{xx} = f_{uu} + 2f_{uv} + f_{vv}$, $f_y = f_u - f_v$, $f_{yy} = f_{uu} - 2f_{uv} + f_{vv}$, $f_{xx} - f_{yy} = 4f_{uv}$.

5. The answer is the second degree Taylor polynomial of $\sqrt{(1+h)/(1+k)}$ at $(0,0)$.

7. $f_{xx} = 2xy/(x^2+y^2)^2 = -f_{yy}$ **9.** $f_{xx} = 20x^3 - 60xy^2 = -f_{yy}$

11. $f_{xx} = 2(y^2-x^2)/(x^2+y^2)^2 = -f_{yy}$

13. $x = r\cos\theta,\ y = r\sin\theta,\quad f_r = f_x\cos\theta + f_y\sin\theta,$

$\quad f_{rr} = f_{xx}\cos^2\theta + 2f_{xy}\cos\theta\sin\theta + f_{yy}\sin^2\theta,\quad f_\theta = r[-f_x\sin\theta + f_y\cos\theta],$

$\quad f_{\theta\theta} = r[-f_x\cos\theta - f_y\sin\theta] + r^2[f_{xx}\sin^2\theta - 2f_{xy}\sin\theta\cos\theta + f_{yy}\cos^2\theta].$

$\quad$ Now combine: $f_{rr} + f_r/r + f_{\theta\theta}/r^2 = f_{xx} + f_{yy}$.

15. $-\frac{7}{32}$ **17.** $\frac{2}{3}\sqrt{3}$, taken at $\pm(\frac{1}{6}\sqrt{3}, \frac{2}{3}\sqrt{3})$ **19.** 6912 at (6, 4)

21. $\frac{1}{8}(\sin\alpha)/(1+\sin\frac{1}{2}\alpha)^2$ **23.** $4(abc)^{1/4}$, taken at $(abc)^{-1/4}(a, b, c)$

25. $z_{max} = (9kP + 8 - 4\sqrt{4 + 9kP})/9k$; the corresponding x is $(-2 + \sqrt{4 + 9kP})/9k$, and $y = 3x$.

CHAPTER 17

Section 2,   page 843

1. $\frac{1}{2}y^3,\quad 20x$ **3.** $\frac{1}{5} + 2y^2 + y^4,\quad \frac{32}{5}[(x-1)^5 - (x-2)^5]$

5. $\frac{1}{4}\pi(1+y^2),\quad \frac{32}{3}/(1+x^2)$ **7.** $\frac{15}{2}$ **9.** $\frac{4}{9}$ **11.** 0   **13.** $-\frac{21}{4}$

15. $\ln\frac{4}{3}$ **17.** 0   **19.** $\frac{1}{2}\pi$ **21.** $\frac{9}{2}\ln 3 + 3\ln 2 - \frac{15}{4}$ **23.** $\frac{16}{3}$ **25.** $\frac{16}{3}$ **27.** $\frac{46}{3}$

29. $\iint = \int_a^b \left(\int_{-c}^c f(x,y)\,dy \right) dx = 0$ because the inner integral is 0 since $f(x,y)$ is an odd function of y

31. $\iint f\,dx\,dy$

Section 3,   page 851

1. $e-1$ **3.** 1   **5.** $\frac{1}{30}$ **7.** $\frac{1}{27}$ **9.** $\frac{16}{3}$ **11.** $\frac{1}{2}\ln\frac{3}{2} + \frac{1}{8}(\frac{1}{81} - \frac{1}{16})$

13. $\frac{1}{2}(e + e^{-1}) - 1 = 2\sinh^2(\frac{1}{2})$ **15.** $\frac{1}{3}$

17. $\int_0^3 \left(\int_x^{2+x/3} f(x,y)\,dy \right) dx$ **19.** $\int_1^3 \left(\int_{(y-9)/2}^{(y-1)/2} f(x,y)\,dx \right) dy$

21. $\int_0^{10} \left(\int_0^{\sqrt{10x - x^2}} f(x,y)\,dy \right) dx = \int_0^5 \left(\int_{5 - \sqrt{25 - y^2}}^{5 + \sqrt{25 - y^2}} f(x,y)\,dx \right) dy$

23. $\int_0^1 \left(\int_{3 - 3x}^{3\sqrt{1 - x^2}} f(x,y)\,dy \right) dx = \int_0^3 \left(\int_{1 - y/3}^{\sqrt{1 - y^2/9}} f(x,y)\,dx \right) dy$

25. $2(e^2 - e)$ **27.** 0   **29.** $\frac{13}{48}$ **31.** $\frac{4}{77}$ **33.** $\frac{5}{4}\sqrt{5}$

35. Integrate over the triangle with vertices $(0, 0)$, $(1, 0)$, $(1, 1)$.

37. $\int_2^3 \left(\int_4^{y^2} f(x,y)\,dx \right) dy$

Section 4,   page 857

1.

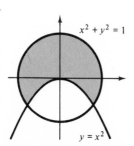

3.

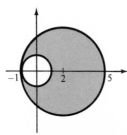

5.

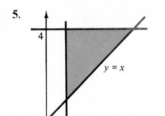

7.

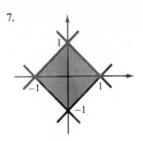

9. $\displaystyle\int_{-\sqrt{3}/2}^{\sqrt{3}/2} \left(\int_{1-\sqrt{1-x^2}}^{\sqrt{1-x^2}} f(x,y)\,dy \right) dx$ **11.** $\displaystyle\int_{0}^{1} \left(\int_{\arcsin x}^{\pi - \arcsin x} f(x,y)\,dy \right) dx$

13. $\displaystyle\int_{-1}^{0} \left(\int_{-4x}^{(11-x)/3} f(x,y)\,dy \right) dx + \int_{0}^{2} \left(\int_{3x/2}^{(11-x)/3} f(x,y)\,dy \right) dx$

15. 1   **17.** 8 ln 2   **19.** $\frac{9}{2}$ **21.** 42   **23.** $e^6 - 3e^2 + 2$ **25.** $\frac{32}{75}$ **27.** $\frac{1}{3}\sqrt{2}$

29. $\displaystyle\int_{0}^{1} \left(\int_{y}^{e^y} f(x,y)\,dx \right) dy$

Section 5, page 867

1. 0   **3.** $\pi a^2[2 \ln(2a) - 1]$ **5.** $\frac{1}{8}\pi a^6$ **7.** $\pi \ln 2$ **9.** $\frac{15}{2}\pi$ **11.** $\frac{15}{16}$ **13.** $\frac{2}{3}\pi a^3$

15. $\frac{5}{12}\pi$ **17.** $\frac{20}{27}$ **19.** $\frac{3}{640}\pi$ **21.** $\frac{1}{64}\pi$ **23.** $AD - BC$ **25.** $-2(u+v)$

27. In terms of u and v, the relations defining D are $u - v \geq 0$, $v \geq 0$, $a \leq (u - v) + v \leq b$, that is, $0 \leq v \leq u$ and $a \leq u \leq b$, the relations defining E. Also $\partial(x,y)/\partial(u,v) = 1$.

29. $\displaystyle\int_{a}^{b} uf(u)\,du$ **31.** $1/v$ **33.** $\frac{1}{2}(ab)^{3/2}(b-a)$

35. $\mathbf{E} = \left\{ (u,v) \ \middle| \ -\sqrt{v^2 + 4ab} \leq u \leq \sqrt{v^2 + 4ab}, \ 0 \leq v \leq b - a \right\}$

Section 6, page 873

1. $F = (1 - e^{-t})/t$, $F' = e^{-t}/t - 1/t^2 + e^{-t}/t^2$

3. $F = [(1+t)^{n+1} - t^{n+1}]/(n+1)$, $F' = (1+t)^n - t^n$

5. $\mathbf{x} = u(\mathbf{b} - \mathbf{a}) + v(\mathbf{c} - \mathbf{a})$ where $u \geq 0$, $v \geq 0$, and $u + v \leq 1$; $\frac{1}{2}|\mathbf{b} \times \mathbf{c} + \mathbf{c} \times \mathbf{a} + \mathbf{a} \times \mathbf{b}|$

7. $\displaystyle\int_{0}^{2\pi} \left(\int_{0}^{h} a\,dz \right) d\theta = 2\pi ah$

9. $\mathbf{x} = ((A + a \cos \alpha) \cos \theta, \ (A + a \cos \alpha) \sin \theta, \ a \sin \alpha)$, $0 \leq \theta \leq 2\pi$, $0 \leq \alpha \leq 2\pi$;

$\displaystyle\iint a(A + a \cos \alpha)\,d\theta\,d\alpha = 4\pi^2 Aa$

11. $\displaystyle\int_{0}^{2\pi} \left(\int_{0}^{\pi} [c^2 \sin^2 u \,(b^2 \cos^2 v + a^2 \sin^2 v) + a^2 b^2 \cos^2 u]^{1/2} \sin u\,du \right) dv$

13. $\displaystyle\iint_{\mathbf{D}} (1 + a^2 + b^2)^{1/2}\,dx\,dy = (1 + a^2 + b^2)^{1/2} \cdot |\mathbf{D}|$

15. $\displaystyle\int_{0}^{2\pi} \left(\int_{0}^{1} \frac{r\,dr}{\sqrt{1 - r^2}} \right) d\theta = 2\pi$ **17.** $4a^2(\pi - 2)$ **19.** $2\pi \displaystyle\int_{a}^{b} x(s)\,ds$

21. $-\frac{1}{2}\pi$ **23.** $+\infty$ **25.** $\pi/(1-p)$ **27.** $\frac{1}{2}\sqrt{\pi/a}$ **29.** $f_{n+1}(t) = -f'_n(t)$

Section 7, page 882

1. $\frac{9}{4}$, $(\frac{5}{9}, \frac{5}{9})$ **3.** 3, $(\frac{4}{9}, 0)$ **5.** $\frac{5}{4}$, $(\frac{7}{15}, \frac{7}{15})$ **7.** 50, $(0, \frac{59}{10})$ **9.** $\frac{4}{3}$, $(0, \frac{2}{5})$

11. $\frac{1}{3}$, $(\frac{3}{4}, \frac{3}{10})$ **13.** $\frac{1}{2}\pi ab$, $(0, 4b/3\pi)$ **15.** $\frac{1}{4}\pi a^2$, $(4a/3\pi, 4a/3\pi)$

17. $\frac{1}{2}a^2\alpha$, $\frac{2}{3}a((\sin \alpha)/\alpha, (1 - \cos \alpha)/\alpha)$ **19.** $\frac{1}{2}\pi a$, $(2a/\pi, 2a/\pi)$

21. $\frac{1}{2}ka\pi^2$, $(-4a/\pi^2, 2a/\pi)$ **23.** 12, $(\frac{3}{2}, 1)$

25. By Example 4, $\bar{x} = 4a/3\pi$, so $V = 2\pi\bar{x}A = 2\pi(4a/3\pi)(\frac{1}{2}\pi a^2) = \frac{4}{3}\pi a^3$.

27. The right triangle of base a revolves about its leg of length h into a right circular cone. $\bar{x} = \frac{1}{3}a$, $A = \frac{1}{2}ah$, and $V = \frac{1}{3}\pi a^2 h = 2\pi\bar{x}A$.

29. A = area of sphere = $4\pi a^2$, L = length of semi-circle = πa, $\bar{x} = A/2\pi L = 2a/\pi$.

31. $\frac{1}{2}Ma^2$ **33.** $\frac{3}{2}Ma^2$ **35.** $\frac{7}{20}M = \frac{7}{30}$

Section 8,   page 887

1. 1.776   **3.** 0.6377   **5.** 0.6888

7. 0.05208, exact 0.05   **9.** 1.1847 ($m = n = 4$: 1.1842)

11. 0.52772; exact $\frac{15}{8} + 4\ln 2 - \frac{15}{4}\ln 3 \approx 0.52779$

13. 1.6502   **15.** 0.064   **17.** 4×10^{-3}

19. $(\frac{1}{3}h\sum B_i p_i)(\frac{1}{3}k\sum C_j q_j) = \frac{1}{9}hk\sum B_i C_j p_i q_j = \frac{1}{9}hk\sum A_{ij}f_{ij}$

21. Both sides equal $A + \frac{1}{2}B + \frac{1}{3}C + \frac{1}{4}D$ **23.** 0.0487; exact 0.04

25. $|\iint| \leq \frac{1}{2}N \iint y(1-y)\,dx\,dy + \frac{1}{2}M\int_0^1 x(1-x)\,dx = \frac{1}{12}(M+N)$.

27. Apply Ex. 26 to $g(u, v) = f(a + hu, c + kv)$, $0 \leq u \leq 1, 0 \leq v \leq 1$, etc.

29. $|\text{error}| \leq \frac{1}{8} = 0.125$

Section 9,   page 888

1. $P = \frac{3}{2}C$ **3.** $\frac{1}{8}a^4$ **5.** $2(e^{2ab} - 1)/a$ **7.** $V = \int_a^b A(z)\,dz$ **9.** $\frac{1}{2}L$

11. $A = \frac{1}{4}a^2(2\cot\alpha - \pi + 2\alpha)$, $m_x = \frac{1}{6}a^3(1 - \sin\alpha)^2/\sin\alpha$

13. $aA^2 - \frac{1}{2}\pi a^2 A + \frac{2}{3}a^3$ **15.** $M = \frac{2}{3}\pi(3 + 4\pi^2)$, $\bar{\mathbf{x}} = 3(3 + 4\pi^2)^{-1}(6\pi, 5 - 4\pi^2)$

17. By Ex. 16, $\left(\int f\right)^2 - \int f^3 = 2\iint$ (non-negative) ≥ 0.

19. The change of variable $s = x - t$ does it.

21. $L(f * g)(s) = \int_0^\infty e^{-sx}\left(\int_0^x f(t)g(x - t)\,dt\right)dx = \int_0^\infty f(t)\left(\int_t^\infty e^{-sx}g(x - t)\,dx\right)dt$

$= \int_0^\infty f(t)e^{-st}\left(\int_t^\infty e^{-s(x-t)}g(x - t)\,dx\right)dt = \int_0^\infty f(t)e^{-st}\left(\int_0^\infty e^{-sy}g(y)\,dy\right)dt$

$= \left(\int_0^\infty f(t)e^{-st}\,dt\right)\left(\int_0^\infty e^{-sy}g(y)\,dy\right) = L(f)(s) \cdot L(g)(s)$.

23. First, $\rho - \rho_0 = kp = k\int_0^z \rho(u)\,du$. Differentiate with respect to z: $d\rho/dz = k\rho(z)$. Therefore $\rho(z)$ is an exponential function, $\rho(z) = \rho_0 e^{kz}$, so $p = k^{-1}(\rho - \rho_0) = k^{-1}\rho_0(e^{kz} - 1)$.

25. $(2\pi\rho_0/k^3)[e^{ak}(ak - 1) + 1 - \frac{1}{2}(ak)^2]$

27. The result follows from $\iint (x - y)[f(x) - f(y)]\,dx\,dy \geq 0$, where the integral is taken over $a \leq x \leq b, a \leq y \leq b$.

29. Pythogorean theorem

CHAPTER 18

Section 1,   page 897

1. $\frac{1}{3}$ **3.** $\frac{9}{2}\ln\frac{4}{3}$ **5.** 2

7. $(a + b + c)^6 - (a + b)^6 - (b + c)^6 - (c + a)^6 + a^6 + b^6 + c^6$ **9.** $\frac{11}{60}$ **11.** $\frac{1}{48}$

13. $\frac{1}{15}$ **15.** $1/6!$ **17.** $\frac{1}{24}$ **19.** $\frac{5}{8}$ **21.** $\frac{33}{20}$ **23.** 54   **25.** $k(\frac{1}{6}a^3)^3$

27. $\frac{1}{2}\int_0^a (a - x)^2 g(x)\,dx$ **29.** $\frac{1}{3}$ **31.** $\frac{5}{16}$ **33.** $2^6/9!$

Section 2, page 904

1. $r\left(\dfrac{\cos\theta}{a}+\dfrac{\sin\theta}{b}\right)+\dfrac{z}{c}=1$ **3.** $r=2a\sin\theta$ **5.** $\frac{1}{16}a^4b^2$ **7.** $\frac{2}{15}a^5$

9. $\frac{1}{2}\pi(e-1)^2$ **11.** $\frac{2}{3}\pi$ **13.** $\frac{25}{8}\pi a^4$ **15.** 2π **17.** $\frac{1}{3}a^3b^2$ **19.** 16π

21. $\dot{\mathbf{x}}=\dot{r}\mathbf{u}+r\dot{\theta}\mathbf{w}+\dot{z}\mathbf{k}$ **23.** $L=\displaystyle\int_a^b\sqrt{\dot{r}^2+r^2\dot{\theta}^2+\dot{z}^2}\,dt$

25. $\mathbf{x}_u=r_u\mathbf{u}+r\theta_u\mathbf{w}+z_u\mathbf{k},\ \mathbf{x}_v=r_v\mathbf{u}+r\theta_v\mathbf{w}+z_v\mathbf{k}$

27. $A=\displaystyle\iint_D\sqrt{r^2\left(\dfrac{\partial(\theta,z)}{\partial(u,v)}\right)^2+\left(\dfrac{\partial(z,r)}{\partial(u,v)}\right)^2+r^2\left(\dfrac{\partial(r,\theta)}{\partial(u,v)}\right)^2}\,du\,dv$

29. Take the parameterization $r=r$, $\theta=\theta$, $z=\sqrt{a^2-r^2}$, where $0\le r\le a$, $0\le\theta\le 2\pi$. The set-up is

$$A=\iint\frac{ar}{z}\,dr\,d\theta=2\pi a\int_0^a\frac{r\,dr}{\sqrt{a^2-r^2}}=2\pi a^2.$$

31. $df=f_r\,dr+f_\theta\,d\theta+f_z\,dz$

Section 3, page 913

1. $\rho=2a\cos\phi$ **3.** $\rho\sin^2\phi=\cos\phi$ **5.** $\rho\sin\phi=2a\cos\theta$ **7.** $\frac{1}{16}\pi a^4$

9. $4\pi a^{n+3}/(n+3)$ **11.** $\frac{1}{36}\pi$ **13.** $\frac{4}{35}\pi a^7$ **15.** $\frac{32}{105}\pi a^7$ **17.** $\frac{1}{15}\pi\sqrt{2}\,a^5$

19. $-\frac{4}{9}\pi$ **21.** $\frac{1}{4}\pi$

23. $\operatorname{grad}f=f_\rho\,\boldsymbol{\lambda}+\dfrac{1}{\rho}f_\phi\,\boldsymbol{\mu}+\dfrac{1}{\rho\sin\phi}f_\theta\,\mathbf{v}$ **25.** $\mathbf{v}=\dot{\rho}\boldsymbol{\lambda}+\rho\dot{\phi}\boldsymbol{\mu}+\rho\sin\phi\,\dot{\theta}\mathbf{v}$

27. $L=\displaystyle\int_a^b\sqrt{(\dot{\rho})^2+\rho^2(\dot{\phi})^2+\rho^2\sin^2\phi\,(\dot{\theta})^2}\,dt$ **29.** $L=\displaystyle\int_a^b\sqrt{1+(\sin^2\alpha)\theta^2}\,d\theta$

31. $\mathbf{x}_u=\rho_u\boldsymbol{\lambda}+\rho\phi_u\boldsymbol{\mu}+\rho\sin\phi\,\theta_u\mathbf{v},\qquad \mathbf{x}_v=\rho_v\boldsymbol{\lambda}+\rho\phi_v\boldsymbol{\mu}+\rho\sin\phi\,\theta_v\mathbf{v}$

33. $A=\displaystyle\iint_D\rho\sqrt{\rho^2\sin^2\phi\left[\dfrac{\partial(\phi,\theta)}{\partial(u,v)}\right]^2+\sin^2\phi\left[\dfrac{\partial(\theta,\rho)}{\partial(u,v)}\right]^2+\left[\dfrac{\partial(\rho,\phi)}{\partial(u,v)}\right]^2}\,du\,dv$

35. $\frac{1}{2}\pi(\sin\alpha)(2a^2+b^2)$ **37.** $\frac{1}{48}a^2b^2c^2$ **39.** $\frac{1}{20}\pi abc^4$

41. $I=\Delta\displaystyle\iiint_S f[\mathbf{x}(\mathbf{u})]\,du\,dv\,dw$, where $\mathbf{S}$ is the tetrahedron with vertices $\mathbf{0}$, $(1,0,0)$, $(0,1,0)$, $(0,0,1)$, and $\Delta=|\det(\mathbf{b}-\mathbf{a},\ \mathbf{c}-\mathbf{a},\ \mathbf{d}-\mathbf{a})|$. Thus $\mathbf{S}$ is described by $u\ge 0$, $v\ge 0$, $w\ge 0$, $u+v+w\le 1$.

Section 4, page 919

1. $(\frac{3}{8}a,\frac{3}{8}a,\frac{3}{8}a)$ **3.** $(0,0,\frac{3}{8}a(1+\cos\alpha))$ **5.** $(0,0,\frac{3}{8}(b^4-a^4)/(b^3-a^3))$

7. $(0,0,\frac{1}{2}a(1+\cos\alpha))$ **9.** $(-\frac{1}{3}a,0)$; use the Addition Law.

11. $(0,0,\frac{2}{3}a\cos\alpha)$ **13.** $(\frac{1}{2}a,\frac{1}{2}a,\frac{1}{2}a)$ **15.** $(\frac{1}{4}\pi a(\sin\alpha)/\alpha,0,0)$

17. The centroid is on the axis of the frustum, at distance $\frac{1}{4}h(b^2+2ab+3a^2)/(b^2+ab+a^2)$ from the larger base.

19. $(\frac{3}{8}a,\frac{3}{8}b,\frac{3}{8}c)$

Section 5, page 923

1. $I_x=\frac{1}{3}M(b^2+c^2)$, etc.   **3.** $I_x=I_y=\frac{1}{12}M(3a^2+4h^2)$, $I_z=\frac{1}{2}Ma^2$

5. $I_x=\frac{1}{12}M(3a^2+4h^2)$, $I_y=\frac{1}{12}M(15a^2+4h^2)$, $I_z=\frac{3}{2}Ma^2$

7. $I_x=I_y=\frac{1}{20}M(3a^2+2h^2)$, $I_z=\frac{3}{10}Ma^2$ **9.** $I_x=I_y=\frac{1}{6}M(a^2+h^2)$, $I_z=\frac{1}{3}Ma^2$

11. $I_x=I_y=I_z=\frac{2}{3}a^2M$ **13.** $I_x=I_y=\frac{1}{4}M(2A^2+5a^2)$, $I_z=\frac{1}{2}M(2A^2+3a^2)$

15. $I_x = I_y = I_z = \frac{1}{3}(I_x + I_y + I_z) = \frac{2}{3}\iiint (x^2 + y^2 + z^2)\, dM$, etc.

17. $I_{xy} = -\frac{1}{4}Mab$, etc.   **19.** $I_{yz} = I_{zx} = -\frac{2}{3}\pi^{-1}Mah$, $I_{xy} = -\frac{1}{2}\pi^{-1}Ma^2$
21. $I_{xy} = I_{yz} = I_{zx} = -\frac{1}{9}Ma^2$

Section 6,   page 930

1. $-\frac{1}{2}\pi$ **3.** $\frac{1}{2}(ab + bc + ca)$ **5.** -2 **7.** 35   **9.** $b - a$ **11.** $\mathbf{F} = \mathrm{grad}(x^3y^2z)$, 1
13. $-y\, dx + x\, dy = -(r\sin\theta)(dr\cos\theta - r\, d\theta\sin\theta) + (r\cos\theta)(dr\sin\theta + r\, d\theta\cos\theta)$
    $= r^2(\sin^2\theta + \cos^2\theta)\, d\theta = (x^2 + y^2)\, d\theta$
15. $\frac{2}{9}$ **17.** $\mathbf{F} = \mathrm{grad}(-\frac{1}{3}|\mathbf{x}|^{-3})$, $|\infty|^{-3} = 0$; $\frac{1}{3}|\mathbf{a}|^{-3}$ **19.** $\mathbf{F} = \mathrm{grad}\,\ln|\mathbf{x}|$
21. $(\frac{3}{2}, 2, \frac{5}{2})$ **23.** $(0, \frac{2}{5}, 0)$ **25.** $(0, -\frac{2}{3}, -2)$
27. $mab(\sin t - t\cos t, -\cos t - t\sin t, a/b)$

Section 7,   page 936

1. 12   **3.** 0   **5.** 2   **7.** $\frac{1}{3}$ **9.** $\frac{3}{4}\pi ab(a^2 + b^2)$ **11.** $-\frac{2}{3}a^5$ **13.** 2π **15.** 0
17. 0
19. $\oint_{\partial D} (-uv_y + vu_y)\, dx + (uv_x - vu_x)\, dy = \iint_D [(uv_x - vu_x)_x - (-uv_y + vu_y)_y]\, dx\, dy$

$$= \iint_D [u(v_{xx} + v_{yy}) - v(u_{xx} + u_{yy})]\, dx\, dy.$$

The other terms cancel each other. The proof requires u and v to have continuous second partials on $\mathbf{D}$.

21. $F_y(x, y) = Q(x, y)$ and $G_x(x, y) = P(x, y)$ **23.** $A = \frac{1}{2}\oint r^2\, d\theta$

Section 8,   page 947

1. 3   **3.** $\frac{1}{4}\pi a^4$ **5.** $2\pi - \frac{16}{3}$ **7.** 8   **9.** 0
11. 0 if n is even, $4\pi a^{n+2}/(n + 2)$ if n is odd
13. $\dfrac{\partial}{\partial x}(fA) + \dfrac{\partial}{\partial y}(fB) + \dfrac{\partial}{\partial z}(fC) = (f_x A + fA_x) + \cdots$
    $= (f_x A + f_y B + f_z C) + f(A_x + B_y + C_z) = $ etc.
15. Choose the z-axis upward, with 0 at the fluid surface (to avoid any funny stuff with orientation). Then $p = -\delta gz$, and the buoyant force is

$$\mathbf{F} = -\iint_{\partial D} p\, d\boldsymbol{\sigma} = -\iiint_D (\mathrm{grad}\, p)\, dV = \delta g\iiint_D \mathbf{k}\, dV = \delta g|\mathbf{D}|\mathbf{k}.$$

Thus the force is directed upward, and its magnitude is the mass of fluid displaced by $\mathbf{D}$ times the constant of gravity, i.e., the weight of the displaced fluid. (The first sign is negative because $\mathbf{n}$ is outward, but pressure acts inward.)

17. $\mathbf{0}$ **19.** $\mathbf{F} = (0, 0, -F_3)$ where $F_3 = 2\pi\,\delta Gc\left(\dfrac{1}{c} - \dfrac{1}{\sqrt{a^2 + c^2}}\right)$.

21. $F_3 = \dfrac{2GMc}{a^2}\left(\dfrac{1}{c} - \dfrac{1}{\sqrt{a^2 + c^2}}\right)$, and $\lim_{a\to 0}\mathbf{F} = (0, 0, -GM/c^2)$.
23. $\mathbf{F}(\mathbf{x}) = -f(\rho)\mathbf{x}$, where $\rho = |\mathbf{x}|$.
25. Let $\mathbf{S}$ be the sphere through $\mathbf{x}$ with center $\mathbf{0}$. Then

$$-4\pi GM = \iint_S \mathbf{F}\cdot d\boldsymbol{\sigma} = -\iint_S f(\rho)\,\mathbf{x}\cdot d\boldsymbol{\sigma} = -f(\rho)\iint_S \mathbf{x}\cdot d\boldsymbol{\sigma}.$$

But $\mathbf{x} \cdot d\boldsymbol{\sigma} = (\rho \mathbf{n}) \cdot (\mathbf{n}\, d\sigma) = \rho\, d\sigma$, so $\displaystyle\iint_S \mathbf{x} \cdot d\boldsymbol{\sigma} = \rho \iint_S d\sigma = 4\pi\rho^3$.

Therefore $-4\pi G M = -4\pi\rho^3 f(\rho)$, $f(\rho) = GM/\rho^3$.

27. $p = \frac{2}{3}\pi G\, \delta^2(a^2 - \rho^2)$, $p(0) = \frac{2}{3}\pi G\, \delta^2 a^2$

29. $\operatorname{curl}(f\mathbf{A}) = \operatorname{curl}(fA, fB, fC) = ((fC)_y - (fB)_z, \,\cdots)$
$= (f_y C - f_z B + f(C_y - B_z), \,\cdots) = (f_y C - f_z B, \,\cdots) + f(C_y - B_z, \,\cdots)$
$= (f_x, f_y, f_z) \times (A, B, C) + f\operatorname{curl}\mathbf{A} = \operatorname{grad} f \times \mathbf{A} + f\operatorname{curl}\mathbf{A}$.

31. By Ex. 29, $\operatorname{curl}[f(\rho)\mathbf{x}] = [\operatorname{grad} f(\rho)] \times \mathbf{x} + f(\rho)\operatorname{curl}\mathbf{x} = [(f'(\rho)/\rho)\mathbf{x}] \times \mathbf{x} + \mathbf{0} = \mathbf{0}$.

33. $\operatorname{curl}(\mathbf{a} \times \mathbf{x}) = \operatorname{curl}(bz - cy, cx - az, ay - bx) = ((ay - bx)_y - (cx - az)_z,$
$(bz - cy)_z - (ay - bx)_x, (cx - az)_x - (bz - cy)_y) = (2a, 2b, 2c) = 2\mathbf{a}$

35. $\Delta f = \operatorname{div}(\operatorname{grad} f)$ **37.** $n(n + 1)\rho^{n-2}$

39. $\mathbf{u} = \boldsymbol{\omega} \times \mathbf{x}$ so $\operatorname{curl}\mathbf{v} = 2\boldsymbol{\omega}$. See p. 709.

41. $\phi = 0$, $\displaystyle\iint = \iint [(x^3 + y^3 + z^3) - 2(x^2 z + y^2 x + z^2 y)]\, d\sigma$

$\displaystyle = \iint 3(z^3 - 2xy^2)\, d\sigma = 3 \iint (\cos^3 \phi \sin \phi - 2\sin^4 \phi \cos \theta \sin^2 \theta)\, d\phi\, d\theta$

$= 3[(\frac{1}{4})(\frac{1}{2}\pi) - 2(\frac{3}{16}\pi)(\frac{1}{3})] = 0$

43. $\displaystyle\phi = \int_0^{2\pi} (\sin^3 \theta - \cos^2 \theta)\, d\theta = -\pi$,

$\displaystyle\iint = \iint [\cos^3 \theta - 2z \cos^2 \theta + \sin^3 \theta - 2\cos \theta \sin \theta]\, d\theta = -2\int_0^{2\pi} \cos^2 \theta\, d\theta \int_0^1 z\, dz = -\pi$

Section 9, page 949

1. $A = 2\pi a h$

3. $f(\mathbf{x}) = f(\rho \mathbf{x}/\rho) = \rho^n f(\mathbf{x}/\rho)$, that is, $f[\rho, \phi, \theta] = \rho^n f[1, \phi, \theta]$.

5. 0 if n is odd, $12\pi a^{n+3}/(n + 1)(n + 3)$ if n is even

7. The level surfaces are toruses, lateral surfaces of right circular cones, and planes on the z-axis. They are mutually orthogonal.

9. It is a pyramid with base a unit square, apex one unit above one vertex of the base; volume $\frac{1}{3}$.

11. $\mathbf{F} = (-4\pi G\, \delta a/b, 0, 0)$

13. By symmetry $\displaystyle\iiint y^{2n}\, dV = \iiint z^{2n}\, dV$, taken over the unit sphere $\rho \le 1$. Now evaluate in spherical coordinates. The required formula falls out.

15. $4a^2 b \arcsin(b/a) - \frac{8}{3}a^3 + \frac{4}{3}(2a^2 + b^2)\sqrt{a^2 - b^2}$

17. Parameterize the surface by $\mathbf{x}(s, v) = \mathbf{x}(s) + v\mathbf{t}(s)$. Then $|\mathbf{x}_s \times \mathbf{x}_v| = |(\mathbf{t} + vk\mathbf{n}) \times \mathbf{t}| = vk$, hence $\displaystyle A = \iint v\, k(s)\, ds\, dv$ over $0 \le s \le L$, $0 \le v \le 1$. The result follows.

19. $\frac{1}{5}\pi$ sr

21. By the Divergence Theorem, $\displaystyle\iiint_D (f - \operatorname{div}\mathbf{A})\, dV = 0$ for such $\mathbf{D}$. If $f - \operatorname{div}\mathbf{A} > 0$ at some point $\mathbf{x}_0$, then by continuity $f - \operatorname{div}\mathbf{A} > 0$ at all $\mathbf{x}$ sufficiently near $\mathbf{x}_0$, hence if $\mathbf{D}$ contains $\mathbf{x}_0$ and is sufficiently small, $\displaystyle\iiint_D (f - \operatorname{div}\mathbf{A}) > 0$, a contradiction, etc.

23. By Ex. 22,

$$\iint_{\partial \mathbf{D}} (A\boldsymbol{\lambda}) \cdot d\boldsymbol{\sigma} = \iint_{\substack{\phi_0 \le \phi \le \phi_1 \\ \theta_0 \le \theta \le \theta_1}} [A(\rho_1, \phi, \theta)\rho_1^2 - A(\rho_0, \phi, \theta)\rho_0^2] \sin \phi \, d\phi \, d\theta$$

$$= \iint_{\substack{\phi_0 \le \phi \le \phi_1 \\ \theta_0 \le \theta \le \theta_1}} \left(\int_{\rho_0}^{\rho_1} \frac{\partial}{\partial \rho} (\rho^2 A) \, d\rho \right) \sin \phi \, d\phi \, d\theta = \iiint_{\mathbf{D}} \frac{\partial}{\partial \rho} (\rho^2 A) \sin \phi \, d\rho \, d\phi \, d\theta$$

$$= \iiint_{\mathbf{D}} \frac{1}{\rho^2} \frac{\partial}{\partial \rho} (\rho^2 A) \, \rho^2 \sin \phi \, d\rho \, d\phi \, d\theta = \iiint_{\mathbf{D}} \frac{1}{\rho^2} \frac{\partial}{\partial \rho} (\rho^2 A) \, dV.$$

25. $\displaystyle \iint_{\partial \mathbf{D}} (C\mathbf{v}) \cdot d\boldsymbol{\sigma} = \iiint_{\mathbf{D}} \frac{1}{\rho \sin \phi} \frac{\partial C}{\partial \theta} \, dV.$

27. If $\mathbf{A} = A\boldsymbol{\lambda} + B\boldsymbol{\mu} + C\mathbf{v}$, then

$$\operatorname{div} \mathbf{A} = \frac{1}{\rho^2} (\rho^2 A)_\rho + \frac{1}{\rho \sin \phi} (B \sin \phi)_\phi + \frac{1}{\rho \sin \phi} C_\theta$$

29. $\Delta f = \rho^{-1}(\rho f)_{\rho\rho} = \rho^{-1}(e^{-k\rho})_{\rho\rho} = k^2 \rho^{-1} e^{-k\rho} = k^2 f$

Index

8. $\displaystyle\int \frac{dx}{\sqrt{x^2 \pm a^2}} = \ln |x + \sqrt{x^2 \pm a^2}|$

9. $\displaystyle\int \sqrt{x^2 \pm a^2}\, dx = \frac{x}{2}\sqrt{x^2 \pm a^2} \pm \frac{a^2}{2}\ln |x + \sqrt{x^2 \pm a^2}|$

10. $\displaystyle\int \sin^2 ax\, dx = \frac{x}{2} - \frac{\sin 2ax}{4a}$

11. $\displaystyle\int \cos^2 ax\, dx = \frac{x}{2} + \frac{\sin 2ax}{4a}$

12. $\displaystyle\int \sin^n ax\, dx = -\frac{\sin^{n-1} ax \cos ax}{na} + \frac{n-1}{n}\int \sin^{n-2} ax\, dx$

13. $\displaystyle\int \cos^n ax\, dx = \frac{\cos^{n-1} ax \sin ax}{na} + \frac{n-1}{n}\int \cos^{n-2} ax\, dx$

14. $\displaystyle\int \sin ax \cos bx\, dx = -\frac{\cos(a+b)x}{2(a+b)} - \frac{\cos(a-b)x}{2(a-b)} \qquad (a \neq \pm b)$

15. $\displaystyle\int \sin ax \sin bx\, dx = \frac{\sin(a-b)x}{2(a-b)} - \frac{\sin(a+b)x}{2(a+b)} \qquad (a \neq \pm b)$

16. $\displaystyle\int \cos ax \cos bx\, dx = \frac{\sin(a-b)x}{2(a-b)} + \frac{\sin(a+b)x}{2(a+b)} \qquad (a \neq \pm b)$

17. $\displaystyle\int x \sin ax\, dx = \frac{\sin ax}{a^2} - \frac{x \cos ax}{a}$

18. $\displaystyle\int x \cos ax\, dx = \frac{\cos ax}{a^2} + \frac{x \sin ax}{a}$

19. $\displaystyle\int x^n \sin ax\, dx = -\frac{x^n \cos ax}{a} + \frac{n}{a}\int x^{n-1} \cos ax\, dx$

20. $\displaystyle\int x^n \cos ax\, dx = \frac{x^n \sin ax}{a} - \frac{n}{a}\int x^{n-1} \sin ax\, dx$

21. $\displaystyle\int \tan ax\, dx = -\frac{1}{a}\ln |\cos ax|$

22. $\displaystyle\int \cot ax\, dx = \frac{1}{a}\ln |\sin ax|$

23. $\displaystyle\int \tan^2 ax\, dx = \frac{1}{a}\tan ax - x$

24. $\displaystyle\int \cot^2 ax\, dx = -\frac{1}{a}\cot ax - x$

25. $\displaystyle\int \tan^n ax\, dx = \frac{\tan^{n-1} ax}{(n-1)a} - \int \tan^{n-2} ax\, dx \qquad (n > 1)$

26. $\displaystyle\int \cot^n ax\, dx = -\frac{\cot^{n-1} ax}{(n-1)a} - \int \cot^{n-2} ax\, dx \qquad (n > 1)$